Biochemistry

BIOCHEMISTRY

N. V. Bhagavan, Ph.D.
Professor of Biochemistry and Medical Technology
John A. Burns School of Medicine
University of Hawaii
Honolulu, Hawaii

Consultant Biochemist to the
Kaiser Foundation Hospital
Honolulu, Hawaii

SECOND EDITION

J. B. Lippincott Company
Philadelphia • Toronto

Library of Congress Cataloging in Publication Data

Bhagavan, N V
  Biochemistry.

  Published in 1974 under title: Biochemistry, a comprehensive
review.
  Bibliography: p.
  Includes index.
  1. Biological chemistry. I. Title.
[DNLM: 1. Biochemistry. QU4.3 B575b]
QP514.2.B45 1978   574.1'92   78-8518
ISBN 0-397-52086-7

Preface

Like the first edition, this book is intended to create an interest and a sustained enthusiasm for the study of biochemistry on the part of students pursuing medical and other health-related studies. Most of these students must continually relate the application of biochemical principles to the art of healing. It is hoped that this book will guide students and practitioners of medicine and related areas in acquiring sufficient information in biochemistry to increase their comprehension of how biochemically determinable constituents vary in normal and abnormal states, and to relate metabolic disorders to biochemical lesions. It is the author's conviction that the acquisition of this knowledge will aid in the alleviation of human suffering.

Every chapter of the first edition has been revised and extended by pertinent basic and clinical information, without disrupting its organization and style. Although the subject of nutrition is not treated as a separate entity, it is one of the underlying themes throughout the book.

In order to provide concise and complete information, detailed discussions of experimental observations have been minimized. In the process, subtle points and various ramifications of the subject material may not appear in the text. With this in mind, the student might refer for greater details to references listed at the end of the book. It should be understood that biochemistry is an extremely rapidly growing field, and in order to keep abreast of new information it is essential to review current literature on a systematic basis.

I am appreciative of the contributions made to various parts of the book by the authors whose names appear on the chapters on which they have worked. Special mention should be made of the invaluable contributions made by Paul S. Takiguchi and John H. Bloor, who reviewed major portions of the manuscript and suggested changes that have enhanced the quality of the work.

I am indebted to Janet Kawata for her painstaking, dedicated help in the typing and preparation of the manuscript for the second edition.

I am grateful to Dr. Alfred G. Scottolini for many fruitful discussions pertaining to the clinical aspects of laboratory medicine.

It has been a pleasure to work with the staff of J. B. Lippincott Company, and in particular I express my gratitude to Mr. J. Stuart Freeman, Jr., for his patience and encouragement.

N. V. Bhagavan

Honolulu, Hawaii
April 1978

Contents

4 Carbohydrates *(Continued)*

Biochemistry

1
Acids, Bases and Buffers

1. Acids are substances that liberate hydrogen ions (protons) in solution (proton donor).

2. Bases are substances that bind hydrogen ions (protons) and thus remove them from solution (proton acceptor). Note: This is the most useful definition of acids and bases for biological work. There are other definitions: Acid is an electron-pair acceptor and base is an electron-pair donor (known as the Lewis Acid Concept).

3. Strong and weak electrolytes.

   a. Strong electrolytes (acids, bases or salts) are completely ionized in aqueous solutions.

   b. Weak electrolytes (acids, bases or salts) are partially ionized in aqueous solutions. The following equation shows the relationship between an undissociated compound and its ions.

$$HA \rightleftharpoons H^+ + A^-$$
$$\text{(weak acid)} \qquad \text{(ions)}$$

   Note that A^- is also known as a conjugate base, because it is a proton acceptor (reaction of A^- with H^+; leftward direction). Similarly, in the dissociation of a weak base, a conjugate acid is produced. The concentration of the conjugate base or acid generated from a weak acid or base is small, since by definition weak acids and bases are only slightly dissociated in aqueous solution.

   c. Applying the Law of Mass Action

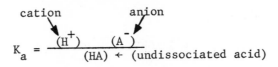

$$K_a = \frac{(H^+)\ (A^-)}{(HA)} \leftarrow \text{(undissociated acid)}$$

   where K_a = ionization constant of the acid and parentheses indicate molar concentrations of the species. Similarly, K_b represents the ionization constant of a base. The higher the value of K_a, the greater the number of H^+ ions liberated per mole of acid in solution and hence the stronger the acid. K_a is thus a measure of the strength of an acid.

$$K_a = \frac{(H^+)\,(A^-)}{(HA)}$$

Rearranging $\qquad H^+ = \dfrac{K_a(HA)}{(A^-)}$

Taking logarithms (to base 10) on both sides

$$\log (H^+) = \log K_a + \log (HA) - \log (A^-)$$

Multiplying by -1

$$-\log (H^+) = -\log K_a - \log (HA) + \log (A^-)$$

Using the definitions

$$-\log (H^+) = pH \text{ and } -\log K_a = pK_a$$

one gets

$$pH = pK_a + \log \frac{(A^-)}{(HA)}$$

or

$$pH = pK_a + \log \frac{salt}{acid}$$

This relationship is commonly known as the <u>Henderson-Hasselbalch Equation</u>.

Dissociation of Water

1. Considering water as a weak acid

$$H_2O \leftrightarrows H^+ + OH^-$$

The dissociation constant

$$K_a^{H_2O} = \frac{(H^+)\,(OH^-)}{(H_2O)}$$

Since in dilute aqueous solutions, the molarity of water is very nearly constant, this is usually rewritten as

$$K_w = K_a^{H_2O}\,(H_2O) = (H^+)(OH^-)$$

The concentration of pure water is 55.6 M so that $K_w = 55.6\ K_a^{H_2O}$. K_w, called the dissociation constant or ion product for water, has the value of 1×10^{-14} at 25°C (room temperature).

2.   In aqueous solution, then $\quad (H^+)(OH^-) = 1 \times 10^{-14}$
So that, in <u>pure water</u> $\qquad\quad (H^+) = 1 \times 10^{-7}$ moles/liter
$\qquad\qquad\qquad\qquad\qquad (OH^-) = 1 \times 10^{-7}$ moles/liter

and: pH = 7
      pOH = 7

For all aqueous solutions $\quad pK_w = pH + pOH = 14$
$\qquad\qquad\qquad$ and $\quad pOH = 14 - pH.$

pOH is not used descriptively (as is pH), but it is useful in calculating the pH of an alkaline solution.

<u>Problem 1</u>   What is the pH of 0.0004 N NaOH solution? Strong electrolyte, completely dissociated.

<u>Method 1</u>: $(OH^-) = 0.0004\ M = 4 \times 10^{-4}\ M$

$$pOH = -\log (OH^-) = -\log (4 \times 10^{-4}) = -\log 4 - \log 10^{-4}$$

$$= -(0.602 - 4) = 3.398$$

pOH + pH = 14 so that pH = 14 - pOH

Then pH = 14 - 3.398 = <u>10.602</u>

<u>Method 2</u>: $(H^+)(OH^-) = 10^{-14}$

$$(H^+) = 10^{-14}/(OH^-) = 10^{-14}/(4 \times 10^{-4}) = 2.5 \times 10^{-11}$$

$$pH = -\log (H^+) = -\log (2.5 \times 10^{-11})$$

$$= -\log 2.5 - \log 10^{-11} = -(0.398 - 11)$$

pH = <u>10.602</u>

<u>Problem 2</u>   Calculate the pH of 0.005 M HCl solution. Strong electrolyte, completely dissociated.

$$(H^+) = 5 \times 10^{-3}\ M$$

$$pH = 3 - 0.7 = \underline{2.3}$$

Problem 3 What is the pH of 0.1 M solution of CH_3COOH? Given $K_a = 1.86 \times 10^{-5}$. Acetic acid is a weak acid and therefore poorly dissociated.

$$CH_3COOH \rightleftarrows (CH_3COO^-) + (H^+)$$

$$K_a = \frac{(CH_3COO^-)\,(H^+)}{(CH_3COOH)}$$

Since $(CH_3COO^-) = (H^+)$, define $(CH_3COO^-) = (H^+) = X$ at equilibrium.

In order to get one CH_3COO^- and one H^+, one CH_3COOH has to dissociate. Therefore, if the concentrations of the ions are X, the loss in CH_3COOH concentration must also be X. Since the initial concentration of CH_3COOH is 0.1 M, the concentration at equilibrium is $(CH_3COOH) = 0.1 - X$. Then

$$K_a = \frac{(X)\,(X)}{(0.1 - X)}$$

To make this easier to solve, assume that the loss of CH_3COOH due to dissociation is small compared to 0.1 M. This is acceptable because

$$\frac{K_a}{(CH_3COOH)}$$ is less than approximately 10^{-3}.

Then $(0.1 - X) \simeq 0.1$ and $K_a = \dfrac{x^2}{0.1}$.

Note: HAc = acetic acid = CH_3COOH.

$$\left[\text{If } \frac{K_a}{(HAc)} \geqslant 10^{-3}, \text{ there will be an error of} \geqslant 3.2\% \text{ in the value of } \left[H^+\right].\right]$$

$$x^2 = 0.1\, K_a$$

$$X = \sqrt{0.1\, K_a}$$

$$= \sqrt{(0.1)(1.86 \times 10^{-5})} = \sqrt{1.86 \times 10^{-6}}$$

$$= 1.36 \times 10^{-3} = (H^+)$$

Note: The concentration of $X = (1.36 \times 10^{-3}$ M) is small compared to the concentration of free acid = (0.1 M).

$$pH = 3 - 0.13$$

$$= \underline{2.87}$$

<u>Problem 4</u> What is the pH of 0.1 N NH_4OH?

$$K_b = 2 \times 10^{-5}$$

Answer: pH = <u>11.15</u>

Buffers

1. These are solutions that minimize changes in pH when acids or bases are added to them.

2. They are usually mixtures of a weak acid and a salt of the same acid with a strong base, although they can also be mixtures of a weak base and its salt with a strong acid.

     Example: CH_3COOH and H_2CO_3 are weak acids. When reacted with NaOH, a strong base, the salts $CH_3COO^-Na^+$ and $Na^+HCO_3^-$ are formed.

      Then: $(CH_3COOH + CH_3COO^-Na^+)$ and $(H_2CO_3 + Na^+HCO_3^-)$ are both buffer systems.

3. A buffer works in the following manner.

    To make a buffer, mix CH_3COOH, $CH_3COO^-Na^+$, and water. The species present in solution are CH_3COOH, CH_3COO^-, Na^+, and H_2O. (There are also small amounts of H^+ and OH^- as there must be in all aqueous solutions. These need not be considered in calculations.)

     Add H^+: $H^+ + CH_3COO^- \rightarrow CH_3COOH$. Almost all of the protons react with acetate ions to produce weakly ionized acetic acid. The H^+ is thereby prevented from changing the pH appreciably.

     Add OH^-: $OH^- + CH_3COOH \rightarrow CH_3COO^- + H_2O$. Almost all of the hydroxyl radicals react with acetic acid molecules to produce more acetate and more water. The OH^- is thus absorbed and does not affect the pH very much.

4. Adding H^+ or OH^- to a buffer causes <u>slight</u> pH changes (since mass action expressions must be obeyed) provided there is enough salt (CH_3COO^-) or acid (CH_3COOH) to react with them. If, for example, all of the acid is converted to a salt by adding a large amount of

OH⁻, then the solution no longer behaves as a buffer. Adding more OH⁻ will cause the pH to rise rapidly, just as if the solution contained only the salt. The maximum buffering is said to occur when the molarities of the salt and acid are equal. Then the buffer has its maximum capacity to absorb either H^+ or OH⁻. (Note also that, mathematically, the pK_a corresponds to an inflection point in the titration curve. Hence, the pK_a is the point of minimum slope or minimum change in pH for a given change in $[H^+]$ or $[OH^-]$. Since a buffer is intended to give only a small pH change with added H^+ or OH⁻, the best buffer is the one which gives the smallest change.) From the Henderson-Hasselbalch Equation, this is seen to occur when the pH of the solution equals the pK_a (or pK_b) of the weak acid (or base) forming the buffer.

$$pH = pK + \log \frac{(salt)}{(acid)}$$

if (salt) = (acid) then $\log \dfrac{(salt)}{(acid)} = \log 1 = 0$ and $\underline{pH = pK}$.

5. Buffers are part of the homeostatic mechanisms whereby the neutrality of the body fluids is regulated.

Blood pH: 7.35-7.45 = normal range

```
    7.8 → death
     ↑ ←――――――― alkalosis
    7.45
        }normal range
    7.35
     ↓ ←――――――― acidosis
    6.8 → death
```

6. Following is a buffer problem worked out in detail to illustrate the principles and calculations.

What volume of 0.1 M NaOH is required to prepare 200 ml of 0.1 M buffer, pH 3.25, using an acid (HA) with a pK of 3.85? (Note: All concentrations are in moles per liter, symbolized by M.)

The final buffer solution will contain a mixture of HA (undissociated acid) and A⁻ (the conjugate base). Since the final buffer molarity is 0.1 M,

HA + A⁻ = 0.1 M (conservation of mass equation).

In addition, since the buffer is an equilibrium solution, the Henderson-Hasselbalch equation must be satisfied:

$$pH = pK + \log \frac{A^-}{HA} \quad OR \quad 3.25 = 3.85 + \log \frac{A^-}{HA}$$

Simplifying:
$$-0.60 = \log \frac{A^-}{HA}$$

The negative sign on the left side of the equation can be removed by inverting the argument of the logarithm ($\log x = -\log 1/x$):

$$0.60 = \log \frac{HA}{A^-} \quad AND \quad \text{taking antilogarithms:} \quad 4.0 = \frac{HA}{A^-}$$

$$\text{or } HA = 4.0 \ A^-.$$

Substituting this into the conservation of mass equation (above) gives:

$$A^- = 0.02 \ M \quad \text{and} \quad HA = 0.08 \ M.$$

The conjugate base is generated by reaction of the acid (HA) with NaOH (HA + NaOH $\rightleftarrows$ Na$^+$ + A$^-$ + H$_2$O). Sufficient NaOH is needed, then, to react with HA to produce a 0.02 M solution of A$^-$. Since one mole of NaOH will react with one mole of HA to produce one mole of A$^-$, the relationship

$$\text{moles of NaOH needed} = \text{moles of A}^- \text{ needed}$$

holds true; and since

$$\text{molarity x volume (liters)} = \text{number of moles,}$$

$$0.1 \ M \text{ x volume of NaOH needed} = 0.02 \ M \text{ x } 0.2 \text{ liter of buffer}$$

$$\text{or volume of NaOH needed} = 0.04 \text{ liter} = 40 \text{ ml.}$$

The buffer solution is prepared by mixing 40 ml of 0.1 M NaOH, 0.02 moles (= 0.2 liter x 0.1 M) of pure HA, and enough water to make 200 ml of solution.

Buffer Systems of the Blood

1. a. The most important buffer of plasma is the bicarbonate-carbonic acid system. Carbonic acid, a diprotic acid, has two pK values of 3.8 and 10.2.

    (I) $H_2CO_3 \rightleftarrows H^+ + HCO_3^-$ $pK_1 = 3.8$

    (II) $HCO_3^- \rightleftarrows CO_3^= + H^+$ $pK_2 = 10.2$

It is apparent from the pK values that neither equilibrium can serve as a buffer system at the physiological pH of 7.4. However, carbonic acid (the proton donor) is in equilibrium with dissolved CO_2 (which in turn is in equilibrium with gaseous CO_2).

$$H_2O + CO_2 \text{ (aqueous)} \rightleftharpoons H_2CO_3$$

The hydration reaction, coupled with the first dissociation of carbonic acid (I, above), produces an apparent pK of 6.1 for bicarbonate formation. (This reaction, however, is very slow to attain equilibrium in the absence of the erythrocyte enzyme carbonic anhydrase.)

This can be summarized as:

$$CO_2 \text{ (aqueous)} + H_2O \rightleftharpoons H^+ + HCO_3^-$$

$$pK_{apparent} = \frac{[HCO_3^-][H^+]}{[H_2CO_3]} = 6.1$$

b. The ratio of HCO_3^- to H_2CO_3 that must exist to maintain the physiologic pH of 7.4 can be calculated by using the Henderson-Hasselbalch relationship as follows:

$$7.4 = 6.1 + \log \frac{HCO_3^-}{H_2CO_3}$$

$$\log \frac{HCO_3^-}{H_2CO_3} = 1.3$$

Taking antilogarithms on both sides:

$$\frac{HCO_3^-}{H_2CO_3} = \frac{20}{1} = \frac{\text{Proton acceptor}}{\text{Proton donor}}$$

Because the ratio is so high (due to $pK_{apparent}$ being much lower than pH), the bicarbonate system at pH 7.4 serves as a good buffer toward acid (can neutralize large amounts of acid) but a poor one for alkali.

In addition, although the amount of H_2CO_3 is small, it is in rapid equilibrium with a relatively large amount of CO_2 (about 1000 times as much) and it can thus function as an effective buffer to increases in alkalinity. The HCO_3^-/H_2CO_3 ratio in the blood is intimately coupled to the production of CO_2 (an end product of substrate oxidation in the tissues) and to loss of CO_2 from the lungs during respiration (see Chapter 12).

c. In the equilibrium expression for the bicarbonate/carbonate buffer system at pH 7.4, the carbonic acid term can be replaced by a pressure term, since the carbonic acid concentration is proportional to the partial pressure of CO_2 (P_{CO_2}) to which the solution (blood) is exposed.

$$pH = 6.1 + \frac{\left[HCO_3^{-} \right]}{\left[aP_{CO_2} \right]}$$

where a, the proportionality constant, is defined by the equation:

$$H_2CO_3 = aP_{CO_2}.$$

The numerical value of a depends on the solvent and temperature. For normal plasma at $37°C$, a = 0.0301 millimoles dissolved CO_2/liter of plasma/torr of CO_2 pressure. This value can be derived from two facts:

i) At $37°C$ and at 760 torr of CO_2 pressure, CO_2 will dissolve in normal plasma to the extent of 521 ml of gas per liter of plasma.

ii) At standard temperature and pressure, one mole of dry CO_2 occupies 22.26 liters (not 22.4 liters, because CO_2 is not an ideal gas).

$$a = \frac{521 \text{ ml } CO_2/\text{liter of plasma}}{760 \text{ torr x } 22.26 \text{ ml/mmole of } CO_2}$$

$$= 0.0301 \frac{\text{mmoles } CO_2/\text{liter of plasma}}{\text{torr of } CO_2 \text{ pressure}}$$

d. The Henderson-Hasselbalch equation can be further modified by substituting for the bicarbonate term. When a strong acid is added to plasma, CO_2 is released, which is derived from dissolved CO_2, carbonic acid, and bicarbonate ions. Thus,

$$\text{Total } CO_2 = HCO_3^{-} + \text{dissolved } CO_2 + H_2CO_3.$$

But $\text{dissolved } CO_2 + H_2CO_3 = 0.0301 \, P_{CO_2}$, so that $HCO_3^- = \text{total } CO_2 - 0.0301\ P_{CO_2}.$

Finally,

$$pH = 6.1 + \log \frac{\text{Total } CO_2 - 0.0301\ P_{CO_2}}{0.0301\ P_{CO_2}}$$

This form is useful for calculating pH from the readily measurable values of total CO_2 and P_{CO_2}.

e. The effectiveness of the HCO_3^-/H_2CO_3 (or CO_2) buffer system in maintaining a constant blood pH of 7.4 depends on the efficient control of bicarbonate and CO_2 concentrations to maintain a ratio of 20/1. The HCO_3^- concentration is regulated by selective excretion and reabsorption by the renal tubular epithelial cell membranes. The P_{CO_2} in the blood can be altered by changes in the rate and depth of respiration. Hypoventilation (slow, shallow breathing) leads to increased blood P_{CO_2} while hyperventilation has an opposite effect. Note that the changes mediated by the lungs (P_{CO_2}) are much more rapid than those effected through the kidneys (HCO_3^- and other anions). These aspects are further discussed in Chapter 12.

2. Serum Protein Buffer System. There are many proteins in serum and they have many weakly acidic (glutamate, aspartate) and weakly basic (lysine, arginine, histidine) amino acid sidechains which act as components of a buffer system. Such effects are insignificant, however, compared to the buffering capacity of hemoglobin (in erythrocytes) and of the bicarbonate system (in the plasma).

3. The Phosphate Buffer System. Phosphoric acid (H_3PO_4) has three dissociable protons (polyprotic acid):

$$H_3PO_4 \rightleftharpoons H_2PO_4^- + H^+ \qquad pK_1 = 2.1$$

$$H_2PO_4^- \rightleftharpoons HPO_4^= + H^+ \qquad pK_2 = 6.7$$

$$HPO_4^= \rightleftharpoons PO_4^{\equiv} + H^+ \qquad pK_3 = 12.3$$

The principal equilibrium functioning at a given pH depends on which pK is closest to the pH. Therefore, at the plasma pH of 7.4, the conjugate pair is $HPO_4^=/H_2PO_4^-$. The Henderson-Hasselbalch equation can then be used to obtain the value of the ratio $HPO_4^=/H_2PO_4^-$ at pH 7.4:

$$7.4 = 6.7 + \log \frac{HPO_4^=}{H_2PO_4^-}$$

$$\log \frac{HPO_4^=}{H_2PO_4^-} = 0.7 \qquad \text{and} \qquad \frac{HPO_4^=}{H_2PO_4^-} = 5.01$$

As was the case with the bicarbonate/carbonic acid system, the salt form ($HPO_4^=$; conjugate base) of the phosphate buffer is present in large (five-fold) excess compared to the acid species ($H_2PO_4^-$). Again, this provides appreciable buffering only for the acid-generating systems. Since the metabolism of a person maintained on an average diet produces more acid than base, this ratio is in the correct direction for neutralizing this acid and maintaining homeostasis. The $HPO_4^=/H_2PO_4^-$ buffering system plays a minor role in the plasma because of the low concentrations of these ions. It is, however, important in raising the plasma pH through the excretion of $H_2PO_4^-$ by the kidney.

4. The hemoglobin buffer system buffers CO_2 (as carbonic acid) produced during metabolic processes. There are two mechanisms for this.

a. Hemoglobin (Hb) exists as oxyhemoglobin ($HHbO_2$), deoxy-hemoglobin (HHb), and as their potassium salts ($KHbO_2$ and KHb). Since HHb and $HHbO_2$ are weak acids, the two buffer systems, (HHb/KHb) and ($HHbO_2/KHbO_2$), are present in the erythrocyte. (This is similar to the buffering of serum proteins in 4 (above). The proton released is not the same one as in b below.)

b. More important than a and a very effective buffer is the acceptance of H^+ ions by histidine in the hemoglobin molecule with the release of O_2 (Bohr proton).

Oxyhemoglobin   Deoxygenated Hemoglobin

or $HbO_2^- + H_2CO_3 \rightleftarrows HHb + O_2 + HCO_3^-$

An increase in P_{CO_2} and a decrease in pH will <u>favor</u> the dissociation of $Hb\bar{O}_2$ and thus permit the imidazole group to accept H^+ ions. In the tissues then

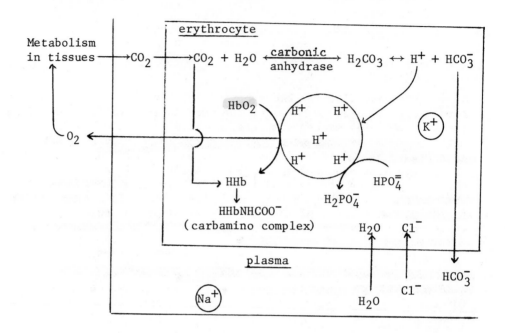

HCO_3^-, produced within the erythrocytes, diffuses into the plasma and in order to maintain electrical neutrality, the principal plasma anion Cl^- goes into the erythrocytes. Note that the concentrations of the principal intracellular cation (K^+) and extracellular (plasma) cation (Na^+) do not undergo significant changes due to the relative impermeability of the erythrocyte membrane. The exact opposite set of changes takes place at the lungs with the release of CO_2.

The bicarbonate/carbonic acid system participates in buffering against acids and bases as follows:

$$Na^+HCO_3^- + HCl \longrightarrow H_2CO_3 + NaCl$$

$$H_2O + CO_2 \text{ (can be removed by the lungs)}$$

$$H_2CO_3 + NaOH \longrightarrow NaHCO_3 + H_2O$$

The above reactions may be summarized as follows:

Hemoglobin actually absorbs about 60% of the hydrogen ions produced by H_2CO_3 formation from CO_2. This is far more than any other buffer system in the blood. However, since hemoglobin and carbonic anhydrase are present only in the erythrocytes, the HCO_3^-/H_2CO_3 system in the plasma is an indispensable intermediary in transporting the acid. This is discussed further in Chapters 7 and 12. Thus, the principal method of CO_2 transport is in the form of HCO_3^- in blood plasma.

Titration Curves

1. The profiles below (Figure 1) are constructed by measuring the pH of an acid solution as it is titrated with a base. Titration curves were prepared using 25 ml of 1 M acid and titrating with 1 M NaOH (or any other strong base). Note that, at any point in the titration profile, the acid remaining = concentration of initial acid − concentration of salt formed.

2. Calculation of pK_a for unknown weak acids and bases (strong acids are completely ionized in H_2O and show no pK_a) is possible with a titration curve for the acid or base. Recall from the Henderson-Hasselbalch equation that pH = pK when the concentration of the weak acid is equal to concentration of its conjugate base (salt). This situation exists when one-half of the initial amount of the weak acid has been titrated with a strong base.

3. One can determine the pK_a of a weak acid used in the preparation of a buffer of known pH by observing the shape of the titration curve of the buffer. Buffer capacity, defined as the mole equivalents of H^+ or OH^- required to change 1 liter of buffer by 1.0 pH unit, is maximal at pH = pK.

(Text continues on p. 16)

Figure 1. Titration Profiles of Some Acids*

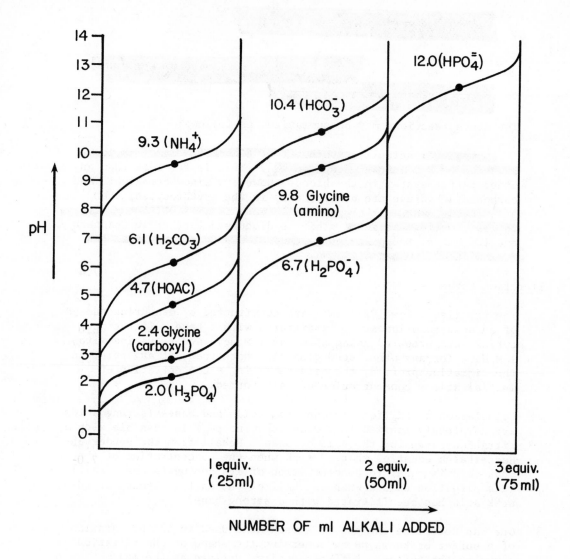

Note: $pK_{apparent}$: $CO_2 (aqueous) + H_2O \rightleftharpoons H^+ + HCO_3^-$ pK_{a_2}: $HCO_3^- \rightleftharpoons CO_3^= + H^+$

Note: The points marked, and the accompanying numbers on the above
diagram, are the pK_a values of the compound indicated.
HOAC = acetic acid.

* Reproduced with permission from Raffelson, Max E., Jr., Brinkley,
Stephen B., and Hayashi, James A.: Basic Biochemistry, ed. 3,
New York, 1971, The MacMillan Company.

Table 1

pH Values of Human Body Fluids and Secretions

Body Fluid or Secretion	pH
Blood	7.4
Milk	6.6-6.9
Hepatic Bile	7.4-8.5
Gallbladder Bile	5.4-6.9
Urine (normal)	6.0
Gastric Juice (parietal secretion)	0.87
Pancreatic Juice	8.0
Intestinal Juice	7.7
Cerebrospinal Fluid	7.4
Saliva	7.2
Aqueous Humor of Eye	7.2
Tears	7.4
Urine (range in various disease states)	4.8-7.5
Feces	7.0-7.5

Intracellular pH of a resting muscle cell
(at 37°C and at an extracellular pH of 7.4)   about 7.1

2
Amino Acids and Proteins

Chemistry of the Amino Acids

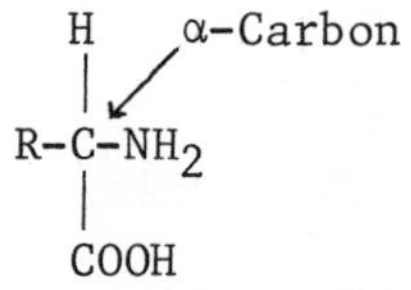

$$
\begin{array}{c}
\text{H} \quad \swarrow \alpha\text{-Carbon} \\
| \\
\text{R-C-NH}_2 \\
| \\
\text{COOH}
\end{array}
$$

1. Almost all of the naturally occurring amino acids are α-amino acids.
    In these, an amino group, a carboxyl group, and an R-group are
    all attached to the α-carbon.

2. They are classified in accordance with the structure of the R-group.

3. With the exception of glycine (R=H) and β-alanine (amino group
    attached to the β-carbon, the second carbon from the carboxyl group),
    the amino acids have at least one asymmetric carbon atom (the
    α-carbon) and hence are optically active.

4. Most amino acids found in living systems are L-α-amino acids.   The
    L indicates the absolute configuration based on the relationship
    of the amino acid to D and L glyceraldehyde. Diagrams showing this
    are given below. Recall, however, that the α-carbon will be
    tetrahedral so that, actually, the -CHO and -CH$_2$OH groups of
    glyceraldehyde are going back into the page and the -H and -OH
    groups are coming out of the page.

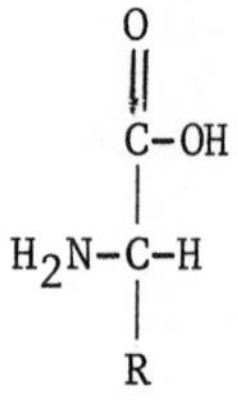

$$
\begin{array}{cccc}
\text{O} & \text{O} & \text{O} & \text{O} \\
\parallel & \parallel & \parallel & \parallel \\
\text{C-OH} & \text{C-H} & \text{C-H} & \text{C-OH} \\
| & | & | & | \\
\text{H}_2\text{N-C-H} & \text{HO-C-H} & \text{H-C-OH} & \text{H-C-NH}_2 \\
| & | & | & | \\
\text{R} & \text{CH}_2\text{OH} & \text{CH}_2\text{OH} & \text{R}
\end{array}
$$

L-amino acid 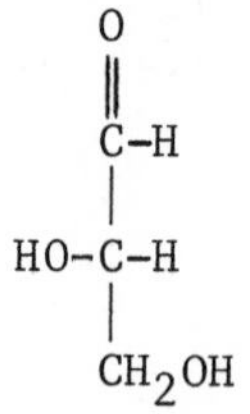   L-glyceraldehyde 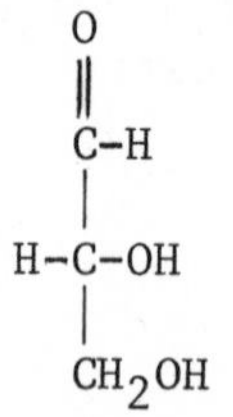   D-glyceraldehyde 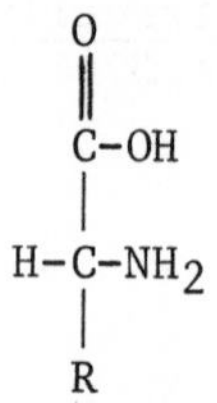   D-amino acid

The designation D or L does not indicate the direction in which a
solution of the compound rotates the plane of polarized light.
This is designated d or l and varies from one amino acid to
another.

16

5. All amino acids isolated from plant and animal protein are L-α-amino acids. β-alanine is found in the vitamin pantothenic acid. Some D-amino acids are found in polypeptide antibiotics such as the gramicidins and bacitracins. D-glutamic acid and D-alanine are found in bacterial cell walls. Carnosine (β-alanyl histidine) and anserine (β-alanyl-1-methyl histidine) are found in skeletal muscles (20-30 millimoles per kg) but their exact function is not known.

6. Threonine, cystine, isoleucine, 3- and 4-hydroxyproline, and hydroxylysine have <u>two</u> asymmetric carbon atoms and therefore have <u>four isomeric forms</u>. General rule: Number of possible isomeric forms of any compound equals 2^n where n represents the number of asymmetric carbon atoms.

Example:

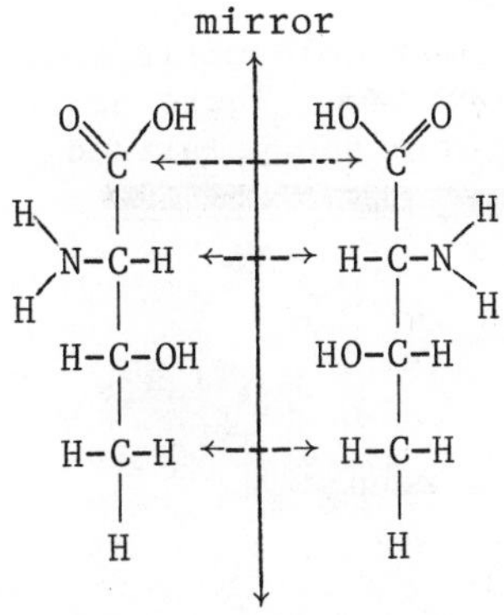

L-threonine   D-threonine

enantiomeric pair

Enantiomeric pairs - similar chemical properties, rotate beam of plane polarized light equally but in opposite directions.

Note: Identical atoms of each molecule are equal distances on either side of the mirror plane, as indicated by the dotted arrows in the figure above. The other two isomers are designated as L-allothreonine and D-allothreonine (Greek-allos=other). The structures of these follow.

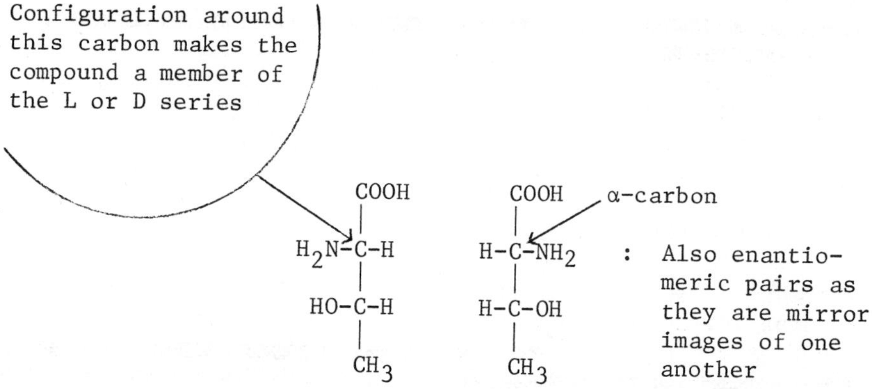

Configuration around
this carbon makes the
compound a member of
the L or D series

$$
\begin{array}{ccc}
\text{COOH} & \text{COOH} & \alpha\text{-carbon} \\
| & | & \\
\text{H}_2\text{N-C-H} & \text{H-C-NH}_2 & : \text{ Also enantio-} \\
| & | & \text{meric pairs as} \\
\text{HO-C-H} & \text{H-C-OH} & \text{they are mirror} \\
| & | & \text{images of one} \\
\text{CH}_3 & \text{CH}_3 & \text{another}
\end{array}
$$

L-allothreonine   D-allothreonine

▽

enantiomeric pair

L-allothreonine and L-threonine are a diastereoisomeric pair
(diastereomers) with different chemical properties. There is no
relationship between the degree of rotation of plane polarized
light by these compounds. They are not mirror images of one
another.

7. Amino acids as electrolytes

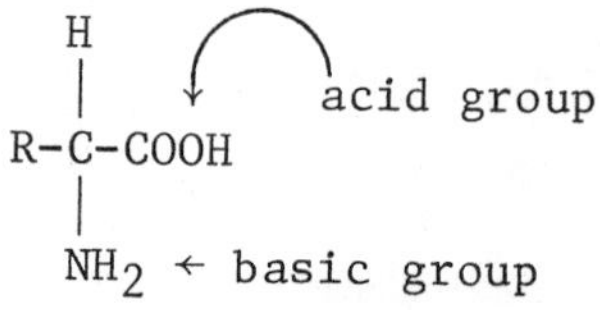

$$
\begin{array}{l}
\text{H} \\
| \quad\quad\text{acid group} \\
\text{R-C-COOH} \\
| \\
\text{NH}_2 \leftarrow \text{basic group}
\end{array}
$$

a. They are ampholytes and contain both acidic and basic groups
    (or proton donating and proton accepting groups).

b. At physiological pH's they exist as dipolar ions or zwitterions

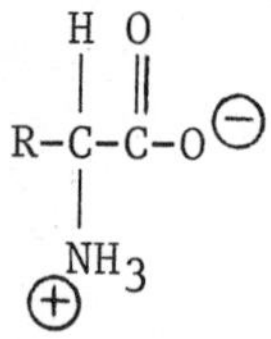

$$
\begin{array}{l}
\text{H} \quad \text{O} \\
| \quad\; \| \\
\text{R-C-C-O}^{\ominus} \\
| \\
\overset{\oplus}{\text{NH}_3}
\end{array}
$$

This form is electrically neutral (does not migrate in an electric
field).

In acidic solution (below pH 2) they exist as

$$\begin{array}{c} H \quad O \\ | \quad \parallel \\ R-C-C-OH \\ | \\ NH_3^+ \end{array}$$

In basic solution (above pH 9.5) they exist as

$$\begin{array}{c} H \quad O \\ | \quad \parallel \\ R-C-C-O^- \\ | \\ NH_2 \end{array}$$

They hardly ever occur as

$$\begin{array}{c} H \quad O \\ | \quad \parallel \\ R-C-C-OH \\ | \\ NH_2 \end{array}$$

Isoelectric point – The pH at which a dipolar ion does not migrate in an electric field. (i.e., the pH at which the dipolar ion is the predominant species in solution; also called pI).

Isoionic point – The isoelectric point when the isoelectric pH is determined in water solution only, in the absence of any other solutes. The isoionic point is also defined as the pH at which the number of cations equals the number of anions.

c. Addition of (H^+) and (OH^-) ions: When H^+ is added, the electrically neutral amino acid becomes positively charged (curves below are for glycine; note the directions of the pH scales).

Figure 2. Titration of Glycine with an Acid

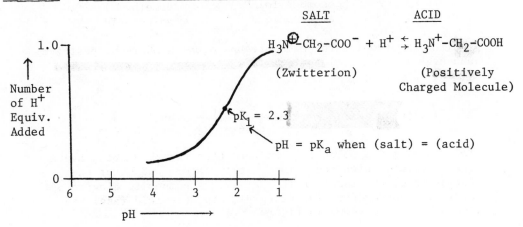

$$\underset{\text{(Zwitterion)}}{H_3\overset{\oplus}{N}-CH_2-COO^-} + H^+ \underset{\rightarrow}{\overset{\leftarrow}{}} \underset{\substack{\text{(Positively} \\ \text{Charged Molecule)}}}{H_3N^+-CH_2-COOH}$$

SALT   ACID

$pK_1 = 2.3$

pH = pK_a when (salt) = (acid)

Number of H^+ Equiv. Added

pH ⟶

When OH⁻ is added, the electrically neutral amino acid becomes negatively charged.

Figure 3.   Titration of Glycine with a Base

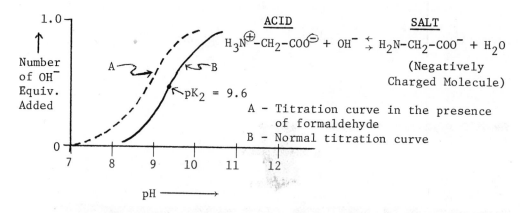

ACID

$H_3\overset{\oplus}{N}-CH_2-COO^{\ominus}$ + OH⁻ $\underset{\to}{\leftarrow}$ $H_2N-CH_2-COO^-$ + H_2O

SALT

(Negatively Charged Molecule)

A – Titration curve in the presence of formaldehyde

B – Normal titration curve

Number of OH⁻ Equiv. Added

pK_2 = 9.6

pH ⟶

Reaction with formaldehyde (HCHO): Formaldehyde combines with the NH_2 groups in neutral or slightly basic solutions and thus alters the pK of the NH_2 group. This is shown by the dashed line in the above graph.

Reaction:

$$R-CH-COO^- \underset{-(CH_2O)}{\overset{+(CH_2O)}{\rightleftharpoons}} R-CH-COO^- \underset{-(CH_2O)}{\overset{+(CH_2O)}{\rightleftharpoons}}$$
with NH_2 group, and $HN-CH_2OH$ group

$$R-CH-COO^-$$
with N bonded to HOH_2C and CH_2OH

monomethylol amino acid

dimethylol amino acid

pI (the pH at which the molecule has equal numbers of positive and negative charges) can be determined from pK_1 and pK_2 as shown.

For Glycine $pI = \dfrac{pK_1 + pK_2}{2} = \dfrac{2.3 + 9.6}{2} = 5.95$

To find pI for a molecule with more than two pK values, average the pK values which occur immediately before and after the formation of the isoelectric species. A composite diagram of titration curves for glycine (gly), aspartic acid (asp) and lysine (lys) is shown in Figure 4 .

Figure 4. Titration Profiles of Glycine, Lysine and Aspartic Acid*

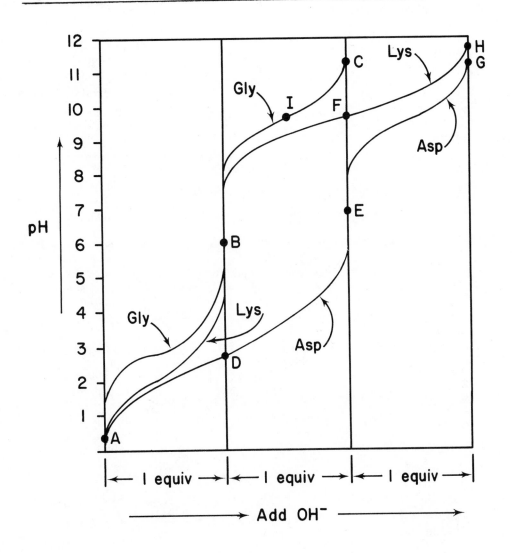

* From Orten, James M. and Neuhaus, Otto W.: <u>Biochemistry</u>, ed. 8,
  St. Louis, 1970, The C.V. Mosby Co. used with permission

The following amino acid species are the predominant forms in the solution at the pH's indicated by the lettered points on the above graph.

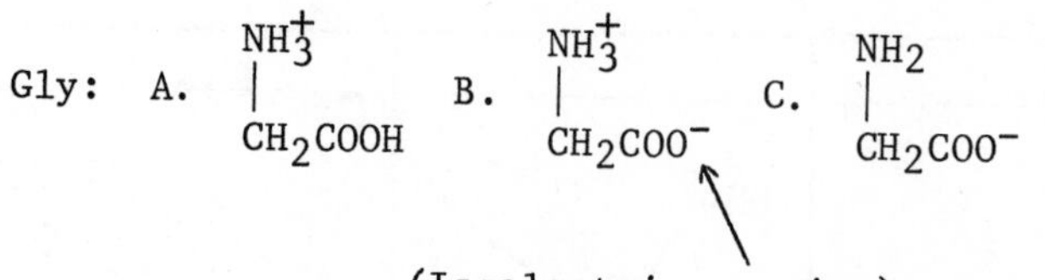

Gly: A.
$$\begin{array}{l} NH_3^+ \\ | \\ CH_2COOH \end{array}$$
B.
$$\begin{array}{l} NH_3^+ \\ | \\ CH_2COO^- \end{array}$$
C.
$$\begin{array}{l} NH_2 \\ | \\ CH_2COO^- \end{array}$$

At point I, there is an equal mixture of B + C; pH = pK_a for $\alpha\text{-}NH_3^+ \rightleftharpoons \alpha\text{-}NH_2 + H^+$

(Isoelectric species)

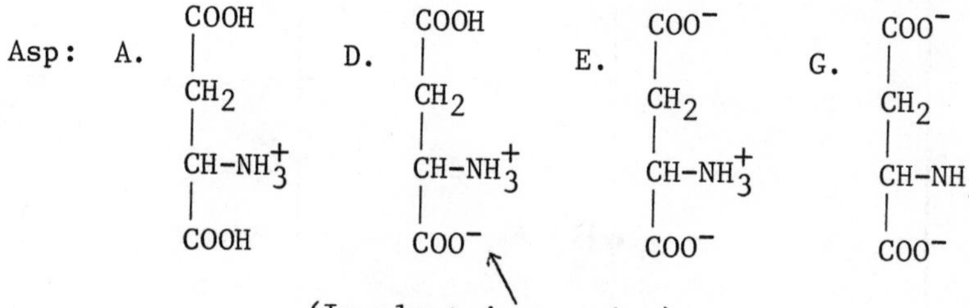

Asp: A.
$$\begin{array}{l} COOH \\ | \\ CH_2 \\ | \\ CH\text{-}NH_3^+ \\ | \\ COOH \end{array}$$
D.
$$\begin{array}{l} COOH \\ | \\ CH_2 \\ | \\ CH\text{-}NH_3^+ \\ | \\ COO^- \end{array}$$
E.
$$\begin{array}{l} COO^- \\ | \\ CH_2 \\ | \\ CH\text{-}NH_3^+ \\ | \\ COO^- \end{array}$$
G.
$$\begin{array}{l} COO^- \\ | \\ CH_2 \\ | \\ CH\text{-}NH_2 \\ | \\ COO^- \end{array}$$

(Isoelectric species)

(pK_1 = 2.0; pK_2 = 4.0; pK_3 = 9.8; pI = $\dfrac{4.0+2.0}{2}$ = 3.0 for aspartate)

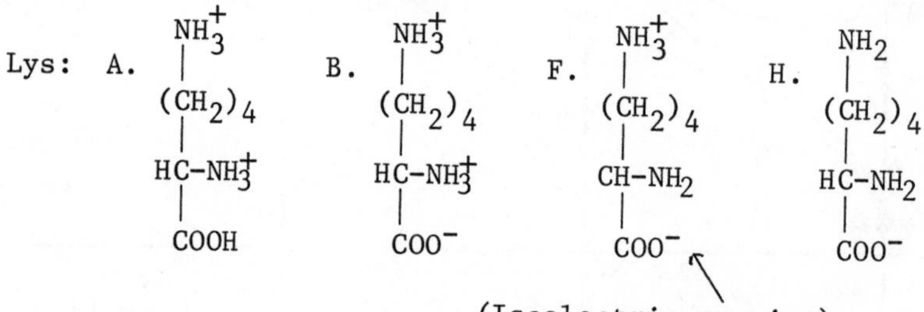

Lys: A.
$$\begin{array}{l} NH_3^+ \\ | \\ (CH_2)_4 \\ | \\ HC\text{-}NH_3^+ \\ | \\ COOH \end{array}$$
B.
$$\begin{array}{l} NH_3^+ \\ | \\ (CH_2)_4 \\ | \\ HC\text{-}NH_3^+ \\ | \\ COO^- \end{array}$$
F.
$$\begin{array}{l} NH_3^+ \\ | \\ (CH_2)_4 \\ | \\ CH\text{-}NH_2 \\ | \\ COO^- \end{array}$$
H.
$$\begin{array}{l} NH_2 \\ | \\ (CH_2)_4 \\ | \\ HC\text{-}NH_2 \\ | \\ COO^- \end{array}$$

(Isoelectric species)

(pK_1 = 2.2; pK_2 = 8.9; pK_3 = 10.5; pI = $\dfrac{8.9+10.5}{2}$ = 9.7 for lysine)

Table 2

pK' and pI Values of Ionizable Groups of Some Selected

Amino Acids at 25°C*

Amino Acid	pK_1' (COOH)	pK_2'	pK_3'	pI**
Alanine	2.34	9.69 (NH_3^+)		6.00
Aspartic acid	2.00	4.00 (COOH)	9.80 (NH_3^+)	3.00
Glutamic acid	2.19	4.25 (COOH)	9.67 (NH_3^+)	3.22
Arginine	2.20	9.04 (NH_3^+)	12.48 (guanidinium)	10.76
Histidine	1.82	6.00 (Imidazolium)	9.17 (NH_3^+)	7.59
Lysine	2.20	8.90 (α-NH_3^+)	10.53 (ϵ-NH_3^+)	9.70
Cysteine	1.71	8.33 (-SH)	10.78 (NH_3^+)	5.02
Tyrosine	2.20	9.11 (NH_3^+)	10.07 (phenolic OH)	5.66
Serine	2.21	9.15 (NH_3^+)		5.68

* For discussion, refer to next chapter.

** pI = pH at the isoelectric point.

<u>Structures of Some Amino Acids</u> (Abbreviations used throughout this text
are indicated in parentheses following the names of the amino acids.)

1. Glycine (Gly). Simplest amino acid (not optically active); used in
    many biosynthetic reactions such as porphyrin and purine
    biosynthesis; has the smallest R-group, and thus can fit into
    crowded regions of the peptide chain. Collagen, a fibrous protein,
    has glycyl residues about every third amino acid residue.

2. **Alanine (Ala)**. Substrate for alanine aminotransferase (also known as glutamic pyruvic transaminase or GPT), an enzyme of clinical significance; other amino acids may be considered derivatives of alanine, with substitutions on the β-carbon. The β-carbon and substitutions attached to it make up the R-group.

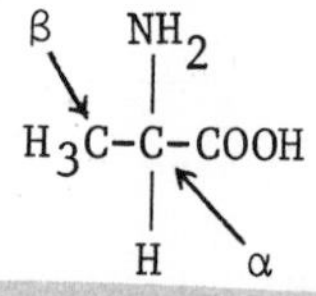

$$\beta \qquad NH_2$$
$$H_3C-C-COOH$$
$$H \qquad \alpha$$

3. **Cysteine (Cys)**. The HS-group is referred to as a sulfhydryl group. It is essential for the activity of many enzymes. Heavy metals inactivate some proteins by combining with HS-groups.

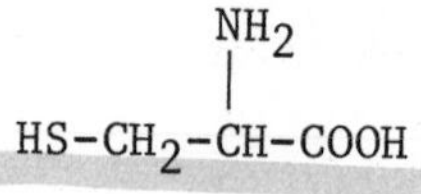

$$NH_2$$
$$HS-CH_2-CH-COOH$$

4. **Cystine.** Oxidized form of cysteine; 2 cysteine $\rightleftarrows$ cystine + 2H; in proteins, this amino acid often links two separate peptide chains together; it is also capable of forming a disulfide bridge within the same peptide chain. This amino acid does not have its own transfer RNA and is not incorporated into a polypeptide during ribosomal protein synthesis. It is, instead, formed after completion of the polypeptide by the oxidation of two cysteine residues. It is responsible for the formation of one kind of kidney stone.

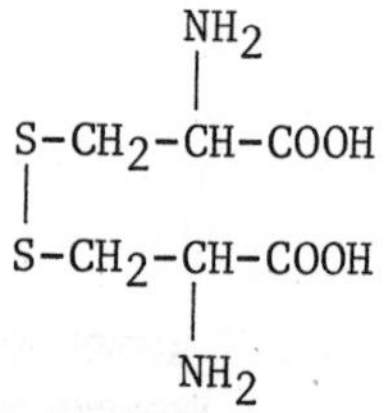

$$NH_2$$
$$S-CH_2-CH-COOH$$
$$S-CH_2-CH-COOH$$
$$NH_2$$

5. **Methionine (Met)**. An essential amino acid for humans; responsible for donating methyl groups in biosynthetic reactions involving transmethylation.

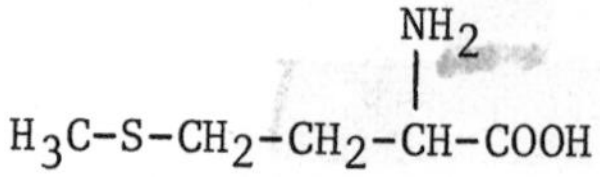

$$NH_2$$
$$H_3C-S-CH_2-CH_2-CH-COOH$$

(Note: 3, 4, and 5 are also referred to as sulfur containing amino acids.)

6. **Leucine (Leu).** One of the branched chain amino acids; the R-group is hydrophobic (water hating) and interacts with other hydrophobic substances (including other hydrophobic amino acid R-groups). It is an essential amino acid for man.

$$
\begin{array}{c}
CH_3 \\
\diagdown \\
CH-CH_2-CH-COOH \\
\diagup \qquad\quad | \\
CH_3 \qquad\quad NH_2
\end{array}
$$

R-Group

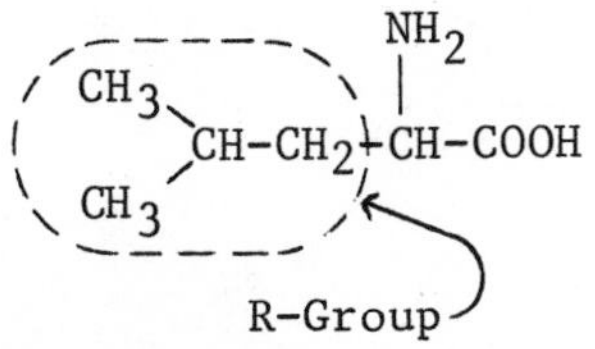

7. **Isoleucine (Ile).** An essential amino acid for man; hydrophobic sidechain.

$$
\begin{array}{c}
CH_3 \\
\diagdown \\
CH_2 \\
\diagdown \quad\quad NH_2 \\
CH-CH-COOH \\
\diagup \\
CH_3
\end{array}
$$

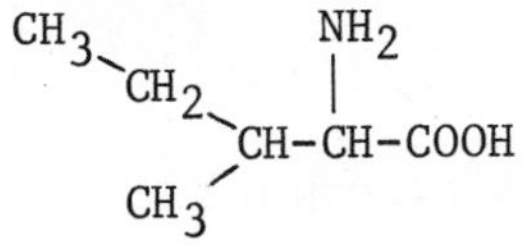

8. **Valine (Val).** An essential amino acid for man; hydrophobic side chain.

$$
\begin{array}{c}
\qquad\qquad NH_2 \\
CH_3 \qquad | \\
\diagdown \\
CH-CH-COOH \\
\diagup \\
CH_3
\end{array}
$$

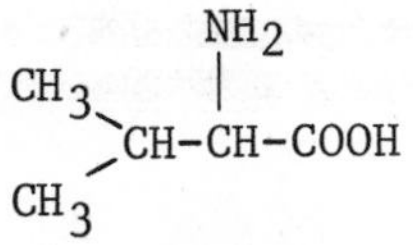

(Note: 6, 7, and 8 are known as branched chain amino acids with hydrophobic R-groups. A defect in the metabolism of these three compounds is the cause of maple syrup urine disease.)

9. **Serine (Ser).** In phosphoproteins, the phosphate is linked through serine; involved in the active center of many enzymes.

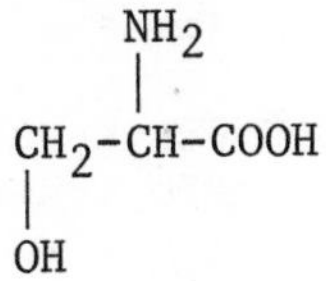

$$
\begin{array}{c}
NH_2 \\
| \\
CH_2-CH-COOH \\
| \\
OH
\end{array}
$$

10. **Threonine (Thr).** An essential amino acid for man.

$$
\begin{array}{c}
NH_2 \\
| \\
H_3C-CH-CH-COOH \\
| \\
OH
\end{array}
$$

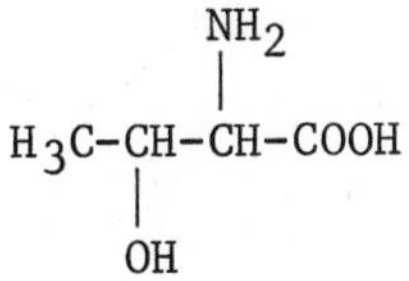

(Note: 9 and 10 are hydroxyl group-containing amino acids.)

11. **Phenylalanine (Phe).** An essential amino acid in man; its metabolism is defective in phenylketonuria; hydrophobic sidechain.

$$\text{C}_6\text{H}_5-\text{CH}_2-\underset{\underset{\text{NH}_2}{|}}{\text{CH}}-\text{COOH}$$

12. **Tyrosine (Tyr).** Accumulates in tissues in tyrosinosis and tyrosinemia; analysis for the phenol ring is used to quantitate proteins by Folin's method. It is involved in the synthesis of thyroxine, catecholamines, and melanin.

$$\text{HO}-\text{C}_6\text{H}_4-\text{CH}_2-\underset{\underset{\text{NH}_2}{|}}{\text{CH}}-\text{COOH}$$

13. **Tryptophan (Trp).** Essential amino acid in man; indole ring system is involved in the formation of serotonin; metabolites of tryptophan are involved in carcinoid disease (an epithelial growth resembling cancer).

$$\text{(indole)}-\text{CH}_2-\underset{\underset{\text{NH}_2}{|}}{\text{CH}}-\text{COOH}$$

(Note: 11, 12, and 13 are known as aromatic amino acids.)

14. **Aspartic Acid (Asp).** Substrate for aspartate aminotransferase (also known as glutamic oxaloacetic transaminase or GOT), an enzyme of clinical significance.

$$\text{HOOC}-\text{CH}_2-\underset{\underset{\text{NH}_2}{|}}{\text{CH}}-\text{COOH}$$

15. **Glutamic Acid (Glu).**   Substrate for both aspartate and alanine
    aminotransferases.   *GOT + GPT*

$$\underset{\displaystyle HOOC-CH_2-CH_2-\overset{\displaystyle NH_2}{\overset{\displaystyle |}{CH}}-COOH}{}$$

Note: The above amino acids (14 and 15) are known as acidic amino
        acids and are often present in proteins as the corresponding
        amides, asparagine and glutamine.

$$\overset{\displaystyle O}{\overset{\displaystyle \|}{(H_2N)-C}}-CH_2-\overset{\displaystyle NH_2}{\overset{\displaystyle |}{CH}}-COOH \qquad\qquad \overset{\displaystyle O}{\overset{\displaystyle \|}{(H_2N)-C}}-CH_2-CH_2-\overset{\displaystyle NH_2}{\overset{\displaystyle |}{CH}}-COOH$$

Asparagine (Asn)   Glutamine (Gln)

Note: Asn and Gln have their own specific transfer RNA carriers.

Glutamine is an intermediate in a number of important reactions.
The γ-amido nitrogen, derived from ammonia, can be utilized in the
de novo synthesis of purine and pyrimidine nucleotides, converted
to urea in the liver, or released as NH_3 in the kidney tubular
epithelial cells. The last reaction, catalyzed by the enzyme
glutaminase, plays an important role in acid-base regulation by
neutralizing acid in the urine (see Chapter 12).

16. **Arginine (Arg).** Involved in urea synthesis; has a guanido group.

$$\overset{\displaystyle H}{\overset{\displaystyle |}{H_2N-C-N}}-(CH_2)_3-\overset{\displaystyle NH_2}{\overset{\displaystyle |}{CH}}-COOH$$
$$\underset{\displaystyle NH}{\overset{\displaystyle \|}{}}$$

Guanido group

17. **Lysine (Lys).**

$$H_2N-\overset{\displaystyle H_2}{\underset{\displaystyle \epsilon}{C}}-\overset{\displaystyle H_2}{\underset{\displaystyle \delta}{C}}-\overset{\displaystyle H_2}{\underset{\displaystyle \gamma}{C}}-\overset{\displaystyle H_2}{\underset{\displaystyle \beta}{C}}-\overset{\displaystyle NH_2}{\underset{\displaystyle \alpha}{\overset{\displaystyle |}{CH}}}-COOH$$

The end NH_2 group is referred to as the epsilon (ε) NH_2 group, since it is attached to the fifth carbon (ε) from the carboxyl group; an essential amino acid in man. δ-Hydroxylysine is found in collagen and is formed postribosomally. Lysine is an essential amino acid.

In some glycoproteins, sugar residues are attached covalently to hydroxylysine residues. Lysine residues also participate in intramolecular cross-links in collagen and elastin. In these molecules, the ε-carbon is oxidized to an aldehyde

$$-CH_2-CH_2-CH_2-CH_2-NH_2 \longrightarrow -CH_2-CH_2-CH_2-C\overset{\displaystyle O}{\underset{\displaystyle H}{<}}$$

by lysyl oxidase. The aldo group can then undergo several types of condensation reactions with other amino acid residues to produce the cross-links. In elastin some of the cross-linking amino acids produced are dehydrolysinonorleucine, lysinonorleucine, merodesmosine, dehydromerodesmosine and desmosine (see elastin synthesis, Chapter 5). Lysine undergoes another covalent reaction in the formation of fibrin, a reaction essential for the clotting of blood. In this case, a peptide bond is formed between glutamine and lysine side chains of two different molecules of fibrin.

$$Fibrin-CH_2-CH_2-\overset{\displaystyle O}{\overset{\displaystyle \|}{C}}-NH_2 \; + \; H_3\overset{+}{N}-CH_2-CH_2-CH_2-CH_2-Fibrin$$

$$\longrightarrow Fibrin-CH_2-CH_2-\overset{\displaystyle O}{\overset{\displaystyle \|}{C}}-\underset{\displaystyle H}{N}-CH_2-CH_2-CH_2-CH_2-Fibrin \; + \; NH_4^+$$

This reaction is catalyzed by the enzyme transamidase. Deficiency of the transamidase leads to bleeding disorders. In a number of proteins the ε-NH_2 group of a lysyl residue is covalently linked (through an amide bond or Schiff base) to compounds that play a role in catalysis. Some examples are:

a. Biotin in carboxylating enzymes (biotinyllysine).

b. Lipoic acid in α-keto acid oxidative decarboxylases (pyruvate, α-ketoglutarate, branched chain keto acid decarboxylases).

c. Opsin, a protein present in the rod cells of the retina, has specific lysyl residues which can form a Schiff-base linkage to 11-cis-retinal (a derivative of vitamin A). This opsin-11-cis-retinal conjugate is called rhodopsin. Light striking these cells triggers a series of conformational changes leading to hydrolysis of the Schiff-base linkage, converting 11-cis-retinal to 11-trans-retinal, and releasing opsin. These reactions are primary events in the process of visual excitation.

d. Many enzymes in amino acid metabolism (transaminases in particular) use pyridoxal phosphate, bound to the ε-NH_2 group of a lysyl residue, as a cofactor. This Schiff-base linkage is usually broken and remade during the functioning of the enzyme in catalysis.

18. Histidine (His):

$$HC \!=\!\!=\!\! C\text{-}CH_2\text{-}\!\!\!\!\overset{\displaystyle NH_2}{C}\!H\text{-}COOH$$

has an imidazole ring attached to the β-carbon of alanine. It functions in the hemoglobin molecule as a proton transporter from the tissues to the lungs as well as in the functioning of the heme group. Histidine participates as a nucleophile or general base in the active center of many enzymes. The decarboxylated product of histidine is histamine, which is produced in several cell types, including the mast cells and non-mast cells. In man, mast cells, found in loose connective tissue and capsules particularly around blood vessels, produce histamine. In blood, histamine is synthesized exclusively in the basophils. In the oxyntic (acid-producing) gland portion of the stomach, histamine appears to be stored in cells which share similarities with polypeptide hormone-producing endocrine cells. Histamine has two sets of actions depending on the receptors (H_1 or H_2) present on the target cells. Constriction of bronchial smooth muscles and dilation of capillaries (resulting in edema and a drop in blood pressure) are examples of H_1 receptor activity, which can be inhibited by classical antihistamines. Stimulation of gastric secretion, inhibition of rat uterine muscle contraction and stimulation of isolated atria are mediated through H_2 receptors. These effects can be antagonized by a new class of inhibitors exemplified by burimamide and metiamide. Note 16, 17, 18 are basic amino acids.

19. Proline (Pro). Present in collagen; a heterocyclic amino acid containing a pyrrolidine ring.

$$H_2C \overset{4}{\rule{2cm}{0.4pt}} \overset{3}{} CH_2$$

(proline structure: ring with H_2C at position 4, CH_2 at position 3, H_2C at position 5, $C\text{-}COOH$ with H at position 2, N with H at position 1)

Proline (and hydroxyproline) residues act as turning or disrupting points in the formation of α-helices in a protein molecule. However, the proline ring does not prevent other types of helical structures. Also note—proline, when present as a part of a peptide, has no additional H on the N to participate in hydrogen bond formation.

20. 4-Hydroxyproline. Present in collagen. The isomeric compound 3-hydroxyproline is also found in collagen. Hydroxylation reactions of some of the proline residues present in collagen are postribosomal modifications. Urine levels reflect bone matrix metabolism, and determination of this amino acid has been suggested as an alternative to alkaline phosphatase isoenzyme fractionation as a way of differentiating between bone and liver diseases in some patients.

(4-hydroxyproline structure: H on top, $HO\text{-}C$ at position 4, CH_2 at position 3, H_2C at position 5, $C\text{-}COOH$ with H at position 2, N with H at position 1)

Note: Although 19 and 20 are sometimes referred to as imino acids this is incorrect since they do not contain an imino ($>C=NH$) group. They are more nearly secondary amines and the term *imino acid* should be reserved for the intermediates formed during amino acid deamination (see pages 574-578).

Miscellaneous Amino Acids

1. β-Alanine. Part of pantotheine and of coenzyme A.

$$\overset{\beta}{CH_2}\text{-}\overset{\alpha}{CH_2}\text{-}COOH$$
$$|$$
$$NH_2$$

2. Taurine. Comes from metabolism of the S-containing amino acids; conjugated with bile acids in the liver.

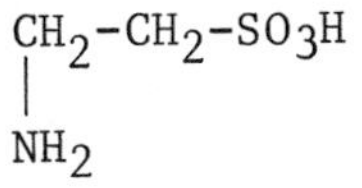

$$CH_2-CH_2-SO_3H$$
$$|$$
$$NH_2$$

3. α-Aminobutyric Acid.

$$NH_2$$
$$|$$
$$H_3C-CH_2-CH-COOH$$

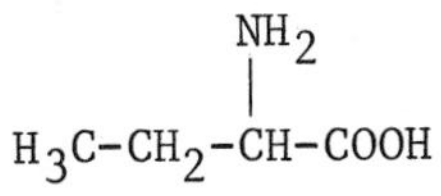

4. γ-Aminobutyric Acid. Present in brain tissue, may be a synaptic transmitter.

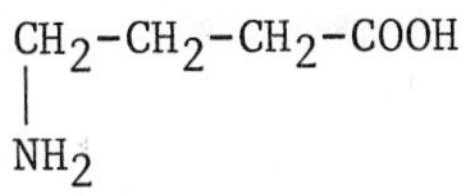

$$CH_2-CH_2-CH_2-COOH$$
$$|$$
$$NH_2$$

5. β-Aminoisobutyric Acid. End-product in pyrimidine metabolism; found in urine of patients with an inherited error in this metabolism.

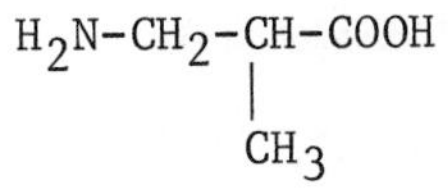

$$H_2N-CH_2-CH-COOH$$
$$|$$
$$CH_3$$

6. Homocysteine. Involved in methionine biosynthesis. Demethylated methionine.

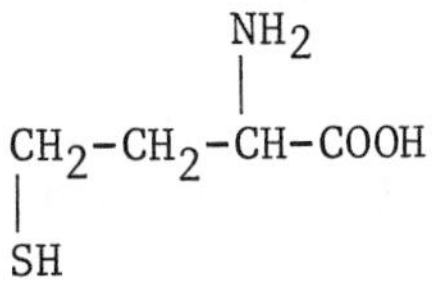

$$NH_2$$
$$|$$
$$CH_2-CH_2-CH-COOH$$
$$|$$
$$SH$$

7. Homoserine. Involved in threonine, aspartate, and methionine metabolism.

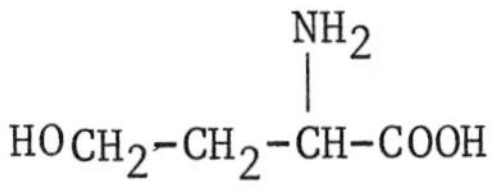

$$NH_2$$
$$|$$
$$HOCH_2-CH_2-CH-COOH$$

8. Cysteinesulfinic Acid. Present in rat brain tissue.

$$\underset{\displaystyle SO_2H}{\overset{\displaystyle NH_2}{CH_2-CH-COOH}}$$

9. Cysteic Acid. Present in wool.

$$\underset{\displaystyle SO_3H}{\overset{\displaystyle NH_2}{CH_2-CH-COOH}}$$

10. Ornithine. A urea cycle intermediate.

$$\underset{\displaystyle NH_2}{\overset{\displaystyle NH_2}{CH_2-CH_2-CH_2-CH-COOH}}$$

11. Citrulline. A urea cycle intermediate.

H$_2$C-CH$_2$-CH$_2$-CH-COOH
| |
HN   NH$_2$
|
C=O
|
NH$_2$

12. Homocitrulline. Present in urine of normal children.

H$_2$C-CH$_2$-CH$_2$-CH$_2$-CH-COOH
| |
HN   NH$_2$
|
C=O
|
NH$_2$

13. 5-Hydroxytryptophan. Decarboxylated product is serotonin, or 5-hydroxytryptamine. It is present in the central nervous system as well as in the intestinal mucosa.

14. Monoiodotyrosine. Present in thyroid tissue and blood serum.

15. 3,5-Diiodotyrosine. Found in association with thyroid globulin.

16. 3,5,3'-Triiodothyronine. Designated T_3; thyroid hormone present in thyroid tissue.

17. Thyroxine (3,5,3',5'-Tetraiodothyronine). Designated T_4; less
     active than T_3, has intrinsic biologic activity although in tissues
     there is evidence that T_4 is converted in part to T_3 by
     monodeiodination of the outer ring. T_3 and T_4 are thyroid hormones.

18. Reverse T_3 (RT_3; 3,3',5'-triiodothyronine). Obtained by the
     monodeiodination of the inner ring of T_4. Appears to show no
     metabolic potency; present in high concentrations in human
     umbilical cord blood.

19. Azaserine. A potent inhibitor of tumor growth. It is not a
     naturally occurring amino acid.

$$N\!\equiv\!N\!=\!CH\!-\!\underset{\underset{O}{\|}}{C}\!-\!O\!-\!CH_2\!-\!\underset{\underset{NH_2}{|}}{CH}\!-\!COOH$$

20. 2-Pyrrolidone-5-Carboxylic Acid (5-oxoproline; pyroglutamic acid).
     A cyclic lactam form of glutamic acid. It is found in several
     peptide hormones including thyrotrophin releasing factor; may be
     involved as an intermediate in amino acid transport via the
     γ-glutamyl cycle.

21. Hypoglycin A (L-α-amino-methylenecyclopropylpropionic acid). This unusual amino acid is present in unripe akees which are fruits from a tree (akee tree) indigenous to West Africa and introduced to Jamaica and other Antilles. When the unripe fruit is eaten, it produces toxic effects, namely vomiting (hence the disorder is called Jamaican Vomiting Sickness), central nervous system depression and hypoglycemia, which can be fatal (for mechanism of action, see hypoglycemia--Chapter 4).

$$H_2C=C—CHCH_2CH-COOH$$

with CH_2 bridging the cyclopropyl ring and NH_2 on the α-carbon.

Hypoglycin B is the γ-glutamyl conjugate of hypoglycin A and is less toxic.

Chemical Properties of Amino Acids Due to Carboxyl Group

1. Salt formation and titration (discussed above under electrolytes).

2. Formation of esters and amides

$$
\underset{\substack{\text{amino}\\\text{nitrogen}}}{R-CH-COOH \atop NH_3^+Cl^-} + C_2H_5OH \underset{+H_2O}{\overset{-H_2O}{\rightleftharpoons}} \underset{NH_3^+Cl^-}{R-CH-\overset{O}{\overset{\|}{C}}-O-C_2H_5} \xrightarrow[\text{NH}_3]{\text{Excess}} \underset{NH_2}{R-CH-\overset{O}{\overset{\|}{C}}-NH_2} + C_2H_5OH
$$

     amino acid + alcohol   Ester   Amide

(amide nitrogen indicated on the amide product)

3. Formation of aminoacyl chlorides

$$
\underset{R'-NH}{R-CH-COOH} \xrightarrow{PCl_5} \underset{R'-NH}{R-CH-COCl}
$$

where R' is one of several groups which can be attached to the α-amino group. This blocks the amino group so that the aminoacyl chloride formed will not react with it.

4. Decarboxylation

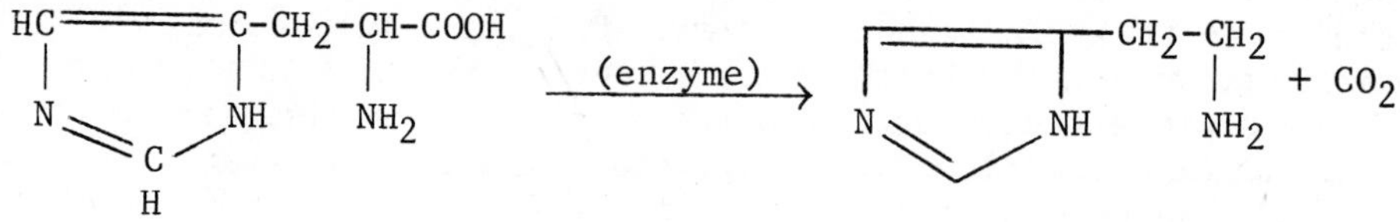

$$HC{=\!=\!=\!=}C-CH_2-CH-COOH \qquad \xrightarrow{\text{(enzyme)}} \qquad \boxed{}-CH_2-CH_2 \quad + \; CO_2$$

Histidine   Histamine

Chemical Properties Due to NH$_2$ Group

1. Acylation

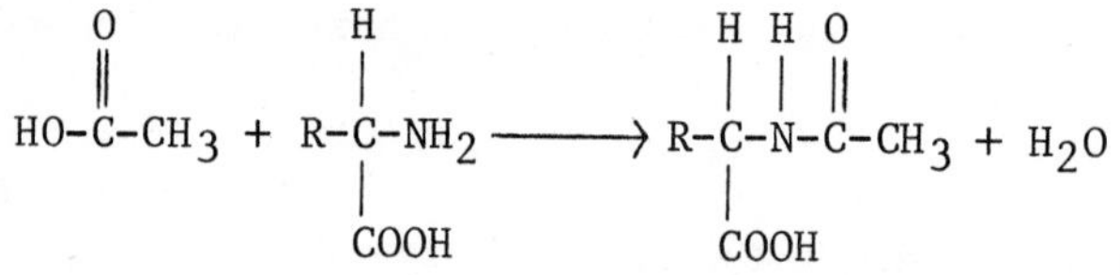

$$HO-\overset{\overset{\displaystyle O}{\|}}{C}-CH_3 \; + \; R-\overset{\overset{\displaystyle H}{|}}{\underset{\underset{\displaystyle COOH}{|}}{C}}-NH_2 \longrightarrow R-\overset{\overset{\displaystyle H}{|}}{\underset{\underset{\displaystyle COOH}{|}}{C}}-\overset{\overset{\displaystyle H}{|}}{N}-\overset{\overset{\displaystyle O}{\|}}{C}-CH_3 \; + \; H_2O$$

2. Benzoylation

$$\bigcirc\!\!-COOH \; + \; H_2N-CH_2-COOH \xrightarrow{-H_2O} \bigcirc\!\!-\overset{\overset{\displaystyle O}{\|}}{C}-\overset{\overset{\displaystyle H}{|}}{N}-CH_2-COOH$$

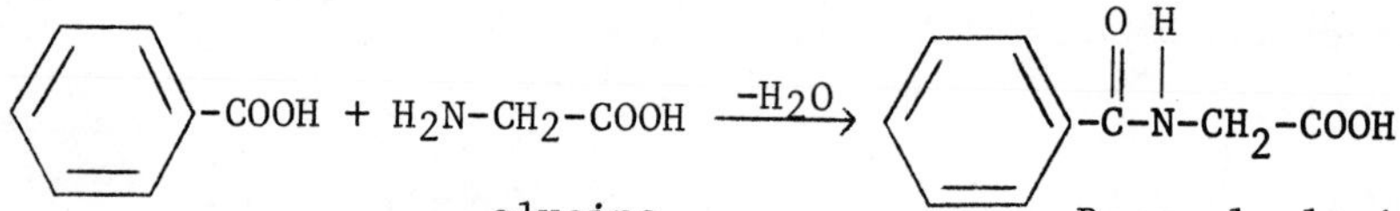

glycine   Benzoyl glycine
(Hippuric acid)

3. Methylation

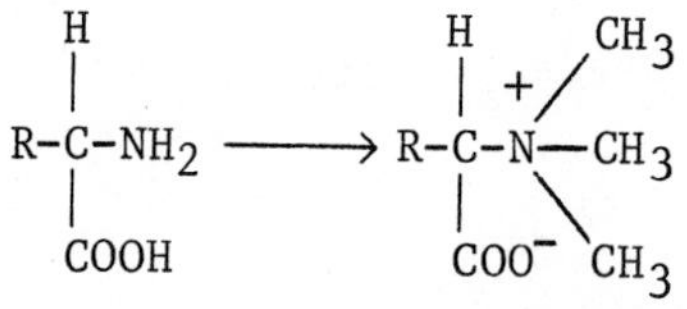

$$R-\overset{\overset{\displaystyle H}{|}}{\underset{\underset{\displaystyle COOH}{|}}{C}}-NH_2 \longrightarrow R-\overset{\overset{\displaystyle H}{|}}{\underset{\underset{\displaystyle COO^-}{|}}{C}}-\overset{+}{N}\!\!\underset{CH_3}{\overset{\diagup CH_3}{\diagdown CH_3}}$$

Betaine of an amino acid

4. Sanger's reagent (1-fluoro-2,4-dinitrobenzene, FDNB) reacts with free amino end groups to form dinitrophenylamino acids (DNP a.a.) which are yellow. This reaction is used to identify terminal amino acids:

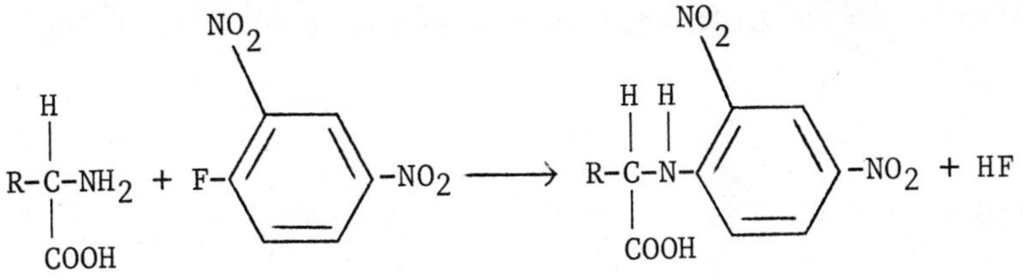

DNP – amino acid

5. Reaction with nitrous acid (HNO_2). This reaction is the basis of the amino nitrogen method of Van Slyke.

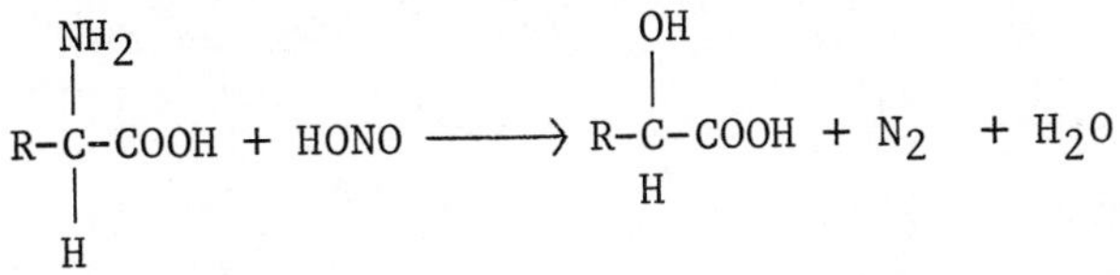

6. Oxidative Deamination

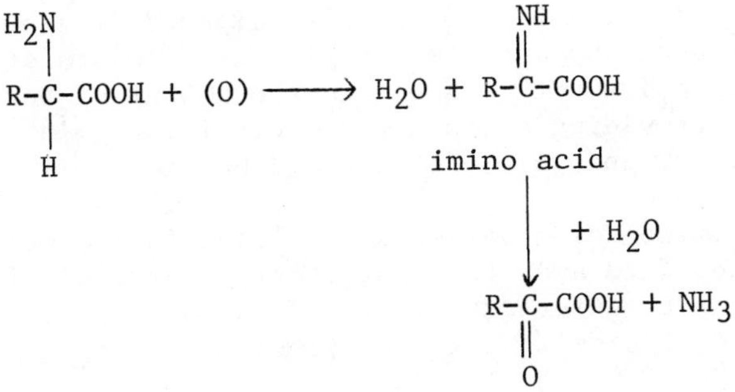

imino acid

$$+ H_2O$$

α–keto acid

7. Reaction with Formaldehyde

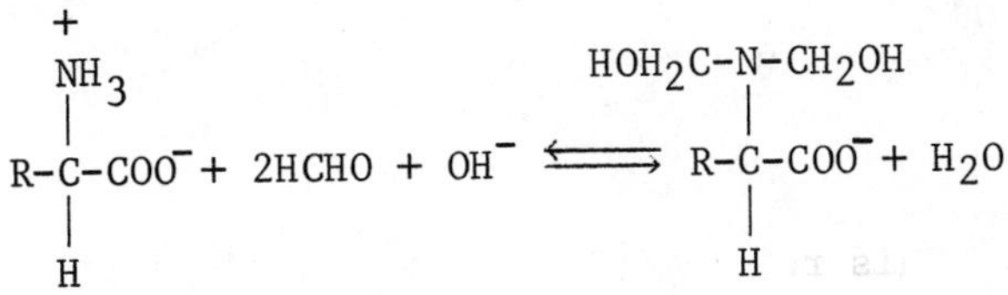

dimethylol amino acid

8. Reaction with aromatic aldehydes in the presence of alkali (Schiff base formation)

Schiff base

9. Ninhydrin (triketohydrindene hydrate) reacts with α-amino acids to produce CO_2, ammonia, an aldehyde one carbon smaller than the amino acid and, in most cases, a blue or purple compound (the exceptions are proline and hydroxy-proline, which give a yellow color). The reaction is quantitative with respect to the amount of both color and CO_2 produced, providing a basis for determining amino acids. Ammonia, some amines, and some proteins and peptides will also give the color reaction (but not the CO_2), so that the method is not specific for amino acids unless CO_2 release is measured or the amino acids are purified away from interfering materials (the usual procedure). The color reaction is used to determine amino acids on some automatic amino acid analyzers.

Ninhydrin α-Amino Acid   Hydrindantin

Diketohydrindylidene-diketohydrindamine
(Purple Compound)

10. Reaction with CO_2 to form a carbamino group

11. Chelation of amino acids with metal ions

Copper diglycinate

Chelates are nonionic. Amino acids and other chelate formers
(e.g., EDTA, penicillamine) form soluble metal complexes. Cu^{2+}
chelates of peptide bonds are the basis of the biuret measurement
of protein. This is the best general method for protein but it
is not as sensitive as other techniques.

12. Peptide Linkage. This is the most common linkage which attaches
one amino acid to another and is the result of a dehydration
reaction between the α-amino group of one a.a. and the carboxyl
group of another:

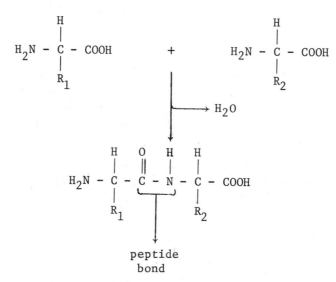

Amino Acid Analysis, Sequence Determination and Chemical Synthesis
of Proteins

1. In order to understand the structure and function of proteins, it
is essential to know

    a. the number and kinds of amino acids present in the protein, and
    b. the order in which the amino acids are connected (called
        sequence or primary structure).

This information is needed, for example, in assessing the effect
of mutations on the amino acid sequence; understanding the
mechanisms of enzyme-catalyzed reactions; synthesis of peptides to
obtain specific therapeutic effects. In the last example, the use
of synthetic, species-specific peptides may eliminate undesirable
hypersensitivity reactions.

2. The first protein sequenced was bovine insulin (M.W. 6,000) in 1955 by Sanger. This molecule is composed of two peptide chains of 21 and 30 amino acids each, linked by two interchain disulfide bonds. This work was significant not only for the methodology which it established, but also because it demonstrated that proteins are characterized by unique primary structures.

3. In 1960, Hirs, Moore, and Stein, and Anfinsen reported the first primary structure of an enzyme. The enzyme was ribonuclease (M.W. 13,700) containing a single peptide chain of 124 amino acid residues with four intrachain disulfide bonds. These investigators established many of the procedures which are currently used in sequence analysis. These include the use of ion exchange resins for the separation of peptides and amino acids, and their quantitation by the ninhydrin reaction.

4. A general approach to the determination of primary protein structure is shown below.

    a. Determine amino acid composition of <u>highly purified</u> protein.

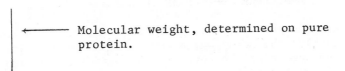

                    Molecular weight, determined on pure protein.

    b. Use composition and molecular weight to determine the number of residues of each amino acid present per protein molecule to the nearest whole number.

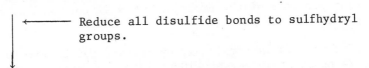

                    Reduce all disulfide bonds to sulfhydryl groups.

    c. Determine N- and C-terminal amino acids. A unique residue for each terminus suggests that the native protein contains only one peptide chain. (If, for example, two different amino acids are found in equimolar quantities when the pure protein is treated with carboxypeptidase, this would imply at least two different chains, present in equimolar amounts, each having a different C-terminal amino acid. In this case, one might expect to find two amino terminal residues also.)

                    Enzymic and chemical hydrolysis of peptide bonds.

d. <u>Different</u>, preferably overlapping sets of smaller peptides are obtained next. Each of these is sequenced and, by looking at the overlapping parts, the sequence of the original protein is assembled (see example next). Most of the actual sequencing is done by the Edman reaction, a procedure which results in the stepwise removal of amino acids from the amino terminus of a peptide.

Example - overlapping peptide mapping

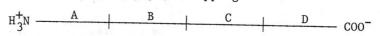

Cleavage of a protein gave four peptides: A, B, C, and D. When these were separated and sequenced, it was not known whether the correct order was ABCD or ACBD. A second hydrolysis (by a different method) was done giving three peptides

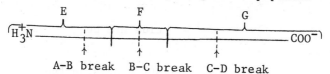

A-B break   B-C break   C-D break

By looking at the sequence of fragments E, F, and G containing, respectively, the bonds which were cleaved in the first hydrolysis, it was deduced that the correct order was ABCD. This was the only way in which the proper sequence could be maintained in going from A to B, B to C, and C to D.

5. Determination of Amino Acid Composition. The a.a. composition of a protein is the number of each of the 20 common amino acids present per mole of the protein. This is obtained by the following procedure:

a. The protein is completely hydrolyzed using acid (6 N HCl, 24 hours or longer at $110^{\circ}C$) into a mixture of its constituent amino acids and the resultant hydrolysate is evaporated to dryness. The lability of some amino acids (notably tryptophan, serine, and threonine) requires alkaline hydrolysis for the determination of these compounds (performed on a separate sample of the protein).

b. The hydrolysate is dissolved in a small volume of an acidic buffer to obtain the protonated form of the amino acids. It is then placed on the top of a cation exchange resin column (Dowex50). The resin is an insoluble synthetic polymer containing $-SO_3^-$ Na^+ groups. It can bind to a protonated amino acid with the release (in this case) of NaCl.

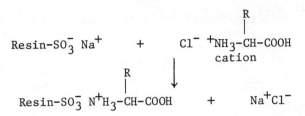

$$\text{Resin-SO}_3^- \text{ Na}^+ \quad + \quad \text{Cl}^- \; ^+\text{NH}_3\text{-CH-COOH}$$
$$\text{cation}$$

$$\text{Resin-SO}_3^- \; \text{N}^+\text{H}_3\text{-CH-COOH} \quad + \quad \text{Na}^+\text{Cl}^-$$

Thus all of the (cationic) amino acids bind to the resin rapidly and stay at the top of the column. The intensity of binding to the resin is dependent upon many factors. These include the number of positive charges on the molecule, the nature and size of the R-group (affecting the adsorption of the amino acids onto resin) and the pK_a of the group involved in the interaction. For example, basic amino acids (lysine, histidine, and arginine) all have more than one positive charge and, hence, bind more tightly than the other amino acids. On the other hand, acidic amino acids are bound less tightly.

c. In addition to the ionic binding to the $-SO_3^-$ groups on the resin, the amino acids are held by hydrophobic and van der Waals interactions. The nature of these bonds helps to determine the resin adsorption characteristics of amino acids having different sidechains. This means that, depending on the R-group, various amino acids bind with different strengths and are released at different times during the elution.

d. The amino acids (cations) from the column are differentially eluted by using a stepwise pH gradient varying between pH 3 and pH 5.

e. As each amino acid appears in the effluent volume, it is reacted with ninhydrin and the color intensity is measured spectrophotometrically to detect and quantitate the amino acid.

f. Using ribonuclease hydrolysate, the amino acids are eluted from the column in the order: Asp, Thr, Ser, Pro, Glu, Gly, Ala, Val, Cys, Met, Ile, Leu, Tyr, Phe, ammonia (produced in the hydrolysis of glutamine and asparagine, each of which have one acid amide $-CONH_2$ group), Lys, His, and Arg.

g. The process of separation and quantitation of amino acid has been automated. In this method, the acidic and neutral amino acids are separated on a "long" column and the basic amino acids on a "short" column. On the short column, the acidic and neutral amino acids emerge unseparated prior to the elution of the basic components. (Recall that the basic amino acids are more tightly bound than the rest of the amino acids.) The analyzer is capable of detecting as little as 10 nanomoles of an amino acid and a complete analysis can be obtained in about four hours.

6. Determination of Sequence

a. Determination of the terminal residues. The terminal residues may be identified by both chemical and enzymatic methods. The C-terminal residue can be determined by hydrazinolysis, a method which liberates the hydrazide derivatives of all amino acids whose carboxyl group is involved in a peptide linkage. The C-terminal residue is liberated as the free amino acid. The carboxypeptidases are enzymes which digest proteins from the carboxyl terminus and are often useful for determining this residue. The amino terminal residue is readily identified by reaction of the α-amino group with reagents such as fluorodinitrobenzene, dansyl chloride or phenylisothiocyanate. The derivatized amino acid is isolated by hydrolysis and may be characterized by thin layer chromatography. Enzymatically, leucine amino peptidase, an exopeptidase which attacks proteins from the amino terminus, is used in determining the N-terminal amino acid.

b. Cleavage of Disulfide Bonds. The disulfide bonds of a protein may be cleaved oxidatively or reduced and alkylated. Treatment of the native protein with performic acid, a powerful oxidizing agent, breaks disulfide bonds and converts cysteine residues to cysteic acid. Reduction of the disulfide linkage by thiols such as β-mercaptoethanol yields reactive sulfhydryl groups. These may be stabilized by alkylation with iodoacetate or ethyleneimine, yielding the carboxymethyl- or aminoethyl- derivatives, respectively.

c. Hydrolysis of Peptide Bonds. Specific hydrolysis of a protein is important in terms of reproducibility of an experiment and yield of the resulting peptides. This may be achieved both chemically and enzymatically. Reagents such as N-bromo-succinimide and cyanogen bromide will hydrolyze proteins at tryptophan and methionine residues, respectively. The most specific protease is trypsin, which hydrolyzes the peptide linkages following the basic residues lysine and arginine. The purification of the products of hydrolysis is often the most challenging aspect of sequence determination. Both anion and cation exchange resins are used extensively, due to the ionic nature of peptides. Preparative paper chromatography and electrophoresis are also useful. The purified peptides are analyzed for amino acid composition and terminal residues. Small peptides may be sequenced directly but large peptides must be further hydrolyzed. Proteases such as chymotrypsin, pepsin, and papain, which are much less specific than trypsin, are often utilized.

d. The sequences of purified peptides and proteins are determined by reaction with phenylisothiocyanate (Edman's reagent). This compound couples with the free alpha amino group of the N-terminal amino acid, giving the phenylthiocarbamyl (PTC) derivative of the peptide or protein. This is then cleaved to a peptide or protein with one less residue, and the thiazolinone of the N-terminal amino acid. This rearranges to the phenylthiohydantoin (PTH) derivative which is extracted from the reaction mixture, converted to a volatile derivative, and identified by gas chromatography. These reactions are illustrated below, using a tripeptide.

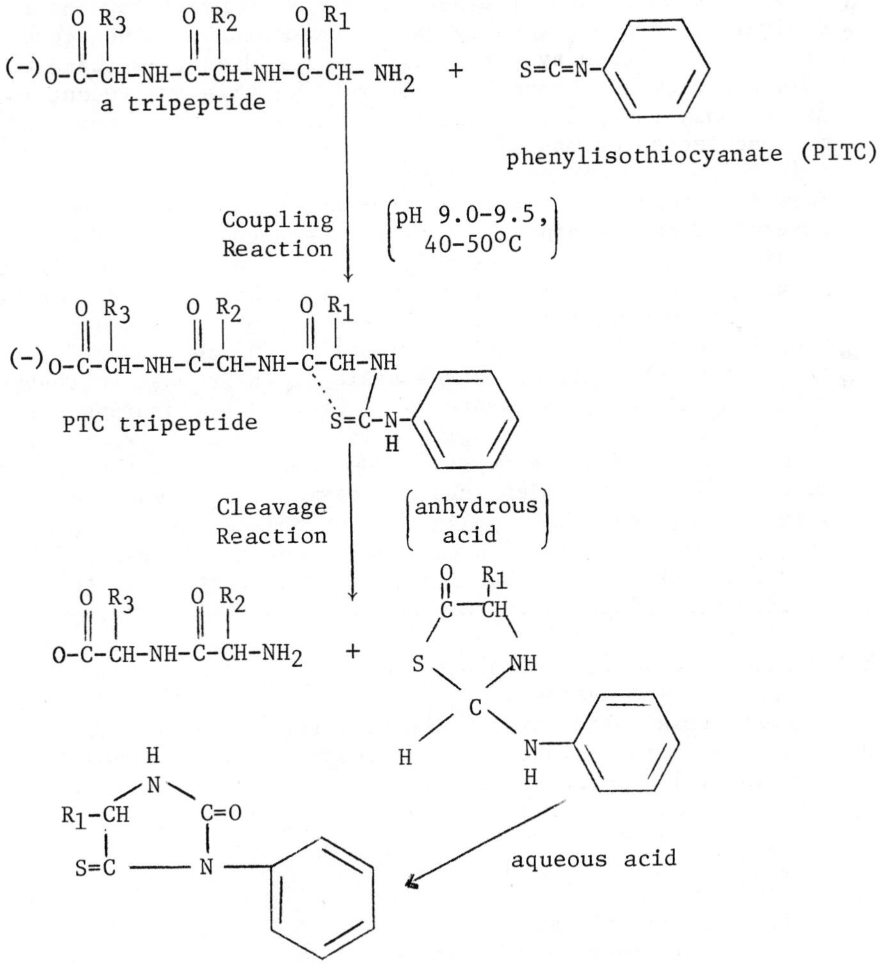

$$(-)O-\overset{\overset{\displaystyle O}{\|}}{C}-\overset{\overset{\displaystyle R_3}{|}}{CH}-NH-\overset{\overset{\displaystyle O}{\|}}{C}-\overset{\overset{\displaystyle R_2}{|}}{CH}-NH-\overset{\overset{\displaystyle O}{\|}}{C}-\overset{\overset{\displaystyle R_1}{|}}{CH}-NH_2 \quad + \quad S=C=N-\langle\ \rangle$$

a tripeptide

phenylisothiocyanate (PITC)

Coupling Reaction $\quad\left(\begin{array}{c}\text{pH } 9.0\text{-}9.5,\\ 40\text{-}50^{\circ}C\end{array}\right)$

$$(-)O-\overset{\overset{\displaystyle O}{\|}}{C}-\overset{\overset{\displaystyle R_3}{|}}{CH}-NH-\overset{\overset{\displaystyle O}{\|}}{C}-\overset{\overset{\displaystyle R_2}{|}}{CH}-NH-\overset{\overset{\displaystyle O}{\|}}{C}-\overset{\overset{\displaystyle R_1}{|}}{CH}-NH$$

PTC tripeptide $\qquad$ $S=\overset{}{C}-\underset{H}{N}-\langle\ \rangle$

Cleavage Reaction $\quad\left(\begin{array}{c}\text{anhydrous}\\ \text{acid}\end{array}\right)$

$$O-\overset{\overset{\displaystyle O}{\|}}{C}-\overset{\overset{\displaystyle R_3}{|}}{CH}-NH-\overset{\overset{\displaystyle O}{\|}}{C}-\overset{\overset{\displaystyle R_2}{|}}{CH}-NH_2 \quad + \quad$$

aqueous acid

phenylthiohydantoin (PTH) of the amino acid

e. The PTH derivative may also be converted to the dansyl derivative which is highly fluorescent and can be detected with greater sensitivity. Alternatively, the amino acid composition of the remaining peptide may be determined and the extracted N-terminal residue surmised by the difference from the original composition (subtractive analysis). The recovered peptide may be coupled again with the Edman reagent and the entire process repeated to identify the penultimate residue of the original peptide. The Edman degradation can be continued for about ten steps, depending on the properties of the particular peptide.

f. The utility of the Edman reaction has led to the development of the protein sequenator. This instrument is programmed to conduct these reactions continuously, delivering the PTH derivative at a rate of about one every 90 minutes. These derivatives are analyzed by gas chromatography. The sequenator has been used to verify the first 60 residues from the amino terminus of whale myoglobin. Currently, this instrument is useful only for proteins and large peptides; small peptides must be sequenced manually.

g. Mass spectrometry provides another approach to sequence determination. Although not widely used currently, the high sensitivity and inherent simplicity of this technique offers much promise. The method consists of bombarding peptides with electrons causing fragmentation at the peptide linkages and yielding a mixture of various pieces of the peptide. The fragments are separated on the basis of their mass to charge ratio and the original sequence is deduced by locating the products of sequential fragmentation. Thus peaks would be observed for the whole peptide, the peptide minus the terminal amino acid, the peptide minus the terminal and penultimate amino acids, etc. The application of this technique to mixtures of peptides, with on line computer analysis of the spectrum, offers the potential for simplification of the entire sequencing procedure.

7. The Chemical Synthesis of Proteins

a. In Merrifield Solid Phase Synthesis, the carboxyl group of the C-terminal amino acid of the peptide to be synthesized is covalently linked to a resin.

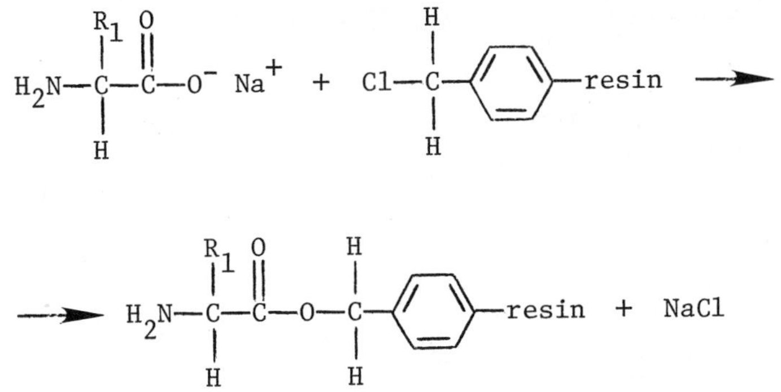

$$H_2N-\underset{\underset{H}{|}}{\overset{\overset{R_1}{|}}{C}}-\overset{\overset{O}{||}}{C}-O^-\ Na^+\ +\ Cl-\underset{\underset{H}{|}}{\overset{\overset{H}{|}}{C}}-\langle\ \rangle-resin\ \longrightarrow$$

$$\longrightarrow\ H_2N-\underset{\underset{H}{|}}{\overset{\overset{R_1}{|}}{C}}-\overset{\overset{O}{||}}{C}-O-\underset{\underset{H}{|}}{\overset{\overset{H}{|}}{C}}-\langle\ \rangle-resin\ +\ NaCl$$

b. The amino group of the next amino acid to be added is protected
with a t-BOC group and the carboxyl group is then coupled to
the amino acid bound to the resin.

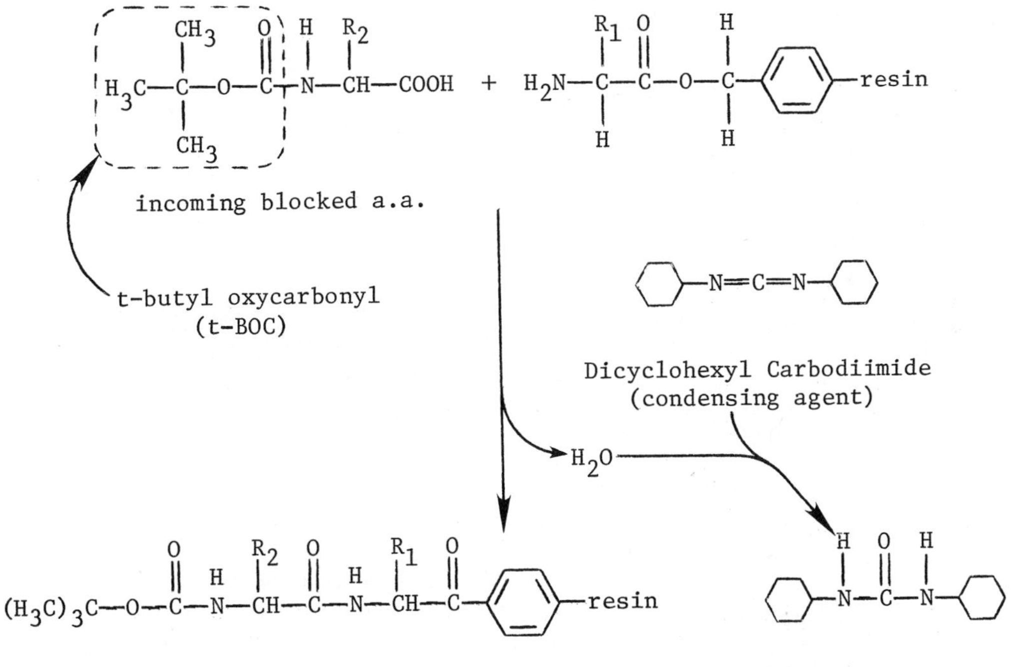

incoming blocked a.a.

t-butyl oxycarbonyl
(t-BOC)

Dicyclohexyl Carbodiimide
(condensing agent)

H_2O

$(H_3C)_3C-O-\overset{\overset{O}{||}}{C}-\underset{\underset{H}{|}}{N}-\underset{\underset{R_2}{|}}{CH}-\overset{\overset{O}{||}}{C}-\underset{\underset{H}{|}}{N}-\underset{\underset{R_1}{|}}{CH}-\overset{\overset{O}{||}}{C}-\langle\ \rangle-resin$

Dicyclohexyl urea

c. The protecting group (t-BOC) is removed by treatment with acid, which converts it to isobutylene and CO_2, both gaseous products. The free amino group is now ready for the next sequential addition.

isobutylene

After the last amino acid has been added, the peptide is cleaved from the resin. The advantages of this method include

1) quantitative yields of products
2) non-purification of intermediate peptides in the synthesis
3) amenable to automation
4) ease and rapidity.

Classical methods of peptide synthesis are also of importance. They can use the same reactions given above, but they are done in a homogeneous solution, rather than bound to insoluble resin beads.

Notable achievements in peptide synthesis include ribonuclease (124 amino acids) and insulin (2 chains of 21 and 30 amino acids, respectively). It is noteworthy that both have been synthesized by homogeneous, classical methods as well as by solid phase techniques, although the solid phase synthesis was much more rapid.

The Structure of Proteins

The forces that determine protein structure include

1. Covalent bonding, the sharing of an electron pair by two atoms, one electron (originally) coming from each atom. Bond energies are about 30–100 kcal/mole of bonds. Important examples in protein structure include peptide, disulfide, ester, and amide bonds. They can be between the reactive groups of amino acids in the same chain (intrachain) or in different polypeptide chains (interchain).

2. Coordinate covalent bonding, the sharing of an electron pair by two atoms, with both electrons (originally) coming from the same atom. These are similar to covalent bonds, but with much lower energies (4–5 kcal/mole of bonds). Consequently, they are much more labile (are made and broken much more readily). The electron pair donor is called a ligand or Lewis base and the acceptor is the central atom (since it frequently can accept more than one pair of electrons) or Lewis acid. These bonds are important in all interactions between transition metals and biological molecules, such as Fe^{+2} in hemoglobin and the cytochromes, and Co^{+3} in vitamin B_{12}.

3. Ionic forces, the coulombic attraction between two groups of opposite charge. The bond energy is less than 10 kcal/mole of bonds and is strongly dependent upon distance. These bonds are found in bonding between positive residues (α-ammonium, ε-ammonium, guanidinium, imidazolium) and negatively charged groups (ionized forms of α-carboxyl, β-carboxyl, γ-carboxyl, phosphate, sulfate).

4. Hydrogen bonding, the sharing of a hydrogen atom between two electronegative atoms which have unbonded electrons. The bond energy is 2–10 kcal/mole of bonds. These bonds are extremely important in water-water interactions and their existence explains many of the unusual properties of water and ice. In proteins, groups having a hydrogen atom which can be shared include ⊃N–H (peptide nitrogen, imidazole, indole), –OH (serine, threonine, tyrosine, hydroxyproline), $-NH_2$ and $-NH_3^+$ (arginine, lysine, α-amino), and –CONH (carbamino). Groups which can accept the sharing of a hydrogen include $-COO^-$ (aspartate, glutamate, α-carboxylate), –S–S– (disulfide), and ⊃C=O (in peptide and ester linkages).

5. Van der Waals' attractive forces, also called London forces. These operate between all atoms, ions, and molecules and are due to a fixed dipole in one molecule inducing an oscillating dipole in

another molecule by distortion of the charge (electron) cloud. The positive end of a fixed dipole will pull an electron cloud toward it, the negative end will push it away. The strength of these interactions is strongly dependent on distance, varying as $1/r^6$, where r is the interatomic separation. These bonds are particularly important in the non-polar interior structure of proteins, providing attractive forces between non-polar sidechains.

6. <u>Hydrophobic interactions</u> cause non-polar sidechains (aromatic rings, hydrocarbon groups) to cling together in polar solvents, especially water. These are not true "bonds" since there is no sharing of electrons between the groups involved. The groups are pushed together by their "expulsion" from the water. This is discussed more fully later in this chapter. The energy of these interactions is about 0.3-1.5 kcal/mole of interactions. These forces are also important in lipid-lipid interactions in membranes.

7. <u>Electrostatic repulsion</u>, between charged groups of like charge. These are just the opposite of ionic (attractive) forces above. They depend on distance and upon the charges of the interacting groups according to Coulomb's law: q_1q_2/r^2, where q_1 and q_2 are the charges and r is the interatomic separation.

8. <u>Van der Waals' repulsive forces</u>, operating between atoms at very short distances. They result from the induction of induced dipoles by the mutual repulsion of electron clouds. Since there is no involvement of a fixed dipole (as there was in Van der Waals' attractive forces), the distance dependence is even greater, $1/r^{12}$ in this case. These forces operate when atoms not actually bonded to each other try to approach more closely than a minimum distance. They are the underlying force in steric hindrance between all atoms.

Types of Protein Structure

There are several levels at which polypeptide structure is considered. The most basic one is the <u>primary structure</u>: the <u>number</u>, <u>kind</u>, and <u>order</u> of the amino acids in the chain. For example:

$$\begin{array}{c} H \\ | \\ H-N^+-Glu-Lys-Ala-Gly-Tyr-His-Ala-\overset{\displaystyle \overset{O}{\|}}{C}-O^- \\ | \\ H \end{array}$$

N-terminal amino acid   C-terminal amino acid

Note that the peptide is written starting with the amino acid having
a free α-amino group (N-terminus) and ending with the residue having
a free α-carboxyl group (C-terminus). This is a convention, and
naming is conducted in the same manner. The proper name for the
above peptide is

glutamyllysylalanylglycyltyrosylhistidylalanine.

The C-terminal residue is always given the name of the free amino
acid while all other residues have the root of the residue name
plus -yl (e.g., glutamate → glutamyl; tyrosine → tyrosyl).

The residues are linked covalently by peptide bonds. This is a very
important linkage and is illustrated below. The bond lengths and
angles are average values and will vary somewhat (but not much)
depending on the amino acids linked and the molecule of which they
are part.

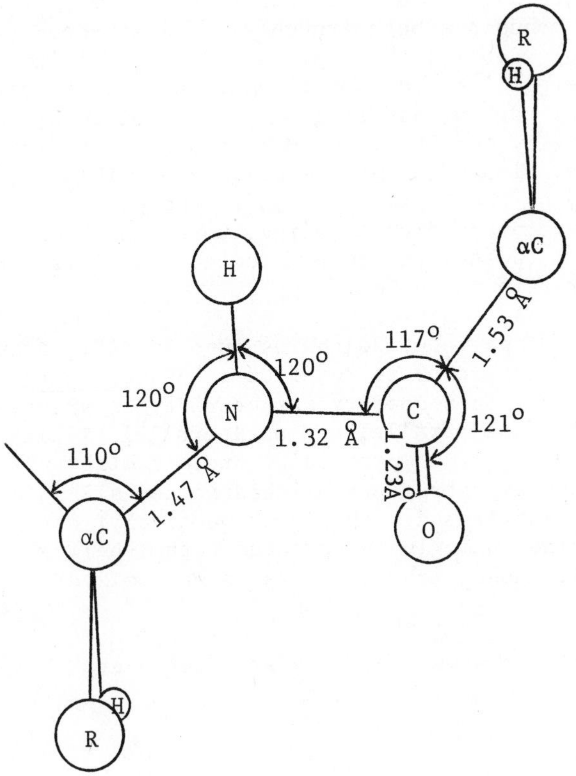

It is also important to notice that the peptide group is in a
planar, trans configuration, with very little rotation or twisting
around the bond linking the α-amino nitrogen of one amino acid and
the carbonyl carbon of the next one (the "peptide" bond). This

is due to an amido-imido tautomerization, lending partial double-bond character to the N-C bond. This is illustrated below, with the transition state, (II), being probably what actually exists in nature.

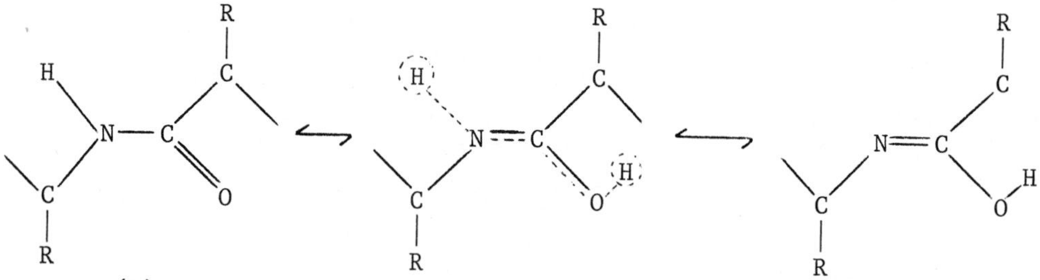

(I)   (II)   (III)
(amido form)   (imido form)

The α-amino proton is shared by the nitrogen and oxygen atoms, and the N-C and C-O bonds are both (roughly) "one-and-one-half" bonds (not single, not double). The planarity and rigidity follow from this, since there is no free rotation around any but single bonds. The nitrogen, carbon, and oxygen atoms involved are all partially sp^2, partially sp^3 hybrids. Many features of protein structure (α-helices, pleated sheets, etc.) are possible, due in part to the geometry of the peptide groups in the protein backbone. Based largely on these properties, Pauling, Corey and Branson in 1951 were the first to postulate the existence of helices and pleated sheets in protein molecules.

The folding of parts of polypeptide chains into specific structures held together by hydrogen bonds is referred to as _secondary structure_. The most common secondary structure types are the **right-handed α-helix, the parallel and antiparallel β-pleated** sheets, and the random coil. A particular protein might possess only one kind of secondary structure (α-keratin and silk protein consist entirely of α-helix and β-sheet, respectively) or it may have more than one kind (hemoglobin contains both α-helical and random coil regions). Most globular proteins have mixed structures.

The essential features of some secondary structures are shown below.

1. Random Coil

This term is misleading. The structure is random in the sense that there is no repeating pattern to the way in which each residue of the peptide chain interacts with other residues (i.e., the Nth residue bonded to the N-3 and N+3 residues, as

in an α-helix). However, given a particular sequence, there
is only one (or at most 2 or 3) ways in which it will coil
itself. This conformation will either have the minimum energy
or will be in a local energy minimum (see diagram below).

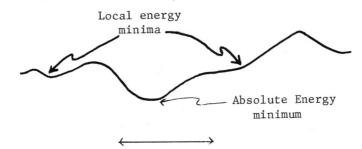

Local energy
minima

Energy of the
conformation
(increasing
upwards)

Absolute Energy
minimum

changes in molecular conformation move
the system along this axis

Since energy must be added to a molecule to make it change
conformation (move from a valley over a hill into another
valley in the diagram), the molecule can remain trapped in a
conformation corresponding to any minimum in the energy map,
even though it would prefer to be at the absolute minimum in
internal energy.

    This concept (of a molecule seeking a preferred, low-energy
shape) is the basis for postulating that the primary structure
(sequence) of a polypeptide determines the secondary and
tertiary structures of the molecule. There are two principal
snags to this hypothesis.

I.   If there is more than one peptide chain in the molecule
     (as in insulin), the way in which the chains associate may
     be determined by factors other than their amino acid
     sequence. (This is different than tertiary structure,
     discussed below, since the insulin chains cannot assume
     their normal secondary and tertiary structures separately
     from each other.)

II. If there are more than two cysteine residues in a
     polypeptide, there is more than one way in which cystine
     (-S-S-) bridges can form (i.e., cys 1 can form a disulfide
     bond with cys 2 or cys 3). Since these bridges are
     covalent bonds and can be formed under conditions where the
     weaker forces determining coiling are not operative, these
     forces will not be powerful enough to break the disulfide
     bonds and pull the molecule back to the true minimum energy
     structure.

2. A right-handed alpha helix is shown below. It is intended to illustrate hydrogen bonding and is <u>not</u> drawn to scale. It is shown wrapped around a cylinder for purposes of presentation.

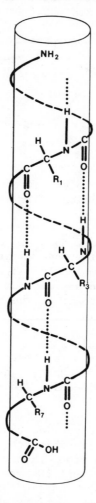

There are 3.6 amino acid residues in each complete turn of the helix (and, hence, $100°$ of turn per residue). Hydrogen bonds between coils of the helix form the "surface" of the structure (see diagram). If the C-terminus is considered down, an amino group always hydrogen bonds to a carboxyl group above it. Since carboxyl and amino groups alternate in the backbone, hydrogen bonds alternate up and down in the helix. The amino proton and carboxyl oxygen of a residue (N) are hydrogen bonded, respectively, to the carboxyl oxygen of residue (N−3) and the amino proton of residue (N+3). This is shown in the uncoiled, schematic diagram as follows.

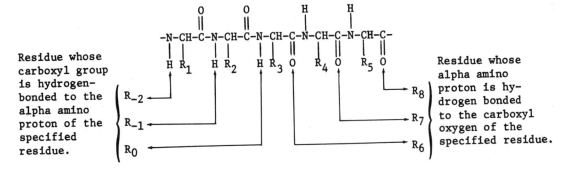

Residue whose carboxyl group is hydrogen-bonded to the alpha amino proton of the specified residue.

Residue whose alpha amino proton is hydrogen bonded to the carboxyl oxygen of the specified residue.

Note also that
a) hydrogen bonds are parallel to the helix axis.
b) the peptide groups are planar and in the trans configuration.
c) the R-groups (sidechains) are roughly perpendicular to the helix axis.

3. The beta (pleated sheet) structure is harder to visualize than is the α-helix. A side (edge-on) view of such a structure would appear as shown below. The hydrogen bonds, between the α-carbonyl oxygens and the α-amino protons, extend into and out of the page in planes perpendicular to the page. One important feature of this structure is that the R-groups (sidechains) are above and below the sheets and nearly perpendicular to them. Amino acids with bulky R-groups tend to not form pleated sheets because their sidechains interfere with each other.

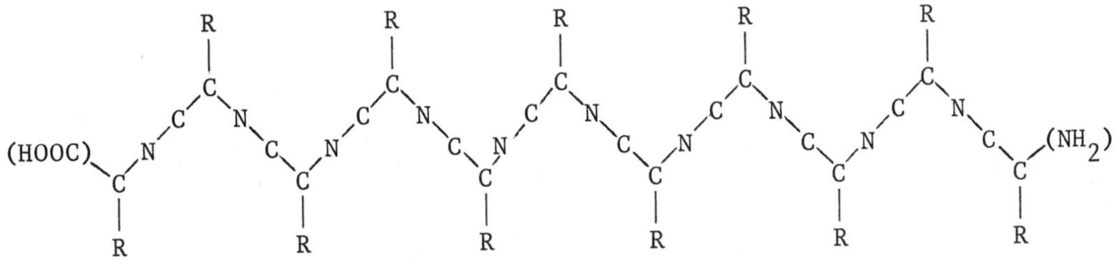

Edge-on View of a Pleated Sheet Structure

The α-carbons always serve as "corners" in the representation. Note, however, that since the α-carbons and the nitrogens are more or less tetrahedral and the carbonyl carbons are trigonal,

some of the atoms shown above to be in the plane of the paper
are actually above or below that plane. The isometric
projection of a pleated sheet (below) will help to clarify
this. Note that the R-groups extend up and down from the
edges of the folds, perpendicular to the edges.

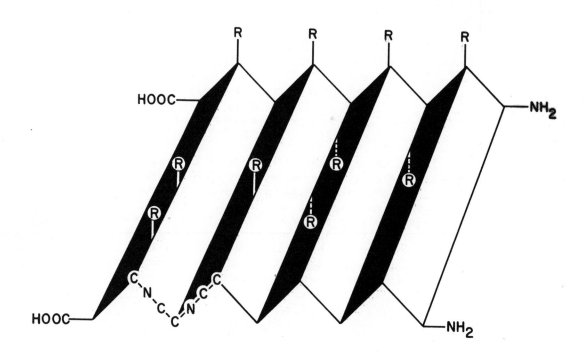

The hydrogen bonds can be arranged two ways, yielding parallel
and antiparallel pleated sheets. The diagrams below illustrate
this. They are top views of the pleated sheets. The hydrogen
bonds (dotted lines) extend into and out of the paper
diagonally. The ones which do not appear to connect to
anything extend to the adjacent chains in the sheet.

```
                          (COOH)    (H₂N)
                                                                              
   R-C                  R-C-H        R-C-H          R-C-H        R-C-H
        \                    \            \              \            \
         N                   N-H•  •  •O=C       •        N-H•         N-H•
          \                                                                 •
           C        •  •  •O=C         N-H•  •  •          •O=C         •O=C      •
            \                \            \          •         \            \
             C-R            H-C-R        H-C-R                 H-C-R        H-C-R
            /                  /            \                      /            \
           N        •  •  •H-N             C=O•  •  •    •        •H-N         •H-N         •
          /                    \            /         •              \            \
         C                    C=O•  •  • •H-N              •          C=O•         C=O•      •
        /                       /            \                          /            \
   R-C                       R-C-H        R-C-H          R-C-H        R-C-H
        \                        \            \              \            \
         N                       N-H        O=C       •        N-H•         N-H•
          \                        /          \                  /            /
           C        •  •  •O=C                 N-H•  •  •    •   •O=C         •O=C      •
            \                \            /                       /            \
             C-R            H-C-R        H-C-R                 H-C-R        H-C-R
            /                  /            \                      /            /
   Edge-on                 (H₂N)         (COOH)              (H₂N)        (H₂N)
   view (same
   as above                        anti-parallel
   diagram)                          chains                        parallel chains
```

The edge-on view at the left is what one would see if the
hydrogen-bonded structures were stood on edge. (This amounts
to rotating them 90° around an axis in the plane of and parallel
to the long edge of the paper.) The hydrogen bonds would then
be in the planes indicated by the lines in the edge view
connecting the R-group (α) carbon atoms. In the parallel and
anti-parallel drawings, an R-group extending to the left is
above the sheet and one to the right is below it.

In the beta structures, the hydrogen bonds are usually said to
be interchain rather than intrachain (as they are in the
α-helix). This is completely correct in, for example, silk.
Frequently, however, a single peptide chain will fold back
upon itself and regions of pleated sheet structure are formed
between different parts of the same chain. Such conformations
have actually been identified in crystal structures of
globular proteins.

4. There are other types of secondary structures which are found in proteins. One important example is present in collagen. Here, the peptide chains are twisted together into a three-stranded helix. The resultant "three-stranded rope" is then twisted into a superhelix. This is similar, in principle, to the double-stranded superhelix found in nuclear DNA.

5. The amino acid sequence (primary structure) strongly influences the types of secondary structure which are present in a protein. Below is a list of amino acids classified according to the sort of secondary structure which they tend to prefer. These divisions are <u>not</u> absolute and have considerable overlap (glutamic acid residues are found in α-helices; alanine residues occur in β-structures).

Table 3.   Preferred Secondary Structure of Some Amino Acids

α-helix formers[a]	non-α-helix[b]	β-structure[c]	Random Coil[d]	α-helix breaking[e]
Ala Trp Gln	Ser	Gly	Glu ⎱ (−)	Pro
Leu Cys Tyr	Thr		Asp ⎰	Hypro
Phe Met Asn	Ile		Lys ⎱ (+)	
	Val		Arg ⎰	

Notes: [a] Gln, Asn   =   glutamine , asparagine; these residues readily form α-helices in aqueous solution.

[b] Thr, Ile , Val - two substitutions on the β carbon causing steric problems in fitting them into an α-helix; (notice that most amino acids have a β-carbon which is a $\geq CH_2$ group). Ability of ser to form sidechain H-bonds causes it to prefer other structures.

[c] Because of its small (−H) sidechain, there is little steric hindrance even though the R-groups in a sheet are very close together.

[d] Ionic repulsion prevents close approximation of glutamic and aspartic sidechains above about pH 4, and of lysyl and arginyl sidechains below about pH 9.

[e] Proline and hydroxyproline , because their α-amino group is part of a ring, are unable to fit into the rigid geometry required by the α-helix. Also, when incorporated into a peptide, they lack an α-amino proton for hydrogen bond formation. Therefore these amino acids are <u>never</u> part of an α-helix. (They do, however, form other helices, as found in collagen.)

The <u>tertiary structure</u> involves the arrangement and interrelationship of the folded chains (secondary structure) of a protein into a specific shape which is maintained by salt bonds, hydrogen bonds, -S-S- bridges, Van der Waals' forces, and hydrophobic interactions. The hydrophobic "bond" is considered to be a major force in maintaining the shape or conformation of proteins. (The word <u>bond</u> in this context is really a misnomer.) Hydrophobic bonds are interactions between non-polar sidechains of alanine, valine, leucine, isoleucine and phenylalanine within H_2O envelopes.

1. Hydrophobic interactions may involve the sidechains of several molecules or may occur between chains on the same molecule (as shown below). They occur because of the inability of non-polar sidechains to interact with water molecules either ionically or through hydrogen bonds.

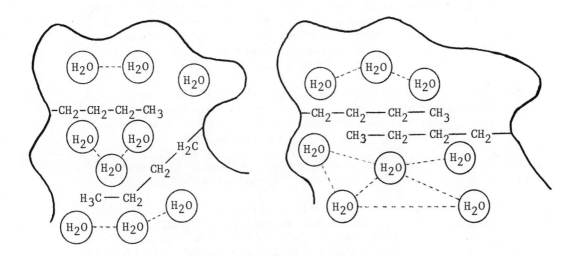

(dashed lines represent
hydrogen bonds)

(I) No hydrophobic interaction between hydrocarbon sidechains; increased non-polar surface area.

(II) Hydrophobic clustering of lipid-like sidechains; water "sees" less lipid surface.

Without going into detail, the general idea is that, when hydrophobic (water-hating; oil or lipid-like) molecules are put into an aqueous solution, they disturb the water structure. To minimize this disturbance (and to return to the lowest possible free energy state), the hydrophobic groups clump together as much as possible so that the surface to volume ratio of the hydrophobic material is minimal. (Two small spheres, each of 1 volume unit, have a greater total surface area than one larger sphere of 2 volume units.) Once the chains are brought close together, van der Waals' attractive forces can operate to assist in holding them there. The van der Waals' forces are quite weak, however, and exactly where most of the "hydrophobic bond energy" comes from is not clear. The probable source is the free energy made available when water is able to achieve a more stable structure. This can be envisioned in two possible ways.

a. The water which surrounds a hydrophobic group is ice-like (highly ordered) because of the limitations imposed on its movement by the presence of the hydrophobic material. A decrease in hydrophobic surface area causes a decrease in order and, hence, an increase in entropy which helps to lower the free energy of the solution.

b. Hydrophobic regions in an aqueous solution break up hydrogen-bond networks. A decrease in the area of hydrophobic surface permits more hydrogen bonds to form, thus lowering the free energy of the solution.

Final answers about the properties and causes of hydrophobic interactions await more definitive experimental data.

2. A schematic diagram of the tertiary structure of lysozyme is shown below. This structure was determined by single crystal x-ray diffraction techniques. This method is an extremely powerful one for elucidating molecular structures, provided suitable crystals of the material can be prepared. This example is given to show a protein molecule which contains several types of secondary structure. It should be noted that, in general, hydrophobic regions are buried inside the molecule while hydrophilic (water-loving; charged) groups appear on the surface, exposed to the aqueous environment. The arrowheads along the backbone indicate the chain direction, from the amino terminus to the carboxyl terminus.

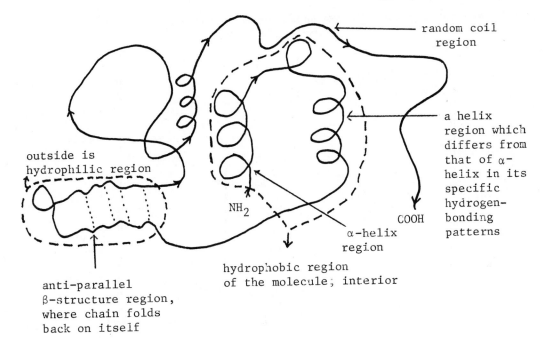

random coil
region

a helix
region which
differs from
that of α-
helix in its
specific
hydrogen-
bonding
patterns

outside is
hydrophilic region

NH₂

α-helix
region

COOH

anti-parallel
β-structure region,
where chain folds
back on itself

hydrophobic region
of the molecule; interior

Quaternary Structure is the association of similar or dissimilar
protein subunits into oligomers (small polymers) or polymers.

1. The subunits are held together by non-covalent forces.
    Consequently, they can be dissociated under relatively mild
    conditions.

2. Many proteins require more than one protein subunit in their
    aggregate structures to perform their biological function
    (hemoglobin, fatty acyl synthetase, actomyosin in muscle, etc.).

3. Enzymes which catalyze the same chemical reaction but which
    have different kinetic properties (ability to bind substrate,
    maximum velocity, etc.) are called isoenzymes or isozymes.
    Important examples include lactic acid dehydrogenase, alkaline
    phosphatase, hexokinase, and hemoglobin (it is somewhat
    incorrect to call hemoglobin an enzyme, though). The isozymes
    of a given enzyme are electrophoretically separable and are
    generally composed of varying proportions of several subunits.

4. The schematic diagram below shows how some of the non-covalent
    forces function in stabilizing quaternary protein structure.

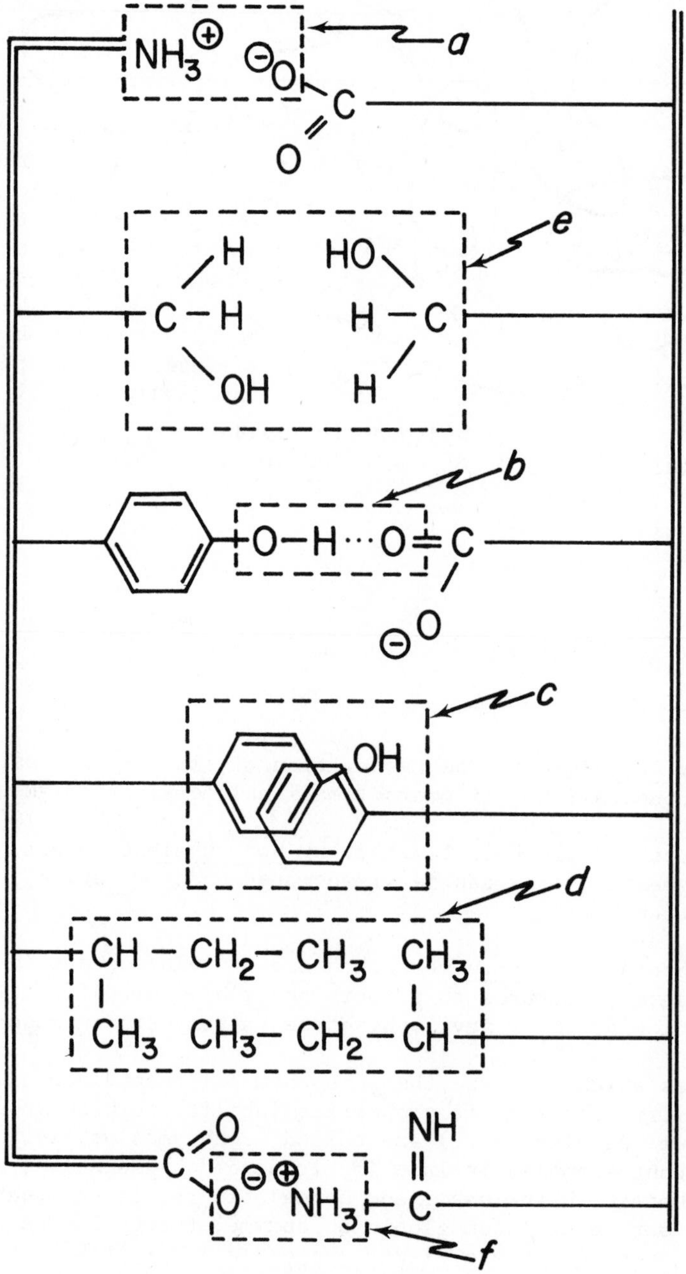

a. Electrostatic attraction between the N-terminal amino group
    and a sidechain carboxyl (from aspartate or glutamate).

b. Hydrogen bond between the phenolic proton of tyrosine and the carbonyl oxygen of a sidechain carboxyl group.

c. Van der Waals' interaction (ring stacking) between the benzene rings of phenylalanine and tyrosine. This is probably also stabilized by hydrophobic forces.

d. Same as (c), between two isoleucyl groups. Here, the hydrophobic forces are probably greater relative to the van der Waals' forces. The residues need not be the same for such interactions.

e. Van der Waals' interaction between two serine residues. Here, the groups are hydrophilic, so that the hydrophobic contribution is minimal.

f. Same as (a), between an arginyl sidechain and the C-terminal carboxyl group.

Denaturation of Proteins

1. Denaturation involves changes in the physical, chemical, and biological properties of a protein molecule.

2. Some of the changes in properties are

   a. **Decreased solubility (often but not invariably).**
   b. Alteration in the internal structure and in the arrangement of peptide chains, which does not involve the breaking of peptide bonds.
   c. Decreased symmetry (e.g., loss of helical structure).
   d. Increased chemical reactivity, particularly of ionizable and sulfhydryl groups.
   e. Increased susceptibility to hydrolysis by proteolytic enzymes.
   f. Decrease or total loss of the original biological activity.

3. The causes of denaturation include

   a. A significant change in the pH of the protein solution.
   b. Temperature changes (particularly high temperature).
   c. Ultraviolet radiation.
   d. Ultrasonic vibration (known as "sonication").
   e. Vigorous shaking or stirring of aqueous solutions which spreads the protein in a thin film over the surfaces of air bubbles (i.e., foaming of the protein solution).
   f. High concentrations of neutral polar compounds such as urea or guanidine. These compounds break hydrogen bonds, allowing new ones to form.

g. Treatment with organic solvents such as ethanol, acetone, etc. This sort of denaturation is minimized at low temperatures (2° to 50°C).

h. Grinding of proteins which results in mechanical deformation and the breakage of peptide chains.

4. The extent and reversibility of denaturation are dependent upon the complexity of the protein and the intensity and duration of the denaturing treatment. Normally, denaturation is an irreversible process, although there are exceptions such as:

a. The denaturation of hemoglobin with acid and renaturation by neutralization under appropriate conditions.

b. The heat denaturation of pancreatic ribonuclease and renaturation on cooling.

It can generally be stated that denaturation is reversible if no disulfide linkages present in the native protein are broken. The greatest block to successful denaturation is the proper reformation of the disulfide bonds.

| specific tertiary structure of α-helix, β-structure, and random coil; all molecules of a given protein have the same shape | chains unfolded and randomly arranged; each chain of slightly different shape | each protein molecule returns to its original shape |

5. A knowledge of the denaturability of proteins is very important in clinical laboratory work. If the activity of an enzyme is being measured to evaluate the health of a patient, a false result can be obtained because of changes in the activity brought about by denaturation. Care must be taken to avoid this during specimen collection, handling, and assay.

Classification of Proteins

Simple Proteins

Fibrous proteins are insoluble in water and resistant to proteolytic enzyme digestion. They are elongated molecules

which may consist of several coiled peptide chains tightly
linked.

1. Collagens

    a. Proteins of connective tissue.
    b. 30% of the total protein in a mammal is collagen.
    c. Contain large amounts of proline, glycine,
        hydroxyproline and hydroxylysine.
    d. Contain no tryptophan.
    e. Collagens, when boiled in water or dilute acid, yield
        gelatins which are soluble and digestible by enzymes.

2. Elastins are present in tendons, arteries and other elastic
    tissues. They cannot be converted to gelatins by boiling
    in water or dilute acid.

3. Keratins are proteins of hair, nails, etc. They contain
    large amounts of cystine. Human hair is 14% cystine.

Globular Proteins are soluble in water and in salt solutions.
In solution these molecules are spheroids or ellipsoids. All
known enzymes, oxygen-carrying proteins, and protein hormones
are members of this class.

1. Albumins: soluble; coagulable by heat.
2. Globulins: insoluble in water; soluble in dilute neutral
    salt solutions; coagulable by heat.
3. Histones: basic proteins found complexed with nucleic
    acids; contain large amounts of arginine and lysine and
    very few aromatic amino acids.
4. Protamines: basic proteins found complexed with nucleic
    acids, particularly in certain fish sperm; rich in
    arginine.

Conjugated Proteins are complex. They are combined with non-amino
acid substances.

Nucleoproteins: DNA⎫
                   RNA⎭ + proteins

Mucoproteins (or mucoids): carbohydrate (more than 4%)+ protein

Glycoproteins: small amount of carbohydrates (less than 4%)
+ protein

Lipoproteins: water soluble   ⎫
                                  ⎬ both contain lipid
Proteolipids: insoluble in water ⎭ and protein

Others: hemoproteins (protein + heme); other metalloproteins;
flavoproteins (protein + flavin); phosphoproteins

Some Properties of Proteins

1. Amphoterism is the ability to behave as an acid or base, depending
    on the conditions.

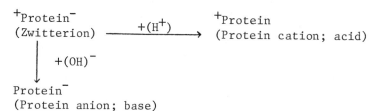

$^+$Protein$^-$
(Zwitterion) $\xrightarrow{\;+(H^+)\;}$ $^+$Protein
(Protein cation; acid)

$+(OH)^-$ ↓

Protein$^-$
(Protein anion; base)

a. Isoelectric pH (pI). The pH at which a protein does not
    migrate in an electric field. The protein exists in the
    zwitterion form with a net charge of 0. There are an equal
    number of cationic and anionic sites on each molecule.

b. Isoelectric precipitation. Many proteins are easily precipitated
    when the pH is adjusted to their isoelectric point (i.e., many
    proteins have their minimum solubility when pH = pI).

c. Electrophoretic mobility is zero at isoelectric pH.

d. Because different proteins have different pI values,
    electrophoresis at different pH's can be used for purification.

2. Solubility can be changed in several ways.

a. Effect of salt concentration (salting-in and salting-out
    phenomena). By adding ionic solutes to a protein solution, the
    affinity of protein molecules for each other compared to the
    affinity of protein molecules for H_2O may be altered, thus
    changing protein solubility.

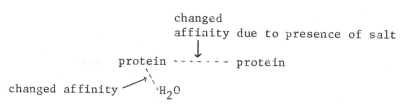

changed
affinity due to presence of salt
↓
protein - - - - - - - protein

changed affinity ⟶ H_2O

<u>Salting-in phenomenon</u>. Adding <u>small</u> amounts of ionic solutes
(NaCl, $(NH_4)_2SO_4$, etc.) decreases the protein-protein interaction
(affinity) but increases protein-H_2O interaction, leading to
solubilization of the protein. The protein-protein affinity is
caused by interactions between oppositely charged groups on
the protein molecules.

$$\underset{(-\overset{\underset{\|}{O}}{C}-O^- \text{ and } {}^+H_3N-)}{}$$

and is decreased by salts (X^+ and Y^-) which provide ions that
bind to the charged protein groups and lower their interaction.

$$(-\overset{\underset{\|}{O}}{C}-O^-\ldots\ldots X^+ \ Y^-\ldots\ldots H_3^+N)$$

<u>Salting-out phenomenon</u>. Adding <u>large</u> amounts of ionic solutes
results in protein precipitation. The mechanism is not well
understood. It is explained as the possible "dehydration of
active water," which increases the interaction between solute
molecules. It is important in one method of protein separation
and purification.

b. <u>pH</u>: As noted earlier, proteins (ampholytes) contain both
positively and negatively charged groups. At its isoelectric
point the ampholyte has no net charge and thus will not
migrate in an electric field. If a mixture of proteins with
different isoelectric points is placed in a pH gradient and
subjected to an electric field, each protein will migrate to
the positions corresponding to its isoelectric points and
will come to rest at that particular pH. This method of
separating proteins according to their isoelectric points is
known as <u>isoelectric focusing</u>. The method consists of
establishing a <u>stable</u> pH gradient in water by subjecting a
mixture of synthetic low molecular weight (300-600 daltons)
polyampholytes (generally mixed polymers of aliphatic amino
and carboxylic acids) to an electric field. When the field
is applied, the ampholytes start to migrate and come to rest
at their respective isoelectric pH values. If proteins with
different pI values are mixed with the ampholytes, they will
also migrate to the position of their respective isoelectric
points, making them separable from one another.

c. <u>Precipitation with water-miscible organic solvents.</u> Aqueous solutions of methanol, acetone, ethanol, and some other organic solvents that are water-soluble and which alter the structure of liquid water can cause denaturation of many proteins. This tends to increase protein-protein interaction and leads to decreased protein solubility. Again, "dehydration of active water" is invoked as a possible cause. Water molecules are occupied with hydrating the organic solvent and are much less available to solubilize the protein molecules through inter-actions with their hydrophilic amino acid residues. Proteins can be separated from one another by varying the concentration of organic solvent.

<u>Purification of Proteins</u> is very important in any attempt to describe exactly the properties of a molecule. Impurities can cause completely erroneous results. Some methods are

1. Differential solubility
2. Specific precipitation
3. Chromatography
4. Preparative electrophoresis
5. Preparative ultracentrifugation
6. Selective enzyme digestion
7. Others

<u>Estimations of the Amount of Protein Present</u>

1. <u>Kjeldahl Procedure</u> (measurement of total nitrogen present)

$$\text{Organic N} \xrightarrow[\substack{\text{catalyst} \\ \text{(Cu, Se, or Hg)}}]{H_2SO_4} CO_2 + H_2O + NH_4HSO_4$$

$$NH_4HSO_4 + 2NaOH \xrightarrow{\text{distillation}} NH_3\uparrow + SO_4^= + 2Na^+ + 2H_2O$$

The NH_3 produced by distillation is passed into a known amount of standard acid and at the end of distillation, the acid remaining is titrated with a standard base. The acid consumed is then a measure of NH_3 produced which in turn is a measure of nitrogen present in the protein. This procedure requires much time and very large amounts of protein. In general, proteins contain about 16% nitrogen. This means that when the weight of nitrogen measured by the above method (taking into account any non-protein nitrogen) is multiplied by the factor 6.25 (100/16) one gets the weight of the total protein present.

2. <u>U.V. absorption at 280 nm</u>, due to tyrosine, phenylalanine, and tryptophan residues (aromatic groups). Since different proteins have differing amounts of these residues (some have none of them),

the method has inherent inaccuracies. It is also not very
successful in the presence of nucleic acids and other substances
which absorb near this wavelength.

3. Biuret Method. peptide + CuSO$_4$ + alkali $\longrightarrow$ violet color

This procedure is specific for peptide bonds, but is not very
sensitive. The violet color is produced by the complex shown below.

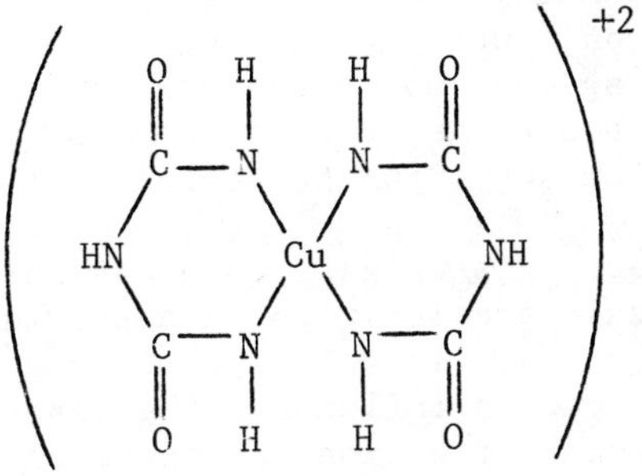

4. Folin - Ciocalteu Reaction

protein + phosphomolybdo $\longrightarrow$ blue color
tungstic acid

The color is due to a reaction with tyrosine. The procedure is
very sensitive but cannot be used in the presence of other phenols.
Also, proteins contain varying amounts of tyrosine, some even
having none. Thus, the method has some of the same drawbacks that
U.V. absorption has.

3
Enzymes

*with John H. Bloor, M.S.**

1. Almost all of the chemical changes (absorption, digestion, metabolism, locomotion, putrefaction, etc.) that take place in a living organism are speeded up by enzymes (catalysts). Without these catalysts the reactions proceed too slowly for biological systems to function at any significant rate.

2. A catalyst is defined by Ostwald as an agent which affects the velocity of a chemical reaction without appearing in the final products of that reaction. In an enzyme catalyzed reaction (A + B $\rightleftharpoons$ C + D), the enzyme influences the reaction velocity of both forward and backward reactions to the same extent. However, the direction in which the reaction proceeds is dependent upon mass-law considerations and the availability of free energy. Enzymes also show the typical features of catalysts such as a) not being consumed in the reaction, b) being needed only in minute quantities, c) reversible and irreversible inhibition, etc. This is discussed much more fully later on, in the section on enzyme kinetics.

3. Enzymes are proteins and are produced by living cells. They possess a high degree of specificity (i.e., they usually catalyze only one type of reaction, frequently acting only on one molecular species) and are classified according to the type of reactions they catalyze.

   a. oxidation-reduction
   b. transfer of groups
   c. hydrolysis of compounds
   d. non-hydrolytic removal of groups
   e. isomerization
   f. joining of two molecules with the breaking of a pyrophosphate bond.

   The following table is a list consisting of examples of enzymes in the above classes. It is by no means complete with respect to classes, subclasses, or examples.

(Text continues on p. 74)

---

*Presently at the Department of Biochemistry, State University of New York at Buffalo.

Table 4. Classification of Selected Enzymes with Clinical Importance According to the Enzyme Commission (E.C.) and Their Common Names

E.C. Code	Systematic Name	Common Name with Some Abbreviations
1	Oxidoreductases	
1.1	Acting on CHOH groups	
	With NAD^+ or $NADP^+$ as hydrogen acceptor	
1.1.1.1	Alcohol:NAD^+ oxidoreductase	Alcohol dehydrogenase (AD)
1.1.1.14	L-iditol:NAD^+ oxidoreductase	Sorbitol dehydrogenase (SD or ID)
1.1.1.27	L-lactate:NAD^+ oxidoreductase	Lactate dehydrogenase (LD or LDH)
1.1.1.37	L-malate:NAD^+ oxidoreductase	Malate dehydrogenase (MD)
1.1.1.42	threo-D_s-isocitrate-NAP^+ oxidoreductase (decarboxylating)	Isocitrate dehydrogenase (ICD)
1.1.1.44	6-phospho-D-gluconate:$NADP^+$ 2-oxidoreductase (decarboxylating)	Phosphogluconate dehydrogenase
1.1.1.49	D-glucose 6-phosphate:$NADP^+$ 1-oxidoreductase	Glucose 6-phosphate dehydrogenase (GPD or G6PD)
1.2	Acting on aldehyde or,keto groups	
1.2.1	With NAD^+ or $NADP^+$ as acceptor.	
1.2.1.12	D-glyceraldehyde 3-phosphate:NAD^+ oxidoreductase (phosphorylating)	Glyceraldehydephosphate (triosephosphate) dehydrogenase
1.4	Acting on $CH.NH_2$ groups	
1.4.1	With NAD^+ or $NADP^+$ as acceptor	
1.4.1.3	L-glutamate:NAD(P) dehydrogenase (deaminating)	Glutamate dehydrogenase
1.6	Acting on NADH or NADPH	
1.6.4	With a disulphide as acceptor	
1.6.4.2	NAD(P)H:glutathione oxidoreductase	Glutathione reductase
2	Transferases	
2.1	Transferring one-carbon groups	
2.1.3	Carboxyl- and carbamoyltransferases	
2.1.3.3	Carbamoylphosphate:L-ornithine carbamoyltransferase	Ornithine carbamoyltransferase (OCT)
2.3	Amino acid transferases	
2.3.2.2	γ-glutamyl-peptide:amino acid γ-glutamyl transferase	γ-Glutamyltransferase (GGT)
2.6	Transferring nitrogenous groups	
2.6.1	Aminotransferases (transaminases)	
2.6.1.1	L-aspartate:2-oxoglutarate aminotransferase	Aspartate transaminase (AST) (also known as Glutamate oxaloacetate transaminase (GOT)
2.6.1.2	L-alanine:2-oxoglutarate aminotransferase	Alanine transaminase (ALT) (also known as Glutamate alanine transaminase (GPT)

E.C. Code	Systematic Name	Common Name with Some Abbreviations
2.7	Transferring phosphorus-containing groups	
2.7.1	Phosphotransferases with alcohol group as acceptor	
2.7.1.1	ATP:D-hexose 6-phosphotransferase	Hexokinase (HK)
2.7.1.40	ATP:Pyruvate 2-0-phosphotransferase	Pyruvate kinase (PK)
2.7.3	Phosphotransferases with nitrogenous group as acceptor	
2.7.3.2	ATP:Creatine N-phosphotransferase	Creatine kinase (CK) (also known as Creatine phosphokinase (CPK)
2.7.4	Phosphotransferases with phospho-group as acceptor	
2.7.4.3	ATP:AMP phosphotransferase	Adenylate kinase
2.7.5	Phosphotransferases catalysing intramolecular transfers	
2.7.5.1	α-D-glucose 1,6-diphosphate:α-D-glucose 1-phosphate phosphotransferase	Phosphoglucomutase
2.7.7	Nucleotidyltransferases	
2.7.7.12	UDP-glucose:α-D-galactose 1-phosphate uridylyltransferase	Hexose 1-phosphate uridylyltransferase
3	Hydrolases	
3.1	Acting on esters	
3.1.1	Carboxylic ester hydrolases	
3.1.1.3	Triacylglycerol acyl-hydrolase	Lipase
3.1.1.7	Acetylcholine hydrolase	Acetylcholinesterase
3.1.1.8	Acylcholine acyl-hydrolase	Cholinesterase
3.1.3	Phosphoric monoester hydrolases	
3.1.3.1	Orthophosphoric monoester phosphohydrolase	Alkaline phosphatase (ALP)
3.1.3.2	Orthophosphoric monoester phosphohydrolase	Acid phosphatase
3.1.3.5	5'-ribonucleotide phosphohydrolase	5'-nucleotidase (5'-NT)
3.1.3.9	D-glucose 6-phosphate phosphohydrolase	Glucose 6-phosphatase
3.2	Acting on glycosyl compounds	
3.2.1	Glucoside hydrolases	
3.2.1.1	1,4-α-D-glucan glucanhydrolase	α-amylase
3.2.1.31	β-D-glucuronide glucuronosohydrolase	β-glucuronidase
3.4	Acting on peptide bonds	
3.4.1	α-Aminopeptide amino acid hydrolases	
3.4.11.1	α-Aminoacyl-peptide hydrolase	Leucine aminopeptidase
3.4.4	Peptide peptidohydrolases	
3.4.23.1	No systematic name	Chymotrypsin
3.4.21.4	"	Trypsin
3.4.21.1	"	Pepsin
3.5	Acting on C-N bonds, other than peptide bonds	
3.5.3	In linear amidines	
3.5.3.1	L-arginine amidinohydrolase	Arginase

E.C. Code	Systematic Name	Common Name with Some Abbreviations
4	Lyases	
4.1	C-C lyases	
4.1.1	Carboxy lyases	Pyruvate decarboxylase
4.1.2	Aldehyde lyases	Aldolase
4.2	C-O lyases	
4.2.1	Hydrolyases	Fumarate hydratase (= fumarase)
4.3	C-N lyases	Histidine-ammonia lyase (= histidase)
5	Isomerases	
5.1	Racemases and epimerases	Ribulose-5-phosphate epimerase
5.1.3	Acting on carbohydrates	Maleylacetoacetate isomerase
5.2	Cis-trans isomerases	
5.3	Intramolecular oxidoreductases	Glucosephosphate isomerase
5.3.1	Interconverting aldoses and ketoses	Methylmalonyl-CoA mutase
5.4	Intramolecular transferases	
6	Ligases	
6.1	Forming C-O bonds	Amino acid-activating enzymes
6.1.1	Amino acid-RNA ligases	
6.3	Forming C-N bonds	Glutamine synthetase
6.3.1	Acid-ammonia ligases	Peptide synthetase, glutathione synthetase
6.3.2	Acid-amino acid ligases	
6.4	Forming C-C bonds	Acetyl-CoA carboxylase
6.4.1	Carboxylases	

Thermodynamics and Kinetics of Biological Reactions

An important aspect of biochemistry is concerned with the elucidation of chemical interconversions in biological systems. Since these reactions primarily involve the orderly release, storage, and utilization of energy, it is logical that a knowledge of chemical energetics (thermodynamics) is required for their discussion. While studies of metabolism, nutrition, and physiological homeostasis may not directly invoke it, an awareness of thermodynamics contributes to the understanding of these subjects. A number of reactions which are important in living systems result in the transfer of electrons, in addition to transfer of energy. To handle these, some background in electrochemistry is necessary. This can be directly related to more classical thermodynamics.

The actual energy transfers, though, are only part of the story. Thermodynamics treats only initial and final states, with no word about how one group of molecules (the reactants) are converted to another, quite different group (the products). Chemical kinetics is needed for this.

Many of the reactions needed by the body would proceed too slowly to be of use, however, were it not for the catalytic abilities of enzymes. To deal with the ways in which enzymes alter the kinetics of chemical reactions, enzyme kinetics has developed as a separate field. As a prelude, then, to talking about specific metabolic processes, this section discusses the fundamental language of thermodynamics and kinetics and the manner in which they are related to biological systems.

Thermodynamics

1. During the conversion of reactant molecules (A + B) to product molecules (C + D), there is a change in the potential energy (called Gibbs free energy by chemists and symbolized by G or F) of the atoms and molecules involved. This is shown below by plotting the variation in the free energy of the participating molecules as the reaction progresses. The phrase "progress of the reaction" was selected to label the abscissa of this diagram because of its generality. It is intended to include all changes in bond lengths and angles (i.e., in the shapes and atomic interactions) of the reactant molecules as they are smoothly converted to product molecules. "Smoothly" is used to imply reversibility. If both reactants and products are present in a reaction mixture, there will be molecules of reactant being converted to product and "product" molecules becoming "reactant" molecules at any given moment. Thus, the equilibrium state is dynamic, not static.

Figure 5.   Potential Energy vs. Progress of Reaction

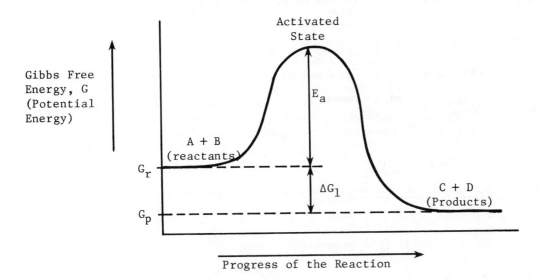

In this diagram E_a = activation energy (free energy of activation; discussed further under kinetics)

G_p = free energy of the products

G_r = free energy of the reactants

$$\Delta G_1 = (G_p - G_r) = (G_{C+D} - G_{A+B}) < 0$$

(The symbol Δ indicates a change or the difference between two quantities.)

Since ΔG is negative, energy is released in converting reactants to products and the reaction is exergonic. If ΔG is positive, energy is absorbed from the environment and the reaction is endergonic.

Thermodynamics deals with (among other things) the size (magnitude) and sign (+ or -) of ΔG, the difference in free energy between the reactants and products. For a given set of reactants and products at constant pressure (for a given reaction), ΔG depends only on the temperature and the concentrations of the reactants and products. Physically, ΔG is the amount of useful work which can be obtained from a chemical reaction at a specified temperature and set of concentrations.

2. Suppose that one starts with (C + D) and wants to convert them
    to (A + B). This is the reverse of the first reaction.
    ΔG is defined as

$$\Delta G_{-1} = (G_{A+B} - G_{C+D}) = -(G_{C+D} - G_{A+B}) = -\Delta G_1 \text{ and } \Delta G_{-1} > 0.$$

(Note that ΔG_1 is negative so that $-\Delta G_1$ is positive.) The (-1)
in the subscript indicates the reverse of reaction designated
by the subscript (1). The reverse of an exergonic reaction
is always an endergonic reaction. Similarly, for (Y + Z) →
(W + X) (the reverse of the reaction portrayed in Figure 6),

$$\Delta G_{-2} = (G_{W+X} - G_{Y+Z}) = -(G_{Y+Z} - G_{W+X}) = -\Delta G_2 \text{ and } \Delta G_{-2} < 0.$$

Figure 6. Potential Energy vs. Progress of Reaction

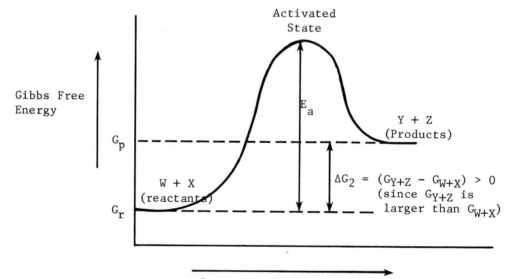

Every reaction is thermodynamically reversible. If, however,
ΔG is very large, so that a great deal of energy must be
supplied for reversal, a reaction may be considered
irreversible for all practical purposes. This means that
under the conditions being used, the back-reaction proceeds to
such a small extent that it need not be considered. (This is
different from saying that a reaction proceeds so slowly that
it can be ignored. That situation will be considered under
kinetics, below.)

3. The dependence of ΔG on temperature and concentration may be written

$$\Delta G = \Delta G^{o} + RT \ln \frac{(Products)}{(Reactants)}$$

where ΔG^{o} = standard free energy change in calories/mole of reactant consumed,

$\quad\quad$ R = gas constant (1.987 Cal. $mole^{-1}$ $deg.^{-1}$),

$\quad\quad$ T = absolute temperature (= ^{o}C + 273), Kelvin

$\quad\;\;$ ln = natural (base e) logarithm,

and (products), (reactants) are molar concentrations (for solvents and solutes) or pressures in atmospheres (for gases).

Since at equilibrium, ΔG = 0, this equation can be rewritten

$$\Delta G^{o} = -RT \ln K_{eq}$$

where $K_{eq} = \dfrac{(Products)}{(Reactants)}$, with the concentrations measured at equilibrium.

The equilibrium constant, K_{eq}, depends only on the temperature of the system. This equation provides a way of determining ΔG^{o} from equilibrium concentration data.

Another commonly used form is

$$\Delta G^{o} = -RT(\ln 10)(\log K_{eq}) = -2.303 \; RT \log K_{eq}$$

where log means base 10 logarithm.

The "standard" state for which ΔG^{o} is defined is arbitrary. By far the most common choice, though, is one in which the temperature is $25^{o}C$, all gases are present at pressures of one atmosphere, all liquids and solids are pure (i.e., not mixtures), and all solutes have unit activity. (Activity is similar to concentration but it takes into account interactions of solute molecules with each other and with the solvent.) Activities are difficult to measure, however, and molar concentrations are commonly used in their place. Since concentration and activity are almost equal in dilute solutions and many biological reactions occur at quite low concentrations, this approximation is acceptable.

A physical interpretation of ΔG^{o} can be obtained by noting that, if

$$\frac{(Products)}{(Reactants)} = 1 \text{ then } \Delta G = \Delta G^{o}.$$

ΔG^O is the amount of useful work which can be obtained by conversion of one mole of each of the reactants in their standard states to one mole of each of the products in their standard states.

The standard free energy change for a reaction is also useful for predicting whether a reaction will occur or not under standard conditions. If $\Delta G^O < 0$, then the reaction will occur spontaneously, provided standard conditions prevail; if $\Delta G^O > 0$, the reaction will not occur by itself. Note, however, that it is really ΔG which determines whether or not a reaction will occur under conditions different from the standard state, such as those existing within a cell. Conclusions based on ΔG^O concerning what reactions will and will not occur in the body can be quite erroneous.

4. In biological systems two other conditions are generally assumed.

    a. Since most solutes are present at fairly low concentrations (less than 0.1 M) and water is present at high concentration (55.6 M for pure water), it is assumed that the water concentration is constant. If water enters into a reaction it is incorporated into K_{eq}.

    b. If H^+ or OH^- participates in a reaction, their concentrations influence ΔG^O. As most biological reactions occur in systems buffered to pH $\simeq$ 7, it is useful to define $\Delta G^{O'}$ as being the value of ΔG^O measured at pH = 7 (sometimes other pH's are used). This corresponds to $(H^+) = 10^{-7}$ M. $\Delta G^{O'}$ values can be compared to each other but not to values of ΔG^O. Although this convention is usually followed, if one sees ΔG^O (without the prime) in a biochemical context, it is advisable to check at what pH the measurement was made.

These two conventions are illustrated by the hydrolysis of ethyl acetate at 25°C, followed by dissociation of the acetic acid produced.

$$CH_3\overset{\overset{\displaystyle O}{\|}}{C}\text{-}O\text{-}CH_2CH_3 + H_2O \overset{\leftarrow}{\rightarrow} CH_3\overset{\overset{\displaystyle O}{\|}}{C}\text{-}O^- + H^+ + HOCH_2CH_3$$

$$K_{eq} = \frac{(CH_3COO^-)(H^+)(HOCH_2CH_3)}{(CH_3COOCH_2CH_3)(H_2O)} = 5.8 \times 10^{-6}$$

ΔG^o = -2.303 RT log K_{eq}

$\quad$ = -(1.987 cal mole^{-1} deg $^{-1}$)(298°K)(log 5.8 x 10^{-6})(2.303)

$\quad$ = -1364 log 5.8 x 10^{-6}

$\quad$ = +7143 cal mole^{-1} = +7.1 kcal mole^{-1}

Alternatively, setting (H$^+$) = 10^{-7} M (pH = 7) and (H$_2$O) = 55.6 M,

$$K' = K_{eq} \frac{(H_2O)}{(H^+)} = \frac{(CH_3COO^-)(HOCH_2CH_3)}{(CH_3COOCH_2CH_3)}$$

$$= 5.8 \times 10^{-6} \frac{55.6}{1 \times 10^{-7}} = 3.22 \times 10^{+3}$$

and $\Delta G^{o\prime}$ = -1364 log 3.22 x 10^{+3}

$\quad$ = -4785 cal mole^{-1} = -4.8 kcal mole^{-1}

Although under standard conditions the reaction does not occur spontaneously (ΔG^o > 0), if more commonly encountered concentrations of water and hydrogen ion are present, the reaction would proceed by itself.

5. a. In order to understand the concept of free energy, an example (described by Baldwin) is useful. Consider a self-operative heat engine which functions by taking in an agent (e.g., steam) at temperature T_1 and releasing it at T_2 ($T_1 > T_2$). The heat extracted (designated Q) is converted as completely as possible into useful work. If W represents the maximum useful work available,

$$W = Q \frac{T_1 - T_2}{T_1} = Q - Q \frac{T_2}{T_1}$$

$\quad$ where T_1 and T_2 are absolute (Kelvin) temperatures. Unless T_2 = 0 or T_1 = ∞, the useful work is always less than the total energy supplied, by a factor of Q (T_2/T_1) = (Q/T_1)T_2.

$\quad$ b. The ratio (Q/T_1) is known as the entropy of the system and is designated by S. The amount of energy unavailable for useful work (= T_2S) is that which is lost in the process of energy transfer. It can be thought of as the amount of randomness or disorder introduced into the system during the transfer.

c. For chemical systems, one can define a quantity G by the
equation $G = H - TS$. Here, G (free energy) is analogous to
W, the amount of useful work available, H (enthalpy) is
analogous to Q, the heat content of the system at constant
pressure, S (entropy) is analogous to (Q/T_1), the wasted
heat energy, and T is the absolute temperature. (Although
it is not shown here, this and the following functions can
all be derived with complete rigor from first principles
of classical thermodynamics.)

This relationship is more commonly used to describe <u>changes</u>
in these quantities. If a system goes from state I (with
$G = G_I$, $H = H_I$, $S = S_I$) to state II ($G = G_{II}$, $H = H_{II}$,
$S = S_{II}$) at constant temperature, one can write

$$G_{II} - G_I = (H_{II} - H_I) - T (S_{II} - S_I)$$

or, in condensed form,

$$\Delta G = \Delta H - T \Delta S$$

where Δ again is read as <u>change</u>.

6. The enthalpy, H, is related to the internal energy of a system
by the equation

$$H = E + PV$$

where E = internal energy,
       P = external pressure on the system, and
       V = volume of the system.

In terms of changes between states, this becomes

$$\Delta H = \Delta E + \Delta(PV) = \Delta E + P\Delta V + V\Delta P.$$

At constant pressure, $\Delta P = 0$ and $\Delta H = \Delta E + P\Delta V$. If, in addition,
the volume is constant, then $\Delta V = 0$ and $\Delta H = \Delta E$.

In many biological reactions $\Delta V = \Delta P = 0$. Consequently, for a
change of state, ΔH, the maximum amount of energy which can be
released as heat is equal to ΔE, the change in the internal
energy of the system.

ΔH can be measured for a particular material by burning the
material in a bomb calorimeter at constant pressure. The heat
released is measured as the temperature rise in a large water
bath surrounding the combustion chamber. The values of ΔH

obtained in this manner are indicative of the total energy available in a compound when it is completely oxidized.   The factors which usually contribute to ΔH under these conditions are the heat of fusion (melting), the heat of vaporization (boiling), and the heat of combustion (bond making and breaking). The first two are important, for example, if at the standard temperature (25°C) a solid is combusted to a combination of liquids and gases, as in the examples below.

Just as with ΔG, ΔH for a process is equal to $-\Delta H$ for the reverse process.   Thus: ΔH (liquid $\rightarrow$ gas) = $-\Delta H$ (gas $\rightarrow$ liquid); ΔH (liquid $\rightarrow$ solid) = $-\Delta H$ (solid $\rightarrow$ liquid); ΔH (making a bond) = $-\Delta H$ (breaking the same bond).

The following reactions were all run at one atmosphere pressure. The subscripts s, l, and g (solid, liquid, and gas, respectively) indicate the phase of the material under one atmosphere pressure at the given temperature.   The ΔH^O values given are standard heats of formation since all reactants are in their standard (natural) state for the temperature given.

a.   Oxidation of one mole of glucose (a carbohydrate) to carbon dioxide and water

$$C_6H_{12}O_{6\,(s)} \;+\; 6O_{2\,(g)} \;\rightarrow\; 6H_2O_{(1)} \;+\; 6CO_{2\,(g)}$$

at 20°C, ΔH^O = -673 kcal /mole

b.   Oxidation of one mole of palmitic acid (a fatty acid) to carbon dioxide and water

$$C_{16}H_{32}O_{2\,(s)} \;+\; 23O_{2\,(g)} \;\rightarrow\; 16CO_{2\,(g)} \;+\; 16H_2O_{(1)}$$

at 20°C, ΔH^O = -2,380 kcal /mole

c.   Oxidation of one mole of glycine (an amino acid) to carbon dioxide, water, and nitrogen

$$C_2H_5O_2N_{(s)} \;+\; 2\text{-}1/4\;O_{2\,(g)} \;\rightarrow\; 2CO_{2\,(g)} \;+\; 2\text{-}1/2\;H_2O_{(1)} \;+\; 1/2\;N_{2\,(g)}$$

at 25°C, ΔH^O = -233 kcal /mole

ΔG and ΔG^O are more useful to biochemists than are ΔH and ΔH^O, because the Gibbs free energy is the useful work (or energy) available from a reaction.   Values of ΔH can be misleading when used to discuss the amount of useable energy which a reaction can supply.

However, ΔH (enthalpic energy or the maximum available energy) is frequently used in calculating the energy yield derived from various foodstuffs. For practical purposes, ΔH is conveniently measured by burning the food substance in an atmosphere of oxygen in a bomb calorimeter. The energy which is released heats a known volume of water surrounding the calorimeter. The product of the increase in temperature and the weight of the water yields the number of calories liberated by the combustion of the food. The following average values obtained from such measurements, expressed in kilocalories per gram of various foodstuffs, are: carbohydrates = 4.1; lipid = 9.4; protein = 5.6. Note that in these studies, heat is measured in kilocalories (kcal = 1000 calories) which is defined as the amount of heat required to raise 1 kg of water from 15°C to 16°C (i.e. an increase of 1°C; also, 1 kcal = 4.184 kilojoules). It should be pointed out that there are variations in the above value within each class of food, depending upon its chemical structure. For example, the value for starch is 4.1 kcal per gram whereas for glucose the value is 3.8. Furthermore, in man, the end product of protein metabolism is urea instead of CO_2 and nitrogen, which are the end products obtained after complete oxidation in the bomb calorimeter. Therefore, a realistic estimate of energy derived from protein in the body, taking into account its incomplete oxidation and specific dynamic action (discussed later), is 4.1 kcal/gm instead of 5.6. In human nutrition, the commonly used values for energy yield (expressed as kcal/gm) from different food substances are as follows: carbohydrates = 4; lipids (fats) = 9; and protein = 4.

During the metabolism of food substances in the body, the volume of oxygen consumed and CO_2 produced can be measured. The molar ratio of CO_2 produced to oxygen consumed, known as the respiratory quotient (RQ), is characteristic of a given substrate. For example, in the complete oxidation of glucose, the RQ is 1:

$$C_6H_{12}O_6 + 6\ O_2 \longrightarrow 6\ CO_2 + 6\ H_2O$$

$$RQ = \frac{\text{Volume of } CO_2 \text{ Produced}}{\text{Volume of } O_2 \text{ Consumed}} = \frac{6\ CO_2}{6\ O_2} = 1.0$$

For a fatty acid:

$$C_{16}H_{32}O_2 + 23\ O_2 \longrightarrow 16\ CO_2 + 16\ H_2O$$

$$RQ = \frac{16\ CO_2}{23\ O_2} = 0.7$$

The RQ for protein is difficult to measure because, as noted earlier, it is not oxidized completely in the body and part of the carbon, hydrogen and oxygen is lost in the urine and feces. Taking all the factors into account, the protein has an RQ of 0.8. The measurement of RQ thus provides a means of assessing the type of food that is being metabolized.

An RQ of 0.85 is obtained in a normal person with a mixed diet, but this value is enhanced with carbohydrate feeding and decreased with lipid feeding.

Heat production can be measured from the oxygen consumption and RQ measurements as follows:

For glucose oxidation:

$$C_6H_{12}O_6 + 6\ O_2 = 6\ CO_2 + 6\ H_2O$$

180gm   192gm
(134 liters)

As noted earlier, the energy yield of 1 g of glucose = 3.8 kcal. Therefore, 180 gm of glucose yields 180 x 3.8 kcal. The energy equivalent per 1 liter of oxygen consumption (with a RQ = 1) is = (180 x 3.8)/134 = 5.1 kcal.

Similarly, for a fatty acid,

$$C_{16}H_{32}O + 23O_2 \qquad 16CO_2 + 16H_2O$$

256gm   736gm
(513.7 liters)

The energy yield of 1 g of fat = 9 kcal. Therefore, 256 gm of fat yields 256 x 9 kcal. The energy equivalent per 1 liter of oxygen consumption (with an RQ = 0.7) is = (256 x 9)/513.7 = 4.49 kcal.

The energy from food in the body is used principally to maintain body heat and synthesize ATP. ATP is then transformed to other forms of energy: chemical (synthesis of new compounds), mechanical (muscle contraction), electrical (nerve activity), electrochemical (various ion pumps), light (light photons) and thermal (maintenance of body temperature). In general, the food energy provides for the maintenance of basal metabolism (determined by basal metabolic rate-- BMR), the additional energy expenditure required for various types of activity, and the specific dynamic action of food (discussed later). The BMR represents the energy that is required for the maintenance of body temperature, muscle tone, circulation, respiration, and other glandular and cellular activity in a person who is awake and at rest with regard to physical, digestive and emotional activities. The BMR is measured in an individual who has been informed of the test procedure and is at rest after a 12-hour fast. The test is performed in surroundings which are warm, comfortable and quiet. Although the BMR can be directly obtained by measuring the amount of heat produced (a cumbersome procedure), the indirect method which measures the volume of oxygen consumed and carbon dioxide evolved per unit time provides acceptable values. An estimate of BMR in an individual can be obtained from the following relationship.

$$BMR \text{ (kcal/day)} = \text{Weight in kg} \times 24 \text{ kcal/day}$$

It should be pointed out that BMR is affected by a number of parameters: size (height and weight relationships); sex (males have higher BMR than females; however, this difference is not noticeable in infants); prior activity; thyroid status, etc. It is increased during growth (due to energy cost of growth) and decreases with age. The BMR is altered in a number of pathological states. Some of the conditions that lead to an increased BMR include: hyperthyroidism, fever (about a 12% elevation in BMR for each degree centigrade above the normal body temperature), Cushing's syndrome, tumors of the adrenal gland, anemia, leukemia, polycythemia, cardiac insufficiency, etc. Some of the disorders that are associated with a decreased BMR include: hypothyroidism, starvation and malnutrition, hypopituitarism, hypoadrenalism (e.g., Addison's disease), etc.

In order to calculate caloric requirements in an individual, it is necessary to take into account two factors: basal metabolic rate and activity (muscular work). The caloric requirements of muscular work can be measured. Following are some examples of the energy required (metabolic cost) for certain types of activity, calculated for a man weighing 70 kg and expressed as calories per hour: sleeping = 75, awake but lying still = 77, sitting at rest = 100, walking slowly (2.6 mph) = 200, active exercise = 290, severe exercise = 450, swimming = 500, running (5.3 mph) = 570, walking upstairs = 1110.

The assimilation of food and its preparation for energy storage or energy yield costs energy which has to be derived from body energy stores. These processes involve oxidative and/or synthetic changes. The increase in metabolic rate resulting from these processes has been termed the <u>specific dynamic action</u> (SDA). The values for SDA vary with the type of food substance ingested. That for protein is the highest (about 30% above its caloric value) as it has to undergo many changes before its carbon skeleton can be used appropriately. One the other hand, the SDA values for carbohydrates and lipids are only 5% and 4% above their caloric value, respectively. The heat produced as a result of SDA is wasted heat and, as noted above, is maximal with protein intake. It should be emphasized that thermodynamically, all energy derived from the metabolism of substrates to CO_2 and H_2O must eventually appear as heat. Following is a generalized relationship between energy intake and output.

Energy Intake = Internal work + obligatory heat production + external
           work + heat production incidental to external work
           $\pm$ energy storage (or body mass changes)

Note that in the above relationship, a number of these processes are under neural and hormonal control. Obesity and cachexia (a disorder characterized by general physical wasting and malnutrition) are extreme examples of problems affecting the energy stores of the body.

7. In living systems, equilibrium is the exception rather than the rule. In fact, life has been defined as the ability to utilize energy from an external source to maintain chemical reactions in a non-equilibrium state. Death corresponds to the attainment of equilibrium by these same reactions.

In a reaction sequence

$$A \rightarrow B \rightarrow C$$

The reaction $A \rightarrow B$ can be prevented from reaching equilibrium by removing B (converting it to C) faster than it can be made from A. This will be discussed further under kinetics. All reactions in the body are interrelated and the system <u>as a whole</u> is in some sort of balance, although individual reactions usually are not in equilibrium. This condition is known as a <u>steady-state</u>. A change in concentration of any one component (product of one reaction which is used as reactant by another reaction) shifts the concentration of all the other components which are linked to it by a sequence of chemical reactions. This results in the attainment of a new steady-state.

Of interest here is the value of ΔG for a non-equilibrium system. This can be determined from ΔG^o and the actual concentrations

of products and reactants in the cell, by use of the equation
(given previously)

$$\Delta G = \Delta G^O + 2.303 \ RT \ \log_{10} \frac{(Products)}{(Reactants)}$$

If ΔG is negative, then under cellular conditions the reaction
proceeds to form products. If ΔG is positive, then the materials
called reactants here are what is actually being made in the
cell.

8. Oxidation and reduction reactions

Oxidation is the loss of electrons or hydride (H^-) ions (but
not hydrogen (H^+) ions) by a molecule, atom, or ion. (The term
oxidation was originally applied to a group of reactions in
which some material combined with oxygen (O_2). It is now
recognized that these reactions belong to the more general
class of conversions which result in electron loss by a compound
and the name is now used for this larger group. In some, but
by no means all such reactions, oxygen is the electron
acceptor and one or more atoms of oxygen become part of the
product molecule.)

Reduction is the gain of electrons or hydride (H^-) ions by a
molecule, atom, or ion.

Note that transfer of one hydride ion results in the transfer
of two electrons.

The amount of work required to add or remove the electrons is
called the electromotive potential or force (emf) and is
designated E or ε. It is measured in volts (joules/coulomb,
where a coulomb is a unit of charge, a quantity of electrons).

The standard emf, E^O (or, at pH 7, $E^{O'}$) is the emf measured
when the temperature is 25°C and the materials being oxidized
or reduced are present at concentrations of 1.0 M. In
biological systems, $E^{O'}$ is most commonly used.

A half-reaction is one in which electrons or hydride ions are
written explicitly. Values of E^O and $E^{O'}$ are tabulated with
the half-reactions for which they are measured. For example,

$$NADH \rightarrow NAD^+ + H^- \qquad E^{O'} = -0.32 \text{ volts}$$

and $H_2O \rightarrow 1/2 \ O_2 + 2H^+ + 2e^-$ $\qquad E^{O'} = -0.816 \text{ volts}$

are half-reactions since they show electrons (e^-) and hydride
ions (H^-).

Alternatively, these half-reactions can be written

$$NAD^+ + H^- \rightarrow NADH \qquad\qquad E^{O\prime} = +0.32 \text{ volts}$$

$$2e^- + 2H^+ + 1/2 \; O_2 \rightarrow H_2O \qquad\qquad E^{O\prime} = +0.816 \text{ volts}$$

Notice that the sign of $E^{O\prime}$ changes. Changing the direction of a reaction reverses the sign of the potential change (just as with ΔG and ΔH).

The oxidation potential is actually a statement of the ease of removing electrons from a material compared to the ease of removing electrons from hydrogen in the half-reaction,

$$H_2 \rightarrow 2H^+ + 2e^- \qquad E^O = 0.00 \text{ volts}$$

(Note that, in this reaction, E^O is defined as zero, thus fixing the scale of E^O value for other reactions.) If it is easier to remove electrons from something, $E^{O\prime}$ will be negative; if it is harder, $E^{O\prime}$ is positive. E^O for the hydrogen half-reaction is used as the zero-point even when talking about $E^{O\prime}$ values. Thus,

$$H_2 \rightarrow 2H^+ + 2e^- \qquad E^{O\prime} = -0.42 \text{ volts}$$

It is easier to remove electrons from hydrogen (and produce H^+) when $(H^+) = 10^{-7}$ M than when $(H^+) = 1.0$ M. The choice of a negative sign for the standard emf to mean "easier to remove electrons" is arbitrary. Although the "negative-easier" choice is used in this book, the International Union of Pure and Applied Chemistry has recommended the opposite convention. The older convention is retained here because it is commonly encountered in medically and biologically related texts and literature.

Since free electrons combine exceedingly rapidly with whatever is at hand, half-reactions never occur by themselves. There must always be something accepting electrons just as fast as they are being released. The substance releasing electrons (or H^-) is the reductant or reducing agent (since it is oxidized); and the substance accepting electrons is the oxidant or oxidizing agent (since it is reduced). Two half-reactions, when combined, give a redox (oxidation-reduction) reaction. When balanced, such reactions never show free (uncombined) electrons. For example,

$$H^+ + NADH + 1/2 \; O_2 \rightarrow NAD^+ + H_2O$$

$$E_T^{O\prime} = -0.32 - (+0.816) = -1.136 \text{ volts}$$

By definition, $E_T^O{}'$, termed the net or total potential or emf, for this reaction is calculated according to

$$E_T^O{}' = E^O{}'\ \text{(reductant)} - E^O{}'\ \text{(oxidant)}.$$

Under standard conditions (and pH = 7), the reaction will occur as written if $\underline{E_T^O{}' < 0}$. Otherwise the reaction will proceed from right to left (the reverse of the way in which it is written).

Notice also that $E^O{}'$ is the amount of energy (work) <u>per coulomb</u> rather than the <u>total</u> energy required for oxidation. Consequently, although the amount of work changes, if the amount (number of moles) of material transformed is altered, $E^O{}'$ -- and $E_T^O{}'$ -- are <u>independent</u> of the number of moles oxidized or reduced.

9. When $E_T^O{}'$ in volts is converted to calories/mole, the result is $\Delta G^O{}'$ for the redox reaction being considered. This is accomplished by means of the equation

$$\Delta G^O{}' = \frac{n\ E_T^O{}'\ F}{4.184} = (23{,}061)\ n\ E_T^O{}'$$

where F = the faraday (= 96,487 coulombs/gram-equivalent),
   n = number of faradays (gram-equivalents) of electrons
       transferred per mole of material oxidized or reduced,
$E_T^O{}'$ = $E^O{}'$ (reductant) - $E^O{}'$(oxidant) in volts, and
$\Delta G^O{}'$ = the standard free energy change in calories per mole
       of material oxidized or reduced (since there are
       4.184 joules/calories).

For illustration, the reaction

$$H^+ + NADH + 1/2\ O_2 \rightarrow NAD^+ + H_2O \qquad E_T^O{}' = -1.136\ \text{volts}$$

is used. For this,

   n = 4 gram-equivalents of electrons per mole O_2 (since
       the half-reaction $O_2 + 4e^- \rightarrow 2O^{-2}$ involves four
       electrons),

   = 2 gram-equivalents of electrons per mole H_2O, NAD^+,
       or NADH (since the half-reactions for one mole of
       each of these materials involves only two electrons).

Then $\Delta G^{O'}$ = 23,061 x (-1.136) x n = -26,197n

$\Delta G^{O'}$ = -26,197 x 2 = -52,394 calories/mole of H_2O, NADH, or NAD^+

= -26,197 x 4 = -104,788 calories/mole of O_2 transformed.

As in all reactions, the actual value of $\Delta G^{O'}$ depends on the reactant or product for which it is calculated.

10. The importance of understanding free energy changes and redox reactions can be seen by noting that all life on earth is based on the redox reaction,

$$mCO_2 + mH_2O + \text{energy} \underset{\text{animals}}{\overset{\text{plants}}{\rightleftarrows}} mO_2 + (CH_2O)_m$$

$E_T^{O'}$ ≃ +1.24 volts

$\Delta G^{O'}$ ≃ +114,300m calories/mole of carbohydrate

In animals, carbohydrate (represented by $(CH_2O)_m$) and oxygen are consumed and energy, water, and carbon dioxide are released. The energy is stored, primarily as adenosine triphosphate (ATP), to be used when needed by cellular processes. In plants, energy from the sun interacts with and is absorbed by the chloroplasts of green plants, causing water to be oxidized and carbon dioxide reduced, producing oxygen and carbohydrate.

In order to understand the many steps between carbohydrate and carbon dioxide and water, a knowledge of the energetics involved is fundamental. Oxidation potentials are of particular use in discussing the initial and terminal steps of the process: the trapping of light energy by chloroplasts during photosynthesis, and the formation of water by the reduction of oxygen via the electron transport chain. After the energy from carbohydrate oxidation is stored as ATP, further transfers of it are usually spoken of in terms of free energy changes.

General Kinetics

Thermodynamics deals with the relative energies and states of

reactants and products, not caring how one travels between these
states. Kinetics, then, is complementary, considering how fast
a reaction occurs and the actual pathway (energetic, structural) --
termed the mechanism -- which it follows. In fact, there is no way,
using kinetics, to decide whether a reaction will or will not occur,
provided an effectively infinite period of time is available. This
will be discussed below in more detail.

1. Reaction Rates

The conversion of $A \xrightarrow{k} B$ (where k is the rate constant,
discussed below) can be represented diagrammatically as the
increase of product, B, with respect to time.

Figure 7. Time Course of a Chemical Reaction

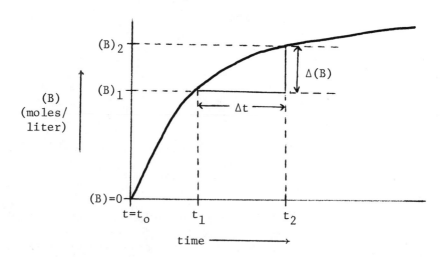

where (B) means the molar concentration of substance B. The
average velocity ($\bar{v}$) for the period t_1 to t_2 is the slope,

$$\bar{v} = \frac{\Delta(B)}{\Delta t} = \frac{(B)_2 - (B)_1}{t_2 - t_1}$$

The instantaneous velocity at some time t' ($\neq t_o$) is

$$v = \frac{d(B(t))}{dt} \qquad \text{evaluated at } t'$$

The expression $\frac{d(B(t))}{dt}$ is the first derivative of (B(t)) with
respect to t.

It is the limit of the slope $\frac{\Delta(B)}{\Delta t}$ as $\Delta t \to 0$, or

$$v = \lim_{\Delta t \to 0} \bar{v} = \lim_{\Delta t \to 0} \frac{\Delta(B)}{\Delta t} = \frac{d(B(t))}{dt}$$

Alternatively, one can consider the change in concentration of the reactants. Thus,

$$v = \frac{d(B(t))}{dt} = -\frac{d(A(t))}{dt}$$

since (A), the reactant concentration, decreases at the same rate as (B), the product concentration, increases. The expressions (B(t)) and (A(t)) are used to imply that one has a detailed knowledge (i.e., an analytic function) of the way in which the concentrations of A and B vary as time (and the reaction) progresses. In reality, however, this is seldom true. To avoid this implication and to make a more compact notation, (B) and (A) will be used from now on instead of, respectively, (B(t)) and (A(t)).

Consideration of the chemistry of the situation provides a way around the need for this analytic function. The rate of formation of B should increase if (A), the reactant concentration, increases. It is reasonable, then, to write

$$\frac{d(B)}{dt} = k(A)^n$$

where k is the <u>rate constant</u> (mentioned above), and $n = 0,1,2...$ is the <u>order</u> of the reaction with respect to A. (The order is zero, first, second, etc., with respect to A if $n = 0,1,2$, etc.)

The rate constant is the proportionality constant between the rate of formation of B and the molar concentration of A. It is characteristic of a particular reaction. The units of k depend on the order of the reaction. For zero-order, they are moles liters^{-1} time^{-1} (time is frequently in seconds). For first-order, they are time^{-1}; second order is liters moles^{-1} time^{-1}, etc. They are whatever is necessary to make $d(B)/dt$ have units of moles liters^{-1} time^{-1}. Notice that, if $n = 0$, $v = k$ (zero-order reaction). This means that (B) changes in a constant way, <u>independent of the concentration of reactants</u>. This is especially important in enzyme kinetics. A plot of (B) versus t for such a reaction is a <u>straight line</u>.

A somewhat more complicated example is

$$A + B \xrightarrow{\quad k \quad} C + D$$

$$v = \frac{d(C)}{dt} = \frac{d(D)}{dt} = \frac{d(Products)}{dt} = k(A)(B)$$

This reaction is first order with respect to A, first-order with respect to B, and second-order overall. Most reactions are zero-, first-, or second-order.

The concept of reaction order can be related to the number of molecules which must collide simultaneously for the reaction to occur. In a first-order reaction, no collisions are required (since only one reactant molecule is involved). Every molecule having sufficient free energy to surmount the activation barrier will spontaneously convert to products. In a second-order reaction, not only must two molecules have enough free energy, they have to collide with each other for the products to form. A third-order reaction requires the simultaneous meeting of three molecules, a very unlikely event. Because of this, reactions of order higher than second are very seldom encountered in simple chemical conversions.

Higher apparent orders are encountered, however, in some cases where an overall rate equation is written for a process which actually proceeds in several steps, via one or more intermediates. In such situations, the individual steps will seldom if ever involve a third-or higher order process. This is sometimes encountered in enzyme catalyzed reactions. The concept of reaction order for such an overall process is somewhat artificial and of little use.

Zero-order reactions can be accounted for in two ways. If a process is truly zero-order, with respect to all reactants (and catalysts, in the case of enzymes), then either the activation energy is zero or every molecule has sufficient energy to overcome the activation barrier. This kind of reaction is seldom if ever found in homogeneous reactions in gases or solutions.

Alternatively, a reaction can be zero-order with respect to one or more (but not all) of the reactants. This is a very important case, especially in enzyme kinetics and clinical assays utilizing enzymes (see the section on the uses of enzymes in the clinical laboratory). This can be explained by noting that one of the reactants (or a catalyst, such as an enzyme) is in limited supply. Increasing the availability (concentrations) of the other reactants can result in no increase in the velocity beyond that dictated by the limiting reagent. This makes the

rate independent of the concentrations of the non-limiting
materials and, hence, zero-order with respect to those materials.

As was pointed out in the section on thermodynamics, all
reactions are theoretically reversible. The correct way to
write $A \rightarrow B$, then, is

$$A \underset{k_{-1}}{\overset{k_1}{\rightleftarrows}} B$$

where k_1 is the rate constant for the conversion of A to B and
k_{-1} is the rate constant for the reverse reaction, conversion
of B to A. The reaction is kinetically reversible if $k_1 \simeq k_{-1}$;
it is irreversible for all practical purposes (proceeds in the
reverse direction so slowly that it can be ignored) if
$k_{-1} \ll k_1$. The rate of a reversible reaction is written

$$v = \frac{d(B)}{dt} = k_1(A) - k_{-1}(B).$$

For $A + B \underset{k_{-1}}{\overset{k_1}{\rightleftarrows}} C + D$

$$v = \frac{d(Products)}{dt} = \frac{d(C)}{dt} = \frac{d(D)}{dt} = k_1(A)(B) - k_{-1}(C)(D).$$

These kinetic schemes (ways of expressing reaction velocities
as functions of concentrations and rate constants) can become
very complicated, if there are many linked (sequential)
reactions and intermediates. They are very important in
biological systems, however, since enzymes make reversible many
otherwise kinetically irreversible reactions. They will be
discussed further in enzyme kinetics.

2. The rate at which a reaction will proceed (measured by k) is
    directly related to the amount of energy which must be supplied
    before reactants and products can be interconverted. This
    activation energy, E_a, comes from the kinetic energy possessed
    by the reactants. This may be translational and rotational
    energy (needed if two molecules must collide to react); or
    vibrational and electronic energy (useful when one molecule
    rearranges itself or eliminates some atoms to form the product).
    The larger the activation energy, the slower the reaction rate,
    and the smaller the rate constant.

    E_a is a free energy, so it has both ΔH and ΔS terms. This means
    that, not only must the reactants have enough energy (ΔH) to make
    and break the requisite bonds, but they must also be properly

oriented (ΔS) for the products to form.   This is useful in thinking about how catalysts function.

Figure 8.   Illustration of Forward and Reverse Activation Energies

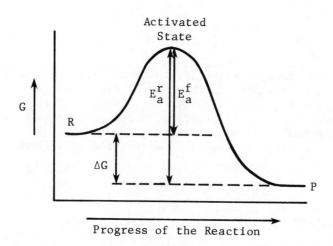

For this free energy diagram

$$R \xrightleftharpoons[k_{-1}]{k_1} P$$

E_a^f = G (activated state) - G (reactants) = activation energy for
$\qquad\qquad$ R → P

E_a^r = G (activated state) - G (products) = activation energy for
$\qquad\qquad$ P → R

$E_a^r = E_a^f - \Delta G$; r and f = reverse and forward, respectively.

Since $\Delta G < 0$ (the reaction as written is exergonic), $E_a^r > E_a^f$.

3.   The <u>Arrhenius Equation</u> interrelates rate constants and activation energies

$$k = Ae^{-E_s/RT}$$

where E_s = Arrhenius activation energy (= $E_a + T\Delta S_a + RT = \Delta H_a + RT$)
$\qquad$ in calories/mole;
$\qquad$ R = gas constant = 1.987 calories mole^{-1} degrees K^{-1};

T = absolute temperature (^{o}K);

A = the Arrhenius pre-exponential factor, a constant;

k = rate constant; and

e = 2.71828, the base of the natural logarithms.

Note that this is an <u>empirical</u> equation, while E_a, the free energy of activation, is a <u>theoretical</u> concept, derived from the kinetic theory of reaction rates. From the Arrhenius equation, it is apparent that the larger E_s (or E_a) is, the smaller is the value of k. The basis for kinetic irreversibility (see page 90, 1, above) is the same, then, as that for thermodynamic irreversibility: the more energy a process takes (the more positive is ΔG or E_a), the more difficult and less likely it is to occur.

Although values for A are not well understood, the Arrhenius equation can be used to calculate E_s by rearranging it to

$$-R \ln k = \frac{E_s}{T} + R \ln A.$$

The slope of a plot of $-R \ln k$ versus $(1/T)$ equals E_s and $\Delta H_a = E_s - RT$. In order to get E_a, ΔS_a (entropy of activation) is needed. (Note: E_s is sometimes designated E_A while E_a is $G\ddagger$, ΔH_a is $H\ddagger$, and ΔS_a is $S\ddagger$.)

4. For a reversible reaction, the equilibrium state is one in which the concentrations of reactants and products are such that the forward and reverse rates are equal. This leads to the conclusion that

$$K_{eq} = \frac{k_1}{k_{-1}},$$

which is demonstrated below.

For a reversible reaction, R $\underset{k_{-1}}{\overset{k_1}{\rightleftharpoons}}$ P,

the rate expression is $\dfrac{d(P)}{dt} = k_1(R) - k_{-1}(P).$

By definition, the concentrations of products and reactants are constant at equilibrium so that

$$\frac{d(P)}{dt} = 0 = k_1(R) - k_{-1}(P)$$

This can be rearranged to $\dfrac{k_1}{k_{-1}} = \dfrac{(P)}{(R)}$

and, since $K_{eq} \equiv \dfrac{(P)}{(R)}$, it is true that $\dfrac{k_1}{k_{-1}} = K_{eq}$.

This is another idea which is important in understanding enzyme kinetics. If some reversible step of an **enzyme-catalyzed** reaction has low enough activation energy, it can be treated as a true equilibrium, simplifying the mathematics of the system.

5. So far, K_{eq} has always been written as a dissociation constant. That is

$$K_{eq} = \frac{(products)}{(reactants)} = K_{diss} = \frac{k_1}{k_{-1}}$$

It is just as correct, though less common, to write it as an association constant:

$$K_{assoc} = \frac{(reactants)}{(products)} = \frac{1}{K_{diss}} = \frac{k_{-1}}{k_1}$$

The more stable the materials termed products are, the larger is K_{diss} and the smaller is K_{assoc}. This follows from the thermodynamic arguments presented previously.

6. Catalysts and catalysis

Although ΔG^o and K_{eq} depend solely on temperature, E_a can be decreased for most reactions by catalysts. Since rate constants are functions of E_a, catalysts also increase them and, consequently, reaction rates. Catalysts do not affect ΔG^o or K_{eq}.

Catalysts are not altered in reactions that they catalyze. If X is a catalyst for $A \underset{\leftarrow}{\rightarrow} B$, then

$$X + A \underset{\leftarrow}{\rightarrow} X + B$$

Although A and B are interconverted, X is unchanged in the net reaction. It may and probably does exist, at some intermediate step, in an altered form, but it always is returned unchanged. For this reason, a very small (catalytic) amount of X can cause the reaction of a great deal of A and B.

Since catalysts function by lowering E_a, they speed up forward and reverse reactions equally. This is essentially saying, in a different way, that catalysts do not affect K_{eq}. This can be seen in the potential energy diagram below, for $A \rightleftarrows B$.

Figure 9. <u>Effect of a Catalyst on the Activation and Free Energies of a Chemical Reaction</u>

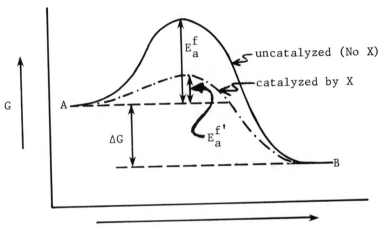

Progress of the Reaction

Catalysts do, however, hasten the <u>attainment</u> of equilibrium. For a given system, ΔG (a function of temperature and concentration) always determines whether or not a reaction will occur.

Since E_a is a free energy change, it is correct to write

$$E_a = \Delta H_a - T\Delta S_a.$$

A catalyst can

i) <u>decrease</u> ΔH_a by changing the electronic distribution in the reactants, making it easier to make or break a bond;

ii) <u>increase</u> ΔS_a by binding and orienting the reactant molecules (decreasing their randomness), making successful collisions more likely.

Both of these effects work to lower E_a. The extent to which each occurs in a particular situation depends on the reaction and the catalyst.

Enzyme Kinetics

Although the thermodynamics of biological systems can be readily
handled in terms of classical ideas and methods, the kinetics of
bodily processes are more clearly understood when discussed in the
language of enzyme kinetics. This is due primarily to the many
types of enzymes involved and the myriad ways in which they can
function. The principles of enzyme kinetics, however, are no
different from those of classical chemical kinetics.

1. Enzymes as catalysts

Most biological reactions would be kinetically irreversible were
it not for the class of catalysts known as enzymes. These are
proteins which can (usually in association with one or more
coenzymes or cofactors) greatly accelerate biochemical
reactions. The diagram below shows schematically what an
enzyme does.

Figure 10.   Effect of an Enzyme on the Activation and Free Energies
           of a Chemical Reactions

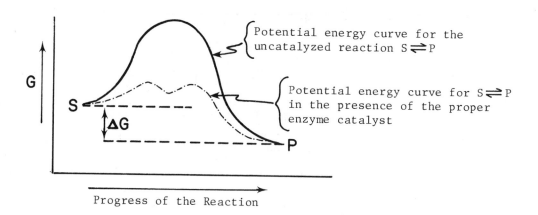

Progress of the Reaction

The catalyzed reaction can be written

$$S \underset{k_{-1}}{\overset{k_1}{\rightleftharpoons}} P$$

where R (for reactant) is replaced by S (for substrate). The
substances which an enzyme can convert to product are termed

substrates. For a freely reversible reaction, the product for
one direction is the substrate for the reverse.

The dip (local energy minimum) in the enzyme catalyzed curve
indicates the presence of a metastable intermediate (an
intermediate having more free energy than either S or P, but
which requires some activation energy for conversion to
either one). This reaction can be written

$$E + S \underset{k_{-1}}{\overset{k_1}{\rightleftharpoons}} ES \underset{k_{-2}}{\overset{k_2}{\rightleftharpoons}} E + P$$

The intermediate is ES, an enzyme-substrate complex. (Notice
that the enzyme (catalyst), E, is regenerated, as it must be.)
It is also apparent from the curve and the rate equation that
there are four activation energies. The original rate constants
and activation energies have no bearing on the rates of the
enzyme-catalyzed reaction.

The scheme given above is the simplest one involving
intermediates. Some enzyme catalyzed reactions may have no
metastable intermediates (as with $X + A \rightleftharpoons X + B$, where the
activated state represents an unstable intermediate); or they
may have many intermediates, with one or more possible kinetic
pathways from S to P. The description of such reactions
requires a knowledge of the mechanisms by which they occur,
and this must be worked out for each reaction and enzyme.
Cases are known where one enzyme can catalyze two different
(though similar) reactions by apparently different mechanisms
(e.g., hydrolysis of amide and ester bonds by chymotrypsin).
The general procedure for obtaining this information is to
measure rates and rate constants for as many processes as
possible, then fit them together into a self-consistent pattern
which is chemically plausible. Certain approximations are
usually possible which greatly simplify this task. The elucidation
of enzyme mechanisms is discussed later in more detail.

2. The rate at which an enzyme-catalyzed reaction occurs is
    influenced by several factors. These include

    a. Temperature. The rates of all chemical reactions are
        increased by a rise in temperature. Such a rise causes the
        average kinetic energy and the average velocities of the
        molecules to increase, resulting in a higher probability
        of effective (reaction-causing) collisions. Enzymes,
        however, become denatured and inactivated at high
        temperatures. Below the denaturation temperature for an

enzyme, it is approximately true that the reaction rate will double for every rise in temperature of 10°C. Since the exact ratio of the new (higher temperature) rate to the old rate is termed Q_n where n is the temperature change in degrees centigrade, this is equivalent to saying that $Q_{10} \simeq 2$. Most enzymes have an optimal temperature, that is, one at which they catalyze at a maximum rate. This is usually close to the normal temperature experienced by the organism of which they are a part. In man, most enzymes have an optimum temperature of 37°C.

b. pH. The activity of most enzymes depends on pH because of the involvement of H^+ or OH^- in the reaction, denaturation at certain pH's, or for other reasons. As with temperature, most enzymes have an optimum pH at which their activity is maximal. It is usually at or near the pH of the fluid in which the enzyme must function. Thus, most enzymes have their highest activity between pH 5 and pH 8 (e.g., pH of human blood is $\sim$7.4). Pepsin, which must operate at the low pH of gastric juice, has maximal activity at pH $\simeq$ 2.

c. Concentration of the Enzyme. The reaction rate increases in direct proportion to the concentration of the enzyme catalyzing it, suggesting that the enzyme-substrate interaction obeys the mass-action law. In fact, at a fixed substrate concentration, the rate of a reaction is frequently used as a measure of the enzyme concentration (see uses of enzymes in the clinical laboratory).

d. Concentration of the Substrate. At a fixed enzyme concentration, the initial velocity (before much substrate is consumed) increases at first with increasing substrate concentration. Eventually a maximum is reached and adding more substrate has no further influence on velocity (v). This is shown in the diagram below. The curve is known as a rectangular hyperbola and, in general shape, is characteristic of all non-allosteric enzymes (allosterism will be discussed later, under enzyme control).

Figure 11.   Effect of Initial Substrate Concentration on the Velocity
of an Enzyme Catalyzed Reaction, where the Concentration
of the Enzyme is Kept Constant

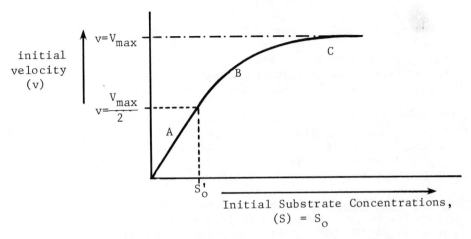

Segment A: linear; first-order with respect to substrate and enzyme
$(v = k(E)S_0)$; initial substrate concentration is low and
limits the velocity.

Segment B: curved; first-order with respect to enzyme; mixed (first
and zero) order with respect to substrate.

Segment C: almost linear; first-order with respect to enzyme; very
close to zero-order with respect to substrate $(v = k(E))$;
at infinite substrate concentration $(S_0 = \infty)$, $v = V_{max}$, a
limiting value for the velocity; as $S_0 \to \infty$, $v \to V_{max}$.

In segment C, effectively all of the enzyme molecules have
substrate bound to them. As soon as one reacts and leaves
the enzyme, another substrate molecule binds. Increasing
(S) cannot speed up the reaction because the enzyme is
being fully utilized already. The value of V_{max} (maximum
velocity) is an important kinetic parameter, characteristic
of a particular enzyme and substrate at specified
temperature and pH.

Another important point is the value of S_o which results in the velocity being half-maximal. In the above diagram,

$$v = \frac{V_{max}}{2} \text{ when } S_o = S_o'$$

The value S_o' (in moles/liter) is called the Michaelis constant, symbolized by K_m. It is a ratio of rate constants, the exact form of which (in terms of rate constants) depends on the kinetic scheme for the reaction. Both V_{max} and K_m are discussed below in more detail.

In the body, substrate depletion is an important control mechanism. When a reaction is no longer needed, the supply of substrate for it is reduced or stopped and, as the existing pool runs out, the enzyme catalyzing the reaction ceases to function for lack of reactant.

e. Presence of Inhibitors. The rates of most enzyme-catalyzed reactions can be decreased by a variety of substances, collectively called inhibitors. They can be general in their action (e.g., urea and other denaturing agents); type specific (acting against phosphorylases, for example); or specific for one enzyme (as are substrate analogues). They are also classified according to the way that they affect the observed kinetics. Inhibitors will be discussed more fully later on.

f. Presence of Allosteric Effectors. A group of enzymes (called allosteric enzymes) exhibits unusual kinetics. Their activity can be increased or decreased by substances termed allosteric effectors. Each allosteric enzyme is influenced by its own set of effectors which can include a wide variety of compounds. The effectors are usually related in some way to the products of the enzyme or to intermediates in some pathway of which the enzyme is part. As the name allosteric (other site) implies, the effectors indirectly influence what happens at the active site by binding to some part of the enzyme separate from the substrate binding site. This group of enzymes will be discussed more thoroughly later on.

It should be noted that paragraphs a-e (above) apply to some extent to all enzymes, while allosteric effects are observed with relatively few of them.

3. Michaelis-Menten Treatment

One procedure for handling enzyme kinetics which has found wide applicability was worked out by L. Michaelis and

M.L. Menten in 1913. Again, consider the reaction

(I)
$$E + S \underset{k_{-1}}{\overset{k_1}{\rightleftharpoons}} ES \xrightarrow{k_2} E + P$$

Here, the release of product is essentially irreversible and the reverse reaction $(E + P \xrightarrow{k_{-2}} ES)$ does not take place to any appreciable extent. This can also be described by saying that k_{-2} is much less than k_2, corresponding to a large, negative ΔG from ES to $(E + P)$. Define

E_o = total concentration of enzyme present (moles/liter)

   = $(E) + (ES)$

S_o = initial substrate concentration (moles/liter).

Assume that
a) $S_o \gg E_o$ so that as substrate is consumed, the substrate concentration (S) changes very little from its initial value of S_o.
b) Only initial velocities are measured. This is necessary to insure that, during the period of measurement, $(S) = S_o$ to a good approximation.
c) (ES) is constant during the period of observation (steady-state condition). This means that
   i) enough time has elapsed since mixing E and S for (ES) to build up,
   ii) not enough time has passed for the rate of formation of ES to decrease due to substrate depletion, and
   iii) $k_2 < k_{-1}$ so that (ES) can build up. If this is not true, ES breaks down as fast as it is formed and a steady-state can never be established.

From this

   rate of formation of ES = $k_1(E)(S) = k_1[E_o - (ES)]S_o$

   rate of breakdown of ES = $k_{-1}(ES) + k_2(ES) = (k_{-1} + k_2)(ES)$

In steady-state

   rate of formation = rate of breakdown so that

   $$k_1[E_o - (ES)]S_o = (k_{-1} + k_2)(ES)$$

(II) $\dfrac{[E_o - (ES)]S_o}{(ES)} = \dfrac{k_{-1} + k_2}{k_1} \equiv K_m$, the Michaelis constant
(mentioned previously)

Notice that K_m is <u>not</u> an equilibrium constant, although it is a ratio of rate constants, because $ES \rightarrow E + P$ is irreversible.

Using the definition of K_m, equation (II) can be rearranged to

$$S_o E_o - S_o(ES) = K_m(ES)$$

$$S_o E_o = (ES)(K_m + S_o)$$

(III) $$S_o = \dfrac{(ES)}{E_o}(K_m + S_o)$$

It is usually difficult to measure (ES) in a reaction mixture. Consequently, equation (III) is not very useful from an experimental point of view. On the other hand, the velocity, v, and the maximum velocity, V_{max}, are readily measured by a variety of methods.

Previously, V_{max} was defined as the value which v approaches as $S_o \rightarrow \infty$. At $S_o = \infty$, all enzyme molecules have substrate bound to them so that $(E) = 0$ and $E_o = (ES)$.

Using this and $v \equiv \dfrac{d(P)}{dt} = k_2(ES)$, one can write $V_{max} = k_2 E_o$.

These expressions can both be solved for k_2, giving

$$k_2 = \dfrac{v}{(ES)} \text{ and } k_2 = \dfrac{V_{max}}{E_o}.$$

Combining these, then, give

$$\dfrac{v}{(ES)} = \dfrac{V_{max}}{E_o} \text{ or } \dfrac{v}{V_{max}} = \dfrac{(ES)}{E_o}$$

Substituting this into equation (III),

$$S_o = \dfrac{v}{V_{max}}(K_m + S_o)$$

(IV) $$v = \dfrac{S_o V_{max}}{K_m + S_o}$$

Equation (IV) is known as the <u>Michaelis-Menten Equation</u>. Regarding this relationship

a) K_m is a constant characteristic of an enzyme and a particular substrate. It is independent of enzyme and substrate concentrations.

b) V_{max} $(= k_2 E_0)$ depends on enzyme concentration. For a particular enzyme, it is largely independent of the specific substrate used.

c) K_m and V_{max} may be influenced by pH, temperature, and other factors.

d) A plot of v versus S_0 (shown before) is a rectangular hyperbola.

e) If an enzyme binds more than one substrate, the K_m values for the various substrates can be used as a relative measure of the affinity of the enzyme for each substrate (the smaller the value of K_m, the higher is the enzyme's affinity for that substrate).

f) In a metabolic pathway, K_m values for enzymes catalyzing the sequential reactions may indicate the rate-limiting step for the pathway (the highest K_m roughly corresponding to the slowest step).

4. It was stated previously that $K_m = S_0$ when $v = (V_{max}/2)$. This can be demonstrated by substituting for v in equation (IV).

$$\frac{V_{max}}{2} = \frac{V_{max} S_0}{K_m + S_0}$$

$$(K_m + S_0)(V_{max}) = 2 V_{max} S_0$$

$$K_m + S_0 = 2 S_0$$

and
$$K_m = S_0, \text{ when } v = \frac{V_{max}}{2}$$

From equation (IV) it can also be seen that

a. v <u>increases</u> if V_{max} <u>increases</u> at constant S_0 and K_m, and

b. v <u>decreases</u> if K_m <u>increases</u> at constant S_0 and V_{max}.

5. Straight lines are more amenable to data evaluation than are curves and it is, therefore, convenient to recast equation (IV)

in some form which can be plotted as straight lines.   There
are two such forms commonly used.

a.   The **Lineweaver-Burk form** is arrived at by taking
     reciprocals of both sides of equation (IV) and rearranging

$$\frac{1}{v} = \frac{K_m + S_o}{S_o V_{max}}$$

(V)

$$\frac{1}{v} = \frac{K_m}{V_{max}} \left(\frac{1}{S_o}\right) + \frac{1}{V_{max}}$$

According to (V), if the system obeys Michaelis-Menten
kinetics, a plot of $1/v$ versus $1/S_o$ will be **a straight line**
of slope (K_m/V_{max}) and a $(1/v)$ intercept of $(1/V_{max})$.
Further, the $(1/S_o)$ intercept, occurring when $(1/v) = 0$,
is $(-1/K_m)$.   This is seen by setting $(1/v)$ equal to zero.

$$0 = \frac{K_M}{V_{max}} \left(\frac{1}{S_o}\right) + \frac{1}{V_{max}}$$

then,

$$\frac{1}{S_o} = \frac{-(1/V_{max})}{(K_m/V_{max})}$$

$$\frac{1}{S_o} = -\frac{1}{K_m} \quad .$$

This is shown in Figure 12 in a **Lineweaver-Burk
plot**.

Figure 12.   Lineweaver-Burk Plot of the Michaelis-Menten Equation

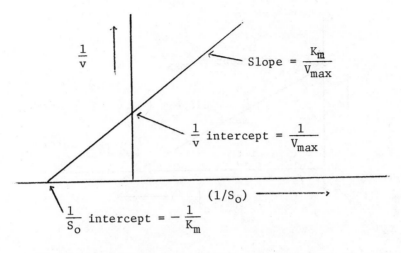

b.   The Eadie-Hofstee form is obtained by multiplying both sides of equation (IV) by $(K_m + S_o)/S_o$ and rearranging

$$v \left(\frac{K_m + S_o}{S_o} \right) = V_{max}$$

$$v \left(\frac{K_m}{S_o} \right) + v = V_{max}$$

(VI) $$v = -K_m \left(\frac{v}{S_o} \right) + V_{max}$$

It can be seen from equation (VI) that a plot of v versus (v/S_o) will have a slope of $-K_m$, a v-intercept of V_{max}, and a (v/S_o)-intercept of (V_{max}/K_m). The diagram below, illustrating this, is an Eadie-Hofstee plot.

Figure 13.   Eadie-Hofstee Plot of the Michaelis-Menten Equation

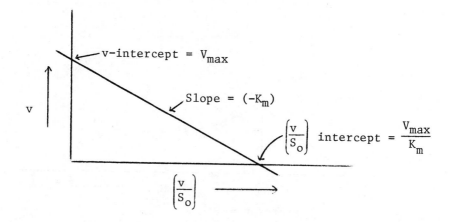

Of these two methods, the Lineweaver-Burk plot is more commonly
used. It must be remembered that all of these conclusions are
based on the premise that the system under study obeys an
equation of the same form as 3(I) **(page 103)** and that the
assumptions (3a, b, c, page 103) are correct. Systems with
different reaction equations can be treated in a similar manner,
but equations II, III, and IV must be rederived for each case.
One example, frequently encountered, involves ES going to EP,
an additional intermediate which then breaks down to products.

6.   Enzyme inhibition is one of the most important ways in which
enzyme activity is both artificially and naturally regulated.
It is the manner in which most therapeutic drugs perform their
function, usually acting on a specific enzyme. Inhibitor
studies have contributed much of the information presently
available about enzyme kinetics and mechanisms. Trypsin, a
highly active proteolytic enzyme distributed in many body
tissues, might rapidly hydrolyze much of the protein in the
body were it not for the presence of specific trypsin
inhibitors in the plasma and many tissues. A correlation has
been demonstrated between a lack of α_1 Trypsin inhibitor
and the occurrence of one form of pulmonary emphysema.

Inhibition can now be considered in terms of the effect that an
inhibitor has on the form of equations 3(IV), 3(V), and 3(VI)
and, consequently, on their graphs. A general scheme for
enzyme inhibition can be drawn:

Figure   14.   General Scheme for Enzyme Inhibition.

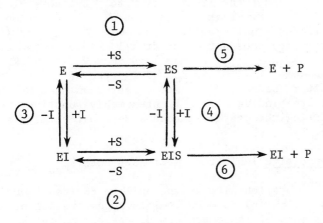

where I represents an inhibitor molecule.

a. The normal enzyme-substrate reaction is ① followed by ⑤.
    Irreversible inhibition occurs if I binds to some form of E
    (E or ES) and the resultant complex (EI or EIS) cannot break
    down to regenerate E. This cannot be treated by Michaelis-
    Menten methods, since the amount of functional enzyme is not
    constant. An irreversible inhibitor usually induces a covalent
    change in the enzyme (by breaking a bond in the enzyme or by
    itself reacting with the enzyme) which destroys or blocks the
    active site.

b. Other reversible inhibitors form EI or EIS (or both) but do so
    reversibly, allowing E to become functional again. There are
    three general ("simple") types of enzyme inhibition:

    i) Competitive inhibition occurs when I binds to the same
       form of the enzyme as does S. That is, in Figure 14,
       reaction ③ occurs but ② and ④ (and hence ⑥) do not.
       This is also called dead-end inhibition. This type of
       inhibition can always be overcome by increasing the
       substrate concentration. Since I and S compete for the
       same binding site, the law of mass action applies and the
       more S present, the greater will be the fraction of E
       bound to S. That is, the ratio (I/S) determines the degree
       of inhibition. Since the actual conversion of S to P is
       not affected, V_{max} is unchanged. $K_{m_{apparent}}$ increases,
       however, since the binding of S to E is impaired.

ii) <u>Noncompetitive</u> inhibition is exhibited by a substance which does not affect the binding of S to E but does block the formation of P. That is, in Figure 14, reactions ① and ② proceed equally well, as do reactions ③ and ④; but reaction ⑥ does not occur at all. This is the reverse of i). $V_{max_{apparent}}$ is now less than V_{max}, but $K_{m_{apparent}}$ is the same as K_m. This type of inhibitor binds at some site other than that of the substrate binding site. It presumably functions by distorting the enzyme in such a way that catalysis cannot occur.

iii) <u>Uncompetitive</u> inhibition decreases both $K_{m_{apparent}}$ and $V_{max_{apparent}}$, relative to K_m and V_{max}. Such inhibitors bind only to ES (reaction ④) and do not allow ⑥ to occur. This sequestration of ES causes more of it to be formed, thereby making $K_{m_{apparent}}$ < K_m. V_{max} is decreased because ESI is a dead-end complex, thereby lowering the amount of product formed.

c. These three simple types of inhibition are primarily useful as limiting cases. That is, most enzymes actually show more complex patterns of inhibition, primarily because they do not fit the Michaelis-Menten single-substrate, single-ES-complex, single-product model, which is also a limiting case. This is not to deny the importance of discussing these systems but to emphasize that the actual problems which a researcher will encounter are usually quite complex; and the interpretation of the symptoms of an enzyme-based disease probably cannot be interpreted in terms of these elementary schemes.

Figure 15. Effects of Inhibitions on Velocity-Substrate Plots and Lineweaver-Burke Double-Reciprocal Plots.

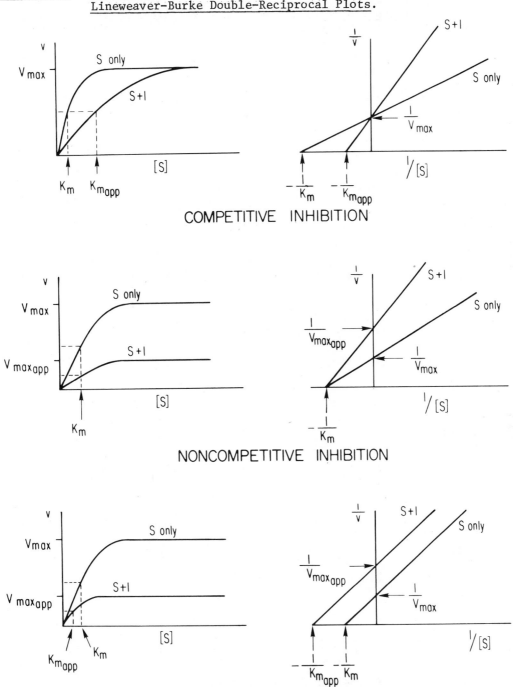

COMPETITIVE INHIBITION

NONCOMPETITIVE INHIBITION

UNCOMPETITIVE INHIBITION

Table   5 .   Effects of Inhibitors on K_m and V_{max}, the Michaelis-Menten Kinetic Parameters.

The Michaelis-Menten equation, in the presence of inhibitors, has the form:

$$v = \frac{S_o V_{max_{app.}}}{K_{m_{app.}} + S_o}$$

where $K_{m_{app.}}$ and $V_{max_{app.}}$ are defined below for the three cases of simple inhibition. K_m and V_{max} are constants for a given enzyme, substrate, and temperature. These values are unaltered by inhibitors. $K_{m_{app.}}$ and $V_{max_{app.}}$ are the observed values in the presence of inhibitor.

Inhibitor Type	V_{max}	K_m
Competitive	None $(V_{max_{app.}} = V_{max})$	$K_{m_{app.}} = K_m(1 + \frac{I}{K_i})$
Noncompetitive	$V_{max_{app.}} = \frac{V_{max}}{(1 + \frac{I}{K_i})}$	None $(K_{m_{app.}} = K_m)$
Uncompetitive	$V_{max_{app.}} = \frac{V_{max}}{(1 + \frac{I}{K_i})}$	$K_{m_{app.}} = \frac{K_m}{(1 + \frac{I}{K_i})}$

Note: $K_i = \frac{(E)(I)}{(EI)}$ or $\frac{(ES)(I)}{(ESI)}$. In uncompetitive inhibition, the inhibitor binds equally well to E and ES.

7. Examples of enzyme inhibition

   a. Irreversible inhibition (covalent linkage of inhibitor to
       enzyme):

      i) Enzymes containing free cysteinyl (sulfhydryl) groups
          can react with alkylating agents.

$$\text{E-SH} + \text{ICH}_2\overset{\overset{\displaystyle O}{\displaystyle \|}}{\text{C}}\text{-NH}_2 \longrightarrow \text{E-S-CH}_2\overset{\overset{\displaystyle O}{\displaystyle \|}}{\text{C}}\text{NH}_2 + \text{HI}$$

       (iodoacetamide)   (inactive covalent
                                 derivative of E-SH)

      ii) Enzymes having seryl hydroxyl groups can be
           phosphorylated by diisopropylfluorophosphate (DFP).

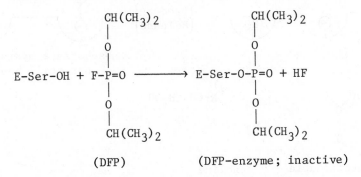

          (DFP)   (DFP-enzyme; inactive)

A specific example is the DFP-inhibition of acetylcholinesterase, the enzyme which inactivates the
neurotransmitter, acetylcholine, by hydrolysis. (Esterases
and other hydrolases frequently have serine as part of
their active site, so DFP inactivates many enzymes of
this type.) The normal reaction is shown below. The
enzyme is represented by

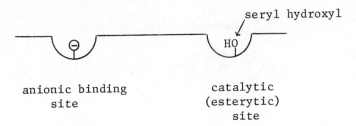

        anionic binding   catalytic
            site   (esterytic)
                                 site

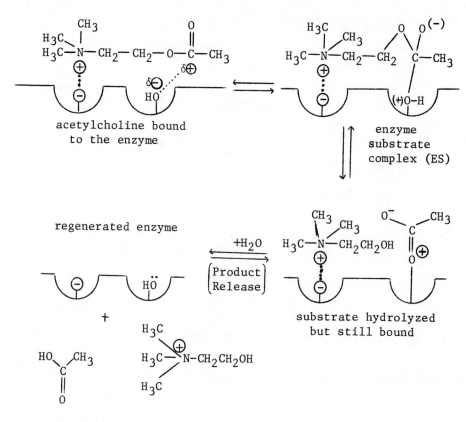

acetylcholine bound
to the enzyme

enzyme
substrate
complex (ES)

regenerated enzyme

+H₂O

[Product Release]

substrate hydrolyzed
but still bound

+

acetic acid   choline

DFP can also bind to the enzyme, form a complex similar to ES, and be hydrolyzed. Hydrolysis of the diisopropyl-phosphate derivative is very much slower than is the release of acetic acid. This is shown below.

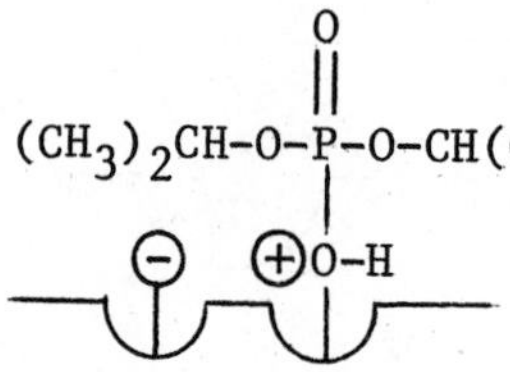

$$(CH_3)_2CH-O-\overset{\overset{\displaystyle O}{\|}}{\underset{\underset{\displaystyle \ominus \quad \oplus O-H}{|}}{P}}-O-CH(CH_3)_2$$

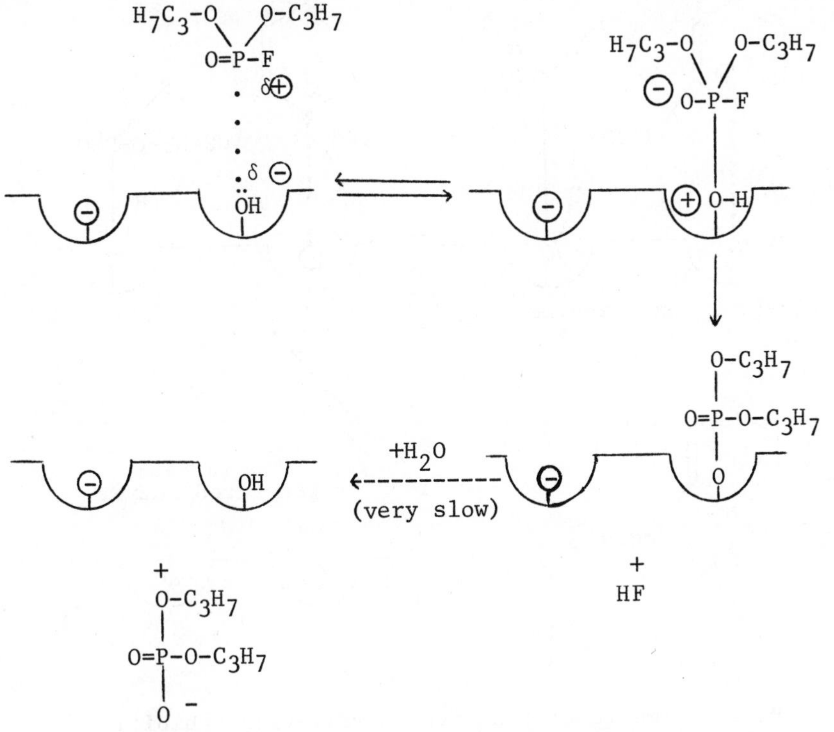
H_7C_3-O O-C_3H_7
O=P-F
δ⊕
δ⊖
OH
H_7C_3-O O-C_3H_7
O-P-F
O-H
O-C_3H_7
O=P-O-C_3H_7
O
$+H_2O$
(very slow)
OH
+
O-C_3H_7
O=P-O-C_3H_7
O -
+
HF

In this particular case, the acetylcholinesterase can be reactivated by hydroxamic acids, and the general formula is

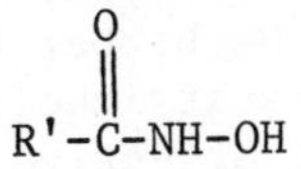
O
R'-C-NH-OH

Pralidoxime (pyridine-2-aldoxime methiodide), a drug used clinically to treat DFP poisoning, increases the rate of phosphoryl enzyme hydrolysis by the mechanism below, where

R = $-CH(CH_3)_2$ and

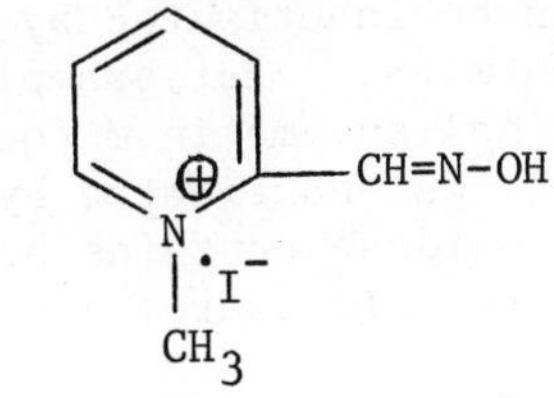
CH=N-OH
N
I⁻
CH_3

is pralidoxime

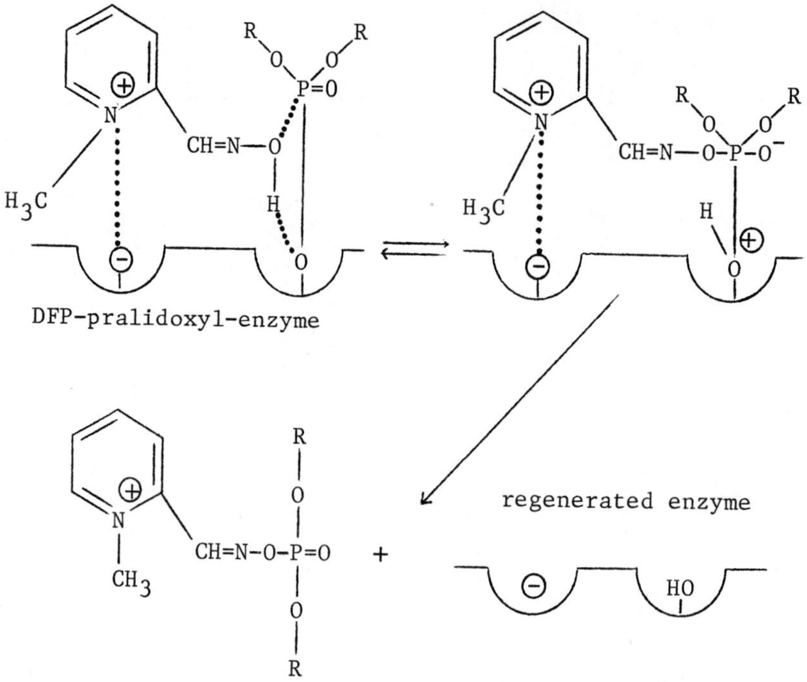

DFP-pralidoxyl-enzyme

$\rightleftharpoons$

regenerated enzyme

Pralidoxime is most active in relieving inhibition of
skeletal muscle acetylcholinesterase. The effects
(increased tracheobronchial and salivary secretion,
bronchoconstriction, central respiratory paralysis and
other CNS disorders) of autonomic system acetylcholinesterase inhibition can be better antagonized
by atropine, which blocks the acetylcholine receptor
sites. Atropine is especially effective in relieving
the excessive secretion and bronchoconstriction.
Notice that it does not relieve the inhibition but
renders ineffective the accumulated acetylcholine.

iii) Metalloenzymes can be inactivated by forming <u>stable
      complexes</u> with the metal. For example, cytochrome
      oxidase, the terminal enzyme in mitochondrial
      electron transport, is inactivated by CN^- (cyanide
      ion). Cytochrome oxidase contains both copper and
      iron. The normal reactions are

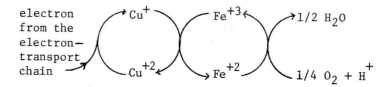

Presumably, CN^- forms a loose complex with Fe^{+2}. Then Fe^{+2} is oxidized to Fe^{+3} which forms a very stable complex with CN^-. This complex cannot be reduced by Cu^+, so the flow of electrons and the uptake of O_2 are stopped. Hydrogen sulfide (H_2S) toxicity operates by the same mechanism.

Cyanide poisoning can result from the ingestion of certain seeds such as peach, cherry, plum, apricot, cassava, bitter almond, etc. The active principle is known as <u>amygdalin</u>. It has the structure:

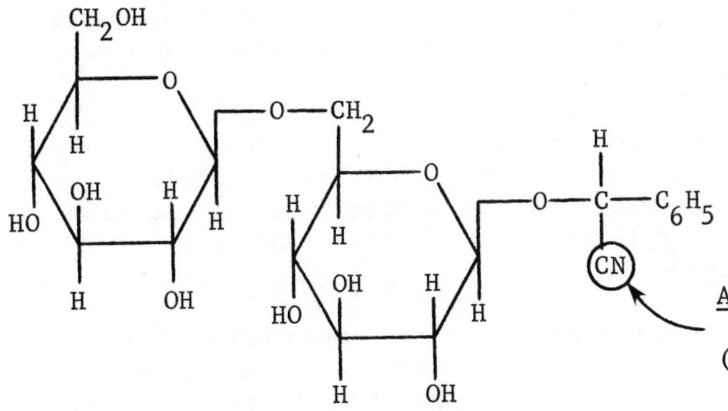

Amygdalin
cyanide
(nitrilo)
group

The cyanide is released when the seed is crushed, hydrated, and the amygdalin enzymatically hydrolyzed according to the reaction:

$$C_{20}H_{27}NO_{11} + 2 H_2O \longrightarrow 2 C_6H_{12}O_6 + C_6H_5CHO + HCN$$

amygdalin   glucose   benzaldehyde hydrogen
                                                                      cyanide

The initial toxic symptoms consist of sudden vomiting and crying followed by fainting, lethargy or coma. If untreated, death eventually ensues. Amygdalin and a similar compound known as laetrile (structure shown below) are currently at the center of a controversy regarding their efficacy as anticancer agents. Although they are used in some parts of the world, they are increasingly proscribed in the United States. The anticancer activity is claimed to depend upon the hydrolysis of glycosides at the tumor site by B-glucuronidase or B-glucosidase, releasing HCN. Presumably the normal cells are not affected by HCN because they contain CN^--inactivating enzymes (rhodanese).

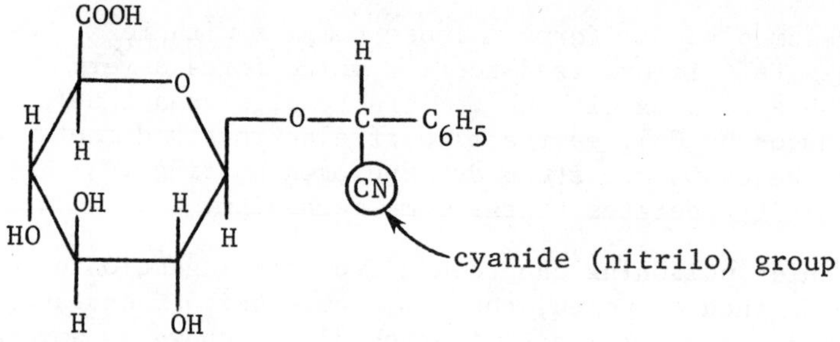

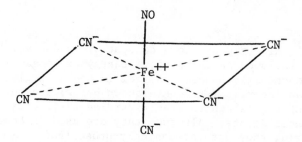

COOH

H—O—C—C_6H_5

(CN) ← cyanide (nitrilo) group

Laetrile

At the time of this writing, there is no objective evidence that these compounds have any therapeutic value as anticancer agents.

Another source of cyanide toxicity is accidental exposure to acrylonitrile ($H_2C=CHC\equiv N$) which is used widely as a starting material in the plastics industry and as a fumigant to kill dry-wood termites, powder-post beetles and wood borers.

Sodium nitroprusside (structure shown below), used therapeutically in the treatment of hypertensive crisis, has cyanide moieties which can be liberated when nitroprusside reacts with sulfhydryl groups in red cells and tissues. The small amounts of released cyanide are inactivated by conversion to thiocyanate (SCN^-) in the liver (see below) and eliminated via the kidney. Although thiocyanate is relatively non-toxic, prolonged exposure to high levels may lead to hypothyroidism, presumably due to thiocyanate's antithyroid action (see thyroid hormone synthesis).

NO

CN^- CN^-

Fe^{++}

CN^- CN^-

CN^-

Iron Coordination Complex Present in Nitroprusside

Cyanide and sulfide toxicity are handled clinically by making use of the ability of methemoglobin (oxidized hemoglobin, $HbFe^{+3}$) to bind CN^- and HS^- as strongly as does cytochrome oxidase. The level of methemoglobin in the erythrocytes is increased by administration of nitrites ($NaNO_2$ solution intravenously; amyl nitrite by inhalation). This causes a shift of the equilibrium.

$$HbFe^{+3} + CytFe^{+3}(CN^-) \rightleftharpoons HbFe^{+3}(CN^-) + CytFe^{+3},$$

to the right, releasing the inhibition of cytochrome oxidase. Cyanomethemoglobin is no more toxic than methemoglobin and can be removed by the normal processes which degrade erythrocytes. Excess, uncomplexed cyanide is removed by administration of sodium thiosulfate.

$$CN^- + Na_2S_2O_3 \xrightleftharpoons[\text{SCN}^- \text{ oxidase}]{\text{sulfurtransferase}} SCN^- + Na_2SO_3.$$

Thiocyanate (SCN^-) is relatively non-toxic and can be eliminated in the urine. The sulfurtransferases, also known as rhodaneses, are mitochondrial enzymes found in liver, kidney, and other tissues. The reverse reaction, catalyzed by thiocyanate oxidase, is slow. Its action can, however, occasionally cause a return of the toxic manifestations. Cyanide is normally found in man complexed to Co^{+3} in vitamin B_{12} (cyanocobalamin).

In experimental animals, it has been shown that hyperbaric oxygen (oxygen under pressure) exerts a protective effect on the cyanide poisoned organism.

b. <u>Competitive inhibition</u> usually or always involves a substrate analogue which can bind in the same place as the substrate but which can only react very slowly or not at all. When used in clinical and biological situations (i.e., with whole cells or tissues), competitive inhibitors are frequently called antagonists or antimetabolites of the substrates with which they compete.

i) The reaction catalyzed by succinate dehydrogenase is shown below.

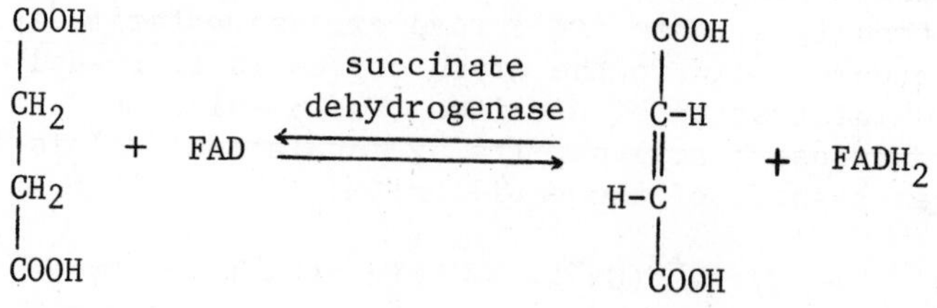

$$
\begin{array}{ccc}
\text{COOH} & & \text{COOH} \\
| & & | \\
\text{CH}_2 & & \text{C-H} \\
| \quad + \text{ FAD} \underset{\text{dehydrogenase}}{\overset{\text{succinate}}{\rightleftharpoons}} & & \| \\
\text{CH}_2 & & \text{H-C} \quad + \text{ FADH}_2 \\
| & & | \\
\text{COOH} & & \text{COOH}
\end{array}
$$

Succinate   Fumarate

(where FAD is a coenzyme which serves as a hydride ion acceptor). This enzyme is competitively inhibited by malonate which acts as an analogue for succinate.

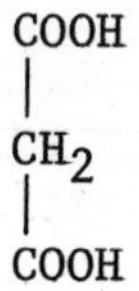

$$
\begin{array}{c}
\text{COOH} \\
| \\
\text{CH}_2 \\
| \\
\text{COOH}
\end{array}
$$

ii) Folic acid (pteroylglutamic acid) is a coenzyme used in certain biosynthetic reactions. Although it is a vitamin in man, some bacteria synthesize it using para-aminobenzoic acid (PABA). Sulfonamides are chemically and physically similar to PABA and can competitively inhibit the synthesis of folic acid, thus leading to inhibition of the growth of susceptible organisms. The structures are shown below.

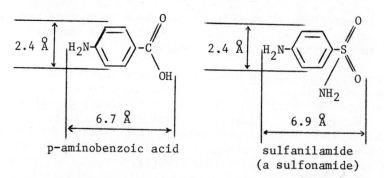

p-aminobenzoic acid   sulfanilamide
                                   (a sulfonamide)

iii) Prior to its functioning as a coenzyme, folic acid must undergo two reductions, first to dihydrofolate

(FH$_2$), then to tetrahydrofolate (FH$_4$). The reductases catalyzing these reactions are competitively inhibited by amethopterin (methotrexate) and aminopterin, making FH$_4$ unavailable. Since FH$_4$ is needed for the synthesis of DNA precursors, a deficiency of it does most harm to those cells synthesizing DNA rapidly. Certain types of malignancies, such as the leukemias, exhibit an extremely high rate of cell division and are consequently particularly susceptible to folate antagonists. These reactions and the structures of the compounds are discussed in the section on one-carbon metabolism.

iv) Xanthine oxidase (X.O.) is an enzyme of purine catabolism which catalyzes the oxidation of hypoxanthine to xanthine and xanthine to uric acid. Allopurinol, an analogue of hypoxanthine, competitively inhibits X.O. Its structure along with that of hypoxanthine is shown below.

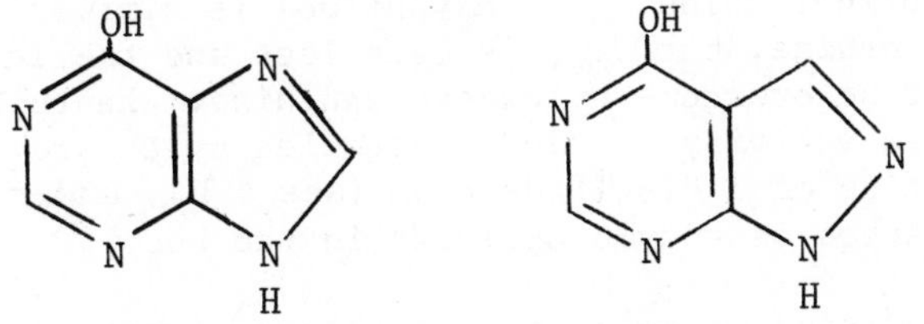

hypoxanthine   allopurinol

Allopurinol is useful for reducing blood uric acid levels in hyperuricemic conditions such as gout and is further discussed under gout. Xanthine oxidase also provides an interesting example of non-competitive inhibition (see below).

c. Non-competitive inhibition

i) It was mentioned in b(iv), above, that xanthine oxidase also can be non-competitively inhibited. In addition to being a competitive inhibitor, allopurinol functions as a substrate for X.O., about 45-65% of an administered dose being oxidized to oxypurinol (alloxanthine) within a few hours. Oxypurinol bears the same relationship to xanthine that allopurinol does to hypoxanthine.

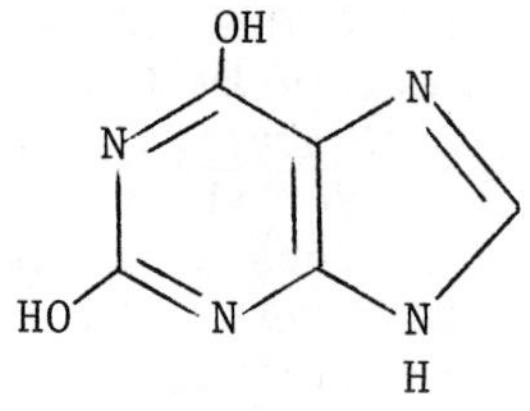

OH

xanthine

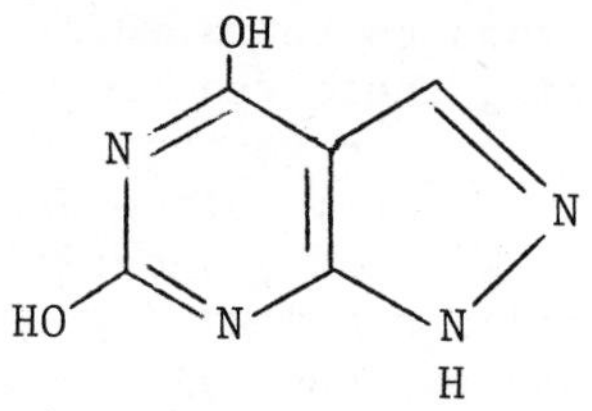

OH

oxypurinol
(alloxanthine)

In the absence of xanthine, allopurinol causes non-competitive inhibition, apparently by combining with a transition state of the enzyme during the oxidation of allopurinol to oxypurinol. Both K_m and V_{max} for allopurinol are less than for xanthine. In the presence of xanthine, allopurinol exhibits only competitive inhibition but oxypurinol behaves non-competitively, probably by the same mechanism as does allopurinol. The K_m of oxypurinol is similar to that for xanthine, but V_{max} is much less and the inhibition cannot be overcome by excess xanthine. Xanthine oxidase activity is also controlled by the reversible reduction of a disulfide bond (see 4.b., under regulation of enzyme activity in the body).

ii) In addition to being susceptible to irreversible inhibition, enzymes having cysteinyl sulfhydryl groups can be non-competitively inhibited by heavy metal ions (Ag^+, Pb^{+2}, Hg^{+2}, etc.). The general sort of reaction is shown below with Hg^{+2}.

$$E\text{-}SH + Hg^{+2} \rightleftharpoons E\text{-}S\text{-}Hg^+ + H^+.$$

Heavy metals can also form covalent bonds with $-COO^-$ and histidyl residues.

Allosteric Control of Enzyme Function

1. Allosterism was mentioned previously in **sections 2.f (p. 102) and 6.b (p. 110)** under enzyme kinetics. It is frequently observed in enzymes that catalyze the first step in a pathway which is unique to that pathway (the committed step). In this way, an end product of the pathway quite unlike the initial substrate can shut off its own synthesis when it has reached an appropriate concentration. This conserves material and shunts the initial substrate into alternate metabolic routes. Well-studied examples are found in the interactions of glycolysis,

gluconeogenesis, fatty acid synthesis and oxidation, and the TCA cycle. These will be discussed later.

2. An allosteric enzyme which has been extensively studied is aspartate transcarbamylase. It catalyzes the reaction:

carbamyl aspartate

Carbamyl aspartate is eventually converted to the pyrimidine nucleotides which are used in a number of processes including nucleic acid synthesis.

Aspartate transcarbamylase has a molecular weight of 310,000 and is made up of twelve polypeptides. Six peptides, designated α, have a molecular weight of 33,000 each; the other six (β-chains) have a molecular weight of 17,000 each. The α-chains combine into two groups of three, forming type A subunits of formula α_3. The β-chains form three B subunits each of formula β_2. The A subunits bind substrate but have no control sites and are termed <u>catalytic subunits</u>. The B subunits are <u>regulatory subunits</u> with control sites but no catalytic activity. The chain molecular weights do not add up exactly to 310,000 due to difficulties in measurement.

When pyrimidine nucleoside phosphates accumulate in the cell, they bind to some of the control sites and allosterically inhibit the enzyme. CTP is the most effective inhibitor. In addition, the <u>purine</u> nucleoside phosphates (adenine and guanine mono-, di-, and triphosphate: AMP, ADP, ATP, GMP, GDP, and GTP) allosterically <u>activate</u> this enzyme. This is logical since, in double-stranded DNA, the ratio (A+G)/(T+C) = 1. For most efficient DNA synthesis, then, purine and pyrimidine nucleotides should be available in equal amounts. Finally, the substrates themselves act as allosteric activators (positive allosteric effectors).

The physical change in the enzyme seems to be one of "swelling" upon binding substrate and releasing CTP (and presumably other negative effectors) and "contracting" into an inactive (or less active) state upon binding CTP. The higher activity form appears to dissociate into free subunits more readily than does the low activity state. This is summarized in the diagram below.

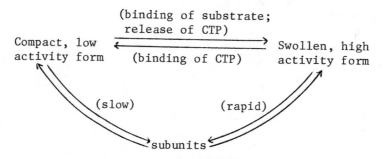

This example illustrates four important general characteristics of allosteric enzymes and their control.

Allosteric enzymes contain multiple subunits. The subunits may be identical or different. Each subunit may or may not be a separate polypeptide chain.

Changes between active and inactive forms are associated with changes in quaternary structure: the way in which the subunits interact with each other.

Control sites and catalytic sites are chemically and spatially separate. They may be on different subunits but are not necessarily so.

For a particular enzyme, allosteric effectors can include both substrate molecules and compounds very different from the substrate. Depending on the enzyme, members of either class may be positive or negative effectors.

3. The kinetics of allosteric systems are rather different from those of enzymes obeying Michaelis-Menten kinetics. For an allosteric enzyme, a plot of v versus S_0 is sigmoidal (S-shaped) rather than rectangular hyperbolic, as shown below.

Figure 16.   Comparison of Allosteric and Non-Allosteric Kinetics

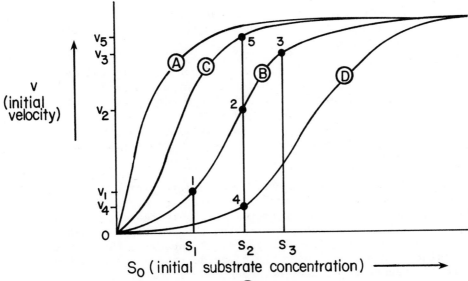

In this diagram: Curve (A) is a rectangular hyperbola,
characteristic of Michaelis-Menten kinetics.

Curve (B) is a typical sigmoid curve for an
allosteric enzyme; the only possible
effector present is substrate.

Curve (C) shows the effect of adding a
positive allosteric effector to the system
of Curve (B).

Curve (D) illustrates the result of adding
an allosteric inhibitor to the system of
Curve (B).

Notice that

i)   Curve (B) is sigmoidal even though only substrate is present.
If an enzyme is allosteric, it will not obey Michaelis-
Menten kinetics even in the absence of a modifier.

ii) At constant S_o (for example at $S_o = S_2$), the activity of
an allosteric enzyme can be controlled by changing the
concentrations of positive and negative effectors.

$$\text{no effector, } v = v_2$$
$$\text{positive effector, } v = v_5 > v_2$$
$$\text{negative effector, } v = v_4 < v_2$$

The important things to consider are the relative affinity of the enzyme for the effectors and the ratio of their concentrations.

iii) Although an allosteric enzyme is insensitive to changes in S_0 at low and high S_0-values, a large change in velocity is observed for small changes in S_0 at intermediate values (e.g. $v_1 \rightarrow v_3$ when $S_1 \rightarrow S_3$). The rate of a non-allosteric enzyme-catalyzed reaction is most sensitive to changes in S_0 at low concentrations of substrate.

The reason for the sigmoidal curve can be interpreted physically. At low S_0, the slope is small because it is difficult for the reaction to occur. As S_0 increases, the enzyme is activated in some way, making the reaction proceed more rapidly, and the slope increases. Eventually, the enzyme is working at a maximal rate, the slope decreases again, and further increases in S_0 have no effect on velocity. Although this appears to be the most common situation, cases are known where a substrate can actually cause a <u>decrease</u> in enzyme activity (for example, the binding of $\overline{NAD^+}$ to rabbit muscle D-glyceraldehyde-3-phosphate dehydrogenase). The v versus S_0 curves for such enzymes will not appear as above, although the enzymes are allosteric. There are many variations on the basic theme of allosterism.

4. There have been several models devised to explain the anomalous (non-Michaelis-Menten) kinetics exhibited by allosteric enzymes. The most important ones are those due to Monod, Wyman, and Changeux (MWC); Koshland, Nemethy, and Filmer (KNF); and Eigen (EIG).

   a. Several things are common to all three models.

      i) Allosteric enzymes are assumed to be oligomers (small polymers) of identical subunits. (It is possible, however, to modify the models to include more than one type of subunit.) A subunit may or may not be a single polypeptide chain.

      ii) There are multiple catalytic and regulatory binding sites on the subunits.

      iii) The change from active to inactive forms involves

non-covalent forces and results in alteration of the
quaternary structure of the enzyme. Changes in the
subunit (tertiary) conformation also generally occur.

Specific aspects of the models are given below.

b. MWC

   i) There are two conformations, designated R and T, of an
       allosteric enzyme. These differ in tertiary and
       quaternary structure; in their affinities for
       substrates and for positive and negative allosteric
       effectors; and in catalytic activity.

  ii) All binding sites in a particular conformation (R or
       T) are completely identical. This means that the only
       differences in affinity that can be observed are
       between R enzyme molecules and T enzyme molecules.
       On a given enzyme molecule, the binding of one
       substrate molecule does not affect the ease of binding
       of a second substrate molecule of the same type.

 iii) In a given enzyme molecule, all subunits are
       either in the R-form or the T-form. There are no
       mixtures of R- and T-subunits in the same enzyme
       molecule.

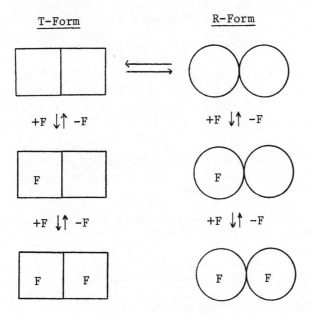

(where F is a substrate or effector molecule)

With nothing bound, the R- and T-forms are in equilibrium, with one form frequently predominating. As soon as a substrate (or effector) molecule binds, the enzyme molecule is "committed" to the form which it had when binding occurred. It cannot change form until all effector, substrate, or product molecules dissociate from it. (For example, the conversion of $\boxed{\text{F}\ \ }$ to $\text{ⓕ}\text{◯}$ cannot occur directly.) The binding of substrate or effector <u>does not</u> <u>cause</u> the conformational change.

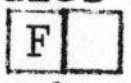

c.   KNF

    i)   In the absence of substrate or of allosteric effectors, the enzyme exists in only one conformation.

    ii) The binding of the substrate or effector to one of the subunits <u>causes</u> that subunit to change conformation (induced-fit hypothesis).

    iii) The basis for cooperativity is the difference in interaction between subunits having substrate bound and those without substrate. That is, there is a difference between circle-circle, circle-square, and square-square interactions in the diagram below.

For example, it would be easier for a square to become a circle if it already had a circle in contact with it. This example again uses a dimer.

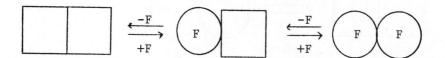

d.   EIG   This is a fusion and extension of the MWC and KNF schemes. The principle features are relaxations of the restraints of these models.

    i)   Changes in subunit conformation can occur whether or not a substrate or effector is bound.

    ii) There can be mixtures of subunits in different conformations in the same enzyme molecule.

This can best be visualized in a diagram.

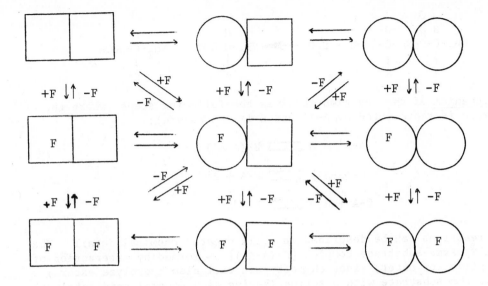

Because of difficulties in drawing, this omits the form
⬭F, which is in equilibrium with all of the structures
above except for those at the very corners. It is apparent
how the MWC and KNF models can be derived from this model
by elimination of certain forms of the enzyme.

e. The testing of these models on actual systems is just
beginning. One major problem is that all three necessarily
predict sigmoid curves for allosteric enzymes. Some other
measurement must therefore be used to differentiate between
the models in a particular system. A comparison of the
values predicted by the theories to the actual experimental
value of such a property will provide some basis on which
to choose between the models. Even then, it will probably
be found that each allosteric enzyme falls at some unique
point intermediate between the extremes.

Other allosteric systems will be pointed out as they are
encountered. Hemoglobin, the classical example, will be
discussed later in detail.

Enzyme Mechanisms

The <u>mechanism</u> of the reaction catalyzed by an enzyme is a detailed
description of the chemical reactions which occur between the substrates,
enzyme, and cofactors. For example, although the overall reaction
catalyzed by chymotrypsin (a pancreatic peptidase involved in digestion;
see Chapter 6) can be written

$$NH_2-\overset{H}{\underset{R}{C}}-\overset{\overset{O}{\|}}{C}-NH-\overset{H}{\underset{R'}{C}}-COOH + H_2O \longrightarrow NH_2-\overset{H}{\underset{R}{C}}-COOH + NH_2-\overset{H}{\underset{R'}{C}}-COOH$$

the <u>mechanism</u> of the reaction involves the following steps (where the substrate is represented by A-A' and E is the enzyme):

$$A\text{-}A' + E \rightleftharpoons E\text{-}(A\text{-}A')$$

$$E\text{-}(A\text{-}A') \rightleftharpoons E\text{-}A + A'$$

$$E\text{-}A + H_2O \rightleftharpoons E + A$$

These reactions can be described in still greater chemical detail. The initial enzyme-substrate complex [E-(A-A')] is formed by interaction of the peptide (or ester, since chymotrypsin will also hydrolyze esters) bond in the substrate with a serine (acting as a general acid catalyst) and a histidine (acting as a general base catalyst) in the active site of the enzyme. (Serine is found in the active site of many other hydrolases.)

An ionized aspartate residue helps to stabilize this complex. The enzyme-bound intermediate (E-A) has the structure

$$E(ser)-O-\overset{\overset{O}{\|}}{\underset{R}{C}}-\overset{H}{C}-NH_2$$

and is termed the acyl-enzyme intermediate (where the hydroxyl sidechain of serine is acylated). Release of the second product (A') involves hydrolysis of the acyl (ester) linkage and restoration of the enzyme to its original form.

This is only a rough outline of the chemical events going on in the active site of chymotrypsin as it is performing its function. The point here is not to give a detailed description of one enzyme but to illustrate the <u>type</u> of information that one needs to describe an enzyme mechanism.

1. What sort of experiments must be done to arrive at this description?

    a. <u>Isotope exchange</u>: If a particular atom of the substrate or product is labeled with an isotope other than the common one, its location at various intermediate steps can be followed. This provides information on exactly what is happening to each part of the molecule.

b. Irreversible-inhibition studies: There are reagents which react with only one type of amino acid residue. If an enzyme is inactivated by one of these reagents, it is reasonable to conclude that one or more of the reagent-specific amino acid residues is involved in enzyme catalysis.

c. pH dependence: If an acidic or basic residue is involved in catalysis, usually only one form (protonated or unprotonated) of the residue will work. By looking at enzyme activity as a function of pH, deductions about the role of such residues can be made.

d. Other kinetic studies, with and without reversible inhibitors of various types, can provide much useful information.

e. Structural studies, particularly ones which elucidate the three-dimensional relationships of the amino acid residues in the enzyme, permit judgments about whether the involvement of one or another sidechain is physically and sterically possible. Such techniques include nuclear magnetic resonance, electron spin resonance, single crystal X-ray diffraction, and crosslinking studies.

2. The real question to be answered by mechanistic studies is how can an enzyme catalyze the reactions that it does with such speed? There seem to be two general factors involved.

a. Because all of the reactants are bound to a small region on the enzyme, their effective concentrations are greatly increased. That is, the adjacent binding makes it easier for an effective collision to occur between reactants.

b. As a reaction proceeds, the conformation of the reacting molecules must change from that of the substrates to that of the products. This involves alterations in bonds, bond lengths, and bond angles. The activation energy represents (at least in part) the work needed to bring about this transformation. The conformation which lies between those of the substrates and products, and which has the maximum amount of energy, is termed the transition state intermediate. The other contribution which an enzyme makes to speed up a reaction is its use of the binding energy to distort the substrates, so that they approach or attain the conformation of the transition state.

3. Chymotrypsin and the serine proteases are among the most thoroughly studied enzymes in terms of mechanism. Other enzymes for which much information has been collected are lysozyme, carboxypeptidase, ribonuclease, lactate dehydrogenase, and aspartate transcarbamylase. This is not intended to be a complete list, but the X-ray structures

are known for all of these, and they represent the range of reaction types which are being studied.

4. Why are these things worth knowing? Aside from curiosity, an understanding of enzyme mechanisms may permit the development of methods for treating enzyme defects. Experiments directed towards the prevention of the hemolytic anemia which occurs in patients with sickle-cell hemoglobin (see Chapter 7) illustrate this possibility. Also, man-made catalysts are much less efficient than enzymes. Observing the strategies employed by nature to do its tasks may provide insight into how we can improve synthetic catalysts.

5. Comments on the Catalysis of Carboxypeptidase A:

Catalysis by carboxypeptidase A, a digestive enzyme which catalyzes the hydrolysis of C-terminal peptide bonds in a polypeptide chain, illustrates the principles of induced fit (substrate binding leads to structural changes at the active site so that the catalytic groups are brought into proper orientation) and electronic strain (a rearrangement in the electronic distribution of the substrate which makes it more susceptible to hydrolytic cleavage).

Figure 17 is a schematic representation of the binding of substrate (glycyltyrosine) to carboxypeptidase A. The enzyme slowly hydrolyzes the substrate to glycine and tyrosine, and thus has been beneficial in the elucidation of the mechanism of reaction. One postulated mechanism which fits chemical and kinetic data involves proton transfer from tyrosine residue 248 to the amide nitrogen and attack by water molecules on the substrate carbonyl carbon. The occurrence of these events is facilitated by polarization of the carbonyl carbon by zinc and hydrogen-bond formation between water and glutamic acid residue 270. The interaction which initiates structural changes at the catalytic site may involve the binding of the negatively charged terminal carboxyl group of the substrate with the positively charged guanidinium group of arginine 145. This leads to the displacement of water from the catalytic cavity, migration of other residues into their proper position around the substrate (induced fit), and an electronic strain in the substrate, leading to hydrolysis.

6. An example of an enzyme mechanism in which there is a steric distortion induced in the substrate molecule is provided by lysozyme. This enzyme catalyzes the cleavage of polysaccharides with an alternating sequence of N-acetylglucosamine and N-acetyl-neuraminic acid joined by $\beta(1\rightarrow4)$ glycosidic linkages. In the polysaccharide substrate, the six-membered hexose ring adjacent to the site of bond cleavage undergoes distortion into a half-chain conformation (an intermediate carbonium ion) in order to fit into

the active site. The energy needed for producing this distortion
may, in part, be derived from the binding energy of enzyme to
substrate.

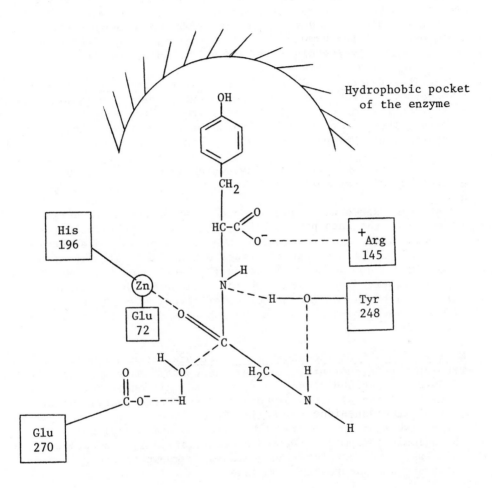

Figure 17. Schematic Representation of the Binding of Substrate to
Carboxypeptidase A.

The numbered amino acid residues placed in boxes are located at the
active site of the enzyme molecule, and the broken lines show their
interactions with the various parts of the substrate molecule.
Notice that the single zinc (Zn) atom which is essential for enzyme
activity is located near the carbonyl bond to be cleaved and the
hydrophobic pocket of the enzyme into which the side chain of the
carboxy-terminal residue fits.

Multiple Enzyme Forms

Enzymes are usually classified according to the reactions that they catalyze rather than by the nature of the enzyme protein. Consequently, there are frequent cases where one reaction can be catalyzed by several enzymes differing with regard to their structure. This even occurs between members of a single species or within individuals. Some of the causes of this multiplicity are discussed below, with examples.

1. a. Genetically-determined differences in the primary (amino acid) sequence. There can be two independent genes (cytoplasmic and mitochondrial malate dehydrogenase) or multiple alleles at one locus (glucose-6-phosphate dehydrogenase variants). The International Union of Biochemistry has recommended that the terms isoenzyme and isozyme be reserved for this type of enzyme multiplicity (see also b, below) which is due to primary structure differences.

    b. Quaternary structure. Two or more types of subunits, derived from separate loci, can combine in varying proportions. For example, lactate dehydrogenase (LDH or LD) in man is a tetramer composed of different combinations of two kinds of subunits, H and M. The five possible forms (H_4, MH_3, M_2H_2, M_3H, M_4) can all be observed upon electrophoresis of normal human serum. This will be discussed further under glycolysis. These are also termed isoenzymes since the subunits differ in amino acid sequence.

    c. Post-translational covalent modifications, both enzymatic and non-enzymatic, not affecting peptide bonds. Many proteins (glycogen phosphorylase and synthetase, histones, hormone-sensitive lipase) are phosphorylated and dephosphorylated during normal functioning of the cell. Other proteins are glycosylated or deglycosylated (liposomal hydrolases, human amylase, hemoglobin) enzymatically and non-enzymatically; and deamidation of asparagine and glutamine residues is known to occur in human salivary and pancreatic α-amylase.

    d. Precursor forms are known for a number of enzymes and other proteins. The active (functional) molecule is derived from them by the selective cleavage of one or more peptide bonds by the action of a protease. The classic examples include the zymogen forms of the digestive hydrolases secreted by the pancreas (chymotrypsinogen, procarboxypeptidase, trypsinogen, proelastase) and proinsulin, from which the C-peptide is cleaved to produce the active hormone. More recent examples are the so-called pre-forms of apparently all secretory enzymes (preamylase, preprocarboxypeptidase, etc.). The pre-prefix indicates the presence of a small N-terminal sequence which causes the growing

peptide chain to enter the cisternae of the endoplasmic
reticulum where it is prepared for secretion from the cell.

e. <u>Non-covalent association</u> occurs between proteins and various
cofactors and small molecules. For example, dihydrofolate
reductase can be isolated with and without bound NADPH; and
the metals in some metallo-enzymes can be removed by chelating
agents.

2. There are other causes for a single enzyme activity to assume
several forms. Polymers occur, either with one type of subunit
(glutamate dehydrogenase of molecular weight 250,000 and 1,000,000)
or several types (the macroamylases seen in some disease states).
Conformational differences, as in allosteric modification, can
produce multiple enzyme forms. In general, though, the five sources
of multiplicity listed above seem to account for the great majority
of such situations.

3. Implicit in the discussion above is the ability to distinguish one
enzyme form from another. The most common methods for doing this
are:

a. Electrophoresis and isoelectric focusing--Any change which
produces a charge difference (gain or loss of an acidic or
basic amino acid; association with or covalent attachment to
a charged molecule) will cause the protein to migrate at a
different rate in an electric field and will alter its
isoelectric point. Electrophoresis is discussed more fully
under LDH and hemoglobin.

b. Ion-exchange chromatography--This method depends on the same
type of change as electrophoresis. Neither technique will
detect changes which do not alter the net charge on the protein.

c. Molecular weight measurements--Gel filtration chromatography
and electrophoresis in a gel containing sodium dodecyl sulfate
(SDS; an anionic detergent) will separate molecules which differ
by at least 2,000 daltons in molecular weight. This has been
used to identify precursor forms and, in some cases, differing
degrees of glycosylation. Since these methods (especially SDS
electrophoresis) will tend to dissociate molecules held together
by non-covalent bonds, association polymers and subunits held
together by non-covalent interactions will be reduced to their
lowest covalently-bonded subunits.

d. Heat denaturation--Many amino acid changes will alter the ability
of an enzyme to maintain its active conformation at an elevated
temperature. This method is very sensitive to small changes in
structure (for example, single residue changes or the association

with small molecules), because of the large (100-150 kcal/mole) energy of activation for the denaturation reaction. This means that the slope of a plot of log denaturation rate versus 1/temperature is very steep and a change of 1^OC can cause a two-fold change in reaction rate. (Compare this to an enzyme-catalyzed reaction where E_{act} typically is about 10-15 kcal/mole and a 10^OC change in temperature is required to double the reaction rate.)

e. Microcomplement fixation--Uses immunological differences as a basis for distinguishing between molecules. It is very sensitive but has not been widely employed.

f. Kinetic properties--Such as inhibitor effects, K_m for natural substrates, actions of allosteric modifiers, and activity toward artificial substrates have all been used to demonstrate heterogeneity in an enzyme preparation.

4. The great majority of cases of multiple enzyme forms have been detected by methods which find only charge differences (particularly electrophoresis). This ignores any situations where the net charges on the molecule are unaltered. In terms of isoenzymes, only about one-third of the possible amino acid changes will produce a change in ionization so that many isoenzymic forms are missed and the genetic variation in an enzyme will be underestimated. It is because of considerations such as this that other techniques (variations in electrophoresis to really find all of the charge variants; heat denaturation; microcomplement fixation) are becoming more popular.

5. The biological significance of multiple enzyme forms is not known and it undoubtedly is different from one instance to another. Some reasons which have been suggested are listed here.

a. Regulation of enzyme activity is the explanation in most cases where the multiple forms seem to exert a significant effect on metabolism.

   i) The K_m may be tailored to meet varying substrate concentrations (hepatic glucokinase and hexokinase) or tissue needs (heart versus skeletal muscle LDH).

   ii) A precursor form of the protein may be useful to prevent premature activity of an enzyme (trypsinogen, proelastase), ensure proper folding of the polypeptide chain (proinsulin) or differentiate between secretory and non-secretory proteins (preamylase, prealbumin).

   iii) Phosphorylation, as an example of reversible covalent

modification of a protein, has been shown to be extremely important in regulation of activity.

b. Some forms may arise simply from aging of the enzyme (with or without the loss of activity). The reactions may be spontaneous (glycosylation of hemoglobin A_1 to produce A_{1c}; deamidation of amylase) or catalyzed by an endogenous or exogenous enzyme (deglycosylation of α-amylase).

c. Some multiple forms are the result of disease states (amylase complexes in macroamylasemia; increase in hemoglobin A_{1c} in diabetics; secretion of a unique form by a malignant tumor).

d. The presence of several loci may provide a buffer against genetic change. If one locus is lost by mutation the reaction can still occur because of the remaining isoenzyme.

e. There may be an adaptive advantage to genetic polymorphism (multiple allelic forms of a protein). The presence of two enzymes with different properties which can catalyze the same reaction may mean that an animal can survive under a wider range of dietary and climatic conditions. It has also been suggested, however, that the observed polymorphism simply represents non-deleterious (but also non-advantageous) mutations which have accumulated.

6. Isoenzymes (and other types of multiple forms, such as proinsulin) are of increasing importance in clinical diagnosis. Changes in isozyme patterns--the per cent of a total enzyme activity which is due to each isozymic form--of several enzymes have been correlated with various diseases and are being actively used in diagnosis. Examples include LDH (discussed later in Chapter 4), CPK, amylase, and alkaline phosphatase (see Chapter 15).

Regulation of Enzyme Activity in the Body

Several methods of enzyme control in biological systems, where temperature and pH are quite constant, have already been mentioned. There are others which are also important. The principal mechanisms are summarized below. A knowledge of them is necessary both to understand the influences of foreign materials (drugs, poisons) on the body and to explain the interdependence of bodily functions. All of the preceding material on kinetics, allosterism, and isozymes is ultimately useful in this way.

1. Inhibition. Competitive, non-competitive, irreversible, others; substrate analogues are of particular importance; mode of action of many drugs.

2. Allosterism. Usually involves multisubunit enzymes; binding
    of a molecule similar to or unlike the substrate to a site
    distinct from the catalytic site; can be positive or negative;
    frequently exhibited by a key control enzyme at the beginning
    of a pathway; examples include the control of hemoglobin
    oxygenation by O_2 and 2,3-diphosphoglycerate (although hemoglobin
    is not truly an enzyme), a)of pyruvate carboxylase by acetyl-
    CoA, b)of phosphofructokinase by AMP, ADP, ATP, and citrate,
    and many others.

3. Isozymes. Multiple isomeric forms of the same enzyme; catalyze
    the same reaction but differ in K_m, V_{max} or both; generally
    separable by electrophoresis; examples include lactic acid
    dehydrogenase, alkaline phosphatase, carbonic anhydrase, and
    many more.

4. Activation of Latent Enzymes

    a. Enzyme is synthesized in an inactive (zymogen, pro-protein,
        proenzyme) form which is activated, frequently by cleavage
        of one or more peptide bonds, before or after secretion.
        Examples include chymotrypsin (chymotrypsinogen), elastase
        (proelastase), pepsin (pepsinogen), and many others.

    b. Enzyme is converted back and forth between active and
        inactive states, depending on the needs of the body,
        usually by a reversible covalent modification such as
        phosphorylation or oxidation and reduction of disulfide
        bonds; mode of control exerted by epinephrine, insulin,
        and other hormones; frequently mediated by cyclic AMP;
        examples include glycogen synthetase and phosphorylase,
        xanthine oxidase (a further control on this enzyme), and
        probably adipose tissue lipase and palmityl-CoA synthetase.

5. Control of mRNA Translation. Similar, in principle, to 4 above;
    less conservative of material but able to respond more rapidly;
    less sensitive than 1, 2, or 4; examples include heme activation
    of globin synthesis in reticulocytes (this and other controls
    of hemoglobin synthesis are discussed under the control of
    protein synthesis and under hemoglobin); ACTH activation of
    desmolase (cholesterol oxidase) synthesis in the adrenal
    cortex (mechanism 4 (b) is also involved); large increase
    in protein synthesis observed upon fertilization of an ovum;
    albumin synthesis may be partly regulated in this manner; see
    also under albumin synthesis and regulation of protein
    synthesis.

6. Control of DNA Transcription. Prevention of mRNA synthesis
    from the DNA template coding for an enzyme; one explanation
    for inducible and repressible enzyme systems in bacteria and
    perhaps in higher animals; conserves energy and raw materials;
    the differentiation of different cell types is at least partly
    controlled in this way; see under regulation of protein synthesis.
    Steroid hormones bind initially to specific cytoplasmic receptors
    and the hormone-receptor complexes then bind to DNA, turning on
    the appropriate genes. The thyroid hormone, triiodothyronine
    (T_3), on the other hand, initiates its action by binding to
    specific receptors located on the chromatin.

7. Five and six (above) are examples of control of the rate of
    synthesis. The rate of degradation may, in some cases, be
    regulated as well. This has been studied very little, but in
    inbred mice a variant is known which has high levels of red-cell
    catalase due to decreased degradation of the enzyme.

Coenzymes

1. Many enzymes, in order to perform their catalytic function,
    require the presence of small, non-protein molecules. These
    are termed coenzymes, cofactors or prosthetic groups more or
    less interchangeably. In this text, these designations will
    be used with complete interchangeability. The enzyme with
    bound cofactor is called a holoenzyme, the protein portion
    being the apoenzyme.

         apoenzyme + cofactor = holoenzyme

    As implied in the above reaction, cofactor binding is
    frequently reversible. In some cases, however, it is
    covalently bound to the apoenzyme. If the linkage is
    non-covalent, the cofactor is said to be dialyzable (i.e.,
    because the cofactor is much smaller than the apoenzyme, the
    two can be separated by dialysis across a semipermeable
    membrane).

    Coenzymes also

    a. Are generally stable towards heat and other agents which
        denature proteins.

    b. Are frequently derived from vitamins. This is one of the
        principle functions for vitamins.

c. Are recycled in many cases. In such situations only small, catalytic amounts of them are needed for the conversion of a great deal of reactant to product. Biotin-requiring carboxylations serve as a good example.

d. Function as cosubstrates. They are chemically altered during the reaction and must be regenerated by another enzyme. The distinction between the substrate and the cofactor in these cases is that the cofactor is returned to its original form while the substrate usually undergoes further chemical modifications. Dehydrogenase reactions provide a good example of this.

2. A list of some important cofactors is given below. It is not complete but includes the principal ones. Their structure, synthesis, and function are discussed with the reactions they catalyze and under the vitamins from which they are derived.

a. Coenzymes involved in hydride (H^-) ion and electron transfer.

   i) Nicotinamide adenine dinucleotide (NAD^+); also known as diphosphopyridine nucleotide (DPN^+); reduced form is NADH (DPNH).

   ii) Nicotinamide adenine dinucleotide phosphate ($NADP^+$); also known as triphosphopyridine nucleotide (TPN^+); reduced form is NADPH (TPNH).

   (i and ii, above, both require niacin (nicotinic acid; nicotinamide) for their synthesis. This is a vitamin in man, although small amounts are also formed from the essential amino acid tryptophan.)

   iii) Flavin mononucleotide (FMN); reduced form is $FMNH_2$.

   iv) Flavin adenine dinucleotide (FAD); reduced form is $FADH_2$.

   (iii and iv, above, are derived from the vitamin riboflavin. They are not truly flavin nucleotides since the flavin ring system is attached to ribitol, a sugar alcohol, rather than ribose.)

   v) Lipoic acid (6,8-dithio-n-octanoic acid); not a vitamin in man, since it apparently can be synthesized in sufficient quantity; involved in acetyl-group transfer during oxidative decarboxylations of amino acids.

vi) <u>Coenzyme Q</u> (CoQ; ubiquinone); a group of closely related compounds, differing only in the length of a sidechain; can be synthesized in man from farnesyl pyrophosphate, an intermediate in cholesterol biosynthesis; structure is similar to that of vitamin K_1 but no interrelationship has been established.

b. Coenzymes participating in group transfer reactions.

    i) <u>Coenzyme A</u> (CoA, CoASH); -acetyl and other acyl group transfers; requires the vitamin pantothenic acid for its synthesis.

    ii) <u>Thiamine pyrophosphate</u> (TPP; Cocarboxylase); used for oxidative and nonoxidative decarboxylations of amino acids and formation of α-ketols (e.g., for the transketolase-catalyzed steps of the HMP shunt); derived from thiamine (vitamin B_1).

    iii) <u>Pyridoxal phosphate</u>; cofactor for an unusually wide variety of reaction types including amino acid racemization, decarboxylation and transamination and the elimination of water and hydrogen sulfide; usually associated with reactions of amino acids; derived from pyridoxine, pyridoxal, and pyridoxamine (collectively called vitamin B_6).

    iv) <u>Tetrahydrofolic acid</u> (FH_4); carrier of one-carbon fragments, such as formyl, methylene, methenyl, and formimino groups; derived from folic acid (folacin).

    v) <u>Biotin</u>; carboxylation reactions; tightly bound to the apoenzyme in an amide linkage to the ε-amino group of lysyl residue; biotin is itself a vitamin.

    vi) <u>Cobamide coenzyme</u> (5,6-dimethylbenzimidazole cobamide; 5'-dehydroadenosyl cobalamine); contains cobalt bound in a porphyrin-like carrier ring system; unusual in that it contains a cobalt-carbon (organometallic) bond; involved in methyl-transfer reactions; very small amounts required by the body; derived from cyanocobalamin (vitamin B_{12}).

    vii) <u>Adenosine triphosphate</u> (ATP); can contribute phosphate, adenosine, and adenosine monophosphate (AMP) for various purposes; has many other functions; does not contain any vitamin portion and hence the complete molecule is synthesized in the body.

viii) Cytidine diphosphate (CDP); carrier of phosphoryl choline, diacylglycerides, and other molecules during phospholipid synthesis; CDP does not contain any vitamin-derived group and can, consequently, be synthesized entirely in the body.

ix) Uridine diphosphate (UDP); carrier of monosaccharides and their derivatives in a variety of reactions; see bilirubin, lactose, galactose and mannose metabolism, glycogen synthesis, and other pathways; as with ATP and CDP, this molecule can be completely synthesized in the human body since it contains no vitamin moiety.

x) Phosphoadenosine phosphosulfate (PAPS; "active sulfate"); sulfate donor in synthesis of sulfur-containing mucopolysaccharides; sulfate donor in detoxification of sterols, steroids and other compounds; see metabolism of the sulfur-containing amino acids; derived from ATP and an inorganic sulfate, hence, contains no vitamin portions and can be synthesized completely by man.

xi) S-adenosyl methionine (SAM; "active methionine"); methyl group donor in biosynthetic reactions; formed from ATP and the essential amino acid methionine.

c. Metalloenzymes. In addition to cobamide (cobalt) requiring enzymes, many others require a variety of metals for activity, maintenance of tertiary and quaternary structure, or both. If these metals are present complexed to a porphyrin ring system, the coenzyme is the metalloporphyrin. A few examples are given below to indicate the diversity of metals involved.

i) Mg^{+2} is required by most enzymes using ATP. The active form of ATP is a Mg^{+2}-ATP^{-4} complex.

ii) Ca^{+2} is involved in a wide variety of processes, notably muscle contraction, blood clotting, nerve impulse transmission, and cAMP-mediated processes. The exact way in which interaction occurs between Ca^{+2} and protein is not known.

iii) Fe^{+2}/Fe^{+3} (ferrous/ferric iron); iron is needed in hemoglobin, the cytochrome chain of oxidative phosphorylation, non-heme iron proteins, and other enzyme systems.

iv) Cu^{+1}/Cu^{+2} (cuprous/cupric copper); cytochrome oxidase in mitochondrial electron transport contains iron and copper. Tyrosinase and other oxidases also require copper.

v) Zn^{+2} (zinc); is used by lactic acid dehydrogenase, the alcohol dehydrogenases, carbonic anhydrase and other enzymes.

vi) Mo^{+6} (molybdenum ion); is part of xanthine oxidase (in addition to iron), and of other oxidases and dehydrogenases.

vii) Mn^{+2} (manganous manganese ion); is required by acetyl-CoA carboxylase, deoxyribonuclease (Mg^{+2} can replace Mn^{+2} in this case) and other enzymes.

viii)   Other metals may be involved in trace amounts as cofactors for some enzymes. Because of the minute quantity of the metal usually required, studies of this sort are difficult.

The Use of Enzymes in the Clinical Laboratory

1.   There are two principal ways in which enzymes are used clinically. They are

   a.   determinations of the amount of an enzyme (or enzyme activity) present in a biological sample,

   b.   the use of enzymes as specific reagents to measure the **concentration of other enzymes or non-enzymic molecules** (e.g., substrates) in biological material.

Examples of these are given below and in other sections of this book.

Most clinical tests involving enzymes make use of kinetics, regardless of the role that the enzyme plays in the procedure. In the section on enzyme kinetics it was pointed out that the velocity (measured as the change in some property with respect to time) of an enzyme-catalyzed reaction depends on, among other things, the concentrations of enzymes and substrate. In assays in which the enzyme activity is being determined in a biological sample, the

substrate is present in great excess and the concentration of the
enzyme determines the velocity. When an enzyme is used as a
reagent, it is present in excess and the concentration of the
compound being assayed (the substrate) is the factor limiting the
velocity.

2. a. Again referring to enzyme kinetics, if the initial substrate
    concentration, S_O, is sufficiently high, the velocity of an
    enzyme-catalyzed reaction will be independent of S_O (zero-order
    with respect to substrate; $v = k(E)$). In addition, if the
    period of measurement is short enough, little substrate is
    consumed. The substrate concentration can then be considered
    constant ($= S_O$) and the reaction will be zero-order during
    the entire assay. For such a system, if X is the property
    being measured, a plot of X <u>v.s.</u> time is a straight line as
    shown in the diagram below.

(optical density,
oxygen pressure,
amount of acid
produced, or any
other property
which can be
monitored)

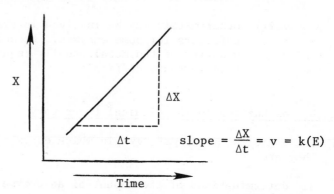

$$\text{slope} = \frac{\Delta X}{\Delta t} = v = k(E)$$

If all other conditions (temperature, pH, S_O, etc.) are constant,
the slope (velocity) depends only on (E). After k is determined
using a sample having a known enzyme activity, this relationship
can be used to quantitate enzyme activities in biological
materials.

This is the principle behind <u>fixed incubation time (two-point)
assays</u>. In such an assay the property being monitored is
measured twice: immediately after mixing the reactants and again
after some time interval. The velocity is the change in the
property over the time interval. If the X <u>v.s.</u> time curve is
linear, with maximum slope this is valid. In some cases, however,
the velocity is not constant initially, although it may become so
later (if a lag period occurs, for example). This results in
a false value for the enzyme activity. To avoid this problem,
methods have been developed for continuously monitoring the
rates of some reactions. These are called <u>kinetic assays</u>.
(This nomenclature is unfortunate, since, as was indicated
previously, <u>all</u> enzyme assays are kinetic in the sense that

they all employ enzyme kinetics.) If such an assay is used, a continuous plot of the property being measured as a function of time is obtained and an appropriate linear portion of the curve can be selected. One major cause for uncertainty in the enzyme assay results can thus be eliminated.

b. Enzymes in plasma can be grouped into two classes: "plasma specific" and "non-plasma specific." The plasma-specific enzymes are those which are normally present in the plasma at higher levels than in most tissue cells and which perform their primary function in the plasma. Examples of these are lipoprotein lipase, cholinesterase, ceruloplasmin, plasmin, and thrombin. These enzymes are synthesized in the liver and released into the circulation at a rate which maintains their optimal steady-state concentrations. Genetic defects or impaired liver function can cause them to fall to suboptimal concentrations.

Thrombin and plasmin deficiency, respectively, cause defective blood coagulation and fibrinolysis; deficiency of ceruloplasmin (a copper-containing oxidase) is associated with hepatolenticular degeneration (Wilson's disease); sensitivity to the muscle relaxant succinylcholine can arise from the absence or deficiency of cholinesterase.

c. Non-plasma-specific enzymes are normally absent from plasma or present at much lower concentrations compared with their concentrations in tissues. They can be further subdivided into secretory enzymes (e.g., amylase, lipase, pepsin, trypsin) and constitutive enzymes associated with the metabolism of the cell (e.g., transaminases, dehydrogenases, kinases, etc.). The constitutive enzymes can be categorized as enzymes present in most cells of the body (e.g., glycolytic enzymes) and those found principally in one or two tissues (tissue- or organ-specific enzymes). In some instances (lactate dehydrogenase, creatine kinase, alkaline and acid phosphatases) one of several isoenzymic forms can be tissue specific (heart and skeletal muscle forms of LDH and CPK; skeletal muscle, bone, liver, prostate, and placental forms of the phosphatases). Elevation of plasma concentrations of the non-plasma-specific enzymes suggests an abnormality (tissue breakdown or ectopic secretion). Examples of this are discussed below and throughout the book.

d. The plasma levels of secretory enzymes are increased when their usual pathways of secretion are obstructed. For example, large amounts of pancreatic amylase and lipase are present in the circulation in patients having pancreatitis. In these cases, the normal secretory route (into the digestive tract) is obstructed, causing release of these enzymes into the circulation.

Salivary amylase is found in the bloodstream in cases of mumps.

e. The factors that affect the release of intracellular enzymes are cell membrane permeability changes and/or destruction. The primary condition that can lead to these changes is the level of ATP in the cell which is required for maintaining the membrane integrity. A decrease in intracellular ATP concentration can result from several causes. These include: a deficiency of one or more of the enzymes needed in ATP synthesis (e.g.,pyruvate kinase in red blood cells); glucose deprivation; anoxia; high extracellular K^+ (ATP depletion in this case is due to the increased activity of the Na^+/K^+ ATPase in the plasma membrane which is required to maintain the proper K^+/Na^+ ratio between the intracellular and extracellular environments).

f. The amount of enzyme released depends upon the degree of cellular damage, the intracellular concentration of the enzyme, and the mass of the affected tissue. The nature of the insult (viral infection, surgical hypoxia, or chemical or mechanical trauma) has no bearing on the enzyme released into the circulation.

g. The type of enzyme released will also reflect the severity of the damage. In general, mild inflammatory conditions are likely to release cytoplasmic enzymes while necrotic conditions yield enzymes from mitochondria as well. Thus, in severe hepatitis and active cirrhosis, the serum aspartate transaminase is raised to extremely high levels (much greater than those of alanine transaminase) because of the mitochondrial damage which yields another isoenzyme of AST in addition to the enzyme released from the cytoplasm.

h. In general, the amount of enzyme released into the plasma from an injured tissue is much greater than can be accounted for on the basis of the tissue enzyme concentration and the magnitude of injury. This has been attributed to the fact that the loss of enzymes from the cells may stimulate their synthesis, which serves to further increase the amount released. For example, in experimental animals, biliary occlusion leads to increased synthesis of alkaline phosphatase. A number of drugs, such as the barbiturates and hydantoins, stimulate synthesis of bilirubin glucuronyl transferase; alcohol leads to increased synthesis of gammaglutamyl transferase. Both of these transferases are microsomal enzymes. It should also be emphasized that, while enzymes in the plasma can become elevated due to tissue injury, the levels may drop (in spite of the continued progress of the injury) to normal or below normal levels when there is a compromised blood circulation or when the functional part of the tissue is replaced with non-functional material (e.g., connective tissue, as in fibrosis).

i. Processes for inactivation or removal of enzymes from the blood include: denaturation of the enzyme due to dilution in the plasma or separation from its natural substrate or coenzyme; presence of enzyme inhibitors (e.g., falsely decreased activity of amylase in acute pancreatitis with hyperlipemia); removal by the reticulo-endothelial system; digestion by circulating proteases; uptake by the tissues and subsequent breakdown by tissue proteases; and clearance in the kidneys of low molecular mass enzymes (amylase and lysozyme). The removal of alkaline phosphatase via the bile has now been discredited.

j. Following is a schematic representation of causes for appearance and disappearance of non-plasma-specific enzymes:

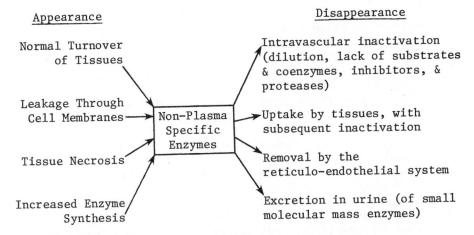

Appearance

Normal Turnover of Tissues

Leakage Through Cell Membranes

Tissue Necrosis

Increased Enzyme Synthesis

Non-Plasma Specific Enzymes

Disappearance

Intravascular inactivation (dilution, lack of substrates & coenzymes, inhibitors, & proteases)

Uptake by tissues, with subsequent inactivation

Removal by the reticulo-endothelial system

Excretion in urine (of small molecular mass enzymes)

k. It is apparent from the above discussion that enzymes have different rates of disappearance, making it important to know when the blood specimen was obtained relative to the time of injury. Additionally, it is essential to know how soon after the occurrence of the injury when various enzyme levels begin to rise. Following is an example of the plasma enzyme changes that take place in myocardial infarction.

Enzyme	Earliest Evidence of Rise Post-Infarction	Half-Life
Creatine kinase (CK-MB)	3-6 hours	1.3 days
Aspartate transaminase (AST)	6-8 hours	2.0 days
Lactate dehydrogenase (LD)	12 hours	4.4 days

LD shows a biphasic half-life profile. This is due to different rates of removal of various isoenzyme fractions. LD5 is removed more rapidly than LD1, the latter having a half-life of 6 days.

In chronic conditions, however, the enzyme levels continue to be elevated until the abnormality has been corrected or relieved (for example, alkaline phosphatase in obstructive jaundice, transaminase in hepatitis).

1. It is important to emphasize the use of appropriate normal ranges in evaluating abnormal levels of plasma enzymes. The former is affected by a variety of factors such as age, sex, race, degree of obesity, pregnancy, alcohol and other drug consumption, and malnutrition. Drugs not only can alter enzyme levels in vivo but may also interfere with their measurement, in vitro. A detailed list of conditions that affect enzymes and other plasma parameters is available (Clinical Chemistry 21(5), April, 1975).

m. Enzymes may also be measured in biological specimens other than plasma, such as urine, cerebrospinal fluid, bone-marrow, amniotic cells or fluid, red blood cells, and leukocytes. Cytochemical localization is also possible in leukocytes and in biopsy specimens from a variety of tissues such as liver and muscle. These aspects are discussed later, in the appropriate sections.

n. In order to understand what an enzyme activity means in terms of normal values, it is necessary to know the units in which it is measured. Thus, alkaline phosphatase activity may be expressed in Bodansky units, King-Armstrong units, or Bessey-Lowry-Brock units, all of which give different numerical values for the normal range. In general, one unit of an enzyme will transform some defined quantity of substrate in a given length of time under specified conditions. To standardize reports of enzyme activity, the Commission on Enzymes of the International Union of Biochemistry has proposed that the unit of enzyme activity be defined as that quantity of enzyme which will catalyze the reaction of one micromole of substrate per minute under standard conditions, and that this unit be called the International Unit (U. or I.U.). The "standard conditions" refer to optimal pH and substrate concentration and, where practical, a temperature of 30°C. (This supersedes the original recommendation of a 25°C temperature.)

o. Under ideal conditions, it is essential to measure both the concentration of the enzyme present and its activity. A low activity may be due either to the absence of an enzyme or to a lack of functional enzyme. The concentration of the enzyme is not commonly measured, however, because of the unavailability of pure human enzyme standards. This is either due to the fact that the enzyme has not yet been isolated in a pure form or that it is not available in sufficient quantities for its use to be

practical. Therefore, the common practice is to measure enzyme
activity. Herein lies one of the main problems in this field,
that is, the lack of an accurate reference method for each
enzyme assayed.

p. Many enzymatic reactions can be monitored at 340 nm if they
are linked (either directly or indirectly) with a pyridine-
nucleotide (NAD^+ or $NADP^+$) dehydrogenase reaction. The coenzyme
absorbs ultraviolet light at 340 nm in the reduced state,
whereas the oxidized coenzyme does not. Hence, changes in
reaction rate in either direction, oxidation or reduction, can
be followed by measuring absorbance at 340 nm. The following
is an example of one such enzyme assay.

Red cell G6PD can be specifically assayed by preparing a red
cell hemolysate. The enzyme catalyzes the reaction:

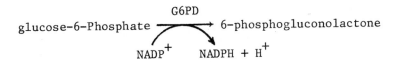

$$\text{glucose-6-Phosphate} \xrightarrow{\text{G6PD}} \text{6-phosphogluconolactone}$$
$$NADP^+ \qquad NADPH + H^+$$

Figure 18. Absorption Spectrum of the Oxidized and Reduced Forms of
NADPH. The equivalent spectra for NADH are essentially
the same.

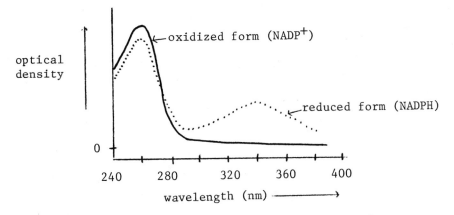

Since NADPH absorbs light at 340 nm while $NADP^+$ does not (see
Figure 18; the rate of the reaction can be followed by measuring
the change in optical density at this wavelength with time. This
assay not only illustrates the measurement of a tissue enzyme,
but it is also an example of a pyridine-nucleotide linked assay.
The difference in optical density between $NADP^+$ and NADPH (also
NAD^+ and NADH) is so readily measured, that a number of assays
make use of it either directly, as here, or indirectly. In the

latter situation, the reaction actually being measured is linked by one or more steps to a reaction in which a pyridine nucleotide is oxidized or reduced. Such an example is seen in the measurement of transaminases (see amino acid metabolism), where one of the products of the primary reaction serves as a substrate for a secondary reaction in which the reduced NADH is required.

$$\begin{array}{c} \text{Aspartate} \\ + \\ \text{α-Ketoglutarate} \end{array} \underset{\longleftarrow}{\overset{\text{Transaminase}}{\longrightarrow}} \begin{array}{c} \text{Oxaloacetate} \\ + \\ \text{Glutamate} \end{array}$$

$$\begin{array}{c} \text{Oxaloacetate} \\ + \\ \text{NADH + H}^+ \end{array} \xrightarrow{\begin{array}{c}\text{Malate}\\\hline\text{Dehydrogenase}\end{array}} \text{Malate + NAD}^+$$

In practice, the molar absorptivity (absorbance or optical density of a one molar solution one centimeter thick) of NADH is used in the following way:

$$A = a\ b\ c$$

where A = observed absorbance (optical density, O.D.),
a = molar absorptivity,
b = pathlength of the spectrophotometer cell in
centimeters, and
c = concentration of the solution in moles/liter
(molarity, M).

This relationship is known as <u>Beer's Law</u> or the Beer-Lambert Law.

For NADH (and (NADPH), at 340 nm., $a = 6.22 \times 10^3$

If b = 1 cm and c = 1 M, then $A = 6.22 \times 10^3$ O.D. units.

If b = 1 cm and c = 1 μmole/ml $(= 10^{-3}$ M), then A = 6.22 O.D. units.

In an enzyme assay in which the amount of NAD^+ (or $NADP^+$) reduced or NADH (or NADPH) oxidized is proportional to enzyme activity, the international unit is related to absorbance change (ΔA) by the equation:

$$\text{I.U.}/\text{ml} = \frac{\Delta A}{\text{min.}} \ \times \ \frac{V_t}{6.22\ V_s}$$

where V_t = total volume of the reaction mixture being measured and V_s = volume of the sample (serum, urine, etc.) being assayed. It is important to emphasize that changes in A can be used to measure concentration because <u>a</u> is a constant for a given substrate and <u>b</u> is made constant.

q. A number of other enzymes are also very important in clinical diagnosis. One example, discussed under glycolysis, is lactic acid dehydrogenase (LDH) and its isozymes. Others include creatine phosphokinase (CPK), alkaline and acid phosphatase, lipase, amylase (discussed in the next section), the transaminases (glutamate-oxaloacetate transaminase (GOT; aspartate transaminase) and glutamate-pyruvate transaminase (GPT; alanine transaminase)), and many more.

3. a. If an enzyme is being used as a reagent, the substrate is the limiting quantity, as was pointed out earlier. The amount of substrate consumed or product formed is proportional to the initial substrate concentration, S_o. The initial enzyme concentration, E_o, is set high enough to avoid saturation by any substrate concentrations which are likely to be encountered (i.e. (E) will never limit the observed velocity).

Unlike the assays discussed so far, the velocity here varies with time since the substrate concentration is not constant. Consequently, a fixed incubation time is used for all standards and unknowns. The average velocity over this time interval, when plotted <u>v.s.</u> S_o, should give a straight line from which unknown values of S_o can be calculated.

b. One example of an assay of this type is the use of glucose oxidase to measure serum glucose levels. This enzyme catalyzes the reaction

$$\beta\text{-D-glucose} + O_2 + H_2O \xrightarrow{\text{glucose oxidase}} H_2O_2 + \text{gluconic acid.}$$

The reaction can be followed by measuring P_{O_2} with a Clark oxygen electrode. If, in addition, peroxidase and ferrocyanide are added, the reactions

$$H_2O_2 \xrightarrow{\text{Peroxidase}} H_2O + \text{Nascent oxygen}$$

$$\text{Nascent oxygen} + \text{ferrocyanide} \longrightarrow \text{ferricyanide}$$

can occur. Since ferrocyanide is colorless while ferricyanide is red, the reaction can now be followed spectrophotometrically. Nascent oxygen is a more reactive form of oxygen than O_2 (molecular oxygen), and is probably atomic oxygen.

c. The serum level of amylase is elevated in acute pancreatitis. Its assay provides an interesting example of the use of linked reactions and of the use of enzymes as reagents in measuring the activity of another enzyme. The serum to be assayed is added to a reaction mixture containing starch, maltase, and glucose oxidase, resulting in the reactions

$$\text{starch} \xrightarrow{\text{amylase}} \text{maltose} + \text{glucose}$$

$$\text{maltose} \xrightarrow{\text{maltase}} \beta\text{-D-glucose}$$

$$\beta\text{-D-glucose} + O_2 \xrightarrow{\text{glucose oxidase}} \text{gluconic acid} + H_2O_2.$$

The reaction is followed the same way as in the measurement of glucose in the preceding example. As in that example, peroxidase and ferrocyanide can be used for a colorimetric procedure.

4. There are several other important factors which pertain to all measurements involving enzymes in the clinical laboratory.

   a. It is important that only S_o or (E) vary in a given enzyme assay. All other reaction conditions (temperature, pH, other molecules present) must be constant. The analyst will otherwise be unable to decide whether an observed value is due to the component being measured or the result of other unrelated effects.

   b. Enzymes are very sensitive to temperature and other denaturing agents, and to inhibition by drugs, other foreign substances, and, in some cases, natural, endogenous inhibitors. One must be very careful then in assessing an enzyme activity value, that the observed activity is not altered by effects such as those mentioned above.

Table   6 . Alteration of Serum Enzymes in Diseases

Tissue/Cell	Disease	Comments
From Myocardium and Related Organs		
LD (LD 1 and 2)	Myocardial infarction (MI)	1>2 (1/2 Flip)
LD (LD 1, 2 and 5)	MI with congestive heart failure	1>2 with 5↑
CK (CPK II or MB fraction)	MI	MB↑
AST	MI	↑

The above enzymes are slightly increased or normal in pericarditis, angina pectoris or myocarditis.

Tissue/Cell	Disease	Comments
From Liver and Biliary Tract		
Aspartate transaminase (AST)	Hepatocellular disease	↑
Alanine transaminase (ALT)	Acute & chronic hepatitis, hepatocellular disease & cirrhosis	↑ usually (ALT>AST)
LD (LD5)	Hepatocellular disease	↑
Alkaline Phosphatase	Cholestasis (an obstructive disorder of the biliary tract)	↑
5' Nucleotidase	"	↑
Gammaglutamyl transferase	"	↑
Leucine aminopeptidase	"	↑

Tissue/Cell	Disease	Comments
From Skeletal Muscle		
CK (MM fraction)	Myopathies	↑
Aldolase	"	↑

Note that in uncomplicated neurogenic myopathies the enzymes are not elevated.

CK	Thyrotoxicosis	↓
CK	Hypothyroidism	↑

Tissue/Cell	Disease	Comments
From Bone		
Alkaline Phosphatase	Paget's disease and other mineralization defects, osteomalacia, rickets	↑ ↑ ↑

Note that in osteoporosis alkaline phosphatase is not elevated.

Tissue/Cell	Disease	Comments
From Prostate		
Acid Phosphatase	Carcinoma	↑

Tissue/Cell	Disease	Comments
From Leukocytes		
Muramidase (lysozyme)	Acute monocytic leukemia, acute & chronic myeloid leukemia lymphocytic leukemia	↑ ↑ ↓

Notes: ↑ = increase; ↓ = decrease; 1>2 refers to relative isoenzyme concentrations.

4
Carbohydrates

<u>Structures and Classes</u>

<u>Carbohydrates</u> (also called sugars or saccharides) are divided into <u>four</u> major groups.

1. <u>Monosaccharides</u> cannot be hydrolyzed into a simpler form. Their general formula is $C_n(H_2O)_n$. The simplest compound having this formula is

$$CH_2O \text{ (or } H-\overset{\displaystyle O}{\overset{\|}{C}}-H)$$

named formaldehyde, but it is not usually termed a carbohydrate. The monosaccharides are called trioses, tetroses, pentoses, hexoses and so on, according to the number of carbon atoms in the molecule. They are further divided into aldo-sugars and keto-sugars. The prefix indicates the presence of an aldehyde or ketone group. The monosaccharides of biological importance in man are

<u>Trioses</u>

D-glyceraldehyde (glycerose; an aldotriose)
dihydroxyacetone (a ketotriose)

<u>Tetroses</u>

D-erythrose (an aldotetrose)

<u>Pentoses</u>

D-xylulose (a ketopentose)
D-ribose (an aldopentose)
D-deoxyribose (an aldopentose)
D-xylose (an aldopentose)
D-lyxose (an aldopentose)

<u>Hexoses</u>

D-glucose (an aldohexose)
D-galactose (an aldohexose)
D-mannose (an aldohexose)
D-fructose (a ketohexose)

154

L-idose (an aldohexose)
L-fucose (6-deoxy-L-galactose)

Heptoses

D-sedoheptulose (a ketoheptose)

2. Disaccharides are composed of two of the same or different
    monosaccharides. The general formula is $C_n(H_2O)_{n-1}$.

    Examples include lactose, maltose, cellobiose (all reducing sugars)
    and sucrose (a non-reducing sugar).

3. Polysaccharides are composed of many monosaccharide units. For
    those of biological importance, the general formula is $(C_6(H_2O)_5)_n$.
    This indicates a polymer of hexoses.

    a. Homopolysaccharides contain only one type of monosaccharide
        unit. Examples include

                cellulose.......... a polymer of glucose
                glycogen........... a polymer of glucose
                starch............. a polymer of glucose
                inulin............. a polymer of fructose

    b. Heteropolysaccharides contain two or more monosaccharide units.
        Examples include: agar-agar, vegetable gums, mucilages, and
        mucin.

    c. Mucopolysaccharides are nitrogen-containing polysaccharides
        such as heparin, chondroitin sulfate, hyaluronic acid and
        chitin.

4. Miscellaneous Compounds (carbohydrate derivatives): Sialic acid,
    vitamin C, streptomycin, inositol, glucuronic acid, sorbitol,
    galactitol, xylitol, and others.

Structural Considerations of Monosaccharides

1. Assignment of absolute configuration, designated as D or L (not to
    be confused with d and l, which indicate the sign of rotation of
    the plane polarized light) is based on the position of substituent
    groups about the penultimate (next to last) carbon. (Refer to
    the material on the optical activity of the amino acids.)

2. If two compounds differ from one another in configuration around
    one carbon, they are termed epimers. Galactose and glucose are
    epimers with respect to carbon 4; mannose and glucose are epimers

with respect to carbon 2. The interconversion of glucose and
galactose is known as epimerization and the enzyme which catalyzes
this reaction is an epimerase. The number of optical isomers of
a compound depends on the number of asymmetric carbon atoms (carbons
with four different substituent groups). If n = the number of
asymmetric carbons, 2^n = the number of optical isomers.

$$
\begin{array}{l}
^1CHO \\
| \\
H-^2C-OH \\
| \\
OH-^3C-H \\
| \\
H-^4C-OH \\
| \\
H-^5C-OH \\
| \\
H_2-^6C-OH
\end{array}
$$

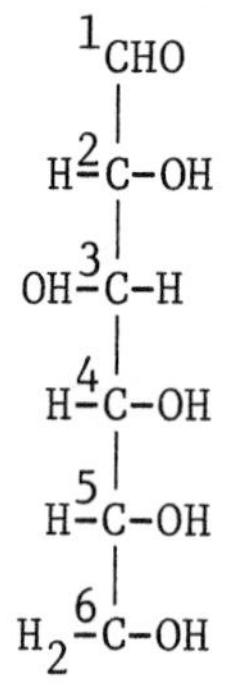

The structure that is written in the
adjacent form has carbons, 2, 3, 4, 5
as asymmetric centers. Therefore the
number of optical isomers expected is
2^4 = 16 isomers (of which D-glucose,
shown here, is one such isomer).
This is a Fischer projection formula.

3. The observed optical activity of glucose, however, indicates that
there were two forms of D-glucose itself. For this to be so, there
must be another asymmetric carbon. This additional asymmetric
center is generated at the carbon atom of an aldehyde (C_1 of
glucose) or keto (C_2 of fructose) group by reaction of either one
of these groups with an alcohol group in the same sugar molecule.
This gives rise to an internal hemiacetal (or hemiketal)
structure. A hemiacetal or hemiketal is a compound formed by
reaction of an aldehyde or ketone with an alcohol.

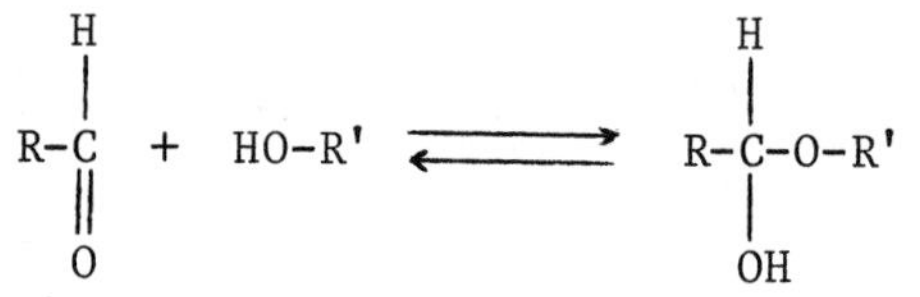

$$
\begin{array}{ccc}
H & & H \\
| & & | \\
R-C \quad + \quad HO-R' & \rightleftarrows & R-C-O-R' \\
\| & & | \\
O & & OH
\end{array}
$$

Note that the reactions are reversible so that the reactants and
products are readily interconvertible.

In the case of glucose, the functional groups of carbon 1 (aldehyde) and carbon 5 (alcohol) will react to give the cyclic hemiacetal form shown below.

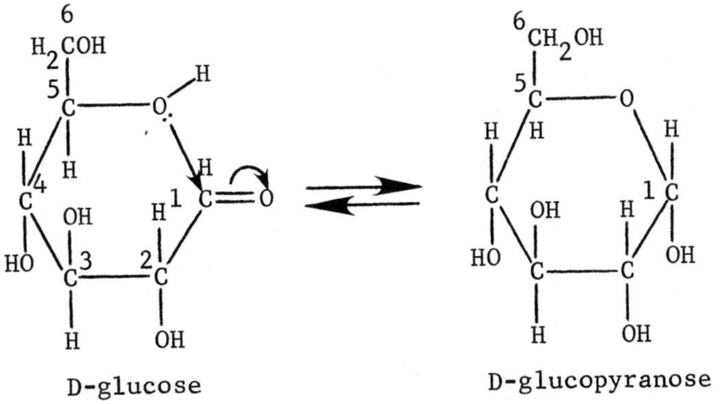

D-glucose   D-glucopyranose

Note: In the above reaction, the proximally located oxygen (a nucleophile) of the hydroxyl group linked to C5 attacks the electron-deficient aldehyde at carbon 1. Carbon 1, known as the anomeric carbon, is an additional, new asymmetric center. The six-membered rings are termed pyranoses because they are derivatives of pyran, a heterocyclic compound. Thus, the systematic name for the ring form of D-glucose is D-glucopyranose.

The formation of the two hemiacetal forms of D-glucopyranose is shown below, using Haworth projection formulae.

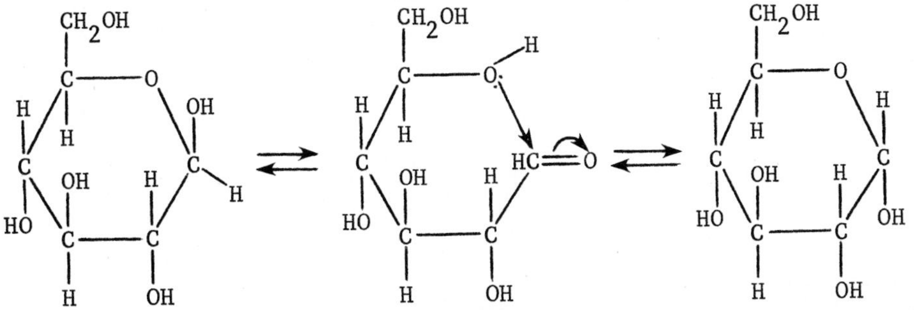

β-D-glucopyranose   D-glucose   α-D-glucopyranose
                           (open form)

The α and β forms of glucose designated in the above structures are known as <u>anomers</u>. Notice the α and β forms differ only in the configuration around carbon 1. Also note that the cyclic hemiacetal forms are in equilibrium with the open chain structure in aqueous systems.

Thus the formation of the hemiacetal structure introduces a <u>new asymmetric center</u> at C_1 (which now has four different groups attached to it) of D-glucose. Therefore the total number of isomers of glucose is $2^5 = 32$.

The α and β cyclic forms of D-glucose (known as anomers) have different optical rotations. These values, however, will not be of equal magnitude with opposite signs as one would expect from enantiomorphs (mirror images). This is because the α and β isomers as a whole are not mirror images of each other. They differ in configuration about the anomeric carbon (1) but have the same configuration at other asymmetric carbons (2,3,4,5).

Crystalline glucose is a mixture of the α and β forms (ordinary shelf crystalline glucose is largely the α-form). When glucose is dissolved in water, the optical rotation of the solution gradually changes and attains an equilibrium value. This change in optical rotation is called <u>mutarotation</u>. It represents the interconversion of α and β forms to an equilibrium mixture of them. This can be represented as

α-D-glucose $\longrightarrow$ Equilibrium mixture containing α, β, & $\longleftarrow$ β-D-glucose
open-chain forms

Specific optical
rotation $[\alpha]_D^{20}$:   +112.2°   +52.7°   +18.7°

$$\begin{bmatrix} \beta \text{ form:} & 63\% \\ \alpha \text{ form:} & 36\% \\ \text{open form:} & 1\% \end{bmatrix}$$

The predominance of the β-form in aqueous solution is due to its more stable conformation relative to the α-form. This can be explained as follows. The pyranose ring is not planar and exists in two conformations, termed boat and chair forms. The chair form is more thermodynamically stable than the boat form because a maximum number of bulky groups (-OH and -CH_2OH) are equatorial rather than axial, thus minimizing steric hindrance. In this respect, the pyranose ring is similar to the ring of cyclohexane because the hemiacetal C-O-C bond angle of 111° is close to that of the C-C-C cyclohexane ring angles of 109°. The boat conformations of the two anomers are shown below.

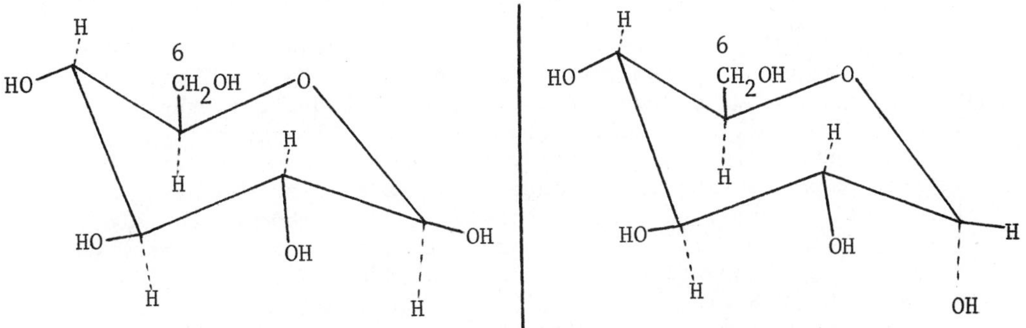

β-D-glucopyranose   axis   α-D-glucopyranose

--- axial bonds   —— equatorial bonds

Note that in β-D-glucopyranose, <u>all</u> of the bulky groups (-OH and -CH$_2$OH) are in equatorial positions whereas in α-D-glucopyranose the anomeric hydroxyl is axial. Thus the β-form is more stable than the α-form and predominates in the solution.

4. Fructose (and other ketoses with 5 or more carbon atoms): These monosaccharides can also occur as α and β anomeric forms. The structures for fructose are shown below:

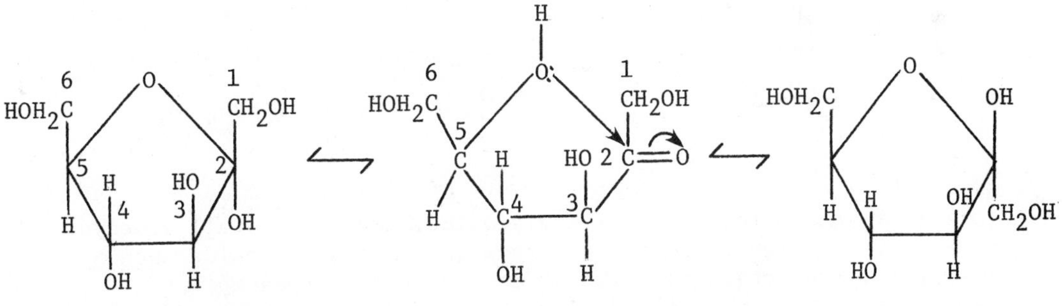

α-D-fructofuranose   D-fructose (open form)   β-D-fructofuranose

Note that in the above structures, the hydroxyl group attached to carbon 5 attacks the electron-deficient carbonyl (#2) carbon yielding a <u>hemiketal</u> in the form of a five-membered <u>furanose</u> ring. The five-membered ring is termed a furanose structure since it is derived from furan, a five-membered oxygen-containing heterocyclic compound.

Some Reactions of Monosaccharides

1. Formation of hemiacetals and hemiketals has already been discussed.

2. <u>Glycoside formation</u>: This is an important reaction which involves the condensation of a hemiacetal (or hemiketal) hydroxyl with a hydroxyl on another carbohydrate or non-carbohydrate (methyl alcohol, glycerol, a sterol, a phenol, etc.). The generic name for the new bond is a <u>glycosidic bond</u>. It can also be named for the specific sugar bearing the hemiacetal or ketal (for example, glucose would form a <u>glucosidic</u> bond). A non-carbohydrate moiety in a glycoside is called an <u>aglycone</u>. If the reaction is between the hemiacetal hydroxyl of one monosaccharide and a hydroxyl of another monosaccharide, the resulting glycoside is a disaccharide. A large number of monosaccharides joined by glycosidic bonds yields a polysaccharide (or polyglycoside). An example of glycoside formation is shown below:

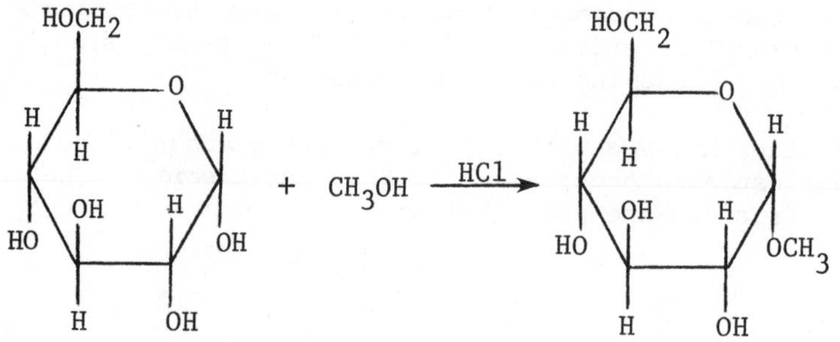

α-D-glucopyranose   Methyl-α-D-glucopyranoside

Note that in the formation of a glycoside, the (potential) aldehyde group is converted to an acetal group and thus a glycoside <u>does not</u> have properties characteristic of an aldehyde (for example the glycoside is no longer a reducing sugar) and does not exhibit mutarotation (because it is not in equilibrium with the open chain form). Most disaccharides and polysaccharides, however, are reducing sugars because any carbohydrate molecule with a free anomeric hydroxyl will confer reducing properties.

Further examples of glycosides:

a.

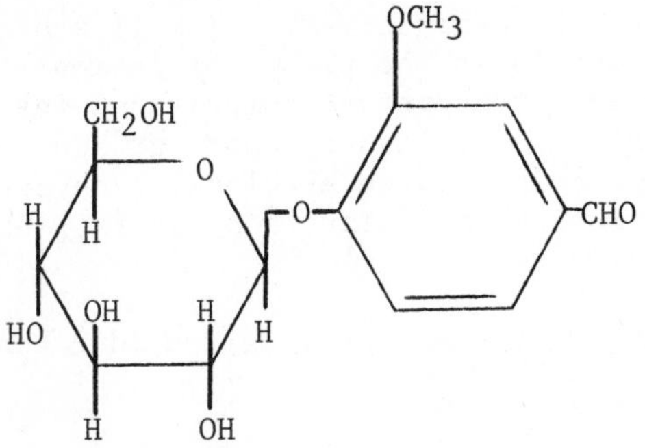

Vanillin-D-Glucoside
(Glucovanillin;
vanilloside): A β-D-
glucoside (natural source
of vanilla flavor)

b. Cardiac glycosides: Digoxin and Digitoxin;

Structure of Digoxin (Digitoxin has the same structure except
it lacks the circled OH group):

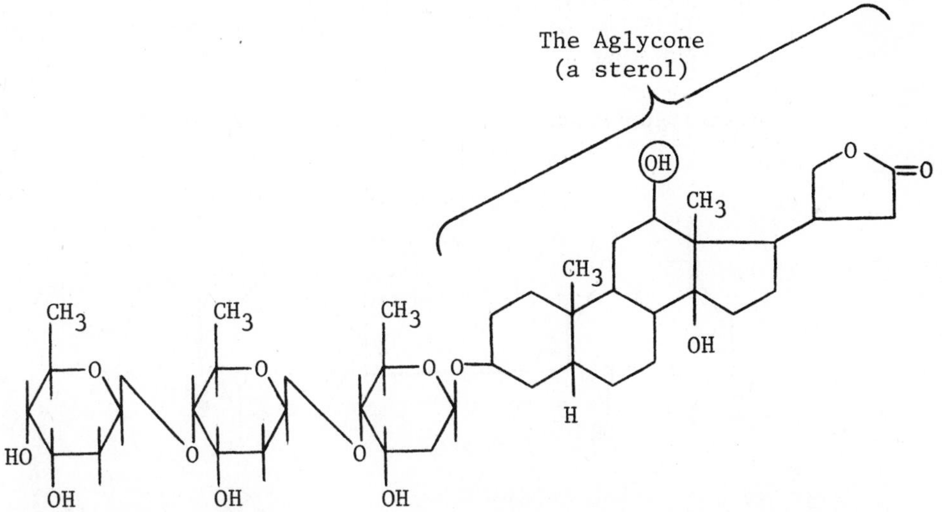

The Aglycone
(a sterol)

Digoxin, a stimulator of cardiac muscle contraction, is obtained
from Digitalis lanata. It is orally effective with the onset of
action occurring 15-30 minutes after administration and a half-
life of 32-45 hours. Digitoxin, another cardiac glycoside,
obtained from Digitalis purpurea, is longer acting with a half-
life of 4-6 days. Digoxin is the principal glycoside now in

clinical use due to its route of administration (oral) and early
onset of action. The drug levels in the plasma of patients
receiving these glycosides must be constantly monitored (by
radioimmunoassay procedures) because either super- or sub-optimal
levels can lead to second or third-degree AV block, tachycardia
associated with block, nodal or ventricular tachycardia, and
other defects in conduction.

3. Formation of disaccharides (many of these contain glycosidic bonds).

Example: Sucrose

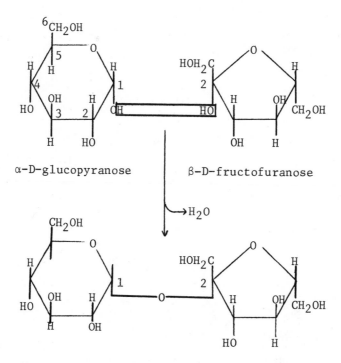

α-D-glucopyranose   β-D-fructofuranose

→H$_2$O

α-D-glucopyranosyl - (1→2)β-D-fructofuranoside

Note: The D-fructose has been "flipped" over.
       The anomeric carbons are joined in sucrose.

Sucrose is a <u>non-reducing sugar</u> since both aldehyde and ketone
carbons participate in the glycosidic bond. Disaccharides can
also be named as glycosides, since they are formed by substituting
a monosaccharide for an aglycone. In a glycosidic bond between
two sugars, the arrow indicates the direction of the bond, from
the non-reducing to the reducing end (free aldehyde or ketone group)
end of the molecule.

4. Formation of phosphoric acid esters

    Example: Formation of D-glucose-6-phosphate

$$H_2C-O-PO_3H_2$$

5. Methylation of hydroxyl groups

6. In solutions of strong mineral acids, sugars are usually dehydrated.

    Example:

$$+ \quad 3 \ H_2O$$

      Any aldopentose   Furfural

7. Reactions in alkaline solution usually involve enol formation

$$\xrightarrow{(OH^-)} \quad + \quad H_2O$$

    Ketone or aldehyde   Enolate

8. There are a number of reactions in which sugar molecules react with various reagents giving products which absorb light in the visible region. Some of these reactions have been used in the determination of sugars or of molecules which contain sugar residues. Two such examples are the reaction of orcinol with pentoses (used in the determination of ribonucleic acid) and the reaction of diphenylamine with 2-deoxypentoses (used in the determination of deoxyribonucleic acid).

9. Reduction of monosaccharides

Example:

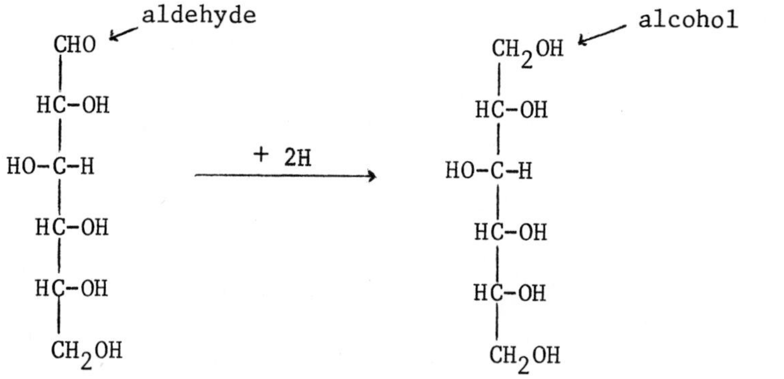

aldehyde

$$\begin{array}{c} CHO \\ | \\ HC-OH \\ | \\ HO-C-H \\ | \\ HC-OH \\ | \\ HC-OH \\ | \\ CH_2OH \end{array} \quad \xrightarrow{+\ 2H} \quad \begin{array}{c} CH_2OH \\ | \\ HC-OH \\ | \\ HO-C-H \\ | \\ HC-OH \\ | \\ HC-OH \\ | \\ CH_2OH \end{array}$$

alcohol

D-glucose   Sorbitol (Sugar alcohol)

Note that, in this reaction, the aldehyde group on carbon 1 has been reduced to an alcohol.

10. Primary oxidation of aldoses

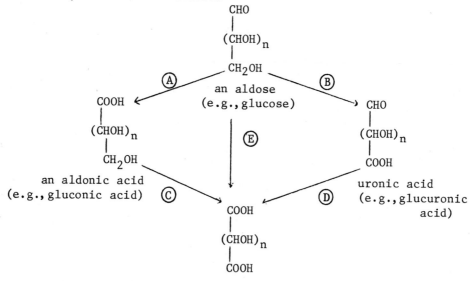

a saccharic or aldaric acid (e.g., glucosaccharic or glucaric acid)

Reactions Ⓐ and Ⓓ are the oxidation of the terminal aldehyde group, reactions Ⓑ and Ⓒ are the oxidation of the terminal alcohol group, and reaction Ⓔ is the oxidation of both groups simultaneously.

Other Disaccharides

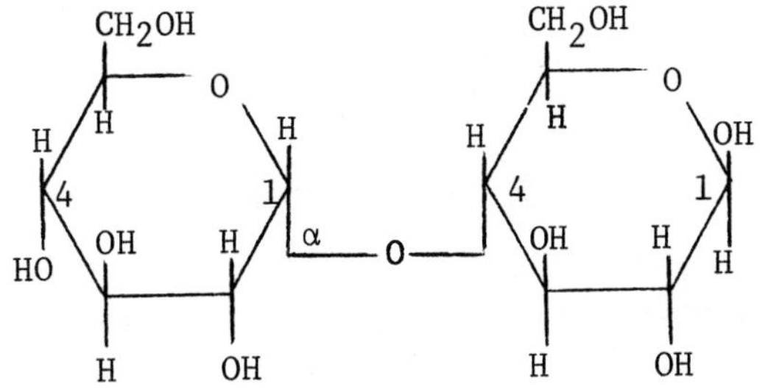

This is a reducing sugar, since at least one of the aldehyde carbons is not involved in a glycosidic bond, and can undergo mutarotation.

β-Maltose: α-D-glucopyranosyl-(1→4)-β-D-glucopyranose

Note: The anomeric carbon of one glucose molecule is joined to a non-anomeric carbon on the other (unlike sucrose). The free anomeric carbon may be in the β-configuration (as shown above), or it may be in the α-configuration, in which case it is called α-maltose (α-D-glucopyranosyl-(1→4)-α-D-glucopyranose).

Lactose

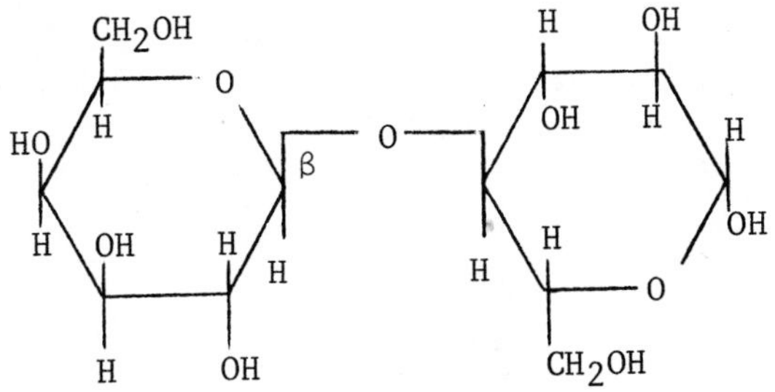

β-D-galactopyranosyl-(1→4)-β-D-glucopyranose.

Note: the second glucose molecule has been flipped over, with the axis of rotation as shown.

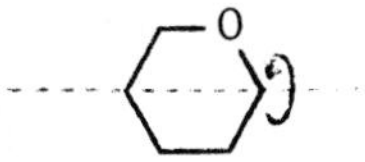

Lactose can also be written

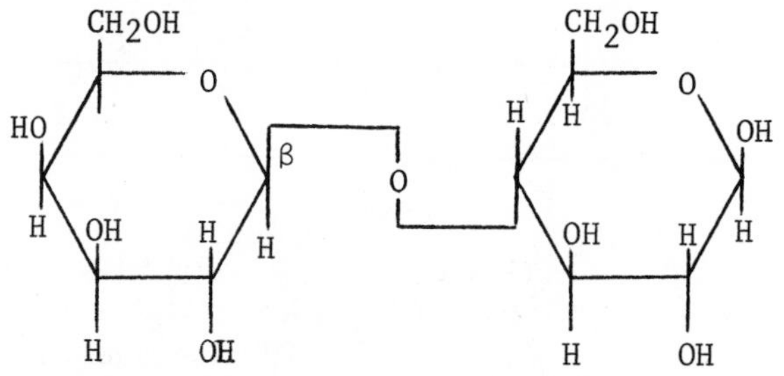

11. Oxidation with periodate and staining for glycogen

Periodic acid (H_5IO_6) will oxidize carbons adjacent to hydroxyl groups, as are found in carbohydrates. The types of reaction are:

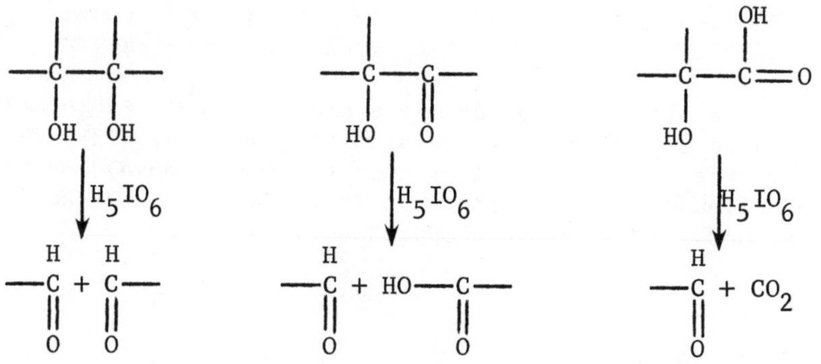

This reaction, followed by determination of the products, can be used to elucidate the structure of carbohydrates. For example, it can distinguish a furanoside from a pyranoside, as shown below.

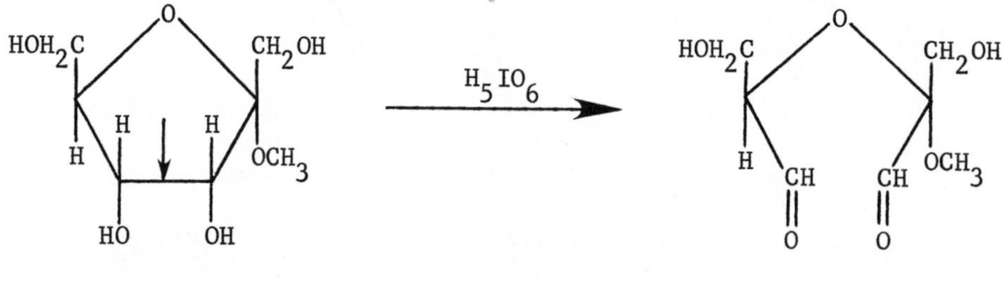

Methyl
α-D-arabinofuranoside

A dialdehyde

Methyl
α-D-glucopyranoside

Periodate cleavage of glycogen (a polyglucose, structure discussed later) yields a polyaldehyde which can be coupled to a dye for visualization purposes. In histology, this is commonly known as periodic acid-Schiff (PAS) staining. The procedure involves the treatment of the tissue slice with periodate followed by staining with Schiff reagent (basic fuchsin bleached with sulfurous acid). The polyaldehyde, if present, will combine with the NH_2 groups of the bleached dye to produce a magenta or purple-colored complex. The specificity of the reaction can be tested by first treating the tissue with α-amylase (which breaks down the glycogen to smaller fragments which can be removed by washing) and then staining with PAS. Absence of the purple-staining substance confirms that the material was glycogen (or another α1,4 linked polysaccharide, such as starch).

Polysaccharides

Cellulose is the most abundant <u>structural</u> polysaccharide of plants and the most abundant organic substance on earth. It occurs almost exclusively in plants.

Structure
1. Total hydrolysis yields only D-glucose.
2. <u>Partial</u> hydrolysis yields the disaccharide cellobiose (β-form): β-D-glucopyranosyl-(1→4)-β-D-glucopyranose.
3. Hydrolysis of fully methylated cellulose yields only 2,3,6-tri-O-methyl glucose, showing a lack of branching in the cellulose molecule.

All of this evidence indicates that cellulose is a linear array of D-glucopyranose residues linked by β-(1→4) glycosidic bonds.

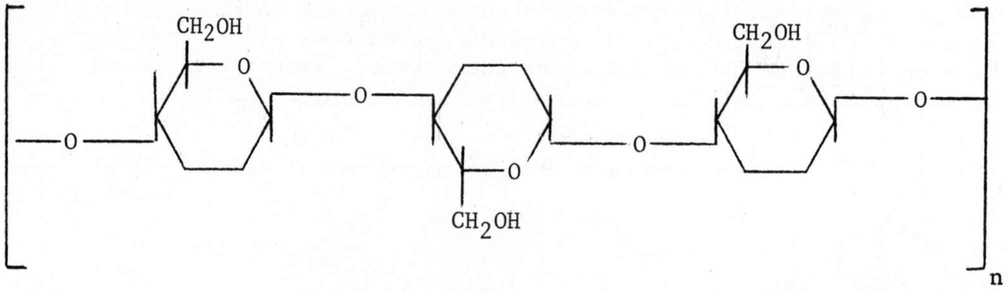

Glucosyl residues are alternately "rotated" 180°, a consequence of the β-glucosyl bond. The above structure can also be written as follows:

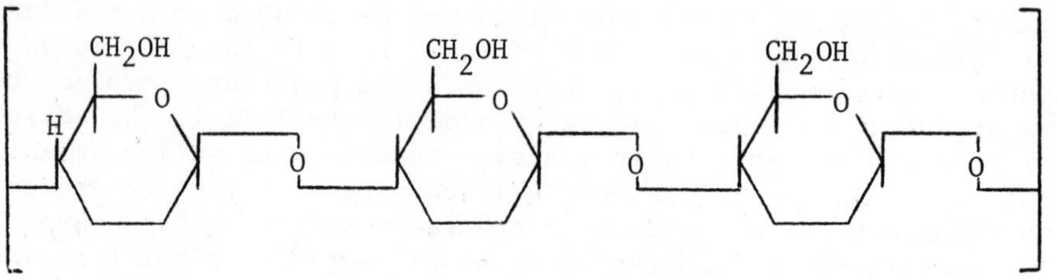

(The hydrogen and hydroxyls on the glucose residues have been omitted for clarity.)

Starches

1. Carbohydrate (energy) reservoir in plants.

2. The basic subunit involves the linkage of successive D-glucose molecules by α-$(1\rightarrow4)$-glycosidic bonds.

3.

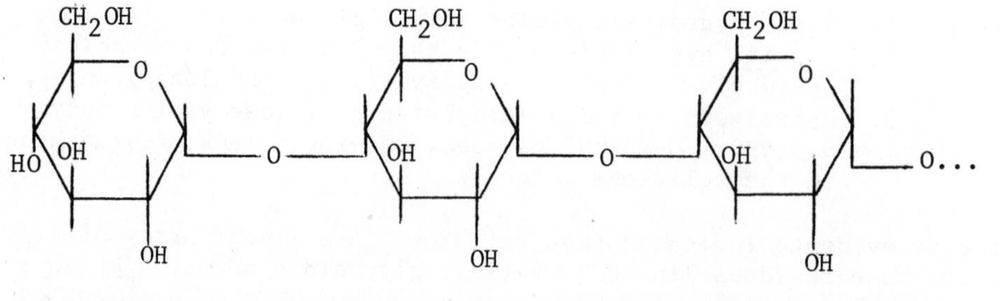

Comparing this structure to that of cellulose and other structural polysaccharides one observes that, in the structural polysaccharides, the monosaccharide units are joined together with β-linkages in contrast to the nutritional carbohydrates (starch, glycogen) which are joined by α-linkages. Mammals do not have the enzymes necessary to digest β-linked compounds. Even cows and horses who live on cellulose do not digest it themselves. Bacteria in their intestinal tracts perform this function.

4. Starch falls into two distinct classes

    a. long unbranched chains (amyloses) having only α(1→4) bonds, and

    b. branched chain polysaccharides (amylopectins) having α(1→6) bonds at the branch-points but α(1→4) linkages everywhere else.

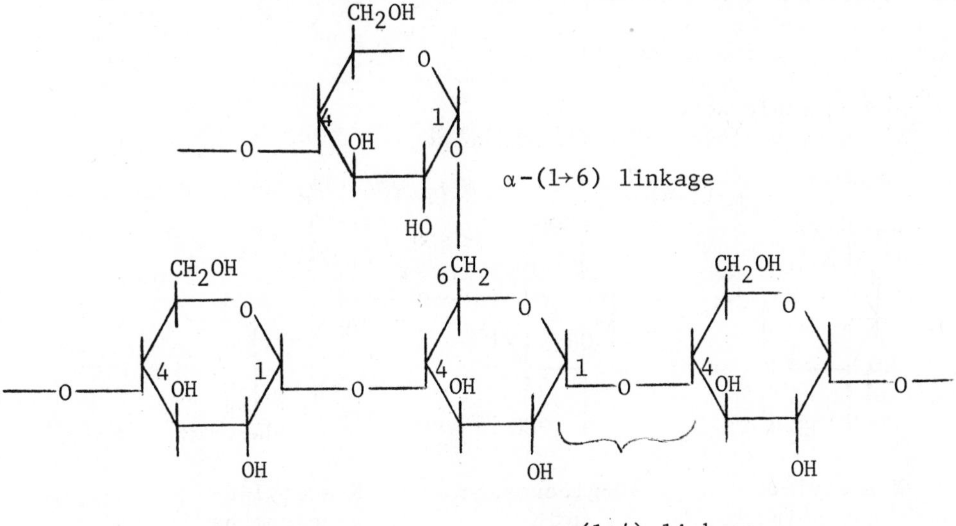

α-(1→6) linkage

α-(1→4) linkage

5. Starches vary widely in their molecular weights and molecular weights of several million are not uncommon.

6. Starches are important in human nutrition. They can be broken down in the alimentary tract to maltose and glucose units and then absorbed. The enzymes that hydrolyze starches are known as amylases. α-amylases (pancreatic and salivary) hydrolyze α(1→4) bonds wherever they occur in polymers of glucose which contain at least three residues. There are also β-amylases which liberate maltose units. They start at the reducing ends of a starch molecule and cleave every other α(1→4) linkage until a branch-point α(1→6) linkage is reached. β-amylases do not hydrolyze β-linkages.

Glycogen (A Polyglucose)

1. Serves as nutritional reservoir in animal tissues.

2. The structure is similar to amylopectin but is more highly branched.

3. Polydisperse (wide range of chain lengths); molecular weights may be as great as 1×10^8.

4. The linear linkages are $\alpha(1{\rightarrow}4)$ and the branch-point linkages are $\alpha(1{\rightarrow}6)$, as in amylopectin.

Hyaluronic Acid (A Mucopolysaccharide)

1. Present in the connective tissues, synovial fluid, and vitreous fluid; usually found associated with proteins. They may act as lubricants and shock absorbants in the joints.

2. Composed of equimolar quantities of

    a. D-glucuronic acid
    b. N-acetyl-D-glucosamine (an amino sugar)

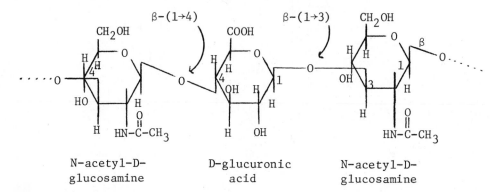

N-acetyl-D-glucosamine	D-glucuronic acid	N-acetyl-D-glucosamine

Chondroitin Sulfates

1. They are widely distributed in connecting tissues and serve as a structural material (cartilage, tendons, and bones).

2. There are three chondroitin sulfates: A, B, and C. Their compositions are

    <u>A</u>: (D-glucuronic acid)-(I)-(N-acetyl-D-galactosamine-4-sulfate)
    -(II)-(D-glucuronic acid)

    <u>B</u>: (L-iduronic acid)-(I)-(N-acetyl-D-galactosamine-4-sulfate)
    -(II)-(L-iduronic acid)

<u>C</u>: (D-glucuronic acid)-Ⓘ-(N-acetyl-D-galactosamine-6-sulfate)

-Ⓘ-(D-glucuronic acid)

Note: Ⓘ indicates a β(1→3) linkage and Ⓘ indicates a β(1→4) linkage.

<u>Kerato Sulfate</u> is a polysaccharide present in the cornea and connective tissue (nucleus pulposus and costal cartilage, for example). Its structure is

(β(1→4))   (β(1→3))
(D-galactose)---(2-N-acetyl-3-deoxy-D-glucose-6-sulfate)---(D-galactose)

Heparin

1. The exact structure and biological function is <u>not</u> clear. It is a strongly acidic substance with a molecular weight of about 17,000.

2. It occurs in the granules of the circulating basophils and in the granules of the mast cells which are present in many tissues (liver, lungs, walls of arteries). Mast cells also contain histamine. In anaphylaxis (undesirable antigen-antibody reactions), heparin is released from the mast cells leading to a lack of coagulation of blood.

3. It functions as an anticoagulant by preventing the activation of factors VIII and IX and, in association with a plasma factor inhibits the action of thrombin.

4. Heparin causes the liberation of lipoprotein lipase. It presumably functions as a cofactor in the action of this enzyme on triglycerides which are associated with chylomicrons and other lipoproteins. This results in the clearing of the turbidity of a lipemic plasma.

5. It is bound to protein.

6. α-(1→4) linkages are involved in its structure. A probable unit of heparin is

Sialic Acids

a. Widely distributed in tissues; particularly present in mucins (mucoproteins of saliva) and blood group substances.

b. Sialic acids are most often N-acetyl derivatives of neuraminic acid. A hydroxyl group may be acetylated in some instances.

c. Neuraminic acid is a 3-deoxy, 5-amino, nine carbon sugar. It is the condensation product of pyruvic acid and mannosamine.

Spontaneous hemiketal ring formation between the keto group on C-2 and the hydroxyl on C-6

Mucopolysaccharides in Connective Tissue

The connective tissues are derived from mesoderm. They are responsible for holding other tissues together as well as providing support and nourishment for the epithelial membranes and glands that arise from the ectoderm and endoderm. Connective tissue, in its broadest definition, encompasses: (a) the loose connective tissue (so named because it is soft, pliable and semi-elastic) which is distributed throughout the body, providing intimate support and nourishment for many other tissues; (b) the hemopoietic tissue (which includes blood cells--erythrocytes, leukocytes, and others--and the myeloid and lymphatic tissues); and (c) the strong supporting types of connective tissue (cartilage, bone, and joints). Loose connective tissue consists of fibers (composed mostly of the proteins collagen and elastin), ground substance (acid mucopolysaccharides, water, and salts), and cells such as fibroblasts, mast cells, fat cells, and macrophages or plasma cells. These cells are surrounded by the intercellular substances (mucopolysaccharides) which are bathed in tissue fluid originating from the capillaries by ultrafiltration. The fibroblasts synthesize collagen and elastin (for details see protein synthesis) while the principal product of the mast cells is heparin. The composition and function of other types of connective tissues are discussed later in the appropriate sections.

The acid mucopolysaccharide composition varies from one connective tissue to another. Hyaluronic acid is the main one in vitreous humor, synovial fluid, and other connective tissues in which there is increased water content. Chondroitin-4-sulfate predominates in cartilage. The nature of the interaction between the polysaccharide and protein parts of these mucopolysaccharides is not well understood. In several cases it appears to be an ether linkage between the 1-hydroxyl group of a xylose or N-acetylgalactosamine and a seryl hydroxyl group on the protein.

The synthesis of the mucopolysaccharides is a subject of active investigation. Chondromucoproteins (those containing chondroitin sulfate) have been the most thoroughly studied. Their elaboration is summarized in the following diagram.

Rough endoplasmic reticulum (polyribosomes); protein portion synthesized and sugar-protein links formed.	⟶ Smooth endoplasmic reticulum: polysaccharide is elongated; sulfate groups are added.
	↓
Completed molecules are secreted by the cell.	⟵ Golgi apparatus: molecules of chondromucoprotein are finished and packaged for secretion.

It is important to notice that, in all of these steps, the enzymes involved are closely associated with (bound to) membranes. The degradation of mucopolysaccharides is accomplished by several enzymes. These include hyaluronidase (degrades hyaluronic acid and the chondroitin sulfates to oligosaccharides); β-glucuronidase (removes

uronic acid residues); several N-acetylhexosaminidases (remove
N-acetylglucosamine, etc.); at least one sulfatase enzyme (removes
sulfate residues from the sulfated polysaccharides).

The mucopolysaccharidoses are a group of inherited lysosomal enzyme
disorders (catabolic defects). They involve, with the exception of the
Morquio syndrome, the accumulation and excretion (in urine) of dermatan
and/or heparin sulfate. The structures of heparin sulfate and dermatan
sulfate are shown below. The circled numbers refer to the biochemical
lesions whose properties are presented in the accompanying table.

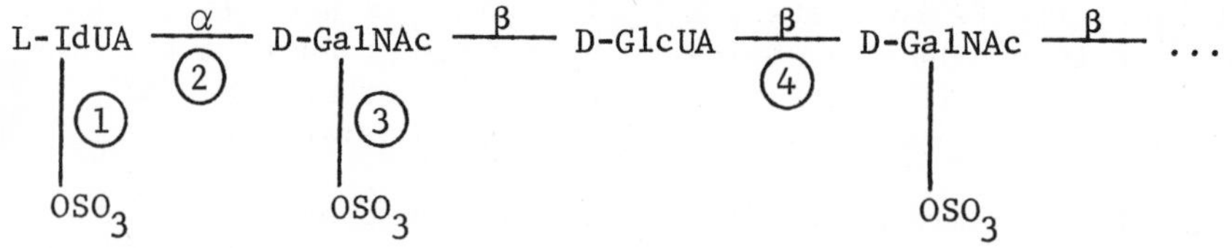

L-IdUA $\xrightarrow{\alpha}$ D-GalNAc $\xrightarrow{\beta}$ D-GlcUA $\xrightarrow{\beta}$ D-GalNAc $\xrightarrow{\beta}$ ...

(2) (1) (3) (4)

OSO$_3$ OSO$_3$ OSO$_3$

Dermatan Sulfate

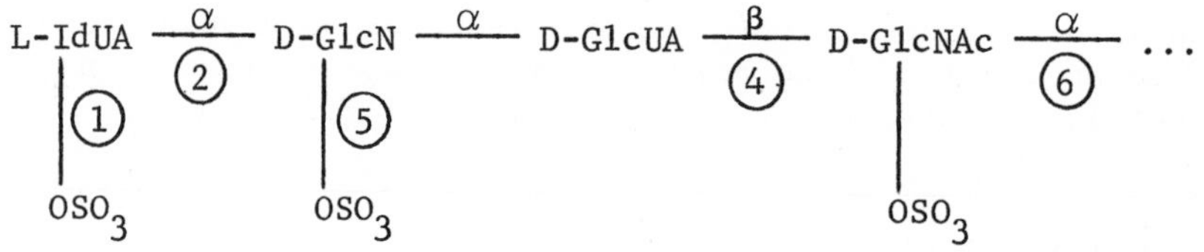

L-IdUA $\xrightarrow{\alpha}$ D-GlcN $\xrightarrow{\alpha}$ D-GlcUA $\xrightarrow{\beta}$ D-GlcNAc $\xrightarrow{\alpha}$ ...

(2) (1) (5) (4) (6)

OSO$_3$ OSO$_3$ OSO$_3$

Heparin Sulfate

IdUA = Iduronic acid
GalNAc = N-acetyl-D-galactosamine
GlcUA = Glucuronic acid
GlcN = Glucosamine
-OSO$_3$ = Sulfate group

The above polymers show microheterogeneity, with variations in the ratio
of iduronic acid to glucuronic acid, of N-sulfated to N-acetylated
glucosamine, and of sulfated to unsulfated iduronic acid residues. In
addition, the places on the polypeptide chain to which the sugar residues
are attached may vary. The relationship of the clinical manifestations
and the biochemical lesion is sometimes not obvious, as can be seen in
Table 7 . For example, deficiency of α-L-iduronidase may lead to
either mild (Scheie Syndrome) or very severe (Hurler Syndrome)
disorders. This lack of correlation between a particular enzyme
deficiency and the clinical presentation of the patient may be due to:
a) differences in the environments of the patients which can alter
expression of the genotype; b) variation in the amount of enzyme

Table 7 . Mucopolysaccharidosis

Syndrome	Enzyme Deficiency	Comments
Hunter	Iduronate sulfatase (1)	X-linked recessive, moderate mental and skeletal deterioration; no corneal involvement; early deafness.
Hurler	α-L-iduronidase (2)	Autosomal recessive; manifests in late infancy (with 'gargoyle' features); severe mental retardation, skeletal deformities; corneal opacities; hepatosplenomegaly.
Scheie	α-L-iduronidase (2)	Autosomal recessive; no mental retardation; mild skeletal changes; corneal opacities.
Maroteaux-Lamy	N-acetylgalactosamine sulfatase (3) (arysulfatase B)	Autosomal recessive; severe skeletal deformities; no mental retardation; corneal opacities, growth retardation; may reach adulthood but often die from progressive heart disease.
Mucopolysaccharidosis	β-glucuronidase (4)	Skeletal abnormalities; hepatomegaly; granular inclusions in leukocytes, mental development not retarded (only one patient described hitherto).
Sanfilippo A-subtype	Heparin N-sulfatase (5)	Autosomal recessive; both subtypes are clinically indistinguishable; severe mental retardation; mild dwarfing; skeletal abnormalities, corneal clouding and hepatosplenomegaly.
Sanfilippo B-subtype	N-acetyl-α-glucosaminidase (6)	
Morquio	Defect in keratan sulfate degradation, enzyme deficiency not known. However, a chondroitin sulfate degrading enzyme (N-acetyl β-galactosamine 6-sulfatase) deficiency has been noted.	Severe skeletal deformities with spondylo-epiphyseal displasia; corneal clouding; no mental retardation.

remaining (the residual activity); c) tissue specificity of the defect (enzyme absent in some tissues but present in others); d) differences in the microheterogeneity of the substrate between individuals; e) variation in the ability to tolerate the large accumulation of substrate; and f) the fact that some mutations are pleiotropic (phenotype exhibits changes in several, apparently unrelated body systems) and that the enzyme deficiency may represent only one facet of the phenotype.

Since the prognosis in most cases of mucopolysaccharidosis is extremely poor, <u>antenatal</u> diagnosis by measuring the enzyme levels in the amniotic fluid and in fibroblast cells cultured from the fluid is presently being investigated. Because amniocentesis is not a totally risk-free procedure, these tests are usually done only in pregnancies where there is reason to suggest the presence of the disease (both parents known heterozygotes; other relatives have the disease; one parent homozygous). If the abnormality is diagnosed in the fetus, the pregnancy can be aborted.

Chemistry of Blood Group Substances (see also the chapter on Immunochemistry)

1. They are composed of polysaccharides and proteins and are found in many places, including erythrocyte membranes, saliva, gastric mucin, cystic fluids, and elsewhere.

2. They are water soluble compounds with molecular weights ranging from about 200,000 to 2 million.

3. They are immunologically distinct from one another, yet they show many similarities in chemical composition. The immunological specificity resides in the oligosaccharide side chains and not in the polypeptide portion of the molecule. The residues contained in the oligosaccharides are L-fucose, D-galactose, N-acetyl-D-galactosamine, and N-acetyl-D-glucosamine (see Chapter 11).

Digestion of Carbohydrates

1. The major ingested carbohydrates are

   a. starch, a polymer of glucose joined by α-linkages. This includes both the branched-chain (amylopectin) and straight-chain (amylose) forms;

   b. lactose (galactose-glucose) and sucrose (fructose-glucose) which are disaccharides;

   c. the monosaccharides glucose and fructose.

2. Digestion of polysaccharides begins in the mouth with the action of salivary amylase (ptyalin). This process is terminated by

swallowing, since the acid gastric juice inactivates salivary
amylase activity.

3.   The principle digestion of polysaccharides takes place in the
     intestines by the action of pancreatic α-amylase. The products of
     hydrolysis of starch and other disaccharides present in the diet
     are further hydrolyzed, into monosaccharides, by specific
     disaccharidases which are present on the mucosal cell surface.

4.   The absorption of glucose (and galactose) into the columnar
     epithelial cells of the villi of the small intestine is greatly
     facilitated by sodium ion. It appears that this process is mediated
     by a membrane-bound carrier which binds glucose and Na^+ at different
     sites. The glucose and Na^+ enter the cell together, the glucose
     being transported up a concentration gradient by the energy obtained
     from the Na^+ moving down a concentration gradient into the cell.
     ATP is not hydrolyzed directly by this process and the potential
     energy stored in the Na^+ gradient is utilized to move the glucose.
     However, to maintain the intracellular sodium ion concentration low
     relative to the extracellular concentration, given this Na^+ influx,
     a Na^+/K^+ ATPase must function, removing Na^+ from the cell and bringing
     in K^+. The energy of this ATP hydrolysis is stored as potential
     energy in the trans-membrane sodium ion gradient for later use by the
     Na^+/glucose carrier. The overall process, then, is considered to be
     active transport. As would be expected, K^+ facilitates the metabolism
     of glucose since it indirectly stimulates glucose uptake by increasing
     the sodium ion gradient (by stimulation of the Na^+/K^+ ATPase). Insulin
     does not play a role in the transport of glucose from the intestinal
     lumen into the mucosal cells.

     Fructose, in contrast to glucose and galactose, is absorbed more
     slowly by a carrier mechanism that is independent both of Na^+ and of
     the glucose-galactose transport mechanism. The monosaccharides are
     then transferred (presumably by diffusion) from the mucosal cells to
     the capillaries of the portal blood circulatory system.

     Figure 19 is a schematic diagram illustrating the salient points
     of carbohydrate digestion and absorption.

     The oligosaccharidases are present only in the mature columnar
     epithelial cells. In man, all of these enzymes are present at birth.
     The life expectancy of these cells is only about 4-6 days so that
     they must be continually regenerated. The functional unit of the
     small intestine is made up of crypt and villus (see Figure 20).
     The epithelial cells, derived by mitosis at the base of the crypt,
     are functionally and morphologically inactive. However, they do show
     DNA and protein synthesis in preparation for their functional
     activity which first appears as they migrate to the lower villus.
     The migration takes about one day after the initial formation.   They

Figure 19. Digestion and Absorption of Carbohydrates

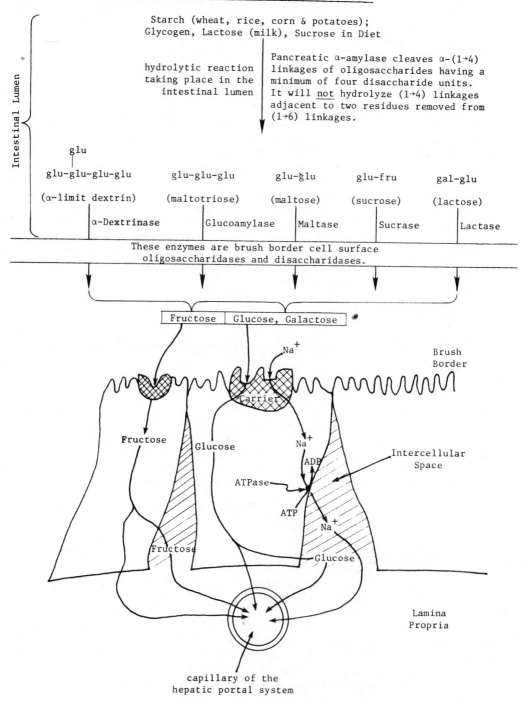

Starch (wheat, rice, corn & potatoes);
Glycogen, Lactose (milk), Sucrose in Diet

hydrolytic reaction taking place in the intestinal lumen

Pancreatic α-amylase cleaves α-(1→4) linkages of oligosaccharides having a minimum of four disaccharide units. It will not hydrolyze (1→4) linkages adjacent to two residues removed from (1→6) linkages.

Intestinal Lumen

glu
|
glu-glu-glu-glu

(α-limit dextrin)

α-Dextrinase

glu-glu-glu

(maltotriose)

Glucoamylase

glu-glu

(maltose)

Maltase

glu-fru

(sucrose)

Sucrase

gal-glu

(lactose)

Lactase

These enzymes are brush border cell surface
oligosaccharidases and disaccharidases.

Fructose | Glucose, Galactose

Na⁺

Brush
Border

Carrier

Fructose

Glucose

Na^+

ADP

ATPase

ATP

Na^+

Intercellular
Space

Fructose

Glucose

capillary of the
hepatic portal system

Lamina
Propria

Figure 20.
A Schematic Representation of a Functional Unit of the Small Intestine.

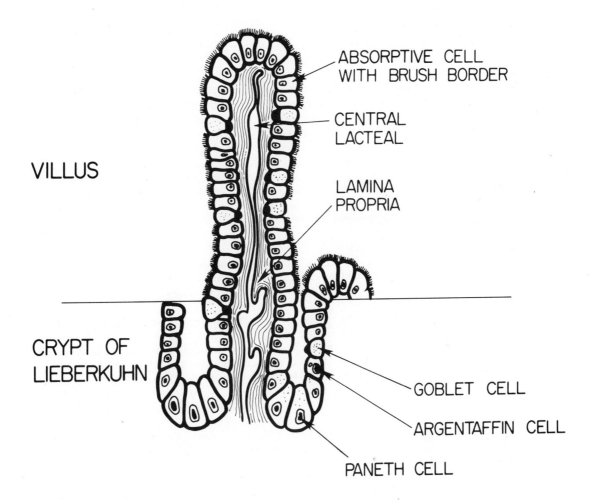

exhibit maximal activity at the middle and upper levels of the villus, 4-5 days after their original formation. At the upper end of the villus, the activity drops and the senescent cells are eventually removed by the continual sloughing off of the intestinal mucosa.

Sucrase and α-dextrinase appear as a hybrid molecule. The glucoamylase present on brush border cells have a different mode of action from pancreatic α-amylase, in that it is an exoglycosidase, sequentially cleaving one glucose at a time from the non-reducing end of a linear $\alpha(1\rightarrow4)$ glucosyl oligosaccharide. The only enzyme present which hydrolyzes a β-glycosidic linkage is lactase. Cellulose, a polyglucose with $\beta(1\rightarrow4)$ linkages, <u>cannot</u> be digested by man for lack of the appropriate enzymes, but it serves as a fiber, adding bulk in the formation of stools.

5. a. A number of problems (intolerances) with the digestion and absorption of disaccharides have been observed in humans. The most common type, <u>lactose intolerance</u>, is quite common in individuals coming from societies where there is traditionally little dairying and milk consumption. For example, in healthy adults of Asia (Indians, Japanese, Formosans, Filipinos, Thais), Africa (Blacks and Arabs), and the Americas (Blacks, American Indians, Mexican Indians, Peruvians, Eskimos), the incidence of lactose intolerance is as high as 90%, while in the Danish, Swiss and white Americans the incidence is relatively low (less than 10%). It is estimated that in the United States, considering the ethnic diversity, there may be 30 million people with lactose intolerance who apparently do not drink milk in order to avoid unpleasant symptoms.

   b. Lactose intolerance is caused by the absence of lactase activity on the brush border cells of the intestine. There are three types of disorders which produce this deficiency.

      i) Primary lactase deficiency is seen in some premature infants. This is similar to other disorders of prematurity, such as transient neonatal tyrosinemia and neonatal hyperbilirubinemia, which are caused by the late maturation of <u>liver</u> enzyme systems. Age-related idiopathic lactose intolerance, as is seen in the various population groups discussed earlier, is also in this category. It is probably genetic in origin.

      ii) Iatrogenic lactase deficiency, secondary to surgery, is seen in some patients with resections of the small bowel, gastrectomy, and pyloroplasty.

      iii) Lactase deficiency due to mucosal damage resulting from treatment with some drugs (neomycin, for example) and some diseases (kwashiorkor; acute gastroenteritis, particularly in infants; infestation with the protozoan <u>Giardia lamblia</u>; celiac disease; tropical sprue; Whipple's disease; intestinal lymphangiectasia; and others).

   c. Lactase activity is maximal in the jejunum and proximal ileum. As we have already seen, the hydrolysis of lactose to glucose and galactose occurs at the surface of the intestinal microvilli and the monosaccharides are absorbed by a carrier-mediated energy and Na^+-dependent process. Lactase activity attains peak values at birth in <u>full-term</u> human infants and it remains high throughout infancy, as milk provides the predominant nutritive support during this period. When the milk intake decreases and is replaced by other types of food, the lactase level drops and lactose intolerance may appear. The <u>extent</u> of the decrease seems to be what distinguishes highly tolerant from intolerant populations. Congenital neonatal lactase deficiency is a rare disease and may prove fatal if not recognized promptly. Acute gastroenteritis in infants frequently produces mucosal damage resulting in lactose intolerance. In both of the above types of lactase deficiency in infants, it is very important to use <u>lactose-free</u> formulas. In infants with gastroenteritis, this type of nutritional support is continued until the infant's gastric mucosa is healed. When there is severe damage to the gastric mucosa, the activities of other disaccharidases (sucrase and maltase) are also affected. However, it appears that lactase is the most sensitive of the disaccharidases so that lactose digestion is the most common problem.

d. After the ingestion of milk, individuals with lactose intolerance exhibit a variety of gastrointestinal symptoms, such as bloating, cramps, flatulence, and loose stools. The severity of the symptoms depends upon the amount of milk consumed and the actual enzyme level. The symptoms disappear with the elimination of the offending disaccharide. An eight-ounce glass of milk contains about 12 grams of lactose and it has been reported that some individuals may show GI symptoms with consumption of as little as 3 ounces of milk. Some lactose intolerant individuals may be able to drink milk if it is consumed along with their meals, and may not show any reaction to milk in cereal or coffee. Lactose-depleted milk or fermented milk products (e.g., yogurt) which have negligible amounts of lactose are good substitutes for milk.

e. The GI problems are due to:

   i) Osmotic effects: The unabsorbed lactose increases the osmolality of the intestinal contents, causing water to be taken up and retained in the GI tract.

   ii) In the colon, lactose gets metabolized by bacterial enzymes to a greater number of smaller molecules (lactate and other short-chain acids), thereby further increasing the osmolality and aggravating the problem of normal fluid reabsorption.

   iii) Fermentation also yields gaseous products (H_2, CO_2, and CH_4) which gives rise to bloating, flatulence and, sometimes, frothy diarrhea.

Diagnosis of Lactase Deficiency

A lactose tolerance test is performed by administering lactose (2 gm/kg up to a total of 100 gm) orally and later measuring blood glucose at various times. The values obtained are compared to those from an oral glucose tolerance test done the day before. If the intestinal lactase is decreased or absent, the ingested lactose will have no effect on the blood glucose and the lactose tolerance profile will be flat, with the peak being less than 20-25 mg/100 ml above the fasting glucose level.

A simple but less reliable test involves the consumption of two glasses of cold milk by the patient, who has been maintained on a milk-free diet for 7-14 days prior to the test. A positive result is indicated by the occurrence of the GI symptoms of lactose intolerance (bloating, cramps, flatulence, loose stools).

Serum Amylase-Clinical Importance

1. As shown in Figure 19, amylase cleaves starch molecules at α-1,4 linkages, releasing maltose, maltotriose, and limit dextrins as the principal products. The enzyme has a molecular mass of about 55,000 daltons, requires chloride and calcium ions for optimum activity, and has a pH optimum of 6.9 (which depends to some extent on chloride ion concentration).

    Amylase is synthesized in the salivary glands and in the acinar cells of the pancreas, which secrete it into, respectively, the saliva and the small intestine. There is evidence for amylase being secreted specifically into human milk and tears, and for synthesis in tissues of the fallopian tubes and ovaries. Serum and urine amylases appear to be a mixture of pancreatic and salivary isoenzymes with the pancreatic one predominating in the urine and the salivary one contributing more to the serum. Both the salivary and pancreatic enzymes show a complex pattern (multiple bands) upon electrophoresis, due to varying degrees of glycosylation and deamidation.

2. Serum amylase is cleared rapidly (half-life is about 130 minutes) by the kidney and the reticuloendothelial system. The pancreatic isozymes seem to be cleared much more rapidly by the kidney than are the salivary types. Measurement of <u>both</u> serum and urinary amylase is beneficial in the diagnosis of disorders <u>which</u> may result in amylase release (see later).

3. Hyperamylasemia (high blood levels) and hyperamylasuria (high urine levels) occur in a number of diseases of diverse origin.

a. Diseases of the pancreas

    i) Acute pancreatitis--commonly caused by alcoholism and gallbladder disease; less frequently by viral hepatitis and by drugs such as steroids, thiazides, sulfonamides, allopurinol, and birth control pills.

    ii) Chronic pancreatitis--most often associated with alcohol abuse.

    iii) Pseudocysts--entrapment of pancreatic juice and inflammatory exudate within a space lined by chronically inflammed, fibrous tissue.

    iv) Ascites--fluid accumulation within the abdominal cavity.

    v) Abscesses of the pancreas; and pancreatic carcinoma.

b. Diseases of non-pancreatic origin (this is a <u>large</u> group of rather unrelated disorders). It includes acute cholecystic disease, diabetic ketoacidosis, disorders of the digestive tract (other than the pancreas), hepatic and renal disorders, salivary gland lesions (mumps), cerebral trauma, burns, pregnancy, pneumonia, prostatic disease, and post-operative situations.

4. Elevated serum and urine amylase levels associated with abdominal pain are strongly suggestive of acute pancreatitis. In this disorder, the renal clearance of amylase is increased, presumably due to an unexplained decrease in renal tubular reabsorption of the enzyme. This observation has led to a clinically useful test which relates the clearance of amylase (C_{am}) to that of creatinine (C_{cr}). The <u>renal clearance</u> of a substance, in milliliters/minute, is expressed as:

$$C = \frac{U}{P} \times V$$

where: U = concentration in urine $\left.\vphantom{\begin{matrix}a\\b\end{matrix}}\right\}$ expressed in same units
            P = concentration in serum or plasma (for example, units/dl)
            V = urine flow, expressed as milliliters/minute.

Therefore, the ratio of amylase clearance to creatinine clearance is:

$$\frac{C_{am}}{C_{cr}} = \frac{\dfrac{\text{urine amylase}}{\text{serum amylase}} \times \text{Volume}}{\dfrac{\text{urine creatinine}}{\text{serum creatinine}} \times \text{Volume}}$$

The volume terms cancel (obviating the need for accurate urine collections) and the expression can be rearranged to:

$$\frac{C_{am}}{C_{cr}} \text{ (expressed as \%)} = \frac{\text{urine amylase}}{\text{serum amylase}} \times \frac{\text{serum creatinine}}{\text{urine creatinine}} \times 100\%$$

To obtain the ratio, all that is required is the determination of amylase and creatinine on <u>simultaneously</u> collected samples of serum and urine. The normal value is 3.1 ± 1.1 (S.D.). In patients with acute pancreatitis, it is elevated to 9.8 ± 3.5. Patients with common duct stones but without pancreatitis, peptic ulcer or intestinal infarction show normal values. However, patients with diabetic ketoacidosis and burns had values similar to those seen in cases of acute pancreatitis. In one study, about 7% of the patients with acute pancreatitis had values below 5.3%. A very low ratio of less than 1.0% suggests macroamylasemia, a poorly understood disease characterized by large molecular aggregates containing both salivary and pancreatic amylases. The "macroamylases" are not cleared by the kidney since they are well above the size cutoff for glomerular filtration.

5. The presence of excessive amounts of lipids in the serum (lipemia) with abdominal pain may result from primary or secondary pancreatic disease. Disorders of primary lipid metabolism (see the section on lipid metabolism) may precipitate an acute attack, and secondary hyperlipemia (lactescent serum) has been observed in 20% of the patients with alcohol-induced pancreatitis. Hyperlipemia in patients with acute pancreatitis interferes with amylase determination, probably due to a non-lipid inhibitor present in both serum and urine. Serial dilution of the lipemic serum or urine with saline (0.15 M NaCl) will unmask the suppressed enzyme activity. Serum amylase elevation is <u>transient</u>, increasing during the first 24 hours following the precipitating event and returning to the normal range during the subsequent 48 hours.

6. Serum amylase measurements are not particularly helpful in diagnosing <u>chronic</u> pancreatitis. In this disorder, progressive functional damage of both the exocrine and endocrine pancreas leads to diabetes and intestinal malabsorption. Measurement of serum lipase (another enzyme secreted by the pancreatic acinar cells in parallel with amylase) and the microscopic examination of stool smears stained for fat with Sudan III or IV are of value in the diagnosis of chronic pancreatitis. If lipase is not secreted into the digestive tract, dietary fat will appear in the feces, undigested. Further information on the pancreatic digestive enzymes can be found in the sections on protein and lipid digestion. Ribonuclease and deoxyribonuclease are two additional digestive enzymes secreted by the exocrine pancreas.

7. Measurement of amylase activity: They can be classified as
    saccharogenic, amyloclastic and those using dyes crosslinked with
    starch.

    a. In the saccharogenic assays, the rate of formation of products
        is measured:

$$\text{Starch} + H_2O \xrightarrow{\text{amylase}} \alpha\text{-limit dextrins} + \text{maltotriose} + \text{maltose}$$

    The reaction can be followed by quantitating the amount of
    reducing sugar formed, since each cut by amylase produces one
    new reducing end. Alternatively, maltose and maltotriose are
    converted to glucose by the addition of α-glucosidase. Glucose
    is then quantitated by an NADH-coupled enzymatic procedure
    which can be easily monitored at 340 nm.

$$\text{Maltose} + H_2O \xrightarrow{\alpha\text{-glucosidase}} \text{Glucose}$$

$$\text{Glucose} + \text{ATP} \xrightarrow{\text{Hexokinase}} \text{Glucose-6-Phosphate (G-6-P)}$$

$$\text{G-6-P} + \text{NAD}^+ \xrightarrow{\text{G-6-P dehydrogenase}} \text{6-Phosphogluconate} + \text{NADH} + H^+$$

    b. In amyloclastic methods, the rate of disappearance of the starch
        substrate is measured. The starch remaining is quantitated at
        various intervals by reacting it with iodine, giving a blue
        starch-iodide complex. This procedure is not recommended because
        it does not follow zero-order kinetics and it is interfered with
        by lipemia. A modification of the amyloclastic method uses
        nephelometry to measure the clearing of a starch suspension.
        Amylase hydrolyzes the starch, making it soluble, and decreasing
        the amount of light scattering by the suspension.

    c. In the dyed-starch method, a water insoluble substrate linked
        to a blue dye is used. When amylase hydrolyzes the substrate,
        fragments bearing the dye molecules are released. This can be
        quantitated by measuring the optical density of the solution.
        The absorbance of the solution is proportional to the amylase
        activity in the test specimen. This procedure is linear,
        follows zero-order kinetics, and is unaffected by lipemia.

Carbohydrate Metabolism

A major function of carbohydrates is to provide energy for body tissues
to use for metabolic processes. The energy is released by oxidation
of the carbohydrates. The first step in this process is the activation
of the carbohydrate by phosphorylation. The carbohydrates directly

used by the cells for fuel are the monosaccharides: glucose, fructose, galactose, and mannose. The latter two are converted (epimerized) to glucose before being metabolized. Glucose is the most extensively used form. The fate of pentose sugars is somewhat uncertain except for one important case. D-ribose and D-deoxyribose are used in the synthesis of nucleotides and, hence, nucleic acids.

The remainder of this chapter will deal largely with pathways of carbohydrate metabolism. It is important to keep in mind the overall purpose of these processes and to note that they are not isolated from each other or from other metabolic pathways. Notice which pathways provide alternate methods for accomplishing the same thing under different metabolic conditions. Pay special attention to the reactions that are capable of responding to changes in metabolic conditions. Often these are the rate-controlling steps. It is also important to see how one cycle or process is connected to others. Below is a list of some of the main subdivisions of carbohydrate metabolism. At the end of the chapter is a diagram illustrating some of the important connecting points.

1. Glycolysis - the oxidation of glucose to pyruvate and lactate (also called the Embden-Meyerhof pathway).

2. Citric Acid Cycle - the oxidation of the acetyl-CoA to carbon dioxide. The acetyl-CoA may come from carbohydrate, fat, or protein metabolism (also called the Krebs Cycle or tricarboxylic acid cycle).

3. Electron Transport and Oxidative Phosphorylation - electron transport is the final process in the oxidation of fuels (not just carbohydrates). Electrons are passed from higher energy molecules to molecular oxygen, giving off much free energy. This energy may be used to phosphorylate ADP, forming a high-energy molecule (ATP) that can be used by the tissues for other metabolic processes. Such ATP synthesis is called oxidative phosphorylation.

4. Gluconeogenesis - the formation of glucose from non-carbohydrate sources such as amino acids and glycerol.

5. Glycogenesis - the synthesis of glycogen from glucose.

6. Glycogenolysis - the breakdown of glycogen to produce glucose (and eventually lactate and pyruvate).

7. Hexose Monophosphate Shunt - an alternative pathway to glycolysis and the citric acid cycle. Again, glucose is oxidized to carbon dioxide and water. (Also called the direct oxidative pathway, the pentose phosphate cycle, the phosphogluconate oxidative pathway).

Figure 21. Glycolysis (Embden-Meyerhof Pathway)

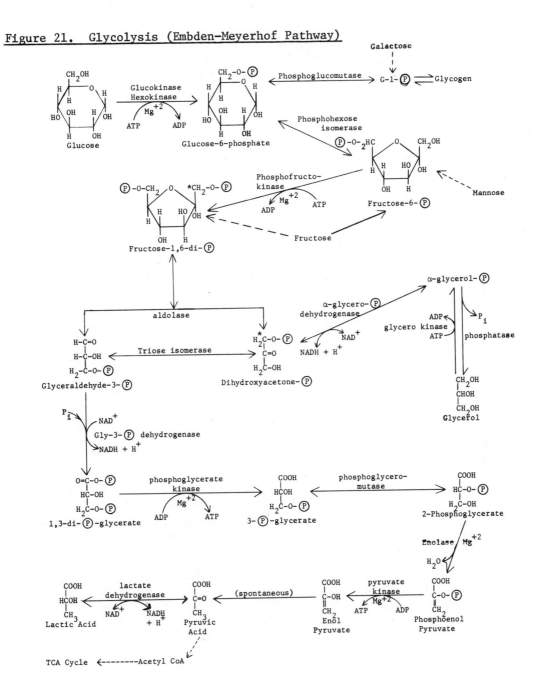

Note: P = phosphate; dotted lines indicate multiple-reaction pathways while solid lines represent single-step
conversions; in fructose-1,6-di P , * indicates the phosphorylated carbon which becomes part of dihydroxyacetone P .

8. Uronic Acid Pathway – another alternative oxidative pathway for glucose. Glucose is converted to glucuronic acid, ascorbic acid (in some animals other than man) and pentoses.

9. Fructose Metabolism – the oxidation of fructose to pyruvate. It is connected to glucose metabolism.

10. Galactose Metabolism – the conversion of galactose to glucose and the synthesis of lactose.

11. Amino Sugar Metabolism – the synthesis of amino sugars from glucose and other sugars and the further synthesis of glycoproteins and mucopolysaccharides.

<u>Comments on E-M Pathway</u>

1. All of the enzymes of E-M Pathway are found in the <u>extra mitochondrial soluble fraction of the cell</u> (in the cytoplasm).

2.

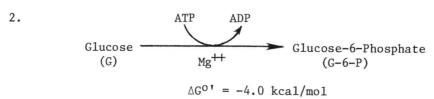

$$\Delta G^{o\prime} = -4.0 \text{ kcal/mol}$$

The equilibrium constant favors the forward reaction (K > 300). This is an important step in regulating the rate at which glucose is metabolized. There are two types of enzymes that can catalyze this reaction.

a. Hexokinase is a non-specific enzyme which acts upon glucose, glucosamine, 2-deoxyglucose, fructose and mannose; it has a higher affinity for aldohexoses than for ketohexoses. In animal tissues, three isoenzymic forms with varying affinities for glucose are present. It is a regulatory enzyme, inhibited by G-6-P, one of its own products. Thus, when G-6-P levels are sufficient to meet the energy demands of the cell, further synthesis of G-6-P is turned off.

b. <u>Glucokinase</u> is a hepatic inducible enzyme and its concentration is affected by changes in the blood glucose levels. Hexokinases of mammalian tissues are present in isoenzymic forms and their activity is inhibited by G-6-P (inhibition may be allosteric). Both enzymes require Mg^{+2} (or Mn^{+2}) for activity. Glucokinase is a hepatic enzyme not present in muscle tissue. It has a higher K_m (10 mM) for D-glucose than does hexokinase (K_m = 100 μM). It is deficient in some patients with diabetes mellitus (see later for further details).

Both enzymes require Mg^{++} or Mn^{++} for activity. These divalent cations form a complex with ATP (which, with glucose, is one of the substrates for these enzymes), discussed later. The hexokinase and glucokinase reactions (phosphorylation of glucose) are <u>irreversible</u> processes, under physiological conditions.

3. Adenosine Triphosphate (ATP) is an important molecule in the cell. It has the structure

where "$\sim$" indicates a "high-energy" bond (one having a large, negative $\Delta G^{o\prime}$ of hydrolysis). (This is a different meaning of high-energy bond than that used by physical chemists.) In the cell, most ATP exists as $(Mg^{+2})(ATP^{-4})$ with the Mg^{+2} chelated to two of the anionic oxygens on the two terminal phosphate groups. In cases where Mn^{+2} replaces the Mg^{+2}, the 7-nitrogen of the adenine ring is also a ligand of the metal ion. Since these complexes are the active forms of ATP, Mg^{+2} (or, in some cases, Mn^{+2}) is essential for practically all reactions involving ATP.

Although ATP has many functions in common with other nucleoside triphosphates (see under nucleic acids), it also serves as the principle "active storage" form of energy within the cell. The principle reaction used in the cell is

$$ATP^{-4} + H_2O \longrightarrow ADP^{-3} + HPO_4^{-2} + H^+ \quad \Delta G^{o\prime} = -7.3 \text{ Kcal/mole}$$

The hydrolysis of the second anhydride bond has the same free energy change

$$ADP^{-3} + H_2O \longrightarrow AMP^{-2} + HPO_4^{-2} + H^+ \quad \Delta G^{o\prime} = -7.3 \text{ Kcal/mole}$$

but it is less frequently used. The reasons for the large, negative $\Delta G^{o\prime}$ of hydrolysis, discussed below in terms of ATP^{-4}, apply equally to ADP^{-3}. They are

a. There are more resonance forms available to ADP^{-3} + Pi than there are for ATP^{-4}. This means that the valence electrons can attain a lower energy state in the hydrolyzed product than in ATP^{-4}. (Note: P_i = inorganic phosphate, HPO_4^{-2} at pH = 7.)

b. ATP^{-4} has 3.8 (since one of the four protons is only about 80% ionized at pH = 7) closely spaced negative charges which repel each other strongly. In the hydrolyzed material (ADP^{-3} + P_i), some of the charges have been carried away by the P_i. This not only favors breaking the bond, it also keeps P_i and ADP^{-3} from coming close enough to each other to recombine.

When fuels are metabolized, the energy released is stored by using it to synthesize the high-energy phosphate anhydride bonds in ATP. These bonds are later hydrolyzed as needed in such a way that the free energy of hydrolysis released is coupled to some bodily process which needs energy. One very important property of ATP is that, despite the large negative $\Delta G^{o'}$ of hydrolysis, this molecule is stable for long periods of time in aqueous solution. This is due to a large activation energy for the hydrolytic reaction.

4. Glucose-6-Phosphate (G-6-P) is an important junction compound.

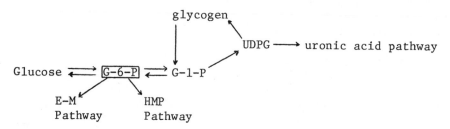

5. Phosphofructokinase

Fructose-6-P $\xrightarrow[\text{Mg}^{++}]{\text{ATP} \quad \text{ADP}}$ Fructose-1,6-diP

$$\Delta G^{o'} = -4.0 \text{ kcal/mol}$$

a. This reaction is essentially irreversible.

b. It is considered to play a major role in the regulation of the rate of glycolysis. The enzyme is an allosteric one, inhibited by high concentrations of ATP, citrate, and long chain fatty acids and activated by AMP or ADP. Thus, glycolysis is turned off when there is an ample supply of non-glycolytic substrates for energy production. It is also shut down when the intra-cellular ATP concentration is high. When the levels of AMP and ADP fall, signalling the depletion of ATP, phosphofructokinase once again becomes active.

6. Fructose-1,6-Diphosphate Aldolase Reaction

F-1,6-diP $\rightleftharpoons$ Dihydroxyacetone Phosphate (DHAP) +

D-glyceraldehyde-3-Phosphate (Gly-3-P)

$\Delta G^{o'}$ = +5.73 kcal/mol

a. The reversal of this reaction (formation of F-1,6-di(P)), also catalyzed by this enzyme, is an aldol condensation, hence the name <u>aldolase</u>.

b. Despite a high positive $\Delta G^{o'}$, the reaction proceeds in the forward direction because of the intracellular concentration of the reactant and the rate of removal of products (that is $\Delta G' < 0$ even though $\Delta G^{o'} > 0$).

c. Mechanism of reaction: A specific lysine residue in the active site of the enzyme ($E-NH_2$) forms a ketimine (an example of a Schiff base) with the keto group of the F-1,6-diP. The steps in the mechanism are as follows:

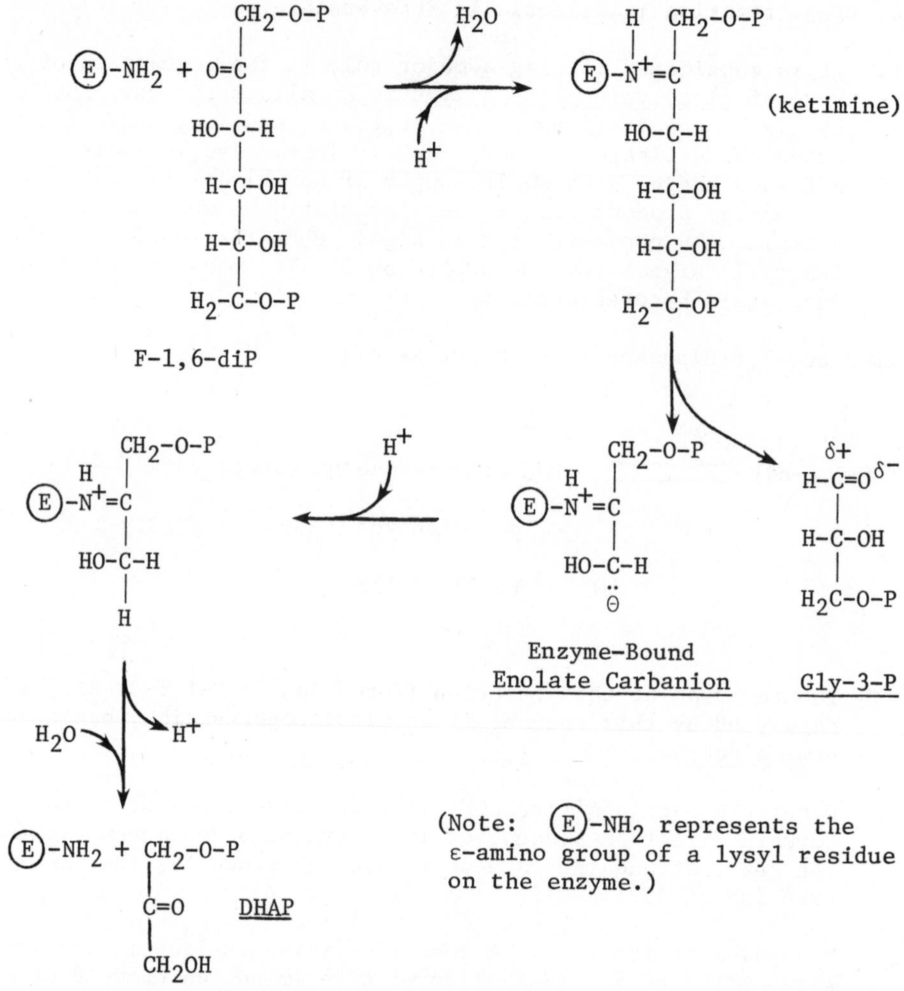

$$
\underbrace{\text{(E)}-NH_2}_{} + \begin{array}{c} CH_2-O-P \\ | \\ O=C \\ | \\ HO-C-H \\ | \\ H-C-OH \\ | \\ H-C-OH \\ | \\ H_2-C-O-P \end{array} \xrightarrow[\text{H}^+]{\text{H}_2\text{O}} \begin{array}{c} H \quad CH_2-O-P \\ \quad | \\ \text{(E)}-N^+=C \\ | \\ HO-C-H \\ | \\ H-C-OH \\ | \\ H-C-OH \\ | \\ H_2-C-OP \end{array} \quad \text{(ketimine)}
$$

F-1,6-diP

$$
\begin{array}{c} CH_2-O-P \\ H \quad | \\ \text{(E)}-N^+=C \\ | \\ HO-C-H \\ | \\ H \end{array} \xleftarrow{\text{H}^+} \begin{array}{c} CH_2-O-P \\ H \quad | \\ \text{(E)}-N^+=C \\ | \\ HO-C-H \\ | \\ \overset{..}{\Theta} \end{array} \qquad \begin{array}{c} \delta^+ \\ H-C=O^{\delta^-} \\ | \\ H-C-OH \\ | \\ H_2C-O-P \end{array}
$$

Enzyme-Bound
<u>Enolate Carbanion</u>   <u>Gly-3-P</u>

$$
\text{(E)}-NH_2 + \begin{array}{c} CH_2-O-P \\ | \\ C=O \\ | \\ CH_2OH \end{array} \quad \underline{DHAP}
$$

(downward arrow) H$_2$O ⟶ H$^+$

(Note: (E)-NH$_2$ represents the
ε-amino group of a lysyl residue
on the enzyme.)

The above reactions are all reversible.

d. Serum aldolase levels are elevated in primary diseases of the muscle fibers (the genetic, inflammatory, endocrine and metabolic myopathies) such as muscular dystrophy, polymyositis, and others. The <u>neurogenic</u> muscular diseases (those secondary to diseases of the nerves supplying the muscle) do not show elevated enzyme levels, except when they have progressed to the point of producing muscle wastage. Examples of this kind of disorder include poliomyelitis, multiple sclerosis, polyneuritis, and myasthenia gravis.

7. Glyceraldehyde-3-P Dehydrogenase Reaction

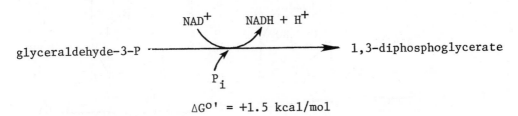

$$\Delta G^{\circ\prime} = +1.5 \text{ kcal/mol}$$

This is the first high energy compound formed in glycolysis.

a. Gly-3-P dehydrogenase is an enzyme made up of four identical subunits. Each subunit contains a sulfhydryl (-SH) group which participates in the formation of an intermediate compound with the aldehyde moiety. The enzyme is inhibited by Hg^+ (non-competitively) and iodoacetamide or iodoacetate (irreversibly). Both of these react with the sulfhydryl group present in the active site.

b. The aldehyde group is oxidized by NAD^+ (a co-substrate) which is converted to NADH.

c. The structure of NAD^+ is

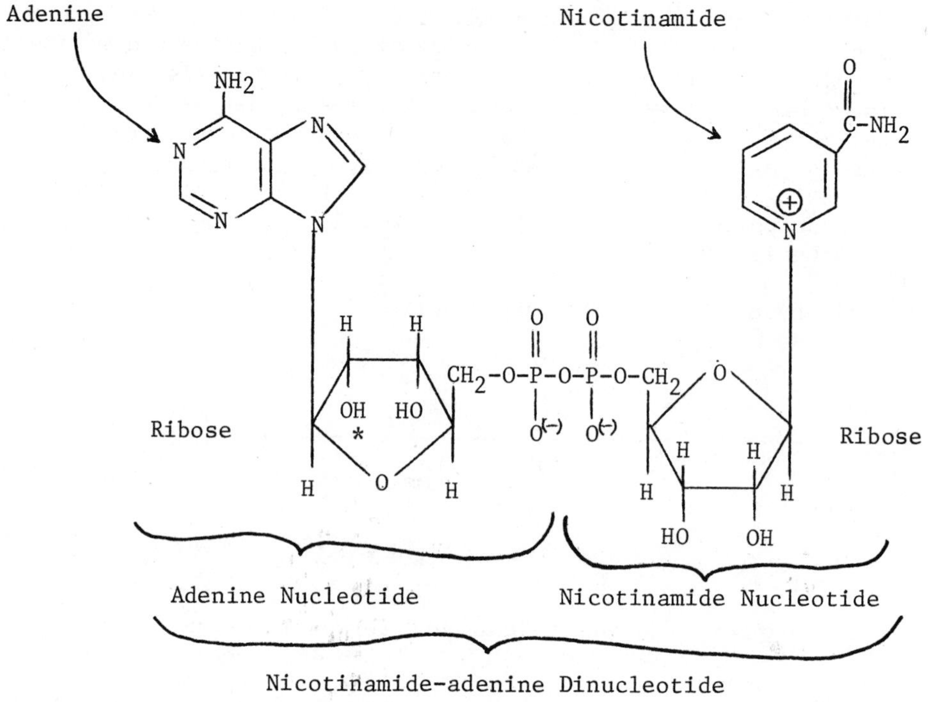

Adenine

Nicotinamide

$$NH_2$$

Ribose

$$CH_2-O-P-O-P-O-CH_2$$

Adenine Nucleotide   Nicotinamide Nucleotide

Nicotinamide-adenine Dinucleotide

* -H is substituted with   O   in $NADP^+$.
$$
-P-OH
$$
OH

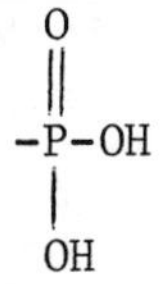

The functional part of NAD^+ and $NADP^+$ is the nicotinamide ring.
Nicotinamide is a vitamin required for the synthesis of NAD^+
and $NADP^+$. It can be made to a limited extent from tryptophan
(see amino acid metabolism).

d. The mechanism of the glyceraldehyde-3-P dehydrogenation
    reaction is shown below. The hydride ion transfer is typical
    of reactions involving NAD^+ and $NADP^+$.

    i) Gly-3-Ⓟ binds to the -SH group of glyceraldehyde-3-Ⓟ
        dehydrogenase.

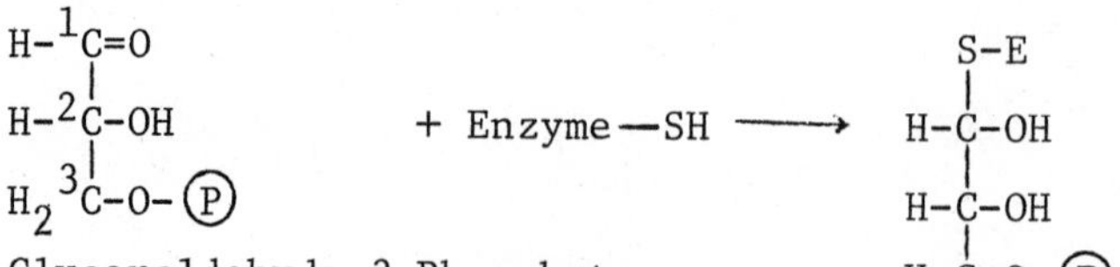

$$H-^1C=O$$
$$H-^2C-OH$$
$$H_2{}^3C-O-Ⓟ$$
Glyceraldehyde-3-Phosphate

+ Enzyme—SH ⟶

$$S-E$$
$$H-C-OH$$
$$H-C-OH$$
$$H_2C-O-Ⓟ$$

ii) A hydride ion (H:⁻) is transferred from the 1 carbon
     of gly-3-P to the para position of the nicotinamide ring.

| Pyridinium ring of NAD⁺ | | Reduced pyridine ring of NADH | Enzyme-bound thiolester intermediate |

Reoxidation by any
of several pathways

iii) Phosphorylation by inorganic phosphate, release of
      1,3-DPG and regeneration of the enzyme.

inorganic phosphate(P$_i$)   1,3-diphospho-glyceric acid

8. The reoxidation of reduced NADH can be accomplished by the following
reactions catalyzed by the appropriate dehydrogenases.

$$\text{Pyruvate} \longrightarrow \text{Lactate (discussed below)}$$

$$\text{Dihydroxyacetone-} \textcircled{P} \longrightarrow \alpha\text{-glycero-} \textcircled{P}$$

$$\text{Oxaloacetate} \longrightarrow \text{Malate}$$

In aerobic cells, the latter two reactions participate in transporting
NADH into mitochondria to be eventually oxidized by oxygen in the
electron transport chain (discussed later in the section on
mitochondria).

9. Fate of 1,3 Diphosphoglycerate (1,3-DPG):

    a. Phosphoglycerate kinase reaction (part of glycolysis)

$$1,3\text{-DPG} + \text{ADP} \rightleftharpoons 3\text{-Phosphoglycerate (3 PG)} + \text{ATP}$$

$$\Delta G^{o\prime} = -4.50 \text{ kcal/mol}$$

    b. Production of 2,3 DPG

$$1,3 \text{ DPG} \xrightarrow{\text{Diphosphoglycerate mutase}} 2,3 \text{ DPG}$$

Large quantities of 2,3-DPG are produced in the human red blood cells (relative to other cell types) where it plays an important role in the regulation of oxygen binding and release by hemoglobin (for more details, see Chapter 7). It is degraded by hydrolysis to 3-PGA by a specific phosphatase. 2,3-DPG also functions as a cofactor in the conversion of 3-phosphoglycerate to 2-phosphoglycerate during glycolysis (see below). Note that 2,3-DPG is not capable of phosphorylating ADP because it has no high energy phosphate bond. Both phosphate linkages are of the ester type whereas in 1,3-DPG one of the phosphates is involved in a mixed acid anhydride linkage (carboxylic acid and phosphoric acid).

10. Phosphoglyceromutase Reaction

$$3 \text{ PG} \rightleftharpoons 2 \text{ PG} \qquad \Delta G^{o\prime} = +1.06 \text{ kcal/mol}$$

In this reaction, 2,3-DPG functions as a cofactor in the following manner:

$$
\begin{array}{c}
\overset{\displaystyle O}{\underset{\displaystyle |}{\|}} \\
C\text{-OH} \\
| \\
HC\text{-O-P} \\
| \\
H_2\text{-C-O-P}
\end{array}
\quad + \ E \qquad \longrightarrow \qquad E\text{-P} \ + \quad
\begin{array}{c}
\overset{\displaystyle O}{\|} \\
C\text{-OH} \\
| \\
H_2\text{-C-O-P} \\
| \\
HC_2\text{-OH}
\end{array}
$$

<u>2,3-DPG</u>   <u>2-PG</u>

11. Enolase Reaction and the Generation of the Second High Energy Compound in Glycolysis

$$2\text{-phosphoglycerate} \rightleftharpoons \text{Phosphoenol pyruvate} + H_2O$$

$$\Delta G^{o\prime} = +0.44 \text{ kcal/mol}$$

a. The enzyme requires Mg^{2+} and is inhibited by fluoride, particularly in the presence of phosphate. This is presumably due to the formation of a magnesium fluorophosphate complex.

b. There appears to be considerable redistribution of energy upon dehydration of the 2-PG molecule. This is indicated by the large difference in the standard free energy of hydrolysis of the phosphate group between 2-PG and PEP (-4.2 kcal/mol versus -14.8 kcal/mol).

12. Pyruvate Kinase Reaction

$$\text{Phosphoenol pyruvate} + \text{ADP} \longrightarrow \text{Pyruvate} + \text{ATP}$$

$$\Delta G^{o\prime} = -7.5 \text{ kcal/mol}$$

a. This is an irreversible reaction under physiologic conditions.

b. It requires Mg^{2+} or Mn^{2+}. The enzyme forms a complex with the divalent cation before it binds the substrate. It also requires K^+ which is apparently needed for the enzyme to attain an active conformation.

c. This enzyme is regulatory, activated by F-1,6-diP and PEP and inhibited by AMP, ATP, citrate, and alanine.

d. Pyruvate kinase deficiency in the red blood cells is one cause of <u>nonspherocytic hemolytic anemia</u>. This disorder is second only to glucose-6-phosphate dehydrogenase deficiency among the enzyme abnormalities of red blood cells.

13. Conversion of Pyruvic Acid to Lactic Acid

a. Under <u>anaerobic</u> conditions, pyruvate is converted to lactic acid by lactate dehydrogenase, LDH. Lactate formation and hence the reoxidation of the NADH formed in the oxidation of glyceraldehyde-3-P allows glycolysis to proceed.

b. LDH exists as several <u>isozymes</u>. Active LDH (M.W. = 130,000) consists of <u>four subunits</u> of <u>two types</u>: M and H. The letters stand for skeletal <u>m</u>uscle and <u>h</u>eart muscle, the tissues in which the respective types of subunit predominate. This is discussed below. Only the tetrameric molecule possesses catalytic activity. Synthesis of the M and H subunits are controlled by distinct genetic loci.

c. At pH 8.6, isozymes of LDH bear different net charges and hence migrate to different regions in an electric field. This is a characteristic property of isozymes. Its usefulness is discussed below.

d. Different tissues have different types of LDH isozymes suited to perform the function of the tissue in which they are present. Heart muscle, which functions under aerobic conditions, has primarily the LDH isozyme made up of four subunits of H (H_4). The tissues that contain LDH-H_4 isozyme produce very little lactic acid because the pyruvic acid formed in glycolysis is channeled into pathways in which it can be oxidized completely (to CO_2 + H_2O), yielding the greater amount of energy needed to sustain a continuous mechanical performance. This happens because pyruvate inhibits the H-isozymes of LDH. As a result, the enzyme is used for the conversion of lactate to pyruvate in these cells

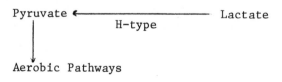

e. On the other hand, in the functioning skeletal muscle, there are times of oxygen deprivation (anaerobic conditions). Under these conditions, muscle (mostly LDH-M$_4$) isozymes (which are only weakly inhibited by pyruvate) convert large amounts of pyruvic acid to lactic acid at a rapid rate. In the relaxation intervals, the oxygen supply increases over oxygen utilization and lactic acid gets reconverted to pyruvate which is completely oxidized via the TCA cycle and oxidative phosphorylation. In vigorous skeletal muscle activity, however, the large amount of lactic acid produced passes into the blood and is transferred to the liver. There it gets metabolized to other products such as pyruvic acid, glucose, etc. This process, known as the Cori cycle is illustrated in the following flow chart:

The Cori Cycle

Muscle	Circulatory System	Liver
muscle glycogen ← ↓ glucose ← anaerobic glycolysis ↓ lactic acid ⟶	blood glucose ← ⟶ blood lactic acid ⟶	glucose ↑ liver glycogen ↑ glucose ↑ pyruvate ↑ ⟶ lactic acid

f. Lactic acid dehydrogenase is an intracellular enzyme and the isozyme profile is related to the type of metabolic activity of the tissue. If there are any changes in the cell permeability or if there is cell damage and destruction, the LDH isozymes are spilled into the blood. The serum LDH levels are then elevated and the serum isozyme pattern is indicative of the tissue that is damaged. In myocardial infarction, for example, LDH levels are increased (primarily due to H$_4$ and H$_3$M) within about 72 hours after infarction. The high levels may be maintained for over a week.

g. LDH Isozyme Profiles in Man

Table 8.   LDH Isozyme Profiles in Man

LDH Isozyme	Subunit Composition	Relative Concentration* (% of total LDH activity)			
		Heart Muscle	Erythro-cytes	Skeletal Muscle	Liver
LD-1	H_4	60	42	4	2
LD-2	H_3M	33	44	7	6
LD-3	H_2M_2	7	10	17	15
LD-4	HM_3	Trace (<1)	4	16	13
LD-5	M_4	Trace (<1)	Trace (<1)	56	64

* These are average values, taken from several reports in the literature.

From these profiles it is apparent that no tissues are "pure" in having only one isozymic form of LDH. Generally two (sometimes three) forms predominate in a particular tissue.

h. The total serum LDH can be determined by an NAD^+-NADH assay (described earlier). Since the reaction is reversible, either pyruvate or lactate can be used as the substrate depending upon the isozyme being measured. Since the heart isozymes are inhibited by pyruvate, lactate is the preferred substrate for these forms.

i. In addition to total activity, it is important to know the amount of activity due to each isozyme (the isozyme profile). There are several ways to do this.

i) The isozymes are separable by electrophoresis (polyacrylamide gel, cellulose acetate membrane, etc.), giving the qualitative pattern shown below at pH = 8.8, 0.05 ionic strength tris-barbital buffer.

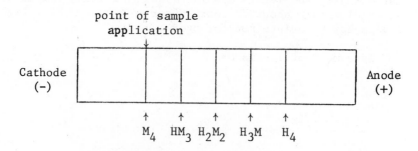

The M-subunits are less negatively charged than the
H-subunits. As the proportion of M in the tetramer
increases, the molecule is less strongly attracted
to the anode. When only M is present, the isozyme
does not move from the point of application or may
even migrate slightly towards the cathode. When
cellulose acetate is used as a support medium, the
M_4 and HM_3 isoenzymes migrate towards the cathode.

ii) The location and amount of LDH activity is determined
following the separation, by putting the gel or membrane in
contact with agar or a solution containing lactate, NAD^+
and a color reagent consisting of phenazine methosulfate
(PMS) and a tetrazolium dye. Wherever LDH is present,
the series of reactions shown below takes place and the
gel or membrane is stained blue. In this diagram,
2,3,5-triphenyltetrazolium chloride (TTC) is used as an
example, although more sensitive dyes are available.
PMS serves as an intermediate electron carrier.

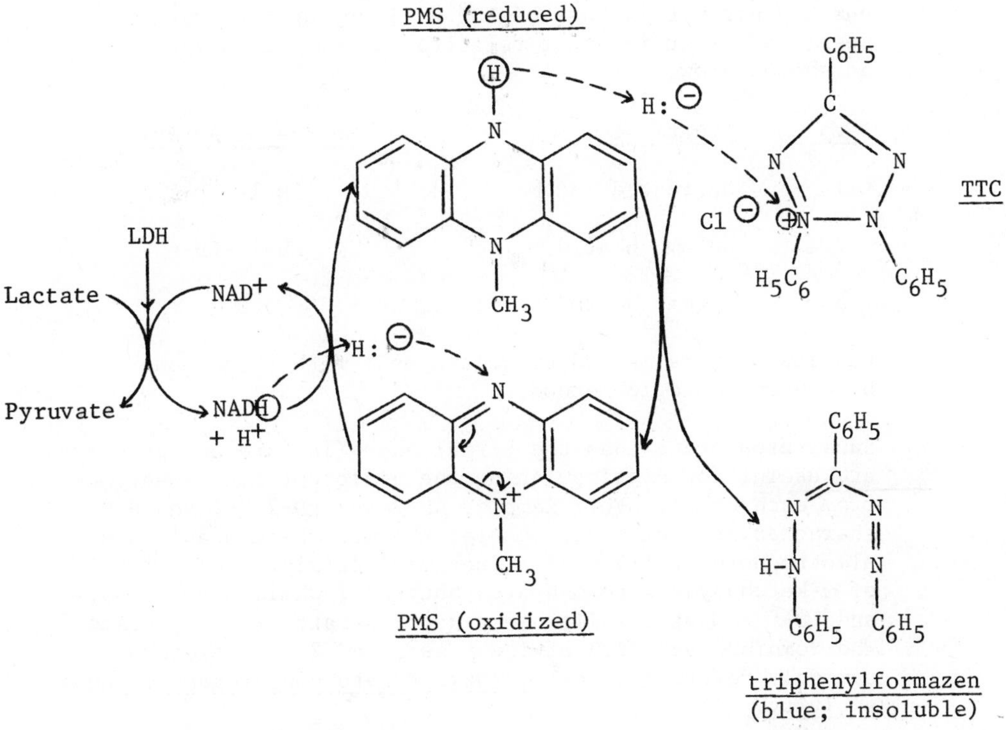

After staining, the gel or membrane can be scanned with
an optical densitometer. The areas under the peaks in
the resulting graph are proportional to the activities of

the different isozymes. This method is highly specific for detecting only LDH activity. It is important to realize (as was pointed out in the section on clinical uses of enzymes) that the activity, and not the amount of enzyme, is being measured.

Any assessment of the activity of the various LDH fractions by electrophoretic methods is only semiquantitative. This is due to differences in the K_m values of the isozymes for the substrate, non-maintenance of zero-order kinetics and differing temperature stabilities of the isozymes. LD4 and LD5 are unstable at cold temperatures. Consequently, sera for isozyme separations should not be refrigerated. In fact, the differences in the relative stabilities of the isozymes to heat has been used in assessing their activities (see later).

iii) M-subunits are much more susceptible to low temperatures, heat, urea, and other denaturing agents than are H-subunits. Treatment of a sample with denaturing agents destroys M_4 and M_3H activity. Loss of LDH activity upon heating has been used as an index of relative isozyme activities, as shown below.

Sample	Treatment	Source of Activity
1	Untreated serum	LD-1--LD-5
2	Serum, heated to 57°	LD-1--LD-4
3	Serum, heated to 65°	LD-1

Results of this method correlate well with those done by electrophoretic methods.

iv) Substrates other than the normal ones (lactate and pyruvate) are useful for distinguishing the different LDH isoenzymes from each other. For example, LD-1 and LD-2 (H_4 and H_3M, characteristic of heart muscle; the most heat stable and electrophoretically mobile isozymes) catalyze the reduction of α-ketobutyrate to α-hydroxybutyrate, while LD-3, LD-4, and LD-5 cannot readily use α-ketobutyrate as a substrate. The combined activity of isozymes 1 and 2 is therefore commonly referred to as hydroxybutyrate dehydrogenase (HBD) activity.

j. Clinical interpretation of LD isoenzyme fractions:

    i) In the clinical interpretation of an isoenzyme analysis, an
       important point to consider is that the total LD activity
       or that of an individual fraction need not be elevated for
       the pattern to be considered abnormal. However, for
       abnormal pattern detection, the essential thing to look for
       is deviation from normal in the <u>relative</u> amounts of the
       fractions. The normal distribution is:

                LD1 < LD2 > LD3 > LD4 <=> LD5.

       Numerical values for each fraction depend on the method and
       electrophoretic support used. An example of a normal LD
       isozyme pattern (carried out on cellulose acetate membrane,
       pH 8.8) expressed as % of total LD activity in the order
       LD1 to LD5 is 25-31, 38-45, 17-22, 5-8, 3-6.

    ii) The actual pattern and the densitometric tracing are both
        examined with the following criteria in mind: Are all the
        bands visible? What is the relationship of LD1 to LD2 and
        their intensity of staining? Is LD2 > LD3? Is LD3 heavily
        stained? What is the intensity of staining of LD4 and LD5?
        Is LD5 deeply stained? Are all fractions heavily stained?
        When LD isoenzyme analyses are performed properly and
        interpreted <u>along with pertinent clinical information</u>, they
        become a powerful tool in the diagnosis and management of a
        number of disorders.

    iii) <u>Serum LD isozyme elevations in various disease states</u>.
         Increased amounts of LD1 and LD2 are seen in <u>myocardial
         infarction</u> (MI) with LD1 usually being greater than LD2
         (referred to as the LD1/2 flip). The total LD is not
         elevated until about 12 hours after infarction and it
         attains a peak value around the third or fourth day. About
         60% of the patients show the LD1/2 flip by 24 hours and 80%
         by 48 hours. CPK (or CK) isoenzyme analysis (discussed
         later in this chapter) provides a remarkably specific and
         sensitive index of myocardial damage when considered along
         with LD isoenzyme analysis, EKG findings, and the clinical
         impressions. It is important to consider the time interval
         between the initial pain episode and the drawing of the
         blood sample for isoenzyme analysis. For example CK_2 (the
         heart-specific isoenzyme; also called MB) is detectable
         4-8 hours after infarction, attains a peak value at about

18 hours, and returns to normal within 48 hours. There is a period during which both the LD1/2 flip and CK_2 are found in the sera of patients with MI. However, due to differences in the rates of disappearance of LD1 and CK_2 from plasma, the CK_2 concentration may be decreasing or absent by the time the LD1/2 flip is apparent, depending on when the sample was drawn. Other causes for elevated LD1 and LD2 (or for the 1/2 flip) are hemolysis, megaloblastic (e.g., pernicious) and hemolytic anemias, renal necrosis, and Duchenne muscular dystrophy. If the LD1/2 flip is due to in vivo hemolysis, this can be confirmed by haptoglobin measurement. Renal tissue contains all of the LD isozymes, but there is considerable variation in isoenzyme distribution within the kidney. LD1 and LD2 predominate in the cortex, tubules and outer medulla while the middle LD fractions are present in large quantities in the inner medulla and the papilla. The glomerulus is rich in LD1, LD3 and LD4. Although normal adult skeletal muscle contains mainly the cathodic fractions (LD4 and LD5), it has been observed in post-natal skeletal muscle diseases (such as the muscular dystrophies) that the total LD levels are increased but that there is a reduction in LD5 with elevations in LD1 and LD2. This type of pattern (anodic and middle fraction predominant) is seen in the normal muscle during the first half of gestation, whereas during the second half of gestation with the differentiation of muscle fibers, the cathodic fractions become predominant. This similarity in the LD isoenzyme pattern between fetal muscle and some types of post-natal muscular dystrophies has been attributed to dedifferentiation of muscle tissue in these pathologic processes.

LD2 and LD3 are elevated in active lung damage (making them useful in differentiating between pulmonary and myocardial infarction), leukemia, collagen diseases, other reticuloendothelial or lymphocyte related processes, pericarditis and viral infections (particularly in children, presumably due to lung involvement). LD5 is elevated in response to skeletal muscle or liver damage. In patients with myocardial infarction complicated with congestive heart failure (due to poor venous return), not only the isoenzyme changes typical to MI (discussed earlier) may appear but also those due to liver necrosis. This is reflected in an elevated LD5 fraction, indicating hypoxic damage of the liver.

11. Reversibility of Glycolysis: Although all reactions are technically reversible (see thermodynamics), the following reactions are physiologically irreversible (i.e. irreversible for all practical purposes in the body). These are processes involving large losses

of free energy (exergonic reactions). They are catalyzed by the enzymes:

a. Hexokinase (glucokinase)
b. Phosphofructokinase
c. Pyruvate Kinase

Alternate routes are available, however, for the synthesis of glucose from lactate, glycerol, and amino acids (gluconeogenesis). Enzymes other than those of glycolysis are used.

Note: The phosphoglycerate kinase reaction is also an exergonic reaction and is not easily reversible, but an alternate route does not exist for this reaction. It is reversed by mass law considerations when 3-(P)-glycerate and ATP accumulate.

12. **Lactate is produced by the E-M pathway (glycolysis) in skeletal** muscle when the muscle is performing under anoxic conditions and in erythrocytes. In the latter case, the enzymes which oxidize pyruvic acid are absent. This lactic acid is carried by the blood to the liver, where it is converted to glucose (Cori cycle already discussed).

13. The net result of anaerobic glycolysis is

$$\text{Glucose} + 2\ \text{ADP} + 2\ P_i \quad \rightarrow 2\ \text{lactic acid} + 2\ \text{ATP}$$

14. Pasteur Effect. In many circumstances glycolysis is decreased by the presence of O_2.

aerobic conditions	anaerobic conditions
low glucose consumption	high glucose consumption
low lactic acid production	high lactic acid production

The mechanism is uncertain but the following are possible explanations:

a. The competition between glycolysis and oxidative phosphorylation for ADP and inorganic phosphate.

b. Inactivation of glycolytic enzymes by oxidation of sulfhydryl groups.

c. Inhibition of phosphofructokinase by ATP (citrate accentuates the effect in the presence of ATP). This may be the major effect.

d. Reversal of glycolysis under <u>aerobic conditions</u> (gluconeogenesis). This would cause an <u>apparent</u> decrease in glycolysis when the glucose concentration is being measured.

e. Aerobic glycolysis is more efficient than anaerobic glycolysis in terms of moles of ATP formed per mole of glucose. Therefore, if the cellular need for ATP does not change, less glucose will be used by each cell, when O_2 is present. As in d above, this would cause an apparent decrease in glycolysis if glucose is being followed.

15. The reciprocal of the Pasteur effect is the <u>Crabtree effect</u>. High concentrations of glucose will inhibit cellular respiration when studied in isolated systems such as ascites tumor cells. The probable mechanism for this effect is more effective competition by glycolysis for inorganic phosphate and NADH. This can lead to a deficiency of materials available to carry on oxidative phosphorylation. These interrelationships are shown in Figure 22.

16. Biochemical Lesions of Glycolysis

a. i) <u>Pyruvate kinase (PK) deficiency</u> in the red cells has been reported. Pyruvate kinase catalyzes the reaction

$$\text{Phosphoenolpyruvate} \longrightarrow \text{Pyruvate}$$
$$\text{ADP} \qquad \text{ATP}$$

ii) The deficiency (classified as a type of congenital nonspherocytic hemolytic anemia) leads to inadequate production of ATP, causing premature destruction of red cells (auto-hemolysis). The lysis is decreased to a small degree by glucose and to a greater extent by ATP, ADP or AMP.

iii) The lesion appears to be transmitted by an autosomal recessive gene. The enzyme is virtually absent in the red cells obtained from the affected individuals. Heterozygous carriers have approximately half of the normal PK activity (compared to 5-25% in homozygotes) but there is a wide range of values in both groups.

Figure 22 .   Competitive Reactions Which Lead to the Pasteur and
Crabtree Effects

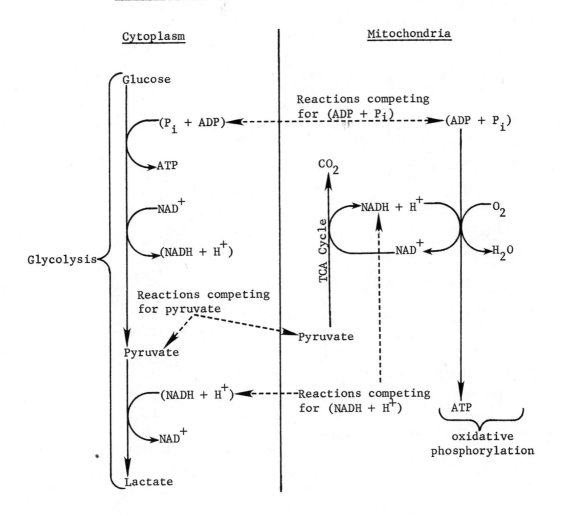

b. In a number of other cases of hereditary hemolytic anemia
specific glycolytic enzymes have been shown to be deficient.
These enzymes include hexokinase, glucose phosphate isomerase,
phosphofructokinase, triosephosphate isomerase,
2,3-diphosphoglyceromutase, and phosphoglycerate kinase. With
the exception of the PK deficiency, discussed above, all such
diseases are rare.

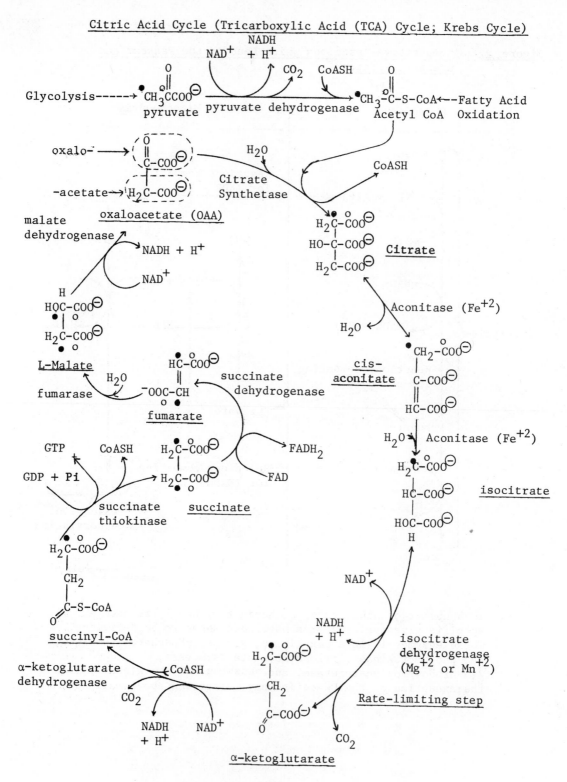

Citric Acid Cycle (Tricarboxylic Acid (TCA) Cycle; Krebs Cycle)

Glycolysis-----→ CH_3CCOO^- pyruvate

$NAD^+ + H^+$ → $NADH$

pyruvate dehydrogenase

CO_2 CoASH

$CH_3-C-S-CoA$ ←--Fatty Acid
Acetyl CoA Oxidation

oxalo- → $C-COO^-$

-acetate→ H_2C-COO^-

oxaloacetate (OAA)

H_2O

Citrate
Synthetase

CoASH

H_2C-COO^-
$HO-C-COO^-$
H_2C-COO^-

Citrate

malate
dehydrogenase

$NADH + H^+$

NAD^+

$HOC-COO^-$
H_2C-COO^-

L-Malate

fumarase

H_2O

$HC-COO^-$
$^-OOC-CH$

fumarate

succinate
dehydrogenase

Aconitase (Fe^{+2})

H_2O

CH_2-COO^-
$C-COO^-$
$HC-COO^-$

**cis-
aconitate**

H_2O Aconitase (Fe^{+2})

H_2C-COO^-
$HC-COO^-$
$HOC-COO^-$
H

isocitrate

GTP CoASH

GDP + Pi

succinate
thiokinase

H_2C-COO^-
H_2C-COO^-

succinate

$FADH_2$

FAD

H_2C-COO^-
CH_2
$C-S-CoA$
O

succinyl-CoA

α-ketoglutarate
dehydrogenase

CO_2

CoASH

$NADH + H^+$ NAD^+

NAD^+

$NADH + H^+$

isocitrate
dehydrogenase
(Mg^{+2} or Mn^{+2})

Rate-limiting step

H_2C-COO^-
CH_2
$C-COO^-$
O

CO_2

α-ketoglutarate

208

<u>Comments on Citric Acid Cycle</u>. The complete oxidation of glucose which
finally produces CO_2 starts with the conversion of pyruvate to acetyl-
CoA with concomitant reduction of $NAD^+ \rightarrow NADH + H^+$. In subsequent
reactions, FAD and additional NAD^+ is reduced, and CO_2 is formed.
Sir Hans Krebs (whose name is frequently associated with the cycle)
received the Nobel Prize in 1953 for his work on intermediary
metabolism.

1. All of the citric acid cycle enzymes are found in the mitochondrial
    fraction of the cell and are in close proximity to the enzymes of
    the <u>respiratory chain,</u> either as components of the cristae membranes
    or as soluble enzymes of the mitochondrial matrix. The enzymes
    aconitase, fumarase and malate dehydrogenase are of the latter type,
    while pyruvic acid oxidodecarboxylase (pyruvate dehydrogenase),
    ketoglutaric oxidodecarboxylase (α–Ketoglutarate dehydrogenase),
    succinate dehydrogenase, and the others are membrane bound.

2. Conversion of pyruvic acid to acetyl-CoA (active acetate)

    a. The five reactions are catalyzed by three different enzymes.
        They function as a multi-enzyme complex and are collectively
        known as pyruvic oxidodecarboxylase. For all practical
        purposes this is an <u>irreversible reaction.</u> Six coenzymes are
        required. They are Mg^{+2}, TPP, lipoic acid, CoASH, FAD and
        NAD^+.

    b. Overall Reaction

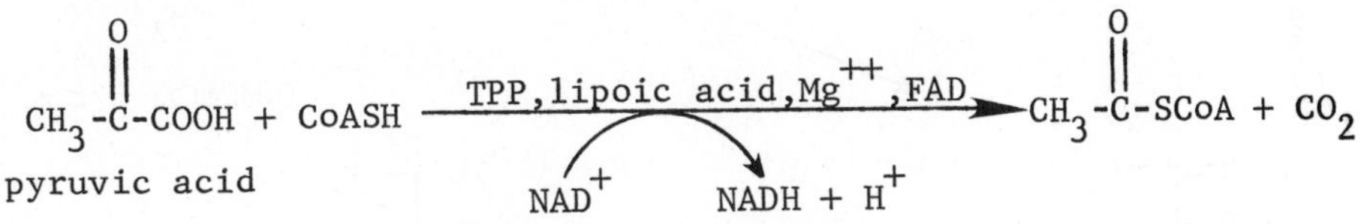

$$CH_3\text{-}\overset{\overset{\displaystyle O}{\|}}{C}\text{-COOH} + CoASH \xrightarrow[\quad NAD^+ \qquad NADH + H^+ \quad]{TPP, \text{lipoic acid}, Mg^{++}, FAD} CH_3\text{-}\overset{\overset{\displaystyle O}{\|}}{C}\text{-SCoA} + CO_2$$

pyruvic acid

$$\Delta G^{o\prime} = -8.0 \text{ kcal/mol}$$

    c. Some details of the reaction

    i)

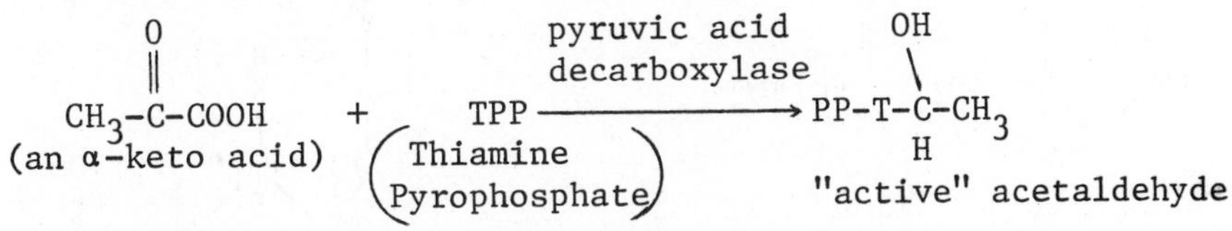

$$CH_3\text{-}\overset{\overset{\displaystyle O}{\|}}{C}\text{-COOH} + \underset{\begin{pmatrix} \text{Thiamine} \\ \text{Pyrophosphate} \end{pmatrix}}{\text{TPP}} \xrightarrow[\text{decarboxylase}]{\text{pyruvic acid}} \overset{\overset{\displaystyle OH}{\diagdown}}{PP\text{-}T\text{-}\underset{H}{C}\text{-}CH_3}$$

(an α-keto acid)   "active" acetaldehyde

In the above reaction the two carbon aldehyde groups of
pyruvate are transferred to thiamine pyrophosphate (TPP). The
details of the reactions are described below. (Note: TPP is
the pyrophosphate ester of Vitamin B_1.)

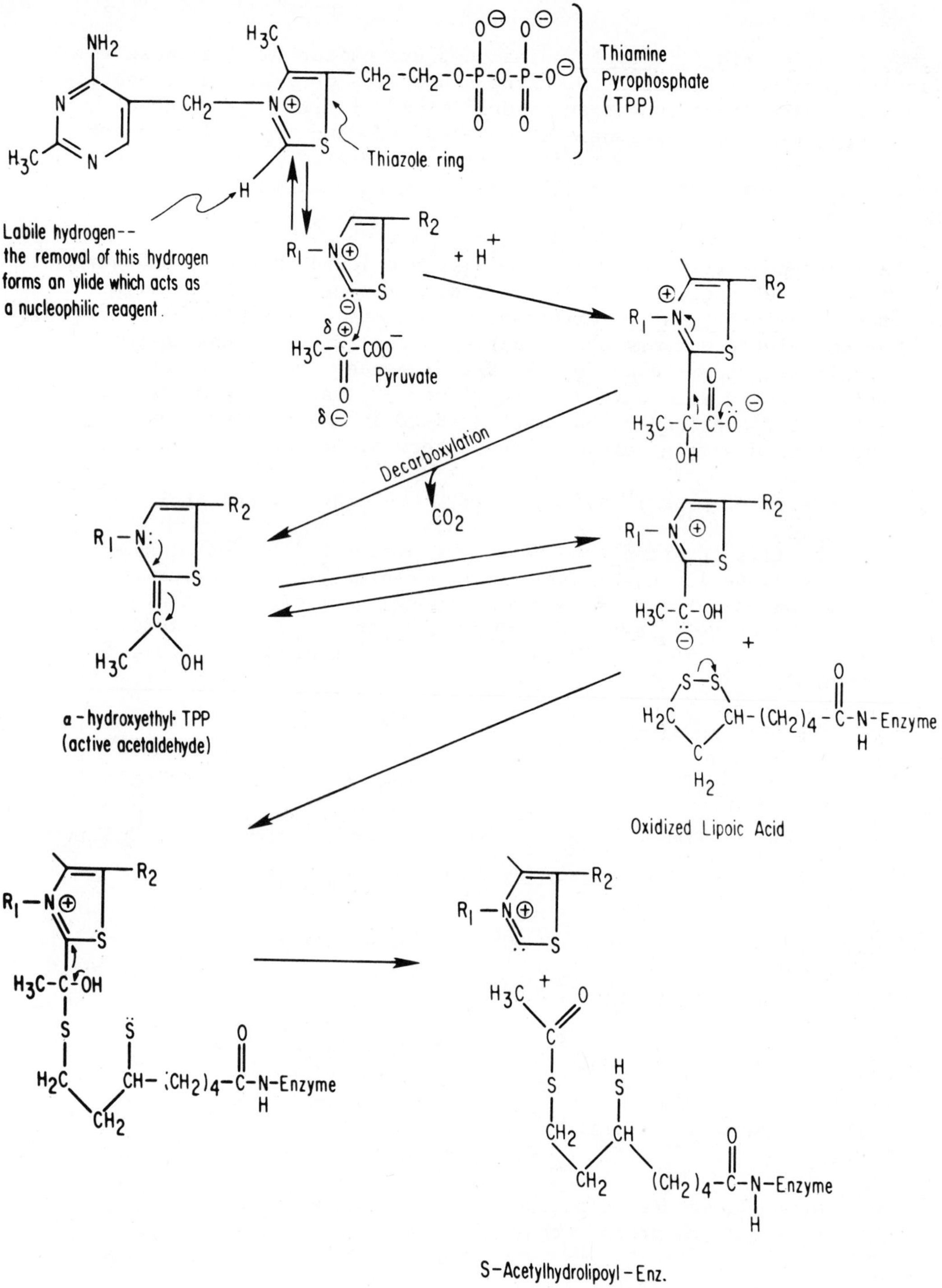

Thiamine Pyrophosphate (TPP)

Thiazole ring

Labile hydrogen-- the removal of this hydrogen forms an ylide which acts as a nucleophilic reagent.

Pyruvate

Decarboxylation

CO_2

α-hydroxyethyl TPP (active acetaldehyde)

Oxidized Lipoic Acid

S-Acetylhydrolipoyl-Enz.

The pyruvate dehydrogenase complex is inhibited (and thus regulated) by high levels of NADH, acetyl-CoA and ATP. The inhibition of the enzyme activity by ATP is mediated through cyclic AMP. These reactions are summarized as follows:

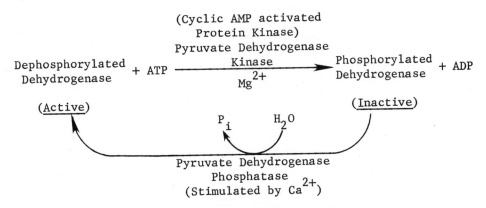

This regulatory mechanism is analogous to that of glycogen synthetase which is discussed under glycogen biosynthesis.

Thiamine pyrophosphate thus functions as a coenzyme for the decarboxylation of pyruvic acid, an α-keto acid. This coenzyme also participates (by a similar mechanism) in the decarboxylation reaction of α-ketoglutarate (another α-keto acid), one of the later reactions in the citric acid cycle. TPP also functions as a coenzyme in the transketolase reactions of the HMP-shunt pathway (discussed later). TPP is obtained from thiamine as follows:

ii) Lipoic acid (considered by some as a vitamin) has the structure

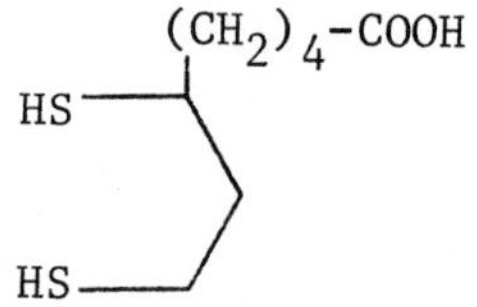

$$(CH_2)_4-COOH$$

HS—

HS—

6,8-dithio-octanoic acid

It is bound to the protein by a peptide linkage involving the epsilon amino (ϵ-NH_2) group of lysine. It is thought to transfer the reaction product from one active center of the enzyme complex to another by virtue of its ability to move over a relatively large area of the enzyme surface.

iii)

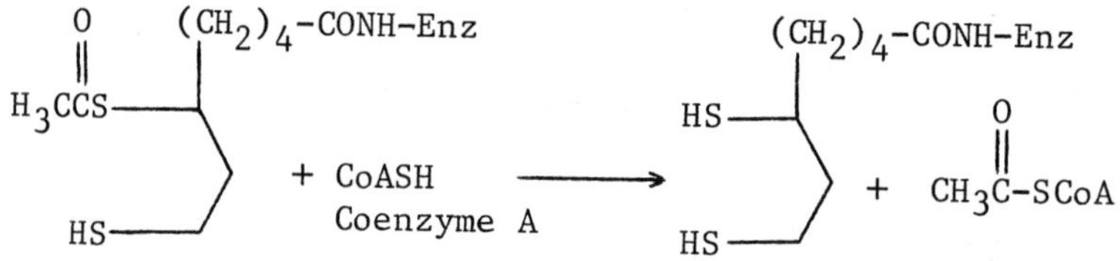

$$\underset{\text{O}}{\overset{\text{O}}{\parallel}}$$
H_3CCS- $(CH_2)_4-CONH-Enz$

+ CoASH
Coenzyme A $\longrightarrow$

HS— $(CH_2)_4-CONH-Enz$

$$\underset{\text{O}}{\overset{\text{O}}{\parallel}}$$
+ $CH_3C-SCoA$

HS—

S-acetylhydro-
lipoyl-enzyme

Reduced
lipoyl-Enz

acetyl-CoA
(one of the
products of the
overall reaction)

The structure of Coenzyme A

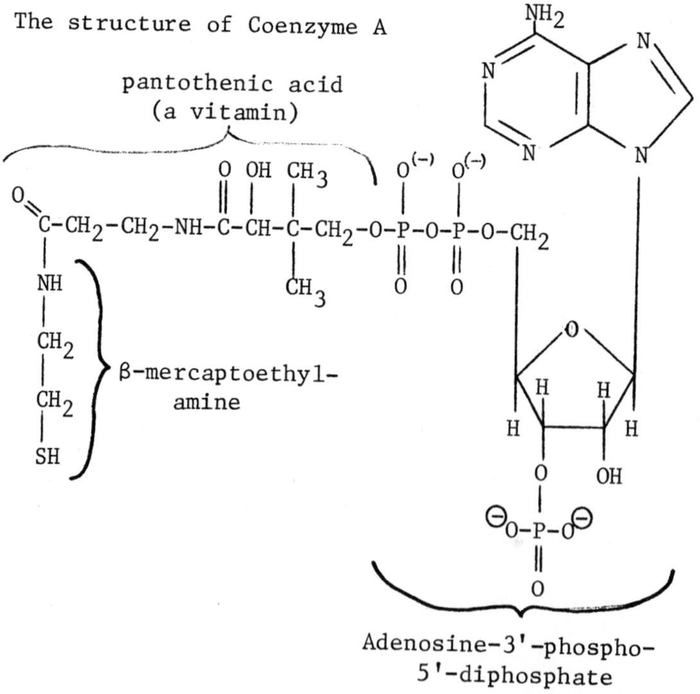

pantothenic acid
(a vitamin)

$$\underset{\text{O}}{\overset{\text{O}}{\parallel}}C-CH_2-CH_2-NH-\overset{\text{O}}{\overset{\parallel}{C}}-\overset{\text{OH}}{\overset{\mid}{CH}}-\overset{CH_3}{\overset{\mid}{C}}-CH_2-O-\overset{O^{(-)}}{\overset{\mid}{\underset{\parallel}{P}}}-O-\overset{O^{(-)}}{\overset{\mid}{\underset{\parallel}{P}}}-O-CH_2$$

NH CH_3 O   O

CH₂

CH₂   } β-mercaptoethyl-
amine

SH

NH_2

H   H

H   H

O   OH

$\overset{\ominus}{O}-P-\overset{\ominus}{O}$

O

Adenosine-3'-phospho-
5'-diphosphate

Acetyl-CoA, a key intermediate, is not only obtained by the oxidation of pyruvic acid, but also from the oxidation of fatty acids and amino acids. In the acetyl-CoA molecule, the acetate group is in an "active" state, and hence can easily function in a variety of synthetic reactions. The acetate is attached to the -SH group in a thioester linkage which has a $\Delta G^{o}{}'$ of -7.5 kcal/mole (similar to that of ATP).

iv) In the next two reactions the oxidized-lipoyl enzyme is regenerated from the reduced-lipoyl enzyme.

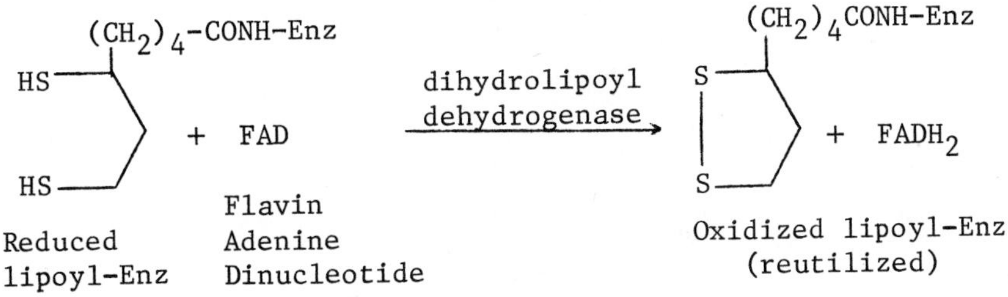

Reduced lipoyl-Enz   Flavin Adenine Dinucleotide   Oxidized lipoyl-Enz (reutilized)

<u>Flavin adenine dinucleotide</u> is the hydrogen acceptor in the above reaction and has the structure

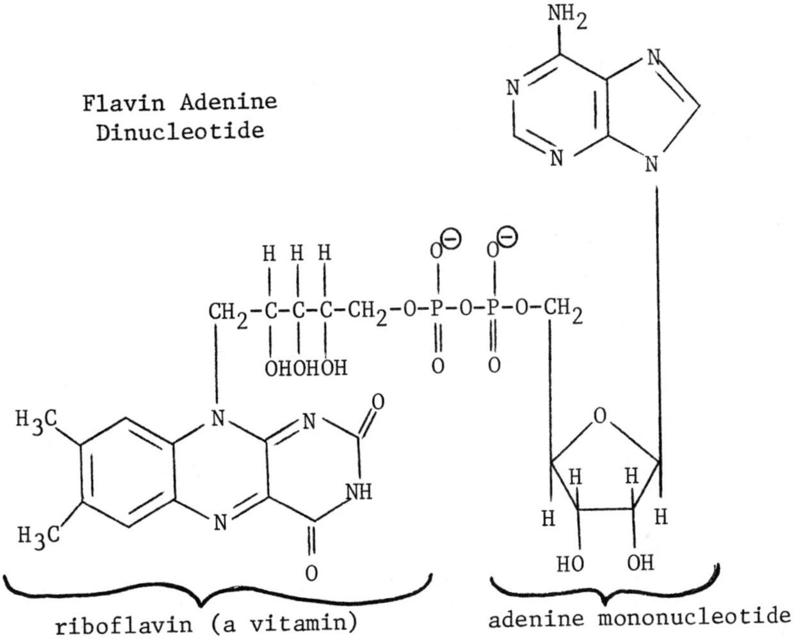

Flavin Adenine Dinucleotide

riboflavin (a vitamin)   adenine mononucleotide

v) In the final reaction, FAD is regenerated with the production of NADH.

$$FADH_2 \ + \ NAD^+ \longrightarrow NADH \ + \ H^+ \ + \ FAD \ (reutilized)$$

d. The pyruvic oxidodecarboxylase complex obtained from E. coli
    consists of three enzymes: decarboxylase (mol wt 183,000),
    transacetylase (mol wt 70,000), and dehydrogenase (mol wt
    112,000) in the ratio of 12:24:6. The complex has also been
    isolated from a variety of other sources, but most of the work
    has been performed using preparations from E. coli. Following
    is a schematic representation of the complex and its reactions.

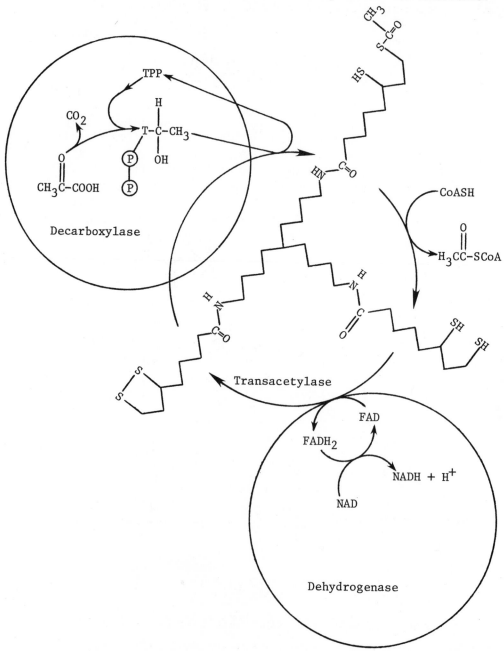

3. Citrate Synthetase and Aconitase Reactions

  a. Acetyl-CoA + Oxaloacetate $\xrightarrow{\text{Citrate Synthetase}}$ Citrate + CoASH

  This is an exergonic reaction which proceeds spontaneously in the direction of citrate synthesis. The energy for the reaction (an aldol condensation between the methyl group of acetyl-CoA and the carbonyl group of oxaloacetate) is derived from the hydrolysis of the thioester bond in acetyl-CoA (presumably with citroyl-CoA as an enzyme-bound intermediate). The enzyme is also known as a condensing enzyme.

  b. Citrate is the first intermediate of the TCA cycle. Its formation is the "committed step" and is subject to regulation. The enzyme is <u>inhibited</u> by high concentrations of NADH, succinyl-CoA (a later product of TCA which competes with acetyl-CoA), ATP and long chain fatty acyl-CoA.

  c. Aconitase, the second enzyme of the TCA cycle proper, catalyzes the reversible addition of water to the double bond of the enzyme-bound intermediate cis-aconitate, yielding either citrate or isocitrate as the product:

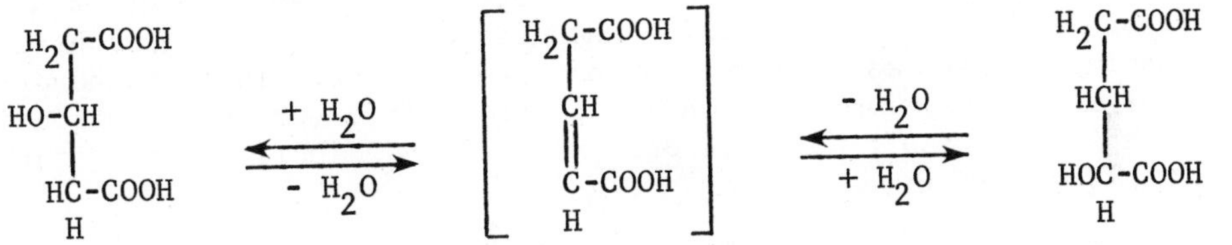

$$
\begin{array}{c}
\text{H}_2\text{C-COOH} \\
| \\
\text{HO-CH} \\
| \\
\text{HC-COOH} \\
\text{H}
\end{array}
\quad
\begin{array}{c}
+ \text{H}_2\text{O} \\
\rightleftharpoons \\
- \text{H}_2\text{O}
\end{array}
\quad
\left[
\begin{array}{c}
\text{H}_2\text{C-COOH} \\
| \\
\text{CH} \\
\| \\
\text{C-COOH} \\
\text{H}
\end{array}
\right]
\quad
\begin{array}{c}
- \text{H}_2\text{O} \\
\rightleftharpoons \\
+ \text{H}_2\text{O}
\end{array}
\quad
\begin{array}{c}
\text{H}_2\text{C-COOH} \\
| \\
\text{HCH} \\
| \\
\text{HOC-COOH} \\
\text{H}
\end{array}
$$

  <u>Citrate</u>   <u>cis-Aconitate</u>   <u>Isocitrate</u>
  (an enzyme-
  bound intermediate)

  Although the ratio of citrate to isocitrate is about 15 at equilibrium, the reaction proceeds towards the formation of isocitrate because of the quick removal of the product in the next reaction. Thus the net reaction catalyzed by aconitase is the isomerization of citrate to isocitrate. Aconitase requires Fe^{2+} and a reduced thiol compound such as glutathione or cysteine.

  d. Although citric acid is a symmetrical molecule, it reacts in an asymmetric manner with the enzyme aconitase (only one of the two $-CH_2-COOH$ groups on the citric acid molecules is modified).

$$\overset{\bullet}{C}H_2 - \overset{o}{C}OOH$$
$$|$$
$$HO-C-COOH$$
$$|$$
$$CH_2-COOH$$
citric acid

: enzyme acts on the oxaloacetate part of the molecule ($\bullet$ and o indicate atoms derived from acetyl-CoA).

<u>Three-point attachment</u> of the enzyme to the substrate molecule (citric acid) accounts for the asymmetrical activity. Condensation of acetyl-CoA with oxaloacetate is also stereospecific with respect to the keto group of oxaloacetate.

4.

(a) isocitrate $\xrightarrow[\text{Mn}^{+2}]{\text{NADP}^+ \quad \text{NADPH} + \text{H}^+}$ (oxalosuccinate) $\xrightarrow{\text{CO}_2}$ α-ketoglutarate

(b) isocitrate $\xrightarrow[\text{Mg}^{+2}]{\text{NAD}^+ \quad \text{NADH} + \text{H}^+ \quad \text{CO}_2}$ α-ketoglutarate

<u>This is the rate-limiting step in the TCA cycle.</u> Two isocitrate dehydrogenases are found in most animal and plant tissues. The cytoplasmic one (which, in most tissues, also occurs in mitochondria but to a much lesser extent) catalyzes reaction 4(a), above. It is <u>non</u>-allosteric, requires NADP$^+$ and Mn^{+2}, and has oxalosuccinate as an enzyme-bound intermediate. The mitochondrial enzyme (catalyzing reaction 4b) is <u>allosteric</u> (positive modulators are ADP and NAD$^+$; negative ones are ATP and NADH), specific for NAD$^+$, and requires Mg^{+2}.

5. a. Conversion of α-ketoglutarate to succinyl-CoA: this oxidative decarboxylation reaction of an α-keto acid, catalyzed by the α-ketoglutarate dehydrogenase (oxidodecarboxylase) multi-enzyme complex is identical to the conversion of pyruvate to acetyl CoA, with respect to the mechanism and cofactors required. It is considered to be the irreversible step in the TCA cycle.

b. Friedreich's ataxia is a hereditary neurological disorder involving spinocerebellar degeneration. The patients suffering from this disorder show high blood levels of alanine (the transamination product of pyruvate), suggesting a defect in the oxidation of pyruvate. By using cultured fibroblasts from these patients, both the pyruvate and α-ketoglutarate

decarboxylase complexes were shown to be defective. Lipoamide
dehydrogenase is the enzyme common to both of these enzyme
complexes. It appears that, in these patients, a mutation in
it has affected the overall activity of both enzyme complexes.
Further, it is of interest to point out that a ketogenic diet
may be useful in the nutritional management of those patients
lacking <u>only</u> pyruvate dehydrogenase complex. Such a diet
consists mainly of lipids and selected amino acids whose
oxidation yields acetyl-CoA directly, thus placing less burden
on the pyruvate decarboxylase complex.

6.

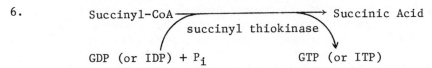

Succinyl-CoA ⟶ Succinic Acid

succinyl thiokinase

GDP (or IDP) + P_i GTP (or ITP)

(GDP = guanosine diphosphate; IDP = inosine diphosphate)

This is the only reaction in the citric acid cycle which directly
generates a high-energy phosphate compound. It is called a
<u>substrate-level phosphorylation</u> as distinguished from oxidative
phosphorylation where oxygen is the terminal electron acceptor.
Note that the glycolytic reactions 1,3-DPG → 3PGA and PEP →
pyruvic acid also are substrate-level phosphorylation reactions.
The GTP formed can be converted to ATP in the following reaction.

nucleosidediphosphate
kinase

GTP   + ADP ⇌ ATP + GDP
(or NTP   (or NDP
in general)   in general)

7.

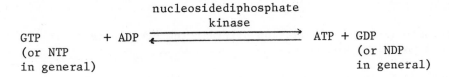

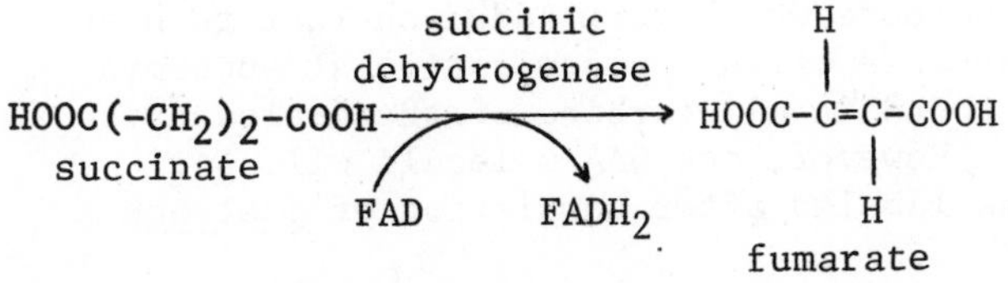

succinic
dehydrogenase

                          H
                          |
HOOC(-CH$_2$)$_2$-COOH ⟶ HOOC-C=C-COOH   Note: The product
    succinate | formed is the
                FAD   FADH$_2$ H   <u>trans</u> isomer.
                                    fumarate

This dehydrogenation is the only reaction in the <u>citric acid cycle</u>
which involves the direct transfer of H to a flavoprotein <u>without</u>
participation of NAD$^+$. The enzyme contains FAD (structure given
previously) and non-heme iron. The oxidation-reduction involves
only the flavin ring system. It proceeds as shown below.
Ⓡ represents the remainder of the FAD molecule.

The two-electron reduction of the flavin ring can also occur in two one-electron steps with the electrons coming from some source other than a hydride ion. In such a case, both hydrogens are derived from H^+ in the solution. This sort of mechanism is typical of all redox reactions involving flavin coenzymes.

8. Labeling experiment: $\bullet CH_3-^oC-S-CoA$: C^{14}-labeled acetyl-CoA.
($\bullet$ and o indicate radioactive methyl and carbonyl carbons, respectively.)

   a. In the first turn of the cycle labeled CO_2 does not appear.

   b. Labeled CO_2 appears in the second and subsequent turns of the cycle.

   c. Succinate is the randomization point in the cycle because it is a symmetrical molecule.

   This labeling is indicated on the diagram of the citric acid cycle. It shows that the carbons which are lost as CO_2 on the initial turn of the cycle are not the carbons which were added on that turn as acetyl-CoA (a, above). Because of the randomization at succinic acid, some label will be found in all carbons of the oxaloacetate formed on the first turn. However, one OAA molecule will not have all four carbon atoms labeled after completion of just one turn.

9. a. At each turn of the TCA cycle, oxaloacetate is regenerated and can combine with the next acetyl-CoA molecule. As can be seen in Figure 23 the TCA cycle is amphibolic, serving both as a catabolic and an anabolic pathway. The reactions that utilize the intermediates of the cycle as precursors for the biosynthesis of other molecules are:

i. α-ketoglutarate + alanine $\rightleftharpoons$ glutamate + pyruvate

ii. oxaloacetate + alanine $\rightleftharpoons$ aspartate + pyruvate

iii. succinyl-CoA + glycine $\longrightarrow$ δ-aminolevulinic acid (ALA)
ALA is then utilized for the synthesis of heme.

iv. citrate + ATP + CoA $\longrightarrow$ acetyl-CoA + oxaloacetate + ADP + Pi
This reaction takes place in the cytoplasm and is a source of
acetyl-CoA for fatty acid biosynthesis.

v. succinyl-CoA + acetoacetate $\longrightarrow$ acetoacetyl-CoA + succinate
This reaction is important in the activation of acetoacetate
(a ketone body) and hence for its utilization in extrahepatic
tissues.

vi. α-ketoglutarate $\longrightarrow$ succinate + CO_2
This reaction is involved in the hydroxylation of proline and
lysine residues of protocollagen, a step in the synthesis of
collagen.

b. However, the utilization of the intermediates for the synthesis of
other molecules leads to the depletion of the intermediates and
hence to a slowdown of oxidation unless they are replenished.
These replacement processes or <u>anaplerotic ("filling-up")
reactions</u>, are listed below:

i. Pyruvate + CO_2 + ATP $\xrightarrow[\text{Biotin, Mg}^{2+}]{\text{Pyruvate Carboxylase}}$ Oxaloacetate + ADP + Pi

This is the most important anaplerotic reaction in animal
tissues. It takes place in the mitochondria. Pyruvate
carboxylase is an allosteric enzyme, requiring acetyl-CoA
for activity (the details of this reaction are discussed
under gluconeogenesis).

ii. Pyruvate + CO_2 + NADPH + H^+ $\underset{\xleftarrow{\hspace{2cm}}}{\xrightarrow[]{\text{Malic Enzyme}}}$ L-Malate + $NADP^+$

iii. Oxaloacetate and α-ketoglutarate may also be obtained from
aspartate and glutamate by transaminase reactions--the
reverse of the reactions mentioned in 9.a.i and ii.

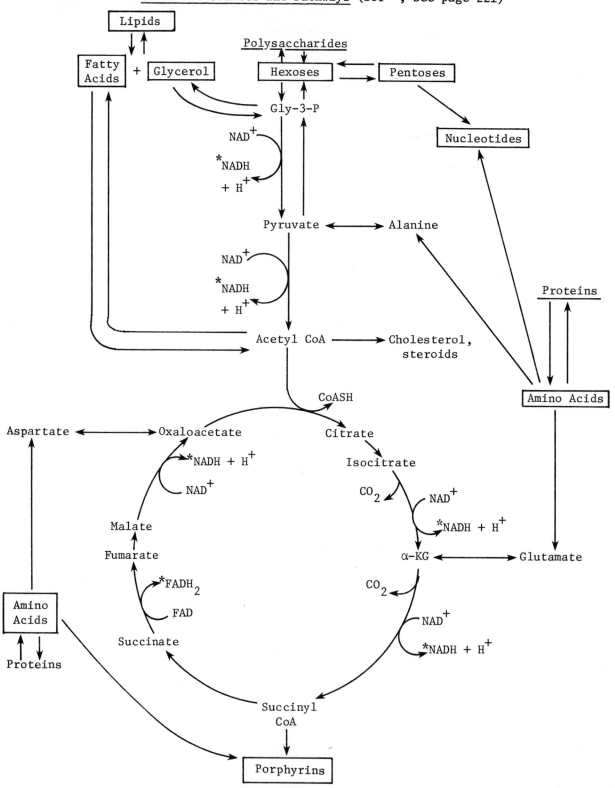

Figure 23. Some of the Interrelationships of Carbohydrate Metabolism with Other Metabolites and Pathways (for *, see page 221)

Note that, since one glucose molecule gives rise to two gly-3-P molecules, 10 NADH molecules and 2 $FADH_2$ molecules (twice the number of stars (*) in Figure 23) are produced per mole of glucose oxidized. These reduced coenzymes are reoxidized in the electron transport chain. Oxidative phosphorylation, when coupled to electron-transport, produces three moles of ATP per mole of mitochondrial NADH and two moles of ATP per mole of mitochondrial $FADH_2$. This is shown schematically below.

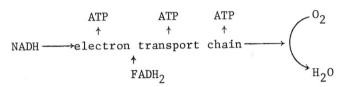

The result is the synthesis of $(3 \times 10) + (2 \times 2) = 34$ moles of ATP per mole of glucose oxidized, excluding substrate-level phosphorylation and ATP consumed. The overall balance of the oxidation of one mole of glucose to CO_2 + H_2O is shown below.

2ATP+glucose ⟶ 2ADP+F-1,6-di Ⓟ   (hexokinase, phosphofructokinase)

F-1,6-di Ⓟ ⟶ 2(gly-3-Ⓟ)   (aldolase)

2(gly-3-Ⓟ)+2P$_i$→2(1,3-DPG)   (gly-3-Ⓟ dehydrogenase)

2ADP+2(1,3-DPG)⟶ 2ATP+2(3-PG)   (phosphoglycerate kinase)

2(3-PG) ⟶ 2PEP   (glycolytic enzymes)

2ADP+2PEP ⟶ 2ATP+2(PA)   (pyruvate kinase)

2GDP+2P$_i$ ⟶ 2GTP   (succinate thiokinase)

2GTP+2ADP ⟶ 2ATP+2GDP   (nucleoside diphosphate kinase)

Glucose+4ADP+4P$_i$→2(PA)+4ATP   total of substrate-level conversions

Adding these four ATP's to the 34 obtained via electron transport-coupled oxidative phosphorylation results in <u>38 ATP synthesized per mole of glucose oxidized</u>. There can be one minor variation in this number. When the NADH reducing-power produced by gly-3-P dehydrogenase in the cytoplasm is transferred to the mitochondria, part of it is used to reduce a flavin coenzyme. When this is subsequently oxidized, only two moles of ATP are produced per mole of flavin. A mole of NADH (worth 3 moles of ATP) has thus been changed to a mole of flavin coenzyme (worth only 2 moles of ATP). Taking this into account, only 36 ATP would be produced per mole of glucose. Actually, the correct number is between 36 and 38. This (and the NADH transfers) are discussed more fully in the section on mitochondria.

10. Glyoxalate Cycle

a. The oxidation of pyruvic acid to acetyl-CoA is an irreversible process. Consequently, the net synthesis of glucose (gluconeogenesis) from acetyl-CoA must come about through bypass reactions. Such reactions are operative in plants and microorganisms but are absent in man and other animals. Therefore, fatty acids, which give acetyl-CoA on oxidation, cannot be converted to glucose in man. The only exceptions to this are the odd-carbon-number fatty acids which give one mole of propionyl-CoA per mole of fatty acid. The propionyl-CoA can be converted to pyruvate or succinate which can subsequently be used to form glucose. The contribution of such routes is very small, though, and is not significant under normal dietary circumstances.

It is important to realize that net synthesis is being discussed here. If labeled acetyl-CoA is used, some of the label will appear in newly made glucose since the carbons added as acetyl-CoA at one turn of the cycle are not the ones lost as CO_2 on the same turn (pointed out before). However, as will be seen in gluconeogenesis, the only route from acetyl-CoA to glucose is through the two CO_2-releasing reactions. The net effect is that, although two carbons are added as acetyl-CoA, two carbons must be lost as CO_2 prior to the formation of glucose. This inability of humans to convert lipids to carbohydrates presents a major problem in diabetes mellitus. This will be discussed later.

b. The key enzymes which make up the bypass in plants and microorganisms are isocitrate lyase (also known as isocitritase) and malate synthetase. They catalyze the reactions shown below.

i)

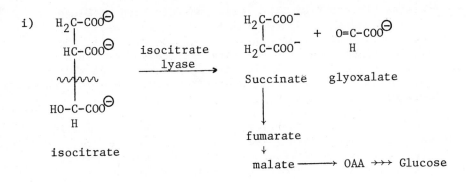

$$H_2C\text{-}COO^{\ominus}$$
$$HC\text{-}COO^{\ominus}$$
$$\overset{}{\underset{}{\wedge\!\wedge\!\wedge}}$$
$$HO\text{-}C\text{-}COO^{\ominus}$$
$$H$$

isocitrate

isocitrate lyase →

$$H_2C\text{-}COO^-$$
$$H_2C\text{-}COO^-$$

Succinate

+ $O=C\text{-}COO^{\ominus}$ H

glyoxalate

Succinate ↓
fumarate ↓
malate ⟶ OAA ⇝ Glucose

ii)

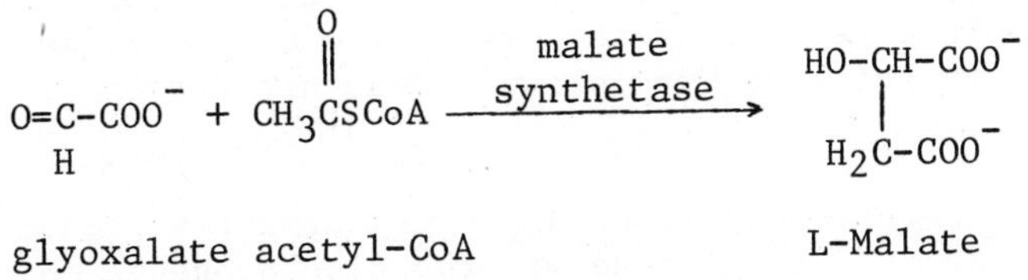

$$O=C\text{-}COO^- + CH_3CSCoA$$
$$\overset{O}{\underset{H}{\|}}$$

glyoxalate acetyl-CoA

malate synthetase →

$$HO\text{-}CH\text{-}COO^-$$
$$H_2C\text{-}COO^-$$

L-Malate

c. The interaction of glyoxalate cycle with the citric acid cycle is shown in the following diagram. Note that the glyoxalate pathway <u>bypasses</u> the CO_2-releasing reactions.

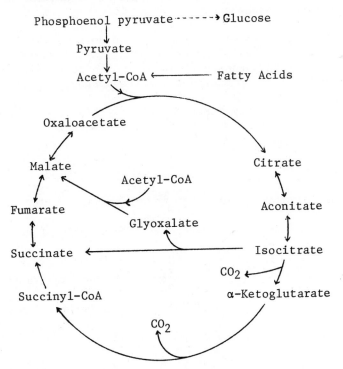

Phosphoenol pyruvate ------> Glucose
↓
Pyruvate
↓
Acetyl-CoA ←—— Fatty Acids

Oxaloacetate

Malate   Citrate

Acetyl-CoA

Fumarate   Aconitate

Glyoxalate

Succinate ←———————— Isocitrate

CO_2

Succinyl-CoA   α-Ketoglutarate

CO_2

<u>Gluconeogenesis</u>.   Reversal of Glycolysis and Formation of Glucose

1. Gluconeogenesis (which means formation of new sugar) is the process
    of formation of glucose from non-carbohydrate sources such as
    amino acids, glycerol and lactic acid. It takes place when the
    energy requirements of the cell are at a minimal level and an energy
    source (ATP) is available.

2. Gluconeogenesis becomes a very important source for supplying
    glucose to various tissues (liver, adipose tissue, brain, etc.)
    when glucose is not otherwise available. The unavailability may
    be due to a number of factors such as starvation.

3. In the human, fatty acids <u>cannot</u> serve as a source of new glucose
    (with the possible exception of the last three carbon atoms of
    odd numbered fatty acids) as was shown in the discussion of the
    glyoxalate cycle.

4. Muscles (under anaerobic conditions) and erythrocytes produce lactic
    acid which is converted to glucose via the gluconeogenic mechanism.
    Glycerol, produced from the hydrolysis of triglycerides in adipose
    tissue, is transported to the liver and can be converted to glucose.
    Glycerol is not metabolized in the adipose tissue due to the lack
    of glycerokinase, the enzyme which catalyzes the conversion of
    glycerol to α-glycerophosphate.

5. The conversion of lactic acid, glycerol, and amino acids into
    glucose and glycogen takes place principally in the liver and
    kidney. A number of hormones play roles in the regulation of these
    reactions. They will be referred to in the appropriate sections.

6. The magnitude of the energy barriers in the following exergonic
    (energy-releasing) reactions prevents their reversal and, hence,
    the reversal of the E-M pathway.

a. phosphoenol pyruvate + ADP $\xrightarrow{\text{pyruvate kinase}}$ pyruvate + ATP

$$\Delta G^{o\prime} = -5.7 \text{ kcal/mole of PEP}$$

b. fructose-6-P + ATP $\xrightarrow{\text{phosphofructokinase}}$ fructose-1,6-DiP + ADP

$$\Delta G^{o\prime} = -3.4 \text{ kcal/mole F-6-P}$$

c. glucose + ATP $\xrightarrow[\text{hexokinase}]{\text{glucokinase or}}$ glucose-6-P + ADP

$$\Delta G^{o\prime} = -3.4 \text{ kcal/mole of glucose}$$

These reactions were mentioned previously under glycolysis.

7. The cell has evolved a series of reactions to circumvent these exergonic reactions and thereby "reverse" glycolysis.

a. Pyruvate is chosen as a starting point to consider gluconeogenesis since the reaction, pyruvate → PEP, is the first one which must be bypassed. Pyruvate produced in the cytoplasm is first transported into the mitochondria, where it is carboxylated to oxaloacetate. The reaction is

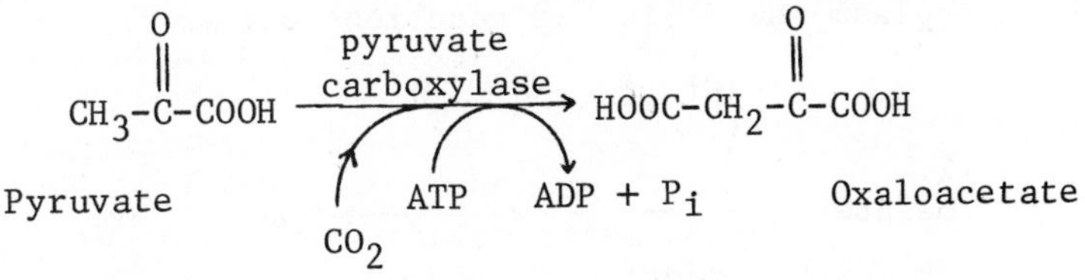

Biotin, a vitamin, is used in the above reaction. It is tightly bound to the carboxylase enzyme by means of the ϵ-NH_2 group of a lysine. It can react with CO_2, forming "active CO_2" in an ATP-requiring reaction. The activated CO_2 can then be transferred to various acceptor molecules. Other CO_2 fixation reactions in the human will be dealt with later. Most of them also use biotin as a coenzyme.

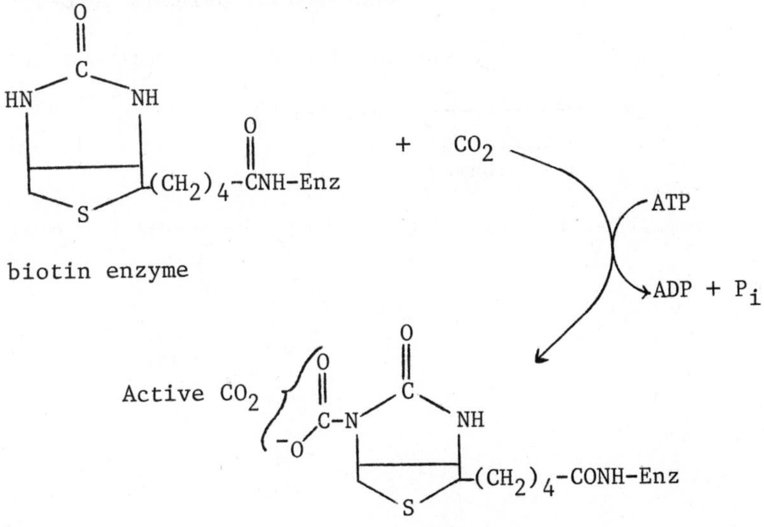

biotin enzyme

Pyruvate carboxylase is found in the mitochondria. It is an allosteric enzyme, inactive in the absence of acetyl-CoA (acetyl-CoA is a positive modulator). This means that pyruvate can be converted back to glucose only in the presence of high acetyl-CoA levels, a condition that occurs when there is a high level of ATP in the cell.

b. OAA (oxaloacetate) does not readily diffuse out of mitochondria so OAA is converted into <u>malate</u>.

$$\text{Oxaloacetate} + \text{NADH} + \text{H}^+ \xrightarrow[\text{malate dehydrogenase}]{\text{mitochondrial}} \text{NAD}^+ + \text{Malate}$$

Malate diffuses out to the cytoplasm by the appropriate carrier system.

c. In the cytoplasm the following reactions occur:

$$\text{Malate} \xrightarrow[\substack{\text{cytoplasmic malate} \\ \text{dehydrogenase}}]{\text{NAD}^+ \quad \text{NADH} + \text{H}^+} \text{OAA}$$

$$\text{OAA} + \text{GTP (or ITP)} \xrightarrow[\substack{\text{pyruvate} \\ \text{carboxykinase}}]{\text{phosphoenol-}} \substack{\text{COOH} \\ | \\ \text{C-O-}\textcircled{P} \\ \| \\ \text{CH}_2} + \text{CO}_2 + \text{GDP (or IDP)}$$

phosphoenol pyruvate (PEP)

8. From PEP to fructose-1,6-di$\textcircled{P}$ the usual glycolytic enzymes are used.

$$\text{PEP} \xrightarrow[\substack{\text{Reversal of} \\ \text{E-M Pathway}}]{} \text{F-1,6-di}\textcircled{P}$$

9. The next irreversible step is F-6-$\textcircled{P}$ $\xrightarrow[\substack{\text{ATP} \quad \text{ADP}}]{}$ F-1,6-di $\textcircled{P}$.

The reverse reaction is catalyzed by fructose-1,6-diphosphatase with the release of inorganic phosphate. The reaction is

$$\text{F-1,6-di}\textcircled{P} + H_2O \xrightarrow{\begin{array}{c}\text{fructose-1,6-}\\ \text{diphosphatase}\end{array}} \text{F-6-P} + P_i$$

$$\Delta G^{o'} = -4.0 \text{ Kcal/mole of F-1,6-di}\textcircled{P}.$$

This is an allosteric enzyme, present in liver, kidney, and striated muscle, but absent in adipose tissue. Low concentrations of AMP and ADP, and high concentrations of ATP and citrate, favor the phosphatase reaction and inhibit the phosphofructokinase reaction.

10.

$$\text{F-6-P} \xrightarrow{\begin{array}{c}\text{Reversal of the}\\ \text{E-M Pathway}\end{array}} \text{G-6-P}$$

11. The third irreversible reaction is bypassed by glucose-6-phosphatase.

$$\text{G-6-P} \xrightarrow{\text{G-6-Phosphatase}} \text{Glucose} + P_i$$

$$\Delta G^{o'} = -4.0 \text{ kcal/mole of G-6-P}$$

G-6-phosphatase is present in the intestine, liver, and kidney but is <u>not</u> found in adipose tissue or muscle tissue. It is a microsomal enzyme (found in the endoplasmic reticulum) and is responsible for providing blood glucose. A deficiency of this enzyme has been implicated as the cause of von Gierke's disease (discussed under glycogenolysis).

Summary of the Reactions of Gluconeogenesis

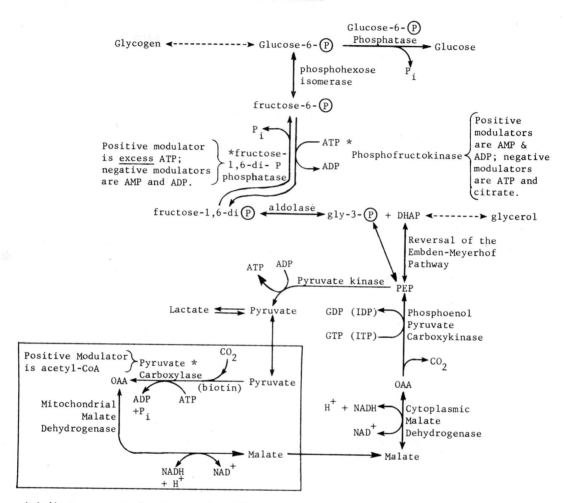

* indicates a control point catalyzed by an allosteric enzyme.
Note that such steps occur at the beginning and the end of the pathways. This avoids
wasteful accumulation of intermediates. The dotted lines in this diagram indicate
reaction pathways for the interconversion of the indicated materials.

228

Summary of the Net Effect of Gluconeogenesis

$$2(CH_3\overset{\overset{\displaystyle O}{\|}}{C}COOH) + 4ATP + 2GTP + 2NADH + 2H^+ + 6H_2O$$

pyruvate

$$\longrightarrow C_6H_{12}O_6 + 2NAD^+ + 4ADP + 2GDP + 6P_i$$

glucose

In contrast, the net effect of glycolysis is

$$C_6H_{12}O_6 + 2ADP + 2P_i + 2NAD^+ \longrightarrow 2CH_3\overset{\overset{\displaystyle O}{\|}}{C}COOH + 2ATP + 2NADH + 2H^+ + 2H_2O$$

glucose   pyruvate

Biochemical Lesions of Gluconeogenesis

1. A deficiency of hepatic fructose-1,6-diphosphatase has been
    observed in humans. This condition is associated with recurrent
    attacks of lactic acidosis and fasting hypoglycemia. The diagnosis
    of the enzyme deficiency is made by either using a specimen of
    liver obtained by laparotomy or by a more recent method using white
    blood cells. The patients show normal levels of liver glycogen
    and of the other enzymes associated with carbohydrate metabolism.

2. Assay of F-1,6-diphosphatase

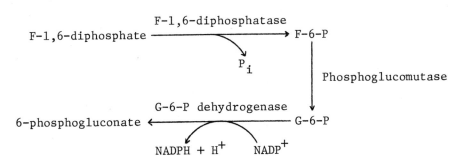

F-1,6-diphosphatase catalyzes the conversion of F-1,6-diP to F-6-P.
The F-6-P produced is converted to G-6-P which is then oxidized to
6-phosphogluconate. The conversion of F-1,6-diP to F-6-P is thus
linked to the reduction of NADP$^+$. This type of assay was discussed
under the use of enzymes in the clinical laboratory.

3. The <u>sudden infant death syndrome</u> (SIDS, also called <u>crib death</u>) is responsible for the death of about 10,000 infants each year in the United States. Numerous mechanisms have been proposed to explain this mysterious disorder. One possibility is a defect in gluconeogenesis. There have been reports of abnormally low levels of phosphoenolpyruvate carboxykinase in liver specimens obtained from victims of crib death. It is postulated that, as infants grow older and begin to sleep through the night without feeding, those in whom gluconeogenesis is defective experience a severe drop in blood glucose levels which can be fatal.

Allosteric Control of Carbohydrate Metabolism

Although no metabolic pathway or group of pathways ever functions alone within the body, the reactions by means of which the body manipulates its energy supply can be discussed as a separate entity. Similarly, although practically every sort of control mechanism can be found in these pathways, the allosteric effects serve as an illustration of the logic of the control processes in the cell. Some of these are summarized below. Several of the paths have already been discussed. Others will be covered later.

Allosteric Modifiers

Enzyme	Positive	Negative
1. Phosphofructokinase	AMP, ADP	ATP, citrate, long chain fatty acids
2. Fructose-1,6-diphosphatase	Citrate, 3-phosphoglycerate, excess ATP	AMP, ADP
3. Pyruvate Kinase	K^+, F-1,6-diP, excess PEP	ATP, AMP, citrate and alanine
4. Pyruvate Dehydrogenase	Dephosphorylation of the phospho-enzyme by phosphatase which is stimulated by Ca^{2+}	ATP-dependent phosphorylation of the enzyme by cyclic-AMP-dependent protein kinase

Allosteric Modifiers, continued

5. Citrate Synthase	---	Succinyl-CoA, NADH, long chain fatty acyl-CoA, ATP
6. Isocitrate Dehydrogenase	ADP, NAD^+	ATP, NADH
7. Succinate Dehydrogenase	Excess succinate, phosphate, ATP, and reduced ubiquinone	Oxaloacetate
8. Pyruvate Carboxylase	Acetyl CoA, ATP	---
9. Acetyl-CoA Carboxylase	Citrate, isocitrate	Long chain fatty acyl-CoA

The following points should be particularly noted concerning these reactions:

a. The end-products of a pathway usually shut off an enzyme at the beginning of that pathway (ATP shuts off glycolysis and the TCA cycle; fatty acyl-CoA shuts off fatty acid synthesis).

b. Buildup of the products from one pathway can turn on an alternate route of metabolism for a precursor (excess citrate switches on fatty acid synthesis and gluconeogenesis (glucose synthesis) to consume acetyl-CoA and pyruvate).

c. A decrease in ATP as a source of energy for biological reactions and an increase in the products ADP and AMP turns off storage routes (fatty acid synthesis and gluconeogenesis) and activates supply pathways (glycolysis, TCA cycle).

d. The enzymes glycogen synthetase and glycogen phosphorylase (reactions 10 and 11 in Figure 24) are _indirectly_ controlled by allosteric modifiers including cyclic adenosine monophosphate (cAMP). This is discussed in detail under glycogen metabolism.

e. Two compounds which are key intermediates in these pathways are glucose-6-phosphate and acetyl-CoA. They serve as branch or junction points for a number of metabolic routes.

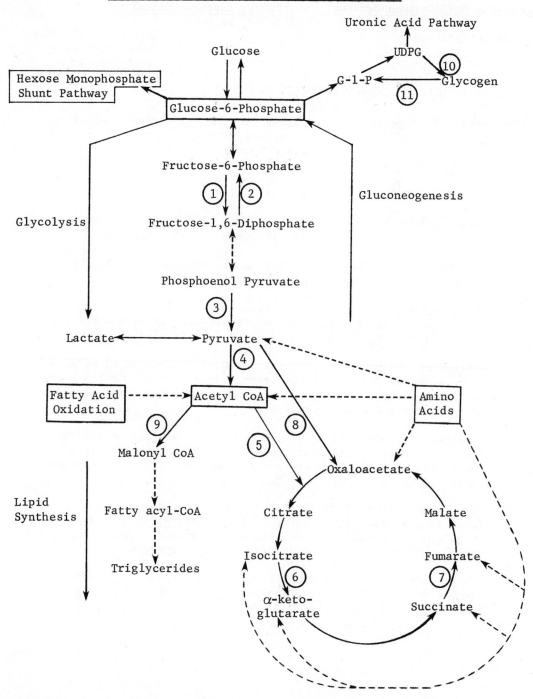

The dotted lines indicate multi-step reaction pathways while solid lines represent single-step conversions.

Mitochondrial Electron Transport and Energy Capture (Oxidative
Phosphorylation) with Richard J. Guillory*, Ph.D.

1. In this section, we shall discuss the capture and storage (in the
   ATP molecule) of the energy released by the oxidation of reduced
   substrates (NADH and $FADH_2$) produced during the catabolism of
   glucose. Because of its involvement in many energy-requiring
   reactions within the cell, ATP is sometimes referred to as the
   "currency of metabolism."

2. The citric acid cycle functions as a common channel for the
   oxidation of metabolites and the production of reducing equivalents
   from a variety of foods including carbohydrates, proteins, and
   lipids. A reducing equivalent is one mole of electrons in the
   form of one gram-equivalent of a reduced electron carrier. Thus
   one mole of NAD^+ gains two moles (two times Avogadro's number) of
   electrons (as well as one mole of H^+) in being reduced to one mole
   of NADH. One mole of NADH is therefore two reducing equivalents.
   Note that the two electrons and one H^+ actually are transferred
   together, in many cases, as a hydride ion ($H:^-$).

3. The following steps are those in which reduced pyridine nucleotide
   (NADH) and reduced flavin-adenine dinucleotide ($FADH_2$) are formed
   during the breakdown of glucose via the Embden-Meyerhof pathway
   and the citric acid cycle.

   a. NADH-producing reactions

      (i)   Glyceraldehyde-3- (P) ⟶ 1,3-diphosphoglycerate

      (ii) Pyruvate ⟶ acetyl-CoA

      (iii) Isocitrate ⟶ α-ketoglutarate

      (iv) α-ketoglutarate ⟶ succinyl-CoA

      (v)   Malate ⟶ oxaloacetate

      The reactions (ii)-(v) take place in the mitochondria and
      reaction (i) takes place in the cytoplasm. Since mitochondrial
      membranes are not directly permeable to NADH, two shuttle pathways
      have been found which transport the reducing equivalents of
      cytoplasmic NADH into the intramitochondrial matrix. These
      reactions are discussed in this section.

---

* Professor, Department of Biochemistry & Biophysics, University of
Hawaii School of Medicine, Honolulu, Hawaii.

b. FADH$_2$-producing reaction

      Succinate ⟶ Fumarate

The above reaction, catalyzed by succinate dehydrogenase, is part of the citric acid cycle and operates within the mitochondria. The enzyme appears to have a mitochondrial structural role as well as a catalytic one.

4. Electrons are catalytically removed together with protons from the reduced substrate (e.g., NADH, FADH$_2$). They are transported over a series of coupled oxidation-reduction reactions to oxygen (O$_2$), which is the ultimate electron acceptor or electron sink. These coupled redox reactions take place in the mitochondria. The flow starts with the electrons having a relatively high potential energy and ends (at oxygen) with them at a low potential. During this electron flow or transport, a portion of the energy lost by the electrons is conserved by a biological energy transducing mechanism (a mechanism which changes the form of the energy from electrical to chemical). Since the energy is stored by using it to phosphorylate ADP to ATP, the overall coupled process is termed oxidative phosphorylation.

Mitochondria

1. Mitochondria are present in the cytoplasm of almost all aerobic eukaryotic cells. They are frequently found in close proximity to the fuel sources and to the structures which require ATP for maintenance and functional activity,(e.g., the contractile mechanisms, energy-dependent transport systems, secretory processes, etc.).

2. The number of mitochondria present within a single cell varies from one type of cell to another. By way of an example, a rat liver cell contains about 1,000 mitochondria, while a giant amoeba (Chaos chaos) has some 10,000 mitochondria. In a given cell, the number of mitochondria may also depend on the cell's stage of development and/or functional activity.

3. The size and shape of mitochondria also varies considerably from one cell type to another. Even within the same cell, mitochondria can undergo rapid changes in volume and shape depending upon the metabolic state of the cell. In general, they are about 1 μ in diameter and 3 to 4 μ in length. Mitochondria have been found to aggregate end to end forming long filamentous structures.

4. Mitochondria consist of about 30% lipid (primarily phospholipids) and 70% protein on a dry weight basis. They also contain specific types of DNA and RNA. The HeLa cell mitochondrial DNA is a circular duplex, made up of a heavy strand and a light strand.

5. Mitochondria undergo self-duplication, with fission taking place following cell division. They have their own protein synthesizing system (details of which are discussed under protein synthesis) but appear to be incapable of synthesizing completely all of the required structural proteins, lipid components, and enzymes. For example, HeLa cell mitochondrial DNA has about 15,000 base pairs corresponding to about 25 genes coding for 25 proteins. Since a mitochondrion has more than 25 different proteins, synthesis of the remaining proteins must require nuclear DNA. In other words, mitochondria are not completely autonomous and are dependent upon other parts of the cell for their synthesis.

6. The protein synthesis system of mitochondria appears to be analogous to that of bacterial systems. This observation, along with the similarity in size and the presence of circular DNA, has been viewed as suggesting that the mitochondria of eucaryotic cells arose from invading bacteria. A probable initial parasitic relationship may have later developed into a beneficial (symbiotic) relationship between the host and the parasite (anaerobic eukaryotic cell and mitochondrion, respectively).

7. Mitochondria consist of two membranes: an outer one which is a smooth, closed system surrounding the inner membrane. These membranes are separated from each other by about 50 to 100 Å. The intermembranal space contains the enzymes responsible for translocation of metabolites, in addition to adenylate kinase and nucleoside diphosphokinase. The inner membrane has many folds directed towards the interior of the mitochondrion. These invaginations, known as cristae, are presumably present to increase the surface area of the inner membrane. This provides more space for the attachment of the multi-enzyme clusters, each about 220 Å long, which carry on electron transport. Attached to these transport complexes are the inner membrane spheres, having a diameter of about 80 Å. These spheres contain ATP-synthesizing enzymes. Under the proper conditions, the spheres also show an ATPase activity, presumably due to a reversal of the normal ATP-synthetic process. It should be pointed out that the inner membrane spheres are not seen in intact mitochondrial preparations. They are observed only in submitochondrial particles prepared by the sonic disruption of mitochondria.

8. The inner membrane surrounds the matrix of the mitochondria and contains

   a. all of the factors essential for oxidative phosphorylation;
   b. the citric acid cycle enzymes;
   c. the fatty acid oxidation enzymes;
   d. the mitochondrial DNA;
   e. ribosomes and other accessories essential for protein synthesis.

9. The membranes, both inner and outer, although performing different functions, have structures similar to that of other biological membranes. Each membrane consists of a middle, non-polar layer (primarily phospholipids), on either side of which are protein layers. Polar regions of the proteins appear to be present on both sides of a single membrane.

10. There is a variable transport of molecules through the mitochondrial membrane. For example, mitochondrial membranes are permeable to pyruvate, while oxaloacetate, ATP and NADH enter with difficulty, if at all. The latter group of molecules is transported only by special carrier mechanisms. There are many specific translocating enzymes (carrier molecules) which participate in this transport. The equimolar exchange of external ADP with internal ATP maintains proper nucleotide pools. A compound isolated from the plant Atractylis gummafera, known as atractyloside, inhibits the exchange of ATP between the inside and outside of mitochondria. Accidental ingestion of this plant has been known to cause death. The active transport mechanism of the mitochondrial membrane are able to accumulate a number of components such as Ca^{++}, phosphate, and other ions, all of which appear to be essential for the proper functioning of the organelle.

   A highly schematic diagram of a mitochondrion is shown in Figure 25.

Electron Transport

1. Electrons are transported in the mitochondria from substrate (NADH, $FADH_2$) to oxygen through intermediate carriers. These carriers are arranged in a definite spatial relationship based upon their individual oxidation-reduction potentials. Electrons flow from a higher (potential) energy level to a lower energy state.

2. At specific locations in the electron-transport chain, part of the energy released by electron transport is conserved in the form of the phosphoric acid anhydride bonds of ATP. This is accomplished by directly coupling the energy-releasing process to the synthesis of ATP from ADP and P_i.

Figure 25.   Schematic Diagram of a Mitochondrion

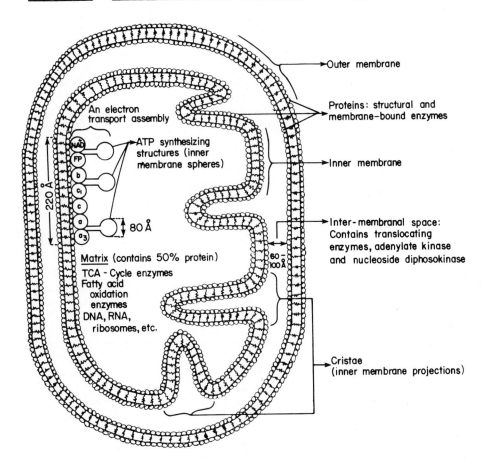

The above diagram is not drawn to scale. It should be pointed out that a liver mitochondrion contains between 15,000 and 30,000 electron-transport assemblies, depending upon the functional activity of the cell.

3. Using NADH as the substrate, in the presence of sufficient amounts of the other required factors, the overall oxidation reaction coupled to phosphorylation can be written:

$$NADH + H^+ + 3ADP + 3P_i + \tfrac{1}{2} O_2 \longrightarrow NAD^+ + 3ATP + 4H_2O$$

The exergonic and endergonic components of the above relationship are, respectively:

$$NADH + H^+ + \tfrac{1}{2} O_2 \longrightarrow NAD^+ + H_2O \qquad \Delta G^{o'} = -52 \text{ kcal/mol}$$

$$3ADP + 3P_i \longrightarrow 3ATP + 3H_2O \qquad \Delta G^{o'} = 3 \times +7.3 = +21.9 \text{ kcal/mol}$$

Before discussing details of the specific reactions and the members of the electron-transport chain, a schematic diagram of the flow of electrons with the relevant energy changes is presented in Figure 26.

Carriers Present Within the Electron-Transport Assembly

1. Nicotinamide Adenine Dinucleotide is the co-substrate which accepts hydride ion (hydrogen with two electrons, $H:^-$) from a suitable donor and is converted to NADH. The structural details of this reaction have been outlined under glycolysis.

2. Flavoproteins. The NADH produced is oxidized by NADH dehydrogenase, an enzyme which contains flavin mononucleotide (FMN) as a coenzyme. The enzyme appears to be associated with non-heme iron (iron bound directly to a protein rather than through a porphyrin group). The role of non-heme iron in the oxidation of NADH by NADH dehydrogenase is not yet completely understood. The structure of FMN was presented in the discussion of the TCA cycle. Other flavin-linked dehydrogenases which are located in the mitochondria and participate in electron transport and respiration are: succinate dehydrogenase (a TCA cycle enzyme), dihydrolipoyl dehydrogenase (one of the enzymes of the pyruvate and α-ketoglutarate dehydrogenase complexes), and acyl-CoA dehydrogenase (the enzyme which catalyzes the first dehydrogenation reaction during fatty acid oxidation). All of these enzymes contain FAD covalently linked to an amino acid residue of the protein by means of the isoalloxazine ring of the FAD. There are many other flavin enzymes which are not a part of the main electron

transport system but which participate in oxidation and reduction reactions. These include: D-amino acid oxidase (see amino acid metabolism); xanthine oxidase (see purine metabolism); aldehyde oxidase (see alcohol metabolism) and orotate reductase (see pyrimidine synthesis).

A significant difference between flavin-linked and NAD^+-linked dehydrogenases is that, in the former, the flavin nucleotide is tightly bound to the enzyme and remains bound during and after catalysis. NAD^+ (and $NADP^+$; and their reduced forms) are not tightly bound, remaining in association with an enzyme only during the actual reaction. As previously noted, NADH dehydrogenase, an FMN-containing protein, also contains iron associated with reactive sulfur atoms. This <u>non-heme iron</u> is oxidized and reduced ($Fe^{+2} \rightleftarrows Fe^{+3}$) during the normal functioning of the enzyme. Four such iron-sulfur centers are essential for catalytic activity but their precise mechanism of action is not clear. It appears that one of them is involved in the transfer of electrons from the NADH to the FMN while the others participate in the transfer of electrons from FMN to CoQ. Three other iron-sulfur centers have been identified in the electron transport chain by the use of electron spin resonance spectroscopy. These are associated with cytochrome b (2) and cytochrome c_1 (1).

3. <u>Coenzyme Q</u> (also known as ubiquinone, a quinone found ubiquitously).

   a. Quinone-coenzymes occur in the lipid fraction of the membrane and are composed of a group having the following basic structure.

   (oxidized form)

The predominant coenzyme Q species isolated from mammalian tissues has ten isoprenoid units ($n = 10$, hence CoQ_{10}).

Figure 26. Mitochondrial Electron-Transport System

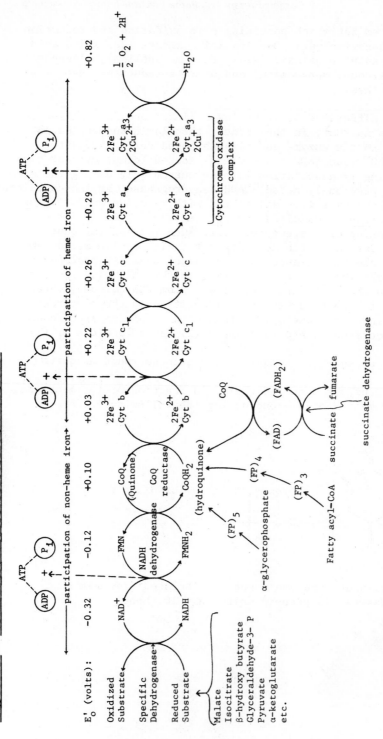

Note: (1) FP = flavoprotein. There are several of these, designated (FP)₃, (FP)₄, etc. All contain flavin or some
derivative of it as a coenzyme.

(2) The relative position of CoQ to cytochrome b is not certain. The E'₀ value for CoQ was measured in 95%
ethanol, a highly artificial situation.

(3) Most of the E'₀ values given are for heart muscle mitochondrial enzymes. There is considerable variation
in reported values due to differing methods of preparation and measurement.

(4) The three dashed vertical arrows leading to ATP indicate the points at which
ATP synthesis is coupled to electron transport.

b. Oxidation-reduction reaction

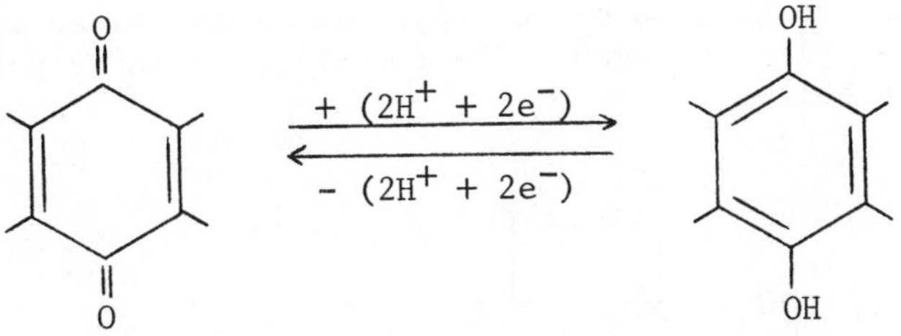

$$+ (2H^+ + 2e^-) \longrightarrow$$
$$- (2H^+ + 2e^-) \longleftarrow$$

quinone   hydroquinone

c. Vitamin K_1 has a structure similar to that of coenzyme Q. In mycobacteria, which do not contain CoQ, vitamin K has a comparable functional role. In animals, however, CoQ has no known vitamin precursors and is presumably synthesized within the body.

Structure of Vitamin K_1

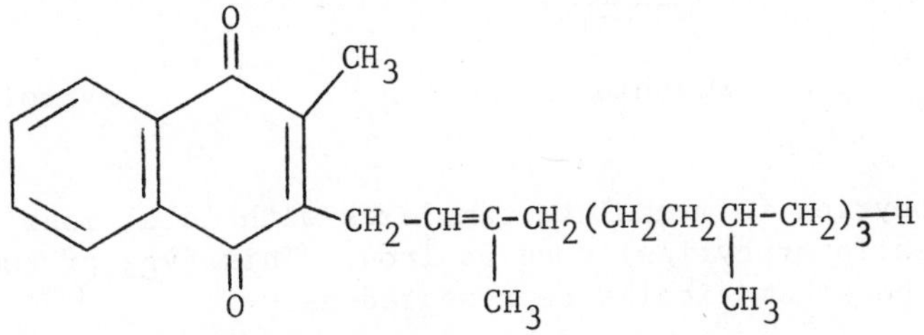

The function of Vitamin K_1 in humans is discussed in the section on vitamins.

4. Cytochromes

a. These are proteins containing iron porphyrins as prosthetic groups (heme-proteins). The iron undergoes repeated oxidations and reductions:

$$Fe^{3+} + e^- \rightleftharpoons Fe^{2+}$$

(Note: This does not happen in hemoglobin, which is also a heme-protein.) Cytochromes can be differentiated by their absorption spectra, as can the oxidation state of any specific cytochrome. The characteristic spectra of the different cytochromes are due to differences in the sidechains of the porphyrin ring.

b. <u>Porphyrin ring systems</u>

   i) These are made up of four pyrrole rings joined together
       by methylene bridges. The parent compound is porphin,
       shown below.

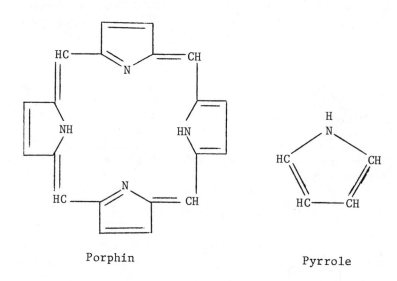

               Porphin   Pyrrole

  ii) Porphyrins form chelate complexes with metal ions
       (metalloporphyrins) such as iron. This type of complex
       can be schematically represented as

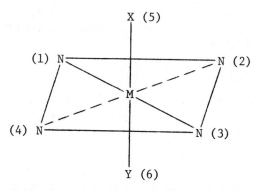

       where M is a metal ion (central metal ion) such as Fe^{+2} or
       Fe^{+3}. Iron can form six coordinate-covalent bonds with
       any six ligands which have pairs of electrons in non-bonding
       orbitals (lone pairs of electrons). These complexes have

a shape or symmetry which is octahedral. In the iron-porphyrin complex, four of the ligands are the four nitrogens of the porphyrin ring (N(1) to N(4), above). The other two positions, X and Y, are occupied by H_2O, an amino acid sidechain or some other ligand. When the metalloporphyrin is bound to a protein, the fifth and sixth positions are usually occupied by amino acid sidechains in the protein. A frequently encountered iron-porphyrin complex is iron-protoporphyrin IX (for porphyrin names, see the section on porphyrin synthesis). If the iron is present as Fe^{+2}, this complex is called <u>heme</u> (or protoheme); if it is Fe^{+3}, the name is changed to <u>hemin</u>. These designations arose because the complex was first found in hemoglobin.

c. <u>Cytochrome b</u>

i) This is the first cytochrome of the electron-transport sequence. As a prosthetic group it contains a heme (an iron-protoporphyrin IX complex). This coenzyme is also present in the heme proteins hemoglobin and myoglobin. There are significant differences in other respects, however, between hemoglobin, myoglobin, and the cytochromes. These are presented below in Table 9. Synthesis of protoporphyrin IX and other aspects of the porphyrins are discussed under the section on hemoglobin.

ii) The prosthetic group of cytochrome b (and hemoglobin and myoglobin) is

In cytochrome b, the (5) and (6) positions of the iron are occupied by protein ligands and are therefore inaccessible for binding with external ligands such as O_2, CO, CN^-, etc.

iii) Cytochrome b accepts electrons from reduced coenzyme Q ($CoQH_2$) and passes them on to cytochromes of higher (more positive) oxidation-reduction potential.

iv) Cytochrome b is tightly bound to the lipoprotein membrane.

v) The table below provides a comparison of some of the properties of the cytochromes with those of hemoglobin and myoglobin.

<u>Table 9.</u> <u>Comparison of Different Types of Metalloporphyrin-Containing Proteins</u>

	Cytochromes*	Hemoglobin & Myoglobin
5 position occupied by	R-groups of specific amino acid residues	Imidazole group of a histidine residue
6 position occupied by	R-groups of specific a.a. residues; unavailable for binding with O_2, CO, CN, etc.	H_2O (in deoxy Hb or deoxy myoglobin) or oxygen (O_2) or CO, etc. (in oxyhemoglobin or carbon monoxyhemoglobin and myoglobin)
Valence State of Iron	$Fe^{2+} \rightleftharpoons Fe^{3+} + e^-$ (oxidation changes back and forth during the normal functioning of the cytochromes)	The normal valence state of Fe^{2+} is unchanged when O_2 is gained or lost. When the reaction $Fe^{2+} \longrightarrow Fe^{3+}$ takes place via an oxidizing agent (e.g., ferricyanide), methemoglobin is produced methemoglobin cannot function as an oxygen carrier

* Cytochrome oxidase differs from the other cytochromes in a number of ways (see later).

d. <u>Cytochrome c</u>

i) Prosthetic Group

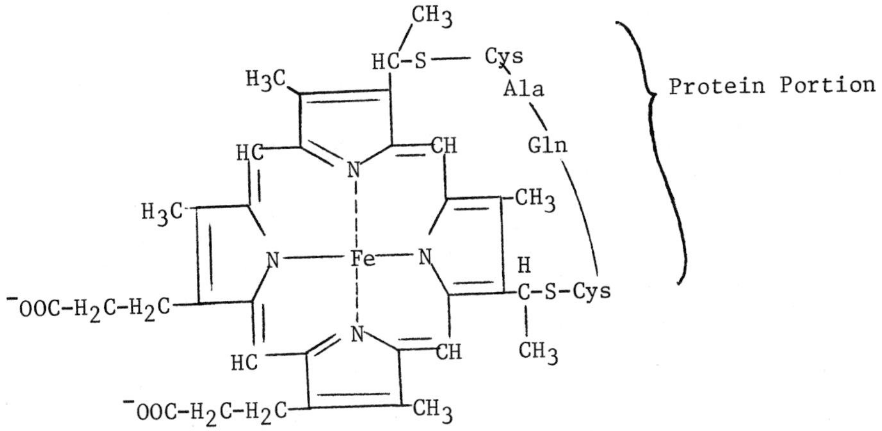

Note the linkage of the heme group to the protein through cysteine residues.

ii) This is the central member of the electron-carrier system and has an intermediate redox potential. It links the oxidation of cytochrome b to the cytochrome oxidase complex. The molecule is relatively loosely bound to the respiratory chain system and is easily extracted in a fairly pure form.

e. <u>Prosthetic Group of Cytochrome *a*</u>

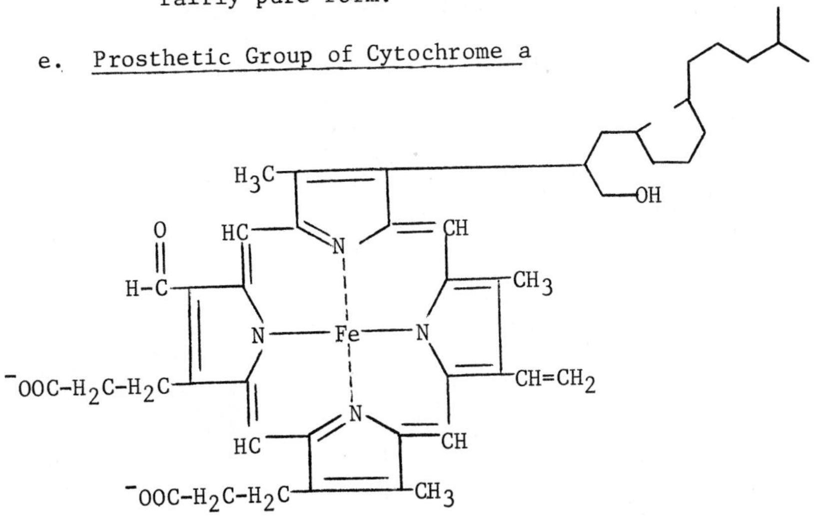

f. Cytochrome a/a$_3$ (collectively termed cytochrome oxidase)

i) This is the terminal member of the electron-transport
assembly. It reoxidizes cytochrome c and reduces oxygen
and has a high (positive) redox potential.

ii) The oxidase contains copper in addition to an iron-
porphyrin system. The copper also undergoes an oxidation-
reduction reaction ($Cu^{2+} + e^- \rightleftharpoons Cu^+$) during electron
transport.

iii) The oxidase is inhibited by CN^-, H_2S, N_3^- (azide), and
CO. The details on CN^- binding have already been
discussed in the section on irreversible inhibitors of
enzymes.

g. It should be pointed out that, with the possible exception of
NAD^+ and cytochrome c, the electron-transport chain members
are intimately associated with the lipids of the membranes.
It is likely that electron transport and energy conservation
take place in the hydrophobic environment of the membrane.

h. Other forms of mitochondrial-type electron-transport systems
occur in microsomes (pieces and vesicles of endoplasmic
reticulum and other intracellular membranous structures
formed during cellular fractionation). Several examples are
given below.

i) In rat liver microsomes, there is a pathway which uses
oxygen for the ω-oxidation of fatty acids (oxidation
at the carbon farthest removed from the carboxyl
group). Three enzymes have been shown to be part of
this system: a unique cytochrome (known as B420 or
P450), an NADPH-cytochrome P450 reductase, and a heat-
stable lipid factor. The numbers 420 and 450 attached
to these cytochromes indicate the wavelengths of
maximum absorption of the cytochrome and its carbonmonoxy
derivative, respectively. The system shows broad
specificity, being able to hydroxylate cyclohexane,
fatty acids, n-alkanes, and a variety of drugs and
other compounds having O- and N-methyl groups.

ii) Other NADPH- and O_2-requiring hydroxylases are present
in microsomes from the adrenal gland. Some of these

are involved in the oxidation of endogenous and exogenous aromatic compounds and of steroids. Details of some of the microsomal oxidase systems are given in appropriate sections.

iii) The enzymes described above are examples of oxygenases – enzymes that catalyze the incorporation of oxygen atoms into organic molecules using molecular oxygen as a substrate. There are <u>monooxygenases</u> (phenylalanine hydroxylase, for example; see amino acid metabolism) which add one atom of an O_2 molecule to the substrate, reducing the other to H_2O; and <u>dioxygenases</u> (e.g., tryptophan-2,3-oxygenase, formerly known as tryptophan pyrrolase; also discussed in the chapter on amino acids) which put both atoms from O_2 onto the organic substrate. The reducing equivalents for monooxygenase frequently come from NADH or NADPH.

Some of these enzymes contain a heme group (tryptophan-2,3-oxygenase) while others do not (phenylalanine hydroxylase uses a pteridine cofactor). The identifying characteristic, though, is the ability to incorporate $^{18}oxygen$ into the product given $^{18}O_2$ as a substrate.

Experiments Used in the Establishment of the Sequence of the Electron-Transport Assembly

1. <u>Spectroscopic evidence</u> has been used for

   a. identification of the presence of suspected or postulated carriers in various subcellular fractions;

   b. determination of the amount of the carrier present;

   c. studies on the kinetics of reduction and reoxidation reactions.

2. <u>Fragmentation and reconstitution</u> of submitochondrial particles

   By means of mechanical disruption and detergent treatment, four different complexes have been obtained from mitochondria. The functions of these complexes and their interrelationships are shown in the following diagram.

Figure 27. Composition, Interrelationship, and Function of Protein Complexes Obtained from Mitochondrial Fractionation

oxidized substrate
NADH + H+
reduced substrate
NAD+
2e-
2H+
Complex I
Flavoprotein
CoQ → CoQH2
Complex II
Flavoprotein
CoQ → CoQH2
succinate
fumarate
Complex III
Cytochromes b and c1
CoQH2 → CoQ
Cyt c (Fe+3) → Cyt c (Fe+2)
Complex IV
Cytochromes a and a3
Cyt c (Fe+2) → Cyt c (Fe+3)
O2 → H2O

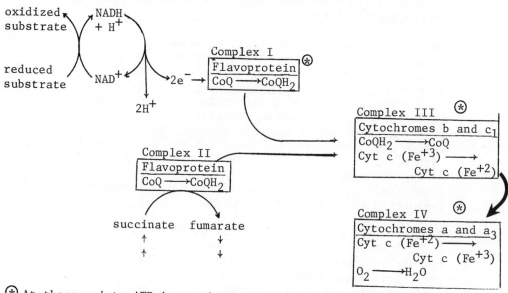

(*) At these points ATP is synthesized by the reaction ADP + $P_i \longrightarrow$ ATP.

3. Inhibitors that block the passage of electrons at selected places have been used in understanding the nature of the carriers, the routes of the flow of electrons and the sites of ATP synthesis.

Figure 28. Inhibitors of Electron Transport and the Specific Transfers Which They Inhibit

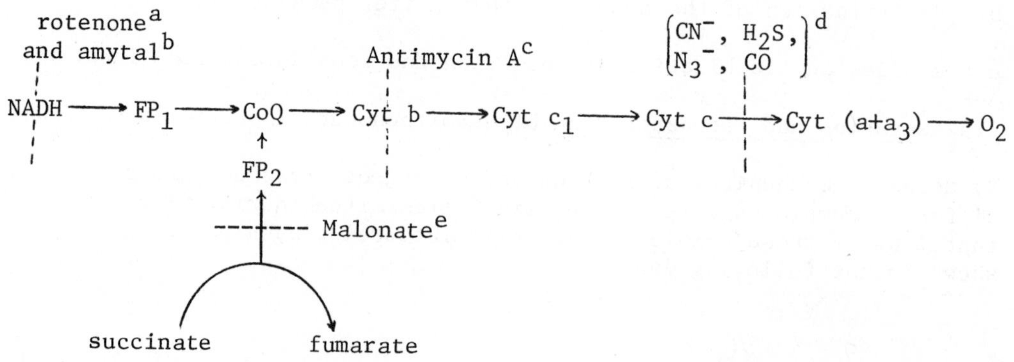

rotenone[a] and amytal[b]
Antimycin A[c]
CN−, H2S, N3−, CO [d]
NADH → FP1 → CoQ → Cyt b → Cyt c1 → Cyt c → Cyt (a+a3) → O2
FP2
Malonate[e]
succinate
fumarate

a. Rotenone is obtained from the roots of the tropical plants <u>Derris elliptica</u> and <u>Lonchocarpus nicou</u>.

b. Amytal is a barbiturate: It is also known as amobarbitol or 5-isoamyl-5-ethylbarbituric acid.

c. Antimycin A is an antibiotic isolated from the bacteria <u>Streptomyces</u>.

d. All of these molecules form stable complexes with Fe^{+3}, preventing its reduction to Fe^{+2}. This prevents further e^- transfer to O_2 and was discussed earlier in greater detail (see enzyme inhibition).

e. Malonate is an analogue of succinate. It competitively inhibits succinic acid dehydrogenase (see TCA cycle).

4. There are other types of inhibitors which have been useful in these studies.

a. <u>Uncoupling agents</u>, such as 2,4-dinitrophenol (DNP), block phosphorylation (ATP synthesis from ADP and P_i) while allowing electron flow to proceed at a maximal rate. Normally, these two processes are tightly coupled--that is, if one cannot occur (due to lack of ADP or oxidizable substrate, for example) the other is also blocked. When an uncoupler is present:

   i) O_2 consumption is stimulated in the absence of ADP;

   ii) ATP <u>hydrolysis</u> occurs, rather than the normal ATP synthesis;

   iii) Glycolytic phosphorylations are not affected.

b. Oligomycin (an antibiotic) and rutamycin block phosphorylation <u>without</u> uncoupling it from electron transport. Both processes are stopped by these agents but they have been shown to affect ATP formation without having any <u>direct</u> affect on the electron carriers of the respiratory chain. Further, addition of an uncoupler, together with oligomycin, allows electron transport to occur but the ATP hydrolysis usually associated with an uncoupler is absent. This allows the study of some of the partial reactions of ATP synthesis by mitochondria.

c. Mitochondria are normally impermeable to monovalent cations (although divalent cations move in and out at a measurable rate). <u>Ionophores</u> are lipid-soluble compounds which can complex with Na^+ and K^+ and carry them across the mitochondrial membrane. They also decrease the steady-state level of ATP in the mitochondria. There are two classes of ionophores.

i) Valinomycin and nonactin cause the transport of K^+ (and Na^+ to a much lesser extent) across the mitochondrial membrane in response to both a chemical and electrical gradient. They do this by forming a complex with the ion in the interior of the molecule. This lipid-soluble complex then carries the K^+ ion inside the mitochondria, causing a collapse of the electrochemical gradient set up by the electron transport chain thereby causing uncoupling of phosphorylation from oxidation.

ii) Gramicidin, on the other hand, allows the passive diffusion of K^+ and Na^+ across the mitochondrial membrane by forming a pore in the membrane itself. This pore disrupts the integrity of the membrane, making the formation of an electrochemical gradient impossible.

5. a. <u>Standard redox-potential data</u> is useful since electrons flow from compounds having a higher $E^{O'}$ (more negative value) to compounds with a lower $E^{O'}$ (more positive value). A knowledge of $E^{O'}$ values for the members of the electron-transport chain permits the placing of the individual carriers in the proper order for concerted electron transport.

b. Some of the $E^{O'}$ and $\Delta G^{O'}$ (= $23,061$ $nE^{O'}_T$) values of the electron transport system components are shown below.

<u>Figure 29</u>. <u>Free Energy Changes Associated with Electron Transport</u>

$$|\leftarrow\!\!-E^{O'}_T = -1.14 \text{ volts}; \ \Delta G^{O'} = -52.6 \text{ kcal/mole NADH} \longrightarrow|$$

$$NADH \rightarrow FP_1(FMN) \rightarrow CoQ \rightarrow Cyt \ b \rightarrow Cyt \ c_1 \rightarrow Cyt \ c \rightarrow Cyt \ (a+a_3) \rightarrow O_2$$

$E^{O'}$ (volts):	−0.32 −0.12	+0.03 +0.22	+0.29 +0.82
$E^{O'}_T$ (volts):	−0.20	−0.19	−0.53
$\Delta G^{O'}$ (kcal/mole):	−9.2	−8.8	−24.4
	Complex I	Complex III	Complex IV

Recall that these complexes are also the sites of ATP synthesis (see Figure 27).

c. The formation of ATP can be described by the reaction

$$ADP + P_i \longrightarrow ATP + H_2O \qquad \Delta G^{o\prime} = +7.3 \text{ Kcal/mole ATP}$$

The $\Delta G^{o\prime}$ values of each of the three complexes in Figure 29 are sufficiently negative (i.e., enough free energy is released) to drive ATP synthesis. This can be seen by noting that the $\Delta G^{o\prime}$ values for the three complexes are all less than -7.3 kcal. Experimentally, it has been found that three molecules of ATP are synthesized for each pair of electrons passed from NADH to oxygen (for each atom of oxygen reduced). The ratio of moles of ATP synthesized to gram atoms of oxygen reduced is called the P/O ratio (or "P to O ratio"). For oxidative phosphorylation under normal conditions, P/O = 3.

d. Since 52.6 kcal/mole of NADH are released in the overall process, it is thermodynamically possible to synthesize

$$\frac{52.6 \text{ kcal/mole NADH}}{7.3 \text{ kcal/mole ATP}} = 7 \text{ moles ATP/mole NADH}$$

Under standard conditions, at pH = 7, then, oxidative phosphorylation is ∿43% (= 3/7 x 100%) efficient. Under actual cellular conditions, however, ΔG for ATP formation is probably about 10–12 kcal/mole. If the values of $\Delta G^{o\prime}$ for the electron carriers are similar to their actual intracellular (ΔG) values, then the efficiency will be better than 43%. The difficulty of discussing intracellular processes in terms of isolated systems (i.e., values of $\Delta G^{o\prime}$ rather than ΔG) is once again apparent. The energy not used to synthesize ATP appears as heat and is important for maintaining the constant body temperature of warm-blooded animals.

e. The P/O ratio for a given substrate is dependent upon the energy level at which the compound transfers its electrons to the electron-transport chain. P/O ratios are given below for some of the electron-transport substrates together with the oxidized products.

	P:O Ratio
Pyruvate ⟶ Acetyl-CoA	3
Isocitrate ⟶ α-Ketoglutarate	3
α-Ketoglutarate ⟶ Succinate	4←(includes 1 substrate-level phosphorylation)
Succinate ⟶ Fumarate	2
Malate ⟶ Oxaloacetate	3

The complete oxidation of glucose by mitochondria yields 36 or 38 ATP, depending upon where the reducing equivalents from the two moles of cytoplasmic NADH enter the electron transport chain (that is, what shuttle is used to transport them into the mitochondria; discussed later). The overall reaction for glucose oxidation is shown below:

$$C_6H_{12}O_6 + 6\ O_2 + 36\ P_i + 36\ ADP \longrightarrow 6\ CO_2 + 36\ ATP + 42\ H_2O.$$

It can be dissected into two parts:

Exergonic component:

$$C_6H_{12}O_6 + 6\ O_2 \longrightarrow 6\ CO_2 + 6\ H_2O,\ \Delta G^{o\prime} = -686\ kcal/mol$$

Endergonic component:

$$36\ P_i + 36\ ADP \longrightarrow 36\ ATP + 36\ H_2O,\ \Delta G^{o\prime} = +263\ kcal/mol$$

The overall efficiency, in terms of energy stored as ATP versus total energy released, is about 38% (263/686 x 100). As noted earlier, under intracellular conditions, the percent efficiency is much greater since the true efficiency

$$= \frac{\Delta G_{ender.}}{\Delta G_{exerg.}}\ x\ 100$$

depends on the actual intracellular concentrations of the substrates and intermediates.

6. It was noted earlier that the inner mitochondrial membrane is the primary site of oxidative phosphorylation. The components of the respiratory chain are situated in and on the membrane in a vectorial fashion, with some components facing the matrix, some facing the outer membrane and some located in the interior of the inner membrane. Much of this information has been gained by working with submitochondrial particles (largely developed by E. Racker and his associates). These particles are prepared by sonication or by mechanical disruption of purified mitochondria. They are vesicles formed from everted pieces of mitochondrial membrane and are capable of electron transport and oxidative phosphorylation. This is shown below:

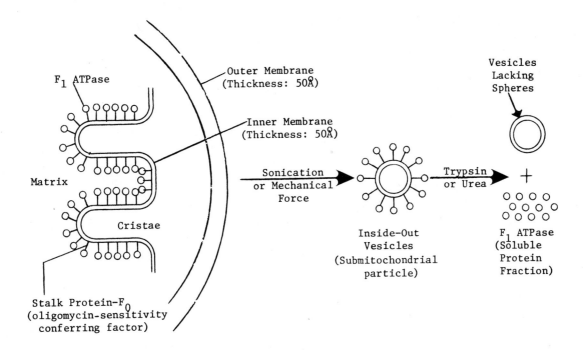

Note that in the above diagram the inside-out particles can be
further dissociated into membranous vesicles (capable of electron
transport but <u>not</u> phosphorylation of ADP) and a soluble protein
component capable of ATP hydrolysis but not electron transport.
The latter protein is called the F_1-ATPase. It is a complex
molecule with a mass of 360,000 daltons. This enzyme complex is
presumed to catalyze the ATP formation from ADP and P_i which
occurs during oxidative phosphorylation. Oligomycin does not
inhibit F_1-ATPase activity. However, when a large protein (with
5 or 6 subunits), known as F_0, is added to F_1, the new complex
becomes sensitive to oligomycin inhibition. The F_0 factor is
consequently known as oligomycin-sensitivity-conferring factor.
These factors, together with the membranous vesicles, can be
reconstituted to yield a system which is capable of both electron
transport and oxidative phosphorylation.

Regulation of Coupled Electron Transport

1. As was mentioned earlier in the discussion of inhibitors of

oxidative phosphorylation, electron transport and ATP synthesis in mitochondria are normally tightly coupled. When respiration (O_2 consumption) occurs, there must be concomitant phosphorylation of ADP.

2. Normally, in a coupled system, once all of the ADP or P_i has been used up, very little respiration takes place. Even if NADH is present, it is not reoxidized. If an uncoupler is added to such a system, O_2 uptake resumes and NAD^+ is produced. In the normal cell, the amount of ADP present is dependent on the expenditure of ATP.

3. There are four other states (types, conditions) of respiratory control in addition to the (ADP + P_i) supply. These are described in the table below, as originally defined by Chance and Williams.

Control State	Conditions which are satisfactory to maintain normal respiration	Condition which controls or limits respiratory rate
1	O_2, respiratory chain	substrate and ADP concentration
2	O_2, ADP, respiratory chain	substrate concentration
3	O_2, ADP, substrate	functional capacity of the respiratory chain
4	O_2, substrate, respiratory chain	ADP concentration
5	ADP, substrate, respiratory chain	O_2 concentration

In this table, substrate refers to NADH, $FMNH_2$, $FADH_2$, or some source of these reduced coenzymes.

Most resting cells are controlled by the availability of ADP (state 4).
The ADP-ATP cycle is shown in the next diagram.

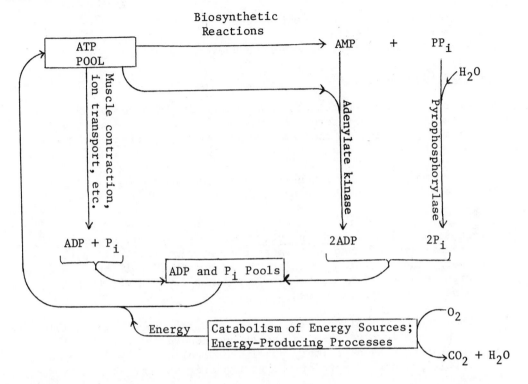

4. Instead of being used for ATP formation, the energy derived from
   electron transport can be used directly to pump certain divalent
   cations (Ca^{2+}, Mn^{2+}, Fe^{2+}) against a concentration gradient into
   the mitochondria. When calcium ion is taken up, an equivalent
   amount of phosphate is also translocated. For every pair of
   electrons transported from NADH to oxygen, about six calcium ions
   are taken up, two each at the three phosphorylation sites.
   Calcium accumulation in mitochondria plays an important role in the
   calcification process.

Possible Mechanisms for the Conservation of Energy During Electron Transport

1. Chemical coupling (biochemical coupling)
2. Chemiosmotic coupling
3. Conformational coupling (mechanochemical coupling)

1. Chemical Coupling

a. This hypothesis postulates that energy is conserved by the formation of high-energy intermediates as electrons are passed from one carrier to the next. The energy is ultimately stored as a phosphoric acid anhydride bond in ATP.

b. After many years of intensive investigation, no high energy intermediates have been isolated.

c. Oxidative phosphorylation occurs only in the presence of reasonably intact membrane structures and hence structural organization of components must play a major role. If high energy intermediates do exist, it is possible that they are tightly bound to a membrane. This might explain why they have not yet been isolated.

d. The name "chemical coupling" is unfortunate since any possible mechanisms involve chemical reactions to a great extent. It is generally used to imply purely chemical as opposed to one involving another sort of energy transfer at some stage.

2. Chemiosmotic Coupling (largely developed by P. Mitchell)

a. It is assumed (and experimentally shown) that the mitochondrial membrane is impermeable to protons (H^+) and hydroxyl ions (OH^-).

b. During electron transport in intact mitochondria, protons are released to the outside of the mitochondria. This results in the establishment of a proton gradient across the membrane, with a high concentration of H^+ (low pH) outside the mitochondria and a low concentration of H^+ (high pH) inside. The energy released during electron transport is stored by using it to maintain this gradient. Gradients of other cations may also be involved.

c. A vectorial (anisotropic) ATPase is located in the membrane.
It is oriented so that, to hydrolyze ATP, it must draw H^+ from
the inside of the mitochondria and OH^- from the outside.
Because of the concentrations which are actually present, the
reverse reaction, ADP + P_i $\longrightarrow$ ATP + H^+ + OH^- occurs. The
H^+ is released to the inside of the membrane, the OH^- goes
to the outside. The ATP synthesis is actually driven, then,
by the rapid reaction of H^+ with the OH^- inside and of the
OH^- with the H^+ outside. The energy for the formation of the
phosphoric acid anhydride bond is supplied by the large,
negative value of ΔG which results from water formation under
the existing H^+ and OH^- concentrations.

This is illustrated as follows:

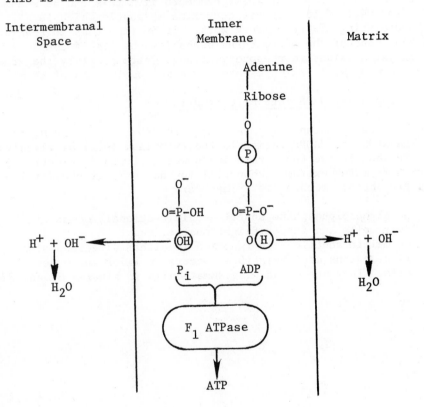

d. Figure 30 is a schematic drawing of a possible arrangement of the enzymes involved. Any spatial association of phosphorylation with the individual components of the electron-transport chain (complexes I, III, and IV for example) has been ignored. FMN and CoQ are actually associated with enzymes and the oxidation of NADH involves a specific dehydrogenase. These enzymes are not explicitly shown.

3. Conformational Coupling

a. Mitochondria undergo changes in the size and shape of their cristae during active respiration stimulated by ADP.

b. These conformational changes may be the driving force for the formation of ATP. At least conceptually, this can be seen as a reversal of the process occurring in muscle which is hydrolyzing ATP to perform work. It is important to realize that none of the above three mechanisms by itself completely explains oxidative phosphorylation. Quite possibly the true mechanism will be represented by some combination of them all.

Entry of Cytoplasmic NADH into Mitochondria

The mitochondrial membrane is impermeable to NADH, NAD^+, NADPH, and $NADP^+$. The NADH and NADPH formed in the cytoplasm (e.g., by glycolysis and the HMP shunt) have their reducing power (i.e., their electrons) transferred from the extramitochondrial to the intramitochondrial space by two shuttle mechanisms. These are

a. the glycerolphosphate shuttle--unidirectional; occurs in liver cells and insect flight muscle, and

b. the malate shuttle--can function in either direction since all reactions are reversible; occurs in liver and other cells. They are shown diagrammatically in Figures 31 and 32.

Figure 30. Schematic Diagram Showing the Possible Coupling of Electron Transport to Phosphorylation by Chemiosmosis

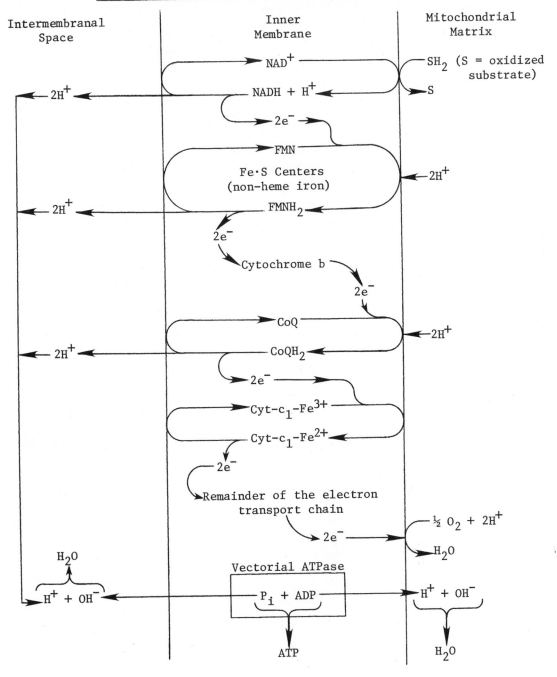

In this diagram, cytochrome b is placed before CoQ to account for the translocation of an appropriate number of protons.

Figure 31. <u>Glycerolphosphate Shuttle for Transporting Cytoplasmic</u>
<u>Reducing Equivalents into the Mitochondria</u>

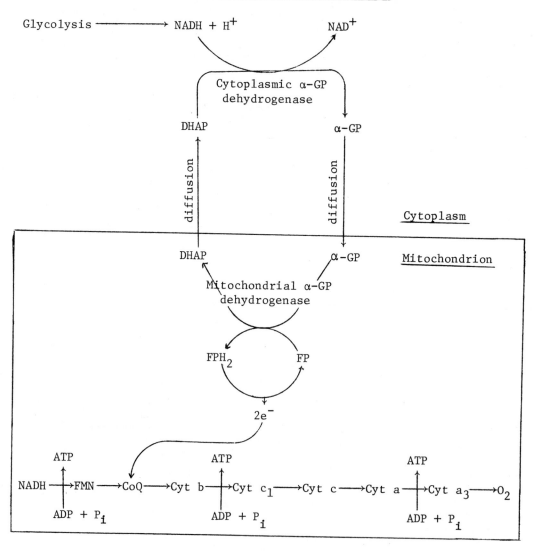

Note: α-GP = α-glycerophosphate; DHAP = dihydroxyacetonephosphate.

1. Since the intramitochondrial α-GP dehydrogenase is linked to a
    flavoprotein, the electrons are put into the electron-transport
    chain at the CoQ level rather than at the NADH dehydrogenase
    level. One phosphorylation site is thus bypassed and only two
    ATP's are synthesized for every two electrons. An NADH molecule
    put through this pathway produces only two ATP's, whereas an
    NADH formed within the mitochondrion (by, say, the TCA cycle) or
    one transported by the malate shuttle (see Figure 32) produces
    three ATP's. (This was mentioned previously in discussing the
    energy output from glycolysis and the TCA cycle.)

    In the presence of oxygen, the intramitochondrial supply of
    reducing substances is depleted and cytoplasmic NADH is needed as
    a source of electrons. Under these conditions, α-GP dehydrogenase
    has a higher affinity for NADH than does lactic acid dehydrogenase
    (LDH), and the α-GP shuttle operates.

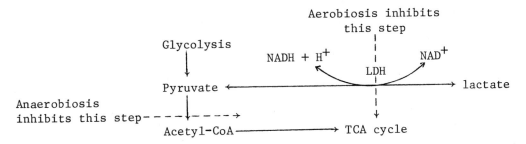

    In the above diagram, the presence or absence of oxygen causes the
    indicated effects by several mechanisms other than the differential
    affinity of these particular enzymes for NADH.

    Most cancer cells accumulate lactate during respiration (i.e., in the
    presence of oxygen). This could be due to a lack of α-GP
    dehydrogenase or a lack of control over glucose transport. Such
    cells show alterations in the structure and function of cell-
    surface membranes relative to normal cells. These changes include
    a loss of contact inhibition and changes in the membrane
    glycoprotein. Similar alterations could also occur in the
    mitochondrial membranes.

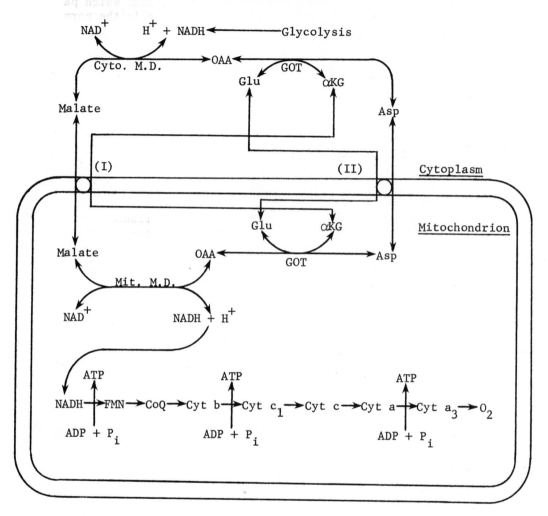

Note:
   OAA = oxaloacetic acid
   Asp = aspartic acid
   Glu = glutamic acid
   α-KG = α-ketoglutarate
   GOT = Glutamate-oxaloacetate transaminase
       (aspartate transaminase)
 Cyto. M.D. = cytoplasmic malate dehydrogenase
  Mit. M.D. = mitochondrial malate dehydrogenase
    (I) = a membrane-bound malate-α-ketoglutarate carrier
   (II) = a membrane-bound aspartate-glutamate carrier.

2. When cytoplasmic NADH is oxidized via this pathway, 3 moles of ATP are produced per mole of NADH. It is not clear which pathway (this or the glycerolphosphate shuttle) operates in the normal cell, but it appears more likely that it is this one.

In adipose cells obtained from obese patients, the activity of both cytoplasmic and mitochondrial α–GP dehydrogenase is lower than that of lean control individuals. This implies that the increased lipogenesis which caused the obesity was due to the availability of greater amounts of α–GP (used in triglyceride synthesis) and to an increased efficiency in the use of metabolites (malate shuttle pathway) by the obese patients.

Water of Oxidation (Metabolic Water)

1. Water that is produced as a metabolite in the electron-transport system is known as water of oxidation. It contributes to the total body-water needs. In humans living in a temperate climate, the average intake and internal production of water is about 2600 ml per day. The water sources with their approximate contributions are indicated below.

Source	Amount (ml/day)
Water and other beverages	1200
Water present in solid foods	1100
Water of oxidation	300
	2600 ml/day

To maintain a balance within the body, water is lost via the lungs, skin (perspiration), feces, and urine.

2. In animals that live under conditions of limited external water supply (e.g.,desert animals), oxidative metabolism becomes the principal source of water. These animals lose very minimal amounts of water through perspiration, urination, and defecation. In addition, they produce more metabolic water by increasing lipid oxidation and decreasing glycolysis.

3. The table below indicates the amount of water produced and the energy liberated in the oxidation of carbohydrates, proteins, and lipids. These values were obtained by burning the respective materials in a bomb calorimeter.

Substance	Water of Oxidation (gm H_2O/gm material)	Energy Released (kcal/gm material)
Carbohydrate	0.55	4.2
Protein	0.41	5.6
Lipid	1.07	9.3

Of the three, lipids have the highest yield/gm of both energy and water. This is because they have a higher ratio of (carbon + hydrogen) to oxygen than do either proteins or carbohydrates. Consider, for example, stearic acid ($C_{18}H_{36}O_2$) and glucose ($C_6H_{12}O_6$).

For stearic acid $(C + H)/O = 54/2 = 27$

For glucose $(C + H)/O = 18/6 = 3$

Because stearic acid contains less oxygen than glucose, more oxygen must be supplied by respiration. This means during the oxidation of stearic acid, a greater number of electrons pass down the electron-transport chain, more ATP molecules are synthesized, and more new water is formed by the reduction of oxygen than during the analogous catabolism of glucose.

Glycogenesis (Glycogen Synthesis)

1. The synthesis of glycogen (the principle storage form of carbohydrate in mammals) occurs in every tissue of the body. It is most active in liver and muscle.

Structure of glycogen:

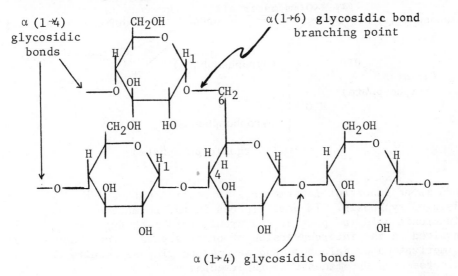

The monomer units which polymerize to form glycogen are glucose molecules. Glycogen is a <u>polyglucose</u>, as are starch and cellulose.

2. The first two steps of the biosynthesis are

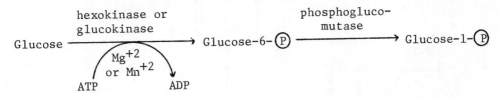

3. The third step involves the formation of a nucleoside diphosphate sugar derivative, uridine diphosphate glucose (UDPG).

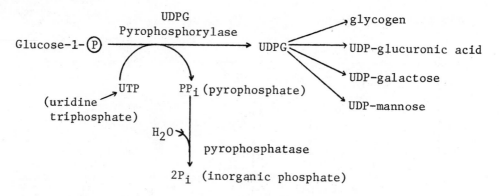

UDPG is an important intermediate in several pathways besides
glycogen synthesis. It participates in the formation of UDP-
glucuronic acid (as part of the uronic acid pathway) and is
involved in the interconversion of other sugars. For example, the
epimerization of galactose and mannose to glucose require UDP-
galactose and UDP-mannose as intermediates.

Structure of UDPG

This reaction is one example of the functioning of nucleoside phosphates as carriers of small molecules during the synthesis of larger ones. This and other roles of nucleoside phosphates are discussed under nucleic acids and in other appropriate sections.

4. The glucosyl group of UDPG is transferred to the terminal glucose residue at the non-reducing end of an amylase chain (primer) to form an α-$(1\rightarrow4)$ glycosidic linkage. The primer may be a polyglucose or an oligoglucosaccharide (small glucose polymer) of at least four residues.

$$\text{UDP-glucose} + (\text{glucose})_n \xrightarrow{\text{Glycogen Synthetase}} \text{UDP} + (\text{glucose})_{n+1}$$

Primer

Amylose [$\alpha(1\rightarrow4)$ polyglucose] chain

Part of the free energy of glucose phosphorylation is conserved in the $\alpha(1\rightarrow4)$ glycosidic bond.

5. Glycogen synthetase cannot make the $\alpha(1\rightarrow6)$ bonds that occur at the branch points of glycogen. For this, another enzyme, amylo $(1\rightarrow4, 1\rightarrow6)$ transglucosylase (branching enzyme) is needed. After the synthetase has added about eight to twelve glucose units to the non-reducing end of an amylose chain, the transglucosylase transfers six or seven of these to the sixth carbon of the glucose residue which was about the 12th residue from the end. The new molecule now has two non-reducing ends and, hence, two growing points. This is shown in the following.

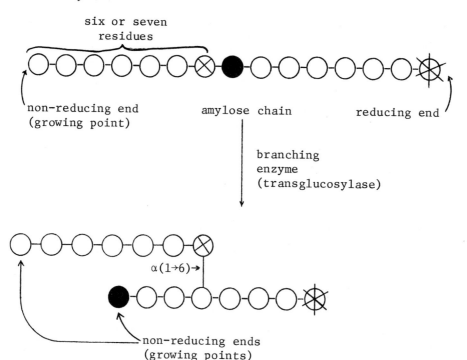

In this diagram, each circle represents a glucose residue.

The α(1→6) bonds occur about every 8-12 glucose units. Since the sidechains can be branched in the same way as the main chain, the result is a very highly branched molecule.

Glycogen Synthetase (Glycogen Synthase; UDP-α-D-glucose-glycogen-glucosyltransferase)

1.   Glycogen synthetase is a key enzyme and control point in glycogenesis. It exists in two forms: (1) a phosphorylated form, called glycogen synthetase-D because it has little or no activity by itself but is greatly stimulated by glucose-6-phosphate (and thus is dependent on G-6-P) and (2) a dephosphorylated form, called glycogen synthetase-I (because it is independent of G-6-P and is active in its absence). The phosphate group is attached to a serine residue by an ester linkage. The activation by G-6-P appears to be of the allosteric type. Glycogen synthetase-D activity is prevented by the presence of UDP. The two forms of glycogen synthetase may be interconverted by the appropriate enzymes. One of these enzymes, glycogen synthetase kinase (GSK),

also exists in an active and inactive form. (<u>Kinase</u> is a general name for any enzyme which catalyzes the phosphorylation of another compound.) The inactive form is transformed to the active form in the presence of the positive modulator, <u>cyclic AMP</u>, which is formed from ATP by the enzyme, <u>adenylate cyclase</u>. This enzyme, present in the plasma membrane of many cells, is stimulated by epinephrine. Insulin depresses the level of cyclic AMP in some tissues. These relationships are shown in Figure 33.

Cyclic AMP may activate GSK by increasing the affinity of the enzyme for Mg^{+2}. It is known that   1) in the absence of Mg^{+2}, the maximum stimulation caused by cyclic AMP is only a 20% increase over that in the absence of the cyclic nucleotide; and 2) at high Mg^{+2} concentration where the kinase is already maximally stimulated, cyclic AMP cannot further increase the activity.

Note: In the glycogen synthesis system, phosphorylation converts the glycogen synthetase enzyme from an active form to a less active form and dephosphorylation transforms the enzyme back to the active form. In the glycogen breakdown (glycogen phosphorylase) system, the opposite effect occurs. Phosphorylation of a key enzyme activates it,while dephosphorylation inactivates it. All of these systems are controlled by cyclic AMP levels. These are discussed below.

Note that the effect of epinephrine is to make glycogen synthesis dependent upon G-6-P. Thus, in a fight-or-flight situation where energy demands are high and G-6-P levels are consequently low (since G-6-P has been used up for energy production),epinephrine release turns off the glycogen synthesis while promoting glycogen breakdown. Note also that glycogen synthetase phosphatase is inhibited by glycogen (negative feedback).

Comments on Cyclic AMP

1. The formation of cyclic-3',5'-adenylic acid (cyclic AMP, cAMP) and its action is implicated in the functioning of many hormones. Some hormones exercise their influence, at least in part, by causing the formation of cAMP in their target tissues. Cyclic AMP thus acts like an intracellular hormone or "second messenger," the first messenger being the hormone or other stimulus which invokes it. Ca^{++} ion is often required for cyclic AMP action. Binding of a hormone at specific sites on target tissue cell membranes causes an increase in cell-wall permeability to Ca^{++}. After the hormone is removed from the binding site or inactivated, the membrane again becomes impermeable to Ca^{++}. An ATP-requiring Ca^{++} pump returns the intracellular Ca^{++} concentration to a low level.

Figure 33.   Regulation of Glycogen Synthetase Activity Within the Cell

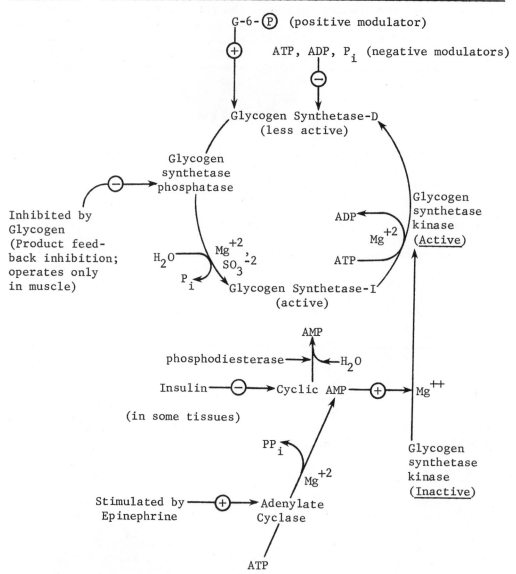

In this diagram, a ⊕ or ⊖ means that a substance increases or decreases respectively the level or activity of the compound or enzyme indicated. Note: At the physiological concentrations of ATP, ADP, P_i, and G-6-Ⓟ in the muscle, the difference in activity between the dependent and independent forms of glycogen synthetase is <u>maximal</u>. That is, the conversion from D to I makes a significant change in activity. The independent form is not activated by G-6-Ⓟ or inhibited by ATP, ADP, or P_i to any great extent.

Adenylate cyclase, the enzyme which catalyzes formation of cyclic AMP, is present in the plasma membrane of many cells. It is especially abundant in liver cells. The structure of cAMP is shown below.

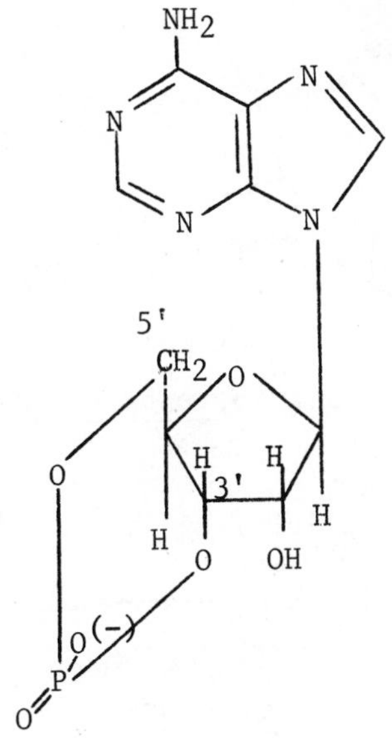

2. Figure 34 illustrates the synthesis and breakdown of cAMP and indicates the general way in which it has been shown to exert its influence on cellular processes.

3. A number of hormones can alter the levels of cAMP, bringing about diverse responses. These responses include substrate mobilization, hormone release, changes in the permeability of the cell membrane, sensory and neural excitation, and melanocyte dispersion. The specificity of the hormone action on a target cell depends upon the presence of unique hormone binding sites (receptor sites) on the cell membrane, as well as the enzymic makeup of the cell. For example, in liver cells both epinephrine and glucagon can increase the level of cAMP, but only epinephrine can bring about such a response in a muscle cell. This is explained on the basis of the absence of receptor sites for glucagon on the muscle cell membrane. In general, it appears that the binding of a hormone molecule to a receptor site sets into action a chain of events which, in one way or another, releases a constraint on the appropriate intracellular systems.

Figure 34. Mechanism by Which Cyclic AMP Mediates the Effect of Hormones on Intracellular Processes

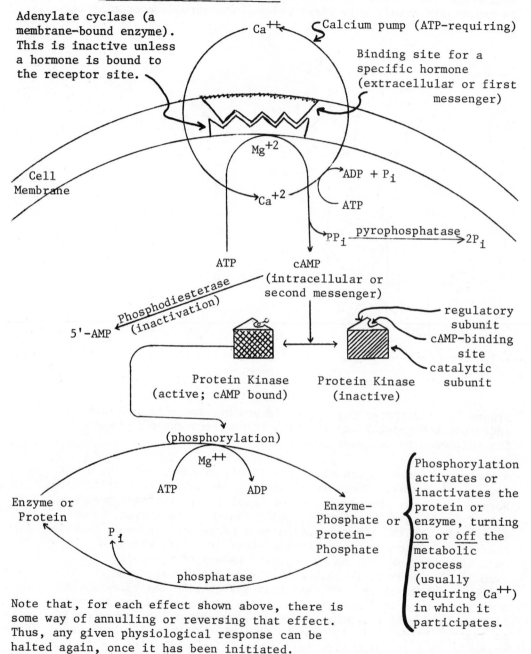

Adenylate cyclase (a membrane-bound enzyme). This is inactive unless a hormone is bound to the receptor site.

Calcium pump (ATP-requiring)

Binding site for a specific hormone (extracellular or first messenger)

Ca^{++}

Mg^{+2}

Cell Membrane

Ca^{+2}

$ADP + P_i$

ATP

$PP_i \xrightarrow{\text{pyrophosphatase}} 2P_i$

ATP

cAMP (intracellular or second messenger)

Phosphodiesterase (inactivation)

5'-AMP

regulatory subunit

cAMP-binding site

catalytic subunit

Protein Kinase (active; cAMP bound)

Protein Kinase (inactive)

(phosphorylation)

Mg^{++}

ATP

ADP

Enzyme or Protein

P_i

Enzyme-Phosphate or Protein-Phosphate

phosphatase

Phosphorylation activates or inactivates the protein or enzyme, turning on or off the metabolic process (usually requiring Ca^{++}) in which it participates.

Note that, for each effect shown above, there is some way of annulling or reversing that effect. Thus, any given physiological response can be halted again, once it has been initiated.

4. Until recently, the concept of receptor sites was largely
    hypothetical. Now, however, receptors have been isolated from
    cell membranes and have been shown to bind appropriate hormones
    in appreciable quantities at physiological concentrations. Some
    examples of hormones bound and tissues from which the receptors
    were isolated (in parenthesis) are: ACTH (adrenal cortex);
    vasopressin (kidney); insulin (liver, adipose tissue, lymphocytes,
    fibroblasts); epinephrine (erythrocyte, spleen capsule); and
    glucagon (liver). Norepinephrine has been shown to bind to
    β-adrenergic receptors isolated from cardiac tissue. This
    binding does not activate the adenyl cyclase known to be present,
    however, unless phosphatidyl inositol (known to be present in
    myocardial cell membranes but removed during receptor preparation)
    is added. Diseases associated with hyper- or hyposensitivity to
    specific hormones may be the result of an increase or decrease
    (respectively) in the number of active hormone receptors in the
    affected tissues. In the genetically obese laboratory mouse, a
    marked decrease in insulin-binding sites in both liver and
    adipose tissue has been demonstrated. Other diseases which are
    likely candidates for similar explanations include
    pseudohypoparathyroidism and vasopressin-resistant diabetes
    insipidus. Catecholamine hypersensitivity appears to occur in
    the aging process, certain cases of hypertension, hyperthyroidism,
    and following denervation. These may eventually be related to an
    increase in catecholamine receptor sites.

5. In addition to the control of cAMP levels through the modulation of
    adenylate cyclase activity, the concentration of cAMP may also be
    controlled by variations in the activity of the phosphodiesterase.
    Recall that this enzyme degrades cAMP to 5'-AMP by hydrolysis of the
    3'-ester bond. The methyl xanthines (caffeine, theophylline, and
    theobromine) are known to inhibit the diesterase. It has also been
    observed in certain tissues (liver, adipose) that insulin depresses
    the level of cAMP; and that in many tissues, prostaglandins are
    antagonists to the action of cAMP although they do not _necessarily_
    alter the cAMP concentration. The mechanisms by which insulin and
    prostaglandins function are still unclear. Prostaglandins have, in
    fact, been shown to stimulate cAMP formation in intact fat pads
    while inhibiting it in isolated fat cells.

    Besides the central role which cyclic AMP plays in many cellular
    functions of higher organisms, it is also found in unicellular
    organisms (e.g., E. coli and cellular slime molds). In these
    organisms (and in a wide variety of species studied, up to and
    including man), an increase in cAMP synthesis is a general response
    to a decrease in glucose supply (i.e., to hunger). In the slime
    molds, if the glucose level remains low for a sufficient time, the
    cAMP eventually causes aggregation of the free-living amoeboid form
    into a multicellular, sporulating form, better able to survive
    the food deprivation.

6. In addition to the well-documented role of cyclic AMP, other cyclic nucleotides (particularly cyclic 3',5'-guanylic acid, cGMP) have been implicated in intracellular control processes.

7. In order to understand the action of cyclic AMP, a convenient, sensitive method must be available for measurement of cAMP concentrations. Several have been developed and will be discussed here.

    a. The original method, used by Sutherland (who received the Nobel Prize for his work on cyclic AMP) and his coworkers, makes use of the activation of phosphorylase kinase by cAMP (see the section on glycogenolysis). The rate of phosphorylation of substrate by the kinase is proportional to the cAMP concentration in the sample. Other enzymatic methods have also been developed.

    b. The dilution of a known amount of a radioactive compound by unlabeled molecules of the same compound present in a sample is the basis for both <u>radioimmunoassays</u> (RIA; see also the measurement of insulin levels) and <u>competitive protein-binding assays</u> (CPB assays). Both types of assays are known for their high specificity (measure <u>only</u> the desired compound) and sensitivity (can, for example, measure 1-2 nanomoles or less of cAMP per gram of tissue), and their relative ease of performance. The methods differ only with respect to the source of the protein used to measure the dilution. The general scheme for both of these is shown in Figure 35. Several radioactive atoms have been used as labels. Frequently, ^{125}I, ^{131}I, and ^{3}H are the ones employed. In both methods, a standard curve is prepared and used to calculate the unknown concentrations. Since these are essentially isotope dilution techniques, the concentration of unknown is <u>inversely</u> proportional to the amount of radioactivity bound in the final separation.

        i) The use of radioimmunoassays began in the late 1950's when Berson and Yalow published a method for insulin. Since then, assays have appeared to measure a wide variety of peptide hormones and other, non-protein substances. In the preparation of the antibody needed for the RIA, a small molecule such as cAMP is not antigenic by itself. In order to induce antibody formation in the animal, the small molecule (termed the hapten, the part of an antigen molecule for which an antibody is specific) must be coupled to a much larger protein molecule such as albumin. The nature of the immune response is discussed more fully in the chapter on immunochemistry.

Figure 35. Flow Chart for Radioimmunoassay and Competitive Protein-Binding Assay

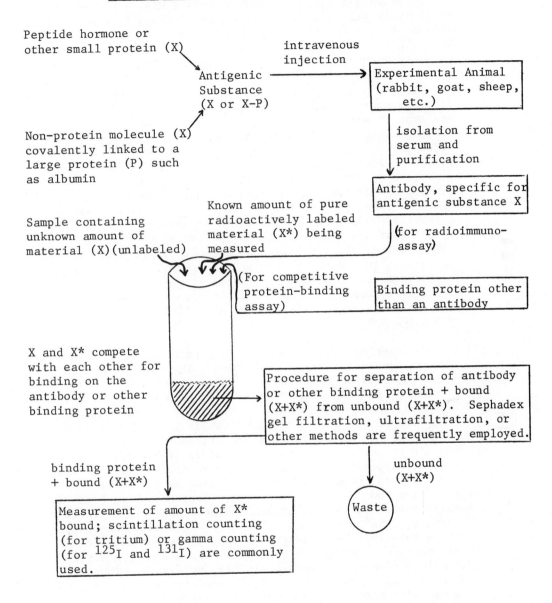

Peptide hormone or other small protein (X)

Non-protein molecule (X) covalently linked to a large protein (P) such as albumin

Antigenic Substance (X or X-P)

intravenous injection

Experimental Animal (rabbit, goat, sheep, etc.)

isolation from serum and purification

Antibody, specific for antigenic substance X

(for radioimmunoassay)

Sample containing unknown amount of material (X)(unlabeled)

Known amount of pure radioactively labeled material (X*) being measured

(For competitive protein-binding assay)

Binding protein other than an antibody

X and X* compete with each other for binding on the antibody or other binding protein

Procedure for separation of antibody or other binding protein + bound (X+X*) from unbound (X+X*). Sephadex gel filtration, ultrafiltration, or other methods are frequently employed.

binding protein + bound (X+X*)

unbound (X+X*)

Waste

Measurement of amount of X* bound; scintillation counting (for tritium) or gamma counting (for ^{125}I and ^{131}I) are commonly used.

Note: X is the hapten, the part of the antigen molecule for which the antibody is specific.

275

In the "double-antibody" modification of RIA, a second antibody is prepared using the first antibody (the one against X, the substance being measured) as an antigen. When the two antibodies are employed together, a more complete separation of free and bound (X+X*) is affected.

ii) Competitive protein-binding assays make use of the ability of certain proteins and enzymes to bind selectively to one type of small molecule. For example, one of the protein kinases involved in the control of glycogen metabolism is activated by the binding of cAMP. Other cyclic nucleotides and related compounds have essentially no effect on the kinase activity. An assay has been developed based on this selective binding which uses protein kinase isolated from muscle. The procedure is essentially identical to the radioimmunoassay for cAMP, except that the kinase replaces the antibody to cyclic AMP. Another similar assay has been published using an uncharacterized cAMP-binding protein fraction from beef adrenal cortex. The use of biological molecules to selectively bind other molecules is also the basis of the purification technique known as affinity chromatography.

iii) One interesting and potentially important source of specific binding agents is hormone receptors. They are found in three subcellular compartments: the plasma membrane, the cytoplasm, and the nucleus. Peptide and glycoprotein hormones, catecholamines and prostaglandins have specific receptors on the plasma membrane, while steroid hormones have receptors in the cytoplasm. Triiodothyronine (a thyroid hormone, T_3) binds to receptors within the cell nucleus. Specific receptors have also been found in other intracellular organellar membranes such as the endoplasmic reticulum and the golgi apparatus. Hormones which bind to these membranes include catecholamines, prolactin, growth hormone, somatomedin, and substances which show non-suppressible insulin-like activity. Assays using the hormone receptors are termed radioreceptor assays (RRA), a major subclass of competitive protein binding assays. They are becoming more popular than RIA because the binding of the hormone is based on the same structural and steric characteristics which determine the biological activity, rather than those which dictate its immunogenicity. RRA's are used clinically for the measurement of ACTH, human chorionic gonadotrophin (HCS), prolactin, and insulin. The cAMP assay described above is another example of a RRA. The detection of estrogen receptors in breast tumors by "inverse RRA" (use of hormone to assay the binding protein) is another important application of the technique. It permits the screening of tumors to find those which will be susceptible to ablative (antiestrogen) treatment (i.e. those with receptors, whose growth is stimulated by estrogens). Receptors are discussed further in Chapter 13, on hormones.

<u>Glycogenolysis</u> (glycogen breakdown) is catalyzed by the following enzymes:

A. <u>Glycogen phosphorylase</u>: forms glucose-1-(P) from glycogen by adding an inorganic phosphate across the α(1→4) linkages; cannot hydrolyze the three α(1→4) bonds adjacent to an α(1→6) branch-point.

B. <u>Oligo-α(1→4)→α(1→4) glucan transferase</u>: transfers the three glucose residues attached to a branch-point residue to the non-reducing end of the main chain, thus exposing the branch-point.

C. <u>Debranching enzyme</u> (amylo-(1→6)glucosidase): hydrolyzes the α(1→6) glucosyl linkages present at branch-points, releasing one molecule of free glucose per branch-point.

1. These enzymes work in a cyclic manner to completely degrade glycogen to glucose and glucose-1-(P). The sequence is A, B, C, A, B, C, etc. The glucose-1-(P) undergoes the transformations shown below prior to entering into other metabolic reactions.

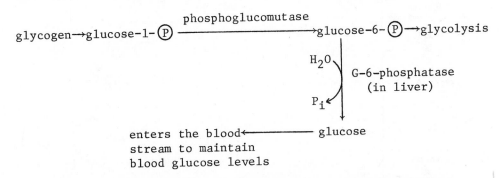

2. The cyclic functioning of enzymes A, B, and C in glycogen

Note that the action of debranching enzyme causes the release of some free glucose. Thus, it is possible for a small (but not significant) rise in blood sugar to occur even in the absence of glucose-6-phosphatase. This happens in von Gierke's disease, discussed later in this section. The rate-limiting step of glycogenolysis is the formation of glucose-1-(P) by the action of glycogen phosphorylase. This enzyme is a key control point and is discussed later in detail. The phosphorylase found in the liver differs significantly from that found in muscle.

Figure 36. **Action of the Glycogen-Breakdown Enzymes on a Typical (Although Small) Glycogen Molecule**

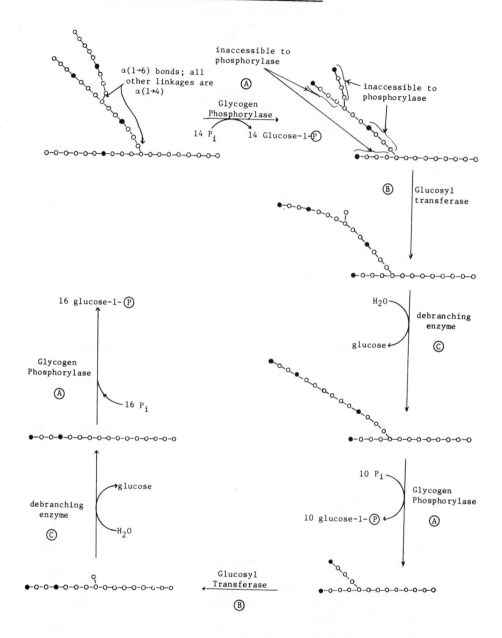

breakdown is shown in Figure 36. Each circle in the chains represents a glucose residue. The solid circles are four residues removed from α (1→6) branch-points and hence are the beginning of stretches inaccessible to the phosphorylase without prior transferase action. The overall process shown can be summarized by the equation

$$(glucose)_{42} + 2\ H_2O + 40\ P_i \longrightarrow 2\ glucose + 40\ glucose\text{-}1\text{-}\textcircled{P} \ .$$

The polymers of glucose which are left at each step, following the action of the three enzymes are called <u>limit dextrins</u>. A dextrin is any polymeric product of the incomplete hydrolysis of a polysaccharide. Since these molecules are degraded as completely as the specific enzymes are capable, they are limit **dextrins**. See limit dextrinosis (Cori's disease, type III glycogen storage disorder) later in this section.

3. Glycogen breakdown is initiated in both liver and muscle by glycogen phosphorylase. This enzyme specifically catalyzes the phosphorolysis of the glucopyranosyl-α(1→4)-glucopyranose linkages in glycogen, releasing glucose-1-$\textcircled{P}$. The phosphorylases from liver and muscle are immunologically different, but they are similar in a number of other ways.

    Both appear to be allosteric enzymes.

    Pyridoxal phosphate is required as a cofactor for both enzymes. In the case of the muscle phosphorylase, there is one molecule bound per subunit. The function of the pyridoxal phosphate is not known.

    Both enzymes undergo an interconversion between a phosphorylated, active form and a dephosphorylated, inactive form. The phosphates are attached to seryl or threonyl residues by ester bonds, as in the synthetase enzyme.

    Epinephrine causes the activation of both the phosphorylases. Its effect is mediated by cyclic AMP.

Of the two enzymes, the muscle phosphorylase has been much more thoroughly studied.

    a. Figure 37 indicates the way in which the activity of muscle glycogen phosphorylase is controlled. In this diagram, PCMB = p-chloromercuricbenzoate. This is a reagent which reacts with sulfhydryl groups, thereby inactivating them (see the section on enzyme inhibition by heavy metals). Although it appears that at sufficiently high (saturating) 5'-AMP levels, the activities of phosphorylases a and b are similar, at the actual substrate and 5'-AMP levels found within the cell, control of phosphorylase

activity results from the a $\rightleftarrows$ b conversion. In addition, in
at least some species, the phosphorylase a dimer is more
active than the tetramer.

Figure 37. Regulation of Muscle Glyocgen Phosphorylase

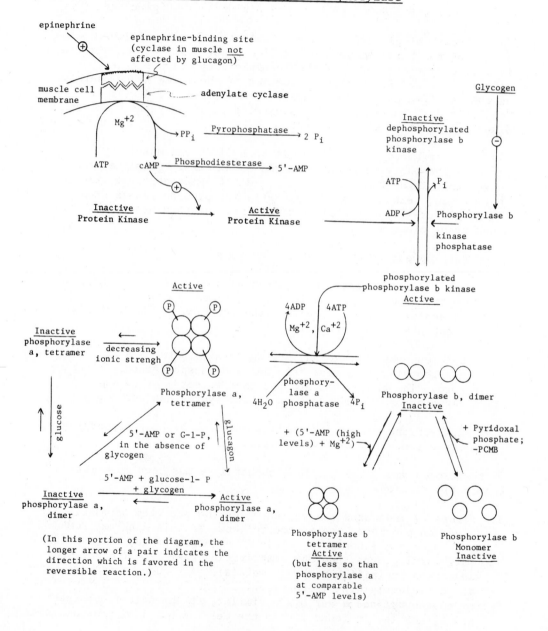

b.   As was pointed out above, the mechanism of the liver glycogen phosphorylase system has been studied less thoroughly than that of muscle. Figure 38 illustrates the general pattern, which is quite similar to that in muscle. Particular points to note about the liver system are

   i)   The depressive effect of insulin on glycogen breakdown.

   ii) The stimulatory effect of glucagon as well as epinephrine.

   iii) The liver dephosphophosphorylase (analogous to muscle phosphorylase b) is not activated by 5'-AMP.

   iv) Although it has not been conclusively demonstrated, liver phosphorylase is probably a dimer in both the phospho- and dephospho-forms.

4. a.   Although the control of glycogen synthesis and breakdown appears very complicated, at least part of the difficulty is that some or perhaps all of the effects (allosteric modulators, dissociation of subunits, etc.) observed in vitro probably play a minor role in vivo. For example, although it appears that at sufficiently high (saturating) levels of 5'-AMP the activities of muscle phosphorylases a and b are similar, at the actual substrate and 5'-AMP concentration found within the cell, this does not occur. In the body, the important controls seem to be

   i)   cAMP stimulates glycogenolysis and inhibits glycogenesis.

   ii) Glycogen accumulation enhances the rate of its own breakdown and inhibits further synthesis.

   iii) Adjustments in the ratios of (phosphorylase a/phosphorylase b) and (synthetase D/synthetase I) are the real sources of control.

   b.   In muscle, the conversion of phosphorylase b to phosphorylase a may be the mechanism by which glycogenolysis is coupled to muscle contraction. The basis for this lies in the activation of the muscle phosphorylase kinase by Ca^{+2} and the involvement of Ca^{+2} in the $a \rightleftarrows b$ interconversion. At present, this is largely speculation.

5. Liver and kidney have glucose-6-phosphatase which catalyzes the last reaction in glycogenolysis.

Figure 38. Regulation of Liver Glycogen Phosphorylase

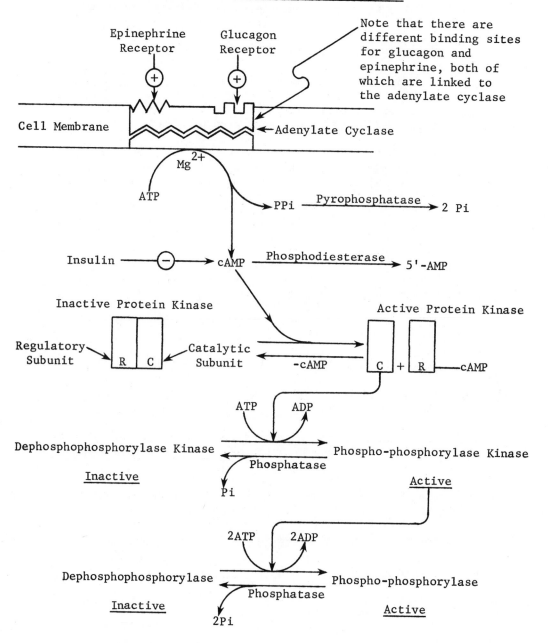

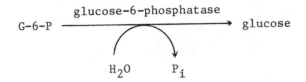

Muscle does not have this enzyme. Thus, glucose can be added to the blood by the former two tissues, especially liver, but muscle glycogen does not participate in the maintenance of blood glucose levels. Muscle glycogen is used primarily to supply the energy needs of the muscle. A small amount of free glucose is released by a debranching enzyme, but it is not significant in terms of the amount in the bloodstream; and most of it is probably utilized without ever leaving the cell.

Glycogen Storage Disorders

These are a group of inherited diseases characterized by the deposition of abnormally large amounts of glycogen in the tissues. These are presented in Tables 10, 11, and Figure 39.

Comments on the HMP Shunt (Figure 40)

1. The reactions of this pathway are catalyzed by enzymes found in the cytoplasm of animal cells. It is therefore termed a soluble (as opposed to membrane-bound) enzyme system. Since the glycolytic enzymes are present in the same cellular compartment, it is important to realize that the following reactions, in addition to those in Figure 40, can also occur.

$$Gly-3-\text{(P)} \xrightleftharpoons[]{\text{triosephosphate isomerase}} DHAP$$

$$Gly-3-\text{(P)} + DHAP \xrightleftharpoons[]{\text{aldolase}} F-1,6-di\text{(P)} \xrightarrow{\text{F-1,6-diphosphatase}} F-6-\text{(P)} + P_i$$

$$F-6-\text{(P)} \xrightleftharpoons[]{\text{hexosephosphate isomerase}} G-6-\text{(P)}$$

To obtain two molecules of Gly-3-(P) for the synthesis of F-6-(P) consider six molecules of glucose passing through the HMP Shunt (i.e. twice the number shown in Figure 40):

$$6 \text{ glucose} \longrightarrow 4 \text{ F-6-(P)} + 2 \text{ Gly-3-(P)} + 6 \text{ CO}_2 \searrow \text{ F-6-(P)} \Bigg] \longrightarrow 5 \text{ G-6-(P)} + 6 \text{ CO}_2$$

(Text continues on p. 289)

Figure 39.   Recapitulation of Glycogenesis and Glycogenolysis

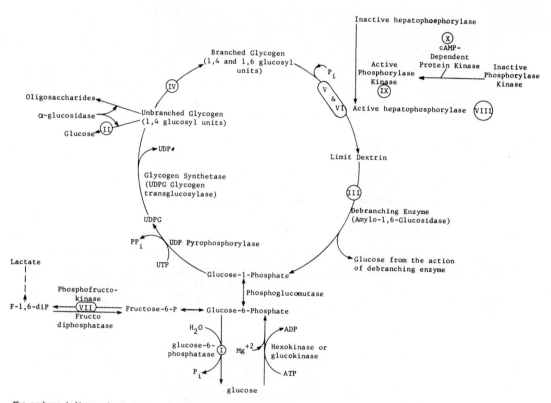

The numbers indicate the Cori-type of the glycogen storage disorder.

Glycogen Storage Disorders

Cori-Type	Name	Deficiency
I	von Gierke's	Glucose-6-phosphatase
II	Pompe's	α-1,2-glucosidase (probably works as shown on diagram)
III	Cori's	Amylo-1,6-glucosidase (debrancher enzyme)
IV	Andersen's	(1,4 1,6)-transglucosylase (brancher enzyme)
V	McArdle's	Myophosphorylase
VI	Hers'	Hepatophosphorylase (+ perhaps others)
VII	---	Phosphofructokinase
VIII	---	Hepatophosphorylase activity reduced
IX	---	Hepatophosphorylase kinase
X	---	cAMP-dependent protein kinase (liver and muscle)

Table 10. Biochemical Lesions Associated with the Glycogen Storage Disorders

Type	Name	Decrease or Deficiency of	Glycogen Structure
I	von Gierke's disease (hepato-renal glycogenosis)	Glucose-6-phosphatase in the cells of liver and of renal convoluted tubules	Normal
II	Pompe's disease (generalized glycogenosis)	Lysosomal α-1,4-glucosidase (acid maltase)	Normal
III	Cori's disease (limit dextrinosis)	Amylo-1,6-glucosidase (debrancher enzyme)	Abnormal; outer chains missing or very short; increased number of branched points
IV	Andersen's disease (brancher deficiency; amylopectinosis)	$(1,4\rightarrow1,6)$-transglucosylase (brancher enzyme)	Abnormal; very long inner and outer unbranched-chains
V	McArdle's Syndrome	Muscle glycogen phosphorylase (myophosphorylase)	Normal
VI	Hers' disease	Liver glycogen phosphorylase (hepatophosphorylase)	Normal
VII	---	Muscle phosphofructokinase	Normal
VIII	---	No enzyme deficiency and normal phosphorylase activating system; liver phosphorylase activity is reduced, however. Brain is also involved.	Normal
IX	---	Liver phosphorylase kinase defect. The phosphorylase is normal but is not activated.	Normal
X	---	cAMP-dependent protein kinase in both liver and muscles	Normal

Table 11. Clinical Characteristics of the Glycogen Storage Disorders

Cori-Type	Name	Comments
I	von Gierke's	Hypoglycemia; lack of glycogenolysis under the stimulus of epinephrine or glucagon; ketosis, hyperlipemia; hepatomegaly; autosomal recessive.
II	Pompe's	How this deficiency leads to glycogen storage is not well understood. In some cases, the heart is the main organ involved; in others, the nervous system is severely affected; autosomal recessive.
III	Cori's	Hypoglycemia; diminished hyperglycemic response to epinephrine or glucagon; normal hyperglycemic response to fructose or galactose; autosomal recessive.
IV	Andersen's	**Rare, or difficult to recognize; cirrhosis** and storage of abnormal glycogen; diminished hyperglycemic response to epinephrine; abnormal liver function; autosomal recessive.
V	McArdle's	High muscle glycogen content (2.5 to 4.1% versus 0.2 to 0.9% normal); fall in blood lactate and pyruvate after exercise (normal is sharp rise); normal hyperglycemic response to epinephrine (thus normal hepatic enzyme); autosomal recessive.
VI	Hers'	Not as serious as G-6-phosphatase deficiency; liver cannot make glucose from glycogen, but can make it from pyruvate; mild hypoglycemia and ketosis; hepatomegaly; probably more than one disease (Type VIII, below, was originally part of this group).
VII	---	Shows properties similar to Type V; autosomal recessive.
VIII	---	Hypoglycemia; enhanced liver glycogen content; hepatomegaly; X-linked trait.
IX	---	Cerebral degeneration.
X	---	Hepatomegaly, hypoglycemia, acidosis.
XI	---	Hepatomegaly.

Figure 40.   <u>Hexose Monophosphate (HMP) Shunt</u> (also known as the pentose
pathway, direct oxidative pathway, and phosphogluconate
oxidative pathway)

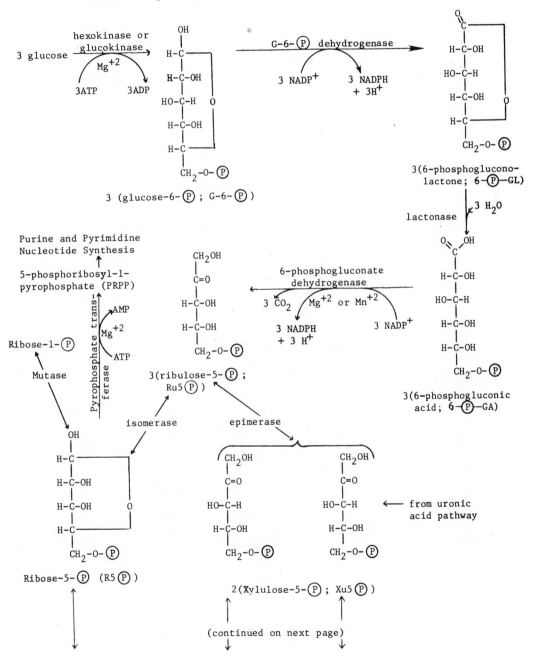

3 (glucose-6-(P) ; G-6-(P))

3(6-phosphoglucono-
lactone; 6-(P)—GL)

3(6-phosphogluconic
acid; 6-(P)—GA)

Purine and Pyrimidine
Nucleotide Synthesis

5-phosphoribosyl-1-
pyrophosphate (PRPP)

Ribose-1-(P)

3(ribulose-5-(P) ;
Ru5(P))

Ribose-5-(P) (R5(P))

2(Xylulose-5-(P) ; Xu5(P))

from uronic
acid pathway

(continued on next page)

Figure 40. Hexose Monophosphate (HMP) Shunt (continued)

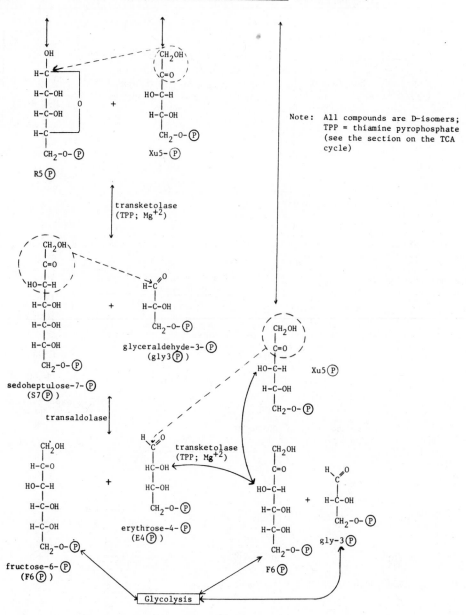

Note: All compounds are D-isomers; TPP = thiamine pyrophosphate (see the section on the TCA cycle)

288

Doubling the entire pathway, then, one arrives at:

6 glucose + 6 H_2O + 12 $NADP^+$ + 6 ATP

$$\longrightarrow 5 \text{ G-6-}\textcircled{P} + 6 \text{ } CO_2 + 12 \text{ NADPH} + 12 \text{ } H^+ + 6 \text{ ADP} + P_i$$

The <u>net</u> conversion, catalyzed by the HMP Shunt enzymes together with the glycolytic enzymes shown above, is the oxidation of one mole of glucose to CO_2 and reduced $NADP^+$, with the hydrolysis of one mole of ATP.

$C_6H_{12}O_6$ + 6 H_2O + ATP + 12 $NADP^+$

$$\longrightarrow 6 \text{ } CO_2 + \text{ ADP} + P_i + 12 \text{ NADPH} + 12 \text{ } H^+$$

2. The pentose shunt has been demonstrated in liver, lactating mammary tissue, adipose tissue, leukocytes, testes, and the adrenal cortex.

3. Due to the absence of glucose-6-$\textcircled{P}$ dehydrogenase, the complete pathway does not function in striated muscle. The 5-phosphoribosyl-1-pyrophosphate (PRPP), which is needed for nucleotide synthesis in the muscle cells, originates from fructose-6-$\textcircled{P}$ and glyceraldehyde-3-$\textcircled{P}$ formed in glycolysis. These compounds are converted to ribose-5-$\textcircled{P}$ by the action of the two transketolases, the transaldolase, and the epimerase and isomerase which are part of the HMP shunt (see nucleotide synthesis).

4. Concerning specific reactions in this sequence

    a. The conversion of 6-phosphogluconic acid to ribulose-5-$\textcircled{P}$ is an oxidative decarboxylation analogous to those catalyzed by the malic enzyme (malate + $NADP^+$ → pyruvate + CO_2 + NADPH + H^+; this is another important source of NADPH) and isocitrate dehydrogenase (see TCA cycle). The reaction presumably involves 6-phospho-3-ketogluconic acid as an enzyme-bound intermediate. The free compound has never been isolated, however.

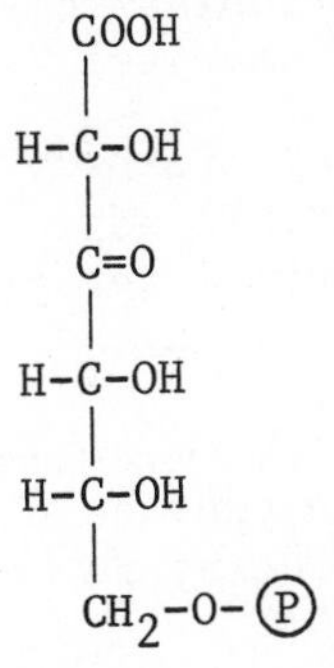

```
        COOH
         |
      H-C-OH
         |
        C=O                        6-phospho-3-ketogluconic acid
         |
      H-C-OH
         |
      H-C-OH
         |
        CH2-O- (P)
```

b. The two reactions catalyzed by transketolases are similar to the formation of acetyl-CoA by the oxidative decarboxylation of pyruvic acid. Thiamine pyrophosphate and magnesium are required for all three conversions.

5. This pathway requires only one ATP molecule for each glucose molecule oxidized, and it is independent of the TCA cycle components. The shunt is energetically comparable to glycolysis and the TCA cycle if the NADPH is transhydrogenated to NADH.

$$NAD^+ + NADPH \xrightleftharpoons{transhydrogenation} NADH + NADP^+$$

There are few, if any, examples of transhydrogenase enzymes which can directly catalyze this conversion. When reducing equivalents are exchanged between NADH and NADPH, it is usually accomplished by two enzymes, one able to use $NADP^+$ or NADPH as a substrate, the other requiring NAD^+ or NADH. The reduced product of one enzyme is a substrate for the other one. The enzymes are usually separated by a membrane, which the product/substrate is able to cross. One example of such a system is shown below.

Mitochondrial:

$$NADH(mito.) + H^+ + \alpha\text{-}KG + CO_2 \xrightleftharpoons[\text{(mito.)}]{ICDH} NAD^+(mito.) + isocitrate$$

Cytoplasmic:

$$NADP^+(cyto.) + isocitrate \xrightleftharpoons[\text{(cyto.)}]{ICDH} NADPH(cyto.) + H^+ + \alpha\text{-}KG + CO_2$$

Net Reaction:

$$NADH(mito) + NADP^+(cyto.) \xrightleftharpoons{} NAD^+(mito.) + NADPH(cyto.)$$

(Abbreviations: mito., cyto. = mitochondrial, cytoplasmic; ICDH = isocitrate dehydrogenase; α-KG = α-ketoglutarate)

The NADH can be oxidized via the electron-transport pathway. For each glucose oxidized, 36 ATP can be produced in this way. The net result is the formation of 36 − 1 = 35 ATP formed per glucose oxidized in the HMP shunt. This is slightly less than the 36 or 38 ATP formed during glycolysis and the TCA cycle.

6. The use of the HMP shunt for glucose oxidation to provide ATP is probably of very minor importance. The NADPH reducing power must

be transferred to the intramitochondrial space, which can be inefficient, depending on the shuttle used (see the section on **mitochondria**). **The real importance of this pathway is more likely:**

    a.   Provision of a mechanism for the formation of pentoses needed in the synthesis of nucleotides.

    b.   A route for the interconversion of pentoses, and of pentoses and hexoses.

    c.   A source for NADPH which serves as a reducing agent in the synthesis of fatty acids, steroids, glutathione (GSH; see G-6-(P) dehydrogenase deficiency), etc. This reducing power is made available to the extramitochondrial part of the cell, where most anabolic (synthetic) processes take place.

7.   It appears that the shunt pathway is stimulated in any tissue which requires NADPH. Thus, many tissues in which this pathway operates specialize in synthesizing fatty acid or steroids, both NADPH-requiring activities.

Biochemical Lesions of Carbohydrate Metabolism and the Red Blood Cells

1.   Circulating erythrocytes (red blood cells, RBC's) are biconcave discs having no nuclei. They are made in the bone marrow and have an average circulatory life expectancy of about 120 days. Their primary functions are the delivery of oxygen to the tissues and participation in the transport of CO_2 to the lungs.

2.   Figure 41 shows the development of mature erythrocytes from the primitive nucleated cells. The pattern of the intracellular changes which eventually provide the metabolism required for the specialized function of the red blood cell is indicated. Also included in the diagram is the development of the other cells **such as white cells, whose metabolisms are discussed later in this section.**

    The mature erythrocytes possess (a) glycolytic pathway to provide energy and 2,3-diphosphoglycerate (DPG), a key intermediate which plays a role in oxygen delivery to the tissues (discussed under hemoglobin); and (b) HMP-shunt pathway to yield NADPH which helps maintain the -SH groups of sulfhydryl-containing proteins in a reduced state. Such proteins function as enzymes (glyceraldehyde-3-(P) dehydrogenase) and as parts of the membrane system in the erythrocytes.

3.   The normal production and release of erythrocytes (normal erythropoiesis) is under the control of a glycoprotein hormone

[292]

Figure 41. Differentiation of Stem Cells and Characteristics of the Developing Red Blood Cells

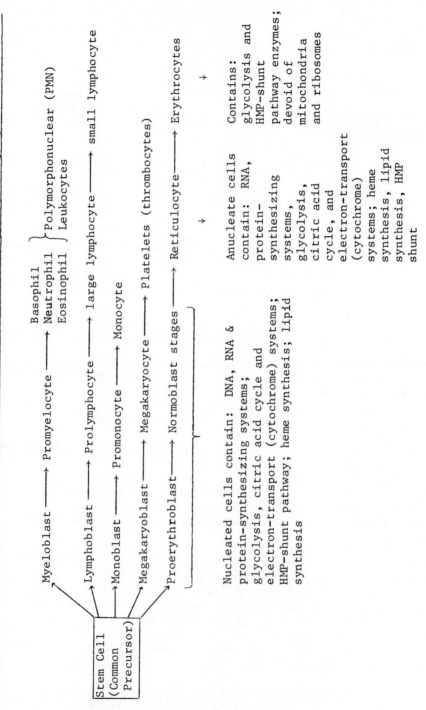

(erythropoietin) which stimulates conversion of some stem cells to proerythroblasts in anemic or hypoxic situations. The mechanism of action of the hormone is presumably through the stimulation of mRNA and protein-synthesizing system (discussed later). Erythropoietin is produced in the plasma by the action of a factor (erythropoietic factor) coming from either kidney or liver on a plasma globulin. The synthesis of the factors is increased by a variety of conditions and agents such as hypoxia, androgens, cobalt salts, etc.

4. Drugs can cause a decrease in the number of circulating erythrocytes either through impaired production (marrow aplasia caused by chloramphenicol) or increased destruction of red cells (hemolysis caused by a variety of chemical compounds). This occurs particularly in individuals with deficiencies of one or more of the enzymes associated with carbohydrate metabolism. Following are some examples of such instances.

Glucose-6-Phosphate Dehydrogenase (G-6-PD) Deficiency

1. When certain drugs are administered to susceptible persons, an acute hemolytic anemia results. Some of these drugs are:

   a. Antimalarials: Primaquine, pentaquine, pamaquine (all are 8-aminoquinoline derivatives).

   b. Sulfonamides and Sulfones: Sulfanilamide, sulfacetamide, sulfapyridine, sulfamethoxypyridazine, salicylazosulfapyridine, thiazosulfone, dapsone.

   c. Analgesics: Acetanilid.

   d. Nonsulfa antibacterials: Nitrofurazone, nitrofurantoin (Furadantin).

   e. Miscellaneous: Ingestion of Vicia fava beans by persons of Mediterranean origin; napthalene, Neoarsphenamine, phenylhydrazine hydrochloride.

2. If primaquine is administered daily to sensitive persons an acute hemolytic crisis develops, beginning on the second or third day. From 30 to 50% of the patient's RBC's are destroyed and the patient becomes jaundiced with dark, often black urine. Older cells (63-76 days old) are destroyed while younger cells (8-21 days) are not affected. If the patient survives the crisis, recovery takes place as new cells become preponderant, despite continued administration of primaquine.

3. Susceptibility to drug-induced hemolytic disease may be due to a
    hereditary deficiency of G-6-℗ dehydrogenase activity in the
    erythrocyte, as explained next.   Normally the following sequence
    of reactions occurs.

    a.   G-6-℗ dehydrogenase catalyzes the reaction

$$G\text{-}6\text{-}℗ \xrightarrow{\hspace{3cm}} 6\text{-Phosphogluconolactone}$$

$$NADP^+ \qquad NADPH + H^+$$

    b.   The NADPH formed is used in the conversion of oxidized
          glutathione (GSSG) to reduced glutathione (GSH).

$$GSSG + NADPH + H^+ \xrightarrow{\text{GSH reductase}} 2\ GSH + NADP^+$$

oxidized glutathione   reduced glutathione

    (GSSG is actually a dimer of GSH, the molecules being linked
    by a disulfide bond.)   GSH plays a very important role in red
    cell metabolism, by "protecting" the -SH groups of different
    substances, such as hemoglobin, catalase, and the lipoprotein
    of cell membranes.   Since GSH is more readily oxidized than
    the protein sulfhydryls, it serves as a source of oxidizing
    power to reduce disulfide bonds which may be deleterious to
    normal cellular function.   Glutathione also helps by destroying
    oxidizing agents which may be present in the cell.   In the red
    cells, the concentration of GSH is 70 mg per 100 cc of red
    cells.   A deficiency of GSH can give rise to hemolysis.   It
    is not yet known exactly how drugs like primaquine affect the
    G-6-℗ dehydrogenase-deficient red cells to cause hemolysis.
    Some of these interrelationships are shown in Figure 42.

    For an explanation of the numbers (i-vi) used in Figure 42,
    see item 5 in this section.   Following are the explanations
    for the abbreviations used in that figure.

    G-6-℗ = glucose-6-phosphate

    Gly-3-℗ = glyceraldehyde-3-phosphate

    DPG(1,3;2,3) = diphosphoglycerate

    3-℗ GA = 3-phosphoglyceric acid

    6-℗ GL = 6-phosphogluconolactone

    6-℗ GLD, G-6-℗ D, Gly-3-℗ D = the respective dehydrogenases

    GSH; GSSG = reduced and oxidized glutathione, respectively.

Figure 42. Interrelationships of Carbohydrate Metabolism in the Erythrocyte

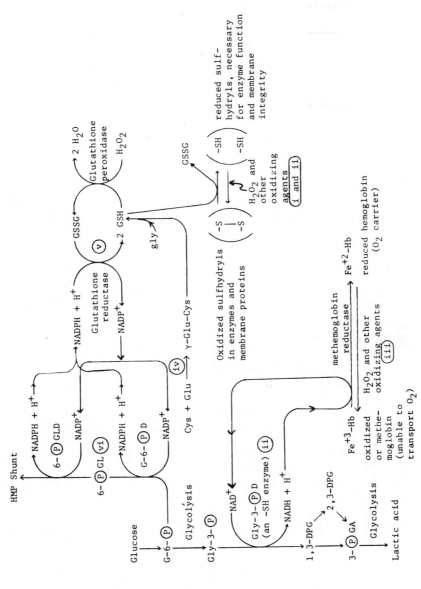

For a key to the numbers (i-vi), see page 297. The abbreviations used in this figure are
explained on page 294.

4. Structure and biosynthesis of glutathione

    i) Structure

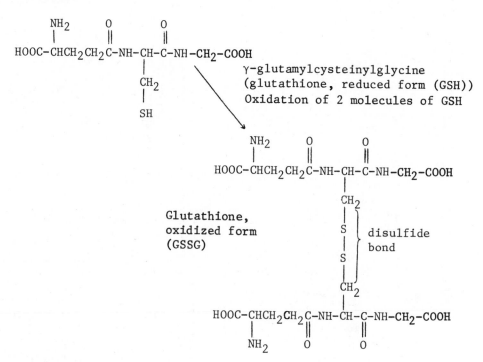

  γ-glutamylcysteinylglycine
  (glutathione, reduced form (GSH))
  Oxidation of 2 molecules of GSH

Glutathione,
oxidized form
(GSSG)

disulfide
bond

    ii) Synthesis

It is important to realize that the synthesis of this
tripeptide takes place in the <u>absence</u> of the normal protein-
synthesizing systems (involving mRNA, tRNA, ribosomes and
appropriate enzyme systems, discussed later). Glutathione
can consequently be formed even in the anucleate, mature
erythrocyte.

L-Cys + L-Glu —— γ-GC Synthetase ——→ γ-Glu-Cys

      Mg^{+2}

  ATP   ADP + P$_i$

                          ATP

                        Mg^{+2};
                        GSH Synthetase

        Gly   ADP + P$_i$

            γ-glutamylcysteinylglycine
        (γ-Glu-Cys-Gly; reduced glutathione)

5. There are several factors that may contribute to the hemolysis (the numbers below refer to Figure 42). These include

   i) Damage to the cell membrane when sulfhydryl groups are oxidized (a process reversed by GSH).

   ii) Interruption of normal pathways of metabolism inside the cell (Gly-3-(P) dehydrogenase, for example, is an important enzyme that requires GSH).

   iii) Degradation of hemoglobin, and the shifting of the equilibrium between oxyhemoglobin and methemoglobin (which can be influenced by a lack of NADH or by the enzyme glutathione peroxidase that requires GSH to degrade H_2O_2).

   iv) Biochemical lesions in the synthesis of GSH.

   v) Glutathione reductase is a flavin enzyme. Deficiency of the vitamin riboflavin results in lowering of the reductase activity.

   vi) Other deficiencies in the production of NADPH.

The above factors are all interrelated. For example, a lack of Gly-3-(P) dehydrogenase would cause both an interruption of glycolysis and a decrease in NADH. The lack of NADH may result in an increase in methemoglobin due to a decrease in the reaction

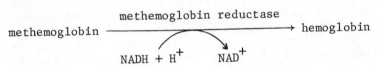

which requires NADH. In addition, the oxyhemoglobin-deoxyhemoglobin equilibrium would be affected, with an increase in the level of oxyhemoglobin at a given oxygen pressure. This would occur because of a deficiency in 2,3-DPG, which lowers the oxygen affinity of hemoglobin by an allosteric mechanism. It is formed from 1,3-DPG, the product of Gly-3-(P) dehydrogenase.

$$\text{Gly-3-(P)} \longrightarrow \text{1,3-DPG} \longrightarrow \text{2,3-DPG}$$

Both of these topics are discussed further in the section on hemoglobin. Notice that the general pattern in the diagram is one in which harmful changes are brought about by H_2O_2 and other oxidizing agents, both exogenous (e.g.,nitrites) and endogenous.

These are prevented or reversed by reducing equivalents derived from carbohydrate oxidation. The reductions are mediated by pyridine nucleotides, glutathione, and specific reductase and dehydrogenase enzymes.

6. a. More than 100 variants of G-6-PD have been reported, differing with respect to the enzyme's kinetic properties, pH optimum, inhibition by NADPH, heat stability, and affinity for normal and abnormal substrates. In some of the variants, specific amino acid changes have been identified. Presumably most of the others also differ in some aspect of their amino acid sequence. Many of these variants produce only a mild enzyme deficiency in the red cells and are not associated with hemolytic anemia. Others show hemolysis only in the presence of exogenous substances which produce peroxide or function directly as oxidants. About 20 of the G-6-PD variants cause a chronic nonspherocytic hemolytic anemia even in the absence of exogenous agents.

    b. There are two clinically important variants of G-6-PD, designated GdA$^-$ and GdM. Although they are no more serious than several other G-6-PD variants, they are of significance because of the number of persons they affect. The normal phenotype is GdB.

        GdA$^-$ is found predominantly in Blacks. The frequency varies with the population, but in the United States about 11% of the males are affected and about 20% of the females are heterozygotes. GdM is found in individuals of Mediterranean and Middle Eastern extraction, from which it takes its name. The incidence is about 5-10% on the average, although one population isolate has been reported with an incidence of 50%. Another common variant among blacks is GdA$^+$. This is catalytically similar to GdB, has the electrophoretic mobility of A$^-$ (which differs from B), and is benign. GdM and GdB are electrophoretically identical.

        Clinically, GdM and GdA$^-$ patients are quite similar. They appear healthy until challenged with one of the substances listed earlier, which can precipitate a hemolytic attack. Biochemically, however, they differ with respect to the enzyme level in their erythrocytes. In GdA$^-$ patients, young red cells have nearly normal G-6-PD levels, which decrease rapidly as the cells age. Thus, older cells are more susceptible to hemolysis. In GdM patients, the G-6-PD level is quite low, even in young cells, although there is a slightly greater activity in reticulocytes. Both, then, appear to be unstable variants, with GdM being much less stable than GdA$^-$. They also differ in a number of kinetic properties.

        The red cells of affected persons are, on the average, much younger than normal, since the cells tend to lyse with a half-life considerably less than the usual 120 days. In addition, a hemolytic episode will subside as the mean age of the surviving cells decreases due to loss of the older ones.

7. G-6-PD deficiency is inherited as a sex-linked trait. It is closely linked to hemophilia and color blindness. The three possible cases of G-6-PD genotype are shown below.

$\overline{X}Y$ = full expression (where $^-$ indicates an X-chromosome carrying the defective enzyme. This is the genotype of affected males (termed hemizygous).

$\overline{XX}$ = full expression (genotype of homozygous females).

$\overline{X}X$ = female heterozygote; unaffected.

Heterozygous females (for example, carriers of GdA$^-$) have two populations of red cells: those with only GdB and those with only GdA$^-$. According to the hypothesis published by Lyon in 1962, this can be explained by the random inactivation of one of the X-chromosomes in each cell at some early stage of embryonic development. Once a cell has committed itself to leaving a particular X-chromosome active, all of its daughter cells will maintain the same one functional. Thus, a female heterozygous for any X-linked characteristic should be a mosaic with respect to that phenotype, with about half of the cells producing protein from one allele, half from the other. Experimental tests, including the G-6-PD system, have provided strong evidence that this idea is correct in most instances. The mechanism of inactivation is not known and there is some evidence that not all of the genes are inactive. The X-chromosome which is not active condenses during interphase and appears as a darkly staining ovoid near the cell membrane known as the nuclear chromatin or <u>Barr body</u>.

Note that, in the heterozygous female, the GdA$^-$ red cells are as likely to hemolyze as those of a male hemizygous for GdA$^-$. The GdA$^-$ male is more severely affected, however, because <u>all</u> of his erythrocytes are susceptible, whereas in the female, half of the cells are normal, containing GdB.

8. G-6-(P) dehydrogenase deficiency can be detected by an NADP-linked assay

Primaquine sensitivity can be detected in three ways.

a. Quantitative determination of G-6-(P) dehydrogenase in red cells.

b. GSH determination after aerobic incubation of cells with acetyl phenylhydrazine (which oxidizes GSH to GSSG).

c. Test for the reduction of methylene blue by the methemoglobin formed upon the addition of $NaNO_2$ to the red cells (qualitative procedure). In G-6-Ⓟ dehydrogenase deficiency, the methemoglobin fails to be reduced and its brown color can be recognized visually.

d. In some instances it may be difficult to detect the G-6-PD deficiency, particularly when hemolysates are used for direct quantitative enzyme assays or dye reduction tests. For example, in GdA⁻ males after a hemolytic episode, the younger cells, relatively rich in the enzyme, predominate and the value may be falsely normal or high. Heterozygous females who have two distinct cell populations, one with normal enzyme activity and the other without any activity (as described above), can present a problem in detection since there are more cells with GdB than with GdA⁻ (due to greater rate of attrition of GdA⁻ cells). In patients with anemia due to sickle cell disease, GdA⁻ may also go undetected. Sickle cell hemoglobin (see the hemoglobin chapter) also causes increasing hemolysis with cell age, biasing the population toward young cells.

9. The geographic distribution of G-6-Ⓟ dehydrogenase deficiency gene could be due to the fact that the condition may confer some protection against malaria. Malarial parasites (<u>Falciparum malaria</u>) require GSH and hence, the HMP-shunt pathway for growth. The G-6-Ⓟ dehydrogenase heterozygotes have enough GSH to carry on metabolism but not enough for the parasites, so the parasites cannot survive. This is similar to the protection conferred by the sickle cell-trait (see hemoglobin diseases).

10. In rats, selenium deficiency has been shown to enhance peroxide-induced hemolysis. This occurs because glutathione peroxidase, an enzyme responsible for destroying peroxides (Figure 42), requires selenium for activity. It has a molecular weight of 84,000 daltons and is made up of four subunits, each containing an atom of selenium. The enzyme is quite specific for GSH as the source of reducing equivalents but is able to use a wide variety of peroxides as substrates. GSH peroxidase is present in a number of other tissues as well, including the liver, kidney, endothelial lining of the blood vessels, and the lens of the eye; and is apparently important in their metabolism.

11. Another group of disorders which cause anemia due to increased hemolysis are the inherited defects in γ-glutamyl-cysteinyl synthetase and GSH synthetase, the enzymes which make glutathione.

In patients with γ-glutamyl-cysteinyl synthetase deficiency, all tissues are affected, including the erythrocytes. They have hemolytic anemia, central nervous system disease, and aminoaciduria.

The last symptom (and perhaps the CNS disorder) is consistent with the postulated role of glutathione in amino acid transport (Meister's γ-glutamyl cycle; see the chapter on amino acid metabolism).

There are two types of GSH synthetase deficiency known. In one group, many tissues are affected, producing hemolytic anemia and CNS disturbances (including mental retardation). The erythrocytes show high concentrations of many amino acids, and pyroglutamic acidemia (5-oxoprolinemia) occurs. The latter two symptoms (and perhaps the CNS disease) can again be attributed to disturbances of the γ-glutamyl cycle and amino acid transport. In the other type of GSH deficiency, only the erythrocytes are affected, producing a mild, usually sub-clinical, hemolytic anemia.

12. a. Except for G-6-PD deficiency, pyruvate kinase deficiency (already discussed under glycolysis) is the most common red blood cell enzyme defect. Reduced activities of erythrocyte hexokinase, glucose-phosphate isomerase, phosphofructokinase, aldolase, triose phosphate isomerase, glyceraldehyde phosphate dehydrogenase, phosphoglycerate kinase, lactate dehydrogenase, adenylate kinase, pyrimidine 5'-nucleotidase and catalase have all been reported, but are found only rarely. All of these disorders (with the exception of phosphoglycerate kinase deficiency which is sex linked) are inherited as autosomal recessive traits. With the exception of lactate dehydrogenase and catalase deficiencies, they all give rise to congenital nonspherocytic hemolytic anemia.

   b. Hereditary spherocytosis, which has an autosomal dominant inheritance, is characterized by abnormal red cell morphology, increased osmotic fragility, and an autohemolysis which is partly corrected by the administration of glucose. The disorder responds well to splenectomy. The specific defect is not yet defined, but a deficiency in the red cell membrane protein kinase, which catalyzes the phosphorylation of spectrin, a membrane protein, has been implicated.

13. Formaldehyde, a potent reducing agent, causes hemolysis by converting NAD^+ to NADH, thereby changing the redox state of the erythrocyte and inhibiting glycolysis at glyceraldehyde-3-phosphate dehydrogenase. An outbreak of hemolysis in a group of patients on hemodialysis was traced to the formaldehyde used as a preservative in the dialysis equipment.

Carbohydrate Metabolism and Phagocytosis

1. Phagocytosis is the internalization of particulate material by single cells. Free-living unicellular organisms (e.g., amoebae and protozoa) obtain nourishment by this process while specialized cells in higher

organisms use it to remove harmful microorganisms and debris from the circulation.

2. In man, the principal cells that perform phagocytosis are the neutrophilic polymorphonuclear leukocytes and the mononuclear phagocytes. Blood monocytes and tissue macrophages (histiocytes) comprise the latter group.

3. The initial steps in the reaction of phagocytes to an invading microorganism--migration directed by chemotaxis and ingestion of the offending cells--are discussed in detail in Chapter 11 (Immunochemistry). Both processes are accomplished by the action of microtubules and microfilaments present in the cytoplasm of the phagocytes. This is discussed at the end of this chapter.

4. Following ingestion, the pinocytic vacuole fuses with cytoplasmic granules which contain enzymes responsible for killing and digesting the vacuole's contents. This process of <u>phagosome</u> formation is also mediated by microfilaments and tubules.

5. The <u>respiratory burst</u> following phagosome formation is marked by a rise in cellular oxygen consumption. Glucose oxidation via the HMP shunt (with consequent production of NADPH) increases the release of hydrogen peroxide and superoxide. These reactions are probably part of the process by which the phagocyte destroys the harmful substance or organism. The hydrogen peroxide formation is insensitive to cyanide, indicating that the oxidase enzyme responsible for H_2O_2 synthesis does not contain heme.

6. The biochemistry of the killing process is not well understood. The three killing agents usually proposed are the compounds produced during the reduction of oxygen to water. This reaction requires four electrons, so there are three intermediates produced:

$$O_2 \xrightarrow{+e^-} O_2^- \xrightarrow{+e^-} H_2O_2 \xrightarrow{+e^-} \cdot OH \xrightarrow{+e^-} H_2O$$

The <u>superoxide anion</u> (O_2^-), hydrogen peroxide (H_2O_2), and the hydroxyl radical $(\cdot OH)$ are all highly reactive. These presumably bring about the demise of the invading organism by oxidizing some components necessary for their survival. Little is known, however, about the actual systems affected.

   a. The superoxide anion is generated by a number of biological reactions, including the autoxidation (not enzyme catalyzed) of hydroquinones, thiols, tetrahydropteridines, and hemoproteins; and the reactions catalyzed by xanthine oxidase, aldehyde oxidase, and some flavoprotein dehydrogenases. The O_2^- is produced as an intermediate in these reactions and is only incidentally released into the cell under most circumstances. If this is actually a major contributor to phagocytic destruction, then a specific, regulated production of O_2^- in white cells must be postulated.

b. Hydrogen peroxide is the normal product of many flavoenzymes, including NADPH oxidase:

$$NADPH + H^+ + O_2 \xrightarrow{\text{NADPH Oxidase}} NADP^+ + H_2O_2$$

NADH may be the electron donor, rather than NADPH. Hydrogen peroxide is further discussed in the section on red cell carbohydrate metabolism.

c. The hydroxyl radical has seldom been observed in biological systems. It may be involved in the functioning of xanthine oxidase. It can be generated as follows:

$$O_2^- + H_2O_2 \longrightarrow OH^- + \cdot OH + O_2$$

in a non-enzymatic reaction. This can occur in the cell since both O_2^- and H_2O_2 are present. The $\cdot OH$ is a much stronger oxidizing agent than either O_2^- or H_2O_2 and may be responsible for some of the destructiveness now attributed to these two compounds.

7. Superoxide can be exceedingly toxic to cells. Consequently, the enzyme <u>superoxide dismutase</u> is present in all cells which can grow in the presence of oxygen (i.e. all but obligate anaerobes). The reaction catalyzed is as follows:

$$2 H^+ + 2 O_2^- \xrightarrow{\text{Superoxide Dismutase}} H_2O_2 + O_2$$

This enzyme contains a heme group and is responsible for protecting the cell from the harmful effects of superoxide anion. The cell is in turn protected from H_2O_2 by catalase, which catalyzes the following decomposition:

$$2 H_2O_2 \xrightarrow{\text{Catalase}} 2 H_2O + O_2$$

a. There are two forms of superoxide dismutase in eukaryotic cells. The extramitochondrial form contains Cu^{+2} and Zn^{+2} while the mitochondrial form, similar to many bacterial dismutases, requires Mn^{+2}.

b. Other readily reducible compounds (glutathione, vitamin E, ascorbic acid) may also serve as protection against oxidizing agents. Peroxidases (such as glutathione peroxidase, discussed earlier in connection with the erythrocyte) also destroy H_2O_2, but use something other than another molecule of H_2O_2 as the electron acceptor (e.g., GSH in the case of GSH peroxidase).

8. Disorders of Phagocytosis

a. A number of disorders have been reported involving defects in the production and maintenance of phagocytes, locomotion and chemotaxis, ingestion and degranulation, and peroxide and superoxide production. Some of these are discussed further in the chapter on Immunochemistry.

b. Chronic granulomatous disease (CGD) is an X-linked recessive genetic disorder in which neutrophils can ingest bacteria but do not show the respiratory burst when they come into contact with a susceptible microorganism. It has been observed in one study that the neutrophils of the affected individuals do not produce superoxide. The identity of the defective enzyme(s) is still unclear, but the lesion appears to be related to the NAD(P)H oxidase systems involved in hydrogen peroxide and superoxide production.

Patients with this disorder suffer from chronic and recurrent suppurative infections, particularly due to Staphylococcus aureus and the coliform bacteria. It is of interest that those bacteria (e.g., Streptococcus haemolyticus) which do not contain catalase and therefore release H_2O_2 to the medium are killed by neutrophils from CGD patients. Furthermore, the killing of Staphylococcus aureus can be accomplished when a hydrogen peroxide generating system is introduced into the phagosome along with the bacteria. Patients with neutrophils which are deficient in glucose-6-phosphate dehydrogenase (with a consequent underproduction of NADPH) show a similar clinical picture to those with CGD. Presumably the lack of NADPH in these patients impairs the production of H_2O_2 and O_2^-.

c. Neutrophil dysfunction in one patient was associated with abnormal locomotion, ingestion, and degranulation. The rates of production of ATP, lactate, and NAD(P)H (as measured by the nitroblue tetrazolium reduction) in the neutrophils were all normal. The lesion appears to be a failure of actin polymerization. The mechanical processes of locomotion, ingestion, and the movement of granules necessary for digestion all are performed by microfilaments, located in the cytoplasm. They are composed of actin and perhaps myosin, both of which have been found in these cells. The functioning of actin and myosin in muscle cells is discussed later.

Figure 43. Uronic Acid Pathway (Glucuronate Pathway)

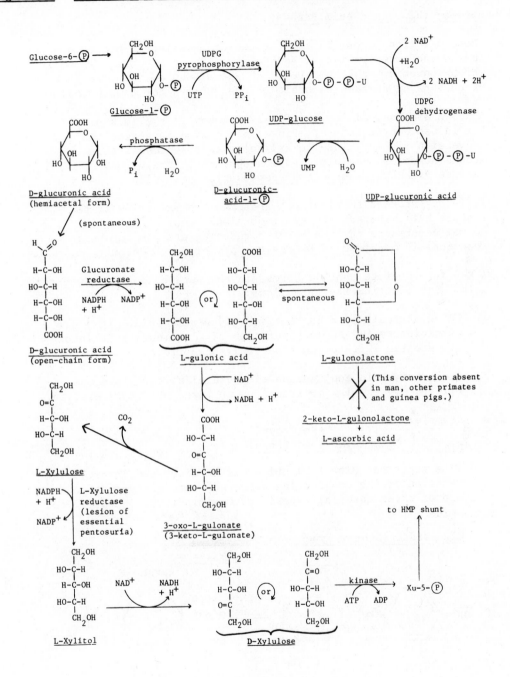

Comments on the Uronic Acid Pathway

1. These reactions occur in the tissues of animals and higher plants. In man and other primates, and guinea pigs, the enzyme is absent which converts L-gulonolactone to 2-keto-L-gulonate. The lack of this pathway makes ascorbic acid a vitamin for these species.

2. A known biochemical lesion of this pathway is the lack of L-xylulose dehydrogenase. Since this enzyme catalyzes the conversion of L-xylulose to xylitol, its absence causes an accumulation of L-xylulose, which appears in the urine. The disease is known as <u>essential pentosuria</u> (see below).

3. It is of interest to point out that the two oxidation steps require NAD^+, whereas the two reductions require NADPH. It is a normal occurrence that whenever a reducing process (anabolic reaction) is required, NADPH is the preferred cofactor.

4. A key intermediate that is formed in the pathway is UDP-glucuronic acid. Its functions are as follows:

    a. UDP-glucuronic acid is "active" glucuronic acid. It participates in the incorporation of glucuronic acid into chondroitin sulfate and other polysaccharides.

    b. Glucuronic acid conjugation with steroids, certain drugs, bilirubin, etc. is important for purposes of detoxification.

    Example:

    $$\text{Bilirubin + 2 UDPG} \xrightarrow{\text{glucuronyl transferase}} \text{Bilirubin diglucuronide + 2 UDP}$$

    (The reactions with bilirubin will be discussed later.) The reaction of the conjugating compounds with glucuronic acid are of three chemical types: glucosidic (ether), ester, and amide

Glucosidic (Ether) Type

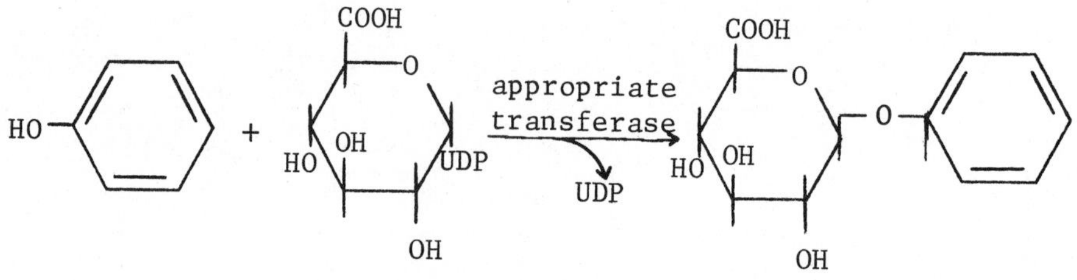

Phenol   UDP-α-glucuronate   phenyl-β-glucuronide

Ester Type (as in bilirubin conjugates)

Benzoic Acid   UDP-α-glu- β-glucuronic acid
                curonate   monobenzoate

Pentosurias

This is a group of diseases characterized by the urinary elevation of one or more of the pentoses. Normally present in the urine are L-xylulose (up to 60 mg/24 hrs), D-ribose (up to 15 mg/24 hrs), and D-ribulose (traces).

1. Essential Pentosuria is characterized by excretion of excessive amounts of the pentose L-xylulose (1-4 gm/24 hrs in the urine).

   a. No apparent disturbance in physiological function has been demonstrated (i.e., the condition is innocuous).

   b. It is often mistakenly diagnosed as diabetes mellitus (due to the presence of a reducing substance in the urine).

   c. It is a recessively inherited biochemical disorder occurring in persons of all age groups.

   d. The biochemical lesion is a deficiency of L-xylulose dehydrogenase.

   e. Diagnostic measures are indicated below. One might suspect this disease if

      i) reducing substances are present in the urine without the other symptoms of diabetes mellitus,

      ii) negative results are found when glucose is tested for by any of the enzyme methods specific for glucose.

   f. More specific methods include

      i) Reduction of Benedict's reagent at low temperature: L-xylulose is a strong reducing agent in contrast to glucose and most other urinary sugars. Benedict's reagent is reduced by L-xylulose at 55°C in 10 minutes or at room temperature in 3 hours. (Caution: Fructose will also reduce this reagent at low temperature.)

ii) Paper chromatography: Using as solvent n-butanol, ethanol, and water (50:10:40), L-xylulose has an R_f = 0.26 which exceeds all other commonly observed urinary sugars. It also gives a red color when treated with orcinol-trichloroacetic acid reagent. This is probably the most convenient method.

iii) Determination of the melting point of the phenylosazone derivative of the sugar.

2. Other types of pentosuria include alimentary pentosuria (L-arabinose and L-xylose are found in the urine following ingestion of large quantities of fruit) and ribosuria (elevation of urinary D-ribose levels associated with muscular dystrophy).

Fructose Metabolism

1. Fructose (a ketohexose) is widely distributed in plants and is an important source of dietary carbohydrate, accounting for 1/6 to 1/3 of the total carbohydrate intake. It is found as free fructose and, linked to glucose, occurs as sucrose. Inulin, a polymer of fructose, is present in chicory and sweet potato. Since inulin is relatively resistant to hydrolysis in the intestine, this is an unimportant source of fructose.

2. Fructose is present in significant quantities in prostate and seminal fluid. It is presumably synthesized in the prostate gland by way of the following reactions:

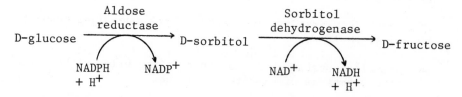

For other aspects of this conversion, see galactose metabolism and diabetes.

3. For the detection of fructose, glucose is first removed with the enzyme glucose oxidase. Fructose is then determined either by chromatographic or colorimetric methods.

4. Fructose can enter the Embden-Meyerhof Pathway by several routes. Notice in particular that fructokinase forms the 1-Ⓟ, unlike hexokinase and glucokinase.

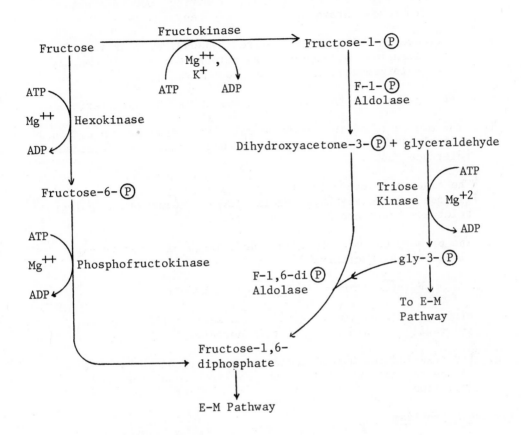

5. In essential fructosuria

    a. Fructose is found in the urine, due to a lack of the enzyme fructokinase.

    b. The condition is harmless and asymptomatic.

    c. Glucose and galactose metabolism are normal.

    d. Insulin has no effect on fructose levels.

    e. The disorder is inherited as an autosomal recessive trait.

6. Hereditary fructose intolerance

    a. Occurs in infants during the weaning period. Symptoms appear
        only after ingestion of fructose. Vomiting occurs shortly
        after fructose-containing foods are eaten.

    b. Occurrence in children gives rise to jaundice, vomiting,
        hyperbilirubinemia, albuminuria, aminoaciduria, hepatomegaly
        (liver enlargement), and failure to thrive.

    c. A strong aversion for fruits and sweets is characteristic.

    d. Fructose is found in the urine only after fructose ingestion.
        High blood fructose levels are reached during oral fructose
        tolerance tests.

    e. The rise in fructose level is accompanied by a fall of blood
        glucose to severely hypoglycemic levels. Glucose and galactose
        tolerance tests are normal, however.

    f. The primary enzyme defect is a deficiency of liver F-1-$\circled{P}$
        aldolase, which catalyzes the reaction

        $$F\text{-}1\text{-}\circled{P} \longrightarrow D\text{-glyceraldehyde} + DHAP$$

        Although this enzyme is virtually absent in the disorder,
        F-1,6-di $\circled{P}$ aldolase activity is normal.

    g. Fructose-induced hypoglycemia is probably caused by a block
        in the release of glucose from the liver resulting in ATP
        depletion.

Galactose Metabolism

1. The principle dietary source of galactose is the disaccharide lactose
    (β-D-galactopyranosyl-(1$\rightarrow$4)-β-D-glucopyranose).

    This molecule is hydrolyzed to galactose and glucose by the enzyme
    lactase, found in the intestinal microvilli.

    $$lactose \xrightarrow[\;H_2O\;]{lactase} galactose + glucose$$

    The metabolic role of lactase has already been discussed.

2. a. After the release of galactose from lactose, the following
        reactions provide for its entry into glycolysis.

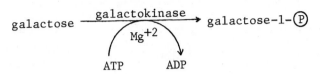

$$\text{galactose} \xrightarrow[\text{Mg}^{+2}]{\text{galactokinase}} \text{galactose-1-} \textcircled{P}$$

$$\text{ATP} \qquad \text{ADP}$$

$$\begin{array}{c} \text{Galactose-1-} \textcircled{P} \\ \text{uridyl} \\ \text{galactose-1-} \textcircled{P} + \text{UDP-glucose} \xrightarrow{\text{transferase}} \text{UDP-galactose} + \text{glucose-1-} \textcircled{P} \\ \text{(block in} \\ \text{galactosemia)} \end{array}$$

$$\begin{array}{c} \text{UDP-Galactose-} \\ \text{4-Epimerase} \\ \text{UDP-galactose} \xleftrightarrow{\hspace{2cm}} \text{UDP-glucose} \longrightarrow \text{glycogen} \longleftrightarrow \text{glycolysis} \end{array}$$

    b. The mechanism of galactose-1-phosphate uridyltransferase is
        an example of a double-displacement or ping-pong reaction.
        The overall reaction is shown above. It proceeds by the
        following steps:

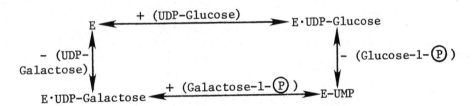

        Note that all of the above reactions are reversible.

3. Galactosemia is the inability to metabolize galactose.

    a. It is a rare congenital disease of infants inherited as an
        autosomal recessive trait and varies tremendously in clinical
        severity.

    b. The biochemical lesion is a lack of galactose-1-$\textcircled{P}$ uridyl
        transferase. This results in the accumulation of galactose-1-$\textcircled{P}$.
        In one study, evidence has been presented to show that the lack
        of enzyme activity in galactosemics is due to a structural gene
        mutation (see protein synthesis for further information).

c. Clinical manifestations include nutritional failure (infants lose weight), hepatosplenomegaly, mental retardation, jaundice, and the development of cataracts. Since the lens tissue is freely permeable to galactose but relatively impermeable to galactitol (the polyalcohol derived from this sugar), the last symptom may be due to the formation of galactitol by the reaction below.

D-galactose   D-galactitol

(See also fructose metabolism and diabetes). The accumulation of the alcohol causes an increase in the osmolarity and water uptake of the aqueous humor, resulting in swelling of the lens. Myopia and tissue damage occur, leading ultimately to the opacity associated with cataracts. It is not clear whether the other symptoms also result from galactitol accumulation, from galactose-1-(P) accumulation, or from some other aspect of the impaired galactose metabolism.

d. Laboratory findings in galactosemia include galactosuria, amino aciduria albuminuria, ketonuria, and impaired galactose tolerance (abnormal galactose tolerance test).

e. A good diagnostic test for this disorder is the measurement of galactose-1-(P) uridyl transferase enzyme activity in red cells. If the diagnosis is made early and milk (lactose) and other galactose sources are withdrawn from the diet, the symptoms and findings recede and normal development appears to ensue. Even when early postnatal diagnosis is made, however, there may be some prenatal damage due to exposure of the fetus to galactose ingested by the mother.

4. Galactokinase Deficiency. Hypergalactosemia and galactosuria have been observed in a few individuals due to the deficiency of galactokinase, the enzyme which catalyzes the reaction

$$\text{Galactose} + \text{ATP} \longrightarrow \text{Galactose-1-}\textcircled{P} + \text{ADP}$$

Unlike galactosemic patients, the individuals with galactokinase deficiency do not develop hepatic and renal problems. Instead, they showed the formation of cataracts at an early age (less than one year after birth) possibly due to the mechanism discussed under galactosemia. Increased incidence of cataracts was also observed in heterozygous individuals, who also exhibited a mild form of galactose intolerance.

5. Galactose is necessary for the synthesis of a number of biomolecules. These include lactose (in the lactating mammary gland; discussed below), glycolipids, cerebrosides, chondromucoids, and mucoproteins. A galactosemic individual is able to synthesize these compounds, even on a galactose-free diet, because of the reversibility of the UDP-galactose-4-epimerase reaction. The UDP-glucose needed for this pathway is readily synthesized from glucose-1-$\textcircled{P}$ and UTP in the reaction catalyzed by UDP-glucose-pyrophosphorylase.

6. Lactose (milk sugar) is synthesized by the following reaction

$$\text{UDP-galactose} + \text{glucose} \xrightarrow[\text{(+}\alpha\text{-lactalbumin)}]{\substack{\text{galactosyl} \\ \text{transferase}}} \text{lactose} + \text{UDP}$$

Galactosyl transferase (also called lactose synthetase) requires alpha-lactalbumin, a protein found in milk, as a cofactor. In the absence of α-lactalbumin, galactosyl transferase catalyzes the reaction

$$\text{UDP-galactose} + \text{N-acetylglucosamine} \xrightarrow{\substack{\text{galactosyl} \\ \text{transferase}}} \text{N-acetyllactosamine} + \text{UDP}.$$

Alpha-lactalbumin synthesis is initiated late in pregnancy by a decrease in the progesterone level.

Metabolism of Amino Sugars

1. These compounds are important components in many complex
    polysaccharides. They are frequently found in glyco- and
    mucoproteins and lipopolysaccharides. The structures of two
    representative compounds are given below.

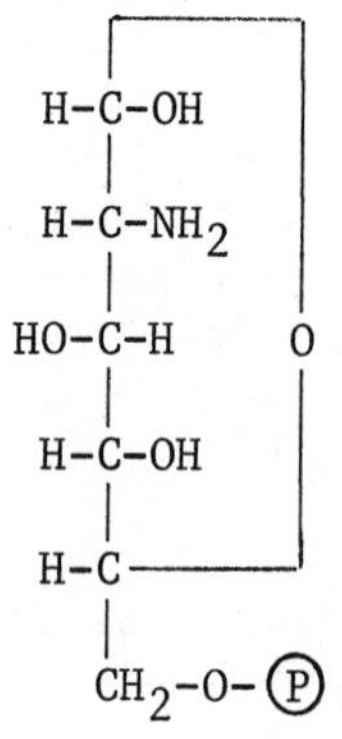

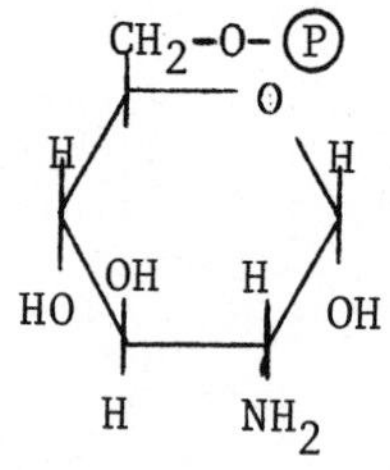

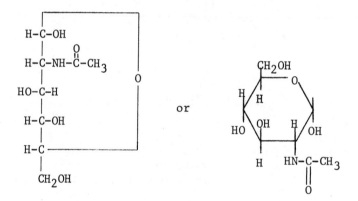

D-glucosamine-6-(P)

N-acetylglucosamine

2. Some of the pathways which involve these amino sugars are shown
    in Figure 44.

Figure 44. Metabolism of Amino Sugars

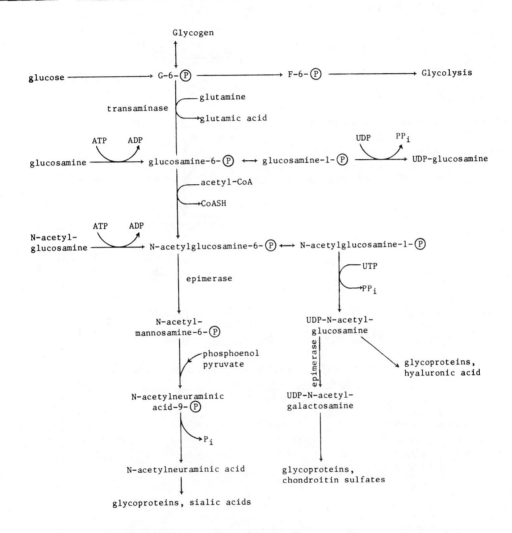

Blood Glucose

1. At any given time, the blood glucose level is dependent upon the
    amount of glucose <u>entering</u> the blood from dietary sources,
    glycogenolysis and gluconeogenesis and the amount that is being
    <u>removed</u> by oxidative and biosynthetic processes. These are
    illustrated in the following diagram.

<u>Figure 45</u>.   Factors Which Affect Blood Glucose Levels

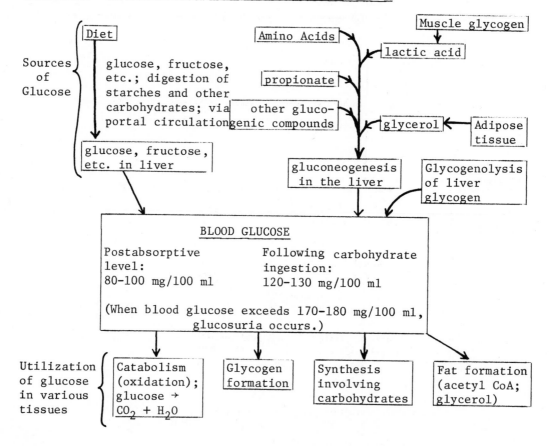

2. The addition of glucose to the blood stream through hepatic
    glycogenolysis can make only a minor contribution to the blood
    glucose level. This is due to the limited supply of glycogen.
    A liver weighing about 1500 grams can provide only about 45-50 grams
    (3-4%) of glucose. This is probably most useful for short periods

of time in emergency situations. Note that epinephrine and glucagon stimulate hepatic glycogenolysis (previously discussed under glycogen). In conditions where there is sustained lack of glucose, however, (e.g., starvation) gluconeogenesis plays a principal role in providing blood glucose. The muscle glycogen cannot furnish glucose to be added to the blood stream. This is due to the absence in muscle tissue of the enzyme glucose-6-phosphatase, which catalyzes the cleavage of G-6-(P) to free glucose and phosphate. The muscle also lacks some of the transaminases needed to convert amino acids to keto acids which then can be used for gluconeogenesis.

3. In the resting body, the brain consumes about two-thirds of the blood glucose. Most of the other one-third is used by the erythrocytes and skeletal muscle. The brain <u>must</u> have an energy source or behavioral changes, confusion, and coma result. In prolonged starvation, the body meets this need in two stages. Initially, muscle protein is broken down and the amino acids are used to synthesize glucose (gluconeogenesis). In particular, alanine is released from the muscle protein and carried in the blood to the liver and kidney where gluconeogenesis occurs. This is known as the <u>alanine cycle</u> and is similar to the Cori cycle for lactate. Both pathways simply recycle a fixed amount of glucose.

The following points need to be emphasized with regard to the alanine cycle and skeletal muscle proteolysis:

a. Alanine (and to a lesser extent other amino acids) is mobilized from the body protein stores particularly during an insult such as starvation, infection, trauma, thermal injury, peritonitis, thyrotoxicosis, prolonged exercise, cold exposure, etc. In the liver, the carbon fragments are converted to glucose (gluconeogenesis). The NH_2 groups are converted to urea (ureagenesis), which is excreted in increased quantities in the urine.

b. These effects are initiated by epinephrine (released from the adrenal medulla-sympathoadrenal discharge) and adrenergic activity and augmented by the action of glucocorticoids, glucagon and growth hormone. The release of excessive amounts of these hormones leads to a hypermetabolic state (see next section). Furthermore, epinephrine and norepinephrine (through their α-receptor effects) suppresses the release of insulin. It should be noted that the β-adrenergic receptor structure augments insulin elaboration; however, during the early phase of injury, the α-receptor effects appear to predominate (discussed later in this section; see Chapter 6

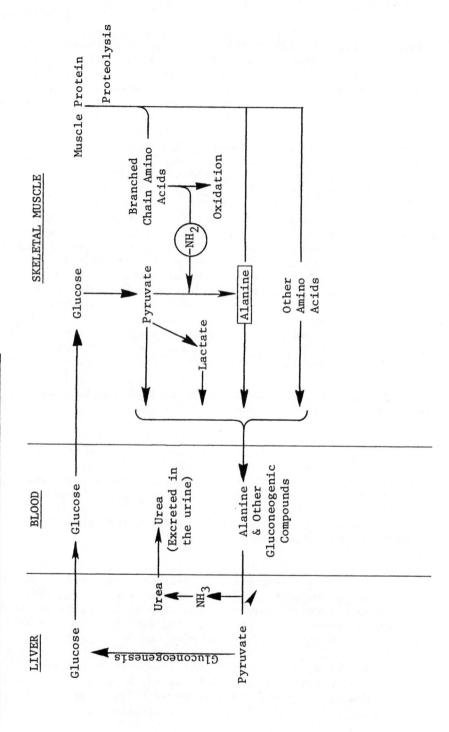

Figure 46. The Alanine Cycle and Muscle Proteolysis

for a discussion on α- and β-receptors of catecholamines).
Thus the net effect is hyperglycemia with increased nitrogen
loss following injury (negative nitrogen balance).

c. Hyperglycemia is maintained primarily at the expense of body
    protein. Recall that liver glycogen stores are limited, and
    fatty acids cannot be converted to glucose.

d. In the nutritional management of the hypermetabolic state,
    it is essential to provide sufficient amounts of carbohydrates,
    proteins and lipids to meet the increased caloric requirements
    as well as to minimize protein loss. The catabolism of
    branched chain amino acids (all of which are essential) is
    increased in skeletal muscle and must be supplemented in
    addition to the other amino acids (in the diet or intravenous
    administration) in order to revert negative nitrogen balance.
    Appropriate maintenance of fluid and electrolyte balance is also
    essential (see Chapter 12).

This is a heavy drain on body protein, however, and if the glucose
supply remains low, the brain adapts to the use of ketone bodies
(acetoacetic acid, acetone, and β-hydroxybutyric acid) as a source
of energy. The most important of these three are acetoacetic acid
and β-hydroxybutyric acid. The synthesis of these compounds from
acetyl-CoA is discussed in the section on ketosis in lipid
metabolism. Normally the body produces only small amounts of the
ketone bodies. In the absence of glycolysis, acetyl-CoA is
primarily derived from fats, however, so ketone body synthesis
directly relieves the need to catabolize muscle protein for energy,
drawing upon the adipose tissue instead. It has been shown that
the brain possesses the enzymes necessary for utilizing this
new energy source. The mechanism by which the body adapts to
ketone body synthesis is not yet known, but it is probable that
hormones (insulin: lower than normal during starvation; glucagon:
higher than normal, relative to insulin, during starvation) are
involved.

4. A number of hormones influence the blood glucose levels. Briefly,
    epinephrine, glucagon, ACTH, cortisol, and growth hormone tend to
    raise blood glucose levels (hyperglycemic substances) while insulin
    lowers the levels (hypoglycemic hormone). These are discussed
    more fully in the section on diabetes.

Relations of the Endocrine Glands to Carbohydrate Metabolism

1. The major control of carbohydrate metabolism is affected through
    hormones excreted by tissues, especially the pancreas, anterior
    pituitary (adenohypophysis), and adrenal cortex.

2. Each endocrine organ generally secretes more than one hormone and these often exert apparently opposing effects.

3. The adenohypophysis produces not only hormones having a direct effect, but also, through tropic hormones, regulates the other endocrine organs (see Chapter 13).

```
                     ,hormones → direct effect on carbohydrate metabolism
   Adenohypophysis
                     `tropic hormones
                              ↘
                          other endocrine organs
                                 ↘
                              hormone
                                  ↘
                              effect on carbohydrate metabolism
```

Some authors use the term trophic (nourishing; functioning in nutrition) rather than tropic (bringing about a change) for those hormones which regulate the levels of other hormones. Despite their rather different roots, the terms have become interchangeable in this context. Because of the interrelations of pathways (fat, protein, carbohydrate), it is difficult to distinguish between primary and secondary hormonal effects.

Synthesis and Secretion of Insulin

1. Evidence has been presented to show that insulin is synthesized as a single polypeptide chain by the ribosomes of the rough endoplasmic reticulum in the classical protein-synthesizing system. The peptide chain then folds on itself making the proper -S-S- bridges to yield proinsulin. Proinsulin is activated by removal of the C-peptide, containing 33 amino acid residues. This is indicated in Figure 47. Notice the presence of three disulfide bonds (two interchain, one intrachain) in both insulin and proinsulin. The amino acid sequence is due to Sanger and his coworkers. For his central role in the work, Sanger was awarded a Nobel Prize in 1957.

2. a. Proinsulin is put into storage granules (β-granules) in the Golgi apparatus. The transformation of proinsulin to insulin, involving proteolysis, takes place presumably during the granule formation and maturation.

Proinsulin (Molecular Weight About 9,000) $\xrightarrow{\text{Trypsin and/or converting enzyme}}$ Insulin (Molecular Weight About 6,000) $+$ C-Peptide (Molecular Weight About 3,000)

Insulin (and some unactivated proinsulin) stored in the granules is secreted under the influence of a suitable stimulus (see below). The granules move toward the cell's plasma membrane, the granule membranes fuse with the plasma membrane, and the contents of the granules are emptied into the blood stream. This process is called exocytosis or emiocytosis and involves no loss of cytoplasm from the cell. It is probably mediated by cAMP. Proinsulin has significantly less biological activity than insulin as measured using several different assay procedures (see measurement of insulin levels).

Physiologically, however, this may not always be true. In one family, eighteen out of fifty individuals were asymptomatic (i.e., clinically normal) yet had proinsulin as the major portion of the insulin-immunoreactive material in their circulation. The inheritance of this hyperproinsulinemia is autosomal dominant. The defect is either in the amino acid sequence of the insulin (making it resistant to the proinsulin converting protease) or in the protease itself. The fact that there is some insulin present in these patients suggest that the protease is normal but that the insulin has become a poorer substrate due (presumably) to an amino acid change.

b. In cell-free mRNA-directed protein synthesizing systems, using proinsulin mRNA to direct the synthesis, a polypeptide larger than proinsulin has been obtained. This precursor, named preproinsulin, has an additional 23 amino acid residues attached to the amino-terminus of the proinsulin molecule. This N-terminal extension contains predominantly amino acids with hydrophobic side groups and is the first part of the polypeptide to be synthesized. As discussed in Chapters 2 and 5, this piece probably binds the insulin polysome to the endoplasmic reticulum and causes the proinsulin molecule to enter the cisternae to be packaged for secretion. Preproinsulin has not been detected in intact β-cells or in the circulation because of proteolytic removal of the hydrophobic "signal peptide" during processing prior to sequestration in the secretory granules.

Figure 47. Conversion of Proinsulin to Insulin*

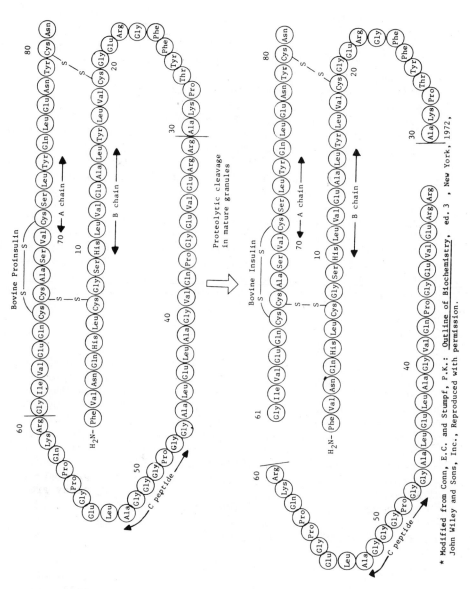

* Modified from Conn, E.C. and Stumpf, P.K.: Outline of Biochemistry, ed. 3 , New York, 1972, John Wiley and Sons, Inc., Reproduced with permission.

c. It is of interest to point out that the synthesis of proinsulin as a precursor to insulin probably occurs to insure the proper folding of the peptide and correct formation of the disulfide bonds. As was pointed out earlier, if there are more than two cysteine residues in a polypeptide (and, in this case, more than one chain, in the active form), the disulfide bonds can form in more than one way, leading to one or more types of incorrect (inactive) molecules. Even in the proinsulin, it is not clear whether the proper disulfide links form in response to a preferred conformation or if there is some more complicated mechanism which specifies the proper alignment. Isolated A and B chains, prepared by chemical synthesis or by reduction of the disulfide bonds in purified native insulin, when subjected to oxidizing conditions, recombine in several inactive forms despite their having the correct amino acid sequence.

3. Insulins isolated from different species show minor variations in amino acid composition and sequence. Although this does not significantly alter their biological activity, prolonged administration of insulin leads to the production of anti-insulin antibodies if the donor and recipient are of different species. These antibodies <u>do</u> interfere with the effectiveness of the insulin. In diabetics who still retain some β-cell function, the antibodies have been shown to react with the patient's own proinsulin and insulin, thereby further reducing the amount of available hormone.

4. The fate of the C-peptide, which is biologically inactive, is unclear. It is secreted by the pancreas in a 1:1 ratio with insulin and is present in the circulation. Recently, radioimmunoassay measurements (see measurement of cAMP and insulin) of C-peptide immunoreactivity (i.e., amount of material which reacts with antibodies against C-peptide) have been used to evaluate pancreatic β-cell function in patients with insulin-treated diabetes. Direct determination of insulin levels is hampered by the presence of antibodies synthesized by the patient in response to the exogenous insulin used in the treatment. Both free C-peptide and proinsulin (which still contains the C-peptide region) are measured in these assays. The contribution of each of these materials to the total immunoreactivity can be determined by performing a preliminary separation by gel filtration. The results of this study are discussed in the section on diabetes.

5. The predominant physiologic stimulator of insulin secretion is glucose. The glucose level sensing device of the β-cells is not well understood. It has been suggested that this process is mediated through the formation of cyclic AMP. Note, however, that insulin <u>reduces</u> the cyclic AMP level in some other cells (e.g., adipose tissue cells) where insulin-mediated processes take place.

Other agents which stimulate the insulin release are the amino acids (leu, arg, lys, and phe), glucagon (a hyperglycemic hormone produced by the α-cells of pancreas which functions by increasing hepatic glycogenolysis; it also has a direct effect on the β-cells); secretin and pancreozymin (hormones which activate the exocrine function of the pancreas); corticotropin; growth hormone; sulfonyl urea compounds such as tolbutamide, used in the treatment of maturity onset diabetes; theophylline (an inhibitor of phosphodiesterase, the enzyme which inactivates cAMP); and stimulation of the right vagal nerve. Examples of <u>inhibitors</u> of insulin secretion include: α-adrenergic stimulating agents (epinephrine and norepinephrine; epinephrine also causes hepatic and muscle glycogenolysis); and β-adrenergic blocking agents (thiazide diuretics, 2-deoxyglucose).

Metabolic Effects of Insulin

1. Insulin is required for the entry of glucose into muscle (skeletal, cardiac, and smooth), adipose tissue, leukocytes, the mammary gland, and the pituitary. Tissues which are freely permeable to glucose include brain, intestinal mucosa, lens, retina, nerve, kidney, erythrocytes, blood vessels, and islet cells. Although the liver is not dependent on insulin for glucose uptake, the hormone still influences this organ's glucose metabolism by stimulating the liver glucokinase activity.

2. Insulin increases glycogen formation and oxidation of carbohydrates in muscle and liver. It suppresses the principal gluconeogenic enzymes: pyruvate carboxylase, PEP carboxykinase, fructose-1,6-diphosphatase and G-6-phosphatase, thus reducing the formation of sugar. It enhances the utilization of glucose by inducing glucokinase (in the liver), phosphofructokinase, and pyruvate kinase. Insulin also stimulates the HMP-shunt pathway and lipogenesis (discussed under lipid metabolism) in the adipose tissue.

3. Insulin has an anabolic effect and affects growth. This is due to

    a. its amino acid sparing action (glucose can be used for energy rather than amino acids),

    b. enhanced transportation of amino acids into cells and their incorporation into proteins. Growth is a complex process and is determined by a number of genetic, nutritional and hormonal growth hormone (thyroxine, insulin and androgens) factors.

4. It is important to note that insulin has none of these effects in

experimental cell-free extracts. This (and other evidence)
suggests that insulin exerts its primary effect at the cell membrane.

Measurement of Insulin Levels

1. A number of bioassays have been developed for the measurement of
    insulin. These are based on the occurrence or magnitude of a
    physiological response evoked by administration of an insulin-
    containing sample to an intact animal or a piece of isolated tissue.
    Such methods include

    a. hypoglycemic response in rabbits;

    b. mouse convulsive assay;

    c. glucose uptake by rat epididymal fat pad;

    d. glucose uptake by isolated rat diaphragm.

    Bioassay procedures are frequently criticized for being inaccurate,
    insensitive, nonspecific, and time consuming. Few studies have
    been published, however, which specifically compare their results
    to those of radioimmunoassays and other new techniques.

2. The measurement of insulin by radioimmunoassay (RIA) was published
    in the late 1950's by Berson and Yallow. It was the first example
    of the method and because of its high sensitivity and specificity,
    it made possible a wide range of new studies and measurements.
    As of 1973, at least 53 different substances could be measured by
    kits which are available commercially. Other methods for
    additional materials are currently being used in research. A
    discussion of this method and the related competitive protein-
    binding technique are presented in the section on cyclic AMP.

    One difficulty in the use of immunologic methods for the
    determination of insulin and other peptides is the partial species-
    specificity in their amino acid sequence. Antibodies produced in
    response to insulin of one species may bind insulin from another
    species to a lesser extent. Thus, if porcine insulin is used to
    prepare antibodies which are then employed to measure human insulin
    levels, care must be taken to standardize the assay against human
    insulin.

    Another problem encountered specifically with peptide hormone RIA's
    is the purification and labeling of the radioactive hormone.
    Hormones are only transiently present in biological fluids and then
    usually at low concentrations; and they are not readily synthesized

by chemical means. These problems are frequently of lesser magnitude with other molecules.

In molecules isolated from living systems, radioactive labeling must usually be done after purification. One method which has found wide application is the iodination of tyrosyl residues in the molecule with ^{131}I (or ^{125}I) in the following reaction:

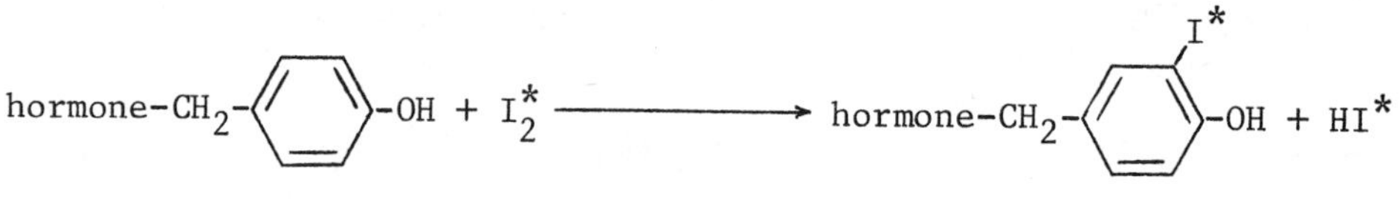

$$\text{hormone-CH}_2\text{-}\langle\bigcirc\rangle\text{-OH} + I_2^* \longrightarrow \text{hormone-CH}_2\text{-}\langle\bigcirc\rangle\text{-OH} + HI^*$$

where * indicates radioactivity. This procedure depends on the presence of a tyrosyl residue in the peptide (phenylalanine will not work). In some non-peptide RIA's, tyrosyl groups are coupled to the molecule which the assay is intended to measure, to provide a site for labeling.

In the insulin assay, the presence of proinsulin complicates matters (as it does in the C-peptide radioimmunoassay). It cross-reacts with the insulin antibodies and is also a contaminant in the insulin used to prepare the antibodies. The assay may also fail to measure all of the physiologically active insulin in the body fluids.

Glucagon

1. This is a peptide hormone, containing 29 amino acid residues. It is synthesized by the α-cells of the pancreas and secreted into the portal blood flow to the liver. Glucagon, along with insulin, is a major hormone in the regulation of carbohydrate metabolism. While insulin tends to lower blood glucose, glucagon is a hyperglycemic factor which works in a number of ways to raise the blood glucose concentration.

2. Glucagon stimulates hepatic glycogenolysis (mechanism already discussed) and gluconeogenesis; lipolysis in the liver and adipose tissue; and the release of insulin, growth hormone, thyrocalcitonin, and catecholamines. It decreases gastrointestinal motility and pancreatic exocrine secretion, produces an increase in heart rate and inotropism, and promotes the excretion of sodium and calcium in the urine.

3. A glucagon-like substance which cross-reacts with antibodies produced against pancreatic glucagon has been identified in the mucosal lining of the gut. This material, as well as other gastrointestinal tract hormones, stimulate the release of insulin.

It has been postulated that this insulin release prepares the liver cells for the proper utilization of the glucose, amino acids, etc., which are transported from the GI tract to the liver by the portal circulation following a meal.

4. a. Glucagon <u>deficiency</u> disorders such as idiopathic α-cell deficiency and chronic pancreatitis give rise to fasting <u>hypoglycemia</u>.

    b. Glucagon <u>excess</u> leads to <u>hyperglycemia</u>. It can be caused by: glucagonoma, multiple endocrine adenomatosis, ectopic glucagon production, and stress (as in starvation, trauma, uncontrolled diabetes). The role of glucagon in diabetes is discussed under diabetes mellitus.

5. Therapeutically, commercially available glucagon has been used in the treatment of hypoglycemia. Glucagon administration has been employed in the diagnosis of insulinoma (causing a <u>drop</u> in blood glucose in patients with insulin-producing tumors); growth hormone deficiency (pituitary dwarfs will fail to show an increase in growth hormone in response to glucagon); pheochromacytoma (stored catecholamines are released when glucagon is administered, leading to a brisk rise in blood pressure, etc.); and some types of glycogen storage disease (there is a lack of the expected rise in blood glucose if the disease affects liver; this is discussed under glycogen storage disorders).

Hypoglycemic Disorders

When the blood sugar drops below 40 mg/dl of whole blood (or below 45-50 mg/dl of serum or plasma), there is an acute release of epinephrine which leads to tachycardia, tremulousness, sweating, weakness, anxiety, nervousness, and hunger. Additional signs may include visual disturbances, faintness, mental confusion, and the inability to concentrate. This <u>hypoglycemia</u> is self-correcting (by glycogenolysis, for example) or if not properly managed, may lead to coma. There are two basic types of hypoglycemia, one occurring in the fed state and the other in the fasting state.

Hypoglycemia in the Fed State

1. From the pancreas, insulin reaches the liver via the portal circulation sooner and in greater quantities than it does any other organ in response to feeding. This primes the hepatic tissue for alimentary glucose utilization (glycogen synthesis) and shuts off hepatic glucose production. When the surge of dietary glucose passes, blood glucose levels return to normal or below normal, insulin secretion ceases and circulating insulin decreases. The liver then resumes glucose synthesis and release.

If however, there is a delay in the resumption of hepatic glucose output, hypoglycemia can occur.

2. Causes of this type of hypoglycemia include:

a. Excessive insulin release

    i. <u>Reactive hypoglycemia</u> can occur in mild, maturity-onset diabetics in whom the initial release of insulin is delayed. Hyperglycemia results and when insulin does come, an excessive amount is released, as the secretory response of the pancreas overshoots. Liver glycogen deposits are also reduced in these patients.

    ii. <u>Alimentary hyperinsulinism</u> is observed in patients with gastrectomy or surgical intestinal bypass. These individuals absorb glucose more rapidly than normal because the carbohydrate is made available more rapidly to the absorptive sites of the small intestine. This can lead to greater than normal insulin secretion. The same effect is seen in persons suddenly removed from oral or parenteral glucose infusion.

    iii. <u>Leucine hypersensitivity</u> of infancy and childhood: Leucine in the diet stimulates the β-cells to secrete insulin. In affected infants the insulin release is excessive resulting in hypoglycemia.

b. Inherited enzyme defects of the liver leading to a lack of production of glucose

    i. Galactose-1-phosphate uridyl transferase deficiency (galactosemia)

    ii. Fructose-1-phosphate aldolase deficiency (hereditary fructose intolerance)

These have both already been discussed.

c. Functional hypoglycemia: Although two-thirds of the cases of hypoglycemia come under this category, the biochemical lesion(s) are not known. It occurs most frequently in young women who are thin, tense, and hyperkinetic; in individuals with vagotonia (hyperirritability of the parasympathetic nervous system); and in individuals with compulsive, conscientious, and intense personality trends.

Fasting Hypoglycemia

This disorder reflects an underproduction of glucose and is associated with many types of pathologic conditions (see Table 12). As it has already been discussed, fasting blood glucose (4 to 8 hours postprandially) is maintained principally by hepatic gluconeogenesis and to a lesser degree by glycogenolysis. Various defects in these processes that lead to hypoglycemia are listed in Table 13. In this table, NSILA-s stands for non-suppressible insulin-like activity soluble in acid alcohol. This material is a polypeptide, synthesized and released by non-pancreatic tumors. It does not cross-react with antibodies produced against insulin. NSILA-s behaves like insulin, however, in that it binds to insulin receptors, making it measurable by the insulin radioreceptor assay.

A common cause of fasting hypoglycemia in adults is the consumption of alcohol in the fasting state. As we will see later, a mole of ethanol oxidation gives rise to the reduction of two moles of NAD^+ to NADH, thereby diverting NAD^+ from glycolytic and citric acid cycle reactions. This leads to a lack of production of substrates required for gluconeogenesis.

Hypoglycin A, the toxic principle of the unripe akee fruit, produces severe hypoglycemia when ingested. Another less toxic compound found in the akee fruit, known as hypoglycin B, is a γ-glutamyl conjugate of hypoglycin A. The akee tree (indigenous to West Africa) grows in Jamaica, other islands of the Antilles, the southern part of Florida and Central America. Hypoglycin also produces vomiting (therefore known as vomiting sickness) and central nervous system depression. In Jamaica, the vomiting sickness apparently occurs in epidemic proportions. This is due to consumption of akee fruit during the colder months of the year when the akees are still unripe and the availability of other food sources is limited. It should be noted,

Table 12. Classification of Fasting Hypoglycemias*

A. Organic hypoglycemia: recognizable anatomic lesion:
  1. Pancreatic islet beta-cell disease with hyperinsulinism in the
     adult:
     (a) Adenoma, single or multiple
     (b) Microadenomatosis, with or without macroscopic adenoma
     (c) Carcinoma, with metastases
     (d) Adenoma(s), microadenomatosis or carcinoma, associated with
         adenomas or hyperplasia of other endocrine gland (MEA)
     (e) Hyperplasia (very rare in adults)
     In infancy and childhood:
         (a) Hyperplasia (leucine-sensitive or -insensitive)
         (b) Nesidioblastosis
         (c) Adenoma
  2. Nonpancreatic tumors associated with hypoglycemia
  3. Anterior pituitary hypofunction
  4. Adrenocortical hypofunction
  5. Acquired diffuse hepatic disease
  6. Severe congestive heart failure
  7. Severe renal insufficiency in non-insulin-dependent diabetes
B. Hypoglycemia due to specific hepatic enzyme deficiency
   (infancy & childhood)
C. Functional hypoglycemia: no recognizable or no persistent
   anatomic lesion:
  1. Deficiency of glucagon
  2. Severe inanition
  3. Ketotic hypoglycemia (childhood)
  4. "Insulin autoimmune syndrome" (no previous insulin
     administration) & hyperinsulinemia
D. Hypoglycemia induced by exogenous agents:
  1. Ethanol (& poor nutrition)
  2. Insulin administration   ⎫
  3. Sulfonylurea administration ⎬ iatrogenic or factitious
  4. Ingestion of ackee fruit (hypoglycin)
  5. Miscellaneous drugs

*Reprinted from the publication, Fasting Hypoglycemia in Adults, with
 the permission of the New England Journal of Medicine (Vol. 294,
 Page 768, 1976) and the authors, S.S. Fajans and J.C. Floyd.

Table 13. Derangements in Mechanisms of Glucose Homeostasis Leading to Fasting Hypoglycemia*

DECREASED GLUCOSE OUTPUT

I. Decreased hepatic gluconeogenesis:
  A. Decrease in supply of substrate for gluconeogenesis:
    1. Decreased mobilization of certain amino acids from muscle
      (a) hormonal excess--insulin, NSILA-s (nonpancreatic tumors), other polypeptides (nonpancreatic tumors)
      (b) hormonal deficiency--cortisol, epinephrine
      (c) idiopathic (ketotic hypoglycemia)
      (d) depletion of muscle stores
    2. Increased uptake of amino acids by extrahepatic tissues
    3. Decrease in lactate & pyruvate originating from formed blood cells, renal medulla, nervous tissue, etc.
      (a) Hormonal deficiency--growth hormone, cortisol, epinephrine
    4. Decreased mobilization of glycerol (& free fatty acids for energy requirements) from adipose tissue:
      (a) Hormonal excess--insulin, NSILA-s
      (b) Hormonal deficiency--glucagon, growth hormone, cortisol, ACTH, epinephrine
      (c) Depletion of adipose stores
  B. Decreased hepatic uptake of substrates:
      (a) Hormonal excess--insulin, NSILA-s
      (b) Hormonal deficiency--cortisol, growth hormone, epinephrine
  C. Deficiency in activity of rate-limiting hepatic enzymes (change in enzyme activation or synthesis--or both) involved in disposition of extracted substrate:
      (a) Hormonal excess--insulin, NSILA-s
      (b) Hormonal deficiency--glucagon, epinephrine, cortisol
      (c) Delay in maturation of enzyme synthesis (newborn of low birth weight)
      (d) Genetic enzyme deficiency (infancy & childhood)
  D. Decrease in substrate oxidation:
    1. Increase in $NADH_2$:NAD (alcohol)
    2. Decreased oxidation of long-chained fatty acids
  E. Extensive, diffuse hepatocellular damage
  F. Interference with hepatic blood flow
II. Decreased hepatic glycogenolysis:
  A. Deficiency in activity of hepatic enzyme:
      (a) Hormonal excess--insulin, NSILA-s
      (b) Hormonal deficiency--glucagon, epinephrine
  B. Genetic enzyme deficiency (infancy & childhood)
  C. Extensive, diffuse, hepatocellular damage

INCREASED GLUCOSE UPTAKE

  I. Liver:
    A. Hormonal excess--insulin, NSILA-s
  II. Muscle:
    A. Hormonal excess--insulin, NSILA-s
    B. Hormonal deficiency--cortisol, growth hormone, epinephrine
    C. Exercise
III. Adipose tissue:
    A. Hormonal excess--insulin
  IV. Conceptus
  V. Tumor (extrapancreatic)

INCREASED UPTAKE OF AMINO ACIDS BY EXTRAHEPATIC TISSUES

  I. Muscle:
    A. Excess--insulin, NSILA-s
    B. Deficiency--cortisol, epinephrine
  II. Conceptus
III. Tumor (extrapancreatic)

---

*Reprinted from the publication, Fasting Hypoglycemia in Adults, with the permission of the New England Journal of Medicine (Vol. 294, Page 766, 1976) and the authors, S.S. Fajans and J.C. Floyd.

however, that the edible portion of the ripe akee fruit contains only small amounts of hypoglycins and is still a main staple in the diet of Jamaica. The biochemical basis for the hypoglycemic action of hypoglycin appears to be related to inhibition of gluconeogenesis. Recall that ATP is required for gluconeogenesis, with the principal source of ATP coming from the oxidation of long chain fatty acids in the mitochondria. The oxidation of fatty acids requires activation by forming a thioester linkage with CoA $(R-\overset{O}{\overset{\|}{C}}-S-CoA)$ and transportation into mitochondria by complexing with carnitine present in the mitochondrial membrane (see fatty acid oxidation, Chapter 8). Hypoglycin A is converted to non-metabolizable CoA and carnitine esters, depleting the limiting cofactors, carnitine and CoA. This leads to inhibition of long chain fatty acid oxidation and gluconeogenesis with consequent hypoglycemia. The formation of a CoA ester from hypoglycin A is shown below.

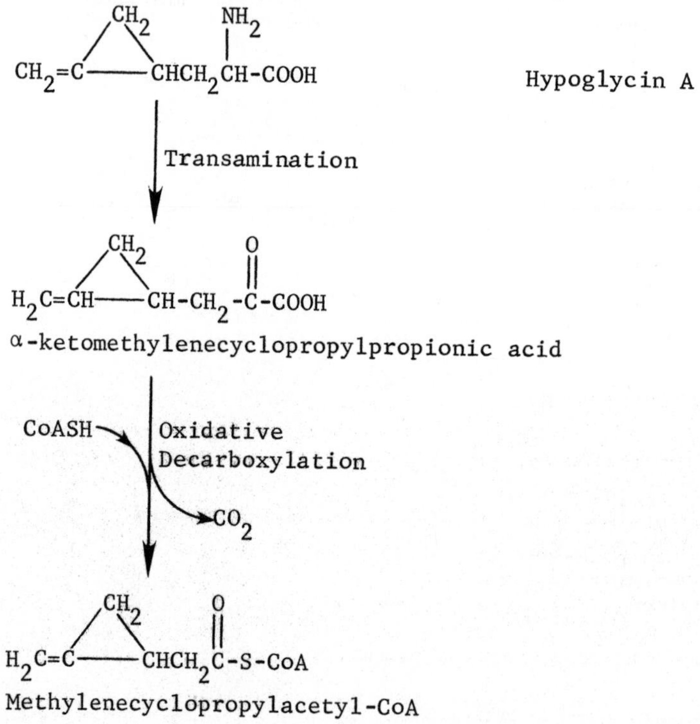
Hypoglycin A
Transamination
α-ketomethylenecyclopropylpropionic acid
CoASH
Oxidative Decarboxylation
CO2
Methylenecyclopropylacetyl-CoA

The hypoglycin A-induced hypoglycemia is relieved by a prompt administration of glucose by an intravenous route.

The identification of the specific cause of hypoglycemia in a given individual involves an extensive laboratory workup. This includes various hormone assays (with appropriate provocative tests) for insulin, proinsulin, glucagon, ACTH, cortisol, NSILA-s, growth hormone, and epinephrine; measurement of alanine and other gluconeogenic substrates; determination of alcohol susceptibility by testing gluconeogenic capability after alcohol infusion following a 12-hour fasting; and a 5-hour glucose tolerance test.

Hyperglycemia

A rise in blood glucose above the level which prevails most of the time is termed hyperglycemia. It is a normal consequence of eating, as carbohydrate absorbed from the digestive tract transiently elevates the concentration of sugar in the circulation. This triggers the release of insulin and suppresses glucagon, working to restore homeostasis. If, however, blood glucose remains above about 110 mg/dl, the hyperglycemia is no longer transient, and ill effects begin to appear. Mild hyperglycemia has few acute effects, but if it is chronic, there may be serious consequences. This is a poorly understood subject at present.

More severe hyperglycemia usually produces diabetes. Although this word is frequently used to indicate a specific disease, it actually refers to any disorder in which polyuria (excretion of excessive amounts of urine) occurs. The most common types of diabetes are those caused by the inability of the proximal convoluted tubule to reabsorb all of the filtered glucose. This increases the osmotic load in the urine and interferes with water resorption in the collecting ducts. A diuresis results, which can be quite massive (as high as 6-8 liters of urine per 24 hours). (Renal physiology and the mechanism of urine production are discussed in more detail in Chapter 12.) In most individuals, glycosuria accompanied by diuresis occurs when the plasma glucose approaches or exceeds about 300 mg/dl. It can be produced experimentally by the administration of phlorizin, a drug which prevents glucose reabsorption by binding to the receptor site of the carriers in the tubular cell membranes.

A number of diseases can produce hyperglycemia with diabetes and glycosuria. The most widely studied one is diabetes mellitus, actually a group of diseases with certain symptoms in common. This is discussed a little later in a separate section. Some other hyperglycemic disorders are described below.

1. Lipoatrophic Diabetes is a rare disease and is inherited as an autosomal recessive trait. It is characterized by an absence of grossly demonstrable adipose tissue, hyperlipidemia, hepatomegaly (leading to cirrhosis), hypermetabolism, polydipsia, polyuria, and glucosuria. These patients sometimes show a mild ketoacidosis and have a marked resistance to insulin, perhaps due to a lack of insulin receptors.

2. The question of insulin resistance has been studied recently in a
    group of patients who share many symptoms with lipoatrophic diabetics.
    These patients are neither obese nor lipoatrophic, are all female,
    and have a skin disorder known as <u>acanthosis nigricans</u>. This group
    of patients appears to consist of two subgroups:

    > Type A--young women; virilization or accelerated growth; primary
    > defect in insulin receptor (decrease in apparent number of
    > receptors).

    > Type B--older women; show many laboratory characteristics of
    > autoimmune diseases, including increased erythrocyte
    > sedimentation rate, proteinuria, and circulating antinuclear
    > and anti-DNA antibodies; circulating antibodies found against
    > insulin receptors.

    Plasma insulin concentrations were normal or high in these patients,
    infusion of large amounts of insulin had little or no effect on the
    polyuria and glucosuria, and no anti-insulin antibodies were
    detectable. The contrast between these types is interesting, one
    being the <u>absence</u> of a <u>receptor</u>, the other an immune disorder,
    <u>producing</u> an unwanted protein.

3. Secondary types of diabetes

    a. <u>Hemochromatosis</u>: This disorder is characterized by deposits of
        iron (due either to excessive iron ingestion or absorption; or
        to multiple transfusions--discussed in hemoglobin metabolism)
        in many tissues. In the pancreas, iron accumulation leads to
        the destruction of the β-cells, causing a deficiency in insulin.
        The accompanying liver disease aggravates the insulin
        insufficiency problem.

    b. <u>Stressful situations</u>: Acute stress as in severe trauma,
        septicemia, peritonitis, pneumonia, burns, stroke, or myocardial
        infarction leads to an increase in circulating catecholamines
        and to stimulation of the sympathetic innervation to the
        endocrine pancreas. This results in the suppression of insulin
        release with enhanced release of glucagon, thus producing
        hyperglycemia. In normal persons, this hyperglycemia is
        transient but in diabetics it can persist and can lead to very
        high elevations of blood glucose. In addition, if parenteral
        glucose is administered, the patient may progress into a
        hyperglycemic nonketotic hyperosmolar coma (glucose levels
        > 400 mg per dl of plasma). This situation is a medical
        emergency. It is treated by insulin administration, fluid and
        electrolytic replacement, and appropriate treatment of
        associated medical problems.

c. <u>Excessive growth hormone production</u>, as that associated with acromegaly, gives rise to an "insulin resistant state" leading to persistent hyperglycemia.

d. <u>Excess glucocorticoids</u>, from an exogenous (e.g., therapeutic) or endogenous source (as in Cushing's syndrome), lead to hyperglycemia due to an increased glucose output by the liver via gluconeogenesis. There is perhaps also a decrease in the capacity of β-cells to release insulin.

e. <u>Pheochromacytoma</u> (a tumor of the adrenal medulla, discussed under amino acid metabolism) secretes large amounts of catecholamines. Hyperglycemia results from increased hepatic glycogenolysis, decreased insulin release by the β-cells, increased glucagon secretion by the α-cells, and peripheral inhibition of glucose uptake. Also, as a consequence, increased free fatty acid mobilization from fat stores occurs.

f. <u>Thyrotoxicosis</u> is a hypermetabolic state with accelerated gastric emptying and absorption which may lead to hyperglycemia. This state may also unmask a latent diabetic (i.e., trigger the appearance of overt diabetes).

g. The <u>hypokalemia</u> (low levels of K^+ in the blood) which results from hyperaldosteronism or diuretic therapy diminishes the insulin release and can cause hyperglycemia.

h. In <u>pregnancy</u>, increased estrogen levels appear to be diabetogenic.

i. <u>Hyperlipidemia</u> leads to hyperglycemia (discussed under lipid metabolism).

j. <u>Diabetes insipidus</u> is a defect in the ability of antidiuretic hormone to function. There is no abnormality of glucose metabolism and no glucosuria accompanies the polyuria.

4. Disorders of endocrine secretion of the pancreas: These diseases, grouped under diabetes mellitus (discussed below), are characterized by abnormalities in carbohydrate, lipid and protein metabolism. They eventually lead to abnormal structural changes in a variety of organs.

Diabetes Mellitus

1. This disorder is complex, affecting the metabolism of lipid, carbohydrate and protein. There are two major types: juvenile-onset and maturity (or adult)-onset. Juvenile diabetics have a greatly decreased production of insulin (about 5% of the normal amount).

Consequently, there is no release of the hormone from the pancreas when there is an elevation of blood glucose. On the other hand, in maturity-onset diabetes, the total amount of insulin present in the pancreas may be normal, but its release in response to hyperglycemia is delayed. The juvenile disease can be produced in animals by pancreatectomy, or by the destruction of the islet tissue with toxic compounds such as <u>alloxan</u> or the antibiotic <u>streptozotocin</u>.

Alloxan

Streptozotocin

(A derivative of
D-glucosamine)

2. Figure 48 shows insulin response to oral glucose administration in various groups of patients. The following points need to be emphasized about these profiles:

a. In patients with severe diabetes (juvenile diabetics) there is clearly a deficiency of insulin.

b. Adequate or increased values for circulating insulin are found in mildly diabetic patients (maturity-onset diabetes). However, there is a delay in insulin <u>release</u>, giving rise to a gradual increase in blood glucose. This hyperglycemia in turn stimulates an exaggerated insulin secretion. <u>Obesity</u> is an aggravating factor in diabetes, apparently causing a decrease in tissue responsiveness to insulin (i.e., insulin <u>resistance</u>), thereby increasing the demand for insulin. Obesity in general is associated with hyperinsulinemia. This chronic, excessive insulin production can lead to pancreatic exhaustion, and, ultimately, diabetes mellitus in genetically susceptible individuals. The basic mechanisms of these abnormalities have not been delineated.

3. As already mentioned, <u>glucagon</u> plays a major role in carbohydrate and lipid metabolism. It stimulates glycogenolysis, gluconeogenesis, lipolysis, and ketogenesis (see later). Glucagon may, consequently, make a significant contribution to the pathogenesis of diabetes

mellitus. It has been suggested that diabetes is a "bihormonal disease" involving insulin deficiency or defects in its utilization (e.g., receptor abnormalities or presence of antibodies to receptor) together with glucagon excess. In fact, glucagon suppression by somatostatin has been beneficial in the short-term management of some of the effects of diabetes (discussed under somatostatin). It appears, however, that hyperglucagonemia without an accompanying insulinopenia has little effect on blood glucose. It does not alter glucose tolerance in normal persons nor upset diabetic control, provided insulin is available.

Figure 48*.   Comparison of Serum Insulin Levels in Various Groups

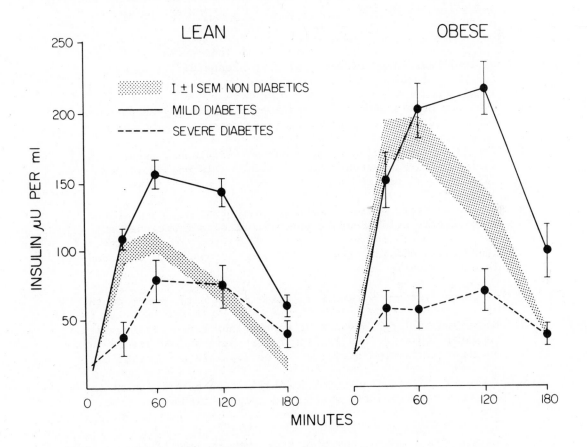

*Reproduced from the article, Current Concepts of Insulin Secretion, in the book Diabetes Mellitus, 4th ed., 1975, page 19, with permission from the American Diabetes Association, Inc. and the authors, M. Tzagournis and S. Cataland.

4. Metabolic consequences of diabetes mellitus: In mild cases, as in maturity-onset diabetes, the major problem appears to be a reduced capacity, particularly of the liver, to utilize material derived from the diet. In severe diabetes, however, the following events take place.

    a. The blood sugar increases to high levels (hyperglycemia) and glucose is found in the urine (glucosuria). The hyperglycemia and glucosuria persist during fasting.

    b. Liver glycogen falls to very low levels. Muscle glycogen is decreased but much less as compared to liver glycogen.

    c. Ingested and endogenous glucose is not taken into muscle and fat cells but is instead excreted to a great extent in the urine.

    d. The respiratory quotient, which is an indication of the type of fuel being metabolized, falls to around 0.71 (the value for fat oxidation) and is not raised by ingestion of glucose.

    e. The rate of tissue-protein breakdown is markedly accelerated (increased urinary nitrogen excretion occurs).

    f. The injection of glucose into normal individuals results in a rise in blood pyruvate and lactate. This does not happen in a diabetic patient.

    g. In the diabetic, large quantities of ketone bodies (acetoacetic acid, β-hydroxybutyric acid, and acetone) are produced due to increased fatty acid metabolism and decreased capacity to oxidize ketone bodies in the muscles. This can lead to diabetic coma.

    Ketone bodies $\longrightarrow$ severe acidosis $\longrightarrow$ coma $\longrightarrow$ death

    h. Excretion of large amounts of glucose and ketone bodies results in water and salt loss leading to dehydration and severe thirst. An abnormally large amount of urine is excreted (polyuria).

    i. There is an increased rate of mobilization of triglycerides and unesterified fatty acids from adipose tissue into the blood.

5. If the diabetes is caused by a lack of insulin, and if the patient is not "insulin-resistant," an injection of insulin promptly corrects all the foregoing metabolic disturbances. An overdose of insulin is very dangerous, producing insulin shock due to hypoglycemia. This condition superficially resembles a diabetic

coma. In the insulin-resistant form of diabetes, even excessive amounts of insulin will not produce insulin shock.

6. A normal adult secretes about 2 mg (45 units) of insulin per day. A diabetic patient being treated with insulin receives about 60 to 70 units per day.

7. As was indicated under the metabolic effects of insulin, a number of tissues do not require the hormone for the entry of glucose into the cells. Consequently, in the hyperglycemia associated with diabetes, the intracellular glucose of these tissues attains a level similar to that of the blood. This condition is unaffected by insulin unless total blood glucose can be reduced. Within these cells, the excess glucose is reduced to sorbitol by aldose reductase (see also galactosemia).

$$
\text{D-glucose} \xrightarrow[\substack{\text{NADPH} \\ + \text{ H}^+}]{\substack{\text{aldose} \\ \text{reductase}} \quad \text{NADP}^+}
$$

$$
\begin{array}{c}
CH_2OH \\
| \\
H-C-OH \\
| \\
HO-C-H \\
| \\
H-C-OH \\
| \\
H-C-OH \\
| \\
CH_2OH
\end{array}
$$

Sorbitol

Part of the sorbitol is oxidized to fructose by sorbitol dehydrogenase.

$$
\text{Sorbitol} \xrightarrow[\substack{\text{NAD}^+ \\ }]{\substack{\text{Sorbitol} \\ \text{dehydrogenase}} \quad \substack{\text{NADH} \\ + \text{ H}^+}} \text{D-fructose}
$$

Although the cell membranes are freely permeable to glucose, fructose and sorbitol do not readily pass through and they are retained within the cells. Even if the hyperglycemia is ultimately controlled by insulin, any sorbitol and fructose which may have accumulated remain until they are removed via other metabolic pathways.

Large amounts of sorbitol and fructose within the cell cause hypertonicity and water retention. Many of the physiological and pathological alterations associated with diabetes can be related to these effects. For example:

a. Sorbitol accumulation in the lens may result in cataract formation (see also galactose metabolism). This is supported by studies of cataract formation in diabetic rats and rats on high-galactose diets.

b. Sorbitol elevation in the peripheral nerves is associated with peripheral neuropathy.

c. In erythrocytes, sorbitol will displace 2,3-DPG, reducing the oxygen-carrying ability of the blood (see hemoglobin).

d. Other changes include vascular problems leading to atherosclerosis, nephropathy, and retinopathy.

A probable mechanism by which sugar alcohols (sorbitol, galactitol) bring about these abnormalities is indicated below.

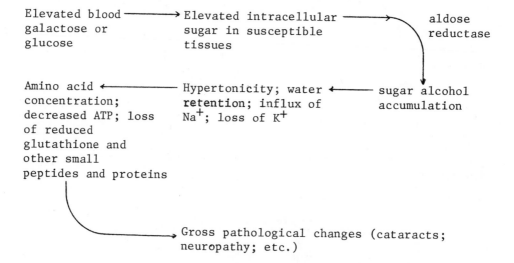

Aldose reductase, the key enzyme in this process, has been found in the affected tissue cells at high concentration.

8. Basement Membrane Changes in Diabetes

One of the clinical manifestations of diabetes which is poorly understood is the change in the basement membrane (BM), leading to its thickening in tissues such as the small blood vessels (microangiopathy) and the renal glomerulus. The BM is made up of a fibrillar network of proteins and its functions are to support cells and to act as a selective filter (particularly in the case of the glomerulus).

The BM-protein is a fibrous glycoprotein closely related to collagen. It is distinguished from collagen by the presence of mannose, hexosamine, sialic acids, and fucose, in addition to the glucose and galactose residues found in collagen. BM-protein also has a large number of half-cystine residues (few of which occur in collagen) and a greater number of hydroxylysine residues than collagen.

The biosynthesis of BM-protein is similar to that of collagen, requiring many post-ribosomal modifications. These include hydroxylation reactions and the addition of sugar residues (see collagen synthesis). In alloxan diabetic rats, increases have been measured in the amount of lysine which is hydroxylated, the number of glucosylgalactose units present (attached to the δ-hydroxyl group of lysine), and the activity of the glycosyl-transferases, needed for the formation of this type of bond. The presence of these extra, bulky sidechains could disrupt the packing of the protein fibers in the membranes, resulting in the observed thickening and increased porosity. If insulin is administered in the early stages of diabetes, the changes in the membranes are reversed. Prolongation of the disease progressively decreases the reversibility, suggesting the value of early diagnosis and treatment.

Although the biochemical lesions in diabetes are still not clear, this work indicates that they do not directly involve coding for the enzymes responsible for the membrane changes (i.e., lysinehydroxylase, glycosyltransferases, etc.). Rather, the observed BM-protein alterations probably result from the decreased insulin supply, the increased intracellular glucose concentration (in insulin-independent tissues such as the kidney and the blood vessels), or both. This is analogous to the proposed mechanism for cataract formation (via the sorbitol pathway) in diabetes mellitus and galactosemia.

9. <u>Clinical diabetes</u> in young persons is frequently due to a lack of insulin (juvenile-onset diabetes). This lack of insulin may be relative rather than absolute (i.e., there may be some insulin produced but not enough). The production of abnormal insulin (mutated amino acid sequence; failure to activate proinsulin) has the same effect as a deficiency in the quantity of insulin secreted. These types of diabetes can usually be treated by administration of an adequate insulin supply.

10. Maturity-onset diabetes, appearing in adults, is usually insulin-resistant. The insulin supply is adequate and even elevated, due to overstimulation of the pancreas by the high blood glucose. In some patients, there may be a delay in the pancreatic response to hyperglycemia followed by an oversecretion of insulin. Other possible causes include

   a. The inability to use or metabolize glucose in the tissues,

   b. The presence of an insulin antagonist (e.g., an insulin antibody or an insulinase),

   c. The inability of normally insulin-sensitive tissues to respond to the hormone.

11. Diabetes is found in association with several rare inherited diseases such as Weiner's, Klinefelter's, Refsum's and Turner's syndromes; Friedreich's ataxia; and others. It is not yet clear whether gout and diabetes are significantly associated. The malfunction of other endocrine glands which influence carbohydrate metabolism may also produce diabetes. Diabetes insipidus is a failure of the kidneys to respond to or a deficiency of, antidiuretic hormone. This results in polyuria, polydipsia (excessive thirst), and persistant hypotonicity of the urine.

12. Morphological changes in the pancreas which are associated with diabetes include

   a. Primary pathologic changes of the pancreas.

   b. Reduction in granulation of the β-cells (this does not necessarily imply a diminution in insulin production).

   c. Anatomic changes in the islets of Langerhans (seen in autopsy cases).

   d. In many cases there are no morphological changes. Furthermore, in an individual whose entire pancreas has been removed (as in carcinoma) the syndrome of diabetes is not as severe as that of many patients with spontaneous diabetes. A severe diabetic needs 80-150 units of insulin per day, whereas patients with pancreatectomy need only about 40 units per day. It appears that the requirement for insulin is greater in the diabetic than in normal persons and that the pancreas of the diabetic is unable to meet this increased requirement.

13. Although it is clear that diabetes mellitus is a familial disorder and that a predisposition for the disease is at least partly

inherited, the mode of genetic transmission is far from clear.
It is hoped that, by further clarification of the genetics of
diabetes, it will become possible to provide accurate information
on the prognosis and treatment in individual cases. This
understanding will also make possible the separation of genetic
and environmental factors in the etiology of diabetes and improve
regimens for the prevention of the disease.

a. Evidence accumulated thus far strongly points to diabetes as a
    genetically heterogeneous group of disorders that share glucose
    intolerance. Thus diabetes can result from a number of different
    genetic defects, with or without environmental factors affecting
    the phenotype. From the data available at the present time, it
    is not clear whether the mode of inheritance of diabetes is auto-
    somal recessive, autosomal dominant with incomplete penetrance, or
    polygenic. That there is a distinct genetic contribution to
    diabetogenesis is shown in the concordance rates for diabetes
    between monozygotic (identical) and dizygotic (fraternal) twins.
    Identical twins show a concordance rate of 45 to 96% for diabetes
    as demonstrated by glucose intolerance, while fraternal dizygotic
    twins show a concordance rate of 3 to 37%. Monozygotic twins result
    from a single fertilization event and consequently are genetically
    identical. Dizygotic twins arise from separate zygotes and are
    genetically distinct, though they may share some traits。

b. As indicated before, more than thirty genetic disorders give
    rise to carbohydrate intolerance and thus contribute to the
    apparent heterogeneity of the disease. Furthermore, juvenile-
    onset and maturity-onset diabetes appear to be clinically,
    physiologically, and genetically distinct.

    Physically, the juvenile diabetic is a thin, ketosis-prone,
    insulin-dependent individual, while the maturity-onset diabetic
    is usually obese, non-ketotic, and "insulin resistant."
    Genetically, histocompatibility studies (see Chapter 11 for a
    discussion of the histocompatibility loci) have shown association
    between the HLA B-8 antigen and antigens at the D locus with
    juvenile-onset diabetes but not with the maturity-onset disorder.
    The HLA B-8 antigen is also found in association with known
    autoimmune diseases such as Addison's and Grave's diseases.

14. Biochemical defects associated with diabetes include

a. The capacity of the liver to deposit and hold glucose as
    glycogen is practically lost in the diabetic. The rate of
    glycogenolysis is also enhanced.

b. The glucokinase reaction is severely impaired.

c. There is a decreased rate of glucose transport into certain tissue cells (particularly muscle and adipose tissue).

d. There is an increase of G-6-phosphatase activity in the liver.

$$\text{G-6-} \textcircled{P} \xrightarrow{\text{Glucose-6-Phosphatase}} \text{glucose} + P_i$$

e. Glucose formation from amino acids is increased (gluconeogenesis).

15. Insulin antagonists, mentioned previously, can cause ineffectiveness of insulin.

a. Insulin antibodies, produced as a result of insulin therapy, are found in the β and γ globulin fractions of plasma. Antibody production is a complexity and must be considered in the treatment of diabetes.

b. Another type of insulin antagonist is distinguishable from antibodies in that it is present without prior administration of exogenous insulin. It occurs in the β lipoprotein fraction of serum.

c. Physiological antagonists include growth hormone, thyroxine, and cortisol.

16. Breakdown of Insulin

a. Insulin released into the portal blood system through the pancreatic vein reaches the liver before it enters the systemic blood circulation to exert its effects on the insulin-dependent tissues such as muscle and adipose. In the liver, in addition to producing its physiologic effect, insulin is degraded by reductive cleavage.

b. The enzyme system that catalyzes the reductive cleavage of the interchain disulfide bonds is shown below.

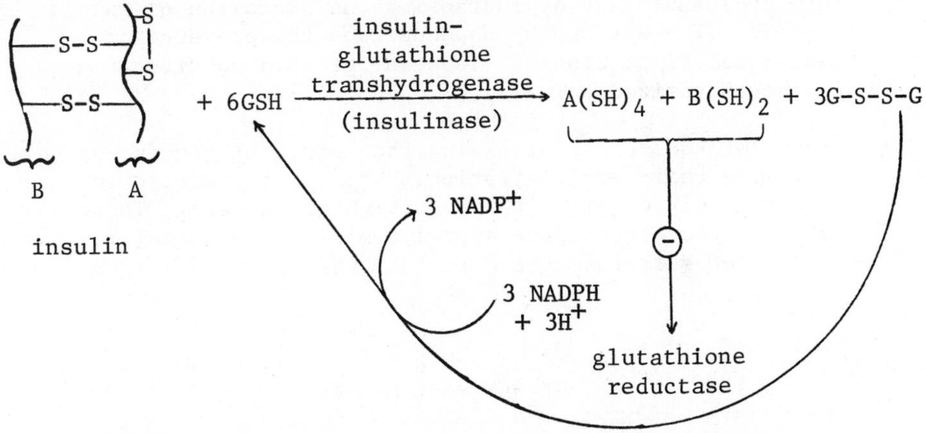

$$\begin{array}{c}
\text{B} \quad \text{A} \\
\text{insulin}
\end{array} + 6\,GSH \xrightarrow[\text{(insulinase)}]{\substack{\text{insulin-}\\ \text{glutathione}\\ \text{transhydrogenase}}} A(SH)_4 + B(SH)_2 + 3G\text{-}S\text{-}S\text{-}G$$

3 NADP$^+$

3 NADPH + 3H$^+$

glutathione reductase

Insulin inactivation is self-limiting because the G-S-S-G reductase is inhibited by the reduced insulin chains so that no more GSH (which is required for the reduction of insulin) is produced.

c. The A and B chains are further degraded by proteolytic enzymes (peptidases).

17. Stimulation of Insulin Secretion

a. Sulfonylureas are drugs used in maturity-onset diabetes.

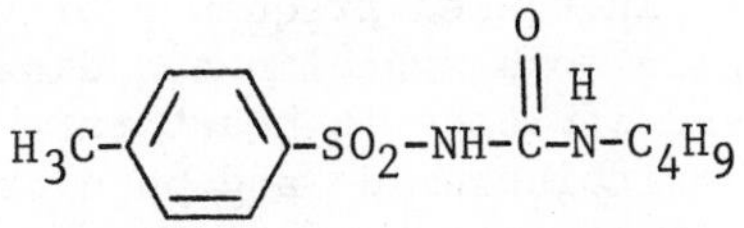

$$H_3C\text{-}\langle\bigcirc\rangle\text{-}SO_2\text{-}NH\text{-}\overset{O}{\overset{\|}{C}}\text{-}\overset{H}{N}\text{-}C_4H_9$$

Tolbutamide (orinase Ⓡ)

$$Cl\text{-}\langle\bigcirc\rangle\text{-}SO_2\text{-}\overset{H}{N}\text{-}\overset{O}{\overset{\|}{C}}\text{-}\overset{H}{N}\text{-}C_3H_7$$

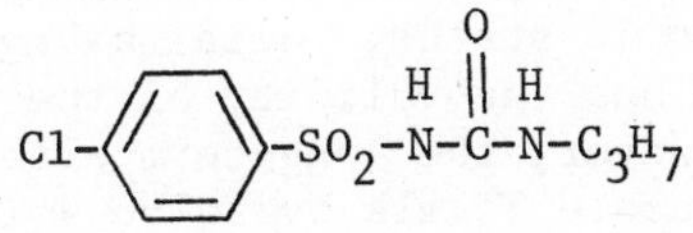

Chlorpropamide (diabinase Ⓡ)

They presumably act by increasing the secretion of insulin by β cells. The efficiency depends upon the presence of functional islet tissue. These drugs have no effect on pancreatectomized animals.

b. Biguanidines act by increasing the uptake of glucose by cells. Following their administration,there is an increase in anaerobic glycolysis with production of lactate. These compounds do not produce hypoglycemia in the normal individuals and do not stimulate the β cells. Example:

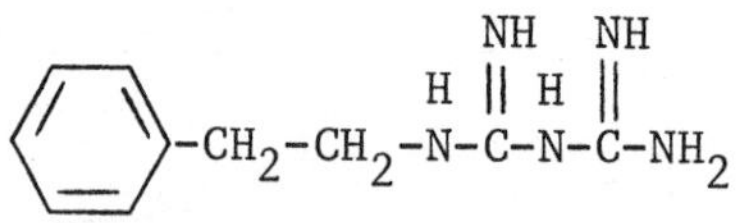

$$\text{(phenyl ring)}-CH_2-CH_2-\underset{\text{H}}{N}-\underset{\underset{NH}{\|}}{C}-\underset{\text{H}}{N}-\underset{\underset{NH}{\|}}{C}-NH_2$$

Phenethylbiguanidine (phenformin; DBI ®)

Biguanidines have been shown to inhibit electron transport in the mitochondria. A side-effect of this compound is lactic acidosis, and this drug has recently been banned by the FDA.

c. A multicenter, long-term study known as the University Group Diabetes Program (UGDP) reported an unexpected increase in cardiovascular mortality in the group of maturity-onset diabetic patients treated with tolbutamide and phenformin, compared with the placebo- or insulin-treated groups. A positive inotropic effect on the heart increased frequency of ventricular fibrillation, a rise in both systolic and diastolic blood pressure, and a marked increase in heart rate were all associated with administration of tolbutamide and phenformin. The mechanism for this cardiotoxicity is not known. In the UGDP study, the distinction between obese and non-obese maturity-onset diabetes was not made in assessing the toxicity of tolbutamide and phenformin. Further, the patient population varied greatly with respect to socioeconomic status, racial background, and initial health conditions. Thus the validity of the UGDP findings have been questioned. However, the long-term effectiveness of controlling blood glucose levels by tolbutamide or phenformin appears to be minimal, and the suggestion has been made that if dietary management alone (see below) is not successful in correcting hyperglycemia in maturity-onset diabetics, therapy with insulin rather than hypoglycemic agents may be more desirable.

18. Role of somatostatin in the treatment of diabetes:

a. Somatostatin, a peptide hormone (14 amino acid residues; contains one disulfide linkage), inhibits growth hormone, glucagon and insulin secretion. The infusion of the hormone into insulin-

dependent diabetics leads to an improvement in carbohydrate
tolerance as well as preventing diabetic ketoacidosis following
the acute withdrawal of insulin from these patients. Although
somatostatin inhibits insulin secretion, which will <u>augment</u>
hyperglycemia, the major effect of this hormone is the <u>suppression</u>
<u>of glucagon release</u> and the consequent amelioration of its
hyperglycemic effects. Further, as we have already seen, the
growth hormone suppression is beneficial in achieving a normal
glucose tolerance. Despite these remarkable effects, the
potential therapeutic use of somatostatin has a number of
limitations such as its short duration of action (half life in
the body is less than 4 minutes), the multiplicity of target
tissues (leading to diverse physiologic effects), and a lack of
data on long-term effectiveness and safety. However, it is
hoped that studies on structural analogs of the hormone will
yield a specific and longer-acting form of the compound which
may be a useful adjunct to insulin in the treatment of diabetes.

It is of interest to point out that, whereas somatostatin can
be used to suppress hyperglycemia in adults, it can also be used
to increase blood glucose levels in idiopathic hypoglycemia
associated with pancreatic nesidioblastosis. Nesidioblastosis
is characterized by the presence of nests of pure β-cells and
small islets arising from the pancreatic ductules. These regions
lack other cell types (α and D) and, therefore, the insulin
secretion may not be regulated, leading to hypoglycemia due to
hyperinsulinism. In a recent report of one case, intravenous
or subcutaneous administration of a long-acting preparation of
synthetic somatostatin (protamine zinc somatostatin) produced
suppression of insulin secretion with concomitant elevation
of blood glucose. The main therapy in this condition, however,
is to reduce the number of β-cells by subtotal pancreatectomy.

b. Somatostatin is present in the hypothalamus and other parts of
the central nervous system (e.g., the spinal cord); in highly
specialized neuro-endocrine cells ontogenically derived from
the neural crest primordia; in the D-cells of the pancreas
(which are located in the islets, in tight-junction to both the
α- and β-cells); in cells of the fundic and antral mucosa of the
stomach; in the duodenum; and in a few cells of the glandular
element of the jejunum and ileum. The release of pancreatic
somatostatin is stimulated by many of the same factors that
cause insulin release: arginine, leucine, glucose, mixed amino
acids; and by pancreozymin, the hormone which stimulates release
of the digestive enzymes from the exocrine pancreas. The true
physiological role of somatostatin is not known. It may be
involved with the regulation of uptake of nutrients from the
digestive tract. It is known to inhibit the secretion of a
number of peptide hormones involved in digestion, but most of
the experiments have been with pharmacological doses.

c. There have been two recent reports of somatostatinomas (malignant tumors of the pancreatic D-cells). Both cases were confirmed histologically. In one of the patients, hypersomatostatinemia was proven by radioimmunoassay. In both cases, insulin and glucagon levels were extremely low, a mild hyperglycemia was present (less than 200 mg/dl), body weight was reduced, and gallbladder disease was present (both cases were found during a cholecystectomy). These symptoms, as well as the steatorrhea and hyperchlorhydria, seem consistent with the postulated role of somatostatin in nutrient utilization.

19. Diagnosis of diabetes:

a. Because of the complexity and heterogeneity of diabetes, its laboratory evaluation is elaborate. It involves measuring a number of endocrine parameters, including:

i. Changes in blood insulin and glucose levels in response to an oral glucose load (oral glucose tolerance test; OGTT);

ii. Insulin release by the pancreas in response to tolbutamide administration;

iii. Leucine tolerance test;

iv. Circulating concentration of glucagon, catecholamines, growth hormone, and cortisol; and

v. Glucagon test.

Of these, the OGTT is the one most frequently performed. It should be emphasized, however, that an abnormal glucose tolerance test may result from anxiety, severe carbohydrate deprivation (see below), old age, infection, acute illness, delayed gastric emptying, certain drugs (e.g., hypoglycemic agents), and other factors, as well as diabetes mellitus.

b. An oral glucose tolerance test consists of the measurement of the fasting blood glucose level, oral administration of glucose to a properly-prepared patient and the subsequent measurement of the blood glucose at intervals of 30 minutes, 1 hour, and 2 hours. Measurements at longer intervals of 3, 4, and 5 hours are also performed in assessing special problems such as hypoglycemia.

i. Preparation of the patient: The subject should have received at least 150 grams of carbohydrate per day for the three days preceding the test (in order to have proper glycogen stores) and should have fasted for at least 8 hours (but no longer

than 16 hours) prior to testing. Consumption of water is
permitted and desirable to prevent dehydration. Coffee
consumption and smoking are not permitted during the test.
If the patient is emotionally upset, the test is postponed.
In general, the test is performed on ambulatory individuals
and <u>not</u> on bed-ridden patients. Many drugs interfere with
the glucose tolerance and therefore should be discontinued
for an appropriate length of time prior to the test,
depending upon a particular drug's disappearance rate. A
few examples of such drugs are: salicylates, diuretics,
oral contraceptive agents, corticosteroids, L-asparaginase,
phenothiazines, thyroid products, sympathomimetics, alcohol,
theophylline, propranolol, and alpha-adrenergic blocking
agents.

ii. The test is performed in the morning as glucose tolerance
is poorer in the afternoon. A general procedure for adults
consists of administering a glucose dose of 75 grams in a
volume of 300 ml over a period of about 5 minutes. The
75 gram load is based upon a desired dose of 40 gram per
square meter of body surface and an average adult surface
area of 1.88 square meters. In children, a dose of 1 gram
per pound up to 75 grams is used.

iii. Interpretation of results: Table 14 and Figure 49 show
currently used values for the diagnosis of diabetes. It
has been pointed out that a confirmed <u>fasting blood sugar</u>
greater than 130 mg/dl establishes the diagnosis of diabetes,
making a tolerance test superfluous. However, the
administration of the loading dose of glucose to an
unidentified diabetic presents no clinical complications.
Note also that values for <u>plasma</u> glucose are about 15%
<u>higher</u> than those for <u>whole blood</u>. A low or flat OGTT
profile is seen in 15 to 20% of healthy individuals and
apparently has no clinical significance.

20. Dietary management of diabetes

a. Diet therapy is the cornerstone in the overall treatment of
diabetes, a point which cannot be overemphasized. Diabetes
mellitus is a <u>chronic condition</u> which requires <u>long-term
management</u>. The dietary strategy for treatment of obese,
maturity-onset diabetics is distinctly different from that for
lean, insulin-dependent diabetics.

i. In fat patients, there is overwhelming evidence that
alleviation of excess <u>adiposity</u> (not just an alleviation of
excess weight, since this may be due to increased muscle-
mass) leads to a decrease in the severity of the diabetes.

This may be due partly to an increased peripheral sensitivity to insulin (see below) and partly to improved beta-cell function. The dietary goals for this group, then, are to restrict total caloric intake, as well as to reduce the fraction of calories derived from fat (to about 30%). Saturated fats are replaced in part with polyunsaturated fats (vegetable oils, special margarines), starch, and protein. Restriction of dietary cholesterol and refined sugar and abstinence from alcohol are also essential. In order to provide a diabetic patient with day-to-day food variety, the standard exchange system, available from the American Diabetic Association or the American Dietetic Association, should be consulted.

ii. In lean, insulin-dependent patients, dietary regulation is more important than is elimination of any specific food. A properly planned diet makes insulin therapy safer and more effective.

Table 14*.   Glucose Tolerance with Age

Age	Fast	1 Hr	2 Hr ***	Sum OGTT/2 Hr
0-30	110**	185	165(185)	460
30-40	112	191	175(195)	478
40-50	114	197	185(205)	496
50-60	116	203	195(215)	514
60-70	118	209	205(235)	532
70-80	120	215	215(245)	550

*Reproduced from the article, The Use of Screening and Diagnostic Procedures: The Oral Glucose Tolerance Test, published in Diabetes Mellitus, 1975, 4th ed., page 62, with permission from The Diabetic Association, Inc. and the author, T.E. Prout.

**The values are expressed as mg/100 ml of plasma.

***In parentheses are given the approximate values for patients above normal but not clearly diagnostic of diabetes. In Figure 48, this range of values has been designated as "probable diabetes."

b. <u>Hyperlipidemia</u> is common in diabetic patients, presenting as hypertriglyceridemia, hypercholesterolemia, or both. This is considered to be one of the factors which increases a person's chances for developing coronary disease. It is discussed further in Chapter 8 under lipid metabolism.

c. In a number of nutritional studies it has been shown that obesity is associated with hyperinsulinemia. In obese individuals there is a net reduction in the numbers of insulin receptors per cell, decreasing the tissue responsiveness to the hormone. This causes the pancreas to secrete more insulin in an attempt to elicit a desired effect. This decrease in the number of receptors appears

<u>Figure 49*</u>. <u>Diagnostic Levels of Plasma Glucose Two Hours After</u> <u>Glucose Challenge for Various Ages</u>

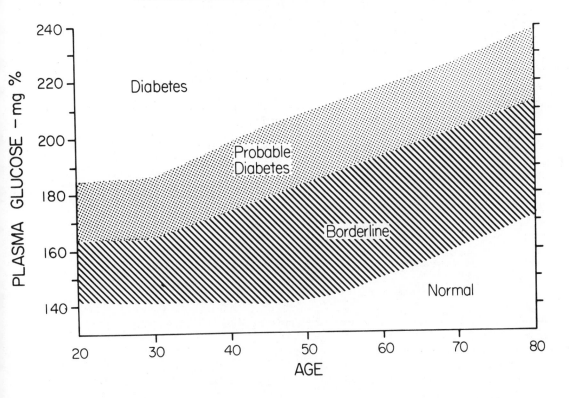

*Reproduced from the article, The Use of Screening and Diagnostic Procedures: The Oral Glucose Tolerance Test, published in Diabetes Mellitus, 1975, 4th ed., page 63, with permission from The Diabetic Association, Inc. and the author, T.E. Prout.

to be reversible. With a reduction in adiposity, the number of receptors returns to normal as do blood glucose and insulin concentrations.

    d.   Alteration in the cell surface insulin receptors, resulting in impaired insulin binding, has been observed in patients with uncontrolled diabetic acidosis, in insulin-resistant diabetics, and in patients with excess glucocorticoids, either exogenous or endogenous. These studies have provided a further theoretical basis for the importance of diet therapy in the treatment of diabetics.

Summary of Action by Hormones on Carbohydrate Metabolism
(Notice that all of these hormones are hyperglycemic, except for insulin.)

1.   Insulin

    a.   Plays a central role in: the removal of glucose from blood; lipid synthesis (in adipose tissue and liver); glycogen formation (principally in muscle); and glucose oxidation (in all insulin-dependent tissues).

    b.   Is secreted into the blood as a direct response to hyperglycemia.

    c.   Has an immediate effect in increasing the glucose uptake by certain tissues; produces hypoglycemia.

    d.   Is produced by the β cells of the pancreas.

2.   Glucagon (29 amino acid residues)

    a.   Release is stimulated by hypoglycemia.

    b.   Accelerates glycogenolysis in the liver.

    c.   Enhances gluconeogenesis from amino acids and lactate.

    d.   Is produced by the α cells of the pancreas.

3.   Anterior Pituitary: GH (growth hormone) and ACTH (adrenocorticotropin)

    a.   Tend to elevate blood glucose levels.

    b.   GH decreases glucose uptake by the tissues.

    c.   GH → adipose tissue → releases free FA → inhibits glucose utilization.

d. Chronic administration of GH leads to diabetes (by producing chronic hyperglycemia which causes β cell exhaustion).

e. ACTH stimulates the adrenal cortex.

4. Adrenal Cortex: glucocorticoids (hyperglycemic agent)

a. Increase gluconeogenesis by the following mechanism:

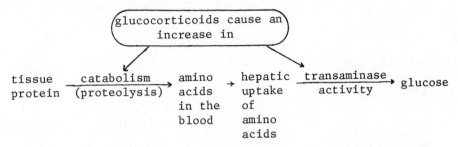

b. Inhibit utilization of glucose in extrahepatic tissues.

5. Adrenal Medulla: epinephrine stimulates glycogenolysis in liver and muscle. It also inhibits glycogenesis.

6. Thyroid hormone (thyroxine, triiodothyronine)

a. Has diabetogenic action.

b. Fasting blood sugar is elevated in hyperthyroid and decreased in hypothyroid patients.

c. Effects on carbohydrate metabolism are not understood.

Carbohydrate Metabolism and Mechanical Functions in Biological Systems

1. Movement in biological systems is coupled to carbohydrate metabolism via ATP. During muscle contraction and other mechanical processes, ATP is hydrolyzed to ADP + P_i and the energy originally stored during oxidation is used to do useful work.

2. The immediate sources of ATP for rapid bursts of muscular contractions are threefold.

a. During times of muscular relaxation, ATP is used by the enzyme creatine phosphokinase (CPK) to phosphorylate creatine (a substituted guanidine; see glycine metabolism). Since the nitrogen-phosphorous bond is unstable (has a high free energy of hydrolysis; $\Delta G^{O'} = -10.3$ kcal/mole), it can be used to phosphorylate ADP, so the reaction (shown below) is freely reversible.

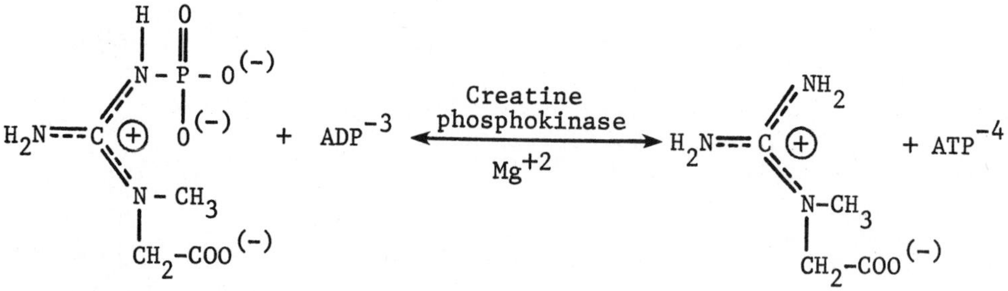

$$
\begin{array}{c}
\text{H} \quad \text{O} \\
\underset{\displaystyle \underset{\displaystyle N-CH_3}{\displaystyle \underset{\displaystyle CH_2-COO^{(-)}}{|}}}{\overset{\displaystyle \overset{\displaystyle |}{\overset{\displaystyle |}{H_2N = C \oplus}}}{}} \quad N-P-O^{(-)} \\
\end{array}
+ \text{ADP}^{-3}
\xrightleftharpoons[\text{Mg}^{+2}]{\substack{\text{Creatine} \\ \text{phosphokinase}}}
\quad H_2N = C \oplus
\begin{array}{c} NH_2 \\ N-CH_3 \\ CH_2-COO^{(-)} \end{array}
+ \text{ATP}^{-4}
$$

Creatine phosphate   Creatine

Creatine phosphate is thus another <u>storage form of high-energy phosphate bonds</u>, especially in muscle tissue.

b. Adenylate kinase (myokinase) reuses ADP to make ATP.

$$
2 \text{ ADP} \xrightarrow{\substack{\text{adenylate} \\ \text{kinase}}} \text{ATP} + \text{AMP}
$$

c. ATP synthesized in glycolysis is also a source of energy. Initially, in the working muscle, oxygen stored in oxymyoglobin is available to maintain aerobic glycolysis. When this supply is used up, anaerobic glycolysis begins with the production of lactic acid. The lactate diffuses from the muscle into the blood, where it is carried to the liver and either further oxidized or used for gluconeogenesis. Glucose formed in this way can return to the muscle, via the circulation. This is known as the <u>Cori cycle</u> (see glycolysis).

3. Muscles which are capable of long-term or sustained activity depend upon the oxidation of free fatty acids, ketone bodies, and pyruvate (either from glucose or lactate) under aerobic conditions to produce ATP. Examples of this are the cardiac and diaphragm muscles.

4. Of the three known systems in plants and animals which transform chemical energy (ATP) into mechanical energy, the best-studied of these is striated muscle.

a. There are <u>three</u> types of muscle tissue: striated, cardiac, and smooth. They differ from one another in structure, function, and innervation.

i. <u>Striated muscle</u> (so named because in light microscopy the fibers appear to have cross striations), also known as

skeletal muscle, is the most abundant type. It is attached
to the bones of the skeleton and is responsible for moving
those bones which are under voluntary control (hence they
are also referred to as voluntary muscles).

ii. <u>Cardiac muscle</u>, microscopically similar to striated muscle,
is present in the walls of the heart and pulmonary veins.

iii. <u>Smooth muscle</u> is found in the walls of the viscera and
blood vessels. Although smooth muscle contains similar
contractile proteins (see below), it differs from the other
two types in being rather slow in contraction, making it
appear similar to some contractile, non-muscle cells.

b. Skeletal muscle fibers are cylindrical (1 to 40 mm in length and
10 to 40 μ in width), multinucleated cells which are formed by
the fusion of many embryonic cells. One fiber, considered as a
single cell, may contain as many as 100-200 nuclei. Striated
muscle can be microscopically distinguished from cardiac muscle
in two ways. First, in skeletal muscle the nuclei are peripherally
located whereas in cardiac (and smooth) muscle the nuclei occupy
a central position. The other difference is that the cardiac
muscle functions as one unit while it is contracting. This is
because the individual cardiac muscle cells are held together by
a junctional complex. In light microscopy this appears as unique
dark-stained bands, called intercalated disks.

c. The naming of the subcellular organelles of a muscle fiber is
different from the non-muscle cell:

> Plasma membrane (plasmalemma): Sarcolemma
> Cytoplasm: Sarcoplasm
> Mitochondria: Sarcosomes
> Endoplasmic reticulum: Sarcoplasmic reticulum

d. <u>Myofibrils</u>, organized bundles of proteins not separated by
membranes from the cytoplasm, are characteristic of muscle cells.
The <u>sarcomere</u> is the functional unit of the contractile apparatus.
It occupies the region between each pair of Z-lines and contains
two half I-bands and one A-band (see Figure 50). In these cells,
the sarcoplasmic reticulum is organized as interconnecting tubules
spaced longitudinally through the myofibrils (the sarcotubular
network). These tubules come into contact at regular intervals
with a <u>transverse tubular system</u> (t-tubules). The t-tubule
network is an extension of the sarcolemma, thus rendering the
extracellular space continuous with the sarcoplasm. In a
myocardial cell, the t-tubules run in a longitudinal as well as
a transverse direction. A nerve impulse entering the t-tubules
is transmitted by an unknown mechanism to the longitudinal tubule

Figure 50. A Schematic Representation of the Structure of Myofibrils

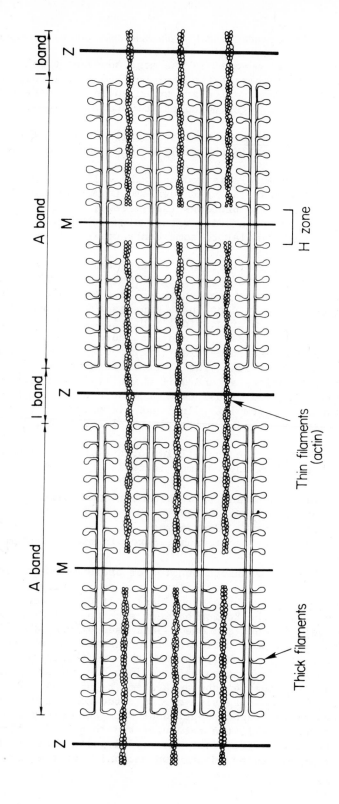

of the sarcoplasmic reticulum where large amounts of calcium ion are sequestered. The nerve impulse causes the rapid release of the calcium ions into the sarcoplasm where they bind to the C subunits of the troponin, eventually leading to contraction (discussed below).

e. In a muscle cell, most aspects of contraction can be explained by examining the properties and <u>cooperative interactions</u> of four proteins: myosin, actin, tropomyosin, and troponin. They are assembled into <u>two</u> multimolecular aggregates known as thick and thin filaments. The thick filament contains all of the myosin of the muscle cell, with the myosin heads projecting along the entire length of the filament except for the bare zone in the middle. Small amounts of other proteins, whose function is not known, are also found in the thick filament. The thin filament contains the other three major proteins: actin, tropomyosin and troponin.

i) Actin (F-actin) is a polymer, probably helical, of subunits (G-actin molecules). The subunits have a molecular weight of about 47,000 daltons. The active form is F-actin, which is able to stimulate the ATPase activity in myosin.

ii) Myosin has properties of both fibrous and globular proteins and a molecular weight of about 5×10^5 daltons. It appears to have the structure shown schematically below.

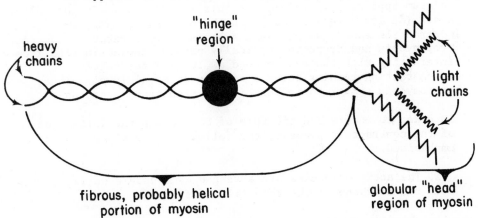

"hinge" region

heavy chains

light chains

fibrous, probably helical portion of myosin

globular "head" region of myosin

Notice that (1) there are four peptide chains, two light and two heavy; (2) the globular part of the molecule contains the ATPase activity and, presumably, is the region which functions during muscle contraction; and (3) myosin is susceptible to hydrolysis by proteolytic enzymes at the "hinge" point. The function of this region is hypothetical only.

f. Figure 50 shows the interdigitated (interpenetrating) arrangement of thick and thin filaments. The thin filaments are linked to adjacent Z-bands and extend toward each other. They are composed of double-helical F-actin polymers with long, helical tropomyosin molecules, having a molecule of troponin attached at the end lying in one groove of the double helix. The thick filaments are made up of myosin molecules having their fibrous portions parallel to each other and their globular regions (containing the ATPase) extending from the bundles like arms or paddles. Contraction is thought to occur by a sliding of thick and thin filaments past one another, propelled by a "pulling" motion of the myosin "arms." During contraction, the H-zone becomes narrower and the Z-lines move closer to each other. In this <u>sliding-filament hypothesis</u>, proposed in 1939 by H.E. Huxley and J. Hanson, it is assumed that the fibers maintain a constant length. It should be emphasized that the details of this model are still hypothetical, supported mostly by indirect evidence. Recently, however, Huxley has observed changes in the X-ray diffraction pattern of contracting muscle which indicate that the myosin heads do move toward the thin filaments during contraction, providing direct support for the theory. The exact nature of the interaction between actin and myosin is not known. The way in which the chemical energy of ATP is used to move the protein molecules is also unclear.

g. <u>Role of Calcium</u>: For muscle contraction to take place, the myosin heads <u>must</u> interact with actin, leading to hydrolysis of ATP. Calcium appears responsible for this interaction. In the relaxed state (at low levels of Ca^{2+}) in the presence of ATP, the contractile apparatus of the muscle is poised, ready to contract. Tropomyosin and troponin physically prevent the actin site from interacting with the myosin head as long as the calcium concentration remains below a certain critical level. Thus, tropomyosin and troponin function as regulatory proteins.

In order to understand the role of calcium in the initiation of the contractile process, the following points need to be emphasized.

i. The actin molecules are arranged as a double helix, forming the backbone of the thin filaments.

ii. The tropomyosin molecules lie end to end in the two grooves of the double helix, one tropomyosin extending over seven actin monomers. The two ends of tropomyosin are different from one another, giving the molecule polarity. Apparently the actin binding sites of tropomyosin are all about the same, as the molecule shows repeats in its sequence.

iii. The troponin molecule is a complex of at least three distinct peptide chains. It is attached about two-thirds of the way from the NH_2 terminus of the tropomyosin molecule. The three troponin subunits are: TnT (largest of the subunits; strongly bound to tropomyosin), TnI (the second largest of the subunits; bound to actin, interfering with the actin-myosin interaction), and TnC (the smallest of the subunits; the calcium-binding protein).

The arriving nerve impulse causes calcium release from the sarcoplasmic reticulum. This Ca^{+2} binds to the troponin C (TnC) subunit. This sets into motion a series of conformational changes which culminates in the movement of tropomyosin deeper into the F-actin groove, allowing the previously inaccessible sites on actin to interact with myosin heads, producing contraction (see Figure 51). During relaxation, the Ca^{2+} is reaccumulated in the storage sites and ATP is regenerated.

h. The calcium-activated troponin-actin system found in vertebrates is replaced by a calcium-activated myosin system (calcium initiates contraction by directly acting on the myosin) in mollusks. Most other animal species, however, contain both myosin-linked and troponin-actin linked contractile tissues.

i. Rigor Mortis is the extreme rigidity and inextensibility that develops in muscle after death. It is due to the depletion of ATP in the muscle. There are two types of myosin-actin complexes. One forms in the presence of ATP (active complex) and the other in its absence (the low-energy or rigor complex). The high-energy complex is generated by the binding of a myosin-ATP intermediate to actin, as described above. Following binding, ATP hydrolysis occurs, the filaments slide, and the low energy complex forms. Binding of another ATP molecule regenerates the high-energy complex and the cycle continues, providing the supply of ATP is constantly replenished. Upon death this replenishment ceases, all of the myosin-actin complexes cycle to the low energy form, and rigor mortis sets in.

In ischemic heart disease there is a rapid decline in myocardial contractility. Note that ischemia occurs due to oxygen deprivation secondary to reduced perfusion (due to occlusion), whereas hypoxia is a result of reduced oxygen supply despite adequate perfusion. Anoxia is due to the complete absence of oxygen supply despite optimal perfusion. It has been postulated that the abnormalities in contraction are caused by the excessive production of H^+ ions which interferes with the interaction of Ca^{++} with the TnC subunit. Another cause may be decreased levels of ATP (see below) and creatine phosphate at certain critical locations, although their overall levels may be normal. The biochemical changes that take place in an ischemic heart are: accumulation of lactate with consequent inhibition of phosphofructokinase,   hexokinase, phosphorylase kinase, glyceraldehyde-3-phosphate dehydrogenase, and carnitine palmitoyl coenzyme-A transferase. The inhibition of the latter enzyme leads to accumulation of acyl-CoA esters which in turn inhibit the ADP/ATP exchange between the cytoplasm and mitochondria, by affecting adenine nucleotide translocase. Carnitine has been shown to reverse the inhibition of adenine nucleotide translocase in the ischemic myocardium.

Figure 51. Schematic Representation of the Arrangement of Tropomyosin and Troponin Molecules in the Thin Filaments and Their Relationship to the Myosin Heads in the Absence and Presence of Ca^{2+}

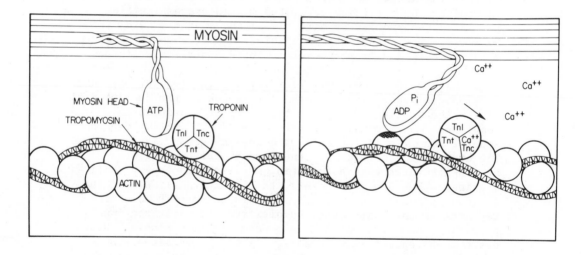

5. Microfilaments are a second energy-transducing system within the cell. They are associated particularly with cytokinesis (cleavage of the cytoplasm during cell division) in animal cells and movement of non-ciliated cells. (Cytokinesis in plant cells seems to involve microtubules only. These are discussed below.) They have been shown to contain an actin-like protein but nothing similar to myosin. Although they have not been chemically well-characterized, they may represent a primitive actomyosin-muscle-like system. The disruption of microfilamentous contractile systems has been postulated as the mechanism by which cytochalasins halt a number of cellular processes (see below).

6. While the two previous systems have been filamentous and best adapted to pulling, <u>microtubules</u> are found where a pushing force is needed. Processes which appear to involve microtubules include the motion of sperm tails and other flagella, ciliary function, and spindle formation and nuclear division during mitosis.

   a. The arrangement of microtubules in cilia and flagella is the "9 + 2" pattern (9 pairs of tubules around the rim with two tubules in the center), illustrated below in a schematic cross-section of a cilium.

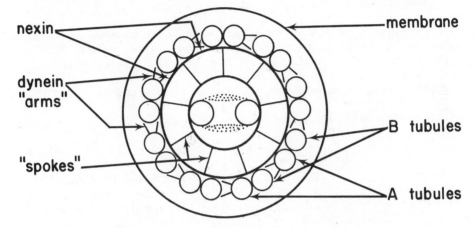

   The basic molecular components of this structure which have been identified are:

   i) <u>tubulin</u> - a globular protein which stacks upon itself to form the walls of the tubules; it appears analogous to actin in muscles; sometimes called flagellin in certain flagella.

   ii) <u>dynein</u> - this protein has the only ATPase activity in the flagellum; extends from the walls of the A tubules; probably equivalent to the myosin "head" region in muscle.

   iii) the <u>spokes</u> and the <u>nexin band</u> are composed of proteins and can be hydrolyzed by proteases. Their function appears structural, maintaining the tubule arrangement during ciliary and flagellar movement.

   b. The A and B tubules of a pair are slightly different, both chemically and morphologically. Theoretical and experimental studies favor a "sliding-tubule" mechanism, similar, perhaps to the sliding-filament mechanism proposed for muscle. In this model, the dynein arms transduce the energy released by ATP hydrolysis into motion by attaching to the adjacent B tubule, bending (thereby sliding one tubule relative to the other), releasing, and returning to their original position. It has been shown that ATP is hydrolyzed in this process.

c. Several compounds, the classical one being colchicine, apparently disrupt microtubular movement. Colchicine is noted for halting mitosis in metaphase, with destruction of the spindle structure which is probably microtubular in nature. Colchicine is used in the treatment of acute gout and the mechanism of its action is discussed under gout.

7. The <u>cytochalasins</u> are an important class of compounds which produce a number of interesting effects when applied to cells. They are proving to be very useful tools in cytological research.

a. The first cytochalasins discovered were A and B, shown below.

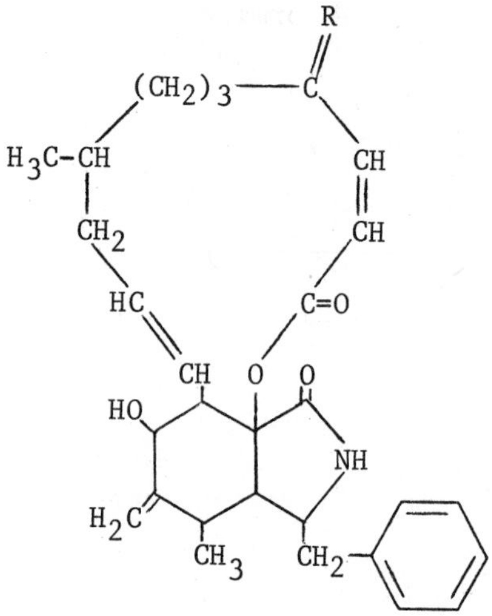

R = O in cytochalasin A

R = H and OH (one bond to the carbon from each) in cytochalasin B

b. General effects which these compounds have on cells include

i)   interference with cytoplasmic cleavage (cytokinesis);

ii) inhibition of cell movement;

iii) cell enucleation (extrusion of the nucleus from a cell).

c. Specific processes inhibited by cytochalasins include:

i)   clot retraction (platelets);

ii) phagocytosis;

    iii) platelet aggregation;

    iv) cardiac muscle contraction;

    v) thyroid secretion and growth hormone release.

  d. On the basis of limited evidence, some authors claim that the cytochalasins function strictly or primarily by disrupting microfilaments and, from this, infer that all of the above processes involve microfilaments. This remains to be proven. Elucidation of the mechanism of action of the cytochalasins will go along with an increased knowledge of the mechanisms for subcellular movement.

Clinical Significance of Creatine Kinase
(CK; also known as creatine phosphokinase, CPK)

1. a. Creatine kinase catalyzes the reversible transfer of the terminal phosphoryl group of ATP to creatine forming creatine phosphate (a N-phosphate):

$$\text{Creatine Phosphate + ADP} \rightleftharpoons \text{Creatine + ATP}$$

Creatine chemistry is discussed under glycine metabolism in Chapter 6.

The optimum pH for the forward reaction is 6.8 while for the reverse reaction it is 9.0. In muscle, there is five times more creatine phosphate than there is ATP. Since the pH of the sarcoplasm is about 6, the equilibrium within the cell lies far to the right, favoring the formation of ATP at the expense of creatine phosphate. Thus, creatine phosphate can only be synthesized when ATP concentrations are high, as in a resting muscle. As soon as contraction begins and the ATP supply is depleted, the stored creatine phosphate is rapidly broken down to make ATP. The energy for muscle contraction is provided by ATP and no other high energy compound can substitute for it. The enzyme CK is activated by divalent cations such as Mg^{2+}, Mn^{2+}, and Ca^{2+} and is inhibited by ions which bind to thiol groups, such as zinc, copper, and mercuric ions. The enzyme contains two reactive sulfhydryl groups; hence thiols such as glutathione, dithiothreitol and cysteine activate the enzyme.

  b. The greatest amount of CK is found in the skeletal muscle. Cardiac muscle, brain, thyroid, and lung also contain appreciable amounts, but the liver and erythrocytes have very little enzyme activity.

c. A number of methods are available for the measurement of CK activity in serum. The one outlined below involves a series of coupled enzyme reactions:

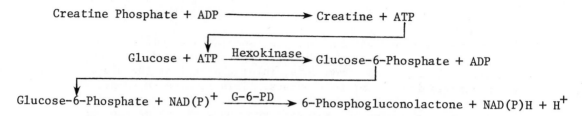

Creatine Phosphate + ADP ⟶ Creatine + ATP

Glucose + ATP $\xrightarrow{\text{Hexokinase}}$ Glucose-6-Phosphate + ADP

Glucose-6-Phosphate + NAD(P)$^+$ $\xrightarrow{\text{G-6-PD}}$ 6-Phosphogluconolactone + NAD(P)H + H$^+$

The NAD(P)H formation is monitored by measuring the increase in absorbance at 340 nm. Adenylate kinase (myokinase) in the serum, also present in the muscle, interferes with this assay. It catalyzes the reaction:

$$2 \text{ ADP} \rightleftharpoons \text{ATP} + \text{AMP}$$

and thus short-circuits the CK assay procedure. If AMP, which inhibits adenylate kinase, is included in the CK assay, the problem is eliminated. Too much AMP will also inhibit CK, though, so no more should be added than is necessary.

d. The major clinical usefulness of the measurement of CK activity in serum is the assessment of diseases of skeletal and cardiac muscles. It should be emphasized that the total CK activity in the blood rises whenever there is the slightest trauma to the skeletal muscle (e.g., brisk walking, running, intramuscular injections). Therefore, separation of the CK isoenzymes is essential to distinguish between skeletal muscle and myocardial muscle injury (discussed later). Muscle disorders can be broadly categorized into two classes.

   i) Diseases of the muscle fiber itself, known as myopathies. Their cause can be genetic (the various muscular dystrophies) or inflammatory (polymyositis); or secondary to endocrine and metabolic problems (alcohol, tetanus toxin).

   ii) Diseases of the muscle secondary to a dysfunction of the nerve supply (neurogenic disorders). Examples include Werdnig-Hoffmann paralysis, peritoneal muscular atrophy, multiple sclerosis, polyneuritis, poliomyelitis, and myasthenia gravis.

Serum CK determinations have proven valuable for distinguishing between these two types, since only in myopathies is the serum CK activity greatly elevated. The neurogenic muscle disorders show only infrequent, modest increases.

2. Multiple Forms (Isoenzymes) of CK

   a. Creatine kinase is a symmetrical dimer. It is made up of two
       subunit types: M (for muscle type) and B (for brain type).
       Thus there are three possible isoenzymes: MM, MB, and BB.
       The isozymes can be separated by electrophoresis or ion
       exchange chromatography under various conditions. The schematic
       diagram below illustrates the electrophoretic separation of CK
       isozymes on an agarose gel, pH 7.5, relative to the major
       serum proteins run under the same conditions. Cellulose acetate
       is also frequently used as the supporting medium.

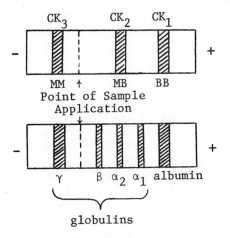

       Following electrophoresis of the serum sample, the CK isoenzymes
       are located by "activity staining." The electrophoresis plate
       is incubated in a solution containing all of the reagents used
       in the total CK assay procedure (discussed above). Wherever CK
       activity is present, NADH is generated. It can be detected and
       quantitated by either of two methods. NADH fluoresces when
       irradiated with long-wavelength (about 350 nm) ultraviolet
       light. The fluorescence intensity of each isoenzyme region can
       be quantitated by fluorescence scanning and detecting equipment.
       Alternatively, the NADH can be reacted with nitroblue
       tetrazolium, which undergoes reduction to a blue-colored
       product known as formazan. (The same procedure is used in LDH
       isoenzyme detection.) The intensity of the blue color, which
       can be measured in a scanning densitometer, is approximately
       proportional to the enzyme level.

   b. Tissue distribution of CK isozymes: CK_1 (BB) is found in the
       brain, thyroid, prostate, kidney, stomach, bladder, lung, small
       bowel, spinal cord, and in normal serum from children and
       umbilical cord blood. CK_2 (MB) is found in the heart, diaphragm,

and tongue, with small amounts being present in umbilical cord serum and serum from children. CK_3 (MM) is found primarily in muscle tissue including the myocardium. Note that heart muscle contains <u>both</u> MM and MB isoenzymes.

c. CK isoenzyme analysis has been most useful in the assessment of myocardial infarction (MI). As early as four hours after a myocardial infarction, CK_2 is detectable in the serum. It reaches a maximum concentration at 16 to 24 hours, then decreases to undetectable levels around 48 hours post-infarction. Studies have shown that the presence in a serum sample of the MB fraction, together with an LDH 1/2 inversion (see the discussion on LDH isoenzymes), virtually confirms the diagnosis of MI. It is important to emphasize that an ECG and an evaluation of the patient's clinical history must always be included in any diagnosis of MI.

Recently, serial determination of CK_2 (by a radioimmunoassay method) has been used to estimate the infarct size. This is possible because the release of CK_2 from the damaged cells and its disappearance from the blood occur at constant rates and follow first-order kinetics. Therefore, by determining the rate of release of CK_2, the volume of distribution of the enzyme and the rate of disappearance from serum, it is possible to calculate the infarct size. This information is of prognostic value. Other disorders which may cause the appearance of MB in the serum are carbon monoxide poisoning, muscular dystrophy, and malignant hyperthermia (see below).

d. In two disorders, malignant hyperthermia and muscular dystrophy, all three CK isoenzymes are found in the serum. It is somewhat puzzling that while the skeletal muscle contains only CK_3 (MM), myopathies show the presence of both MB and BB fractions in the serum. It has been suggested that in these disorders, a dedifferentiation of the muscle causes a reversion to fetal isoenzyme production. This is based on the observation that early embryonic muscle contains predominantly a BB fraction, which is gradually replaced during maturation with MB and MM forms. The adult isozyme pattern is reached at about the fifth to sixth month of gestation.

<u>Metabolism of Alcohols</u>

1. <u>Ethyl Alcohol</u>

a. Ethyl alcohol is rapidly absorbed from the stomach, small intestine, and colon. It is oxidized in the liver to acetyl-CoA in the following steps:

$$CH_3CH_2OH \xrightarrow[\substack{NAD^+ \quad NADH+H^+}]{\substack{\text{alcohol} \\ \text{dehydrogenase}}} CH_3CHO \xrightarrow[\substack{NAD^+ \quad H_2O \quad NADH+H^+}]{\substack{\text{aldehyde} \\ \text{dehydrogenase}}} \underset{\text{acetate}}{CH_3\overset{\overset{\textstyle O}{\|}}{C}-O^-}$$

$$CH_3-\overset{\overset{\textstyle O}{\|}}{C}-O^- \xrightarrow[\substack{CoASH \quad ATP \quad AMP+PP_i}]{\text{Activating Enzyme}} CH_3-\overset{\overset{\textstyle O}{\|}}{C}-SCoA$$

Alcohol dehydrogenase is a cytoplasmic, zinc-containing enzyme. The conversion of acetaldehyde to acetate and acetyl-CoA takes place in the mitochondria. It is not clear whether the acetaldehyde produced is converted to acetyl-CoA via the production of acetate (as shown in the reactions above) or via direct reaction, involving both oxidation and coupling. The rate of oxidation of the alcohol is independent of the blood alcohol concentration (that is, the limiting or rate-controlling factor in the reaction is not the circulating alcohol concentration).

b. It has been reported that alcohol is also oxidzed to acetaldehyde by hepatic microsome (that is, by enzymes located in the smooth endoplasmic reticulum). This has been termed the microsomal ethanol oxidizing system. The reaction requires NADPH, O_2, microsomal hemoprotein P_{450}, and other components (see the section on microsomes, under mitochondria). Other drugs, such as barbiturates, are also metabolized by microsomal oxidases and other enzymes. The observation that alcohol potentiates the effect of barbiturates is <u>partly</u> explained by the fact that both compounds are competing for the same metabolic pathway. <u>Chronic</u> alcohol consumption, on the other hand, causes proliferation of the smooth endoplasmic reticulum, <u>increasing</u> the rate of detoxification of drugs (phenobarbital, warfarin, etc.) as well as enhancing the catabolism of natural compounds such as testosterone. This latter effect may contribute to the feminization observed in chronically alcoholic males. One other pathway implicated in the oxidation of alcohol to acetaldehyde involves catalase and hydrogen peroxide. The <u>predominant</u> reaction of alcohol oxidation, however, is the step catalyzed by alcohol dehydrogenase.

c. Many of the metabolic effects of alcohol center around the production of NADH, leading to an increased NADH/NAD$^+$ ratio in the hepatic cell. We have already seen that NADH affects a number of regulatory sites. Some of the effects of an increased

$NADH/NAD^+$ ratio are: inhibition of citric acid cycle activity, increased lactate production (leading to lactic acidemia and hyperuricemia; see uric acid metabolism for discussion), stimulation of δ-aminolevulinic acid synthesis causing increased synthesis of porphobilinogen, and inhibition of gluconeogenesis (presumably by inhibition of pyruvate carboxylase).

d. Alcohol-induced hypoglycemia is mostly due to decreased stores of liver glycogen (caused by poor nutrition) and to inhibition of gluconeogenesis, as mentioned above. Alcohol inhibits the hepatic synthesis of albumin, transferrin, and complement, while stimulating the synthesis of lipoproteins. Alcohol ingestion gives rise to hypertriglyceridemia and fatty liver. The triglyceride synthesis is increased due to augmented production of α-glycerol phosphate from dihydroxyacetone phosphate (a reaction which requires NADH), and of fatty acids from acetyl-CoA. Note that both NADH and acetyl-CoA are products of alcohol metabolism. Alcoholics show various coagulapathies (due to liver cell injury as well as thrombocytopenia) and leukopenia.

e. The pathogenesis of alcoholic ketoacidosis is not clear but appears to be related to hormonal factors and excessive generation of NADH. The serum of a patient with alcoholic ketoacidosis exhibits an elevated concentration of ketone bodies (see lipid metabolism, Chapter 8, for a discussion on ketone bodies); a pH below 7.4; low, normal or very mildly elevated glucose; a marked anion gap (discussed in acid-base balance, Chapter 12); and a mild to moderate increase in lactate. Growth hormone, epinephrine, glucagon, and cortisol are all elevated causing a mild hyperglycemia and the transfer of free fatty acids from adipose tissue to the liver. The catabolism (β-oxidation) of these fatty acids yields acetyl-CoA which is used to synthesize acetoacetate and (in turn) β-hydroxybutyrate using NADH (discussed in fatty acid metabolism). It is important to distinguish between diabetic ketoacidosis and alcoholic ketoacidosis. In diabetic ketoacidosis the blood glucose levels are very high and insulin is required in the management. This is not true of the alcoholic disorder. Both require varying degrees of treatment with fluid and electrolytes (Na^+, K^+, HCO_3^---refer to Chapter 12, on water and electrolyte balance). In addition, the management of alcoholic ketoacidosis requires glucose.

f. The major clinical problems with alcoholism are alcoholic hepatitis and cirrhosis, both of which can eventually lead to liver failure (hepatic coma) and death. Other complications include intestinal disorders leading to several nutrient absorption problems, variceal bleeding, hepatorenal syndrome, chronic pancreatitis, and infections. As already discussed,

alcohol produces hepatocellular injury which is reflected in a
number of biochemical parameters such as impaired secretory
function (perhaps because of microtubular failure). This
interferes with the export of proteins such as albumin,
transferrin, and others. The hepatocytes also contain <u>Mallory
bodies</u> (alcoholic hyaline) which may be due to antitubular
action of alcohol. Mallory bodies, under the electron microscope,
appear to be fibrillar and tubular aggregations of "intermediate
filaments."

The biochemical abnormalities of alcoholism also lead to
morphological changes in the liver cells which can be detected
microscopically. These include the presence of Mallory bodies,
ballooning of cells, maldistribution of subcellular organelles,
accumulation of triglyceride, and necrosis of the centrolobular
hepatocytes. The centrolobular region normally has a limited
supply of oxygen. Alcohol induces a hypermetabolism which
probably further depletes the supply. The necrosis is then due
to the resultant hypoxia. Lymphocyte infiltration is also seen
and is implicated in the pathogenesis of alcoholic hepatitis.

g. Alcoholic cirrhosis is an end-stage (irreversible) manifestation
of chronic, excessive alcohol consumption. The term <u>cirrhosis</u>
is a morphologic one. It applies when there is widespread hepatic
fibrosis accompanied by the formation of nodules of parenchymal
cells. Alcoholic cirrhosis is a micronodular type. The fibrosis
causes separation of the lobules (the functional units of the
liver--see Chapter 15, on liver function) and leads to the
gradual destruction of hepatocytes, impaired bile flow and
portal circulation, and eventual hepatic failure. The pathways
that lead to alcoholic cirrhosis are not clearly defined at this
time. Collagen (Types I and III) synthesis, as well as
fibrogenesis, is stimulated by alcohol.

h. Disulfiram is a compound used in the treatment of chronic
alcoholism. It inhibits the aldehyde dehydrogenase by competing
with NAD^+ for binding sites on the enzyme molecule. This leads
to accumulation of acetaldehyde giving rise to such undesirable
effects as nausea, vomiting, thirst, sweating, headache, etc.

Structure

$$H_5C_2 \diagdown C_2H_5$$
$$N-C-S-S-C-N$$
$$H_5C_2 \diagup \underset{S}{\Vert} \underset{S}{\Vert} \diagdown C_2H_5$$

Disulfiram

2. Methyl Alcohol Degradation

    a. Methyl alcohol is usually introduced into the body
    accidentally and is metabolized in the liver as follows:

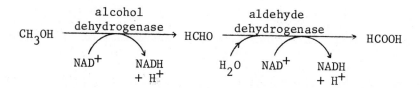

    The initial oxidation produces formaldehyde which damages the
    retinal cells, leading to blindness. The second oxidation
    step generates formic acid which produces acidosis.

    b. The initial dehydrogenase which acts on methyl alcohol to
    produce formaldehyde is the same enzyme that oxidizes the
    ethyl alcohol to acetaldehyde. This observation has been used
    in the treatment of methanol toxicity. Ethanol is administered
    to ameliorate the toxic effects of methanol. The ethanol
    competes with the methanol for enzyme and NAD^+, thereby
    reducing the rate of formaldehyde synthesis to the point where
    the aldehyde dehydrogenase can remove it rapidly enough to
    prevent any buildup. The acidosis is corrected by
    administering bicarbonates, etc.

5
Nucleic Acids

Nucleic Acid Composition

1. Nucleic Acids. There are two types of nucleic acids: ribonucleic
    acids (RNA) and deoxyribonucleic acids (DNA). The "deoxy" refers
    to the absence of an -OH in the 2-position of the ribose molecules.
    They are called acids because, due to the presence of the phosphate
    groups, they behave chemically as acids. The diagram below
    illustrates their components.

Constituents of Nucleic Acids

NUCLEIC ACIDS
(polynucleotide)
↓
NUCLEOTIDES
(base-sugar-phosphate)

NUCLEOSIDES   PHOSPHORIC ACID
(base-sugar)

PURINES & PYRIMIDINES   SUGARS
("bases")

*Adenine   Thymine (in DNA only)   D-Ribose (in RNA only)
*Guanine   *Cytosine   D-Deoxyribose (in DNA only)
           Uracil (in RNA only)

(* means that the base occurs in both DNA and RNA)

2. Mononucleotides (or just nucleotides) are structural units of
    nucleic acids. They consist of

          Purine or pyrimidine (nitrogenous base)

          D-ribose or 2-deoxy-D-ribose (sugar)

          Phosphoric acid

    a. The nitrogenous bases found in most nucleotides are indicated
        next. The numbering systems are those used by Chemical
        Abstracts.

Purines

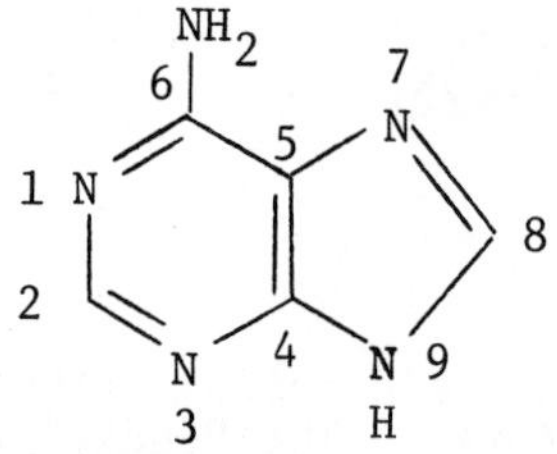

NH$_2$

6 | 7
5
1 N   N
8
2
N   4 N 9
3   H

Adenine: 6-aminopurine

O

HN   N

H$_2$N   N   N
H

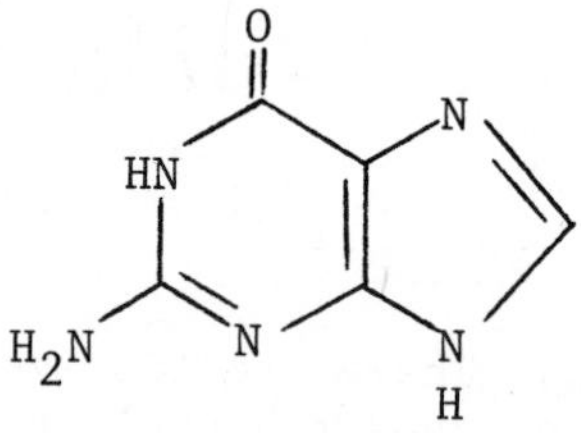

Guanine: 2-amino-6-oxypurine

Pyrimidines

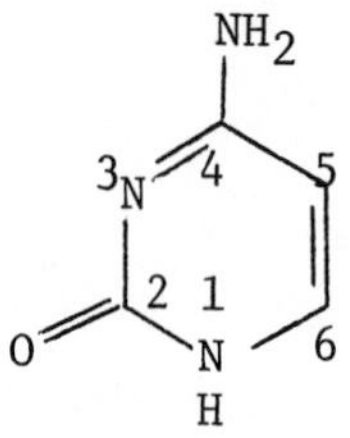

NH$_2$

3 N 4 5
2 1
O   N   6
H

Cytosine: 2 oxy-
4 amino-
pyrimidine

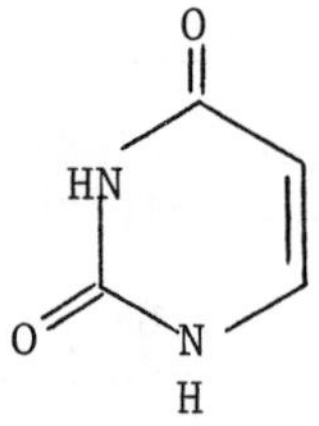

O

HN

O   N
H

Uracil: 2,4-
dioxy-
pyrimidine

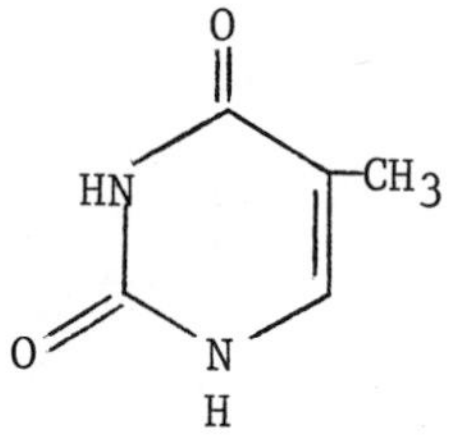

O

HN   CH$_3$

O   N
H

Thymine: 5-methyl-
2,4-dioxy-
pyrimidine

Other bases, particularly methylated derivatives of the
five indicated above, are found in some instances.
For example, many (probably all) transfer RNA (tRNA)
molecules contain inosine (a purine), pseudouridine
(a pyrimidine),and others (see under RNA).

b. In oxypurines and oxypyrimidines,an equilibrium exists between
keto and enol forms. This interconversion, known as
tautomerization, involves the exchange of two protons between
different positions on the molecule. This is illustrated
below for uracil. This property is very important in nucleic
acid structure and the genetic code, since the ability of the
keto and enol forms to hydrogen bond differs. The tautomers
favored in the equilibrium are indicated for uracil
and this form will be used throughout the book. The
constant presence of at least small amounts of the other
form may be a contributory factor in the rate of spontaneous
mutation.

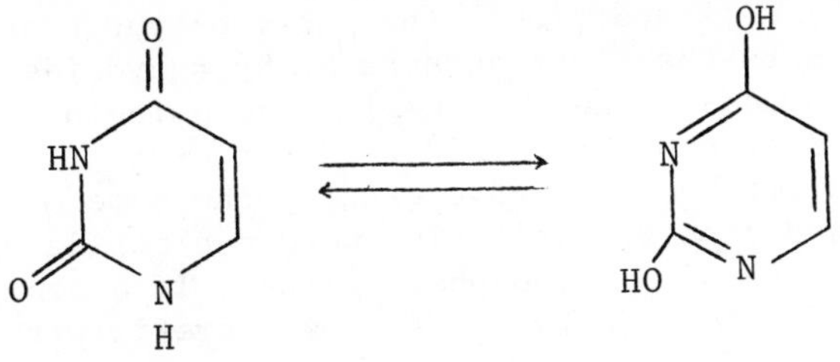

Lactam or keto form
of Uracil

Lactim or enol form
of Uracil

c. Nucleosides are β-N-glycosides of pyrimidine and purine bases.
    Examples are shown below.

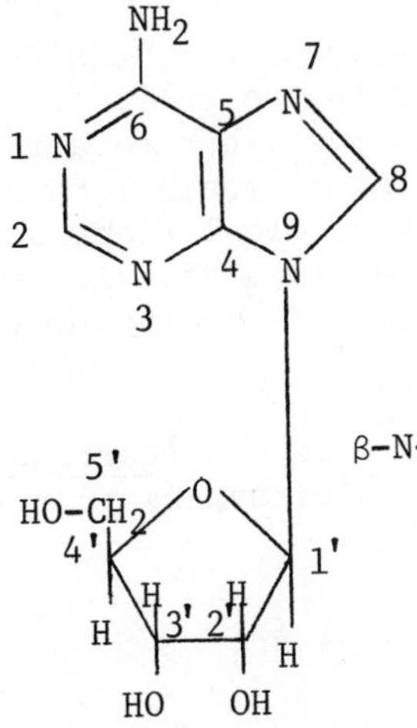

β-N-glycosidic linkage

This compound is <u>adenosine</u> or <u>adenine ribonucleoside</u>
(9-β-D-ribofuranosyladenine). If the -OH on the 2'-position
of the ribose is replaced by -H, the corresponding
2'-deoxynucleoside (in this case 2'-deoxyadenosine) is formed.
Note that it is the N-9 position on the purine ring which is
involved in the N-glycosidic bond. In pyrimidine nucleosides,
the N-1 nitrogen forms the linkage.

d. <u>Nucleotides</u> are formed from nucleosides by addition of a
    phosphate group. The phosphate is attached in an ester
    linkage, usually to the 5'-hydroxyl but sometimes to the
    3'-hydroxyl of the ribose. If the position (5' or 3') is not
    specified, the 5'-ester can generally be assumed. For example,
    adenine-5'-nucleotide is more usually known as adenylic acid
    (since the phosphate group is acidic) or adenosine

monophosphate (AMP). If additional (one or two) phosphate
groups are attached to the first phosphate (by anhydride
linkages), the compounds formed are called adenosine diphosphate
(ADP) and adenosine triphosphate (ATP). Notice that
nucleoside di- and triphosphates are exactly the same as
nucleotide mono- and diphosphates. The word nucleotide already
indicates the presence of one phosphate group. If a deoxy
sugar is present, the term deoxy is used as a prefix with
the nucleotide name. When bases other than adenine are
involved, the corresponding names (guanosine, guanylic acid;
cytosine, cytidylic acid; uridine, urydylic acid; thymidine,
thymidylic acid) are used. A summary of these names is given
with purine and pyrimidine nucleotide biosynthesis.

Functions of Nucleotides

1. By means of the ATP $\rightleftarrows$ ADP + P_i interconversion, ATP serves as the
    <u>primary carrier of chemical energy</u> in the cell. (Other nucleoside
    triphosphates (NTP's) are also involved, but to a much smaller
    extent.) The energy can be used for mechanical function (as in
    the ATPase of muscle) and to drive chemical reactions (as in the
    synthesis of glutathione). For reactions of this sort,
    only the terminal phosphate anhydride is usually hydrolyzed,
    giving ADP and P_i.

2. Different parts of NTP molecules are used as <u>carriers and
    activators</u> of a variety of groups in biosynthetic reactions.
    Several examples are given below.

    a. <u>Adenosine Triphosphate</u>

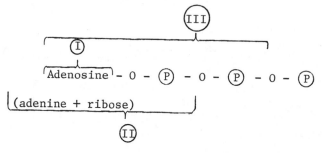

    I . adenosine, as <u>S</u>-adenosylmethionine, carries "active
         methyl" groups for biological methylations (see
         metabolism of the sulfur-containing amino acids).

    II . AMP, as part of 3'-phosphoadenosine-5'-phosphosulfate
          (PAPS; see metabolism of sulfur-containing amino acids)
          provides "active sulfate" for synthesis of the

chondroitin sulfates, etc. AMP is also the carrier of amino acids during the loading (activation) of the tRNA's used in protein synthesis.

III . ADP is part of coenzyme A. In this and other molecules containing NDP's (see below), even though ATP is the immediate source of adenosine, usually only one phosphate comes along with it. The other phosphate was already part of the molecule to which the adenosine monophosphate is being added. For example:

$$\text{4'-phosphopantetheine} + \text{ATP} \longrightarrow \text{dephospho-CoA} + PP_i .$$

This also illustrates the release of inorganic pyrophosphate (PP_i), another feature of many reactions in which a NMP is incorporated into some other molecule. The pyrophosphate is rapidly hydrolyzed to inorganic phosphate by pyrophosphatase.

$$PP_i + H_2O \xrightarrow{\text{pyrophosphatase}} 2\ P_i$$

About 7.3 kcal are released per mole of PP_i hydrolyzed. This helps to make the resynthesis of ATP by the reversal of these reactions quite difficult and the desired product (here, dephospho-CoA) much more stable.

b. Uridine triphosphate is used to synthesize UDP-glucose, UDP-galactose, UDP-glucuronic acid, and other UDP-sugars. These compounds are used in the biosynthesis of polysaccharides and for other reactions. See for example glycogen synthesis, galactose metabolism, the uronic acid pathway, amino sugar metabolism, and bilirubin (hemoglobin) metabolism.

c. Cytidine triphosphate serves as a source of CMP for CDP-choline and CDP-diglyceride in lipid and phospholipid biosynthesis.

Cytidine diphosphate choline (CDP-choline)

3.   NTP's, especially ATP, <u>provide inorganic phosphate and
     pyrophosphate for biosynthesis</u>.

   a.   Phosphorylation reactions include those catalyzed by hexokinase,
        glucokinase, fructokinase, the protein kinases, and others.
        These are discussed in a number of places. If X is a molecule
        which can be phosphorylated, a general reaction is

$$X + NTP \xrightarrow{\text{kinase}} X\text{-}P + NDP$$

   b.   Pyrophosphate transfers (pyrophosphorylations) are less common.
        One important example is the formation of 1-phosphoribosyl-5-
        pyrophosphate (PRPP), the form of ribose used in nucleotide
        synthesis. The reaction is

$$\text{Ribose-1-P} + ATP \longrightarrow PRPP + AMP$$

     This is discussed further under purine and pyrimidine synthesis.

4.   A number of <u>coenzymes</u> involved in electron transfer reactions
     contain nucleotides. These include NAD^+, $NADP^+$, and FAD. FMN's
     (and part of FAD) are not true nucleotides since they contain
     D-ribitol, a sugar alcohol, in place of D-ribose. The nitrogenous
     base in FMN (and part of FAD) is 6,7-dimethylisoalloxazine (flavin),
     which is not one of the bases found in the nucleic acid nucleotides.
     Coenzyme A, mentioned above, is another nucleotide coenzyme.

5.   ATP, CTP, GTP, and UTP are <u>precursors for the nucleotides in RNA</u>.
     The deoxynucleotides (dATP, dCTP, dGTP, and dTTP) probably
     function similarly in DNA synthesis. Some evidence exists, however,
     that the deoxynucleotide triphosphates are not the <u>immediate</u>
     precursors of DNA (i.e., they are not substrates for DNA polymerase),
     at least in <u>E. coli</u>. There may also be two or more intracellular
     pools of nucleotides which provide material for different
     processes. In tRNA, the "unusual" nucleotides (those containing
     bases other than A, C, G, or U) are usually if not always the
     result of modifications performed on A, C, G, or U <u>after</u> these
     "usual" nucleotides are already part of the polynucleotide. This
     is equivalent to postribosomal modification in protein synthesis.

6.   The most recently described function of nucleotides is their
     role as <u>intracellular hormones or "second messengers."</u> This is
     best typified by cyclic AMP, discussed previously. Other cyclic
     nucleotides may perform a similar function.

Structure and Properties of DNA

1. A portion of a polynucleotide chain is shown below.

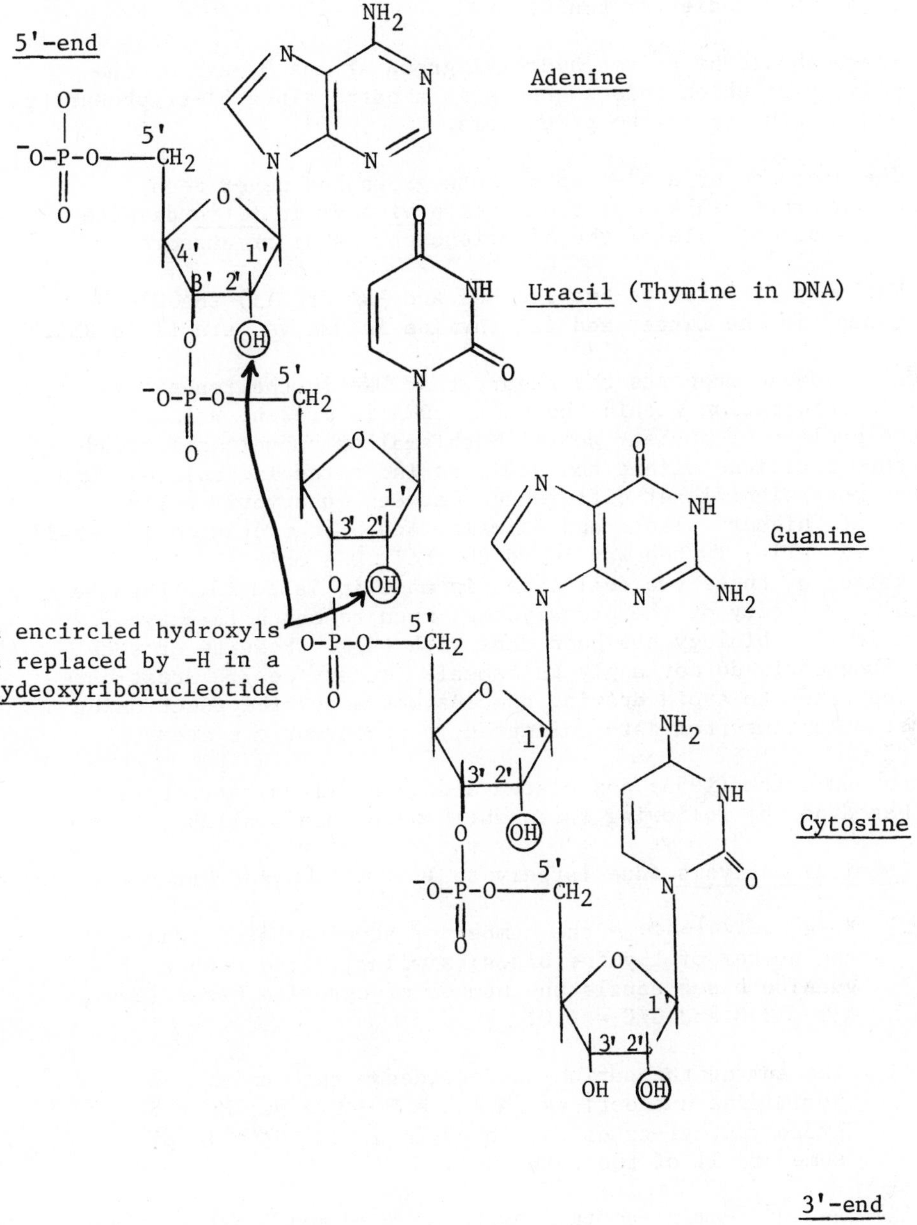

5'-end

Adenine

Uracil (Thymine in DNA)

Guanine

The encircled hydroxyls are replaced by -H in a polydeoxyribonucleotide

Cytosine

3'-end

Regarding this structure, notice that

a. In both DNA and RNA the nucleotides are connected by 3',5'-phosphodiester bonds.

b. There should be a free hydroxyl group at the 3'-end of the polymer of which this sequence is a part, since 5'-triphosphates are thought to be the precursors.

c. The presence of a 5'-triphosphate group has never been demonstrated in DNA synthesized *in vivo* or *in vitro* despite the apparent role of the 5'-triphosphates as precursors.

d. The only differences between DNA and RNA are (1) the 2'-OH groups in the latter and (2) thymine in DNA *vs*. uracil in RNA.

2. Strong evidence supports the theory that DNA is the repository of **genetic information within the cell.** DNA is present primarily in the nucleus or nuclear zone. Much smaller amounts are found in other locations within the cell, as indicated in Table 8. One of the most significant differences between eucaryotic cells (those of "higher" plants and animals) and procaryotic cells (small, primitive cells) is the way in which their DNA exists. A comparison of these two cell types is made in Table 15. Because of the simplicity of the procaryotes, much research in enzymology and molecular biology has been done with them. Results of such work frequently do not apply to mammals (eucaryotes). Great care must be taken to avoid drawing conclusions about the functioning of higher organisms from data collected in procaryotic systems.

3. Watson and Crick (1953) postulated a double-helical structure for DNA based on the following facts which were then available to them.

    a. <u>Chemical Analysis</u> (due largely to I. Chargaff and his coworkers)

        i) Base equivalence - the number of adenine bases equals the number of thymine bases; similarly, the number of guanine bases equals the number of cytosine bases (i.e., A/T = 1.0 and G/C = 1.0).

        ii) The sum of the purine nucleotides = the sum of the pyrimidine nucleotides (A + G = T + C + MC; MC = 5-hydroxymethyl-cytosine, found in a few DNA's in place of some or all of the cytosine).

        iii) 4- and 6-amino-containing bases = 4- and 6-keto-containing bases (A + C + MC = G + T).

Table 15.   Comparison of Eucaryotic and Procaryotic Cells

Procaryotic Cells	Eucaryotic Cells
(Eubacteria (including Escherichia coli and Bacillus subtilis); the spirochetes and rickettsiae; and others)	(All higher plants and animals, including man)
I) Small cells	I) Cells are 1000–10,000 times larger
II) Cytoplasm is structureless except for storage granules, ribosomes, and the nuclear zone; only membrane present is the cell membrane, surrounding the entire cell	II) In addition to the outer cell-membrane, many structured, membrane-bounded organelles occur within the cytoplasm. These include mitochondria, chloroplasts (in photosynthetic cells), Golgi bodies, rough and smooth endoplasmic reticulum, peroxisomes, lysosomes, and the nucleus
III) Ribosomes have a sedimentation rate (in the ultracentrifuge) of 70S; composed of two subunits (30S and 50S)(see under RNA)	III) Larger ribosomes (sediment at 80S); subunits are 40S and 60S (varies somewhat between plants and animals)
IV) Most DNA exists as a single macromolecule (chromosome); it is thought that no protein is associated with the DNA (although this is not well established; in B. subtilis there is some evidence for DNA association with a basic protein); small amounts of cytoplasmic DNA occur and are called plasmids or episomes	IV) DNA is distributed among multiple chromosomes (8 in Drosophila (fruit fly); 46 in man; 78 in chickens); one or more molecules of DNA per chromosome; DNA associated with large quantities of basic proteins (protamines and histones; DNA + protein = chromatin); nucleus surrounded by membrane; nucleolus (site of ribosomal synthesis) associated with nucleus; some DNA (0.1–0.2% of total cellular DNA) found in mitochondria and chloroplasts; very small amounts of DNA (satellite DNA) found free in the cytoplasm

iv) The base ratio $(A + T)/(G + C + MC)$, known as the
dissymmetry ratio, is a characteristic ratio for a given
species and shows species variation.

b. <u>X-ray diffraction analysis</u> of DNA fibers by Franklin and
Wilkins showed a regularity in the physical structure of
the DNA molecule. (Wilkins shared the Nobel Prize in medicine
and physiology with Watson and Crick in 1962.)

c. The development, by Cochran, Crick and Vand (1952), of the
theory of diffraction of helical molecules and the
postulated (Pauling, Corey, and Branson, 1951) α-helical
structure for some polypeptides   (Pauling received a Nobel
Prize for his work on protein structure) were important also.

4. Details of their proposed structure are given below.

a. The basic structure is two chains wound around a common
(hypothetical) axis, with the ribose-phosphate backbone on the
outside (away from the center) and the bases pointing in,
towards the axis. The diameter of the helical cylinder is 20 Å.

b. The bases are planar and in the keto tautomeric form. The
planes of the bases are all perpendicular to the central axis
and the distance from one base to the next is 3.4 Å. This
results in the stacking of the bases, one upon the other,
not unlike a pile of plates (albeit a twisted one, since the
stack must follow the curve of the helix). There are ten
nucleotides in each complete turn of the helix. The
van der Waal's and hydrophobic interactions resulting from
the base stacking provide at least part of the energy needed
to stabilize the structure.

c. The two chains are complementary and are of opposite polarity
to each other. This is shown schematically below. In the
drawing, the four bases are represented by their initial
letters (A, C, T, G) and the deoxyribose-phosphate backbone is
indicated by the dashes between the letters; the dotted
lines between bases represent hydrogen bonding.

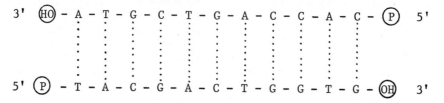

Note the <u>opposite polarity</u> of the strands (one runs 3' → 5' while the adjacent (complementary) one runs 5' → 3') and the complementarity of the bases (A always opposite T; G always opposite C). Notice in each case, a purine is paired with a pyrimidine.

d.   The complementarity indicated above is necessary to explain the observed base ratios (chemical composition). The origin of the specificity (A-T and G-C) lies in the ability of these pairs to form the most stable hydrogen-bonded dimers. The hydrogen-bonding patterns thought to be significant are indicated below. As in the case of the α-helix and peptide-group geometry (mentioned before),knowledge of this pattern and the base geometries contributed to the prediction of the double helix.

Other hydrogen-bonding patterns (notably those proposed by Hoogsteen) can also be drawn, but they are probably of lesser significance (however, see under tRNA).

5. The two strands and the specificity of the base pairing in this structure account for accurate replication of the DNA molecule during mitosis.

6. The DNA duplexes undergo unwinding (denaturation) upon heating, changes in pH, and other denaturing influences (see protein denaturation). Upon denaturation, the optical absorbance at 260 nm increases due to the unstacking of the bases. This change in optical density is a very useful method for measuring the degree of denaturation of a DNA solution under various circumstances. Other changes (decreased viscosity, increase in optical rotation, increase in buoyant density) also occur upon denaturation.

7. Experiments in bacterial transformation (transfer of genetic information from one bacterial cell to another bacterial cell; discovered by Griffith and used extensively by Avery and others), viral transduction (virus-mediated transfer of parts of the bacterial genome from one cell to another), other viral techniques (notably the Hershey-Chase experiments) have helped greatly to establish the role of DNA as the genetic material. Through these experiments, the genetic map (location of the genes relative to one another in the DNA) of E. coli has been determined. It has been shown that the single chromosome (DNA molecule) in E. coli is a closed loop (circular). Certain other DNA's, isolated from several sources (some viruses, mitochondria, other bacteria), appear also to be circular.

8. It has been observed, in general, that the amount of DNA per cell increases with the increase in complexity of a cell's function. This is indicated in Table 16.

Table 16.   The DNA Content of Some Selected Cells

	DNA in picograms per cell or per virion
Bacteriophage T	0.00024
Bacterium (E. coli)	0.009
Fungi (Neurospora crassa)	0.017
Higher Plant (tobacco)	2.5
Sponge (Tube sponge; diploid)	0.12
Bird (chicken erythrocytes)	2.5
Reptile (alligator)	4.98
Fish (carp)	3.0
Amphibia (Xenopus laevis)	8.4
Mammal (human)	6.0

9. In the genomes of eucaryotic cells (but most probably not in viruses and bacteria), a given nucleotide sequence may be repeated many times (gene amplification). In other words, many extra copies of the same gene are present. This observation is widespread and may be true in all higher organisms. It may help to explain why amphibians, for example, have more DNA per cell than do mammals.

    The occurrence of multiple copies of some genes was concluded largely through the use of DNA-DNA hybridization experiments. In doing these, one first isolates DNA from the cells to be tested, denatures it to obtain single-stranded DNA, and immobilizes it by bonding it to some material (e.g., resin beads) which can be used to pack a chromatography column. Labeled (usually tritiated) DNA is then prepared by growing more of the cells on a medium containing radioactive thymidine. This DNA is similarly isolated and denatured, then passed over the beads containing the bound "reference" DNA. Renaturation (annealing) of the DNA is a second-order process, depending on the frequency with which two genes capable of pairing meet each other. Consequently, the more copies there are present of one gene, the greater is the probability that a particular meeting (collision) will be successful (result in a stable, base-paired hybrid). The amount of pairing which occurs in a given length of time is therefore directly related to the number of copies of each gene present. Alternatively, one can say that the rate of hybridization is proportional to the number of copies of the same DNA sequence which are present.

    Nucleic acid hybridizations (DNA-DNA, DNA-RNA and RNA-RNA) are also used to measure the degree of homology between the genomes of different species and hence to evaluate how closely related two species are (molecular taxonomy), to measure the number of copies of RNA made from a particular gene (DNA molecule), and for other purposes. These methods generally measure the maximum amount of DNA which can be hybridized, rather than the rate of hybridization. Such experiments are easier with procaryotic nucleic acids than with eucaryotic ones, due to the much greater complexity of the genetic materials in the latter cells. Some very interesting data have resulted from these experiments, despite the many pitfalls in interpretation of the results.

10. There are at least three reasons why multiple copies of one gene might be necessary.

    a) The need for synthesis of large quantities of one type of protein or nucleic acid. Thus, many copies of the genes for collagen, keratin, and structural proteins of membranes would be useful. That this actually occurs has been demonstrated in the case of ribosomal RNA.

b. The occurrence, in some proteins, of repeating homologous amino acid sequences. Examples are the immunoglobulins and hemoglobin. In at least the latter case, however, gene amplification has been shown to <u>not</u> occur. Rather, <u>many copies of the mRNA</u> for the hemoglobin peptides are transcribed from a few (2-3) copies of the appropriate gene.

c. A mutation in one copy, leading to the production of an inactive enzyme, will not completely eliminate that enzyme from the cell if there is another good gene for it somewhere else on the DNA. This is also cited as a reason for the occurrence of isozymes (see under proteins and enzymes).

11. Among higher organisms, all of the somatic diploid cells of a given species contain the same amount of DNA. This is independent of the metabolic state of the cell (not modified by diet or environmental circumstances), unlike the RNA content of the same cells. This indicates again the permanence of the DNA complement, a necessary feature for anything which must store and transmit information.

12. The base compositions of DNA specimens vary from one species to another, but the complementarity (A=T; G=C) is present regardless of species.

Replication of DNA

1. Double-stranded DNA undergoes semiconservative replication. This means that the two complementary antiparallel chains separate and each serves as a template for the synthesis of its complement, so that the two daughter double-stranded molecules are identical to the parent, each containing one strand from the parent and one "new" strand. This was postulated by Watson and Crick and experimentally verified by Meselson and Stahl. Although the details are far from clear, it appears that the replication proceeds as the unwinding is occurring, the new synthesis taking place just behind the unwinding-point and moving along with it.

2. There is still a great deal of controversy surrounding the replicative mechanism. Despite the discovery of at least three DNA-dependent DNA polymerases, it is not clear which enzyme actually performs the polymerization <u>in vivo</u>. It is also not known for certain exactly what the immediate precursors to the deoxynucleotides in the DNA are.

3. Kornberg and his coworkers were the first to isolate a DNA-dependent DNA polymerase [(DNA polymerase I), Kornberg enzyme; from <u>E. coli</u>]. For this work Kornberg received the Nobel Prize.

As a polymerase, this enzyme requires

a. The four deoxynucleoside triphosphates (dATP, dTTP, dGTP, dCTP) and Mg^{+2}. All four must be present for synthesis to proceed.

b. Preformed DNA called a <u>template-primer</u>. This consists of a template, which is a long, single strand of DNA or a double-stranded DNA with a single-strand nick or gap. Base-paired to the template is the primer strand (need not be long) which has a free 3'-OH group.

Under these conditions, the enzyme starts adding deoxynucleotides to the free 3'-OH end of the primer. The deoxynucleotides are complementary to the template strand. The growing point of the new strand is at its 3' terminus. Total synthesis of biologically active DNA of the ØX 174 bacteriophage has been achieved using Kornberg's DNA polymerase and a DNA ligase (connects the 5' end to the 3' end of a strand of DNA, making a circle). This work was published by Goulian, Kornberg and Sinsheimer.

4. The reactions involved in the functioning of DNA polymerase I are shown below.

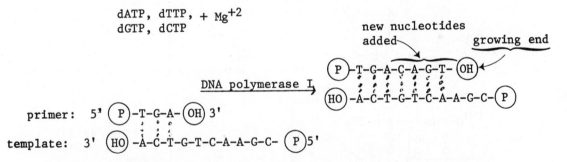

In this reaction, one pyrophosphate (PP_i) is released for each nucleotide added to the polymer. The polymerization is made more irreversible by the action of a pyrophosphatase, which catalyzes the highly exergonic reaction.

$$PP_i + H_2O \xrightarrow{\text{pyrophosphatase}} 2\ P_i$$

This was described previously in the section on functions of the nucleotides. The structural relationship of the growing chain to the nucleotides being added is shown in the following diagram.

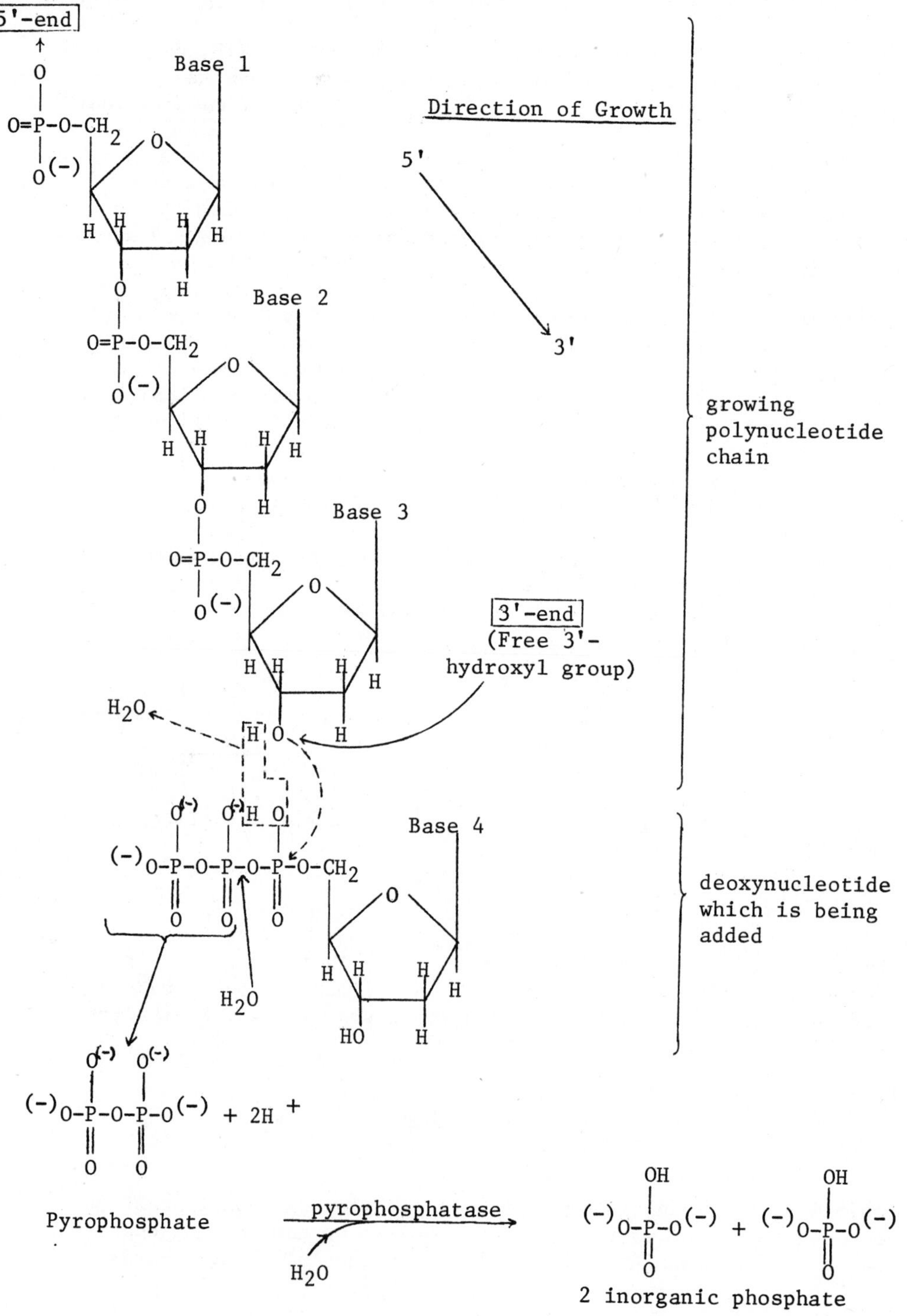

5'-end

Base 1

Direction of Growth

5'

3'

growing
polynucleotide
chain

Base 2

Base 3

3'-end
(Free 3'-
hydroxyl group)

H_2O

$O=P-O-CH_2$

deoxynucleotide
which is being
added

Base 4

H_2O

HO   H

Pyrophosphate $+ 2H^+$

pyrophosphatase

H_2O

2 inorganic phosphate

5. a. The Kornberg enzyme possesses two types of exonuclease activity
        in addition to the polymerase activity described above. It can
        remove, by hydrolysis, nucleotide residues from either the
        3'-(3' → 5' exonuclease) or 5'-(5' → 3' exonuclease) end of a
        single strand of DNA. This ability may aid in the elimination
        of mismatched nucleotide pairs so that DNA synthesis proceeds
        correctly; or it may be needed by the enzyme in performing DNA
        repair, as described below.

    b. The biggest setback to the theory that DNA polymerase I (Pol I)
        was the enzyme responsible for the replication of DNA came from
        the discovery that E. coli mutants lacking this enzyme were
        capable of normal growth rates. These mutants contain two
        other DNA polymerases, designated II and III. Later, it was
        shown that "normal" (wild-type) E. coli also had these enzymes.

        These two enzymes differ with respect to template and primer
        requirements and exonuclease activities, when compared with
        Pol I. DNA polymerase II (Pol II) and III (Pol III) have only
        3' → 5' exonuclease activity while Pol I possesses both
        exonuclease activities. The polymerases II and III have
        stringent template-primer requirements in that they only use
        double-stranded DNA with short gaps. Polymerase I, on the
        other hand, requires long, single-stranded DNA or duplex DNA
        with a single-strand nick or gap (template) base-paired with
        another DNA strand that has a free 3'-OH group (primer).

        DNA polymerase III is the most active of the three enzymes and
        it is thought that it may be the replicative enzyme. The
        biologically active form of polymerase III (Pol III*) is
        believed to be a dimer. The enzyme complex which appears to
        initiate the growth of a DNA-strand is composed of Pol III*
        and a protein known as copolymerase III* (Copol III*). In
        order to initiate the replication, the complex (Pol III*-
        Copol III*) combines with ATP, a DNA template, and an RNA
        primer (discussed later).

    c. The E. coli mutants which do not contain DNA polymerase I and
        yet undergo normal growth are extremely sensitive to ultraviolet
        and x-ray irradiation and show an increased number of deletion
        mutants (see under mutations--this chapter). Furthermore, the
        repair of nicks and gaps in the DNA of these mutants is
        seriously impaired. All of these observations suggest that
        the primary function of polymerase I is repair and not
        DNA replication.

6. As mentioned earlier, DNA replication is bidirectional (both
    strands are duplicated simultaneously), yet the direction of
    synthesis is always 5' → 3'. That is, all of the known DNA
    polymerases can add nucleotides only to a free 3'-OH group.

a. Before synthesis begins, the double-helical DNA must be unwound, exposing the strands. This is brought about by specific proteins (mol. mass 10,000–75,000 daltons) which have oligonucleotide binding sites with a capacity of about eight nucleotides. The binding of these proteins to DNA is a cooperative phenomenon (binding of the first one makes binding of the succeeding ones easier). They attach in succession to one strand, causing a "bubble" to form in the duplex in advance of the replicating fork (the Y-shaped region at the end of a bubble, where active DNA polymerization is taking place). This is illustrated below.

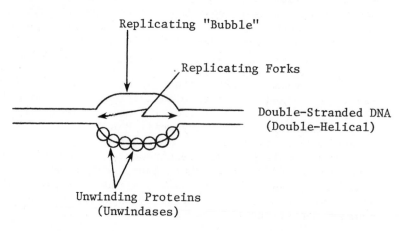

These proteins have been isolated from both virus-infected bacterial cells and eukaryotic cells. From a mechanical viewpoint, this progressive unwinding during the replication of double-stranded circular DNA leads to supercoiling. The torsional stress is probably relieved by creating an opening (nick) in one of the strands, permitting relaxation of the supercoils. The factor which produces and afterwards seals the nicks is a protein, presumably an enzyme.

b. The problem of 3' → 5' synthesis is explained in the following way. DNA synthesis takes place in "bursts," producing small fragments (known as Okazaki fragments) of 1000 to 2000 residues. Each fragment is preceded by a primer RNA (see below) which is subsequently excised. The Okazaki fragments are then joined by the enzyme DNA ligase. As can be seen from the accompanying diagram, this type of replication permits copying of both (antiparallel) strands of DNA, without requiring a 3' → 5' polymerase.

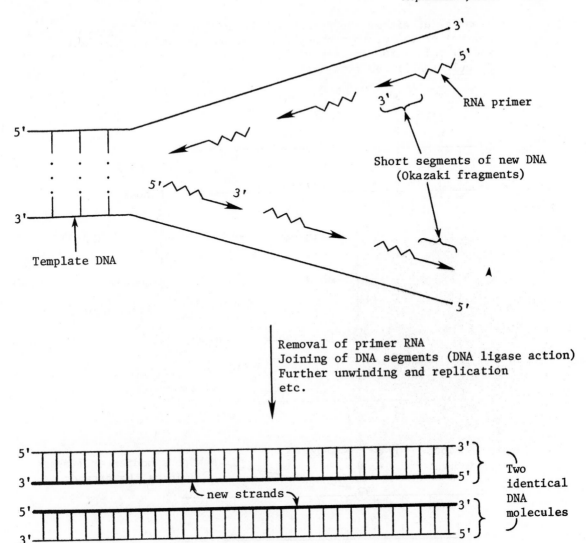

RNA primer

Short segments of new DNA
(Okazaki fragments)

Template DNA

Removal of primer RNA
Joining of DNA segments (DNA ligase action)
Further unwinding and replication
etc.

new strands

Two
identical
DNA
molecules

c.   DNA ligase action: This enzyme catalyzes the joining of two
     pieces of DNA by forming a single phosphodiester bond between
     the 3'-OH and 5'-phosphate ends of the two pieces. The
     catalysis occurs only when the nucleotides to be linked are
     paired to adjacent bases in the other strand and when the
     strands themselves are part of a double-stranded DNA. The
     energy for the formation of the bond is derived from the

cleavage of either the pyrophosphate bond of NAD^+ (E. coli
enzyme) or the α,β-pyrophosphate bond of ATP (eukaryotic cells
and E. coli infected with bacteriophage T4). The steps of the
E. coli (NAD^+) reaction are shown below.

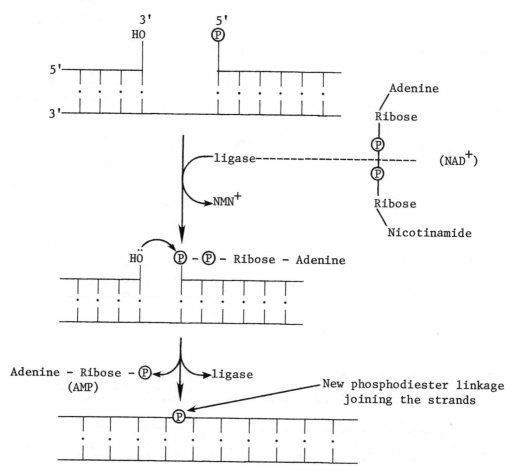

Note: NMN^+ = nicotinamide mononucleotide

7. In vitro experiments have shown that none of the known DNA
    polymerases can use native double-stranded DNA as a primer. It has
    been demonstrated, however, that DNA-polymerase can use an RNA-DNA
    hybrid. It is also known that DNA-dependent RNA polymerase can
    synthesize RNA using double-stranded DNA as a template (discussed
    later). All of this information has led to the formulation of the
    following probable sequence of events that take place during DNA
    synthesis in vivo.

a. The synthesis is begun at a specific initiation site on the DNA
molecule by a <u>DNA-dependent RNA polymerase</u>. This enzyme
generates a short, complementary RNA primer using ribonucleoside
5'-triphosphates.

b. The Pol III*-Copol III* complex, in the presence of ATP,
initiates the addition of deoxyribonucleotides to the 3'-end
of the primer RNA. After initiation, the Copol III* is detached
from the enzyme complex and replication of the template is
continued by Pol III*.

c. Extension of the replication bubble occurs with the help of
unwinding proteins.

d. Removal of the RNA primer from the 5'-end, one residue at a
time, is catalyzed by the 5' → 3' exonuclease activity of DNA
polymerase or by a specific ribonuclease.

e. Finally, the oligo-polydeoxyribonucleotides are joined by DNA
ligase.

8. The accurate transmission of genetic information requires great
stability of the structure of DNA in order to guarantee precise
replication. To assist in this, cells have developed certain
error-correcting mechanisms. One common type of damage which
most cells can reverse is thymine-dimer formation. Adjacent
thymine residues in the DNA molecule, upon exposure to ultraviolet
light, undergo the reaction shown below:

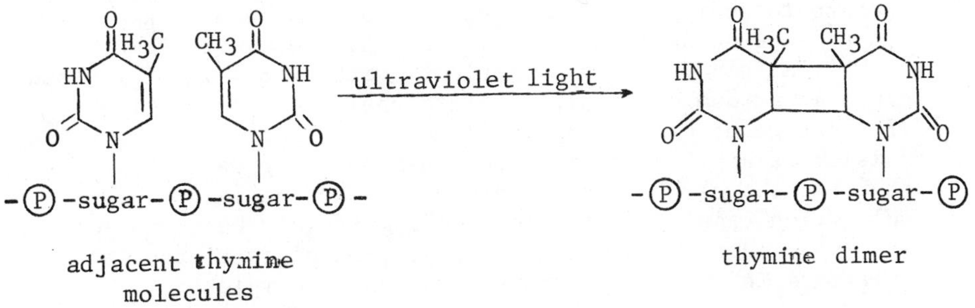

adjacent thymine molecules   thymine dimer

These dimers form "lumps" in the DNA and prevent replication. Two
mechanisms are known for dimer removal.

a. <u>Photo-reactivation</u> is just the reverse of the above reaction.
There is thought to be an enzyme which binds selectively to
thymine dimers. Upon exposure to visible light the dimers
are cleaved and the photoreactivating enzyme is released.

b. Thymine dimers can also be reactivated in the dark in a more
complex process probably involving excision of the dimer and
resynthesis of a short stretch of DNA. An endonuclease
(enzyme which hydrolyzes nucleic acids at points other than
the ends) specific for regions containing thymine dimers has
been found in Micrococcus luteus. This enzyme makes a nick
(single cut) in the defective strand and exposes a free 3'-OH
for a polymerase to act on. The resynthesis of the defective
region (complementary to the undamaged strand) and final
excision of the dimer-region are probably performed by DNA
polymerase I. Evidence in support of this includes the
discovery of a strain of E. coli mutants which are more
sensitive to ultraviolet light than is the wild type and which
lack the DNA polymerase I enzyme and the fact that the requirements for thymine-dimer repair synthesis are the same as those
for DNA polymerase I synthesis of DNA. For these and other
reasons, Kornberg's enzyme is probably not directly involved in
DNA replication in vivo. The use of the complementarity of the
two strands of DNA to direct the repair process may indicate an
evolutionary reason for the development of double-stranded DNA.
Another type of repair process consists of replacing a large
portion of a double-stranded DNA with an intact corresponding
DNA segment obtained from another DNA molecule. This process
is known as recombination repair.

c. In humans, xeroderma pigmentosum is an autosomal recessive
disease in which the mechanism of repair synthesis is defective
in the skin fibroblasts. Persons so afflicted are more likely
to develop skin cancer and show an abnormal sensitivity to
exposure to sunlight. The defect appears to be a lack of the
endonuclease needed to recognize the thymine dimers and make
the initial nick. However, patients with xeroderma pigmentosum
show a great deal of genetic heterogeneity (at least five
different genetic loci are involved). The patients have normal
excision repair (the process for thymine dimer removal,
described above). The problem may be a defect in the post-
replication repair ("proof-reading") process. Post-replication
repair, as opposed to excision repair, involves correction of
any errors in the DNA after replication is completed.

DNA repair processes are of fundamental importance in the
protection of the genetic material against structural changes
brought about by intrinsic metabolites or environmental
carcinogens and mutagens. Defects in DNA repair can lead to a
broad range of problems including malignant transformation,
immunodeficiency, neurological abnormalities, and premature
senescence. It appears that the origin of malignant transformation may be caused by somatic mutation in some cases, so that
the absence of a repair process may increase the frequency of

tumor formation. Immunological impairment and premature aging
are possibly due to an abnormal rate of cell destruction.

Other inherited human disorders in which DNA repair is defective
are Fanconi's anemia, Bloom's syndrome, and ataxia telangiectasia.
All of these disorders are associated with a predisposition to
malignancy and chromosomal aberrations. The incidence of
chromosome breaks after treatment of cells in vitro with
alkylating agents, DNA cross-linking agents, and radiation is
also increased. Disorders associated with premature aging which
show defective DNA repair include progeria and Cockayne's
syndrome. The affected children appear normal at birth but
they age rapidly and have a shortened life span. In patients
with Down's syndrome, there is an increased incidence of leukemia
and their cells exhibit a reduced DNA polymerase activity.
Subjects with actinic keratosis, a premalignant lesion of the
epidermis, develop squamous cell carcinomas more often than
normal. These cells are defective in the system for UV-induced
DNA repair.

9. The first complete nucleotide sequence of a DNA genome was deciphered
by Sanger and associates. They used the single chromosome from a
bacteriophage, ϕX174, which contains single-stranded DNA with 5375
nucleotides. The methods employed were quite ingenious. They
included cleavage of the molecule into smaller fragments with
restriction endonucleases (discussed later); labeling of the
strands at the 5'-end with ^{32}P (using nucleotide phosphorylase);
digestion of the strands with different types of DNAse's; separation
of the digestion products by electrophoresis and homochromatography;
and identification of the fragments by autoradiography.

10. DNA replication has been much more thoroughly studied in viruses
and procaryotic cells than in eucaryotic ones, as is the case with
most processes involving nucleic acids. Much work has been devoted
to explaining functions which are largely unique to these "simple"
systems, such as replication of circular chromosomes. Bacterial
and viral studies are also simplified by the apparent absence of
the histones and other basic proteins present in eucaryotic chromosomes. It still remains to be seen just how many of the findings
in such systems really apply to man and the other mammals. There
is no doubt, however, that much valuable insight and many novel
techniques have arisen from the work on primitive genomes.

Recombinant DNA

1. This is a new, exciting, rapidly developing area of DNA research.
Most of the present work in the field deals with the transfer of
genetic information from one species to another species. This is
accomplished by splicing segments of DNA from one organism onto

extrachromosomal pieces of DNA from another organism. The resulting recombinant DNA (DNA chimera) is then inserted into cells of the species from which the extrachromosomal DNA originally came. Recombinant DNA molecules undergo propagation within the host cells, permitting the cloning (production of many identical copies from one original) of foreign DNA.

2. Some of the details of these processes are as follows:

    a. The host cells most frequently used are E. coli which contain, besides the DNA of the circular chromosome, small circular pieces of extrachromosomal DNA known as plasmids. The plasmids replicate independently of the chromosome of the bacterial cell. Among other things, antibiotic-resistance genes (or R-factors) are coded for by plasmids. Plasmids can be transferred between bacteria by a cell to cell contact so that R-factors and other plasmid-born genes have become widespread.

        The plasmid DNA is separated from the chromosomal DNA by using cesium chloride gradient equilibrium ultracentrifugation in the presence of ethidium bromide. The ethidium bromide binds preferentially to plasmid DNA, making it denser, so that it sediments to a different region of the gradient. The antibiotic resistance property is used in selecting those bacteria which have acquired the hybrid DNA containing the resistance factor, among those bacteria that have not been transformed. Therefore, the transformed bacteria can be isolated as colonies growing on a media containing the antibiotic.

    b. The next step is the preparation of the host DNA (plasmid) and the donor DNA for splicing. This is accomplished with restriction endonucleases.

        i) This is a group of site-specific enzymes whose members are isolated from a number of bacterial strains. They recognize regions of the DNA which possess a two-fold axis of rotational symmetry. (Such regions are frequently called palindromes, but this is not really correct.) An example is shown below.

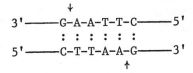

```
              ↓
3'————G-A-A-T-T-C————5'
         : : : : : :
5'————C-T-T-A-A-G————3'
                 ↑
```

The sequence on the upper strand, read 3' → 5', is the
same as that on the lower strand, also read 3' → 5'. The
restriction endonuclease binds to sites such as this and
cleaves both strands of the DNA either at random intervals
throughout the DNA (Type I) or between specific bases
within the site (Type II; of interest in DNA sequencing
and recombination experiments).

Each restriction endonuclease has its own recognition
site. The one shown is for EcoRI (isolated from a strain
of E. coli, from which it derives its name). This is a
type II enzyme which cleaves within the site at the
places shown by the arrows (between the G and the A, on
each strand). The products of this digestion have, at
their 3'-end, short pieces of single-stranded DNA:

```
3'————G-OH 5'                3'-OH-A-A-T-T-C————5'
                    +
5'————C-T-T-A-A-OH 3'              5' HO-G————3'
```

These "sticky ends" (cohesive termini) are complementary
to each other and will base-pair correctly with any other
EcoRI-generated sticky end that they encounter.

ii) The host-organism plasmid is treated with a restriction
     endonuclease under conditions where singly-cut forms
     predominate (accomplished by limited digestion or
     selection of a nuclease for which there is only one
     recognition site). The nuclease selected must not
     recognize (and cleave) a site within a gene essential for
     the expression or replication of the plasmid. The donor
     DNA is completely digested, producing a number of pieces
     with two sticky ends. The linear plasmid and the donor
     pieces are mixed and allowed to anneal to once again form
     circles. The remaining nicks are closed by using DNA
     ligase (discussed earlier). This is illustrated below.

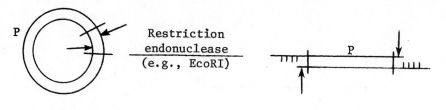

P   Restriction
               endonuclease
               (e.g., EcoRI)

<u>intact circular</u>   <u>cleaved, linear plasmid</u>
<u>plasmid</u>

(Note: ╪══╪ defines the symmetrical recognition site;
The small arrows show the points of cleavage;
P ≡ plasmid DNA)

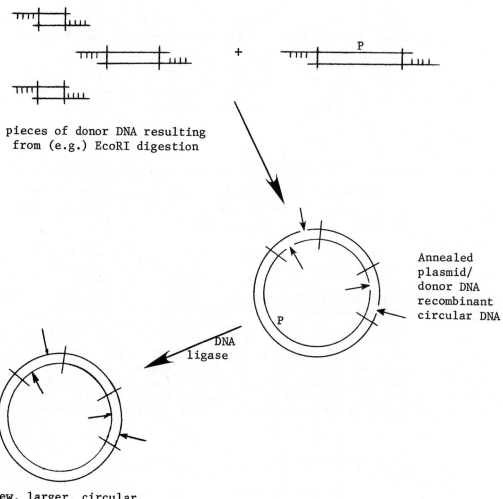

pieces of donor DNA resulting
from (e.g.) EcoRI digestion

                                              Annealed
                                              plasmid/
                                              donor DNA
                                              recombinant
                                              circular DNA

                    DNA
                    ligase

New, larger, circular
plasmid carrying the
donor DNA sequence

396

iii) Although restriction endonucleases have access to the
DNA of the bacteria from which they are isolated, their
DNA is not degraded. Another group of enzymes known as
modification methylases prevent this. They recognize
the same site as the nuclease and attach methyl groups
to either the 5'-position of a cytosine or the 7'-amino
group of an adenosine within the site, preventing cleavage
by the restriction enzyme.

c. The new, circular plasmid carrying the donor DNA is added to an
E. coli culture in the presence of $CaCl_2$. The $CaCl_2$ alters the
bacterial cell wall, making it permeable to the DNA. Since
good plasmids are picked up by only a few cells, these
successful transfers are selected for by growing the E. coli on
a medium containing the appropriate antibiotic. Cells which
took up plasmids and, hence, possess the resistance factor will
form colonies; all others will die.

d. DNA from a variety of sources has been spliced onto plasmid DNA
and inserted into bacterial cells. These hybrid plasmids
undergo successful replication. E. coli undergoes division
about once every 25 to 30 minutes so that this technique
provides a method for the synthesis of a large amount of a
particular DNA, provided that it can be incorporated into a
plasmid. Recombinant techniques are being used in the mapping
of eucaryotic genes and for the study of their fundamental
properties. In higher organisms with multiple chromosomes and
many more genes than procaryotes, the recombinant-DNA technique
provides an avenue by which single genes can be isolated and
synthesized in large quantities. A particular eucaryotic gene
can be isolated by the same selection techniques that have
been applied to bacterial genes for years. Other potential
benefits of this research include: i) cures for genetic
defects by replacement of the missing gene (referred to as
genetic engineering); ii) synthesis in large quantities of a
gene product used therapeutically, such as human insulin, human
growth hormone, clotting factors, antibiotics, etc. (there is
a recent report that the rat insulin gene has been cloned
successfully); iii) development of nitrogen-fixing strains of
bacteria capable of growing in the roots of plants which must
now depend on synthetic nitrogen fertilizers--or even the
development of strains of plants which can themselves fix
nitrogen; iv) bacteria which can break down hydrocarbons into
useful products; and there are other possibilities.

But these techniques also present potential hazards. It is
easy to imagine the inadvertent (or intentional?) creation and
escape of new, virulent, antibiotic-resistant organisms, or of
organisms containing genes that can cause tumors. At the present

time this problem is receiving considerable attention from many
sources: scientists, government, press, and the public. All
are working to develop appropriate safeguards against the
potential risks of recombinant-DNA research. One major step
has been the development of mutant, enfeebled, attenuated
strains of E. coli (EK1, EK2, EK3) which cannot survive in its
natural environment (that is, outside of the laboratory).
Presumably, if these strains did escape it would rapidly die,
so that any hazardous genes which it might carry would not be
propagated. Standards for facilities with special safety
features have also been drawn up, to permit the handling of
these organisms with a minimum of risk to the researcher and
the public.

Types and Properties of RNA

1. RNA (ribonucleic acid) is the second type of nucleotide polymer in
    the cell. The principal differences between DNA and RNA are the
    presence of an -OH group in the 2' position of the ribose in RNA,
    and the occurrence of uracil in place of thymine as one of the
    four bases in RNA.

    There are four major types of RNA in the cell.

    a. Ribosomal RNA (rRNA)

    b. Transfer RNA (tRNA) also called soluble or sRNA

    c. Messenger RNA (mRNA)

    found in all cells; function discussed below

    d. Nuclear RNA (nRNA) - found only in eucaryotes; function not
        known but may be involved in the control of transcription (see
        under regulation of enzyme synthesis); not well characterized

2. Properties of E. coli RNA's

Type	Sedimentation Coefficient	Approximate Molecular Weight (daltons)	No. of Nucleotide Residues	% of Total Cellular RNA
ribosomal RNA (3 species)	5S	35,000	~120	82
	16S	550,000	~1,500	
	23S	1.1×10^6	~3,000	
transfer RNA (60 species)	4S	23,000 to 30,000	75-90	16
messenger RNA (many species)	6S-25S	25,000 to 1×10^6	75-3000	~2

3. In bacterial cells, all RNA is found in the cytoplasm but in the mammalian cells the RNA is distributed among the various organelles. For example, in a liver cell the nucleus contains 11% of the total cellular RNA, the mitochondria have 15%, the ribosomes have 50%, and 24% is free in the cytoplasm. The most stable RNA's are tRNA and rRNA. The least stable is mRNA.

4. RNA has been shown to be the genetic material in most (if not all) plant viruses (such as the wound tumor and tobacco mosaic viruses); some bacterial viruses or bacteriophages (the $Q\beta$ and R^{17} viruses of E. coli, for example); and some animal viruses (poliomyelitis and type A influenza; leukemia and solid tumor viruses of mice and chickens; others). Some of these viruses have been shown to possess an RNA-dependent DNA polymerase (an enzyme which can form DNA polymers by reading an RNA template; reverse of DNA-dependent RNA polymerase) and to require active DNA replication for infection. The presence of this polymerase in some human leukemic cells has been interpreted as suggestive evidence for a viral etiology of some human cancers. However, RNA-dependent DNA polymerase, also called reverse transcriptase, has been found in "normal" human cells, in animal cells, and in the wild type of E. coli. The role of the enzyme in these cells is not understood at this time. Reverse transcriptases have been isolated from several different RNA tumor viruses. Replication of these viruses is quite similar to DNA-dependent DNA polymerase-directed replication (already discussed).

Some of the properties of the enzyme are:

   i) The molecular mass ranges from 70,000-160,000 daltons (the enzyme from avian myeloblastosis virus consists of two subunits of molecular masses 70,000 and 110,000 daltons and contains two atoms of Zn^{2+});

  ii) The direction of synthesis is $5' \rightarrow 3'$;

 iii) They require deoxyribonucleoside-5'-triphosphates;

  iv) They require a short primer, as well as an RNA template strand with a 3'-hydroxyl terminus;

   v) They are very active when using natural or very large synthetic RNA templates.

The enzyme first synthesizes a DNA-RNA hybrid. After removal of the RNA strand by the action of ribonucleases, the DNA strand is replicated to give a double-stranded DNA. The newly formed DNA strand then serves as a template for the synthesis of many copies of the viral RNA and, ultimately, of more virus particles. In addition, the DNA strand can lysogenize (be incorporated into) the host DNA, in some cases transforming the host cell into a neoplastic cell. Experimentally, the reverse transcriptase has been used in synthesizing DNA, known as cDNA (complementary DNA) from unique RNA molecules (such as the mRNA which directs hemoglobin synthesis). This complementary DNA, if synthesized using labeled nucleotide triphosphates, can be used for hybridization studies (described earlier in this chapter).

5. <u>Ribosomal RNA</u> (rRNA) is found associated with protein in ribosomes. These are roughly spherical bodies found in most cells. Their role in protein synthesis is discussed later. There are three principal types of ribosomes: procaryotic ribosomes, eucaryotic cytoribosomes (found in the cytoplasm and bound to the rough endoplasmic reticulum), and mitoribosomes (present in the mitochondria and chloroplasts of eucaryotic cells).

    a. The most thoroughly studied ribosomes are those of <u>E. coli</u>. They are composed of two subunits, whose composition is given in Figure 52.

<u>Figure 52</u>. <u>Molecular Composition of Ribosomes in E. coli</u>

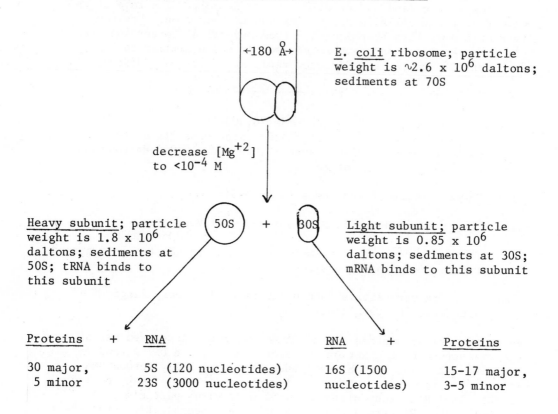

$\leftarrow$180 Å$\rightarrow$

E. coli ribosome; particle weight is $\sim 2.6 \times 10^6$ daltons; sediments at 70S

decrease $[Mg^{+2}]$ to $<10^{-4}$ M

50S   +   30S

Heavy subunit; particle weight is 1.8×10^6 daltons; sediments at 50S; tRNA binds to this subunit

Light subunit; particle weight is 0.85×10^6 daltons; sediments at 30S; mRNA binds to this subunit

Proteins	+	RNA	RNA	+	Proteins
30 major, 5 minor		5S (120 nucleotides) 23S (3000 nucleotides)	16S (1500 nucleotides)		15-17 major, 3-5 minor

b. Eucaryotic cytoribosomes, while basically similar from species to species, show much more heterogeneity than do procaryotic ribosomes. A comparison of the two types is presented below. Considering the other major differences between eucaryotes and procaryotes, the similarities in functions and properties of the ribosomes are probably more important than the differences.

Procaryotic Ribosomes	Eucaryotic Cytoribosomes
60-65% RNA	About 55% RNA
35-40% protein	About 45% protein
Particle weight is about 2.6×10^6 daltons; diameter is 180 Å; sedimentation coefficient of 70S	Somewhat larger, with a particle weight of about 3.6×10^6 daltons and a sedimentation coefficient of 77-80S
Two subunits of sedimentation coefficients 30S and 50S	Two subunits of sedimentation coefficients 38-45S and 58-60S (particle weights of 1.2×10^6 and 2.4×10^6 daltons, respectively)
Large subunit has two rRNA molecules of size 5S and 23S; small subunit has one rRNA of size 16S	Large subunit has 2 or 3 rRNA molecules; small subunit has one rRNA molecule; size of rRNA is somewhat species-dependent
Genes (cistrons) for rRNA synthesis are part of the one bacterial chromosome	rRNA is synthesized on DNA contained in the nucleolar organizer; this becomes part of the nucleolus during mitosis
Ribosomes occur in the cytoplasm either free or associated with mRNA (polysomes)	Ribosomes occur free in the cytoplasm or bound to the rough endoplasmic reticulum; when associated with mRNA, they are called polysomes

c. Mitoribosomes are found in the chloroplasts and mitochondria present in the cytoplasm of eucaryotic cells. These organelles also have their own supply of DNA which is used for the synthesis of some (but by no means all) of the proteins and enzymes in these structures. Although these ribosomes resemble procaryotic ribosomes with respect to inhibitor sensitivity, details of peptide-chain initiation, and other properties,

they are distinct from them in size and RNA composition. In
some ways they appear to be intermediate between the
cytoribosomes and the bacterial ribosomes. For this and other
reasons, it has been suggested that mitochondria and
chloroplasts originated as bacterial invaders of eucaryotic
cells. Eventually they lost the ability to exist outside of
their hosts and a symbiotic relationship developed. There is
little or no experimental data to substantiate this
other than circumstantial evidence.

d. The proteins associated with the ribosomes are basic in nature
and it appears that the major ones are present only once in
a particular ribosome. Although their exact purposes are not
known, they probably function at least partly as structural
proteins to maintain proper conformation of the RNA for
protein synthesis. Other postulated functions for the
ribosomal proteins include

    i) Control of mRNA translation (e.g., selection of the proper
        mRNA molecules).

    ii) The fulfillment of some of the enzymatic functions needed
        for protein synthesis, such as movement of the ribosome
        along the message, transfer of the amino acids from the
        tRNA to the growing chain (peptidyl transferase), GTPase
        activity, etc.

Few experimental data are available to either confirm or deny
these roles.

6. <u>Transfer RNA</u> (tRNA; also called soluble RNA or sRNA)

a. These RNA molecules are carriers of specific amino acids
during protein synthesis on the ribosomes.

b. Each amino acid has at least one specific corresponding tRNA.
Some have multiple tRNA's (leucine and serine each have
5 corresponding tRNA's in <u>E. coli</u>).

c. RNA contains rather large numbers of minor or "unusual" bases
which are usually methylated forms of the four "usual" bases.
(Any base other than A, G, C, or U is considered "unusual"
in RNA.)

d.   Examples of these "unusual" nucleotides are

Pseudo-Uridylic Acid   Ribothymidylic Acid
                                       (5-methyl-uridylic acid)

Other "unusual" bases found in tRNA include 4-thiouracil,
N^6-isopentyladenine, 2'-O-methyladenosine (a nucleoside),
hypoxanthine, 5-methylcytosine, and several more. These
bases are formed by specific enzymes (e.g., tRNA methylases)
following polymerization of the tRNA on the DNA template.

e.   The structure in Figure 53 is a schematic representation of
yeast alanyl tRNA. The general features are quite similar
to those of the other tRNA's whose structures have been
determined. (These include phenylalanyl, tyrosyl, and two
species of seryl tRNA, all from yeast.)

Note that the codon on the mRNA (5' → 3') is oriented in the
opposite direction relative to the anticodon on the tRNA
(3' → 5'). This is analogous to the relationship between the
two strands of double-stranded DNA. The fact that the third
letter of the anticodon (here I, for inosine) need not always
be the base which is usually complementary to the third letter
of the codon (here C) is explained by the wobble hypothesis.
Basically this states that due to a certain flexibility or
"wobble" in the recognition of the anticodon bases by those
of the codon, the same tRNA can be recognized by codons which
differ by only the last letter. This is not necessarily true
of all codons or all tRNA molecules, as is indicated by the
occurrence of five seryl tRNA's in E. coli for six codons;
and by the fact that, while asparagine is coded for by AAU and
AAC, the codons AAA and AAG specify lysine (see under genetic

Figure 53.   Schematic Structure of Alanine Transfer RNA from Yeast

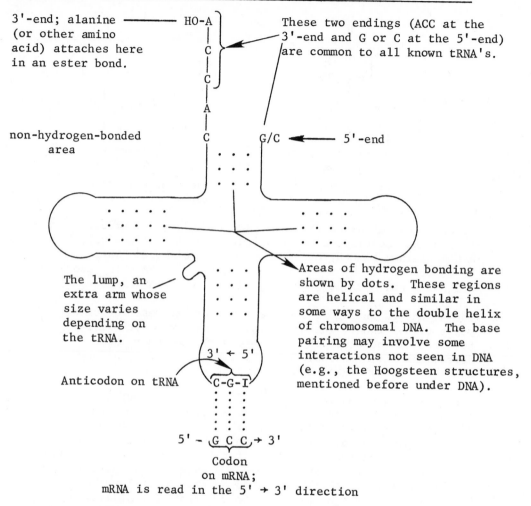

3'-end; alanine ———— HO-A   These two endings (ACC at the
(or other amino   3'-end and G or C at the 5'-end)
acid) attaches here   are common to all known tRNA's.
in an ester bond.

non-hydrogen-bonded
area   5'-end

The lump, an   Areas of hydrogen bonding are
extra arm whose   shown by dots. These regions
size varies   are helical and similar in
depending on   some ways to the double helix
the tRNA.   of chromosomal DNA. The base
                                   pairing may involve some
                                   interactions not seen in DNA
Anticodon on tRNA   (e.g., the Hoogsteen structures,
                                   mentioned before under DNA).

Codon
on mRNA;
mRNA is read in the 5' → 3' direction

code). In the first case, the six serine codons fall into
two groups.

   i) UCU, UCC, UCA, UCG and

   ii) AGU, AGC.

If the wobble hypothesis was always used, these two groups would
require two seryl tRNA's at most (since the codons are all
either UC- or AG- where the "-" indicates more than one letter).
In the instance of asparagine and lysine, if the last letter

of the codon were totally unimportant, then there would be frequent errors observed where Asp and Lys were put in the wrong places in proteins. Moreover, if the third base of a codon really did not matter, then the three-letter code would be reduced to an effective two-letter code. For more on this, refer to the section on the genetic code.

The unusual nucleosides found in yeast alanyl tRNA are pseudouridine, inosine, dehydrouridine, ribothymidine, methyl guanosine, dimethyl guanosine, and methyl inosine. They occur largely (but not exclusively) in or adjacent to the non-hydrogen-bonded loops and the lump.

f. The general reaction of tRNA with an amino acid is shown below. The tRNA, amino acid, and the aminoacyl-tRNA synthetase enzyme must all match for the loading of the tRNA to take place.

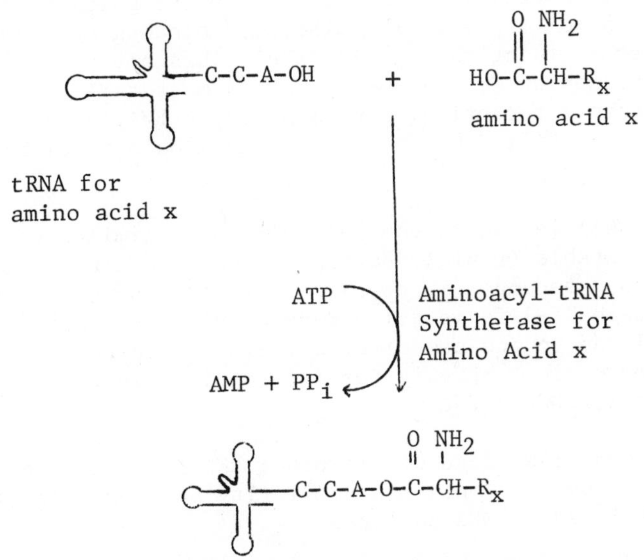

$$\text{tRNA}-\text{C-C-A-OH} \quad + \quad \underset{\text{amino acid x}}{\overset{\overset{\displaystyle O \quad NH_2}{\displaystyle \|\quad |}}{HO-C-CH-R_x}}$$

tRNA for
amino acid x

$$\text{ATP} \longrightarrow \quad \text{Aminoacyl-tRNA Synthetase for Amino Acid x}$$

$$\text{AMP} + \text{PP}_i \longleftarrow$$

$$\text{tRNA}-\underset{}{C-C-A-O-\overset{\overset{\displaystyle O \quad NH_2}{\displaystyle \|\quad |}}{C}-CH-R_x}$$

Aminoacyl-tRNA$_x$
(loaded or charged tRNA)

The reaction appears to involve an enzyme-bound intermediate of the form Aminoacyl$_x$ ∿ AMP. Of the twenty common amino acids, all are attached to tRNA by this reaction except glutamine. It appears that glu-tRNA (glutamyl-tRNA) is transamidated with glutamine or asparagine which transfers their amide nitrogen to the γ-carboxyl of glu to form gln-tRNA (glutamine-tRNA) (see also protein biosynthesis, below).

g. The first tRNA sequenced was yeast tRNA^{ala} (R.W. Holley and coworkers, 1965; Holley received a Nobel Prize for this work in 1968). As of 1969,the complete nucleotide sequences of twenty tRNA species had been worked out. Work is also progressing on the elucidation of the three-dimensional structure of tRNA molecules by x-ray diffraction. These results indicate a high degree of similarity in the structures of many if not all of the different tRNA species.

h. A tRNA molecule must be able to uniquely recognize

   i)   the proper aminoacyl-tRNA synthetase;

   ii) the binding site for tRNA on the large ribosomal subunit;

   iii) the codon on the mRNA.

It has been clearly established that the codon recognition site is the anticodon base triplet. The ribosomal binding site, which should be similar or identical on all tRNA's, has not yet been isolated. The sequence -G-T-ψ-C-G-, found in most tRNA's, may be involved in this interaction. The aminoacyl-tRNA synthetase recognition site is still unknown, but the anticodon is clearly not directly involved.

7. Messenger RNA (mRNA) is the ribonucleic acid most similar to DNA. It is the most unstable (rapidly degraded) RNA with a half-life varying from 4 to 6 seconds (E. coli) to 8 to 12 hours in rat liver cells. The half-life is quite dependent on the generation time of the cell and on the number of proteins which must be made from each mRNA molecule. mRNA is the intermediate in the information transfer DNA → RNA → Protein.

a. The synthesis of mRNA (and of the other RNA molecules in the cell) is shown below. Note the similarity of these reactions to those catalyzed by DNA polymerase I.

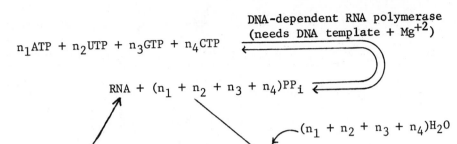

$$n_1 ATP + n_2 UTP + n_3 GTP + n_4 CTP \quad \xrightarrow{\text{DNA-dependent RNA polymerase (needs DNA template + Mg}^{+2})}$$

$$RNA + (n_1 + n_2 + n_3 + n_4)PP_i$$

$$(n_1 + n_2 + n_3 + n_4)H_2O$$

pyrophosphatase

$$2(n_1 + n_2 + n_3 + n_4)P_i$$

Polymer of ribonucleotides;
per molecule of RNA – there
are n_1 adenylic acid residues,
n_2 uridylic acid residues,
n_3 guanylic acid residues,
and n_4 cytidylic acid residues.

b. Either Mg^{+2} or a 4:1 mixture of $(Mg^{+2} + Mn^{+2})$ is required for the reaction. If Mn^{+2} alone is used, some deoxyribonucleotides are incorporated.

c. All four ribonucleoside-5'-triphosphates are required.

d. The enzyme adds mononucleotide units to the 3'-hydroxyl end of the RNA chain so that the direction of RNA synthesis is 5' → 3'. This is the same as the direction in which the DNA is synthesized.

e. The reaction can be reversed by increasing the concentration of pyrophosphate. In the cell a pyrophosphatase $(PP_i \rightarrow 2 P_i)$ ensures that the reaction proceeds toward the synthesis of RNA.

f. The polymerase enzyme is found in bacterial, plant and animal cells. In the higher organism, it is found in the nuclei, in the nucleolus and in the mitochondria. Regardless of the source, the same reaction is catalyzed with the same chemical requirements.

g. Unlike DNA polymerase I, RNA polymerase has a complex structure containing five subunits. One of the subunits, designated sigma (or σ), is necessary in order for the RNA polymerase to recognize the DNA start signals for RNA polymerization. There appear to be many different sigmas, each perhaps capable of recognizing different initiator regions on the DNA. It is thought that when viruses invade cells, they provide their own unique sigmas to permit transcription of their genomes by the cells' RNA polymerase molecules. E. coli DNA-dependent RNA polymerase, an extensively investigated enzyme, has a molecular mass of about 490,000 daltons with a subunit composition of $\alpha_2\beta\beta'\sigma$. The sigma subunit in this case brings about initiation by keeping the entire enzyme ($\alpha_2\beta\beta'\sigma$ or holoenzyme) in the proper conformation for binding to the initiation site (pyrimidine-rich areas of 10 or more residues) in the DNA template. Catalytic function is accomplished by the $\alpha_2\beta\beta'$ portion of the enzyme (core polymerase).

h. Double-stranded DNA is most active as a template; single-stranded DNA shows less activity.

i. The RNA formed has a base composition complementary to that of the template DNA. The frequencies of occurrence of the bases A, T, G, and C in the DNA template are equal, respectively, to the frequencies of occurrence of U, A, C, and G in the RNA formed.

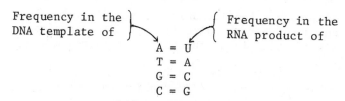

Frequency in the DNA template of ⎫⎬⎭   ⎧⎨⎩ Frequency in the RNA product of

$$A = U$$
$$T = A$$
$$G = C$$
$$C = G$$

j. By using highly purified E. coli enzyme in the presence of double-stranded DNA as a template, it has been shown that only one of the two strands is transcribed. The mechanism is not understood. The strand being transcribed can switch, depending on the region of the DNA.

k. In E. coli the other RNA's (transfer-RNA and ribosomal-RNA) are also made by the enzyme using DNA as template.

1. The product of the RNA polymerase <u>does</u> carry a 5'-triphosphate group, unlike the product of Kornberg's enzyme (DNA polymerase I). Most RNA chains found <u>in vivo</u> start with either pppA or pppG at the 5' end.

m. The antibiotic actinomycin D inhibits RNA synthesis by binding to the DNA, probably adjacent to guanine residues. This prevents transcription by the RNA polymerase. Rifampin (also rifampicin, rifamycin B, and other closely related compounds of the rifamycin family) blocks RNA synthesis by binding to the polymerase and preventing chain initiation. They are specific for polymerases of procaryotic origin and do not affect those of nuclear, eucaryotic origin. The rifamycins are isolated from <u>Streptomyces mediterranei</u> and are currently used in the treatment of tuberculosis. Rifampicin is administered orally. The structures of actinomycin D and rifampin are shown below.

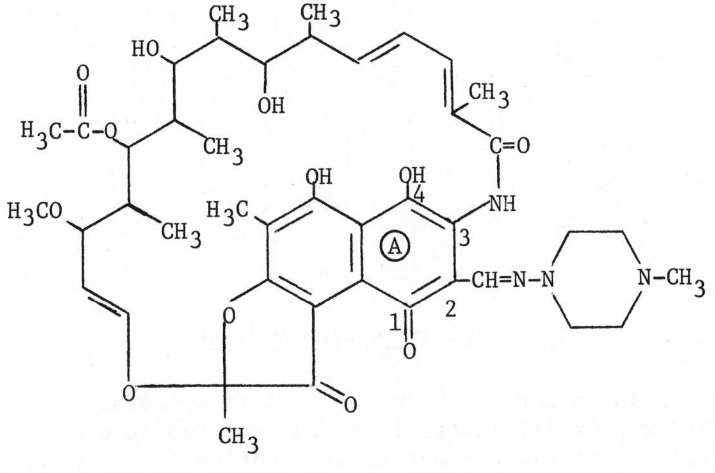

<u>Structure of Rifampin</u> ($C_{43}H_{58}N_4O_{23}$)

(Rifamycin B and rifampicin differ in the substituents in the 1 and 2 positions of the (A) ring.)

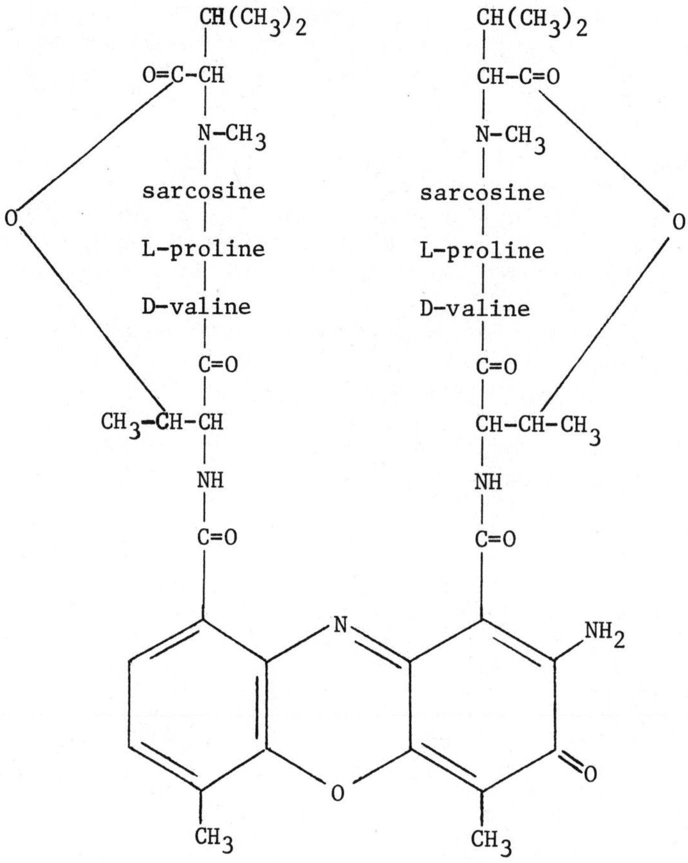

$$CH(CH_3)_2 \qquad CH(CH_3)_2$$

O=C-CH   CH-C=O

N-CH$_3$ N-CH$_3$

sarcosine   sarcosine

L-proline   L-proline

D-valine   D-valine

C=O   C=O

CH$_3$-CH-CH   CH-CH-CH$_3$

NH   NH

C=O   C=O

Structure of Actinomycin D

Some of the other agents which inhibit RNA synthesis include
ethidium bromide (intercalates between two successive DNA base
pairs), aflatoxin, 2-acetylaminofluorene (acts on the DNA),
streptolydigin (blocks RNA chain elongation), and α-amanitin
(blocks eukaryotic nuclear polymerase but not bacterial,
mitochondria, or chloroplast RNA polymerases). Both aflatoxin
(synthesized by the fungus _Aspergillus flavus_ which grows on
peanuts) and 2-acetylaminofluorene (a synthetic compound) are
extremely potent carcinogens and can produce liver cancer.
These two compounds, in addition to inhibiting transcription,
also cause the inhibition of DNA replication. Alpha-amanitin
is produced by the toxic mushroom _Amanita phalloides_. Following
are the structures of aflatoxin, ethidium bromide and
2-acetylaminofluorene.

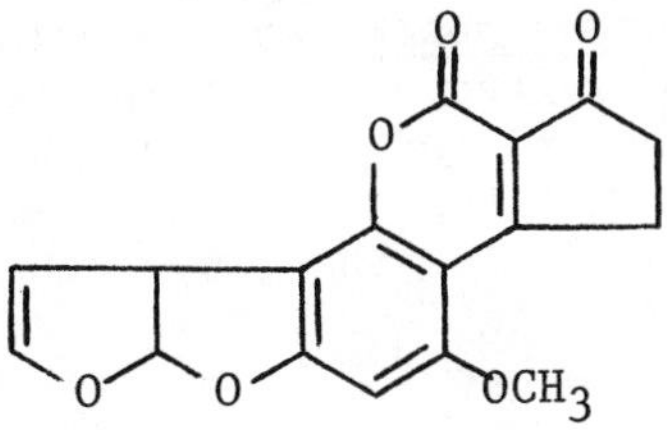

Structure of Aflatoxin B

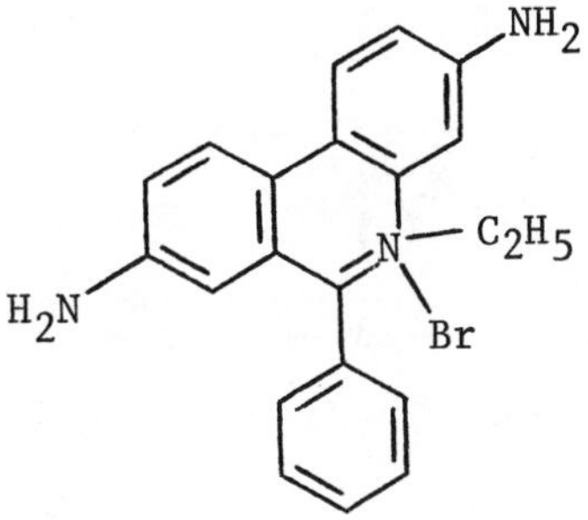

Structure of Ethidium Bromide

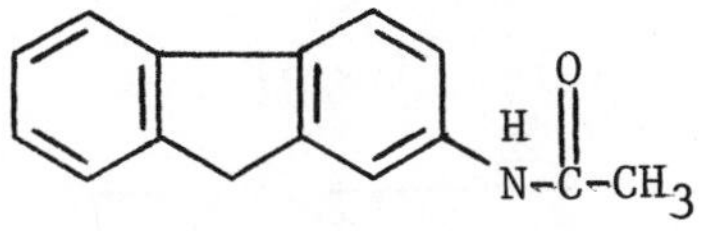

Structure of 2-Acetylaminofluorene

Nucleases

1. These are enzymes which catalyze the hydrolysis of nucleic acids. They have varying degrees of specificity and have proved quite useful in the determination of sequence and other properties of these compounds. The use of their specificity is the basis of the nearest-neighbor analysis, used to prove complementarity of various nucleic acid strands.

2. In addition to the obvious difference in specificity between ribonucleases and deoxyribonucleases, there are phosphodiesterases which are non-specific, hydrolyzing both RNA and DNA. There is also the distinction between exonucleases (requiring a free

3' or 5' end to start) and <u>endonucleases</u> (severing polynucleotides internally; do not require a free 3'- or 5'-hydroxyl group). Some of the known nucleases are indicated in Table 17, with the characteristics of their reactions. A polynucleotide chain is shown schematically below. Notice that, by successive application of the nucleases described in Table 17, a wide variety of highly specific products can be obtained.

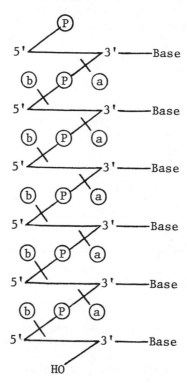

The Genetic Code

1. The sequence of the purines and pyrimidines of mRNA (originally transcribed from DNA) provides the information for the sequence of the amino acids of proteins (primary structure).

2. The information present in DNA is transcribed into the form of RNA. mRNA is formed from the DNA template, catalyzed by RNA-polymerase (Jacob and Monod, 1961).

Table 17. Sources and Specificities of Some Nucleases

a. Exonucleases

	Substrate	Cleavage Site and Products
i) Snake venom phosphodiesterase	DNA or RNA	All (a) linkages, starting at the end with a free 3'-OH group and working toward the 5'-end; releases nucleoside-5'-phosphates.
ii) Bovine spleen phosphodiesterase	DNA or RNA	All (b) linkages, starting at the end with a free 5'-OH group and working toward the 3'-end; releases nucleoside-3'-phosphates.

b. Endonucleases

i) Bovine pancreatic deoxyribonuclease (DNAse I)	DNA	Cleaves all (a) linkages, but prefers those between purine and pyrimidine bases; releases two smaller polynucleotides, one with a free 3'-OH group, the other with a 5'-phosphate group.
ii) Deoxyribonuclease II (DNAse II); from mammalian spleen and thymus, some bacteria)	DNA	Randomly cleaves all (b) linkages; releases two smaller polynucleotides, one with a free 5'-OH group, the other with a 3'-phosphate group.
iii) Pancreatic ribonuclease	RNA	Cleaves all (b) linkages in which the phosphate is also bound to the 3'-OH of a pyrimidine nucleoside; releases two smaller polynucleotides, one with a free 5'-OH group, the other with a pyrimidine-3'-phosphate at the 3'-terminus.

iv) Taka-diastase (isolated from the Mold *Aspergillus oryzae*)

(I) Ribonuclease T_1	RNA	Same as pancreatic ribonuclease except that guanine is required in place of a pyrimidine; cleaves (b) linkages.
(II) Ribonuclease T_2	RNA	Same as ribonuclease T_1, except that adenine must be on the 3' side, rather than guanine; cleaves (b) linkages.

3. The genetic code for a particular amino acid consists of a <u>triplet</u>
    of <u>non-overlapping</u> bases which act as a code word (called a <u>codon</u>).
    For example:

5'-end --U-U-U-U-U-A-C-C-U-- 3'-end ← messenger RNA strand

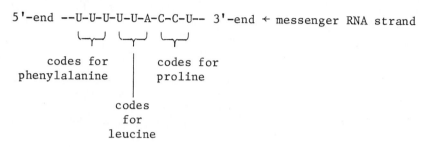

| codes for | codes for |
| phenylalanine | proline |

codes
for
leucine

4. Elucidation of the Genetic Code:

    a. If the code words were each made up of two nucleotides and since
        there are only four bases, there would be only 16 (= 4^2 possible
        dinucleotide combinations) codons which is not enough to code
        for the 20 amino acids. There are, however, 64 (= 4^3)
        trinucleotide combinations, which is more than enough to code
        for twenty amino acids. This suggested that the code must be
        a <u>triplet</u> code, with each amino acid specified by a group of
        <u>three nucleotide bases</u>.

    b. The existence of a triplet code was proved experimentally by
        using synthetic systems and analyzing the peptide synthesized;
        and by binding studies of aminoacyl-tRNA's to mRNA triplets.

5. There may be more than one codon for a given amino acid. For
    example: <u>GUU</u>, <u>GUC</u>, and <u>GUA</u> all code for valine. This
    characteristic is referred to as the <u>degeneracy</u> of the code.

6. Note that the first two nucleotides in each of these triplets are
    the same in all codons for a given amino acid, but that the third
    nucleotide is variable. This suggests that the third base is less
    important in specifying the incorporation of an amino acid. The
    wobble hypothesis (see under tRNA) is partly based on this
    observation.

7. Of the 64 possible triplets, 61 have been shown to code for amino
    acids. The 3 triplets which do not code for any amino acid are
    chain-terminating triplets (also called nonsense triplets) and
    they signal completion of a polypeptide chain. They are UAA, UAG,
    and UGA. The codons UAG and UAA are also called amber and ochre
    codons, respectively.

8. There may be initiating codons as well. In E. coli for example, the chain-initiating codon is AUG which codes for N-formylmethionyl-tRNA (fMet-tRNA).

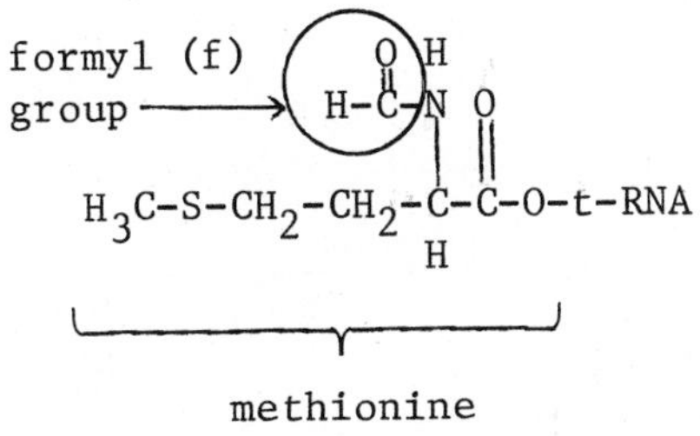

methionine

One of the valine codons (GUG) can also bind fMet tRNA (bearing anticodon UAC). This codon may also be involved in chain initiation.

9. AUG also codes for methionine, raising the question of how one codon can code for two amino acids. After the synthesis of a peptide chain is terminated, the formyl group and, in some cases, the methionine are cleaved from the N-terminus of the polypeptide.

10. In E. coli, there are two types of tRNA which can have methionine attached to them. Only one of these can be recognized by the enzyme which attaches a formyl group to the Met residue. They are therefore designated tRNA^{Met} and tRNA^{fMet}. Both of these can bind to the AUG codon but only tRNA^{fMet} can be recognized by the valine codon GUG. In eucaryotic cells, which use Met rather than fMet as an initiating amino acid, there are also two distinct tRNA^{Met} species. One of these can be formylated by the enzyme from E. coli, the other cannot. While the significance of all this is not yet completely clear, it suggests that there is something more than the genetic code which is used for the recognition of the tRNA needed for peptide-chain initiation.

11. The genetic code is universal, the same code words being used to designate the same amino acids in all organisms (see also under loading of tRNA in the section on protein synthesis).

12. The genetic code is shown in Table 18. It is important to realize that these codons are the bases found in the mRNA and that the sequence in the DNA is complementary to this. The relationship of DNA, mRNA, and protein sequence is shown below. Notice also that the direction of reading of both DNA and RNA is important.

<u>Table 18.</u> <u>The Genetic Code</u>

First Position (5' end)	Second Position				Third Position (3' end)
	U	C	A	G	
U	Phe	Ser	Tyr	Cys	U
	Phe	Ser	Tyr	Cys	C
	Leu	Ser	Term[a]	Term[a]	A
	Leu	Ser	Term[a]	Trp	G
C	Leu	Pro	His	Arg	U
	Leu	Pro	His	Arg	C
	Leu	Pro	Gln	Arg	A
	Leu	Pro	Gln	Arg	G
A	Ile	Thr	Asn	Ser	U
	Ile	Thr	Asn	Ser	C
	Ile	Thr	Lys	Arg	A
	Met[b]	Thr	Lys	Arg	G
G	Val	Ala	Asp	Gly	U
	Val	Ala	Asp	Gly	C
	Val	Ala	Glu	Gly	A
	Val[b]	Ala	Glu	Gly	G

[a] Chain-terminating codons

[b] Chain-initiating codons

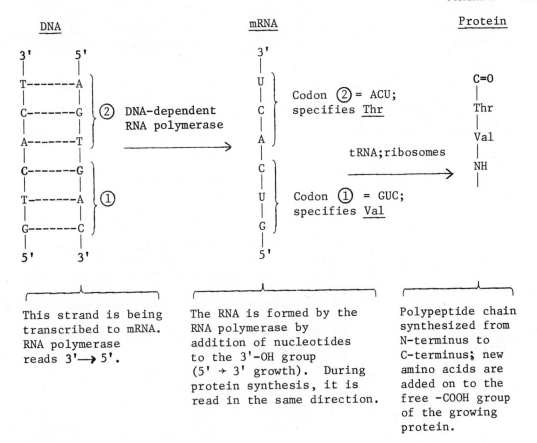

| DNA | mRNA | Protein |

This strand is being transcribed to mRNA. RNA polymerase reads 3' → 5'.

The RNA is formed by the RNA polymerase by addition of nucleotides to the 3'-OH group (5' → 3' growth). During protein synthesis, it is read in the same direction.

Polypeptide chain synthesized from N-terminus to C-terminus; new amino acids are added on to the free -COOH group of the growing protein.

Mutations

Mutations have been known for a long time in terms of their phenotypic or outward expression. In some cases, variation in a specific protein has been implicated as the cause of the modified phenotype. In others, such as an alteration in eye color or in height, the new phenotype appears to involve a more complex change in the molecular makeup. Diseases which can be shown to be inherited are called "inborn errors of metabolism," a term coined by Garrod in 1909. Now that DNA has been established as the genetic material, research has been concentrated on relating heritable characteristics to changes in the chromosomes and in DNA.

A major breakthrough in this work was the discovery of the genetic code (see above). Changes in the amino acid sequence of a protein could be related to a specific alteration in the composition of the DNA. Certain general types of mutations (changes in the base sequence of the DNA) were then recognized and a start could be made toward attributing them to particular chemical events. The mechanisms of action of some of the known mutagens (things which increase the number of mutations over that found normally) were elucidated in terms of the chemistry of DNA. These basic types and the way in which mutagens may cause them are shown below with examples. The changes indicated are based on the wild-type DNA sequence ("normal" or reference sequence; the one found in the unmutated genome) below.

Wild-type
DNA sequence: -G-C- $\boxed{\text{A}}$ -C-
                   . .   .   .
                   . .   .   .
                 -C-G- $\boxed{\text{T}}$ -G-
                         ↑

The examples will indicate changes (mutations) involving this base pair. Note that the
<u>pair</u> must be considered even though the protein is specified by just one strand, because the DNA in the chromosomes is double stranded.

1. <u>Transition</u>. One purine-pyrimidine base pair is replaced by another.
Example: $\boxed{\text{A}\cdot\cdot\text{T}}$ is replaced by $\boxed{\text{G}\cdot\cdot\text{C}}$.

-G-C- $\boxed{\text{A}}$ -C-   -G-C- $\boxed{\text{G}}$ -C-
  . .   .   .   ⟶   . .   .   .
  . .   .   .
-C-G- $\boxed{\text{T}}$ -G-   -C-G- $\boxed{\text{C}}$ -G-

These mutations can occur spontaneously. One possible mechanism involves tautomerization of the adenine to the enol form during DNA replication. In this form, A can pair with C, resulting in insertion of a cytosine in place of thymine in the strand being synthesized. In the next replicative cycle, the C will specify a G in its complementary strand and one of the resultant cells will now have a G-C pair in place of the original A-T pair. These are called <u>copy-errors</u> and require DNA replication for their appearance.

Mutations of this type may also be induced by 5-bromouracil (5BU) which resembles thymine and which may be incorporated into one DNA strand in place of thymine during replication. Since 5BU pairs with G (unlike T, which pairs with A), T and A are replaced with C and G. Similarly, 2-aminopurine may be read as either A or G. Nitrous acid (HNO_2) deaminates A, forming hypoxanthine which pairs with C.

2. <u>Transversion</u>. A purine-pyrimidine base pair is replaced by a pyrimidine-purine pair.

Example: A··T is replaced by T··A .

```
-G-C- A -C-                    -G-C- T -C-
   .   .  .                       .   .  .
   .   .  .        ⟶              .   .  .
-C-G- T -G-                    -C-G- A -G-
```

This type of mutation is commonly observed in spontaneous mutations in some species. About half of the mutations of the α-and β-chains of hemoglobin are transversions. The mechanism is unknown and no agents that specifically cause this type of mutation have been found.

3. <u>Insertion</u>. The insertion of one or more extra nucleotides

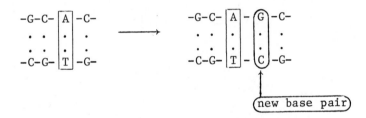

This is one type of frame-shift mutation (see below). It may be caused by acridine orange and proflavin . These molecules can become inserted between two DNA bases (a process called intercalation). They spread the bases farther apart than normal and cause addition of an extra base in the complementary strand during replication. Insertion of <u>more</u> than one base probably occurs during lysogeny by a virus.

Acridine

4. <u>Deletion</u>. The deletion of one or more nucleotides

```
-G-C- A -C-                -G-C-C-
   .   .  .                   . . .           A,T    deleted bases
   .   .  .        ⟶          . . .
-C-G- T -G-                -C-G-G-
```

This is also a frame-shift mutation. It may be caused by hydrolysis (and then loss) of a base due to high temperature or pH change, by covalent cross-linking agents, or by alkylating or deaminating agents. The latter three cause the altering of bases so that pairing cannot occur.

5. Transition and transversion mutations affect at most one codon in the DNA. Due to the existence of chain initiation and termination codons and the degeneracy of the code, a variety of results can occur. It is important to remember that a <u>mutation</u> occurs in the DNA whereas the <u>codon change</u> actually is manifest in the mRNA. The box around one letter of the codon indicates the letter affected.

<u>Table 19</u>. <u>Changes in Amino Acid Sequence Brought About by Mutation</u>

<u>Mutation</u>	<u>mRNA Change</u>	<u>Old Codon</u>	<u>New Codon</u>	<u>Change in Protein</u>
T → C (transition)	A → G	CU [A]	CU [G]	Leu → Leu (no change)
		U [A] G	U [G] G	term → Trp (no termination; next gene read)
		[A] UG	[G] UG	Met → Val (may result in loss of initiation site or incorrect amino acid in protein)
A → T (transversion)	U → A	U [U] A	U [A] A	Leu → term (premature chain termination)
		AG [U]	AG [A]	Ser → Arg (if Ser is active site, as in many enzymes, loss of activity occurs)
		[U] UA	[A] UA	Leu → Ile (probably no change in structure)

Mutations which produce a new chain termination codon are called <u>nonsense</u> mutations while those which replace one amino acid codon by another amino acid codon are called <u>missense</u> mutations. These are all hypothetical examples. Some actual mutations are shown in the section on abnormal hemoglobins.

6. Insertion and deletion mutations result in a shift of the reading frame of the DNA molecule (frame-shift mutations). In some organisms, the majority of the spontaneous mutations observed appear to be of these types.

A frame-shift mutation affects all of the information in the strand from the point of mutation up to the next initiation or control site. Consider the loss of -G-C-C-A- from the sequence shown below.

region deleted in the deletion mutant

DNA: 3' -T-A-T-(G-C-C-A)-T-G-G-A-A-A-C-G- 5'

RNA polymerase

mRNA: 5' -A-U-A-C-G-G-U-A-C-C-U-U-U-G-C- 3'

Protein: —Ile—Arg—Tyr—Leu—Cys—

Normal (wild-type)

DNA: 3' -T-A-T-T-G-G-A-A-A-C-G- 5'

RNA polymerase

mRNA: 5' -A-U-A-A-C-C-U-U-U-G-C- 3'

Protein: —Ile—Thr—Phe—Ala—

deletion mutant

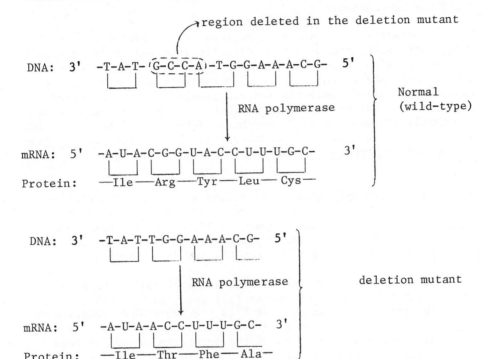

The first amino acid (Ile) is unaffected, since it precedes the deleted region. The succeeding amino acids are all altered and the chain is shortened by one amino acid. (Notice that GC- is enough information to specify alanine, since all four codons (GCU, GCC, GCA, GCG) are for the same amino acid.) If, by chance, the deletion or addition starts between codons and deletes or adds a number of bases which is a multiple of three, then there is no shift in the reading frame. A net loss or gain of codons is all that occurs.

The actual deletion and insertion events only affect one strand of the DNA, initially. During replication, a new strand is made, complementary to the affected sequence. This new, double-stranded DNA is shorter in both strands.

7. Transition mutations resulting from copy-errors (above) cannot occur unless DNA replication is in process. Other mutations can occur in the absence of replication, but require replication for their expression. One can also imagine a mutagenic event which affects a gene necessary for normal cell function. This could both occur and influence the cell in the absence of any DNA replication. If such a mutation affected a process vital to the cell's survival, the cell would probably die prior to mitosis and the mutation would not be transmitted.

8. Errors (mutations) can occur in regions of the DNA which code for rRNA, tRNA, mRNA, and sequences directly involved in the control of transcription (discussed later). These mutations may have a more drastic effect than a change in the sequence of one protein. Loss of control of transcription can result in over- or under-production of one or more gene products. Defects in rRNA or tRNA can completely upset protein synthesis. Regions of the DNA which are used for the control of transcription are called control genes, as opposed to structural genes which specify functional polypeptides.

9. Mutagens usually only affect one strand at a time of a double-stranded DNA. If a mutation occurs in an organism, in most cases half of the progeny of the organism will be normal, the other half mutant. This is further evidence in support of a semi-conservative mechanism for DNA replication (see above).

10. Mutations are frequently deleterious and even lethal to an organism, but they are not necessarily so. They provide the basis for genetic variations which is necessary for evolution to occur. Scientifically, they have been extremely valuable as a tool for gaining knowledge about the mechanism of inheritance and cell function.

11. Our understanding of the etiology of cancer has been advanced by the discovery that certain organic compounds (natural and synthetic) which are widely disseminated (in air, water, soil, food, clothing, etc.) are carcinogenic. In order to prevent cancers caused by such substances, the carcinogenicity of organic compounds present in our environment is being extensively tested.

    a. Investigations of this sort have traditionally required the use of large numbers of experimental animals to which high doses of the compound in question are administered. The validity of this type of testing has been questioned because of the use of animals (usually rodents) and of the large doses given. Four points must be made in refutation of this criticism and in support of these test procedures.

   i) It is necessary to use animals for economic and moral reasons. The test species are mammals and are similar to humans in many (although not all) of their responses to various substances.

   ii) The large dosages are economically necessary. If low dosages were used, similar to those encountered by people consuming normal amounts of food, hundreds of thousands of animals would be required in order to see the small percentage of them which contracted cancer.

  iii) The high dosages increase the probability that a given animal will be affected. The significant fact, however, is whether <u>any</u> of the animals are affected. Most chemicals, in large <u>or</u> small amounts, simply are not carcinogenic.

   iv) There seems to be no safe level for carcinogens. Carcinogens, while causing fewer tumors at low concentrations, still appear to be harmful even at the lowest testable dosages. Thus, drawing conclusions from high-dosage studies about the tumorigenicity of smaller dosages seems qualitatively justifiable.

b. A correlation between the mutagenicity of a compound and its carcinogenicity has been known for some time. B. Ames has used this fact to develop a new assay for carcinogenicity which avoids many of the problems cited above. The Ames test uses certain selected strains of <u>Salmonella typhimurium</u> which carry a mutation that prevents them from synthesizing histidine. They are normally unable to grow unless the medium contains histidine, and are called histidine-negative strains. If, however, these mutant bacteria are exposed to a mutagen, then spread on a histidine-free medium, colonies will appear. The mutagen will cause an <u>additional</u> mutation in some of the bacteria which will <u>correct</u> the original one which made them histidine-dependent (a back-mutation to wild type). If a compound is <u>not</u> mutagenic, none of the bacteria will revert to wild type and no colonies will grow. The number of colonies is a measure of the mutagenicity of a substance. The revertant

colonies can be tested for the type of back-mutation produced
(single base changes, frame-shifts) by selection of the proper
test strain. All of the test strains also lack the DNA
excision repair system so that they are unable to repair the
mutations.

The major drawback to this test is the assumption that a
mutagen is a carcinogen, and vice versa. Although, as was
stated earlier, these assumptions <u>seem</u> valid, based on many
comparisons between this and animal tests (which look directly
at carcinogenicity), there are bound to be exceptions. The
speed, low cost, and effectiveness of the Ames test, however,
make it an important new tool in the search for carcinogens.

One example of the sensitivity of the Ames test is its ability
to differentiate between 2-acetylaminofluorene (AAF) and its
active tumorigenic metabolite, N-hydroxyacetylaminofluorene
(HAAF). AAF is non-mutagenic in the test system, whereas AAF
preparations which have been incubated with liver microsomes
(which convert it to HAAF) show a high degree of mutagenicity.

Protein Biosynthesis

The overall process is one in which mRNA is read by the ribosomes,
one codon at a time. The proper tRNA binds at each codon and adds its
amino acid to the growing chain.

There are five major stages.

1. Activation of the Amino Acids
2. Initiation of the Polypeptide Chain
3. The Elongation Process
4. Termination of the Polypeptide Chain
5. Post mRNA-Ribosomal Modifications of Proteins

Although steps 1 and 5 have been studied extensively in both eucaryotic
and procaryotic systems, steps 2, 3, and 4 are much more clearly worked
out for procaryotes. Steps 1-4 are discussed below in terms of the
mechanism elucidated in <u>E. coli</u> and other procaryotes. As more data
accumulate on eucaryotic protein synthesis, it appears to be similar
in many ways to the process described here.

1. <u>Activation of the Amino Acids</u>. This stage requires ATP, amino
    acids, tRNA, aminoacyl-tRNA synthetases, and Mg^{++}. It occurs
    in two stages:

Note: In this and the other diagrams in this section, tRNA is
       represented schematically by   C -C-A-OH   , where

       -C-C-A-OH is the nucleotide sequence and the free 3'-OH
       group found in all known tRNA molecules. It should be
       understood that this drawing actually represents a
       nucleotide polymer, containing 75-90 nucleotide bases.
       Another representation might therefore be

$$(G/C) \text{---} (N)_{75-90} \text{---} C \text{---} C \text{---} A \text{---} OH.$$

a. The aminoacyl-tRNA synthetases are highly specific for both
    the amino acid and the corresponding tRNA. The enzyme possesses
    three sites for binding: (i) the specific amino acid,
    (ii) the specific tRNA, and (iii) ATP.

b. In the eucaryotic cells, mitochondria contain their own tRNA
    and aminoacyl synthetases which are different from those
    found in non-mitochondrial systems.

c. The mRNA-ribosome complex does not recognize the aminoacyl portion of the loaded tRNA. The specificity of the tRNA-mRNA interaction is entirely due to the relationship between the tRNA anticodon and the mRNA codon. This was demonstrated by the following experiment.

Cysteine, attached to cysteinyl-specific tRNA was converted to alanine in the reaction

$$
\text{(Cys-specific tRNA)}-O-\overset{\overset{\displaystyle O}{\|}}{C}-CH-CH_2-SH
$$

$$
\underbrace{\quad NH_2 \quad}
$$

cysteine

hydrogenation in the presence of Raney Nickel

$$H_2$$

$$\rightarrow H_2S$$

$$
\text{(Cys-specific tRNA)}-O-\overset{\overset{\displaystyle O}{\|}}{C}-CH-CH_3
$$

$$
\underbrace{\quad NH_2 \quad}
$$

alanine

When this tRNA, having the anticodon for cysteine but carrying the amino acid alanine, was used in a protein synthesizing system, alanine was incorporated into the polypeptide product where cysteine should have been.

d. The anticodon region (recognition site) of the tRNA consists of a triplet of bases which aligns with the codon on the mRNA molecule (see previously under the structure of tRNA). The matching between the anticodon region of the tRNA and the codon region of the mRNA presumably occurs by standard base-pairing (A with U and G with C). It appears that the requirements for base pairing in the positions of the first two nucleotides (starting from the 3' end of the anticodon) is more stringent than in the third nucleotide position (at the 5' end of the anticodon) (see the wobble hypothesis, under the structure of tRNA).

e. The anticodons can contain unusual bases as do other parts of tRNA. For this reason, and because the base-pairing of the 5' end of the anticodon with the 3' end of the codon is not as rigidly controlled, the anticodons are often different from what would be predicted by base pairing. This is illustrated in Table 20.

Table 20. Anticodon Assignments for Some Transfer RNA's

Amino Acid	mRNA Codon	Anticodon expected in tRNA on the basis of standard base pairing	Literature Reported Anticodons in tRNA
	(5' → 3')	(3' → 5')	(3' → 5')
Tyrosine	UAC	AUG	AψG
	UAU	AUA	
Phenylalanine	UUC	AAG	AAGm
	UUU	AAA	
Alanine	GCA	CGU	CGI
	GCG	CGC	
	GCC	CGG	
	GCU	CGA	
Methionine (fMet)	AUG	UAC	UAC (for tRNA^{fMet})
Isoleucine	AUU	UAA	UAI
	AUC	UAG	
	AUA	UAU	
Valine	GUU	CAA	CAI
	GUC	CAG	
	GUA	CAU	
	GUG	CAC	

Note: The methionine anticodon is the one predicted by standard base pairing; ψ = pseudouridine; Gm = 2'-O-methyl guanosine; all anticodons are from yeast tRNA molecules except that for fMet.

f. Although the genetic code is universal (the code word for a particular amino acid is the same in all organisms), the tRNA of one organism may not function in another organism. Presumably this is due to the presence of a different nucleotide sequence (at regions other than the anticodon) in some of the tRNA's of different species. These nucleotide sequences recognize the activating enzymes of the particular species.

g. There may be more than one tRNA specific for the same amino acid (just as there may be more than once codon for an amino

acid). However, there need not be a separate activating enzyme
for each of the several tRNA's for one amino acid (since the
tRNA binds to the enzyme and the mRNA template at different
sites, the enzyme-binding sites may be identical even though
the mRNA template-binding sites are different).

2. Initiation of the Polypeptide Chain. Requirements are the
initiating aminoacyl-tRNA (formyl Met-tRNA in bacteria), mRNA, GTP,
initiating factors (F_1, F_2, F_3), 30S ribosomal subunit and 50S
ribosomal subunit. In eucaryotic systems, it appears that a special
Met-tRNA is used for chain initiation. This tRNA is similar in some
respects to the bacterial fMet-tRNA and it cannot be used to
insert a Met residue into a growing peptide chain except at the
N-terminus.

Probable Sequence of Events. (Note that the shape of the ribosomal
subunits in the diagram below is highly schematic):
a. The initiating aminoacyl-tRNA (in bacteria) is formed from
methionine tRNA by formylation of the α-amino nitrogen (for
structure, see under tRNA) in the reaction:

Met-tRNA + N^{10}-formyl tetrahydrofolate

$\longrightarrow$ fMet-tRNA + tetrahydrofolate

b. The procaryotic ribosome (70S), composed of two subunits (30S,
50S), undergoes continuous dissociation and reassociation.
The activation of the ribosome begins with its dissociation.
The 30S subunit interacts with three specific proteins called
initiation factors (IF-1, IF-2 and IF-3). These factors have
been isolated and purified from E. coli. They have molecular
masses of 9,000, 65,000 and 21,000 daltons, respectively.
The probable steps in the formation of the initiation complex
are shown in Figure 54. Note in the sequence of reactions
that:

i) the initiating aminoacyl-tRNA is bound to the peptidyl
site;

ii) in the process of formation of the functional ribosome,
the GTP is hydrolyzed to GDP + P_i, providing the energy
required to make the complex;

iii) the initiation factors that dissociate from the 30S
subunit are reutilized; and

iv) the translation of the codons on mRNA occurs in the
5' → 3' direction with the anticodon of tRNA positioned
in the 3' → 5' direction.

Figure 54. Steps Involved in the Initiation Complex

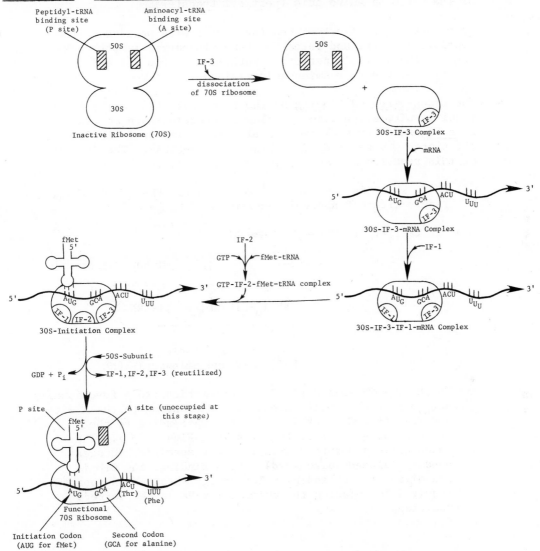

Initiation factors have also been isolated and characterized from eukaryotic cells. In general, they are similar to the prokaryotic ones.

3. The <u>elongation process</u> consists of three steps.

   <u>Codon-directed binding</u> of the proper aminoacyl-tRNA to the aminoacyl binding site (A) of the 50S subunit.

Transfer of the growing peptide chain to the α-amino group of the incoming amino acid (<u>peptidyl transfer</u>).

<u>Translocation</u> of the tRNA bearing the growing peptide to the peptidyl site (P) and release of the discharged (empty) tRNA located there. This permits repetition of step (a) and addition of the next amino acid residue.

a. The <u>codon-directed binding</u> of the aminoacyl-tRNA to the A site requires GTP, a specific cytoplasmic protein known as (in *E. coli* and other prokaryotes) elongation factor T (EF-T) and the proper charged tRNA (in this case Ala-tRNA). The details of this reaction are as follows:

   i) EF-T, which is composed of two subunits, $EF-T_s$ and $EF-T_u$, reacts with GTP and aminoacyl-tRNA.

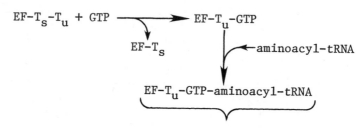

a ternary complex of
$EF-T_u$, GTP and aminoacyl-tRNA

   ii) The $EF-T_u$-GTP-aminoacyl-tRNA complex binds to a functionally active ribosome, with aminoacyl-tRNA binding to the A site and the anticodon portion of the tRNA binding to the corresponding codon of the mRNA (see Figure 55). The codon-anticodon interaction involves specific hydrogen bonding (already discussed). Upon binding, the GTP is hydrolyzed to GDP and P_i. The energy obtained is apparently required for placing the aminoacyl-tRNA in the correct position.

Figure 55. EF-T$_u$-GTP-Aminoacyl-tRNA Binding to Functionally Active Ribosome

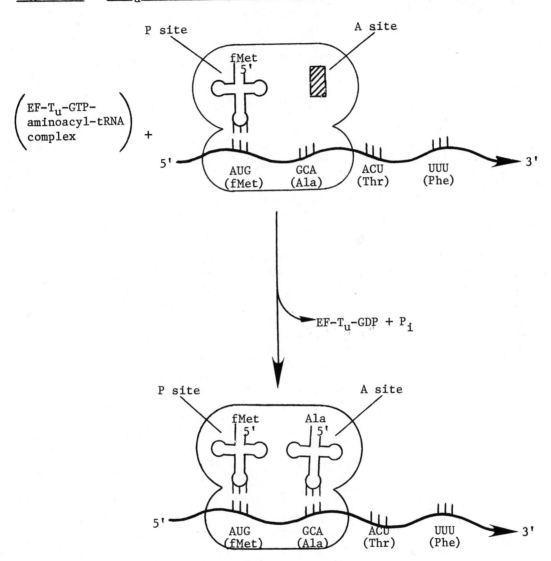

In eukaryotic cells, factors known as EF-1 and EF-2 perform functions quite similar to that of factor EF-T of prokaryotes. One important difference between the factors from eukaryotes and prokaryotes is that the eukaryotic factor (EF-2) is <u>inactivated</u> by <u>diphtheria toxin</u>, an enzyme secreted by the diphtheria bacterium. The inactivation involves a reaction between EF-2 and NAD^+, catalyzed by diphtheria toxin.

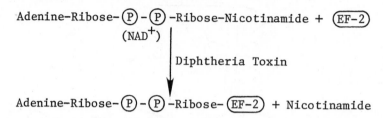

The EF-2 is covalently bonded to the ribose which originally bore the nicotinamide, inactivating the EF-2. Diphtheria toxin thus inhibits eukaryotic protein synthesis at the translocation stage.

b. Formation of the peptide bond requires <u>peptidyl transferase</u>, an enzyme which is part of the 50S ribosome. This step requires no ATP or GTP. The energy needed to form the peptide bond probably comes from hydrolysis of the ester linkage between the peptide chain and the tRNA.

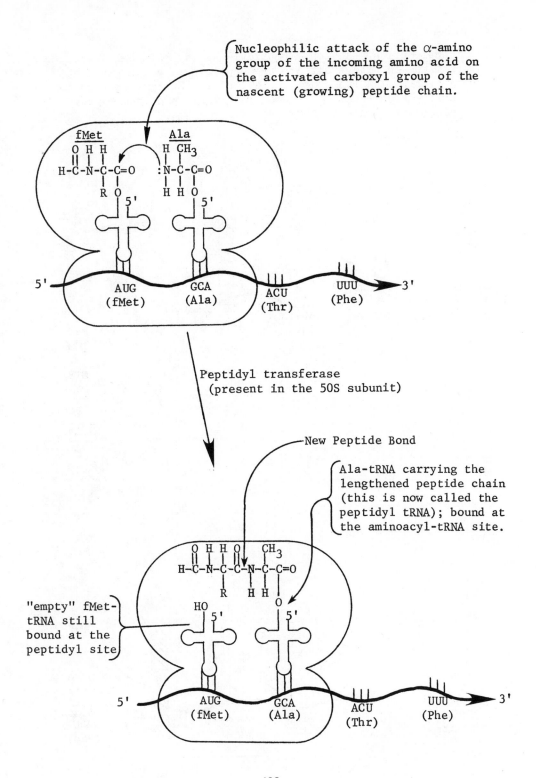

Nucleophilic attack of the α-amino group of the incoming amino acid on the activated carboxyl group of the nascent (growing) peptide chain.

Peptidyl transferase (present in the 50S subunit)

New Peptide Bond

Ala-tRNA carrying the lengthened peptide chain (this is now called the peptidyl tRNA); bound at the aminoacyl-tRNA site.

"empty" fMet-tRNA still bound at the peptidyl site

c.   <u>Translocation Reaction</u>.   Peptidyl-tRNA is physically shifted
from the aminoacyl site to the peptidyl site, displacing
the empty tRNA from the peptidyl site.   Energy for the
conformational change in the ribosome is provided by the
hydrolysis of GTP to GDP + P_i

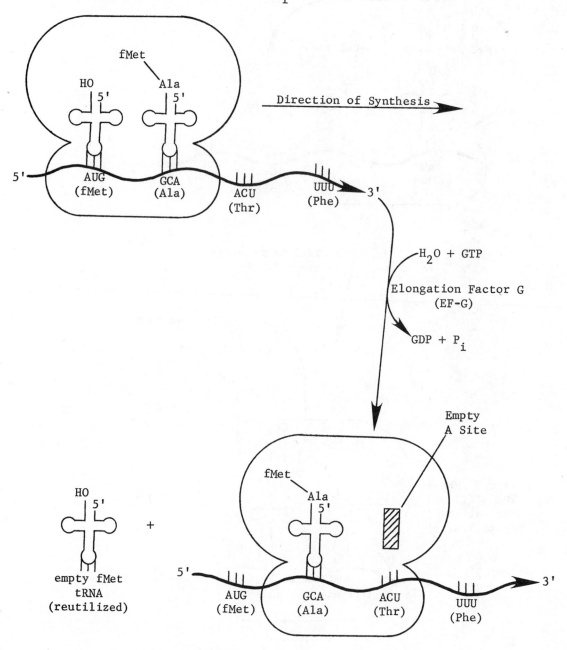

d. These three reactions are cyclically repeated until one of the termination codons is reached. Note that in the process of elongation, two GTP molecules are hydrolyzed for each amino acid added: one when the aminoacyl-tRNA binds to the A site of the ribosome and the other during translocation. Thus, the total amount of energy expended in the synthesis of <u>one</u> peptide bond comes from the hydrolysis of four high energy phosphate bonds (equal to about 30 kcal = 7.5 x 4). Recall that the synthesis of each molecule of aminoacyl-tRNA consumes two high energy phosphate bonds (ATP $\rightarrow$ AMP + PP_i; $PP_i \rightarrow 2 P_i$). (Note in these calculations that the one GTP molecule expended in the formation of the initiation complex is not included, because of its small contribution in the overall synthesis of a polypeptide.)

Despite the fact that a large amount of energy (30 kcal) is used to synthesize a peptide bond, the amount of energy conserved in the single peptide bond is only about 5.0 kcal (standard free energy of hydrolysis of a peptide bond $\Delta G^{o'}$ = -5.0 kcal). The net $\Delta G^{o'}$ for peptide bond formation is about -25 kcal $\left[-30 - (-5) \right]$. Therefore, this process is overwhelmingly exergonic and is, for all practical purposes, irreversible. It appears that a large fraction of a cell's energy is expended in the synthesis of proteins. This process of energy expenditure may have evolved to assure perfect fidelity in the translation of mRNA.

e. Ribosomes undergo complex changes in shape during each bond-making cycle. It has been shown that the mRNA is bound to the 30S subunit at a specific site which is about 30 nucleotide units long. Although it is not completely clear just what role the rRNA plays in this binding, it has been suggested that Mg^{+2} ions form bridges between the rRNA and mRNA bases. Whatever the exact mechanism is, there is definitely a portion of the mRNA molecule which cannot be hydrolyzed by RNAse as long as ribosomes are bound to it.

4. Termination of the Polypeptide Chain

a. Termination is signaled by a termination codon in the mRNA after the last amino acid has been added. To terminate polypeptide synthesis, two interactions must be ended: i) the tRNA (anticodon) and mRNA (codon) interaction, and ii) that between the tRNA and the nascent polypeptide chain.

b. The detachment process requires three specific proteins known as release factors (R_1, R_2 and R_3, in E. coli and other prokaryotes) which bind to the ribosome, bringing about a shift of the polypeptidyl-tRNA from the A site to the P site. At the

P site the polypeptide chain is hydrolytically cleaved from the tRNA. The enzyme that catalyzes this hydrolytic reaction is probably peptidyl transferase, the same enzyme which is responsible for peptide bond synthesis. The altered specificity of the transferase is apparently brought about by the binding of the releasing factors. The sequence of the termination steps are shown in Figure 56.

5. Post-Ribosomal Modifications of Proteins

    a. After their synthesis on mRNA-ribosome complexes, many proteins are modified by hydroxylation (collagen); phosphorylation (glycogen phosphorylase, ribosomal proteins, etc.); acetylation (histones); methylation (histones, muscle proteins, cytochrome c, brain protein); addition of various carbohydrate groups (glycoproteins); or removal of a peptide (activation of insulin, proteolytic enzymes).

    b. The exact biological significance of these post-ribosomal modifications is not known, but some of the possibilities include

        i) protection against proteolytic enzymes (e.g., in the case of collagen and elastin);

        ii) self-assembly of single proteins into an active, multisubunit form (as in the formation of the collagen superhelix from collagen monomers; discussed later);

        iii) initiation or termination of a biological function (methylation and acetylation of histones; activation of insulin and proteolytic enzymes; phosphorylation and dephosphorylation of certain enzymes in glycogen metabolism);

        iv) identification of proteins intended for export from the cell (e.g., the glycoproteins); and

        v) introduction of molecular heterogeneity following synthesis on the mRNA ribosome complex (see the immunoglobulins in the chapter on immunochemistry).

    c. Methylation predominantly occurs on the sidechains of the amino acids lysine and arginine, and on free carboxyl groups. Phosphorylation usually occurs at serine hydroxyls.

    d. Extensive hydroxylation of preformed protein occurs in collagen (discussed later), primarily at lysyl and prolyl residues.

Figure 56. The Sequence of the Termination Process

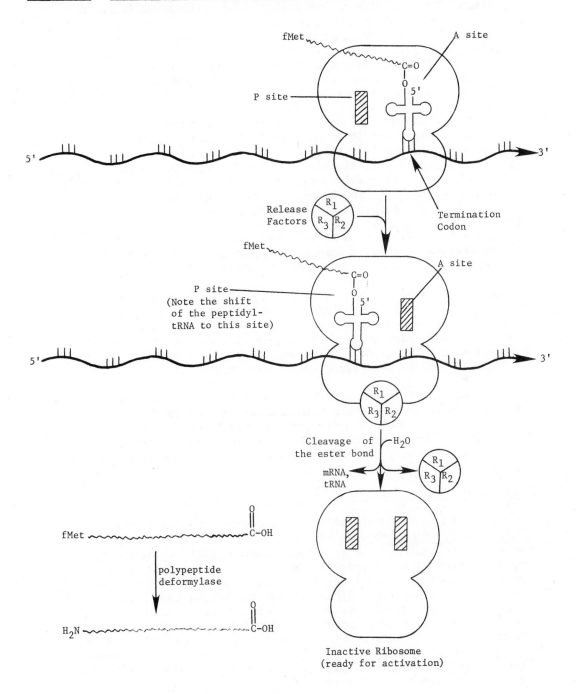

e. Trimethylation of a lysine residue ($-NH_2 \rightarrow -N(CH_3)_3^+$) with S—adenosylmethionine as the methyl donor (discussed in amino acid metabolism) is involved in the synthesis of carnitine (see fatty acid metabolism). This methylation does not take place with free lysine but only with protein-bound lysine. The free trimethyllysine derived from the breakdown of proteins is then converted into carnitine by a series of reactions that are not completely known.

f. Vitamin K-dependent γ-carboxylations of glutamic acid residues in prothrombin and other coagulation factors (VII, IX, and X) are post-ribosomal modifications that are required to make these molecules functional. The γ-carboxyglutamic acid residues of prothrombin serve as binding sites for calcium, which is essential in the conversion of prothrombin to thrombin, its physiologically active form.

g. Reversible adenylation at tyrosine residues of glutamine synthetase (see amino acid metabolism) is involved in the metabolic control of the reaction catalyzed by this enzyme.

h. Hexose binding to the NH_2-terminal valine residues of the β chains of hemoglobin changes the binding of 2,3-diphospho-glycerate to hemoglobin (see Chapter 7). The reaction is not enzyme-catalyzed and occurs to a greater extent in diabetics, who have an elevated blood glucose. This modification may be involved in the regulation of oxygen affinity.

i. Deamidation of amino acid residues (Gln → Glu, Asn → Asp) is also observed. Since this changes the net charge on the protein, it produces additional electrophoretic forms. The multiple banding pattern of human salivary and pancreatic amylase has been demonstrated to result, in fact, from this type of modification.

Inhibitors of Protein Synthesis

1. Puromycin has a structure very similar to that of the terminal adenosine of aminoacyl-tRNA, so that it is able to enter the aminoacyl (A) site on the ribosome. Another part of the puromycin molecule resembles the amino acids tyrosine and phenylalanine. Peptidyl transferase will catalyze bond formation between the previously synthesized polypeptide and the puromycin. Because this bond is not easily broken and because the puromycin is not really attached at the A site, the peptidyl-puromycin is released. Thus puromycin affects peptide chain elongation, causing premature termination of synthesis. The released chains are shorter than they should be and have a puromycin molecule covalently attached to the C-terminus (see Figure 57).

Figure 57. Schematic Diagram Illustrating the Action of Puromycin in Its Inhibition of Protein Biosynthesis

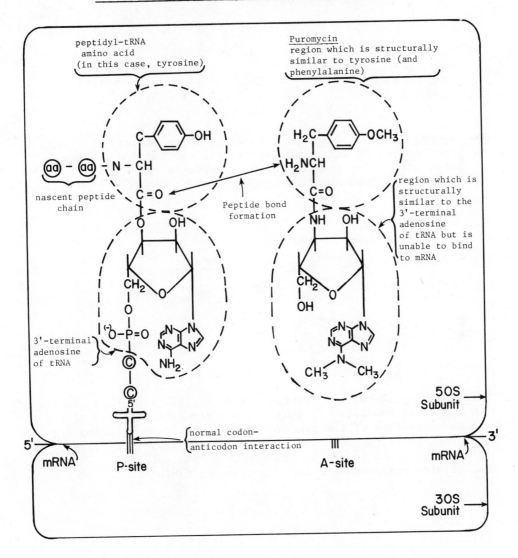

439

2. <u>Chloramphenicol</u> probably interferes with the peptide-forming steps by binding to the 50S subunit. It inhibits protein synthesis by the 70S ribosomes in procaryotic cells and in the mitochondria of eucaryotic cells,but not by the 80S ribosomes of eucaryotic cells. In susceptible organisms, chloramphenicol prevents chain elongation beyond the first peptide bond.

<u>Cycloheximide</u> (also called actidione) inhibits protein synthesis in 80S ribosomes of eucaryotic cells, but not the 70S ribosomes of procaryotes. The results which it produces are similar to those induced by chloramphenicol.

3. <u>Streptomycin</u> (one of the aminoglycoside antibiotics) binds to the 30S ribosomal subunit. It inhibits protein synthesis and also causes misreading of the genetic code, probably by altering the conformation of the 30S subunit so that aminoacyl-tRNA's are less firmly bound to codons and are thus less specific. This and related compounds are important antimicrobials.

Antibiotics which are related to streptomycin and which produce similar effects are the neomycins and kanamycins. Bacterial cells can mutate and become resistant to streptomycin (able to grow in its presence). They have been shown to have an altered 30S subunit which is not affected by this antibiotic.

4. Tetracyclines inhibit protein synthesis by preventing the binding of aminoacyl-tRNA to the 30S subunit.

Tetracycline

5. Diphtheria toxin (already discussed) inhibits protein synthesis by inactivating EF-2 in eukaryotic cells and thus interferes with translocation.

6. Some other inhibitors of protein synthesis are emitine (inhibits binding of aminoacyl-tRNA to the ribosome), and abrine and ricin (plant proteins which interfere with elongation on eukaryotic ribosomes by binding to the 60S subunit).

7. Agents which inhibit DNA transcription and RNA synthesis thereby also prevent protein synthesis. Actinomycin D binds to DNA and prevents transcription by blocking movement of the RNA polymerase. Rifampicin (and other members of the rifamycin family of antibiotics) binds directly to the DNA-dependent RNA polymerase and inhibits transcription in this manner. The rifamycins affect procaryotic polymerase but not the polymerase found in eucaryotic cells. Rifamycin B occurs naturally and rifampicin is a semi-synthetic derivative of it. The structures of actinomycin D and rifampin (a typical rifamycin) are given in the section on messenger RNA.

Polyribosomes

1. Clusters of ribosomes attached to one another by a strand of mRNA are known as polyribosomes (polysomes; ergosomes). In electron microscopic studies, the connecting fiber can actually be seen.

2. Polysomes result from the reading of one mRNA molecule by more than one ribosome at a time. One of the four peptide subunits of hemoglobin contains about 150 amino acids. The corresponding mRNA needed to synthesize this peptide has $3 \times 150 = 450$ nucleotide bases. Since each base occupies about 3.4 Å of the mRNA, 450 bases is about 1500 Å long. An 80S reticulocyte ribosome is about 220 Å in diameter and it has been found that polysomal ribosomes are spaced about 50-100 Å apart. Based on these values, there should be 5-6 ribosomes in each hemoglobin polysome. This is illustrated in Figure 58.

3. Polysomes of many sizes have been found. In E. coli they have been found with up to 40 ribosomes per mRNA molecule. From liver, polysomes containing up to 20 ribosomes have been reported.

4. Such synthesis increases the efficiency of utilization of mRNA, since several polypeptide chains can be made simultaneously from one message. Prior to the discovery of polysomes, measurements of the amount of mRNA in a cell gave values which were too low to explain the observed rate of protein synthesis.

5. The direction of translation is 5' → 3' and it is not reversible. mRNA is always read from the 5' → 3' end.

Figure 58. Polysome for Hemoglobin Subunit Synthesis Containing Five
Ribosomes Per Message

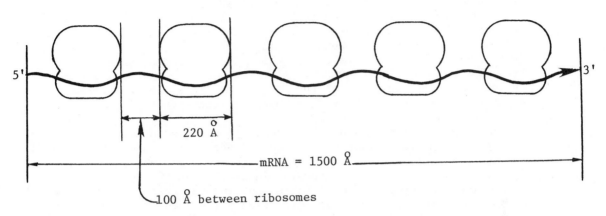

If there were only 50 Å between ribosomes, there would be just about
enough room for a sixth ribosome on the message. The drawing is on
a scale of 100 Å to the centimeter.

Protein Synthesis in Mitochondria

1. Protein synthesis does take place in mitochondria.

2. Mitochondria possess circular DNA and have a DNA-directed RNA
    polymerase as well as specific forms of tRNA and activating
    enzymes.

3. Mitochondrial ribosomes are roughly 70S, similar to those of
    bacteria (procaryotic cells; see under rRNA). Thus, eucaryotic
    mitochondria are affected by drugs like chloramphenicol.

4. Experiments strongly indicate that the cytochromes and other
    mitochondrial enzymes are specified by nuclear chromosomes and are
    synthesized extramitochondrially. Mitochondrial DNA may code for
    only a few membrane-located proteins. Similarly, chloroplast DNA
    may also code for some membrane-bound proteins but information
    contained in the nuclear DNA is necessary for the formation of
    functional chloroplasts.

Protein Synthesis and Secretion in Eucaryotic Cells

1. As was mentioned previously, procaryotic ribosomes and polysomes
    are found free in the cytoplasm. In eucaryotic cells, proteins
    which are to be used <u>within</u> the cell are also synthesized by
    ribosomes (polysomes) present in the cytoplasm, unattached to any
    of the other subcellular organelles. An example of this type of
    synthesis is the formation of hemoglobin in the reticulocytes.

2. Some cells synthesize a large amount of protein which is intended for export (excretion). This protein is made by ribosomes which are bound in rows to the outer surface of parts of the endoplasmic reticulum (ER). The rough endoplasmic reticulum appears rough in the electron microscope because of the ribosomes present as bumps on the surface. It is thought that the large (60S) ribosomal subunit contains the point of attachment to the endoplasmic reticulum and that the growing peptide extends through the 60S subunit into the cisternae of the reticulum. Thus, the completed peptide is released into the cisternae for transport.

3. a. Examples of this include the pancreatic exocrine cells (which synthesize proteolytic enzymes for secretion into the intestinal lumen) and the liver (which makes albumin and other serum proteins). The synthesis, intracellular transport, and ultimate discharge of exportable proteins have been investigated in the pancreatic exocrine cell using pulse chase radioautography. Tissue slices are cultured in a medium containing radioactive amino acids which are incorporated into newly synthesized protein. The slices are then washed free of unincorporated labeled amino acids and placed in a medium containing unlabeled amino acids. Previously synthesized labeled protein is "chased" by the unlabeled protein that is made subsequently. The "pulse" of labeled protein can be followed as it moves through the cell by removing tissue slices at various intervals and subjecting them to fixing, radioautography, and electron microscopy.

    b. The secretory cells of the pancreas (acinar cells) are arranged in roughly ellipsoid structures known as acini. Each acinus contains about a dozen cells. The central region of this aggregation is the lumen, which is the beginning of the pancreatic duct system. Each acinus contains a centro-acinar cell which secretes bicarbonate solution. The acinar cells are joined to one another by tight junctions, desmosomes, and gap junctions. The tight junctions probably prevent leakage of pancreatic secretions into the extracellular spaces. These cells are pyramidal shaped and are polar with the nucleus on the side away from the lumen (the base) and the secretory granules at the lumenal side (the apex). The centro-acinar and acinar cells show microvilli on their free borders. These aspects are schematically shown in Figure 59.

    c. Synthesis of secretory proteins begins on the polysomes attached to the exterior membrane of the endoplasmic reticulum (the rough ER) of the acinar cell. The larger subunit of the ribosome makes contact with the membrane. As discussed earlier, the protein is synthesized and transferred into the cisternal space of ER, through pores in the ER membrane. This is illustrated in Figure 60.

Figure 59. Schematic Diagram of an Acinus

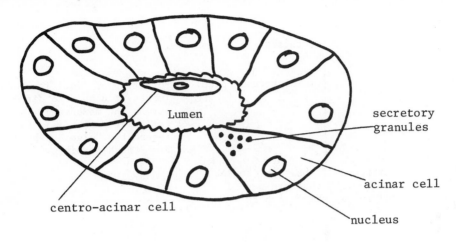

Figure 60. Schematic Diagram of the Synthesis of Secretory Protein by Polysomes on the Rough Endoplasmic Reticulum

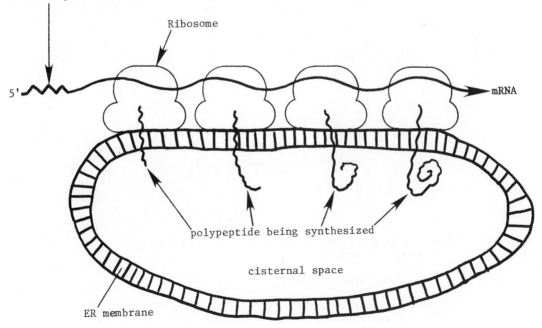

d. The migration of the secretory proteins can be summarized as follows (also see Figure 61).

Site of Synthesis
(Rough ER)

| Via cisternal space: 1. Formation of disulfide bonds (makes it unable to diffuse across the membrane)
2. Selective proteolysis to remove the "signal peptide" from the N-terminus.

Passage through "Transition" Region
(portions of ER that are devoid
of polysomes, the smooth ER)

Budding from the Transition Region

Transporting Vesicles

Passage through Golgi Complex
(series of flattened, smooth-surfaced
saccules lying one on top of the other)

Secretory proteins undergo extensive modifications--
carbohydrate additions, inorganic sulfate addition,
selective proteolysis, etc.

Condensing Vacuole
(immature granule formation)

Mature Storage Granule
(zymogen granule)

Appropriate external stimuli

Crowding of Granules at the Apex

Fusion with the Plasmalemma
(plasma cell membrane)

Exocytosis

Ejection of the Contents
of the Granule into the
Acinar Lumen

Figure 61.[*] <u>A Detailed View of the Pancreatic Exocrine Cell Showing</u>
<u>the Events Leading to Secretion of Proteins into the</u>
<u>Acinar Lumen</u>

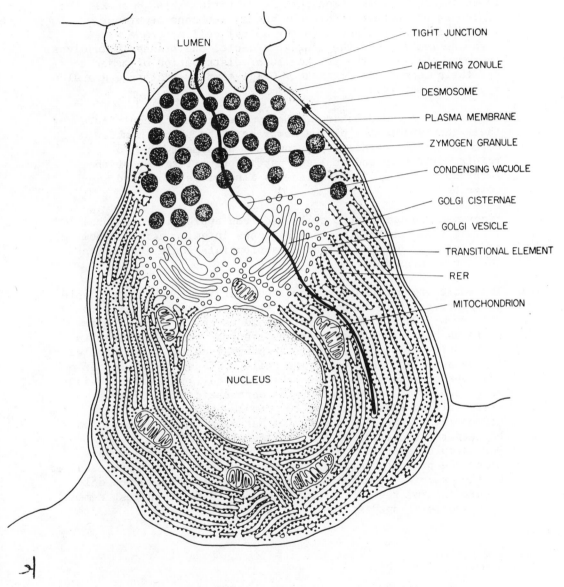

*Reprinted with permission from J.D. Jamieson, "Membranes and
Secretion," <u>Hospital Practice</u>, Vol. 8, No. 12, and from "Cell
Membranes: Biochemistry, Cell Biology & Pathology," G. Weissman
and R. Claiborne, eds., HP Publishing Co., Inc., 1975.

e. Some characteristics of this system of transport are: protein for export is sequestered within the cisternal space of the ER; transport always occurs within a membrane-bound container; the contents cross a membrane only once; the protein is transported in an insoluble form; transport is vectorial (base → apex); the synthesis of protein is discontinuous; membrane-membrane interaction takes place when the apical cell membrane fuses with the membrane of the storage granule; and transport involves enormous change in the intracellular distribution of membranes and their components (from ER → Golgi complex → storage granules → plasma membrane).

f. During fusion and discharge, the zymogen granules contribute their membrane to the plasma membrane. The plasma membrane does not expand permanently, however, presumably due to reinternalization and at least partial reutilization of the membrane. Some of the membrane is also broken down and resynthesized.

g. Exocytosis in pancreatic cells requires appropriate hormonal stimuli (see gastrointestinal hormones in amino acid metabolism), continued ATP synthesis, and an appropriate ionic environment (Ca^{+2} concentration).

h. The secretory process described thus far is probably applicable to a variety of cells possessing storage granules ranging in size from 250 to 1500 mμ. This includes cells which export proteins but do not show storage granules such as are seen in pancreatic cells (e.g., liver parenchymal cells, plasma cells, fibroblasts, and some endocrine cells). It has been assumed that the secretory process in the liver cell, for example, is in general similar to that of the pancreatic cell. They presumably contain much smaller granules (about 50 mμ in diameter). There are, however, other significant differences between the secretion by pancreatic cells and liver cells. For example, secretion by the liver cell may be continuous, since the products (albumin, clotting factors, etc.) are required by the organism at all times, while pancreatic secretion only occurs intermittently and only when needed (as to digest a meal), in response to specific hormonal signals.

Recapitulation of Some Chemical Reactions Which Occur in Protein
Biosynthesis. An overall scheme of protein synthesis is shown in
Figure 62.

1. Activation of the amino acid by aminoacyl-tRNA synthetase.
    E, below, is the synthetase enzyme.

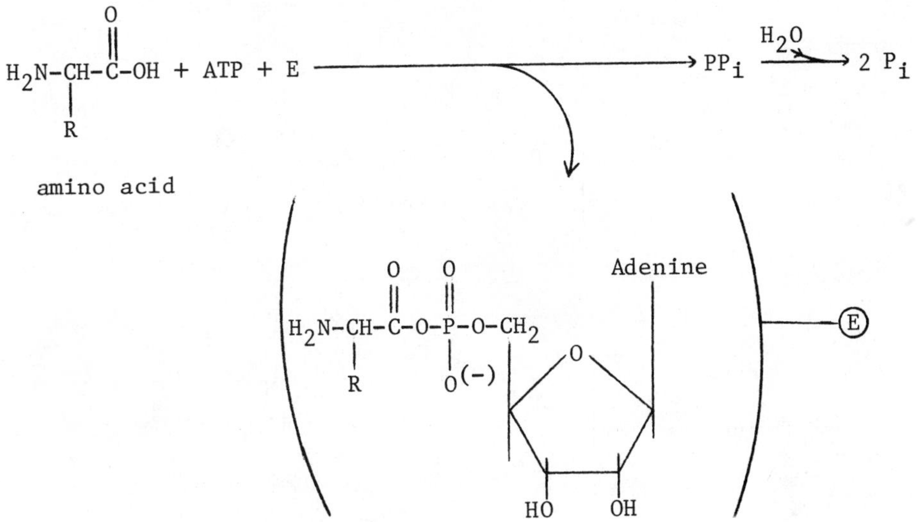

$$H_2N-CH-\overset{\overset{O}{\|}}{C}-OH \ + \ ATP \ + \ E \ \longrightarrow \ PP_i \ \overset{H_2O}{\longrightarrow} \ 2 \ P_i$$

amino acid

mixed acid anhydride, between
an amino acid and a phosphate
(a.a.-AMP- Ⓔ)

2. Loading of the tRNA by transfer of the activated amino acid to the
    3'-OH group of the tRNA. This reaction is also catalyzed by
    the synthetase enzyme, E. Ⓒ and Ⓟ represent, respectively,
    cytosine and phosphate.

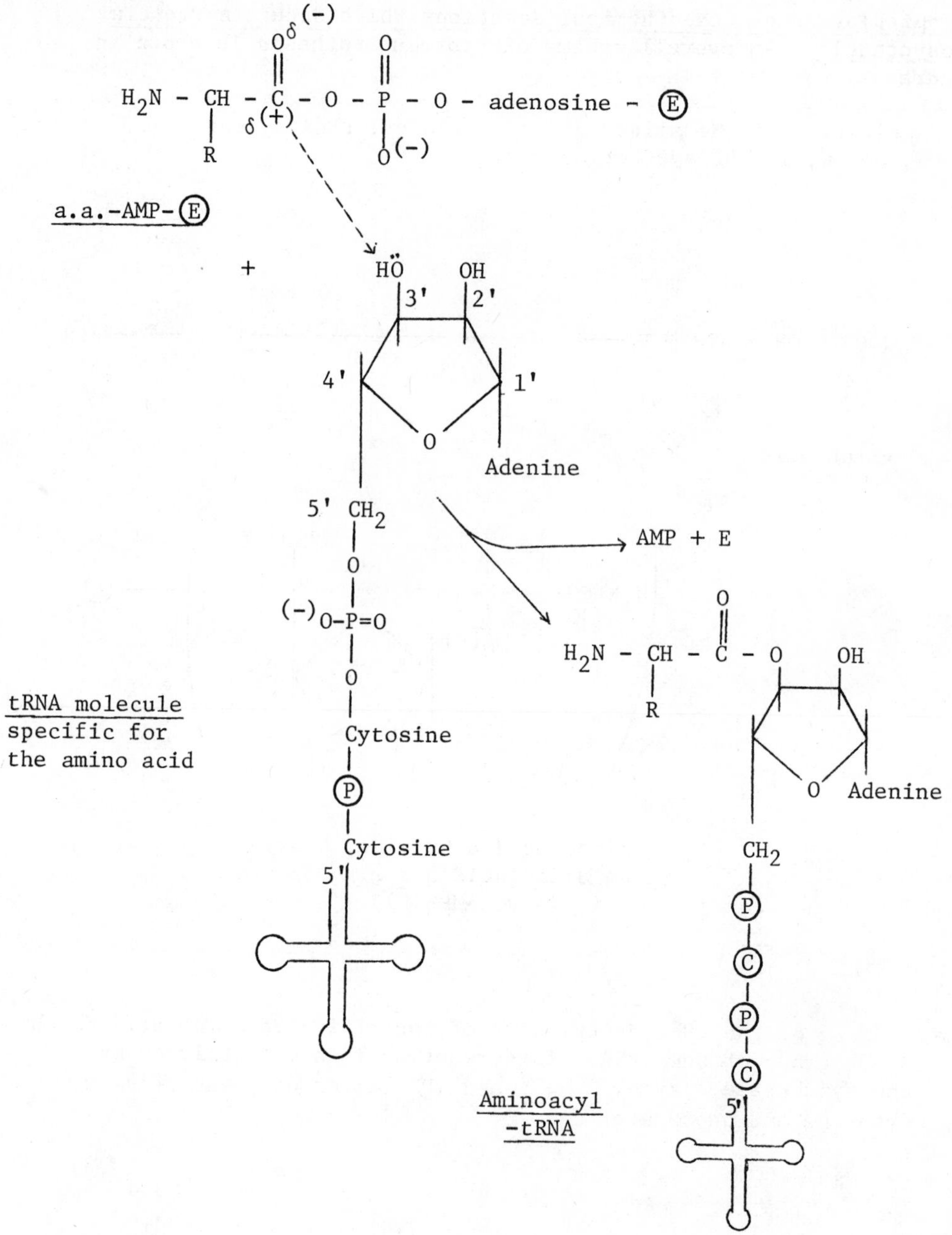

$$H_2N - CH - \overset{O^{\delta}\,(-)}{\underset{\delta\,(+)}{C}} - O - \overset{O}{\underset{O\,(-)}{P}} - O - \text{adenosine} - \textcircled{E}$$

with R below CH.

a.a.−AMP−(E)

+

HO   OH
3'   2'

4'   1'

O

Adenine

5' CH$_2$

O

$(-)$ O−P=O

O

tRNA molecule
specific for
the amino acid

Cytosine

(P)

Cytosine

5'

→ AMP + E

$$H_2N - CH - \overset{O}{C} - O \qquad OH$$

with R below CH.

O   Adenine

CH$_2$

(P)

(C)

(P)

(C)

5'

Aminoacyl
−tRNA

3. Transfer of the growing peptide chain from the tRNA bound to the peptidyl site on the ribosome to the α-amino group of the aminoacyl tRNA bound at the aminoacyl site. The reaction is catalyzed by peptidyl transferase.

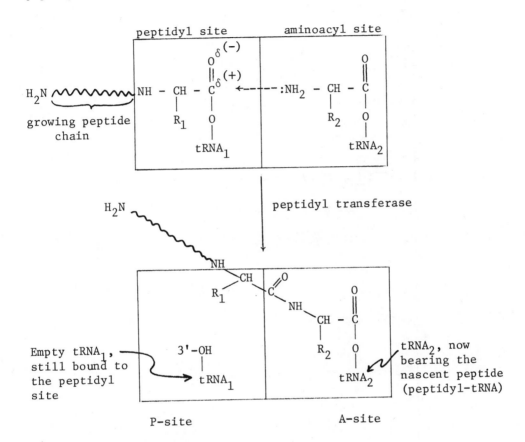

Synthesis of Short Peptides

1. Short peptides such as gramicidin S, glutathione, and others are not synthesized on free ribosomes. Some short peptides, however, are derived from larger polypeptides (which <u>are</u> synthesized on ribosomes) by cleavage (e.g., peptide hormones). Glutathione, synthesized by a non-ribosomal pathway, has been discussed in carbohydrate metabolism in red blood cells.

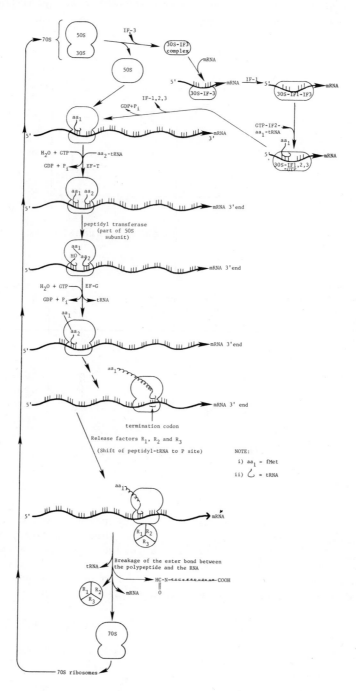

452

2. Gramicidin S, an antibiotic produced by certain strains of
    <u>Bacillus</u> <u>brevis</u>, is a cyclic decapeptide consisting of two
    identical pentapeptides. It is synthesized by an enzymatic
    process which does not involve ribosomes or mRNA. The structure
    of gramicidin S is:

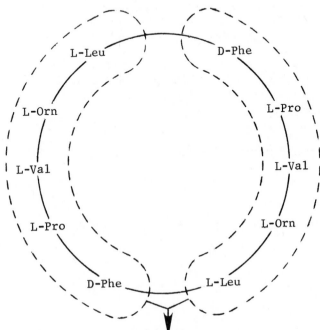

Two identical pentapeptides linked head to tail.

3. The synthesis involves two enzymes, E_I and E_{II}. The reactions are
    as follows:

    Activation of Amino Acids (aa): All of the aa's except D-Phe are
    activated by E_I in the presence of ATP and each aa is bound in a
    thioester linkage at a <u>specific site</u> on the enzyme. D-Phenylalanine
    complexes with E_{II} in a thioester linkage, with ATP providing the
    energy.

$$aa + ATP \longrightarrow aa\text{-}AMP \quad + PP_i$$
$$(\text{aminoacyl-AMP})$$

$$aa\text{-}AMP + E_I\text{-}SH \longrightarrow E_I\text{-}S\text{-}\overset{\overset{O}{\|}}{C}\text{-}\overset{\overset{H}{|}}{\underset{|}{C}}\text{-}NH_2$$

(with "$- AMP$" leaving)

Thioester linkage — amino acid (R)

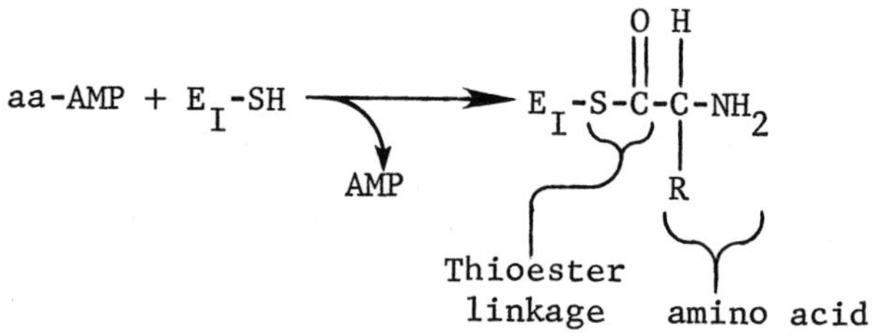

Similarly,

$$D\text{-}Phe + ATP \longrightarrow Phe\text{-}AMP + PP_i$$

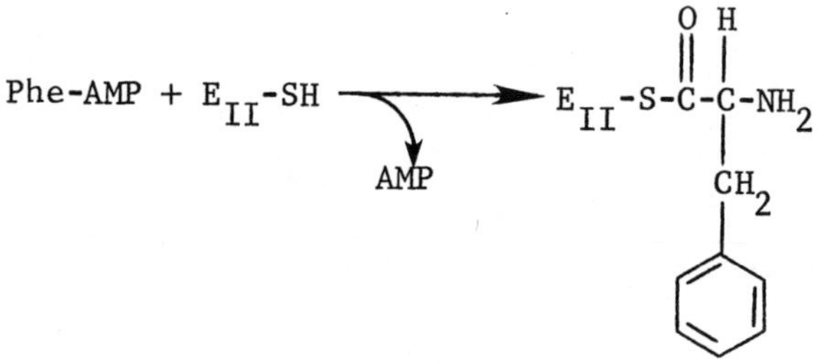

$$Phe\text{-}AMP + E_{II}\text{-}SH \longrightarrow E_{II}\text{-}S\text{-}\overset{\overset{O}{\|}}{C}\text{-}\overset{\overset{H}{|}}{\underset{|}{C}}\text{-}NH_2$$

(with "$- AMP$" leaving; side chain CH_2—phenyl)

Peptide synthesis begins from the N-terminus with the interaction of E_{II}-Phe and E_I-Pro.

$$E_{II}\text{-}S\text{-}\overset{\overset{O}{\|}}{C}\text{-}Phe + E_I\text{-}S\text{-}\overset{\overset{O}{\|}}{C}\text{-}Pro \longrightarrow E_{II}\text{-}SH + E_I\text{-}S\text{-}\overset{\overset{O}{\|}}{C}\text{-}Pro\text{-}Phe$$

The remainder of the synthesis proceeds one residue at a time by transfer of the growing N-terminal sequence from the peptidyl E_I to the α-amino group of the next E_I-amino acid. This leads to the production of

$$E_I\text{-}S\text{-}\overset{\overset{O}{\|}}{C}\text{-}Leu\text{-}Orn\text{-}Val\text{-}Pro\text{-}Phe\text{-}NH_2$$

The cyclic decapeptide is obtained by interaction of two
E_I-pentapeptide molecules:

H₂N-Phe-Pro-Val-Orn-Leu-C-S-E_I

(with O double-bonded to C, arrows connecting the two peptides)

E_I-S-C-Leu-Orn-Val-Pro-Phe-NH₂

The specificity for the synthesis of the gramicidin S sequence
resides in the specificity of E_I and E_{II}.

Some Selected Examples of Protein Synthesis in Mammalian Systems

1. Comments on Collagen Biosynthesis

    a. Collagen is the principal protein of the connective tissue.
        Its major function is to provide strength and maintain the
        structural integrity of tissues and organs. In its final
        form, in the absence of any denaturation or degradation,
        it is insoluble.

    b. Collagen fibrils are made up of subunits called <u>tropocollagen</u>.
        These subunits (molecular mass of about 285,000 daltons) are
        in turn composed of three polypeptide chains. There are at
        least four genetically distinct types of collagen, based on
        these three chains of tropocollagen.

        <u>Type I</u>--consists of two α1-type-I chains and one α2 chain--
        [α1(I)]₂α2. This collagen is present in a number of connective
        tissues such as mature skin (dermis), tendon, bone, and dentin.

        <u>Type II</u>--consists of three identical α1-type-II chains--
        [α1(II)]₃. This type of collagen is found mostly in
        cartilagenous tissues. The α1-type-II chains have a molecular
        mass similar to α1-type-I chains. However, they differ in
        primary sequence and the former has increased hydroxylysine
        and glycosylated hydroxylysine residues.

        <u>Type III</u>--is made up of three identical chains known as
        α1-type-III chains--[α1(III)]₃. Unlike α1(I), α1(II) and α2
        chains, the α1(III) peptide chains contain cysteine residues
        which participate in the formation of intramolecular disulfide
        linkages. Presumably this type of collagen is found in tissues
        associated with epithelium or endothelium (e.g., rat lung,
        human placental membrane, human aorta) and absent in tissues
        which do not support an epithelium (e.g., tendon).

Type IV--is made up of genetically distinct α1 chains (amino acid composition is different from the other α1 chains). Type IV collagen is primarily found in various basement membranes (e.g., the glomerulus, Descemet's membrane, the basement membrane surrounding the smooth muscle cells of arteries, etc.). The various collagen peptides (α1(I), α1(II), α1(III), α2 and basement membrane peptide) have different amino acid compositions and unique sequences.

In solution, collagen molecules exist as long rods. Consequently, collagen solutions have a high viscosity. They also exhibit a strong, negative optical rotation. When a solution of collagen is heated, the molecules undergo denaturation. This process can be followed by measuring changes in viscosity (more viscous to less viscous state) and optical rotation. These changes are due to melting of the collagen structure from a helix to a random coil conformation (a process quite analogous to the melting of double-helical polynucleotides). The midpoint of the denaturation curve of collagen is termed the melting temperature (T_m). For a fully hydroxylated native collagen, T_m is about 37°C.

c. The α1 (type I) peptide contains a repeating tripeptide unit made up of glycine with two other amino acids: Gly-aa_1-aa_2. A large number of the tripeptides contain proline in the aa_1 position, yielding Gly-Pro-aa_2. In the aa_2 position, the amino acids most frequently found are hydroxyproline and alanine. Other residues, such as lysine, hydroxylysine, and serine, are also present, but less frequently. The amino acid sequence of other collagen peptides is currently being investigated.

d. The hydroxylation of selected lysine and proline residues occurs after the translational process (that is, it is a post-ribosomal modification of the peptide chain). The hydroxylation is catalyzed by a mixed-function oxidase. It requires α-ketoglutarate, a reducing agent (ascorbic acid), Fe^{++}, and oxygen. Hydroxylation of lysine does not occur in elastin (discussed later), but some of the prolyl residues are hydroxylated. The Clq component of complement, which shares certain features with the collagen peptides, also undergoes similar hydroxylations. Although the exact function of the hydroxylations is not clear at this time, they play a role in helical stability, affect the carbohydrate content (since carbohydrates are attached to the hydroxylysyl residues), and are important in the formation of crosslinks (residues of hydroxylysine appear to be the major precursors of crosslinks in collagen, particularly at certain unique sites). As we will see later, all of the crosslinks in elastin are derived from residues of lysine and modified lysine. In collagen, however, lysine, hydroxylysine and histidine are involved in the crosslinking process.   In general, collagen has fewer crosslinks than elastin.

e.   In the tropocollagen molecule, each of the three peptide chains are present in a left-handed polyproline-type helix with about three residues per turn.   Furthermore, the three helical chains form a gradual right-handed superhelix by winding around a central axis (coiled-coil).   The individual chains are interconnected by hydrogen bonds.   Note that the collagen helix is not an α-helix.   In the α-helix, there is only one peptide chain and all of the hydrogen bonds are intrachain (between parts of the same chain).   Uncertainty exists at this time regarding how the helices fit together as well as the number of interchain hydrogen bonds that are formed.   As mentioned earlier, every third residue in the three peptide chains is glycine, an amino acid which contains the smallest of the R-groups (hydrogen).   Glycine can therefore be accommodated in the interior of the triple-stranded superhelix whereas the two amino acid residues present on either side of glycine (proline and hydroxyproline) are present on the outside of the helix.

f.   Tropocollagen molecules (length of about 3000 $\overset{\circ}{A}$) are packed in parallel linear arrays to form collagen fibrils.   Each tropocollagen N-terminus is displaced by about one-quarter the length of the molecule from that on either side ("quarter-staggered").   The resulting fibers show the characteristic 670 $\overset{\circ}{A}$ spacing seen in electron micrographs.   There is a gap or hole of about 400 $\overset{\circ}{A}$ between the end of one tropocollagen molecule and the start of another.   It has been suggested that these gaps may function in bone formation.   Bone consists of an organic component which is almost entirely collagen and an inorganic phase of calcium phosphate.   It is known that collagen is necessary for the deposition of calcium phosphate in the formation of bone.   It is thought that the gaps are nucleation sites for this crystal formation.   Supporting this is the finding of crystals in the initial stages of bone formation at intervals of about 670 $\overset{\circ}{A}$.   Figure 63 is a schematic illustration of the packing of tropocollagen molecules to yield a collagen fibril.

g.   Collagen contains small amounts of tyrosine and very little tryptophan and cysteine.   The absence of cysteine in most collagens eliminates the possibility of formation of disulfide bonds crosslinking the peptide chains.   Covalent crosslinking in collagen takes place almost entirely through aldehyde and amino groups.

The aldehyde groups are formed by the action of lysyloxidase on the ε-amino groups of lysine and hydroxylysine.   The general reaction is

$$R-CH_2-NH_2 \xrightarrow{\text{lysyl oxidase}} R-\overset{\overset{\displaystyle O}{\|}}{C}H$$

The specific products are

$$\text{lysine} \xrightarrow{\text{lysyl oxidase}} \alpha\text{-amino adipic acid } \delta\text{-semialdehyde (AL)}$$

$$\delta\text{-hydroxylysine} \xrightarrow{\text{lysyl oxidase}} \delta\text{-hydroxy }\alpha\text{-amino adipic acid} \\ -\delta\text{-semialdehyde (HAL)}$$

Lysyl oxidase, probably present in multiple forms, requires copper for its activity. Copper deficiency leads to the lack of enzyme activity, thereby leading to the inhibition of collagen maturation (discussed later).

<u>Figure 63.</u> <u>A Schematic Representation of the Collagen Microfibril.</u>
     N = amino terminus, C = carboxy terminus.

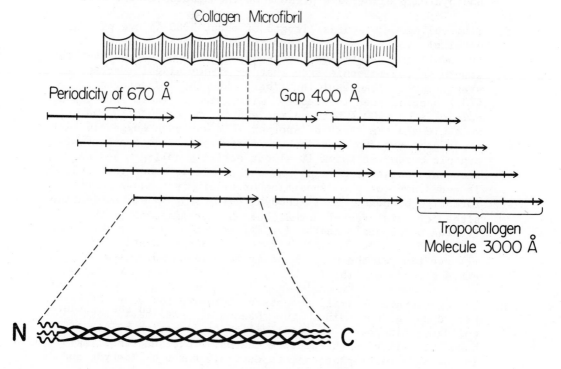

The cross-links can form by either of two basic reactions:

i) Schiff base formation

$$R_1-\overset{\overset{\displaystyle H}{|}}{C}=O \; + \; H_2N-CH_2-R_2 \; \underset{\longleftarrow}{\longrightarrow} \; R_1-\overset{\overset{\displaystyle H}{|}}{C}=N-CH_2-R_2 \; + \; H_2O$$

The residues involved can be any combination of δ-semialdehyde and ε-amino group. This is not a very stable linkage. R_1 and R_2 represent two separate collagen polypeptides.

ii) Aldol condensation

$$R_1-CH_2-\overset{\overset{\displaystyle H}{|}}{C}=O \; + \; \underset{\overset{\displaystyle |}{\underset{\displaystyle \overset{C=O}{\underset{|}{H}}}{}}}{CH_2-R_2} \longrightarrow R_1-CH_2-\overset{\overset{\displaystyle H}{|}}{\underset{\overset{\displaystyle |}{OH}}{C}}-\underset{\overset{\displaystyle |}{\underset{\displaystyle \overset{C=O}{\underset{|}{H}}}{}}}{CH-R_2}$$

$$\searrow H_2O$$

$$R_1-CH_2-\overset{\overset{\displaystyle H}{|}}{C}=\underset{\overset{\displaystyle |}{\underset{\displaystyle \overset{C=O}{\underset{|}{H}}}{}}}{C}-R_2$$

The reaction shown is between two AL residues, although two HAL residues or an HAL and an AL residue could undergo an aldol condensation also. This cross-linkage, particularly the product formed after dehydration, is very stable.

The overall process of cross-linking (oxidation and coupling) is called maturation of the protein. The final cross-linked product is insoluble under most conditions unless degradation and denaturation occur. Cross-links similar to these have been found in elastin, another important connective tissue protein found predominantly in elastic tissues (e.g.,walls of the large blood vessels).

h. The chronological sequence of events that lead to the synthesis of mature collagen. These events that take place within the fibroblasts, osteoblasts and chondroblasts are summarized below.

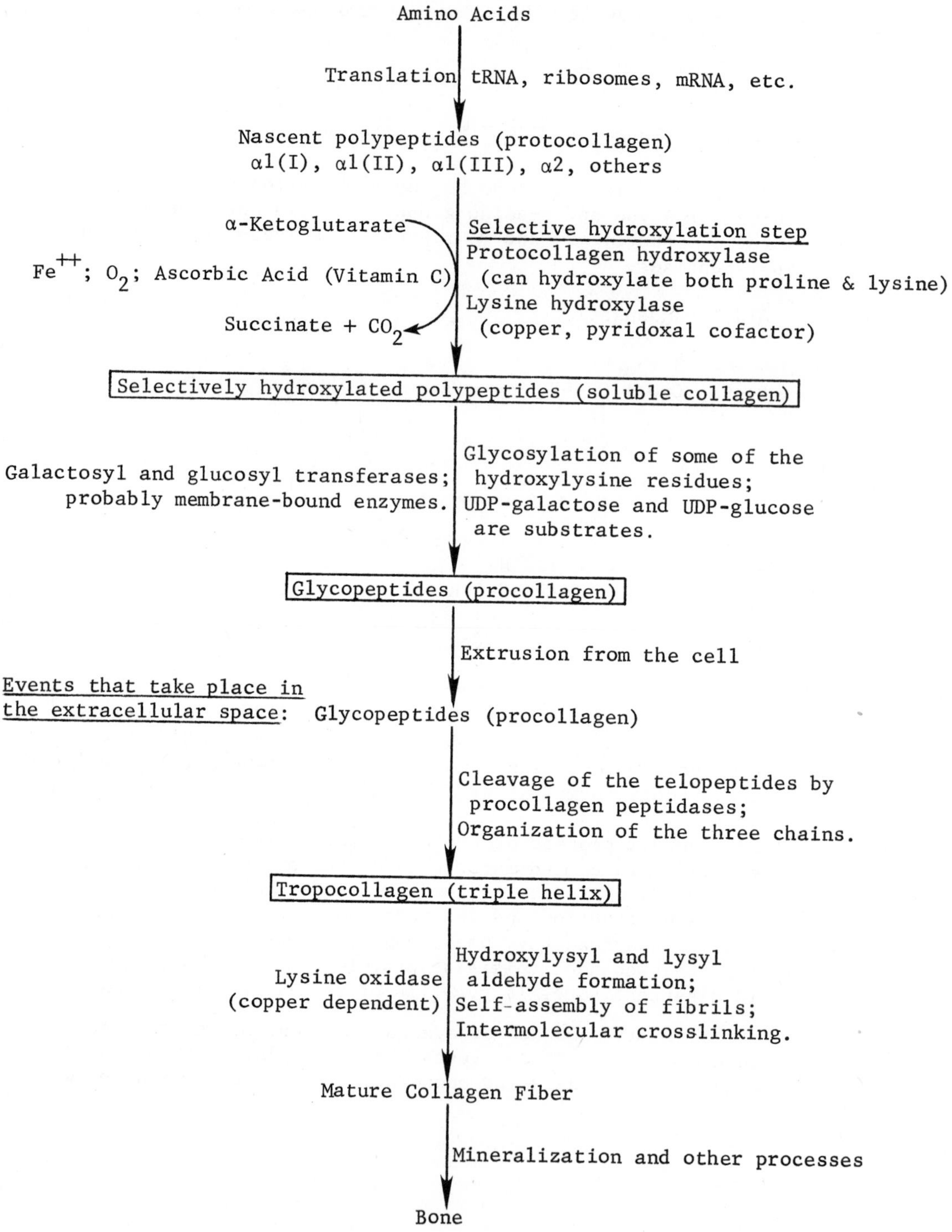

Amino Acids

Translation | tRNA, ribosomes, mRNA, etc.

Nascent polypeptides (protocollagen)
$\alpha 1(I)$, $\alpha 1(II)$, $\alpha 1(III)$, $\alpha 2$, others

α-Ketoglutarate

Fe^{++}; O_2; Ascorbic Acid (Vitamin C)

Succinate + CO_2

<u>Selective hydroxylation step</u>
Protocollagen hydroxylase
  (can hydroxylate both proline & lysine)
Lysine hydroxylase
  (copper, pyridoxal cofactor)

┌───┐
│ Selectively hydroxylated polypeptides (soluble collagen) │
└───┘

Galactosyl and glucosyl transferases;
  probably membrane-bound enzymes.

Glycosylation of some of the
  hydroxylysine residues;
UDP-galactose and UDP-glucose
  are substrates.

┌──────────────────────────────┐
│ Glycopeptides (procollagen) │
└──────────────────────────────┘

Extrusion from the cell

<u>Events that take place in
the extracellular space</u>: Glycopeptides (procollagen)

Cleavage of the telopeptides by
  procollagen peptidases;
Organization of the three chains.

┌──────────────────────────────┐
│ Tropocollagen (triple helix) │
└──────────────────────────────┘

Lysine oxidase
(copper dependent)

Hydroxylysyl and lysyl
  aldehyde formation;
Self-assembly of fibrils;
Intermolecular crosslinking.

Mature Collagen Fiber

Mineralization and other processes

Bone

i) Note that one of the hydroxylation reactions requires the concomitant conversion of α-ketoglutarate to succinate and carbon dioxide. This conversion appears to be different from the decarboxylation reaction of α-KG that takes place in the mitochondria which involves CoA, thiamine pyrophosphate, lipoic acid, etc. The other lysine-specific hydroxylase has not yet been completely characterized, but it does appear different from the better known protocollagen hydroxylase.

ii) The determining factors that establish the extent of hydroxylation and glycosylation of the polypeptide are not understood. Certainly those are methods by which a cell can introduce heterogeneity into peptides with identical amino acid sequence made on the mRNA-ribosome complex.

iii) It should be pointed out that types of collagen differ from one species to another and between tissues in the same species. The differences **include chain type, amino** acid composition, and carbohydrate content.

iv) The distinction between procollagen and tropocollagen is that the peptide chains in the former are longer than in tropocollagen. The purpose of this extra peptide (the telopeptide) is not completely clear. It is possibly involved in the self-assembly of the collagen monomers into the three-stranded helix present in tropocollagen. Following this assembly, the telopeptide is no longer needed and is removed by a protease.

i. The rate of collagen synthesis is stimulated by growth hormone at the level of the mRNA-ribosome—mediated protein synthesis. It is inhibited at the hydroxylation step by ascorbic acid (vitamin C) deficiency. At least *in vitro*, however, the vitamin C requirement of protocollagen hydroxylase can be practically replaced by reduced pteridines.

j. Proteases that specifically break down collagen are known as collagenases. Collagenolytic activity has been demonstrated in a number of tissues. These enzymes are inhibited by two protease inhibitors: α_1 antitrypsin and α_2 macroglobulin. Consequently, collagenase activity is not readily detectable in tissue homogenates. The exact role of collagenases is not clear at this time. However, these enzymes play a role in a number of processes such as wound healing and bone growth. These remodelling processes require continual synthesis and breakdown of collagen. Collagenase activity is detected in polymorphonuclear leukocytes where the enzyme may play a role

in the inflammatory process. It is presumed that the enzyme
is normally present in a precursor form (procollagenase) which
is activated under certain physiologic conditions. Elucidation
of the functions of the collagenases may help in understanding
many pathologic states such as rheumatoid arthritis, bone
resorption, corneal ulcers, and local invasion by tumors.

2. <u>Comments on Elastin Synthesis</u>

    a. Elastin, like collagen, is found in connective tissue but,
unlike collagen, it has elastic properties and is present in
the walls of the major arteries (up to 10-15% of the fresh
weight of the tissue). Mature elastin is apolar and yet can
bind calcium. It has a long half-life (indicating that its
turnover rate is negligible). Lysyl residues in elastin are
not hydroxylated and the protein contains little or no
carbohydrate (recall that OH-lys is the principal point of
carbohydrate attachment in collagen). The elastin fibers are
held together by intra- and intermolecular crosslinkages
derived primarily from lysine and modified lysine residues
(discussed below). The structure of elastin can be altered
(over long periods of time) by changes in diet or by certain
drugs.

    b. Like collagen, elastin synthesis begins inside the cell (smooth
muscle cells and fibroblasts). After peptide synthesis and
some post-translational modifications, the peptides (at this
stage known as soluble elastin) are extruded from the cell.
Mature fibers are formed by proper registration of peptides
(self-assembly), organization around microfibrillar proteins
(which are also products of the smooth muscle cells and
fibroblasts), selective proteolysis, modification of some lysyl
residues, and crosslinking. All of these processes take place
extracellularly and the finished product, mature elastin, is
an insoluble fibrous substance.

    c. The amino acids in soluble elastin are predominantly of the
type which contain non-polar side chains. There is little or
no histidine or methionine and other sulfur-containing amino
acids. It contains small amounts of arginine. Only 5% of the
proline residues are hydroxylated (about 40% of proline residues
in collagen are hydroxylated) and it contains no hydroxylysine.

    d. A major portion of elastin appears to be made of repeating
sequences of <u>tetra</u>-, <u>penta</u>- and <u>hexa</u>peptide units (recall that
collagen is largely made up of repeating <u>tri</u>peptide units).
These are:

```
Tetrapeptide:   -Gly-Gly-Val-Pro-
Pentapeptide:   -Pro-Gly-Val-Gly-Val-
Hexapeptide:    -Pro-Gly-Val-Gly-Val-Ala-
```

e. Crosslinking in Elastin: The crosslinks are derived from
    residues of lysine. Some of these residues undergo oxidative
    deamination to yield allysine. This reaction is catalyzed by
    lysyl oxidase, a copper-dependent enzyme. This was discussed
    earlier, under collagen synthesis. The chemical linkages occur
    between specific lysyl or allysyl residues by aldol condensation
    and Schiff-base reactions. Some of the resulting crosslinks
    are shown below.

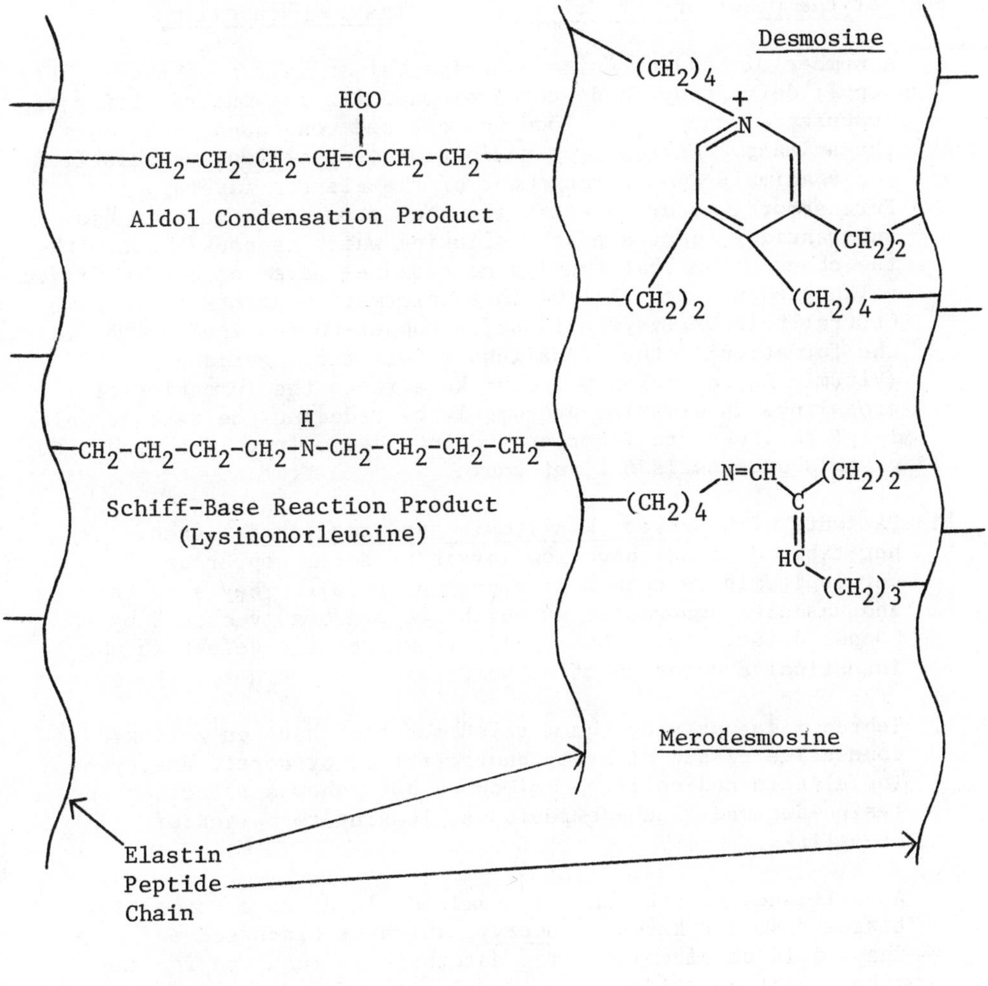

f. Various models have been suggested for the macromolecular structure of elastin. They include: crosslinked globular elastin subunits, crosslinked random elastin chains, the "coiled-coil" model, and the fibrillar model. The fibrillar model incorporates a unique conformation due to the presence of a residue of glycine followed by an amino acid residue with a bulky side group (e.g., valine) in a peptide chain. This type of sequence gives rise to near right-angle turns in the polypeptide backbone. These are known as "β-turns." The various models that are proposed for elastin differ significantly. It is not known at this time which of the models best accounts for all of the properties of elastin.

3. Some of the Disorders of Collagen and Elastin Metabolism

a. A number of studies using experimental animals have shown that copper deficiency leads to cardiovascular lesions resulting in ruptures of the major blood vessels and consequent death of the animals. Histologic examination of the blood vessels from these animals shows derangement of the elastin fibers. Furthermore, mature elastin isolated from these animals has substantially decreased crosslinking, which is consistent with the observation that the lysine residues have not been modified as they should be. All of this information points to the central role of lysyl oxidase, a copper-containing enzyme, in the formation of the crosslinks in elastin. Pyridoxine (Vitamin B$_6$) deficiency in chicks affects the formation of crosslinks in elastin, presumably by reducing the rate at which lysyl residues are incorporated into desmosine. The mechanism of this process is not yet known.

b. Patients with Menkes' kinky hair syndrome (an X-linked, heritable disease) have low levels of serum copper and ceruloplasmin (a copper transport protein). They show hair and vascular abnormalities which are presumably caused by the copper deficiency. These patients may have a defect in the intestinal absorption of copper.

c. Inbred mice carrying the mottled mutation have an X-linked connective tissue disorder characterized by aortic aneurysms. The elastin and collagen molecules both show a reduction in lysine-derived aldehyde residues, leading to a lack of crosslinks.

d. A deficiency of ascorbate (vitamin C) leads to a connective tissue disorder known as Scurvy, which is discussed in Chapter 10 on vitamins. The ascorbate is required for the mixed-function oxidative hydroxylation of residues of lysine and proline (discussed earlier). The metabolism of ascorbate

and copper are interrelated. Ascorbic acid oxidase, an enzyme
which plays a role in the maintenance of ascorbate concentrations
in the tissues, is a copper-dependent enzyme. Dopamine
β-hydroxylase, an enzyme involved in the synthesis of
catecholamines (see tyrosine metabolism in the amino acid
chapter), is a copper-containing protein that also requires
ascorbate for catalytic activity.

e. Lathyrism, a disease associated with the eating of certain
peas of the genus Lathyrus, occurs in two forms. In
osteolathyrism, the abnormalities involve bone and connective
tissue. The active principle (from L. odoratus) responsible
for the deleterious effects is β-aminopropionitrile (BAPN):

$$N \equiv C-CH_2 - \overset{\overset{\displaystyle H}{|}}{\underset{\underset{\displaystyle H}{|}}{C}} - NH_2$$

This compound is a noncompetitive inhibitor of lysyl oxidase,
thereby preventing the formation of aldehyde derivatives from
lysyl residues. The neurotoxic agent responsible for
neurolathyrism is β-N-oxalyl-L-α,β-diaminopropionic acid
(isolated from L. sativus). The mode of action of this agent
is not known. However, it is speculated that its oxidative
decarboxylation product, 2-imidazolidone-4-carboxylic acid,
is the toxic agent.

β-N-oxalyl-L-α,β-
diamino-propionic acid

2-imidazolidone-
4-carboxylic acid

The latter compound is a potent, competitive inhibitor of
5-oxoprolinase, an enzyme which is part of the γ-glutamyl
cycle. The γ-glutamyl pathway (see amino acid metabolism) is
involved in amino acid transport. Interference with this
process can thus lead to serious metabolic consequences.

f. The Ehlers-Danlos syndromes, a group of inherited connective
tissue disorders, exhibit a wide variety of abnormalities,
including hyperextensibility of the joints, fragility, skin
bruisability ("cigarette paper" scarring), and friability of
the tissues. Some of the other problems observed in these
disorders are pneumothorax, friability of the intestine

(leading to spontaneous rupture), gastrointestinal diverticula, hernias, ruptures of the great vessels, ocular problems, mediastinal emphysema, gingival bleeding, and periodontosis with early loss of teeth.

In type VI Ehlers-Danlos syndrome, which shows severe ocular abnormalities, a defective collagen has been demonstrated in which the lysyl residues are not adequately hydroxylated. The biochemical lesion appears to be diminished activity of lysyl hydroxylase (due either to an abnormal structure or a defect in the activation process). This has been verified by assaying for lysyl hydroxylase activity in fibroblast cultures from the patient's skin. Patients with another type of Ehlers-Danlos syndrome (Type VII), characterized by generalized connective tissue disorder, are deficient in procollagen peptidase, preventing the formation of mature collagen. Dermatosparaxis, a disorder found in cattle, is also characterized by a defect in the procollagen peptidase. The cattle with this autosomal recessive disorder show extremely fragile skin.

g. Marfan's syndrome, an autosomal dominant disorder of the connective tissue, is characterized by abnormalities in the lens of the eye, skeletal abnormalities with loose-jointedness and excessive length of the extremities, and cardiovascular abnormalities with weakness of the aortal media leading to diffuse or dissecting aneurysms. The biochemical lesion in this disorder is yet to be identified.

h. Homocystinuria (discussed in amino acid metabolism), an autosomal recessive disease, is due to a deficiency of cystathionine synthetase. It is accompanied by homocystinemia. The homocysteine interferes with the maturation of both elastin and collagen, probably by binding the copper necessary for lysyl oxidase activity and perhaps by combining with the aldehyde groups on the oxidized lysines. The connective tissue abnormalities found in homocystinuria and Marfan's syndrome are very similar.

i. Osteogenesis imperfecta, an inherited disease, is characterized by fragility of the bones, skin, scleras and teeth. Using skin fibroblasts from these patients, it has been shown that the cells have substantially reduced levels of Type I collagen and increased levels of Type III collagen. Recall that bone contains only type I collagen. Scleras and dentin contain predominantly the same type of collagen. Therefore, if the pattern seen in the skin fibroblasts with regard to collagen synthesis is generalized, it may account for the abnormalities seen in bone, sclera and dentin.

j. Drugs which have chelating ability (and can thus bind Cu^{2+} and Fe^{2+}) can interfere with collagen synthesis and maturation. <u>Hydralazine</u>, a hypotensive drug, affects the maturation of collagen in a number of different ways: It binds to aldehydes, forming protein-bound hydralazones which may be responsible for causing autoimmune reactions to collagen and elastin; it inhibits lysyl oxidase by binding to copper; it inhibits lysyl hydroxylase by binding to Fe^{2+}. The structure of hydralazine is shown below:

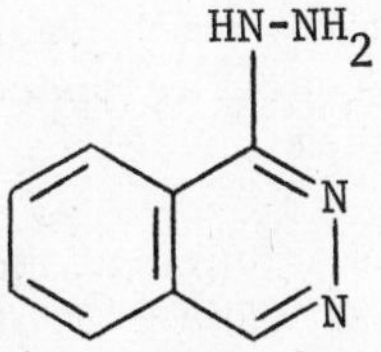

<u>Penicillamine</u>, a copper-chelating drug used in the treatment of Wilson's disease (see copper metabolism), also interferes with the lysyl oxidase reaction giving rise to connective tissue abnormalities.

4. Comments on Albumin Synthesis and Edema

a. Albumin, the principal serum protein, has a molecular weight of about 65,000 daltons and contains about 575 amino acids. In humans, the normal blood plasma range is 3.5-5.0 gm of albumin/100 ml. Albumin is also found in the extravascular spaces, the lymph, and in other biological fluids including amniotic fluid, bile, gastric juice, sweat, tears, etc. In kidney diseases, it is found in the urine and is a major component of edema fluid (see later, under edema).

b. Plasma albumin has two primary functions.

   i) Maintenance of osmotic pressure (see later, under edema).

   ii) Transport of a variety of substances which are bound non-covalently to albumin. These include metals and other ions, bilirubin, amino acids, fatty acids, enzymes, hormones, drugs, and others.

c. Albumin is synthesized by the liver cells (hepatocytes) and is the principal protein made in the liver. No other sites of synthesis are known at present. In a 70 kg human, the liver synthesizes about 12-14 gm of albumin per day. The half-life of an albumin molecule in man is about 20 days. The albumin synthetic system is the classical mRNA-ribosome mediated process discussed earlier in detail. The mRNA for albumin (575 amino acids) is composed of (575 x 3) + 3 nucleotides for initiation + 3 nucleotides for termination = 1731 nucleotides. Allowing about 90 nucleotides (310 Å) from the center of one ribosome to the center of the next, the polysomes should contain about 19-20 ribosomes per albumin message. This has been found experimentally by the use of sucrose gradient centrifugation. Since albumin is an <u>export</u> protein, it travels to the Golgi apparatus where it is packaged preparatory to secretion. The exact biochemical transport mechanism is not known but it requires K^+ and is inhibited by ouabain (see previously under proteins for export). Albumin which leaks through the blood vessel walls into the extravascular spaces is returned to the blood via the lymphatic system.

d. Control of albumin synthesis occurs at both the transcriptional (DNA → RNA) and translational (RNA → Protein) levels.

   i) Adequate nutrition (protein-nitrogen intake) is basic to the regulation of all protein synthesis. Although protein deficiency and overall malnutrition (starvation) differ somewhat in their molecular effects, they both result in a decreased rate of protein synthesis due to alterations in RNA metabolism. Refeeding of starved animals results in an immediate, threefold increase in the amount of albumin present in the newly synthesized protein, implying that the albumin mRNA is more stable (has a longer half-life) than other mRNA.

   ii) Although all of the amino acids must be present to support protein synthesis, albumin synthesis seems especially sensitive to tryptophan. This amino acid appears to stimulate mRNA synthesis and to increase the activity of the RNA polymerase which makes rRNA. Tryptophan may also increase the stability of the ribosome-mRNA-ER complex on which albumin is synthesized. It is far from clear, however, why this specific amino acid plays such an important role in this regulation.

   iii) Albumin synthesis is stimulated <u>in vivo</u> by both cortisone and the thyroid hormones (thyroxine and triiodothyronine) perhaps synergistically. Cortisone is known to increase the synthesis of hepatic rRNA, tRNA, and mRNA and it may promote ribosomal binding to the endoplasmic reticulum. Thyroid hormones have a similar effect, stimulating mRNA and rRNA synthesis as well as increasing the binding of rRNA to the ER and perhaps stimulating the ribosomes. Testosterone causes the same changes as do the thyroid hormones. Increased albumin synthesis has been noted in Cushing's Syndrome (involving hyperactivity of the adrenal cortex). In hyperthyroidism the rates of both albumin synthesis and breakdown are increased. In hypothyroid individuals, these processes are decreased but, more important, edema can result in this state from the accumulation of albumin in extravascular spaces. It should be noted that despite the increase in albumin synthesis associated with cortisol administration, the hormone has been considered antianabolic because it causes an overall negative nitrogen balance.

iv) Insulin may also be needed for maximal albumin synthesis, although hypoalbuminemia and decreased albumin synthesis are not usually associated with diabetes mellitus. Growth hormone enhances albumin synthesis by stimulating rRNA polymerase activity.

v) In vitro studies have shown that albumin synthesis in hepatic tissue is influenced by an osmotic regulatory mechanism. This is a significant observation in view of albumin's role in maintaining colloidal osmotic pressure. Albumin synthesis is increased in blood hypotonia and decreased in hypertonic situations.

e. Albumin levels in disease states

i) Despite earlier studies, it now appears that although serum albumin levels may decrease in cirrhosis, this observation has no direct bearing on total liver function, prognosis, or serum albumin synthesis.

Analbuminemia is a rare, probably inherited lack of serum albumin. The symptoms associated with it are surprisingly mild, considering the protein's central role in osmotic regulation. This is at least partly explained by an increase in serum levels of cholesterol, phospholipid, gamma globulins, fibrinogen, and transferrin. These replace albumin to some extent in maintaining the blood osmolarity. As yet, no specific metabolic defect has been demonstrated in patients having this disorder.

ii) Variations in albumin synthesis may be related in a more complex fashion to cirrhosis and gastrointestinal disease. In the latter case, and in nephrosis, an increased rate of destruction contributes to the hypoalbuminemia. In patients suffering from severe and extensive burns, there is extensive loss of serum, hence water and albumin. This loss is too great to be met by increased synthesis.

f. Edema means swelling and is characterized by an increase in the fluid in intercellular and extravascular spaces. This can be visibly observed in the subcutaneous tissue. The basic mechanisms are related to alterations in the permeability of the blood vessels and to changes in the osmotic and hydrostatic pressures of the blood and extravascular fluids. Figure 64 illustrates Starling's hypothesis and shows the interrelationships of these factors in a normal capillary.

From Figure 64 it is apparent that a shift in blood pressure can cause a change in the direction of filtration (water movement). An increase or decrease in the other pressures involved may result in a similar reversal. The blood and tissue osmotic pressures are the pressure which would be measured across a semipermeable membrane with water on one side and blood or extravascular fluid, respectively, on the other side. The arrows indicate the direction of net water flow under the influence of the individual pressures.

g. Based on the foregoing discussion, four types of edema can be differentiated by their etiologies.

i) If the serum protein concentration (especially the albumin concentration) is decreased, the blood osmotic pressure also decreases. When that pressure falls below its normal minimum (about 23 torr), the net filtration pressure (and the filtration rate) increases at the arterial end of the capillary and the absorption rate decreases at the venous end. The net result is a shifting of water into the extravascular space. This type of edema can occur in severe albuminuria (which may be associated with lipoid nephrosis, the nephrotic stage of glomerulonephritis, and amyloid nephrosis). It can also result from the decreased albumin synthesis present in protein starvation and malnutrition.

ii) A decrease in the blood osmotic pressure and an increase in the tissue osmotic pressure occurs if the capillary endothelium is damaged sufficiently to permit passage of protein into the interstitial fluid. This amounts to simultaneously decreasing serum osmolarity and increasing the extravascular fluid osmolarity. Once again, the filtration rate (arterial capillary) increases and the absorption rate (venous capillary) decreases. This mechanism is involved in the production of blisters, hives, and in other localized inflammation. Toxins, trauma, and histamine-like substances function similarly (see further under immunochemistry).

Figure 64. Pressure Interrelationships Between Intravascular and Extravascular Fluids

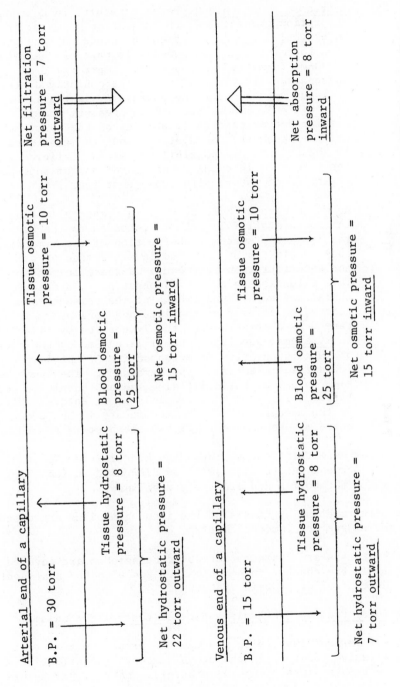

Arterial end of a capillary

B.P. = 30 torr

Tissue hydrostatic pressure = 8 torr

Net hydrostatic pressure = 22 torr outward

Blood osmotic pressure = 25 torr

Tissue osmotic pressure = 10 torr

Net osmotic pressure = 15 torr inward

Net filtration pressure = 7 torr outward

Venous end of a capillary

B.P. = 15 torr

Tissue hydrostatic pressure = 8 torr

Net hydrostatic pressure = 7 torr outward

Blood osmotic pressure = 25 torr

Tissue osmotic pressure = 10 torr

Net osmotic pressure = 15 torr inward

Net absorption pressure = 8 torr inward

Note: B.P. = blood pressure; torr = mm of Hg, a unit of pressure; inward refers to from the tissue into the capillary lumen; outward refers to into the tissue from the capillary lumen.

[471]

iii)   <u>Increased venous blood pressure</u> results in decreased net
absorption pressure and the retention of water in the
interstitial fluid. This can be caused by the presence
of a tourniquet and supposedly is part of the mechanism
of edema formation in congestive heart failure. It is
associated with varicose veins (permanently distended,
tortuous veins, occurring especially in the lower
extremities). Increased venous pressure occurs in
cardiac decompensation (inability of the heart to
maintain adequate circulation), but the exact mechanism
is not clear. One possibility is that failure of the
heart results in renal retention of sodium and, hence,
retention of water, with subsequent edema.

iv)   As was mentioned before, protein (especially albumin) lost
to the interstitial fluid is returned to the blood via
the lymphatic system. Obstruction of the lymphatic vessels
thus results in an <u>increase</u> in extravascular osmolarity
and net filtration rate and a decrease in reabsorption.
This situation occurs in <u>elephantiasis</u>, a disease
characterized by hypertrophy of the skin and subcutaneous
tissue, especially of the legs and genitals. It is
caused by obstruction of the lymphatics by filarial worms.
In <u>myxedema</u>, swelling (edema) is caused by the presence
of large quantities of mucoproteins in the lymph vessels.
It is associated with hypothyroidism. The lymph vessels
seem to be metabolically sluggish and unable to remove
interstitial protein as rapidly as it enters. Since it
is the mucoproteins which accumulate in the edema fluid,
there may be a selectivity in the protein removal process.

h.   Other aspects of edema, including electrolyte (Na^+, K^+) levels
and hormone interactions,will be discussed in the chapter on
water metabolism and electrolyte balance.

5.   The structure and synthesis of the immunoglobulins are discussed
in the chapter on immunochemistry.

6.   The synthesis and characteristics of hemoglobin are covered in
detail in the chapter on hemoglobin. See also the next section
on the control of protein synthesis.

Regulation of Protein Synthesis and Other Control Mechanisms

1.   It is important that cells produce the proper substances
(qualitative control) in the proper amounts (quantitative control)
at the proper times (temporal control). Energy, raw materials,
and space would be wasted if there **was** no way to control production.
Cellular development and differentiation are based on the cells'

ability to control their molecular makeup and function. There are three major levels of control within cells:

a. Regulation of enzyme activity has been discussed previously in the chapter on proteins and enzymes. One of the most important examples of this type of control is allosterism. It is exemplified by the reaction below.

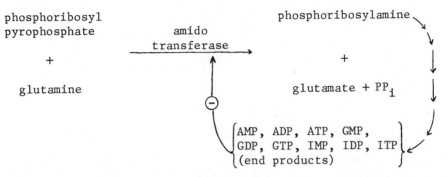

phosphoribosyl pyrophosphate

+

glutamine

amido transferase

phosphoribosylamine

+

glutamate + PP$_i$

$\ominus$

AMP, ADP, ATP, GMP, GDP, GTP, IMP, IDP, ITP (end products)

Glutamine phosphoribosyl pyrophosphate amidotransferase catalyzes the first step unique to purine biosynthesis (see under nucleotide biosynthesis). The end-products of this pathway are the purine nucleotides, all of which are quite unlike the substrates or products of the amidotransferase-catalyzed reaction. They interact allosterically with the enzyme, decreasing its activity. Thus, accumulation of the products of a pathway causes a decrease in the activity of that pathway. This is an illustration of one mechanism of feedback or end-product inhibition.

b. Regulation of mRNA translation may be a factor in the biosynthesis of hemoglobin and other proteins. With respect to hemoglobin, this mechanism will be discussed later in this section. At least in some cells, the presence of unusual tRNA's may be a control mechanism. The occurrence of these tRNA's is influenced by factors exterior to the cell. This type of control apparently exists in many plants. Cortisone and the thyroid hormones appear to regulate albumin synthesis at least partly at this level.

The ability of ribosomes to initiate protein synthesis may also be modulated by reversible phosphorylation of the subunits. As was pointed out earlier, ribosomes must dissociate into subunits before they are able to bind to an initiation codon. The phosphorylation may inhibit this dissociation.

c. Regulation of gene expression (transcription) is the "highest" level of control. It permits conservation both of protein-synthetic materials and of the RNA precursors. Induction and repression in bacteria; differential expression of the genome during development of higher animals; and control of the synthesis of hemoglobin, albumin, and perhaps other molecules, operate at least partially by this mechanism.

The first and third types of control (a and c) are recognized to be of major importance in most cells. Using these two levels, cells can function optimally. The second level is less well studied but it appears to be important in at least some situations.

Coarse control ─────────────→ regulation of gene expression

Intermediate control ─────────→ regulation of mRNA translation
(protein synthesis)

Fine control ─────────────────→ allosteric control of enzymes

Examples of the control of translation and transcription are discussed below.

2. A further example of the modulation of translation via changes in ribosomal aggregation has been **described elsewhere in text.** In bacterial systems, an initiation factor (F_3, see under protein synthesis) is necessary to prevent association of the 30S and 50S subunits and thereby to permit initiation of protein synthesis. An entirely analogous protein factor (tentatively designated IF-3rr) has been demonstrated to be necessary for globin mRNA initiation in rabbit reticulocytes. IF-3rr appears to prevent subunit aggregation following release from the polysomes, rather than actively causing dissociation of 80S ribosomes. Heme (Fe^{+2}-protoporphyrin III complex, see under hemoglobin) is involved in some way, being responsible either for IF-3rr synthesis or activity. In the absence of IF-3rr, all protein synthesis would cease in the reticulocyte. Thus, this factor may represent one step in a coordinated mechanism for regulation of both heme and globin synthesis during erythroid development. This larger role is still only circumstantial. These findings are especially significant because they support the generality of the protein synthetic process originally described in procaryotes.

3. In eucaryotes, all RNA, except that made in the mitochondria, is synthesized within the nuclear membrane (which includes the nucleolus). Recent work has indicated that variable release of this RNA to the cytoplasm and endoplasmic reticulum may be a potential source of control over gene expression. Stretches of polyadenylic acid (poly A) have been found associated with some of the nuclear RNA and there is increasing evidence that the mRNA's of eucaryotic cells are terminated by regions of poly A, probably

at the 3'-OH end. This poly A may be involved in selection of
the mRNA to be released from the nucleus or in the ability of mRNA
to be used for translation. This whole process is still very new
and requires much more investigation.

4. Regulation of gene expression is important not only in the cell's
   economy, but also plays a major role in cell differentiation. For
   example, the development of a neuron and a pancreatic acinar cell
   from the same zygote are both controlled by the same DNA. The
   way in which this DNA is used (how much and what part of it is
   used for RNA transcription and when this use occurs) must be
   controlled for the differential development of cells.

5. Enzyme Induction

   Enzymes may be classified according to the conditions under which
   they are present in a cell. There are two basic types of enzymes.

   a. Constitutive enzymes are formed at constant rates and in
      constant amounts. Their presence in a cell is not related to
      the presence or absence of their substrates. They are
      considered to be part of the permanent enzymatic makeup of the
      cell. The enzymes of the glycolytic pathway are examples of
      this group.

   b. Inducible enzymes (adaptive enzymes) are always present in at
      least trace amounts, but their concentrations vary in
      proportion to the concentrations of their substrates. The
      classical example is β-galactosidase, which is present in small
      amounts in wild type E. coli and whose concentration increases
      in the presence of lactose.

$$\text{lactose} \xrightarrow{\substack{\text{Induction of} \\ \text{β-galactosidase}}} \text{glucose + galactose}$$

lactose
(β-D-galactopyranosyl-
(1→4)-β-D-glucopyranoside)

   In this example, the inducer (lactose; however see later, under
   induction mechanisms and the lac operon) is a substrate for the
   enzyme induced. This is the usual case, but compounds similar
   to the substrate may also be inducers and there are substrates
   which are not inducers. The enzyme β-galactosidase is coded
   for by part of the lac (for lactose) operon. The other enzymes
   which are also part of the lac operon and whose formation is
   also induced by lactose (a β-galactoside) are β-galactoside
   permease and β-thiogalactoside acetyltransferase. The permease
   participates in the transport of β-galactosides into the cell.
   The exact role of the β-thiogalactoside acetyltransferase is

not known at this time. The induction of more than one enzyme
by a single inducer is known as <u>coordinate induction</u>. In the
case of the lac operon it has been shown that the genes for
these three enzymes are contiguous to each other. This is also
true of other operons in bacteria, such as the his operon in
<u>S</u>. <u>typhimurium</u> and <u>E</u>. <u>coli</u> and the arabinose operon in <u>E</u>. <u>coli</u>.

   c. A given enzyme may be constitutive in one organism or in one
strain of an organism, inducible in another, and absent in a
third. In addition, the magnitude of the response to an
inducer varies from organism to organism and from strain to
strain, and is genetically determined. Mutations are known in
bacteria which cause inducible enzymes to become constitutive.

6. a. <u>Enzyme Repression</u>: If a bacterial strain capable of synthesizing
a particular amino acid is placed in a culture medium containing
that amino acid, synthesis of the amino acid ceases, due to
repression. The amino acid acts as a corepressor of the apparatus
(operon) that produces the enzymes for its synthesis. This is
another type of feedback system very similar to the end-product
inhibition of an enzyme. For example, if <u>E</u>. <u>coli</u> cells are
provided with sources of carbon and NH_4^+, they will synthesize
all 20 amino acids. If histidine is added to the culture medium,
the entire sequence of enzymes that participates in histidine
biosynthesis is no longer produced, but the production of enzymes
required for the synthesis of other amino acids is not affected.
This type of system, where a single substance (e.g., histidine)
represses the synthesis of a group of enzymes catalyzing the
consecutive reactions of a metabolic pathway leading to the
synthesis of the repressor, is known as <u>coordinate repression</u>.

   b. Another type of repression of enzyme synthesis, known as
<u>catabolite repression</u>, occurs when bacteria are allowed to grow
in the simultaneous presence of several food sources where one
of the sources is glucose. In this situation, the bacteria
will use glucose preferentially over other sources such as
lactose. This is because glucose or some catabolite of glucose
represses the enzyme synthesis that is necessary for the
metabolism of other fuel sources (e.g., β-galactosidase).
Furthermore, in facultative bacteria, glucose can suppress the

enzymes involved in electron transport and respiration, because in the presence of adequate amounts of a fermentable sugar, the bacteria prefer to use primitive catabolic pathways (e.g., glycolysis) rather than more complex ones.

7. In 1961 and 1962, Jacob and Monod published a genetic mechanism for the induction and repression of enzyme systems in bacteria. They postulated that in addition to the structural genes which direct the synthesis of specific enzymes and other proteins using a mRNA template (previously described), there is an operator gene which precedes and controls the structural genes. If the operator gene is repressed, the structural genes cannot function. The sequence of DNA, composed of an operator gene and one or more structural genes, is called an operon. It is transcribed as one continuous piece of mRNA (a polycistronic messenger). For example, when the lac operon is induced, β-galactoside permease and β-thiogalactoside transacetylase are synthesized in addition to β-galactosidase. All of these enzymes are needed for metabolism of the galactose released from lactose.

There is also a regulator gene on the chromosome, not necessarily adjacent to the operon. This gene codes for the mRNA for a repressor protein. In a repressible enzyme system (such as the histidine synthesis operon), the repressor protein is termed an aporepressor, since it is unable to cause repression without the binding of the corepressor. The inducer and corepressor substances probably cause a conformational change in the repressor or aporepressor.

A promotor region (P) has also been identified as a part of the lac operon and its regulatory gene system. This region consists of two sub-areas, catabolite gene activator protein binding site (CAP binding site) and the DNA-dependent RNA polymerase interaction site. The promotor region is contiguous with the operator gene and precedes both the operator and structural genes. The role of the promotor region will be discussed further under the mechanism of catabolite repression. In the following diagrams, the various aspects of repression and induction using the lac operon have been illustrated.

a. Enzyme Induction

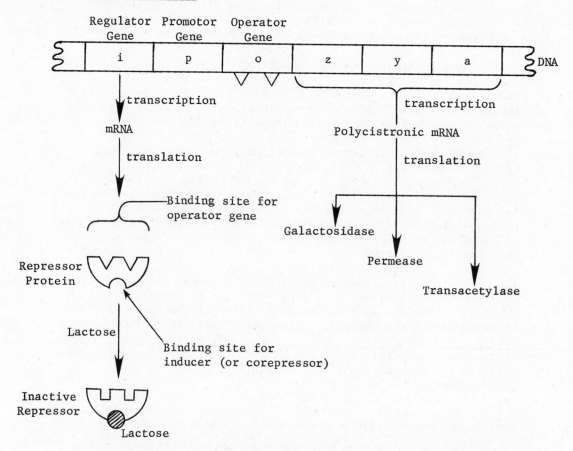

In the above schematic illustration, note that the repressor protein is inactive when bound to the inducer (lactose), so that the structural genes can be transcribed. The three structural genes are transcribed sequentially, to produce a single, polycistronic mRNA from which all three enzymes are translated.

b. Enzyme repression: In the absence of lactose, the repressor protein (the product of the i gene) binds to the operator and blocks transcription.

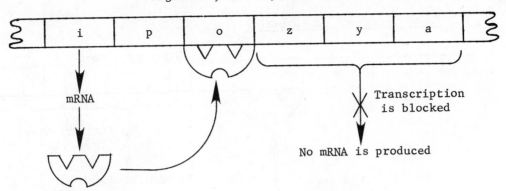

Repressor Protein
(active in the
absence of inducer)

c. <u>End-product repression</u>: Recall that the addition of histidine
to an <u>E. coli</u> culture represses the formation of all the
enzymes required to synthesize this amino acid. In this type
of repression, the end-product (histidine) is responsible for
blocking the transcription of the structural genes. These
reactions are illustrated as follows.

<u>In the absence of histidine</u>

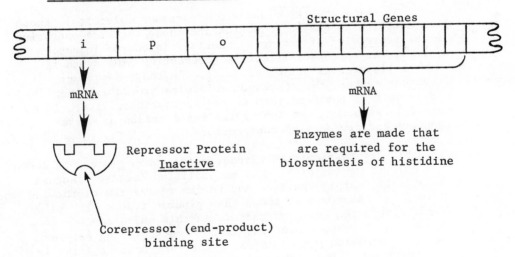

In the presence of histidine

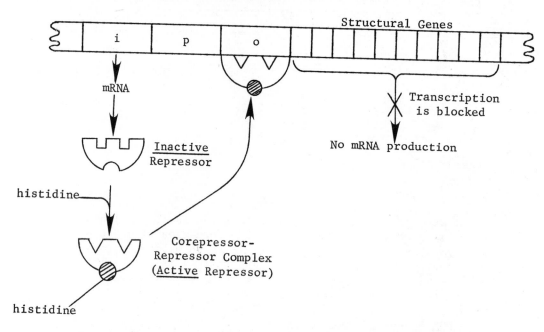

Note that there are two classes of repressor molecules: those that function in enzyme induction and those that are operative in the repression of enzyme synthesis. In enzyme induction systems, the repressor is active (preventing mRNA synthesis) when the inducer is not bound to it. In the end-product repressor system, the repressor molecules are active when the end-product is bound to them at the aporepressor site. In a metabolic sequence, the substrates can function as enzyme inducers and products as corepressors.

d. Catabolite repression: As already noted, when glucose is added to a culture medium of E. coli metabolizing lactose, lactose utilization sharply declines due to the repression of the lac operon. It has now been shown that glucose reduces the cAMP levels within the cell and that it is this which causes the repression of synthesis of the lactose-metabolizing enzymes. The means by which glucose affects the cAMP concentration is not understood. The mechanism is specific, however, in that only cyclic AMP is effective. Elevated levels of cAMP cause derepression and transcription of the structural genes. The sequence of events can be summarized as follows: cAMP combines with a protein known as catabolite gene activator protein (CAP), a dimer of two identical subunits each with a molecular mass of 22,000 daltons. The cAMP-CAP complex binds to a specific CAP binding site, within the promotor site. This interaction leads to a destabilization of DNA in the adjacent region, permitting the binding of RNA polymerase, initiating transcription. A schematic illustration of these reactions is shown below.

<u>In the absence of glucose</u>

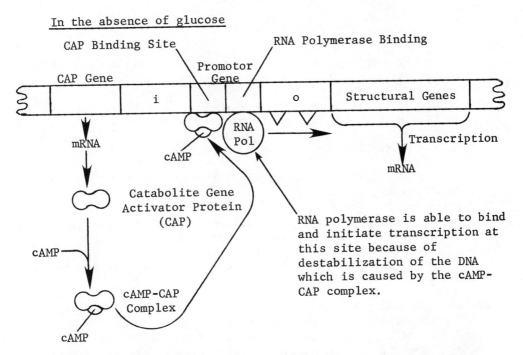

In the presence of glucose, the cAMP level is reduced below a certain critical value and CAP is inactive. RNA polymerases then cannot bind to DNA and transcription is prevented. The transcription of DNA, then, is modulated by the interplay between the repressor protein, a substrate or a product, cAMP, and CAP.

e. The lac operon is the most thoroughly studied example of an inducible enzyme system. The lac repressor has been isolated and shown to be a protein composed of four subunits and having a molecular weight of about 1.5×10^5 daltons. Its isolation was quite a remarkable feat since it is estimated that there are only about 10 molecules of the repressor per E. coli cell (about 10^{-8} M). Constitutive mutants (ones in which lac operon transcription is maximal in the absence of any inducer) have been found with errors in the regulatory and operator genes. In the regulator mutant, the repressor is unable to recognize the operator binding site; in the operator mutant, normal repressor is present but the operator is abnormal. It is interesting that all known operator mutants are deletions. Regulator mutants are also known in which the inducer substance cannot bind to the repressor molecule and the lac operon is not transcribed, even in the presence of an inducer. These are called super-repressed strains. A curious fact is that the true intracellular inducer of the lac operon is known not to be lactose. Rather, certain galactosides appear to function in this role.

f.   As has been pointed out before, the experimental work on which the preceding discussion is based was obtained exclusively in bacterial systems. It is an example of how far our knowledge of procaryotic control mechanisms has advanced but it is not clear to what degree similar thinking can be applied to eucaryotic cells. Some indication of this is given in the following sections on control in higher organisms.

8.   The control of enzymes at the transcriptional and translational levels in higher, eucaryotic organisms is less well understood. Several inducible liver enzymes (e.g., glucokinase) have been found. Steroids are known to penetrate into the cell nucleus and bind to the DNA, as are certain antibiotics, carcinogens, and other molecules. This may cause changes in the DNA transcription, both qualitatively and quantitatively. The role of steroids is discussed further under steroid hormones.

a.   As mentioned previously, selective derepression of the nuclear DNA is probably the basis for cellular differentiation during development. The concept of genetic totipotency (possession of the entire genome by all diploid cells of higher organisms and, under proper conditions, the expression of these genes) is supported by many observations and experiments. Histones (a class of basic proteins associated with DNA; see under proteins and enzymes) may function as permanent (or long-term) repressors within differentiated cells. Histones can be displaced from DNA by polyanions such as pieces of RNA and phospho- and lipoproteins and they can be phosphorylated by protein kinases which are activated by cyclic AMP. Phosphorylation of the histones decreases their affinity for DNA. This effect provides a further mechanism for hormonal regulation of protein synthesis. Enzymatic methylation of certain bases in DNA presents still another mechanism for permanent gene repression in differentiation.

b.   One important characteristic of malignant cells is their dedifferentiation, with the concomitant derepression of large parts of their genomes. This might be due to a lack of production of (functional) repressors or production of antibodies to the repressors (assuming that the repressors are proteins). The antibodies could bind to the repressor, thereby inactivating it. Another approach to carcinogenesis is

based on the observation that histones can be displaced from
DNA by nuclear RNA (which is a polyanion as are all
polynucleotides). Perhaps viral RNA (or DNA) can similarly
bind to nuclear DNA, remove the histones, and cause derepression.
At the present, this is largely either speculative or based
only on circumstantial evidence.

c. When RNA binds to double-helical DNA, histones are displaced
    and the duplex appears to unwind. Since single-stranded DNA's
    (or single-stranded regions within duplex DNA) are required for
    transcription, this suggests an additional mechanism for
    transcriptional control.

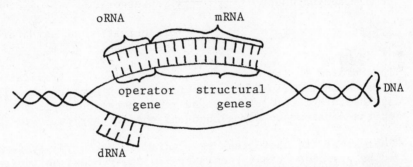

For transcription to occur, derepressor RNA (dRNA) must bind
to the DNA strand complementary to the one to be transcribed.
This causes local unwinding of the DNA duplex. In the next
step, during transcription, the entire operon (operator gene +
structural genes) is transcribed. After release of this large
RNA, it is cleaved by an endonuclease into operator RNA (oRNA,
complementary to the operator gene) and messenger RNA (mRNA,
complementary to the structural genes). Suppose (as shown
above) that the dRNA binds to the DNA region which is
complementary to the operator gene. This provides a self-
regulation of the transcription. The oRNA formed can bind to
the dRNA, giving a double-stranded RNA. The dRNA-oRNA duplex
is unable to bind to the DNA and initiate transcription.
Removal of the dRNA by complexation with oRNA results in
renaturation of the single-stranded region and cessation of
its transcription. Although there are some suggestive
experimental data, this system has yet to be actually
demonstrated either <u>in vivo</u> or <u>in vitro</u>.

d. Although the repressor-inducer model of Jacob and Monod has
    been substantiated in bacterial systems, evidence for it in
    eucaryotic systems is weaker. The clearest demonstration has
    come from the studies of Granick and his coworkers on the
    control of heme synthesis in mammalian reticulocytes. Heme is
    the prosthetic group (cofactor) of hemoglobin, myoglobin, and

some of the cytochromes. It consists of a porphyrin ring
(protoporphyrin III; see under porphyrin synthesis) and an atom
of iron. Mention has already been made (see the section on
enzyme regulation) of the control of translation by heme and
initiation factor 3rr (IF-3rr). It appears that heme also
functions as a corepressor for transcription of the gene for
δ-aminolevulinic acid synthetase (ALA synthetase). This enzyme
catalyzes the first unique step in the synthetic pathway for
protoporphyrin III. This is another example of end-product
inhibition of the committed step in a synthetic pathway. Just
as in procaryotes, one can summarize this by the equations

heme + aporepressor ⟶ repressor

repressor + operator gene ⟶ inactivation of structural
                                     gene transcription

In addition to its effect on transcription and translation, heme
is a negative allosteric effector for ALA synthetase. The
synthesis and role of heme and related compounds is presented
in the sections on hemoglobin and the porphyrins.

e. Repressors and aporepressors studied in bacterial systems seem
to be proteins rather than RNA. In eucaryotes, as indicated
above, protein and RNA may be involved in transcriptional
regulation as repressor and aporepressor molecules. Currently,
however, even the existence of operons (or their equivalent)
in eucaryotic cells has not been established with any certainty.
For reasons such as this, all of the preceding discussion of
regulation in eucaryotes should be regarded as a tentative
explanation of certain experimental data.

Purine and Pyrimidine Nucleotide Biosynthesis

Before we proceed with the synthesis, it is essential to be familiar
with   (1) the names of the nucleotides and their relationship to
nucleosides and bases; (2) one-carbon transfer reactions which are
utilized in the nucleotide synthesis; and (3) synthesis of 5-phospho-
ribosyl-1-pyrophosphate (PRPP), a compound utilized in the de novo
and salvage pathways of nucleotide synthesis.

Nucleotide Nomenclature

Following is a list of nucleotides whose synthesis will be discussed
in this chapter. Also given in the list are the corresponding base
and nucleosides. Note that in the deoxyribonucleotides, the sugar
present is D-deoxyribose rather than D-ribose. This is not specifically
shown in the diagram. The deoxyribose is indicated with a "d", as for
example dAMP. The only exception to this is thymidylic acid,
abbreviated TMP even though it contains deoxyribose.

Base	Nucleoside (Base+sugar)	Nucleotide (Base+sugar+phosphate)

Purines

Adenine	Adenosine	Adenylic acid (AMP, adenosine monophosphate)
Guanine	Guanosine	Guanylic acid (GMP, guanosine monophosphate)
Hypoxanthine	Inosine	Inosinic acid (IMP, inosine monophosphate)

Pyrimidines

Uracil	Uridine	Uridylic acid (UMP, uridine monophosphate)
Cytosine	Cytidine	Cytidylic acid (CMP, cytidine monophosphate)
Thymine	*Thymidine	*Thymidylic acid (TMP, thymidine monophosphate)

*Note: In these molecules, the usual sugar present is deoxyribose.

One-Carbon Metabolism

1. One-carbon compounds such as formate and formaldehyde are utilized
   in the biosynthesis of the purines, thymine, serine, and
   methionine. The addition of a one-carbon compound is accomplished
   by a group of carriers known as the folic acid coenzymes.
   (S-adenosyl methionine, discussed with the sulfur-containing amino
   acids, donates methyl groups for biosynthetic reactions. These
   are known as biological methylation reactions and constitute a
   group of one-carbon transfers distinct from those involving folate
   derivatives.) As their name implies, these carriers are derived
   from folic acid (a vitamin in man). A vitamin is an "accessory
   food factor," a compound which must be present in the diet for
   normal metabolism. A vitamin cannot be synthesized by the
   organism for which it is a vitamin.

2. Structure of folic acid (F)

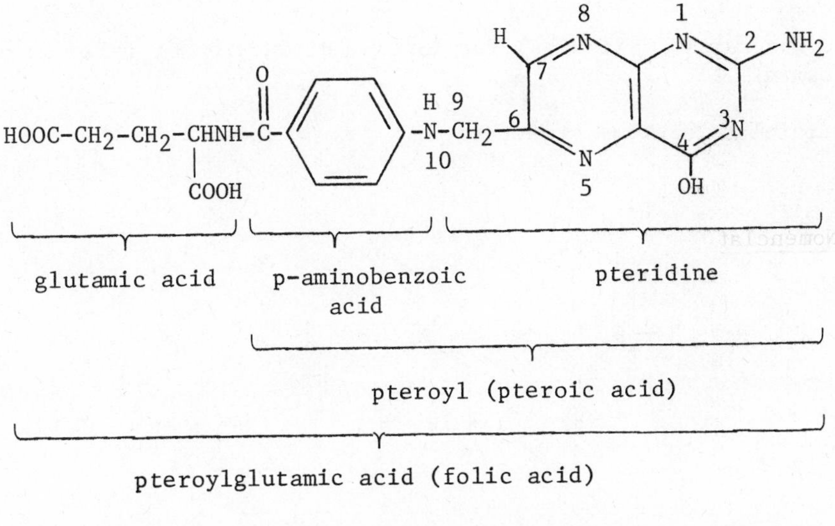

glutamic acid   p-aminobenzoic acid   pteridine

pteroyl (pteroic acid)

pteroylglutamic acid (folic acid)

3. Folic acid (shown above) is not active as a coenzyme. <u>It must go</u> <u>through two reductions</u>, first to dihydrofolic acid (FH_2), and then to tetrahydrofolic acid (FH_4) before it may act as a carrier.

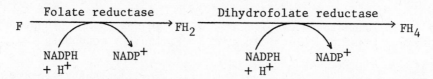

It is not completely clear whether one enzyme catalyzes both reductions or, as indicated above, there are two separate enzymes. Natural folates usually occur as polyglutamates (oligoglutamyl folates). The additional glutamyl residues (one to six, or more) are attached through gamma-carboxyl linkages. During absorption in the intestinal tract, which appears to be a carrier-mediated, energy-dependent process, the polyglutamyl chain is cleaved by enzymes, termed "conjugases," to monoglutamyl folate and glutamic acid. Conjugase activity is detected predominantly in the lysosomes of the intestinal epithelial cells. The predominant folate derivatives in tissues and extracellular fluid are:

Serum: Monoglutamyl form: N^5-methyl tetrahydrofolate
(abbreviated 5-methyl FH_4 <u>or</u> 5-methyl-THF)

Tissues in general: N^5-methyl-FH_4 and small amounts of other forms (discussed later)

Liver (storage form): oligoglutamyl folates

Red Cells: oligoglutamyl folates

Urine: monoglutamyl forms

Refer to Chapter 10 (Vitamins) for further discussion of folic acid as a vitamin.

4. Tetrahydrofolate has the structure

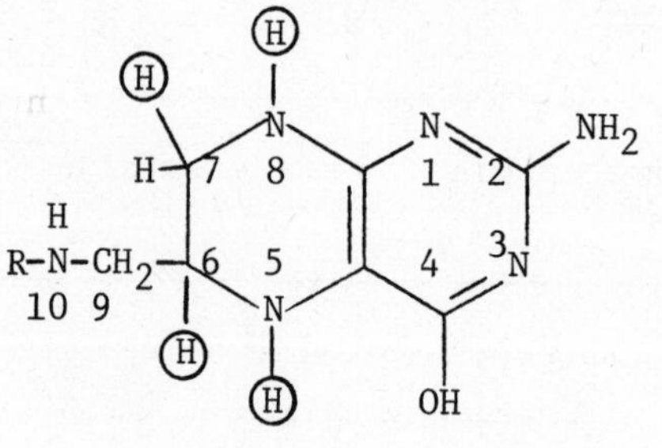

where R is the remainder of the molecule, identical to the equivalent portion of folic acid. Thus, FH_4 is 5, 6, 7, 8-tetrahydrofolic acid. The partially reduced dihydrofolate (FH_2) found <u>in vivo</u> appears to be 7,8-dihydrofolate. The other possible isomers (5,6- and 6,7-) have never been unequivocally demonstrated as products of either the chemical or enzymatic reduction.

5. Dihydrofolic acid reductase is **very strongly** inhibited competitively by methotrexate (amethopterin) which is an analogue of FH_2 and thus is an important "folic acid antagonist." Folic acid antagonists are used to halt the growth of certain fast-growing cancer cells (especially the leukemias). They function by preventing the formation of FH_4. This slows down the synthesis of nucleic acids, which requires FH_4. The sulfonamides (sulfa drugs) interfere with folic acid activity in a different way. They are very similar in structure to a p-aminobenzoic acid (PABA) which is used in the biosynthesis of folic acid. These analogues of PABA interfere with folic acid synthesis in microorganisms which have this synthetic capacity. Since man cannot synthesize folic acid (it is provided in the diet as a vitamin), sulfonamides do not interfere with human metabolism.

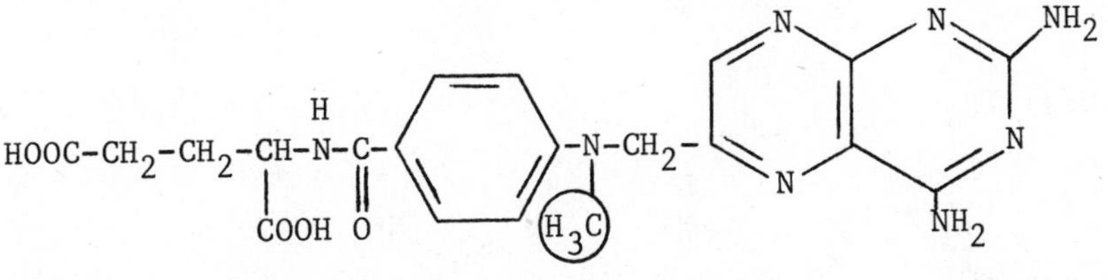

<u>Methotrexate</u> (amethopterin) (in aminopterin, another antineoplastic agent, the circled methyl group is replaced by a hydrogen).

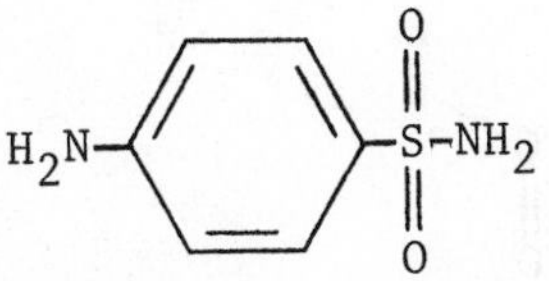

<u>Sulfanilamide</u> (one of the sulfonamides)

6. There are several active species within the folic acid group. Different species function as one-carbon carriers in different metabolic processes. However, in all biosynthetic reactions for which a folic acid derivative provides a one-carbon fragment, the carbon is carried in a covalent linkage to one or both of the nitrogen atoms at the 5 and 10 positions of the pteroic acid portion of tetrahydrofolate. The five known forms of carriers are shown below. The portions of the structure which are not drawn are identical to the corresponding portion of FH_4.

a. N^5, N^{10}-methylene FH_4

b. N^5, N^{10}-methenyl FH_4 (also called N^5, N^{10}-methylidynel FH_4)

c. N^{10}-formyl FH_4

d. N^5-formyl FH_4 (folinic acid)

This form, also called <u>leucovorin</u> or <u>citrovorum</u> <u>factor</u>, is one of the most chemically stable of the tetrahydrofolate derivatives. It is therefore used clinically to prevent or reverse ("rescue") the toxic effect of folate antimetabolites (methotrexate, aminopterin; discussed later).

e. N^5-formimino FH_4

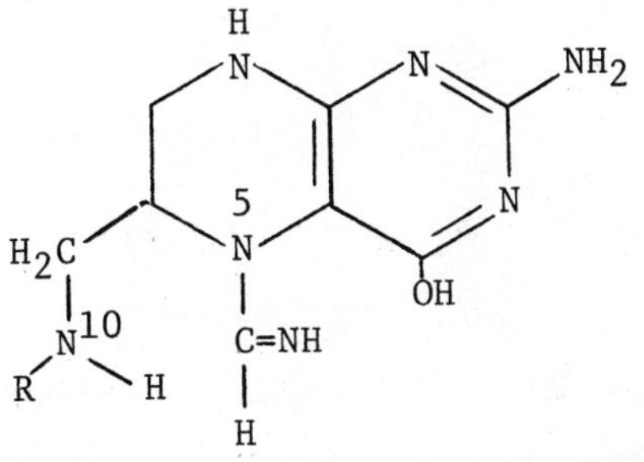

7. The reactions which produce some of the active species of the folate coenzymes are shown here. These interconversions are summarized in Figure 65.

a. Formation of N^5,N^{10}-methylene FH_4 (a <u>formaldehyde</u> derivative of FH_4) may occur in several ways.

i) FH_4 + H-C-H $\xrightarrow{\text{(non-enzymatic)}}$ N^5,N^{10}-methylene FH_4

  formaldehyde

ii) FH_4 + NH$_2$-CH-COOH $\xrightarrow[\text{(pyridoxal phosphate)}]{\text{serine hydroxymethylase}}$ N^5,N^{10}-methylene FH_4
      |
      CH$_2$-OH
                                                                          +
    serine   NH$_2$-CH$_2$-COOH

                                                                glycine

iii) N^5,N^{10}-methenyl FH_4 $\xrightarrow[\text{NADPH + H}^+ \quad \text{NADP}^+]{\substack{N^5,N^{10}\text{-methylene-}FH_4 \\ \text{dehydrogenase}}}$ N^5,N^{10}-methylene FH_4

The last two reactions (above) are the most important sources of this form of the coenzyme <u>in vivo</u>.

b. Formation of N^5,N^{10}-methenyl FH_4 (a <u>formate</u> derivative of FH_4) may also occur in several ways.

i)

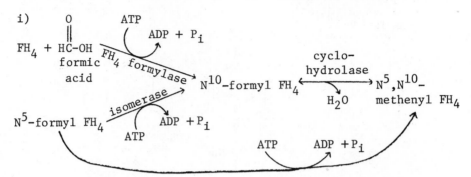

The two reactions above also show synthetic routes for N^{10}-formyl FH_4. Note that N^5-formyl FH_4 can be converted to N^5,N^{10}-methenyl FH_4 either directly or <u>via</u> N^{10}-formyl FH_4.

ii) Reversal of the tetrahydrofolate dehydrogenase reaction (a, iii, above).

iii) The action of cyclodeaminase on N^5-formimino FH_4 releases NH_3 and N^5,N^{10}-methenyl FH_4.

c. Formation of N^5-formyl FH_4 may occur from N-formyl glutamate and FH_4.

$$\text{N-formyl glutamate} + FH_4 \longrightarrow N^5\text{-formyl } FH_4 + \text{glutamate}$$

d. The N^5 position of FH_4 can also carry a formimino group. The formimino group of N^5-formimino FH_4 can be transferred to other compounds, producing formimino derivatives which are needed in biosynthesis. For example, formimino glycine, a precursor of the purines, can be formed by the reversal of reaction (i) below. The two principal synthetic routes for this FH_4 compound are

i)

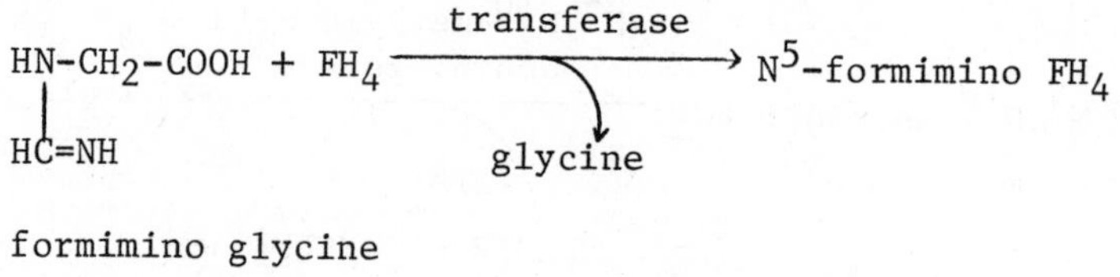

$$\text{HN-CH}_2\text{-COOH} + FH_4 \xrightarrow[\text{glycine}]{\text{transferase}} N^5\text{-formimino } FH_4$$
$$|$$
$$\text{HC=NH}$$

formimino glycine

Figure 65. Some Interrelationships of One–Carbon Transfer Reactions Involving Folate–Derived Carriers

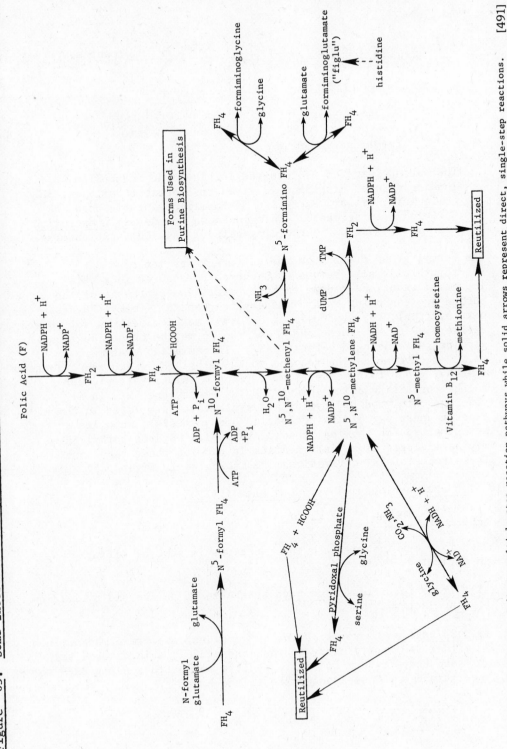

Note: Dashed arrows indicate multiple–step reaction pathways while solid arrows represent direct, single–step reactions.

[491]

ii)

$$\text{HOOC-CH}_2\text{-CH}_2\text{-CH-COO}^{(-)} + \text{FH}_4 \xrightarrow[\text{glutamate}]{\text{transferase}} \text{N}^5\text{-formimino FH}_4$$

$$\underset{\displaystyle \text{HC=NH}}{\overset{\displaystyle |}{\underset{\displaystyle |}{\text{NH}}}}$$

formiminoglutamate

The second reaction is involved in histidine metabolism. If a loading dose of histidine is given to a patient deficient in folic acid, urinary excretion of formiminoglutamic acid ("figlu") is increased. This is known as the figlu excretion test and is useful in the diagnosis of megaloblastic anemias.

8. The formyl group present on the different species of tetrahydrofolic acid serves as a one-carbon source in several important reactions. Some examples are: the synthesis of the purine nucleus using N^5,N^{10}-methenyl FH_4 and N^{10}-formyl FH_4 (see next section); the synthesis of N-formyl methionine-tRNA (discussed earlier) using N^{10}-formyl FH_4; and the formation of serine from glycine, also using N^{10}-formyl FH_4. In mammalian systems, N^5,N^{10}-methylene can be reduced to N^5-methyl FH_4 which can transfer a methyl group to homocysteine, forming methionine (see under sulfur-containing amino acids). N^5,N^{10}-methylene FH_4, in a reaction catalyzed by thymidylate synthetase (see later in this chapter under pyrimidine nucleotide biosynthesis), supplies a methyl group for thymidylic acid synthesis. Hydroxymethyl FH_4 may participate as an intermediate in some of these reactions. These roles of FH_4 derivatives are indicated in Figure 65.

Role of 5-Phosphoribosyl-1-Pyrophosphate (PRPP)

1. ### Functions of PRPP

   PRPP is a key intermediate and functions in several different ways in nucleotide synthesis.

   i) It is required for de novo synthesis of purine and pyrimidine nucleotides.

   ii) Salvage pathways of purine nucleotides take preformed purines (derived from dietary sources and nucleic acid catabolism) and convert them to the respective nucleotides using PRPP to attach a ribose-1-phosphate to the base.

$$\text{(P)}-\text{Ribose}-\text{(P)}-\text{(P)} \longrightarrow \text{Base}-\text{Ribose}-\text{(P)} + PP_i$$
$$\qquad\quad \text{(PRPP)} \qquad\qquad\qquad \text{(nucleotide)}$$

2. <u>Synthesis of PRPP</u>

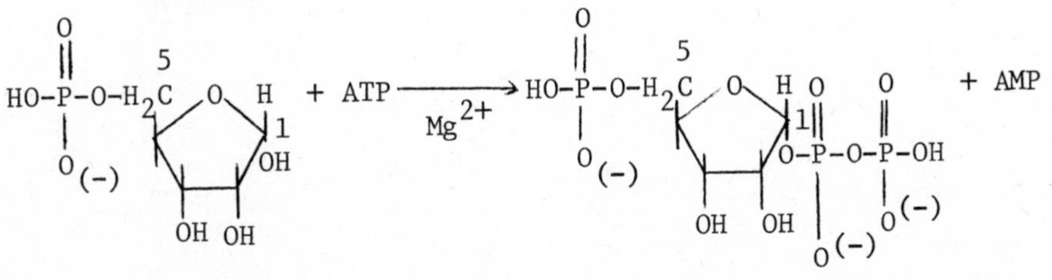

(R-5-P)   (PRPP)

Note: This reaction is catalyzed by a kinase which transfers the terminal pyrophosphate group of ATP to R-5-P. This is unusual, since kinases generally transfer only phosphate groups.

3. <u>Origin of Ribose-5-P</u>. R-5-P which is utilized in the above synthesis of PRPP can come from three possible pathways.

    a. The hexose monophosphate shunt (phosphogluconate pathway) is probably the major source of R-5-P in tissues (liver, bone marrow) in which glucose can be oxidized aerobically.

    b. Uronic acid pathway. Xylulose-5-P produced in this pathway is converted to R-5-P as follows:

$$\text{D-Xylulose-5-P} \xrightarrow{\text{Epimerase}} \text{D-Ribulose-5-P} \xrightarrow{\text{Isomerase}} \text{D-Ribose-5-P}$$

The contribution made by this pathway to the supply of R-5-P may be a minor one. The epimerase and isomerase enzymes are the same ones which participate in the HMP shunt.

    c. Ribose-5-phosphate can also be produced in tissues (such as skeletal muscle) in which glucose is metabolized principally by anaerobic glycolysis. This is shown in the following diagram, starting from fructose-6-phosphate (F-6-P).

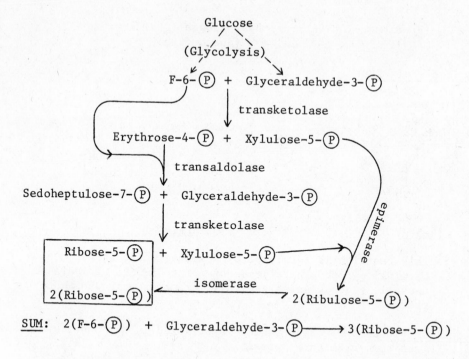

Glucose

(Glycolysis)

F-6-(P) + Glyceraldehyde-3-(P)

transketolase

Erythrose-4-(P) + Xylulose-5-(P)

transaldolase

Sedoheptulose-7-(P) + Glyceraldehyde-3-(P)

transketolase

Ribose-5-(P) + Xylulose-5-(P)

isomerase

2(Ribose-5-(P)) ⟵ 2(Ribulose-5-(P))

epimerase

SUM: 2(F-6-(P)) + Glyceraldehyde-3-(P) ⟶ 3(Ribose-5-(P))

Notice that all of the enzymes in this pathway are also part of the HMP shunt. In the muscle, however, complete operation of the HMP pathway does not occur, due to the absence of glucose-6-phosphate dehydrogenase. These interconversions can occur under either aerobic or anaerobic conditions but they are probably more important in the absence of oxygen.

Purine Nucleotide Biosynthesis

In the diagrams below, the portion of the molecule enclosed by a dashed line was added in the preceding reaction.

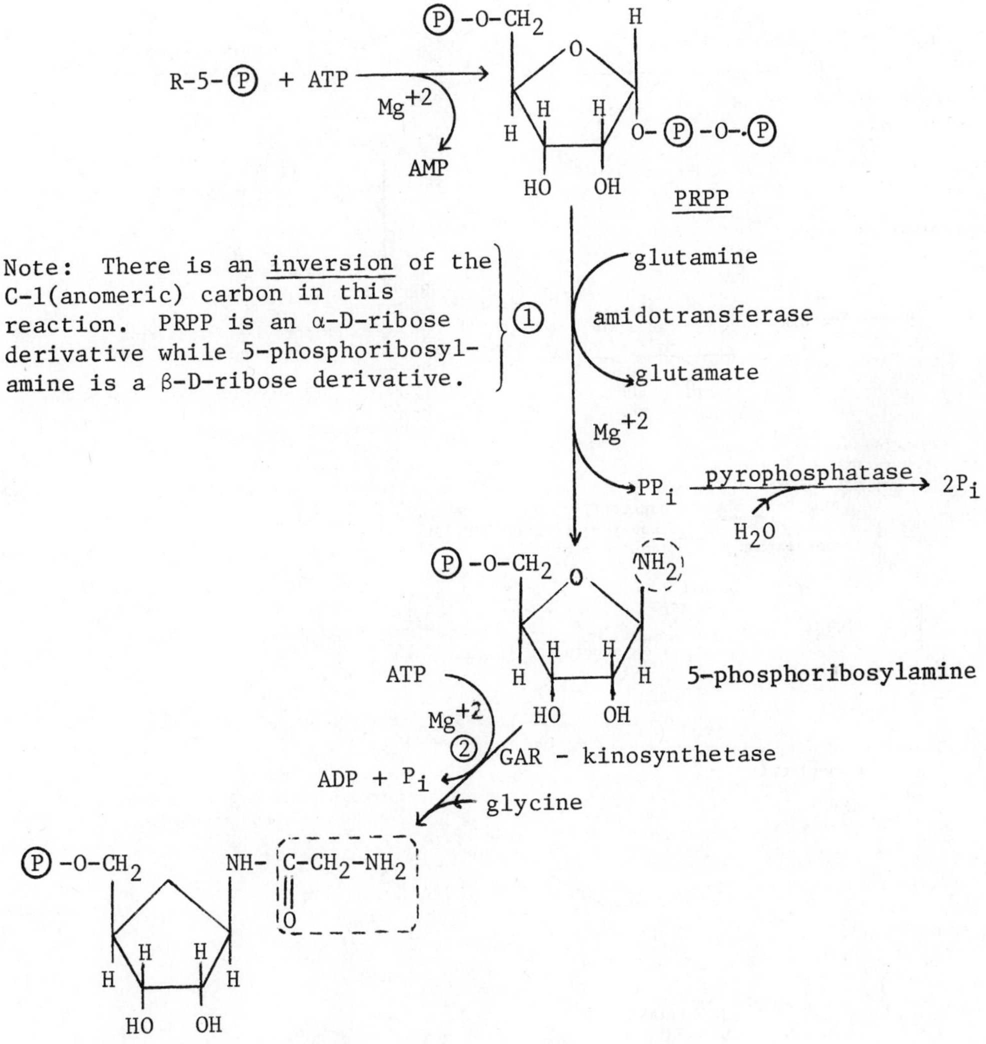

R-5-(P) + ATP $\xrightarrow[\text{Mg}^{+2}]{}$

(P)-O-CH$_2$ ... H ... O ... H ... H ... H ... O-(P)-O-(P)

AMP

HO   OH   **PRPP**

Note: There is an <u>inversion</u> of the C-1(anomeric) carbon in this reaction. PRPP is an α-D-ribose derivative while 5-phosphoribosyl-amine is a β-D-ribose derivative.

① glutamine

amidotransferase

glutamate

Mg^{+2}

PP$_i$ $\xrightarrow{\text{pyrophosphatase}}$ 2P$_i$

H$_2$O

(P)-O-CH$_2$ ... O ... (NH$_2$)

H   H

ATP

H   H   **5-phosphoribosylamine**

Mg^{+2}

HO   OH

② GAR - kinosynthetase

ADP + P$_i$

glycine

(P)-O-CH$_2$ ... NH-C-CH$_2$-NH$_2$ ... ‖ O

H   H

H   H

HO   OH

<u>Glycinamide ribosyl-5-(P) (GAR)</u>

GAR can be written in a different form for the sake of convenience. Note that the ribosyl phosphate group is now represented by "R-5-P."

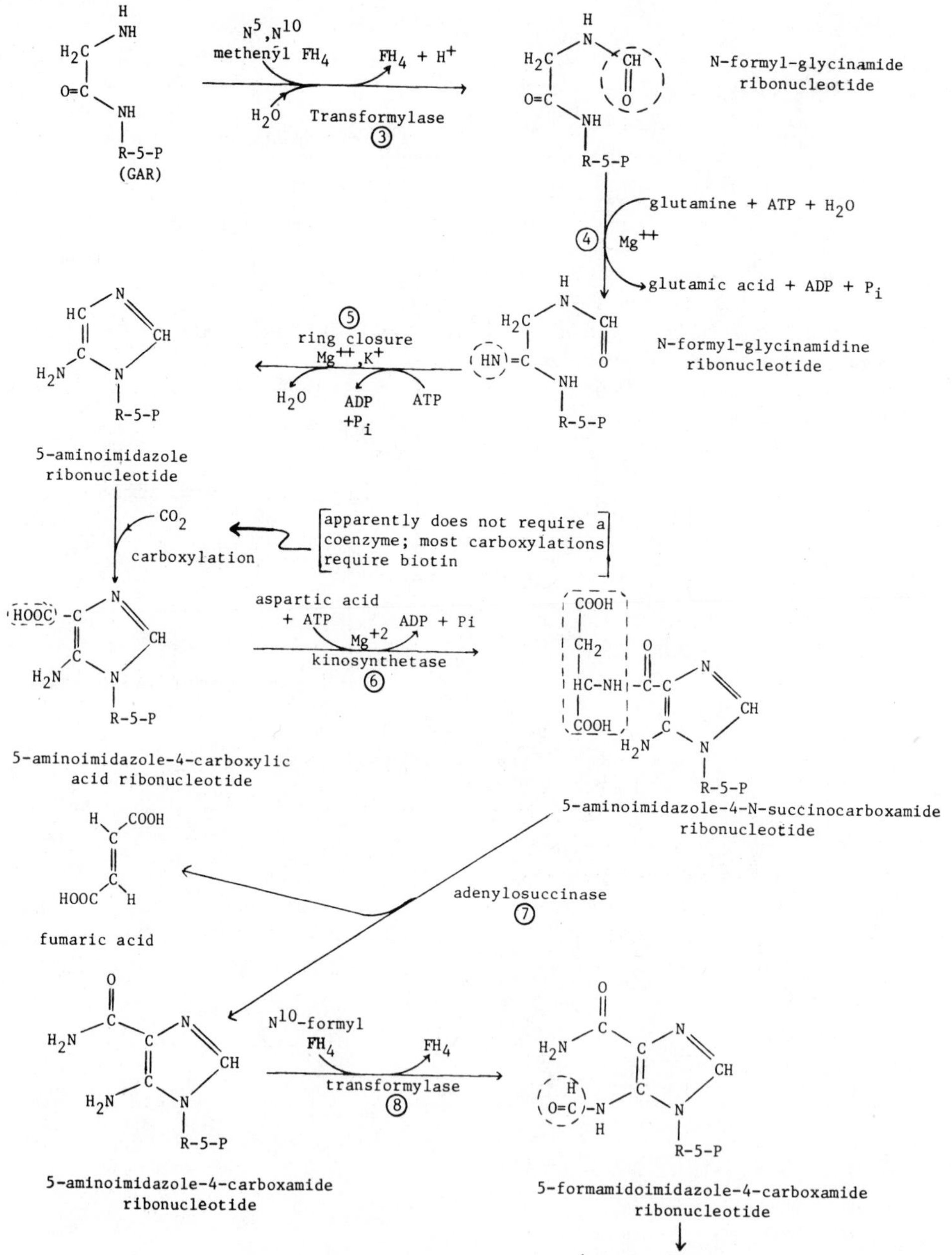

N^5, N^{10} methenyl FH_4 $FH_4 + H^+$

H_2O Transformylase ③

(GAR)

N-formyl-glycinamide ribonucleotide

glutamine + ATP + H_2O

④ Mg^{++}

glutamic acid + ADP + P_i

N-formyl-glycinamidine ribonucleotide

⑤ ring closure Mg^{++}, K^+

H_2O ADP $+P_i$ ATP

5-aminoimidazole ribonucleotide

CO_2

carboxylation

apparently does not require a coenzyme; most carboxylations require biotin

aspartic acid + ATP   ADP + Pi

Mg^{+2} kinosynthetase ⑥

5-aminoimidazole-4-carboxylic acid ribonucleotide

5-aminoimidazole-4-N-succinocarboxamide ribonucleotide

fumaric acid

adenylosuccinase ⑦

N^{10}-formyl FH_4 FH_4

transformylase ⑧

5-aminoimidazole-4-carboxamide ribonucleotide

5-formamidoimidazole-4-carboxamide ribonucleotide

(to top of next page)

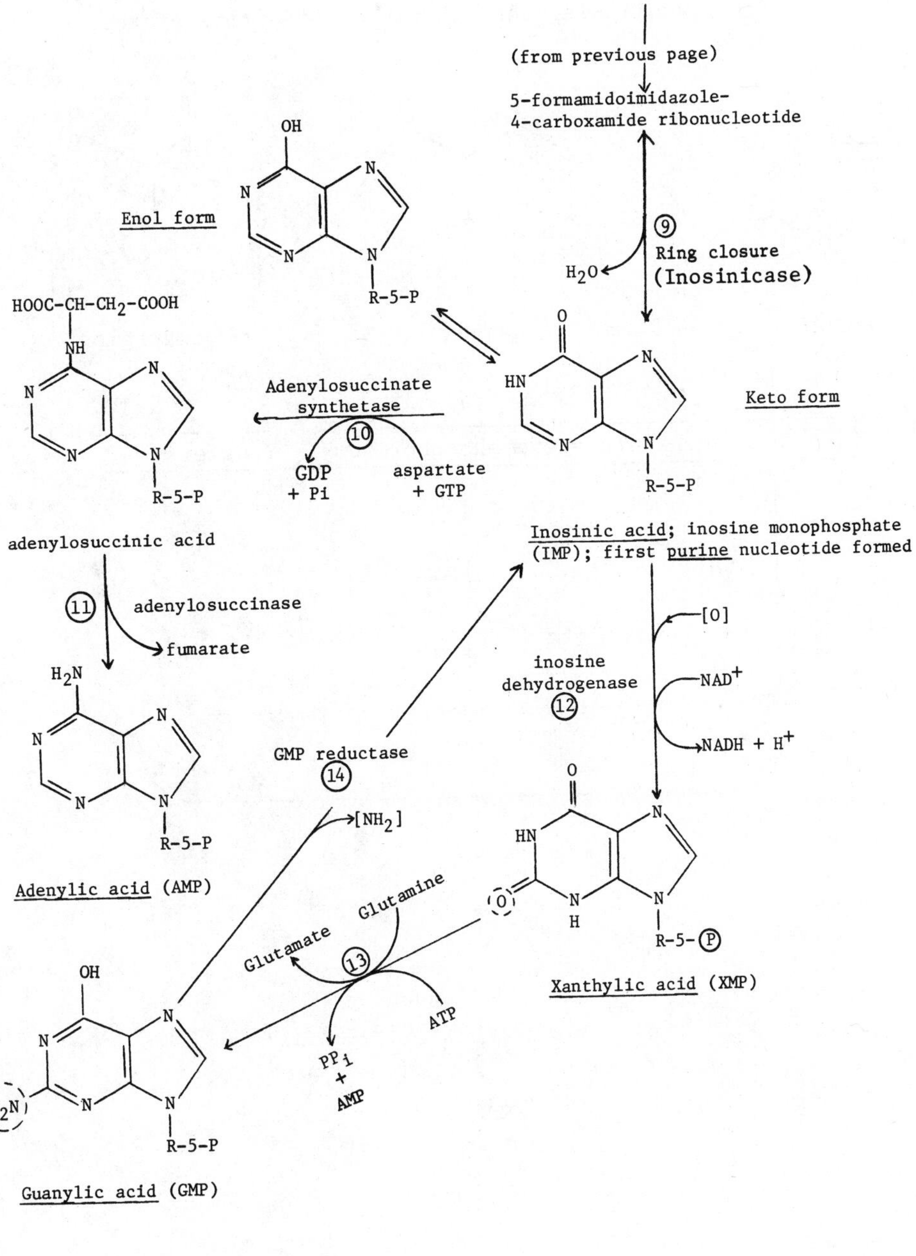

(from previous page)

5-formamidoimidazole-
4-carboxamide ribonucleotide

Enol form

⑨ Ring closure
(Inosinicase)

H_2O

HOOC-CH-CH₂-COOH

Adenylosuccinate
synthetase
⑩

GDP
+ Pi

aspartate
+ GTP

Keto form

adenylosuccinic acid

Inosinic acid; inosine monophosphate
(IMP); first purine nucleotide formed

⑪ adenylosuccinase

fumarate

inosine
dehydrogenase
⑫

[O]

NAD^+

$NADH + H^+$

GMP reductase
⑭

$[NH_2]$

Adenylic acid (AMP)

Glutamate   Glutamine

⑬

Xanthylic acid (XMP)

PPi
+
AMP

ATP

Guanylic acid (GMP)

497

Other reactions of importance in the synthesis of purine nucleotides are the <u>salvage pathways</u>, summarized below. The significance of these is discussed later, following the inhibition of purine synthesis.

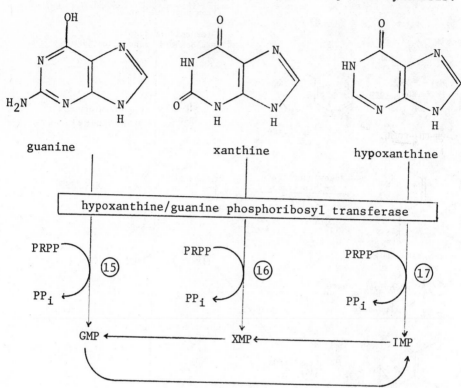

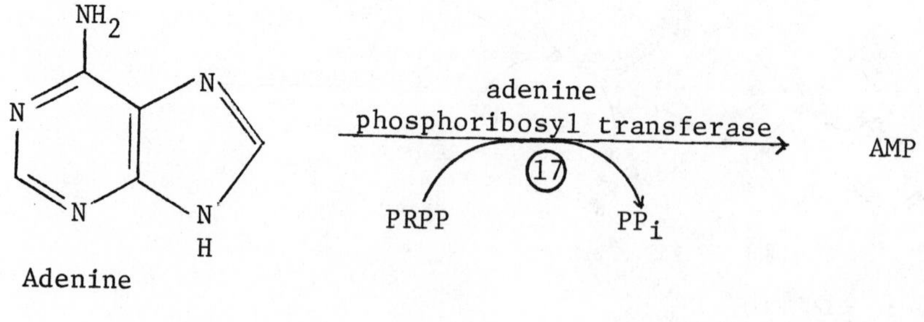

guanine   xanthine   hypoxanthine

Adenine

AMP

General Comments on Antineoplastic Agents

This includes antimetabolites, alkylating agents, antibiotics, steroidal hormones, specific mitotic inhibitors, and miscellaneous drugs. We have already mentioned, in the section on nucleic acid synthesis, chemicals that are effective against cancer. In this section, additional compounds in this category will be discussed. Following are a few general comments concerning cancer chemotherapy.

1. Chemicals directed against DNA, RNA, and protein synthesis have made major contributions to cancer therapy. Surgery and radiation are still, however, the major modes of treatment for cancer. For example, surgery is extremely beneficial in the treatment of cancers of the skin, breast, colon, and uterus. Radiation is a particularly useful procedure for treating solid tumors in their early stages of development. It is used against seminoma (a type of cancer of the testis), retinoblastoma (a cancer of the eye), Hodgkin's disease (a cancer of the lymph system), and cancer of the nasopharynx. In disseminated malignancies, though, (leukemias, lymphomas, hemopoietic cancers) where surgery and radiation are not particularly effective, chemotherapy has proven quite useful.

2. Some of the principles involved in chemotherapy are:

   a. Aggressive treatment against tumors that are small and in the early stages of growth is most effective.

   b. A combination of drugs is significantly more effective than are the same drugs used individually.

   c. Chemotherapy is frequently effective against metastatic tumors (those arising from neoplastic cells which became detached from the primary tumor and are transported by the circulatory and lymphatic system to other parts of the body), particularly after eradication of the primary tumor by surgery or radiation. It should be noted that tumors (primary or metastatic) are not clinically detectable until they contain at least 10^9 cells.

3. All of the antitumor agents used at the present time have serious toxic effects on normal cells. Furthermore, a number of these agents themselves are mutagenic and can be carcinogenic. They can either directly affect the DNA or, through immunosuppression, encourage the growth of other tumors. The latter possibility is supported by the increased incidence of tumors among transplant patients who have been treated with immunosuppressive drugs.

4. Following is a list of some of the compounds used as antineoplastic agents, together with their mode of action. Additional details are given at the appropriate places.

L-Asparaginase: Blocks protein synthesis by depleting the amino acid asparagine (neoplastic cells require a high rate of protein synthesis).

Methotrexate: Inhibits dihydrofolate reductase, prevents availability of (thereby limiting the supply of) single-carbon fragments and interferes with purine and pyrimidine synthesis and certain amino acid interconversions.

6-Mercaptopurine and 6-Thioguanine: Inhibit purine biosynthesis and the interconversion of purines.

5-Fluorouracil: Inhibits thymidylic acid (a pyrimidine nucleotide) synthesis by blocking the thymidylate synthetase reaction.

Arabinosyl Cytosine: Blocks the conversion of cytidylic to deoxycytidylic acid; inhibits DNA polymerase.

Hydroxyurea: Inhibits ribonucleoside diphosphate reductase and thus blocks the conversion of cytidylic to deoxycytidylic acid.

Adriamycin (and Daunomycin): Anthracyclenes; fermentation products of the fungus _Streptomyces_ peuceticus var. caesius;

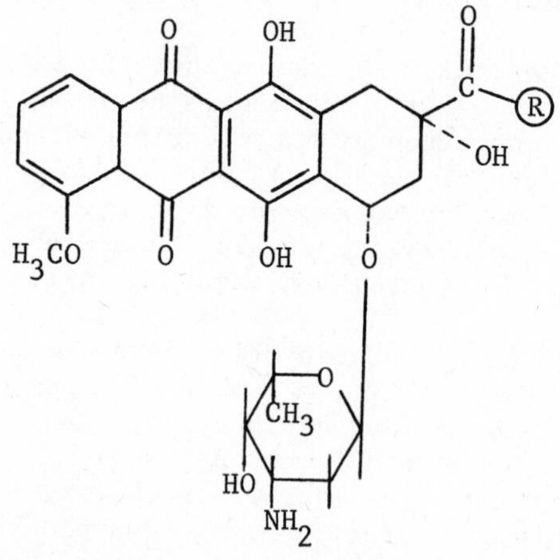

R = CH_3 in Daunomycin

R = CH_2OH in Adriamycin

intercalates between adjacent base pairs on the DNA strand, leading to uncoiling of the DNA helix with inhibition of DNA-directed RNA polymerase and DNA polymerase activity. Adriamycin is rapidly accumulated in cardiac tissue and is a cardiotoxic drug.

Bleomycin: A mixture of several polypeptide antibiotics which are fermentation products of Streptomyces verticillus. It inhibits the progression of cells through the premitotic (G_2) and mitotic (M) phases of the cell cycle. Other effects include in vitro scission and fragmentation of DNA, chelation of heavy metals, and the inhibition of incorporation of thymidine into DNA. Bleomycin appears to have minimal cytotoxic action on rapidly proliferating normal tissue such as bone marrow (a property useful in combination chemotherapy). It is particularly effective against squamous cell carcinomas and the malignant lymphomas.

Nitrosoureas: Structures of some nitrosoureas are shown below.

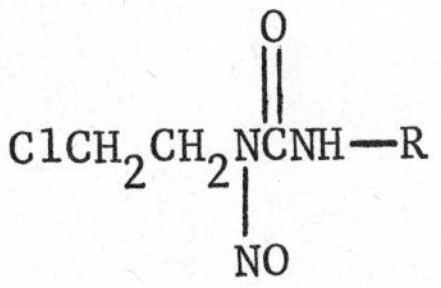

$$ClCH_2CH_2\overset{\displaystyle O}{\overset{\|}{\underset{\underset{\displaystyle NO}{|}}{N}}}CNH{-\!}R$$

Chloroethylnitrosourea

R = CH_2CH_2Cl; BCNU: 1,3-bis(2-chloroethyl)-1-nitrosourea

R = ⟨⟩ ; CCNU: 1-(2-chloroethyl)-3-cyclohexyl-1-nitrosourea

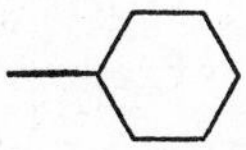

R = ⟨⟩—CH_3; methy CCNU: 1-(2-chloroethyl)-3-(4-methylcyclohexyl)-1-nitrosourea

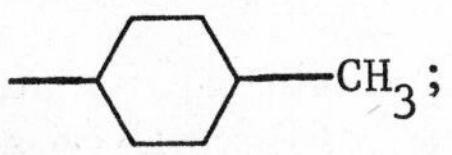

These compounds act by: i) The alkylation of DNA (with the chloroethyl portion of the molecule; BCNU can crosslink DNA because it contains two chloroethyl groups); ii) Carbomylation by the isocyanate portion of the molecule; iii) Inhibition of DNA repair by the isocyanate metabolite of the parent compound.

Cyclophosphamide: Functions as an alkylating and crosslinking agent.

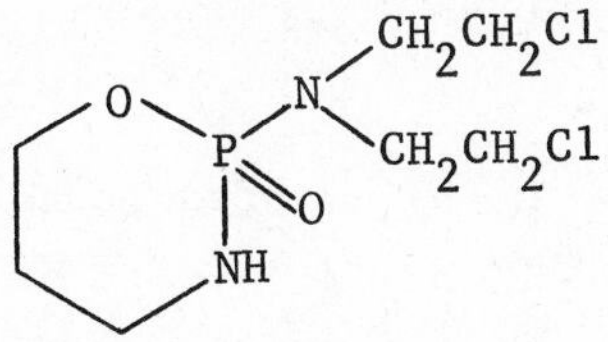

The active toxic metabolite appears to be derived from the action of the liver microsomal mixed-function oxidase system. The reaction is shown below.

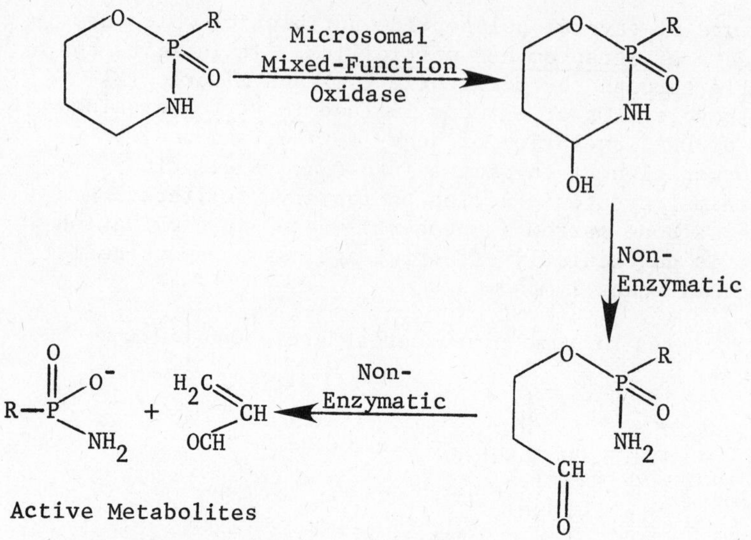

$$R-\overset{\displaystyle O}{\underset{\displaystyle NH_2}{\overset{\|}{P}}}-O^{-} \quad + \quad \overset{H_2C}{\underset{OCH}{>}}CH$$

Active Metabolites

Vinka Alkaloids: Vincristine and vinblastine: These compounds prevent the assembly of the protein precursor of microtubules and mitotic spindles, thereby arresting cell division. Colchicine, a drug used for acute gout, has a similar mode of action.

Actinomycin D: Discussed earlier; binds to GC-rich regions and interferes with transcription. It is particularly effective in Wilm's tumor, a cancer of the kidney in children.

Prednisone: A corticosteroid with little salt-retaining activity (see Chapter 13 on hormones). It is an anti-inflammatory agent having its major effects on lymphoid tissue. It causes the nuclei to become pyknotic and disintegrate. It also inhibits DNA and RNA polymerases in lymphocytes.

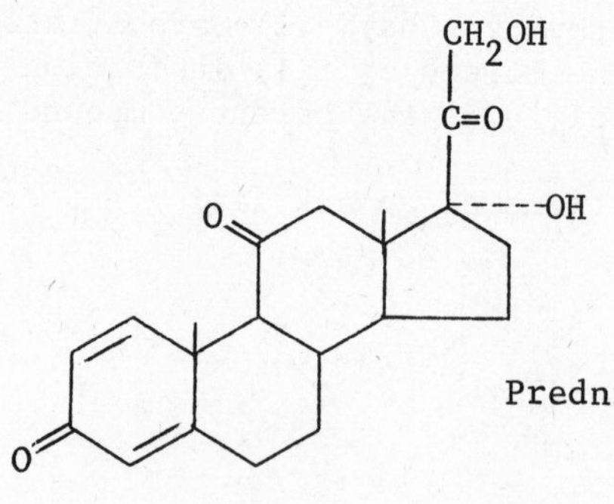

Prednisone

Comments on Purine Biosynthesis

1. Sources of atoms in the purine molecule

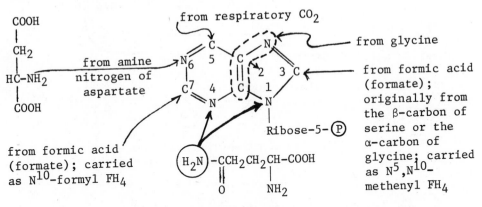

The numbers in this diagram indicate the <u>order of addition of the</u> <u>atoms</u> in the purine nucleus. Note that the three atoms of glycine are added as a group and that glycine is the only amino acid which provides <u>both</u> carbon and nitrogen atoms to the purine ring. (In pyrimidines, aspartate plays an equivalent role.)

2. In purine biosynthesis, the glycosidic bond is formed when the first atom of the purine ring is incorporated. This is in contrast to pyrimidine biosynthesis, where the pyrimidine ring is completely formed prior to addition of ribose-5- P .

3. In <u>E. coli</u> cells, there are several control mechanisms present to regulate purine synthesis.

    a.

    <u>Amidotransferase</u> is a multivalent (many binding sites) regulatory enzyme. Feedback inhibition is by ATP, ADP, or AMP, or by GTP, GDP, GMP; each group of nucleotides acting at a separate control site on the enzyme. IMP, IDP, and ITP also inhibit this reaction (see the section on enzyme regulation for more information about this reaction).

b.

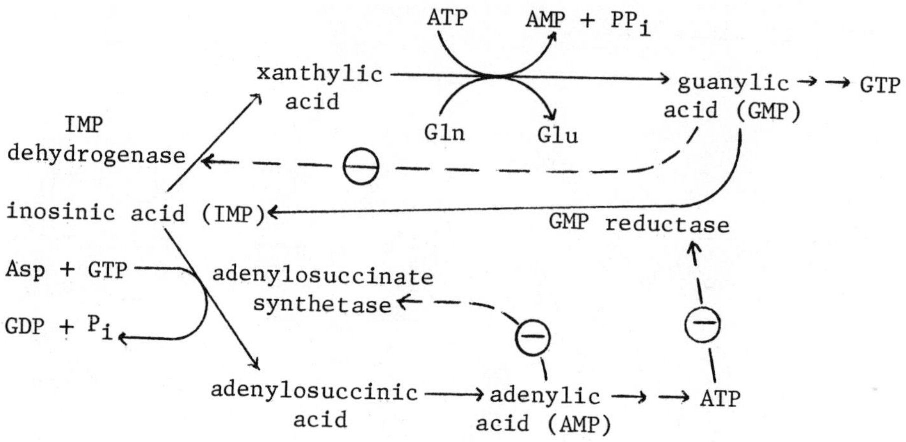

i) ATP is required as a cofactor for GTP synthesis and, reciprocally, GTP is needed to form ATP. This regulates the synthesis of one nucleotide relative to the other.

ii) End-product inhibition also occurs in both pathways above. Adenylosuccinate synthetase is inhibited by AMP and IMP dehydrogenase is inhibited by GMP.

iii) GMP reductase, which converts GMP to IMP, is inhibited by ATP. Thus, an adequate supply of adenine nucleotides prevents their further synthesis at the expense of guanine nucleotides.

c. There is also evidence that guanosine and adenosine suppress the synthesis of some of the enzymes which catalyze steps in purine nucleotide biosynthesis.

d. Some interrelationships and control points in purine biosynthesis are summarized in the next diagram. Heavy lines indicate reactions while dashed lines show control pathways.

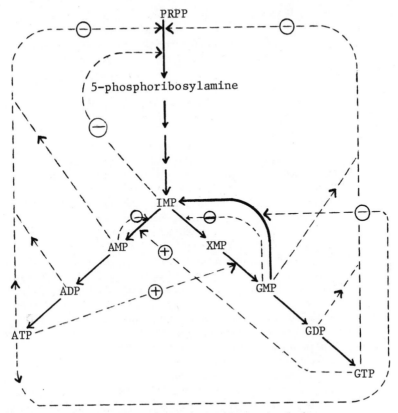

4. Inhibitors of purine nucleotide biosynthesis include

    a. Glutamine Analogues. Azaserine, an antibiotic which has been
        isolated from a species of Streptomyces, is a structural
        analogue of glutamine.

$$N \equiv N = CH - \overset{\overset{\displaystyle O}{\|}}{C} - O - CH_2 - \underset{\underset{\displaystyle NH_2}{|}}{CH} - COOH$$

Azaserine

Azaserine inhibits steps
1, 4, and 13 of purine
synthesis (see diagram of
purine synthetic pathway).

$$H_2N - \overset{\overset{\displaystyle O}{\|}}{C} - CH_2 - CH_2 - \underset{\underset{\displaystyle NH_2}{|}}{CH} - COOH$$

Glutamine

b. i. Folic acid analogues, such as methotrexate, act as folate antagonists. They block the production of functional folate derivatives by inhibiting dihydrofolate reductase. Methotrexate functions as an antagonist without any preliminary metabolic transformation. The reduction in size of the active folate pools leads to diminished purine biosynthesis (inhibition of reactions 3 and 8 in purine synthesis). The diminution in functional folate derivatives also affects amino acid and pyrimidine metabolism (inhibition of thymidylate synthesis--discussed under pyrimidine biosynthesis). Thus, methotrexate inhibits DNA, RNA, and protein synthesis. It has been found effective in the treatment of a number of neoplastic diseases, including breast cancer, head and neck cancer, choriocarcinoma, osteogenic sarcoma, and acute forms of leukemia. In the treatment of these neoplastic diseases, it has been observed that the administration of high doses of methotrexate can be tolerated provided that the patient also receives <u>citrovorum factor</u> (N^5-formyl tetrahydrofolate; folinic acid; its synthesis requires dihydrofolate reductase) after the drug therapy. This ameliorates the damage to bone marrow caused by methotrexate (myelosuppression leading to leukopenia and thrombocytopenia). Some of the other toxic effects of the drug include mucositis, gastrointestinal symptoms, and liver toxicity. The drug has direct toxic effects on the hepatic cells and chronic oral administration of methotrexate in psoriasis has been associated with an increased incidence of postnecrotic cirrhosis.

The "rescue" of normal but not tumor cells from methotrexate toxicity by citrovorum factor is partly explained by differences in membrane transport. For example, osteogenic sarcoma cells (which do not respond to conventional doses of methotrexate, making high doses of the drug necessary for effective treatment) are not rescued by citrovorum factor when it is administered after a treatment with methotrexate. This is presumably due to the absence of transport sites for citrovorum factor in the neoplastic cells. The proper administration of methotrexate along with a citrovorum factor rescue program has been found superior to that of therapy with methotrexate alone. However, the combined treatment has been reported to be associated with new toxicities such as cutaneous vasculitis and renal impairment.

ii.   L-asparaginase, another antineoplastic agent (asparaginase hydrolyzes asparagine to aspartic acid and NH_3, thus making it unavailable for protein synthesis in tumor cells), when given at appropriate intervals, limits methotrexate toxicity on the bone marrow. This is thought to result from the inhibition of protein synthesis by L-asparaginase. Absence of protein synthesis prevents the cells from entering the DNA synthetic phase of the cell cycle, where the cells are most susceptible to methotrexate. In patients with acute myelocytic and lymphocytic leukemia, the combination of L-asparaginase and methotrexate has yielded very promising results in cases where the two drugs used individually are not effective. The value of the combined drug regimen is limited in its applicability, though, because the usefulness of L-asparaginase is limited to the treatment of leukemias.

iii. Resistance to methotrexate can develop for several reasons. Dihydrofolate reductase activity may actually increase after therapy, or an enzyme which has a lower affinity for the inhibitor may be synthesized, or the ability of the tumor cells to transport the drug may decrease.

c. Purine Analogues

i) 6-mercaptopurine

(I) Mechanism of action. 6-MP is a structural analogue of hypoxanthine and can be converted to a nucleotide by one of the salvage enzymes.

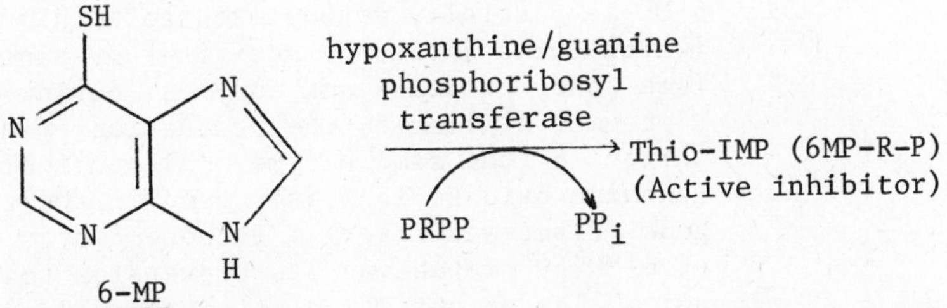

This reaction is similar to salvage pathway reactions 15, 16, and 17. Thio-IMP prevents the production of both AMP and GMP by inhibiting the following reactions:

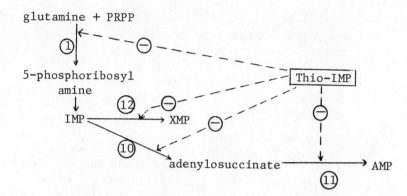

Although unmetabolized 6-MP, thio-ITP, deoxythio-ITP, and thio-IDP are biologically active in several ways (which may account for some of the side-effects of 6-MP treatment), thio-IMP is the most active metabolic form of 6-MP.

(II)   6-MP is used in the treatment of acute leukemias. Eventually, however, 6-MP-resistant tumor cells do become preponderant. The probable mechanisms of resistance include development of altered specificity or a lack of phosphoribosyl transferases so that thio-IMP (the active inhibitor) is not formed. In support of this mechanism, the resistant cells do respond to 6-methylmercaptopurine ribonucleoside, which is converted to the corresponding nucleotide. Other means for 6-MP-resistance may be alterations in the permeability of the cell and increased rate of destruction of 6-MP.

(III) 6-MP is partially metabolized to 6-thiouric acid (which lacks antitumor activity) by xanthine oxidase (see under proteins and enzymes; purine catabolism), a process similar to the production of uric acid from purines by the same enzyme. Allopurinol is a xanthine oxidase inhibitor used in the treatment of gout (discussed later). It potentiates the action of 6-MP by preventing its conversion to 6-thiouric acid. The potentiation effect of allopurinol is one of the properties that is taken into consideration in 6-MP treatment. 6-MP is also degraded by methylation of the sulfhydryl group and subsequent oxidation of the methyl group. In order to prevent this type of degradation, S-substituted derivatives of 6-MP have been synthesized such as azathioprine.

ii)   Azathioprine (imuran) is a derivative of 6-MP and functions as an antiproliferative agent. The compound has its principal use as an immunosuppressive agent. Azathioprine presumably releases 6-MP in the body by reacting with sulfhydryl compounds such as GSH.

iii)   6-Thioguanine is similar in its action to 6-MP. Again, the most active form is 6-thio-GMP, the ribonucleotide of the drug.

6-Thioguanine

d.   Sulfonamides prevent formation of functional folic acid in bacteria. They are structural analogues of PABA (see also under one-carbon metabolism).

a sulfonamide; if R = -H, the compound is sulfonilamide

5.   The primary pathway of purine nucleotide synthesis has already been discussed. There are, however, other secondary reactions by which purine nucleotides are formed. These are called "salvage (reutilization) pathways" because they involve the synthesis of purine nucleotides from pre-formed purines which are "salvaged" from dietary sources and from tissue breakdown. The reutilization reactions are as follows:

a.

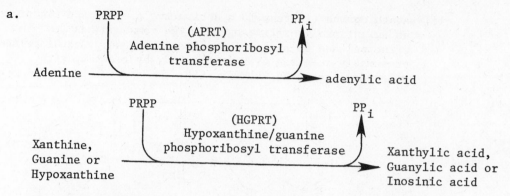

The <u>Lesch-Nyhan Syndrome</u> is an X-linked recessive neurological disorder associated with an overproduction of uric acid (see also gout; hyperuricemia), mental retardation, and a tendency for self-mutilation. In red cells, skin fibroblasts, and brain and liver cells of patients having this disease, an almost complete lack of hypoxanthine/guanine phosphoribosyl transferase activity has been demonstrated. The absence of this salvage enzyme is presumably the reason why 6-MP and imuran are ineffective in controlling purine synthesis in such patients. The connection between the observed symptoms and the lack of the transferase is not clear.

The clinical picture that is observed in the complete deficiency of APRT is in <u>contrast</u> with the dramatic symptoms seen in the Lesch-Nyhan syndrome. The major clinical manifestation due to the absence of adenine phosphoribosyl transferase is the formation and excretion of 2,8-dihydroxyadenine as insoluble material (gravel) in the urine. 2,8-Dihydroxyadenine is highly insoluble (solubility of 1 to 3 mg/l, which is 50 times less soluble than uric acid, the normal end-product of purine metabolism in man--discussed later) and leads to urolithiasis (formation of stones in the urine). In the routine urine stone analysis, 2,8-dihydroxyadenine stones can be confused with those of urate. Thus, appropriate chemical analyses of the stones are required, particularly in the <u>pediatric age group</u>, to identify APRT-deficient individuals. The formation of 2,8-dihydroxyadenine in APRT deficiency is shown below.

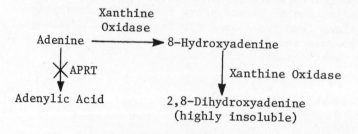

The mode of transmission of this disorder in one family study showed that the defect is inherited as an autosomal recessive trait.

The neurologic disorders characteristic of HGPRT deficiency are not found with APRT deficiency. This difference is presumably related to the levels of these two enzymes in the brain, where APRT is much less active than HGPRT. In a patient with APRT deficiency, HGPRT levels in erythrocytes were found to be normal. APRT deficiency is treated with a low purine diet and the inhibition of xanthine oxidase with allopurinol (a structural analogue of hypoxanthine--discussed later). Recall that xanthine oxidase is responsible for the formation of insoluble 2,8-dihydroxyadenine.

b. Purine nucleosides may be formed from free purines and ribose-1-phosphate

$$\text{free purine} + \text{R-1-P} \xrightarrow{\begin{array}{c}\text{purine}\\\text{nucleoside}\\\text{phosphorylase}\end{array}} \text{nucleoside} + P_i$$

The phosphorylase may also be involved in degradative reactions (see purine catabolism, below). Nucleosides may then be converted to nucleotides by nucleoside kinases as in the example here.

$$\text{Adenosine} + \text{ATP} \xrightarrow{\begin{array}{c}\text{Adenosine}\\\text{kinase}\end{array}} \text{Adenylic acid} + \text{ADP}$$
$$(\text{AMP})$$

6. Guanylic acid (GMP) and adenylic acid (AMP) are converted to GTP and ATP respectively in several ways.

a. A group of nucleoside monophosphokinases, of which adenylate kinase is one, catalyze the formation of nucleoside diphosphates from nucleoside monophosphates using ATP. If Z is the purine group, then the reaction is as follows:

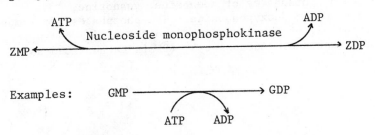

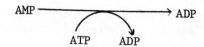

b. Nucleoside diphosphokinases catalyze the formation of a
    nucleoside triphosphate from a nucleoside diphosphate.

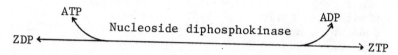

Examples:

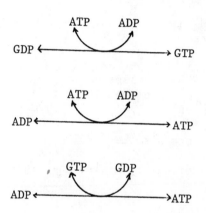

The reactions above are used in the formation of pyrimidine
nucleoside triphosphates as well.

7. Formation of deoxyribonucleotides will be discussed later in this
    chapter. Purine and pyrimidine deoxynucleotides will be covered
    together.

8. Purine Catabolism

    a. The purine nucleotide products of RNA and DNA catabolism are
        the 3'- and 5'-phosphates of adenosine, guanosine,
        deoxyadenosine and deoxyguanosine. The phosphates are removed
        (as P_i) from all of these except adenosine-5'-℗ by appropriate
        phosphatase enzymes. Adenosine-5'-℗ is catabolized as
        shown next.

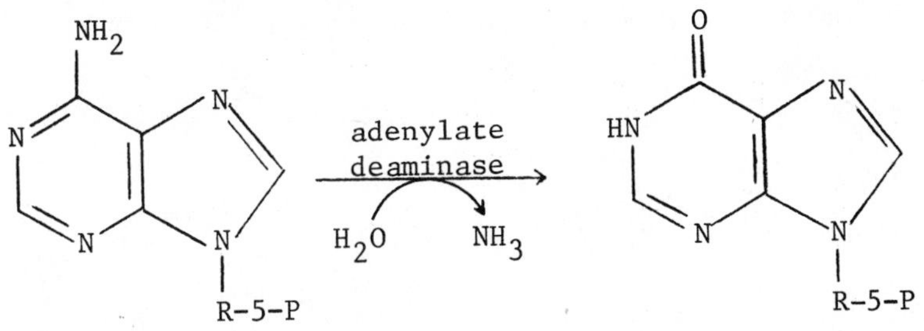

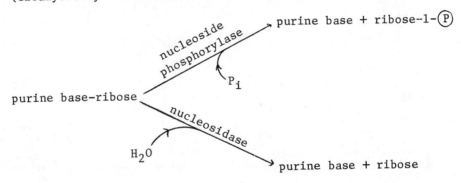

adenosine-5'-$\text{\textcircled{P}}$ inosine-5'-$\text{\textcircled{P}}$

The inosine-5'-$\text{\textcircled{P}}$ is converted to inosine and P_i by a phosphatase.

b.   The purine nucleosides produced by the above reactions next have the N-glycosidic linkage broken. This can be done by addition of P_i (catalyzed by a nucleoside phosphorylase) or of H_2O (catalyzed by a nucleosidase). The general reactions are

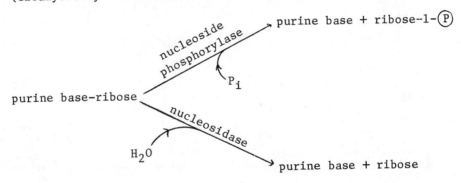

The analogous reactions occur for the deoxyribonucleosides, producing deoxyribose and deoxyribose-1-$\text{\textcircled{P}}$. The sugars are metabolized further or excreted. Note that the nucleoside phosphorylases may also be involved in synthetic (salvage) pathways.

c.   The purine bases (adenine, guanine, and hypoxanthine) are further catabolized in the reactions below.

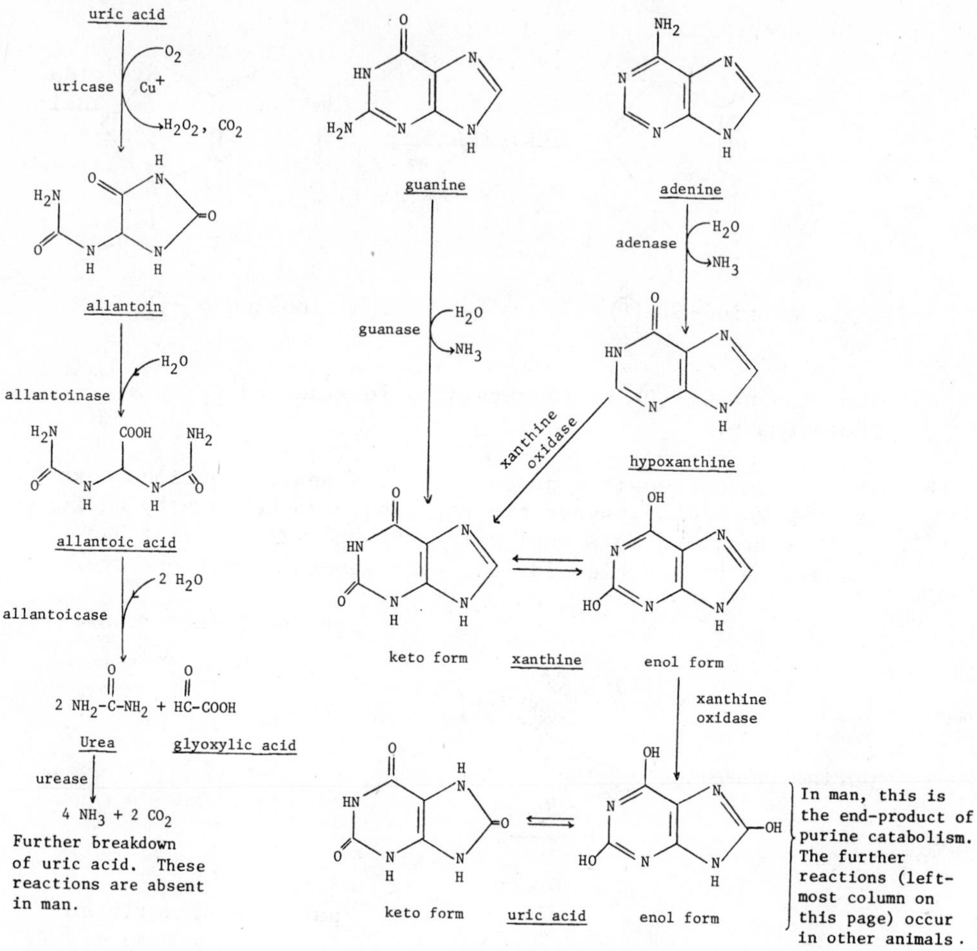

uric acid

uricase ⟨O₂, Cu⁺ ⟩ → H₂O₂, CO₂

allantoin

allantoinase ← H₂O

allantoic acid

allantoicase ← 2 H₂O

$$2\ NH_2\text{-}C\text{-}NH_2 + HC\text{-}COOH$$

Urea   glyoxylic acid

urease

$$4\ NH_3 + 2\ CO_2$$

Further breakdown
of uric acid. These
reactions are absent
in man.

guanine

guanase ← H₂O → NH₃

adenine

adenase ← H₂O → NH₃

hypoxanthine

xanthine oxidase

keto form   xanthine   enol form

xanthine
oxidase

keto form   uric acid   enol form

Steps in purine degradation
which are common to all
animals.

In man, this is
the end-product of
purine catabolism.
The further
reactions (left-
most column on
this page) occur
in other animals.

d. **Xanthine oxidase (X.O.)** is an important and interesting enzyme.
The reactions catalyzed by this complex flavoprotein oxidase
are:

$$Hypoxanthine + H_2O + O_2 \longrightarrow Xanthine + O_2^-$$

superoxide anion

$$Xanthine + H_2O + O_2 \longrightarrow Uric\ Acid + O_2^-$$

The superoxide radical undergoes conversion to hydrogen peroxide by the action of superoxide dismutase. The hydrogen peroxide is then inactivated either by catalase or peroxidase. Xanthine oxidase is inhibited by allopurinal (a purine analog), a drug used in the treatment of gout and discussed in the next section, on gout.

9. Final Excretory Products of Purine Metabolism in Some Animals

Animal Group (examples)	Product Excreted
1. **Some** marine invertebrates, crustaceans	Ammonia
2. Most fish, amphibia (ureotelic animals)	Urea
3. Some teleost fish	Allantoic acid
4. Most mammals other than primates; turtles; insects	Allantoin
5. Primates (including man); dalmatian dogs; birds; some reptiles (uricotelic animals)	Uric acid

10. Certain specific immunodeficiency diseases are associated with defects in enzymes of purine metabolism.

a. Adenosine deaminase deficiency: A relation between deficiency of this enzyme (as assayed in the erythrocytes) and a severe form of combined immunodeficiency (CID) has been observed. The disease is characterized by the onset of severe infections early in life. The deficiency involves both B and T cell-mediated immunity (hence the disease is termed as combined immunodeficiency; refer to Chapter 11 on immunochemistry for discussion of T and B cells). Adenosine deaminase deficiency appears to be transmitted as an autosomal recessive trait. It should be noted that there are disorders of primary immunodeficiency other than CID, which show normal levels of adenosine deaminase. As discussed earlier, adenosine deaminse catalyzes the deamination of adenosine to inosine:

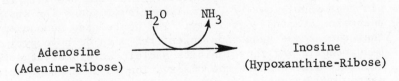

Adenosine
(Adenine-Ribose)

Inosine
(Hypoxanthine-Ribose)

The inosine generated is either reutilized in the purine salvage pathways or converted to uric acid. Evidence for a causal relationship between the enzyme deficiency and the pathogenesis of the immune defect was obtained from a study which showed restoration of immunologic function in lymphocyte cultures of a patient with CID by the addition of exogenous adenosine deaminase. Transfusion of normal erythrocytes to the same patient yielded positive results. The biochemical relationship between the failure to deaminate adenosine and the inhibition of lymphocyte proliferation is not clear at this time. However, it has been reported that high concentrations of adenosine impair in vitro cellular proliferation from lymphoid lines by inhibition of pyrmidine synthesis, with resulting pyrimidine starvation.

b. Purine nucleoside phosphorylase deficiency: This enzyme deficiency has been observed in association with a disorder known as selective cellular immunodeficiency in which there is a severe defect in T-cell immunity but normal B-cell immunity. The inheritance pattern of the disorder appears to be autosomal recessive. As we noted earlier, nucleoside phosphorylase catalyzes the reversible conversion of a nucleoside and inorganic phosphate to free base and ribose-1-phosphate:

$$\text{Inosine} + P_i \rightleftharpoons R\text{-}5\text{-}P + \text{Hypoxanthine}$$
(Hypoxanthine-Ribose)

$$\text{Guanosine} + P_i \rightleftharpoons R\text{-}5\text{-}P + \text{Guanine}$$
(Guanine-Ribose)

$$\text{Xanthosine} + P_i \rightleftharpoons R\text{-}5\text{-}P + \text{Xanthine}$$
(Xanthine-Ribose)

In one patient with this enzyme deficiency, it has been noted that there is overproduction and consequently overexcretion of inosine and guanosine. The patient also showed hypouricemia and hypouricosuria, which are of diagnostic significance. This is due to a lack of production of the free purine bases required for uric acid synthesis. The patients with this disorder show, however, normal levels of adenosine and adenosine deaminase activity. The postulated mechanism is that the excess guanosine interferes with de novo pyrimidine biosynthesis and thus inhibits nucleic acid biosynthesis. This may also explain the observation of a megaloblastic bone marrow and hypochromic, microcytic, peripheral blood.

Gout

1. Characteristics of gout

    a. Gout is characterized by elevated levels of urates (salts of
        uric acid) in the blood (hyperuricemia) and uric acid in the
        urine (uricosuria). Virtually all patients with gout (even those
        who are asymptomatic) have hyperuricemia (urate levels above
        7.0 mg/100 ml of plasma). Significant differences in plasma
        urate levels have been noted between different racial groups.
        For example, among the Maoris (Polynesians native to New Zealand),
        the mean serum urate level in the adult male population is
        greater than 7 mg/100 ml. Although there is a higher incidence
        of gout in the adult male Maori population, hyperuricemia is
        not always sufficient, in itself, to cause gout (see below,
        under primary gout). The disease may remain asymptomatic or
        it may become clinically manifest with the characteristic
        attacks of gouty arthritis, the formation of uric acid stones
        in the kidney, or both.

    b. Many of the clinical symptoms of gout arise from two things.

        i) The relative insolubility of uric acid (the end-product of
            purine metabolism in humans) in aqueous solutions (sodium
            urate is more soluble than the acid itself);

        ii) the absence, in man, of the enzyme uricase, which degrades
            uric acid to the more-soluble compound allantoin
            (mentioned above).

        Thus, the accumulation of uric acid and urates above normal
        levels results in precipitation and the formation of sodium
        urate crystals in the joints, kidney, etc. This leads to the
        typical gout syndrome by the mechanism discussed below.

    c. Primary gout is a genetically determined disorder of purine
        metabolism, seen predominantly in men. The occurrence of gout
        in women is uncommon and when it does occur it is usually
        found in postmenopausal individuals. There may be a metabolic
        or hormonal sexual factor involved. Men normally have a blood
        urate concentration which is about 1 mg/100 ml higher than
        women, a difference which disappears after menopause. The
        different hormonal makeup of men and women may directly
        influence the onset of gouty attacks. In the proposed mechanism
        for episodes of gouty inflammation, small amounts of sodium
        urate initially precipitate and are phagocytized by poly-
        morphonuclear (PMN) leucocytes. After engulfment, the urate

crystals interact with the membranes of the intracellular lysosomes and with the PMN leucocyte cell membrane. This interaction occurs due to the ability of urate to hydrogen bond with groups in the membrane, and it leads to destruction of the membrane and release of the cytoplasmic and lysosomal components into the surrounding fluid. The lysosomal enzymes (proteases, lipases, etc.) proceed to indiscriminately destroy the tissue and inflammation results (see the section on immunochemistry). It is reported that 17-β-estradiol (present in premenopausal women but largely absent in men and postmenopausal women) confers a greater stability to the lysosomal membranes than does testosterone.

Primary gout may be due to an overproduction or an under-excretion of uric acid, or, more likely, to a combination of both. Frequently, siblings and other close relatives of afflicted individuals have high levels of uric acid but do not develop gout, indicating that hyperuricemia is not the only factor involved. Primary renal gout is due to an underexcretion caused by an enzymatic deficiency that affects the uric acid transport. Primary metabolic gout is due to an overproduction of uric acid and its precursors (purines).

d. Secondary gout is an acquired form of the disease, in which gouty arthritis develops as a complication of hyperuricemia caused by another disorder (such as leukemia, chronic nephritis, polycythemia, etc.). This type of hyperuricemia is usually due to high levels of uric acid caused by abnormally rapid turnover of nucleic acids. Notice that here hyperuricemia is sufficient to produce the clinical symptoms of gout.

e. Gout is very rare in children and adolescents. When it does occur, it may represent a unique form of secondary gout. In fact, it is uncommon before the thirties, the greatest incidence of initial attacks being seen in men in their forties or fifties. Such patients have probably had hyperuricemia for many years but did not experience acute attacks of gout until later in life.

f. In the usual development of the (untreated) disease, acute attacks follow the initial appearance with increasing frequency and usually with increasing severity. There is often a build-up of chalky deposits of sodium monourate crystals, either in nodules (called tophi; singular tophus) or diffusely in cartilage and in tissues around the joints. The arthritic attacks of gout are caused by the deposition of large amounts

of urate crystals in the joints, probably in the synovial
fluid leukocytes. This chronic gouty arthritis may be
grotesquely deforming, totally disabling, and excruciatingly
painful. The syndrome has been known since the early Greek
and Roman physicians. The modern history of gout dates from
1683 when T. Sydenham, himself a sufferer from the disease,
chronicled its symptoms in great detail.

g. A diagnosis of gout is strongly indicated by the following
    parameters:

    i.   In males or postmenopausal females over the age of 40,
         the occurrence of acute distal-joint pain (frequently
         in the big toe), with attacks brought on by stress or
         trauma. The pain can be relieved by treatment with
         colchicine.

    ii. Hyperuricemia (> 7 mg/dl in male plasma)

    iii. Demonstration in the laboratory of negative, birefringent
         monosodium urate crystals in tophi or synovial fluid.
         These can be observed in the polarizing microscope.
         Calcium pyrophosphate crystals, which occur in cases of
         pseudogout, can be distinguished from monosodium urate
         crystals because the latter appear yellow when they are
         parallel to the axis of the polarizing lens, whereas
         the pyrophosphate crystals appear blue.

2. Drugs used to treat gout

    a. Colchicine (see also the section on microtubules) is used to
        treat the pain and inflammation of an attack of acute gouty
        arthritis. Other anti-inflammatory drugs such as
        adrenocorticotropin, phenylbutazone, corticoids, oxyphenbutazone
        and indomethacin are also helpful.

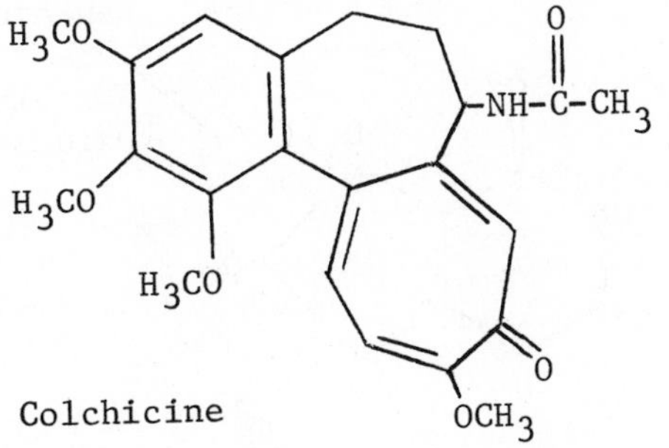

Colchicine

The mechanism by which colchicine relieves acute gouty attacks is obscure. Microtubules are involved in leukocytic motion and it is colchicine which apparently disrupts such motile systems. Although this is a possible mode of action, it remains to be demonstrated that this occurs in vivo.

b. **Allopurinol**, which is structurally similar to hypoxanthine, inhibits the enzyme xanthine oxidase and thus reduces the formation of xanthine and of uric acid. Xanthine oxidase **slowly oxidizes the allopurinol to alloxanthine (oxypurinol)** which is also a xanthine oxidase inhibitor. These interactions are shown diagrammatically below and are discussed further in the section on enzyme inhibition.

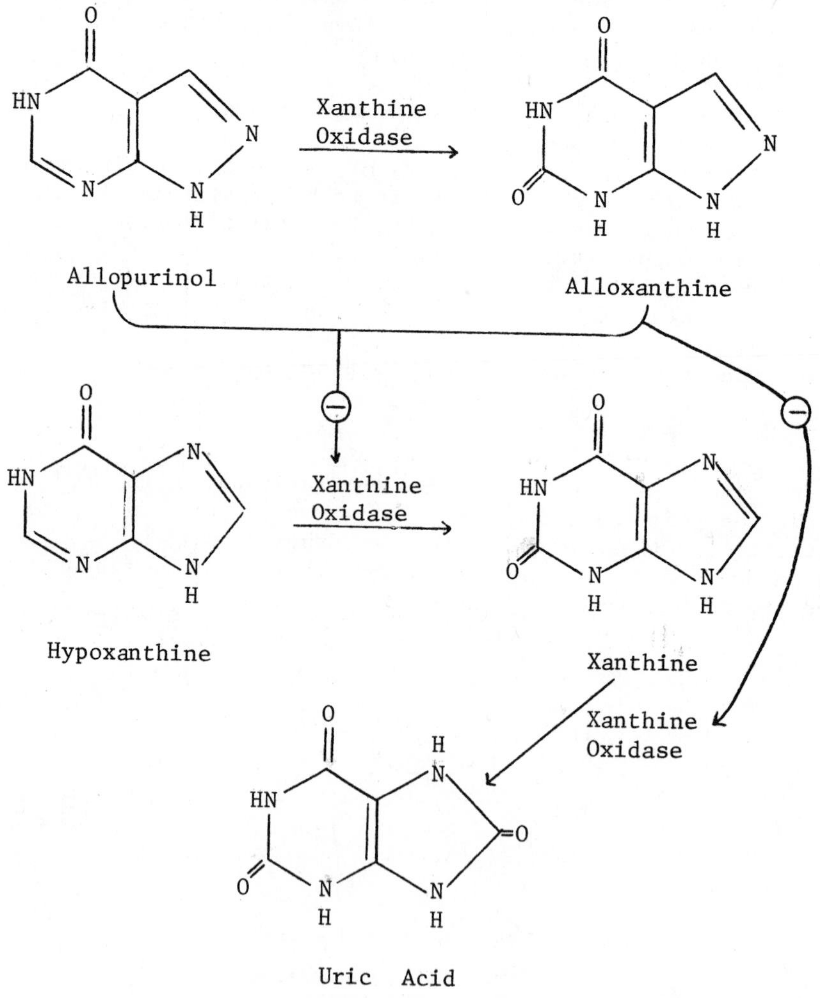

Allopurinol

Alloxanthine

Hypoxanthine

Xanthine

Xanthine Oxidase

Uric Acid

Figure 66. A Proposed Mechanism for the Pathogenesis of Acute Gouty Arthritis*

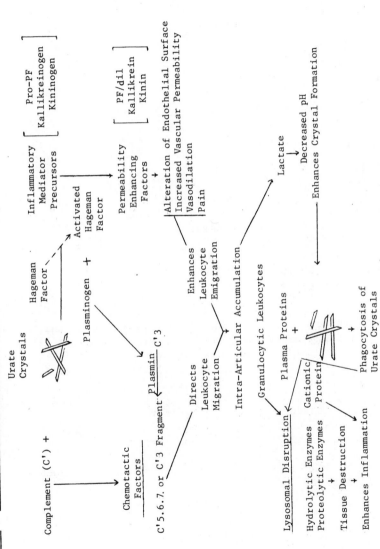

Pro-PF = Propermeability factor; PF/dil = permeability factor/diluted serum. Some of the factors indicated above (complement system, kallikrein, Hageman factor, leukocytic function) are discussed in the chapter on immunochemistry. Note that the urate crystals appear to function in two primary ways: (1) by directly and indirectly activating the complement systems; and (2) through activation of the Hageman factor. Both routes lead to the same ultimate effect.

* Reproduced with permission from "Hageman Factor and Gouty Arthritis", R.W. Kellermeyer, Arthritis and Rheumatism 11, 453, 1968).

Allopurinol is also converted to allopurinol ribonucleotide. This decreases uric acid production by decreasing overall purine synthesis through depletion of PRPP, a substrate used in the initial step of this pathway. Additionally, the allopurinol ribonucleotide allosterically inhibits PRPP-amidotransferase, the first enzyme needed for de novo purine biosynthesis, by a feedback mechanism. These effects are illustrated below.

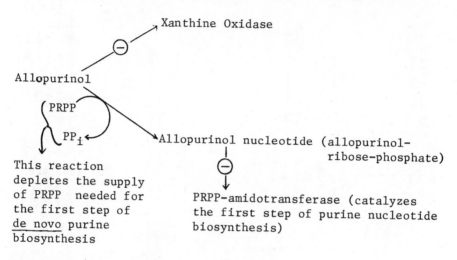

(PRPP + Gln $\longrightarrow$ 5-phosphoribosylamine + Glu + PP_i)

Since xanthine is both a product and a substrate for xanthine oxidase, allopurinol treatment could be expected to cause its accumulation in the body. Since this compound is only sparingly soluble in urine (but more soluble than uric acid), this could lead to the development of urinary xanthine crystalluria or lithiasis. This complication has not been observed in patients receiving the drugs for treatment of gout or uric acid stones, but it has occurred in two patients with Lesch-Nyhan Syndrome and one with lymphosarcoma.

c. A large number of drugs produce uricosuria in man. One such compound is probenacid which is effective both in the

regulation of hyperuricemia and for the resolution and
prevention of tophi. Another drug which produces similar
effects is <u>sulfinpyrazone</u>. Both of these agents are weak
organic acids like uric acid. They probably act as competitive
inhibitors of tubular reabsorption of uric acid, thus increasing
its excretion.

d. Acute gouty arthritic attacks may sometimes be precipitated by
sudden alterations of uric acid levels by antihyperuricemic
agents. This apparently paradoxical effect of a sudden
<u>lowering</u> of uric acid levels leading to an increase in the
likelihood of developing gouty arthritis is well documented
but not at all understood.

3. <u>Hyperuricemia</u>, which can lead to gout, is produced by a variety of
causes and only in a few instances have the biochemical lesions
been delineated. The presence of high levels of uric acid in a
majority of the individuals who have gout as a clinical disorder
is not explained by an established defect in a biochemical process.
Lesions that may result in the overproduction of uric acid are
listed below.

a. <u>Reactions leading to an overproduction of PRPP</u>

i) Glutathione reductase variant. The enzyme glutathione
reductase catalyzes the following reaction (see also under
erythrocyte metabolism):

$$GSSG + NADPH + H^+ \longrightarrow 2\ GSH + NADP^+$$

The NADPH required in this reaction is produced in the
hexose monophosphate shunt pathway (already discussed,
see page 245). In a few patients with gout, it has been
observed that there is an increased activity of glutathione
reductase. This puts a drain on the supply of NADPH and,
in a cascade process, stimulates the HMP shunt, increases
R-5-P production (R-5-P is a product of the HMP shunt), and
provides more PRPP for purine nucleotide biosynthesis.
This leads to an overproduction of purine nucleotides and
eventually to hyperuricemia. These interrelationships can
be diagrammatically shown as below.

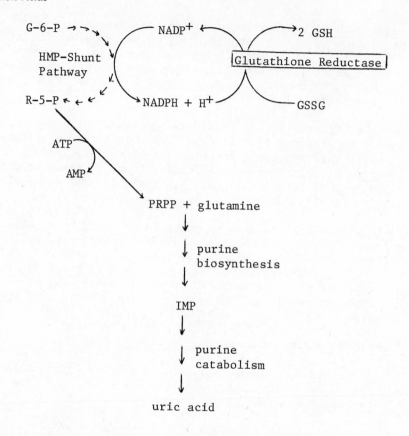

b. <u>Reactions leading to the overproduction of glutamine</u>. Glutamine is produced as follows:

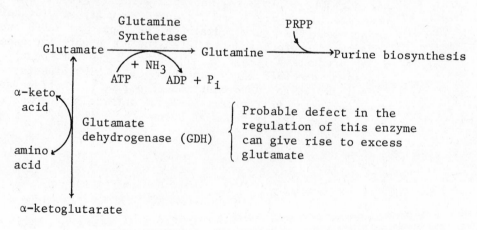

(The details of these reactions will be discussed under amino acid metabolism)

In the above relationships, the GDH-catalyzed reaction is favored in the direction of synthesis of α-ketoglutarate from glutamic acid. Elevated plasma glutamic acid levels have been observed in gouty patients, suggesting the possibility of a defect in the control of GDH activity. It is presumed that the increased amount of glutamic acid is diverted to glutamine synthesis and purine production.

c. Abnormalities in the function of glutamine-PRPP-amidotransferase. This enzyme, an allosteric protein, catalyzes one of the rate controlling steps in purine biosynthesis. Hence, changes in the level of this enzyme or defects in its control can lead to enhanced production of uric acid.

d. Deficiency of Hypoxanthine/Guanine Phosphoribosyl-Transferase (HGPRT). This enzyme participates in the salvage (reutilization) pathway of nucleotide formation.

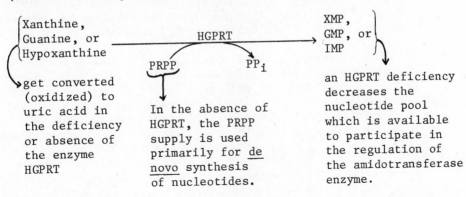

Xanthine, Guanine, or Hypoxanthine $\xrightarrow[\text{PRPP} \quad \text{PP}_i]{\text{HGPRT}}$ XMP, GMP, or IMP

get converted (oxidized) to uric acid in the deficiency or absence of the enzyme HGPRT

In the absence of HGPRT, the PRPP supply is used primarily for de novo synthesis of nucleotides.

an HGPRT deficiency decreases the nucleotide pool which is available to participate in the regulation of the amidotransferase enzyme.

As is indicated above, in the discussion of salvage pathways, an HGPRT deficiency or absence can lead to elevated levels of uric acid. There are two categories of HGPRT enzyme defects which have been observed.

i) Virtually complete deficiency of the enzyme occurs in the Lesch-Nyhan Syndrome. This is an X-linked disorder, affecting only males. It is characterized by excessive production of uric acid leading to gouty arthritis, urate overexcretion, and the formation of renal stones. Neurological disorders such as mental retardation, a tendency towards self-mutilation, spasticity, and choreoathetosis are also apparent to some degree in all cases. The relationship between the neurological symptoms and the deficiency of the salvage enzyme is not clear at present.

ii) A partial deficiency of the enzyme has been observed in some adult hyperuricemic individuals. However, extensive epidemiological studies correlating the enzyme activity with uric acid levels in patients with gout is not available at this time.

e. <u>Excessive production of organic acids leading to elevated levels of uric acid.</u> Lactate, acetoacetate, and β-hydroxy-butyrate (the latter two are known as ketone bodies) compete with uric acid for secretion in the kidneys. Thus, elevated levels of these organic acids reduce the uric acid excretion and can increase its levels in the body fluids. Excess lactic acid production (lactic acidosis) can occur in G-6-phosphatase deficiency and alcohol ingestion. Excess ketone bodies are formed in untreated diabetes mellitus, starvation, G-6-phosphatase deficiency, etc. Additionally, since alcohol oxidation produces NADH and acetyl-CoA, glycolytic activity is decreased and glucose is metabolized to R-5-P in the HMP shunt under conditions of excessive ethanol ingestion.

Although untreated diabetes mellitus, starvation, and G-6-P phosphatase deficiency all cause an elevation of the ketone bodies in the blood, in diabetes, the hyperglycemia actually can increase the glomerular filtration rate and thereby improve uric acid elimination. As was pointed out in the discussion of diabetes, its relationship to gout is not yet clearly understood.

f. A summary of the known biochemical defects which can result in hyperuricemia is presented in Figure 67. For more details, refer to the preceding material.

g. Several studies, particularly the one conducted over a period of 25 years in Framingham, Massachusetts, have shown that hyperuricemia falls into several groups.

   i. Gouty arthritis occurs in 90% of persons with serum urate levels of 9 mg/dl or higher (2 mg/dl above the normal level). The incidence drops to 25% in those with values between 8 and 9 mg/dl. These patients were also at risk for the development of urate nephropathy, a renal disorder frequently associated with gout. This observation strongly suggests a relationship between sustained serum urate values in excess of 9 mg/dl and the risk of developing gout.

   ii. Five to eight per cent of the general population have asymptomatic hyperuricemia. The importance of this is not clear. Uric acid can be decreased to normal amounts by drug treatment, although in some obese hyperuricemics, caloric restriction and weight reduction may obviate the need for drugs.

   iii. Regardless of the magnitude of the hyperuricemia, if a person excretes more than 700 mg of uric acid per day in the urine, there is a greater than 34% chance that he will suffer from renal stones. These patients are also at risk for gouty arthritis.

   iv. There are, then, two groups for whom antihyperuricemic drug therapy may be appropriate: Those with serum urate levels in excess of 9 mg/dl; and those who excrete more than 700 mg urate per day in the urine.

Figure 67. A Composite Diagram Depicting the Biochemical Lesions that May Lead to Hyperuricemia

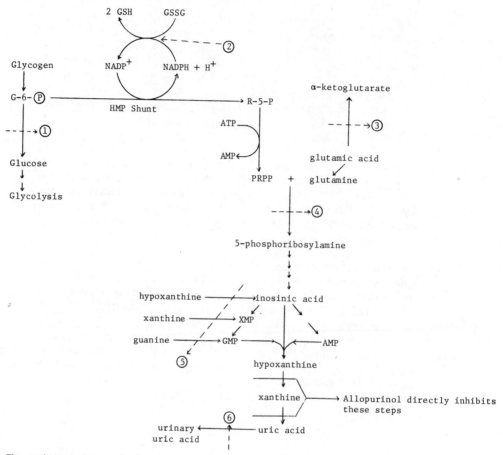

The numbers refer to the key below

Number	Enzyme	Comments
1	glucose-6-phosphatase deficiency	leads to excess PRPP production
2	glutathione reductase overactivity	leads to drain on NADPH and consequent overproduction of PRPP
3	glutamate dehydrogenase deficiency	oversupply of glutamine for purine biosynthesis
4	PRPP-amidotransferase overactivity (loss of control)	increased rate of purine biosynthesis
5	hypoxanthine-guanine phosphoribosyl transferase decrease or absence	degradation of nucleotide bases to uric acid rather than salvage; enzyme is completely absent in Lesch-Nyhan Syndrome patients
6	overproduction of other organic acids	decreased secretion of uric acid due to competition with other organic acids in the kidneys

528

Biosynthesis of Pyrimidine Nucleotides

1. The first step in the pathway is the formation of carbamyl phosphate:

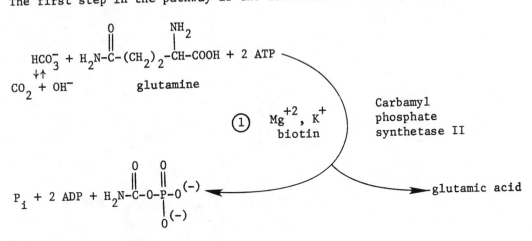

carbamyl phosphate

2. Formation of the pyrimidine ring:

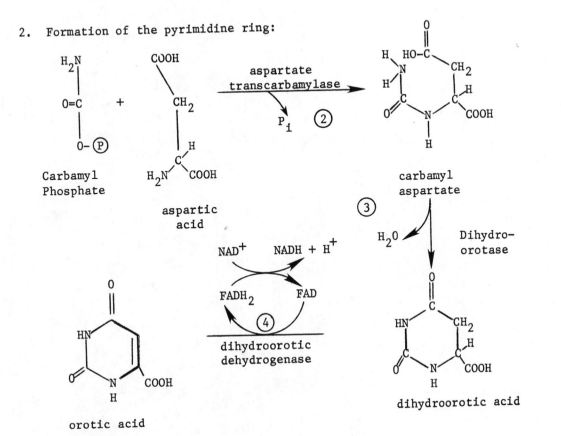

3. Formation of UMP

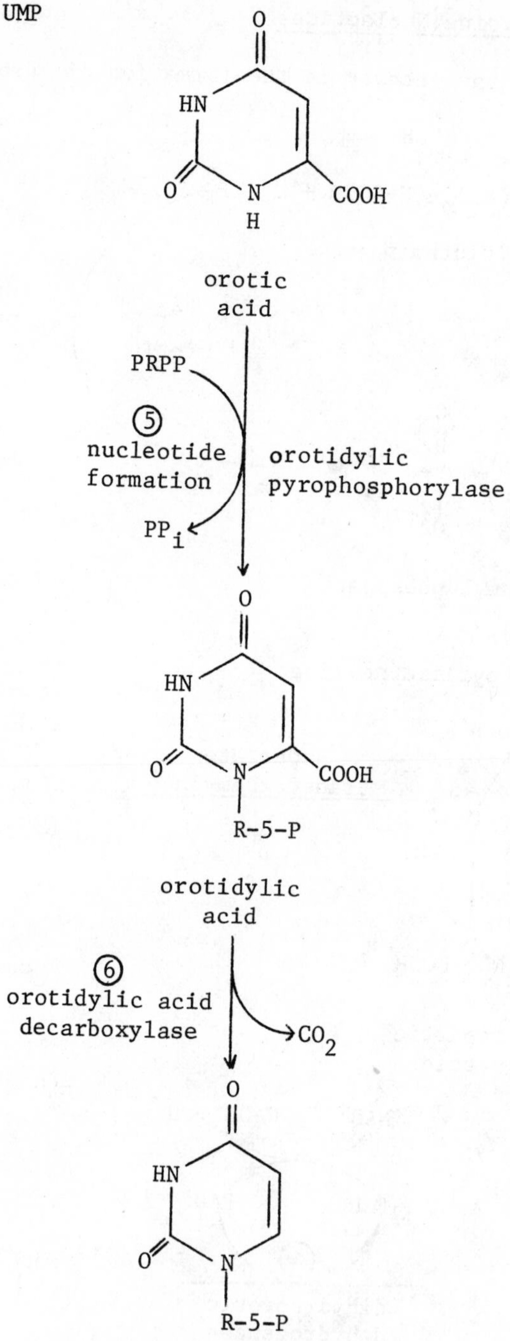
O
HN
O
N
H
COOH
orotic acid
PRPP
5
nucleotide formation
PP$_i$
orotidylic pyrophosphorylase
O
HN
O
N
COOH
R-5-P
orotidylic acid
6
orotidylic acid decarboxylase
CO_2
O
HN
O
N
R-5-P

uridylic acid (UMP)

4. Conversion of UMP to CTP, dUMP, TMP, TTP, dCMP, dCTP, and 5-hydroxymethyldeoxycytidylic acid

Uridine-5'-$\textcircled{P}$ (UMP)

deoxyuridine-5'-pyrophosphate (dUDP)

Ribonucleoside diphosphate reductase

UMP kinase Mg^{+2} ATP → ADP

$NADP^+ + H_2O$

$NADPH + H^+$

UDP

phosphatase P_i

dUMP

nucleoside diphosphate kinase Mg^{+2} ATP → ADP

N^5, N^{10}-methylene FH_4 Mg^{+2} FH_2

thymidylate synthetase

UTP

CTP synthetase Gln → Glu

ATP + H_2O → ADP + P_i

thymidylic acid (TMP)

deoxyribose-5-$\textcircled{P}$

N^5, N^{10}-methylene FH_4 FH_2

thymidylate synthetase

2ATP kinases 2ADP

TTP

cytidine triphosphate (CTP)

Ribose-5-$\textcircled{P}$-$\textcircled{P}$-$\textcircled{P}$

dUMP

AMP → ADP

NH_3 H_2O

dCMP aminohydrolase

CDP

ribonucleoside diphosphate reductase Mg^{+2}; ATP $NADPH + H^+$ → $NADP^+ + H_2O$

dCDP

phosphatase → dCMP P_i

dHMP synthetase

N^5, N^{10}-methylene-FH_4 FH_4

5-hydroxymethyldeoxycytidylic acid (dHMP)

deoxyribose-5-$\textcircled{P}$

ATP → ADP

dCTP

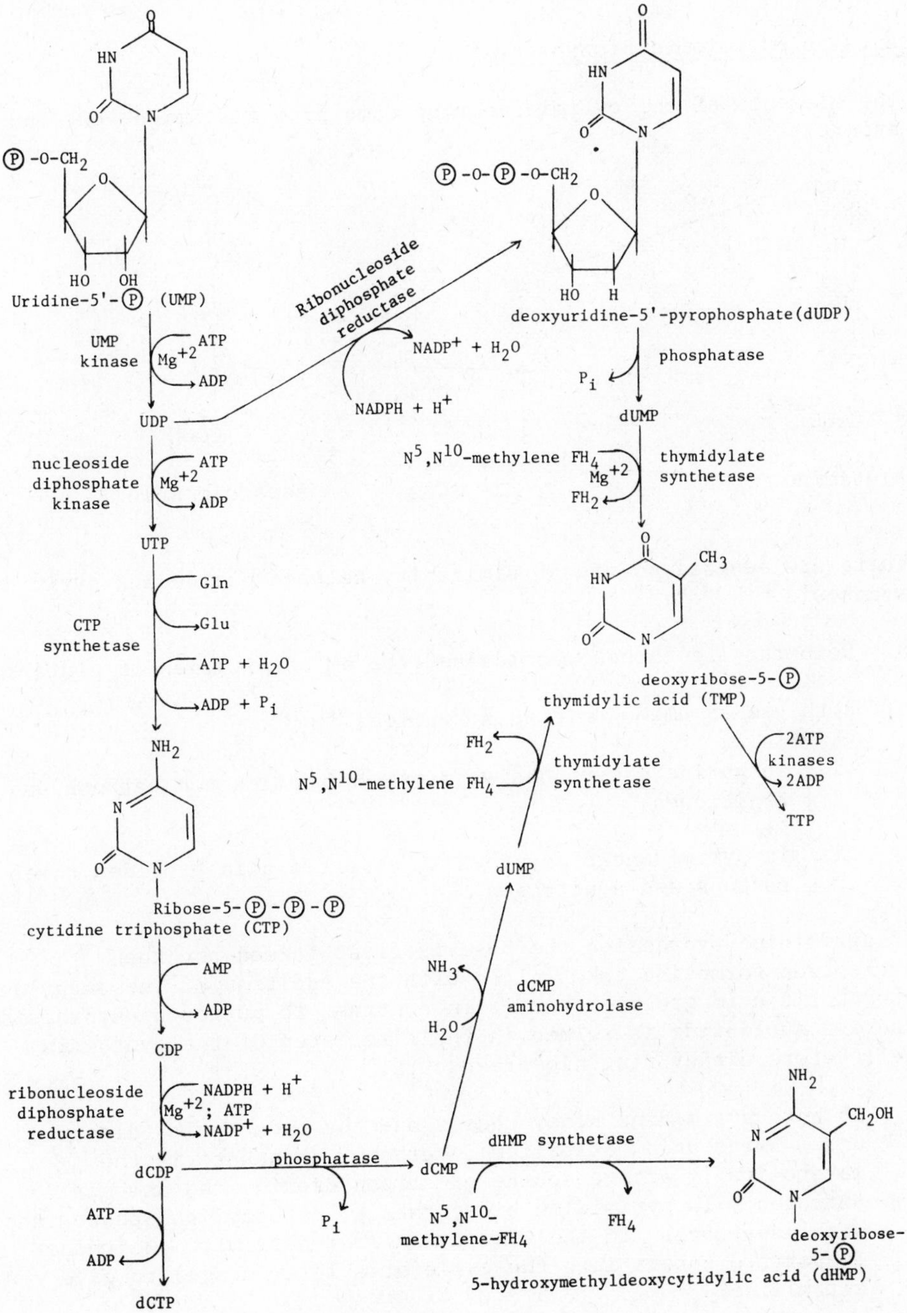

5. The nucleoside and deoxynucleoside triphosphates, needed for nucleic acid synthesis, are formed from the mono- and diphosphates by kinase enzymes in reactions entirely analogous to those shown for the purine nucleotides.

Comments on Pyrimidine Biosynthesis

1. The elements of the pyrimidine ring come from glutamine, CO_2 and aspartate.

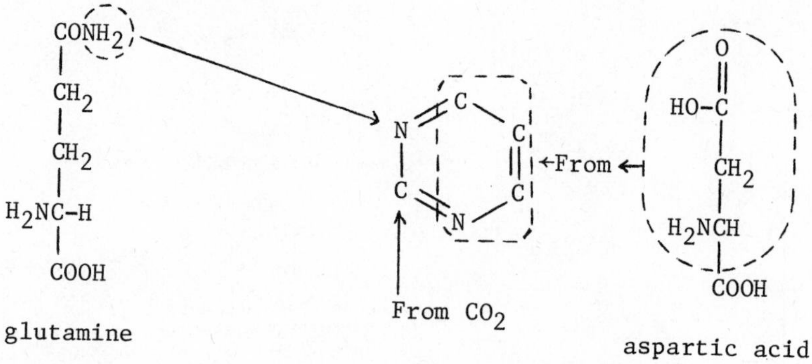

$CONH_2$
|
CH_2
|
CH_2
|
H_2NC-H
|
$COOH$

glutamine

From CO_2

O
||
HO-C
|
CH_2
|
H_2NCH
|
COOH

←From←

aspartic acid

There are several points of similarity between purine and pyrimidine synthesis.

a. Both require carbon dioxide and the amide nitrogen of glutamine.

b. Both use an amino acid as a "nucleus."

    i) In purine biosynthesis, glycine supplies two carbons and a nitrogen.

    ii) In pyrimidine biosynthesis, aspartic acid provides three carbons and a nitrogen.

2. In pyrimidine synthesis, the base is first formed and then nucleotide formation takes place with the addition of the sugar and phosphoric acid group. This is in contrast to purine biosynthesis, where a nucleotide is formed in the first step of the synthesis well before either ring closure occurs.

3. a. In both pyrimidine biosynthesis and the urea cycle (discussed under amino acid metabolism), carbamyl phosphate is the metabolically active source of carbon dioxide and ammonia nitrogen. In pyrimidine biosynthesis, carbamyl phosphate reacts with aspartate. In the urea cycle, it reacts with ornithine (aspartate enters into the cycle at a later stage) to give citrulline.

b. Within the eucaryotic cell there are two separate pools of carbamyl phosphate and two separate enzymes, designated carbamyl phosphate synthetase (CPS) (I) and (II). CPS (I) provides carbamyl phosphate for the urea cycle. It is strictly a mitochondrial enzyme, requires N-acetylglutamate as a positive allosteric activator, and can use either NH_3 or the amide nitrogen of glutamine as a nitrogen source. CPS (II) is found in the cytoplasm and maintains the carbamyl phosphate pool used in pyrimidine nucleotide biosynthesis. Its activity is not influenced by N-acetylglutamate but, unlike CPS (I), it is inhibited by glutamine analogues such as D-carbamyl-L-serine. CPS (II) can only use the amide nitrogen of glutamine as a nitrogen source.

c. The presence of these two physically separated enzymes, each with its own properties, probably reflects the need for the cell to be able to control pyrimidine biosynthesis and amino acid catabolism independently of each other, despite the fact that both pathways require carbamyl phosphate. E. coli is unusual in this respect, having apparently only one carbamyl phosphate synthetase to supply both pathways. This may be reflected in a difference in the control steps in procaryotes and eucaryotes (see below).

d. Although there is an absolute biotin requirement in the CPS reaction, a biotin deficiency is difficult to produce, presumably due to biotin synthesis by the intestinal bacteria. Biotin can, however, be inactivated by avidin, a protein (molecular weight about 45,000 daltons) found in raw egg white. Avidin is readily inactivated by heating. The nature of its interaction with biotin is not known.

4. The conversion of dihydroorotic acid to orotic acid (reaction ④) is catalyzed by a metalloflavoprotein dehydrogenase. This enzyme contains FAD as a prosthetic group. Following reduction of the protein-bound flavin to $FADH_2$, the complex is reoxidized by NAD^+, forming $NADH + H^+$.

5. a. Thymidylate synthetase, the enzyme which catalyzes formation of TMP from dUMP, has been best studied in L. coli. The enzyme isolated from this bacterium is similar to that from L. casei, in that both use N^5,N^{10}-methylene FH_4 as a methyl source and as a reducing agent (note that FH_2 is released in the reaction, rather than FH_4).

b. One substrate for thymidylate synthetase, dUMP, comes either from dUDP (by action of a phosphatase) or from dCMP (in a reaction catalyzed by dCMP aminohydrolase). dCMP aminohydrolase is activated by dCTP and inhibited by dTTP, presumably by an allosteric mechanism. In regenerating rat liver and other rapidly growing tissues, the aminohydrolase is very active while in normal tissue it is almost undetectable. This, and its inhibition by dTTP, suggest that it may be a key control point in DNA synthesis, along with thymidine kinase (discussed below under salvage pathways).

6. In a group of viruses known as the T-even bacteriophages, 5-hydroxymethyldeoxycytidylic acid (5-HMP) replaces thymine in the DNA. Note that it differs from thymine only by the presence of a hydroxyl group on the methyl substituent at the 5-position of the cytidine ring.

7. Regulation of pyrimidine biosynthesis

   a. Control of pyrimidine biosynthesis has been much more thoroughly studied in bacteria than in eucaryotes. In bacteria, several important examples of negative allosteric feedback control have been elucidated. For example, aspartate transcarbamylase (reaction ② in the synthetic pathway) and orotidylic acid decarboxylase (reaction ⑥) are both inhibited by UMP. This is negative feedback, since a product (UMP) controls its own production. Aspartate transcarbamylase is a particularly well-studied example of an allosteric enzyme which has both control and catalytic capabilities. Substrates and purine nucleotide act as positive modulators of this enzyme while pyrimidine nucleotides (products of the pathway) are negative effectors.

   b. In mammalian systems, there is evidence that carbamyl phosphate synthetase (II) is the site of feedback inhibition by the pyrimidine nucleotides. Because there are two CPS enzymes in mammals, the formation of pyrimidine-channeled carbamyl phosphate is the first unique (committed) step in mammalian pyrimidine biosynthesis. If this was the control point in bacteria which have only one CPS (see above), both pyrimidine biosynthesis and the urea cycle would have to be regulated simultaneously, which is not necessarily desirable. The committed step (first unique reaction) in bacterial pyrimidine biosynthesis is catalyzed by aspartate transcarbamylase. As described above, this is the feedback inhibition point of this pathway in bacteria. The logic of these processes, to conserve material and energy at the earliest feasible step, is very important to recognize.

   c. Also in mammals, dihydroorotic dehydrogenase (reaction ④) is inhibited by several purines and pyrimidines; and orotidylic acid decarboxylase (reaction ⑥) is inhibited by UMP as it is in E. coli.

d. In cultured human lymphocytes during phytohemagglutinin-induced blastogenesis, carbamyl phosphate synthetase II (glutamine-utilizing; CPSII) and aspartate transcarbamylase, the initial enzymes of the de novo pyrimidine biosynthetic pathway, are co-induced in an associated form. Addition of adenine, guanine, and guanosine to this system leads to inhibition of the synthesis of these enzymes. The toxic effect of adenosine is attributed to the inhibition of de novo pyrimidine synthesis by the phosphorylated derivatives of adenosine at orotidylic acid formation. The immunodeficiency diseases (already discussed in purine metabolism) associated with defects in enzymes of purine metabolism show elevated levels of purine nucleosides. The toxicity may be due to an inhibition of pyrimidine biosynthesis by these nucleosides.

It should be emphasized that a balanced synthesis of pyrimidine and purine nucleotides is essential for the biosynthesis of the nucleic acids in growing cells. As noted above, purine nucleotides regulate the biosynthesis of pyrimidine nucleotides at orotidylic acid formation. Phosphoribosyl-1-pyrophosphate (PRPP) also plays a significant role in maintaining a balance between the rate of purine and pyrimidine biosynthesis. It is an allosteric activator of carbamyl phosphate synthetase II and a precursor for the synthesis of 5-phosphoribosyl-1-amine (the first step in de novo purine biosynthesis).

8. a. While salvage pathways for purines and purine nucleosides are well established, reutilization of pyrimidines by similar pathways has not been described in any detail. That pyrimidine salvage does occur is evident in the treatment of orotic aciduria. This disease (discussed below) results from a defect in the pyrimidine-synthetic pathway. Many of its symptoms can be relieved by treatment with uridine which must be utilized via a salvage route.

b. The phosphorylation of nucleosides may also be regarded as a salvage pathway. An important example is the reaction catalyzed by thymidine kinase.

$$\text{deoxythymidine} + \text{ATP} \longrightarrow \text{thymidylic acid} + \text{ADP}$$

This enzyme has very low activity in normal rat livers. In partially hepatectomized animals, however, its activity is greatly elevated. There is evidence that it may be one of the rate-controlling enzymes in DNA synthesis. It is inhibited by dTTP (a feedback-process) and by dCTP.

9. Certain pyrimidine derivatives serve as powerful antimetabolites:

a. The metabolism of 5-Fluorouracil (5FU) is shown below.

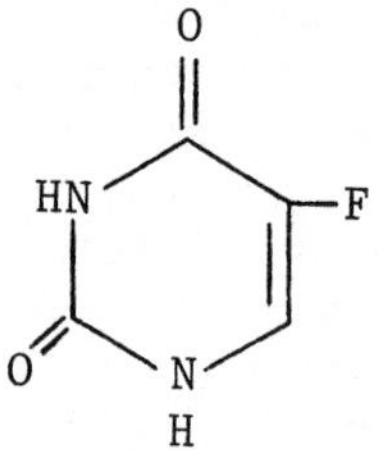

5-Fluorouracil

(i)   5-FU → 5-FU-Ribonucleoside
             ↓
      5-FU-Ribonucleotide
             ↓
      FUdR → 5'-fluoro-2'-deoxyuridine-5'- (P)
                        (F-dUMP)
             (inhibits thymidylate synthetase)

(ii) F-dUMP inhibits (both competitively and non-competitively) the <u>conversion of dUMP to TMP</u>. Note that, in the above reaction, the active, selectively toxic agent is F-dUMP which can also be directly derived from fluorodeoxyuridine (5-fluorouracil deoxyribose; FUdR) by the action of deoxyuridine kinase and ATP.

(iii) Both compounds (5FU and FUdR) are used in the treatment of carcinomas of the ovary, breast, colon, etc.

(iv) 5-Fluoro-deoxyuridylate (F-dUMP) inhibits thymidylate synthetase by binding strongly to the active center of the enzyme in the presence of the cofactor N^5,N^{10}-methylene FH_4.

b. 5-Iododeoxyuridine (IRdR). This compound has no significant action on thymidylate synthetase but the phosphorylated derivative of IUdR can be incorporated into DNA in the place of thymidylic acid and thus interferes with the growth of the cell. <u>IUdR is used in viral infections</u> such as vaccinia and herpes.

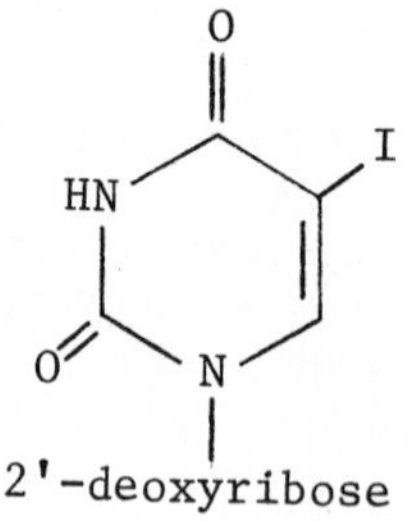

5-iododeoxyuridine

2'-deoxyribose

c. 1-β-D-arabinofuranosyl cytosine (cytosine arabinoside,
    cytarabine) blocks the formation of dCMP and inhibits
    biosynthesis of DNA. It is suggested that cytarabine <u>inhibits
    the conversion of ribonucleotides to deoxyribonucleotides.</u>
    The drug is used to treat acute lymphocytic and acute
    myelocytic leukemias.

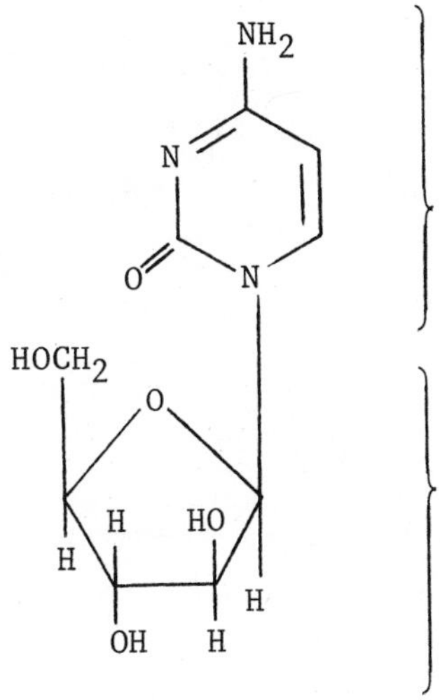

NH_2

cytosine

$HOCH_2$

arabinose: note the
similarity of this sugar
to 2'-deoxyribose

d. Azacytidine, an analogue of cytidine used in the treatment of
    acute granulocytic leukemia, is thought to affect DNA, RNA,
    and protein synthesis.

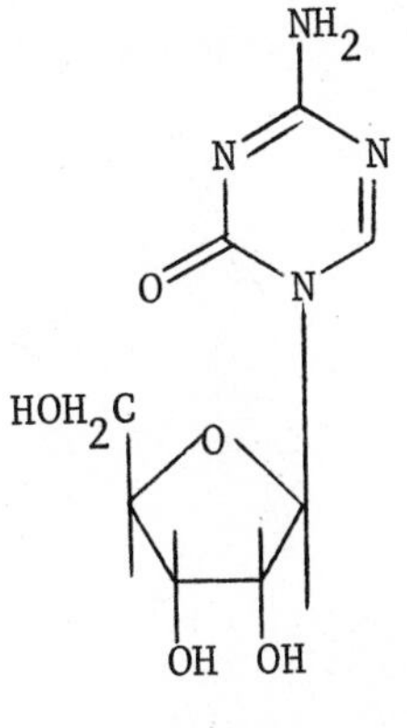

NH_2

Structure of 5-Aza-Cytidine

HOH_2C

e. We have already noted that methotrexate and its related
    analogues, aminopterin-4-amino folic acid and 3'-5'-dichloro-
    methotrexate, are potent inhibitors of dihydrofolate reductase

in purine biosynthesis. In pyrimidine biosynthesis, the thymidylate synthetase reaction is inhibited by methotrexate, since this conversion requires N^5,N^{10}-methylene FH_4:

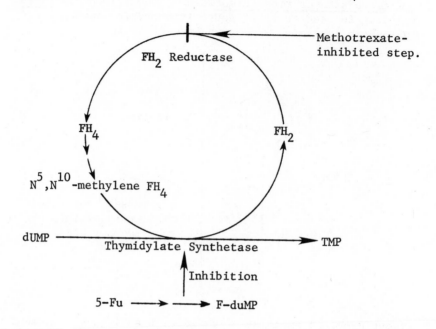

5-Fluorouracil's action on thymidylate synthetase is also indicated.

10. <u>Orotic aciduria</u> is an inherited metabolic disease caused by biochemical lesions in pyrimidine metabolism. In the affected child, it is characterized by an increase in the excretion of orotic acid due to a defect in pyrimidine biosynthesis. In the type I disease, both orotidylic acid pyrophosphorylase (reaction 5) and orotidylic acid decarboxylase (reaction 6) are present but decreased markedly in activity. In the type II disorder, pyrophosphorylase activity is normal or elevated, but the decarboxylase is almost entirely absent. The single patient observed so far to have the type II disease had both orotic aciduria and orotidinuria. Patients also display hypochromic erythrocytes and megaloblastic marrow which are unrelieved by iron, pyridoxine, B_{12}, folic acid or ascorbic acid. Leukopenia is also present. Treatment with uridine (2 to 4 gm/day) results in an excellent reticulocyte response, rise in hemoglobin, an increase in growth, etc. It is important to emphasize that these individuals are <u>dependent</u> upon exogenous pyrimidine supply in order to maintain normal growth and development. This is identical to the need for vitamins and essential amino and fatty acids in all humans. Antagonists to orotidylic acid decarboxylase (6-azauridine and allopurinol) will produce transient orotic aciduria.

11. Since folate and vitamin B_{12} are involved in thymidylic acid synthesis, their deficiencies can also cause megaloblastic anemia.

12. The catabolic pathways for the pyrimidine bases derived from DNA and RNA are shown next. Note that cytosine is first converted to uracil, then further degraded. Presumably the free bases are obtained from the nucleotides in reactions similar to those shown for the purine nucleotides.

Pyrimidine Catabolism

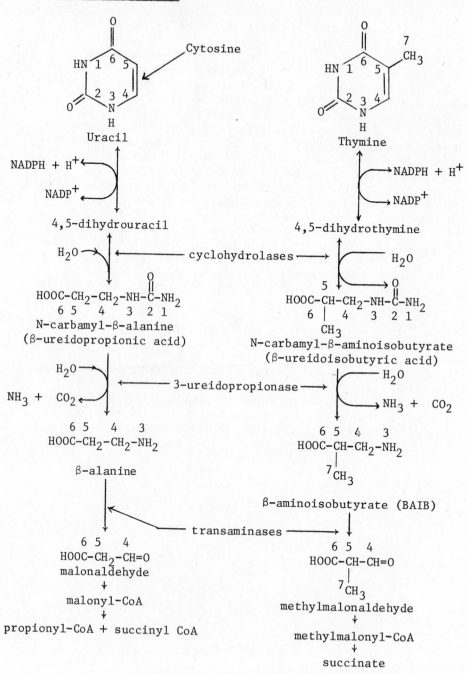

540

High levels of BAIB are found in urine when excessive tissue destruction takes place. Also, in the Oriental population, familial high-excretors often have been reported, the significance of which is not understood.

Biosynthesis of Deoxyribonucleotides

1. The pathways for TTP synthesis have been discussed and the formation of dCTP has been mentioned previously. Both can be found in the section on pyrimidine nucleotide biosynthesis.

2. In E. coli, the nucleoside diphosphates are the substrates for the reductase; in Lactobacillus leichmanii the triphosphates function in this capacity and, in addition, a cobamide coenzyme is required. Although much work remains to be done, the mammalian system seems most similar to that occurring in E. coli. The enzyme system shown below is from E. coli.

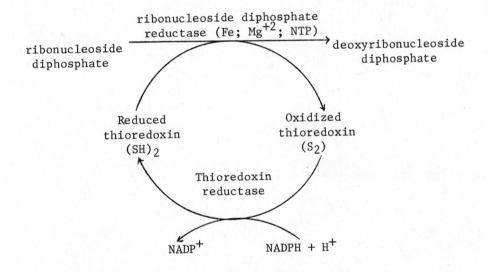

Different nucleoside triphosphates (NTP's) are required for the reduction of the various ribonucleoside diphosphates, implying that they are needed as allosteric effectors rather than as an energy source. For example, ATP must combine with the reductase for it to catalyze the conversion of CDP to dCDP.

3. Control of deoxyribonucleotide biosynthesis appears to be complex.
    Ribonucleoside diphosphate reductase (RDR) is inhibited by
    hydroxyurea and similar compounds, apparently through their
    interactions with the non-heme iron present in RDR. Other effects,
    probably all allosteric, are summarized below. Further information
    on some of these conversions is given under pyrimidine biosynthesis.

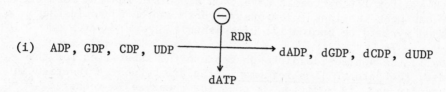

(i)   ADP, GDP, CDP, UDP ———————→ dADP, dGDP, dCDP, dUDP

               RDR; ⊖ ; dATP

(ii) TdR ———————→ TMP

       TK; ⊖ ; dTTP, dCTP

(iii) dCMP ———————→ dUMP

       dCMPAH; ⊖ ; dTTP

(iv) CDP, UDP ———————→ dCDP, dUDP

       ATP; RDR; ⊕ ; TMP ; TTP

(v)   ADP, GDP ———————→ dADP, dGDP

       TTP; RDR; ⊕

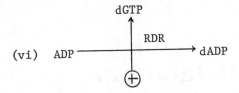

The enzyme abbreviations used in these reactions are: RDR = ribonucleotide diphosphate reductase; TK = thymidine kinase; and dCMPAH = dCMP amino hydrolase. Regarding these reactions, notice that (i-iii) all involve feedback inhibition and tend to prevent the accumulation of unneeded nucleoside triphosphate in the cell. Reactions (iv-vi) serve to maintain a balance between purines and pyrimidines and between the two members of each group. It should be remembered, however, that these control properties were all worked out using in vitro systems. The extent to which they apply in vivo remains to be demonstrated. The controls on reactions (ii) and (iii) have been demonstrated in mammalian systems while those in reactions (iv-vi) have been shown only in E. coli. (The work on the synthesis and control of deoxynucleosides in E. coli has been largely worked out by P. Reichard and his collaborators in Stockholm.) In reaction (i), dATP inhibits all four reductions whereas it has been demonstrated only for GDP → dGDP in mammals and for CDP → dCDP in birds. Thus, once again, caution should be exercised in asserting that regulation of these pathways as described here functions in mammals.

6
Amino Acid Metabolism

General Scheme of Amino Acid Metabolism

1. Generalized picture of protein metabolism

    Sources (a–c)   Utilization (d–f)

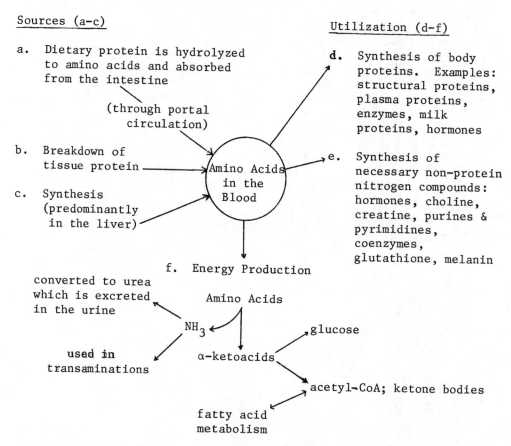

    a. Dietary protein is hydrolyzed
        to amino acids and absorbed
        from the intestine

                    (through portal
                    circulation)

    b. Breakdown of
        tissue protein

    c. Synthesis
        (predominantly
        in the liver)

                    Amino Acids
                    in the
                    Blood

    d. Synthesis of body
        proteins. Examples:
        structural proteins,
        plasma proteins,
        enzymes, milk
        proteins, hormones

    e. Synthesis of
        necessary non-protein
        nitrogen compounds:
        hormones, choline,
        creatine, purines &
        pyrimidines,
        coenzymes,
        glutathione, melanin

    f. Energy Production

    converted to urea
    which is excreted
    in the urine

                    NH$_3$ Amino Acids

                                                              glucose

        used in
    transaminations   α-ketoacids

                                                  acetyl-CoA; ketone bodies

                            fatty acid
                            metabolism

2. Cellular proteins and their constituent amino acids are in a
    dynamic equilibrium. Degradation and synthesis (to replace the
    existing molecules) are in constant operation.

3. The turnover rate (quantity of a substance synthesized or degraded
    per unit of time) of different proteins in different tissues varies
    enormously. Example:

Proteins	Half-life Times in Days
From blood, liver, other internal organs	2.5-10
muscle protein	180
collagen	greater than 1000

Therefore, the turnover rate of body protein depends upon the tissue. Intestinal epithelial cells are sloughed off every 4 to 6 days which contributes about 70 g/day (endogenously derived protein) in the intestinal lumen to be absorbed. The liver secretes about 20 g/day of plasma proteins. Muscle tissue, due to its mass and rate of metabolism, makes a significant contribution to the total body protein turnover rate. Urinary output of creatinine provides a rough estimate of total muscle mass (creatinine, derived from creatine phosphate in a non-enzymatic process, is excreted in the kidneys without significant reabsorption--discussed under creatine metabolism) and 3-methylhistidine urinary output provides an index of muscle protein turnover. 3-Methylhistidine (a post-ribosomal modified residue of histidine), an amino acid present in muscle contractile proteins and released upon muscle protein breakdown, is not reutilized and is quantitatively excreted in the urine.

4. Some mechanisms that contribute to protein turnover are tissue replacement, breakdown and synthesis of individual tissue proteins, and utilization and replacement of proteins such as digestive enzymes, hormones, serum proteins, etc.

5. Amino acids derived from protein turnover may make a significant contribution toward the energy requirements of the body.

6. Nitrogen Balance

   a. When nitrogen intake (the amount of nitrogen in the diet) equals nitrogen losses (the amount of nitrogen excreted in urine plus feces), the body is said to be in nitrogen balance. In such a situation, anabolism = catabolism. This is the condition in normal adults.

   b. When nitrogen intake exceeds nitrogen losses, then the body is retaining nitrogen as tissue protein. This is called positive nitrogen balance and anabolism exceeds catabolism. This condition is characteristic of growing children, patients recovering from emaciating illnesses, and pregnant women.

c. <u>Kwashiorkor</u> (which means displaced children in the Bantu
    language) is a disease caused by malnutrition (specifically, by
    a prolonged insufficient intake of necessary proteins) in
    children. Some of its characteristics are lack of appropriate
    cellular development (because of failure to synthesize normal
    amounts of protein); edema; diarrhea (may be with fatty stools);
    atrophy of the pancreas (which requires a high rate of protein
    synthesis) and the intestinal mucosa; gray and scaly skin (due
    to lack of melanin) which may develop ulcerating patches, etc.
    <u>Marasmus</u> is another disease associated with malnutrition. It
    is due, however, to generalized starvation rather than an
    insufficiency specifically of protein.

7. Amino acids are used as a source of energy. They enter the
    tricarboxylic acid (TCA) cycle for ultimate oxidation. The
    relationship of amino acid metabolism to the TCA cycle is shown in
    Figure 68.

8. Amino acids may be classified according to the types of products
    that are formed when the amino acids are metabolized. Some amino
    acids give rise only to acetoacetic acid and other ketone bodies
    and are called <u>ketogenic amino acids</u>. Others give rise only to
    glucose or glycogen (or compounds which can be converted to these
    compounds) and are called <u>glucogenic amino acids</u>. Some amino
    acids are <u>both</u> glucogenic and ketogenic. A list is shown below.
    The assignment of an amino acid to one group or another is based
    on actual experiments in intact animals and not on considerations
    of isolated metabolic pathways. For example, any of the amino
    acids which are metabolized to pyruvate can be glucogenic (via
    pyruvate carboxylase) or ketogenic (via pyruvate oxidodecarboxylase).
    When these amino acids were fed to animals, however, predominantly
    glucose was synthesized from them.

Classification	Amino Acid
Ketogenic only	leucine
Ketogenic and Glucogenic	isoleucine, lysine, phenylalanine, tyrosine, tryptophan, threonine
Glucogenic Only	all other amino acids

9. Essential and Non-Essential Amino Acids

a. The body can synthesize some amino acids. These are known as
    non-essential amino acids (because it is not essential that
    they be available in the diet for normal growth and development).

Figure 68.   Interrelationships of Amino Acid Metabolism to the TCA
Cycle and Energy Production

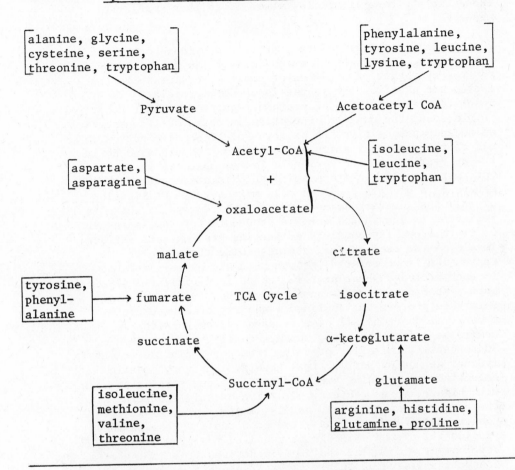

These amino acids are synthesized by the transamination of
keto acids which have the appropriate carbon skeletons.
Transamination is discussed in detail later in this chapter.
Note that the terms essential and non-essential have no meaning
with respect to the relative importance of an amino acid in
metabolism. They are strictly <u>dietary</u> terms.

b.   Essential amino acids are those which the body cannot synthesize
and which must thus be obtained from an external source, the
diet. Adequate amounts of each of these essential amino acids
are required to maintain proper nitrogen balance. The essential
L-amino acids (indispensable) for adult man are <u>tryptophan</u>,
<u>phenylalanine</u>, <u>lysine</u>, <u>threonine</u>, <u>valine</u>, <u>leucine</u>, <u>isoleucine</u>
and <u>methionine</u>. In infants, histidine is required for proper

development and growth and is thus considered essential. In
adults, the requirement of histidine is probably negligible
except in the case of uremia, where it is then considered
essential.

c. Protein Requirements: Dietary protein provides both organic
nitrogen and the essential amino acids. The former is used
for the synthesis of a number of small molecules while the
latter are used for the synthesis of body proteins. The
quantitative estimation of protein requirements must take
into account the quality of protein in terms of its essential
amino acid composition. A protein should provide not only
all of the essential amino acids but they should also be in
the appropriate concentrations. For example, if the
concentration of a certain amino acid is greater than the
others in a protein or amino acid mixture, the utilization
of the other dietary amino acids can be depressed, which is
reflected in growth failure. A procedure for the assessment
of protein quality consists of feeding growing rats various
levels of the test protein and assessing the slope of
regression lines relating growth rate and protein intake.
In such a study, wheat protein, which is deficient in lysine
when compared with lactalbumin or egg protein (used as a
reference protein and assigned a quality of 100 percent
because it contains all of the essential amino acids in
desirable concentrations), shows a quality of less than
20 percent. Similarly, proteins from corn, also deficient in
lysine, do not support optimal growth. However, genetic
selection and breeding programs have yielded strains of corn
with higher lysine content. Proteins of animal origin (with
the exception of gelatin), namely, meats, eggs, milk, cheese,
poultry and fish, are considered good quality because they
provide all of the amino acids required for growth and
development. Gelatin, the protein derived from collagen (a
connective tissue protein), lacks tryptophan, which is an
essential amino acid. It is thus termed an incomplete
protein. As noted above with the proteins of wheat and corn,
the plant proteins in general are of poor quality because
they are deficient in one or more of the essential amino
acids. Of the plants, the best quality proteins are found in
the legumes (beans, peas and peanuts) and in nuts. Therefore,
the diet of a person who is a pure vegetarian (all vegetable
diet without any animal foods, dairy products, or eggs)
requires careful planning in order to achieve a combination
of proteins which provide all of the necessary amounts of
essential amino acids. Following are examples of such
combinations of complementary vegetable proteins: rice and
black-eyed peas; whole wheat or parched crushed wheat (bulgur)

with soybeans and sesame seeds; cornmeal and kidney beans; and soybeans, peanuts, brown rice and bulgur wheat. A deficiency of vitamin B_{12} (see Chapter 10) is a problem in persons who are on a pure vegetarian diet. Although plant proteins used singly do not provide all of the essential amino acids, their inclusion in the diet provides non-essential amino acids which may otherwise have to be derived from essential amino acids. The following are estimates of daily protein needs for human subjects who derive their proteins from a western-type diet: Adults--0.8 g/kg; infants at birth--2.2 g/kg; infants (0.5 to 1 year)--2.0 g/kg. During pregnancy an increased protein intake over the normal adult level is recommended. It is important to emphasize that the above protein requirements are valid only when energy needs are appropriately met by non-protein sources. If energy intake in the form of carbohydrates and lipids is insufficient to meet the energy expenditure, the protein is utilized to meet the energy deficit, resulting in a negative nitrogen balance.

10. Amino acids and their derivatives have been shown to influence nerve impulse transmission, both in the central and peripheral nervous systems. The post synaptic effect of these materials may involve cyclic AMP and, hence, adenylate cyclase.

   a. In the brain, compounds of the formula $HOOC-CH_2-(CH_2)_n-NH_2$ (where n = 0-4) exert an inhibitory influence on synaptic transmission. One member of this group is γ-aminobutyric acid (GABA), obtained by α-decarboxylation of glutamic acid (see proline and phenylalanine metabolism). Taurine (see metabolism of sulfur-containing amino acids) and alanine have similar depressive effects. Glutamic and aspartic acids, on the other hand, display potent excitatory effects on central neurons. Note that the inhibitors are neutral amino acids while the stimulatory substances are acidic.

   b. Glycine appears to have an inhibitory effect on neural transmission within the spinal cord. On the other hand, acetylcholine, which is metabolically related to glycine, is strongly implicated as a neurotransmitter in the central nervous system.

   c. Aromatic amines, derived from phenylalanine, tyrosine, histidine, and tryptophan, have for some time been considered as neurotransmitter substances. The most thoroughly studied of these materials are the catecholamines, epinephrine and norepinephrine (see phenylalanine); serotonin (see tryptophan); and histamine (from histidine).

d. Since the 1930's, a group of polypeptides, designated
    "substance P," have been known for their ability to cause
    smooth muscle contraction and vasodilation. Although this
    appears to be truly a peptide effect and not simply due to
    small-molecule impurities, much work remains to be done
    in this area.

11. Protein Synthesis and Storage

$$\text{amino acids} \underset{\text{catabolism}}{\overset{\text{anabolism}}{\rightleftarrows}} \text{proteins}$$

The mechanism of protein synthesis has already been discussed.
Each tissue of the body has its own characteristic proteins and
most of these proteins are synthesized within their own tissue
cells. However, some tissues also make proteins for export.
For example:

a. The liver synthesizes fibrinogen, albumin, and most of the
    plasma globulins.

b. The lymphocytes of the lymphoid tissue synthesize plasma
    γ-globulins.

c. Protein hormones are synthesized by various organs.

d. Trypsin, chymotrypsin, lipase, and amylase (all digestive
    enzymes) are made in the pancreas.

The body apparently does not store free amino acids and has little
capacity even for the storage of proteins (unlike carbohydrates and
fats). The minimal reserves that are available are proteins that
are incorporated into the structure of the liver (and perhaps
muscle).

Effects of Some Hormones on Protein Metabolism

1. Thyroxine (produced by the thyroid gland) affects the metabolism
    of carbohydrates, proteins and fats. The overall metabolic rate
    of the entire body reflects changes in the amount of thyroxine
    available.

    a. Severe deficiency ⟶ low metabolic rate characterized by

        i) failure to grow, imbecility
        ii) decrease in protein synthesis

   b. Excessive production ———→ high metabolic rate characterized by

       i)   increased rate of tissue breakdown
       ii) marked emaciation, nervousness
       iii) increased rate of oxidation of food; more amino acids
            are destroyed in oxidation, and thus fewer go into
            protein synthesis

2. <u>Growth hormone</u> (produced by the anterior pituitary) fosters protein
   anabolism, possibly by facilitating amino acid uptake by the cells.
   Underproduction leads to dwarfism and overproduction leads to
   gigantism and acromegaly.

3. <u>Insulin</u> promotes increased utilization of glucose and hence
   increased production of ATP as a source for peptide-bond formation.
   This indirectly affects protein metabolism because there follows
   an increased incorporation of amino acids into proteins, and
   probably an increased rate of translation of messenger RNA.

4. The <u>male sex hormones</u> (androgens) stimulate protein synthesis.

5. The <u>hormones of the adrenal cortex</u> promote breakdown of proteins
   and amino acids.

6. <u>Epinephrine</u> lowers the plasma levels of free amino acids, probably
   by facilitating the uptake of amino acids by the cells.

Protein Digestion

1. a. Protein that is to be absorbed in the intestinal tract is
       derived from both the diet and endogenous sources such as
       digestive enzymes and mucosal cell protein. Recall that the
       mucosal cells are sloughed off from the villar tip which are
       replaced by cells which divide in the crypts at the base of
       the villi. The endogenous contribution of protein made in a
       person is about 70 g per day, which is mostly derived from sloughed-
       off mucosal cells.

   b. Digestion in the stomach requires gastric secretion, which is
       usually divided into three phases. The first, called the
       nervous stimulation phase (cephalic phase), is mediated
       through the vagus nerve. With the anticipation of food or
       the presence of food in the mouth, the vagus is thought to
       act directly on acid-secreting parietal cells, mucin-secreting
       gastric chief cells and gastrin-secreting (small amount)
       antral cells. Central stimulation has been localized to the
       ventromedial nuclei of the hypothalamus, with ultimate control
       resting in the frontal lobes.

c.   The second or gastric phase begins with the introduction of
food into the gastric antrum and consequent gastrin production
(a peptide hormone). Stimulation of the gastric juice can
occur as a result of the presence of food in the stomach
(local chemical stimulation), psychic stimulation of the
vagus, or by reflexive stimulation of the vagus with distension
of the stomach wall. Hypoglycemia, mediated through the vagal
reflexes, also stimulates the parietal cell to produce HCl
either directly or indirectly through gastrin. This
observation is supported by the fact that atropine (a
parasympatholytic agent) blocks the gastric secretion while
increasing the serum gastrin levels. Local reflexes (in
addition to vagal reflexes) are present along the cholinergic
axons in the wall of the stomach that function in the oxyntic
(acid-producing) gland area to stimulate the release of
gastrin. These reflexes are stimulated by the presence of
chemicals in the stomach as well as by distension. These
stimuli can be blocked by atropine. The final mediator of
gastric secretion appears to be histamine, for the following
reasons: (i) Histamine is found in large amounts in the
stomach, particularly in the oxyntic gland location;
(ii) Histamine acts directly on the parietal cells to produce
HCl (as shown in the denervated stomach); (iii) The depletion
of histamine and induction of histidine decarboxylase occurs
upon feeding, cholinergic stimulation, or gastrin
administration; and (iv) Metiamide (a histamine H_2-receptor
blocking agent) prevents gastric acid secretion induced by
parasympathomimetics and gastrin. Thus, it appears that
gastric acid secretion involves acetylcholine, gastrin and
histamine.

Gastrin, present in the pyloric antral mucosal cells (G-cells),
exists as two molecular species termed gastrin I and II. The
hormone is stored in granules within these cells. Both
species are 17-residue peptides, but they differ from one
another in that tyrosine (the 12th residue) in gastrin II is
sulfated. The occurrence of larger molecular forms of gastrin
(big gastrin) have been reported. However, the structure
responsible for biological activity is the C-terminal
tetrapeptide (a pentapeptide known as pentagastrin, which
includes this tetrapeptide sequence, is available commercially)
which is identical in all species studied to date. Another
interesting feature in the structure of the gastrins is their
c-terminal pyroglutamyl residue. This is the gamma lactam of
glutamic acid, which is also found in some of the hypothalamic
hormones, (e.g., thyrotropin=releasing hormone).

By means of a negative feedback process, H^+ regulates the
secretion of gastrin. Gastric fluid at pH 1, produced in

response to a protein meal, brings about a maximal suppression of gastrin (at pH 2.5 there is about 80% suppression).

   d. The third or intestinal phase of gastric secretion serves chiefly in an inhibitory capacity. As food and acidified fluid enter the duodenum, secretin and other gastrointestinal hormones (such as bulbogastrone and gastric inhibitory polypeptide) are emitted into the circulation which produces a feedback inhibition of gastric secretion.

2. Gastrin is released into the blood stream where it is carried to its target cells through which it affects several functions of the stomach.

   a. It stimulates motility of the stomach, bringing about an intensification of the churning action which breaks up food and prepares it for intestinal digestion.

   b. It causes the secretion of pepsin, an endopeptidase present in the gastric juice, which hydrolyzes proteins into polypeptides and amino acids. Although the enzyme preferentially hydrolyzes peptide bonds in which tyrosine and phenylalanine participate, its action is not highly specific. Pepsinogen, the zymogen form of pepsin, is secreted by the chief cells of the gastric mucosa. Pepsinogen is converted to the active enzyme by H^+ at pH 1.0 to 1.5 (pH of normal gastric content) and also by autocatalysis of pepsin itself. A milk-curdling enzyme known as rennin, also produced by the chief cells (of the fundic glands), assists in the proteolysis of milk proteins by pepsin. The clotting of milk by rennin slows down its passage through the alimentary tract and thus provides time for the action of pepsin in the stomach.

   c. Hydrochloric acid is produced by the parietal cells of the gastric mucosa under the influence of gastrin. As indicated above, this is important because pepsinogen is activated at very low pH's. The hydrochloric acid is produced by the action of carbonic anhydrase in the parietal cells.

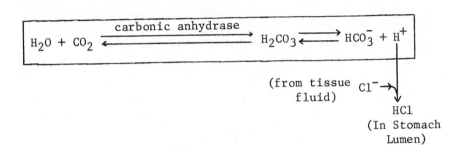

Chloride ions which are present in the gastric juice are
derived from the tissue fluids and transported across the
gastric epithelial cells into the stomach lumen, presumably by
a carrier-mediated ATP-dependent process. Gastric acidity is
the chief factor of concern when one performs a gastric
analysis in the clinical laboratory. The gastric analysis is
performed on individuals who have fasted for a period of
8-12 hours. The procedure consists of: a basal collection
of gastric juice by means of a tube placed in the stomach for
a 30-minute period (or a 12-hour nocturnal collection), the
administration of a stimulant of gastric secretion (Histalog--
an analogue of histamine with fewer side-effects than histamine
or pentagastrin--a stimulant of choice), and collection of
15-minute specimens until the highest acid output is reached
(usually takes about 6-8 specimens). The free hydrochloric
acid is determined by titration with 0.10 M NaOH to a pH of
3.5 using a pH meter. An indicator known as Toepfer's reagent
(0.5 g diethylaminoazobenzene/100 ml ethanol) may be used
instead of a pH meter, with the endpoint being indicated by a
salmon color. From the titration values, the basal acid
secretion rate (normal value = 0-5 mmol/h) and maximal acid
secretion rate (normal value = 1-20 mmol/h) are calculated.
A high acid output is associated with duodenal and prepyloric
ulcer or the Zollinger-Ellison syndrome (see later). True
achylia (no hydrochloric acid or pepsinogen excretion) with a
decreased volume of gastric secretion is found in pernicious
anemia (note that pernicious anemia is a type of megaloblastic
anemia--discussed in Chapter 10). Patients with carcinoma of
the stomach show variable findings with regard to gastric
acidity: they may exhibit achlorhydria (50%), hypochlorhydria,
or hyperacidity. The Zollinger-Ellison (Z-E) syndrome is
caused by either non-beta-cell neoplasms of the pancreas
(adenoma or carcinoma) or certain tumors of the duodenum.
Both of these cause a continuous or intermittent release of
gastrin. In these patients serum gastrin levels are highly
elevated with values ranging from 500 to 3200 (or higher) pg/ml
(the normal range of serum gastrin measured in subjects with
an overnight fast is 50-155 pg/ml; higher values of 500 ± 300
are found in individuals with an age of 80 or over). Patients
with the Z-E syndrome usually have intractable ulcer disease
due to massive gastric hypersecretion (2-3 liters) rich in
hydrochloric acid. Treatment consists of total gastrectomy
(removal of the target organ). The serum gastrin may also be
elevated in patients with pernicious anemia.

Other methods of gastric analysis include the tubeless gastric
analysis and a test used to confirm complete vagotomy
(Hollander test).

Tubeless gastric analysis is used to rapidly screen large numbers of patients without the need for intubation. Caffeine sodium benzoate is usually used to stimulate acid secretion. A resin, Amberlite XE 96, coupled with an indicator dye, Azure A, is taken orally. In gastric juice of pH less than 3, hydrogen ion displaces the Azure A dye from the resin. The dye is then absorbed from the stomach into the bloodstream where it is excreted into the urine, turning it blue. The excreted dye is estimated by comparing the color against a standard using either a colorimeter or a visual comparative block. In positive tests, excretion of greater than 0.6 mg indicates the presence of acid. Excretion of lower values is not a reliable indicator of achlorhydria and must be confirmed by intubation gastric analysis. No quantitative data can be derived, and there are many problems with the test. The main difficulty is getting a complete and accurately timed urine specimen. Color interference by other excretory products such as pyridium can be seen. Individuals with altered gastric motility, such as with subtotal gastrectomy or pyloric obstruction, have falsely abnormal test results. Marked congestive heart failure, severe liver or kidney malfunction and urinary retention also obscure test results.

The principle of the Hollander test is based upon the observation that hypoglycemia (45 mg/100 ml or less), produced by administration of insulin, stimulates gastric secretion via the vagus nerve (initially sensed by hypothalamic cells with subsequent stimulation of the vagal medullary nuclei). Thus, patients who have undergone complete vagotomy do not show any significant gastric acid secretion, confirming the completeness of the vagotomy procedure.

d. Another component of gastric juice is mucus which consists of at least two glycoproteins. Mucus not only coats the stomach lining, protecting it against hydrochloric acid, but it is also essential in the absorption of vitamin B_{12} (see vitamin metabolism).

3. From the stomach, the chyme (partially digested stomach contents) moves into the upper part or bulb of the duodenum. Entry of the acid chyme stimulates the release of <u>bulbogastrone</u> from the mucosal cells of this region. The target cells of this hormone are the G-cells in the pyloric antrum, where it inhibits release of further gastrin. This, and the drop in pH associated with the departure of food from the stomach, shut down the digestive processes in the stomach. The chemical nature of bulbogastrone has not yet been elucidated.

a. As the chyme moves further into the duodenum and jejunum, other hormones are released which perform various functions in aiding the digestive process. Gastric inhibitory polypeptide (GIP), one such hormone, is released into the blood circulation primarily from the cells lining the duodenum and to a lesser extent from the cells lining the jejunum. GIP appears to be released in a biphasic fashion in response to food: the initial peak is reached 30-60 minutes after stimulation by glucose and the second peak two to three hours after stimulation by fat. GIP contains 43 amino acid residues and has a molecular mass of 5104 daltons. There is a strong sequential homology between GIP, secretin, and glucagon. The actions of GIP include: inhibition of gastric secretion, pepsin secretion, food-stimulated gastrin secretion and stimulation of insulin secretion (an enteric hormone affecting insulin secretion--known as the enteroinsular axis). The latter effect of GIP apparently plays a significant role in maintaining blood glucose levels. Hypoglycemia (see hypoglycemia--Chapter 4) observed in some patients with the dumping syndrome may in part be due to excessive release of GIP with consequent increases in insulin secretion.

b. Cholecystokinin-pancreozymin (CCK-PZ) and secretin are also released from cells in the jejunum and lower duodenum.

   i) CCK-PZ release is stimulated by the presence of fats and protein digests (especially long-chain fatty acids and essential amino acids) in the duodenal lumen. It causes contraction of the gallbladder (hence, release of bile needed for fat absorption) and stimulates the exocrine pancreatic cells to secrete pancreatic juice rich in enzymes (amylase, lipase, proteases). It was originally thought that two separate hormones caused these effects, hence the double name for one hormone. CCK-PZ is a polypeptide having 33 amino acid residues. The C-terminal sequence of five amino acids is identical to the same regions of the gastrins.

       Recent studies have indicated that CCK-PZ (specifically its C-terminal octapeptide) stimulates small-bowel motility--a property with potential application in the treatment of postoperative ileus (intestinal obstruction) in order to stimulate peristalsis. Another interesting observation is that the elevated CCK-PZ levels in the blood after a meal may be responsible for producing a state of satiety through a negative feedback process by affecting the central nervous system's electrical activity. These results suggest that morbid obesity may be due to a hormone deficiency--which needs further careful investigation.

ii) Secretin release is stimulated by the low pH of the chyme as it comes from the duodenal bulb. This hormone influences the pancreas to secrete a watery pancreatic fluid, rich in bicarbonate, into the small intestine. The alkalinity of this secretion neutralizes the acid chyme, bringing it to a pH of about 7-8, suitable for the activity of the pancreatic enzymes induced by CCK-PZ. Secretin appears to be controlled by a feedback mechanism (as is gastrin). The increased pH which results from the bicarbonate inhibits secretin release. Porcine secretin is a polypeptide of 27 amino acid residues. It shows a great deal of sequence homology with porcine glucagon.

In the secretin molecule, there is a helical portion between amino acid residues 5 and 13 which is apparently necessary for the full biologic activity of the hormone. The existence of larger molecular variants of secretin have been reported.

4. The foregoing material is summarized in Table 21. It should be pointed out that the action of these hormones is not as singular as it is indicated to be in this table and in the preceding discussion. For example, although CCK-PZ has a strong effect on the gallbladder and the pancreatic enzyme secretion, it can also, especially in large doses, bring about gastric and intestinal motility and secretion of bicarbonate (from both the pancreas and the liver), pepsin, gastric acid, and even insulin. Similarly, while gastrin has a group of primary functions, it can also bring about all of the effects which CCK-PZ induces. There are other examples of multiple functions of the gastrointestinal hormones, some of them overlapping each other as do these.

5. Intestinal glucagon (enteroglucagon; glucagon-like immunoreactive factor), mentioned in Table 21, appears to have properties in common with both pancreatic glucagon and secretin. For example, it stimulates the release of insulin and pancreatic bicarbonate. Intestinal glucagon, however, is _induced_ by ingested glucose, whereas pancreatic glucagon is _suppressed_ by elevated blood glucose levels.

6. Although the six compounds discussed above are the best-characterized of the gastrointestinal (G.I.) hormones, a number of other substances have also been implicated in the regulation of gastric function. Some of these have been chemically identified, but parts of the physiological data needed to fully characterize them are missing. Members of this group include

i) enterocrinin (released by upper small intestine by the presence of chyme; increases the secretion of intestinal digestive juice);

Table 21. Actions and Properties of Some Gastrointestinal Hormones

Hormone	Produced by	In response to stimulation from	Acts on (target organ)	Primary effects produced
I. Gastrin	Cells in the pyloric antral region of the stomach	Ingestion of food; antral distension; small-molecular components of diet	Stomach	Secretion of pepsin and HCl; gastric motility increased
II. Bulbogastrone	Mucosal cells of the duodenal bulb	Acid chyme entering the duodenum	Gastrin-secreting cells of the stomach	Decreased gastrin secretion
III. Gastric Inhibitory Polypeptide (GIP)	Cells in the jejunum and lower duodenum	Glucose and fat entering these regions	Gastrin-secreting cells of the stomach; pancreas	Decreased gastrin secretion; stimulation of insulin secretion
IV. Cholecystokinin-pancreozymin (CCK-PZ)	Cells in the jejunum and lower duodenum	Fats and partially digested proteins entering these regions	Gallbladder and exocrine pancreas	Contraction of the gallbladder and release of bile; secretion of pancreatic enzymes
V. Secretin	Cells in the jejunum and lower duodenum	Low pH of the chyme	Pancreas	Secretion of bicarbonate
VI. Intestinal Glucagon	Cells of the walls of the stomach and small intestine	Ingestion of glucose	Pancreas	Stimulation of insulin secretion

  ii)   hepatocrinin (released by upper small intestine by the presence of chyme; increases hepatic bile secretion);

 iii)   motilin (polypeptide containing 22 amino acid residues obtained as by-product in secretin isolation; stimulates gastric motility in both fundic and antral pouches; release stimulus not known);

  iv)   vasoactive intestinal peptide (VIP) (polypeptide of 28 residues isolated from porcine-upper-intestinal wall; smooth muscle relaxant which thereby increases splanchnic and peripheral blood flow; resembles secretin and CCK-PZ in this respect and in structure; release stimulus and true physiological role unknown).

   v)   Somatostatin (see Chapter 4--Diabetes Mellitus), a hormone of the hypothalamus which inhibits release of growth hormone, has been shown to be present in substantial amounts in the D-cells of the upper gastrointestinal tract and endocrine pancreas. The precise role of somatostatin with regard to gastrointestinal function is not clear at this time. However, when somatostatin is administered exogenously, it is capable of producing the following extrahypophyseal actions: inhibition of insulin and glucagon secretion by direct action on the pancreatic cells, inhibition of fasting- and food-stimulated gastrin release, and inhibition of oxyntic gland secretion by direct action. These findings suggest that somatostatin plays a regulatory role in the endocrine and exocrine secretory activities of the gastrointestinal tract.

7. a. Certain proteolytic enzymes are secreted as inactive enzyme precursors called pro-enzymes or zymogens (discussed previously). These are then converted to the active form by hydrolysis of one or more peptide bonds. The pancreatic pro-enzymes listed below are activated as indicated.

$$\text{Trypsinogen} \xrightarrow{\text{trypsin or enterokinase}} \text{Trypsin}$$

$$\text{Chymotrypsinogen} \xrightarrow{\text{trypsin}} \text{Chymotrypsin}$$

$$\text{Procarboxypeptidases A \& B} \xrightarrow{\text{trypsin}} \text{Carboxypeptidases A \& B}$$

$$\text{Proelastase} \xrightarrow{\text{Trypsin}} \text{Elastase}$$

This system, in which digestive enzymes exist as inactive precursors until they are at a site where their activity is desirable, has obvious value. Trypsin, chymotrypsin and elastase are endopeptidases. Carboxypeptidases (and amino peptidases, mentioned below) are exopeptidases (see Table 22).

b. Among the enzymes produced by the small intestine itself are several which act on proteins:

   i)   Enterokinase: converts trypsinogen to trypsin

   ii) Aminopeptidases: cleave terminal amino acids from the N-terminal end of peptides

   iii) Carboxypeptidases: cleave terminal amino acids from the C-terminal end of peptides

   iv) Dipeptidases: hydrolyze dipeptides into individual amino acids; require cobalt or manganese ions for activity

   v)   Tripeptidases: hydrolyze tripeptides, releasing a free amino acid and a dipeptide

c. The preferred sites of cleavage (specificities) of some of the proteolytic digestive enzymes are indicated in Table 22.

d. Cystic fibrosis (CF), an inherited, multifaceted disease of childhood for which a specific biochemical lesion is not known, affects the lungs, pancreas, salivary glands and mucous glands. The disorder is transmitted as an autosomal recessive trait and is the most common inherited chronic disease in Caucasian children. The frequency of occurrence of CF is between 1/1500 and 1/2500. Therefore, approximately 1 in 20 persons is heterozygous for the disease. CF seldom occurs in Orientals and Blacks, unless they are intermixed with Caucasians. The heterozygotes show no clinical characteristics of CF (completely asymptomatic) and are thus undetectable until they have children stricken with CF. The manifestations of the disease are chronic obstructive pulmonary involvement (with chronic infections), steatorrhea and malabsorption. The gastrointestinal abnormalities are caused by blockage of the pancreatic ducts, due to excessive secretion of mucous (leading to the formation of mucous plugs) as well as insufficient production of digestive enzymes (trypsinogen, lipase, amylase, etc.). Enzyme replacement via the diet appears to ameliorate the severity of the disease.

Table 22. Specificities of Some of the Gastrointestinal Proteolytic Enzymes

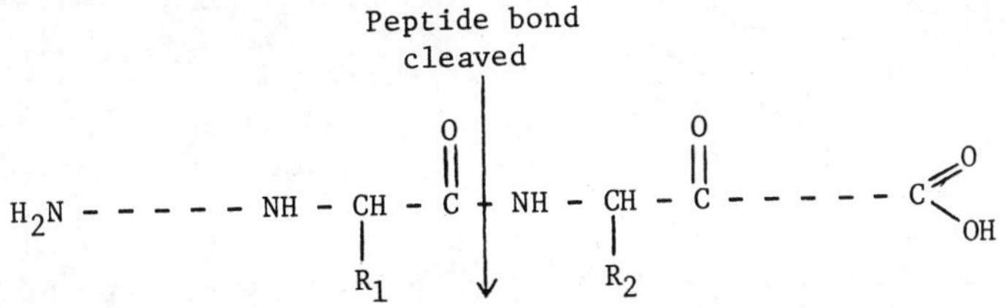
Peptide bond cleaved
H_2N - - - - - NH - CH - C(=O) ǀ NH - CH - C(=O) - - - - - - C(=O)OH
R_1
R_2

Enzyme	Prefers to cleave peptide bonds in which
a. Trypsin	R_1 = Arg or Lys; R_2 = any amino acid residue
b. Chymotrypsin	R_1 = Aromatic amino acid (Phe, Tyr, Trp); R_2 = any amino acid residue
c. Carboxypeptidase A	R_1 = any amino acid residue; R_2 = any C-terminal residue except Arg, Lys, or Pro
d. Carboxypeptidase B	R_1 = any amino acid residue; R_2 = Arg or Lys at the C-terminus of a polypeptide
e. Elastase	R_1 = Neutral (uncharged) residues; R_2 = any amino acid residue
f. Aminopeptidase	R_1 = Most N-terminal residues of a polypeptide chain; R_2 = any residue except Pro
g. Pepsin	R_1 = Trp, Phe, Tyr, Met, Leu; R_2 = any amino acid residue

Dipeptidases

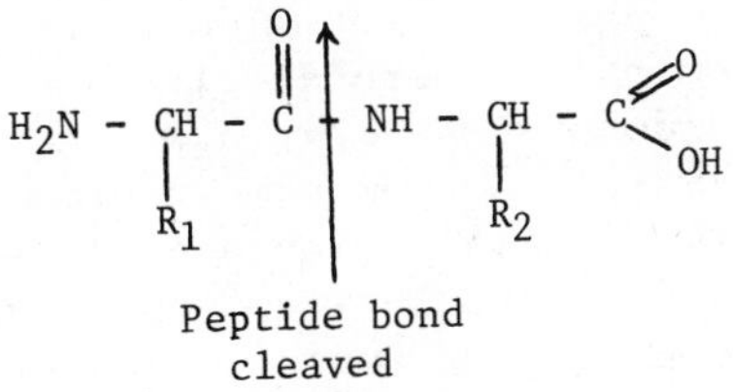
H_2N - CH - C(=O) ǀ NH - CH - C(=O)OH
R_1
R_2
Peptide bond cleaved

Tripeptidases

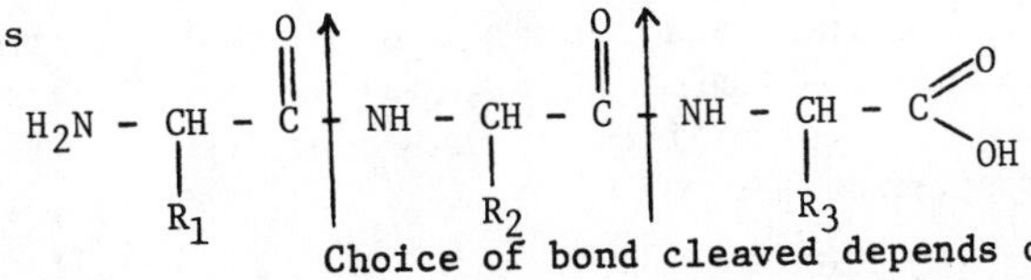
H_2N - CH - C(=O) ǀ NH - CH - C(=O) ǀ NH - CH - C(=O)OH
R_1
R_2
R_3

Choice of bond cleaved depends on specific enzyme. In a given molecule, only one bond is cleaved by a tripeptidase.

R_1, R_2, R_3 = **any amino acid residue**

Note: Of the first 7 enzymes, a, b, e, and g are endopeptidases while c, d, and f are exopeptidases; trypsin is the most specific endopeptidase; pepsin is the least specific endopeptidase.

The objective of dietary management of cystic fibrosis is to provide high calories (infants, 200 cal/kg/day; children, 50-100% more calories than normal children of comparative ages), high protein (6-8 gm/kg/day), and modified fat (medium-chain triglycerides used as a source of dietary fat).

The diagnosis of CF consists of detecting higher levels of sodium in the sweat (quantitative pilocarpine iontophoretic sweat test). A positive sweat test (repeated two or more times) and the presence of one of the clinical characteristics mentioned earlier is strongly suggestive of CF. A neonatal screening procedure which consists of measuring albumin concentration in the meconium is being evaluated. Neonates with cystic fibrosis have increased levels of albumin in the meconium without manifestation of meconium ileus.

e. Pancreatitis also can result in blockage of the pancreatic secretions. Some of the enzymes accumulate and spill over into the bloodstream. Measurement of serum lipase and amylase are useful in the diagnosis of this disorder (see the section on uses of enzymes in the clinical laboratory).

8. a. A number of mechanisms have been proposed to explain the active transport of amino acids into the intestinal mucosal cell. The amino acid transport process shows a number of similarities with the active transport of glucose which is a carrier-mediated, Na^+-dependent, ATP-requiring process (discussed in carbohydrate absorption, Chapter 4). Similar processes also take place in the kidney, where amino acids are reabsorbed from the glomerular filtrate into the tubular epithelial cells and subsequently into the peritubular capillaries. The carriers show some degree of specificity in that structurally similar amino acids are transported by the same transport system. There appears to be at least five sterospecific transport systems present in the kidney and intestine. The transport systems that have been identified for amino acid groups are: small neutral amino acids, large neutral amino acids, acidic amino acids, and basic amino acids. Proline, an imino acid, has its own transport system.

b. The γ-glutamyl cycle developed by A. Meister and his associates provides another mechanism for the translocation of amino acids. The operation of the γ-glutamyl cycle requires: six enzymes (one of which is membrane-bound with the rest being cytoplasmic enzymes); glutathione (GSH, γ-glutamylcysteinylglycine found in all animal tissue cells); and ATP (3 ATP's are required for each amino acid translocation).

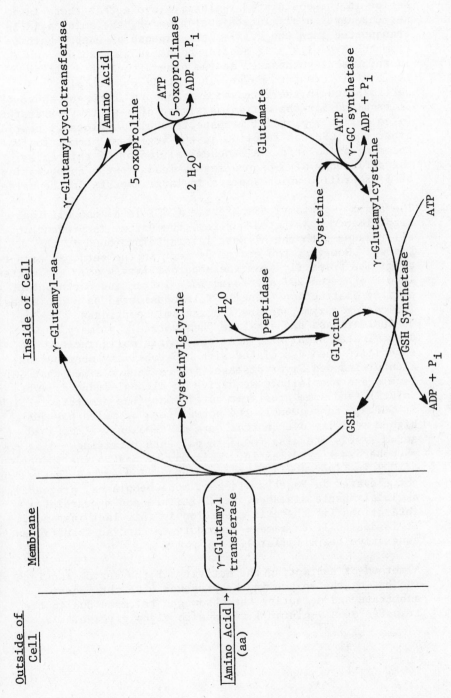

Figure 69. The γ-Glutamyl Cycle for the Transport of Amino Acids.

Notice in the postulated γ-glutamyl cycle that there is no net consumption of glutathione; however, the amino acid is transported into the cell at the expense of energy gained from the hydrolysis of peptide bonds of glutathione, which in turn is resynthesized at the expenditure of 3 ATP's. It is of interest to point out that Na^+, which also participates in the amino acid transport discussed above (8a), is required for the optimal activity of γ-glutamyl transferase (also called γ-glutamyl transpeptidase). γ-Glutamyl peptides other than GSH can function as the γ-glutamyl donor for the incoming amino acid. A biochemical lesion of this cycle, the deficiency of 5-oxoprolinase, leads to the accumulation of 5-oxoproline which appears in large amounts in the urine.

The serum γ-glutamyl transferase (GGT) level has clinical significance. As noted earlier, the enzyme is present in all tissues with the highest levels being found in the kidney. However, the enzyme present in the serum appears to originate primarily from the hepatobiliary system. Elevated levels of serum GGT are found in: intra- and posthepatic biliary obstruction (hence GGT is considered as a cholestasis-indicating enzyme, as are leucine aminopeptidase, 5'-nucleotidase and alkaline phosphatase), primary or disseminated neoplasms, some pancreatic malignancies especially when associated with hepatobiliary obstruction, alcohol-induced liver disease (it has been claimed that GGT levels are exquisitely sensitive to alcohol-induced liver injury), and some prostatic carcinomas (the prostate is rich in GGT and consequently the normal sera of males has 50% higher activity over normal sera of females). Increased serum levels are also found in patients receiving phenobarbital or dilantin. This may be due to the induction of GGT in liver cells by these drugs. The enzyme levels are not elevated in Paget's disease, bone neoplasms, pregnancy, skeletal muscle diseases, renal failure and myocardial infarctions (if there is elevation in this last instance, it is due to liver damage secondary to cardiac insufficiency-congestive heart failure).

A method of estimating the activity of the enzyme consists of incubating the enzyme in the presence of a chromogenic substrate and measuring the chromogen released due to the transfer of a γ-glutamyl residue to glycylglycine.

γ-glutamyl-p-
nitroanilide   Glycylglycine   (chromogen)

The reaction is monitored by measuring the change in absorption
at 405 nm.

c. Peptide and, notably, dipeptide transport systems in the
    intestinal lumen appear to play a significant role in the
    absorption of proteins. The dipeptides absorbed by stereo-
    specific transport systems into the mucosal cells are
    hydrolyzed to constituent amino acids. Thus, the final
    products of protein digestion that are transferred from the
    intestinal lumen to the portal vein are mostly free amino
    acids. The quantitative significance and mechanism of
    peptide transport is not clear at this time. However, it is
    known that the mechanism of uptake for free amino acids is
    different from that used for dipeptides and that there is
    no competition between the two processes. The biological
    importance of peptide absorption comes from the study of
    patients with Hartnup's disease, who have an inborn error
    in the transport of neutral and aromatic amino acids
    (tryptophan in particular). Nevertheless, these patients
    show normal development on a normal diet, because they are
    able to absorb the essential neutral and aromatic amino acids
    (tryptophan) in the dipeptide forms.

Plasma Proteins: Diagnostic Significance

1. Plasma contains more than 100 different proteins, each under
    separate genetic control. These include: transport proteins for
    hormones, vitamins, lipids, metals, pigments, drugs, etc.;
    enzymes; enzyme inhibitors (proteinase inhibitors); hormones;

antibodies; clotting factors; complement components; and kinin precursors. Quantitation (by radial immunodiffusion, electroimmunoassay, nephelometric methods and radioimmunoassay procedures) of the various constituents of plasma is of value when following the course of, and diagnosing, a variety of diseases.

2. One simple and useful technique involves the separation of serum proteins by an electric field, using cellulose acetate as a support medium (electrophoresis). The separation of these proteins is possible because they each carry different charges and hence migrate at differing rates when subjected to an electric potential. Serum is generally used instead of plasma, because the fibrinogen found in plasma appears as a narrow band in the β-region, which may be mistaken for the sign of monoclonal paraproteinemia (discussed later). The support medium, cellulose acetate, possesses several advantages over paper, these are: the time required to separate the major proteins is short, albumin trailing is absent, and rapid quantitative determination of protein fractions by photoelectric scanning (after a suitable staining procedure) is possible. Figure 70 shows normal and abnormal patterns of serum electrophoresis. It should be emphasized that the electrophoretic patterns obtained are not indicative of any one disease or class of diseases. Also, a characteristic pattern may be obscured or not found in a disease entity where normally such a pattern is expected. Therefore, serum electrophoretic patterns provide only a general impression of the disorder, requiring confirmation by other procedures. Furthermore, an alteration (depression or elevation) in a given fraction should be quantitated by more sensitive and specific methods.

3. Normal Values: Adult normal values expressed as g/100 ml are: 3.2-5.6 for albumin, 0.1-0.4 for α_1-globulin, 0.4-0.9 for α_2-globulin, 0.5-1.1 for β-globulin, and 0.5-1.6 for γ-globulin. With the exception of γ-globulin, adult values are attained by three months of age for all the fractions. Cord blood γ-globulin is largely of maternal origin and undergoes catabolism, reaching its lowest value at about three months. Adult levels for γ-globulin are not reached until about 2-7 years. After the age of 40, there is a gradual decline of albumin with an increase in the β-globulin fraction.

4. Albumin

    a. It is predominantly synthesized in the liver.

    b. Elevation of albumin is very uncommon, but many types of abnormalities lead to diminution of albumin (hypoalbuminemia)

<u>Figure 70.</u>   <u>The Serum Protein Electrophoretic Patterns</u>

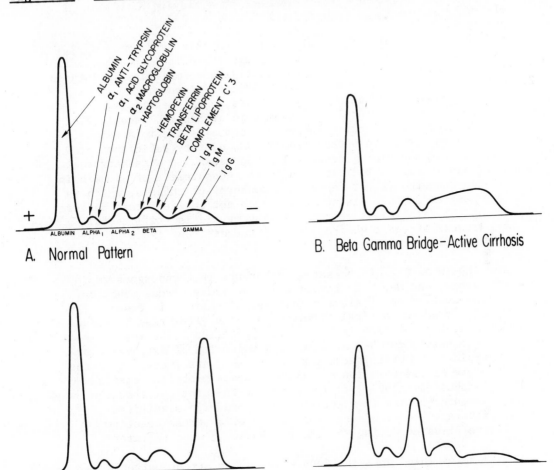

A. Normal Pattern

B. Beta Gamma Bridge–Active Cirrhosis

C. Monoclonal Gammopathy

D. Low Albumin: High α_2, Low γ
(Nephrotic Syndrome)

Pattern A is reproduced with permission from Helena Laboratories, Beaumont, Texas.

with marked depression occurring in hepatic, renal, and systemic disease.

c. Functions: Albumin accounts for about 75% of the osmotic pressure in blood and is thus responsible for the stabilization of blood volume and the regulation of vascular fluid exchange. Therefore, hypoalbuminemia can give rise to edema. It also functions as a transport protein for drugs, pigments, dyes, and some hormones, thus playing a role in their metabolism.

d. <u>Some Hypoalbuminemic Conditions</u>: Nephrotic syndrome, protein-losing enteropathies, cystic fibrosis (hypoalbuminemia with edema may be part of the initial abnormalities in infants with this disease), glomerulonephritis, cirrhosis, carcinomatosis, bacterial infections, viral hepatitis, congestive heart failure, rheumatoid arthritis, uncontrolled diabetes, I.V. feeding (hypoalbuminemia is due to a deficiency of amino acids in portal blood), and dietary deficiency of proteins containing essential amino acids. A screening test employed for cystic fibrosis is the measurement of albumin in the meconium (first feces of a newborn) of infants, which is elevated in this disease. Although not an ideal test because of 15% false negative results, the method is still useful because of the high frequency of the disorder (about 1 in 2000 births in Caucasians).

e. Albumin is one of the serum components that undergoes rapid changes and hence is involved in the acute stress reaction. The acute drop of albumin in these situations is essentially due to adrenocortical stimulation which gives rise to enhanced catabolism of albumin and sodium retention. The latter is responsible for hemodilution and expansion of extracellular fluid. The other serum components that are increased in the acute stress reaction are the α_1 and α_2 globulin fractions (discussed later). It is important to point out that the immunoglobulins (γ-fraction) do not undergo significant alterations during the acute phase of a stress reaction but are elevated during the chronic phase (discussed later).

f. Some examples of situations where acute stress patterns may be seen are: acute infections in the early states, tissue necrosis (myocardium, renal, tumor), severe burns, rheumatoid diseases with acute onset, surgery and active collagen disease.

g. Analbuminemia and Bisalbuminemia:

  i) Analbuminemia, a very rare autosomal recessive disorder, is characterized by a deficiency of albumin. Surprisingly, these patients show a lack of clinical edema, presumably due to an osmotic compensation by the globulins which are mildly elevated. Females will exhibit minimal pretibial edema, mild anemia, normal liver function tests, absence of proteinuria, lowered blood pressure and elevated serum cholesterol levels.

ii) Bisalbuminemia is due to albumin polymorphism. Upon serum electrophoresis, a double albumin band is seen. These double peaks are due to either differences in electrical charge (eleven electrophoretically distinct forms have been reported thus far) or albumin dimers. Both forms are expressed as autosomal dominant traits and apparently present no significant clinical abnormalities. However, an acquired bisalbuminemia may be associated with diabetes mellitus, the nephrotic syndrome, hyperamylasemia and penicillin therapy. In these instances the bisalbuminemia disappears after the disease in question is corrected.

5. α_1-Globulins

a. The proteins that migrate in this region are: α_1-antitrypsin (accounts for 70-90% of α_1-peak), α_1-acid glycoprotein also known as orosomucoid (accounts for 10-20% of α_1-peak), α_1-lipoprotein, prothrombin, transcortin and thyroxine-binding globulin. Note that in acute phase reactions the elevation of the α_1-peak is due to increases in α_1-antitrypsin and α_1-acid glycoprotein, the two principal constituents of the peak. α_1-Peak elevations are also seen in chronic inflammatory and degenerative diseases, being highly elevated in a number of malignant diseases. When the α_1-peak is depressed it is important to rule out α_1-antitrypsin deficiency (see below) by direct measurements using radial immunodiffusion methods or other suitable techniques. The presence of α_1-fetoprotein in non-fetal serum is of clinical significance because of its close association with primary carcinoma of the liver. However, the routine serum electrophoresis seldom gives information in this regard and specific immunochemical methods should be employed in detecting the α_1-fetoprotein in the serum.

b. α_1-Antitrypsin (normal range: 160-350 mg per 100 ml of serum by radial immunodiffusion procedure), one of the principal protease inhibitors (abbreviated as P_i), is synthesized in the hepatocytes. It exhibits inhibitory activity against a broad spectrum of proteolytic enzymes: trypsin, chymotrypsin, elastase and neutral protease. The latter two enzymes released from polymorphonuclear granulocytes can destroy pulmonary tissues unless inhibited. Thus the major role of α_1-antitrypsin in man appears to be the protection of pulmonary tissue from the granulocytic proteases. It is synthesized exclusively in the liver and has a half-life of three to six days. α_1-Antitrypsin is a small-size glycoprotein (molecular mass of 50,000 daltons; consists of a single peptide chain with 12% carbohydrate including sialic acid residue) and is able to penetrate into a wide variety of body fluids. Thus, α_1-antitrypsin may have a widespread function in the protection of tissues against the destructive action of proteolytic enzymes. α_1-Antitrypsin also inhibits collagenases (from skin and granulocytes), plasmin and thrombin, but the principal inhibitor of these enzymes appear to be α_2-macroglobulin. α_1-Antichymotrypsin, Cl esterase inhibitor (deficiency in hereditary angioedema) and antithrombin III (most important inhibitor of thrombin) are other protease inhibitors that are present in the plasma.

α_1-Antitrypsin has been shown to be a polymorphic protein. At least 21 alleles of the gene are known. The concentration of this protease inhibitor in serum is determined by a series of co-dominant alleles. The common normal genotype is designated $P_iM(MM)$. $P_iZ(ZZ)$ is the predominant and clinically significant genotype associated with α_1-antitrypsin deficiency and shows serum values which are 10-20% that of normal mean adult values. P_iS, P_iW and P_iP genotypes have intermediate values which are 60% that of normal. Two disorders, familial emphysema and hepatitis syndrome in infancy and childhood, have been associated with α_1-antitrypsin deficiency, mostly of type P_iZZ and a few with types SZ, SS or PZ. The lung disease, as noted earlier, presumably results from the gradual destruction of elastin, due to the decreased inhibition of elastase released by granulocytes. A variety of secondary factors (smoking, air pollutants, pulmonary infection) contribute towards the age of onset and severity of the disease. It should be pointed out, however, that not all familial emphysemas can be attributed to severe α_1-antitrypsin deficiency. Furthermore, the risk of developing pulmonary emphysema among the heterozygotes (P_iMZ, MS, MZ) is apparently no different from normal individuals with the homozygous genotype P_iM.

The association of hepatitis-like syndrome (sometimes progressing to cirrhosis with liver failure and death) with α_1-antitrypsin deficiency in infancy and childhood is well established, although the pathogenesis of this disorder is not known at the present time. Furthermore, not all infants with severe deficiency of α_1-antitrypsin go on to develop liver disease. An interesting observation, which may have bearing on the mechanism of the disease, is that individuals with genotype P_iZ are deficient in α_1-antitrypsin due to their inability to secrete α_1-antitrypsin from the liver to serum. Electron microscopic studies have revealed that there is an accumulation of α_1-antitrypsin which is sialic acid-deficient in the cisternae of rough endoplasmic reticulum of affected liver cells.

6. α_2-Globulin

The α_2-fraction includes haptoglobin, α_2-macroglobulin, α_2-microglobulin, ceruloplasmin, erythropoietin and cholinesterase. Haptoglobin is nonspecifically increased in the acute stress reaction in the presence of inflammation, tissue necrosis or destruction. A function of haptoglobin is to combine with hemoglobin in order to remove it from the circulation. Thus, in an episode of intravascular hemolysis, there is a depletion of haptoglobin which may require a week or more to return to normal serum levels. The haptoglobin levels in these instances should be quantitated by immunochemical procedures or techniques which involve saturation of haptoglobin with hemoglobin and separation of the uncomplexed hemoglobin by electrophoresis. As noted earlier, α_2-macroglobulin functions as a protease inhibitor. In the nephrotic syndrome there is a characteristic elevation of the α_2-peak with hypoalbuminemia (see Figure 70).

7. β-Globulins

a. Quantitatively significant β-globulins are composed of transferrin, β-lipoproteins, complement C3 and hemopexin. Transferrin, synthesized in the liver, is an iron transport protein in serum which accounts for about 60% of the β-peak. It is increased in iron-deficiency anemia and pregnancy (see iron metabolism, Chapter 7). Transferrin levels are decreased in metabolic (e.g., liver, kidney) and neoplastic diseases.

b. A pattern often observed in advanced cirrhosis is the lack of separation of the β-peak with the gamma region, described as β-γ bridging (see Figure 67). This is presumably due to elevated levels of IgA. In addition, hypoalbuminemia is also found in this disease.

c. Complement levels (C3 and/or C4) are decreased (due to its consumption) in diseases associated with the formation of immune complexes. Examples include glomerulonephritis (acute and membrano-proliferative), systemic lupus erythematosis, and others.

d. β_2-Microglobulin is present in small quantities. Thus, alterations in its level is not reflected in the cellulose acetate electrophoresis (normal range: 155-225 µg/100 ml of serum--determined by a radioimmunoassay procedure). It is a small protein with a molecular mass of 11,700 daltons and is a constituent of many types of cell membranes including those of lymphoid cells. The molecule consists of 57 amino acid residues with one intrachain disulfide linkage. The amino acid sequence of β_2-microglobulin shows a striking homology with immunoglobulin (IgG-C_H3 and C_L regions) polypeptide

chains and is part of the histocompatability antigens. The
exact function of β_2-microglobulin is not known, but it is
thought to be involved in some immunologic process.
β_2-Microglobulin levels in serum and other biologic fluids
are elevated in conditions where there is increased growth
and augmented cell turnover, such as normal growth of the
fetus, neoplasms, inflammation, and immunologic diseases.
In renal failure its level is enhanced due to a depression
of catabolism by the renal tissue. The measurements of
salivary β_2-microglobulin levels have been found to be
useful in following the lymphoid infiltration of the
salivary glands of patients with Sjögren's syndrome, which
is an autoimmune disease characterized by keratoconjuctivitis
sicca (dry eyes), xerostomia (dry mouth) and an associated
connective tissue disorder.

8. <u>Gamma Globulins</u> (Also refer to Chapter 11 for further discussion)

The five major classes of immunoglobulins in descending order of
quantity are IgG, IgA, IgM, IgD and IgE. C-Reactive protein
migrates with the gamma globulins and is increased in trauma and
acute inflammatory processes. During an intense acute
inflammatory process, its concentration may be highly elevated,
giving rise to a sharp protein band which may be mistaken for
a true monoclonal gammopathy (see 8C). The variations in the
electrophoretic pattern due to the gamma globulins can be
categorized into the following three groups.

a. Agammaglobulinemia and hypogammaglobulinemia: May be primary
or secondary. The secondary forms may be found in chronic
lymphocytic leukemia, lymphosarcoma, multiple myeloma, the
nephrotic syndrome, long-term steroid treatment and,
occasionally, in overwhelming infection.

b. Polyclonal Gammopathy: A diffuse polyclonal increase in
gamma globulin, primarily in the IgG fraction. Recall that
in chronic stress there is an elevation of the gamma globulins.
The major categories that give rise to polyclonal gammopathy
are:

i) Chronic Liver Disease: An increase in polyclonal IgG
and IgA with a decrease in albumin levels is a
characteristic finding irrespective of disease etiology.
Recall that the β-γ bridging observed in cirrhosis is
presumably due to elevated levels of IgA which migrates
to a position between the β and γ region, increasing
the trough between these bands.

ii) Sarcoidosis: The electrophoretic pattern shows a stepwise descent of globulin fractions starting from the gamma globulins.

iii) Autoimmune Diseases: Rheumatoid arthritis (increased IgA levels), systemic lupus erythematosus (increased IgG and IgM levels), etc.

iv) Chronic Infectious Disease: Bronchiectasis, chronic pyelonephritis, malaria, chronic osteomyelitis, Kala-azar, leprosy, etc.

c. Monoclonal Gammopathies: Upon densitometric tracing, a protein band shows a thin, spike-like configuration which is due to the presence of a homogeneous class of protein. The pattern with a spike should be further investigated using a number of other laboratory and clinical parameters to rule out multiple myeloma. Although the monoclonal spike is usually observed in the γ-region, it can occasionally be seen in the β or α_2 region. Note that in pregnancy a monoclonal spike in the β region may be due to elevated transferrin levels. The diagnosis of multiple myeloma consists of the presence of monoclonal protein in serum or urine (in the light chain disease, a monoclonal spike in the serum electrophoresis may not be seen, however, the urine electrophoresis shows a monoclonal spike due to Bence-Jones proteins--K or λ light chains; also, in the serum electrophoresis, total gamma globulins are often decreased), immunoelectrophoresis, bone marrow plasmacytosis with plasma cell nodules, radiologic bone lesions and other clinical symptoms. It should also be emphasized that monoclonal proteins without any significant clinical illness have been found with the incidence of occurrence increasing with age. For example, as many as 3% of those over the age of 70 show the presence of monoclonal protein (benign monoclonal proteinemia). The secondary paraproteinemias may be seen in association with hematopoietic malignancy (lymphomas and leukemias), other types of neoplasms (e.g., colon carcinoma), long-standing chronic urinary or biliary tract infection, rheumatoid factor related to IgM monoclonal protein, and amyloidosis.

9. In summary, the alterations (acute phase response) that occur after tissue injury or inflammation are as follows. Constituents that are elevated (positive acute phase reactants) include haptoglobin, α_1-antitrypsin, α_1-antichymotrypsin, α_1-acid glycoprotein, ceruloplasmin, complement components C2, C3 and C4,

antihemophilic globulin, fibrinogen (reflected in the accelerated erythrocyte sedimentation rate), etc. Proteins (negative acute phase reactants) that are reduced consist of albumin, thyroxine-binding prealbumin, and transferrin. These proteins return to normal levels after about ten days to two weeks. However, if the injury persists and there is antigenic exposure (endogenous or exogenous), the immunoglobulins and the negative acute phase reactants begin to rise with a slight elevation in the positive reactants.

Some General Reactions of Amino Acids

1. Synthesis and Interconversion of Amino Acids

    As has been mentioned, essential amino acids must be obtained preformed in the diet. The other amino acids may be synthesized by the conversion of certain α-keto acids to amino acids (the α-keto group is replaced by an amino group). Note that this synthesis by conversion of α-keto acids occurs in several of the reactions discussed below.

2. Deamination of Amino Acids

    Removal of the α-amino group of amino acids is the first step in their catabolism. There are two general categories of deamination: oxidative and non-oxidative.

    a. Oxidative deamination is catalyzed by amino acid oxidases, of which there are two major types: L-amino acid oxidase, and D-amino acid oxidase.

        i) L-amino acid oxidase

α-amino acid   α-imino acid   α-keto acid

L-amino acid oxidase occurs in the liver and kidney (it is
not widely distributed) and is a flavoprotein containing
either FMN or FAD as a coenzyme. The enzyme does not attack
glycine, dicarboxylic amino acids, or β-hydroxy amino acids.
The activity of this enzyme is very low and the metabolism
of L-amino acids by this pathway is not of major importance
for the above reasons.

ii) High levels of D-amino acid oxidase (also called glycine
     oxidase) activity is found in the liver and kidney. This
     enzyme contains FAD and deaminates many amino acids,
     including glycine (L-amino acid oxidase is inactive towards
     glycine). The oxidation of the glycine is shown below.
     The reactions of the other amino acids (D-isomers) are
     entirely analogous to this.

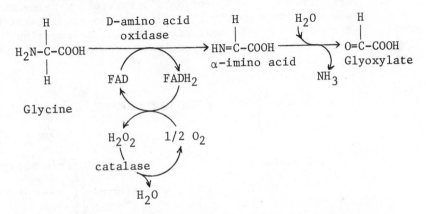

D-amino acid oxidases are found in <u>peroxisomes</u>, also
known as microbodies. Peroxisomes are single membrane
sacs which bud off from the smooth endoplasmic reticulum.
They contain enzymes which are responsible for the
production of H_2O_2 such as D-amino acid oxidase,
L-α-hydroxy acid oxidase, citrate dehydrogenase and
presumably L-amino acid oxidase. They also contain
catalase and peroxidase which <u>destroy</u> hydrogen peroxide.
In the white cells, the killing of bacteria involves the
hydrolytic activity of the lysosomes and the production
of H_2O_2 by the peroxisomes (see under red cell metabolism).

iii) D-amino acid oxidase is very important because it
     catalyzes the conversion of a D-amino acid (which is
     normally not used by the body) to an α-keto acid which
     no longer has an asymmetrical α-carbon. The keto acid
     may be reaminated to form an L-amino acid which can then
     be used by the body. Because this conversion from D- to
     L-amino acids is possible, the body can use D-amino acids
     in the diet as a source of essential amino acids.

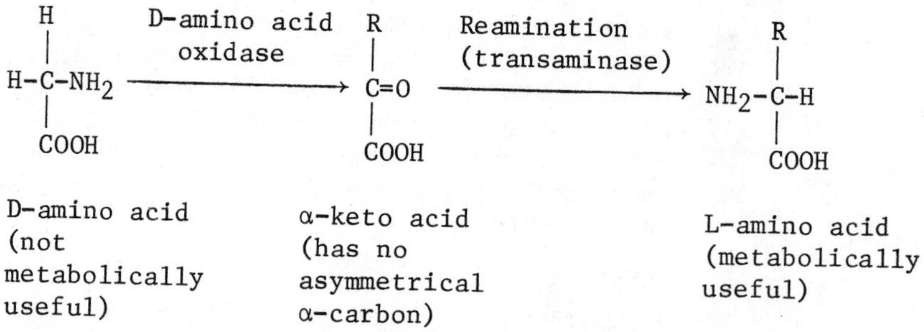

$$\underset{\substack{\text{H}\\|\\\text{H-C-NH}_2\\|\\\text{COOH}}}{} \quad\xrightarrow[\text{oxidase}]{\text{D-amino acid}}\quad \underset{\substack{\text{R}\\|\\\text{C=O}\\|\\\text{COOH}}}{} \quad\xrightarrow[\text{(transaminase)}]{\text{Reamination}}\quad \underset{\substack{\text{R}\\|\\\text{NH}_2\text{-C-H}\\|\\\text{COOH}}}{}$$

D-amino acid   α-keto acid   L-amino acid
(not   (has no   (metabolically
metabolically   asymmetrical   useful)
useful)   α-carbon)

iv) Glutamate dehydrogenase plays a major role in the metabolism of amino acids. It is present in both the cytoplasm and mitochondria of the liver. The enzyme catalyzes the reversible oxidative deamination of L-glutamate by NAD^+ (or $NADP^+$) to yield α-ketoglutarate and NH_3. Presumably the initial step consists of the formation of α-iminoglutarate by dehydrogenation, which is followed by hydrolysis of the imino acid to a ketoacid and NH_3:

$$\underset{\text{L-Glutamate}}{\text{HOOC-CH}_2\text{-CH}_2\text{-}\underset{\substack{|\\\text{NH}_2}}{\overset{\substack{\text{H}\\|}}{\text{C}}}\text{-COOH}} + NAD^+ \rightleftharpoons \text{HOOC-CH}_2\text{-CH}_2\text{-}\underset{\substack{||\\\text{NH}}}{\text{C}}\text{-COOH} + NADH + H^+$$

α-iminoglutaric acid

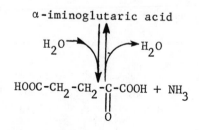

$$\text{HOOC-CH}_2\text{-CH}_2\text{-}\underset{\substack{||\\\text{O}}}{\text{C}}\text{-COOH} + NH_3$$

α-ketoglutarate

The net reaction is:

L-glutamate + $NAD(P)^+$ + $H_2O \rightleftharpoons$ α-ketoglutarate + $NAD(P)H$ + NH_4^+

Glutamate dehydrogenase is an allosteric protein. The enzyme from beef liver has a molecular mass of 336,000 daltons and contains six identical subunits. Its positive modulators (stimulators) are ADP, GDP and some amino acids. Its negative modulators (inhibitors) are ATP, GTP and NADH. The enzyme's activity is also affected by thyroxine and some steroid hormones.

The central role of glutamate dehydrogenase in amino acid metabolism is due to the fact that it is the only active amino acid dehydrogenase present in most cells and participates along with appropriate transaminases (or aminotransferases) in the deamination of all other amino acids.

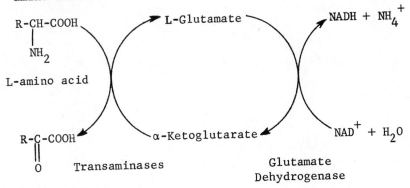

The NH_3 produced in the above reactions is toxic and must be detoxified by converting it to the harmless amides glutamine and asparagine and, eventually, to urea. Urea is then excreted in the urine (these aspects are discussed later in this chapter). The NADH generated is ultimately oxidized by the electron transport chain.

b. <u>Non-oxidative deamination</u> is accomplished by several specific
   enzymes, some of which are shown here.

   i) Amino acid dehydrases

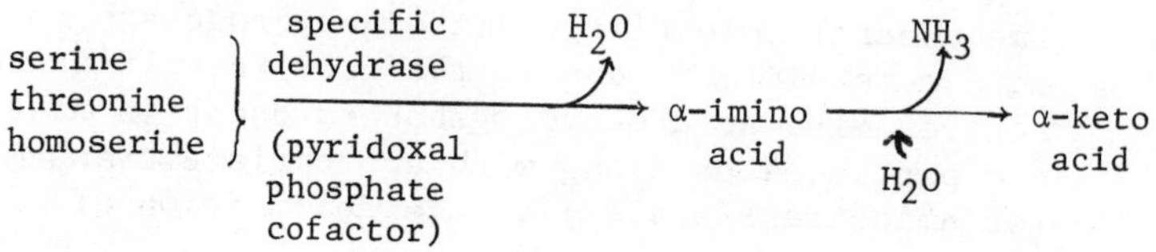

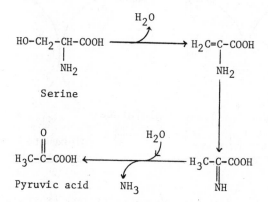

   Example:

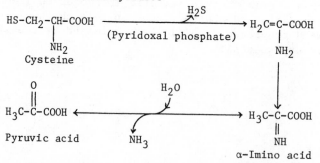

   ii) Amino acid desulfhydrases

3. Reamination

In these reactions, NH_3 is assimilated. Following are some examples.

a. Amination of α-keto acids to give the respective α-amino acids (this has already been mentioned as a mode of synthesis of amino acids).

b. Synthesis of carbamyl phosphate (used in pyrimidine and urea biosynthesis).

c. Glutamine and asparagine synthesis.

4. Transamination of Amino Acids

This combines both deamination and amination reactions. Transaminations are reversible and catalyzed by transaminase enzymes (also called aminotransferases) which are intracellular. These enzymes are found in practically all tissues but are especially numerous in the heart, brain, kidney, testicle, and liver. The general process of transamination is shown below.

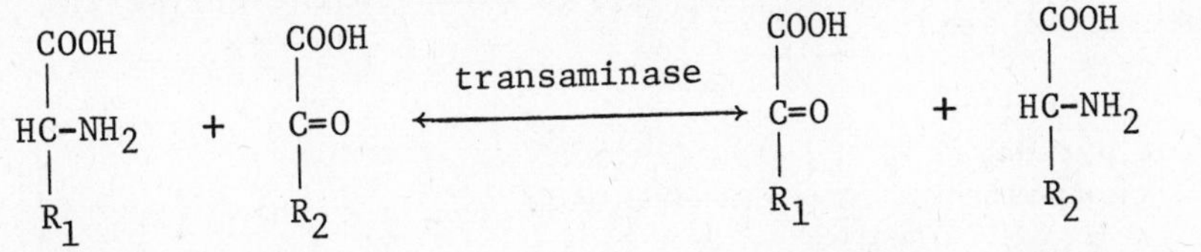

$$
\begin{array}{c}
\text{COOH} \\
| \\
\text{HC-NH}_2 \\
| \\
\text{R}_1
\end{array}
\;+\;
\begin{array}{c}
\text{COOH} \\
| \\
\text{C=O} \\
| \\
\text{R}_2
\end{array}
\;\xrightleftharpoons[\text{transaminase}]{}\;
\begin{array}{c}
\text{COOH} \\
| \\
\text{C=O} \\
| \\
\text{R}_1
\end{array}
\;+\;
\begin{array}{c}
\text{COOH} \\
| \\
\text{HC-NH}_2 \\
| \\
\text{R}_2
\end{array}
$$

amino acid   keto acid   keto acid   amino acid
   (1)   (2)   (1)   (2)

All naturally occurring amino acids participate in transaminase reactions.

Examples:

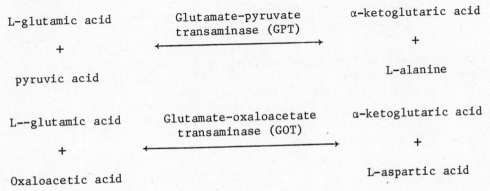

L-glutamic acid $\xrightarrow{\text{Glutamate-pyruvate transaminase (GPT)}}$ α-ketoglutaric acid

+

pyruvic acid

L-alanine

L--glutamic acid $\xrightarrow{\text{Glutamate-oxaloacetate transaminase (GOT)}}$ α-ketoglutaric acid

+

Oxaloacetic acid

L-aspartic acid

The most common general type of reaction found in animals, plants, and microorganisms is transamination with α-ketoglutaric acid.

L-amino acid

+   ←————————————→

α-ketoglutaric acid

α-keto acid

+

L-glutamic acid

5. All transaminase reactions appear to have the same mechanism and use the same coenzyme (pyridoxal phosphate, a derivative of vitamin B_6 which is important in many reactions involving α-amino acids). Pyridoxal phosphate serves as a carrier of amino groups or amino acids. The coenzyme is reversibly transformed from its free aldehyde form, pyridoxal phosphate, to its aminated form, pyridoxamine phosphate.

Pyridoxal phosphate is bound tightly to the enzyme by the formation of a Schiff base with the ε-amino group of a specific lysine residue at the active site and noncovalently through the positively charged nitrogen atom of the ring. Following is a schematic diagram of pyridoxal phosphate bound to the enzyme in the absence of substrate.

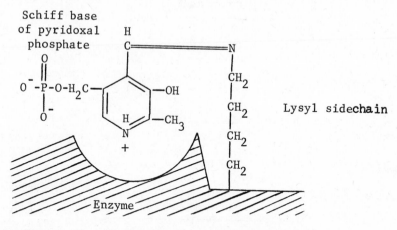

During catalysis, the amino acid substrate displaces the lysyl ε-amino group of the enzyme from the pyridoxal phosphate and forms a Schiff base with the pyridoxal phosphate.

a. Role of pyridoxal phosphate in the conversion of an α-amino acid to an α-keto acid

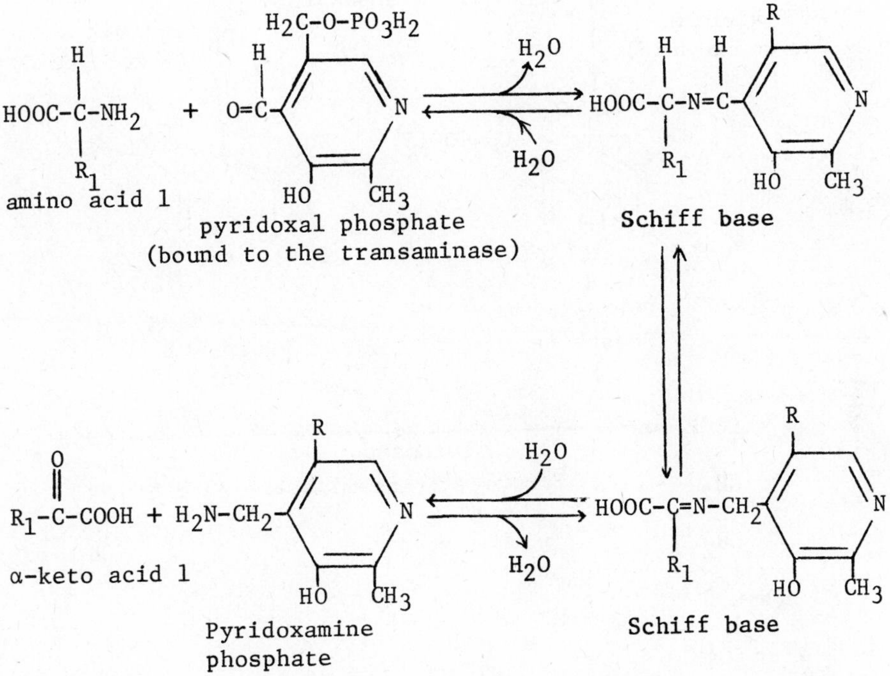

$$HOOC-\underset{\underset{R_1}{|}}{\overset{\overset{H}{|}}{C}}-NH_2 \;+\; \text{pyridoxal phosphate (bound to the transaminase)} \;\underset{H_2O}{\overset{H_2O}{\rightleftharpoons}}\; \text{Schiff base}$$

amino acid 1

pyridoxal phosphate
(bound to the transaminase)

Schiff base

$$R_1-\overset{\overset{O}{\|}}{C}-COOH \;+\; H_2N-CH_2-(\text{pyridoxamine ring}) \;\underset{H_2O}{\overset{H_2O}{\rightleftharpoons}}\; \text{Schiff base}$$

α-keto acid 1

Pyridoxamine phosphate

Schiff base

$$R = -H_2C-O-PO_3H_2$$

b. Conversion of an -keto acid to an -amino acid requires pyridoxaine phosphate and reversal of the above reactions.

$$\begin{array}{c}\text{α-keto acid 2} \\ + \\ \text{pyridoxamine phosphate}\end{array} \rightleftharpoons \begin{array}{c}\text{α-amino acid 2} \\ + \\ \text{pyridoxal phosphate}\end{array}$$

The two reactions above (a and b) are coupled as shown below.

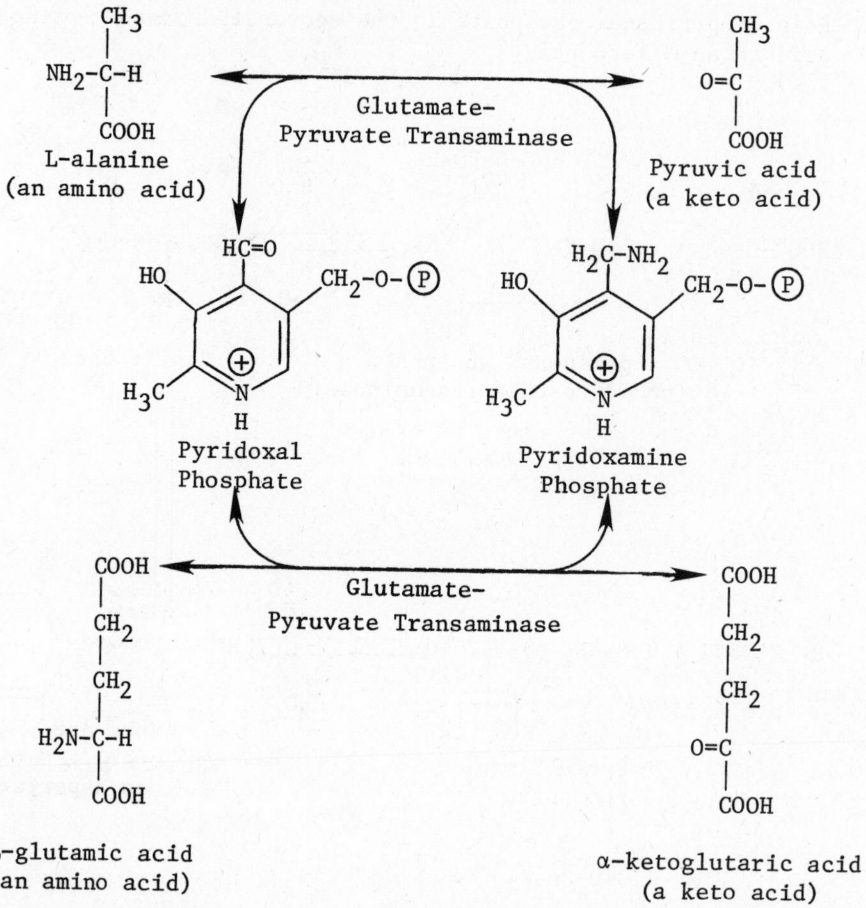
CH3
NH2-C-H
COOH
L-alanine
(an amino acid)
Glutamate-
Pyruvate Transaminase
CH3
O=C
COOH
Pyruvic acid
(a keto acid)
HC=O
HO
CH2-O-P
H3C
N
H
Pyridoxal
Phosphate
H2C-NH2
HO
CH2-O-P
H3C
N
H
Pyridoxamine
Phosphate
COOH
CH2
CH2
H2N-C-H
COOH
Glutamate-
Pyruvate Transaminase
COOH
CH2
CH2
O=C
COOH
L-glutamic acid
(an amino acid)
α-ketoglutaric acid
(a keto acid)

c. As already noted, pyridoxal phosphate is not only the prosthetic group of all transaminases, but also of other enzymes such as decarboxylases, dehydrases, desulfhydrases, racemases and amino acid aldolases. The diverse reactions that pyridoxal phosphate participates in are due to its ability to labilize various bonds of the amino acid when complexed with the prosthetic group.

In the interaction between pyridoxal phosphate and amino
acid, an electron pair is removed from the α-carbon of the
amino acid and transferred to the positively charged
pyridine ring, which is subsequently returned to the
α-carbon of the keto acid. Thus in transaminase reactions,
pyridoxal phosphate functions both as a carrier of amino
groups as well as electron pairs.

6. Physiological Significance of Transaminases

a. Two transaminases of particular clinical importance are
glutamic oxaloacetic transaminase (GOT) and glutamic pyruvic
transaminase (GPT). GOT and GPT are also known as aspartate
aminotransferase (AST) and alanine aminotransferase (ALT),
respectively.

b. Because these are intracellular enzymes, the serum levels of
GOT and GPT are normally very low. Any significant tissue
breakdown gives rise to high serum transaminase levels. For
example, in myocardial infarction, there is an increase in the
serum level of GOT (heart muscle contains high concentrations
of this enzyme). There are alterations in serum GOT and GPT
levels in some of the liver diseases (infectious hepatitis,
infectious mononucleosis). One also finds high serum levels
of these enzymes in conditions where there is damage to
skeletal muscle. When these enzymes are measured in the serum,
the values obtained are referred to as SGOT and SGPT
activities (where the S refers to serum).

Table 23.   Tissue Distribution of Transaminases in Man*

	GOT	GPT
Heart	156	7.1
Liver	142	44
Skeletal Muscle	99	4.8
Kidney	91	19
Pancreas	28	2
Spleen	14	1.2
Lung	10	0.7
Serum	0.02	0.016

Values expressed in terms of units x 10^{-4}/gm wet tissue homogenate.

---

* From Wróblewski, F. and La Due, J., Proc. Soc. Exp. Biol. Med., 91, 569 (1956).

c.   Assay for Transaminases

As was discussed in the section on clinical enzymes, there are two basic methods available for measuring enzyme activity: continuous-monitoring procedures which consist of measuring initial velocities during zero-order kinetics with respect to substrate, and fixed incubation (or two-point assays) in which a cumulative change is measured. In general, the preferred methods are the continuously monitored procedures (due to maintenance of zero-order kinetics and the prevention of product inhibition--oxaloacetate is a potent inhibitor of GOT activity). Since direct monitoring of the transaminase reaction is not possible, it is coupled with the oxidation of NADH, which is catalyzed by appropriate dehydrogenases:

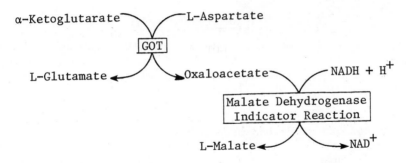

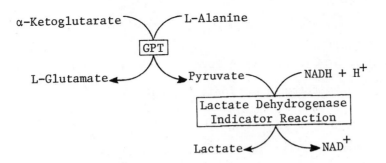

The reactions are monitored continuously by measuring decrease in absorbance at 340 nm. The enzyme activity is calculated by measuring the change in absorbance per unit time ($\Delta A/t$) which is directly related to micromoles of NADH oxidized and, in turn, to micromoles of substrate consumed per unit time.

Specific comments on the SGOT assay: Side reactions which involve direct or indirect oxidations of NADH, by endogenous or exogenous substrates and enzymes, constitute a source of error in the assay procedure. Some of these reactions are:

i)   Pyruvate + NADH + H$^+$ $\xrightarrow{\text{LDH}}$ Lactate + NAD$^+$

ii) α-Ketoglutarate + NH$_4^+$ + NADH $\xrightarrow{\text{GDH}}$ L-Glutamate + NAD$^+$

iii) Alanine + α-Ketoglutarate $\xrightarrow{\text{GPT}}$ L-Glutamate + Pyruvate

Pyruvate generated in reaction (iii) is again reduced in reaction (i). Attempts have been made to minimize or eliminate the side reactions. In one procedure it requires preincubation of the serum specimen with lactate dehydrogenase which converts endogenous pyruvate to lactate. The primary reaction is initiated by the addition of α-ketoglutarate. Since this technique does not eliminate other side reactions, a better method consists of double-beam analysis in which side reactions are allowed to take place in one cuvette, while in the other cuvette all reactions including transaminases are allowed to occur. The difference in change in absorption between these two reactions will reflect true transaminase activity.

In order to obtain optimal activity, addition of pyridoxal phosphate to serum has been found to be necessary.

d. Normal causes for variations in transaminase activity include

 i) Dietary factors: increased (or decreased) pyridoxine intake is followed by increased (or decreased) transaminase levels.

 ii) Physical activity: rigorous physical activity produces high SGOT resulting from skeletal muscle breakdown.

 iii) Sex differences: insignificant.

 iv) Pregnancy: slight increase in both SGOT and SGPT levels.

 v) Age differences: neonatal SGOT levels are normally very high.

e. Conditions which result in hyper serum transaminase levels

 i) Myocardial infarction. Degree and duration of serum GOT elevation depend upon the magnitude of infarct but SGOT can, at its peak, be as much as ten times the normal value. GOT is high within the first 12 hours following the attack, reaching a maximum at 24-48 hours and returning to normal in 4-7 days (post infarction). In myocardial infarction, the increase of SGOT is greater than the increase in SGPT. The serum levels of LDH isoenzymes (1 & 2) and CPK-2, however, are more specific indicators of myocardial infarction.

 ii) Infectious hepatitis (viral). The elevation of GPT is generally more than that of GOT and the elevation may persist for many months after the onset. During the recovery, there is no correlation of symptoms with transaminase levels. The continued high transaminase levels during recovery from hepatitis indicate continuing hepatocellular necrosis.

 iii) Infectious mononucleosis. Alteration in serum GPT and GOT levels appear to parallel the subjective manifestations. Elevations of GPT are more striking than GOT, probably indicating hepatic involvement.

iv) Renal infarction. High levels of GOT are seen in renal arterial infarction.

v) Severe burns. The skin contains a relatively low concentration of transaminase, but extensive burns may liberate quite a large quantity of these enzymes.

vi) Trauma. Transaminase elevations here are usually more transient than those stemming from myocardial infarction, but the diagnosis of myocardial infarction in accident victims and surgical patients is complicated by these elevations.

vii) Poliomyelitis. In acute paralytic poliomyelitis there is a rise in transaminase activity in cerebrospinal fluid, probably a result of nervous tissue damage. Poliomyelitis during the stage of viremia sometimes produces myocarditis and subclinical liver damage, and thus an increase in serum transaminase levels. These complications probably do not cause the elevated cerebrospinal fluid levels, however.

7. Transamidation

The amide nitrogen of glutamine may serve as a source of amino groups. This is illustrated by a step in the synthesis of glucosamine, in which the amide nitrogen of glutamine is transferred to the keto group of fructose-6-phosphate, a reaction catalyzed by a transamidase.

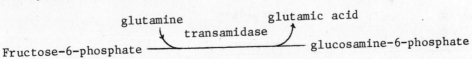

8. Decarboxylation

Decarboxylation of amino acids to produce amines is catalyzed by pyridoxal-dependent bacterial decarboxylases. Toxic amines (ptomaines) are produced in this way during putrefaction of food protein. Examples of this sort of reaction are shown below.

a.

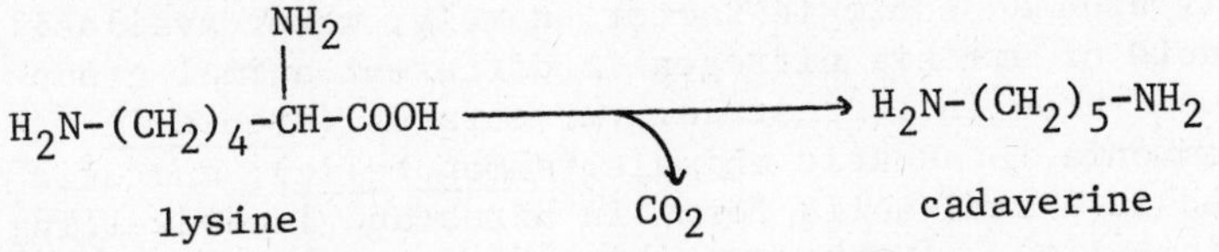

$$H_2N-(CH_2)_4-\underset{\underset{lysine}{}}{\overset{\overset{NH_2}{|}}{CH}}-COOH \xrightarrow[\;CO_2\;]{} \underset{cadaverine}{H_2N-(CH_2)_5-NH_2}$$

b.

$$\underset{\text{ornithine}}{\text{H}_2\text{N}-(\text{CH}_2)_3-\overset{\overset{\displaystyle \text{NH}_2}{|}}{\text{CH}}-\text{COOH}} \longrightarrow \underset{\text{putrescine}}{\text{H}_2\text{N}-(\text{CH}_2)_3-\text{NH}_2}$$

with CO_2 released

c.

$$\underset{\text{arginine}}{\text{H}_2\text{N}-\overset{\overset{\displaystyle \text{NH}}{||}}{\text{C}}-\text{NH}-(\text{CH}_2)_3-\overset{\overset{\displaystyle \text{NH}_2}{|}}{\text{CH}}-\text{COOH}} \longrightarrow \underset{\text{agmatine}}{\text{H}_2\text{N}-\overset{\overset{\displaystyle \text{NH}}{||}}{\text{C}}-\text{NH}-(\text{CH}_2)_4-\text{NH}_2}$$

with CO_2 released

d. The formation of tyramine and histamine from tyrosine and histidine respectively, is shown along with the metabolism of these amino acids (see later in this section under histidine and tyrosine metabolism).

e. Two other amines synthesized from complex acid decarboxylation products are spermine and spermidine.

$$\text{H}_2\text{N}-(\text{CH}_2)_3-\text{NH}-(\text{CH}_2)_4-\text{NH}-(\text{CH}_2)_3-\text{NH}_2$$

Spermine

$$\text{H}_2\text{N}-(\text{CH}_2)_3-\text{NH}-(\text{CH}_2)_4-\text{NH}_2$$

Spermidine

Metabolism of Ammonia and Urea Formation

1. Ammonia (NH_3; at physiological pH, 98.5% exists as ammonium ion, NH_4^+), the highly toxic waste product of protein catabolism, is rapidly inactivated by a variety of reactions. Some of the detoxified metabolites are utilized (thus salvaging a portion of the amino nitrogen) and some are excreted. The excreted detoxified form of ammonia varies quite widely among different vertebrate and invertebrate groups. The development of a pathway for nitrogenous waste disposal in a given species appears to depend chiefly upon an ecologic factor, namely, water availability. The end products of ammonia nitrogen in different animal groups are as follows: urea in terrestrial vertebrates (ureotelic organisms); ammonia in aquatic animals (ammonotelic); and uric acid (excreted in a semi-solid form) in birds and land-dwelling reptiles (uricotelic). In the amphibia, during their aquatic phase of development, ammonia is excreted.

After metamorphosis, the adult frog excretes ammonia nitrogen as urea. During metamorphosis, the liver develops the capability to produce the enzymes required for the synthesis of urea. <u>The principal end-product of protein (or amino nitrogen) metabolism in man is urea.</u>

2. Normally, only faint traces of NH_3 are present in the blood. This is because NH_3 is rapidly removed by the following reactions.

    a. Glutamate dehydrogenase reaction:

$$\alpha\text{-Ketoglutarate} + NH_3 + NAD(P)H + H^+ \rightarrow \text{Glutamate} + NAD(P)^+ + H_2O$$

    b. Amidation of glutamic acid to form glutamine:

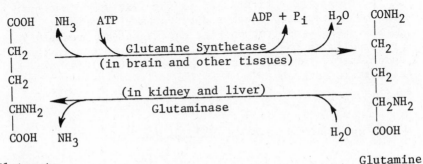

Glutamate   Glutamine

Glutamine functions in the temporary non-toxic storage and transport of NH_3, thus maintaining NH_3 concentrations below toxic levels in the tissues. This is particularly important in the brain which is extremely sensitive to NH_3. Glutamine also functions as a donor of the amide nitrogen in the synthesis of purines, pyrimidines and of amino sugars found in the structural polysaccharides. The reverse reaction catalyzed by glutaminase produces NH_3 in the liver and kidney. In the liver, NH_3 is converted to urea; in the kidney (as ammonium ions), it plays a role in the maintenance of acid-base balance (renal ammonia formation is increased markedly in metabolic acidosis).

    c. By far the most important route for the detoxification of NH_3 is the urea cycle (Krebs-Henseleit cycle) which is discussed later.

3. Sources of NH_3: The endogenous production is due to amino acid oxidation (quantitatively not significant) and deamination of glutamine, glutamate and adenylate. In muscle, deamination of adenylate is responsible for the major production of ammonia. As noted earlier, glutamine yields glutamic acid and NH_3 in the

liver and kidney. Furthermore, ammonia is converted to urea in the liver, which is transported via the blood to the kidneys and ultimately excreted in the urine. In addition, NH_3 in the kidney (derived from glutamine by the glutaminase reaction and dependent upon the acid-base balance) is excreted as ammonium ions. A considerable quantity of NH_3 is derived from the action of bacterial enzymes in the lower intestine on exogenous sources such as urea and other nitrogenous compounds. The liberated ammonia diffuses across the intestinal mucosa to enter the portal blood. It is taken up by the liver and converted to urea.

4. Urea formation (urea cycle; Krebs-Henseleit cycle)

a. Urea formation takes place in the liver by a cyclic series of reactions. The first step is the fixation of ammonia as carbamyl phosphate, catalyzed by carbamyl phosphate synthetase (CPSI). CPSI, a mitochondrial matrix enzyme present principally in the liver (and to a small extent in the gut and kidney), requires N-acetylglutamate (AGA) as an allosteric activator.

$$CO_2 + NH_3 + H_2O + 2\ ATP \xrightarrow[\text{Biotin}]{\text{AGA}} \underset{\text{Carbamyl Phosphate}}{H_2N-\overset{\overset{O}{\|}}{C}-O-\overset{\overset{O^-}{|}}{\underset{\underset{O}{\|}}{P}}-O^-} + 2\ ADP + P_i$$

In the above reaction (a complex and irreversible process), NH_3 is the preferred substrate which is obtained from portal circulation (exogenously derived source) and the glutamate dehydrogenase and glutaminase reactions.

Note that the carbamyl phosphate is also synthesized in the cytoplasm and is used as a precursor for pyrimidine biosynthesis (see pyrimidine metabolism). The cytoplasmic carbamyl phosphate synthetase (designated as CPSII) differs from its mitochondrial counterpart in a number of different ways: CPSII preferentially utilizes glutamine as a nitrogen source; does not require N-acetylglutamate; is inhibited by uridine triphosphate; is activated by 5-phosphoribosyl-1-pyrophosphate; and is much less active. Thus, in ureotelic organisms (e.g., man), there are two separate pools of carbamyl phosphate synthesized by two different enzymes. These enzymes are compartmentalized and used for different purposes with very little exchange taking place between them. However, it has been observed that when there is an accumulation of carbamyl phosphate in the mitochondria due to biochemical lesions in the pathway for urea synthesis, small amounts of mitochondrial carbamyl phosphate serve as a substrate for pyrimidine biosynthesis (discussed later).

b. The second reaction of the urea cycle consists of the transfer of the carbamyl group (of carbamyl phosphate) to ornithine, yielding citrulline. The reaction is catalyzed by ornithine transcarbamylase (also known as ornithine carbamyl transferase), a mitochondrial matrix enzyme:

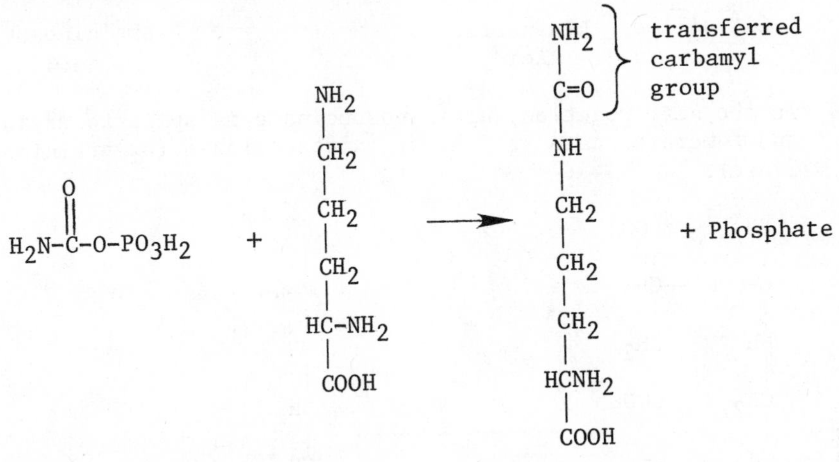

$$
H_2N-\overset{\overset{\displaystyle O}{\|}}{C}-O-PO_3H_2 \quad + \quad
\begin{matrix} NH_2 \\ | \\ CH_2 \\ | \\ CH_2 \\ | \\ CH_2 \\ | \\ HC-NH_2 \\ | \\ COOH \end{matrix}
\quad \longrightarrow \quad
\begin{matrix} \left.\begin{matrix} NH_2 \\ | \\ C{=}O \end{matrix}\right\} \begin{matrix}\text{transferred}\\ \text{carbamyl}\\ \text{group}\end{matrix} \\ NH \\ | \\ CH_2 \\ | \\ CH_2 \\ | \\ CH_2 \\ | \\ HCNH_2 \\ | \\ COOH \end{matrix}
\quad + \text{ Phosphate}
$$

Carbamyl Phosphate   L-Ornithine   L-Citrulline

Ornithine, required in the above reaction, is transported from the cytoplasm to the mitochondria by means of a specific inner-membrane transport system.

c. The remainder of the reactions in urea biosynthesis take place in the cytoplasm. Citrulline, derived from mitochondria, combines with aspartate to form argininosuccinic acid, catalyzed by arginosuccinate synthetase and driven by the hydrolysis of ATP to AMP and PP_i.

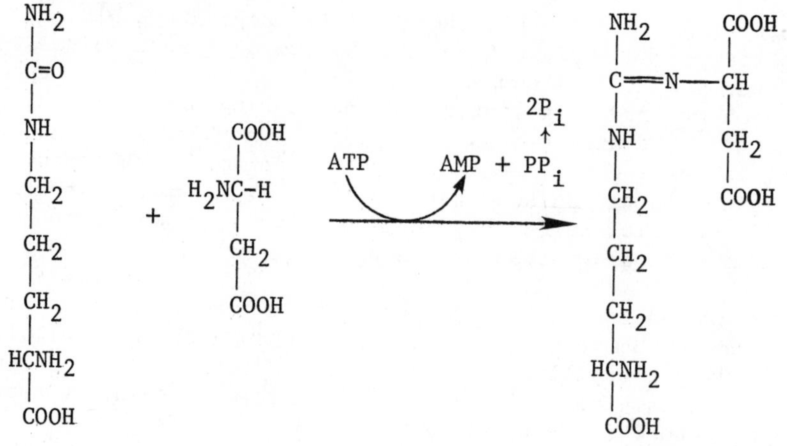

$$
\begin{array}{c}
\text{NH}_2 \\
| \\
\text{C=O} \\
| \\
\text{NH} \\
| \\
\text{CH}_2 \\
| \\
\text{CH}_2 \\
| \\
\text{CH}_2 \\
| \\
\text{HCNH}_2 \\
| \\
\text{COOH}
\end{array}
\quad + \quad
\begin{array}{c}
\text{COOH} \\
| \\
\text{H}_2\text{NC-H} \\
| \\
\text{CH}_2 \\
| \\
\text{COOH}
\end{array}
\quad
\xrightarrow[\text{ATP} \quad \text{AMP + PP}_i]{\quad\quad\quad 2\text{P}_i \quad\quad\quad}
\quad
\begin{array}{cc}
\text{NH}_2 & \text{COOH} \\
| & | \\
\text{C}=\!=\!\text{N---CH} \\
| & | \\
\text{NH} & \text{CH}_2 \\
| & | \\
\text{CH}_2 & \text{COOH} \\
| \\
\text{CH}_2 \\
| \\
\text{CH}_2 \\
| \\
\text{HCNH}_2 \\
| \\
\text{COOH}
\end{array}
$$

L-Citrulline   L-Aspartic   L-Argininosuccinic
                   Acid   Acid

d. In the next reaction, argininosuccinate is split to arginine
    and fumarate, catalyzed by argininosuccinase (or argininosuccinate
    lyase).

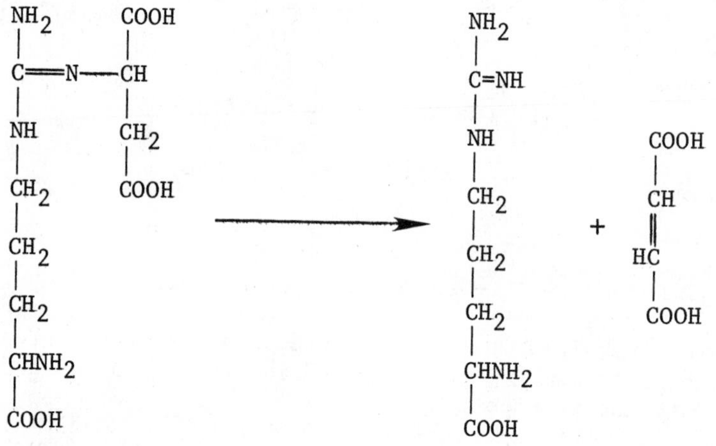

$$
\begin{array}{cc}
\text{NH}_2 & \text{COOH} \\
| & | \\
\text{C}=\!=\!\text{N---CH} \\
| & | \\
\text{NH} & \text{CH}_2 \\
| & | \\
\text{CH}_2 & \text{COOH} \\
| \\
\text{CH}_2 \\
| \\
\text{CH}_2 \\
| \\
\text{CHNH}_2 \\
| \\
\text{COOH}
\end{array}
\quad \longrightarrow \quad
\begin{array}{c}
\text{NH}_2 \\
| \\
\text{C=NH} \\
| \\
\text{NH} \\
| \\
\text{CH}_2 \\
| \\
\text{CH}_2 \\
| \\
\text{CH}_2 \\
| \\
\text{CHNH}_2 \\
| \\
\text{COOH}
\end{array}
\quad + \quad
\begin{array}{c}
\text{COOH} \\
| \\
\text{CH} \\
\| \\
\text{HC} \\
| \\
\text{COOH}
\end{array}
$$

L-Argininosuccinic   L-Arginine   Fumaric
      Acid   Acid

Note that this is the normal pathway for the synthesis of
arginine, a non-essential amino acid. Fumarate can be
converted to oxaloacetate via the tricarboxylic acid cycle

and the latter to aspartate by transamination (fumarate → malate → oxaloacetate → aspartate), which is then reutilized in the urea cycle.

e. In the last reaction the arginine is cleaved to urea with the regeneration of ornithine. The ornithine is transported into the mitochondrial matrix to initiate a new cycle in the synthesis of urea.

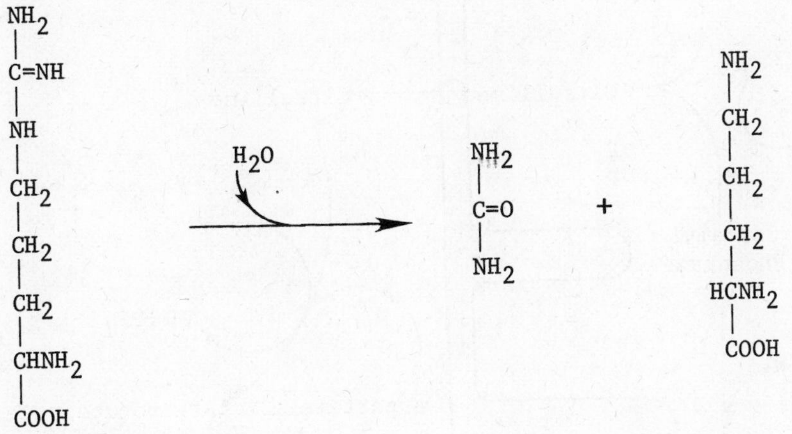

$$
\begin{array}{c}
NH_2 \\
| \\
C{=}NH \\
| \\
NH \\
| \\
CH_2 \\
| \\
CH_2 \\
| \\
CH_2 \\
| \\
CHNH_2 \\
| \\
COOH
\end{array}
\quad
\xrightarrow{\;H_2O\;}
\quad
\begin{array}{c}
NH_2 \\
| \\
C{=}O \\
| \\
NH_2
\end{array}
\quad + \quad
\begin{array}{c}
NH_2 \\
| \\
CH_2 \\
| \\
CH_2 \\
| \\
CH_2 \\
| \\
HCNH_2 \\
| \\
COOH
\end{array}
$$

L-Arginine   Urea   L-Ornithine

Ornithine generated in the above reaction combines with another molecule of carbamyl phosphate, thus initiating the next cyclic process. Note that of the two nitrogen atoms of urea, one arises from carbamyl phosphate, being ultimately derived from ammonia. The other nitrogen is derived from the α-amino group of aspartate which in turn may be obtained from the transamination of oxaloacetate with glutamate.

5. The complete urea cycle with compartmentalization is shown in Figure 71. The formation of one molecule of urea requires the hydrolysis of four high-energy phosphate groups of ATP. The overall reaction is as follows:

$$2\,NH_3 + CO_2 + 3\,ATP + 3\,H_2O \longrightarrow NH_2\overset{\overset{\displaystyle O}{\|}}{-C}-NH_2 + 2\,ADP + AMP + 4\,P_i$$

Urea

Figure   71.   The Krebs-Henseleit Cycle for Urea Biosynthesis Showing the
Steps Involved in the Mitochondrial Matrix and Cytoplasm of
the Liver Cell

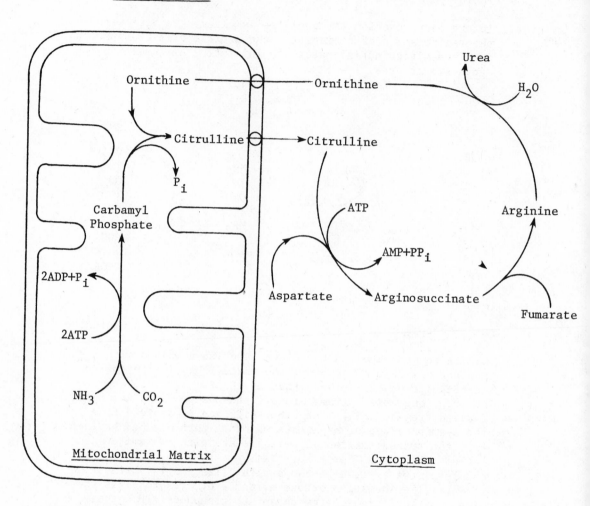

6.   The toxicity of NH_3 stems from several effects.

   a. NH_3 is basic ($NH_3 + H^+ \rightleftarrows NH_4^+$) and, thus, excess ammonia
   raises the intracellular pH. Metabolic alkalosis occurs in
   patients with biochemical lesions in urea synthesis.

   b.   In mitochondria, excess NH_3 drives the reductive amination
   of α-ketoglutarate catalyzed by glutamate dehydrogenase.
   This depletes the key intermediate of the tricarboxylic acid
   cycle, leading to its severe impairment. There results a
   severe inhibition of respiration in the brain with considerable

stimulation of glucose consumption via glycolysis. In the liver, excess ketone body formation (acetoacetate, β-hydroxybutyrate and acetone) from acetyl-CoA takes place.

c. Ammonia interferes with membrane functions, particularly the active transport of monovalent cations. In experimental animals, ammonia intoxication leads to the loss of cations through the mitochondrial membrane, with swelling and loss of matrix. The major clinical manifestation of ammonia toxicity involves the central nervous system and the neurologic sequelae include weakness, unsteadiness, tremors, decerebrate rigidity, seizures, coma and death.

The hyperammonemic conditions (the first three are congenital disorders) are caused by:

a. Inborn errors of urea cycle enzymes.

b. Some disorders of ornithine and lysine metabolism.

c. Some inborn deficiencies in the catabolism of branched-chain amino acids.

d. Chronic or acute liver failure (hepatic encephalopathy).

7. Inborn errors in each of the five enzymes of urea synthesis have been described. These are carbamyl phosphate synthetase deficiency, ornithine transcarbamylase deficiency, citrullinemia, arginosuccinic aciduria and argininemia. All of these disorders are characterized by hyperammonemia but are most pronounced in patients with defects in carbamyl phosphate synthetase and ornithine transcarbamylase. The toxic effects of hyperammonemia in these syndromes include episodes of vomiting, lethargy, and coma after ingestion of protein. In general, the treatment of these disorders consists of decreasing NH_3 production in the body which is accomplished by restricting protein intake and the administration of α-keto analogues of essential amino acids (in particular, valine, isoleucine, leucine, methionine, and phenylalanine). The severe variants of these disorders, particularly in the neonate and infant, lead to death due to hyperammonemia. It is of interest to point out that in patients with ornithine transcarbamylase deficiency, orotic aciduria is observed. This is presumably due to carbamyl phosphate accumulation behind the block, which is transported into the cytoplasm and serves as a substrate for pyrimidine biosynthesis. Recall that orotic acid is an intermediate in pyrimidine biosynthesis.

8. The exact biochemical lesions of the hyperammonemic disorders, which involve other than urea cycle enzymes, are not understood.

As noted earlier, these disorders are due to inborn errors of ornithine and lysine metabolism and catabolism of branched-chain amino acids. Some of these syndromes (the latter in particular), are associated with metabolic acidosis with or without ketosis, whereas urea cycle enzyme deficiency gives rise to metabolic alkalosis.

9.   In diseases of the liver (acute and chronic), the ability of the hepatocytes to detoxify ammonia is decreased, depending upon the severity of the damage. In portal-systemic encephalopathy (caused by cirrhosis), one of the important goals in the management of the disease is to bring about the reduction of ammonia levels.

10.   Patients with <u>Reye's syndrome</u>, presumably caused by a virus, show <u>transient</u> elevation in serum ammonia which has been ascribed to reduced activities of hepatic mitochondrial carbamyl phosphate synthetase and ornithine transcarbamylase. Ultrastructure studies have shown an acute mitochondrial damage. In the hepatocyte, the reesterification of mobilized fatty acids, due to active lipolysis in adipose and other tissues, produces the microvesicular fatty infiltration which appears as lipid droplets uniformly distributed throughout the cell. In this disorder, however, all organs with the exception of the brain appear to recover from the transient dysfunction of mitochondria. Death in this disorder is due to massive swelling of the brain (cerebral edema) caused by disturbances in the maintenance of water and electrolyte balance. The management in this syndrome centers around the reduction of intracranial pressure by using mannitol and glycerol. Thus, in this disorder, acute therapeutic lowering of blood ammonia has been shown not to be very effective in improving the neurologic status, whereas in the management of hepatic encephalopathy, reduction in ammonia levels is essential.

11.   The treatment of hyperammonemia includes:

a.   Reduction in protein intake while maintaining adequate amounts of total calories (caloric deprivation may lead to tissue protein catabolism, thus aggravating the hyperammonemic condition).

b.   Division of daily protein intake into smaller portions in order to prevent massive buildup of ammonia.

c.   Suppression of colonic bacteria by antibiotics (e.g., neomycin) which leads to the lowering of ammonia produced in the gastrointestinal tract.

d. Administration of lactulose, a non-assimilable disaccharide, brings about the lowering of ammonia levels by two mechanisms. Enteric bacteria catabolize lactulose to organic acids. These acids combine with NH_3 to produce NH_4^+ ions, thus preventing its absorption into the portal circulation. In addition, the catabolism of lactulose leads to the formation of an increased number of osmotically active particles. This draws water into the colon and produces loose acid stools, thereby permitting the loss of ammonia as ammonium ions.

e. Reduction of dietary nitrogen intake by substituting some or all essential amino acids with their respective keto acid analogues.

f. Exchange transfusion and peritoneal dialysis in acutely ill patients.

12. Measurement of blood ammonia levels: In one type of procedure, the principle consists of the conversion of NH_4^+ to NH_3 by alkalinization of blood and the quantification of NH_3 either by titration or colorimetric methods. The ammonia liberated in the above methods is separated from blood by either isothermal diffusion or a cation exchange resin. Blood or plasma ammonia levels can also be quantitated by an enzymatic procedure.

$$NH_4^+ + \alpha\text{-Ketoglutarate} + NADH \longrightarrow L\text{-Glutamate} + NAD^+$$

The decrease in absorbance at 340 nm due to the oxidation of NADH to NAD^+ is proportional to the ammonia concentration. NH_3 levels can also be measured by a conductometric method or by use of a specific ammonia electrode. Normal values of ammonia are: venous whole blood = 75–196 μg/100 ml; plasma = 56–122 μg/100 ml.

13. As noted earlier, urea, almost solely synthesized in the liver from NH_3, is excreted by the kidney. Thus, the blood levels of urea represent a balance between urea synthesized in the liver and excretion by the kidney. Its measurement is widely used (along with creatinine--discussed later) as a screening test for the evaluation of kidney function, although it lacks sensitivity. For example, it has been shown that the serum levels of urea may remain normal with significant kidney damage. Elevation of blood-urea nitrogen may occur due to the following causes:

a. Prerenal: Enhanced protein catabolism associated with negative nitrogen balance (e.g., fevers, thyrotoxicosis, major surgery, wasting diseases, diabetic coma), cardiac decompensation, water depletion due to decreased intake or excessive loss, excessive catabolism of blood proteins (e.g., leukemias, gastrointestinal bleeding disorders), etc. The term azotemia is used to characterize the above conditions in which blood urea and/or other nitrogen metabolites is elevated with or without renal disease, whereas the term uremia signifies a clinical syndrome with marked retention of urea in the blood due to renal failure (see below).

b. Renal: Acute glomerulonephritis, chronic nephritis, polycystic kidney, tubular necrosis, hepatorenal syndrome, and nephrosclerosis. In the uncomplicated nephrotic syndrome, the blood urea is normal.

c. Postrenal: Any obstruction of the urinary tract, due to stones, enlarged prostate gland or tumors, gives rise to increased reabsorption of urea through the tubules, diminished filtration, and a consequent high blood urea.

14. Urea can be measured by its conversion to ammonia by the enzyme urease and estimation of ammonia by titration or colorimetric methods. It is also measured by a direct colorimetric procedure using diacetyl monoxime:

$$CH_3-\overset{\overset{O}{\|}}{C}-\overset{\overset{NOH}{\|}}{C}-CH_3 + H_2O \xrightarrow{\ H^+\ } CH_3-\overset{\overset{O}{\|}}{C}-\overset{\overset{O}{\|}}{C}-CH_3 + NH_2OH$$

Diacetyl Monoxime   Diacetyl   Hydroxyl
                                                        amine

$$\begin{matrix} CH_3 \\ | \\ C=O \\ | \\ C=O \\ | \\ CH_3 \end{matrix} \quad + \quad \begin{matrix} H_2N \\ \diagdown \\ \quad C=O \\ \diagup \\ H_2N \end{matrix} \quad \xrightarrow[2\ H_2O]{\ H^+\ } \quad \begin{matrix} CH_3 \\ | \\ C=N \\ | \quad \diagdown \\ \quad\quad C=O \\ | \quad \diagup \\ C=N \\ | \\ CH_3 \end{matrix}$$

Diacetyl   Urea   Diazine derivative
                                            (yellow)

In the United States, the concentration of urea is expressed as urea nitrogen. The urea nitrogen values can be converted to urea by multiplying by 60/14 or 4.28 (note that the molecular mass of urea,

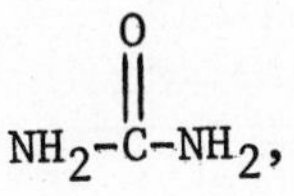

is 60 daltons with each nitrogen atom contributing a mass
of 14). The normal range of urea nitrogen is 8-26 mg per 100 ml
of serum or plasma.

Metabolism of Some Individual Amino Acids

Amino Acid Biosynthesis

Mammalian tissues are capable of synthesizing more than half of the
amino acids needed by the body. These amino acids, known as <u>non-essential</u>
amino acids, are made from carbon skeletons derived from lipid and
carbohydrate metabolites or from transformations of essential amino
acids. The nitrogen of the amino groups is obtained from either NH_4^+
ions or from the $-NH_2$ groups of other amino acids. The amino groups
are transferred by means of transaminase reactions. Following is a
list of some non-essential amino acids with their precursors indicated
in parentheses.

> Glutamic acid (α-ketoglutaric acid)
> Aspartic acid (oxaloacetic acid)
> Serine (3-phosphoglyceric acid)
> Glycine (Serine)
> Tyrosine (phenylalanine)
> Proline (glutamic acid → Δ^1 - pyrroline-5-carboxylic acid)
> Hydroxyproline (proline)
> Alanine (pyruvic acid)
> Cysteine & cystine (sulfur atom comes from methionine and carbon
> atoms from serine)

Some details of the above reactions as well as catabolism of some
individual amino acids are discussed below in the pathways for the
synthesis of the non-essential amino acids.

Glycine Metabolism

1. Metabolic Pathways

    a. Glycine may be formed from serine (using tetrahydrofolate), from
        glyoxylate (by transamination), and from choline.

    b. Glycine breakdown may involve conversion to serine; to
        glyoxylate (which is further broken down to form oxalate or
        formate); or by decarboxylation to N^5,N^{10}-methylene FH_4.

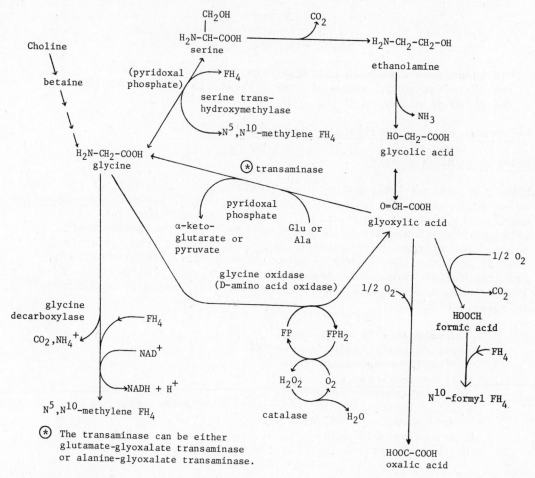

2. Hyperoxaluria

   a. Biochemical lesions.

      i) impairment of oxidation of glyoxylic acid to formic acid.

      ii) inactive glutamic-glyoxylic transaminase or alanine-glyoxylic transaminase.

      If there is impairment of conversion of glyoxylate to formate and to glycine, excess glyoxylate is oxidized to oxalate.

   b. The disease is characterized by high urinary excretion (15 to 60 mg per 24 hours) of oxalic acid (hyperoxaluria) and oxalates (e.g., calcium oxalate). These are poorly soluble and are precipitated, leading to the formation of kidney stones (nephrolithiasis). Hyperoxaluria may also occur in pyridoxine deficiency (required for glutamic-glyoxylic transaminase and alanine-glyoxylic transaminase), excessive ingestion of dietary oxalates (e.g., rhubarb), ingestion of ethylene glycol (an antifreeze component) due to its conversion to oxalate, methoxyflurane anesthesia (due to its conversion to oxalate) and hyperoxaluria secondary to ileal disease (caused by hyperabsorption of dietary oxalates and is probably related to degree of steatorrhea). In patients with steatorrhea in various disease states, a presumed mechanism for the occurrence of hyperoxaluria is that the unabsorbed fatty acids combine with calcium, thereby preventing the formation of insoluble calcium oxalates. This allows the dietary oxalate to remain in solution in the gastrointestinal tract and thus makes it available for absorption. In addition, it is also possible that the unabsorbed fatty acids and bile acids (see lipid metabolism) increase the permeability of the colon to oxalate, leading to its increased absorption. Ascorbic acid, when ingested in large quantities (>4 g/day), may lead to hyperoxaluria.

In addition to hyperoxaluria, the other causes that are associated with kidney stone formation are hypercalciuria (hyperparathyroidism--see calcium metabolism--Chapter 9, renal tubular acidosis, vitamin D intoxication, hyperthyroidism and idiopathic hypercalciuria), hyperuricosuria (see purine catabolism--Chapter 5), cystinuria (genetically determined defect in the renal transport of cystine, lysine, arginine and ornithine) and magnesium ammonium phosphate stone formers (caused by recurrent urinary tract infections). In the latter disorder, bacteria break down the urea to ammonia, which alkalinizes the urine and brings about the precipitation of magnesium ammonium phosphate.

c. Although the exact mechanisms that lead to the formation and growth of renal calculi are not known, the basic physiochemical process appears to involve supersaturation of urine with crystalloid material, thus initiating the precipitation and consequent stone formation. Untreated nephrolthiasis can lead to renal failure (with associated complications) and death.

3. Glycinuria (rare disease)

a. The syndrome is characterized by high excretion of glycine in the urine and a tendency to form oxalate renal stones.

b. The disease is probably due to a defect in renal tubular transport of glycine (decreased reabsorption of glycine).

4. <u>Hyperglycinemia</u> has been observed as a secondary feature in a number of diseases including carbamylphosphate synthetase deficiency (mentioned above), methylmalonic acidemia, and isovaleric acidemia. There are also cases which appear to be due to specific errors in glycine metabolism.

    a.   <u>Ketotic hyperglycinemia</u> may be due to a defect in the enzyme system which oxidizes glycine or, more likely, in the biosynthesis of a cofactor for this system.

    b.   <u>Non-ketotic hyperglycinemia</u> appears to be caused by a defect in the glycine decarboxylase reaction in which glycine is converted to CO_2, NH_3, and N^5,N^{10}-methylene tetrahydrofolate.

5.   Some Special Functions of Glycine

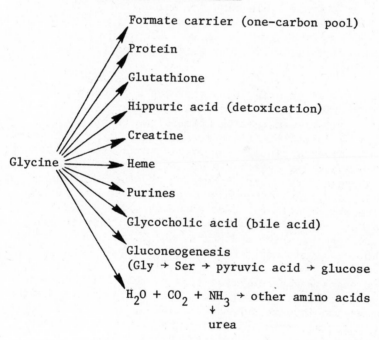

Glycine

- Formate carrier (one-carbon pool)
- Protein
- Glutathione
- Hippuric acid (detoxication)
- Creatine
- Heme
- Purines
- Glycocholic acid (bile acid)
- Gluconeogenesis (Gly → Ser → pyruvic acid → glucose

$H_2O + CO_2 + NH_3$ → other amino acids
↓
urea

Metabolism of Creatine and Creatinine

1. Synthesis of creatine, creatine phosphate, and creatinine.

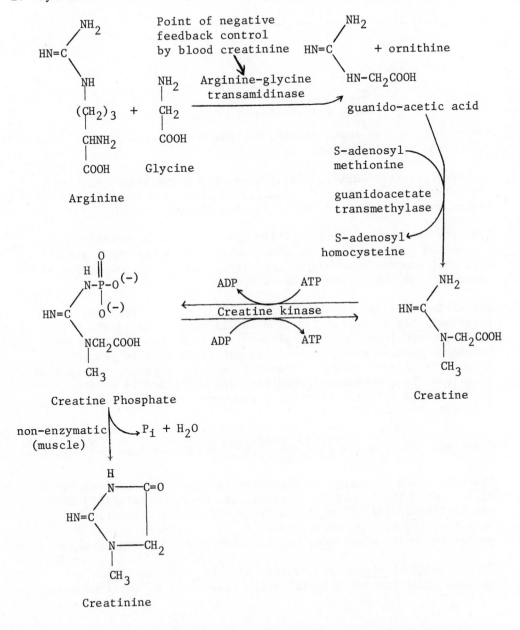

2. Creatine phosphate (also known as phosphocreatine) is a reservoir of high energy phosphate which can readily convert ADP to ATP in muscle and other tissues

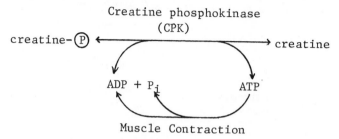

Creatine phosphokinase
(CPK)

creatine-P ⟷ creatine

ADP + P_i ATP

Muscle Contraction

Note: CPK is also known as ATP-creatine phosphotransferase and creatinekinase.

3. Striated muscle, heart muscle, testes, liver, and kidney have high concentrations of creatine. Small quantities are found in the brain and still smaller quantities are found in blood.

4. The rate of creatine synthesis (in the liver) is apparently controlled by blood creatine levels which can alter the arginine-glycine transamidinase enzyme activity. The control mechanism is recognized as one of the feedback type.

5. Creatinine, derived from the dephosphorylation of creatine phosphate, has no useful function and is eliminated in the urine. It is removed from plasma by glomerular filtration and to a small extent by tubular secretion. It is not reabsorbed by the tubules to any significant extent. Thus creatinine clearance approximately parallels the glomerular filtration rate (GFR) and is used as a kidney function test. Creatinine clearance is obtained from the following relationship:

$$\text{Creatinine Clearance} = \frac{\text{Urine Creatinine}}{\text{Serum Creatinine}} \times \text{Urine Volume per Unit Time.}$$

The calculated clearance is corrected to a standard body surface of 1.73 m^2. The normal adult ranges are 84-162 ml/min/1.73 m^2 and 82-146 ml/min/1.73 m^2 for males and females, respectively. Note that the most accurate procedure of estimating GFR is by measuring inulin clearance, which is administered exogenously.

Measurements of <u>plasma</u> creatinine and urea nitrogen levels are used as gross screening tests in the assessment of impaired renal

excretory function. In this regard, plasma creatinine values
provide a better guide than urea nitrogen because creatinine
levels are generally unaffected by endogenous and exogenous
factors and its excretion is fairly constant from day to day in
a given individual. On the other hand, recall that urea levels
are elevated not only by defective glomerular filtration but also
by the quantity of protein intake, protein catabolism due to
hemorrhage (e.g., gastrointestinal) and tissue breakdown following
infection or steroid therapy. Decreased levels of plasma urea
nitrogen are found in liver disease (inability of the liver to
convert NH_3 to urea), starvation, and with low protein diets.
However, a drawback in using plasma creatinine levels as a kidney
function test is that it lacks sensitivity. The plasma creatinine
levels are generally not enhanced until renal function is
substantially impaired. As noted earlier, this is due to the
removal of creatinine by tubular secretion, particularly when
creatinine in the plasma exceeds normal levels. Normal adult
ranges for serum (or plasma) creatinine are 0.9-1.4 mg/100 ml
and 0.8-1.2 mg/100 ml for males and females, respectively.
Plasma creatinine values of 2-4 mg/100 ml with normal renal blood
flow are suggestive of moderate to severe renal damage.

Creatinine excretion is dependent upon skeletal muscle mass and
is constant in a healthy individual. The creatinine coefficient
is defined as: milligrams of (creatinine + creatine) excreted
per kilogram of body weight per day. Normal values are
24-26 for men and 20-22 for women.

6. Creatinuria is characterized by excessive excretion of creatine
    in urine. It may be present in the following conditions: the
    process of growth, fevers, starvation, diabetes mellitus, extensive
    tissue destruction, muscular dystrophy, and hyperthyroidism.

7. Plasma CPK levels are elevated in myocardial infarction as well
    as in any muscle trauma (already discussed in Chapter 4).

8. A common clinical laboratory procedure of creatinine determination
    consists of reacting protein-free filtrates of plasma or serum
    with an alkaline picrate solution which gives rise to a red
    pigment (creatinine-picrate-Jaffe reaction). This substance is
    quantitated by measuring absorbance at 500 nm. This reaction is
    not specific for creatinine, and other chromogens (present in red
    cells) may react with the alkaline picrate solution giving
    red-colored products.

Creatine can be converted to creatinine by heating at an acid pH:

Creatine   Creatinine

Therefore, creatine is measured as the difference between pre-existing creatinine and the total that is formed after the creatine has been converted to creatinine by the above procedure. The adult normal ranges for serum (or plasma) creatine for men and women are 0.1-0.4 and 0.2-0.7 mg/100 ml, respectively.

Serine Metabolism

1. Serine is <u>not</u> an essential amino acid in man. It is synthesized from 3-phosphoglycerate which is formed from glucose as an intermediate in glycolysis.

2. The pathway of serine synthesis is as follows:

3. Some of the other reactions in which serine participates are:
    serine ↔ glycine interconversions, synthesis of cephalins,
    formation of ethanolamine and choline and the synthesis of cysteine.

Interconversion of Proline, Hydroxyproline, and Glutamic Acid

1. All of these amino acids are <u>non-essential</u> (dietarily dispensable).

2. The pathways of interconversion are shown below.

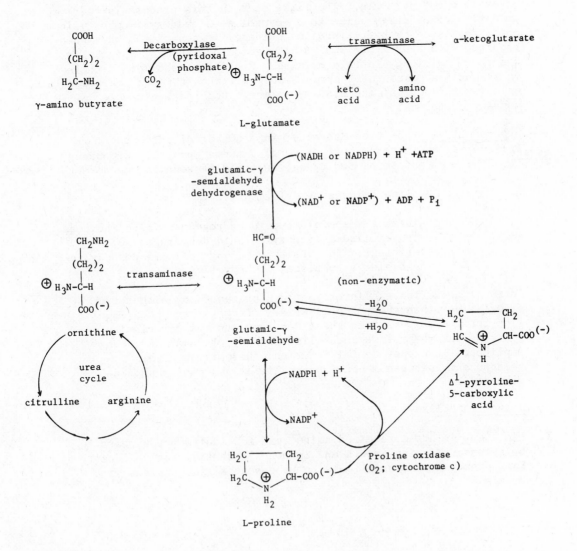

3. Proline is converted to 4-hydroxyproline in a reaction catalyzed by a hydroxylase and requiring molecular oxygen. It takes place predominantly as a post-ribosomal modification of proline residues in collagen polypeptides (see collagen synthesis).

4. GABA is produced by decarboxylation of glutamate. Brain tissue is rich in GABA and its exact role is not understood, although it is implicated in the transmission of nerve impulses.

5. Hydroxyproline undergoes catabolism principally in the liver to glyoxylate and pyruvate. The first step in this process is the oxidation of hydroxyproline to Δ^1-pyrroline-3-hydroxy-5-carboxylic acid, catalyzed by hydroxyproline oxidase. Two patients have been reported with hydroxyprolinuria and hydroxyprolinemia, apparently due to a deficiency in this oxidase. Proline metabolism in these individuals appears to be normal, suggesting that there are separate oxidases for proline and hydroxyproline. Both patients were mentally retarded.

6. Hyperprolinemia has been described in a number of patients. Prolinuria, hydroxyprolinuria, and glycinuria generally accompany it. There appear to be two types, due to different enzyme defects.

   a. Type I: proline oxidase is deficient.

   b. Type II: glutamic-γ-semialdehyde dehydrogenase (also known as Δ^1-pyrroline-5-carboxylic acid dehydrogenase) is deficient. In this type, Δ^1-pyrroline-5-carboxylic acid also accumulates in body fluids.

There is no definite clinical pattern presented by either type of defect other than the aminoaciduria. Attempts have been made to link renal disorders with these diseases but the correlation is poor. Although proline metabolism (by way of GABA) has been implicated in proper neural function, there is no clear relationship between mental abnormalities and faulty proline metabolism.

Histidine Metabolism

1. Histidine is not a required amino acid for adults but appears to be necessary during growth and is hence essential for children. Histidine has also been found to be essential for persons with uremia.

2. Histidine catabolism

a. Histidinemia, an inborn error of histidine metabolism, is due to a deficiency of histidine-ammonia lyase (also known as histidase) which severely impairs the conversion of histidine to urocanate. With a normal diet, this leads to accumulation of histidine (and certain transaminated products such as imidazole-pyruvate, imidazole-lactate, imidazole-acetate) in the blood, cerebrospinal fluid and urine. The disorder is transmitted in an autosomal recessive manner with an incidence of 1/20,000. The clinical manifestation of this disease varies from a totally benign disorder to mental subnormality and speech deficiencies.

b. Note in the catabolism of histidine that the imidazole ring is cleaved to yield N-formiminoglutamate (figlu) from which the formimino group (HN=CH-), a one-carbon fragment, is transferred to tetrahydrofolate (see one-carbon metabolism-- Chapter 5). A folic acid deficiency can thus lead to a lack of figlu utilization, causing it to accumulate with large amounts being excreted in the urine. The figlu excretion is very pronounced when folic acid-deficient individuals are given a loading dose of histidine. This procedure, known as the figlu excretion test, can be used in the assessment of folic acid deficiency. However, more sensitive radioisotopic assays have been developed using folate binders for the measurement of folic acid. It should be pointed out that under certain situations, high urinary figlu levels may co-exist with elevated levels of serum folic acid. Such paradoxical situations may be encountered in the following states:

Vitamin B_{12} deficiency

$$N^5\text{-methyl FH}_4 \xrightarrow[\text{Homocysteine} \quad \text{Methionine}]{B_{12}\text{-Enzyme}} FH_4$$

In the absence of Vitamin B_{12}, folic acid is trapped as N^5-methyl FH_4 and is unavailable for participation as a carrier for one-carbon fragments (e.g., figlu).

c. Inborn errors of metabolism in one or more of the steps involved in folate metabolism can lead to increased excretion of figlu with elevated levels of serum folate. For example, a deficiency of glutamate-formimino transferase leads to the accumulation of figlu with high urinary levels co-existing with high levels of serum folic acid. Thus the serum folate measurement along with the figlu excretion test may be helpful in identifying patients with B_{12} deficiency or inborn errors of metabolism of folate metabolism.

3. Metabolism of histamine and function

a. Histamine is the decarboxylated product of histidine.

Histidine $\xrightarrow[\substack{\text{decarboxylase} \\ B_6}]{\text{Histidine}}$ Histamine $+ CO_2$

b. Histamine is found in three types of cells ("histaminocytes"):
    basophils of the blood, mast cells of the tissues, and certain
    cells of the gastric mucosa and other parts of the body.
    Basophils containing histamine belong to the granular series
    of bone marrow. In patients with myelogenous leukemia, blood
    levels of histamine are increased, whereas in patients with
    lymphatic leukemia, the histamine is not elevated. Mast cells
    are found in loose connective tissue and capsules, especially
    around blood vessels. In mast cells and basophils, histamine
    is contained in granules and is released (by degranulation,
    vacuolization and depletion) under a wide variety of conditions:
    immediate hypersensitivity reactions, trauma, non-specific
    injuries (infection, burns), etc. The histamine release
    appears to be a two-step process--expulsion of granules and
    the release of histamine in the extracellular fluid from the
    heparin-protein complex by a cation exchange process. The
    degranulation process is affected by oxygen, temperature,
    metabolic inhibitors, etc. The regulation of histamine
    release is further discussed in Chapter 11. The release of
    histamine from the gastric mucosal cells is presumably mediated
    by acetylcholine (released due to parasympathetic nervous
    stimulation) and perhaps gastrin. Histamine stimulates
    parietal cells of the gastric mucosa to produce hydrochloric
    acid (discussed earlier in this chapter).

c. The action of histamine can be grouped into two general
    categories:

    i) Contraction of smooth muscle in various organs (gut,
        bronchi). The receptors involved in this action have
        been designated H_1. The conventional antihistamine
        drugs such as diphenhydramine and pyrilamine are
        H_1-receptor antagonists. These compounds are useful in
        the management of various allergic manifestations. Note,
        however, that in acute anaphylaxis, bronchiolar
        constriction is rapidly relieved by a prompt administration
        of epinephrine (a physiologic antagonist of histamine).

    ii) Secretion of hydrochloric acid by the stomach and an
        increase in heart rate are mediated by a different type
        of receptor designated H_2. Some examples of H_2-receptor
        antagonists are burimamide, metiamide and cimetidine.
        These agents have been found effective in the control of
        gastric acid secretion and are thus useful in the treatment
        of gastric ulcers. The effect of histamine on the heart
        involves both H_1 and H_2 receptor sites. H_2 receptors
        mediate the positive chronotropic (rate of contraction)
        and ventricular inotropic (force of contraction) effects,
        whereas H_1 receptors mediate the negative dromotropic
        (property of affecting the conductivity of a nerve fiber)
        effect and possibly the atrial inotropic effect.

d. Structures of some histamine antagonists:

H_1-Antagonists:

Diphenhydramine (Benadryl)

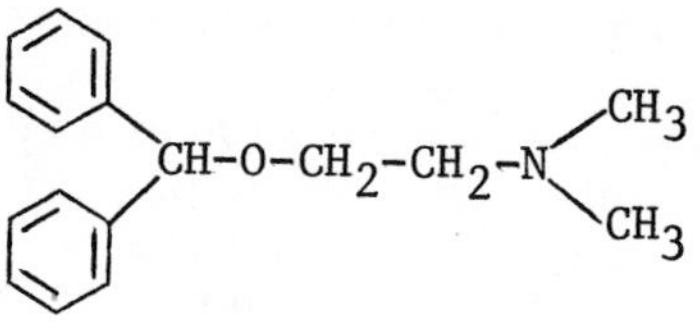

$$CH-O-CH_2-CH_2-N \begin{matrix} CH_3 \\ CH_3 \end{matrix}$$

Pyrilamine (Mepyramine)

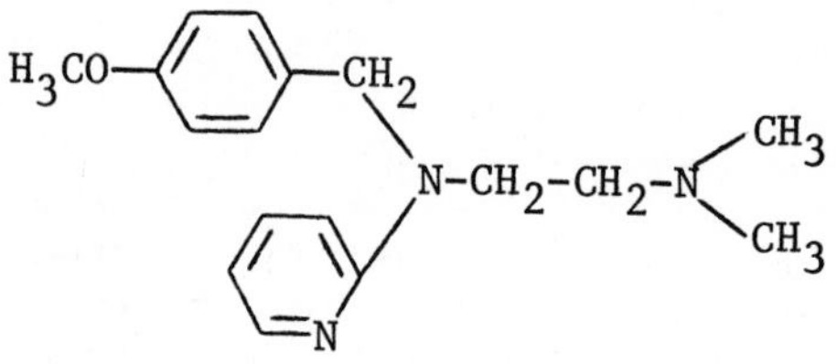

$$H_3CO-C_6H_4-CH_2 \\ N-CH_2-CH_2-N \begin{matrix} CH_3 \\ CH_3 \end{matrix}$$

H_2-Antagonists:

Burimamide

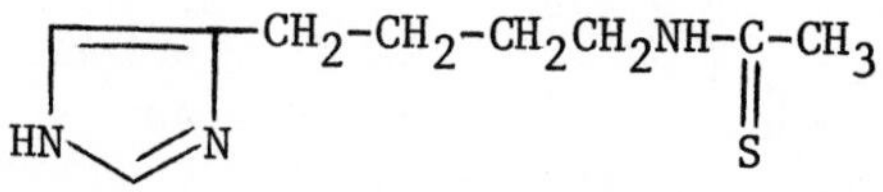

$$CH_2-CH_2-CH_2CH_2NH-\overset{\underset{\|}{S}}{C}-CH_3$$

Metiamide

$$H_3C \qquad CH_2-S-CH_2-CH_2NH-\overset{\underset{\|}{S}}{C}-NHCH_3$$

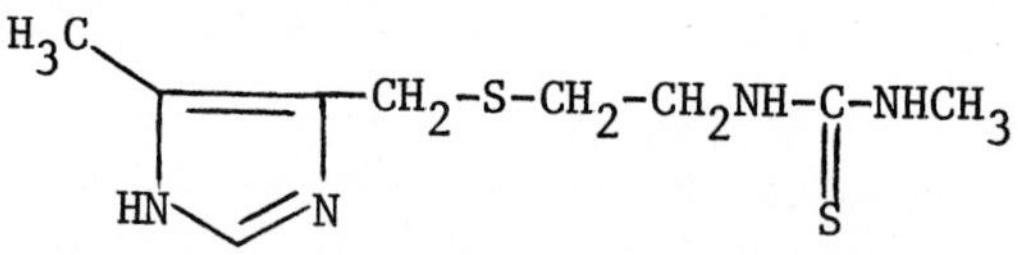

Cimetidine

$$H_3C \qquad CH_2-S-CH_2CH_2NH-\overset{\underset{\|}{NCH_3}}{C}-NHCH_3$$

612

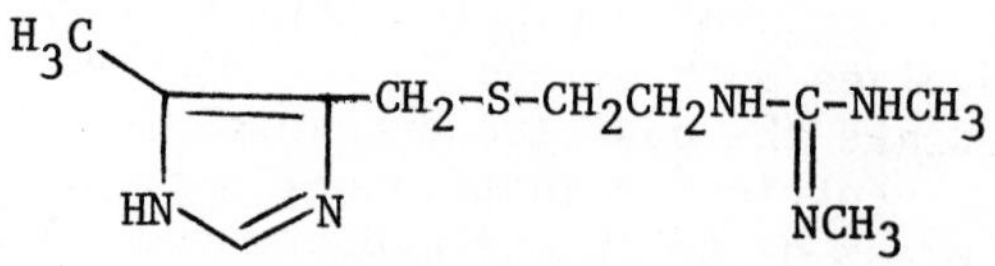

e. Inactivation of histamine

Histamine is rapidly inactivated through one of two mechanisms.
One of these involves methylation by S-adenosylmethionine of
one of the nitrogen atoms of the imidazole ring (catalyzed by
the enzyme N-methyl transferase) or the terminal nitrogen atom
of histamine's sidechain (catalyzed by methyl transferases).
Methyl histamine is subsequently deaminated by monoamine
oxidase to form methyl imidazole acetic acid which is readily
excreted. In the other route, the sidechain of histamine is
deaminated by diamine oxidase to form imidazole acetic acid.
This is then conjugated with ribose and excreted as
1-ribosyl-imidazole-4-acetic acid. The reaction is unique in
that it is the only known reaction in which ribose is used
for conjugation.

Catabolism of Branched-Chain Amino Acids

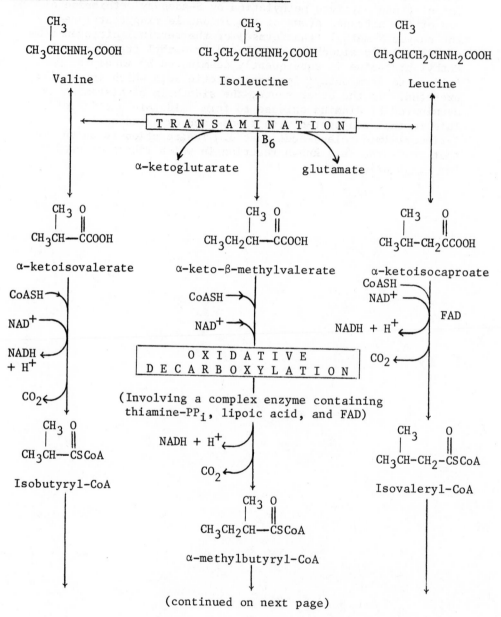

$$CH_3CHCHNH_2COOH$$
(with CH_3 branch)

Valine

$$CH_3CH_2CHCHNH_2COOH$$
(with CH_3 branch)

Isoleucine

$$CH_3CHCH_2CHNH_2COOH$$
(with CH_3 branch)

Leucine

T R A N S A M I N A T I O N
B_6

α-ketoglutarate   glutamate

$$CH_3CH-CCOOH$$
(with CH_3 and O)

α-ketoisovalerate

$$CH_3CH_2CH-CCOCH$$
(with CH_3 and O)

α-keto-β-methylvalerate

$$CH_3CH-CH_2CCOOH$$
(with CH_3 and O)

α-ketoisocaproate

CoASH

NAD^+

NADH + H^+

CO_2

CoASH

NAD^+

CoASH
NAD^+

NADH + H^+

CO_2

FAD

O X I D A T I V E
D E C A R B O X Y L A T I O N

(Involving a complex enzyme containing
thiamine-PP_i, lipoic acid, and FAD)

NADH + H^+

CO_2

$$CH_3CH-CSCoA$$
(with CH_3 and O)

Isobutyryl-CoA

$$CH_3CH_2CH-CSCoA$$
(with CH_3 and O)

α-methylbutyryl-CoA

$$CH_3CH-CH_2-CSCoA$$
(with CH_3 and O)

Isovaleryl-CoA

(continued on next page)

614

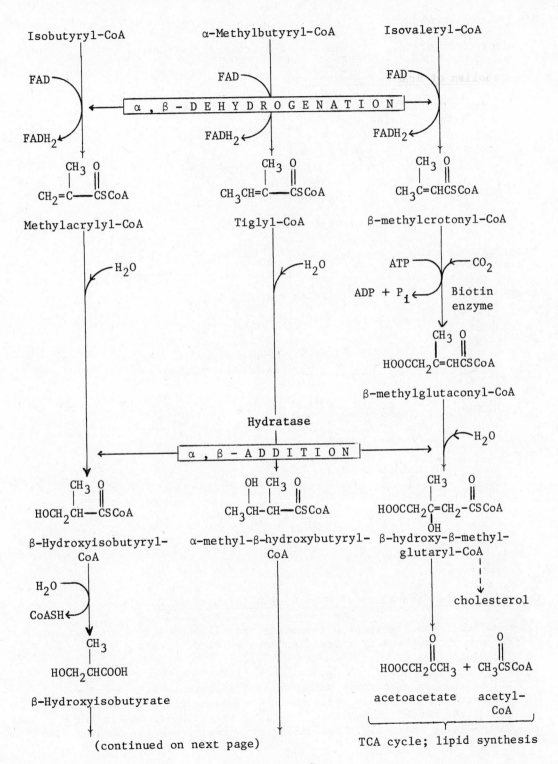

Isobutyryl-CoA   α-Methylbutyryl-CoA   Isovaleryl-CoA

α,β-DEHYDROGENATION

$$CH_2=C\overset{\underset{|}{CH_3}}{}\!\!-\!\!C\overset{O}{\underset{}{\parallel}}SCoA$$

Methylacryl yl-CoA

$$CH_3CH=C\overset{\underset{|}{CH_3}}{}\!\!-\!\!C\overset{O}{\underset{}{\parallel}}SCoA$$

Tiglyl-CoA

$$CH_3C=CHCSCoA$$

β-methylcrotonyl-CoA

ATP → CO₂
ADP + P$_i$ ← Biotin enzyme

$$HOOCCH_2C=CHCSCoA$$

β-methylglutaconyl-CoA

Hydratase

α,β-ADDITION

$$HOCH_2CH\!-\!CSCoA$$

β-Hydroxyisobutyryl-CoA

$$CH_3CH\!-\!CH\!-\!CSCoA$$

α-methyl-β-hydroxybutyryl-CoA

$$HOOCCH_2C=CH_2-CSCoA$$

β-hydroxy-β-methyl-glutaryl-CoA

$$\downarrow cholesterol$$

H_2O
CoASH ←

$$HOCH_2CHCOOH$$

β-Hydroxyisobutyrate

$$HOOCCH_2CCH_3 + CH_3CSCoA$$

acetoacetate   acetyl-CoA

(continued on next page)   TCA cycle; lipid synthesis

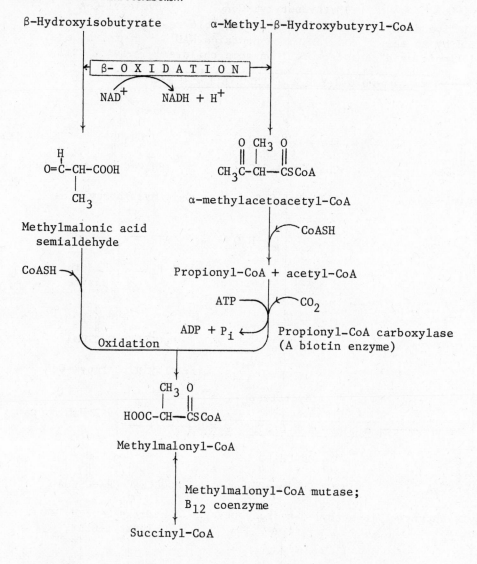

Comments on the Catabolism of Branched-Chain Amino Acids

1. L-valine, L-isoleucine, and L-leucine are all essential amino acids for man although their respective keto acids can replace them in the diet.

2. All of the branched-chain amino acids transaminate with α-keto-glutarate in reversible reactions. One enzyme probably catalyzes all of the three transaminase reactions in the heart muscle but in the liver there are three different enzymes, one for each reaction.

3. The oxidative decarboxylations are irreversible reactions and are catalyzed by an enzyme complex which contains NAD^+, FAD, thiamine PP, and lipoic acid as cofactors. In some children, non-functional decarboxylase prevents further catabolism of all three of the α-keto acids. These branched-chain keto acid derivatives accumulate in the urine. This inborn error of metabolism is known as <u>maple syrup urine disease</u> because the characteristic odor of the urine resembles that of maple syrup. The disease involves severe impairment of the central nervous system, mental retardation, and often death.

4. A biochemical lesion in the metabolism of methylmalonic acid, which is produced from valine and isoleucine, has been noted. (Note: methylmalonic acid can also be obtained from other metabolic sources such as methionine, threonine, etc.) This hereditary condition is characterized by the presence of large amounts of methymalonic acid in the urine, metabolic acidosis, etc.

5. A hereditary deficiency of α-ketoisovaleryl CoA-dehydrogenase has also been reported. This enzyme catalyzes the conversion of α-ketoisovalerate (formed from valine) to isobutyryl-CoA. Its absence causes α-ketoisovaleric acid levels to rise in the blood and urine.

6. The pathways of metabolism of these amino acids have similarities with carbohydrate and fatty acid metabolism. Examples: the oxidative decarboxylation step is similar to pyruvic acid and α-ketoglutarate oxidations; the α, β-dehydrogenation and α, β-addition are similar to steps in fatty acid oxidation.

7. Note that in the pathway for leucine oxidation, β-hydroxy-β-methylglutaryl-CoA is formed. This is an intermediate in the biosynthesis of cholesterol. Also, leucine yields acetoacetate and acetyl-CoA which are both ketogenic and hence it is the only purely ketogenic amino acid. On the other hand, isoleucine is both ketogenic and glucogenic, and valine is only glucogenic.

Summary of Overall Catabolism

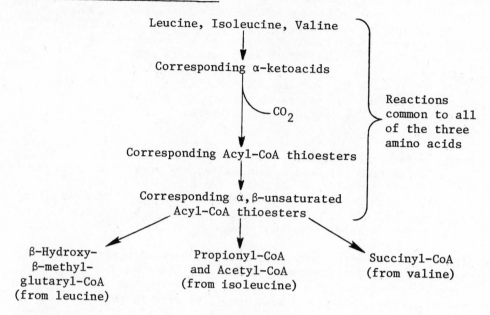

Leucine, Isoleucine, Valine

↓

Corresponding α-ketoacids

↓ —CO₂

Corresponding Acyl-CoA thioesters

↓

Corresponding α,β-unsaturated Acyl-CoA thioesters

Reactions common to all of the three amino acids

β-Hydroxy-β-methyl-glutaryl-CoA (from leucine)

Propionyl-CoA and Acetyl-CoA (from isoleucine)

Succinyl-CoA (from valine)

Metabolism of the Sulfur-Containing Amino Acids

Methionine and cysteine are the principal sources of the body's organic sulfur. Methionine is an essential amino acid (unless adequate homocysteine and a source of methyl groups are available), but cysteine is not essential since it can be synthesized from methionine. Cystine, another sulfur-containing amino acid, is synthesized from cysteine.

Metabolism of Methionine

1. In the body, methionine primarily is

    a. utilized in protein synthesis,

    b. the principal methyl donor (methyl groups are transferred from methionine to other compounds for detoxication processes, biosynthesis, etc.),

    c. a precursor of cysteine.

2. Formation of S-adenosylmethionine (active methionine), the methyl-
    donating form of methionine

a.

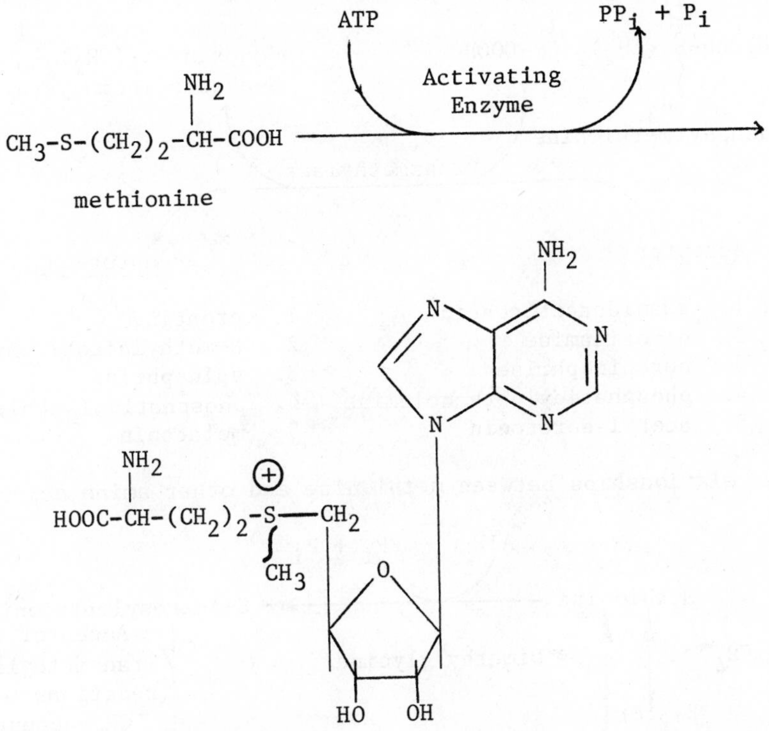

$$CH_3-S-(CH_2)_2-\underset{\underset{NH_2}{|}}{CH}-COOH \xrightarrow{\quad\quad\quad\quad\quad}$$

methionine

(ATP → Activating Enzyme → $PP_i + P_i$)

S-adenosylmethionine (SAM)

b. S-adenosylmethionine is a sulfonium compound. The carbon-sulfur
    bond is a "high-energy" bond, having a free energy of hydrolysis
    of 8 kcal/mole. This energy is available for biosynthetic
    reactions (transmethylation).

A generalized transmethylation reaction is shown below.

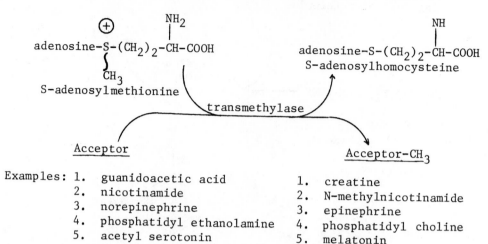

Examples: 1. guanidoacetic acid   1. creatine
          2. nicotinamide   2. N-methylnicotinamide
          3. norepinephrine   3. epinephrine
          4. phosphatidyl ethanolamine 4. phosphatidyl choline
          5. acetyl serotonin   5. melatonin

3. Some interrelationships between methionine and other amino acids:

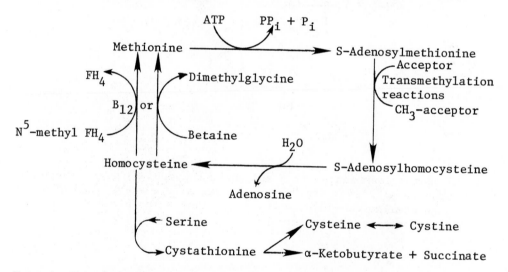

Note in the above reactions that homocysteine can be remethylated to methionine in the presence of suitable methyl donors (and enzymes):

$$
\begin{array}{c}
CH_3 \\
| \\
H_3C\text{-}N^+\text{------}CH_2\text{-}COOH \\
| \\
CH_3
\end{array}
\qquad \text{(Betaine-homocysteine methyl transferase)}
$$

Betaine

or N^5-methyl FH_4 (Methyl transferase--a B_{12}-dependent enzyme)

4.   Formation of Cystathionine and Cysteine

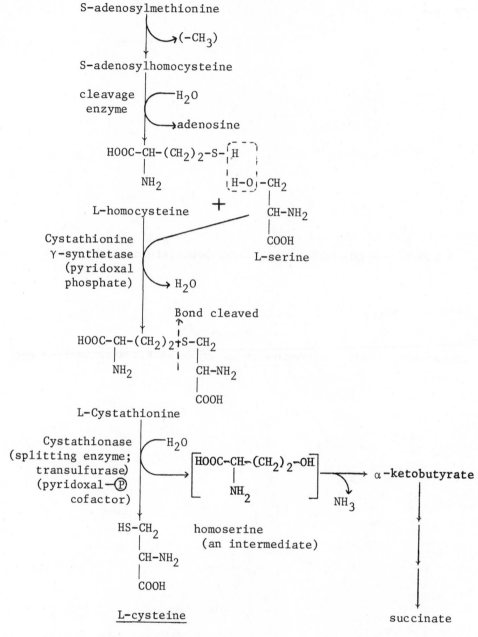

S-adenosylmethionine

$\searrow$ (-CH$_3$)

S-adenosylhomocysteine

cleavage enzyme $\quad$ -H$_2$O

$\rightarrow$ adenosine

HOOC-CH-(CH$_2$)$_2$-S- H
   |
   NH$_2$ $\qquad$ H-O -CH$_2$
                        |
L-homocysteine $\quad + \quad$ CH-NH$_2$
                        |
                        COOH

L-serine

Cystathionine
γ-synthetase
(pyridoxal
phosphate) $\quad \searrow$ H$_2$O

Bond cleaved $\uparrow$

HOOC-CH-(CH$_2$)$_2$-S-CH$_2$
   | |
   NH$_2$ CH-NH$_2$
                        |
                        COOH

L-Cystathionine

Cystathionase
(splitting enzyme;
transulfurase)
(pyridoxal—Ⓟ
cofactor) $\quad$ -H$_2$O

$\rightarrow$ [HOOC-CH-(CH$_2$)$_2$-OH] $\rightarrow$ **α-ketobutyrate**
              |
              NH$_2$ $\qquad$ NH$_3$

HS-CH$_2$ $\qquad$ homoserine
   | (an intermediate)
CH-NH$_2$
   |
   COOH

<u>L-cysteine</u> $\qquad\qquad\qquad\qquad$ succinate

Note: In the above synthesis of cysteine, only the sulfur atom is
       provided by methionine. The carbon chain and the amino group
       are obtained from serine.

5. Catabolism of Cystine & Cysteine

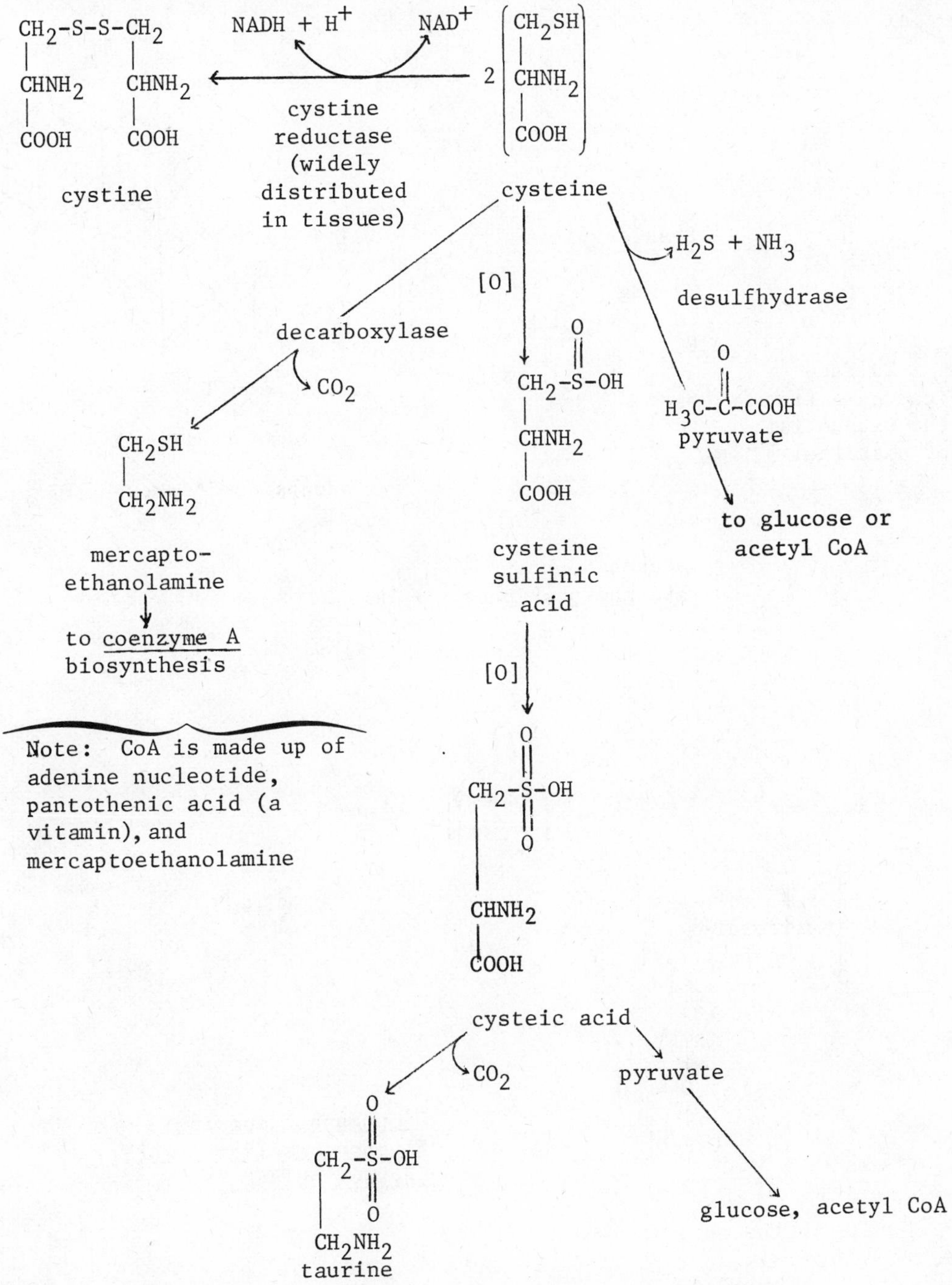

$$CH_2-S-S-CH_2$$ cystine
$$|\qquad\qquad|$$
$$CHNH_2\quad CHNH_2$$
$$|\qquad\qquad|$$
$$COOH\qquad COOH$$

cystine

$\xleftarrow{\text{NADH + H}^+ \quad \text{NAD}^+}$ cystine reductase (widely distributed in tissues)

$$2\begin{pmatrix} CH_2SH \\ | \\ CHNH_2 \\ | \\ COOH \end{pmatrix}$$

cysteine

$\nearrow$ H$_2$S + NH$_3$

desulfhydrase

$$H_3C-\overset{O}{\overset{||}{C}}-COOH$$
pyruvate

to glucose or acetyl CoA

decarboxylase
$\rightarrow$ CO$_2$

$$CH_2SH$$
$$|$$
$$CH_2NH_2$$

mercapto-
ethanolamine

↓

to coenzyme A
biosynthesis

Note: CoA is made up of adenine nucleotide, pantothenic acid (a vitamin), and mercaptoethanolamine

[O]

$$CH_2-\overset{O}{\overset{||}{S}}-OH$$
$$|$$
$$CHNH_2$$
$$|$$
$$COOH$$

cysteine
sulfinic
acid

[O] ↓

$$CH_2-\overset{O}{\underset{O}{\overset{||}{\underset{||}{S}}}}-OH$$
$$|$$
$$CHNH_2$$
$$|$$
$$COOH$$

cysteic acid
$\rightarrow$ CO$_2$
pyruvate
→ glucose, acetyl CoA

$$CH_2-\overset{O}{\underset{O}{\overset{||}{\underset{||}{S}}}}-OH$$
$$|$$
$$CH_2NH_2$$
taurine

Note: Taurine forms conjugates with bile acids to produce taurocholic acids (see Chapter 8).

6. Formation of "Active Sulfate"

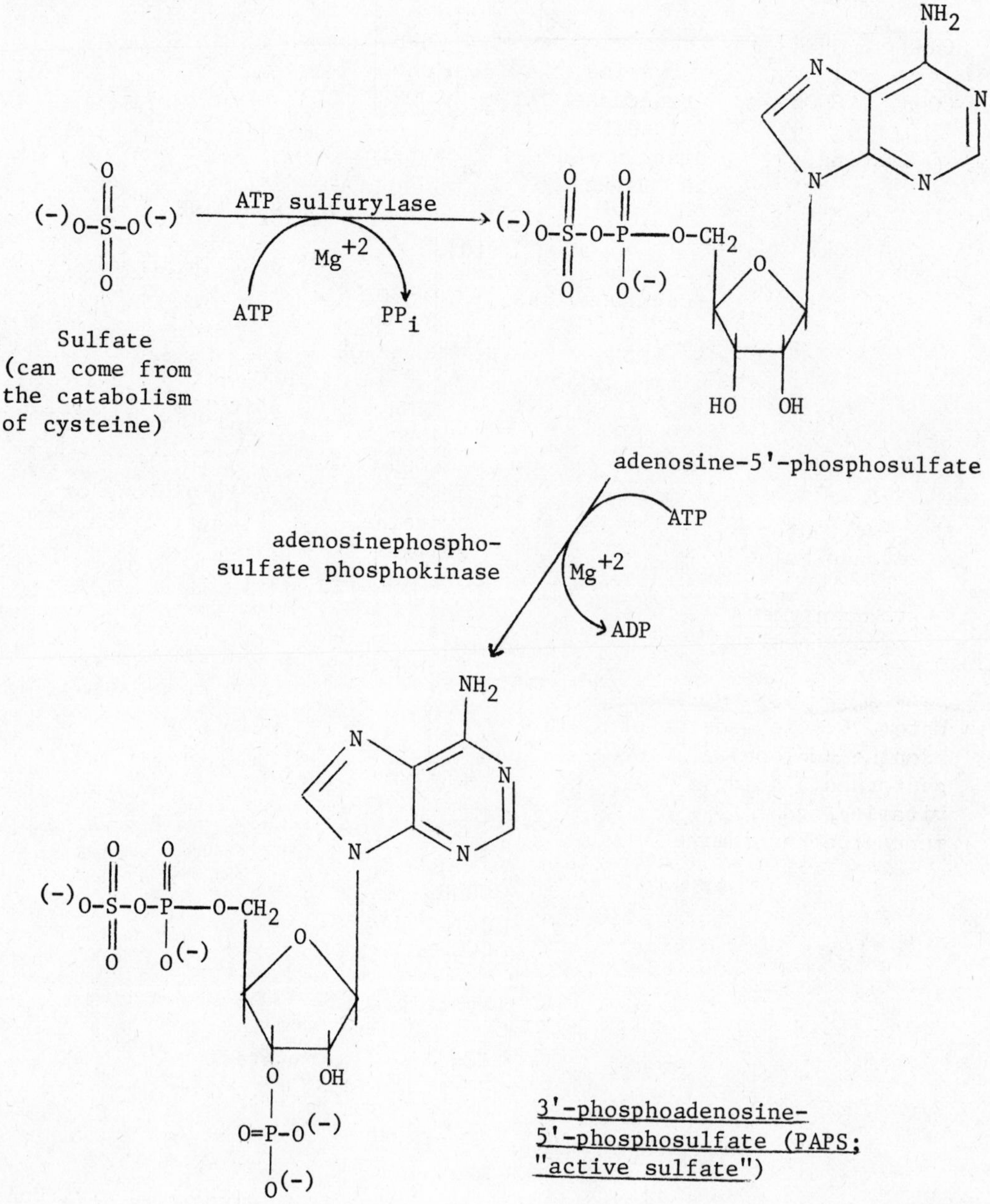

Sulfate
(can come from
the catabolism
of cysteine)

ATP sulfurylase

Mg^{+2}

ATP PP_i

adenosine-5'-phosphosulfate

adenosinephospho-
sulfate phosphokinase

ATP

Mg^{+2}

ADP

3'-phosphoadenosine-
5'-phosphosulfate (PAPS;
"active sulfate")

PAPS is the donor of the sulfate moiety which occurs as an ester
of galactose in compounds such as the chondroitin sulfates and
other sulfate-containing mucopolysaccharides. Phenols, steroids,
etc. also react with PAPS giving the respective sulfate derivatives
which are then eliminated in the urine. This is one of the
detoxication mechanisms which takes place in the liver. These
reactions are summarized as follows:

$$\text{PAPS} + \text{R-OH} \xrightarrow{\text{transferring enzymes}} \underset{\text{sulfate ester}}{\text{R-O-}\overset{\displaystyle O}{\underset{\displaystyle O}{\overset{\|}{\underset{\|}{S}}}}\text{-OH}}$$

PAPS + R-OH   (phenols steroids etc.)

Abnormalities Involving Sulfur-Containing Amino Acids

1. Cystinuria

   a. Excessive excretion of cystine in urine occurs in this
      inherited disease.

   b. There are also defects in renal reabsorptive mechanisms for
      lysine, arginine, and ornithine.

2. Cystinosis

   a. Cystine crystals are deposited in many tissues and organs.
      This disease is also inherited.

   b. General aminoacidura is also present.

   c. Other severe renal impairments occur, and the patients usually
      die at an early age, apparently from renal failure.

3. Homocystinuria

   a. The inherited deficiency of cystathionine synthetase leads
      to elevated plasma levels of methionine and homocysteine and
      an increased urinary homocysteine excretion. The disorder
      is characterized by one or more of the following clinical
      manifestations: skeletal abnormalities (see collagen and
      elastin metabolism--Chapter 5), mental retardation, ectopia
      lentis, malar flush, and susceptibility to arterial and
      venous thromboembolism. The latter is a major cause of
      mortality and morbidity and, hence, patients with cystathionine
      deficiency are prophylactically managed with antithrombotic
      agents (e.g., dipyridamole and aspirin). The mechanism of

thrombosis in homocystinuria is not understood. Some patients
show a reduction in plasma methionine and homocysteine
concentrations as well as urinary homocysteine excretion upon
pyridoxine (B_6) administration (recall that cystathionine
synthetase and the next enzyme in the pathway, cystathionase,
require pyridoxal phosphate as a coenzyme).

b. Homocystinuria has also been observed as a consequence of
defects in the methylation of homocysteine to methionine
which involves both vitamin B_{12} and folate (see folate
metabolism--Chapter 5 and vitamins--Chapter 10).

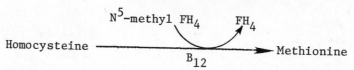

The above reaction is catalyzed by methyl transferase.

4. The Fanconi Syndrome

a. This is a genetically determined abnormality involving renal
tubular defects.

b. There is a generalized decrease in the capacity to reabsorb all
amino acids, glucose, calcium, phosphate, Na^+, K^+, uric acid,
H_2O, etc.

c. The survival of the patient depends on the severity of the
disease.

d. Clinical findings: rickets; acidosis (in severe cases); and
dehydration.

e. In children with Fanconi's Syndrome, cystinosis is almost
always present.

f. Microdissection of the nephrons of infants and adults who have died of the Fanconi Syndrome has demonstrated the presence in nephrons of a very thin, apparently non-functioning initial segment of the proximal convoluted tubule. The remaining portion of the proximal tubules seems to be normal. Whether this anatomical abnormality can account for all reabsorptive problems remains to be proven. The relationship of renal defects to cystinosis is not understood.

g. Treatment: replacement of lost electrolytes and restoration of bone by high vitamin D and calcium intake; administration of alkaline salts to correct the acidosis. There is no known treatment for cystinosis. If untreated, the severe condition (involving acidosis and dehydration) is usually fatal.

5. Cystathionase deficiency which leads to cystathioninuria has also been reported. Most patients with this disorder show some physical or mental abnormality. A defect in the oxidation of sulfite to sulfate, due to the deficiency of sulfite oxidase, has been observed in one patient with severe brain damage, mental retardation and ectopia lentis.

Metabolism of Phenylalanine and Tyrosine

1. Phenylalanine is an essential amino acid. If adequate amounts of phenylalanine are available, tyrosine is not essential; otherwise it is. These two amino acids are involved in the synthesis of a variety of important compounds, including thyroxine, melanin, epinephrine and norepinephrine.

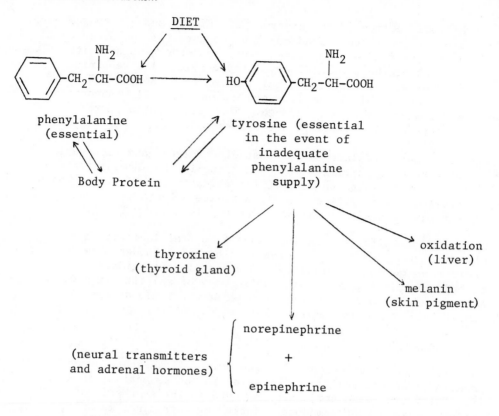

A number of diseases are related to the lesions that are present in the metabolism of phenylalanine and tyrosine. These are phenylketonuria (PKU), alcaptonuria, tyrosinosis, albinism, certain defects in thyroid hormone production, and defects in the production of norepinephrine and epinephrine.

2. Oxidation of Phenylalanine (normal catabolic route):

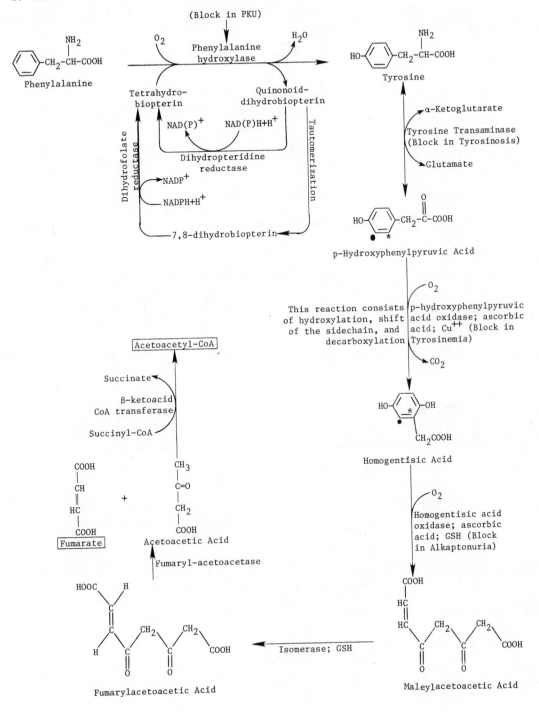

3. Structures of biopterins:

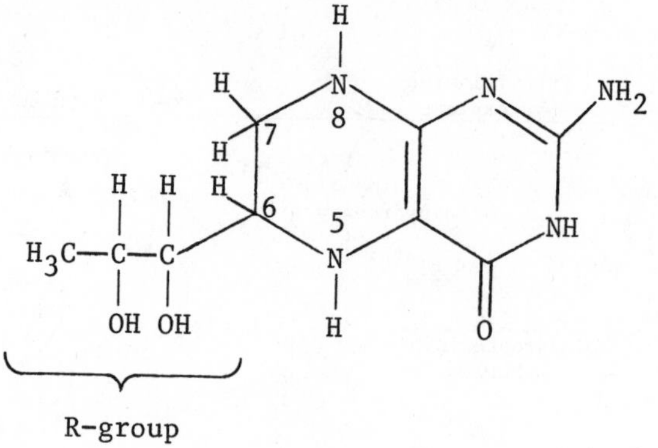

R-group

5,6,7,8-tetrahydrobiopterin

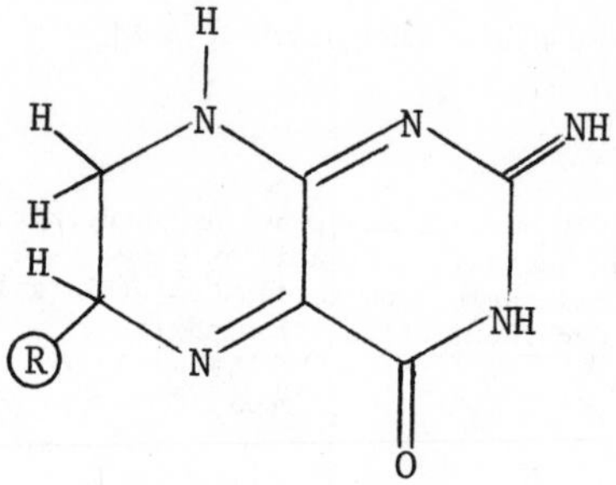

Para-Quinonoid
Dihydropteridine

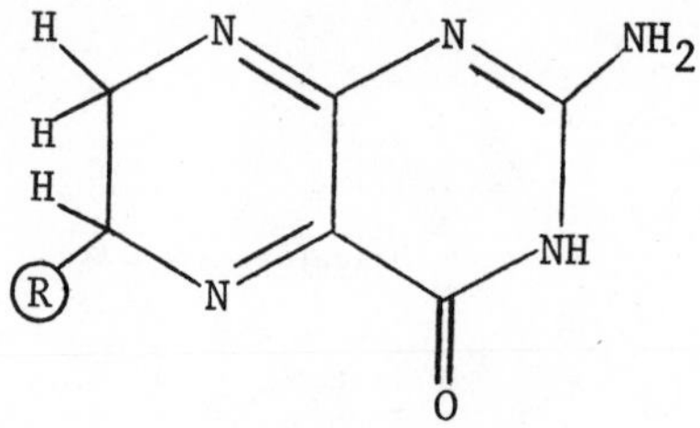

Ortho-Quinonoid
Dihydropteridine

"Quinonoid" – Dihydrobiopterins

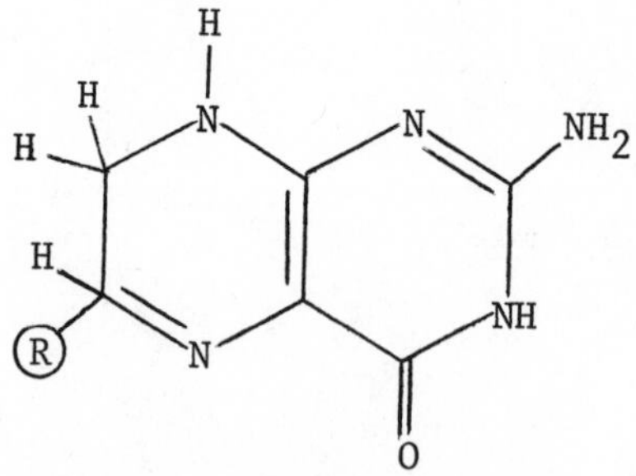

7,8-dihydrobiopterin

4. The phenylalanine hydroxylase system: The phenylalanine reaction is _irreversible_ and occurs primarily in the soluble fraction of liver homogenates in several mammalian species. The reaction is a coupled oxidation in which phenylalanine is oxidized to tyrosine and the tetrahydrobiopterin to a quinonoid dihydro-derivative. Molecular oxygen is the electron acceptor; one atom yields the p-hydroxyl group while the other is reduced to water. The enzyme complex consists of three enzymes and an essential coenzyme, tetrahydrobiopterin. The individual reactions are as follows:

a. Phenylalanine + O_2 + tetrahydrobiopterin $\xrightarrow{\text{phenylalanine hydroxylase}}$ tyrosine + H_2O + quinonoid-dihydrobiopterin

b. Quinonoid-dihydrobiopterin + NAD(P)H + H^+ $\xrightarrow{\text{dihydropteridine reductase}}$ NAD(P)$^+$ + tetrahydrobiopterin

Quinonoid-dihydrobiopterin is an extremely unstable compound and can rapidly rearrange (tautomerization) to 7,8-dihydrobiopterin. This compound is reduced to the tetrahydro-isomer by a NADPH-dependent enzyme, dihydrofolate reductase. Recall that the same enzyme also catalyzes the conversion of dihydrofolate (FH_2) to tetrahydrofolate (FH_4) and that folic acid contains a pteridine ring system (see one-carbon metabolism--Chapter 5).

c. 7,8-dihydrobiopterin + NADPH + H^+ $\xrightarrow{\text{dihydrofolate reductase}}$ NADP$^+$ + tetrahydrobiopterin

Thus, the reductases (b and c) ensure that adequate tetrahydrobiopterin levels are maintained at the expense of reduced pyridine nucleotides. The conversion of phenylalanine to tyrosine can be interfered with if any of the essential components of the system are lacking: phenylalanine hydroxylase, dihydropteridine reductase, dihydrofolate reductase or tetrahydrobiopterin. The deficiency of tetrahydrobiopterin can also affect other important hydroxylations involved in the synthesis of neurotransmitters, namely tyrosine to 3,4-dihydroxyphenylalanine (L-DOPA) and tryptophan to 5-hydroxytryptophan. The former reaction is involved in the synthesis of catecholamines and the latter in the production of serotonin (these conversions are discussed later in this chapter). The origin of pteridines in the human body is unclear. They may be derived from the folic acid breakdown.

Phenylketonuria (PKU) and Its Variants

1. Classical PKU

   a. This disease, inherited as an autosomal recessive trait, is
      due to the hepatic deficiency of the enzyme phenylalanine
      hydroxylase. Thus, the patient is unable to convert
      phenylalanine to tyrosine, resulting in hyperphenylalaninemia
      and the accumulation of metabolites of phenylalanine with
      their excretion in the urine. The homozygous state occurs
      with a frequency of about one in 10,000 among the white
      population. It is estimated that in the United States there
      are about 3.5 million heterozygotes. These numbers are
      rising due to early diagnosis of the disease and the
      institution of a low phenylalanine diet (discussed later),
      leading to an increased survival rate of infants with PKU.
      This has resulted in a greater number of fertile adults with
      the disorder.

b. The formation of various metabolites that accumulate in PKU:

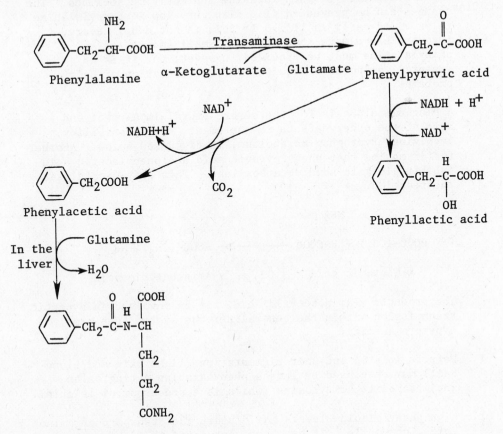

Phenylacetylglutamine

Note that the name phenylketonuria is derived from the presence of the keto acid, phenylpyruvate, in the urine. All of the above metabolites are excreted in the urine.

c. Pathological changes of the skin and brain in PKU are due to the accumulation of phenylalanine and their metabolites. The changes can be prevented if a diet very low in phenylalanine is provided early in life. If this is not done, severe, irreversible damage soon occurs. Small amounts of phenylalanine must be provided, however, for it is an essential amino acid required for the formation of body proteins.

d. A biochemical basis for the severe mental impairment and some of the other defects is not clear. One possibility, involving tryptophan metabolism, is discussed below. Another factor may be that phenylpyruvic acid and phenylacetic acid inhibit glutamic acid decarboxylase. This enzyme catalyzes the reaction.

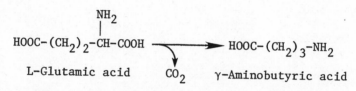

$$\underset{\text{L-Glutamic acid}}{\text{HOOC-}(CH_2)_2\text{-CH-COOH}} \xrightarrow[]{} \underset{\text{γ-Aminobutyric acid}}{\text{HOOC-}(CH_2)_3\text{-NH}_2}$$

The product, γ-aminobutyric acid, may be involved in synaptic transmission within the central nervous system (see also hyperprolinemia).

e. Defects in skin and hair pigmentation (light skin and blond hair) may be caused by excess phenylalanine acting as an antimetabolite to tyrosine (which is a precursor of melanin).

f. High phenylalanine levels may disturb the transport of amino acids into cells. It has been postulated that the variation in the overt pattern of PKU may reflect differences in the efficiency of this transport in the presence of elevated phenylalanine concentrations.

g. It appears that in all infants, phenylalanine hydroxylase matures postnatally in the liver. Only when it fails to mature (during the biochemical differentiation of the liver), as in PKU, does the brain damage occur. However, it has been reported that hepatic hydroxylase activity appears early in the mid-trimester (eleventh to twelfth week) of the human fetus and this activity changes with age during fetal and perinatal development. In premature infants, due to immaturity of phenylalanine hydroxylase, hyperphenylalaninemia may be seen. Therefore, the promptness and consistency of treatment with the proper dietary regimen usually determines the magnitude of irreversible intellectual impairment. An effective screening program consists of measuring blood phenylalanine levels in infants between 6 and 14 days of age. If the value is 4 mg/100 ml of serum or higher, it is generally accepted that a follow-up test is required. If screening is done too early, the PKU infants may not show elevated levels. The follow-up test is performed before 21 days of age because it has been suggested that in order to prevent mental retardation, formulas low in phenylalanine should be instituted before three weeks of age. If the levels remain elevated (20 mg/100 ml or greater), it is strongly suggestive of hyperphenylalaninemia being a manifestation of PKU. The normal range for phenylalanine levels in the newborn is 1.1 to 3.1 mg/100 ml. In a PKU infant, the blood phenylalanine levels are measured weekly until one year of age and once a month thereafter. During this period the phenylalanine level is maintained between 5-10 mg/100 ml.

The low phenylalanine dietary regimen can be eased at about the age of six. This is possible due to the fact that the completion of brain differentiation takes place by that age. An enigmatic aspect of this problem is that in some cases, restricted phenylalanine intake does not correct the defect (see later).

h. In PKU, defective myelination occurs in the brain, there is an increased incidence of epileptic seizures, and abnormal EEG's are common.

i. A defect in tryptophan metabolism also appears to be involved in PKU, as evidenced by the occurrence of abnormal indole derivatives in the urine and low levels of serotonin (a product of tryptophan metabolism) in the blood and brain. There are several points where phenylalanine and tryptophan metabolism are interrelated.

The enzyme 5-hydroxytryptophan decarboxylase, which catalyzes the conversion of 5-hydroxytryptophan to serotonin, is inhibited in vitro by some of the metabolites of phenylalanine.

Phenylalanine hydroxylase is similar to the enzyme that catalyzes the hydroxylation of tryptophan to 5-hydroxytryptophan, a precursor of serotonin. Also, in vitro, phenylalanine is found to inhibit the hydroxylation of tryptophan. There has been some speculation that the mental defects related to PKU may be caused by the decreased production of serotonin.

j. Hyperphenylalaninemia in "mentally normal" adults can occur in homozygous (treated with low phenylalanine in infancy and childhood) or unrecognized heterozygous PKU individuals or in other types of disorders. This can be a problem in pregnant women because high blood levels of phenylalanine (greater than 10 mg/100 ml) have been shown to cause intrauterine growth abnormalities and mental retardation of the offspring. The fetus is unable to convert phenylalanine to tyrosine (tyrosine then becomes an essential amino acid for the fetus), leading to an accumulation of phenylalanine with its consequent disturbances (discussed earlier). It has been suggested that every woman in early pregnancy should be tested for hyperphenylalaninemia. In positive cases, the pregnant woman may choose either therapeutic abortion or a properly supervised, low phenylalanine diet throughout the entire pregnancy. It should be emphasized, however, that if the intake of phenylalanine is too severely restricted, it may affect the nutrition of the fetus. Furthermore, it is not known at this time whether such low phenylalanine diets completely prevent fetal damage.

In mentally ill patients, hyperphenylalaninemia should be ruled out before any form of treatment (e.g., antipsychotic drugs, electroconvulsive therapy) is begun, because in unidentified PKU patients, psychoses may be ameliorated by instituting a diet regimen low in phenylalanine.

2. Variant Forms of PKU

As noted earlier, hydroxylation of phenylalanine to tyrosine is a complex process and involves three enzymes and an essential coenzyme, tetrahydrobiopterin. Recently, variant forms of PKU have been reported which are due to deficiencies of dihydropteridine reductase and tetrahydrobiopterin. These disorders are not amenable to treatment with the low phenylalanine diets. The abnormal neurological manifestations observed in these variants appear to be due to decreased synthesis of serotonin, dopamine and norepinephrine. Recall that dihydropteridine reductase and tetrahydrobiopterin are required in the biosynthesis of these neurotransmitters. In the management of these patients, administration of tetrahydrobiopterin may not be of significant value because of its inability to be transported into the brain from the blood. Suggestions have been made that administration of 5-hydroxytryptophan and/or dihydroxyphenylalanine, which are both products after the metabolic block, may be beneficial.

Methods of Assay for Phenylalanine and Its Metabolites

1. Automated Method

$$\text{Phe + Ninhydrin} \xrightarrow[\text{alkaline Cu-tartrate buffer}]{\text{L-leucyl-L-alanine; heat;}} \text{fluorescent complex}$$

The Leu-Ala dipeptide enhances the fluorescence of the Phe-ninhydrin complex.

2. Ferric chloride ($FeCl_3$) methods are based on the reaction α-keto acid in PKU urine:

$$\text{Urine + FeCl}_3 \text{ (soln.)} \longrightarrow \text{green or blue-green color}$$

This reaction is the basis of the "diaper test" for PKU in newborn infants. It is not sensitive enough, however, to detect the disease until several days after birth, when phenylalanine levels have risen to over 50 mg/100 ml of urine. Also, as was pointed out above, phenylalanine hydroxylase activity matures postnatally and it is the failure to mature which causes PKU. A number of substances interfere with this test, yielding confusing or erroneous results. The reagents have now been incorporated into a dried preparation used to coat test strips which can be employed in place of the ferric chloride solution.

3. Microbiological method. A sensitive assay using a bacterial-inhibition method has been developed by Guthrie. The test is based on the fact that phenylalanine can overcome the inhibitory effect of thionylphenylalanine on the growth of Bacillus subtilis.

4. Measurement of plasma phenylalanine is the most specific diagnostic test for PKU.

Tyrosinosis: This is a very rare disease. The biochemical lesion appears to be in the tyrosine transaminase, leading to a marked elevation of plasma tyrosine.

Tyrosinemia: This is due to p-hydroxyphenylpyruvic acid oxidase deficiency. Heterozygous carriers are found with a prevalence of one in 20-30 in the French-Canadian population. The clinical manifestations of the acute form of the disorder include cirrhosis, renal Fanconi syndrome (generalized aminoaciduria), and early death due to hepatic failure. Tyrosine levels in the plasma are elevated, and some patients show a modest elevation of plasma methionine. A diet low in tyrosine and phenylalanine has been found to be beneficial in the management of this disease. However, it is not known whether the diet therapy offers protection against chronic liver disease.

Alcaptonuria

1. This is a rare, metabolic, hereditary disease in which there is an accumulation and elimination in urine of homogentisic acid. The urine darkens upon exposure to air (due to the oxidation of homogentisic acid).

2. The biochemical lesion is a lack of homogentisic acid oxidase.

3. Homogentisic acid is a good reducing agent and causes positive reactions in reducing tests for glucose which may lead to an erroneous diagnosis of diabetes or renal glycosuria.

4. The clinical features include pigmentation of cartilage and other connective tissues due to the deposition of oxidized homogentisic acid. This generalized pigmentation is called ochronosis. Ocular ochronotic pigmentation has also been observed which can be mistaken for melanosarcoma. Patients nearly always develop arthritis in later years. The exact relationship between the pigment deposition and arthritis is not yet understood.

Biosynthesis of Melanin

Tyrosine →(tyrosinase, [O])→ 3,4-Dihydroxyphenylalanine (DOPA) →(tyrosinase, [O], H_2O)→ DOPA Quinone →([O])→ Dopachrome (hallochrome)

Note: Zn^{+2} is present in melanocytes at a high concentration.

Dopachrome →(Zn^{++}, CO_2)→ 5,6-Dihydroxyindole →([O], H_2O)→ Indole – 5,6-Quinone →(Polymerization)→ Melanin (a complex substance of high molecular weight) →(Protein)→ Melanoprotein

Comments on the Biosynthesis of Melanin

1. Melanin is a polymer of indole nuclei. Its exact structure is not known. Melanin is produced by melanocytes which are normally present in the skin (in hair bulbs and in the dermis), the mucous membranes, the eye, and the nervous system. It is a highly insoluble compound which presents problems in its study.

2. The varying degrees of coloration of skin (white-brown-black) are due to the amount of melanin synthesized in the melanocytes rather than the number of melanocytes. Melanin formation is apparently under both hormonal and neural regulation.

   a. The adenohypophysis secretes melanocyte-stimulating hormone (MSH) which affects skin coloration in fish, reptiles, amphibians, etc., but whose role in mammals is uncertain. It is a polypeptide hormone.

   b. Certain hormonal disturbances are accompanied by changes in pigmentation. For example, in Addison's disease, in which there is an underproduction of cortisol (a hormone produced by the adrenal glands), increased pigmentation of the skin is a frequent clinical finding. Another disease is panhypopituitarism in which there is a deficiency in the production of all of the pituitary hormones. It is characterized by skin pallor. The exact causes of these pigmentation changes are not yet known. A decrease (or increase) in MSH production may be involved or the MSH-like activity of ACTH may account for some of the pigmentation changes.

   c. ACTH has a small amount of MSH-like activity. (ACTH and MSH have overlapping amino acid sequences in their structure.)

   d. Melatonin (N-acetyl-5-methoxy tryptamine) is a hormone present in the pineal body and peripheral nerves. A metabolite of tryptophan, it has been shown to reduce the deposition of melanin in frog melanocytes. In mammals, its synthesis appears to be sensitive to light and it has some inhibitory effect on the ovary. All of this is suggestive of involvement in cyclic or rhythmic processes.

3. A lack of melanin production in the melanocytes gives rise to several hereditary diseases collectively called albinism.

   a. Among the mechanisms that have been investigated as possible causes of albinism are

      (i) deficiency or lack of tyrosinase,

(ii)   lack of melanin polymerization,

(iii) lack of synthesis of the protein matrix of the melanin granule,

(iv)   lack of availability of the substrate tyrosine,

(v)   presence of inhibitors of tyrosinase.

Of these, the basic defect appears to be (i).

b. Clinical features of albinism involving the skin include increased susceptibility to various types of carcinoma (melanin's major function is as a screen which gives protection from solar radiation). When the eyes are involved, photophobia, subnormal visual acuity, strabismus nystagmus (involuntary rapid movement of the eyeball; vertical, horizontal, or rotatory), absence of some of the anatomical components of the eye (such as the sphincter muscle) are some of the aspects observed. The inheritance pattern of albinism varies from type to type. For example, universal albinism (characterized by absence of melanin in the hair bulb, retinal pigment epithelium, skin and ureal tract) is autosomal recessive; Piebald albinism (characterized by isolated and scattered patches or absence of melanin in the skin and hair) is autosomal dominant; and ocular albinism (characterized by absence of melanin in the retinal pigment epithelium) is a sex-linked recessive trait.

The overall frequency of albinism in Europe is about one in 20,000. However, the Indian populations of Central and North America (the Cura Indians of San Blas Province, Lower Panama, the Hopi Indians of Arizona, and the Jemez and Zuni Indians of New Mexico) show a frequency of one in 200. The reason for the unusually high frequency of a deleterious recessive gene has been investigated in the Hopi Indians. In the traditional Hopi society, the albinos are well-accepted and a positive attitude is shown towards them. As an extension of this positive attitude, the male albinos are not expected to labor in the fields and are therefore spared from the sun and hostile nomads. Thus, the adult Hopi albino male is left in the village during the day while others work in the fields, increasing the albino's opportunity for sexual activity. This causes a significant effect on the gene frequencies of the next generation. Furthermore, pre- and extramarital relations are accepted as a way of life in the traditional Hopi society. Maintenance of a deleterious recessive gene at a high frequency in this instance appears to be related to a reproductive advantage conferred on the affected homozygous individuals. This type of cultural selection will disappear with "westernization."

Metabolism of Catecholamines

The principal catecholamines are dihydroxyphenylethylamine (dopamine), epinephrine (E) and norepinephrine (NE). Chemically, these compounds are dihydroxylated phenyl (catechol) ethyl amine derivatives, synthesized in postganglionic sympathetic nerve endings, the brain, and in cells from the embryonic neural crest. The latter includes the adrenal medulla, organ of Zuckerkandl and other visceral and non-visceral paraganglia. Catecholamines present in the brain are synthesized locally because circulating catecholamines in the blood do not cross the blood-brain barrier.

1. Biosynthesis

    The synthesis consists of an amine precursor uptake (tyrosine) and decarboxylation, a unifying characteristic described by Pearse and abbreviated the APUD system, which is present in many organ systems. The major pathway of catecholamine synthesis is as follows: Tyrosine → dihydroxyphenylalanine (DOPA) → dopamine → norepinephrine → epinephrine (see Figure 72).

    The main points to be emphasized in this pathway are:

    a. Tyrosine hydroxylase, which catalyzes the conversion of tyrosine to DOPA, is the major rate-limiting step. This step, similar to the reaction catalyzed by phenylalanine hydroxylase, requires molecular oxygen, tetrahydrobiopterin and the associated enzymes that are necessary for the regeneration of tetrahydrobiopterin from reduced pyridine nucleotides. This reaction is inhibited by norepinephrine, phenylalanine, and the drug α-methyltyrosine.

    b. Dopamine-β-hydroxylase, a copper-containing enzyme present in granules (storage vesicles), catalyzes the conversion of dopamine to norepinephrine. NE is the principal end-product in sympathetic nerve endings.

    c. Phenylethanolamine-N-methyl transferase (present in adrenal medulla) is the enzyme responsible for the conversion of norepinephrine to epinephrine which is stored in the chromaffin granules. The methyl donor is S-adenosylmethionine.

    The synthesis, storage and release of catecholamines are regulated by preganglionic sympathetic nerve impulses as well as by the stored end-products on the rate-limiting steps.

Figure 72.   Biosynthesis of Catecholamines

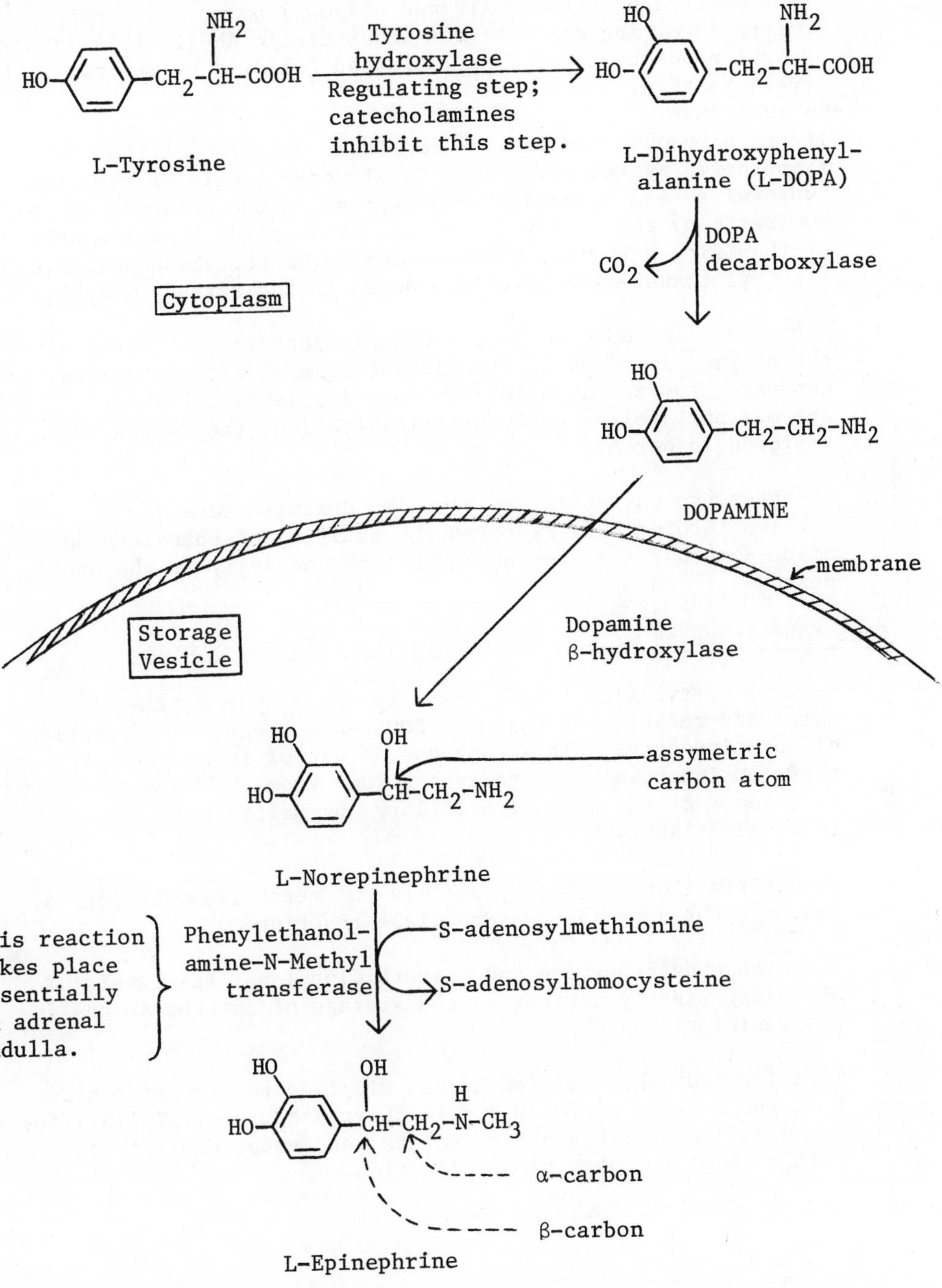

The principal isomers synthesized are L-isomers, which are many times more active than D-isomers.

643

d. The catecholamines are released from the granules along with other materials (such as ATP and enzymes) by an exocytotic process involving calcium (see under cyclic AMP). This is brought about by chemical stimulation followed by membrane depolarization.

e. It has been observed that prolonged stimulation of the sympathetic system gives rise to elevated levels of tyrosine hydroxylase and dopamine-β-hydroxylase in the norepinephrine-synthetic system of nerve endings. Similarly, in epinephrine synthesizing pathways, elevated levels of phenylethanolamine-N-methyl transferase have been noted in the adrenal medulla.

f. Adrenocorticotropic hormone (ACTH) influences the levels of the enzymes involved in the biosynthesis of catecholamines. Hypophysectomized animals show very low levels of these enzymes but, following ACTH administration, the levels are restored to normal.

g. In these hypophysectomized animals, dexamethasone (a corticosteroid) only restores the activity of phenylethanol-amine-N-methyl transferase and not the activity of the other enzymes.

2. Some Inhibitors of Catecholamine Biosynthesis and Storage

a. α-Methyl-p-tyrosine inhibits the tyrosine hydroxylase-catalyzed reaction--Tyrosine → DOPA--and hence, the formation of norepinephrine. This compound is useful in treating the hypertension resulting from pheochromocytoma. Pheochromocytoma is a tumor of the adrenal medullary chromaffin cells (discussed later).

b. Disulfiram (for structure, see alcohol metabolism, Chapter 4) inhibits the dopamine-β-hydroxylase reaction.

c. The compounds reserpine and guanethidine lower the levels of norepinephrine by preventing the storage of catecholamines in the vesicles.

d. Tyramine and other related amines displace the norepinephrine from the storage vesicles due to their structural similarities. The displaced norepinephrine enters the synaptic cleft and brings about its physiological action.

3. **Functions**

a. The functions of the catecholamines are varied and widespread depending upon their concentrations and the target tissue. The action (excitatory or inhibitory) of a particular catecholamine in a given tissue is dependent upon the integration of α-adrenergic, β-adrenergic, and dopaminergic receptor responses, mediated by cyclic nucleotides (see later). The relative potency of adrenergic compounds in eliciting α-receptor effects is E > NE > I (isoproterenol). For β-receptor response, it is I > E > NE. The receptors are macromolecules located in the plasma membranes of target cells.

b. Some examples of the effects caused by α-receptor stimulation are: constriction of blood vessels, sphincter contraction in the stomach, contraction of trigone and sphincter of the urinary bladder, and contraction of uterine smooth muscle. An antagonist of α-receptor action is phentolamine:

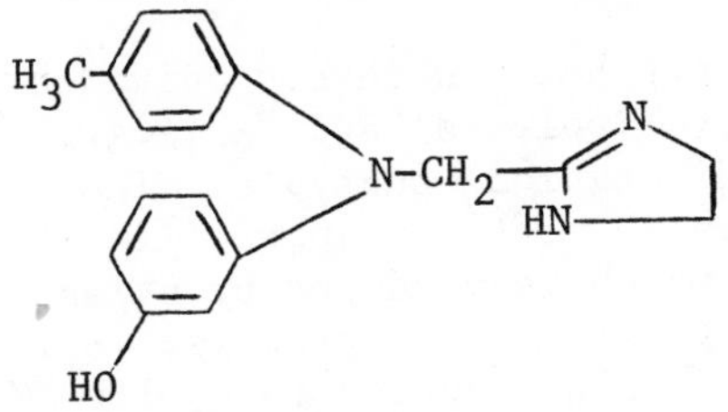

c. Some examples of the effects caused by β-receptor stimulation are: increased heart rate, increased contractility and conduction velocity in the heart, diminished refractory period of the atrioventricular node, dilation of the blood vessels, decreased mobility and tone of the stomach, detrusor relaxation in the urinary bladder, relaxation of uterine smooth muscle and bronchodilatation. The β-adrenergic effects are most potently stimulated by isoproterenol (agonist) and inhibited by propranolol (antagonist).

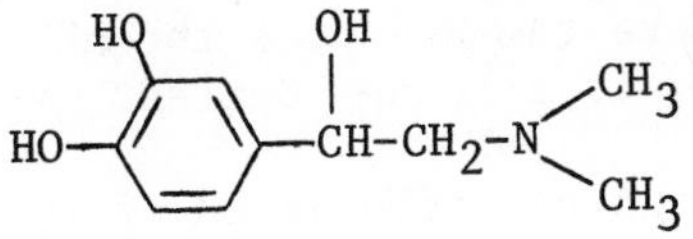

Isoproterenol

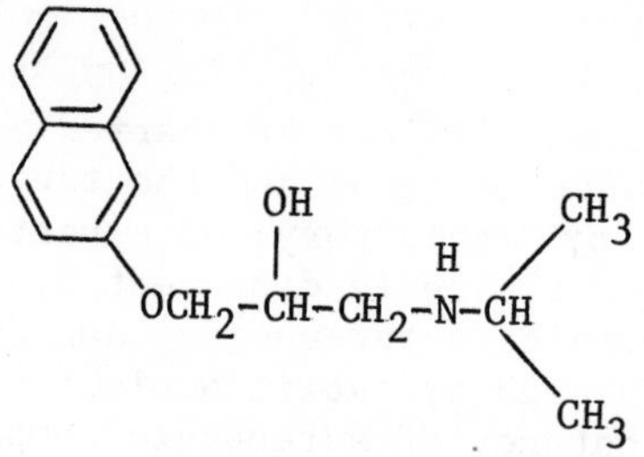

Propranolol

It can be seen from the structures of the agonists of adrenergic receptors that the ability to bind to β-receptors is enhanced by a bulky substituent on the amino nitrogen (compare the structure of isoproterenol and norepinephrine) while such bulky groups cause a decreased affinity for α-receptors. Substitution of the hydroxy group on the β-carbon atom promotes binding to both α- and β-adrenergic receptors while decreasing affinity for dopaminergic receptors.

β-Adrenergic receptors have been further subdivided into β_1 (e.g., cardiac inotrophy, lipolysis) and β_2 (e.g., bronchodilatation, vasodilatation) subtypes. The β-adrenergic effects of catecholamines in most instances (if not all) are mediated by cyclic AMP, which is produced by stimulation of adenylate cyclase (see cyclic AMP--Chapter 4). Similarly, dopaminergic receptors are also coupled to cyclic AMP production by the stimulation of adenylate cyclase. In contrast, it has been speculated that the α-adrenergic effects may be associated with increased levels of cyclic GMP (cyclic 3',5'-guanosine monophosphate) due to stimulation of guanylate cyclase by α-receptor agonists. Keeping the above aspects in consideration, the following is a summary of the action of epinephrine and norepinephrine:

i)   Norepinephrine is a generalized vasoconstrictor, while epinephrine's vasoconstriction is restricted to the arterioles of skin, mucous membrane, and splanchnic viscera. Both stimulate the heart, although norepinephrine does so to a lesser degree.

ii) Epinephrine brings about the relaxation of smooth muscles such as those of the stomach, intestine, bronchioles, and urinary bladder, while contracting the sphincters of the stomach and bladder.

iii) Epinephrine enhances the BMR (basal metabolic rate), glycogenolysis in liver and muscle, and the release of free fatty acids from adipose tissue (thus increasing the fatty acids in the blood). Norepinephrine brings about a similar metabolic effect, but to a lesser extent.

d. Dopaminergic receptors are found in certain brain regions involved in coordinating motor activity. These receptors are also found in the renal vasculature where they apparently cause vasodilatation. In the brain, dopamine functions as a neurotransmitter in dopaminergic neurons. It also is a precursor for the synthesis of other catecholamines. L-DOPA, the immediate precursor of dopamine, is used therapeutically to treat a neurological disorder known as paralysis agitans (Parkinson's disease), possibly by stimulating dopaminergic fibers. L-DOPA is used instead of dopamine, because the latter does not cross the blood-brain barrier. An <u>antagonist</u> of dopamine is haloperidol:

$$F-\langle\bigcirc\rangle-\overset{\overset{\displaystyle O}{\|}}{C}-CH_2-CH_2-CH_2-N\langle\bigcirc\rangle\overset{OH}{\underset{\langle\bigcirc\rangle}{}}Cl$$

4. <u>Catabolism</u>

Figure 73 is a diagram of the inactivation processes of epinephrine and norepinephrine.

The major points to be emphasized are:

a. Metabolites are present both in free and conjugated (with glucuronic acid or sulfate) forms. The conjugation reactions take place in the liver.

b. The catecholamines are almost completely metabolized in the body and, therefore, only small quantities are measurable in the urine.

c. The O-methylated metabolites metanephrine and normetanephrine, produced by the action of catechol-O-methyl transferase, are found in urine in slightly higher concentrations than the catecholamines. The methyl donor is S-adenosylmethionine.

d. The <u>major metabolites</u> produced by the action of monoamine oxidase are vanillylmandelic acid (VMA) and 3-methoxy-4-hydroxyphenylglycol (HPMG), which are excreted in the urine.

Figure 73. Catabolism of Epinephrine and Norepinephrine

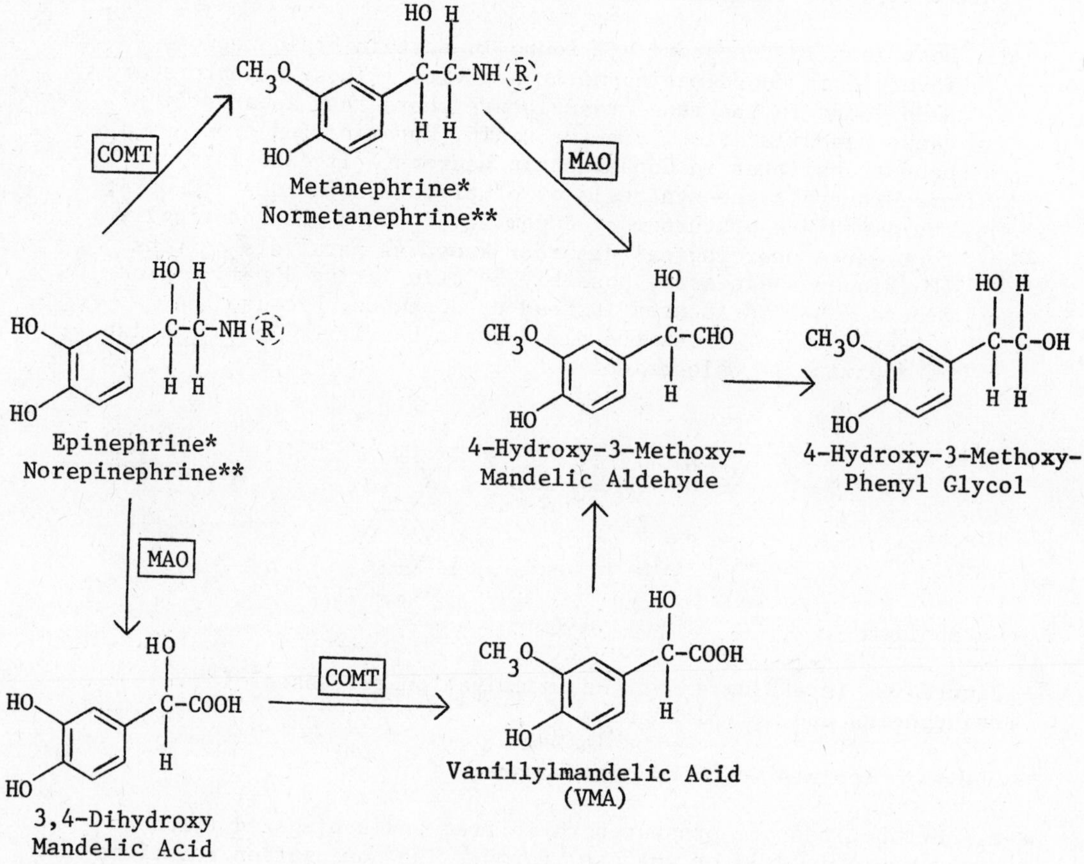
COMT
MAO
MAO
COMT
Metanephrine*
Normetanephrine**
Epinephrine*
Norepinephrine**
4-Hydroxy-3-Methoxy-
Mandelic Aldehyde
4-Hydroxy-3-Methoxy-
Phenyl Glycol
3,4-Dihydroxy
Mandelic Acid
Vanillylmandelic Acid
(VMA)

* (R) = CH_3 COMT = Catechol-O-methyl transferase

** (R) = H MAO = Monoamine oxidase

e. Dopamine's major and minor metabolites are homovanillic acid and methoxytyramine, respectively.

5. Pheochromocytoma: This rare disorder is a tumor of the chromaffin tissues. When present, it is found in the adrenal medulla with an incidence of 90% in adults and 10% in children. The tumor

secretes excessive but varying amounts of norepinephrine and epinephrine. As indicated earlier, norepinephrine is a relatively pure α-receptor agonist. Epinephrine, however, is both a potent α-receptor stimulant and a strong β-receptor agonist. Therefore, the clinical manifestations of pheochromocytoma are not simply dependent on the amount of hormones synthesized by the tumor, but rather on the amounts of norepinephrine and/or epinephrine released. Accordingly, the signs and symptoms of pheochromocytoma, though predictable from the pathophysiology, are extremely varied. In general, the clinical manifestations can be divided into three "hyper" categories: Hypertension, hypermetabolism, and hyperglycemia. Although hypertension is an almost constant manifestation, 60% of the cases are not of the paroxysmal variety but are instead sustained hypertension. In addition, there are signs and symptoms associated with hypertension which are more suggestive of pheochromocytoma, namely, orthostatic hypotension, childhood hypertension crises provoked by minor physical activity, and insulin administration, failure to respond to anti-hypertensive drugs, etc. There are certain neurocutaneous syndromes more commonly associated with pheochromocytoma such as Von Recklinghausen's Disease, Von Hipple-Lindau Disease, Sturge-Weber syndrome and tuberous sclerosis.

Since the hypertension due to pheochromocytoma is treatable by surgery, a number of biochemical tests have been developed to help diagnose this disorder. Concerning these tests, the following points are to be emphasized. Since some tumors preferentially secrete catecholamines and others metabolites, the best chemical screen for the laboratory diagnosis of pheochromocytoma includes urinary catecholamines in addition to at least one of the urinary metabolites.

a. Determination of VMA alone (utilizing extraction and oxidation) would effectively diagnose approximately 85% of the cases.

b. Determination of metanephrines alone would permit diagnosis of approximately 95% of the cases.

c. Determination of catecholamines alone would result in diagnosis of 90% of the cases (virtually 100% if specimen is collected during a hypertensive crisis--seldom possible).

d. Determinations of total urinary catecholamines and metanephrines virtually assure an almost 100% diagnostic accuracy and are the recommended parameters for chemical screening of pheochromocytoma.

Comments on the Biosynthesis of Thyroxine (T_4) and Triiodothyronine (T_3) (Figure 74)

1. The thyroid gland weighs about 30 g in an adult and is made up of many spherical follicles which are well vascularized with blood and lymphatic vessels. The gland's total activity is relegated to endocrine function. Each follicle is surrounded by a single layer of low, cuboidal epithelial cells. The follicles contain the thyroid hormones, T_3 (triiodothyronine) and T_4 (thyroxine), bound to thyroglobulin (TG) which, in histological studies, appears as a colloidal substance.

2. In addition to providing T_3 and T_4, the thyroid gland is also the source of calcitonin, a polypeptide hormone involved in calcium homeostasis (see calcium metabolism, Chapter 9).

3. The structure of a follicle is shown in Figure 75.

Figure 75.   Diagrammatic Representation of a Thyroid Follicle

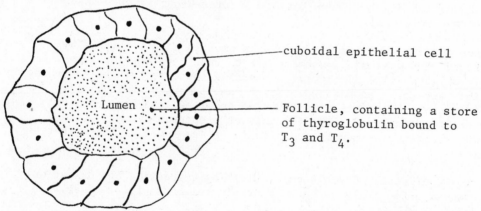

cuboidal epithelial cell

Lumen

Follicle, containing a store of thyroglobulin bound to T_3 and T_4.

Note: In a hypothyroid state (e.g., that produced by I^- deficiency), the follicles are enlarged and the epithelial cells are flattened (squamous type). In a hyperthyroid state (caused, for example, by excess TSH secretion),the follicles are small, contain little colloidal substance, and are linked with columnar cells.

4. The pathway of the synthesis of T_3 and T_4 is shown in Figure 74. Some of the details of these steps are:

    a. The normal intake of iodine as iodide (about 100-200 mg/day) is absorbed from the small intestine into the portal blood circulatory system. About 1/3 of this is taken up from the plasma by the thyroid gland, using an iodide pumping mechanism which works against a concentration gradient. The remaining

Figure 74. Biosynthesis of the Thyroid Hormones (Thyroxine and Triiodothyronine)

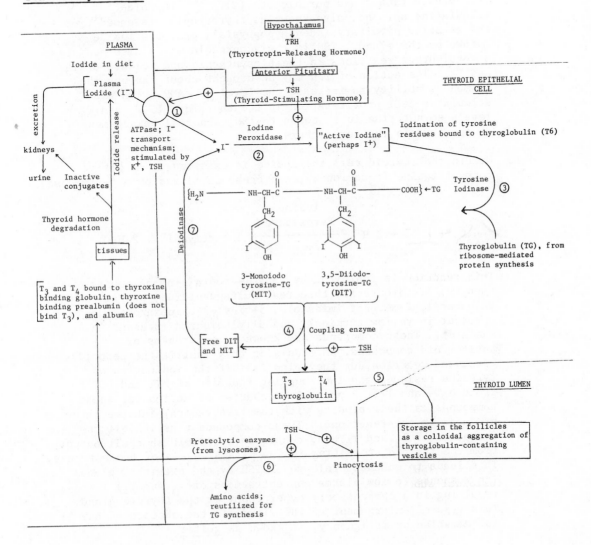

iodide (about 2/3) is eliminated in the urine. The trapping or concentrating of iodide by the thyroid cells is an energy-dependent process and respiratory inhibitors such as cyanide and azide, or uncouplers of oxidative phosphorylation such as dinitrophenol, inhibit the process. Transport of I^- (reaction 1, in Figure 74) is competitively inhibited by anions of thiocyanate (SCN^-) and perchlorate (ClO_4^-). Thyroid-stimulating hormone, also known as thyrotropin, released from the anterior pituitary (adenohypophysis), promotes iodide uptake by the gland. TSH is a glycoprotein which also mediates several other reactions in the biosynthesis and release of T_3 and T_4. The action of TSH may be brought about through the hormone's ability to increase the level of cAMP. TSH also stimulates cell proliferation in the thyroid. If T_3 and T_4 levels do not rise in response to TSH, abnormal enlargement of the thyroid (goiter) occurs.

b. Within the thyroid cell the iodide is oxidized to an active form of iodine (reaction 2), a process catalyzed by a peroxidase.

$$H_2O_2 + 2\ I^- + 2\ H^+ \xrightarrow[\text{TSH}]{\substack{\text{Iodine}\\\text{Peroxidase}}} 2\ \text{"active I"} + 2\ H_2O$$

This reaction is inhibited by many HS-containing compounds such as: 2-thiouracil, thiourea, sulfaguanidine, propyl-thiouracil, 2-mercaptoimidazole, 5-vinyl-2-thiooxazolidone (present in yellow turnips), and allylthiourea (present in mustard). These compounds are known as goitrogens or antithyroid compounds. Not only do they inhibit the peroxidase-catalyzed reaction, but they also inhibit the iodination of tyrosine residues and the coupling reaction of MIT and DIT to produce T_3 and T_4. A probable mechanism of action for these compounds is their binding with the free-radical intermediates produced in the reactions. These compounds thus inhibit the formation of T_3 and T_4 and, consequently, diminish their output from the thyroid, resulting in low plasma levels of the hormones. This leads to enhanced release of TSH by the pituitary gland (in response to low plasma concentrations of T_3 and T_4) resulting in a compensatory hypertrophy of the thyroid gland. This type of enlargement of the gland, in the absence of any inflammation or malignancy, is known as goiter.

c. Iodination of tyrosine residues present on the thyroglobulin molecule. Thyroglobulin is a glycoprotein (molecular weight 660,000 daltons) containing about 140 tyrosine residues. It is synthesized on the ribosomes in the classical protein-synthesizing pathway and transferred to small vesicles where the tyrosine gets iodinated (reaction 3). The vesicles are subsequently stored in the follicles (reaction 5) until their release is stimulated by TSH.

"active I"	monoiodotyrosine-TG (MIT)
+	+
Tyrosine residues of Thyroglobulin	diiodotyrosine-TG (DIT)

d. The coupling reaction of DIT with DIT and DIT with MIT, giving rise to T_4 and T_3 (respectively), involves condensation of two substituted phenols to give a corresponding diphenyl ether and a three-carbon compound (possibly serine or alanine) (reaction 4). This coupling process takes place on the thyroglobulin molecule and both the hormones and the three-carbon residue remain attached to the TG molecule until the hormones are required by the body. The exact mechanisms of iodination and coupling are not known. In the presence of adequate amounts of iodide, thyroglobulin contains considerably more T_4 than T_3, with the ratio being about 10-20 to 1.

e. When there is a need for T_3 and T_4, the thyroglobulin reenters the thyroid cells by pinocytosis (a TSH-mediated step). There it undergoes proteolysis catalyzed by lysosomal proteases (reaction 6). T_3 and T_4 are released into the blood where they are transported, bound to specific carrier proteins and to albumins.

f. In the coupling reaction and proteolysis of TG, small amounts of MIT and DIT are released. The iodine contained in these molecules is recovered by the action of deiodinase and made available to the iodide pool of the cell (reaction 7). This reutilization process is quantitatively significant in that the deficiency of the enzyme deiodinase can lead to hypothyroidism due to inadequate iodine supply.

5. The structures of the thyroid hormones and some of their metabolites:

a. T_4, T_3, and reverse T_3 are derivatives of thyronine:

Thyroxine (T_4) is 3,5,3',5'-<u>tetra</u>iodothyronine. Triiodothyronine (T_3) is 3,5,3'-triiodothyronine. Reverse T_3 (RT_3) is 3,3',5'-triiodothyronine. Note that RT_3 lacks an iodo group at the 5-position of the inner ring, whereas T_3 lacks an iodo group at the 5'-position of the outer ring. This difference in substitution has importance when considering the metabolic potency of these two molecules (discussed later).

b. A metabolite of T_4 is 3,5,3',5'-tetraiodothyroacetic acid (Tetrac).

Tetrac is produced via deamination and oxidative decarboxylation.

c. Transport of thyronine hormones: T_3 and T_4 are carried in the plasma in both bound and free forms. The predominant serum binding protein is thyroxine binding globulin (TBG). The other binding proteins are albumin and thyroxine binding prealbumin. It should be emphasized that it is the concentration of unbound or free T_4 or T_3 which determines the extent of the tissue response (thyrometabolism). T_4 has a higher affinity than T_3 for the binding proteins. For example, T_4 binds to TBG thirty times stronger than T_3 and, consequently, a relatively higher proportion of T_3 exists in the unbound form. The differences in binding of T_3 and T_4 to proteins account for their differences in turnover rate and biologic activity. The turnover rate of T_3 is about 50-70%/day, whereas that for T_4 is 10%/day. In serum from normal adults, the unbound T_4 constitutes about 0.03% of the total, while the unbound T_3 ranges from 0.18-0.46% (for numerical values--see later).

d. Functions of thyroid hormones: These include the regulation of development and metabolic rates in the body by their effects on oxidative reactions. The thyroid hormones have an anabolic effect and are essential for normal growth and maturation. It has been shown in experimental animals that pulmonary maturation during intrauterine life is sensitive to thyroid hormones and that low levels delay pulmonary development. A consequence of this is the inadequate production of pulmonary surfactant (lecithin--see Chapter 8). The function of surfactant is to reduce the surface tension of the alveolar lining, which prevents complete loss of air from the alveoli

during expiration and thereby preventing alveolar collapse.
The latter effect is important as it reduces the work involved
in subsequent breathing. Thus, an inadequate amount of
surfactant may lead to the development of the respiratory-
distress syndrome. In addition to thyroid hormones, several
other factors also affect surfactant synthesis and these
include gestational age, acidosis, pulmonary hypofunction and
plasma glucocorticoids. The observations made in humans are
in accord with the animal studies. The cord-blood thyroid
hormone concentrations (T_3 levels in particular) are
significantly lower (with some increase in TSH) in premature
infants in whom the respiratory-distress syndrome subsequently
develops. However, it should be pointed out that the mechanism
that leads to the low levels of thyroid hormones (in the
presence of increased levels of TSH) and its relationship to
the development of the respiratory-distress syndrome is not
clear.

The hypothyroid state is associated with low basal metabolic
rate, lowered systolic blood pressure, slow pulse, diminished
physical and mental vigor, and obesity in some instances.
The hypothyroid individuals also show elevated serum
cholesterol levels, while free fatty acid levels are decreased
due to tissue lipolysis (see lipid metabolism--Chapter 8).
The tissues obtained from hypothyroid animals show decreased
oxygen consumption, consistent with a decrease in the rates
of oxidative reactions.

Congenital forms of hypothyroidism (cretinism), if untreated,
lead to stunted physical development, deformation, and idiocy
(irreversible damage to central nervous system). With prompt
diagnosis (see later) during the first month of a child's
life and with proper supplementation of thyroid hormones, the
disastrous consequences of cretinism may be prevented.
Placental transfer of thyroid hormones is exceptionally
limited. It has been recommended that there be the establishment
of a mandatory detection program for neonatal hypothyroidism
similar to the one set up for the detection of phenylketonuria
(PKU). The incidence of neonatal hypothyroidism has been
estimated to be one in 5,000-10,000.

Some of the possible causes of congenital hypothyroidism are
aplasia, hypoplasia, maldescent of thyroid (an embryonic
defect of development), inborn errors in the transport of
iodide or in the enzymes involved in hormone synthesis,
autoimmune thyroditis (antibodies produced against one's own
thyroid antigens, the most common one being thyroglobulin,
produce a range of disorders extending from stimulation to
destruction of the thyroid gland), defects in the intrathyroidal

proteolysis of thyroglobulin, abnormal plasma binding of thyroid hormones, and inability to convert T_4 to T_3. The causes of acquired forms of hypothyroidism include iodide deficiency, ingestion of goitrogens, thyroidectomy or radioiodine therapy for thyrotoxicosis, cancer, isolated midline thyroid, lingual thyroid, destruction of the gland by X-ray, thyrotropin deficiency due to diseases of the anterior pituitary gland, and thyrotropin releasing hormone deficiency due to hypothalamic injury, disease, or chronic infections.

In the hyperthyroid state, the symptoms are opposite to that of hypothyroidism. These include systolic hypertension with increased pulse pressure, tachycardia, palpitation, increased irritability, loss of weight, weakness, dyspnea, emotional instability, etc. In hyperthyroid children, exophthalmos (abnormal protrusion of the eye ball) is common. In addition, a majority of the cases exhibit goiter which is characteristically diffuse and usually firm. The causes of hyperthyroidism include tumors of the thyroid, other tumors which produce thyrotropin-like substances with consequent exogenous thyroid hormone excess, and the presence of a long-acting thyroid stimulator (LATS) which is an antibody. Hyperthyroidism is treated by drugs such as propylthiouracil (mechanism already discussed), methimazole, subtotal thyroidectomy, or radiation therapy. It should be mentioned that the latter two modalities of treatment may render the patient hypothyroid, thus requiring thyroid replacement therapy.

6. Active form of the thyroid hormone

In terms of biologic potency (measured as calorigenic activity and capacity to restore euthyroid metabolic status in hypothyroid patients), T_3 is about four times more active than T_4. There is significant conversion of T_4 to T_3 by 5'-monodeiodination in the peripheral tissues (particularly in the liver), resulting in amplification of the biologic activity of the parent molecule. The above observations lead to the consideration of T_4 as a prohormone. It should be noted, however, that a number of studies have indicated that T_4 possesses significant intrinsic biologic activity without the need for conversion to T_3. Another monodeiodination product of T_4 is reverse T_3 (the removal of iodine from the 5-position of the inner ring of T_4). This process also takes place in the peripheral tissues (principally in the extrahepatic tissues) and the product (RT_3) is biologically <u>inactive</u>. It is of interest to point out that regarding the structure-action relationships of T_3, 4'-hydroxylation is necessary (or a group capable of metabolic conversion to a hydroxyl), the ether linkage is not essential (replacement of the ether linkage with a methylene or sulfur group does not result in

substantial loss of activity), iodine or other halogen groups
are not essential for activity and may be substituted with methyl
groups (for example 3,5-dimethyl-3'-isopropyl-L-thyronine is
active; thus halogen or methyl groups are necessary on position 3
and 5), and a carboxyl-containing aliphatic side chain is
essential (L-alanine provides the maximal activity).

Animal studies have revealed that RT_3 in high doses antagonizes
the effects of T_4. In experiments with liver homogenates and
kidney slices, it has been shown that RT_3 competitively inhibits
the formation of T_3 from T_4. Therefore, reverse-T_3 may have a
regulatory control over the process by which T_3 is formed in the
tissues. The conversion of thyroid hormones in a normal adult
is shown in Figure 76.

The following additional aspects should be noted with regard to
the conversion of T_4 to T_3 and reverse-T_3.

a. A major portion of T_3 (35 nmoles/day) is obtained from T_4 by
    5'-deiodination in the peripheral (hepatic) tissues while
    only small amounts (5 nmoles/day) are secreted by the thyroid
    gland. A similar situation exists for reverse-T_3, except
    that 5-deiodination may take place predominantly in peripheral
    tissues other than the liver.

b. Although RT_3 competitively inhibits the conversion of T_4 to
    T_3 in vitro, it is not clear at this time whether RT_3 plays
    a regulatory role in vivo.

c. About 20% of the daily disposal of T_4 through the feces
    involves the formation of tetraiodothyroacetic acid (Tetrac)
    and its conjugation (and unmodified T_4) with glucuronic acid
    and sulfate. Inactivation of active thyroid hormones also
    take place by additional deiodination, oxidative deamination
    and conjugation.

7. The factors which affect T_4 to T_3 conversion.

    a. In the fetus and newborn, the rate conversion of T_4 to RT_3
        is greater than the conversion of T_4 to T_3 when compared
        with adults. Thus, umbilical cord blood and amniotic fluid
        in pregnancy show high levels of RT_3. However, several hours
        after birth the serum T_3 level increases with decreasing RT_3.
        Normal adult levels are reached after a few days. The cause
        for the early post-natal enhancement of T_3 may be due to
        direct thyroidal secretion of T_3 in response to TSH release.

    b. T_3 levels decline with advancing age while T_4 levels remain
        the same.

Figure 76.   Conversion of T_4 to T_3 and Reverse-T_3

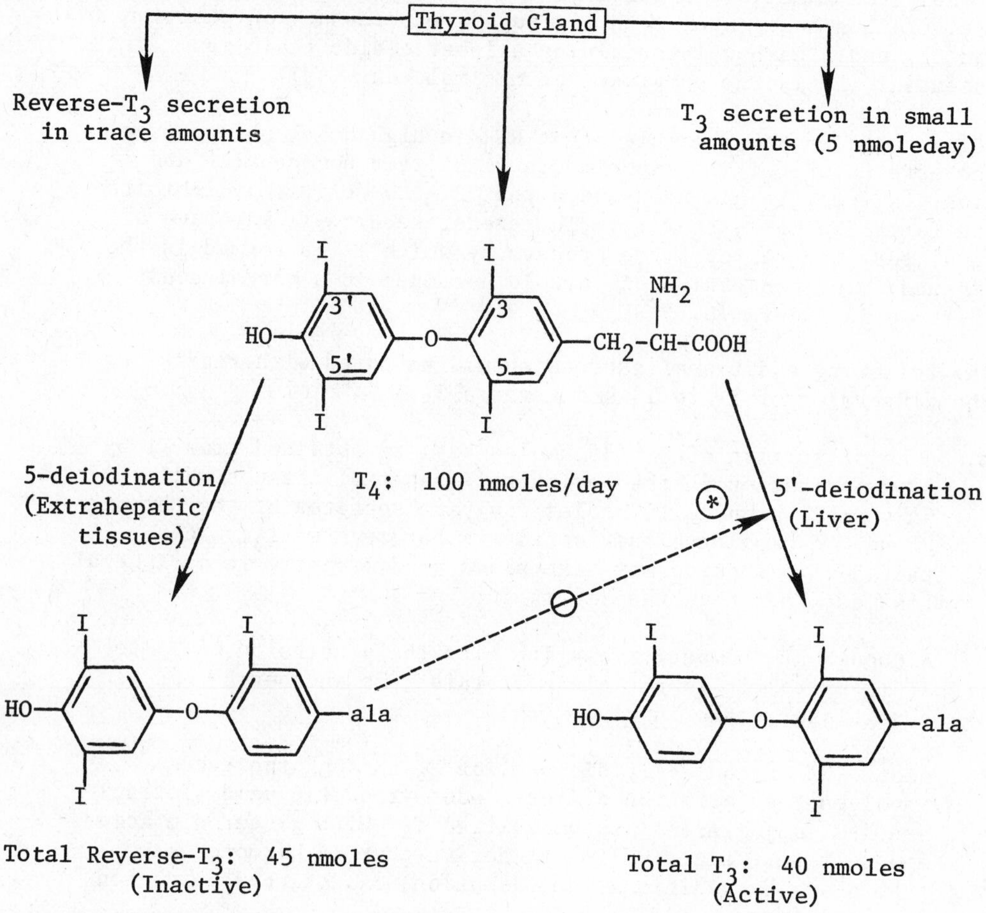

Reverse-T_3 secretion
in trace amounts

Thyroid Gland

T_3 secretion in small
amounts (5 nmoleday)

5-deiodination
(Extrahepatic
tissues)

T_4:   100 nmoles/day

(*)

5'-deiodination
(Liver)

⊖

Total Reverse-T_3:   45 nmoles
(Inactive)

Total T_3:   40 nmoles
(Active)

* Competitive inhibition by reverse-T_3 on $T_4 \rightarrow T_3$ conversion--
an *in vivo* regulatory process?

c.   Hyperthyroidism may result from an enhanced conversion rate
of T_4 to T_3 and/or increased production and secretion of T_3
by the thyroid gland.

d.   Serum T_3 and RT_3 concentrations undergo reciprocal changes
in normal subjects during caloric deprivation, especially
when they are derived from carbohydrate sources, protein-
calorie malnutrition and anorexia nervosa. Fasting leads to

a decrease in T_3 with an increase in RT_3 levels, suggesting
inhibition of the generation of calorigenically active product
and promotion of the synthesis of an inactive product from T_4.
Upon refeeding, the opposite reactions occur, namely, an
increase in T_3 and a decrease in RT_3. During these periods
of fasting and refeeding, total T_4 remains relatively
unchanged (with slightly elevated free T_4 concentrations
during fasting). The decline of T_3 giving rise to a
hypometabolism in caloric deprivation may be viewed as a
protective mechanism, because slowing down the metabolic
rate provides distinct advantages during starvation and
systemic illness associated with catabolism.

e. In patients with liver disease (cirrhosis), the T_3 levels
are decreased with elevated total T_4 and free T_4. This
observation suggests the importance of the liver in the
conversion of T_4 to T_3. It is of interest to point out that
individuals with nonthyroidal systemic illness, fasting, or
protein-calorie malnutrition show low serum T_3 levels and
yet they do not show any symptoms associated with
hypothyroidism. Furthermore, the TSH levels are not elevated
(significance of this is discussed later) in the above cases,
suggesting the absence of hypothyroidism. It is possible,
therefore, that the euthyroid state is maintained by
circulating T_4 which possess intrinsic biologic activity.

8. Mechanism of thyroid hormone action

The present experimental evidence suggests that thyroid hormones
initiate their effects by augmentation of the transcription of
genetic information. The presumed steps involved in the
initiation of hormonal action are as follows: Transport of T_3
into the target cells, binding with cytoplasmic proteins,
penetration (probably in the free form) through the nuclear
membrane, binding to specific nuclear sites attached to chromatin
(acidic proteins of chromatin), augmentation of the synthesis of
specific types of RNA, translation of mRNA and synthesis of
appropriate proteins.

The specificity of the thyroid hormone action presumably resides
in the presence or absence of <u>nuclear</u> binding sites. Nonresponsive
tissues such as the spleen and testis (in rat) have significantly
less nuclear binding sites when compared with responsive tissues.
The cytoplasmic binding proteins of thyroid hormones appear to
lack specificity and probably do not have any directive role in
the initiation of hormone activity. Furthermore, experiments
<u>in vitro</u> have shown that nuclear binding occurs in the absence of
cytoplasmic proteins. In contrast, for the transport of steroid
hormones to nuclei, cytoplasmic binding with a specific protein

is a prerequisite (see Chapter 13). In a study to identify the specific proteins and metabolic processes elaborated due to thyroid hormone action, the sodium pump was found to be augmented. The extrusion of Na^+ from inside to outside of the cell requires energy which is derived from the hydrolysis of ATP to ADP and P_i. The latter two compounds stimulate mitochondrial respiration. Thus, the hormone-mediated Na^+ transport accounts for thermogenesis and increased oxygen consumption.

9. Regulation of thyroid function

a. Control of thyroid hormone secretion is shown in Figure 77. Note in the figure that T_4 and T_3 regulate their own synthesis by acting at both pituitary and hypothalamic sites. The major regulatory site, however, is at the pituitary gland, where thyroid stimulating hormone (TSH, also known as thyrotropin) is released. Thyrotropin is a glycoprotein with a molecular mass of about 30,000 daltons. It consists of two subunits, α and β. The α subunits of thyrotropin, leutinizing hormone, follicle-stimulating hormone, and human chorionic gonadotropin have nearly identical sequences. The biologic specificity in each of the above hormones reside in the β-subunits (see Chapter 13--for further discussion). Thyrotropin's action on the thyroid gland involves all phases of hormone synthesis and release, namely, iodide uptake and organification, hormone synthesis, endocytosis, and proteolysis of the thyroglobulin bearing the T_4 and T_3. The primary action of thyrotropin appears to be mediated through the activation of adenylate cyclase and increased glandular concentrations of cyclic AMP.

As noted earlier, the plasma levels of T_3 and T_4 regulate TSH secretion. When T_3 and T_4 levels decrease (e.g., hypothyroidism), the TSH secretion goes up and vice versa. In an analogous manner, thyrotropin releasing hormone (TRH) and TSH levels show a similar reciprocal relationship to each other.

Figure 77.   Control of Thyroid Hormone Secretion

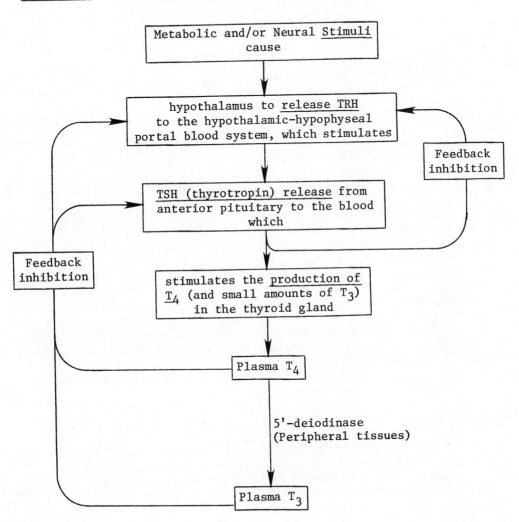

b. TRH is a neutral tripeptide and consists of pyroglutamate, histidine and prolinamide:

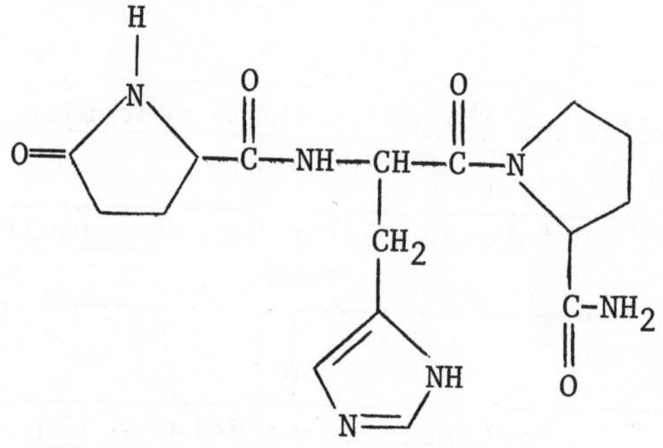

Pyroglutamylhistidylprolinamide

Pyroglutamate is an internal cyclic amide of glutamic acid. It is also found in gastrin, one of the GI hormones. The peptide does not appear to be species-specific and has been chemically synthesized (available commercially). TRH, secreted into the portal venous system, travels from the hypothalamus directly to the pituitary gland where it interacts with specific receptors on the plasma membrane of the thyrotropin-releasing cells, causing the activation of adenylate cyclase with intracellular elevation of cyclic AMP. This causes a prompt release of thyrotropin (a calcium-dependent process). TRH promotes thyrotropin synthesis as well. TRH also stimulates prolactin secretion (see Chapter 13). Exogenous administration of TRH has been found to be useful in assessing the pituitary reserve of thyrotropin and prolactin in patients with pituitary lesions, differentiation of hypothalamic hypothyroidism from secondary hypothyroidism due to pituitary destruction, etc.

10. Laboratory evaluation of thyroid function

a. It is generally accepted that the concentration of <u>unbound</u> (or free) thyroid hormones determines the thyrometabolic status. The equilibrium reaction between unbound T_4, thyroxine binding protein (TBP), and T_4 bound to TBP is shown below:

$$T_4 + TBP \rightleftarrows T_4\text{-TBP}$$

Applying the Law of Mass Action,

$$K = \frac{\left[T_4\text{-}TBP\right]}{\left[T_4\right]\left[TBP\right]} \qquad \ldots\ldots\text{Equation 1}$$

where K = equilibrium constant; T_4 = concentration of unbound T_4; TBP = concentration of thyroxine binding protein (mostly TBG); T_4-TBP = concentration of T_4 bound to TBP.

The concentration of unbound T_4 is obtained by rearranging the above relationship.

$$T_4 = \frac{1}{K}\frac{\left[T_4\text{-}TBP\right]}{\left[TBP\right]} \qquad \ldots\ldots\text{Equation 2}$$

It is apparent from the above equation that the concentration of free T_4 can be determined from the ratio between the concentrations of bound thyroxine to thyroxine binding protein. It is important to emphasize that this ratio will remain a <u>constant</u>, irrespective of alterations in the concentration of hormone binding protein, as long as the individual remains euthyroid. In other words, variations in the concentration of binding protein in itself will not affect metabolism mediated by thyroid hormones, for it is the concentration of free thyroxine which is responsible for maintaining a normal metabolism. It follows, therefore, that the ratio between the concentrations of bound T_4 to binding protein in equation 2 decreases and increases, respectively, in hypo- and hyperthyroidism. The biologically active form of thyroxine (i.e., unbound) is only about 0.05% of the total thyroxine. It is therefore possible to estimate the free thyroxine by making approximations in equation 2.

i) Total T_4 is approximately equal to bound T_4 (numerator term) because 99.95% of T_4 is in the bound form. Total T_4 in a serum sample can be measured either by a competitive protein displacement method (T_4D) or by a radioimmunoassay procedure (T_4RIA). The T_4D method (Murphy-Pattee principle) consists of adding an aliquot of the patient's T_4 (obtained by extracting serum with alcohol and centrifuging to remove the precipitated proteins) to a reference sample of TBG saturated with radioactive T_4 (^{125}I-T_4). The patient's T_4 (unlabeled) displaces the radioactive T_4 bound to TBG by a competitive process, which is dependent upon the concentration of unlabeled T_4. After reaching equilibrium, the unbound T_4 (both radioactive and

non-radioactive forms) is removed by resin absorption
and the radioactivity remaining bound to TBG is measured.
The patient's T_4 is then quantitated by running T_4
standards. T_4RIA methodology utilizes the reaction
between an antigen and its specific antibody (for
details, see Chapter 4).

ii) An estimate of the concentration of TBP (the denominator
term) is arrived at indirectly by measuring the unbound
sites of TBP with radioactive T_3. This test is known
as the T_3-uptake test (abbreviated as T_3U). It does not
depend upon the concentration of T_3 in the serum nor
does it measure T_3 levels. The test is performed by
creating a partition of added ^{125}I-T_3 between the
patient's TBP and resin beads which also bind to labeled
T_3. Thus the unsaturated sites on TBP compete with the
resin for radioactive-T_3. Once equilibrium is reached,
the radioactivity taken up by the resin beads is
measured. If there is a large excess of unsaturated
sites on TBP (e.g., hypothyroidism), the resin uptake
of radioactive T_3 is less, yielding a low value for the
test. Conversely, if there is a decrease in the unbound
sites (e.g., hyperthyroidism), the test will show an
increase in the binding of radioactive T_3 to the resin.
Hence, there is an inverse relationship between the
unsaturated binding sites (an indirect estimate of TBP)
and the uptake of T_3 by the resin. Note that for all
practical purposes, any displacement of T_4 by T_3 can be
ignored, because T_4 is bound much more tightly than T_3
to TBP. Furthermore, the concentration of endogenous
T_3 is considerably less than T_4 and thus the sites that
T_3 occupy are insignificant in relation to the total
unoccupied sites. It has been recommended that the T_3U
results be expressed as a ratio between the patient's
value and the value from a pool of normal reference
serum, included in the same assay procedure.

$$T_3U = \frac{T_3U \text{ of patient's serum}}{T_3U \text{ of pooled normal reference serum}}$$

The normal range for the T_3U ratio is around one.
Therefore, when the T_3U ratio is multiplied by T_4 (D or
RIA) values to obtain the free thyroxine index (see
below), the same numerical values as that of T_4 (D or
RIA), are obtained for normal specimens. Thus, the
modified equation 2 which yields an estimate of free
thyroxine is as follows.

Table 24. T_4, T_3U, FT_4 and Thyroid Status in a Variety of Clinical Situations

Clinical Situation	T_4(RIA or D)	T_3U Ratio	FT_4 Index	Thyroid Status
Hyperthyroidism	↑	↑	↑	Hyperthyroid
Hypothyroidism	↓	↓	↓	Hypothyroid
Pregnancy	↑	↓	↔	Euthyroid
Nephrosis	↓	↑	↔	Euthyroid
Cirrhosis	↓	↑	↔	Euthyroid
TBG Deficiency*	↓	↑	↔	Euthyroid
TBG Excess*	↑	↓	↔	Euthyroid
Administration of Drugs				
Dilantin	↓	↑	↔	Euthyroid
Salicylates (large doses)	↓	↑	↔	Euthyroid
Anabolic Steroids	↓	↑	↔	Euthyroid
Estrogens	↑	↓	↔	Euthyroid

↔ = Normal

↑ = Increased

↓ = Decreased

* These problems are genetic abnormalities affecting thyroxine-binding-globulin synthesis. [665]

Free thyroxine index (FT_4 index) = T_4(D or RIA) x T_3U ratio

Alterations of the above parameters and the thyroid status in a few conditions are shown in Table 24 . Note that in pregnancy, estrogen administration and TBG excess (due to genetic abnormality), T_4 is elevated while the T_3U ratio is decreased, giving rise to a normal thyroid status. The reverse is true in nephrosis, cirrhosis, TBG deficiency, and anabolic steroid administration. Dilantin and salicylates reduce bound T_4 levels because they bind to TBG, thus displacing T_4.

As was shown, the determination of the FT_4 index requires the measurement of two parameters independently, namely, total T_4 and T_3-uptake. However, recently introduced methods for estimating the free thyroxine index consist of dual sequential competitive protein binding assays.

This test combines the results of T_3U in the procedure. The basis for a test known as effective thyroxine ratio, ETR, uses the above principle and consists of estimating the serum thyroxine in the presence of thyroxine binding proteins from the same serum; the results are expressed as a ratio of serum thyroxine similarly determined in normal serum. This compensates for variations in T_4 due to differences in the thyroxine binding protein content of serum.

b. In the diagnosis of hypothyroidism, the FT_4 index is measured. If it is below normal, the disorder is confirmed by measuring TSH (by a RIA procedure). An elevated TSH level is definite evidence of primary thyroid failure, indicating a need for replacement therapy. If the TSH is not elevated, a TRH test (see below) is performed for further clarification.

c. In the diagnosis of hyperthyroidism, the initial screening test is the measurement of the FT_4-index. If it is elevated, the hyperthyroidism is confirmed, if necessary, by determining thyroidal iodide turnover using radioactive iodine or technitium. In rare instances where the FT_4 index is normal but the patient is clinically hyperthyroid, measurement of T_3 (not T_3U) may clarify the problem. These situations are given the name of T_3-thyrotoxicosis.

d. TRH test: When TRH is administered intravenously in a normal individual, there is a doubling of the serum TSH levels at about 15 to 30 minutes after the injection, followed by a steady decline to basal levels in one to four hours. The response of TSH levels to TRH administration in different clinical conditions are as follows:

   i) Primary hypothyroidism--TSH levels are usually high to begin with and show an exaggerated rise with TRH administration.

   ii) Hyperthyroidism (or in the case of overtreatment of hypothyroid states with T_3 or T_4)--the excess T_3 and T_4 will inhibit the release of TSH and thus no rise will be observed.

   iii) Hypopituitarism--the TSH levels are low to begin with, and may remain low depending upon the pituitary TSH reserve.

   iv) Hypothalamic hypothyroidism--shows a normal magnitude of response, but with a delayed peak, usually 45-60 min after the administration of TRH.

e. As noted above, in primary hypothyroidism the TSH is elevated with a low FT_4 index. However, there are situations in which TSH may be elevated with a normal FT_4 index. This can occur in circumstances where there is a reduction in the amount of functioning thyroid gland, because in order to maintain a euthyroid status, a higher level of TSH is required to drive the remaining portion of the gland to produce adequate amounts of the thyroid hormones. Examples of such conditions are chronic thyroiditis, colloid goiter, ^{131}I-therapy, partial thyroidectomy and inadequately supplemented hypothyroid states.

f. Normal ranges (adult values obtained from serum):

   T_4(RIA) = 5-10 µg/100 ml

   T_3U ratio = 0.82-1.18

   FT_4 index (T_4 x T_3U) = 5-10

T_3(RIA) = 60–145 ng/100 ml

FT_3 index (T_3 x T_3U) = 60–145 (determined in a similar manner
as FT_4 index)

TSH (thyrotropin) = 0–14 µU/ml.

Effective Thyroxine Ratio (ETR) = 0.86–1.13

Metabolism of Tryptophan

Comments on Tryptophan Metabolism

1. Tryptophan is an essential amino acid and is required for the
   synthesis of body proteins. It undergoes reversible deamination,
   producing indolepyruvic acid, which accounts for the fact that
   L-tryptophan can be replaced with either the keto acid (indole
   pyruvic acid) or D-tryptophan (D-tryptophan may be converted to
   indole pyruvic acid which in turn may be converted to L-tryptophan).

2. Tryptophan is involved in the synthesis of several important
   compounds.

   a. Nicotinic acid (amide) is a vitamin, required in the synthesis
      of NAD^+ and $NADP^+$, the cofactors of many dehydrogenase reactions.
      It has been found in humans that 60 mg of tryptophan gives rise
      to 1 mg of nicotinamide.

      i) Synthesis of NAD^+ (and some related reactions)

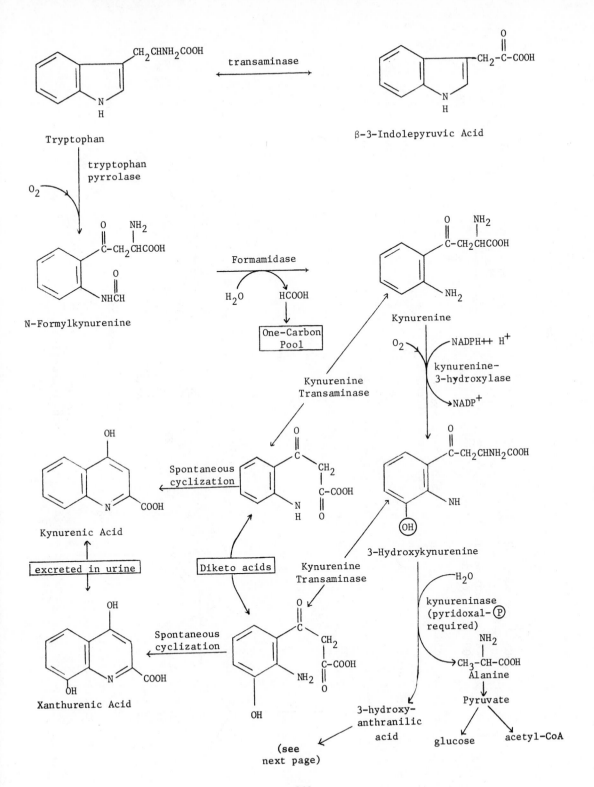

Tryptophan

transaminase

β-3-Indolepyruvic Acid

tryptophan
pyrrolase

O_2

N-Formylkynurenine

Formamidase

H_2O → HCOOH

One-Carbon
Pool

Kynurenine

O_2 → NADPH + H^+

kynurenine-
3-hydroxylase

NADP$^+$

Kynurenine
Transaminase

3-Hydroxykynurenine

Spontaneous
cyclization

Kynurenic Acid

excreted in urine

Diketo acids

Kynurenine
Transaminase

H_2O

kynureninase
(pyridoxal-℗
required)

Alanine

Spontaneous
cyclization

Xanthurenic Acid

3-hydroxy-
anthranilic
acid

Pyruvate

glucose   acetyl-CoA

(see
next page)

669

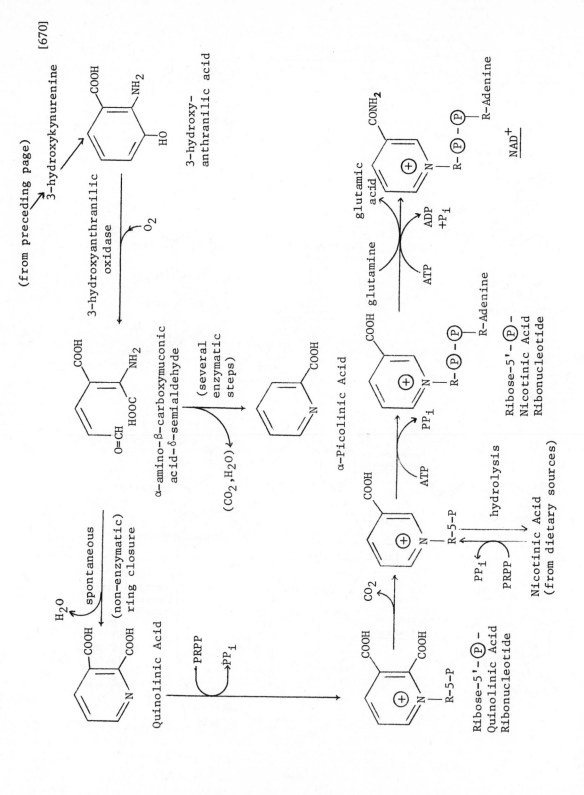

(from preceding page)

[670]

ii) In the synthesis of nicotinic acid from tryptophan, the
first reaction is tryptophan $\rightarrow$ N-formylkynurenine. This
reaction is catalyzed by tryptophan pyrrolase, an
inducible iron-porphyrin enzyme found in the liver.
Molecular oxygen is incorporated into the product in this
reaction.

iii) Deficiency of vitamin B_6 (pyridoxine) results in a
deficiency in the production of nicotinic acid. This
happens because kynureninase which catalyzes the
reaction:

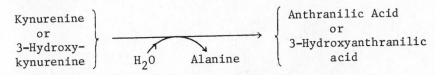

$$\left.\begin{array}{c} \text{Kynurenine} \\ \text{or} \\ \text{3-Hydroxy-} \\ \text{kynurenine} \end{array}\right\} \xrightarrow[\ \ H_2O \qquad \text{Alanine}\ \ ]{} \left\{\begin{array}{c} \text{Anthranilic Acid} \\ \text{or} \\ \text{3-Hydroxyanthranilic} \\ \text{acid} \end{array}\right.$$

is dependent upon B_6-PO_4 and is very sensitive to its
deficiency. In the absence of this vitamin, the above
reaction will not take place and kynurenine and
3-hydroxykynurenine are converted to the corresponding
diketo acids by transaminase reactions. These are
eliminated in the urine. Note: Although the transaminase
reactions are also B_6-PO_4 dependent, it appears that
kynureninase is more sensitive to B_6 deprivation.

iv) In the absence of B_6 and tryptophan in the diet, normal
synthesis of NAD^+ and $NADP^+$ can be achieved if there is
an adequate dietary supplement of nicotinic acid.
However, a diet which is deficient in protein is also
likely to be deficient in vitamins (including niacin and
nicotinamide). In the nutritional deficiency disease
pellagra, there is a lack of both protein (and hence
tryptophan) and the vitamins niacin and nicotinamide.

b. Serotonin

i) Serotonin is synthesized from dietary tryptophan. It
is hydroxylated to 5-hydroxytryptophan and then
decarboxylated to 5-hydroxytryptamine, which is serotonin.

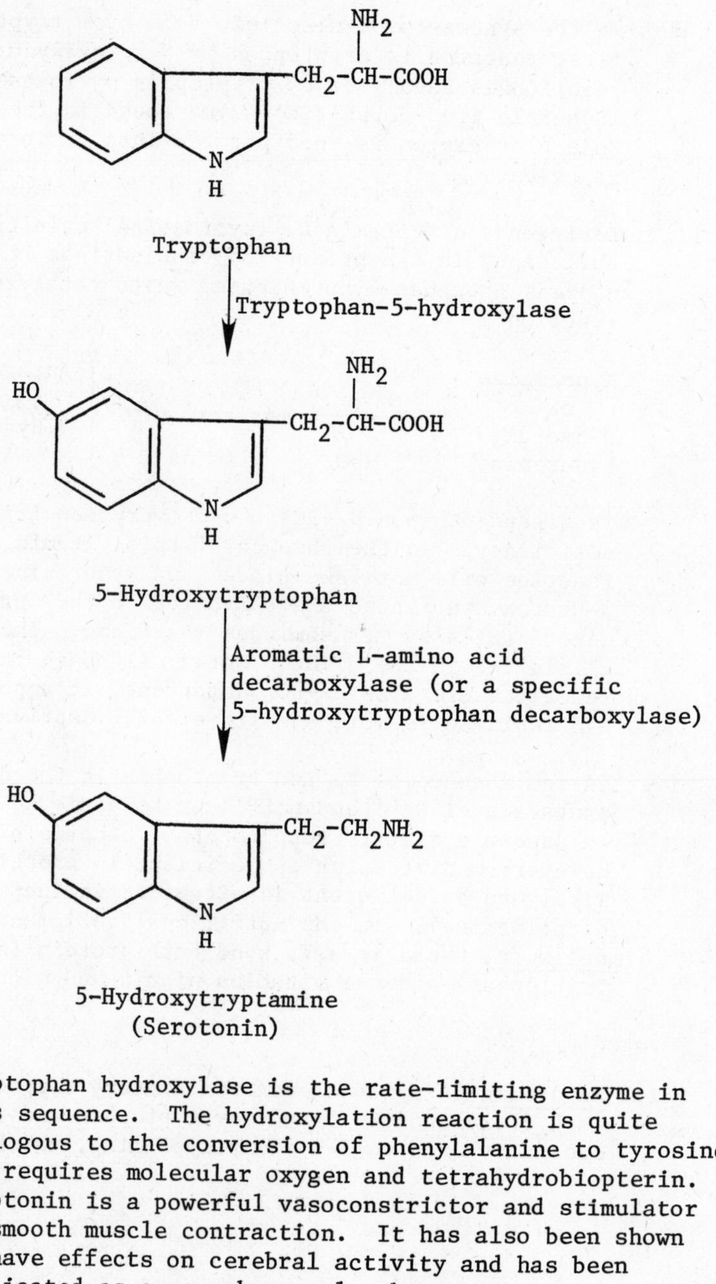

Tryptophan

Tryptophan-5-hydroxylase

5-Hydroxytryptophan

Aromatic L-amino acid
decarboxylase (or a specific
5-hydroxytryptophan decarboxylase)

5-Hydroxytryptamine
(Serotonin)

Tryptophan hydroxylase is the rate-limiting enzyme in
this sequence. The hydroxylation reaction is quite
analogous to the conversion of phenylalanine to tyrosine
and requires molecular oxygen and tetrahydrobiopterin.
Serotonin is a powerful vasoconstrictor and stimulator
of smooth muscle contraction. It has also been shown
to have effects on cerebral activity and has been
implicated as a neurohormonal substance.

ii) Serotonin undergoes oxidative deamination as follows:

$$\text{Serotonin} \xrightarrow{\text{monoamine oxidase}} \text{5-hydroxyindoleacetic acid}$$
$$\text{(5-HIAA)}$$

5-HIAA, the detoxified product of serotonin, is excreted in the urine. Other derivatives of 5-HIAA, such as glycine, acetyl and glucuronide conjugates, have been identified in the urine, particularly in individuals with carcinoid disease.

iii) Many inhibitors of monoamine oxidase have been reported, an example of which is the isopropyl derivative of isonicotinic acid hydrazide (iproniazid). These inhibitors lengthen the action of serotonin by stopping or delaying its inactivation, thus presumably giving rise to a stimulation of cerebral activity (see also comments on catecholamine biosynthesis).

c. Melatonin is a hormone of the pineal gland whose synthesis is shown in Figure 78. Its presumed function has been mentioned under albinism (phenylalanine-tyrosine metabolism).

3. The conversion of tryptophan to skatole and indole takes place predominantly in the large intestines due to the action of bacterial enzymes. The odor of the feces is partly attributed to these molecules.

4. Hartnup Disease

a. This condition is characterized by a defect in the intestinal and renal transport of tryptophan and other amino acids which are excreted in large quantities in urine and feces.

b. In this hereditary disorder, the nature of the biochemical lesions are not known, but a deficiency of tryptophan pyrrolase may be involved.

c. Individuals with this deficiency also excrete indolylacetic acid and indican (mostly 3-indoxylsulfuric acid; some indoxyl-β-glucuronide may also appear). These individuals also have decreased ability to produce kynurenine and nicotinic acid. The major clinical manifestations of this disease include a pellagra-like rash, mental retardation, and other aberrations.

Figure 78. Summary of Some Pathways in Tryptophan Metabolism

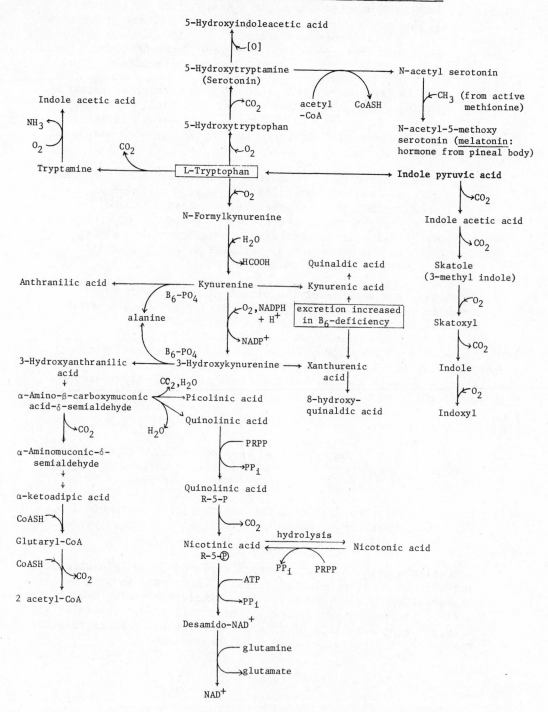

5. Pathways for the synthesis of melatonin, serotonin, and nicotinamide, as well as other reactions of tryptophan metabolism, are summarized in Figure 78. Note that tryptophan is <u>both</u> ketogenic and glucogenic since products of its catabolism include acetyl-CoA (which is the precursor of the ketone bodies) and alanine (which transaminates to pyruvate). Pyruvate can either contribute to gluconeogenesis or be used to form acetyl-CoA.

7

Hemoglobin and Porphyrin Metabolism

with John H. Bloor, M.S.

Hemoglobin (Hb) is the protein which, in vertebrates, functions as the principal carrier of oxygen from the lungs to the tissues. It is normally found within the erythrocytes, annucleate cells which are present in the plasma of the circulatory system. The presence of Hb increases the oxygen-carrying capacity of the blood 70-fold.

Some aspects of erythrocytes have already been discussed (see the chapter on carbohydrate metabolism). Hemoglobin will be considered here in detail. To start with, hemoglobin contains iron and in order to understand its structure and function, a knowledge of the chemistry and metabolism of iron is necessary.

A Brief Review of the Coordination Chemistry of Iron Complexes

1. In the formation of a coordinate covalent bond, both electrons of the bonding pair are supplied by one of the atoms. Transition metal ions (for example, iron ions) accept electrons and form bonds to give rise to complex ions. In a general sense, the transition metal ions may be considered as Lewis acids. In the Lewis sense, acids are electron-pair acceptors and bases are electron-pair donors. The atom, ion, or molecule that donates the electron pair to form a coordinate covalent bond with a transition metal ion is known as a <u>ligand</u>. Thus, the ligand is a Lewis base and the metal is a Lewis acid.

2. Metal atoms (or ions), in forming complex ions, try to reach a noble-gas electron configuration by accepting electron pairs from ligands. The iron atom, which has 26 electrons, can attain the noble-gas configuration of krypton by accepting 10 electrons.

Examples:

Form of Metal	Number of e^-	Complex molecule or ion	Number of e^-
Fe atom	26	$Fe(CO)_5$	$26 + 10 = 36$
Fe^{+2}	24	$Fe(CN)_6^{-4}$	$24 + 12 = 36$
Fe^{+3}	23	$Fe(CN)_6^{-3}$	$23 + 12 = 35$

Note that when Fe^{+3} forms $Fe(CN)_6^{-3}$, the complex ion does not have the noble-gas configuration of krypton.

676

3. It is assumed that the electron pairs donated by the ligands enter hybrid orbitals associated with the iron atom. Iron has a coordination number of 6 (can form a complex with six ligands). The orbital hybridization on the iron is d^2sp^3 which gives rise to octahedral complex ions having six bonds. The bonding orbitals are formed from two 3d, one 4s, and three 4p orbitals. These orbitals all have about the same energy, this being necessary for the hybridization to take place. Distribution and arrangement of electrons in the d^2sp^3 orbitals of iron is shown in Table 25. Note that in each atom of iron in the table, there is an inner core of 18 electrons having the configuration $1s^2$, $2s^2$, $2p^6$, $3s^2$, $3p^6$. Thus, the total number of electrons in each iron is 18 plus the number indicated by the small, vertical arrows in the table.

    Note that in Table 25, the ferrocyanide ion ($Fe(CN)_6^{-4}$, containing Fe^{+2}) is diamagnetic (all electrons are paired) and that ferricyanide ($Fe(CN)_6^{-3}$, containing Fe^{+3}) is paramagnetic (having an unpaired electron).

    The octahedral complex $Fe(H_2O)_6^{+2}$ contains four unpaired electrons and presumably forms sp^3d^2 hybrid orbitals (known as an outer complex; involves 4d orbitals) instead of d^2sp^3 hybrid orbitals (known as an inner complex; uses 3d orbitals), as in the case of $Fe(CN)_6^{-4}$. Apparently the type of hybrid orbitals formed depends upon the strength of the Lewis base (the ligand). For example, CN^- and O_2 are stronger Lewis bases than H_2O. Strong Lewis bases tend to repel electrons more strongly. This causes the electrons to pair up (when possible) and results in inner complex formation. Because of this, outer complexes are also known as high-spin complexes and inner complexes are called low-spin complexes. These terms both refer to the number of unpaired electron spins.

Metabolism of Iron

1. An adult human (70 kg) contains about 4.0-4.5 g of iron which is distributed approximately as follows:

    in hemoglobin---about 2.6-3.0 g

    in ferritin and hemosiderin-----------------------1.0-1.5 g
    (storage forms of iron)

    in myoglobin, transferrin (specific ------------0.1-0.2 g
    iron-binding plasma β_1-globulin),
    cytochromes, peroxidase, catalase,
    tryptophan pyrrolase, etc.

Table 25. Electronic Configurations of Some Oxidation States and Complexes of Iron

	3d	4s	4p	4d
Fe atom	(↑↓)(↑)( )(↑)(↑)	(↑↓)	( )( )( )	( )( )( )( )( )
Fe^{+2} (gas; high-spin complexes)	(↑↓)(↑)( )(↑)(↑)	( )	( )( )( )	( )( )( )( )( )
Fe^{+2} (low-spin complexes)	(↑↓)(↑↓)(↑↓)( )( )	( )	( )( )( )	( )( )( )( )( )
Fe(CN)$_6$$^{-4}$	(↑↓)(↑↓)(↑↓)[(↑↓)(↑↓)	(↑↓)	(↑↓)(↑↓)(↑↓)]	( )( )( )( )( )
		d^2sp^3 hybrid orbitals		
Fe^{+3} (gas; high-spin complexes)	(↑)(↑)(↑)(↑)(↑)	( )	( )( )( )	( )( )( )( )( )
Fe^{+3} (low-spin complexes)	(↑↓)(↑↓)(↑)( )( )	( )	( )( )( )	( )( )( )( )( )
Fe(CN)$_6$$^{-3}$	(↑↓)(↑↓)(↑)[(↑↓)(↑↓)	(↑↓)	(↑↓)(↑↓)(↑↓)]	( )( )( )( )( )
	↑ unpaired electron	d^2sp^3 hybrid orbitals		
Fe(H$_2$O)$_6$$^{+2}$	(↑↓)(↑)( )(↑)(↑)	[(↑↓)	(↑↓)(↑↓)(↑↓)	(↑↓)]( )( )( )( )
		sp^3d^2 hybrid orbitals		
*Fe(His)(Py)$_4$(O$_2$)$^{+2}$	(↑↓)(↑↓)(↑↓)[(↑↓)(↑↓)	(↑↓)	(↑↓)(↑↓)(↑↓)]	( )( )( )( )( )
		d^2sp^3 hybrid orbitals		

*This is the configuration of iron in oxyhemoglobin. His = histidyl nitrogen; Py = pyrrole ring nitrogen (part of porphyrin).

2. Concerning iron absorption, the following points should be emphasized.

    a. Ionic iron is chiefly absorbed in the ferrous state (Fe^{2+}). Absorption occurs principally in the duodenum and progressively decreases to low levels in the ileum. The small intestine also functions as an excretory organ for iron, because the absorptive epithelial cells are replaced by new ones at the end of their lifespan (3-5 days) and the iron stored as ferritin in these cells is lost.

    b. Iron in the ionic form is carried across the intestinal epithelial cell by an active transport system. However, iron complexed with protoporphyrin (heme iron) appears to be transported intact into the cell where iron is removed for further processing.

    c. The iron in food exists mainly in the ferric (Fe^{3+}) state, complexed to proteins, amino acids, organic acids or heme. Since it is absorbed in the ferrous state, the ferric form must be reduced. This is accomplished by ascorbate, succinate and amino acids. The gastric acids potentiate iron absorption by aiding in the formation of soluble ferrous iron chelates which are absorbable. In subjects with achlorhydria or partial gastrectomy, the absorption is below normal and can be increased by the administration of hydrochloric acid.

    d. A number of factors influence the absorption of inorganic iron. Phosphates, phytates, and oxalates can cause decreased iron absorption by forming insoluble iron complexes. The absorption of heme iron, however, is not affected by these agents. Other factors that hinder iron absorption include malabsorption syndromes, severe infection and gastrectomy (or achlorohydria-- discussed above). Factors which facilitate iron absorption include ascorbic acid and the presence of other metabolic reducing agents, normal gastric acid production, and an adequate amount of calcium (which complexes with oxalates, phytates and phosphates and thus makes them unavailable for the formation of insoluble complexes with iron). Therefore, the nutritive value of iron depends upon the relative concentrations in food of the substances mentioned above. Antacids such as aluminum hydroxide and magnesium hydroxide decrease iron absorption. In general, foods of animal origin provide a considerably greater amount of iron which can be assimilated than do foods of vegetable origin. This is due to the fact that the vegetables (on a weight basis) contain less iron while also containing substances (e.g., phytates) which bind to iron that produce insoluble complexes. Foods containing greater than 5 mg of iron per 100 g include organ meats (liver, heart), wheat germ, brewer's yeast, oysters, and certain dried beans. Foods that contain 1 to 5 mg of iron per 100 g include muscle meats, fish, fowl, some fruits (prunes, raisins), most green vegetables, and most cereals. Foods containing less than 1 mg per 100 g (foods low in iron) include milk and milk products and most nongreen vegetables. Cooking in iron utensils enhances the iron content of the food. One type of iron storage disorder, known as hemosiderosis, can result when iron is present in excessive quantities in a diet which is also suited for maximum iron absorption. This is seen in the African Bantu population (Bantu siderosis) who ingest a diet high in corn, which is low in phosphorous and cooked in iron utensils. In addition, they also consume an alcoholic beverage fermented in iron pots. The combination of a low phosphate diet with a high intake of iron increases the iron absorption, leading to this disorder. A general increase in the iron levels of the tissues without any parenchymal cell (functional cells of a tissue) damage is known as hemosiderosis (note: hemosiderin is a storage form of iron in which ferric hydroxide is present as micelles). Hemosiderosis can lead to hemochromatosis (see later) which is associated with liver damage leading to cirrhosis, diabetes mellitus, pancreatic fibrosis, and brown pigmentation, presumably due to the toxic effects of iron accumulation in the cells far in excess of need.

e. The iron entering the mucosal cell is distributed into two pools, a ferrous iron pool, which transports iron to the portal blood from the serosal surface, and a ferric iron pool associated with ferritin, which has a considerably slower turnover rate. Ferritin formed by the combination of iron with apoferritin is a storage form of iron found throughout the reticuloendothelial system and in the parenchyma of organs (liver, muscle, etc.). In ferritin, iron is present as micelles of hydrated ferric oxide-phosphate complexes surrounded by a protein shell (apoferritin). The molecular mass of ferritin is 465,000 daltons and about 20% of the mass consists of iron. Note that much of ferritin present in the mucosal cells is lost through exfoliation.

f. The exact mechanisms that are involved in the regulation of iron absorption are not clear. On the basis of observations that oral administration of a large dose of iron reduced (or inhibited for a period) the absorption of a second dose of administered iron, ferritin has been implicated in the regulation of iron absorption. It has been proposed that the quantity of ferritin does this by preventing iron absorption when it is high and favoring absorption when it is low. This

has been termed the "mucosal block" or "ferritin curtain."
It should be pointed out, however, that this block is
susceptible to disruption at more than one point, and unneeded
iron can be stored in the body.

g. At normal levels of iron intake, absorption of iron into the
    body is comprised of at least two steps: mucosal uptake
    from the intestinal lumen and mucosal transfer of iron to
    the body. Both of these are affected inversely by the state
    of <u>body iron stores</u>; that is, a decrease in the iron stores
    is accompanied by an increase in absorption and vice versa.
    It is of interest to point out that in iron-deficiency states,
    non-ferrous metals which have a similar ionic radius and which
    form octahedral complexes with six coordinate covalent bonds
    are also absorbed at an increased rate. These include cobalt
    and manganese.

    Another factor which affects iron absorption is the rate of
    <u>erythropoiesis</u>. Accelerated erythrocyte production caused by
    bleeding, hemolysis, or hypoxia leads to increased iron
    absorption. Conversely, diminished erythropoiesis caused by
    starvation, blood transfusion or return to sea level from
    high altitudes decreases the iron absorption. It is not
    understood how the magnitude of body iron stores and the rate
    of erythropoiesis transmit information to the duodenum so as
    to transfer appropriate amounts of iron into the plasma.
    This process of information exchange between the duodenum and
    the rate of erythropoiesis appears to be a complex one. For
    example, in iron-deficient subjects, the enhanced iron
    absorption continues long after the hemoglobin is restored to
    normal levels. Furthermore, in chronic hemolytic anemias,
    iron absorption is increased and persists for prolonged
    periods, leading to iron overload.

h. Dietary requirements of iron. The dietary needs depend upon
    several factors: amount of food-iron ingested, composition
    of the food (already discussed), amount of iron lost
    (excretion) from the body and variations in physiological
    states (growth, onset of menses, pregnancy, etc.).

    i) Intake of iron: A normal individual absorbs about 8-10%
        of the ingested iron.

    ii) An average American diet consisting of reasonably varied
        foodstuffs contains about 6 mg of iron per 1000 calories
        and, therefore, contains about 10-15 mg of iron per day.
        Therefore, with 10% absorption, the amount of iron
        absorbed is about 1.0 to 1.5 mg per day.

iii) Iron losses from the body: Iron is tenaciously conserved and the obligatory loss in an adult male is about 1 mg (0.5 to 1 mg) per day. Most of this iron loss occurs due to the normal exfoliation of gastrointestinal mucosa, loss of blood due to gastrointestinal bleeding, loss of cells due to normal shedding from the surfaces of the body (dermal, intestinal, pulmonary, urinary) and loss in bile and sweat. Only insignificant amounts of iron are lost in the urine because iron in the plasma is complexed with proteins which are too large to pass through the kidney glomerular membrane. Iron in the feces is primarily unabsorbed dietary iron. Thus, an adult male can adequately maintain iron balance on a daily basis by ingesting a diet which contains 10-15 mg of iron with 10% of it being absorbed. However, in the case of growing individuals (who have enlarging needs for iron in terms of storage and function), females of child-bearing age (who have superimposed demands for iron due to loss of iron in the menstrual blood), and females in pregnancy (who have to meet the iron demand of the fetus, maintain the increased total number of red blood cells in an expanded circulating blood plasma and compensate for loss of blood at delivery), further considerations must be taken into account in meeting the iron requirements. Thus, the amounts of iron which must be absorbed per day in menstruating and pregnant females are about 2 and 2.5 mg, respectively. In order to meet the appropriate requirements of absorption, the following amounts of dietary iron consumption are recommended during various physiologic states and growth periods. As noted earlier, when more iron is required by the body, the intestine releases its "mucosal barrier," permitting more of the dietary iron to enter.

	Age (years)	Recommended daily dietary iron intake (mg)
Infants	0.0-0.5	10
	0.5-1.0	15
Children	1-3	15
	4-10	10
Males	11-18*	18
	19 and up	10
Females	11-50**	18
Pregnant		greater than 18
Lactating		18

* Adolescent growth spurt.
**Adolescent growth spurt and/or menstrual blood loss.

Note that in pregnancy the increased iron requirement may not be met by ordinary diets or increased absorption and, therefore, the use of supplemental iron is recommended. The newborn infant, which has about a 3-6 months supply of iron stored in the liver during fetal development, requires supplementation of iron-rich foods starting from the third month because milk is poor in iron.

The principal source of iron loss in non-pregnant females in the fertile age group is the menstrual loss of iron. In one study, it has been shown that the mean value of the menstrual blood loss is 43.4 ± 2.3 ml. Since each ml of blood from a normal female contains about 0.5 mg of iron, the amount of iron lost every 27 days is about 20-23 mg. Increases in menstrual flow (known as menorrhagia) can significantly augment iron loss, eventually leading to iron-deficiency anemia (discussed later).

3. Plasma iron transport

a. Iron is carried in the trivalent state (Fe^{3+}) bound to a specific iron-transport protein known as transferrin. It is a β_1-globulin (mucoprotein) and has a molecular mass of 74,000 daltons. Each molecule of transferrin can bind two atoms of iron. Electrophoretic studies have revealed the presence of at least 18 genetically controlled variants of transferrin. In some transferrin molecules, single amino acid substitutions can account for the variations in the electrophoretic mobility. Transferrin is synthesized primarily in the liver and appears as early as 29 days during fetal development.

b. The central role of transferrin in iron metabolism is exemplified by studies of patients with congenital atransferrinemia. These individuals, who lack the iron-carrier protein, show a severe refractory hypochromic anemia in the presence of excess iron stores in many body sites, susceptibility to infection (transferrin has been shown to inhibit bacterial, viral and fungal growth, presumably due to its capacity to bind the iron required for the growth of these organisms) and retardation of growth. These observations have led to the elaboration of the following aspects concerning the role of transferrin, that

    i) it is the proper and specific transport vehicle for iron,

ii) it selectively distributes iron to those tissues which have demands for its utilization (e.g., hemoglobin synthesis), and

iii) it is not required for iron absorption.

A schematic diagram of iron absorption and distribution in the body is shown in Figure 79.

c. Determination of serum iron and transferrin levels

i) The principle used in one method of serum iron determination consists of the precipitation of the iron-protein complex and release of iron from the complex (accomplished by an acid such as hot trichloroacetic acid or acetate buffer at pH 5.0), reduction of iron from the ferric to ferrous state (using hydroxylamine or ascorbic acid), reaction of the iron with a chromogen (α,α'-dipyridyl, 2,2',2"-tripyridine, bathophenanthroline), and measurement of absorbance of the iron-chromogen complex (at 552 nm, when the chromogen used is tripyridine).

Normal ranges: Iron levels are high at birth (110-270 μg/100 ml of serum), fall rapidly in the first 4 to 6 months and rise by age 3 to a range of 61-175 μg/100 ml. The adult normal range is 60-150 μg/100 ml with no differences between sexes. It is important to emphasize that the normal range of a given parameter depends upon the method employed and that each laboratory must establish its own normal range derived from its population base. There is a considerable diurnal variation in serum iron levels, being high in the early morning and low in the evening. The magnitude of this variation in a given individual can be as large as fivefold. Thus, an individual with normal serum iron levels may be mistakenly diagnosed as having iron deficiency if the iron values were determined in a specimen obtained in the evening. It has been reported that the diurnal variation of serum iron is reversed in those who sleep by day and work by night. Furthermore, it appears that there are minimal diurnal variations in diseases such as iron depletion or iron overload, hemolysis, infection, megaloblastic anemias, etc. It should be emphasized, however, that in order to obtain meaningful and reproducible results, the blood for iron determinations must be obtained at standard times.

Decreased serum iron levels may be seen in chronic blood loss, acute and chronic infections, chronic inflammatory

Figure 79. A Schematic Diagram of Iron Absorption and Distribution in the Body

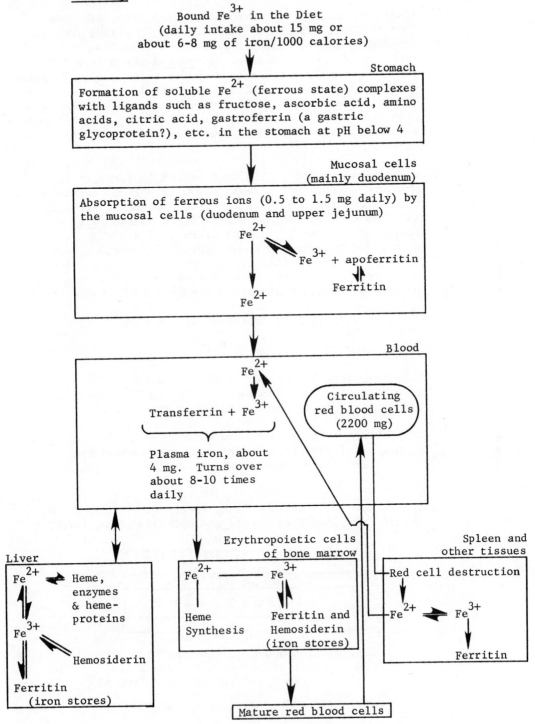

685

disorders, malignancy, extensive dermatitis, etc. In some of these disorders, the decrease in plasma iron is presumably due to improper release of iron from the reticuloendothelial tissue. Increased serum iron levels are, in general, associated with decreased (or ineffective) erythropoiesis (e.g., refractory, hypoplastic anemias, pernicious anemia, thalassemia) or increased release of iron from body stores (e.g., hemosiderosis, hemochromatosis and acute liver disease). These are summarized in Table 26.

ii) Transferrin is not routinely measured in terms of its absolute concentration (mass per unit volume) but is instead quantitated in terms of the amount of iron it is capable of binding. This is known as the total iron-binding capacity (TIBC) of the serum. TIBC is determined by adding a known excess of iron to a serum sample, separating the unbound iron, and either measuring unbound iron or protein-bound iron by the procedure described above. The normal adult ranges are:

TIBC = 250-410 µg Fe/100 ml of serum

Percent Saturation = 20-55%

The difference between the TIBC and serum iron is known as unsaturated iron-binding capacity (UIBC), and its normal range is 140-280 µg Fe/100 ml serum.

In contrast to serum iron and UIBC levels, TIBC does not show diurnal variation. TIBC values are about the same in both sexes, and they show some decline between the ages of 60 to 80.

The relationship between TIBC, amount of iron bound to transferrin, and UIBC is diagrammatically shown below.

Total iron-binding capacity (TIBC)

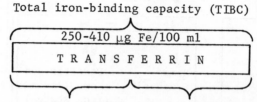

Amount of iron bound   Unsaturated (or latent)
in a normal serum   iron binding capacity (UIBC)
(20-55%)   140-280 µg Fe/100 ml

Table 26. Alterations of Serum Iron, Percent Saturation and TIBC in Various Disorders of Iron Metabolism and Physiologic States

Disorder or Physiologic State	Fe	Saturation	TIBC
Normal Adult	60-150 µg/100 ml	20-55 %	250-410 µg Fe/100 ml
Late Pregnancy	→	↓	↑
Fetus at Parturition (Values are compared with adult range)	↑	↑	↓
Chronic Iron Deficiency (Leads to iron-deficiency anemia)	↓	↓	↑
Anemias (hemolytic, pernicious, aplastic, myelophthisic)	↑	↑	→ or ↓
Hemolytic and pernicious anemias during remission	↓	↓	increase to achieve normal levels
Hemochromatosis and transfusion siderosis	↑	↑	→ or ↓
Acute infections (Hypercatabolic states)	↓	→ or ↓	↓
Chronic infections, malignant tumors, myelomatosis, uremia, leukemia, cirrhosis, acute or subacute liver atrophy	→ or ↓	→ or ↓	↓
Acute liver disease (e.g., acute hepatitis)	↑	↑	→

→ = normal ; ↑ = increase ; ↓ = decrease

iii) Alterations of serum iron and transferrin (TIBC) levels in various disorders of iron metabolism and physiologic states are shown in Table 26. Note that elevated levels of TIBC are consistently seen in chronic iron deficiency and pregnancy. Decreased levels of TIBC are due to impaired transferrin synthesis (e.g., liver diseases, malnutrition) or increased loss (e.g., nephrosis). Disorders associated with iron overload (e.g., hemochromatosis, sideroblastic anemias, pernicious anemias, chronic hemolytic anemias, transfusional siderosis) show a significant decrease in TIBC, which may be due to both hepatic damage and malnutrition.

4. <u>Iron-deficiency anemia</u>

a. Iron deficiency is widespread throughout the world. In general, the factors that are responsible for the iron deficiency include lack of availability of iron and/or availability in an absorbable form (bioavailability) in the diet, loss of iron exceeding absorption both in normal and abnormal physiologic states, increased demands in many normal physiologic states, abnormalities involving sites of absorption, abnormalities concerned with storage and release of iron, and its proper utilization. Various categories are listed below.

i) Dietary iron deficiency

ii) Lack of substances that favor iron absorption (ascorbate, amino acids, succinate, etc.)

iii) Presence of compounds that limit iron absorption (phytates, oxalates, excess phosphates, etc.)

iv) Deficiency of iron absorption due to gastrointestinal disorders (gastrectomy, etc.)

v) Loss of iron in females due to menstruation, pregnancy, puerperium (the period of confinement after labor) and lactation. This group makes up the largest proportion of patients seen with iron deficiency and anemia.

vi) Chronic loss of iron due to abnormal blood loss from the gastrointestinal tract (peptic ulceration, hemorrhoids, malignancy, colonic ulceration, hook worm infestation, etc.) or the genitourinary tract (uterine fibroids)

vii) Enhanced, inadequately met iron demands in growth and/or new blood formation

viii) Deficiency in iron transport from mother to fetus

ix) Abnormalities in iron storage and deficiencies in the release of iron from reticuloendothelial system (infection, malignancy)

x) Inhibition of the incorporation of iron into hemoglobin (lead toxicity)

b. In states of negative iron balance, the iron stores maintain normal levels of hemoglobin and other iron proteins. As noted earlier, storage iron is composed primarily of ferritin and hemosiderin and is found mainly in the liver, spleen and bone marrow. Storage iron can thus be exhausted without any onset of significant anemia (a manifestation of iron deficiency). There is exhibited a <u>late development</u> of hypochromic (erythrocytes deficient in hemoglobin), microcytic (low corpuscular volume) anemia (low hemoglobin level) in the iron-deficiency state. The overlapping sequence of events of iron deficiency, with the eventual development of hypochromic, microcytic anemia, is as follows:

i) Occurrence of iron deficiency--diminished iron stores (measured in bone marrow aspirates), normal erythrocyte morphology, increased iron absorption, absence of anemia; serum iron, TIBC and percent saturation are normal; no clinical features

ii) Continued iron deficiency--exhaustion of iron stores, normal erythrocyte morphology, increased food iron absorption, normal serum iron parameters

iii) Plasma iron levels and percent saturation are reduced, TIBC is expanded, erythrocyte morphology is normal, but anemia may be present. This is a <u>normocytic, normochromic anemic phase</u>, before the red cell morphologic changes become evident in the peripheral blood. It should be pointed out that there are other hematologic disorders which can give rise to normocytic anemias, which include hemolytic anemias, anemias due to acute blood loss, aplastic anemias, myelophthisic anemias, and anemias of infection, pregnancy, uremic or endocrine disturbances.

iv) Progressive lowering of plasma iron and percent saturation, expanded TIBC, hypochromic, microcytic anemia. This stage can progress until it affects and involves many organs and their functions.

c. The clinical characteristics of iron-deficiency anemia which are relatively non-specific include pallor, rapid exhaustion, muscular weakness, anorexia, lassitude, difficulty in concentrating, headache, palpitations, dyspnea on exertion, angina on effort, peculiar craving for unnatural foods (pica), ankle edema, abnormalities involving all proliferating tissues--especially mucous membranes (rhagades at the angles of the mouth, atrophy of the glossal papillae, esophageal ulceration and spasm--Plummer-Vinson syndrome, atrophy of the gastric mucosa, and atrophy of the nasal mucosa). Nail growth is affected with both transverse and longitudinal ridges and fissures, eventually leading to spoon-shaped nails (koilonychia). It should be emphasized that the onset (hypochromic-microcytic anemia and all of the other clinical characteristics) is insidious with a very slow progression (many months to years). There are continual physiologic adjustments that are taking place during the gradual progression of the disorder, so that even a severe hemoglobin deficiency may be accompanied by a remarkable paucity of symptoms.

d. Laboratory evaluation of iron-deficiency anemia:

i) A fully developed iron-deficiency state is characterized by anemia (low hemoglobin levels) and red blood cells which are poor in hemoglobin content (low mean corpuscular concentration) and small in volume (low mean corpuscular volume). The red blood cells may show considerable variation both in size and shape. In a pronounced anemic state with many small and poorly filled small cells, the red cell count may be within normal limits and the hematocrit may only be lowered moderately. As noted earlier, the plasma iron content is low (usually less than 35 µg/100 ml) with an expanded TIBC (350-500 µg/100 ml) and decreased percent saturation (see Table 26).

At present, none of the plasma parameters can detect early iron deficiency. As already indicated, during early stages of iron lack, the iron stores (liver, bone marrow) are decreased,and thus the measurement of these stores may be used. However, liver biopsy and bone marrow aspiration are special procedures and do not serve as routine screening tests in the assessment of iron stores to evaluate negative iron balance.

ii)   Erythrocyte protoporphyrin (due to lack of utilization)
      is often elevated in iron deficiency and has been
      suggested as a possible screening test.

iii) The estimation of the number of sideroblasts in the
      bone marrow can be utilized in the assessment of iron
      stores. Sideroblasts are precursors of erythrocytes
      (normoblasts) that contain ferritin-iron granules in
      the cytoplasm which are unassociated with any cytoplasmic
      organelles and are usually not surrounded by a membrane.
      The iron in these cells is identified by staining with
      prussian blue reagent and appears as a bluish material.
      The other siderotic cells containing cytoplasmic
      aggregates of ferritin are reticulated siderocytes and
      mature siderocytes. There is a close correlation among
      plasma iron levels, transferrin saturation (TIBC) and
      the proportion of sideroblasts found in the bone marrow.
      In hemolytic anemias, pernicious anemia and hemochromatosis,
      the serum iron levels are increased and sideroblast levels
      reach as high as 70% (normal range: 30-50% of total
      cells). In iron deficiency, the sideroblasts are decreased
      or absent.

      It is of interest to point out that sideroblasts loaded
      with excessive amounts of iron in their mitochondria
      (unlike the normal sideroblast with <u>cytoplasmic ferritin
      granules</u>) are observed only under <u>abnormal conditions</u>.
      Mitochondrial iron accumulation is due to defects in
      the proper utilization of iron in the mitochondria,
      thereby leading to sideroblastic anemia. In protoporphyrin
      synthesis (discussed later in this chapter), two or three
      reactions that take place in the mitochondria can be
      inhibited, which will give rise to iron accumulation.
      Delta-aminolevulinic acid (ALA) synthetase catalyzes
      the first step in protoporphyrin synthesis, which consists
      of the condensation of glycine and succinyl-CoA to form
      ALA. This is a rate-limiting enzyme and requires
      pyridoxal phosphate (B_6) as a coenzyme. Either a defect
      in the enzyme or a deficiency of B_6 can lead to
      sideroblastic anemia. B_6 deficiency can occur due to
      the following causes: dietary deficiency of pyridoxine;
      inhibition of pyridoxal kinase which catalyzes the
      conversion of pyridoxine to pyridoxal phosphate; and an
      increased inactivation of pyridoxal phosphate by the
      enhanced activity of pyridoxal phosphate phosphatase.
      In alcoholics, the serum pyridoxal phosphate levels are
      low. This may be due to the inhibition of pyridoxal

kinase or the enhanced activity of pyridoxal phosphate phosphatase. Drugs such as isoniazid, cycloserine, and pyrazinamide can bring about iron accumulation in mitochondria by inhibiting the pyridoxal phosphate-catalyzed reaction, namely, ALA synthetase. Chloramphenicol (an inhibitor of protein synthesis occurring on 70S ribosomes) causes iron accumulation by inhibiting mitochondrial protein synthesis, whose proteins include ALA synthetase, heme synthetase and cytochrome oxidase. Lead inhibits the biosynthesis of heme at a number of steps (e.g., ALA dehydrase, heme synthetase) and therefore prevents iron utilization in mitochondria. Heme synthetase, which inserts iron into the protoporphyrin, may be deficient or inhibited (discussed above). The role of these two enzymes in mitochondrial iron utilization is diagrammatically shown below.

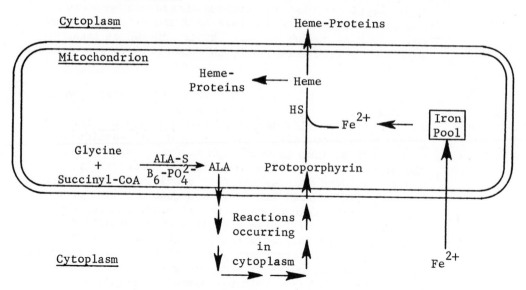

ALA-S = δ-Aminolevulinic acid synthetase;

ALA = δ-Aminolevulinic acid;

HS = Heme synthetase.

Mitochondrial iron overload can also result from impaired globin synthesis (thalassemia). All of these disorders associated with intramitochondrial defects in heme synthesis are associated with a high rate of ineffective erythropoiesis (known as sideroachrestic

<u>anemia</u>). It should be pointed out that the accumulation of iron due to some biochemical lesion in its utilization may eventually cause damage to mitochondria by peroxidation of mitochondrial lipids. This may lead to the impairment of enzymes such as heme synthetase. Mitochondria obtained from patients with sideroblastic anemia (e.g., due to pyridoxine deficiency) are both morphologically and functionally (swelling, lysis, disintegration, peroxidation of lipids) abnormal and, hence, the sideroblasts with many iron-loaded mitochondria will not survive. However, some sideroblasts with a few functionally normal mitochondria can mature into <u>reticulated siderocytes</u> and reach the circulation. These cells are unable to lose their iron or mitochondria and undergo maturation to form siderocytes. The spleen is responsible for the removal of iron-loaded mitochondria from these cells. In the hereditary or congenital form of sideroblastic anemia, which has a sex-linked, partially recessive character, the exact nature of the biochemical lesion is not understood. These disorders are characterized by hypochromasia, microcytosis, hyperferremia (discussed later) and tissue iron overload. "<u>Ringed sideroblasts</u>," seen in light microscopy of bone marrow smears stained with prussian blue, are of diagnostic value in identifying the iron-loaded mitochondria. The distinctive ring is due to the distribution of mitochondria around the periphery of the nucleus. Thus, it should be emphasized that there are other hematologic disorders, namely sideroachrestic anemias, thalassemias, anemia of infection, hemoglobin variants, and others, which exhibit hypochromic, microcytic anemia in addition to iron-deficiency anemia. However, the hypochromic, microcytic anemia due to causes other than iron deficiency can be properly identified. For example, in sideroblastic anemia, the serum iron is normal or increased, TIBC values are normal with a higher per cent saturation of TIBC, there are increased stores of iron in the bone marrow, etc. (see section on thalassemia for differential diagnosis between thalassemia and iron-deficiency anemia). Recall that the late stages of iron deficiency show, in addition to hypochromic, microcytic anemia, a deficiency of plasma iron with an expanded TIBC, and all of these respond only to iron therapy.

iv) Serum ferritin concentration: In normal adults the serum ferritin levels are presumably directly related to the available reticuloendothelial and parenchymal iron stores. The normal adult range for serum ferritin determined by a radioimmunoassay procedure is 12-250 ng/ml, with men

having higher mean concentrations than women. In iron-
deficiency anemia, low values of ferritin are observed,
whereas patients with iron-storage diseases such as
idiopathic hemochromatosis and transfusion siderosis
(discussed later) show strikingly high levels. Furthermore,
in the latter disorders, repeated therapeutic phlebotomy
(see below) has been shown to reduce the initially elevated
ferritin levels to normal or even to the iron-deficient
range. These observations support the assumption that
serum levels of ferritin correlate with total body iron
stores. However, in a number of disorders, the <u>elevated</u>
serum ferritin levels do not parallel the reticuloendothelial
iron stores. This is found in anemia associated with
chronic inflammatory diseases (e.g., rheumatoid arthritis),
acute and chronic liver disease, leukemia, and Hodgkin's
disease. Furthermore, in asymptomatic (e.g., precirrhotic)
persons with increased iron stores due to familial
hemochromatosis, the estimation of serum ferritin levels
has not been helpful.

5. Treatment of iron deficiency: It is essential to establish the
cause of the negative iron balance which has resulted in iron
deficiency. The treatment should include correction of the
underlying cause of anemia while at the same time improving the
iron balance. In general, oral iron therapy is satisfactory in
correcting the iron deficiency. However, parenteral iron therapy
is preferred under certain specific situations, such as patients
with proven malabsorption problems, some post-gastrectomy patients,
patients with gastrointestinal diseases (e.g., ulcerative colitis
or Crohn's disease), patients with excessive blood loss in whom
oral iron may be insufficient to compensate for the loss
(hemorrhagic telangiectasia, hiatus hernia, or menorrhagia), and
patients who cannot be relied upon to take oral iron. In general,
the oral iron preparations are ferrous salts because, as noted
earlier, the iron is transported in the ferrous state across the
mucosal membrane.

Any mandatory fortification of certain foods (e.g., bread and
flour) with iron, in order to prevent iron-deficiency anemia in
young women at risk because of menstrual blood loss (who make up
2 per cent or less of the entire population), exposes the rest of
the 98 per cent of the population to the risk of developing iron-
storage disease. In this regard, it should be mentioned that in
a town in Sweden, where for the past 30 years mandatory iron
supplementation has been practiced, there is an increased number
of cases of men with iron storage problems. Whether or not the
iron overload in these individuals is a consequence of the iron
fortification is not yet established; nevertheless, this aspect
requires further careful study.

6. Some normal ranges for circulating red blood cells, hematocrit, hemoglobin and blood indices are presented in Table 27. The definition of the various parameters are as follows:

Hematocrit = Volume (%) of packed red cells per 100 ml of blood.

Mean Corpuscular Volume (MCV) = Volume of the average red blood cell of a given sample of blood expressed as μ^3. MCV is calculated as follows:

$$\frac{\text{Vol of packed RBC in ml/1000 ml blood}}{\text{RBC in millions/mm}^3} = \text{MCV in } \mu^3$$

Mean Corpuscular Hemoglobin (MCH) = Amount of hemoglobin, by weight, in the average red blood cell in a given sample of blood. MCH is calculated as follows:

$$\frac{\text{Hemoglobin (Hb) in gm/1000 ml of blood}}{\text{RBC in millions/mm}^3} = \text{MCH in pg/cell}$$

Mean Corpuscular Hemoglobin Concentration (MCHC) = Proportion of hemoglobin (W/V) contained in the average red blood cell in a given sample of blood. MCHC is calculated as follows:

$$\frac{\text{Hemoglobin in gm/100 ml blood}}{\text{Volume of packed cells in ml/100 ml blood}} \times 100 = \text{MCHC in g/100 ml RBC's}$$

Iron-Storage Disorders (Hemochromatosis)

1. An excessive accumulation of iron in the body (chronic iron overload) can result from the following conditions.

    a. Defective erythropoiesis (dyserythropoiesis): impaired hemoglobin synthesis leading to the lack of utilization of iron and its accumulation in the mitochondria. Examples of these are the inhibition of ALA synthetase activity due to dietary B_6 deficiency (and hence a B_6-responsive disorder); inhibition of heme synthesis by lead; impairment of pyridoxine metabolism in alcoholic patients (note that the varieties of erythropoietic abnormalities and red cell defects observed in alcoholism are usually related to the duration of alcohol intake and the degree of hepatic damage; the prerequisites for sideroblastic changes that occur in alcoholism are chronic malnutrition and folate deficiency with megaloblastic bone marrow); a heterogeneous group of familial sideroblastic anemias; and thalassemia major (homozygous) syndromes.

Table 27. Normal Ranges for Red Blood Cell, Hematocrit, Hemoglobin and Blood Indices

	Red Cell Count million/mm^3	Hemoglobin (g/100 ml of blood)	Hematocrit %	MCV (μ^3)	MCH (pg/cell)	MCHC (g/dl RBC's)
Newborn Infants	6.0-8.0	15-20	44-62	103-106	36-38	35-36
Children	4.5-5.5	11-12	35-37	77-80	25-28	33-35
Adult Males	4.5-5.5	14-16	40-45	80-94	27-32	33-38
Adult Females	4.1-5.0	12-14	36-45	80-94	27-32	33-38

b. Transfusion-dependent iron overload (transfusion hemosiderosis) occurs in patients with thalassemia major, sickle-cell disease or any other hematologic disorder requiring chronic, multiple blood transfusions.

c. Familial hemochromatosis: The underlying cause predisposing to tissue iron accumulation is not clear. The disorder is expressed as an autosomal dominant trait and is associated with normal hematopoiesis. The identification of preclinical states of familial hemochromatosis by measuring ferritin levels has not been uniformly successful (discussed earlier).

d. Accumulation of iron due to high dietary iron and the presence of substances that enhance its absorption. The Bantu of South Africa ingest up to 200 mg of iron per day, which is obtained from a diet low in phosphate and from an alcoholic beverage.

e. Atransferrinemia, an inherited disorder, results in the accumulation of unneeded iron in the tissues. The defect can be temporarily treated with intravenous administrations of transferrin (the role of transferrin in iron metabolism has already been discussed).

It is important to emphasize that in all of the above disorders the gastrointestinal tract is <u>unable</u> to limit the absorption of unneeded iron to any significant extent. Thus, the mucosal block (see iron absorption), which is responsible for keeping out unneeded iron on a daily basis in a normal person, appears to be susceptible to disruption, perhaps at more than one point.

2. Iron overload leads to parenchymal iron accumulation with progressive deterioration in pancreatic, hepatic, and cardiac function. The clinical manifestations include cirrhosis, diabetes mellitus, impotence, and congestive heart failure. The patients usually succumb to life-threatening arrhythmias or intractable heart failure. Removal of excess iron brings about clinical improvement, particularly of diabetes and congestive heart failure. Therefore, in the management of iron-storage disorders, the cornerstone of therapy is iron removal from the body. This is accomplished in one of two ways depending upon the nature of the disease.

a. In iron-storage diseases accompanied by normal erythropoiesis (e.g., familial hemochromatosis), removal of excessive iron is accomplished by aggressive phlebotomy (incision of a vein for intentional bloodletting). Depending upon the patient's ability to mobilize storage iron and synthesize hemoglobin, the therapeutic phlebotomy of a unit of blood may be performed weekly, twice weekly or even three times per week. One unit of blood contains about 250 mg of iron, so that as much as 750 mg of iron can be removed in a week's period. When the iron stores have been depleted, the patient is maintained free of any significant stores by performing four to six phlebotomies per year.

b. The removal of iron in disorders associated with dyserythropoiesis (e.g., β-thalassemia major – Cooley's anemia, sickle-cell anemia) calls for an alternate mode of therapy. This is due to the fact that these patients <u>require repeated blood transfusions</u> to live through childhood and maintain adulthood. This mode of therapy consists of using iron-chelating agents. <u>Deferoxamine</u> is an iron chelator isolated from <u>Streptomyces pilosus</u> and has the following structure:

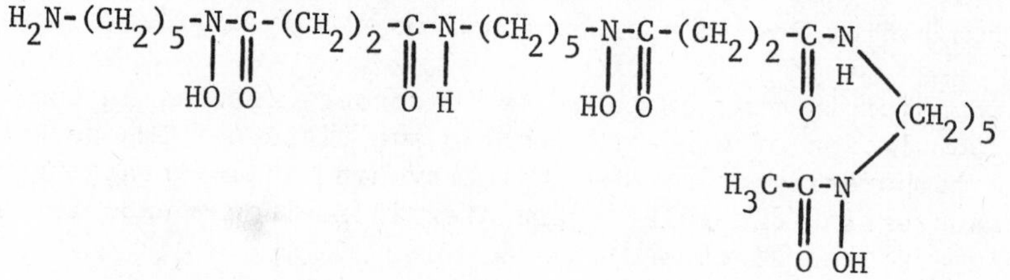

$$H_2N-(CH_2)_5-\underset{\substack{|\\HO}}{N}-\underset{\substack{\|\\O}}{C}-(CH_2)_2-\underset{\substack{\|\\O}}{C}-\underset{\substack{|\\H}}{N}-(CH_2)_5-\underset{\substack{|\\HO}}{N}-\underset{\substack{\|\\O}}{C}-(CH_2)_2-\underset{\substack{\|\\O}}{C}-\underset{\substack{|\\H}}{N}$$

Notice that this molecule has six nitrogens separated by fairly long, flexible stretches of methylene groups. Since iron forms octahedral complexes (i.e., each iron atom has six ligands), one molecule of deferoxamine is probably capable of occupying all six coordination sites. This would produce an iron-deferoxamine complex having a stoichiometry of 1:1. On this basis, the chelator might be compared to a non-cyclic, stretched-out porphyrin with two extra nitrogens.

c. Deferoxamine is a highly selective complexing agent. For <u>ferric</u> iron, K_{assoc} is about 10^{30}, while K_{assoc} for Ca^{+2} is only 10^2. The iron in hemoproteins is not affected by this agent, because these proteins do not come in contact with the chelator (e.g., the cytochromes in mitochondria) or have only ferrous iron (e.g., hemoglobin). Such selectivity makes the compound very useful in the treatment of iron-storage problems (and acute iron poisoning) as its administration does not affect the vital functioning of hemoglobin, the cytochromes, and calcium metabolism. In addition, the ferric iron of ferritin and hemosiderin is chelated in preference to that found in transferrin.

d. The deferoxamine-iron complex is small enough to be filtered by the glomerulus so that the main route of iron excretion during deferoxamine therapy is the urine. Recall that iron is normally not excreted by this route.

e. When deferoxamine is given orally it complexes with dietary
    iron, making both the drug and the iron unavailable for
    absorption. The preferred route is <u>parenteral</u>, usually by
    intramuscular injection (see later). Irritation and pain at
    the site of administration and the need for daily injections
    make the treatment unpopular. In addition, even with
    co-administration of large amounts of ascorbic acid, the
    effected iron loss is far below that necessary to remove the
    amount accumulated during intensive, chronic transfusion
    therapy.

    Recently, the use of a slow, continuous intravenous infusion
    or continuous subcutaneous administration has been reported
    to be much more effective in establishing negative iron
    balance and eliminating stored iron. It is postulated that
    a small, labile (chelatable) iron pool is in slow equilibrium
    with a much larger (storage) pool. When deferoxamine is
    administered, the labile pool is rapidly emptied. Any
    deferoxamine which remains in the body, or which is administered
    shortly thereafter, finds no more iron to bind. Thus, most of
    a single intramuscular dose is simply excreted unchanged. If
    this same amount of chelator (or a larger dose), however, is
    given as a continuous infusion, the labile pool is initially
    depleted and then <u>kept</u> empty. As more iron is released from
    the storage sites, it is immediately chelated and removed.
    Removal of as much as 180 mg of iron per day in a
    β-thalassemia-major patient has been accomplished by this
    method, making it comparable to the use of phlebotomy.
    Massive intravenous doses of deferoxamine have also been
    reported to effect the urinary excretion of large amounts of
    iron in iron-overloaded patients.

f. Some of the other chelators that are presently undergoing
    preliminary animal screening tests or clinical trials are as
    follows:

Rhodotorulic Acid

A hydroxamic acid similar to deferoxamine, it is produced by
a microorganism in high amounts and is potentially less
expensive but equally as effective as deferoxamine.

2,3-Dihydroxybenzoic Acid

This compound is a fragment of iron-chelators synthesized by several microorganisms.

Ethylenediamine-N,N' bis[2-hydroxyphenylacetic acid]

This compound resembles the common chelating agent, ethylenediamine tetraacetic acid (EDTA).

Structural Aspects of Hemoglobin

1. Mammalian hemoglobins are tetramers, made up of four polypeptide subunits: two α-subunits and two other subunits (usually β, γ or δ). These polypeptide chains differ in the type of amino acids present at specific positions (i.e., in their sequence or primary structure). The normal hemoglobin types are: HbA_1: $\alpha_2\beta_2$ (adult human hemoglobin; most common); HbF: $\alpha_2\gamma_2$ (human fetal hemoglobin); HbA_2: $\alpha_2\delta_2$ (a minor component of normal adult hemoglobin). They are discussed later in somewhat greater detail. The structural and functional aspects of hemoglobin have been worked out almost entirely through studies of HbA_1 and its mutants.

2. Hemoglobin has a molecular weight of about 64,500 daltons and is spherical, due to the remarkable fit of the four subunits. The α-subunits have a molecular weight of about 15,750 daltons and that of the β-subunits is about 16,500 daltons. This arrangement of the subunits is known as the quaternary structure. A useful way of visualizing their orientation is to consider the four

polypeptides as being at the four corners of a regular tetrahedron. One should recognize that this is somewhat idealized, since the subunits are not truly spherical and the tetrahedron which their centers describe is not exactly regular. In addition, during binding and release of O_2, the subunits move about, as is discussed later.

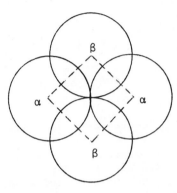

3. In the assembled hemoglobin, contacts between like subunits ($\alpha\alpha$ and $\beta\beta$) are few and limited primarily to six polar interactions between the C-terminal residues and other amino acids in the peptides. These interactions are mediated by salt bridges which can be pictured as (short) chains of salt (e.g., Na^+Cl^- , $(NH_4{}^+)_2SO_4{}^=$) molecules (really, ion pairs) between two groups of opposite charge. For example:

$$-NH_3{}^+ \; (\overline{Cl^-Na^+}) \; \overline{O}OC-$$

(This is an oversimplification, since there will be many other ions present in the solvent which will also interact with the charged groups. A more accurate view might be that of a <u>charged</u> "atmosphere" of ions surrounding both groups with the charge cloud polarized by the charges on the polypeptide.) Despite this paucity of interactions, the importance of those which do exist will be seen later, in the discussion of the mechanism by which hemoglobin functions.

In deoxyhemoglobin, there is no β-β contact (this is significant in explaining the role of 2,3-DPG). Unlike chains are more strongly joined, mostly via van der Waal's (hydrophobic) interactions. There are probably a few hydrogen bonds which also contribute to these attachments.

These facts are derived principally from the x-ray crystallographic results of Perutz, Kendrew and their coworkers at the Medical Research Council, Laboratory of Molecular Biology in Cambridge, England. They are also, however, supported by other data. Under certain conditions (for example, dilute solutions), normal hemoglobin dissociates in the following way:

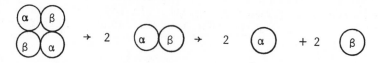

There is no evidence for either or $\beta\beta$ forming although, in the complete absence of α-chains, β_4 is known. Hemoglobin does not dissociate in red blood cells due to its high concentration.

(Note: M.F. Perutz and J.C. Kendrew received the 1962 Nobel Prize in Chemistry for this work.)

4.   In order to understand the mechanism of oxygenation, it is important to realize that hemoglobin is not as symmetrical as it appears to be. If the two α-and two β-subunits are distinguished from each other:

$$\alpha_1 \quad \beta_2$$
$$\beta_1 \quad \alpha_2$$

The asymmetry can be described by saying that the $\alpha_1\beta_1$ contact differs from the $\alpha_1\beta_2$ contact. The contacts $\alpha_1\beta_1$ and $\alpha_2\beta_2$ are the same, however, as are the contacts $\alpha_1\beta_2$ and $\alpha_2\beta_1$. It has been shown that when hemoglobin dissociates into dimers, the contact broken is of the type $\alpha_1\beta_2$ and the one remaining in the dimer is of the $\alpha_1\beta_1$ sort. This is a mildly confusing notation since $\alpha_2\beta_2$ in one context means HbA_1 (consisting of two α-and two β-subunits), while in this situation it could mean the second α subunit plus the second β subunit (although it is not usually used in this way). The best way to keep this straight is by carefully watching the context in which the notation is used.

5.   a.   Figures 80 and 81 are schematic diagrams of the α- and β-chains, respectively, of hemoglobin. Each peptide chain (subunit) is made up of helical and non-helical segments, and each surrounds a heme group. Hemoglobin is unusual among globular, soluble proteins in that almost 80% of its amino acids exist in an α-helical conformation. There are two notations used to specify the locations of amino acids in the hemoglobin peptides. One numbers the residues from the N-terminus of each chain. For example, residue $\alpha42$ of normal adult human hemoglobin is tyrosine, while $\beta42$ is phenylalanine. The other, devised by

Figure 80. Secondary Structure of the α-Chain of Human Hemoglobin

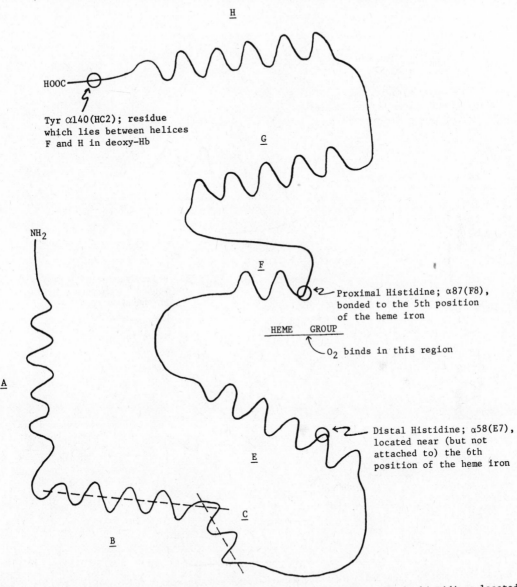

The helical regions (labeled A-H, after Kendrew), N- and C-termini, and the histidines located near the heme group are indicated. The axes of the B and C helices are indicated by dashed lines.

Figure 81. Secondary Structure of the β-Chain of Human Hemoglobin

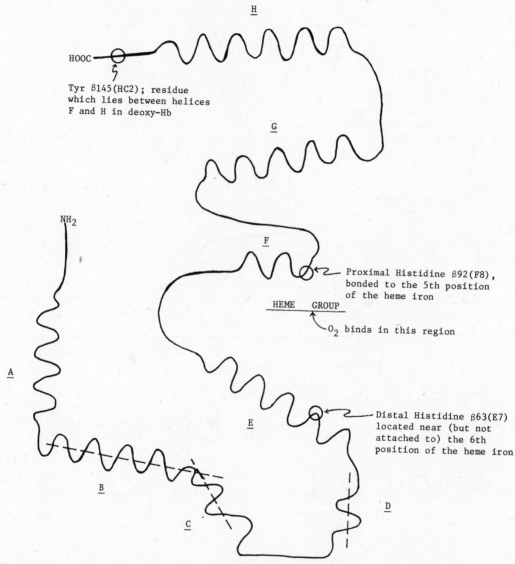

H

HOOC ——○

Tyr β145(HC2); residue
which lies between helices
F and H in deoxy-Hb

G

NH₂

F

Proximal Histidine β92(F8),
bonded to the 5th position
of the heme iron

HEME   GROUP

O₂ binds in this region

A

E

Distal Histidine β63(E7)
located near (but not
attached to) the 6th
position of the heme iron

B

C

D

The helical regions (labeled A-H, after Kendrew), N- and C-termini, and the histidines located
near the heme group are indicated. The axes of the B, C, and D helices are indicated by
dashed lines.

Kendrew, numbers residues within α-helices from the N-terminal amino acid of each of the helices. This is also known as the "helical notation." Residues in non-helical regions are designated by the letters of the helices at either end of the region. Thus, residue α42 would be C7 and β42 is CD1. These notations are often combined, yielding α42(C7) or Tyrα42(C7) and β42(CD1) or Pheβ42(CD1).

b. Heme consists of a porphyrin ring system with an Fe^{+2} fixed in the center through complexation to the nitrogens of the four pyrrole rings. A porphyrin is a planar, resonating (aromatic) ring formed from four pyrrole rings linked by =CH-(methene) groups. The pyrrole rings may have sidechains, the different porphyrins being distinguished on the basis of variations in these sidechains. The structure and properties of porphyrins are discussed more fully later in this chapter.

c. Iron in the ferrous state has a coordination number of six (each atom of iron can bind with six electron pairs). Therefore, in the heme molecule, two of iron's coordination positions are still unoccupied. When heme is associated with the peptide, a histidyl nitrogen (from the so-called proximal histidine) of the peptide bonds with the fifth coordination position of iron, leaving the sixth position open for combination with oxygen, water, carbon monoxide, or other ligands.

6. a. The peptide chain folds around and protects the heme groups, which are located in a crevice near the surface of the subunit structure. As indicated in Figures 80 and 81, the heme lies between helices E and F. One side of the heme group (the distal side) is open for reversible combination with oxygen. This open part of the molecule is surrounded by hydrophobic groups, resulting in a microchemical environment of low dielectric constant. This prevents O_2 from permanently changing Fe^{+2} to Fe^{+3} and is responsible for the ability of hemoglobin to bind oxygen reversibly. Normally, when Fe^{+2} combines with O_2, the Fe^{+2} is oxidized to Fe^{+3} and Fe^{+3} is not capable of binding reversibly with O_2. There is some evidence (from electron paramagnetic resonance) that, in oxyhemoglobin, the oxygen actually does oxidize the Fe^{+2} to Fe^{+3} <u>as long as the O_2 is bound</u>. When the O_2 dissociates, it takes all of its electrons with it, causing the iron to return at once to Fe^{+2}. This transient oxidation does not seriously affect Perutz's model (discussed later) for Hb function, since the dimensions for the Fe^{+2} radius changes were experimental values. What is unique about hemoglobin is not the ability to bind oxygen; it is the ability to bind oxygen <u>reversibly</u>. If the Fe^{+2} in

hemoglobin is permanently oxidized to the ferric state (Fe^{+3}), the Fe^{+3} becomes bound tightly to an OH^- group and O_2 will not bind.

b. The heme groups can be removed from hemoglobin by dialysis. Hence, they are not covalently attached. Not counting the proximal histidine (Hisβ92(F8) or α87(F8)), about 60 amino acids contact (come within 4 Å of) one or more of the atoms of the heme group. This is roughly the maximum length of an effective hydrogen bond or hydrophobic interaction. Of these sixty contacts, all but one in the α-subunit and two in the β-subunit are non-polar. This emphasizes the highly hydrophobic environment of the hemes. These polar interactions all involve the carboxyl groups of the propionic acid sidechains on the hemes.

Functional Aspects of Hemoglobin

1. The primary function of Hb is to transport oxygen from the lungs to the tissues where it is utilized. Hemoglobin forms a dissociable hemoglobin-oxygen complex which can be written as follows:

$$\text{deoxyhemoglobin} + O_2 \rightleftharpoons \text{oxyhemoglobin}$$

This relationship follows the law of mass action in that the reaction is shifted to the right when there is an increase in oxygen pressure (as in the lungs) and to the left when there is a decrease in oxygen pressure (as in the tissues). The sketch below shows the flow of oxygen and some approximate partial oxygen pressures. Recall that a torr (named for Torricelli, the inventor of the barometer) is a unit of pressure equal to 1 mm of Hg.

inspired air $\longrightarrow$ alveolar air $\longrightarrow$ arterial blood in lungs and capillaries

P_{O_2} = 158 torr P_{O_2} = 100 torr P_{O_2} = 90 torr

capillary beds $\longrightarrow$ interstitial fluid $\longrightarrow$ inside tissue cells

P_{O_2} = 40 torr P_{O_2} = 30 torr or less (estimated) P_{O_2} = 10 torr or less (estimated)

It is useful to consider this as the flow of oxygen down a continuous pressure (concentration) gradient. One function of hemoglobin, then, is to increase the solubility of oxygen in the blood, which it does 70-fold.   Table 27 summarizes some facts concerning erythrocytes and hemoglobin.

2. a.   In Figure 82, the sequential change from curve C to B to A (termed a "rightward shift" of the oxygen dissociation curve) could have been caused by <u>an increase in temperature</u> (as in strenuous exercise), <u>a decrease in pH</u> (as in acidosis or exercise), or <u>an increase in the 2,3-DPG concentration</u> (discussed later), as well as by an increase in P_{CO_2}.   As a result of such a change, at a fixed P_{O_2} the per cent saturation of the hemoglobin in the blood stream decreases.   If the blood had initially been saturated to the same extent in all three cases (for example, by oxygen at a pressure of 95 torr, indicated in the figure as P_{O_2} (arterial)), such a shift would mean, physiologically, that the tissues would receive an increasing amount of oxygen in terms of number of moles delivered in a given time.   On the other hand, if the arterial pressure was only about 45 torr (indicated in the figure), a rightward shift would decrease <u>both</u> the initial <u>and</u> final per cent saturations to roughly the same extent and, on a molar basis, the tissues would receive almost the same amount of oxygen regardless of the location of the curve.   It is important to consider these factors when evaluating the significance with respect to altered oxygen transport (i.e., oxygen delivery to the tissues) of any shifts of the oxygen dissociation curve. This can be better seen as follows.   The values discussed below are indicated on Figure 82.

Suppose that a person's hemoglobin is capable of binding 100 moles of O_2 per minute at P_{O_2} = 95 torr.   At this P_{O_2} the per cent saturation of hemoglobin is essentially <u>independent</u> of the pH, temperature, and organic phosphate concentration in the red cell, as can be seen in Figure 82. If this same individual was breathing air containing oxygen at a P_{O_2} of only 45 torr, however, the number of moles of O_2 bound would depend heavily on the dissociation curve and, hence, on factors such as pH, temperature, and 2,3-DPG concentration.   If curve B was in effect while the blood was passing through the vascular beds of the pulmonary alveoli, about 83 moles of O_2 per minute would be picked up, whereas, if curve A was being followed, only about 70 moles of O_2 per minute would be absorbed.

On the other hand, the amount of oxygen released by the blood in the capillary beds of the tissues is almost always determined by the position of the dissociation curve.   Taking the tissue P_{O_2} to be 40 torr, hemoglobin will be 59 or 72 per cent saturated, depending on whether curve A or curve B

Figure 82. Dissociation Curves of Hemoglobin at Several CO_2 Pressures and of Myoglobin

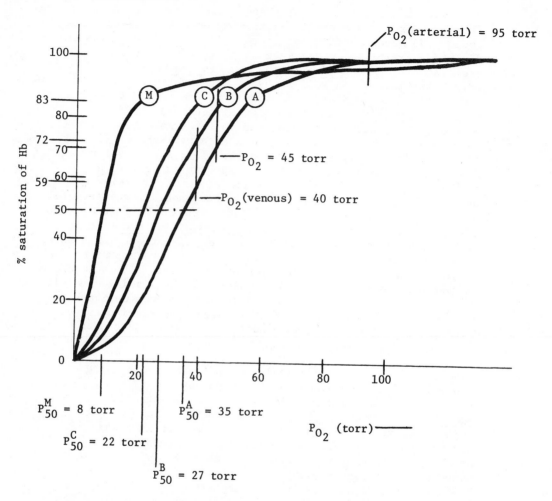

Curve A: P_{CO_2} = 80 torr; Curve B: P_{CO_2} = 40 torr; Curve C: P_{CO_2} = 20 torr; Curve M is for Myoglobin. P_{O_2} (arterial) and P_{O_2} (venous) are normal physiological values for, respectively, arterial and venous oxygen tensions. During exercise, P_{O_2} (venous) will normally be lower, around 20–25 torr.

describes the physiological state. This means that with a P_{O_2} of 95 torr in the lungs and curve A being applicable, $100-59 = 41$ moles O_2 per minute would be delivered to the tissues in this hypothetical individual. With curve B, $100-72 = 28$ moles O_2 per minute would be available. Similarly, for a P_{O_2} of 45 torr in the lungs, curve A gives $70-59 = 11$ and curve B $83-72 = 11$ moles O_2 per minute supplied to the tissues. Thus, at low loading pressures, a rightward shift of the oxygenation curve makes much less difference in the amount of oxygen available to the tissues than it does at higher loading pressures, given that the unloading pressure is the same or nearly so.

It should be recognized, though, that this is an over-simplification of the actual events <u>in vivo</u>. As the blood travels from the lungs to the tissues and back again, pH, P_{CO_2}, temperature, and many other factors vary constantly. Consequently, the curve which accurately describes the affinity of hemoglobin for oxygen differs from one moment to the next. The situation above, wherein the dissociation curve did not change from the time that oxygen was picked up to when it was released, is greatly idealized for the purpose of clarity.

b.   For example, it was observed that in going from a low altitude (10 meters above sea level) to a higher altitude (4509 meters above sea level), the 2,3-DPG concentration within the erythrocytes increased by 25-30% within 24 hours after the altitude change and the oxygen dissociation curve shifted to the right (P_{50} increased). About 50% of the change in 2,3-DPG concentration and P_{50} occurred within 15 hours of the change in altitude. This was interpreted as evidence that 2,3-DPG plays a central role in the body's adaptation to lowered arterial P_{O_2} caused by the inspiration of air with lowered O_2 tension of high altitude (a form of anoxic anoxia). Further evaluation of the experiment revealed, however, that the arterial P_{O_2} also changed, from about 94 torr at 10 meters to about 45 torr at 4509 meters, due to a decrease in atmospheric oxygen pressure from 149 torr to 83 torr. Thus, due to the altitude increase, the arterial pressure decreased to a point where the supply of oxygen to the tissues was largely independent of the position of the curve. Once this was recognized, it was shown that the principal adaptive factor was an increased respiratory rate. Increases in hematocrit and mean corpuscular hemoglobin concentration were also observed.

c.   There are, however, situations where a shift in the curve may prove useful. For example, in anemia or cardiac insufficiency, where the supply of oxygen to the lungs is normal but the ability of the blood to deliver it to the tissues is impaired, the intraerythrocytic 2,3-DPG concentration has been observed to increase. A rightward shift of the dissociation curve will not affect the per cent saturation during loading in the lungs (since the upper part of the curve is quite flat), but it will decrease the per cent saturation in the tissues, thereby providing an increase in the number of moles of oxygen available per unit time for metabolism. Respiration in a normal individual provides another example. Since P_{CO_2} is higher in the tissues than it is in the lungs (and, consequently, pH is lower in the tissues than in the lungs), oxygen unloading is facilitated at the tissue level and loading occurs more readily in the lungs. The decrease in hemoglobin saturation with increasing P_{CO_2} (and hence, decreasing pH) is known as the alkaline Bohr effect (since it occurs maximally at pH 7.4) and the molecular basis of it will be discussed later. The acid Bohr effect, occurring below pH 6.0, involves the increase in Hb saturation with decreasing pH. It perhaps involves a major conformational change of the Hb subunits which occurs about pH 5.9. It is not important in vivo.

3. There are several other important features of Figure 82.

a. The pressure at which the blood is 50% saturated with oxygen, designated P_{50}, depends on the position of the oxygen dissociation curve of the hemoglobin. Since the shape of these curves depends only on the molecular nature of the hemoglobin, all curves for a given hemoglobin belong to the same family of curves and they cannot cross each other. Consequently, P_{50} values are a useful measure of the position of the dissociation curve for a particular hemoglobin: the larger the P_{50} value, the further "right" the curve is.

b. Notice that the hemoglobin curves are all sigmoid (S-shaped) while that for myoglobin is a rectangular hyperbola. The relationship of hemoglobin to myoglobin is the same as that of an allosteric enzyme to a "normal" (non-allosteric) enzyme (see the section on enzyme kinetics). In fact, although hemoglobin is not truly an enzyme, it is the classic and best understood example of allosteric control.

  i) At P_{O_2} values found in tissues, the myoglobin curve lies to the left of the hemoglobin curves. Because of this, hemoglobin can oxygenate myoglobin quite readily. This can also be expressed by saying that in the equation

$$Hb(O_2)_4 + 4Mb \xrightarrow{\longleftarrow} Hb + 4MbO_2$$

the equilibrium lies far to the right. Myoglobin is a heme protein which is quite similar to a single subunit of hemoglobin. It occurs at high concentrations in muscle, where it functions as a storage site for oxygen. As will be seen later in this section, the difference in its oxygen dissociation curve is due primarily to the fact that it has but a single subunit, compared to the four found in hemoglobin.

ii) The sigmoidicity of the Hb curve can be interpreted in exactly the same way as were the curves for other allosteric enzymes. The binding of one molecule of oxygen to a hemoglobin tetramer makes the binding of successive molecules to the same tetramer easier, up to the limit of four molecules of oxygen per molecule of hemoglobin. This "cooperative binding" of oxygen by hemoglobin is the basis, as indicated above, for the regulation of oxygen (and indirectly carbon dioxide) levels in the body. The description of a molecular basis for this phenomenon is one of the major triumphs of x-ray crystallographic studies of protein structure.

4. The mechanism of oxygenation of hemoglobin (as proposed by Perutz) may be explained as follows.

a. Hemoglobin appears to have two quaternary structures corresponding to the deoxygenated (deoxy; five-coordinate iron) and the oxygenated (oxy; six-coordinate iron) forms. X-ray crystallographic results have demonstrated that the number of ligands attached to (coordination number of) the iron is the significant difference between these forms. Thus, all six-coordinate hemoglobins (including oxy- and methemoglobin) have one form and deoxyhemoglobin has the other.

b. Binding of one or more oxygens causes a change in the tertiary structure of the binding subunits, resulting in a quaternary structure alteration which enhances the ability of oxygen to cause tertiary changes in the remaining, unoxygenated (unliganded) subunits. This makes it much easier to bind more oxygen (up to 4 molecules O_2/molecule Hb) after the first one is attached. The best available data indicate that there are two <u>quaternary</u> structures (deoxy and oxy; T and R in the notation of Monod, Wyman, and Changeux) for the tetramer and that, in addition, each subunit can exist in one of two <u>tertiary</u> structures (again deoxy or oxy; sometimes designated t and r). The transition (t → r) is caused by O_2 binding to a subunit and the transition (T → R) occurs when one or more (the exact number is not clear) t → r transitions have occurred.

Thus, there may be structures in which a subunit in the t-state exists as part of a tetramer which is in the R-state, and vice-versa. These may be the partially oxygenated "intermediates" detected by certain studies. It appears clear, however, that the $T \rightleftharpoons R$ equilibrium controls the position of the individual $t \rightleftharpoons r$ equilibria. That is, the quaternary (tetrameric) structure determines which tertiary (monomeric) structure is preferred.

c. This mechanism has some features of both the Monod, Wyman, and Changeux (MWC) and the Koshland, Nemethy, and Filmer (KNF) models (see the section on allosterism).

   i) The tertiary structure of the subunits is altered only upon binding oxygen, as in the KNF model. Unlike this model, however, the liganding of one subunit does not appear to <u>directly</u> affect the ability of other subunits to bind oxygen (i.e., no "induced fit").

   ii) The change in subunit affinity occurs instead by the initial ligands causing a <u>quaternary</u> structure change of the sort described by the <u>MWC</u> model. Unlike the MWC model though, oxygen apparently binds to Hb molecules in the T-state as well as to ones in the R-state. In their original paper, Monod, Wyman, and Changeux state that their assumption of total non-reactivity of the T-state may have been an oversimplification.

5. Although some aspects are still not rigorously proven, a number of molecular details are becoming clear. Perutz has assembled the pertinent information into several articles, reference to which is given at the end of this book. They are very interesting and provide an excellent view into the way in which molecular data can be used to deduce the mechanism by which an enzyme (or protein) functions.

   a. The trigger for the tertiary structure change (and ultimately the quaternary alteration) appears to be the iron atoms in the heme groups. Upon binding oxygen, the iron changes from a high-spin, 5-coordinate ferrous ion having a radius of about 2.24 Å to a low-spin, 6-coordinate ferrous ion with a radius of about 1.99 Å. The radius change occurs because the electrons are nearer the nucleus in the low-spin state. As is indicated in Figure 83, this permits the Fe^{+2} to fit more closely into the hole in the porphyrin ring, resulting in a movement which brings the nitrogen of the proximal histidine (the one bound to the fifth position of the iron) closer to the plane of the porphyrin ring by about 0.85 Å. As was indicated before, the porphyrin is in contact with about

Figure 83. <u>Comparison of the Fe^{+2}-Porphyrin Relationship in Oxy- and</u> <u>Deoxy-Hemoglobin.</u> The large circle represents the iron atom in the heme. The drawings below are roughly to scale, although the iron atom is drawn to the same diameter in both cases.

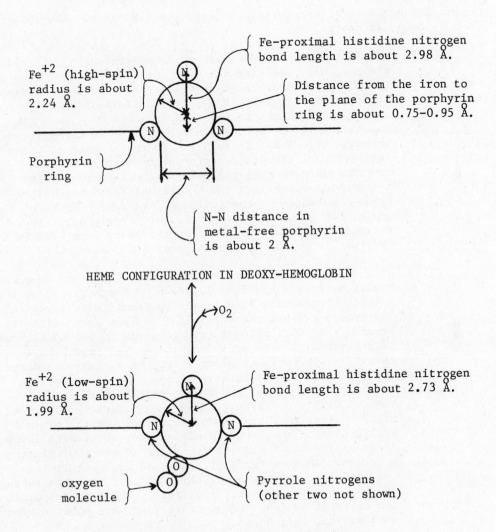

HEME CONFIGURATION IN DEOXY-HEMOGLOBIN

HEME CONFIGURATION IN OXY-HEMOGLOBIN

sixty residues in the globin chain. Such a change in the position of either the histidine, bound rigidly in the peptide backbone, the heme group itself, or both, causes drastic modifications in the positions of the residues to which they are attached. In particular, certain salt bridges which are instrumental in maintaining the deoxy (T) conformation are broken.

b. The sequence of events triggered by the high-spin → low-spin change is as follows:

i) The quaternary deoxy structure is maintained by six salt bridges, all involving the C-terminal residues of each of the four subunits. ArgHC3(141)α_1 forms two salt bridges with residues in subunit α_2; ArgHC3(141)α_2 forms corresponding links to residues in subunit α_1; and the two HisHC3(146)β residues interact with the opposite α-chains (β_1 to α_2; β_2 to α_1). In the absence of these salt bridges, the deoxy conformation is unstable and spontaneously reverts to the oxy structure (e.g., in hemoglobins which lack the C-terminal residues, either through mutation or chemical modification). In each chain, the penultimate (next to last) residue at the C-terminus is tyrosine which in the deoxy form lies in a hydrophobic pocket between helices F and H.

ii) When a subunit binds oxygen, the consequent heme movement causes helices F and H to approach each other, narrow the pocket between them, and expel the penultimate tyrosine. The tyrosine drags the C-terminal residue along with it, breaking the salt bridges (two if it is an α-subunit, one if a β-subunit). In addition, the strain increases in the $\alpha_1\beta_2$-type contacts along which the ultimate movement will occur in the quaternary change. The word "strain" refers to "long" hydrogen bonds, salt bridges, and van der Waal's contacts. These long interactions are of higher potential energy than the normal length, minimum energy ones. Thus, when possible, they shorten to lower their energy and thereby bring about movement within the molecule. Another tertiary change is apparent in the β-subunits. In deoxy-hemoglobin, helix E in these subunits is too close to the sixth position of the iron to permit the entry of oxygen. Consequently, at some point during oxygenation, this pocket opens or is opened by the movement of helix E away from the heme and helix F. This does not occur in the α-subunits since the distance therein between helices E and F is greater.

iii) When a sufficient number of salt bridges have broken and the $\alpha_1\beta_2$ strain has built up, the remaining salt bridges are inadequate to maintain the deoxy conformation. All of the restraining salt bridges now break and the tetramer "clicks over" into the oxy form. The exact number of salt bridges which must be initially broken by oxygen binding is not known and probably depends on factors such as pH, temperature, CO_2 concentration, and concentration of organic phosphates in the solution. This change is illustrated in Figure 84. The motion can be considered more or less to be a rotation of $\alpha\beta$ pairs, joined by $\alpha_1\beta_1$-type interactions, around the α-α axis. The effect is to move the β-subunits closer to each other in a pincer-like motion, narrowing the gap between them. It is clear then why tetramers are <u>required</u> for this cooperative mechanism to work. Dimers do not possess the $\alpha_1\beta_2$-type contacts which undergo the change during quaternary structure alteration. Notice that a change in <u>tertiary</u> structure brings about a <u>quaternary</u> structure change.

In the upper drawing of Figure 84, the dashed lines indicate the six salt bridges which break during quaternary change, while the solid lines indicate "permanent" (non-covalent) interactions between subunits. Note that in the oxy-form, there is probably a β-β link. The upper pair of drawings is more or less a top view of the tetramer, while the lower pair is a perspective, looking roughly along the α-α axis. It should be recognized that these drawings are highly schematized and that the changes are considerably more detailed and complex than indicated.

iv) The tyrosines in unoxygenated (t) subunits are still held in position in the R-state, but only about half as strongly as they were in the T-state. Consequently, the binding of oxygen to these subunits requires about half the original (deoxy) activation energy and occurs with greatly increased ease. This is the source of the "cooperativity" exhibited by hemoglobin.

v) For some time it has been known that at pH 7.4, hemoglobin releases 0.7 moles of H^+ for each mole of O_2 it binds and that the process is reversible. Since this is considered to be the mechanism by which CO_2 decreases the O_2-affinity of Hb (alkaline Bohr effect), these are referred to as the Bohr protons. This is equivalent to saying that

Figure 84. Diagram of the Gross Quaternary Differences Between Oxy- and Deoxy-Hemoglobin

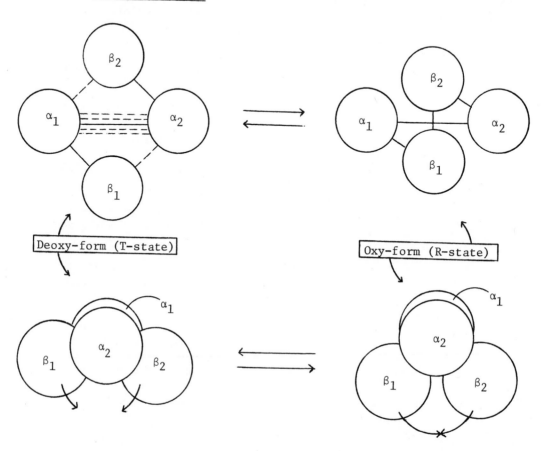

oxy-hemoglobin is more acid than deoxy-hemoglobin. The pK_a for oxy-Hb is 6.62 and the pK_a for deoxy-Hb is 8.18. It appears likely that this change in pK results from the breaking of the six salt bridges and that the groups which release the Bohr protons are those which were involved in the bridges in oxy-hemoglobin. In this way, an increase in P_{CO_2} which causes a decrease in pH can favor the deoxy structure by stabilizing the salt bridges which maintain that conformation.

vi) Deoxy-hemoglobin is able to bind one mole of organic phosphate (in man, notably 2,3-diphosphoglyceric acid (2,3-DPG) and ATP) per mole of Hb, while oxy-hemoglobin cannot. As was indicated earlier, this is significant in controlling the degree of oxygenation of Hb. From a

number of studies, evidence is quite good that at least
2,3-DPG can bind to a group of six positively charged
residues on the β-chains located in the central cavity
between the β-subunits. Upon oxygenation, some of the
positive groups move away and the cavity closes, so that
the 2,3-DPG is expelled. (The role of 2,3-DPG is discussed
in more detail later in this chapter.) Because the
binding of 2,3-DPG is abolished by the high salt
concentrations generally needed to form single crystals
of Hb for x-ray studies, the actual location of its
interaction has not yet been determined. There may,
in fact, be more than one site. It is significant that
despite its apparent interaction with protonated
nitrogens (Arg, His, Lys, and α-amino groups), 2,3-DPG
does not have any influence over the number of Bohr
protons released during oxygenation.

c. Perutz himself succinctly summarizes these steps (Nature, $\underline{228}$,
726 (1970)):

> "... The haem group is so constructed that it amplifies
> a small change in atomic radius, undergone by the iron atom
> in the transition from the high spin to the low spin state,
> into a large movement of the haem-linked histidine
> relative to the porphyrin ring. This movement, and the
> widening of the haem pocket required in the β-chain,
> triggers small changes in the tertiary structure of the
> reacting subunits, which include a movement of helix F
> towards helix H and the centre of the molecule. ... this
> movement can be observed to narrow the pocket between
> helices F and H so that the penultimate tyrosine is
> expelled. The expelled tyrosine pulls the C-terminal
> residue with it, rupturing the salt-bridges that had held
> the reacting subunit to its neighbours in the deoxy
> tetramer. The rupture of each salt-bridge removes one of
> the constraints holding the molecule in the deoxy
> conformation and tips the equilibrium between the two
> alternative quaternary structures some way in favour of
> the oxy conformation. In this conformation the oxygen
> affinity is raised because the subunits are no longer
> constrained by the salt-bridges to maintain the tertiary
> deoxy structure ... [this] also leads to the liberation
> of the Bohr protons, in agreement with the finding that
> their release is exactly proportional to the amount of
> oxygen taken up, and that the two reactions are
> synchronous. I have suggested that the subunits react in
> the order α_1, α_2, β_1, β_2, but this is not certain and my
> cooperative mechanism would work equally well in whichever
> sequence the individual subunits react."

6. Since the discovery in the late 1960's that 2,3-DPG, ATP, inositol
    hexaphosphate (IHP; the principal intraerythrocytic organic
    phosphate in birds), and, to a much lesser extent other organic
    phosphates, bind to Hb and decrease its affinity for oxygen, a
    burgeoning literature on this topic has developed.

    a. The ATP level in erythrocytes is unusually high for a cell
        which does not carry on oxidative phosphorylation. 2,3-
        diphosphoglycerate (2,3-DPG) is present at even higher
        concentrations (about 1:1 mole ratio with Hb) than ATP. The
        2,3-DPG is formed by the rearrangement of 1,3-DPG catalyzed by
        the enzyme diphosphoglycerate mutase. 3-phosphoglyceric acid
        is required as a cofactor and the reaction is allosterically
        stimulated by 2-phosphoglyceric acid. Inorganic phosphate
        appears to be a negative allosteric modifier. 2,3-DPG also
        functions as a cofactor for monophosphoglycerate mutase, an
        enzyme in the glycolytic pathway. In this reaction, it is
        needed only in catalytic amounts. The enzyme 2,3-DPG
        phosphatase catalyzes the reaction

$$2,3\text{-DPG} + H_2O \longrightarrow 3\text{-PG} + P_i$$

    which is the degradative route for 2,3-DPG.

    b. The effect of 2,3-DPG on the oxygen dissociation curve of Hb
        and a possible binding site for 2,3-DPG to Hb were discussed
        earlier. It is appropriate to reemphasize here the <u>many</u>
        factors (pH, hematocrit, amount of Hb per red cell,
        respiratory rate, P_{O_2} in the inspired air, etc.), in addition
        to the concentration of 2,3-DPG, which work together to
        control oxygen delivery to the tissues.

        Regarding the proposed binding site on Hb for the organic
        phosphates, it appears that 2,3-DPG can be readily accommodated
        in the cleft between the β-chains. There is some doubt,
        however, as to whether inositol hexaphosphate (IHP; much more
        effective than 2,3-DPG) and ATP (whose binding constant is
        an order of magnitude less than that for 2,3-DPG) will fit
        into the same area. Recent studies have indicated, moreover,
        that two binding sites on hemoglobin may be involved. At one
        site, on deoxy-hemoglobin, 2,3-DPG and ATP compete for binding
        while at the other site, on oxy-hemoglobin, ATP <u>enhances</u> the
        binding of 2,3-DPG. This also emphasizes the finding that
        oxy-hemoglobin <u>does</u> apparently bind 2,3-DPG, contrary to earlier
        reports. Clearly there is more to be learned about the
        interactions of organic phosphates with hemoglobin molecules.

    c. Another area of investigation has been the mechanism by which
        2,3-DPG concentration can be altered inside the erythrocyte.
        One or more of these controls must be able to act within a

period of less than twelve hours (perhaps much shorter; see later discussion), since significant (15-25%) changes in 2,3-DPG concentration have been noted in such a timespan. The most probable effects involved are summarized below.

i) As was mentioned in the section on thermodynamics, the binding of 2,3-DPG to Hb decreases the amount of 2,3-DPG which can participate in other reactions, notably the 1,3-DPG $\rightleftarrows$ 2,3-DPG equilibrium. This causes increased 2,3-DPG synthesis at the expense of 1,3-DPG.

ii) The intraerythrocytic pH appears to have several effects on the 2,3-DPG concentration. First, a decrease in pH increases the binding of 2,3-DPG to Hb, bringing the mechanism above into operation. Second, an increase in pH is known to stimulate glycolysis, which tends to increase the levels of all glycolytic intermediates such as 2,3-DPG. Finally, pH changes affect the activities of diphosphoglycerate mutase and phosphatase, the enzymes responsible for 2,3-DPG synthesis and degradation, respectively.

iii) The age of the erythrocytes may or may not have some influence on intracellular organic phosphate levels. Although reports differ somewhat, it appears that either the age has no effect or increasing age corresponds to decreasing concentration.

iv) There may also be genetic control over the 2,3-DPG levels. There is evidence that ATP concentrations within the erythrocyte are under hereditary influence and in the hooded strain of rats, levels of both of these organic phosphates appear to be genetically influenced. Obviously this mechanism is quite different from the previous three in being of no importance in rapid, short-term adaptations.

Whatever the actual regulator(s) is, it is apparent that red cell metabolism is intimately involved in the transport of oxygen by hemoglobin.

d. The processes described in the preceding section might be termed final-step controls. The primary stimuli which trigger these terminal steps are many. A few of them are

i) Decreased O_2-delivery to the tissues due to anemia, altitude, cardiac insufficiency, etc. (mentioned before).

ii) The presence of certain hormones, such as thyroxine (which may directly stimulate diphosphoglycerate mutase) and androgens (which also increase erythropoiesis).

iii) Polycythemia (which decreases the intraerythrocytic
      concentration of 2,3-DPG).

iv) Some cases of pulmonary disease.

Whether the shift in the O_2-dissociation curve which accompanies
2,3-DPG changes is beneficial or not depends largely on the
partial pressure of oxygen in the lungs (see previously). There
is wide variation in the degree of change in the 2,3-DPG
concentration exhibited by patients having the same disease.
For example, in severe pulmonary disease the increase may be
from 0-100%; in leukemia (a non-hemolytic anemia), rises from
20-150% have been observed; and patients with iron deficiency
display increases ranging from 40-75%.

e. Two other aspects of 2,3-DPG are important.

    i) It has been reported that the 2,3-DPG level is about 45%
        higher in venous blood than in arterial blood, corresponding
        to the decreased oxygen saturation in venous blood. This
        is important because it demonstrates the extreme rapidity
        with which the 2,3-DPG regulatory mechanism can act. It
        also indicates a possible mechanism by which elevation
        of the 2,3-DPG (useful to unload oxygen) might be prevented
        from impairing oxygen uptake by the pulmonary blood.

    ii) Ion transport across red cell membranes is stimulated by
         deoxygenation of the Hb in the cell. This may be due to
         the binding of 2,3-DPG to the deoxy-Hb, causing enhanced
         glycolysis, glucose uptake, and hence ion transport.

f. An important clinical problem involving 2,3-DPG is the
    observation that 2,3-DPG and, to a lesser extent, ATP
    concentrations decrease rapidly in blood which is stored for
    even a few days in the acid-citrate-dextrose (ACD) medium
    used for some time by many blood banks. The result of this is
    that oxygen affinity is increased and the ability to supply
    oxygen to the tissues is thereby decreased when such blood is
    used for transfusions. In patients who receive this blood,
    there is an initial increase in oxygen affinity which does not
    decrease to normal for six hours or more. Traditionally,
    red cell survival has been used as the main criterion of the
    quality of stored blood. Survival does not necessarily
    correspond to maintenance of adequate organic phosphate (OP)
    levels, however, so this must be considered as a supplementary
    criterion.

    A number of studies have investigated how best to vary the
    composition of the storage medium to prevent this loss of

organic phosphates. (Actually, these compounds do not really leave the cell since red cell membranes are generally quite impermeable to organic phosphates. Instead, they are metabolized to other forms.) Some of these findings are summarized below.

i) Citrate-phosphate-dextrose (CPD) medium is somewhat better than ACD for maintaining OP levels.

ii) Storage in the deoxy-form at a pH of 7.2 is more satisfactory than oxygenation and lower pH. However, at an alkaline pH, ATP is lost, which decreases cell survival time.

iii) ACD supplemented with the nucleoside adenosine maintains ATP but not 2,3-DPG. In fact, the rate of oxygen affinity elevation is mildly increased in this medium.

iv) ACD supplemented with the nucleoside inosine seems to be the most satisfactory medium at present. Inosine maintains both ATP and 2,3-DPG in vitro, as well as retarding significantly the increase in oxygen affinity. It is not known exactly how inosine brings about this change, but it may partly function by supplying ribose for 2,3-DPG formation (via the HMP shunt).

So far, the actual therapeutic significance of these changes is not clear. The greatest effects should probably be looked for in patients who have received repeated, massive transfusions over a period of, say, six hours, so that a significant fraction of their circulating erythrocytes have increased O_2 affinity.

g. As more is learned about the functioning and control of organic phosphates within the red cell, ways of altering this level artificially in particular situations can be developed. The role of hormones (e.g., androgens and thyroxine) in controlling tissue oxygen supply may be better understood. The physician should also be aware that some drugs may, as a side-effect, change the metabolism of the red cell in a manner which will affect the oxygen-carrying ability of the blood.

7. Although hemoglobin is worth understanding in its own right, one must remember that most regulatory enzymes are also allosteric, multisubunit proteins. That they are allosteric means that the rate at which they catalyze a reaction or bind substrate to one subunit is controlled by the binding of another molecule (a positive or negative allosteric effector) either on the same or another subunit. By understanding how hemoglobin changes its oxygen affinity through quaternary structure alterations, other (regulatory)

enzymes can be better understood. Four especially important concepts to arise from the hemoglobin work are

   a.   The change in Fe^{+2} from high spin to low spin acting as a trigger;

   b.   The use of salt bridges in maintaining one structure or another;

   c.   The indication that hemoglobin's allosterism can best be described by the Eigen model, i.e., there are features intermediate between the MWC and KNF models. This opportunity to directly test the allosteric models is very important;

   d.   The description of how an allosteric effector (O_2 and organic phosphates) may function in terms of specific molecular changes.

The approach to the elucidation of hemoglobin's structure and function has been interdisciplinary, involving physical chemistry (x-ray crystallography, magnetic resonance), physiology (respiratory studies, etc.), biochemistry (Bohr effect, Hb derivatives, amino acid sequence), and genetics and clinical medicine (hemoglobinopathies, discussed later). Without all of these contributions, the problem would have been solved much more slowly or not at all.

8.   Another function of hemoglobin is to participate in the transport of CO_2 from the tissues to the lungs with very little change in pH. This buffering action of hemoglobin was already discussed in Chapter 1 and is illustrated in Figure 85. Note that it is primarily the Bohr protons which participate in this.

In Figure 85, note:

   a.   The chloride shift which accompanies the movement of HCO_3^-;

   b.   The role of H^+ and hence CO_2 and carbonic anhydrase, in unloading oxygen from oxy-hemoglobin;

   c.   The intermediary role of the plasma as a carrier of CO_2 (as HCO_3^-);

   d.   That, in addition to the O_2 bound to Hb, there is always some oxygen physically dissolved in the plasma;

   e.   That some CO_2 is chemically combined with Hb, as carbamino compounds.

9.   Hemoglobin concentrations in the erythrocyte are controlled by the levels of <u>erythropoietin</u>. The precise role of erythropoietin, a hormone largely derived from the kidney, is not understood. Its action involves the production of several species of RNA in

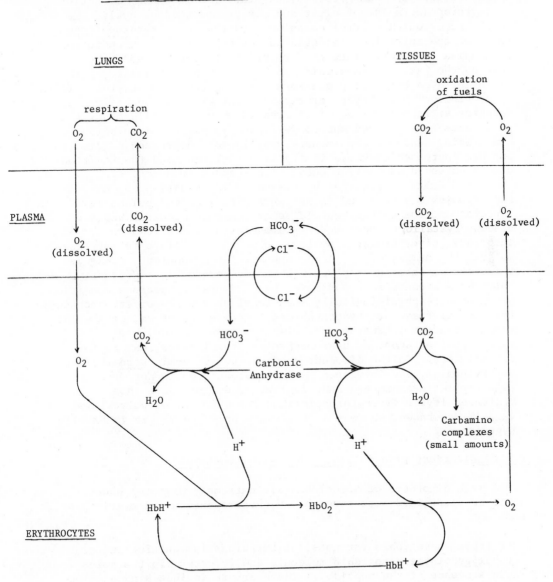

A similar figure is presented in Chapter 1 of this book. Note that the reactions on the right side of this diagram occur in the plasma and erythrocytes in the capillaries, during oxygen delivery to the tissues; while the reactions on the left side occur in the lungs, during CO_2 release and O_2 uptake. The cyclic nature of these changes, in going from lungs to capillaries back to the lungs should be recognized.

the erythroblast which are probably concerned with increased uptake of iron and hemoglobin synthesis. The mechanism of action of erythropoietin is not clear and may be mediated by cyclic AMP or the genes which control hemoglobin synthesis. Erythropoietin triggers the initial differentiation of committed erythropoietin-responsive stem cells into pro-erythroblasts and, in addition, governs the rate of hemoglobin synthesis in the partially differentiated red cell precursors. The synthesis of erythropoietin is controlled by the state of tissue oxygenation at specific receptor sites. Its level is increased when the supply of oxygen to tissues is decreased due to decreased levels of circulating hemoglobin, inadequate pulmonary ventilation, decreased cardiac output, increased oxygen affinity of hemoglobin, and the position of the oxygen dissociation curve with respect to normal or increased demands of oxygen by tissues. In conditions of chronic anoxia which may be found in patients with impaired pulmonary ventilation, congenital heart disease, chronic acquired heart disease, etc., the increased erythropoietin levels lead to secondary polycythemia. The term polycythemia refers to increases in both the number of red cells and hemoglobin content. This is reflected in the elevated hemoglobin and hematocrit. Secondary polycythemia can also occur in certain tumors or tumorous conditions which produce physiologically inappropriate amounts of erythropoietin. Examples of these are hypernephroma, uterine leiomyoma, cerebellar hemangioblastoma, hepatoma, ovarian carcinoma, adrenal cortical hyperplasia or adenoma, hydronephrosis, renal cysts, hamartoma (liver), etc. In the disorder known as polycythemia vera (primary polycythemia), the increased red cell production is not erythropoietin-augmented and, in fact, these patients show decreased levels of erythropoietin. The etiology of polycythemia vera is not known, whereas in secondary polycythemia it is usually hypoxia.

Inherited Disorders of Hemoglobin (Hemoglobinopathies)

The inherited disorders of hemoglobin are extremely diverse; however, they can be classified into three broad, heterogeneous and overlapping groups.

1. Structural hemoglobin variants: Alterations in function and/or stability of the hemoglobin molecule due to errors in the amino acid sequence of the peptides. These errors include single amino acid substitutions at a specific site, the insertion or deletion of one or more residues, or the combination of different pieces of two normal chains resulting in hybrid polypeptides. The hybrid-peptide variants (known as hemoglobin Lepore) and chain-termination mutants are discussed with the thalassemias. More than 150 abnormal hemoglobins have been described, a number of which cause no physiologic abnormalities.

2. Thalassemia syndromes: This group of disorders is characterized by a <u>decreased</u> rate of synthesis of one or more peptide chains of hemoglobin. The clinical problem in this disorder is a result of imbalanced chain synthesis.

3. Hereditary persistence of fetal hemoglobin (HPFH): These conditions result because there is a failure to switch from fetal hemoglobin (HbF) to adult hemoglobin (HbA). Thus, these individuals will continue to have high levels of HbF instead of HbA beyond their neonatal period. It should be pointed out, however, that HPFH does not present any major hematologic abnormalities and is compatible with normal life.

    HPFH and the thalassemias are not two entirely independent entities but are two groups which form a continuum. In addition, thalassemias can also result from the inheritance of structural hemoglobin variants which are inefficiently produced.

Normal Hemoglobins

As a prelude to discussing the hemoglobinopathies, a brief discussion on the composition of various normal hemoglobins and their temporal sequence in development is useful.

1. Each molecule of human hemoglobin is made up of four globin polypeptides. These are shown below with respect to their temporal sequence in development.

    Embryonic hemoglobins:

    $$\text{Gower}_1 = \zeta_2\epsilon_2 \quad (\text{zeta}_2 \text{ and epsilon}_2)$$

    $$\text{Gower}_2 = \alpha_2\epsilon_2 \quad (\text{alpha}_2 \text{ and epsilon}_2)$$

    $$\text{Portland} = \zeta_2\gamma_2 \quad (\text{zeta}_2 \text{ and gamma}_2)$$

    Fetal hemoglobin:

    $$\text{Hemoglobin F} = \alpha_2\gamma_2$$

    The major oxygen carrier in fetal life consists of two α-chains and two non-α-chains.

    Adult hemoglobins:

    $$\text{Hemoglobin A} = \alpha_2\beta_2 \quad (\text{alpha}_2 \text{ and beta}_2)$$

    The major adult type, it constitutes about 95% of the total hemoglobin.

Hemoglobin $A_2 = \alpha_2\delta_2$

The minor adult type, it comprises less than 3.5% of the total hemoglobin because the output of δ-chains is comparatively low at all times.

Hemoglobin $F = \alpha_2\gamma_2$

It comprises less than 2% of the total hemoglobin in the normal adult.

Note that hemoglobin F predominates in fetal life because the output from γ-genes before birth greatly exceeds that of the β-genes. After birth, however, the γ-genes are almost completely suppressed while the β-genes are fully activated. The relative amounts of the hemoglobin polypeptide chains in different stages of human development are shown in Figure 86. In this figure, data concerning the ζ-chain are not included. Note that the ε-chains, present only in the early embryonic fetus, are replaced by γ-chains in the fetal period of development, which in turn are almost completely replaced by β-chains and small amounts of δ-chains. The mechanisms that control these replacements in development are not known.

2. Synthesis of the different hemoglobin chains is controlled by six qualitatively different genes (which have not as yet been assigned to specific chromosomes), one coding for each of the six (α, β, γ, δ, ε and ζ) polypeptides. It is speculated that these evolved from one another by gene duplication followed by mutation. The α-chain gene is considered to be the oldest. At some time long ago, duplication of the α-gene occurred. One copy continued to code for α-chains while the other gradually mutated into what is today the gene for the γ-chain. Some time after this initial duplication event, but still quite a while ago, the γ-chain gene was duplicated. One copy began diverging from the γ-gene, becoming what now codes for the β-chain. The δ-chain gene presumably arose in a similar fashion from the β-gene. The principal evidence for this developmental path comes from a comparison of the amino acid sequences of these chains. The further apart two chains are in evolution, the fewer identical residues (homologies) will presumably occur between them. This sort of study shows that there is a 39% homology between α- and γ-chains, a 42% homology between α- and β-chains, a 71% homology between β- and γ-chains, and a 96% homology between β- and δ-chains. These evolutionary relationships would be very difficult to actually prove and should be considered quite speculative.

3. There is now fairly good evidence that in Caucasian and Oriental populations the α-chain genes are duplicated; therefore each diploid cell has four copies of the α-gene. This is apparently not the case in Blacks and Melanesians, who may have a variable number of α-genes ranging from 2 to 4 per diploid cell. As noted

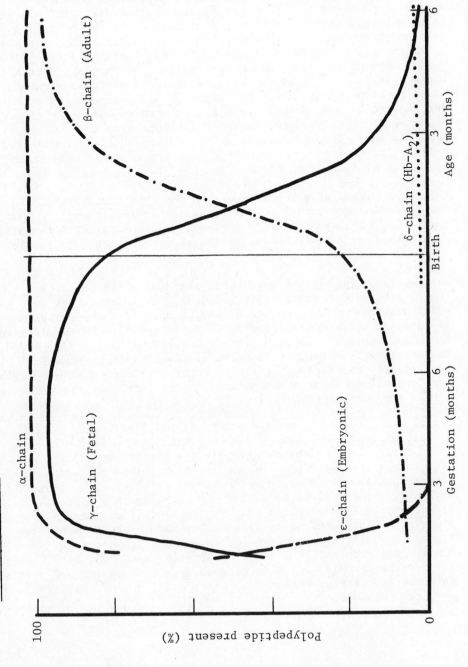

Figure 86. Relation of the Amounts of Hemoglobin Polypeptide Chains to Different Stages of Human Development

From E.R. Huehns, N. Dance, S. Hecht, A.G. Motulsky, Cold Spring Harbor Symp. Quant. Biol. 29, 327 (1964)

earlier, the β-, γ- and δ-genes are closely linked and may be present as a cluster on a single chromosome. These are not linked to the α-genes, which are thought to be located on an entirely different chromosome. Since the β- and δ-genes are not duplicated, there are only two copies each per diploid cell. Each haploid gene complement contains two structural genes responsible for the synthesis of two different γ-chains. These γ-chains differ from one another at the 136th amino acid residue of the polypeptide chain. ^{G}γ contains a glycyl residue while ^{A}γ contains alanine. These γ-polypeptides cannot be separated by electrophoresis or chromatography and, hence, are distinguished on the basis of amino acid composition and sequence determination. The ratio of ^{G}γ/^{A}γ is about 3 in cord blood and less than 1 in adult red blood cells. The ratios of the resulting two types of HbF vary under different clinical conditions.

HbF has a higher affinity for O_2 than does HbA, facilitating O_2 transfer from the mother to the fetus. This is, in part, due to the weaker binding of 2,3-DPG to γ-chains as compared to β-chains and, thus, the HbF oxygen saturation curve is shifted to the left of the HbA-saturation curve.

4. The α-chain contains 141 amino acid residues while the others (β, δ, and γ) each contain 146 residues. It is essential to point out that each chain has a heme group. In the α-chain, it is inserted between histidine residues 58 and 87. In the β-chain, the heme is present between histidine residues 63 and 97. Any changes (due to mutation) in these histidine residues have a particularly deleterious effect on the functioning of the hemoglobin due to problems in maintaining iron in the ferrous state (discussed later; see methemoglobin).

5. Hemoglobin A complexes with a number of small molecules (e.g., glutathione, hexoses), and these complexes can be separated from other hemoglobins by chromatographic and electrophoretic procedures. Hemoglobin A_3 (also called $HbA_1α$) is a complex between HbA and glutathione which occurs in older erythrocytes. In HbA_{Ic} (which appears as a fast-moving hemoglobin upon electrophoresis in alkaline pH), the N-terminal residues of both β-chains are attached to an aldehyde or keto group of a hexose through a Schiff-base linkage (making Hb a glycoprotein). Hemolysates prepared from normal adults contain about 5 to 7 per cent hemoglobin$_{Ic}$. It is decreased in iron-deficiency anemia; in diabetes mellitus either HbA_{Ic} or a similar component is <u>elevated</u> two-fold.

The Thalassemias

1. This is a heterogeneous group of hypochromic, microcytic anemias resulting from the reduced rate of synthesis of one or more globin chains of hemoglobin. In certain geographic regions (e.g., Southeast Asia, the Philippines, China, the Hawaiian Islands, the Mediterranean countries), the thalassemia syndromes are common and may constitute one of the significant public health problems. The clinical severity of these disorders varies considerably, ranging from incompatibility with life or a severe form of anemia to mild or totally asymptomatic forms. Most of the clinical problems are due to deficient production of specific polypeptide chains, leading to ineffective erythropoiesis and the precipitation of unused polypeptide chains (which are unstable) in the red cell precursors. The red blood cells containing these intracellular precipitates of unused chains (called Heinz bodies) escape from the bone marrow into the circulation. They are then sequestered either in the spleen or other parts of the reticuloendothelial system, causing shortened red cell survival in the circulating blood. In some cells, the inclusion bodies may be "pitted" out with consequent membrane damage. These bizarre-shaped red cells may be seen in the peripheral blood. Furthermore, the trapping of an increased number of red blood cells in the spleen leads to its enlargement.

   In brief, the anemia seen in thalassemia is due to ineffective erythropoiesis, decreased red cell survival, and hemodilution. The other pertinent clinical aspects of the thalassemias are discussed further under specific disorders.

2. The thalassemias are classified broadly into α- and β-types. α-Thalassemia refers to disorders associated with deficiencies in α-chain production; similarly, β-thalassemia is associated with deficiencies in β-chain synthesis. These two types are further subdivided into several distinct, genetic subtypes. In general, it should be emphasized that the clinical symptoms in a given thalassemic syndrome may not be uniformly predictable. This can be explained by variations within and among races, acquired conditions modifying the expression of the thalassemic gene, and inheritance of genes for the structural variants of hemoglobin.

α-Thalassemia

1. The different types of α-thalassemias (α-thal) arise from the interaction of three α-thal genes: α-thal 1 results in complete abolition of α-chain synthesis; α-thal 2 partially reduces α-chain synthesis; and the Hemoglobin Constant Spring (Hb-CS) gene, which

is similar to an α-thal gene since its product is produced in a greatly diminished quantity, codes for an elongated α-gene product.

2. Two major clinical disorders of α-thal seen in Oriental and Mediterranean populations are the lethal Hemoglobin Bart's (γ_4) hydrops fetalis condition and the hemoglobin H (β_4) disease. As mentioned earlier, the genes for the α-chain are duplicated; that is, a diploid cell contains four copies of the gene. The broad clinical array of α-thal extending from a lethal condition to a totally benign form is directly related to the number of α-genes deleted. From the genetic evidence available, the homozygous state of the α-thal 1 gene (deletion of all four copies of the α-gene) represents Hb-Bart's hydrops syndrome. The homozygous state of the α-thal 2 gene represents deletion of only two genes and, therefore, is an α-thal trait. Hemoglobin H disease, which is a result of the deletion of three α-genes, results from the heterozygous states for both the α-thal 1 and α-thal 2 genes, or α-thal 1 and Hb-CS genes. Table 28 summarizes these and other aspects of the α-thalassemias. Following are some additional points which need to be emphasized.

a. The presence of a large amount of Hb-Bart's (γ_4) in hydrops fetalis is not compatible with life. In this case the fetuses are delivered stillborn. In Southeast Asia, this appears to be one of the major causes of stillbirth. The γ_4 form of hemoglobin is useless because of its marked increase in oxygen affinity and its inability to deliver oxygen to the tissues.

b. In the hemoglobin H disease, which is due to the absence of three α-genes, there are substantially decreased amounts of α-chains available for combining with either γ-chains in the fetus or β-chains in the adult. Thus, there is a marked accumulation of γ_4 in the fetus and β_4 (hemoglobin H) in the adult (recall that γ-chain synthesis is suppressed immediately after birth while β-chain synthesis is augmented).

Hemoglobin H disease is characterized by the presence of 5-30% HbH in the red blood cells of adults. The tetrameric β-chain hemoglobin is useless for the same reasons given above for γ_4-hemoglobin. Furthermore, because of its instability, HbH tends to precipitate within the red blood cells, causing rigidity and damage to the membrane. As a result, the flexibility needed to negotiate the passage through the reticuloendothelial tissue of the spleen and other organs is lost. During this passage the inclusions are removed ("pitted") and some of the damaged cells escape into the circulation. The survival of these abnormal cells is markedly reduced. The bone marrow responds to this situation by producing small red blood cells which are poorly filled

Table 28. α-Thalassemia Syndromes in Asian Populations

Syndrome (or phenotype)	Genotype	No. of α-genes present	Parental Genotypes*	Frequency	Clinical severity	Hemoglobin pattern
Hydrops fetalis with Hb Bart's	Homozygous α-thal 1	0	α-thal trait x α-thal trait	0.25	Lethal (death in-utero)	Cord blood: Mostly Hb Bart's (γ_4); small amounts of HbH (β_4) and Hb Portland.
Hemoglobin H disease	a. Heterozygous α-thal 1 (α-thal trait) and heterozygous α-thal 2 (silent carrier)	1	α-thal trait x silent carrier	0.25	Severe to moderate microcytosis; hemolysis.	Cord blood: About 25% Hb Bart's; Adult blood: 5-30% HbH.
	b. α-thal trait and heterozygous Hb-CS	1 (and 1 Hb-CS gene)	α-thal trait x Hb-CS-heterozygote	0.25	"	Same as above + 2-3% Hb-CS
α-thalassemia trait	Heterozygous α-thal 1 or Homozygous α-thal 2	2	α-thal trait x normal	0.5	Mild microcytosis	Cord blood: About 5% Hb Bart's; Adult blood: normal composition
Silent Carrier	Heterozygous α-thal 2	3	Silent carrier x normal	0.5	None	Cord blood: About 1-2% Hb Bart's; Adult blood: normal composition
Hemoglobin Constant Spring trait	Heterozygous Hb-CS	3 (and Hb-CS gene)	Hb-CS heterozygous x normal	0.5	None	About 0.5-1% of Hb-CS both in cord and adult blood

*The parental types shown are only representative examples and do not include matings of HbH genotypes with other genotypes.

[731]

with hemoglobin (hypochromic microcytosis). Other clinical characteristics of HbH disease include splenomegaly and hepatomegaly (the degree of enlargement is not as striking as in the case of β-thalassemia major--discussed later), cholelithiasis, erythroid hyperplasia, etc.

The diagnosis of hemoglobin H disease is established by finding: (i) a rapidly moving minor hemoglobin (HbH) on electrophoresis at pH 8.4 on cellulose acetate membrane (see hemoglobin electrophoresis), and (ii) hemoglobin H inclusions (Heinz bodies) in the red blood cells. Hemoglobin H inclusions are demonstrated by incubating one volume of 1% isotonic brilliant cresyl blue with two volumes of freshly drawn blood for one hour. It should be pointed out that any hemoglobin which is unstable, either due to an enzyme defect (e.g., G-6-PD deficiency--see Chapter 4) or structural changes (discussed later), will yield inclusions similar to those seen in the HbH disease. However, it appears from a diagnostic point of view that the one-hour incubation procedure gives rise to positive results only with HbH. Other unstable hemoglobins usually require a longer period of incubation to undergo denaturation and precipitation to the same extent. Some of the other abnormal laboratory parameters may include: elevated unconjugated (indirect) bilirubin, increased LDH isoenzyme fractions of red cell origin due to hemolysis (LDH 1 & 2), and moderate osteoporosis. The hypochromic, microcytic anemia of HbH patients (and other thalassemia patients) is difficult to distinguish from that caused by other disorders (the most frequent cause being iron deficiency), and a scheme for differential diagnosis is presented in Table 32. Patients with both HbH disease and iron deficiency may not show the presence of hemoglobin H. It should be noted that the finding of hemoglobin H is not absolutely diagnostic of α-thalassemia-hemoglobin H disease. In a number of instances of erythroleukemia (or Di Guglielmo syndrome), hemoglobin H has been observed as a result of unbalanced chain synthesis which is presumably due to chromosomal aberrations found in leukemic states. In a recent report, erythrocytes from a patient with acute myeloblastic leukemia have been shown to contain 50-65% HbH. The neoplastic red cells showed very little α-chain synthesis and yet contained both sets of α-chain genes as shown by hybridization techniques. A proposed explanation for this defect consists of activation of specific repressors affecting both sets of α-chain genes, which are normally active only in embryonic life. Chromosomal analysis of the peripheral blood and bone marrow was normal in this patient. Hemoglobin Bart's has been observed in the D_1-trisomy syndrome. Unlike the β-thalassemias (see later), hemoglobin H disease and other α-thal syndromes (α-thal trait and silent carrier) show no change in the <u>relative proportions</u> of hemoglobin A, A_2 and F. This is because these hemoglobins are composed of two α-chains and two non-α-chains and thus are affected equally. In the Black population, the α-thalassemia problem differs significantly from that of the Oriental and Caucasian populations. In Blacks, HbH disease is seen very rarely, and hydrops fetalis with Hb Bart's has not been reported. This may be due to a number of α-genes present in the population. In addition, it is speculated that in homozygous α-thalassemia, α-gene deletion may extend into the α-like gene (ζ), causing early death during the embryonic period due to deficiency of $\zeta_2\varepsilon_2$ or $\zeta_2\gamma_2$ (Hemoglobin Portland). On the other hand, Hb Portland may be responsible for carrying the homozygous Oriental α-thal infant close to term.

The molecular defects in homozygous α-thalassemia and hemoglobin H disease have been shown to be due to, respectively, the complete deletion of α-globin genes and the deletion of a majority of α-globin genes (3 out of 4). The assessment of the number of copies of α-globin genes in the DNA of an individual is carried out by hybridization of host DNA with radioactive complementary DNA (cDNA) which is used as a probe and prepared from a normal α-globin mRNA by the action of reverse transcriptase. Some of the details of the hybridization procedure are described below. To recapitulate, DNA contains two complementary strands (i.e., adenine on one strand is paired with thymine on the other strand and guanine with cytosine) of polydeoxynucleotides wound around each other to form a double helix. One strand is arbitrarily designated as plus and the other as minus. The template DNA strand for mRNA synthesis is a minus strand and since the mRNA is complementary to DNA, it is designated as a plus strand.

Synthesis of cDNA

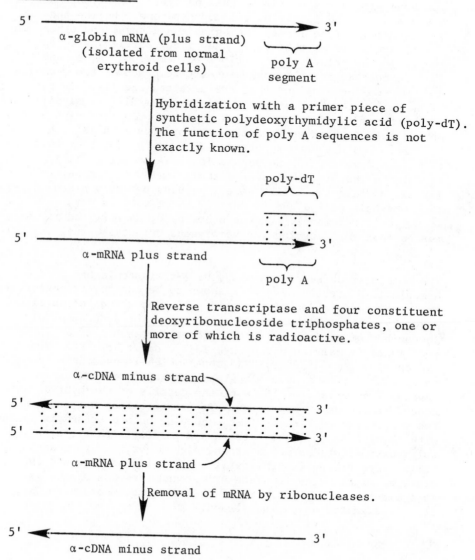

5' ————————————————————► 3'
　α-globin mRNA (plus strand)
　(isolated from normal
　　erythroid cells)

poly A
segment

Hybridization with a primer piece of
synthetic polydeoxythymidylic acid (poly-dT).
The function of poly A sequences is not
exactly known.

poly-dT

5' ————————————————————► 3'
　　α-mRNA plus strand

poly A

Reverse transcriptase and four constituent
deoxyribonucleoside triphosphates, one or
more of which is radioactive.

α-cDNA minus strand

5' ◄———————————————————— 3'
5' ————————————————————► 3'

α-mRNA plus strand

Removal of mRNA by ribonucleases.

5' ◄———————————————————— 3'
　　α-cDNA minus strand

Hybridization Procedure
_____

The labeled cDNA can now be used as a probe in determining
whether or not its complementary sequence exists in the DNA
isolated from a diploid cell.

Diploid Cell
(e.g., fibroblasts cultured from amniotic
fluid--used in prenatal diagnosis)

isolation of DNA, which is sheared by pressure
into lengths approximately equal to α-cDNA

heated to effect complete
dissociation of strands

Single-Stranded DNA
(Since α-gene is duplicated, there are
four single strands of DNA made up of two plus
strands and two minus strands.)

addition of labeled α-cDNA
minus strand; cooling under
controlled conditions of salt
concentration and pH to achieve
reassociation of plus strands
with minus strands (anneal or
hybridize) and separation of
hybridized DNA from single
strands either by hydroxyapatite
(calcium phosphate) chromatography
or enzymatic methods

Double-Stranded-DNA Radioactivity is measured.
(Labeled, double-stranded molecules are due to
hybridization with α-cDNA minus strands.)

In the complete absence of α-genes, there is very little
incorporation of the label in the hybridized DNA. Furthermore,
under appropriate experimental conditions, it can be shown
that the radioactivity in the hybridized DNA is proportional
to the number of α-genes present. These studies have
demonstrated that α-thal syndromes are the result of deletions
of α-structural genes and, similarly, in the conditions of
hereditary persistence of fetal hemoglobin and δβ-thalassemia
(discussed later), β-gene deletion. However, in most of the
β-thalassemias, despite the fact that the β-genes are present,

the erythroid cells are incapable of maintaining an appropriate concentration of mRNA in the cytoplasm to support optimal amounts of β-chain synthesis.

c. The Hb Constant Spring gene gives rise to an elongated α-chain product, which is produced in greatly reduced amounts. Thus, the Hb-CS gene mimics an α-gene deletion. As noted earlier, one of the genetic causes of HbH disease is due to the interaction of the α-thal 1 gene and the Hb-CS gene.

Hb-CS contains 31 additional amino acid residues at the C-terminal end of the α-chain. This is due to a change in the normal termination codon of UAA to CAA as a result of single-base exchange. CAA codes for glutamine, thus allowing the α-chain mRNA to be read until another termination codon at position 173 is reached. Other chain-termination mutant hemoglobins have been reported. Base substitutions in these along with Hb-CS are shown in Table 29. It should be pointed out that the chain-termination mutant hemoglobins are synthesized only in small amounts (about 1%) and the reasons for this are not clear (probably related to stability of mRNA).

Table 29. α-Chain-Termination Mutant Hemoglobins

Hemoglobin	Codon Change at α-142
Hb-Constant Spring	UAA → CAA (Gln)
Hb-Icaria	UAA → AAA (Lys)
Hb-Koya Dora	UAA → UCA (Ser)
Hb-Seal Rock	UAA → GAA (Glu)

3. The β and δβ-Thalassemias:

a. These disorders show considerable heterogeneity, both at the clinical and molecular levels. These are summarized in Table 30. The types of β-thalassemia mutations include β⁺, β⁰, δβ⁰ and Hb Lepore. In homozygous β⁺-thal, there are small amounts of β-chains produced, whereas in β⁰, δβ⁰ and Hb Lepore, the β-chain is totally absent. Thus, clinically, homozygous β-thalassemia (Cooley's anemia) can occur due to many different genotypes (β⁺/β⁺, β⁰/β⁰, β⁺/β⁰, β⁺/δβ⁰, β⁰/δβ⁰, Hb Lepore/Hb Lepore). Hemoglobin Lepore is made up of normal α-chains combined with non-α-chains which have the N-terminal end of the δ-chain fused with the C-terminal end of the

Table 30. The β-Thalassemias

Type	Severity	Hemoglobin Pattern			Molecular Defect	
		A	A$_2$	F	mRNA	Genes
Homozygous States						
β$^+$-thalassemia	Cooley's anemia* (Thalassemia major)	↓	↑ (usually not more than 12-15%)	↑	β-mRNA markedly reduced	β-genes present
β^0-thalassemia	"	none	↑	↑	absent β-mRNA or non-functional β-mRNA	β-genes present
δβ^0-thalassemia	Milder disease (Thalassemia intermedia)	none	none	100%	absent β- and δ-mRNA	deleted β- and δ-genes
Hb Lepore	Cooley's anemia	none	none and 25% Hb Lepore	75%	absent normal β- and δ-mRNA	absent normal β- and δ-genes. δβ-fusion gene present
Heterozygous States						
β$^+$-thalassemia	Mild disease (β-thal trait or β-thal minor)	↓	↑ (usually not more than 6.5%)	→ or slightly ↑	deficiency of β-mRNA	β-genes present
β^0-thalassemia	Mild disease	"	"	"	deficiency of β-mRNA or non-functional β-mRNA	"
δβ-thalassemia	Mild disease	↓	↓	5-20%	deficiency of β- and δ-mRNA	deletion of β- and δ-genes on one homologous chromosome
Hb Lepore	"	↓ and 5-15% Hb-Lepore	↓	↑	deficiency of normal β- and δ-mRNA	replacement of normal β- and δ-genes by δβ-fusion gene on homologous chromosome

*Severe disease with massive dyserythropoiesis. ↑ = increased; ↓ = decreased; → = normal.

[737]

β-chain. The Hemoglobin Lepore genes have arisen due to chromosomal misalignment and an unequal crossing-over between the δ- and β-genes, resulting in δβ-fusion. Thus the Hb-Lepore gene, a fusion product of δ- and β-genes, directs the synthesis of the non-α-chains (the hybrid chains) of Hb-Lepore. It mimics δβ°-thalassemia because, in both instances, normal δ- and β-chains are completely absent. Differing points of crossover between the δ- and β-genes yield different δβ-fusion genes and, therefore, result in distinct variants of hemoglobin-Lepore, namely, Washington, Hollandia and Baltimore. In Hemoglobin Miyada, the normal α-chains are combined with βδ-chains. As noted earlier, δβ°-thalassemia is due to the deletion of δβ-structural genes. Other β-thalassemic conditions are not the result of β-structural gene deletions but are the result of the inability to maintain optimal levels of functionally active mRNA to support adequate amounts of β-chain synthesis.

b. Clinically homozygous β-thalassemia is a severe disease with massive ineffective erythropoiesis. Microcytic anemia is present from the first few months of life. The disorder is also characterized by severe skeletal deformities (due to excessive and abnormal erythroid proliferation), weight loss and a hypermetabolic state. Regular blood transfusions are required to reverse the above mentioned abnormalities. However, ineffective erythropoiesis combined with repeated blood transfusions leads to iron overload. This is the most frequent cause of death (cardiac death--see under iron-storage disorders), usually occurring during the second decade of life. Attempts are being made to prolong the life of these patients by using suitable iron chelators (already discussed). Recall that in this situation removal of iron by phlebotomy defeats the primary purpose of blood transfusion and, thus, iron chelators alone are important in reducing the toxicity due to iron accumulation. As seen in α-thalassemia, the unused chains in the red blood cells are mainly responsible for initiating abnormalities seen in homozygous β-thalassemia. The uncomplexed α-chains are highly unstable and are precipitated, initiating the ineffective erythropoiesis.

c. In β-thalassemia, HbF synthesis persists beyond the neonatal period and HbF is distributed heterogeneously among the red cells (this means that some red cells contain HbF while others are devoid of HbF). The cells that contain HbF have a longer survival time. Therefore, if it were possible to stimulate γ-chain synthesis <u>in vivo</u>, the clinical severity of homozygous β-thalassemia could be ameliorated. The marked stimulation of erythropoietin in this disorder is due to the anemia and tissue anoxia produced as a result of HbF (present in longer-surviving red blood cells), which has a higher oxygen affinity.

d. Quantitation of globin-chain synthesis for the prenatal
diagnosis of α- and β-thalassemias has been accomplished in
fetuses of less than 20 weeks gestation by using techniques
in the fields of obstetrics and molecular biology.

4. <u>Hereditary Persistence of Fetal Hemoglobin (HPFH)</u>

a. In order to understand HPFH disorders, the following aspects
of HbF are summarized.

    i) It is predominantly a hemoglobin of fetal erythrocytes
and is composed of two α- and two γ-polypeptide chains
($\alpha_2\gamma_2$). The γ-chains of HbF and the β-chains of HbA
(the predominant hemoglobin of adult cells) both contain
146 amino acid residues and are similar in primary
structure, differing at only 39 residue positions.
However, HbF and HbA differ in certain physical and
chemical properties, as well as immunological
specificity. HbF and A can be separated by
electrophoretic or chromatographic procedures. HbF is
alkali resistant and has a lower affinity for 2,3-
diphosphoglycerate.

    ii) The amount of HbF present in adult red blood cells is
less than 2% and is distributed <u>heterogeneously</u> among
the red cells (that is, uneven distribution of HbF--
discussed later).

    iii) As noted earlier, there is structural heterogeneity in
the γ-chains; glycine or alanine is found at residue 136.
Thus man has two or more genetic loci, specifying
γ-chains.

b. HPFH conditions are characterized by persistent HbF production
beyond the neonatal period and the absence of any major
hematologic abnormalities. These disorders may be considered
as mild forms of β-thalassemia. Although there is a deficiency
of β-chains (due to β-gene deletion and probable deletion of
the δ-gene), there is minimal chain imbalance due to almost
complete compensation by the increased synthesis of γ-chains
to yield HbF. Homozygous $\delta\beta^0$-thalassemia and homozygous HPFH
both lack β- and δ-structural genes, yet the former is a
severe disorder while the latter is very mild. The molecular
basis for these differences is not yet known. It has been
speculated that the severity of $\delta\beta^0$-thalassemia is probably
due to the deletion of the genetic material extending beyond
the δβ-region to the γ-gene region, thus reducing the production
of γ-chains. Note that the presence of γ-chains markedly
diminishes the clinical severity of the disorder (as in the
case of HPFH). Stimulation of γ-chain production in red blood

cells of patients with homozygous β-thalassemia has <u>potential</u> therapeutic value by decreasing the excess α-chains in these cells. It is of interest to point out that sickle-cell anemia in Saudi Arabians presents itself as an unusually mild disease (while in Blacks this is a severe disease--discussed later) because these patients' red blood cells contain 10-30% HbF.

c. A great deal of heterogeneity is found in HPFH. The heterozygous individuals do not show abnormalities in red cell morphology or in any other red cell indices (this is in contrast with the thalassemic traits which are characterized by hypochromic microcytosis). Distribution of HbF among erythrocytes has been used to distinguish between HPFH and other disorders. In some forms of HPFH the HbF is distributed equally in all of the erythrocytes and is known as a homogeneous pattern. In δβ-thalassemia, however, the HbF is heterogeneously distributed (that is, a few erythrocytes contain most of the HbF). It should be pointed out, however, that there are some HPFH conditions (of European and English origin) without any evidence of thalassemia that show a <u>heterogeneous</u> distribution of HbF among the red blood cells. Thus, the homogeneous distribution of HbF among erythrocytes is no longer the hallmark of HPFH. The commonly observed HPFH in Blacks is subdivided into three groups based upon the ratio of the two types of γ-chains: $^G\gamma^A\gamma$ (2:3), $^G\gamma^A\gamma$ (1:10), $^G\gamma$. The heterozygotes in the above disorders have 15-35% HbF which is evenly distributed in the erythrocytes (homogeneous pattern).

d. The procedure (Kleihauer technique) for assessing the distribution of HbF among red blood cells is based on the differential resistance of HbF and HbA to elution under mild acidic conditions (acid-phosphate buffer pH 3.3-3.5). The procedure consists of initially precipitating hemoglobin in the red blood cells by fixing and drying the blood smear with alcohol followed by exposing the cells to an acidic buffer solution. Under these conditions, HbA is solubilized and is eluted from the cell. The blood smear is stained (hematoxylin-eosin) and examined under the microscope for HbF distribution. The cells which retain the HbF are stained pink.

e. HbF is elevated in the adult red blood cells under a variety of abnormal conditions in addition to HPFH and δβ-thalassemia syndromes. These are presented in Table 31. In some pregnant women, the rise and fall in HbF has been observed.

Table 31.   Conditions Associated with Alteration of Fetal Hemoglobin Levels*

1. Anemias
    Aplastic anemia (congenital and acquired)
    Pernicious anemia
    Hereditary spherocytosis
    Hereditary elliptocytosis
    Congenital nonspherocytic hemolytic anemia
    Anemia of chronic infection
    Anemia of blood loss
    Erythropoietic porphyria
    Paroxysmal nocturnal hemoglobinuria
2. Myelofibrosis
3. Postirradiation fibrosis of marrow
4. Malignancies involving marrow
    Tumor metastasis
    Leukemia (especially those with an erythroleukemia component)
    Histiocytosis
    Plasmacytic myeloma
5. Hemoglobinopathies
    Quantitative (thalassemias, major and minor)
    Qualitative
    HbS
    HbC
    HbD
    HbE
    Hb Lepore
    Hb Pylos (Lepore Boston)
    Mixed syndromes
6. Hereditary persistence of fetal hemoglobin
7. Chromosomal abnormalities
    D_1 – trisomy
    D_1 – trisomy with associated red cell carbonic anhydrase deficiency
    D/D translocation with D_1 – trisomy
    C/D translocation
8. Syndrome of high HbF, low HbA_2, and carbonic anhydrase deficiency
9. Thyrotoxicosis with erythrocyte carbonic anhydrase deficiency
10. Bronchial asthma
11. Familial HbF elevation

---

* Reproduced from Cooper, H.A. and Hoagland, H.C. Fetal Hemoglobin,
Mayo Clin Proc 47(6):409, 1972.

Table 32. Some Laboratory Parameters of Various Hypochromic, Microcytic Anemias

	Serum			Hemoglobin			Other Distinguishing Parameters
	Fe	TIBC	% Saturation	A2	F	Other	
Iron deficiency	↓	↑	↓	→	→ or ↑		Sideroblasts in the bone marrow decreased. Administration of iron corrects anemia
Thalassemia traits							
1. α-thalassemia							
a. α-thal trait	→	→	→	→	→		Require family study and α/β chain synthesis ratio
b. Hemoglobin H disease	→	→	→	→	→	5–30% HbH*	"
2. β-thalassemia							
a. β-thal trait	→	→	→	↑**	→ or ↑		"
b. δβ-thal trait	→	→	→	↓	↑		"
Sideroblastic anemia	→ or ↑	→	→ or ↑	→	→ or ↑		Presence of ringed sideroblasts in the bone marrow
Anemia of chronic disease	↓	↓	↓	→	→ or ↑		

→ = normal   ↑ = increased   ↓ = decreased

* HbH is identified by its rapid migration on electrophoresis at pH 8.6 and formation of inclusion bodies in the erythrocytes when incubated with brilliant cresyl blue.

** Existence of iron deficiency in the thalassemic states may mask the expected hemoglobin pattern.

Hemoglobinopathies Caused by Abnormal Subunits

As was indicated previously, these can be divided into two groups

   a. point mutations, involving substitution of one amino acid
       residue for another;

   b. addition and deletion mutants, which have, respectively, more
       or fewer residues than the normal chain from which they are
       derived.

1. Table 33 summarizes the clinical and molecular characteristics
    of a number of point-mutated hemoglobins. The mechanisms by which
    the codon change could occur were discussed in the chapter on
    nucleic acids. Note that the mRNA codon is listed in the table
    rather than the DNA sequence. Mutatory influences function at
    the DNA level.

2. Mutations of this sort (single residue mutations) are the most
    common and important ones. The occurrence or severity of a
    clinical disorder associated with an amino acid substitution
    generally can be said to depend on the location of the residue in
    the chain, the most critical regions seeming to be those near the
    heme group or ones involved in $\alpha_1\beta_1$ and $\alpha_1\beta_2$ contacts.

   a. Mutants which involve the heme contacts or any other aspect of
       the heme pocket can cause weak or absent heme binding, resulting
       in loss of the heme group from the Hb (Hb Sydney, Hb Hammersmith);
       or stabilization of the heme iron in the ferric (Fe^{+3}) state, so
       that O_2 cannot be reversibly bound (the hemoglobins M).
       Methemoglobin (Hb M) also loses its heme group more readily
       than normal Hb.

   b. Mutants in the $\alpha_1\beta_1$ contacts are known which weaken the contacts
       and thereby increase the subunit dissociation (Hb E), and which
       alter the oxygen affinity, either increasing it (Hb Tacoma) or
       decreasing it (Hb Yoshizuka; Hb E).

   c. The $\alpha_1\beta_2$ contacts are involved in combining the $\alpha\beta$ dimers to
       form tetramers and these are the contacts which undergo the
       greatest change in the quaternary structure ($T \rightleftarrows R$)
       transformation. Some of these show increased (Hb Chesapeake)
       or decreased (Hb Kansas) oxygen affinity. Others, such as
       Hb-G-Georgia, completely dissociate into dimers upon
       oxygenation, then reassociate to tetramers when oxygen is
       removed. The degree of cooperativity (heme-heme interaction)
       is decreased in most of the $\alpha_1\beta_2$-contact mutants, presumably
       due to the tendency to form dimers upon oxygenation.
       (Text continues on p. 747)

Table 33. Molecular and Clinical Aspects of Some Abnormal Hemoglobins

Entries in this table are all mentioned in the text. This is not intended to be anywhere near a complete listing of all hemoglobin variants which are known.

Hemoglobin	Chain and Position*	Amino Acid Change	mRNA Codon Alteration**	Remarks
I-Philadelphia	α16(A14)	Lys → Glu	AA(A,G) → GA(A,G)	Increased alkali resistance
G-Honolulu	α30(B11)	Glu → Gln	GA(A,G) → CA(A,G)	Also called G-Chinese; benign
Chesapeake	α92(FG4)	Arg → Leu	CGZ → CUZ	Increased O_2 affinity; $\alpha_1\beta_2$-contact mutant
G-Georgia	α95(G2)	Pro → Leu	CCZ → CUZ	Oxy-form dissoc. to dimers; incr. O_2 aff.; $\alpha_1\beta_2$-contact mutant
S	β6(A3)	Glu → Val	GA(A,G) → GU(A,G)	See sickle-cell anemia
G-Makassar	β6(A3)	Glu → Ala	GA(A,G) → GC(A,G)	Compare to HbS; benign
C	β6(A3)	Glu → Lys	GA(A,G) → AA(A,G)	Xtalliz. in RBC; mild hemolytic anemia; compare to HbS
C-Harlem	β6(A3) β73(E17)	Glu → Val Asp → Asn	GA(A,G) → GU(A,G) GA(U,C) → AA(U,C)	Double mutant; see HbS and Hb Korle-Bu
G-San Jose	β7(A4)	Glu → Gly	GA(A,G) → GG(A,G)	Compare to HbS; benign

Name	Position	Substitution	Codon change	Notes
E	β26(B8)	Glu → Lys	GA(A,G) → AA(A,G)	Mild hemolytic anemia; incr. subunit dissoc.; decr. O_2 affinity; $\alpha_1\beta_1$-contact mutant
Genova	β28(B10)	Leu → Pro	CUZ → CCZ	B helix disrupted; unstable
Tacoma	β30(B12)	Arg → Ser	AG(A,G) → AG(U,C)	Increased O_2 aff.; unstable; $\alpha_1\beta_1$-contact mutant
Hammersmith	β42(CD1)	Phe → Ser	UU(U,C) → UC(U,C)	Decr. O_2 aff.; cyanosis; poor heme binding; inclusion bodies; unstable
Zürich	β63(E7)	His → Arg	CA(U,C) → CG(U,C)	Compare to Hb-M-Saskatoon; incr. O_2 aff.; sulfa drugs cause hemolysis
Sydney	β67(E11)	Val → Ala	GUZ → GCZ	Compare to Hb Hammersmith; inclusion bodies, unstable; cyanosis; poor heme binding
Korle Bu	β73(E17)	Asp → Asn	GA(U,C) → AA(U,C)	Compare to Hb-C-Harlem; benign
Köln	β98(FG5)	Val → Met	GUG → AUG	Unstable; incr. O_2 aff.; $\alpha_1\beta_2$-contact mutant

[746]

Table 33 (continued)

Hemoglobin	Chain and Position*	Amino Acid Change	mRNA Codon Alteration**	Remarks
Kansas	β102(G2)	Asn → Thr	AA(U,C) → AC(U,C)	Unstable; cyanosis; incr. O₂ aff.; α₁β₂-contact mutant
Yoshizuka	β108(G9)	Asn → Asp	AA(U,C) → GA(U,C)	Unstable; incr. O₂ aff.; α₁β₁-contact mutant
D-Los Angeles	β121(GH4)	Glu → Gln	GA(A,G) → CA(A,G)	Benign; can occur with HbS; also called D-Punjab
M-Boston	α58(E7)	His → Tyr	CA(U,C) → UA(U,C)	
M-Iwate	α87(F8)	His → Tyr	CA(U,C) → UA(U,C)	
M-Saskatoon	β63(E7)	His → Tyr	CA(U,C) → UA(U,C)	Methemoglobins; cause cyanosis
M-Hyde Park	β92(F8)	His → Tyr	CA(U,C) → UA(U,C)	
M-Freiburg	β23(B5)	Val → (deleted)	GUZ GUZ → ---	
M-Milwaukee	β67(E11)	Val → Glu	GU(Z,G) → GA(A,G)	

* Kendrew's helical notation is given in parentheses.

** Because of redundancy in the genetic code, the observed amino acid in the normal and mutant peptides can be coded for by more than one codon (the only exception is the change in Hb Köln, since Met has only one codon). For example, Glu can be coded for by GAA or GAG. When it is replaced by Val (as in HbS), the valine can be coded for by GUA or GUG where the mutation was A→U in the second base of the codon. A Z occurring in a codon (such as CGZ for Arg in Hb Chesapeake) means that the third letter of the codon can be any one of the four bases A, U, C, or G.

Glossary: benign means no observable symptoms.
dissoc. means dissociates.
aff. means affinity.
xtalliz. means crystallization.
incr., decr. mean increased or decreased.

d.   A great number of mutations lead to hemolytic anemia (increased rate of lysis (shortening of the lifespan of erythrocytes)). These are called <u>unstable hemoglobin variants</u> and are usually associated with some change which decreases the solubility of the hemoglobin in the red cell. The lowered solubility results in intraerythrocytic precipitation of hemoglobin (Heinz body formation; Hb appears as intracellular inclusion bodies) which apparently triggers any of several lytic processes. Other factors, such as increased interaction between the sulfhydryl groups of Hb and membrane proteins resulting in weakening of the cell membrane, may also cause increased hemolysis.

Sickle-cell hemoglobin (Hb S) is the best-known example of this and was the first mutant hemoglobin to be described in molecular detail (by L. Pauling, H. Itano, S. Singer, and I. Wells, in 1949). Hb S is discussed more fully later on. Since the presence of heme greatly stabilizes the hemoglobin, mutations which result in impaired heme binding lead to hemolysis due to precipitation of the heme-free globin chains. Similarly, α-subunits by themselves tend to precipitate, so that mutations which decrease intersubunit binding can also result in lytic anemias. This probably also occurs in β-thalassemia. Deletion mutants (Hb Gun Hill, and others; discussed later), which distort the chain conformation have a similar effect. Hemoglobin Köln is an $\alpha_1\beta_2$-contact mutant with an increased positive charge relative to normal Hb. It is unstable and shows increased O_2 affinity. Disruption of a helical region not in contact with the heme group can also apparently distort the molecule sufficiently to cause major changes in its properties. For example, in Hb Genova, a proline is inserted into the B helix in place of a leucine, causing hemolytic anemia even in heterozygotes.

e.   Most hemoglobin mutants are very rare, since they tend to confer negative adaptive advantage to the bearer and his progeny. Exceptions to this are Hb S, Hb C, Hb-D-Los Angeles, Hb E, and some of the thalassemias, which occur in certain populations with a relatively high frequency. Since the appearance of even one mutation is quite uncommon, the occurrence of <u>two</u> abnormalities in one individual is exceedingly rare in most cases. The only known instance of two amino acid substitutions in one chain is Hb-C-Harlem. Since one change is the same as that in Hb S and the other change is the one occurring in Hb Korle-Bu, the double mutant may have resulted from a second point mutation in a carrier of either allele, but most likely in a Hb S heterozygote. It could also have occurred by homologous crossing over in a person heterozygous for both Hb Korle-Bu and Hb S. It is significant that one of the mutations (Hb S) is known to have a high frequency of occurrence. Another double mutant (Hb-C-Georgetown) has been reported but it may be identical to Hb-C-Harlem.

f. Much more common than the doubly mutant hemoglobins is the occurrence of heterozygotes for two abnormal alleles. Examples of this include Hb S-Hb D disease (similar to sickle-cell trait; mild hemolytic anemia) and Hb S-Hb C disease. It is also not uncommon for an abnormal globin chain to be inherited along with the gene for a β-thalassemia. Reports have appeared of Hb S, Hb C, Hb D, Hb E, Hb J, and Hb G occurring in such conjunction. It is important to note that, in general, these "double inborn errors" only happen when one or both of the separate defects occurs with high frequency. In Hb S-Hb C disease, it is significant that not only are both alleles fairly common, but they both appear primarily in the same population.

g. A number of mutants are known which are either completely benign, even in the homozygote, or which result in very mild symptoms of little or no clinical significance. Examples of this include Hb San Jose, Hb-G-Honolulu (also called Hb-G-Chinese; an $\alpha_1\beta_1$-contact mutant), Hb Korle-Bu, and Hb-D-Los Angeles (also called Hb-D-Punjab).

h. Seemingly minor variations between mutations can result in very different clinical and molecular pictures. For example, sickle-cell anemia is a serious disease, brought about by the mutation β6(Glu → Val), while Hb San Jose, which is β7(Glu → Gly), is a harmless variant. Also compare Hb Zürich (α63 His → Arg) and Hb-M-Saskatoon (α63 His → Tyr); and Hb S, Hb C (β6 Glu → Lys), and Hb-G-Makassar (β6 Glu → Ala).

i. Throughout the preceding discussion, reference has been occasionally made to homozygotes for a mutation versus heterozygotes. Persons heterozygous for a trait are usually termed carriers since they frequently display few if any of the symptoms of the homozygous disease. This is probably due to the presence of the normal gene on the other chromosome which can supply normal gene products for use by the cell. (This is similar to complementation in bacterial genetics.) If one carrier has offspring by another carrier however, the children are likely to be homozygous for the trait and have the characteristic symptoms. In some cases, the homozygous state is lethal neonatally and only the heterozygote is ever seen.

3. One of the most common, most severe, and best **studied hemoglobin** mutants is Hb S, which causes sickle-cell anemia (or disease) in homozygotes and **sickle-cell trait** in heterozygotes. The name arises from the characteristic change of shape of the patient's erythrocytes (from the normal biconcave disc to a curved, sickle-like morphology) under anaerobic conditions. Possessors of the trait appear to be largely asymptomatic, although there are some reports of death occurring among carriers when they are subjected to severe tissue hypoxia (high altitude, strenuous exercise). The oxygen pressure probably must be below about 10 torr before sickling occurs in heterozygotes. This could occur during the administration of an anesthetic or during **recovery from anesthesia.**

    a. As indicated in Table 33, the mutation is a replacement of Glu (a polar amino acid) by Val (a non-polar residue) at position six of the beta chains. The substitution decreases the net negative charge (at pH 8.6) of the tetramer by two. Exactly how this causes the aggregation is not clear despite active speculation. One theory is that the substitution of a neutral residue for a charged residue permits a hydrophobic interaction between tetramers which results in a non-covalent polymerization. Oxy-Hb S is not affected but the solubility of deoxy-Hb S is greatly decreased as indicated below.

    $$\frac{\text{Solubility of oxy-Hb A}}{\text{Solubility of deoxy-Hb A}} = 2; \qquad \frac{\text{Solubility of oxy-Hb S}}{\text{Solubility of deoxy-Hb S}} = 50$$

    This is explained by the need for a binding site which is present in deoxy-Hb S but absent in oxy-Hb S. The correctness of this hypothesis will not be known until the 3-dimensional structure of deoxy-Hb S is elucidated.

    As a consequence of this change, upon deoxygenation Hb S precipitates within the cell forming a semisolid gel. Under the microscope, tactoids (a type of liquid crystal) are visible, approximately 1-15 microns long. This precipitation causes the characteristic shape-change of the erythrocytes. The mildness of the heterozygous condition can be explained by the presence in the erythrocytes of roughly a 1:1 mixture of Hb S and normal Hb A. Thus, it requires much more complete deoxygenation to form enough solid to cause sickling. The homozygote has only Hb S within his erythrocytes.

    b. As implied above, almost all of the clinical findings in sickle-cell anemia can be explained by the precipitation and cell shape-change. The sickled cells are recognized as abnormal by the reticuloendothelial system and phagocytized by the RE cells, resulting in a hemolytic anemia. The "crises" which occur in homozygotes (and perhaps occasionally in heterozygotes) are caused by the inability of sickled cells to pass through the finest capillaries. Once a capillary is blocked and the blood flow through it ceases, deoxygenation of the static erythrocytes increases, more cells sickle, and the blockage becomes larger. An infarct of this sort can occur anywhere within the body. The normal circulatory rate does not keep cells deoxygenated for the 15 seconds required for sickling to begin however, and crises occur only when circulation slows or hypoxia is present.

As noted earlier, sickle-cell anemia in Saudi Arabians is a relatively mild disorder. The red blood cells of these patients contain 10-30% HbF which is presumed to be responsible for ameliorating the severity of the disease (see thalassemia syndromes).

c. The sickle-cell allele is transmitted as an autosomal recessive characteristic. The trait is present in about 8% of the Afro-American population and to a much greater extent in some Black African populations. The homozygous condition causes about 60,000-80,000 deaths per year among African children and incapacitates many more. Hb S also occurs in India (primarily among primitive tribes) and in the Mediterranean area. Although there is some disagreement, it seems probable that the gene has remained in these populations because of resistance to the type of malaria caused by the parasite Plasmodium falciparum which heterozygosity confers. This "immunity" is present even in young children (less than two years) who have not developed more conventional forms of resistance and to whom the disease is quite lethal. A possible mechanism for this resistance involves infestation by P. falciparum and growth of the parasite within the red cells. Erythrocytes so infected tend to adhere to the vessel walls where they become deoxygenated, sickled, and phagocytized by the RE cells. Another explanation is that for some reason, P. falciparum does not thrive on Hb S. This appears unlikely since sickle cells are as capable as normal cells of supporting P. falciparum in culture.

d. Recent attempts at treating sickle-cell disease with urea and cyanate (CNO⁻) have been highly controversial. Particularly regarding the use of urea, published results have been contradictory. Urea, being a well-known protein denaturant, might be able to decrease the presumed hydrophobic interactions which maintain deoxy Hb S in the semisolid state. However,

the in vivo levels which appear to be necessary to produce such
an effect are difficult to attain clinically and produce
severe diuresis. It has also been reported that certain
African tribes imbibe cow's urine (presumably high in urea) to
ameliorate sickle cell disease. Cyanate, through its ability
to carbamylate free amino groups, could also decrease this
sort of interaction. Urea in solution is known to be in
equilibrium with small amounts of cyanate

$$H_2N\text{-}\overset{\overset{\textstyle O}{\|}}{C}\text{-}NH_2 \;\Longleftrightarrow\; NH_4^{\oplus} + N\equiv C\text{-}O^{(-)},$$

raising the possibility that the two substances act by the
same mechanism.

Alternatively, Perutz's x-ray work has demonstrated that two
of the six salt bridges needed to maintain the deoxy-conformation
are between the α-carboxyl of Arg α141(HC3) and the α-amino
group of Val α1(NA1) on the other α-subunit. Any α-amino group
carbamylation should decrease the stability of the deoxy- form,
thereby increasing O_2-affinity and making it more difficult to
deoxygenate the cells. Carbamylation of these N-terminal
valines has been shown to occur with oral and intravenous
cyanate administration and the oxygenation curve is shifted to
the right. Studies are currently being conducted to determine
the usefulness of cyanate therapy in the chronic treatment of
sickle-cell anemia. The N-terminal carbamylation also prevents
the formation of carbamino compounds (discussed later) but
does not cause any significant drop in pH which might be expected
to result from an increase in free CO_2. This appears to be
caused by the buffering provided by a slight increase in plasma
bicarbonate which also occurs. There are still many problems
to be solved and much yet to be understood about this disease.

A new class of compounds, tri- or tetrapeptides, has been
found to inhibit the aggregation or the gelation of HbS, in
in vitro experiments performed in reconstituted erythrocytes
containing the inhibitors and HbS. The normal cells are
impermeable to these inhibitors. To be effective, these
peptides require a hydrophobic amino acid at one end and
one with a sidechain capable of hydrogen bond formation at
the other end (e.g., L-phenylalanine-L-phenylalanine-L-
arginine). Presumably, these small molecules bind to HbS
at specific sites and prevent the HbS aggregation, the
primary cause of sickling. In order for this class of
compounds to be therapeutically useful, it needs to be
modified to pass through the erythrocyte membrane.

e. Hemoglobin C is present primarily in American Blacks and in inhabitants of Accra in Ghana, West Africa. The frequency of Hb C and Hb S and their occurrence in the same population probably contributes to the frequency of Hb S-Hb C heterozygotes (mentioned previously). Hb C-Hb A heterozygotes are asymptomatic, but Hb C homozygotes show a decreased erythrocyte lifespan and splenomegaly. The Hb C mutation involves β6(A3) Glu → Lys and shows a decreased Hb solubility (not nearly as severe as in Hb S disease) which results in the appearance of Hb crystals in blood smears from these patients.

4. Although α- and β-mutants have been discussed so far and are much more widely known, γ- and δ-chain mutants have also been found. These are much harder to study because of poorer availability of Hb F and Hb A$_2$. Another reason that such abnormalities are rarely seen is because these are minor hemoglobins postnatally. Consequently, the occurrence of overt clinical symptoms associated with γ- and δ-chain variants is rare or non-existent. At least six δ-chain mutants are known, all of which are stable. None involve residues which contact the heme or are involved in interchain contacts. In addition to the known heterogeneity of fetal Hb (G$_γ$-and A$_γ$-chains, mentioned earlier), six mutant γ-sequences have been determined (involving both A-and G-chains) and several others have been reported but not sequenced.

5. Deletion and addition mutants of hemoglobin are known. They are listed in Table 34. Note that since Hb-M-Freiburg is one of the methemoglobins, it is also listed in Table 34. Mutants of these types can result from unequal crossing over within the same gene producing a deletion or between different genes producing either a deletion or addition; mutation of a termination codon to one which codes for an amino acid (causing an increase in chain length); and perhaps in other ways. Hb Constant Spring (an α-chain extension mutant) probably resulted from the mutation

$$\text{UA(A,G)} \quad\text{———}\quad \text{CA(A,G)}$$
$$\text{Term} \qquad\qquad\qquad \text{Gln}$$

since residue α142 (the first one after the normal C-terminus) is Gln. Recall that "Term" designates a termination codon (see Chapter 5, Table 18). It could also be the product of a cross-over. It is difficult to explain Hb Tak in this way, however, because residue β147 (the first new one added to the normal C-terminus) is Thr. None of the codons for Thr can be obtained from the three usual Term codons by a single base change.

$$\text{Term codons} \begin{cases} \text{UAA} \\ \text{UAG} \\ \text{UGA} \end{cases} \qquad \left. \begin{array}{l} \text{ACU} \\ \text{ACC} \\ \text{ACA} \\ \text{ACG} \end{array} \right\} \text{Thr codons}$$

Table 34. Deletion and Addition Mutants of Hemoglobin

Hemoglobin	Mutation	Remarks
Hb-Gun Hill	Deletion of residues 93-97 in the β-chain	Deletion near the heme position; no heme binding to β-chains
Hb-Leiden	Deletion of residue 6 or 7 (both Glu) in the β-chain	No clear clinical symptoms; mildly abnormal red cell morphology
Hb-Tochigi	Deletion of residues 56-59 in the β-chain	Unstable hemoglobin
Hb-M-Freiburg	Deletion of residue 23 in the β-chain	One of the methemoglobins, discussed later
Hb-Constant Spring	Addition of 31 residues to the C-terminus of the α-chain	Frequently inherited with α-thalassemia; occurs mostly in persons of Cantonese extraction; symptomless by itself
Hb-Tak	Addition of 10 residues to the C-terminus of the β-chain	Asymptomatic

6. In the chapter on carbohydrate metabolism, the sensitivity to
    oxidizing substances, especially primaquine, of persons lacking red
    cell glucose-6-phosphate dehydrogenase (G6PD) was discussed. In
    those patients, the drugs can trigger hemolytic episodes because
    of a lack of reducing power (i.e., glutathione) in the erythrocytes.
    Many, perhaps all, of the unstable Hb variants are similarly
    unstable to oxidants. In particular, sulfonamides precipitate
    hemolytic episodes in persons with Hb Zürich. The α-thalassemia
    known as hemoglobin-H disease confers sensitivity to sulfisoxazole,
    another drug which affects persons deficient in G6PD. These and
    other situations in which an inherited characteristic causes
    persons having the characteristic to respond adversely to a
    medication are part of the field of pharmacogenetics (see also
    the chapter on drug metabolism).

Laboratory Evaluation of Hemoglobinopathies

The differential diagnosis of the hemoglobinopathies is made by using
a number of laboratory procedures. Some of the methods are described
below.

1. <u>Electrophoretic Procedures</u>: Most of the common hemoglobin mutants
    can be identified by carrying out electrophoretic separation of
    the hemoglobins on cellulose acetate at pH 8.4-8.6
    (tris-ethylenediaminetetraacetic acid-boric acid buffer; ionic
    strength 0.01-0.1 M) and citrate agar at pH 6.0-6.2 (0.005 M
    sodium citrate buffer). At pH 8.4-8.6 with cellulose acetate as
    the support medium, the hemoglobins are negatively charged and
    migrate toward the anode to varying degrees in an electric field,
    due to their charge differences (see Figure 87 and Table 35).
    The separation of various hemoglobins with citrate agar (1% agar)
    as the support medium is based on the electrophoretic charge and
    on adsorption (which may be related to hemoglobin solubility) of
    the hemoglobin to the agar (or some impurity contained in the
    agar). The patterns obtained from citrate agar with various
    hemoglobin mutants are shown in Figures 88-90.

The following aspects should be noted concerning the electrophoretic
patterns of the hemoglobins on cellulose acetate and citrate agar.

a. HbS, D and G migrate about half-way between A and A_2 (or C, E)
    upon cellulose acetate electrophoresis. However, with citrate
    agar electrophoresis, HbS can be distinguished from D and G,
    both of which move along with HbA.

b. Similarly, HbC, which migrates with A_2, E, O and C(Harlem)
    upon cellulose acetate electrophoresis, can be confirmed by
    citrate agar electrophoresis.

(Text continues on p. 758)

Figure 87. Relative Mobilities of Some Hemoglobins on Cellulose Acetate (pH 8.4)

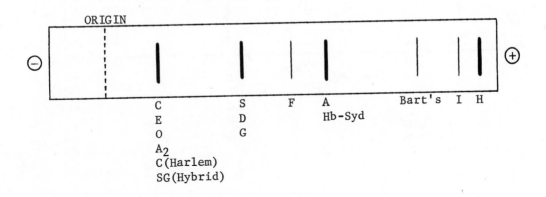

Table 35. Amino Acid Substitutions and Net Charge Alterations in Some Hemoglobins Whose Relative Mobilities are Shown in Figure 87

Hemoglobin	Amino Acid Change*	Charge Alteration in Tetramer*
C	Glu → Lys	+4
E	Glu → Lys	+4
C(Harlem)	Glu → Val	+4
	Asp → Asn	
S	Glu → Val	+2
D-Los Angeles	Glu → Gln	+2
Sydney	Val → Ala	0
I-Philadelphia	Lys → Glu	-4

* As compared to HbA.

Figure 88. Relative Mobilities of Some Hemoglobins on Citrate Agar (pH 6.0)

ORIGIN
+
-
C
S
C (Harlem)
O Arab
A
D
E
G
H
Bart's
F

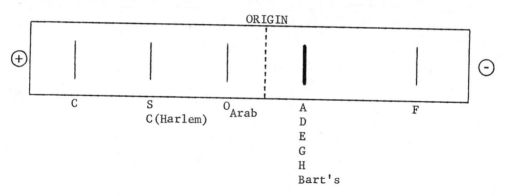

Figure 89. Relative Mobilities of α-Chain Mutants on Citrate Agar Electrophoresis*

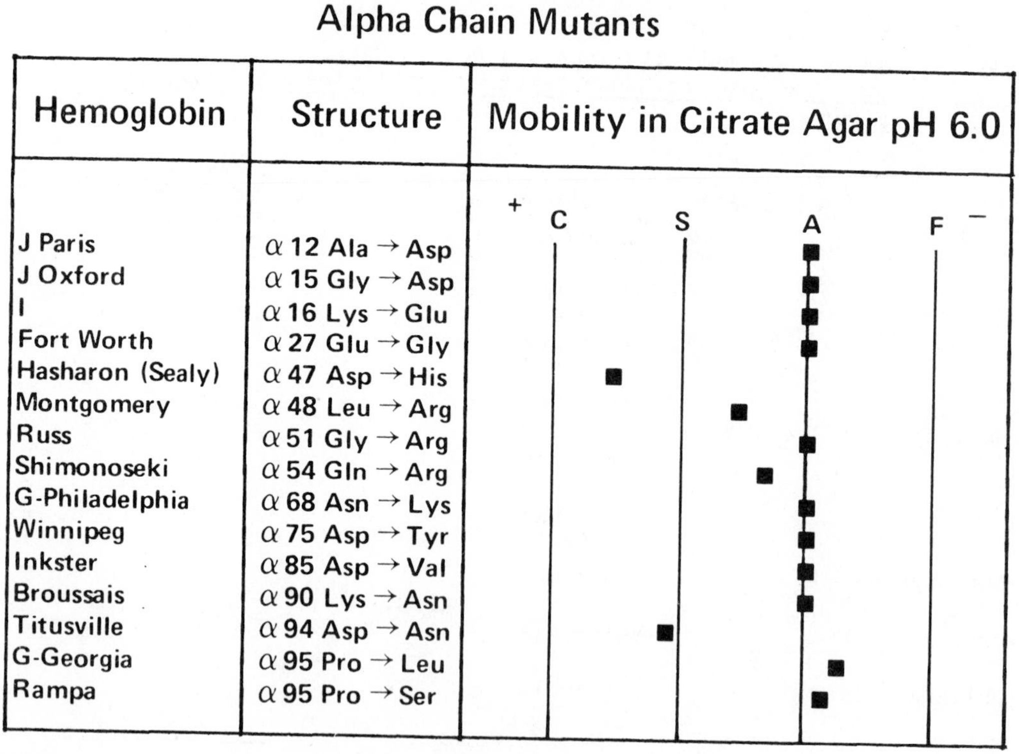

Alpha Chain Mutants
Hemoglobin
Structure
Mobility in Citrate Agar pH 6.0
+ C S A F −
J Paris α 12 Ala → Asp
J Oxford α 15 Gly → Asp
I α 16 Lys → Glu
Fort Worth α 27 Glu → Gly
Hasharon (Sealy) α 47 Asp → His
Montgomery α 48 Leu → Arg
Russ α 51 Gly → Arg
Shimonoseki α 54 Gln → Arg
G-Philadelphia α 68 Asn → Lys
Winnipeg α 75 Asp → Tyr
Inkster α 85 Asp → Val
Broussais α 90 Lys → Asn
Titusville α 94 Asp → Asn
G-Georgia α 95 Pro → Leu
Rampa α 95 Pro → Ser

*Courtesy of Dr. Rose Schnieder, UTMB Galveston, Texas. Reproduced with permission from Helena Laboratories, Beaumont, Texas.

Figure 90. Relative Mobilities of β-Chain Mutants on Citrate Agar Electrophoresis*

Beta Chain Mutants

Hemoglobin	Structure	Mobility in Citrate Agar pH 6.0
S	β 6 Glu → Val	
C	β 6 Glu → Lys	
C Harlem	β 6 Glu → Val β 73 Asp → Asn	
G-San Jose	β 7 Glu → Gly	
J-Baltimore	β 16 Gly → Asp	
G-Coushatta	β 22 Glu → Ala	
E	β 26 Glu → Lys	
Alabama	β 39 Gln → Lys	
G-Galveston	β 43 Glu → Ala	
Williamette	β 51 Pro → Arg	
Osu Christiansborg	β 52 Asp → Asn	
N-Seattle	β 61 Lys → Glu	
Korle Bu	β 73 Asp → Asn	
Mobile	β 73 Asp → Val	
D-Ibadan	β 87 Thr → Lys	
Gun Hill	β 91-95 deleted	
N-Baltimore	β 95 Lys → Glu	
Malmo	β 97 His → Gln	
Koln	β 98 Val → Met	
Kempsey	β 99 Asp → Asn	
Richmond	β 102 Asn → Lys	
Burke	β 107 Gly → Arg	
P	β 117 His → Arg	
D L.A. (Punjab)	β 121 Glu → Gln	
O-Arab	β 121 Glu → Lys	
Camden	β 131 Gln → Glu	
Deaconess	β 131 Gln → 0	
K-Woolwich	β 132 Lys → Gln	
Hope	β 136 Gly → Asp	
Bethesda	β 145 Tyr → His	
Cochin Port Royal	β 146 His → Arg	

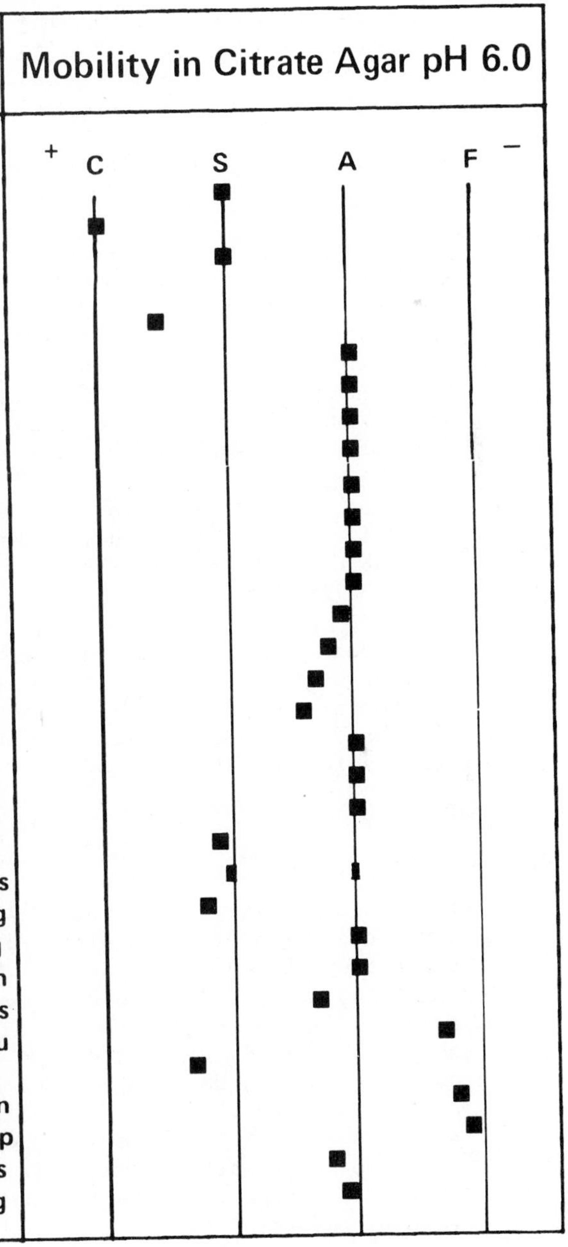

Mobility axis labels (left + to right −): C, S, A, F

*Courtesy of Dr. Rose Schnieder, UTMB Galveston, Texas. Reproduced with permission from Helena Laboratories, Beaumont, Texas.

    c. HbC(Harlem) moves like HbC (Glu → Lys) at pH 8.4 on cellulose acetate due to two mutations, both of which result in the loss of negative charges. There is a β6(Glu → Val) substitution (also found in HbS, thus giving HbC(Harlem) some of the properties of HbS) and a β73(Asp → Asn) substitution. However, with citrate agar electrophoresis, HbC(Harlem) is separated from HbC.

2. Two modifications of classical electrophoresis have also proved useful in studying hemoglobins.

    a. <u>Isoelectric focusing</u> (see Chapter 2) is based on the observation that in electrophoresis along a pH gradient, a molecule will migrate to the pH which corresponds to its isoelectric point (pI = the pH at which a zwitterion has zero mobility in an electric field). This method requires a pI difference of 0.02 pH units between two molecules in order to separate them. It is reported to separate minor hemoglobins which are too similar to be resolved by ordinary electrophoresis.

    b. Electrophoresis of globins and individual chains can be used to confirm and identify hemoglobins which are not normally separable in the electrophoretic systems involving complete hemoglobins.

3. It is apparent from the above discussion that many hemoglobins with different amino acid substitutions demonstrate identical electrophoretic mobility. Therefore, methods which rely on differences other than the electrophoretic charge need to be used in establishing the identity of these hemoglobins.

    a. Solubility differences: HbS, as noted earlier, is insoluble in the deoxygenated state (reduced form). This property of insolubility serves as a basis for differentiating HbS from other hemoglobins. The test is carried out in the presence of a reducing agent (sodium dithionite or sodium metabisulfite) and appropriate buffer systems. When HbS is added to such a solution, it becomes turbid. The degree of turbidity can be assessed in a spectrophotometer and related to the amount of hemoglobin S. By using the same reagents (to create low oxygen tension), HbS can be precipitated within the red blood cells. This results in these cells acquiring a characteristic sickle shape which is viewed under a microscope. HbC(Harlem) and Hemoglobin Bart's behave like HbS in the solubility test. Therefore, these two hemoglobins may give rise to false-positive results when testing for HbS.

    b. HbF is identified and quantitated on the basis of its resistance to denaturation by a strong alkali while all other hemoglobins undergo denaturation. The method consists of exposing the hemolysate to an alkaline solution of pH 12.8 for a specified time, precipitating the denatured hemoglobin by adding ammonium sulfate, and measuring the concentration of undenatured Hb (HbF) which is present in the clear supernatant by spectrophotometric procedures. Using acidic conditions, only HbF can be detected in the red blood cells (see thalassemia syndromes) because other hemoglobins are solubilized and eluted from the cell under these conditions.

c. Hemoglobin A_2 can be quantitated by an ion-exchange chromatographic (using DE52-diethylamino-ethylcellulose, anion exchanger) procedure. Separation in this system is due to differences in the interactions of the charged groups of hemoglobin with the charged groups on the ion-exchange resin at different pH values. Recall that HbA_2 measurement is used in the diagnosis of β-thal trait, in which the HbA_2 is elevated to about twice the normal levels (the upper normal limit is about 3.0-3.5%). It should be noted, however, that in this procedure other hemoglobins (e.g., C, E, O, D) are also eluted along with HbA_2.

d. Some hemoglobin mutants (e.g., Hemoglobins E, Zurich, Sydney, Köln, Gun Hill, Seattle, M-Hyde Park, etc.) and homotetramers (Hemoglobin H-β_4) denature readily within the red blood cell to form precipitates known as inclusion bodies. After inducing the formation of the inclusion bodies, staining with a redox dye (brilliant cresyl blue) produces multiple, greenish-blue bodies in red blood cells containing unstable hemoglobins.

e. Another result of the general instability of the abnormal hemoglobins is that they are much more readily denatured by heat than normal ones. Heating a mutant Hb for 30 minutes at $60^{\circ}C$ usually causes complete denaturation, while HbA is hardly precipitated under the same conditions. The sulfhydryl groups of the mutant chains are more exposed than those in normal Hb. They are, therefore, more reactive towards parachlormercuric benzoate (PCMB; see the section on enzyme inhibitors). Treatment with PCMB for several hours precipitates many mutant peptides, whereas it only causes dissociation of HbA.

f. One of the most useful methods which provides information about the primary structure of hemoglobins is <u>peptide mapping</u> or <u>fingerprinting</u>. In this technique, the chains are separated and partially digested by one of several proteolytic processes (see the section on protein sequence determination), yielding smaller oligopeptides. These are then chromatographed in one dimension by standard methods on paper or silica gel. After the chromatographic separation, the peptides are subjected to electrophoresis in the second dimension. The peptides are located by spraying the chromatogram with ninhydrin or some other reagent which reacts with the peptides to give a color. The application of two-dimensional separation enhances the possibility of separating peptides with similar charge or hydrophobic properties. When digested, chromatographed and electrophoresced under specified conditions, a given peptide will always give the same peptide map. By comparison of a known map (e.g., the β-subunits present in HbA) with the map of an unknown variant, the difference in spot pattern (usually amounting to the absence of one of the standard spots in the mutant chromatogram and the appearance of a new spot) can be used to locate the altered residue. In addition, since the composition of the peptides which cause the spots in the standard is known, the specific residue change can be pinpointed. Even when such results seem to clearly indicate the mutation, however, a new hemoglobin variant is not unequivocally characterized until its entire amino acid sequence is known.

4. The growing knowledge of hemoglobin mutations and their importance in the pathology of certain diseases has led to the development of screening programs to detect some of the most common and most harmful alleles. The primary aim of such screening programs is to locate carriers (heterozygotes). For example, there is increasing evidence that heterozygotes for HbS can, under severely hypoxic circumstances, suffer acute sickling crises. Not only will such screening aid these persons in avoiding this danger, but it should provide a broader base for genetic counseling and thereby decrease the number of homozygotes conceived. One such program which has been established makes use of electrophoresis and differentiates between sickle-cell trait, sickle-cell disease, and sickle-cell-HbC disease, as well as detecting patients with thalassemia minor, hereditary persistence of HbF, and hemoglobin types AC and CC. A method has also been reported for the detection of HbS homozygosity in fetuses as early as the first trimester of gestation. A similar screening program for the detection of thalassemia has been proposed. After sickle-cell disease, β-thalassemia is probably the most serious hereditary hemoglobinopathy in the United States.

Derivatives of Hemoglobin

1. Carboxyhemoglobin (carbon monoxide hemoglobin)

    a. Carbon monoxide (CO) is a relatively inert (chemically non-reactive), odorless gas. The affinity of Hb for CO is 210 times as great as the affinity of Hb for O_2. This means that in the equilibrium reaction

$$HbO_2 + CO \xrightleftharpoons{} HbCO + O_2 \qquad K_{eq} = 2.1 \times 10^2$$

the equilibrium lies far to the right. The type of bond between Hb and CO and the binding site, are the same as those for the Hb-oxygen complex. That is, CO binds to the sixth position of the heme iron with a coordinate-covalent bond.

b. Because of the identical binding sites and the greater binding strength of CO, when both CO and O_2 are present in appreciable quantities, the CO is bound preferentially and O_2 is excluded.

c. CO poisoning may occur in the presence of automobile exhaust, poorly oxygenated coal fires in stoves and furnaces, or any other situation where incomplete combustion of a carbon-containing compound occurs.

d. For clinical diagnosis, history, unconsciousness, cherry-red discoloration of the nailbeds and mucous membranes, and spectrophotometric analysis of the blood are used.

e. Treatment for carbon monoxide poisoning consists of purging the body of CO. Although the equilibrium

$$CO + HbO_2 \xrightleftharpoons{} HbCO + O_2$$

has a large equilibrium constant (is predominantly to the right when CO and O_2 are present in similar amounts), a large excess of O_2 will favor the formation of HbO_2 and the release of CO by the law of mass action. Breathing an atmosphere of 95% O_2 and 5% CO_2 will usually eliminate CO from the body in 30-90 minutes.

f. It is interesting that the oxidation of heme to biliverdin produces CO in amounts equimolar to the amount of biliverdin (and, hence, bilirubin) formed (see later, under Hb catabolism). The CO is transported via the blood to the lungs, where it is released. Although no cases are known wherein endogenous CO proved toxic, it has been suggested that this carbon monoxide may contribute significantly to air pollution.

2. Carbaminohemoglobin

a. As was mentioned previously (see Chapter 1 and earlier in this chapter), some of the CO_2 in the blood stream is carried as carbamino compounds. These form spontaneously, in a readily reversible reaction, with the free α-amino groups of the N-terminal amino acids of the Hb chains

Hemoglobin   Carbaminohemoglobin

b. Although this means that Hb can directly carry as many moles of
    CO_2 as it can O_2, the HCO_3^- system seems to be a much more
    important way of transporting CO_2 in the blood. Evidence for
    this comes from the absence of any marked degree of acidosis
    when the N-terminal amino groups are blocked by carbamylation
    with CNO^- (see sickle-cell anemia), forming carbamylhemoglobin.

c. The presence of carbamino groups decreases the affinity of
    Hb for O_2. This is a mechanism whereby the P_{CO_2} can affect
    oxygen affinity independent of any effect which the carbon
    dioxide may have on the pH and it provides an additional
    process which may influence the position of the oxygen
    dissociation curve under various conditions.

3. Methemoglobin (see also the section on red cell metabolism in
    Chapter 4)

    a. Hemoglobin in solution, on standing in the presence of oxygen,
        is slowly oxidized to methemoglobin, a derivative of hemoglobin
        in which the iron is present in the ferric (Fe^{+3}) state. In
        metHb, the ferric ion cannot accept oxygen because the iron is
        bound tightly to a hydroxyl group or to some other anion.
        The heme porphyrin (protoporphyrin III) containing an Fe^{+3} ion
        (instead of Fe^{+2}) is known as a hemin.

    b. In the body, there is a small but constant amount of
        methemoglobin produced. It is re-reduced by specific enzymes
        (methemoglobin reductases). The electrons are furnished
        primarily by NADH which is generated in glycolysis (refer to
        carbohydrate metabolism). The reductases isolated from
        human red cells are also capable of using NADPH, but to a
        lesser extent. Because of this continuous formation of metHb,
        an inability to re-reduce it is pathological. The resultant
        condition is called methemoglobinemia and causes tissue anoxia.

c. There are two basic types of hereditary methemoglobinemia.

   i) Congenital methemoglobinemia may be due to a deficiency
       in the erythrocytes of one or more of the reducing
       enzymes which convert metHb back to normal Hb. This is
       usually a recessive trait. MetHb values may range from
       10 to 40% of the total Hb (normal = 0.5%). Treatment
       involves administering an agent that will reduce the
       metHb. This is usually ascorbic acid or in severe cases,
       methylene blue. The reactions involved are shown later.

   ii) There may also be a defect in the hemoglobin molecule,
       making it resistant to re-reduction by both the metHb
       reductases and exogenous agents such as ascorbate and
       methylene blue. These abnormal hemoglobins are
       collectively called the hemoglobins M (HbM). The defects
       of the six known types are summarized in Table 33. The
       disease is inherited as a dominant trait. In the four
       types involving a His → Tyr mutation, the problem seems to
       be that the phenolic hydroxyl group of Tyr forms a stable
       complex with Fe^{+3}, making it resistant to conversion back
       to Fe^{+2}. Hb-M-Milwaukee has a similar problem, having
       a glutamic acid residue substituted for a valine at
       position β67(E11) near the distal histidine. The
       γ-carboxyl group of the Glu is able to form a stable
       complex with Fe^{+3}. Although a brownish cyanosis is
       characteristic of blood containing Hb M, diagnosis should
       be confirmed by examination of the absorption spectrum,
       either of the Hb or of its CN^- derivative. The cyanide
       derivative spectrum is preferred, since, of the two, it
       differs from the Hb A spectrum to a greater extent. Some
       of the M hemoglobins are unable to form CN^- derivatives,
       presumably due to a Tyr or Glu blocking the sixth iron
       positions where CN^- must bind. In these cases, the
       absorption spectrum of the Hb M itself must be examined.

d. A third type of hereditary methemoglobinemia could occur, in
    which the hemoglobin is mutated in the region to which the
    methemoglobin reductase binds to perform its function. In the
    mutant, the reductase could not attach to the Hb, thereby
    leaving the iron in a ferric state. This type has not been
    reported. When more is known about the mechanism of
    interaction between metHb and metHb reductase, perhaps reasons
    will become apparent why this is so.

e. Acquired Acute Methemoglobinemia

   i) This is a relatively common condition caused by
       introduction into the body of a variety of drugs (all

oxidizing agents or capable of being converted to
oxidizing agents in the blood) such as phenacetin, aniline,
nitrophenol, aminophenol, sulfanilamide, and inorganic
as well as organic nitrites and nitrates. The condition
is less commonly produced by chlorates, ferricyanide,
pyrogallol, sulfonal, and hydrogen peroxide. These
compounds appear to act as catalysts for the oxidation
of Hb by oxygen, producing metHb.

ii) The symptoms of this condition include brownish cyanosis,
headache, vertigo, and somnolence (sleepiness; unnatural
drowsiness).

iii) Diagnosis is based on history, occurrence of the brownish
cyanosis, and the presence of excessive amounts of metHb
(measured spectrophotometrically).

iv) Treatment is usually not difficult, consisting of removal
of the offending substance and the administration of
ascorbic acid (in mild cases) or methylene blue (in
severe cases). These reducing agents function according
to the reactions below (MB = methylene blue, oxidized;
MBH_2 = methylene blue, reduced).

$$MBH_2 \rightleftharpoons MB + H_2 \text{ (gas)}$$

$$2\ H_{2\,(gas)} + Hb(Fe^{+3})_4 \longrightarrow Hb(Fe^{+2})_4 + 4H^+$$

(The release of H_2 by MBH_2 is spontaneous since methylene
blue is autoxidizable.)

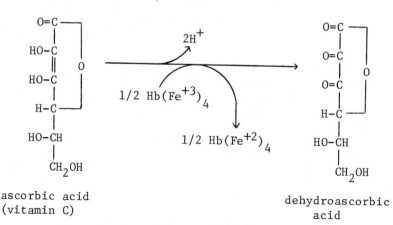

ascorbic acid
(vitamin C)

dehydroascorbic
acid

4. Sulfhemoglobin

    a. Many of the drugs that are responsible for formation of
        methemoglobin also produce sulf-Hb, a greenish Hb derivative
        apparently formed only by oxy-Hb. The two appear together in
        poisoning by sulfanilamide, acetanilid, and sulfanol.

    b. The structure of sulf-Hb is unknown, but the sulfur appears
        associated with the pyrrole nucleus rather than with the iron.
        Symptoms are the same as those for metHb.

    c. There is no specific treatment. The sulfhemoglobin will
        eventually disappear (in mild cases) as the cells containing
        it are destroyed.

5. Cyanmethemoglobin

    a. Cyanide poisoning <u>does not</u> produce cyanohemoglobinemia, nor
        does it produce cyanosis. Cyanosis (bluish coloration,
        especially of the skin and mucous membranes) is caused by an
        excessive concentration (over 5 gm/100 ml) of deoxygenated
        (sometimes called reduced) hemoglobin in the capillary blood.
        Cyanosis may occur:

        i)   when there is a diminished uptake of oxygen in the lungs
             as in extensive pulmonary disease, strangulation,
             pulmonary arterial constriction and certain types of
             congenital heart disease;

        ii) when there is a circulatory inefficiency as in heart
             failure, shock, and localized impairment of venous return;

        iii) when the blood is abnormal (e.g., Hb M) and cannot carry
             enough oxygen.

    b. Cyanide poisoning produces histotoxic anoxia by poisoning
        cytochrome oxidase and other respiratory enzymes. This
        prevents the utilization of O_2 by tissues. It is detected by
        the characteristic odor of HCN gas (bitter almonds) and by
        laboratory tests (absorption spectra, tests for CN^-).

    c. Treatment of cyanide poisoning (discussed with enzyme inhibition
        in Chapter 3) provides an example of selective toxicity. It
        consists of diverting the cyanide into the production of
        cyanmethemoglobin. First, some of the normal hemoglobin is
        converted to methemoglobin by intravenous infusion of a solution
        of $NaNO_2$ or inhalation of amyl nitrite. Once the methemoglobin
        is formed, CN^- can replace the OH^- at position 6 of the iron.

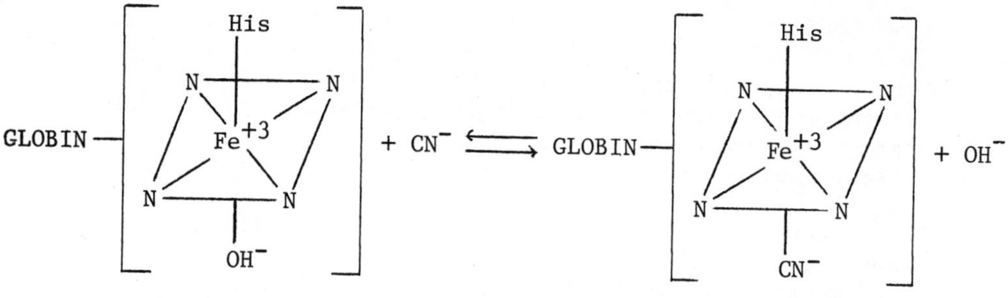

Methemoglobin   Cyanmethemoglobin

The four nitrogens complexed to the iron in the drawing above
are from the pyrrole rings of the porphyrin. The histidine
in position 5 of the iron complex is the proximal His,
α87(F8) or β92(F8). Cyanmethemoglobin is no more toxic than
methemoglobin and cells containing it can be eliminated by
normal body processes. The cyanide bound to the metHb is
always in equilibrium with free CN^- and this uncomplexed
cyanide is converted to thiocyanate (SCN^-; non-toxic) by
administration of thiosulfate (see also pages 116-119).

Synthesis of Porphyrin and Heme

1. Porphyrin consists of four pyrrole molecules joined by methene
    (-CH=) groups. If methylene ($-CH_2-$) bridges are used instead, the
    ring system is called a porphyrinogen. Specific porphyrins and
    porphyrinogens differ from each other in the groups attached to
    the numbered positions (1-8) below. The bridges are designated
    α-δ and the pyrrole rings are A-D. Porphin, the example given,
    has hydrogens at all 8 substituent positions.

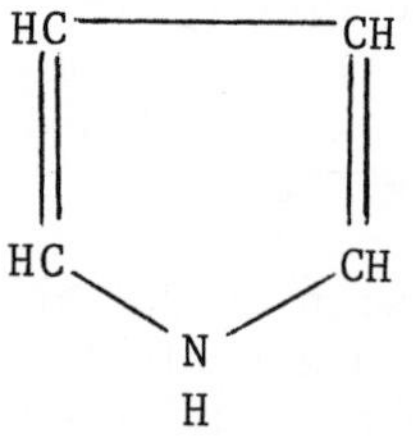

Pyrrole

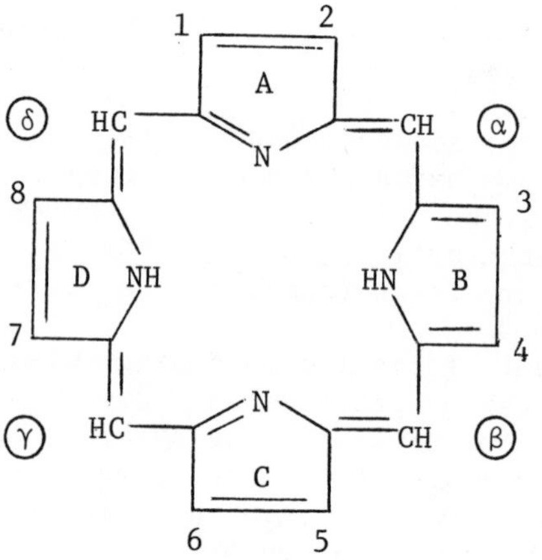

Porphin
($C_{20}H_{14}N_4$)

2. Porphyrins can form complexes with metal ions. This property is very important in their functioning in biological systems. Examples: Heme is an iron porphyrin; chlorophyll is a magnesium porphyrin; vitamin B_{12} is a cobalt porphyrin.

3. These <u>metalloporphyrins</u> are conjugated with proteins to form a number of biologically important macromolecules. Some of these are

    a. Hemoglobin: iron porphyrin + globin; functions as a reversible oxygen carrier; molecular weight 64,500; contains four Fe^{+2} atoms.

    b. Erythrocruorins: iron porphyrin + protein; their function in some invertebrates is similar to that of hemoglobin in higher animals.

    c. Myoglobins: iron porphyrin + protein; present in muscle cells of vertebrates and invertebrates; participate in respiration.

    d. Cytochromes: iron porphyrin + protein; participate in electron transfer reactions (oxidation and reduction reactions).

    e. Catalases and Peroxidases: iron porphyrin enzymes; catalyze the reactions

$$2 \ H_2O_2 \xrightarrow{\text{Catalase}} 2 \ H_2O + O_2$$

$$H_2O_2 + XH \xrightarrow{\text{Peroxidase}} 2 \ H_2O + X$$

    f. Tryptophan pyrrolase: iron porphyrin enzyme; catalyzes the reaction

$$\text{Tryptophan} \xrightarrow{\text{[O]}} \text{Formyl Kynurenine}$$

Biosynthesis of Porphyrins

1. Formation of δ-aminolevulinic acid (ALA)

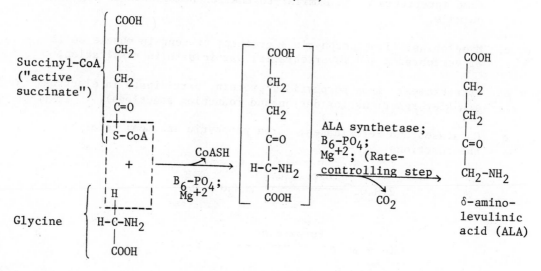

Note: B_6-PO_4 = pyridoxal phosphate; both of the above steps are catalyzed by the same enzyme.

2. Formation of Porphobilinogen (PBG)

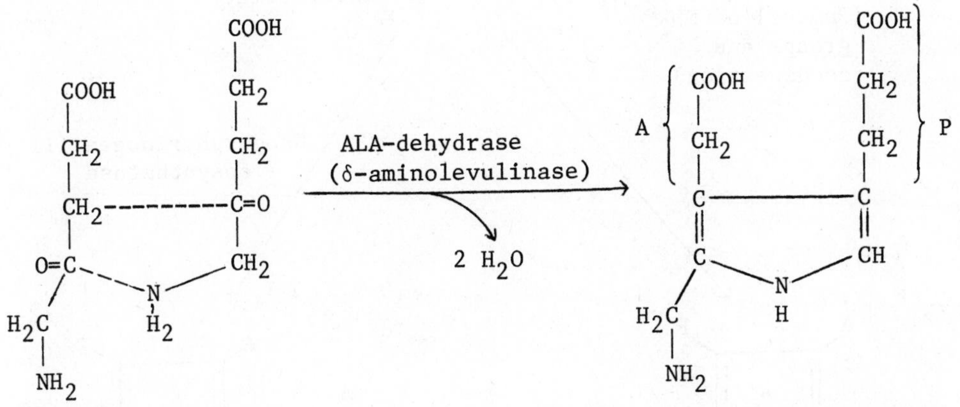

2 molecules of ALA   ALA-dehydrase (δ-aminolevulinase) $\rightarrow$ $2\ H_2O$ Porphobilinogen (PBG)

Note: A = acetate; P = propionate. PBG is more conveniently written as:

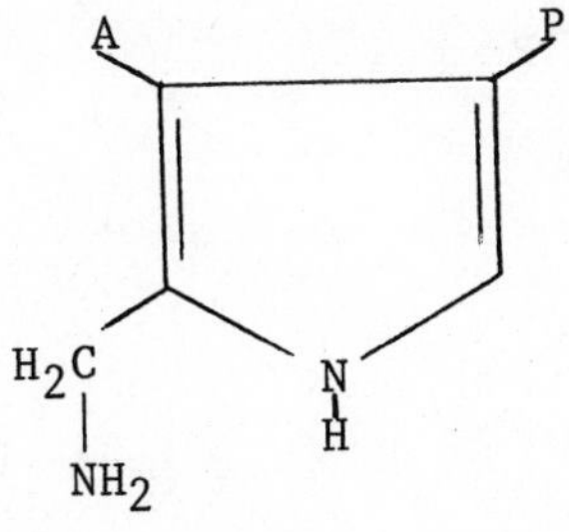

PBG

769

3. Synthesis of Uroporphyrinogens Types I and III

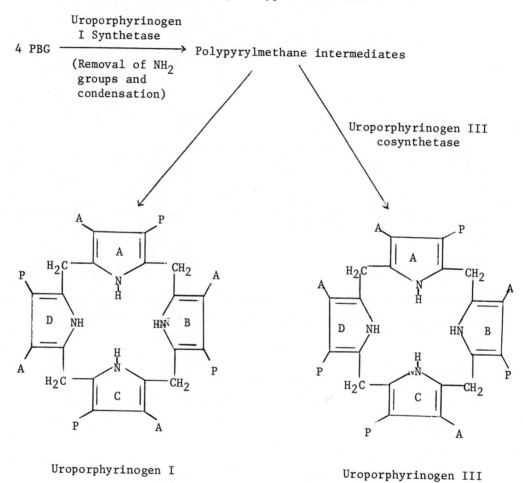

Uroporphyrinogen I   Uroporphyrinogen III

Note: The only difference between types I and III is the orientation of the substituents on ring D.

4. Synthesis of Protoporphyrin III and Heme

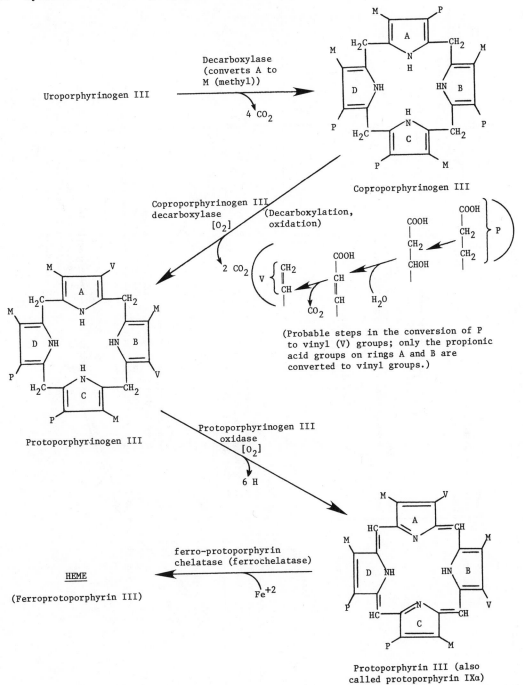

Uroporphyrinogen III

Decarboxylase
(converts A to
M (methyl))

$4 \ CO_2$

Coproporphyrinogen III

Coproporphyrinogen III
decarboxylase
$[O_2]$

(Decarboxylation,
oxidation)

$2 \ CO_2$

(Probable steps in the conversion of P
to vinyl (V) groups; only the propionic
acid groups on rings A and B are
converted to vinyl groups.)

Protoporphyrinogen III

Protoporphyrinogen III
oxidase
$[O_2]$

6 H

HEME

(Ferroprotoporphyrin III)

ferro-protoporphyrin
chelatase (ferrochelatase)

Fe^{+2}

Protoporphyrin III (also
called protoporphyrin IXα)

Figure 91. Summary of the Biosynthesis of Porphyrins and Heme

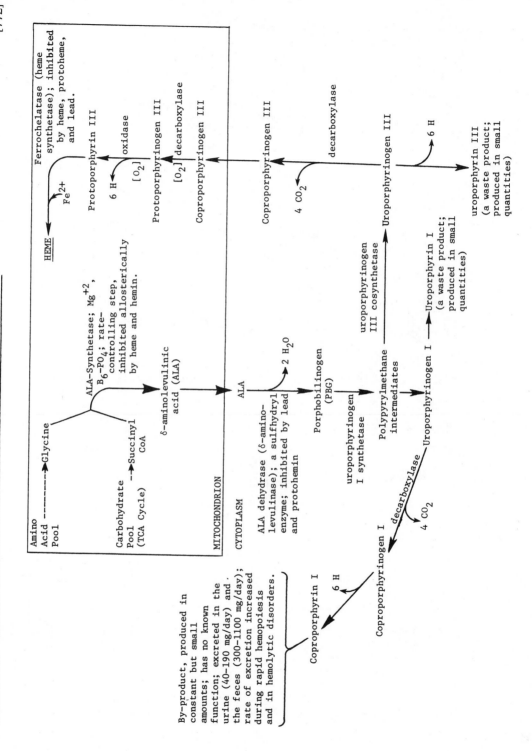

Comments on Porphyrin and Heme Synthesis

1. The ALA-Synthetase-catalyzed reaction:

   a. The first reaction unique to porphyrin synthesis is the
      formation of δ-aminolevulinic acid (ALA) from one mole each
      of succinyl-CoA (a TCA cycle intermediate) and glycine. It
      requires B_6-PO_4 (pyridoxal phosphate) and Mg^{+2} as cofactors.
      A proposed mechanism for ALA-synthetase consists of the
      following steps: formation of a Schiff base by pyridoxal
      phosphate with a reactive amino group of the enzyme; entry of
      glycine and the formation of an enzyme-pyridoxal-phosphate-
      glycine-Schiff-base complex; loss of a proton from the
      α-carbon of glycine with the generation of a carbanion;
      condensation of the carbanion with succinyl-CoA to yield an
      enzyme-bound intermediate (α-amino-β-ketoadipic acid);
      decarboxylation of the enzyme-bound intermediate to ALA and
      its liberation from the enzyme by hydrolysis. Since ALA-
      synthetase is found in the mitochondrial matrix, it does not
      occur in mature erythrocytes which lack functional mitochondria.
      Although the enzyme is found in the cytosol of rat hepatocytes,
      it appears to be functional only when present within the
      mitochondria.

   b. In experimental animals, a deficiency of pantothenic acid (needed
      for CoASH and, hence, succinyl-CoA synthesis) can prevent
      heme synthesis and cause anemia. Similarly, a lack of
      vitamin B_6 or the presence of compounds which block the
      functioning of this cofactor (isonicotinic hydrazide, used in
      the treatment of tuberculosis; L-penicillamine, a chelating
      agent used in the treatment of Wilson's disease, discussed
      under copper metabolism) causes decreased production of ALA
      and heme, leading eventually to anemia. Heme synthesis also
      requires that the TCA cycle be functional and that an oxygen
      supply be available (active respiration is needed).

   c. The primary regulatory step (in the liver) is apparently the
      synthesis of ALA from succinyl-CoA and glycine, catalyzed by
      the enzyme ALA synthetase. The normal end product, heme,
      when accumulated in excess and not utilized in the synthesis
      of conjugated proteins (the principal one being hemoglobin),
      is oxidized to hematin. Hematin contains a hydroxyl group
      attached to the trivalent iron. When the hydroxyl group is
      replaced by a chloride ion, the product is known as hemin
      (or hemin chloride). Hemin and heme allosterically inhibit
      the ALA synthetase reaction (an example of hepatic feedback
      or end-product control of an enzyme). It has also been reported
      that heme, functioning as a corepressor, blocks the formation
      of ALA-synthetase protein at the DNA level. The repressor,

which binds to the DNA, is made of a protein part (aporepressor) and heme. This mechanism controls the rate of structural gene transcription and hence the rate of synthesis of the enzyme protein. It and other aspects of the control of heme synthesis are discussed in greater detail in the chapter on nucleic acids.

Presumably, individuals with a defect (loss of control) in this system can show the condition of porphyria (discussed later in this section) or can develop hepatic porphyria when certain drugs (such as barbiturates) are administered. The postulated mechanism is that these persons produce defective repressor molecules which combine with barbiturates (or sex steroids) at the corepressor site, giving rise to an inactive complex. The result is the repressor does not function and, consequently, overproduction of ALA synthetase occurs, resulting in porphyria.

2. The ALA-Dehydrase catalyzed reaction

a. In this step, two molecules of ALA are condensed to form one molecule of porphobilinogen with the release of two molecules of water. Since ALA dehydrase is present in the cytosol, the ALA must leave the mitochondria prior to this reaction.

The mechanism of ALA-dehydrase consists of Schiff-base formation by the keto group of one molecule of ALA with an amino group of the enzyme, followed by a nucleophilic attack by the enzyme-ALA anion on the carbonyl group of a second ALA molecule. This results in the elimination of water. The free amino group of this second ALA molecule displaces the amino group of the enzyme by transamination or transaldimination to form porphobilinogen.

b. This enzyme may play a role in the regulation of heme synthesis. ALA dehydrase isolated from Rhodopseudomonas spheroides appears to be an allosteric enzyme, activated by K^+ and inhibited by heme.

c. ALA dehydrase is a sulfhydryl enzyme and reduced glutathione is essential for its activity. Agents which bind to -SH groups, such as iodoacetamide, p-chloromercuribenzoate, heavy metal ions, etc. inhibit the reaction (see also page 122). ALA dehydrase is apparently very sensitive to lead and in lead poisoning, there is a significant increase in the urinary excretion of ALA (normal is 2 mg/24 hr). The decrease in ALA-dehydrase activity in blood has been used as a measure of the extent of lead poisoning in clinical situations.

3. Synthesis of Uroporphyrinogens I and III

    a. The nature of these reactions is still not completely clear but certain things are becoming known. There appear to be two enzymes involved: uroporphyrinogen I synthetase and uroporphyrinogen III cosynthetase. When an enzyme preparation originally capable of making both types I and III is heated to 60°C, it ceases to make type III but continues synthesis of type I. This indicates the presence of two different enzymes, of which the cosynthetase is heat labile.

    b. A likely sequence of events for this synthesis is indicated below.

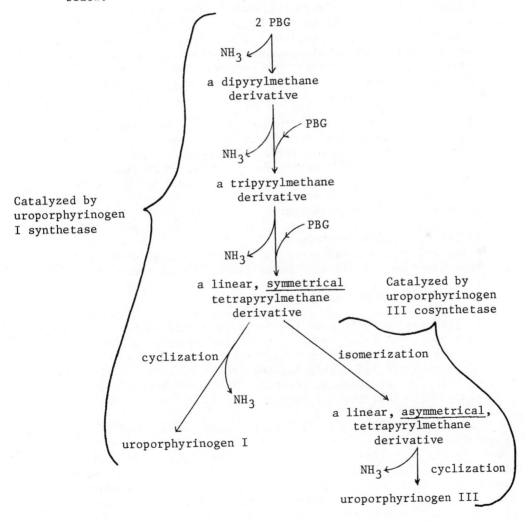

Although these steps remain to be proven, a dipyrrylmethane derivative has been tentatively identified. It has also been demonstrated that only four molecules of PBG are needed for the synthesis of type III isomers, that formaldehyde is neither consumed nor produced in the reaction, and that dipyrylmethenes do not participate as reactants.

4. Formation of Heme

    a. The end-product of uroporphyrinogen I is coproporphyrin I which has no known function and is eliminated (see Figure 91). The type III uroporphyrinogen is ultimately converted to heme.

    b. Uroporphyrinogen III is converted in the cytosol to coproporphyrinogen III in a reaction catalyzed by uroporphyrinogen decarboxylase. The enzyme is highly specific for uroporphyrinogen and is inhibited strongly by mercury, copper, manganese, and oxygen. The oxygen probably functions by promoting the oxidation of uroporphyrinogen III to uroporphyrin III, a waste product.

    c. The coproporphyrinogen III then moves into the mitochondria for further conversions. It is converted to protoporphyrinogen III by coproporphyrinogen III decarboxylase. This mitochondrial enzyme catalyzes the oxidative decarboxylation of two propionic sidechains of coprogen III to vinyl groups with the elimination of two molecules of CO_2. The enzyme appears to have an absolute requirement for molecular oxygen. In the penultimate reaction of heme biosynthesis, protoporphyrinogen III is oxidized to protoporphyrin III (also known as IX), catalyzed by a membrane-associated-mitochondrial enzyme, protophyrinogen oxidase. The primary electron acceptor in this reaction in eukaryotes is not known, but oxygen is required for enzyme activity.

    d. The terminal step in heme synthesis is catalyzed by ferrochelatase (also known as ferro-protoporphyrin chelatase or heme synthetase) which incorporates ferrous iron into protoporphyrin III. Reducing substances such as ascorbic acid, cysteine, or glutathione are required by the enzyme. The reaction is inhibited by heme as well as by lead.

5. Of the many possible porphyrin isomers, only types I and III are found in biological systems. Type I isomers have no known useful function and are eliminated in the urine and feces. The type III isomer, when chelated with iron to form heme or hemin, serves as the prosthetic group in hemoglobin, myoglobin, the cytochromes c, and other enzymes. The magnesium-porphyrin known as chlorophyll a, found in many photosynthesizing plants, is formed by a pathway

very similar to that of heme up to the synthesis of magnesium-protoporphyrin III. It is then further modified for its role in photosynthesis. The synthesis of vitamin B_{12}, a chelate of cobalt with a porphyrin-derived nucleus, is also similar to heme synthesis up to a certain point. It too is finally modified by pathways unique to the bacteria in which it is made.

6. Biosynthesis of Hemoglobin

   The protein portion of hemoglobin is called <u>globin</u>. It is synthesized primarily in the nucleated erythrocytes of the bone marrow and in the reticulocytes, using the labile amino acid pool of the body. The synthesis occurs via the regular protein biosynthetic reactions involving DNA, mRNA, ribosomes, tRNA, amino acid activating enzymes, and so on (see the section on protein synthesis in Chapter 5). The α-and β-chains of hemoglobin are apparently under the control of different genes and yet are synthesized at the same rate. It is also of interest that hemin and protoporphyrin III, while inhibiting or decreasing the rate of synthesis of heme (discussed above), promote the synthesis of globin in an apparant attempt to maintain the proper ratio (1:1) of heme to globin. There are genetic defects in the globin production which give rise to abnormal hemoglobins. These were discussed earlier in this chapter.

7. Some Chemical Properties of Porphyrins

   a. The tertiary nitrogens of the pyrrole rings function as weak bases. The carboxyl groups function as acids. The iso-electric pH range is 3.0-4.5.

   b. Porphyrin<u>ogens</u> are colorless while porphyr<u>ins</u> are colored compounds, due to the conjugation introduced when the methylene bridges are converted to methenes. Porphyrins have characteristic absorption spectra in both UV and visible regions. All porphyrins show an absorption band at about 400 nm known as the Soret band.

   c. When porphyrin complexes with iron or other metals, its spectrum in the visible region changes. For example, in alkaline solutions, protoporphyrin gives rise to several sharp absorption bands at 540, 591, and 645 nm, while heme shows only a broad band with a plateau extending from 540 to 580 nm.

   d. Heme can complex with Lewis bases. A hemochrome (hemochromogen) is a compound of heme with any two other ligands to give hexacoordinate Fe^{+2}.

8. As we shall see later in this section, breakdown (catabolism) of heme does not yield porphyrins but gives rise to products (bile pigments) which are toxic to the body and need to be eliminated. In general, ALA, PBG, and uroporphyrin are excreted in the urine while coproporphyrin (to a large extent) and protoporphyrin (solely) are excreted in the bile after appropriate transformations (discussed later).

9. Effect of lead on heme biosynthesis: Lead may affect the biosynthesis of heme at several sites: intracellular iron transport, ALA-dehydrase, coproporphyrinogen oxidase, and ferrochelatase (heme synthetase). Impairment at these sites results in increased ALA and coproporphyrin excretion, and increased erythrocyte protoporphyrin concentration. Measurements of these parameters, including erythrocyte ALA-dehydrase activity, have been used as diagnostic tests for lead intoxication. Among these tests, the measurement of erythrocyte protoporphyrin (or its zinc chelate) may turn out to be the test of choice. It should be noted, however, that the erythrocyte protoporphyrin is also elevated in other disorders, including iron-deficiency anemia, erythropoietic protoporphyria, and others. Lead toxicity is also established by determining the concentration of lead in whole blood with atomic absorption spectroscopy. In addition to hematopoietic tissue, lead has adverse effects on the kidney and nervous system. The effects of lead on the nervous system are long-lasting and only minimally reversed by therapeutic measures (chelating therapy). Thus, detection in the subclinical stage is essential and consists of measuring whole blood lead and erythrocyte protoporphyrin.

The Porphyrias

1. The porphyrias are groups of diseases caused by abnormalities in heme biosynthesis and are characterized by excessive accumulation and/or excretion of porphyrins or their precursors. The porphyrins are excreted by different routes, depending on their water solubility. For example, uroporphyrin with eight carboxyl substituents is more water soluble than the other subsequent porphyrins and is eliminated in the urine. On the other extreme, protoporphyrin, which contains two carboxyl groups (making it the least water soluble), is found exclusively in the bile. Coproporphyrin, which has four carboxyl groups and intermediate solubility, is found both in the bile and urine.

2. These disorders are associated with acute and/or cutaneous manifestations. In the acute state the clinical presentation may include abdominal pain, constipation, hypertension, tachycardia and neuropsychiatric symptoms. The cutaneous

problem consists of photosensitivity (itching, burning, redness, swelling and scarring), hyperpigmentation and, sometimes, hypertrichosis (an abnormally excessive growth of hair).

3. The porphyrias may be classified according to the location of the metabolic lesion. It may be in the liver (hepatic porphyrias) or bone marrow (erythropoietic porphyrias). It should be noted, however, that there is growing evidence that different porphyrias classified as hepatic or erythropoietic may include enzyme defects common to many tissues. The various enzyme deficiencies are shown in Figure 92. There is a third type of porphyria which is induced by drugs, chemicals and endogenous compounds such as steroids. The major hepatic porphyrias include acute intermittent porphyria, variegate porphyria, hereditary coproporphyria and cutaneous hepatic porphyria. The principal erythropoietic porphyrias are hereditary erythropoietic porphyria and erythropoietic protoporphyria.

4. Hepatic Porphyrias

   a. Acute intermittent porphyria (AIP) is associated with excessive urinary excretion of δ-aminolevulinic acid (ALA) and porphobilinogen (PBG), with little or no corresponding rise in either type I or type III porphyrinogens. Although ALA-synthetase activity is reportedly elevated in these patients, the lack of PBG polymerization suggests that the primary enzymic defect may be a lack of uroporphyrinogen I synthetase. In fact, it has now been observed that the uroporphyrinogen I synthetase activity is depressed in the liver and erythrocytes from AIP patients. A family study has indicated this is probably the primary defect and that the elevation of ALA-synthetase activity either reflects the lack of feedback control by heme and hemin or is an associated enzymic error. Clinical symptoms include central and peripheral nervous disorders, psychosis, and hypertension (although these can vary a great deal). Death can occur due to respiratory paralysis. Since these individuals are unable to make porphyrins to any great extent, they are not photosensitive. This disorder is inherited as an autosomal dominant trait.

   b. Variegate porphyria: Patients with this disease excrete excessive amounts of both uroporphyrin and coproporphyrin (but not porphobilinogen). These individuals are photosensitive (the presence of porphyrins near the surface of the body results in light sensitization because the porphyrins have the ability to concentrate radiant energy) but other symptoms vary widely, hence the term variegate. This disease appears

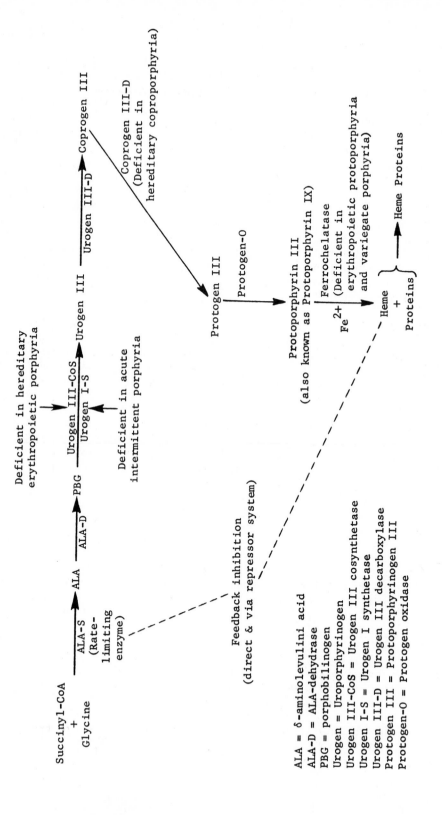

Figure 92. Porphyrin Biosynthesis and the Putative Defects in Various Porphyrias

ALA = δ-aminolevulini acid
ALA-D = ALA-dehydrase
PBG = porphobilinogen
Urogen = Uroporphyrinogen
Urogen III-CoS = Urogen III cosynthetase
Urogen I-S = Urogen I synthetase
Urogen III-D = Urogen III decarboxylase
Protogen III = Protoporphyrinogen III
Protogen-O = Protogen oxidase

to be inherited as an <u>autosomal dominant trait</u>. The primary lesion is not known; however, hepatic ALA synthetase is elevated. It has been suggested that <u>ferrochelatase</u> is defective in these patients.

c. <u>Hereditary coproporphyria</u>: It is inherited as an <u>autosomal dominant trait</u>. In this disorder there appears to be a block in the conversion of coprogen III to protogen III, presumably due to an impairment in the activity of coproporphyrinogen III decarboxylase.

In all of the above disorders, there is an increase in hepatic ALA-synthetase activity. This is due to a deficiency of heme synthesis caused by genetic blocks in its synthetic pathway. There are increased amounts of the porphyrin precursors ALA and PBG in liver, plasma and urine. Recall that ALA synthetase is regulated by the heme feedback process and the repressor system (discussed already). Recently, hematin has been used to treat the acute attacks of these porphyrias with marked clinical improvement. Although it is often inadequate, carbohydrate feeding has also been associated with improvement in acute intermittent porphyria. This has been termed the "glucose effect" and presumably depends upon repression of the induction of ALA-synthetase, the limiting enzyme of porphyrin biosynthesis. The mechanism of the glucose effect is not known.

5. Erythropoietic Porphyrias

a. A defect in the synthesis of type III isomers from polypyrylmethane intermediates, possibly due to a deficiency of uroporphyrinogen III cosynthetase, produces <u>hereditary erythropoietic porphyria</u>. In this case, the type I porphyrin (principally uroporphyrin I) is formed, accumulates in the tissues, and is excreted in the urine. This deficiency in the production of the normal (type III) isomer gives rise to further increases in the levels of the type I isomers by reducing the regulatory action on ALA synthetase. Red cells may also respond to excessive amounts of porphyrins by episodes of hemolysis. The compensatory increase in hemoglobin formation can then exaggerate the already increased production of type I porphyrins. The accumulation of type I porphyrins produces a pink to dark red color in the teeth, bones, and urine. Red-brown teeth and urine are pathognomonic of this disorder. These individuals are also sensitive to long-wave ultraviolet light and sunlight. This disease appears to be inherited as an <u>autosomal recessive trait</u>.

   b. Erythropoietic protoporphyria (EPP) results from a <u>deficiency of ferrochelatase</u> in the reticulocytes of the bone marrow. It is transmitted as an <u>autosomal dominant trait</u> with variable penetrance and expressivity. In general, this disorder is a benign disease and the most prominent symptom is photosensitivity. Occasionally, EPP may lead to liver disease. Reduced levels of ferrochelatase activity in EPP results in the accumulation of protoporphyrin in the maturing reticulocytes and young erythrocytes. When a smear of these cells is exposed to fluorescent light, they appear as red-fluorescing erythrocytes. The unused protoporphyrin in the red blood cells appear in the plasma, which is then picked up by the liver and excreted into the bile. If the protoporphyrin accumulates in the liver it can lead to severe liver disease. Patients with EPP, in contrast to individuals with other porphyrias, have normal urinary porphyrin levels. Abnormally high levels of protoporphyrin are found in erythrocytes, plasma and feces.

The photosensitization in EPP patients is presumably caused by visible light stimulating protoporphyrin present in the dermal capillaries to an excited (triplet) state. This in turn converts molecular oxygen to singlet oxygen (see later under photoinactivation of bilirubin) which can produce cellular damage. The oral administration of β-carotene ameliorates the photosensitivity reaction in EPP. The action of β-carotene may be due to its quenching effect on singlet oxygen and free-radical intermediates.

6. The induced (non-hereditary) porphyrias and porphyrinurias can be caused by the following conditions. The chief urinary porphyrin appearing in each type is mentioned in the parenthesis. Toxic agents: (coproporphyrin III in some cases of lead poisoning, 8000 mg/24 hrs) chemicals, heavy metals, acute alcoholism. Liver diseases: (coproporphyrin I, 100-600 mg/24 hrs) infectious hepatitis, mononucleosis, and cirrhosis. Certain blood diseases: (coproporphyrin I) hemolytic anemias, pernicious anemias, leukemias. Miscellaneous: (coproporphyrin III) poliomyelitis, aplastic anemias, Hodgkin's disease. In pregnancy, especially when associated with hepatic disease and in women on oral contraceptives, a significant elevation of urinary coproporphyrin I and ALA has been found.

Bilirubin Metabolism

Since iron and protein metabolism have already been covered, bilirubin is the only hemoglobin-derived material still to be discussed.

1. Sources and Formation of Bilirubin

   a. The heme moiety derived from the destruction of senescent red blood cells is the predominant source of bilirubin, supplying about 80% of the total (250-300 mg of bilirubin is normally formed per 24 hours). The remaining 20% comes from the catabolism of the other heme proteins listed previously.

b. Bilirubin is produced in the reticuloendothelial cells of the liver, spleen, and bone marrow. These cells engulf older red blood cells, cause them to lyse and release hemoglobin, then catabolize the hemoglobin and form the pigment.

c. Although red blood cells contribute 80% of the bilirubin formed, the metabolic turnover time of hemoglobin (about 120 days) is slow compared to that of some of the hepatic hemoproteins. For example, the microsomal cytochromes, P-450 and b_5, that participate in the detoxication reactions of hormones, toxins, drugs, and of bilirubin itself, have a turnover time of between one and two days.

The relationship between hemoglobin synthesis and bilirubin production has been studied in humans by administering a labeled heme precursor (e.g., ^{15}N-labeled glycine) and measuring the concentration of the label in both the circulating hemoglobin and the end-products of heme appearing in the feces. Results from stool samples show that the labeled material appears in two peaks: an early, small peak (representing about 20% of the total label) at about 8 days after administration of the label which is largely accounted for by the metabolism of non-hemoglobin heme (obtained from cytochrome P-450, cytochrome b_5, etc.) and a large peak appearing between 100-160 days after injection, accounting for the physiological lifespan of red blood cells.

d. Epithelial cells of the renal tubules and the liver parenchymal cells play a role in the conversion of hemoglobin heme to bilirubin when the hemoglobin is produced due to intravascular hemolysis. Bile pigment can also be formed in the macrophages when they catabolize hemoglobin from the phagocytized red blood cells appearing due to blood extravasations.

e. The conversion of heme to bilirubin consists of two steps: the microsomal heme oxygenase system and the biliverdin reductase reaction (see Figure 93). The substrate, heme, effects its own catabolism by activating the heme oxygenase system, presumably by enzyme induction. In hemoglobinuria, the hemoglobin that is present in the glomerular filtrate is taken up by the kidney epithelial cells and stimulates the heme oxygenase system.

f. Note that the natural bile pigments are formed by the opening of the heme ring at the α-position, giving rise to the IXα series. The mechanism for the observed α-selectivity in the cleavage of the heme ring is not yet clear. Heme oxygenase catalyzes the conversion of those heme groups which are easily dissociable from their protein counterpart to equimolar amounts of biliverdin IXα and carbon monoxide (CO). Carbon monoxide, an unusual metabolite of mammals, is derived from the oxidation of the methene-bridge α-carbon. Since the endogenous CO production is only coupled to heme catabolism (or bilirubin formation), its measurement can be used in the assessment of total heme turnover.

g. The formation of bilirubin from biliverdin is a <u>reductive process</u> and occurs rapidly so that no appreciable levels of biliverdin normally accumulate. It is of interest to point out that in birds, amphibians and reptiles, the biliverdin produced in the first step of heme catabolism is excreted directly in the bile in an unconjugated form without any further modification. Since bilirubin is a highly toxic and insoluble product which requires elaborate energy-requiring transport and inactivation systems, the purpose for its production from an apparently non-toxic metabolite poses an intriguing question.

h. The role of haptoglobins (see also serum proteins in Chapter 6) in hemoglobin catabolism: The major serum α_2-globulins are haptoglobins, a polymorphic series of glycoproteins which have the unique property of forming stable complexes with hemoglobin. Normal levels of human plasma haptoglobin can bind approximately 50-150 mg of free hemoglobin per 100 ml of plasma. The haptoglobins appear to be involved with the catabolism of any <u>free hemoglobin</u> that escapes into the plasma. The extracorpuscular hemoglobin in the intravascular spaces (due to hemolysis) binds with haptoglobin. The complexes are sequestered and catabolized by the reticuloendothelial cells (liver > bone-marrow > spleen). Thus, haptoglobin complexes with hemoglobin to prevent accumulation of free hemoglobin in plasma, which in the uncomplexed state is filtered through the glomeruli resulting in the loss of iron via the kidney. Furthermore, the formation of large molecular haptoglobin-hemoglobin complexes prevents deposition of hemoglobin in the renal tubules during intravascular hemolysis. It should be noted that haptoglobin appears to play a minor role in normal hemoglobin catabolism. Four to six per cent of American Blacks normally have no detectable serum haptoglobin.

Plasma haptoglobin levels are elevated in inflammatory diseases and depressed following episodes of intravascular hemolysis. Haptoglobins are presumably synthesized by the liver and, therefore, parenchymal liver disease may be

associated with low haptoglobin levels. However, in biliary
obstruction, elevated levels of haptoglobin have been noted.
Numerous methods are available for the determination of
haptoglobins, including: measurement of the peroxidase
activity of the haptoglobin-hemoglobin complex, electrophoresis
of haptoglobin and the haptoglobin-hemoglobin complex,
immunologic methods, gel filtration, and differential acid
denaturation of hemoglobin and the complex. The normal range,
as determined by electrophoretic procedures, is about 20-200 mg
per 100 ml of serum.

2. Some Properties of Bilirubin

a. Bilirubin is a yellow pigment with an absorption maximum at
450 nm. This property can be utilized in the direct measurement
of bilirubin in the serum. However, carotene also absorbs near
the same wavelength (440 nm), so that the ingestion of
carotene (which is present in carrots, tomatoes, and mangoes)
seriously limits the direct spectrophotometric measurement
of bilirubin in adult sera. In neonatal sera, the direct
measurement is of value in emergency situations. The more
reliable methods involve coupling of the pigment with diazotized
sulfanilic acid (discussed later).

b. Bilirubin is soluble in lipids (and in organic solvents) and
consequently can diffuse freely across cell membranes. Inside
the cell, bilirubin interferes with many crucial metabolic
functions and most of the specific interactions are not
understood. The brain damage which it causes has been
ascribed to the ability of the pigment to uncouple oxidative
phosphorylation on the basis of in vitro experiments.
In vivo results, with levels of bilirubin which actually
occur in certain diseases, do not support this mechanism.

c. Bilirubin is insoluble in aqueous solutions, particularly at
physiological pH. It is soluble in a basic solution because
of the weak acidity conferred by the presence of two propionic
groups whose pK is around 8.

3. Transport and Uptake of Bilirubin by the Hepatocytes

a. Since bilirubin is insoluble in aqueous systems, it is
carried in the blood complexed to albumin. This complex
formation prevents the indiscriminate passage of bilirubin
into tissue cells other than hepatocytes. The unique ability
of hepatocytes to take up, concentrate, and eliminate bilirubin
is presumably due to (i) the presence of a carrier mechanism
available for the transport of bilirubin; (ii) the presence
of bilirubin-binding proteins in the cytoplasm of hepatocytes;

Figure 93.   Microsomal Oxidation of Heme to Biliverdin and the Reduction of Biliverdin to Bilirubin

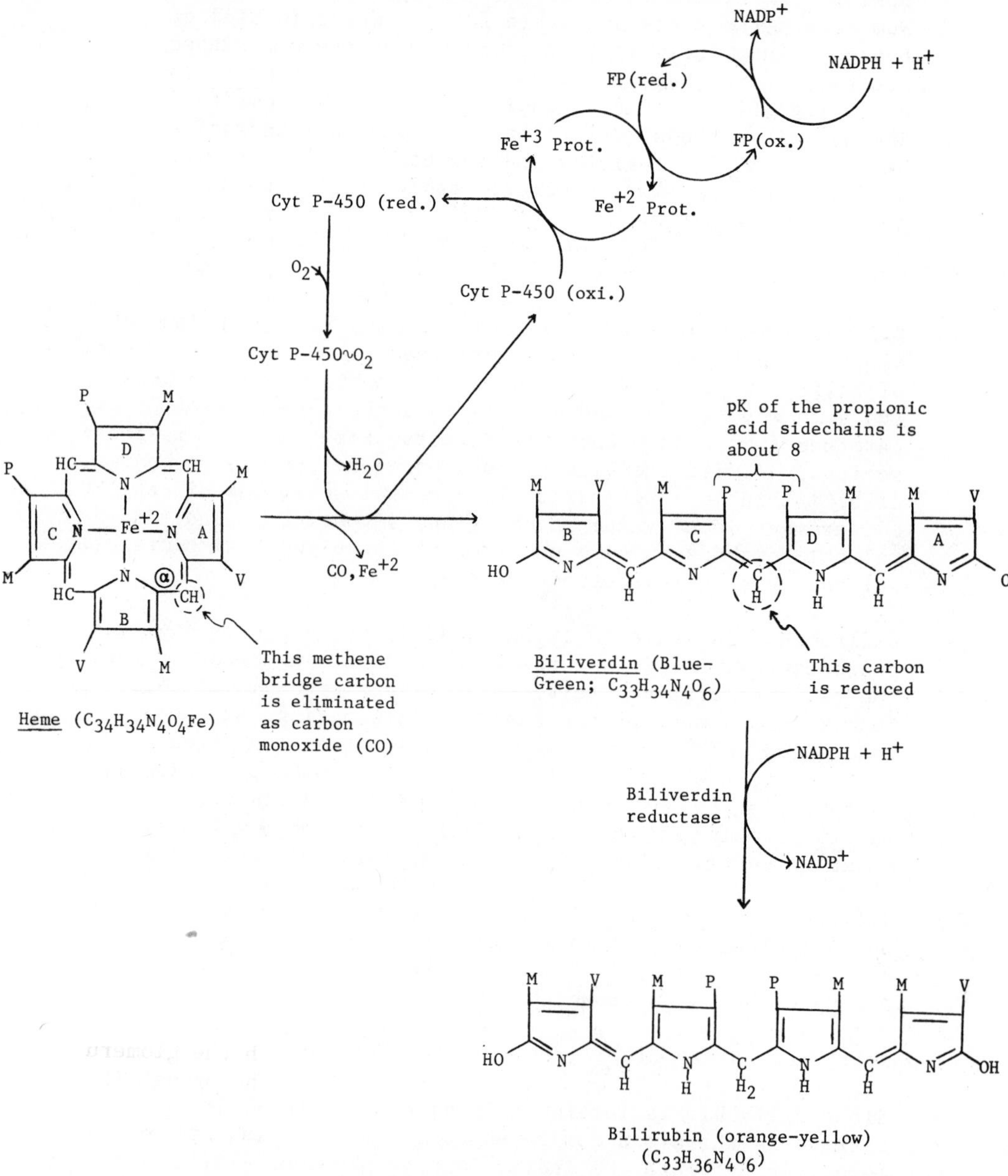

FP = flavoprotein (perhaps containing FAD); Fe^{+2}/Fe^{+3} Prot. is a non-heme iron protein; M = methyl; V = vinyl; P = propionic acid. A similar (probably identical) microsomal oxidase system is described later under the metabolism of drugs.

and (iii) conversion of bilirubin within the cells to water-soluble conjugates which normally cannot reenter the blood but are eliminated through the bile. Two bilirubin-binding proteins, named Y and Z, have been isolated from the hepatocytes and characterized. The Y and Z proteins are soluble and found in the cytoplasm. The binding specificity of these proteins is not restricted to bilirubin as they can bind other organic anions (e.g., sulfobromophthalein, indocyanine green). Recent studies have shown that, in particular, the Y protein participates in the transport of a number of organic anions and hence has been termed <u>ligandin</u>. Furthermore, it has been shown that the ligandins possess glutathione-S transferase activity which is responsible for the conjugation of GSH with compounds bearing an electrophilic center, a process used in the detoxification of potentially harmful agents. Ligandin has also been identified in the proximal tubule cells of the kidney. This renal ligandin may be important in organic anion transport.

The Y protein has greater affinity for bilirubin than Z but its binding capacity is less. Therefore, at low concentrations of bilirubin, the preferential binding agent is Y, but as the concentration of bilirubin saturates the binding sites on Y, the binding to Z increases. Adult levels of the Y protein (and glucuronyl transferase--discussed later) are not reached until several weeks after birth; the Z-protein levels are the same for both neonate and adult. Physiologic, neonatal nonhemolytic jaundice (discussed later) is due to the relative deficiency of Y protein during this period and manifests as a problem in premature infants. Phenobarbital administration in the prenatal period has been shown to enhance the synthesis of ligandin, glucuronyl transferase and, perhaps, other enzymes involved in intrahepatic bilirubin metabolism in the neonate. In the <u>Gilbert Syndrome</u> (nonhemolytic, unconjugated hyperbilirubinemia), the biochemical lesion may in part reside in the deficiency of intrahepatic bilirubin-conjugating proteins.

Bilirubin is not eliminated by the kidney because it is complexed to albumin which does not pass through the glomerular capillary walls. If bilirubin is detected in the urine, it is conjugated bilirubin (which is water soluble) and signifies a pathologic process. Conjugated bilirubin is normally confined to the liver cells where it is eliminated via the biliary system.

b. Since bilirubin is an anion (due to the presence of the two propionic acid residues) and is transported bound to albumin (1 molecule of albumin binds more than 2 molecules of bilirubin), other anions such as sulfonamides, thyroxine and

triiodothyronine, fatty acids, and acetyl-salicylate can
compete for the **bilirubin-binding** sites on albumin and
displace bilirubin. If the plasma concentration of displaced
(free, unbound) bilirubin increases due to a reduction in
albumin-binding capacity (normal binding capacity: 20-25 mg/
100 ml), it can enter the brain (giving rise to brain
encephalopathy) and other tissues, producing cytotoxic effects.
This aspect (cytotoxicity) assumes a particular significance
in the newborn because the liver is not yet mature at birth
(apparently about 10-15 days are required for the liver to
attain its full biochemical potential) and is unable to
metabolize bilirubin. As a result, neonatal bilirubin levels
can surpass the binding capacity of the plasma proteins and
produce cytotoxic effects if the situation is not corrected.
This problem is accentuated in erythroblastosis.

c. As mentioned above, hepatocytes take up unbound bilirubin
from the sinusoidal plasma, presumably in a carrier-mediated
step. Once the bilirubin enters the cell, it is immediately
bound by the acceptor proteins so that it cannot escape back
across the membrane. The uptake of bilirubin by the
hepatocytes is shown next.

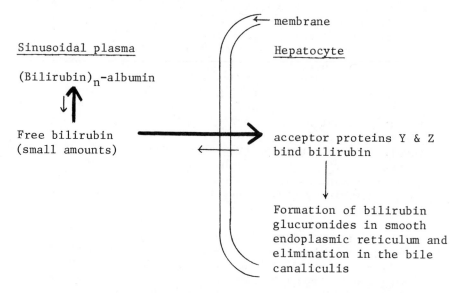

d. Compounds such as phenobarbital and other anionic drugs
undergo metabolic transformations similar to that of bilirubin
before they are eliminated. Studies with **phenobarbital** have
shown that the drug enhances the hepatic uptake, conjugation,
and excretion of bilirubin, presumably by stimulating the
overall process.

4. Formation and Excretion of Bilirubin Glucuronides

    a. The general reaction for glucuronide formation is shown below.

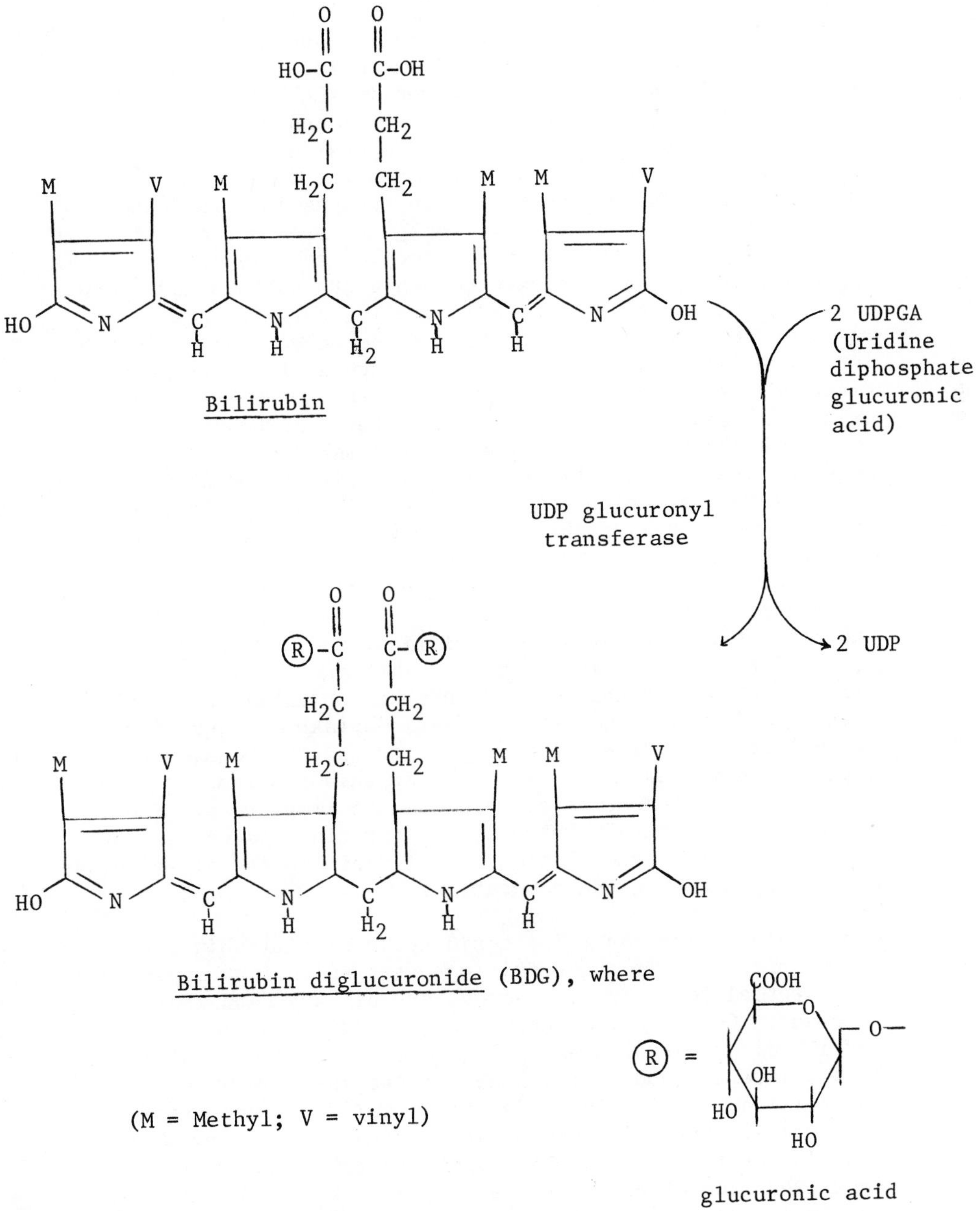

Bilirubin

2 UDPGA
(Uridine
diphosphate
glucuronic
acid)

UDP glucuronyl
transferase

2 UDP

Bilirubin diglucuronide (BDG), where

(M = Methyl; V = vinyl)

$R =$

glucuronic acid

b. Conjugation of bilirubin with glucuronic acid to form
    bilirubin diglucuronide (BDG) takes place within the smooth
    endoplasmic reticulum, catalyzed by the enzyme glucuronyl
    transferase. This transformation brings about significant
    changes in the properties of the bilirubin molecule. It
    changes from a lipid-soluble molecule to one that is freely
    soluble in an aqueous medium, and the molecular mass changes
    from 585 to 937 daltons. BDG formed in the hepatocytes is
    eliminated via the bile excretory system and under normal
    circumstances does not cross the cytoplasmic membrane to
    appear in sinusoidal plasma. With hepatic damage, the
    water-soluble, conjugated bilirubin escapes into the blood
    and is excreted in the urine. Thus, the presence of bilirubin
    in the urine indicates hepatic injury of some type.
    Diglucuronide is the predominant conjugate synthesized. Other
    bilirubin conjugates detected in small quantities are
    monoglucuronides, sulfates, etc. As noted earlier, at birth
    (and especially in prematurity) the levels of Y protein and
    glucuronyl transferase are subnormal with adult levels not
    being attained until several weeks later. Although a
    potential hazard at birth, the delayed maturation of these
    proteins protects the fetal liver by limiting the entry of
    bilirubin and a variety of other organic anions into the
    hepatocytes. On a weight basis, the fetus probably produces
    as much bilirubin as the adult and utilizes the placenta as
    an efficient method of eliminating unconjugated bilirubin by
    transferring it to the maternal circulation. Birth interrupts
    the placental route of pigment elimination and the "immature"
    liver of the neonate is confronted with the elimination of
    bilirubin. During this period there is an imbalance in the
    rate of bilirubin production, hepatic uptake, conjugation,
    and biliary excretion, resulting in transient unconjugated
    hyperbilirubinemia. The magnitude of this imbalance depends
    upon the degree of hemolysis and hepatic immaturity. The
    biochemical nature of the hepatic maturation process is not
    understood. The accumulation of bilirubin in the plasma plays
    a role in hastening the maturation.

c. Canalicular secretion: The conjugation of bilirubin with
    glucuronic acid and the resulting water solubility appears to
    be essential for secretion across the biliary canalicular
    membrane. Only small amounts (concentrations up to 2.0 mg
    per 100 ml of normal human gallbladder bile) of unconjugated
    bilirubin are found in the bile. These small amounts are
    solubilized by the bile acids (or salts) which are also
    secreted into the bile and by increasing the pH (from 6.2 to
    7.8) in the gallbladder. Any problems resulting in a
    decreased production of bile acids, a lack of alkalinization
    or the presence of increased amounts of unconjugated bilirubin

lead to precipitation of bilirubin in the bile with the formation of gallstones. These concretions are known as <u>pigment gallstones</u> as opposed to the more common cholesterol type (see Chapter 8). The hepatobiliary secretory mechanism for bilirubin, which is shared by many other endogenous and exogenous organic compounds, is not understood. The secretory process exhibits active transport against a large concentration gradient and is subject to competitive inhibition by other compounds, thus being saturable.

d. In the clinical laboratory, bilirubin and its glucuronides are distinguished on the basis of the conditions under which they react with diazotized sulfanilic acid (van den Bergh reaction; discussed later). The water-soluble bilirubin glucuronides react <u>directly</u>, without any need for an "accelerator" and are known as <u>direct-reacting bilirubin</u>. Free bilirubin, on the other hand, requires an accelerator such as alcohol or a caffeine-sodium benzoate reagent to make it react under ordinary conditions. It is therefore called <u>indirect-reacting bilirubin</u>. These properties will be discussed more, both in terms of diseases and chemical analysis, later in this section.

5. Metabolism of Bilirubin in the Gastrointestinal Tract

a. Bilirubin in the conjugated form is virtually unabsorbed from the small intestine. In the large intestine the fecal flora converts it by stepwise reduction to a group of colorless compounds collectively known as urobilinogen. During the process of reduction, part of the glucuronides may undergo deconjugation while a portion remains in the conjugated form. Most of the urobilinogen formed in the colon is excreted in the feces. A minor fraction of the urobilinogen absorbed via the portal circulation is taken up by the liver and re-excreted into the bile. The small portion of urobilinogen which escapes into the systemic circulation is subsequently cleared by the liver or the kidneys (up to about 4 mg per 24 hours in the urine). Part of the urobilinogen which is re-excreted in the bile appears to be reabsorbed from the small intestine. Thus, the enterohepatic circulation of urobilinogen involves both the small and large intestines. In abnormal conditions, the urobilinogen levels are altered. For example, in obstructive jaundice, where the normal entry of bile into the intestines is interrupted and urobilinogen production is thereby diminished, there is an almost total absence of urobilinogen in the urine as well as in the stool. In this condition, the stool does not have its characteristic color. In hemolytic states, urobilinogen levels (fecal and urinary) are increased due to the enhanced overall production of pigments. In hepatic disorders, urinary urobilinogen values may be increased due to the inability of the hepatocytes to reabsorb urobilinogens from the portal blood and return them to the bile. This gives rise to elevated urobilinogen levels in the systemic circulation and increases urinary excretion of these compounds.

b. Laboratory evaluation of urinary urobilinogen is frequently made on a two-hour urine specimen (collected particularly in the early afternoon because, during this period, the urobilinogen clearance has been observed to be maximal and relatively constant). Urobilinogen is quantitated by treating the sample with p-dimethylaminobenzaldehyde in HCl (one of two "Ehrlich's reagents"; the other is diazosulfanilic acid - Ehrlich's diazo reagent - used in bilirubin determinations) and measuring the pink color produced. The color is stabilized and intensified by adding sodium acetate. Sodium acetate appears to provide the specificity for the reaction too because it inhibits the color formed by indole and skatole which also react with Ehrlich's reagent. A reducing agent such as ascorbate is needed in this reaction in order to maintain urobilinogens in the reduced state. The results are expressed in Ehrlich units, where each unit represents the amount of color produced by 1 mg of urobilinogen.

c. Urobilinogens in the intestine are oxidized to urobilins in reactions catalyzed by bacterial enzymes. Other decomposition products include dipyrroles (e.g., bilifuscins). These are the compounds which give the characteristic color to the stools. A schematic diagram of heme catabolism is shown in Figure 94.

6. Disposition of Bilirubin by Other Than Hepatic Routes

a. Under normal circumstances, the liver is capable of handling the excretion of bilirubin. Under abnormal conditions, when bilirubin glucuronide accumulates, it can be excreted through the urine.

b. Bilirubin is photolabile, that is, it can be decomposed upon exposure to light. The reaction requires oxygen and gives water-soluble products of unknown structure which can be eliminated in the urine. In neonatal jaundice and unconjugated hyperbilirubinemia in adults, serum levels of bilirubin have been reduced by exposure of the patient to intense light. Presumably, bilirubin in the skin and in the cutaneous capillary blood is destroyed.

The photolability of bilirubin is dependent upon oxygen, which is most likely in the singlet state. The proposed mechanism of photodegradation involves the following steps:

Figure 94. Catabolism of Hemoproteins

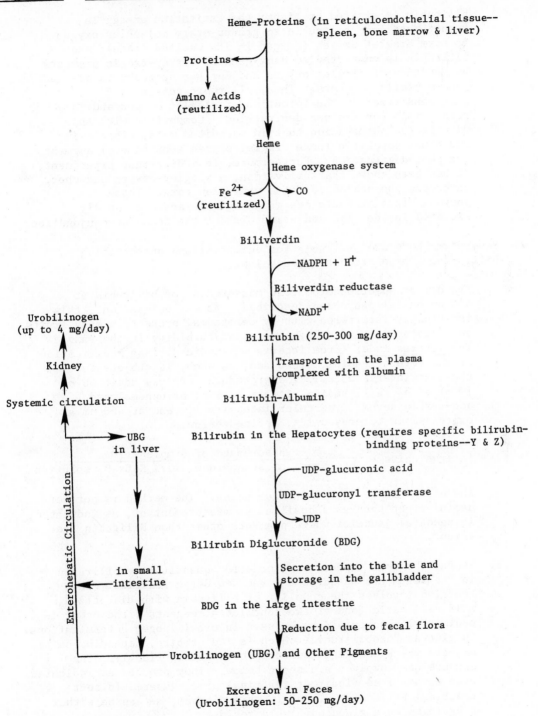

793

bilirubin is excited by light; the excitation energy is transferred from bilirubin to ground-state molecular oxygen to form singlet oxygen (oxygen in its excited state); and bilirubin is converted to dioxygenated tetrapyrrolic pigments by addition of singlet oxygen and further degraded to di- and mono-pyrrolic products. Thus, bilirubin acts as a photosensitizer in the formation of singlet oxygen and is responsible for its own destruction. Compatible with this view is the observation that in jaundiced rats, riboflavin and hematoporphyrin (both singlet-oxygen sensitizers) augment the photodegradation. Furthermore, in a different experiment, it has been shown that biliverdin, a singlet-oxygen quencher, undergoes photooxidation at a very slow rate. These photooxidized products are water soluble and are rapidly excreted in the bile and urine without the need for conjugation.

7. Several methods are available for detecting and quantitating bilirubin. These are discussed below.

   a. The direct spectrophotometric measurement of bilirubin at 450 nm has already been described. As was pointed out, this is strongly interfered with by carotenes, present in many fruits and vegetables; and by hemolysis and turbidity in the sample. The icterus index (from icteric = jaundiced) is a modification of the direct reading method and, as such, is subject to the same interferences. The procedure involves addition of 10 ml of 5% (W/V) sodium citrate to 1 ml of unhemolysed, non-turbid serum. The color intensity is read at 460 nm and expressed in icterus units, defined below.

$$\text{icterus units} = \frac{\text{absorbance of sample} \times 10}{\text{absorbance of standard } 0.0157\% \ K_2Cr_2O_7 \text{ solution}}$$

   The normal range is 3-8 icterus units. The method is not very useful other than as a qualitative measurement or, as indicated, in neonatal jaundice where pigments other than bilirubin are absent.

   b. By far the most important methods for quantitating bilirubin in the serum are based on the van den Bergh reaction. This reaction involves the coupling of bilirubin with diazotized sulfanilic acid (one of two "Ehrlich's reagents"; the other is p-dimethylaminobenzaldehyde, used in urobilinogen determinations) to give an "azobilirubin" which is red or blue, depending on pH, and which can be measured spectrophotometrically. The conversions involved are shown below. They proceed as indicated with either free (indirect reacting) or conjugated (direct reacting) bilirubin. Thus, in this diagram, R- can be either a hydrogen or a glucuronic acid residue.

Diazotization of Sulfanilic Acid

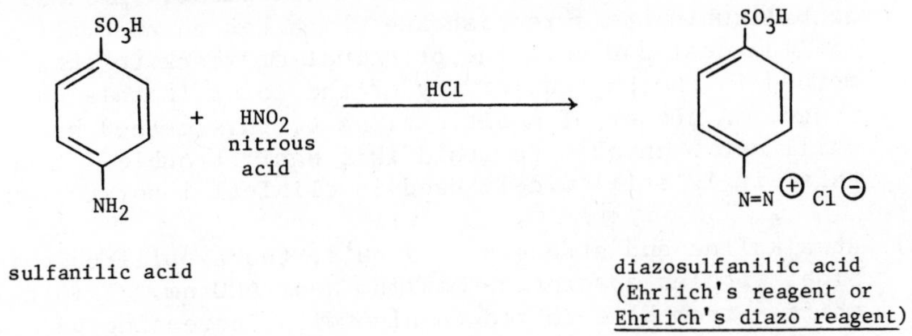
SO_3H
NH_2
+ HNO_2
nitrous
acid
HCl
SO_3H
$N{=}N^{\oplus}$ $Cl^{\ominus}$
sulfanilic acid
diazosulfanilic acid
(Ehrlich's reagent or
Ehrlich's diazo reagent)

Van den Bergh Reaction

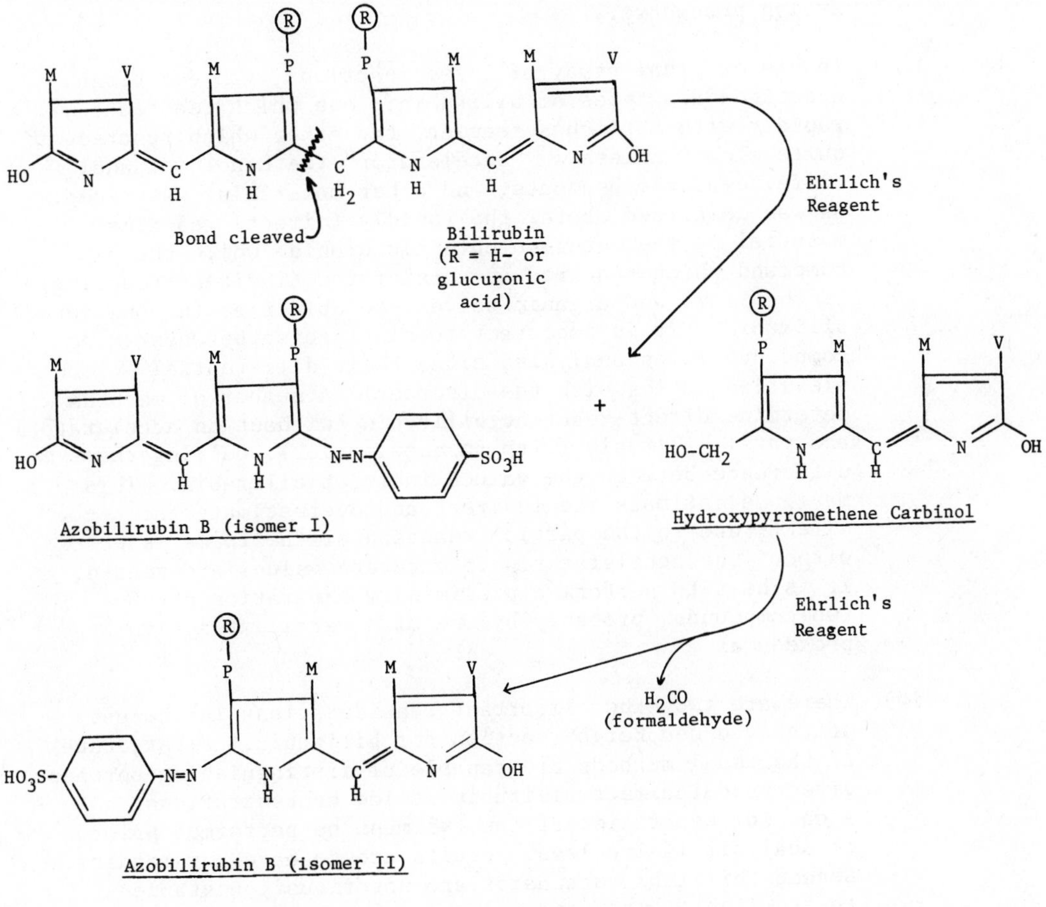
Bond cleaved
Bilirubin
(R = H- or
glucuronic
acid)
Ehrlich's
Reagent
+
Azobilirubin B (isomer I)
Hydroxypyrromethene Carbinol
Ehrlich's
Reagent
H_2CO
(formaldehyde)
Azobilirubin B (isomer II)

i) In the method of Malloy and Evelyn, a pH of 2-2.5 (weakly acid) is used for color development. At this pH, azobilirubin has a reddish hue which has an absorption maximum near 540 nm. The principal drawback to this method is the pH sensitivity of the color in this pH range. A number of modifications to this method have still not been able to avoid this basic trouble. Despite this, it is still widely used in clinical laboratories.

ii) At alkaline and strongly acid pH's, the azobilirubin is blue, with an absorption maximum near 600 nm. This color shift - from blue to red to blue with increasing pH - makes the pigment a **double-range indicator**. Jendrasic and Grof, and Nosslin, developed a method using an alkaline pH. It has the advantages of greater color intensity and greatly decreased pH sensitivity, compared to the Malloy-Evelyn procedure.

iii) In his original study of these reactions, van den Bergh described two types of bilirubin: one which reacted rapidly with Ehrlich's reagent, the other which reacted quite slowly unless an "accelerator" (methanol, ethanol, caffeine-sodium benzoate, and other materials) was added. As was described above, the rapidly (direct) reacting material is the **water-soluble** glucuronide while the compound which required an accelerator (indirect reacting) is the lipid and organic solvent-soluble free (unconjugated) bilirubin. It is important to distinguish between these compounds in an analysis, since their differential elevation varies with the disorder. A number of methods determine direct-reacting bilirubin (without an accelerator) and total bilirubin (with an accelerator), calling the difference between the values indirect bilirubin. This may underestimate the indirect and overestimate the direct, due to the partial reaction of the former even without the accelerator. If accurate values are needed, it is best to perform a preliminary separation of the two compounds, probably by one of several extractive procedures.

iv) There are two other important considerations in the use of the van den Bergh reaction for bilirubin. First, none of the usual methods are capable of distinguishing between direct and indirect bilirubin at low concentrations. A careful separation of the two must be performed prior to analysis if low level results are to have any meaning. Second, bilirubin standards are notoriously unstable toward light, heat, and solvent. Great care must be taken in standard preparation and storage. The College

of American Pathologists, the American Association of
Clinical Chemists, and the American Academy of
Pediatrics have jointly recommended that bilirubin
standards be prepared by the addition of acceptable
bilirubin preparations to pooled, normal human serum,
followed by lyophilization.

c. Bilirubin is not normally present in the urine and, when it is,
only conjugated bilirubin is found. Qualitative methods are
generally used due to the instability of the pigment toward
O_2 and because any bilirubinuria is considered abnormal. These
tests are based on oxidation of yellow bilirubin to green
biliverdin or blue bilicyanin. The oldest one is the Gmelin
or iodine-ring test where I_2 is the oxidizing agent. It is
now considered relatively insensitive and can fail to reveal
bilirubinuria. The best tests are modifications of the
Harrison test, which uses Fouchet's reagent (1% $FeCl_3$ in 25%
trichloroacetic acid solution). Addition of this reagent to
urine containing barium chloride will produce a noticeable
green color if as little as 0.05 mg of bilirubin/100 ml is
present. The Huppert-Cole method is also used.

Hyperbilirubinemias

1. As the name implies, these disorders are characterized by elevated
serum levels of bilirubin, bilirubin conjugates, and other bile
pigment metabolites. When the level of the bile pigments exceeds
the capacity of albumin to bind them, the pigments escape and
accumulate in the tissues. Note that the term bile pigments refers
to catabolites of heme, whereas bile salts (or acids) are derived
from cholesterol catabolism. Both of these molecules are secreted
into the bile. The bile pigments are waste products and are
largely eliminated by further conversion to other products by the
fecal flora. The bile salts are essential for lipid absorption
and are modified in the gastrointestinal tract where most of them
are reabsorbed. The retention of bilirubin, particularly in the
skin and eyes, gives rise to a yellow discoloration. This is
known as jaundice. Accumulation of carotene and lycopene can also
cause yellow discoloration of the skin; however, they do not cause
yellowing of the sclerae. Excessive production of bilirubin in
newborn infants (discussed later) leads to pigmentation of the
brain nuclei ("kernicterus") and may result in serious brain
damage. In a jaundiced adult, on the other hand, there may not
be any potential neurotoxicity but there may be derangements in
the organs that are involved in the formation, uptake, and
excretion of bilirubin. In general, hyperbilirubinemia can arise
from an excessive rate of breakdown of red blood cells, obstruction
of the bile duct, or because of pathological changes in the liver
cells.

2. Unconjugated hyperbilirubinemia (indirect-reacting bilirubin) can result from the three principal causes outlined below. It is important to note that in these disorders, bilirubinuria does not occur.

    a. Excessive hemolysis from a variety of disorders (e.g., erythroblastosis fetalis, pernicious anemia, thalassemia): In these situations, the liver is presented with a much greater than normal load of bilirubin. If the liver is functioning normally, it is quite capable of handling the increase. Note that with the normal daily bilirubin production of 250 mg, plasma levels remain below 1 mg per 100 ml. A new equilibrium is apparently established across the hepatocyte membrane, with a consequent elevation in circulating bilirubin (4 mg per 100 ml). Regardless of the reason for red cell destruction, if the pigment load (plasma levels greater than 4 mg per 100 ml) overwhelms the capacity of the liver to metabolize bilirubin, jaundice may result.

    b. Defects in the uptake of bilirubin by the hepatocytes: In Gilbert's syndrome, the hyperbilirubinemia is presumably due to a deficiency or defect(s) in the intrahepatic bilirubin-binding proteins (Y and Z). The compound flavaspidic acid (an active constituent of male-fern extract), which is used in the treatment of tapeworm infestations, competes with bilirubin in binding with intrahepatic acceptor proteins and thus causes increased levels of unconjugated bilirubin in the blood.

    c. Defects in the conjugation reaction: There are two genetic types. The Crigler-Najjar syndrome (constitutional non-hemolytic hyperbilirubinemia), transmitted as an autosomal recessive trait, is characterized by a complete lack of the conjugating enzyme, glucuronyl transferase. Severe neonatal jaundice may result in brain encephalopathy if not appropriately treated. The second type (type II--Arias syndrome), transmitted as an autosomal dominant trait, is a milder form where the conjugation takes place to a lesser degree, giving rise to moderate elevations of unconjugated bilirubin. Administration of phenobarbital (also used in the treatment of cholestasis) to these patients produces beneficial results by decreasing blood bilirubin levels. As discussed earlier, phenobarbital stimulates the hepatic uptake, conjugation, and secretion of bilirubin. It is important to point out that phenobarbital serves no useful purpose when administered to individuals completely lacking the conjugating enzyme.

3. Types of conjugated hyperbilirubinemia (direct-reacting bilirubin)

    a. In these disorders, the defect lies in the hepatic and/or
        biliary secretory mechanisms. An abnormal flow of bile may be
        due to mechanical obstruction occurring either intra- or
        extrahepatically (obstructive jaundice) or it can be caused by
        non-mechanical factors. This syndrome is known as cholestasis.
        Biliary tract obstruction can be caused by a variety of
        conditions such as tumors, stones, hepatitis, cirrhosis,
        stricture, pancreatitis, or cholecystitis. Non-mechanical
        cholestasis is observed in bacterial infections; after
        administration of steroids, oral contraceptives or other
        drugs; in pregnancy (rarely); and in familial, chronic
        idiopathic jaundice (Dubin-Johnson syndrome). In cholestasis
        there is retention of bile pigments as well as bile salts
        (products of cholesterol catabolism--see Chapter 8). The
        bile salt retention leads to pruritis by its action on sensory
        nerve endings in the skin. The other effects due to retention
        of bile salts are steatorrhea (fatty stool) with
        hypoprothrombinemia and osteomalacia (bile salts are essential
        for the absorption of lipids and lipid-soluble molecules such
        as vitamin K--required for prothrombin synthesis in the liver,
        and vitamin D--required for calcium absorption). Urinary
        urobilinogen is absent only when there is complete obstruction.
        Recall that the formation of urobilinogen is dependent upon
        excretion of bilirubin diglucuronide into the intestine and
        its modification there by fecal flora. The stool specimens
        of patients with cholestasis show a characteristic clay color,
        due to the absence of pigments.

    b. The Dubin-Johnson syndrome is characterized by deposition of
        a brown pigment (possibly related to melanin) in the liver and
        by a fairly mild, chronic, nonhemolytic hyperbilirubinemia.
        Roughly equal amounts of conjugated and unconjugated pigment
        appear in the plasma with the former usually predominating
        slightly. The bilirubin levels and a number of other features
        found in Rotor's syndrome are quite similar to those of the
        Dubin-Johnson syndrome. Rotor, however, did not observe any
        liver pigmentation in his patients and, unlike Dubin and
        Johnson, was able to visualize the gallbladder by
        cholecystography. Because Rotor described only three patients,
        it is difficult to draw any real conclusions. There is a
        great deal of variation in the severity of symptoms in Dubin-
        Johnson patients, however, and it seems possible that the two
        syndromes are actually different manifestations of one
        biochemical lesion. It is interesting that in the Dubin-
        Johnson disorder, the ratio of coproporphyrinogen I to

coproporphyrinogen III excreted in the urine is elevated over that of normal individuals and bromosulphthalein retention is 10-20% (variable) over normal (see liver function tests). The disease appears to be inherited as an autosomal recessive trait.

c. In the above disorders, the mechanism by which conjugated bilirubin appears in the plasma is not clear. It has been suggested that there may be injury to the endothelial lining of the bile ductules (cholangioles) leading to the leakage of the conjugated bilirubin into the plasma. In addition, reverse pinocytosis has been implicated as a mechanism for transfer of bilirubin glucuronides to sinusoidal plasma. Electron microscopic studies have shown a multiplicity of structural derangements suggesting that intrahepatic cholestasis may be due to several independent factors.

d. It is important to point out that when conjugated bilirubin begins to have access to the blood, its appearance in the urine also becomes apparent. In liver disorders (viral, hepatitis, cirrhosis, etc.), plasma levels of both conjugated and unconjugated bilirubin levels are often increased. This is indicative of general damage to the hepatocytes.

4. Jaundice in the neonate

a. Occurrence of some degree of jaundice in the neonatal period (physiologic jaundice) is not uncommon. It is usually due to the immaturity of some aspect of bilirubin conjugation in the liver. As noted earlier, the liver is apparently not "mature" at birth and requires one to two weeks postnatally to complete development of the uptake and conjugating system. Ordinarily, there is a rise in the levels of bilirubin on the first day of life, reaching a maximum on the third or fourth day (at which time it rarely exceeds 10 mg per 100 ml; see Table 36). In erythroblastosis fetalis, high levels of bilirubin occur (giving rise to generalized jaundice, brain jaundice, disturbances of the nervous system, and death). Erythroblastosis fetalis is also characterized by the presence of increased numbers of nucleated red cells in the peripheral blood and extra-medullary hematopoiesis. One cause of erythroblastosis fetalis is Rh incompatibility.

b. Neonatal, nonhemolytic jaundice can be subdivided into two groups: unconjugated and conjugated hyperbilirubinemia. As already mentioned, the most common form of unconjugated, nonhemolytic hyperbilirubinemia is physiologic jaundice. Other disorders falling under this category are as follows.

i)   Hereditary glucuronyl transferase deficiency: Two
     types--recessive and dominant (discussed earlier).

ii) Breast-milk jaundice: In the milk of some mothers,
     there is a relatively high level of a hormone,
     presumably pregnanediol, which inhibits bilirubin
     conjugation. This leads to unconjugated hyperbilirubinemia,
     but the levels do not usually rise so as to result in
     neurotoxicity. Therefore, the mothers of infants with
     this syndrome, in the absence of other complications, may
     not be asked to discontinue breastfeeding.

iii) Intestinal obstruction: The intestine of the newborn
     contains a high activity of β-glucuronidase. This
     enzyme converts conjugated bilirubin to its free form
     which can be absorbed into the portal and, eventually,
     systemic blood circulation. Obstruction and decreased
     intestinal mobility leads to increased enterohepatic
     circulation of bilirubin.

iv) Hypothyroidism.

v)   Jaundice associated with certain drugs (e.g., novobiocin)
     which can inhibit glucuronide conjugation in the neonatal
     period.

c. Conjugated hyperbilirubinemia (direct-reacting bilirubin),
    although rare during the neonatal period, can occur due to
    many causes. In general, the etiology of this disorder is
    due either to impaired hepatocellular function or extrahepatic
    obstruction. The impaired hepatocellular function can be due
    to:

i)   Infection (congenital or acquired): Hepatitis A and B
     viruses, bacterial infection, syphilis, rubella,
     cytomegalovirus, toxoplasmosis, herpes simplex,
     coxsackie virus, and listeria.

ii) Genetic diseases: Cystic fibrosis, galactosemia,
     tyrosinemia, α_1-antitrypsin deficiency, infantile
     Gaucher's disease, Dubin-Johnson syndrome, and Rotor's
     syndrome.

iii) Miscellaneous: Cholangitis, hepatic necrosis, and
     deficiency of intrahepatic ducts.

The extrahepatic biliary obstruction may be caused by biliary
atresia (absence or closure of the biliary tree), bile-plug

Table 36. Serum Bilirubin Levels in Some Normal and Abnormal Conditions

Condition	Bilirubin Values in mg/100 ml of Plasma or Serum		
	Total	Unconjugated (indirect reacting)	Conjugated (direct reacting)
Newborn, cord blood	0.2 – 2.9		
1 day	0 – 6.0		
3 days	0.25–11		
7 days	0.14– 9.9		
Normal (adult)	0.2 – 1.0	0.2 – 1.0	0 – 0.20
Hemolytic Disorders (in adults)	2.2 – 3.4	2.0 – 3.0	0.2 – 0.4
Glucuronyl Transferase Deficiency			
Type I: Crigler–Najjar Syndrome—complete deficiency	15 –48	15 –48	negligible
Type II: Arias Syndrome—partial deficiency	6 –22	6 –22	trace
Dubin–Johnson Syndrome (perhaps identical to Rotor's syndrome—defect in the bile secretory mechanism)	2.49–19 (ave. of 46 cases = 5.6)	2.2 (ave. of 34 cases; ranged from 74%–14% of total)	3.4 (ave. of 34 cases; ranged from 26%–86% of total)
Gilbert's Syndrome (defects in hepatic bilirubin uptake)	2.2 (may occasionally be as high as 8 mg)	2.0 (may occasionally be as high as 8 mg)	0.2
Cholestasis (severe form)	10.0	1.0	9.0
Cirrhosis (severe)	11.0	5.0	6.0
Hepatitis (acute–severe)	10.0	1.5	8.5

syndrome, choledochal cyst, or bile ascites. The liver disorders associated with infection or hereditary metabolic disease can be distinguished from problems with extrahepatic biliary obstruction by identifying the organism causing the infection or measurement of the appropriate enzyme. However, the differentiation between neonatal hepatitis (not due either to infection or hereditary metabolic disease) and biliary atresia is a problem as they both present similar clinical manifestations (conjugated hyperbilirubinemia with jaundice appearing at 1 to 8 weeks, dark urine and pale stools, elevated serum transaminases and alkaline phosphatase, and hepatosplenomegaly). It is essential to distinguish between these two conditions because biliary atresia is correctable by surgery after an exploratory laparotomy (abdominal section), whereas neonatal hepatitis may lead to cirrhosis in patients who have undergone exploratory laparotomy. Some of the diagnostic procedures used in the differentiation of these two disorders are: ^{131}I-rose-bengal excretion after phenobarbital treatment (increased excretion in hepatitis; no increase in biliary atresia), vitamin E administration test (serum vitamin E increases in hepatitis but not in atresia--this is because vitamin E is a lipid-soluble vitamin and requires bile acids (or salts) for its absorption (see Chapter 8) ; in biliary atresia the bile entry into the intestine is interrupted), red cell peroxide hemolysis test (this is an indirect test which measures the intestinal absorption of vitamin E, and its deficiency results in enhanced sensitivity of erythrocytes to hemolysis in the presence of hydrogen peroxide), and measurement of serum bile acids (in cholestasis the serum bile acids are elevated).

5. Jaundice in older children and adults: Algorithms developed by J.D. Ostrow for the diagnosis of jaundice in older children and adults are presented in Figure 95. It should be noted that in developing any algorithms, certain arbitrary decisions have to be made. It is recommended that the reference listed with the figure be consulted.

(Text continues on p. 807)

Figure 95--Algorithm 1. Classification of Jaundice

Circled numbers refer to passages in the original publication.
PSP indicates phenolsulfonphthalein; BSP indicates sulfobromophthalein.

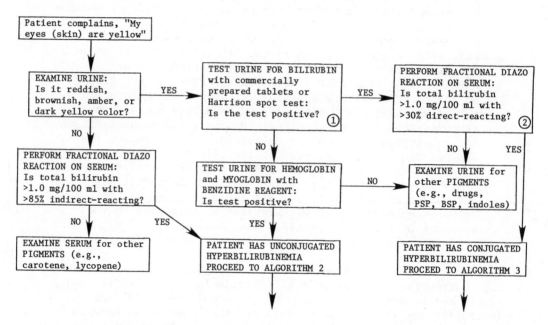

Reproduced with permission from J. Donald Ostrow, JAMA, 234, 522-526, 1975, Copyright 1975, American Medical Association.

Figure 95--Algorithm 2. Unconjugated Hyperbilirubinemia

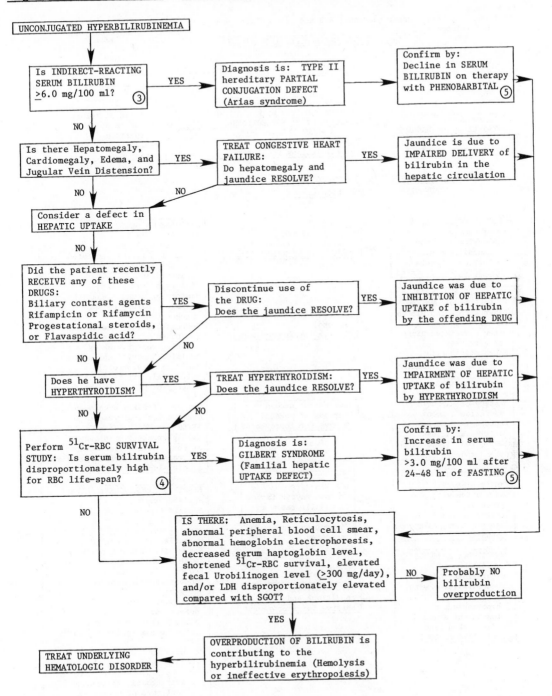

Reproduced with permission from J. Donald Ostrow, JAMA, 234, 522-526, 1975, Copyright 1975, American Medical Association.

Figure 95--Algorithm 3. Conjugated Hyperbilirubinemia

RUQ indicates right upper quadrant; KUB indicates plain x-ray film of abdomen; GI, gastrointestinal.

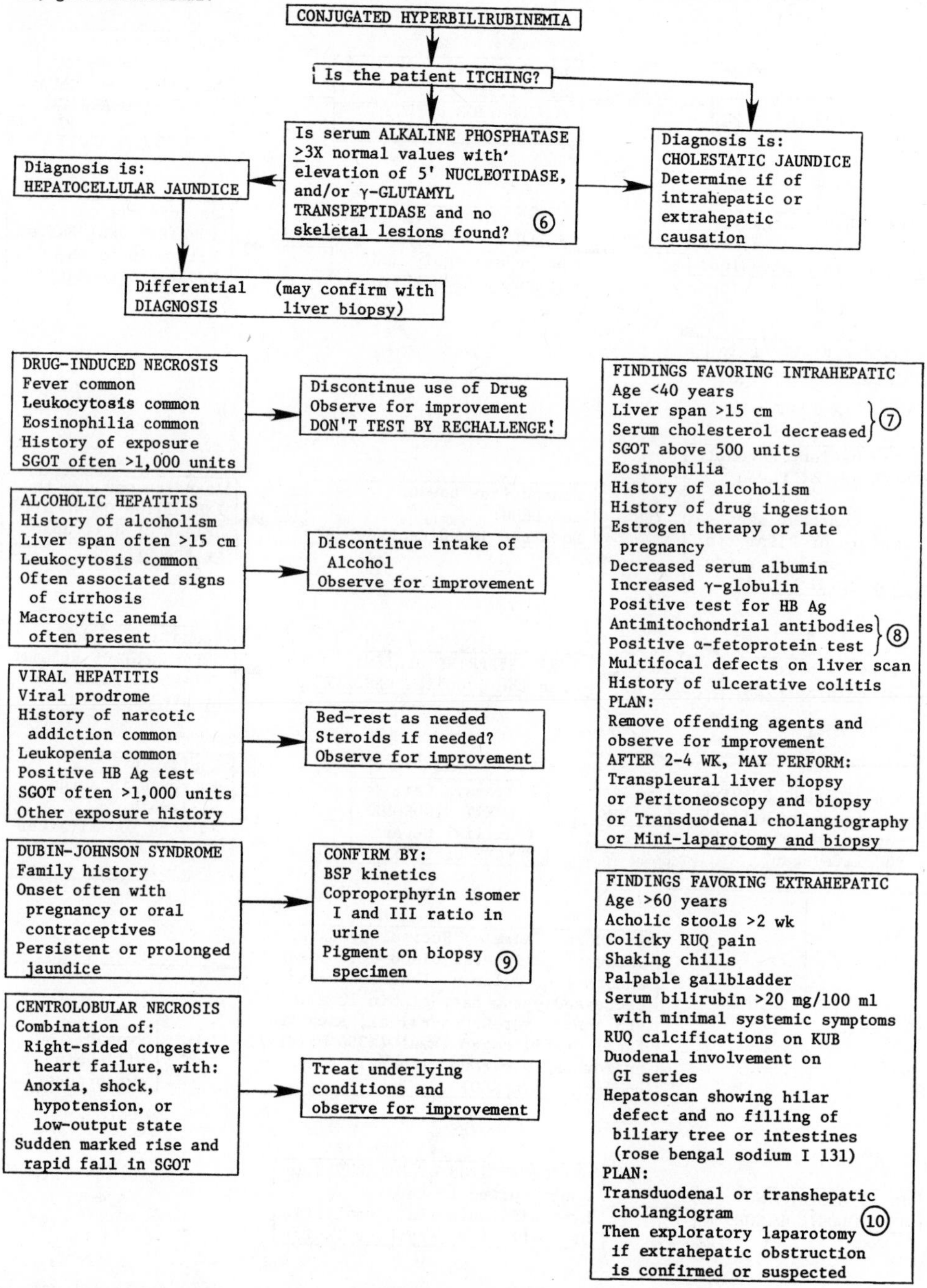

CONJUGATED HYPERBILIRUBINEMIA

Is the patient ITCHING?

Is serum ALKALINE PHOSPHATASE >3X normal values with' elevation of 5' NUCLEOTIDASE, and/or γ-GLUTAMYL TRANSPEPTIDASE and no skeletal lesions found? ⑥

Diagnosis is: HEPATOCELLULAR JAUNDICE

Diagnosis is: CHOLESTATIC JAUNDICE Determine if of intrahepatic or extrahepatic causation

Differential DIAGNOSIS (may confirm with liver biopsy)

DRUG-INDUCED NECROSIS
Fever common
Leukocytosis common
Eosinophilia common
History of exposure
SGOT often >1,000 units

→ Discontinue use of Drug
Observe for improvement
DON'T TEST BY RECHALLENGE!

ALCOHOLIC HEPATITIS
History of alcoholism
Liver span often >15 cm
Leukocytosis common
Often associated signs of cirrhosis
Macrocytic anemia often present

→ Discontinue intake of Alcohol
Observe for improvement

VIRAL HEPATITIS
Viral prodrome
History of narcotic addiction common
Leukopenia common
Positive HB Ag test
SGOT often >1,000 units
Other exposure history

→ Bed-rest as needed
Steroids if needed?
Observe for improvement

DUBIN-JOHNSON SYNDROME
Family history
Onset often with pregnancy or oral contraceptives
Persistent or prolonged jaundice

→ CONFIRM BY:
BSP kinetics
Coproporphyrin isomer I and III ratio in urine
Pigment on biopsy specimen ⑨

CENTROLOBULAR NECROSIS
Combination of:
Right-sided congestive heart failure, with:
Anoxia, shock, hypotension, or low-output state
Sudden marked rise and rapid fall in SGOT

→ Treat underlying conditions and observe for improvement

FINDINGS FAVORING INTRAHEPATIC
Age <40 years
Liver span >15 cm
Serum cholesterol decreased } ⑦
SGOT above 500 units
Eosinophilia
History of alcoholism
History of drug ingestion
Estrogen therapy or late pregnancy
Decreased serum albumin
Increased γ-globulin
Positive test for HB Ag
Antimitochondrial antibodies } ⑧
Positive α-fetoprotein test
Multifocal defects on liver scan
History of ulcerative colitis
PLAN:
Remove offending agents and observe for improvement
AFTER 2-4 WK, MAY PERFORM:
Transpleural liver biopsy
or Peritoneoscopy and biopsy
or Transduodenal cholangiography
or Mini-laparotomy and biopsy

FINDINGS FAVORING EXTRAHEPATIC
Age >60 years
Acholic stools >2 wk
Colicky RUQ pain
Shaking chills
Palpable gallbladder
Serum bilirubin >20 mg/100 ml with minimal systemic symptoms
RUQ calcifications on KUB
Duodenal involvement on GI series
Hepatoscan showing hilar defect and no filling of biliary tree or intestines (rose bengal sodium I 131)
PLAN:
Transduodenal or transhepatic cholangiogram
Then exploratory laparotomy ⑩ if extrahepatic obstruction is confirmed or suspected

8
Lipids

1. Lipids are insoluble in water but are soluble in organic solvents such as chloroform, ether, and benzene. The list of these compounds includes fats, oils (which are liquid at room temperature), waxes, and related compounds.

2. Fats serve as

   a. Insulating material (in subcutaneous tissue and as "padding" around certain organs).

   b. A major source of caloric energy, used directly or stored. Fat is an efficient and concentrated form of stored energy. In the human body, the oxidation of one gram of fat yields about 9 kcal, whereas one gram of protein or carbohydrate gives rise to about 4 kcal. This is partly because the majority of the carbons in triglycerides (the major form of stored lipid) are in a lower oxidation state than are those in carbohydrates. Consequently, the oxidation of fats to CO_2 and water makes possible the synthesis of more moles of ATP per carbon atom than does carbohydrate oxidation (see Chapter 4 and the section on β-oxidation in this chapter). In addition, stored lipid has very little water associated with it, unlike glycogen (the principal storage form of carbohydrate). This means that storage of lipid takes up much less volume than does the storage of an equal mass of carbohydrate.

   Fats and oils provide an economical source of calories to the diet of western society. It is estimated that, in the diet of the United States, fat accounts for 40-45 percent of the total caloric intake. However, the type of fat consumed varies with the different income groups. In low-income families the intake of lard is high, while in high-income families salad and cooking oils predominate.

   c. Combinations of fats with proteins (lipoproteins) are important cellular constituents. They occur in cell membranes and in membranes of intracellular organelles such as nuclei, microsomes, mitochondria, etc. Lipid derivatives are part of the components of the electron-transport system in mitochondria.

   d. Unsaturated fatty acids containing two or more double bonds, (linoleic, linolenic, and arachidonic) are not synthesized in the body and, hence, must be furnished in the diet. The

807

functions of these essential fatty acids are not well
understood, but they are required for normal growth and
function. One recently described role may be in the synthesis
of prostaglandins.

3. Metabolism of Fats   (a few general statements)

  a. The body fat is in a dynamic state.

  b. Much of the carbohydrate of the diet is converted to fat before
      it is utilized to provide energy.

  c. Fat may serve as a major source of energy in many tissues.
      Except for the brain, tissues may use free fatty acids (FFA)
      as a fuel in preference to carbohydrates.

  d. Fat's caloric value is 9 kcal/gm as mentioned above and it is
      associated with less water in storage than either protein or
      carbohydrate.

  e. Fats are the major source of calories in the postabsorptive
      state.

Types of Lipids

1. Simple lipids: esters of fatty acids with various alcohols.

  a. Fats: esters of fatty acids with glycerol (glycerides).

  b. Waxes: esters of fatty acids with monohydroxy aliphatic
      alcohols.

      i) True waxes: esters of cetyl alcohol ($CH_3(CH_2)_{14}CH_2OH$) or
          other higher straight-chain alcohols with palmitic,
          stearic, oleic or other higher fatty acids.

      ii) Cholesterol esters: esters of cholesterol with fatty
           acids.

      iii) Vitamin A esters: esters of vitamin A with palmitic or
            stearic acid.

      iv) Vitamin D esters.

2. Compound lipids: esters of alcohols with fatty acids which also
    contain other groups.

  a. Phospholipids: contain fatty acids, glycerol, phosphoric acid,
      and in most cases, a nitrogenous base.

b. Glycolipids or cerebrosides: contain both carbohydrate and nitrogen but not phosphate or glycerol.

c. Sulfolipids: contain sulfate groups.

d. Lipoproteins: lipid and protein complexes.

e. Lipopolysaccharides: lipid and polysaccharide complexes.

3. <u>Derived lipids</u>: substances derived from the above group by hydrolysis; they include

a. Fatty acids (saturated & unsaturated). Any aliphatic carboxylic acid of about six carbons or more is generally considered a fatty acid. This definition is flexible.

b. Mono-and diglycerides (derived from triglycerides by saponification of, respectively, two or one of the ester linkages).

c. Glycerol, sterols and other steroids (vitamin D), and alcohols containing β-ionone rings (vitamin A).

d. Fatty aldehydes.

e. The lipid portion of lipoproteins.

4. <u>Miscellaneous lipids</u>: aliphatic hydrocarbons (iso-octadecane), carotenoids, squalene (an intermediate in the biosynthesis of cholesterol) and other terpenes (named for turpentine), vitamins E and K, glycerol ethers (diesters of these compounds occur in tissues from various sources including human neoplasms), glycosylglycerols (in plants), etc.

Fatty Acids

1. Saturated fatty acids   General Formula: $C_nH_{2n+1}COOH$
   (No double bonds)

   acetic acid CH_3COOH

   butyric acid C_3H_7COOH – found in fats from butter

   palmitic acid $C_{15}H_{31}COOH$ ⎱ present in all animal and plant

   stearic acid $C_{17}H_{35}COOH$ ⎰ fats

   lignoceric acid $C_{23}H_{47}COOH$ – present in cerebrosides

2. Unsaturated fatty acids (general formula: $C_nH_{2n+1-2m}COOH$ where m = number of double bonds + number of rings in molecules).

   a. One double bond: $C_nH_{2n-1}COOH$

      Examples: Palmitoleic – $C_{15}H_{29}COOH$ (C_{16}; $\Delta 9$)

         cis-9-Hexadecanoate

         Oleic – $C_{17}H_{33}COOH$ (C_{18}; $\Delta 9$)

         cis-9-Octadecanoate

      Note: The Δ indicates the location of the double bond. For example, $\Delta 9$ means that the double bond is between carbons 9 and 10. The carboxyl carbon is carbon number 1. The location of the double bond in the naturally occurring fatty acids can also be related to the ω-end of the fatty acid molecule (i.e., the last methyl carbon of the fatty acid starting from the carboxyl end); the monosaturated oleic acid is an ω-9 acid and the polyunsaturated (listed below) linoleic acid has double bonds at ω-6 and ω-9 carbons.

   b. Two double bonds: $(C)_n(H)_{2n-3}COOH$ – linoleic; all cis-9,12-Octadecadienoic acid (C_{18}; $\Delta 9,12$).

   c. Three double bonds: $(C)_n(H)_{2n-5}COOH$ – linolenic; all cis-9,12,15-Octadecadienoic acid (C_{18}; $\Delta 9,12,15$).

   d. Four double bonds: $(C)_n(H)_{2n-7}COOH$ – arachidonic; all cis-5,8,11,14-eicosatetraenoic acid (C_{20}; $\Delta 5,8,11,14$).

The presence of a double bond in the hydrocarbon chain gives rise to geometric isomerism. This type of isomerism is due to restricted rotation of carbon-carbon double bonds and is exemplified by fumaric acid and maleic acid:

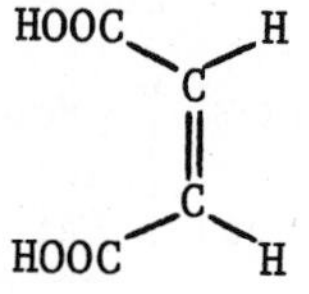

Maleic Acid
(Cis-form)

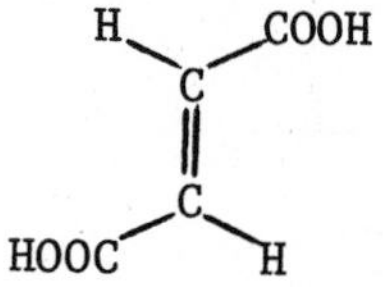

Fumaric Acid
(Trans-form)

It is of interest to point out that almost all of the naturally
occurring, unsaturated long-chain fatty acids are found as the
less stable cis-isomers instead of the more stable trans-isomers.
The cis-configuration in the molecule leads to bending at the
position of the double bond as shown for oleic acid and linoleic
acid:

$$HOOC(H_2C)_7 - C \begin{matrix} H \\ \| cis \\ \end{matrix} C \begin{matrix} \\ H \end{matrix}$$

$$H_3C(H_2C)_7 - C$$

Oleic Acid

$$H - C \begin{matrix} (CH_2)_7COOH \\ \| cis \end{matrix}$$

$$H - C \begin{matrix} CH_2 \\ C \begin{matrix} \| cis \end{matrix} \end{matrix} H$$

$$H_3C(CH_2)_4 - C \begin{matrix} H \end{matrix}$$

Linoleic Acid

The geometries of a saturated hydrocarbon chain and similar
monounsaturated cis- and trans-chains are shown in Figure 96.
Note that the cis-configuration introduces a bend (of about 30°),
whereas the trans-isomer resembles the extended form of the
saturated chain.

Arachidonic acid with four cis-double bonds is a U-shaped molecule.
It is the cis-isomers that are biologically active in terms of
their function as essential fatty acids (discussed later), and
the trans-polyunsaturated fatty acids cannot substitute for the
cis-isomers. The trans-isomers, however, are metabolized in the
body similar to the saturated fatty acids.

Fatty acids with one to eight carbons are liquids at room
temperature while those with more carbon atoms are solids.
However, with long chain fatty acids, unsaturation lowers the
melting point below that of the corresponding saturated fatty
acids. For example, stearic acid (C18) has a melting point of
70°C whereas oleic acid (C18; Δ9) has a melting point of 14°C
and, therefore, is a liquid at room temperature.

3. a. Quantitatively, the largest group of fatty acids in biological
      systems is composed of acyclic, unbranched, non-hydroxylated
      molecules containing an even number of carbon atoms. Appreciable
      amounts of other types of fatty acids do occur in certain
      tissues, however, and may perform important roles in the life
      of various organisms including man. The prostaglandins,
      discussed later, provide one example. Another is the membrane
      phospholipids, many of which have an unsaturated fatty acyl
      group as one of their sidechains. The metabolism of some of

Figure 96. Geometry of Saturated, Trans-Monounsaturated and Cis-Monounsaturated Chains

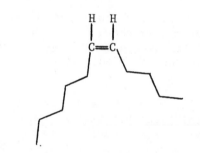

Saturated Chain

Trans-Monounsaturated Chain
(uncommon in naturally
occurring fatty acids)

Cis-Monounsaturated Chain
(common in naturally
occurring fatty acids)

these compounds is discussed later, along with the more common types of fatty acid metabolism.

b. Some of these "unusual" fatty acids are produced only by plants or microorganisms. Others are synthesized by mammals including man. By means of gas chromatography and mass spectroscopy, human earwax (cerumen) has been shown to contain varying amounts of over fifty different types of fatty acids of all of the kinds mentioned above. Sebum (the secretion of the sebaceous glands in the skin) also contains a wide variety of fatty acids in both free and esterified forms. Skin microorganisms and spontaneous oxidative and hydrolytic reactions may be partly responsible for the diversity which is found.

c. Some specific examples of these compounds include

    i) Tuberculostearic acid, part of the lipids of the tubercle bacillus, is 10-methylstearic acid. Thus it is both an odd-chain-length and a branched acid.

ii)   Cerebronic acid, mentioned later under cerebrosides, is 2-hydroxylignoceric acid, an α-hydroxy fatty acid.

iii) Ricinoleic acid (12-hydroxyoleic acid) is a hydroxy-unsaturated compound characteristic of castor oil.

iv)   Chaulmoogric acid, isolated from chaulmoogra oil and previously used in the treatment of Hansen's disease (leprosy), has the structure

Chaulmoogric Acid

It should be noted that the drug of choice for the treatment of all forms of leprosy belongs to a class of compounds known as sulfones. Dapsone is a sulfone, and all the other clinically useful sulfones are derived from it. Its structure is as follows.

4,4'-Diaminodiphenylsulfone
(Dapsone)

v)   Phytanic acid (3,7,11,15-tetramethylhexadecanoic acid; note that hexadecanoic acid is palmitic acid) is the oxidation product of phytol, an open-chain terpene derived from chlorophyll. It is present in foods derived from herbivorous animals, notably cow's milk and animal fat. Because of the 3-methyl group, it is resistant to β-oxidation and is normally catabolized by the phytanic acid α-oxidase system to pristanic acid which can be β-oxidized. In persons with Refsum's syndrome, a nervous disorder, up to about 20% of the serum fatty acids and 50% of the hepatic fatty acids are composed of phytanic acid, whereas the normal amount in serum is less than 1 μg/ml. This is due to the (heritable) absence of phytanic acid α-hydroxylase in affected individuals. This enzyme catalyzes one of the initial reactions in the α-oxidation (discussed later in this chapter).

vi) Bacteria are able to synthesize cyclopropyl fatty acids of general formula

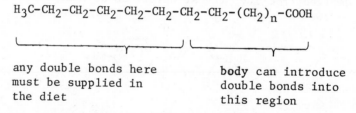

where x and y are integers. The substrate is a cis-unsaturated fatty acid, usually in the β-position of phosphatidylethanolamine. The extra methylene group is supplied by S-adenosylmethionine.

4. There is a group of fatty acids that mammals need but are unable to synthesize. These <u>essential fatty acids</u> must be supplied in the diet. Linoleic, linolenic, and arachidonic acids are generally considered to comprise this group although man can convert linoleic acid to arachidonic acid. These are essential because, in each of them, <u>one or more of the double bonds lies somewhere between the</u> <u>terminal methyl group and the sixth carbon from that group</u>. In this region, indicated in the drawing below, mammals are unable to introduce a double bond.

$$H_3C-CH_2-CH_2-CH_2-CH_2-CH_2-CH_2-CH_2-(CH_2)_n-COOH$$

any double bonds here must be supplied in the diet   body can introduce double bonds into this region

a. The biologic functions of essential fatty acids (EFA) have not been fully defined. They are constituents of cellular membranes and precursors in the synthesis of prostaglandins, which are involved in many physiologic processes (see below).

b. Palmitoleic and oleic acids (both possess a cis Δ9 double bond) are <u>not essential fatty acids</u>, because these compounds can be synthesized from the respective saturated fatty acid coenzyme-A esters. The enzyme system which catalyzes this reaction is a monooxygenase system and is found in the endoplasmic reticulum of the liver and adipose tissue. The <u>overall</u> reaction for palmitoleic acid synthesis is:

$$Palmitoyl\text{-}CoA + NADPH + H^+ + O_2 \longrightarrow Palmitoleyl\text{-}CoA + NADP^+ + 2H_2O$$

Note in the above reaction that one mole of the molecular oxygen accepts two pairs of electrons, one pair derived from palmitoyl-CoA and the other from NADPH. The electrons from NADPH are transported via cytochrome-b_5 reductase to

cytochrome b_5 (microsomal electron transport). The terminal
reaction consists of the activation of palmitoyl-CoA and
oxygen by a protein which is sensitive to cyanide and known
as cyanide-sensitive factor. Oleic acid can be further
dehydrogenated to yield polyunsaturated fatty acids (PUFA)
that are <u>nonessential</u>. These PUFA will <u>not</u> replace the
requirement for the essential fatty acids mentioned above.
A PUFA that is derived from oleic acid is 5,8,11-eicosatrienoic
acid (a 20-carbon fatty acid with three double bonds; the
final double bond is nine carbons removed from the terminal
methyl group--a 9ω fatty acid). It is of interest to point
out that this trienoic acid is found in significant amounts
in the heart, liver, adipose tissue and erythrocytes of
animals fed diets low in essential fatty acids. Its level
is decreased when the diet is supplemented with linoleic or
linolenic acids. The appearance of the 9ω-trienoic acid
(C20; Δ5,8,11) in tissues and plasma has been used in the
assessment of EFA deficiency. One method consists of measuring
the ratio of trienoic acid to arachidonic acid (C20; Δ5,8,11,14;
a tetraene), and it has been observed that in rats, a triene-
to-tetraene ratio of 0.4 or less indicates normal EFA status.
In humans the recommended dietary allowance for EFA is 1 to
2% of the total energy intake, which also maintains a normal
ratio of triene to tetraene.

c. Linoleic acid plays an important role in the prevention of
essential fatty acid deficiency (discussed below). Arachidonic
acid can be synthesized from linoleic acid, but it should be
emphasized that all vertebrates <u>lack</u> the ability to synthesize
linoleic acid <u>de novo</u>. Thus, the requirement of arachidonate
may be dispensed with if there are adequate amounts of
linoleate in the diet.

d. The conversion of linoleic acid to arachidonic acid:

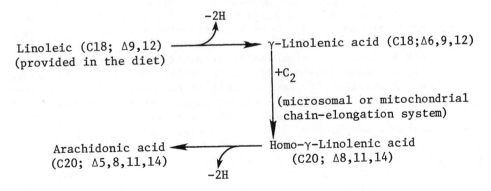

The desaturation steps occur in the microsomes, utilizing the cytochrome-b_5-oxygenase system with NADPH functioning as a coreductant of oxygen. The fatty acid chain elongation is accomplished by mitochondrial or microsomal enzyme systems. The source of carbons for fatty acid elongation differs between microsomal and mitochondrial pathways. In the former, malonyl-CoA is used, whereas in the latter, acetyl-CoA provides the carbons:

Microsomal System

$$C18(6,9,12) \xrightarrow[\substack{\text{NADPH} \\ \searrow \\ CO_2}]{\text{Malonyl-CoA}} C20(8,11,14)$$

Mitochondrial System

$$C18(6,9,12) \xrightarrow[\text{NADPH}]{\text{Acetyl-CoA}} C20(8,11,14)$$

e. Essential fatty acid deficiency. The clinical manifestations of EFA deficiency in humans closely resemble those seen in animals. These include dry, scaly skin, usually erythematous eruptions (generalized or localized and affecting the trunk, legs, and intertriginous areas), diffuse hair loss (seen frequently in infants), poor wound healing, failure of growth and an increase in the metabolic rate. Abnormalities in ECG patterns have been reported and may be due to membrane alterations that occur in EFA-deficient states. Similar changes may be the basis for structural and functional abnormalities observed in the mitochondria obtained from EFA-deficient animals. Surgical patients who are maintained on glucose-amino acid solutions for prolonged periods develop EFA deficiency, and the reported signs and symptoms in these cases include anemia, thrombocytopenia, hair loss and sparse hair growth, increased capillary permeability, dry scaly skin, desquamating dermatitis and a shift in the oxygen dissociation curve to the left. An oral or intravenous administration of linoleic acid is necessary to correct these problems. Commercially available fat emulsions which contain linoleic acid are available for intravenous use. An adult patient needs about 10 g of linoleic acid per day.

EFA deficiency can also occur in infants with highly restricted unsaturated diets (e.g., primarily skim milk intake), in patients with total parenteral hyperalimentation without the appropriate supplements of unsaturated lipids, and in patients with severe malabsorptive defects.

f. Trans fatty acids. As noted earlier, almost all double bonds in the naturally occurring fatty acids are in the cis-configuration. Furthermore, in the EFA the double bonds are located farthest from the carboxyl group on the sixth carbon (ω6) relative to the methyl end of the molecule. Trans-polyunsaturated fatty acids <u>do not</u> replace their cis-counterparts in the requirement for essential fatty acid activity and seem to possess a hypercholesterolemic effect. During the process of <u>partial</u> dehydrogenation of vegetable oils (e.g., margarine), the cis-fatty acids are isomerized to trans-compounds. The various "hydrogenated" margarines contain 15 to 40 percent of unnatural trans fatty acids. The hypercholesterolemic effect of trans fatty acids is presumably due to its impairment of the first rate-limiting step in the formation of bile acid from cholesterol. Since the steady state level of cholesterol is dependent upon its conversion to bile acid (see bile acid metabolism--this chapter), any perturbation in this process affects the cholesterol levels. These aspects are receiving considerable attention because there is evidence that hypercholesterolemia is contributory (a risk factor) to the development of coronary heart disease and that a severe reduction of cholesterol levels may bring about amelioration of the clinical events (see lipoprotein metabolism in this chapter for a discussion on the relationship between cholesterol levels in various lipoproteins and the incidence of coronary heart disease). EFA substitution for saturated fatty acids in the diet lowers the plasma cholesterol levels and the reason for this is not clear. Some of the explanations that have been offered are that EFA's: stimulate the secretion of cholesterol into the intestine, enhance the oxidation of cholesterol to bile acids, esterify with cholesterol and cause an enhanced rate of turnover and excretion, and shift the distribution of cholesterol from the plasma into the tissues, perhaps by decreased secretion of lipoproteins from the liver. It is apparent from the above discussion that the conversion of naturally occurring cis-fatty acids to trans isomers by artificial means changes a beneficial substance into a deleterious product. Most vegetable oils are relatively rich in EFA (except for coconut oil), low in saturated fatty acids and lacking in cholesterol. Animal fats, on the other hand, are generally low in EFA (except for fish), high in saturated fats and contain cholesterol. It is of interest to point out that the EFA content of the body and milk fat of ruminants can be increased by the feeding of EFA that are encased in formalin-treated casein. The EFA is only released at the site of absorption by dissolution of the capsules. These dietary manipulations in the ruminants are accompanied by an increase in carcass EFA, a decrease in saturated fatty acids and no change or an increase in the

cholesterol content. Table 37 summarizes the fatty acid
composition of some fats of animal and plant origin. The
prudent diet recommended by various health organizations
with respect to fat intake is that the fat consumption should
not exceed 30–35% (current average of American consumption
is 40–45%) of the total calories per day, with equal amounts
of saturated and polyunsaturated fats and a cholesterol
intake of not more than 300 mg/day (current average of
American consumption is about 600 mg/day). Note that all
animal fat contains cholesterol, whereas vegetable oils
have no cholesterol.

5. The roles of free fatty acids in the plasma, as well as methods
for their analysis, are discussed in the section on plasma lipids.
They are also important constituents of the acylglycerides,
phospholipids, and cholesterol esters. The synthesis and catabolism
of the fatty acids are also covered later in this chapter.

Prostaglandins (PG)

1. Prostaglandins, discovered initially in the seminal plasma, are
synthesized in every tissue of the body. They are ubiquitous.
The seminal plasma PG's are formed in the prostate gland from
which the name prostaglandins is derived. PG's are rapidly
synthesized as well as inactivated and the catabolizing enzymes
are distributed throughout the body. In particular, the lungs
not only participate in the active synthesis of PG's, but also
play a major role in the catabolism of PG's released into the
circulation from other organs.

2. a. The parent compound of PG's is prostanoic acid. It is a
C20 fatty acid containing a five-membered (cyclopentane)
ring. The dotted bond between carbons 7 and 8 is aimed
back into the page from carbon 8.

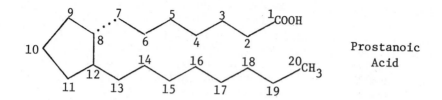

Prostanoic
Acid

b. Differences between various PG's are due to substituents and
their configuration on the five-membered ring (see
Figure 97). Note that the names given to the various PG's
in order to differentiate them from one another consists of
PG followed by a letter (e.g., PGE, PGF, etc.) and a

Table 37. Fatty Acid Composition of Some Fats of Animal and Plant Origin

	Saturated* C14; C16; C18 (predominantly C16 & C18)	Monosaturated* C16:1; C18:1 (predominantly C18:1)	Polyunsaturated* Mostly as linoleic acid - C18:2
Animal Fats			
Butter	59	37	4
Beef	54	44	2
Chicken	40	38	22
Pork	40	46	14
Fish (Salmon & Tuna)	28	29	23 + 20**
Vegetable Oils			
Safflower	11	11	78
Corn	14	26	60
Sesame	14	43	43
Soybean	15	27	58
Peanut	20	45	35
Cotton Seed	29	19	52
Coconut	92	6	2
Palm	53	37	10
Olive	16	69	15
Sunflower	12	18	70

* All values are expressed as weight percentages of total fatty acids.
** Other polyenoic acids.

Figure 97. Structures of Prostaglandins

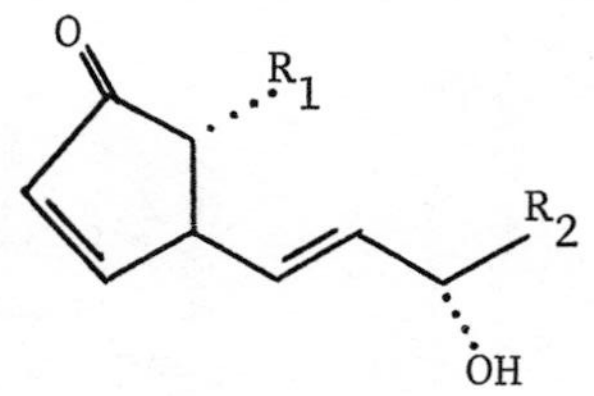

PGA

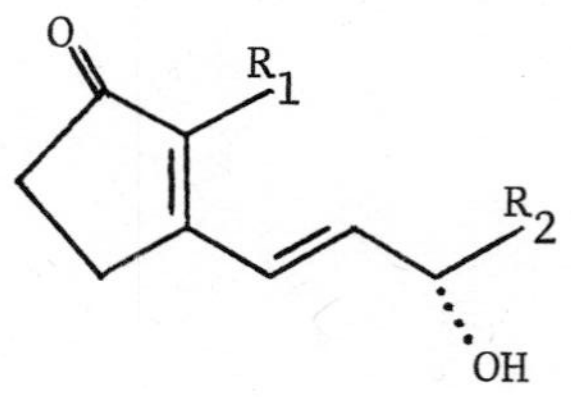

PGB

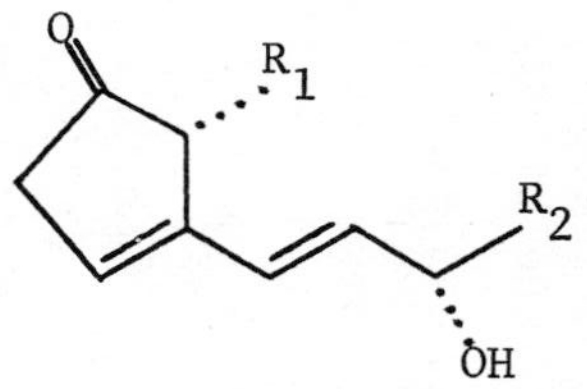

PGC

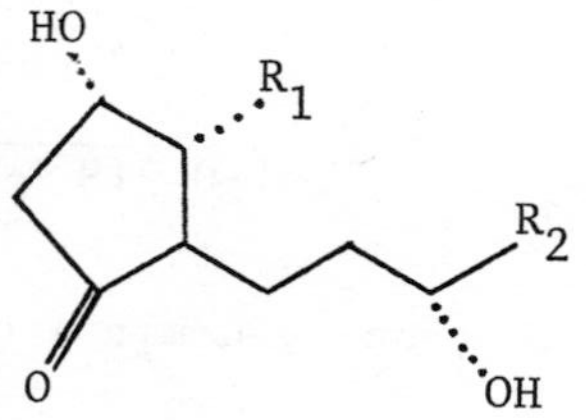

PGD

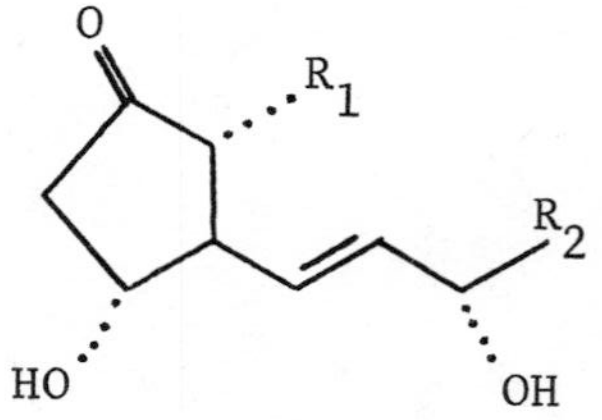

PGE

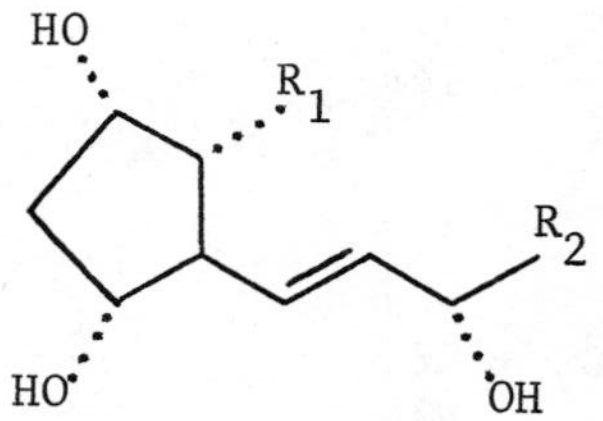

PGF_{α}

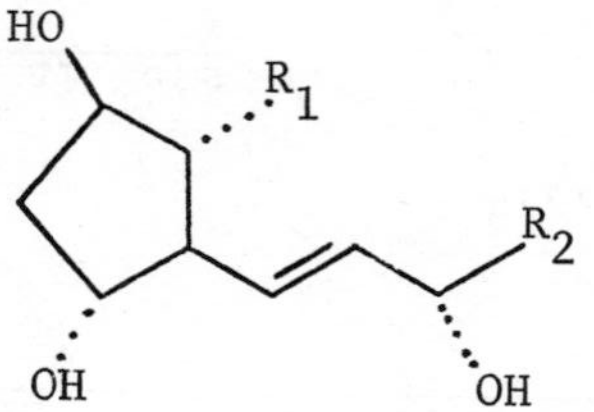

PGF_{β}

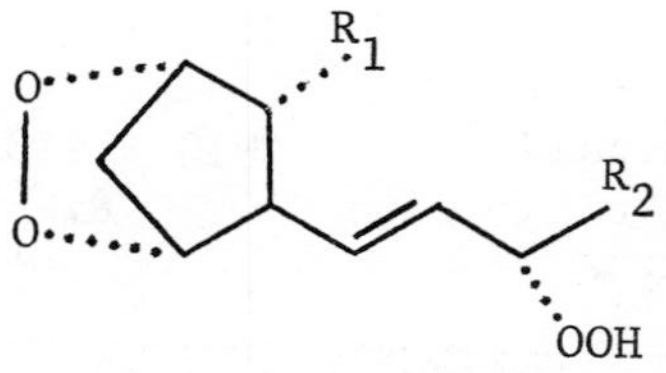

PGG

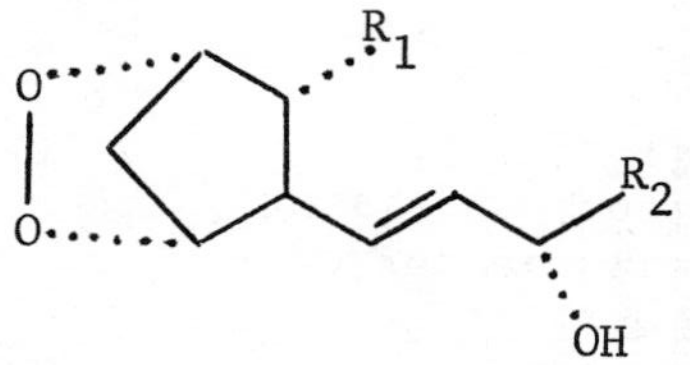

PGH

numerical subscript (e.g., PGE_1, PGF_2) which indicates the
number of specifically located double bonds. The location
and the type of double bonds are as follows: PG_1, trans C-13;
PG_2, trans C-13 and cis C-5; PG_3, trans C-13, cis C-5 and
cis C-17. All PG's have a hydroxy group at C-15 except for
PGG which has a hydroperoxy group (-OOH). The hydroxy group
at C-15 is in the S configuration in the naturally occurring
prostaglandins and is shown by a dotted line. The α and β
notations of the PG abbreviations (e.g., $PGF_{2\alpha}$) designate
the configuration of the substituent at C-9 on the cyclopentane
moiety, similar to that used for steroid chemistry (α = down,
shown by dotted line; β = up, shown by solid line). The
natural compounds are α-derivatives.

3. Biosynthesis

   a. PG's are synthesized from unsaturated fatty acids.
      Arachidonic acid is the predominant precursor, which gives
      rise to PG's with two double bonds (PG_2-series). PG's of
      the 1 and 3 series are derived from 8,11,14-eicosatrienoic
      and 5,8,11,14,17-eicosapentaenoic acids, respectively.
      It is thought that phospholipids may be the precursors of
      20-carbon polyenoic acids:

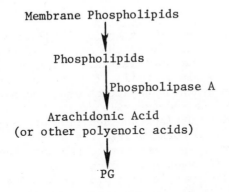

The structural relationships between the precursors and
PG's synthesized are shown in the following examples:

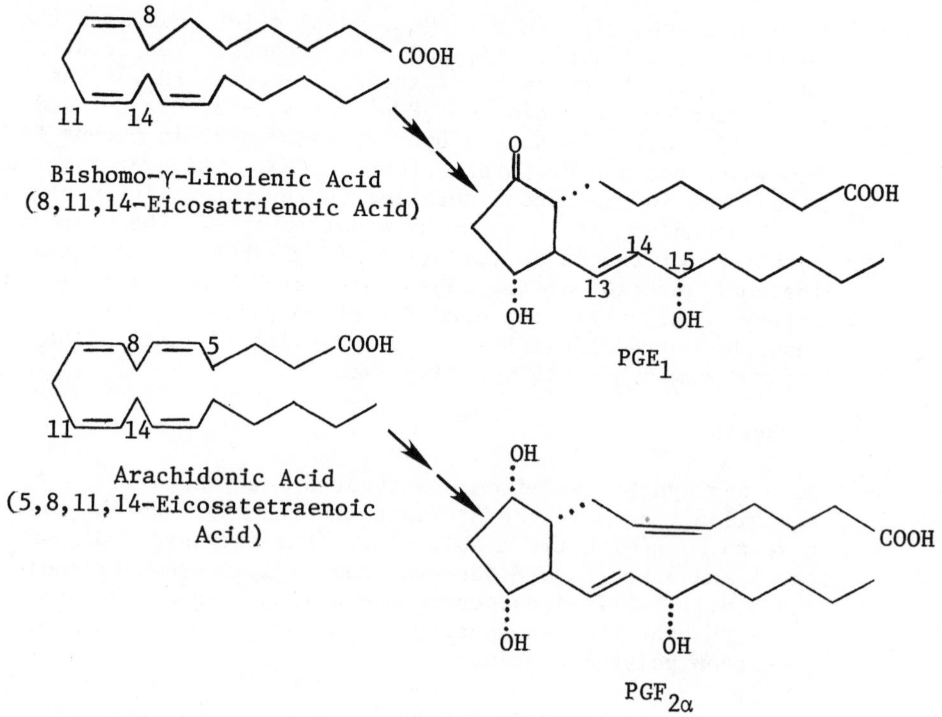

Bishomo-γ-Linolenic Acid
(8,11,14-Eicosatrienoic Acid)

PGE_1

Arachidonic Acid
(5,8,11,14-Eicosatetraenoic
Acid)

$PGF_{2\alpha}$

b. A typical pathway for the synthesis of PGE_2 from arachidonic
    acid is shown in Figure 98.   The fatty acid cyclo-oxygenase
    (or prostaglandin synthase) initially catalyzes the addition
    of one oxygen molecule at C-11.   The product formed undergoes
    cyclization with synchronous attack by a second oxygen
    molecule to form the endoperoxide PGG_2.   The enzyme is a
    dioxygenase because during each addition of oxygen, both
    atoms of O_2 are incorporated into the product.   Other
    examples of dioxygenases are tryptophan pyrrolase
    (tryptophan-2,3-dioxygenase), homogentisic acid-1,2-dioxygenase
    and 4-hydroxyphenylpyruvic acid dioxygenase (see amino acid
    metabolism--Chapter 6).   Dioxygenases contain iron either as
    labile iron-sulfur centers or as heme.   Some also contain
    copper.   Reduced glutathione (GSH) is also required in the
    formation of PGE_2.   Acetylsalicylic acid (aspirin) and
    indomethacin, the anti-inflammatory agents, inhibit the
    prostaglandin synthase enzyme.

c. PGA_2 is obtained from PGE_2 by dehydration. PGC_2 and B_2 are
    isomers of PGA_2 and thus can be synthesized by isomerases
    which are present in plasma.   The formation of these
    compounds are shown in Figure 99.

Figure 98. Biosynthesis of PGE₂ From Arachidonic Acid

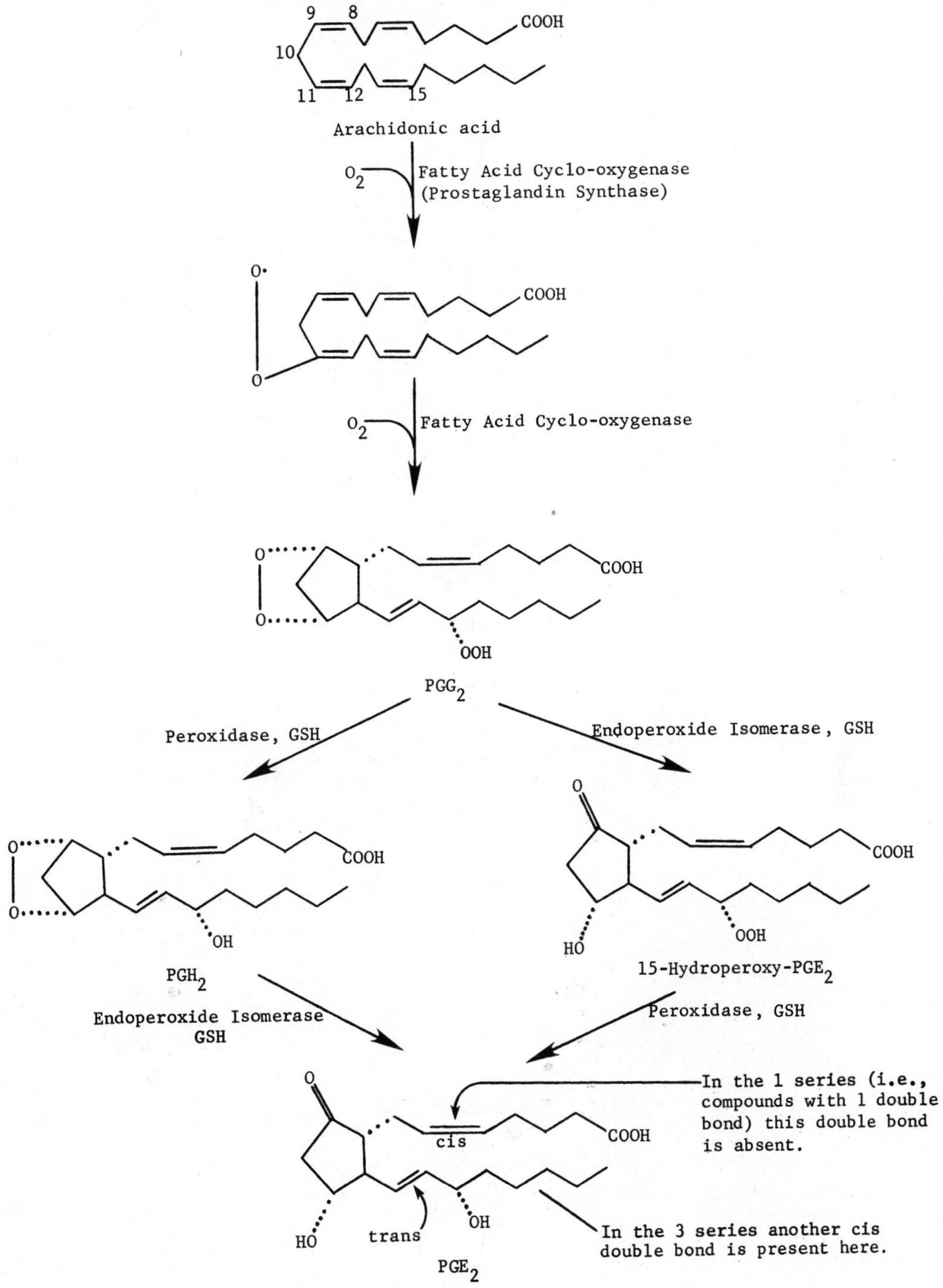

Figure 99. Conversion of PGE₂ to PGA₂, C₂ and B₂

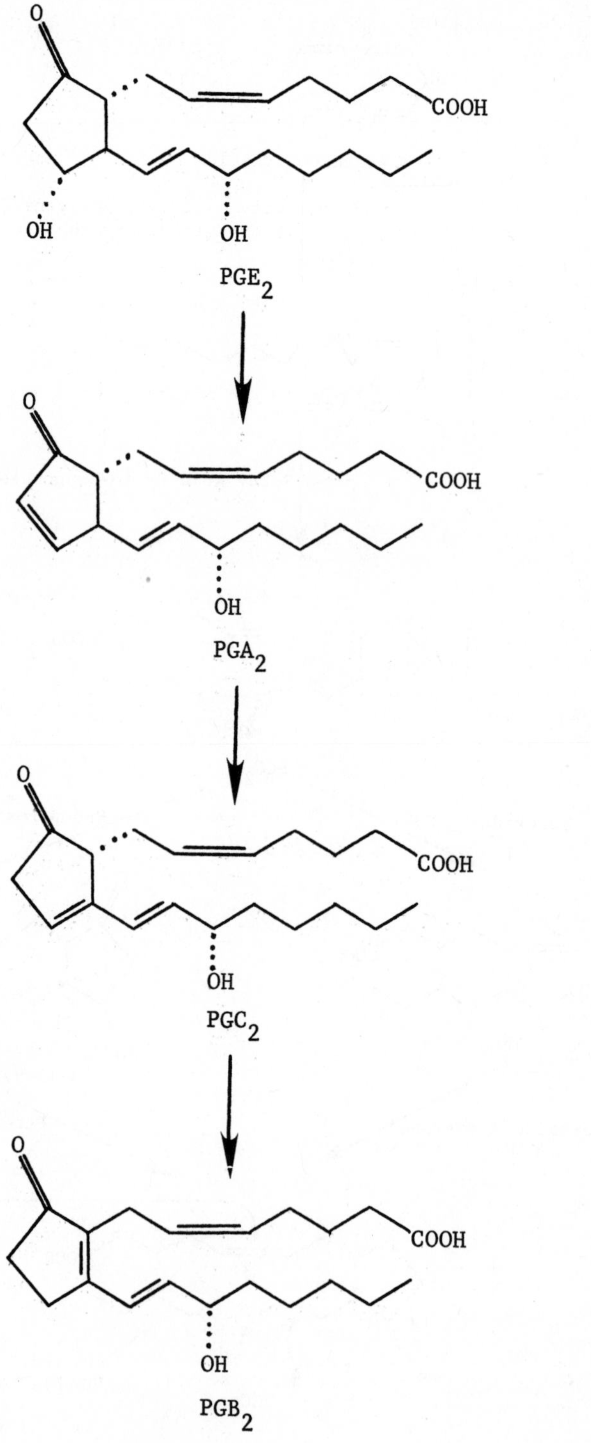

PGE₂

PGA₂

PGC₂

PGB₂

d. The interconversion of PGE_2 and $PGF_{2\alpha}$:

$$NAD(P)H + H^+ \qquad NAD(P)^+$$

9-keto-PG-Reductase
(9K-PGR)

PGE_2 $\qquad\qquad\qquad\qquad\qquad\qquad\qquad$ $PGF_{2\alpha}$

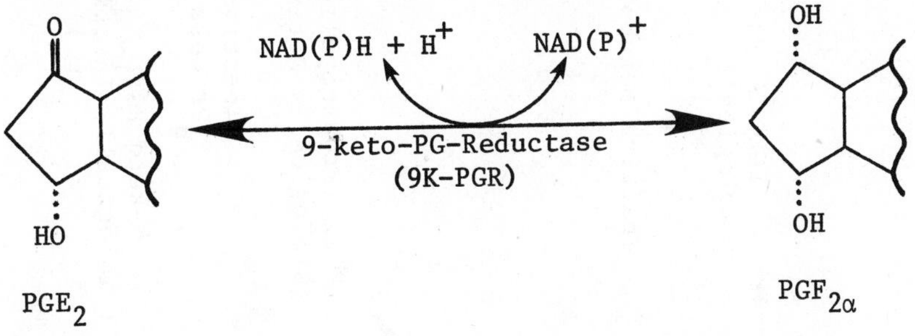

The above reaction takes place mainly in the cytoplasm (lungs). The oxidized pyridine nucleotides ($NAD(P)^+$) inhibit the conversion of PGE_2 to $PGF_{2\alpha}$. In the initial <u>catabolic</u> reaction of both of these compounds (discussed below) by 15-hydroxy-PG-dehydrogenase, the reduced pyridine nucleotides formed in that reaction inhibit the first step. Thus, the ratio of reduced to oxidized pyridine nucleotides may control both the interconversion of PGE_2 and $PGF_{2\alpha}$ and the first step in their catabolism. This is an important consideration with regard to the PG's biological actions because in many tissues PGE and PGF have opposing effects (see later).

e. PG endoperoxides (PGG_2 and PGH_2) are key intermediates in the synthesis of several prostaglandins, thromboxanes (TX), and the newest entry into this group of compounds, the prostacyclins (PGI). The latter compounds are highly <u>unstable</u> and <u>very active</u>. In general, the biologic action of PG's may be due to the production of these highly active and short-lived intermediates at specific receptor sites. The conversion of PG endoperoxides to various derivatives is shown in Figure 100. The structures of PGI_2, TXA_2 and TXB_2 are shown in Figure 102.

4. Catabolism: The catabolism of PG's occurs throughout the body but one of the most efficient processes is the uptake and catabolism of the circulating E and F compounds by the lungs. In fact, the lungs are capable of removing PG's during a single circulatory cycle. Despite this rapid removal, the PG's have adequate access to the target organs to bring about their physiologic effects. The catabolizing processes start with reactions catalyzed by the enzymes PG 15-hydroxy dehydrogenase (15-PGDH; oxidation of allylic -OH group at C-15) and PG reductase (reduction of the Δ^{13}-double bond). These transformations are followed by β-oxidation, ω-oxidation of the alkyl sidechain, ω-hydroxylations, and the eventual elimination of the products of these reactions. (The ω-carbon is the one farthest from the

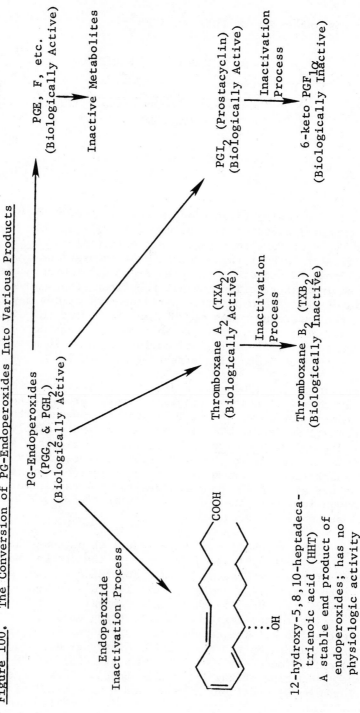

[826]

Figure 100. The Conversion of PG-Endoperoxides Into Various Products

PG-Endoperoxides
(PGG$_2$ & PGH$_2$)
(Biologically Active)

PGE, F, etc.
(Biologically Active)

Inactive Metabolites

PGI$_2$ (Prostacyclin)
(Biologically Active)

Inactivation
Process

6-keto-PGF$_{1\alpha}$
(Biologically Inactive)

Thromboxane A$_2$ (TXA$_2$)
(Biologically Active)

Inactivation
Process

Thromboxane B$_2$ (TXB$_2$)
(Biologically Inactive)

Endoperoxide
Inactivation Process

COOH

OH

12-hydroxy-5,8,10-heptadeca-
trienoic acid (HHT)
A stable end product of
endoperoxides; has no
physiologic activity

number 1 (carboxyl) carbon in a fatty acid.) The catabolism of PGE_2 and $PGF_{2\alpha}$ is shown in Figure 101.

The first step in the catabolism of PG's catalyzed by 15-PGDH appears to be the rate-limiting step. The enzyme is found in the cytoplasm (lungs), requires NAD^+ and is specific for the C-15(S) alcohol group. The substrates for the enzyme include PGA_2, PGE_2, $PGF_{2\alpha}$, and PGH_2, whereas PGB_2, PGD_2 and PGG_2 do not function as substrates. Using monkey lung preparations it has been shown that the enzyme is activated by thyroid hormones and is inhibited by reduced pyridine nucleotide (NADH) and PGB_2. The enzyme activity is not altered by steroids, calcium and cyclic AMP.

5. Biological Actions of Prostaglandins

   a. This is a rapidly growing area of research and results
       continue to pour in. A complete listing of all the actions
       of PG's in various animal systems is beyond the scope of this
       work. Some of the problems in delineating the primary actions
       of PG's are that they give rise to opposing effects and that
       it is difficult to distinguish between normal physiologic
       functions and pharmacological actions. In general, PGE_2 and
       $PGF_{2\alpha}$ have opposite effects on smooth muscle tone, release of
       mediators of immediate hypersensitivity, cyclic nucleotide
       levels, etc. Thus, the ratio between E and F compounds (due
       to changes in oxidized and reduced forms of pyridine
       nucleotides--discussed earlier) may be one of the crucial
       factors involved in controlling a given physiological
       response in the body. Other factors which are contributory
       in determining a biologic action are the relative proportions
       of endoperoxides, thromboxanes and prostacyclin.

   b. Cardiovascular Effects

       i) PG's cause increases in the cardiac output and enhance
           myocardial contractility. PGE causes reduction in
           arterial blood pressure accompanied by dilation of
           blood vessels and increased vascular permeability.
           This is part of the inflammatory response (wheal and
           flare response) brought about by PG's which can be
           blocked by anti-inflammatory agents such as aspirin
           (acetylsalicylic acid). This anti-inflammatory action
           may be due to aspirin's inhibition of the enzyme
           dioxygenase, which catalyzes one step in prostaglandin
           synthesis. It has been suggested that the endoperoxide
           PGG_2 probably plays a pivotal role in eliciting pain,
           swelling and other inflammatory effects.

Figure 101. Catabolism of PGE_2 and $PGF_{2\alpha}$

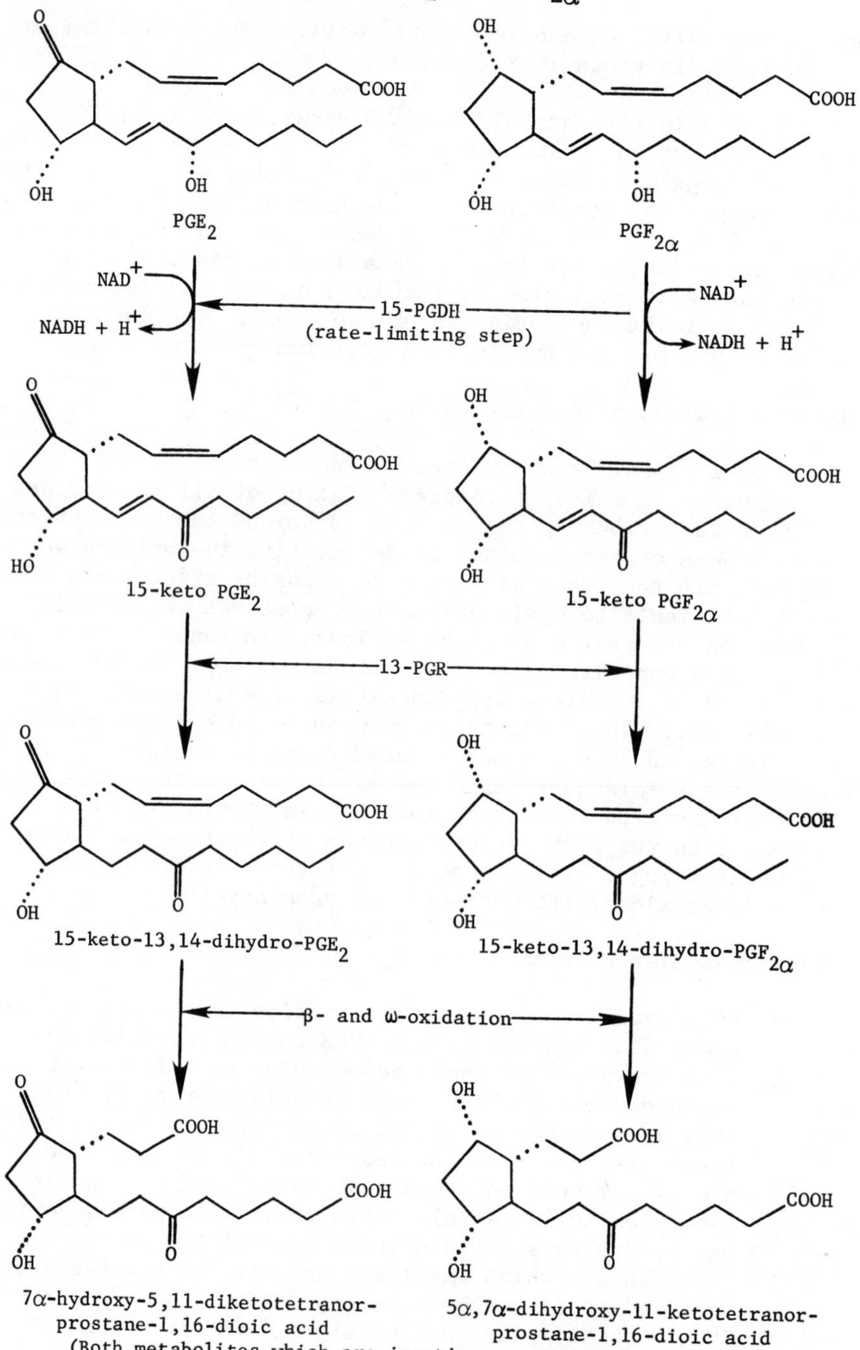

(Both metabolites which are inactive are excreted in urine.)

15-PGDH = 15-hydroxy-PG-dehydrogenase
13-PGR = 13-PG-reductase

ii) The effects of PGF on the cardiovascular system are opposite to those described above for PGE. Thus PGF causes hypertension, constricts blood vessels, and decreases the capillary permeability.

iii) PGE_1 inhibits platelet clumping by stimulation of cAMP synthesis.

c. Renal Effects: PGE_1 (and PGA) increases plasma flow, urinary volume, and the excretion of Na^+, K^+, and Cl^-.

d. Nervous System Effects: PGE in the cerebral cortex produces sedation and tranquilizing effects. In the medulla, PGE and PGF can produce either excitation or inhibition. PGE and PGF can inhibit spinal reflexes in the spinal cord. A wide variety of more specific actions in the nervous system have also been attributed to prostaglandins, namely PGE's inhibit and $PGF_{2\alpha}$ enhances the adrenergic response, whereas the cholinergic responses appear to be enhanced by both PG's.

e. Effects on Reproductive Glands: PGE and PGF cause increased synthesis of steroids. PGE in low concentrations causes lowering of uterine peristaltic action but, in higher concentrations, PGE and PGF can stimulate uterine activity leading to the expulsion of the fetus and placenta. This property makes the prostaglandins potentially useful as abortifacients and labor inducers. In the United States, $PGF_{2\alpha}$ is used for terminating a second trimester pregnancy by the intra-amniotic administration of the drug. PG's may also play a role in fertility by their action on the Fallopian tubes. PG's that are present in human semen can contract the proximal end of the Fallopian tube while dilating the distal end. This leads to a trapping of the ovum, making it more accessible for fertilization by spermatozoa. Since PG's have been shown to produce opposing effects in the fertilization process, it may be the concentration ratio of E over F which decides whether or not fertilization will occur.

f. In the lungs, PGE dilates bronchi and pulmonary vasculature while PGF has the opposite effect. The inhibition of mediators of hypersensitivity (histamine, SRS-A--see Chapter 11) and bronchodilation elicited by the PGE's makes them a potentially useful therapeutic agent (used as aerosol preparations) in the treatment of allergy. However, their clinical use at the present time is precluded due to an anomalous effect of PGE, that is, at very low concentrations it enhances the anaphylactic release of mediators.

Furthermore, PGE causes irritation of the upper respiratory tract (when administered as aerosols) and the effects are quite variable.

g. PGE inhibits gastric secretions and both forms (E and F) cause increased peristaltic action in the intestines, with a corresponding decrease in transit time. They also cause inhibition of Na^+ uptake (by stimulation of adenylate cyclase) by mucosal cells accompanied by increased water flow, thereby producing diarrhea.

h. Metabolic Effects: Many of PG's actions are linked with the production of cAMP by adenylate cyclase or its degradation by phosphodiesterase. In many tissues, PG's stimulate the formulation of cAMP. In the thyroid gland, PGE has a similar action to that produced by TSH. In the adipose tissue cells, however, PGE_1 prevents lipolysis (discussed later in this chapter) by decreasing cAMP levels. Insulin also has an antilipolytic effect in the adipose tissue cells. It has been observed that PGE_1 produces insulin-like responses in the peripheral tissues.

i. Patients with cancers of the lungs, kidney or pancreas often exhibit hypercalcemia, which has been linked to abnormalities in the PGE_2 metabolism. The increased mobilization of calcium in the blood is derived from bone marrow mediated by PGE_2 by an unknown mechanism. Consistent with this, patients with hypercalcemia show abnormally high urinary concentrations of the PGE_2 catabolite, 7α-hydroxy-5,11-diketotetranor-prostanedioic acid (see Figure 101), which is reduced along with serum calcium by aspirin and indomethacin (antagonists of PG synthesis) administration. However, in hypercalcemic cancer patients who have normal levels of PG metabolite and in patients with metastasis to bone marrow, the PG antagonists do not affect significantly the serum calcium level. Hypercalcemia, a lethal complication of cancer, can produce weight loss, nausea, vomiting and brain damage resulting in coma and death.

The potential possibilities for the clinical use of PG's may lie in the treatment of peptic ulcers, hypertension, and in obstetrics and gynecology (fertility control, induction of labor, etc.). At the present time, there is a great need for a drawing together of the proliferating observations concerning the multitudinous actions of the prostaglandins. Until PG research fully enters into the analytic stage, it is difficult to understand the true role of these ubiquitous compounds. In this regard, two of the most powerful techniques which have been developed to date in the assay of PG's are gas chromatography in conjunction with mass spectrometry and radioimmunoassay.

6. Thromboxanes, Endoperoxides, and Prostacyclin in Platelet Function

Since the discovery of prostaglandins, a great deal of work has gone into the investigation of their biosynthesis. Recently, several new, very unstable compounds (their biological half-lives are on the order of minutes or seconds) have been discovered which have the same (or opposite) effects as prostaglandins, are chemically related to them, and are frequently more potent. A general idea of the best-understood parts of this emerging system is shown in Figure 102.

These pathways are particularly significant because they are related to platelet aggregation (blood clotting). Defects in them may partly be the cause of heart attacks and strokes.

a. Prostaglandin I_2 (PGI$_2$; also named prostacyclin because it contains a second five-membered ring in addition to the one common to all PG's) is synthesized in the innermost lining of human arteries and veins, including the coronary arteries, by the enzyme prostacyclin synthetase. Chemically, PGI$_2$ is (5Z)-9-deoxy-6,9α-epoxy-Δ^5-prostaglandin $F_{1\alpha}$. The compound has been abbreviated as PGI$_2$, following the alphabetical nomenclature scheme applied to previously known PGs. It prevents platelet aggregation.

b. Thromboxane A_2 (TXA$_2$) is synthesized by thromboxane synthetase, located in platelets. The name thromboxane was given to these compounds because they were first isolated from thrombocytes (platelets) and they contain an oxane ring. It causes the aggregation of platelets and the formation of blood clots (see Chapter 11 for details on blood clotting).

c. Thromboxane B_2 and 6-keto-prostaglandin $F_{1\alpha}$ are biologically inert degradation products of these compounds.

d. It is known that, although platelets adhere to many surfaces, they do not do so when they contact healthy arterial walls. Workers in this field speculate that this non-adhesion is caused by PGI$_2$, formed from PGG$_2$ which is released by the platelets when they contact the walls. If the arterial intima is damaged by the formation of an atherosclerotic plaque (which may itself be promoted by a lack of prostacyclin synthetase activity) it is unable to make PGI$_2$ and thereby prevent the initiation of a clot. Moreover, when the PGG$_2$ is

Figure 102. Prostaglandins and Related Compounds

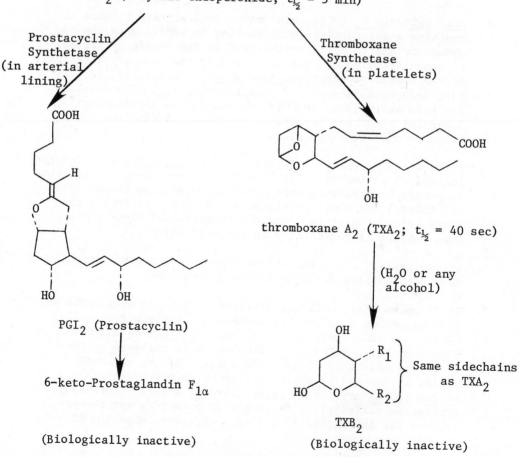

arachidonic acid

inhibited by
aspirin and ⟶ cyclooxygenase
indomethacin

PGD$_2$, PGE$_2$,
PGF$_{2\alpha}$

PGG$_2$ (a cyclic endoperoxide; $t_{\frac{1}{2}}$ = 5 min)

Prostacyclin
Synthetase
(in arterial
lining)

Thromboxane
Synthetase
(in platelets)

thromboxane A$_2$ (TXA$_2$; $t_{\frac{1}{2}}$ = 40 sec)

PGI$_2$ (Prostacyclin)

(H$_2$O or any
alcohol)

6-keto-Prostaglandin F$_{1\alpha}$

Same sidechains
as TXA$_2$

(Biologically inactive)

TXB$_2$
(Biologically inactive)

not converted to PGI_2, it is used as a substrate for thromboxane synthetase in the platelets. The TXA_2 which is formed thereby <u>promotes</u> clot formation, compounding the problem. A third factor which may increase the likelihood of arterial blockage is that PGI_2 causes smooth muscle relaxation (and hence vasodilation), an effect opposed by TXA_2. This information opens up the possibility of synthesizing a stable compound which will prevent clot formation and decrease the risk of coronary thrombosis.

e. Although these new compounds have been primarily studied in relation to blood clotting, they are found in many tissues and may serve a multitude of functions. Thromboxanes, for example, have been identified in platelets, leukocytes, lung, spleen, brain, kidney, umbilical artery, bone marrow, and placenta. There has also been a report of a patient who was lacking cyclooxygenase activity, presumably a genetic defect. This is a new and exciting area which, in the future, will undoubtedly contribute another measure of information concerning bodily processes and their regulation.

f. Aspirin, a well known analgesic and anti-inflammatory drug, has been used in clinical trials to protect against heart attacks. The results obtained from these studies are equivocal. This may be due, in part, to aspirin inhibiting the first step in prostaglandin synthesis (PG-cyclooxygenase), thus blocking the formation of both PGI_2 (vasodilates and prevents platelet aggregation) and TXA_2 (vasoconstricts and promotes platelet aggregation). However, aspirin may still be beneficial in patients with severely diseased coronary arteries where the inhibition of TXA_2 synthesis may be quantitatively more significant, particularly when lacking PGI_2 synthetase activity.

g. Increased platelet aggregation has been noted in vitamin E-deficient older pediatric patients with biliary cirrhosis secondary to congenital extrahepatic biliary atresia and cystic fibrosis. In these patients the platelet hyperaggregability has been reversed with vitamin E supplementation. The presumed mechanism for this observation is that the vitamin E inhibits the formation of platelet-aggregating endoperoxide intermediates. In support of this mechanism, <u>in vitro</u> experiments have shown that vitamin E and other phenolic antioxidants inhibit the peroxidation of arachidonic acid and thus prevent PG synthesis.

Steroids

This group includes cholesterol, ergosterol, the bile acids, the sex
hormones, the D-vitamins, and the cardiac glycosides. The biological
properties of these respective groups are discussed in appropriate
sections of this text.

1.   Steroids are often found in association with fat.

2.   They occur in the "unsaponifiable residue" following saponification.

3.   The basic steroid nucleus, cyclopentanoperhydrophenanthrene,
     consists of four fused rings.

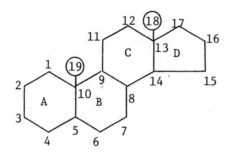

The conformation of these ring systems is discussed under
cholesterol.

a.   Carbons 18 and 19 are angular methyl groups attached at positions
     10 and 13, respectively. They are not part of the steroid
     nucleus but occur in a number of derivatives.

b.   A sidechain at position 17 is common.

c.   If the compound has one or more OH groups and no $>$C=O or -COOH
     group, it is ster<u>ol</u>. If it has one or more $>$C=O or -COOH
     groups, it is ster<u>oid</u>.

Plasma Lipids

1. This is a heterogeneous group of compounds which is seldom
   strongly polar, despite the frequent occurrence of phosphorous and
   to a lesser extent nitrogen. When they are extracted with suitable
   lipid solvents and subsequently separated into the various lipid
   classes, the presence of the following compounds can be demonstrated.

   a. Phospholipids comprise the largest lipid component of
      plasma but their clinical significance is not completely
      understood. One role is as a detergent, to increase the
      solubility of other lipids (e.g., in bile secretions).
      This is possible because many of the phospholipids have
      both polar and non-polar regions. The importance of
      phospholipids in amniotic fluid and lipoprotein metabolism
      is discussed later in this chapter.

   b. Cholesterol is the second largest fraction of the plasma lipids.

   c. The triglycerides are third quantitatively among lipids in
      the plasma.

   d. Unesterified (free) long-chain fatty acids make up most or
      all of the remaining material.

2. Free fatty acids (FFA) are also called NEFA or non-esterified fatty
   acids. Quantitatively, the straight-chain, saturated fatty acids
   of formula $CH_3(CH_2)_n COOH$ are the most important, with stearic (C_{18})
   and palmitic (C_{16}) comprising at least 85% of the total.

   a. Although the FFA account for less than 5% of the total fatty
      acid present in the plasma, they are the most metabolically
      active of the plasma lipids.

   b. FFA is released from the adipose tissue and is transported
      in the plasma bound to albumin.

   c. Plasma FFA levels are increased in diabetes mellitus and are
      often elevated in obese individuals.

d.   Measurement of free fatty acids in serum

serum + $CHCl_3$ $\longrightarrow$ FFA in $CHCl_3$ solution

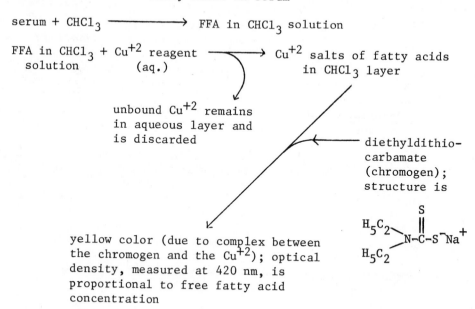

FFA in $CHCl_3$ + Cu^{+2} reagent $\longrightarrow$ Cu^{+2} salts of fatty acids
   solution   (aq.)   in $CHCl_3$ layer

unbound Cu^{+2} remains
in aqueous layer and
is discarded

diethyldithio-
carbamate
(chromogen);
structure is

yellow color (due to complex between
the chromogen and the Cu^{+2}); optical
density, measured at 420 nm, is
proportional to free fatty acid
concentration

Diethyldithiocarbamate is also used in quantitation of copper
in body fluids (see copper metabolism).

Another procedure for free fatty acids in serum and plasma
involves an initial extraction of these molecules from the
sample with a 40:10:1 (V/V/V) mixture of isopropanol, heptane,
and 1 N H_2SO_4. Water and more heptane are then added to
separate the unionized acids (non-aqueous layer) from the other
serum components (aqueous layer). The extracted NEFA are
titrated with dilute alkali to a thymolpthalein end-point.

3. Triglycerides are triple esters of glycerol with three molecules
   of fatty acid. According to the recommendations of IUPAC
   (International Union of Pure and Applied Chemistry) and IUBC
   (International Union of Biochemistry), the new terminology to
   be used for triglycerides, diglycerides and monoglycerides are
   triacylglycerols, diacylglycerols and monoacylglycerols,
   respectively. In this text, the older and the newer

terminologies are used interchangeably. Diglycerides and monoglycerides do not occur in appreciable amounts in nature but they are intermediates in many biosynthetic reactions.

At room temperature triglycerides exist in the solid or liquid form based on the nature of the constituent fatty acids. Most vegetable oils contain unsaturated fatty acids (e.g., oleic, linoleic, linolenic--see discussion on essential fatty acids in this chapter) and are liquids (low melting point). In contrast, triglycerides of animal origin contain higher proportions of saturated fatty acids (e.g., palmitic and stearic) resulting in higher melting points. Therefore, they are semisolid or solid at room temperature. The fatty acids may or may not be identical to each other. The general formula of a triglyceride is indicated below.

$$
\begin{array}{c}
\quad\quad\quad\quad\quad\quad\quad\quad O \\
\quad\quad\quad\quad\quad\quad\quad\quad \parallel \\
\quad\quad\quad\quad CH_2-O-C-R_1 \\
\quad\quad O \quad\quad | \\
\quad\quad \parallel \quad\quad | \\
R_2-C-O-CH \quad O \\
\quad\quad\quad\quad | \quad\quad \parallel \\
\quad\quad\quad\quad CH_2-O-C-R_3
\end{array}
$$

a. Triglycerides derived from the intestinal absorption (details discussed later) of fat are transported in the blood as chylomicrons. Those triglycerides synthesized in the liver and intestines from carbohydrates and FFA are carried as prebetalipoproteins.

b. Triglycerides are stored in adipose tissue. They function as an energy reservoir and participate in the transport of FFA.

c. Triglycerides (TG's) are usually quantitated by saponifying them to release glycerol and free fatty acids, then measuring the amount of glycerol formed. The glycerol may also be released from the triglyceride by the action of triglyceride lipases. The triglycerides are isolated from the serum by an extractive and/or chromatographic procedure. Ways of obtaining and quantitating glycerol are indicated next.

i) Triglycerides

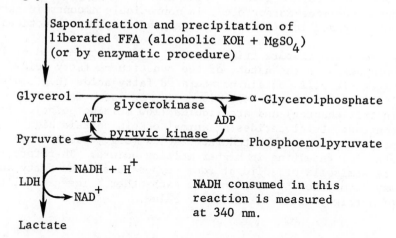

Saponification and precipitation of
liberated FFA (alcoholic KOH + MgSO$_4$)
(or by enzymatic procedure)

Glycerol ——————————→ α-Glycerolphosphate
            glycerokinase
        ATP   ADP

Pyruvate ←————————— Phosphoenolpyruvate
            pyruvic kinase

LDH   — NADH + H$^+$ NADH consumed in this
        → NAD$^+$ reaction is measured
                                at 340 nm.

Lactate

ii) Alternatively, the α-glycerolphosphate can generate
     NADH in the reaction:

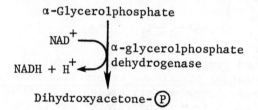

α-Glycerolphosphate

NAD$^+$
                α-glycerolphosphate
NADH + H$^+$ dehydrogenase

Dihydroxyacetone-Ⓟ

The increase in optical density at 340 nm is
proportional to the α-glycerolphosphate concentration
and, hence, to the triglyceride concentration.

iii) A chemical method for glycerol involves its oxidation
      to formaldehyde with sodium periodate followed by
      colorimetric determination of the formaldehyde at
      570 nm with chromotropic acid (4,5-dihydroxy-2,7-
      napthalene sulfonic acid).

4. Cholesterol

a. This compound is distributed in all cells of the body,
    especially in the nervous tissue. It occurs in animal
    fats but not in plant fats. Structure of cholesterol:

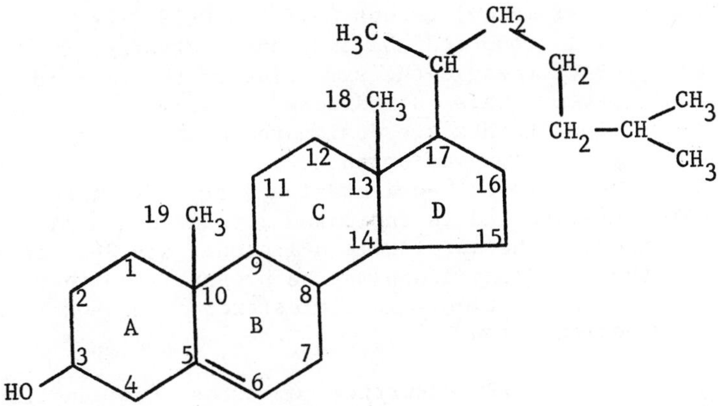

Cholesterol contains three six-membered rings and one
five-membered ring. Note that the cholesterol has one
double bond between carbons 5 and 6. Dihydrocholesterol
is known as <u>cholestanol</u>, which has the following conformation.

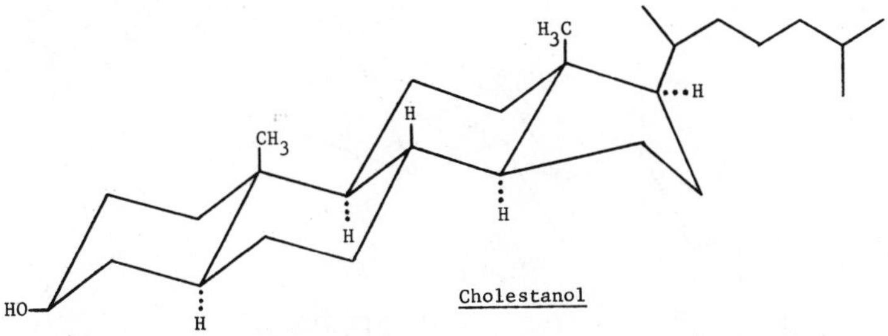

Cholestanol

The following aspects should be noted with respect to the
<u>cholestanol</u> structure.

i) All the ring fusions (between A and B, B and C, and
    C and D) are trans and, therefore, the hydrogen atoms
    or methyl groups attached to the bridgehead carbon
    atoms project on opposite sides of the rings. The
    rings are in the more stable chair conformations.
    However, in <u>cholesterol</u>, the double bond between
    carbons 5 and 6 distorts the conformation of the
    first and second rings which leads to the conformation
    shown below.

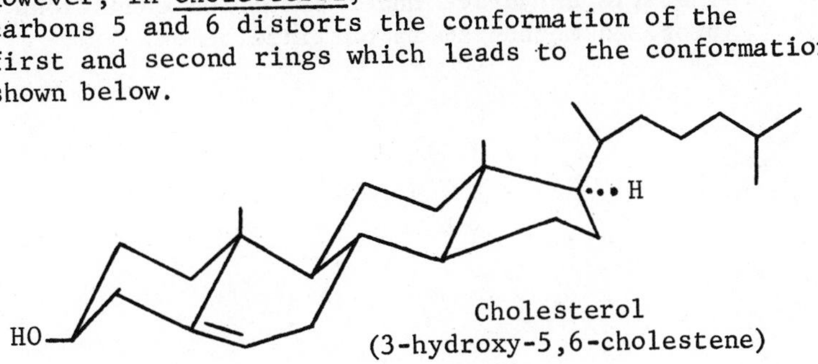

Cholesterol
(3-hydroxy-5,6-cholestene)

ii) The angular methyl groups C-18 and C-19, the
3-hydroxyl group and the side chain attached to C-17
all project towards the same side of the steroid
ring system. This is indicated by solid lines, and
these substituents are designated as β. If a
substituent group is oriented below the plane of the
ring (opposite to β-orientation), it is in the
α-orientation and is indicated with a dotted line.
In general, the angular methyl groups are β-oriented,
but the 3-hydroxy group may be present in either the
α- or β-orientation. In cholesterol, the 3-OH is in
the β-orientation.

iii) In some naturally occurring compounds, the junction
between A and B is <u>cis</u>. An example of this type of
compound is β-coprostanol. This compound occurs in
large quantities in the feces and is synthesized
from cholesterol by the action of fecal flora. The
structure of β-coprostanol is shown below.

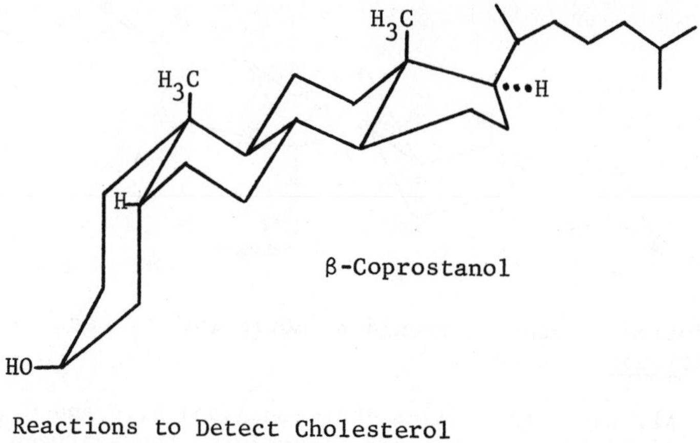

β-Coprostanol

b. Color Reactions to Detect Cholesterol

i) The Liebermann–Burchard reaction and the Salkowski test
are based on the colors produced (green and red,
respectively) when cholesterol is reacted with H_2SO_4
+ acetic anhydride under specified conditions. A
proposed mechanism is outlined below.

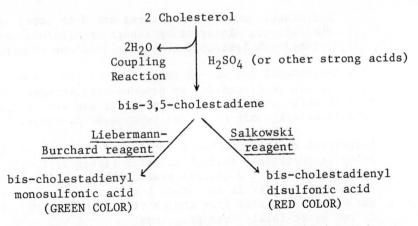

2 Cholesterol

$2H_2O$ ← Coupling Reaction

H_2SO_4 (or other strong acids)

bis-3,5-cholestadiene

Liebermann–Burchard reagent

Salkowski reagent

bis-cholestadienyl monosulfonic acid (GREEN COLOR)

bis-cholestadienyl disulfonic acid (RED COLOR)

The Salkowski reagent contains a relatively higher concentration of H_2SO_4, thereby favoring disulfonate formation. Because the purity of the red or green color is difficult to control, the use of these reactions is not highly reproducible. Other sterols are also known to react, producing a green color. Despite these difficulties, well over 150 different procedures for the determination of cholesterol have been published based on these reactions.

ii) The Zak reaction of cholesterol with $FeCl_3$, glacial acetic acid, and concentrated H_2SO_4, produces a pink to purple color which is much more stable than the colors formed in the preceding methods. This procedure is also more sensitive and reproducible and has been adapted for and widely used in auto-analyzers in the clinical laboratory. The basic reaction is again sulfonic acid formation.

iii) The newest method for cholesterol determination uses cholesterol oxidase, isolated from a species of the bacterium Nocardia. This is quite similar to the determination of glucose by glucose oxidase. The basic reaction is shown below.

$$\text{Cholesterol} + O_2 \xrightarrow{\text{Cholesterol oxidase}}$$

$$\text{Cholest-4-ene-3-one} + H_2O_2$$

The H_2O_2 then reacts with a dye precursor in the presence of a peroxidase to form a colored product. A third enzyme in the reaction mixture, cholesterol ester hydrolase, permits the determination of total serum cholesterol. Interfering substances include ergosterol, 7-dehydrocholesterol, and other $3\beta-\Delta^4$ or Δ^5 sterols.

c. Free cholesterol has an -OH group in position 3 and is precipitable by digitonin (a glycoside) to produce digitonides. This is the form in which cholesterol is biosynthesized and is the form present as a structural unit of animal membranes (discussed later).

d. Esterified cholesterol has a fatty acid attached to the hydroxyl group in position 3 through an ester linkage. This type of cholesterol is not digitonin precipitable. The role of the cholesterol esters is not known but they are an important catabolic product of free cholesterol. Cholesterol esters are formed by at least three processes.

    i) In the liver, a fatty acid can be transferred from fatty acyl CoA to free cholesterol.

   ii) Free fatty acids can be esterified with cholesterol in the intestinal mucosa in the presence of bile salts.

  iii) The enzyme lecithin-cholesterol acyl transferase (LCAT), a plasma-specific enzyme (see uses of enzymes in the clinical laboratory), catalyzes the reaction:

lecithin + free cholesterol $\longrightarrow$

                                       lysolecithin + cholesterol ester

This is a transesterification process involving transfer of an unsaturated fatty acyl group from the β-carbon of lecithin to the 3-position of cholesterol. A deficiency of LCAT (sometimes called Norum's disease), has been described in seven members of three Scandinavian families. These individuals also show high plasma levels of lecithin and unesterified cholesterol, a complete absence of α_1- and pre-β-lipoproteins, proteinuria, anemia, corneal infiltration, and the presence of foam cells in the glomerular tufts. Blood smears of the peripheral blood show the presence of target cells and the erythrocytes contain increased amounts of cholesterol and lecithin. The disease appears to be transmitted as an autosomal recessive trait.

e. In Wolman's disease, cholesteryl esters (discussed below) and triglycerides (all neutral lipids) accumulate in a number of organs and tissue levels of di- and monoglycerides and free fatty acids are usually elevated, accompanied by steatorrhea. Hepatosplenomegaly and calcification of the adrenals are always present, due to the excessive lipid storage in these organs. The disease usually manifests itself within the first week of life and is invariably fatal, normally by the age of six months. Although the disorder is clearly inherited, probably as an autosomal recessive trait, it remains to be clearly shown whether the biochemical lesion in Wolman's disease is one of synthesis, transport, or degradation. The best evidence, obtained from three patients, is the total deficiency of acid lipase activity in the liver and spleen. This enzyme hydrolyzes triglyceride and cholesteryl esters.

Phospholipids

Phospholipids, the fourth group of plasma lipids, are present in all cells as well as in the plasma. They are also known as phosphorized fats, phospholipins, and phosphatides.

1. All but the sphingomyelins contain a molecule of glycerol as a backbone. The middle (β) carbon of the glycerol, as part of the phospholipid molecule, is asymmetric. It usually has an L-configuration.

D-α-phosphatidic acid   L-α-phosphatidic acid

2. There are five types of phospholipids.

   a. Phosphatidic acid is an important intermediate. A mole of it contains one mole of phosphate, one mole of glycerol, and two moles of one type of fatty acid or one mole each of two different types. The general structure of these compounds is given above. Phosphatidic acids are also named phosphatidylglycerol.

b. <u>Lecithins</u> are phosphatidyl cholines (esters of choline with
    phosphatidic acid). They are zwitterions over a wide pH range
    (including physiological pH) due to the presence of the choline
    (+ charge) and the phosphoric acid (- charge; strong acid).
    The fatty acids present usually are palmitic, stearic, oleic,
    linoleic, and arachidonic acids. Lecithins are the most
    abundant phosphoglycerides in animal tissues and are soluble
    in all of the usual fat solvents except for acetone.

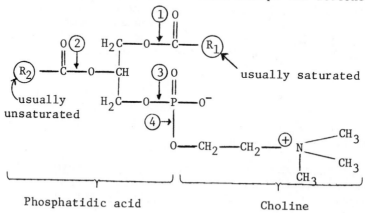

Phosphatidic acid   Choline

Note that lecithin has a quaternary nitrogen (a nitrogen
with four valencies).

Enzymes that degrade lecithins cleave the molecule at the
numbered (1-4) bonds. They are

  i)   Phospholipase A (lecithinase A): hydrolyzes bonds
       of type 1.

  ii) Phospholipase A_2: hydrolyzes bonds of type 2; releases
       a fatty acid and <u>lysolecithin</u> or <u>lysophosphatidylcholine</u>;
       secreted by the pancreas as a proenzyme; some snake venoms
       contain a Ca^{+2}-requiring phospholipase of this type;
       lysolecithins are powerful detergents and hemolytic
       agents (due to their ability to solubilize membrane
       components).

  iii) Lysophospholipase: hydrolyzes type 1 bonds in
       lysolecithins; releases a free fatty acid and
       glycerylphosphorylcholine.

  iv) Phospholipase B (lecithinase B): hydrolyzes bonds of
       types 1 and 2; secreted by the pancreas.

    v) Phospholipase C: hydrolyzes bonds of type 3, releasing a diglyceride and phosphoryl choline; also acts on phosphatidyl serine and ethanolamine (cephalins); found in the α-toxin of <u>Clostridium welchii</u> and certain other bacteria and in some plant and animal tissues.

    vi) Phospholipase D: hydrolyzes bonds of type 4, releasing phosphatidic acid and free choline; can simultaneously remove the choline and add an alcohol in its place (transesterification); found only in plants.

c. <u>Cephalins</u> are phosphatides containing ethanolamine, serine, or inositol in place of choline. They are found in all tissues but are particularly abundant in brain and other nerve tissues. They are probably involved in blood clotting. Cephalins are separated on the basis of their differential solubilities in mixtures of chloroform and alcohol. The structures of these compounds are indicated next.

$$
\begin{array}{l}
\quad\quad\quad\quad\ \ \overset{\displaystyle O}{\overset{\displaystyle \|}{}} \\
H_2C\text{-}O\text{-}C\text{-}R_1 \\
\ \ \ \ \ \ \ \ | \\
\overset{\displaystyle O}{\overset{\displaystyle \|}{}} \\
R_2\text{-}C\text{-}O\text{-}CH \\
\ \ \ \ \ \ \ \ | \quad\quad \overset{\displaystyle O}{\overset{\displaystyle \|}{}} \\
H_2C\text{-}O\text{-}P\text{-}O^{-} \\
\ \ \ \ \ \ \ \ \ \ \ \ \ \ |
\end{array}
$$

$O\text{-}CH_2CH_2\overset{+}{N}H_3 \quad \longleftarrow \quad$ <u>ethanolamine</u>

$O\text{-}CH_2CH(\overset{+}{N}H_3)COO^{-} \quad \longleftarrow \quad$ <u>serine</u>

<u>inositol</u>

d. <u>Plasmalogens</u> (phosphatidyl ethanolamines, serines, and cholines) are found in cardiac and skeletal muscle, brain, and liver. Their function is not known. The general structure is shown below.

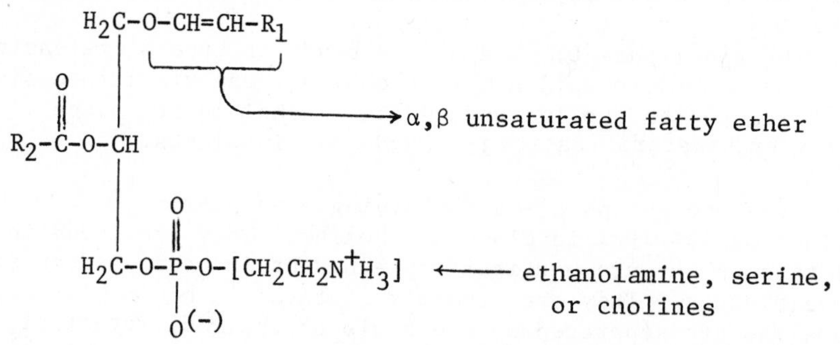

$$H_2C-O-CH=CH-R_1$$

$$R_2-\overset{O}{\overset{\|}{C}}-O-CH$$

→ α, β unsaturated fatty ether

$$H_2C-O-\overset{O}{\overset{\|}{P}}-O-[CH_2CH_2N^+H_3]$$
$$\underset{O^{(-)}}{\big|}$$

← ethanolamine, serine, or cholines

The α, β-unsaturated fatty ether is sometimes called an aldehydogenic group since hydrolysis of the ester releases an α, β-unsaturated primary alcohol which readily tautomerizes to an aldehyde.

e. <u>Sphingomyelins</u> are found in all tissues, especially in brain and other nervous tissue. On complete hydrolysis they yield

1 mole of fatty acid

1 mole of H_3PO_4

1 mole of choline

1 mole of sphingosine

The structure of a **sphingomyelin is shown below**

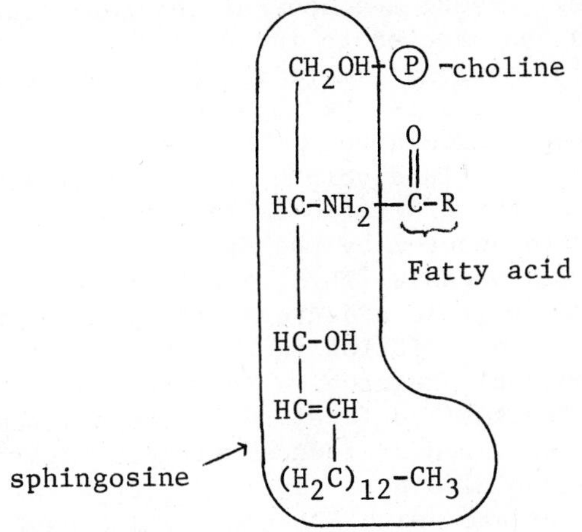

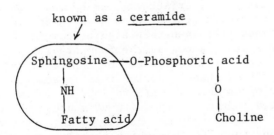

$$CH_2OH \text{--} \textcircled{P} \text{--choline}$$

$$HC\text{-}NH_2 \text{--} \overset{\overset{\displaystyle O}{\|}}{C}\text{-}R$$

Fatty acid

$$HC\text{-}OH$$

$$HC\text{=}CH$$

sphingosine $\longrightarrow$ $(H_2C)_{12}\text{-}CH_3$

OR

known as a <u>ceramide</u>

Sphingosine $\longrightarrow$ O-Phosphoric acid

NH   O

Fatty acid   Choline

In <u>Niemann-Pick</u> disease, large amounts of sphingomyelins accumulate in brain, liver, and spleen, accompanied by mental retardation. The probable biochemical lesion is a deficiency of sphingomyelinase, the enzyme which catalyzes the cleavage of the molecule to phosphorylcholine and ceramide. There is no known treatment.

Measurement of Lecithin and Sphingomyelin in Amniotic Fluid

1. One of the principal causes of death in premature infants is respiratory distress syndrome (RDS) with hyaline membrane disease. This occasionally occurs even in full-term babies. The syndrome is apparently due to immaturity of the lung, specifically with respect to its ability to synthesize lecithin and excrete it into the alveolar air spaces. Consequently, ways of measuring fetal lung maturity are helpful in judging whether a fetus is capable of surviving postnatally.

2. In a normal pregnancy, the lung is adequately developed by about
    the 36th or 37th week. There are several changes occurring at
    about this age which can be used to indirectly evaluate fetal lung
    maturity. A recently developed, more direct method which has
    proved useful in this regard is the measurement of lecithin
    (or more commonly, the lecithin to sphingomyelin (L/S) ratio,
    to correct for changes in fluid volume) in the amniotic fluid.
    In normal pregnancies, the L/S ratio is less than 1 prior to about
    the 31st week, rises to about 2 by the 34th week, 4 at the 36th
    week, and 8 at term (39 weeks). This results from an increase in
    lecithin synthesis rather than a decrease in sphingomyelin
    formation. There is some variation in these values for normal
    gestation and, in abnormal pregnancies (due to many causes
    including maternal diseases and fetal and placental conditions),
    the L/S ratio may be elevated or reduced without regard to
    gestational age. It has also been reported that a low L/S ratio
    is not <u>inevitably</u> associated with RDS and hyaline membrane disease.

While a L/S ratio of greater than 2 is associated with the
absence of serious RDS, a ratio lower than 2 is not uniformly
predictive of the development of RDS. However, the pulmonary
surface-active material (surfactant) can be measured in the
amniotic fluid by a different procedure which depends on the
property of surfactant to generate stable foam in the presence
of ethanol. This test, known as the <u>foam stability test</u> (FST)
or <u>shake test</u>, correlates well with the L/S ratio and fetal
lung maturity. It has been reported that in some instances,
in the presence of an immature L/S ratio (less than 2), the
FST has indicated lung maturity (without subsequent respiratory
distress). This discrepancy may be due to the presence of
surfactants other than lecithin which stabilize the neonatal
alveoli, namely phosphatidyl glycerol (PG) and phosphatidyl
inositol (PI). These acidic phospholipids are synthesized in
a stepwise fashion during the last trimester of normal
pregnancy (PI parallels the L/S ratio up to a value of 2 and
PG appears when the ratio exceeds 2). Thus, the measurement
of PI and PG may be used as additional indices of lung maturity.
It is noteworthy that PI and PG are present in negligible
quantities in blood products whereas lecithin and sphingomyelin
exist in sufficient quantities so as to interfere with the L/S
ratio measurement when the amniotic fluid is contaminated with
blood. Thus, in specimens of amniotic fluid mixed with blood,
the measurement of PI and PG may prove to be of value in the
assessment of lung maturity. It should be emphasized that a
number of factors such as hypoxia and acidosis depress
phospholipid synthesis. The role of thyroid hormones in RDS
is discussed under phenylalanine and tyrosine metabolism (see
amino acid metabolism in Chapter 6). Glucocorticoids (e.g.,
cortisol) have been shown to participate in the normal process

of fetal lung maturation and there appears to exist a positive correlation between the L/S ratio and the cortisol levels in the amniotic fluid. However, it has been reported that the measurement of the levels of cortisol in the amniotic fluid may not be used in the assessment of fetal lung maturation with confidence.

In <u>unanticipated births</u> of premature infants, the risk of RDS can be assessed by measuring surfactants in gastric aspirates of the newborn using the foam stability test. The infant's gastric content of surfactants is taken as an indicator of pulmonary surfactant, since the newborn swallows amniotic fluid <u>in utero</u>.

3. As noted above, lecithin is a powerful detergent (surface-active agent, capable of reducing the surface tension of water) which presumably functions by stabilizing the alveolar air spaces, preventing them from closing due to surface tension upon exhalation. Although the formation of lecithin by trimethylation of α-palmityl, β-myristyl phosphatidyl ethanolamine is detectable from week 22-24 onward, this particular lecithin has only marginal ability to stabilize the alveoli. In the mature mammalian lung, the principal pathway for lecithin synthesis is a salvage reaction:

$$\text{CDP-choline + D-1,2-diacylglycerol} \xrightarrow{} \begin{array}{l}\text{Phosphatidyl choline}\\ \text{(a lecithin)}\end{array}$$
$$\downarrow$$
$$\text{CMP}$$

Apparently the lecithins formed by this route are much more effective at stabilizing the alveolar air spaces.

4. In one method of L/S ratio determination, chloroform is used to extract the phospholipids from the amniotic fluid. The phospholipids are then separated by thin-layer chromatography of a concentrated sample of the chloroform extract. The phospholipids are located by spraying with sulfuric acid and are identified by comparison with known standards. The spots are quantitated by scanning the plates with a suitable densitometer.

Other Complex Lipids

1. <u>Polyglycerol phospholipids</u> are present in large amounts in heart muscle, liver, and brain (in decreasing order). They also occur in plants.

   a. An important example of this group of compounds in animal tissue is cardiolipin (diphosphatidylglycerol). Its structure is shown below.

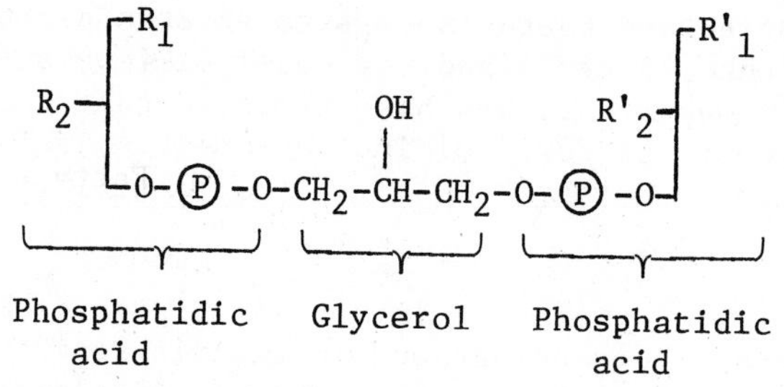

$$
R_2 \left[\begin{array}{c} R_1 \\ \\ O - \textcircled{P} - O - CH_2 - \underset{\underset{OH}{|}}{CH} - CH_2 - O - \textcircled{P} - O \end{array} \right] \begin{array}{c} R'_1 \\ \\ R'_2 \end{array}
$$

Phosphatidic   Glycerol   Phosphatidic
  acid   acid

b. In the mitochondrial membrane, 10% or more of the lipid is
    cardiolipin. It is apparently linked in the manner shown below.

> cytochrome —cardiolipin— structural protein

c. Cardiolipin is the only phosphatidate which is known to be
    antigenic. This property is used in the serologic test for
    syphilis.

2. Cerebrosides (glycolipids) are <u>not</u> phospholipids. On hydrolysis
    they yield

    1 mole of sphingosine

    1 mole of fatty acid

    1 mole of a hexose (usually D-galactose)

    The general structure is

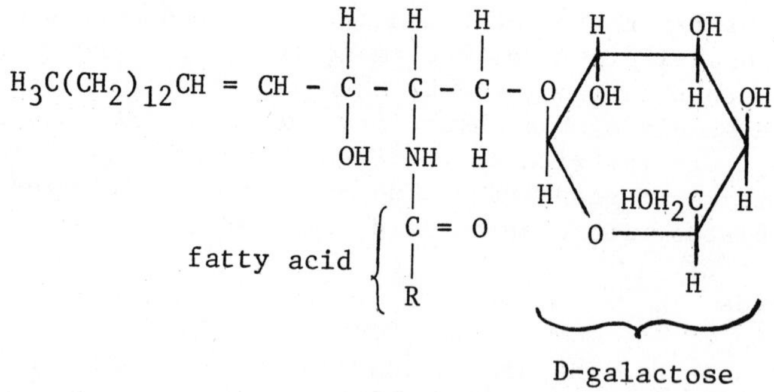

$$
H_3C(CH_2)_{12}CH = CH - \underset{\underset{OH}{|}}{\overset{\overset{H}{|}}{C}} - \underset{\underset{NH}{|}}{\overset{\overset{H}{|}}{C}} - \underset{\underset{H}{|}}{\overset{\overset{H}{|}}{C}} - O
$$

fatty acid $\left\{ \begin{array}{c} | \\ C = O \\ | \\ R \end{array} \right.$

D-galactose

They are found in brain and other tissues and can be considered
as monosaccharide derivatives of ceramides.

In addition to gluco- and galactocerebrosides as general names, different fatty acyl groups confer specific names to the cerebrosides.

Cerebroside	Fatty Acid	Formula of Fatty Acid
kerasin	lignoceric acid	$CH_3(CH_2)_{22}COOH$
phrenosin (cerebron)	cerebronic acid	$CH_3(CH_2)_{21}CHOHCOOH$
nervon	nervonic acid	$CH_3(CH_2)_7CH=CH(CH_2)_{13}COOH$

3. <u>Sulfatides</u> (or sulfatidates) are cerebrosides having a sulfate group attached in an ester linkage to the galactosyl residue. In the white matter of the brain, an analogue of phrenosin, having a sulfate ester at C-3 of the galactose, is present in large quantities.

4. <u>Gangliosides</u> (also called ceramide oligosaccharides), the most complex glycosphingolipids, are found in ganglion cells, spleen, and erythrocytes. They generally consist of a ceramide (N-acyl sphingosine) in an ether linkage to a hexose (usually galactose or glucose). To this cerebroside nucleus are commonly attached at least one mole of N-acetylglucosamine or N-acetylgalactosamine and at least one mole of N-acetylneuraminic acid (see Chapter 4).

   a. One notation for gangliosides, which is used later in the discussion of sphingolipidoses, is composed of an initial capital G (for ganglioside); a subscripted M, D, or T (indicating one, two, or three residues, respectively, of N-acetyl neuraminic acid in the molecule); and a further subscript, either a number or a number and a lower case letter. Thus G_{M1}, G_{D1a}, and G_{T1} designate, respectively, members of the class of gangliosides containing one, two, or three residues of N-acetyl neuraminic acid per molecule of ganglioside.

   b. Globoside, the most abundant glycosphingolipid in erythrocyte stroma, is given below as an example.

      ceramide-(←1β)-glucose-(4←1β)-galactose-(4←1α)

      galactose-(3←1β)→N-acetyl-galactosamine

   Another important ganglioside is ganglioside G_{M1}, a degradation product of which (G_{M2}) accumulates in Tay-Sachs disease. The structure of G_{M1} is shown below.

```
ceramide-(←1β)-glucose-(4←1β)-galactose-(4←1β)
                              |
                              |                    N-acetylgalactosamine
                              |
                              |                    (3←1β)-galactose
                              |
                       (3←2)-N-acetyl neuraminic acid
```

Comments on the Biosynthesis of Phospholipids and Triglycerides

1. The importance of cytidine diphosphate (CDP) and coenzyme A (CoASH) as carriers of intermediates in lipid synthesis is apparent from Figure 103. Further examples will be given throughout this chapter. The structure of CDP is given in the chapter on nucleic acids, while that of CoASH is with glycolysis. The structure of CDP-diacylglyceride is shown below. These biosynthetic reactions take place largely on the endoplasmic reticulum.

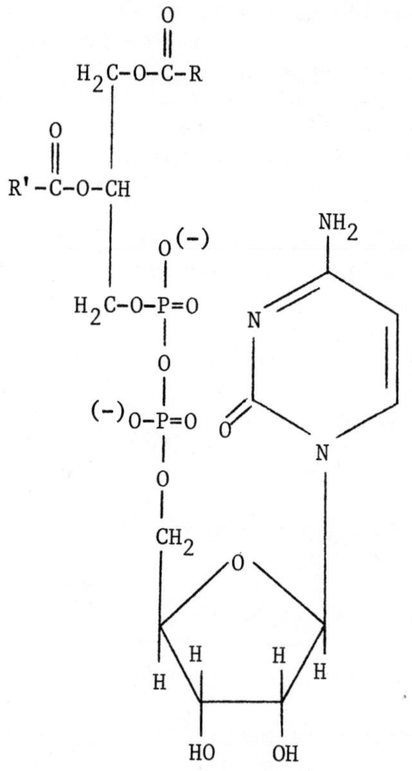

Cytidine
diphosphate
diacylglyceride

Figure 103. Biosynthesis of Di- and Triglycerides, Cephalins, Cardiolipin (A Polyglycerol Phospholipid), and Lecithin

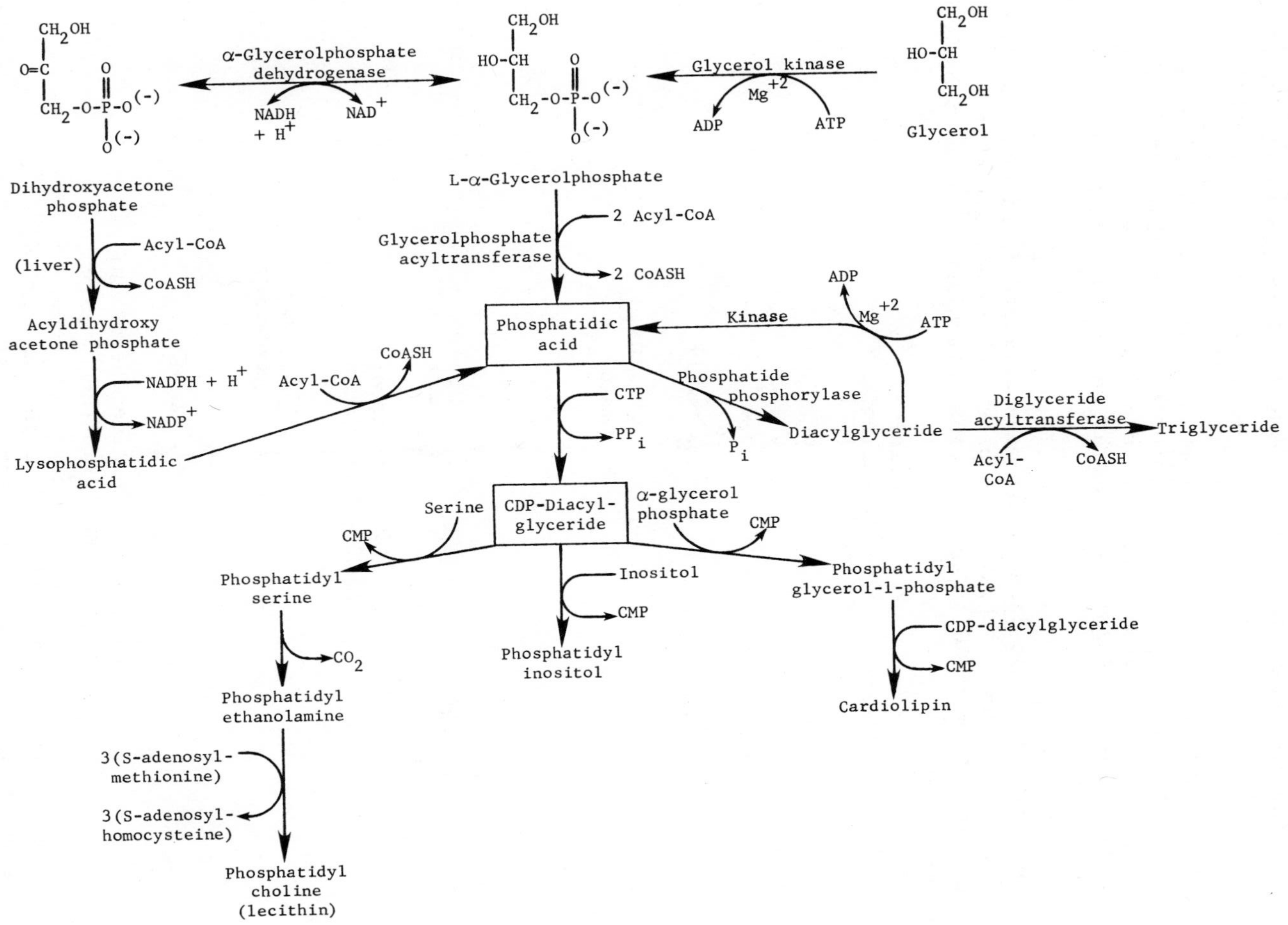

[853]

2. In Figure 103, due to the specificities of the enzyme involved,
    the pathway from dihydroxyacetone phosphate through lysophosphatidyl-
    choline produces phosphatidic acids in which the terminal acyl group
    is saturated and the middle (β) acyl group is usually unsaturated.
    Triglycerides and phosphatides from many sources show this pattern.
    The direct acylation of α-glycerolphosphate, on the other hand,
    produces a random mixture of saturated and unsaturated acyl groups
    in phosphatidic acid. This could produce the lecithins of lung
    tissue which contain predominantly saturated acyl groups to both
    positions.

3. In addition to the <u>de novo</u> pathways outlined in Figure 103, there
    exists a salvage pathway which uses choline and ethanolamine
    obtained from the diet and from the catabolism of other phospholipids.

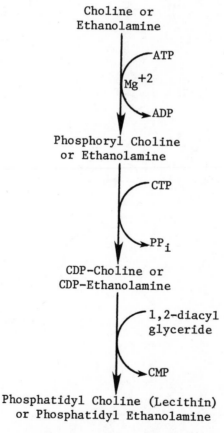

4. Plasmalogens are synthesized by a pathway similar to the
    lysophosphatidic acid route shown in Figure 103. The α,β-unsaturated
    ether is derived from the long-chain alcohol initially coupled to
    dihydroxyacetone phosphate. The reactions are indicated below.

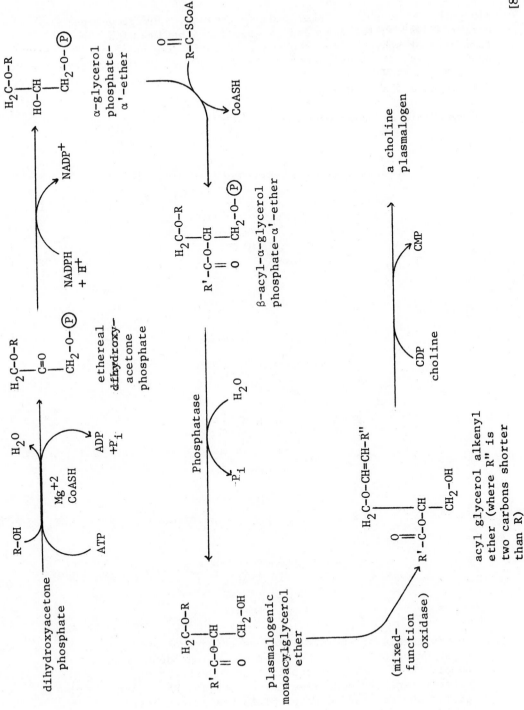

5. Biosynthesis of the sphingolipids (ceramides, sphingomyelins, cerebrosides, sulfatides, and gangliosides) begins with the formation of sphingosine, the characteristic component of sphingolipids (also called trans-D-sphingenine) from palmitoyl-CoA and L-serine. These reactions require pyridoxal phosphate, NADPH, and a flavin coenzyme.

    a. Synthesis of Sphingosine

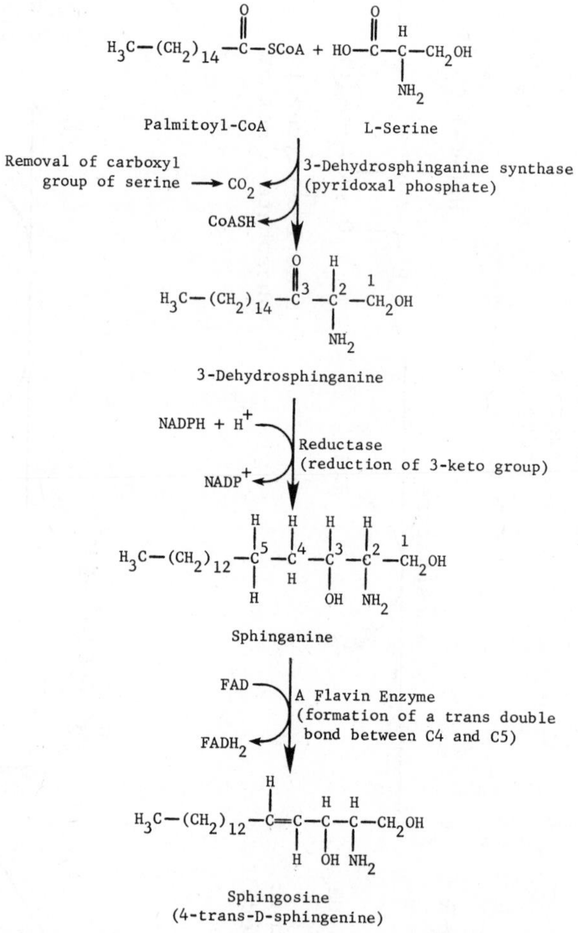

$$H_3C-(CH_2)_{14}-\overset{O}{\overset{\|}{C}}-SCoA + HO-\overset{O}{\overset{\|}{C}}-\overset{H}{\underset{NH_2}{\overset{|}{C}}}-CH_2OH$$

Palmitoyl-CoA   L-Serine

Removal of carboxyl group of serine → CO_2 ← 3-Dehydrosphinganine synthase (pyridoxal phosphate)

CoASH ←

$$H_3C-(CH_2)_{14}-\overset{O}{\overset{\|}{\underset{}{C}}}{}^3-\overset{H}{\underset{NH_2}{\overset{|}{C}}}{}^2-{}^1CH_2OH$$

3-Dehydrosphinganine

NADPH + H$^+$ → Reductase (reduction of 3-keto group)

NADP$^+$ ←

Sphinganine

FAD → A Flavin Enzyme (formation of a trans double bond between C4 and C5)

FADH$_2$ ←

Sphingosine
(4-trans-D-sphingenine)

    b. The synthesis of sphingolipids starting from sphingosine is shown below. Note the roles of nucleotide carriers, namely CDP (carrier of choline), UDP (carrier of carbohydrate residues--see Chapter 4) and PAPS (carrier of active sulfate--see Chapter 6) in the biosynthesis of these complex lipids.

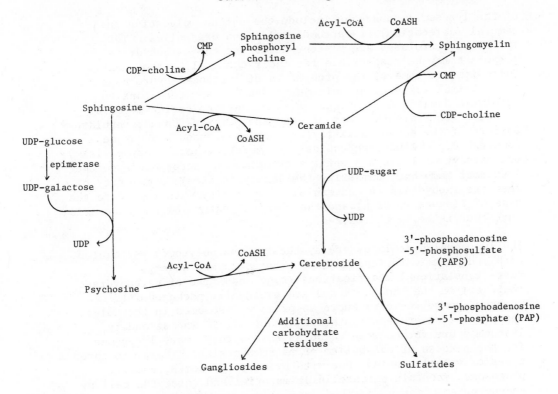

Catabolism and Storage Disorders of Glycosphingolipids

1. An important group of inherited diseases exists in which large
    quantities of different glycosphingolipids accumulate within the
    lysosomes of various tissues. These belong to the group of inborn
    metabolic errors known as lysosomal disorders and all involve
    decreased or absent activity of catabolic enzymes rather than
    overactivity of synthetic pathways. Lysosomes contain over forty
    different hydrolases with specificities for different types of
    molecules. Although there are several criteria for classifying
    a disorder as lysosomal, it is significant that in the majority
    of such diseases, the absence of a specific hydrolase can be
    demonstrated. Other lysosomal disorders include the
    mucopolysaccharidoses (See chapter 4);
    mucolipidoses (deposition of both mucopolysaccharide and lipid;
    most genetic defects not known); Niemann-Pick disease; I-cell
    disease (may be due to a "leaky" lysosomal membrane); cystinosis;
    acid phosphatase deficiency; Wolman's disease; and others.

    The lysosome is the major site of intracellular digestion in all
    types of cells, and their enzymes are capable of degrading both
    intra- and extracellular macromolecules. The overall functions

of the lysosomal apparatus include the uptake, digestion and removal (defecation) of macromolecules and organelles. The lysosomal hydrolases are glycoproteins and the carbohydrate component may be responsible for their chemical heterogeneity. Following are some of the properties of lysosomal hydrolases: show maximal activity at acid pH (3.5-6.4); are resistant to autolytic digestion--a unique feature; have no strict metal or cofactor requirements; most of them are membrane bound; regulatory aspects of the enzyme activity are not understood and may be complex (e.g., glucocerebrosidase, a membrane-bound enzyme, is not catalytically active until a cytoplasmic factor binds to the lysosomal membrane, unmasking the catalytic site); a majority of them are exohydrolases and exhibit specificity with regard to the type of linkage (α or β) and the chemical nature of the monomeric unit that is being hydrolyzed.

The lysosomal hydrolases are synthesized on polysomes associated with endoplasmic reticulum (ER). These enzymes undergo post-translational modifications (e.g., addition of carbohydrate residues) in the smooth ER and are eventually packaged into membrane-bound vesicles known as primary lysosomes in the Golgi apparatus (see protein synthesis--Chapter 5). The secondary lysosomes are formed when vacuoles fuse with primary lysosomes for the purpose of catabolism of macromolecules. There are three types of vacuoles which fuse with primary lysosomes, namely, phagosomes (contain extracellular material and enter the cell by endocytosis; membrane derived from plasma membrane), autophagosomes (contain intracellular macromolecules and organelles; membrane derived from ER) and secretory vesicles (sequestration of granules also packaged in the Golgi apparatus; in some cell types the vesicles containing secretory granules release their contents into extracellular fluid--exocytosis). Catabolism in the secondary lysosomes results in the production of diffusable small molecules as well as undigestable residues. The latter is either defecated by the process of exocytosis or retained in the cell as residual bodies. In some lysosomal disorders, in addition to enzyme deficiency, the defects may lie either in the fusion of the primary lysosome with the vacuole and/or the process of exocytosis.

Defects of the lysosomal hydrolases may be due to mutations in the structural gene (structural gene defects may result in changes of the kinetic properties, physical stability, and/or binding sites other than the catalytic site, which is required for normal catalysis), mutations in the genes responsible for post-translational alterations of the enzyme (note that the lysosomal hydrolases are glycoproteins, and defective glycosylation may yield defective gene products), or mutations in the regulatory genes which control the synthesis of the enzyme. In general, the lysosomal enzyme deficiencies are characterized by the accumulation of unmetabolized substrates which may be derived from both endogenous and exogenous sources, depending on the type of cell. The excessive accumulation of a given substrate interferes with normal physiologic functions. Furthermore, the substrate deposition in different cell types (which can be demonstrated both chemically and morphologically) correlates with the onset of various clinical manifestations.

2. The particular group of glycosphingolipids to be discussed here all contain sphingosine in the form of a ceramide. The catabolism of these compounds involves removal of successive glycosyl residues until the ceramide is released. The errors usually involve the enzymes which catalyze these hydrolyses of glycosidic bonds. The one exception to this is metachromatic leukodystrophy, due to a deficiency in a sulfatidase enzyme.

3. The catabolic pathways for the glycosphingolipids are given in Figure 104 and their disorders are summarized in Tables 38 and 39. Certain aspects of these diseases warrant further comment and it is given in succeeding paragraphs. Accumulation of a specific lipid in these disorders is frequently accompanied by deposition of one or more polysaccharides which are structurally related to the lipid. Their clinical treatment is generally palliative or non-existent. Recently, however, enzyme replacement therapy (administration of an exogenous preparation of the missing enzyme) has proven useful in the treatment of some of these disorders, which is discussed later in this section. Because the exogenous enzymes are apparently unable to cross the blood-brain barrier, the efficacy of this type of treatment in the majority of the glycosphingolipidoses (which have neurological involvement) is doubtful. Attempts are being made to modify the enzymes to overcome this difficulty. Considerable progress has also been made in identifying carriers of these diseases and in prenatal diagnosis of homozygotes. In this regard, laboratory tests have been developed for the assay of enzyme activity in leukocytes or cultured skin cells using chromogenic or fluorogenic synthetic substrates. This will, presumably, improve the prospects for genetic counseling.

Figure 104. Degradative Pathways for Glycosphingolipids

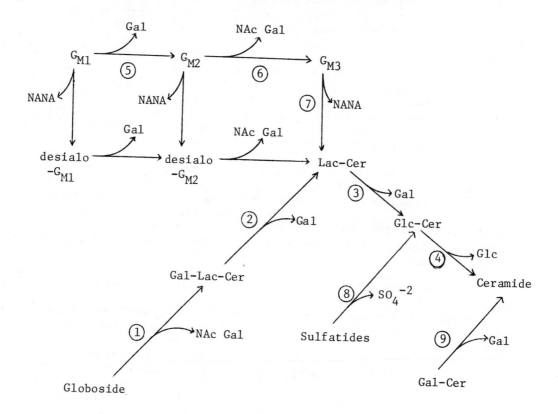

$G_{M1,2,3}$ are gangliosides. The structures of ganglioside G_{M1} and globoside were given previously in the section on gangliosides. The other compounds in this diagram are derived from them by successive removal of the indicated carbohydrate residues.

Abbreviations
    Gal = galactose
    Glc = glucose
    NAc Gal = N-Acetyl-Galactose-2-Amine
    NANA = N-Acetyl Neuraminic Acid (a sialic acid)
    Lac = lactose (galactosyl(β-1$\rightarrow$4) glucose)
    Cer = ceramide
    desialo = without a sialic acid (NANA) residue

The circled numbers correspond to the metabolic lesions listed in Tables 38 and 39.

Table 38. Characteristics of Glycosphingolipid Storage Disorders

Disorders	Major Lipids Accumulated	Other Compounds Affected	Enzyme Lacking*	Remarks
G_{M2} gangliosidosis, Type II (Tay-Sachs Variant; Sandhoff's disease)	Globoside and G_{M2} ganglioside both accumulate		Hexosaminidases A and B (①) and (⑥)	Same clinical picture as Tay-Sachs disease but progresses more rapidly; no Jewish predilection; hepatosplenomegaly, cardiomyopathy
Fabry's disease (Glycosphingolipid Lipidosis)	Gal−(4←1α)−Lac−Cer	Gal−(4←1α)−Gal−Cer accumulates	α-galactosidase (②)	X-linked recessive; hemizygous males have a characteristic skin lesion usually lacking in heterozygous females; pain in the extremities; death usually in the 4th decade results from renal failure or cerebral or cardiovascular disease
Ceramide Lactoside Lipidosis	Gal−(4←1β)−Glc−Cer		Ceramide Lactoside β-galactosidase (③)	Liver and spleen enlargement; slowly progressive brain damage; neurological impairment
Gaucher's disease (glucosyl ceramide lipidoses; three types; see text)	Glc−Cer	Accumulation of G_{M3} ganglioside most frequently; other compounds occasionally	β-glucosidase (glucocerebrosidase; ④)	Hepatosplenomegaly; frequently fatal; no known treatment; occurrence of "Gaucher cells" (RE cells which contain accumulations of erythrocyte-derived glucocerebroside)
G_{M1} gangliosidosis (two types; see text)	G_{M1}− and desialo-G_{M1}-gangliosides	Keratin-sulfate-related polysaccharide accumulates	G_{M1}-β-galactosidase (⑤)	Mental and motor deterioration; accumulation of mucopoly-saccharides is as significant as accumulation of gangliosides; invariably fatal; autosomal recessive; blindness, cherry red macula (30%); hepatosplenomegaly; vacuolated lymphocytes; startle response to sound, dysostosis multiplex

G_{M2} gangliosidosis, Type I (Tay-Sachs disease; see text)	G_{M2}- and desialo-G_{M2}-gangliosides	Other desialo hexosyl ceramides; occasionally other compounds accumulate	Hexosaminidase A (6)	Red spot in retina; mental retardation; severe psychomotor retardation; seizures; blindness; startle response to sound; invariably fatal; autosomal recessive; panracial but especially prevalent among Northern European Jews
G_{M3} gangliosidosis	G_{M3}- and desialo-G_{M3}-ganglioside		Not known (7)?	Sample stored long in formalin prior to analysis; lipid changes could be artifactual; included here primarily for completeness
Metachromatic leukodystrophy (MLD; sulfatide lipidoses; at least three types; see text)	3-sulfato-galacto-sylcerebroside	Cerebrosides other than sulfatides are decreased; ceramide dihexoside sulfate accumulates	Sulfatidases (8); arylsulfatases)	Demyelination; progressive paralysis and dementia; death usually occurs within the first decade; autosomal recessive inheritance
Krabbe's disease (Globoid cell leukodystrophy; galactosyl ceramide lipidosis)	Galactocerebroside	Sulfatides are also greatly decreased, probably as a secondary feature	Galactocerebroside-β-galactosidase (9)	Mental retardation; demyelination; psychomotor retardation; failure to thrive; progressive spasticity; globoid cells in brain white matter; invariably fatal; autosomal recessive inheritance

* The circled numbers following the enzyme names refer to reactions indicated in Figure 104. The abbreviations used here are also the same as those in that figure.

Table 39. Biochemical Lesions and the Sphingolipids Accumulated in Sphingolipidoses

Disease	Major sphingolipid accumulated	Enzyme defect
Gaucher's disease	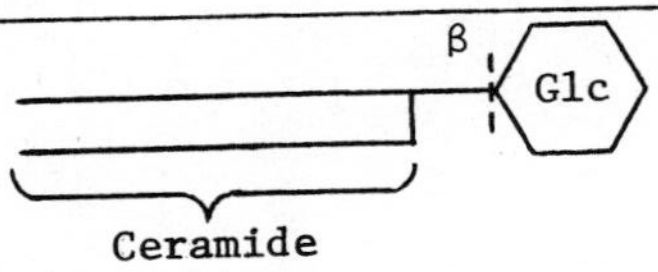 Ceramide glucoside (glucocerebroside)	Glucocerebrosidase (β-Glucosidase)
Niemann-Pick disease	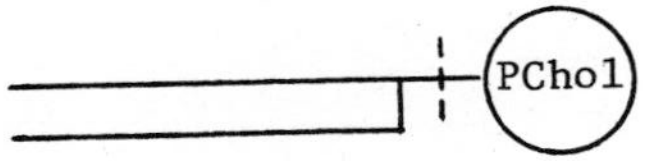 Sphingomyelin	Sphingomyelinase
Krabbe's disease	Ceramide galactoside (galactocerebroside)	Galactocerebroside-β-Galactosidase
Metachromatic leucodystrophy	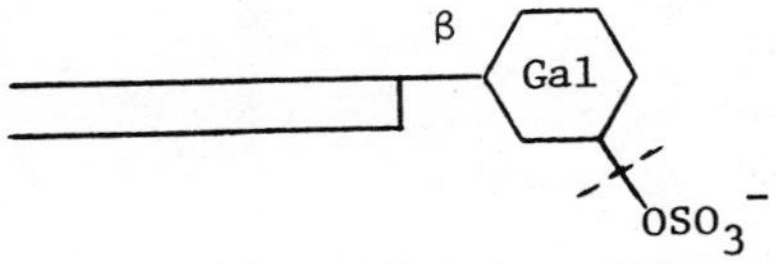 Ceramide galactose 3-sulfate (sulfatide)	Sulfatidase
Ceramide lactoside lipidosis	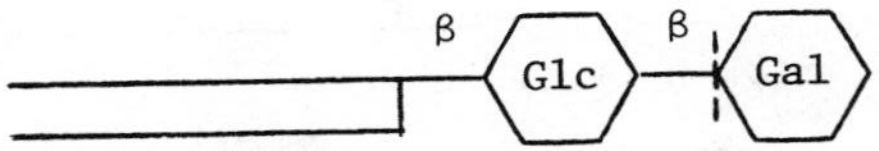 Ceramide lactoside	Ceramide Lactoside β-Galactosidase

Gaucher's disease: β — Glc; Ceramide

Niemann-Pick disease: PChol

Krabbe's disease: β — Gal

Metachromatic leucodystrophy: β — Gal — OSO$_3^-$

Ceramide lactoside lipidosis: β — Glc; β — Gal

[863]

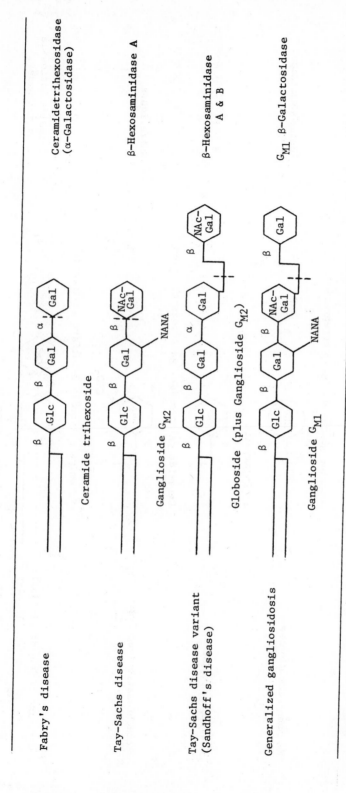

Fabry's disease

Ceramidetrihexosidase
(α-Galactosidase)

Ceramide trihexoside

Tay-Sachs disease

β-Hexosaminidase A

Ganglioside G_{M2}

Tay-Sachs disease variant
(Sandhoff's disease)

β-Hexosaminidase
A & B

Globoside (plus Ganglioside G_{M2})

Generalized gangliosidosis

G_{M1} β-Galactosidase

Ganglioside G_{M1}

Ceramide: N-acylsphingosine
Glc: glucose
PChol: phosphorylcholine
Gal: galactose
NANA: N-acetylneuraminic acid
NAcGal: N-acetylgalactosamine

4. Patients exhibiting glucosyl ceramide lipidosis (Gaucher's disease) fall into one of three groups. The genetic defects in these three types appear to be due to similar errors in the same or related genetic loci.

    a. Type I--Chronic non-neuronopathic (adult): The most common variety of Gaucher's disease. A somewhat heterogeneous group of patients characterized by a lack of cerebral involvement, presence of hematological abnormalities (anemia, thrombocytopenia) and erosion of the cortices of the long bones.

    b. Type II--Acute neuronopathic: Usually appears prior to six months of age and is fatal by two years. Mental damage is a primary characteristic and the disease progresses rapidly from its onset.

    c. Type III--Subacute neuronopathic (juvenile): Another heterogeneous group in which death occurs anywhere from infancy to about 30 years of age. The cerebral abnormalities usually appear at least two years postnatally.

    All of the above types of disorders, despite their diverse and variable organ involvement, share the following common features: hepatosplenomegaly, Gaucher cells in the bone marrow (accumulation of glucocerebroside in reticuloendothelial cells in liver, spleen and bone marrow) and autosomal recessive inheritance. Some studies have shown a correlation between the residual β-glucocerebrosidase activity and the clinical severity. However, other studies have failed to find similar correlations. In any event, the precise molecular basis for the genetic heterogeneity in Gaucher's disease is not known. The Gaucher cells obtained from bone marrow aspirates exhibit a characteristic "wrinkled tissue paper" appearance of the cytoplasm, due to rod-shaped striated inclusion bodies. The deposits are composed primarily of glucocerebroside. It has been shown that patients with Gaucher's disease have elevated levels of acid phosphatase activity in the serum and spleen, increased iron stores, increased angiotensin-converting enzyme activity and a relative deficiency of factor IX (a factor involved in blood coagulation). The β-glucocerebrosidase deficiency in Gaucher's disease is established by assaying the enzyme activity in leukocytes or fibroblasts, using either the labelled natural substrate or water-soluble artificial substrates such as 4-methyllumbelliliferyl-β-D-glucopyranoside.

5. Studies of a number of cases of G_{M1} gangliosidosis have revealed two distinct types.

    a. In generalized gangliosidosis, G_{M1} and desialo-G_{M1}-gangliosides accumulate in both the brain and viscera. Three

β-galactosidase activities have been isolated from normal
human liver and all three are absent in this disorder. The
progress of the disease is rapid, beginning at or near birth
and ending fatally, usually by two years of age.

b. <u>Juvenile G_{M1} gangliosidosis</u> progresses much more slowly.
Psychomotor abnormalities usually begin at about one year and
death ensues at 3-10 years. Only two of the three liver
β-galactosidase activities are absent in this variant,
presumably accounting for the lack of lipid accumulation in
this organ. This enzymatic finding also supports the genetic
separation of the two forms.

6. G_{M2} gangliosidosis has two types.

a. Type 1--Tay-Sachs disease: β-hexosaminidase A (hex-A)
deficiency.

b. Type 2--Sandhoff disease: deficient activities of two enzymes,
β-hexosaminidase A and B (hex-A, hex-B).

By immunologic studies it has been shown that hex-A and hex-B
share a common subunit (β). Based on these studies the tentative
composition of hex-A is $(\alpha\beta)_n$ and hex-B is $(\beta\beta)_n$.

7. The reason for the accumulation of desialo-G_{M1} and desialo-G_{M2}
gangliosides in G_{M1} gangliosidosis and Tay-Sachs disease (TSD),
respectively, is not clear. It has been demonstrated that, <u>in vitro</u>,
tissue from patients with TSD <u>is</u> capable of metabolizing
desialo-G_{M2} ganglioside. The reaction is inhibited by G_{M2}
ganglioside but it remains to be demonstrated that this mechanism
operates <u>in vivo</u>. Although similar studies have apparently not
been done with G_{M1} gangliosidosis material, a parallel explanation
seems plausible.

8. The name metachromatic leukodystrophy (MLD) refers to a
dysfunction of the white matter of the brain which is characterized
by metachromasia of the nervous tissue. Metachromasia is the
ability of certain compounds (notably organic anions such as
sulfatides which either have a high molecular weight or are capable
of polymerizing or forming micelles) to shift the absorption
wavelength of some cationic dyes to shorter wavelengths. This
causes toluidine blue, for example, to appear red or pink. Other
dyes affected include cresyl violet (stains MLD tissue brown),
methylene blue, and thionine.

It is important to distinguish between MLD and other disorders of
the white matter. One common disease of this sort is multiple
sclerosis, which probably has a viral or autoimmune etiology.

9. As in several other of the diseases discussed here, metachromatic leukodystrophy can be subdivided into several disorders which show distinct clinical patterns.

    a. Late infantile form. Most common type; usually appears 1-4 years postnatally; associated with loss of motor and mental functions and ultimately death; due to lack of arylsulfatase A and cerebroside sulfatase deficiency.

    b. Adult. Rare; onset is not until age 21 years or later but difficult to diagnose; dementia and psychosis; later, loss of motor function usually followed by death; deficiency (but not absence) of arylsulfatase A.

    c. MLD variant with multiple sulfatase deficiency. Rare; appears at age 1-3 years; progresses slowly with psychomotor deterioration and other changes; development of these patients is retarded compared to those with late infantile MLD despite similar age of onset; the disorder is fatal and is apparently due to a lack of arylsulfatases A and C and of steroid sulfatase, and decreased activity of arylsulfatase B.

10. One other disorder of lipid catabolism is alpha-fucosidosis. In this disease, a glycoceramide having the antigenicity of H-isoantigen (see the section on ABO blood groups in the chapter on immunochemistry) and mucopolysaccharides having an α-L-fucose residue at the reducing end accumulate in liver, brain, and perhaps other tissues.

Cer-(←1β)-Glc-(4←1β)-Gal-(4←1β)-NAcGlcam
$$\qquad\qquad\qquad\qquad\qquad\qquad | $$
$$\qquad\qquad\qquad\qquad\qquad (3←1β)-Gal$$
$$\qquad\qquad\qquad\qquad\qquad\qquad | $$
$$\qquad\qquad\qquad\qquad\qquad (2←1α)-Fuc$$

α-L-fucose
(6-deoxy-L-galactose)

Structure of H-isoantigenic substance which accumulates in α-fucosidosis. Cer = ceramide; Glc = glucose; Gal = galactose; NAcGlcam = N-acetyl-glucosamine; Fuc = α-L-fucose.

Because of the mucopolysaccharide deposition, this is also classified as a mucolipidosis or even a mucopolysaccharidosis. Clinical features include cerebral degeneration, thick skin, electrolyte abnormalities, and muscle spasticity. In the livers of the patients studied, α-fucosidase was found to be absent. This enzyme hydrolyzes glycosidic bonds between α-fucose and other sugars. This enzyme deficiency is apparently the cause of this disorder.

11. <u>Approaches to the therapy of sphingolipidoses</u>: As noted earlier, the treatment is primarily symptomatic and supportive. For example, in patients with anemia due to Gaucher's disease, thrombocytopenia associated with hypersplenism is relieved by splenectomy. Recently, infusion of appropriate purified human (obtained from placental tissue) enzymes in Gaucher and Fabry patients has resulted in the reduction of accumulated glycolipids in the circulation and liver. This approach of enzyme replacement may prove to be beneficial only in the clinical treatment of the <u>non-neuronal</u> types, because of the blood-brain barriers. One of the major problems in this area is to obtain large amounts of highly purified enzyme from human tissues in order to minimize immunological complications. The recent advances made in the cloning and amplification of the appropriate human DNA segment in a bacterial plasmid and subsequent production of the gene product (somatostatin synthesis) may yield large amounts of the enzyme required for treatment (see Chapter 5). In an attempt to minimize the immunological complications, the use of the recipient's own erythrocytes entrapped with the enzyme is under investigation. When erythrocytes are exposed to hypotonic conditions in the presence of the enzyme, the pores formed in the membrane allow rapid exchange of the enzyme with its intracellular contents. Restoration to isotonicity reseals the erythrocyte membrane, entrapping a certain amount of the enzyme. The other enzyme carriers are <u>liposomes</u>, which are concentric lipid bilayers prepared from cholesterol, lecithin and phosphatidic acid. It should be mentioned that the ideal cure for these disorders would be the addition or replacement of genetic material coding for the missing gene product.

<u>Biological Membranes</u>

1. As was pointed out earlier (Chapter 3), one of the factors which distinguishes living matter (cells) from non-living material is the ability to employ external energy to maintain chemical reactions in a non-equilibrium steady-state. This would be practically impossible without the various types of membranes which permit the cell to establish a "local" (intracellular) environment which differs from whatever surrounds the cell regarding concentrations and even the particular molecules present. In addition to this important physical isolation, membranes provide a support for many metabolic enzymes and an electrical barrier (as in nerve conduction). Membranes cannot be totally impermeable, though, since nutrients must enter the cell and waste products have to be eliminated. Specific carrier enzymes (permeases), apparently mounted in the membrane, perform important

roles in the transport of various molecules. Other compounds can freely diffuse through the membranes. The particular function of the cell dictates which molecules can and cannot cross the membrane barrier and under what conditions they can do so. Additionally, membranes must be physically flexible (plastic). This is evident from observations of cell packing and motility and from the swelling and shrinking which occurs when cells are placed in a medium having an osmolarity different from that within the cell.

2. The preceding discussion also applies to the use of membranes for intracellular localization seen in eucaryotic cells. Examples of such subcellular organelles include mitochondria, chloroplasts, the nucleus, the endoplasmic reticulum and the Golgi apparatus, lysosomes, peroxisomes, and others. Some of these are discussed elsewhere in this book. Recall the role of conformational change in one theory of the mechanism by which electron flow is coupled to phosphorylation in the mitochondria (page 258).

3. Biological membranes are generally composed of proteins, neutral lipids (especially cholesterol), and phospholipids. The specific ratios of these compounds vary considerably among cell types and between plasma membranes and membranes of subcellular organelles. This is to be expected, since many of the properties which make different cells and organelles unique reside in their membranes. For example, in the myelin nerve sheath membrane, the ratio of lipid (grams) to protein (grams) is 4, the molar ratio of cholesterol to phospholipid is 0.7–1.2, and the principal phospholipids present are phosphatidyl ethanolamine and choline, and cerebrosides. For a liver cell membrane (plasma membrane, plasmalemma), the corresponding values are 0.7–1.0 (lipid/protein), 0.3–0.5 (cholesterol/polar lipids), while phosphatidyl ethanolamine, choline, and serine predominate among the phospholipids. A third example is the ribosome-free endoplasmic reticulum, having values of 0.8–1.4 (lipid/protein), 0.03–0.08 (cholesterol/phospholipid), with phosphatidyl choline and serine, and sphingomyelin as the principal phospholipids. The endoplasmic reticulum is typical of the membranes of other subcellular organelles in containing much smaller amounts of cholesterol than the plasma membranes. Animal cell membranes also contain oligosaccharides, composed principally of fucose, galactose, amino sugars, and (in some cell membranes) sialic acid. Some of these apparently provide the antigenic specificity possessed by most cells.

4. In addition to varying considerably in composition from one cell to another, biological membranes apparently are constantly gaining and losing molecules (plasticity in time). Thus, different membranes are constantly exchanging proteins and lipids, and as a cell matures, the protein and (to a lesser extent) lipid content

of its membranes alters. This may affect composition and
structural studies of, say, erythrocyte membranes, where any sample
obtained from fresh blood will contain cells of all ages. Exogenous
substances can change membrane composition. In the rat, phenobarbital
causes proliferation of smooth endoplasmic reticulum (SER) membranes
in which the activities of certain enzymes (including the P-450
cytochromes involved in phenobarbital detoxication) are ten times
higher than the activity in normal SER membranes. These "abnormal"
membranes persist after the phenobarbital has been metabolized and
may explain why this drug is capable of stimulating heme catabolism.

5. Despite intensive study and much speculation, the structure of
    biological membranes is still not well understood. The "lipid
    bilayer" model was originally proposed by Danielli and Davson in
    1935. It has been modified extensively since then but remains
    an integral part of most theories of membrane structure. A
    schematic representation of the fluid mosaic model (proposed by
    Singer and Nicholson) is shown in Figure 105. Cholesterol, which
    is an important part of animal membranes, probably lies in the
    hydrophobic region with the plane of its ring parallel to the
    fatty acids of the phospholipids.

<u>Figure 105.</u>   <u>A Schematic Representation of the Fluid Mosaic Model of</u>
<u>Membranes</u>

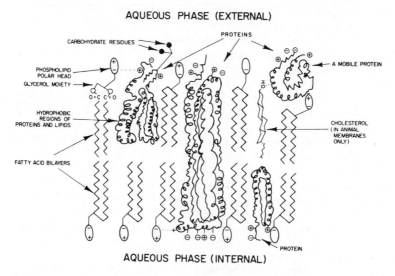

The phospholipids can be any of the phosphatidates (e.g.,
phosphatidyl choline, ethanolamine or serine) or even some of the
sphingolipids. Differences in lipid (and protein) composition
give differing physical and chemical properties to membranes from

various sources. Note that the Danielli-Davson lipid bilayer model is still providing the basic framework for this membrane.

6. A number of studies indicate that there are membrane proteins attached to three places: the inner surface, the outer surface, and those that penetrate the membrane so that they are both inside and outside. The forces involved in the protein-lipid interaction are probably non-covalent in most cases. Ionic groups of the proteins may associate with the polar surfaces of the membrane lipids while hydrophobic portions contact the lipid-like interior of the bilayer.

This leads to the concept of mosaicism or inhomogeneity in biological membranes. The nature and arrangement of the proteins in a particular membrane may well depend on the region of the membrane being studied. The "fluid" part of the fluid-mosaic model means that the proteins on (and in) the membrane may migrate from one part of the surface to another, either randomly or in response to stimuli. In addition, it seems that lipid molecules are constantly being added to and removed from the membrane, although there is, in most cases, no net change in its composition. There is still much work to be done before these crucial structures can be completely described and understood.

7. The central role of membranes in enzyme function is one of the most important ideas to arise from membrane studies. As was indicated previously, membranes are far more than passive envelopes for the cytoplasm or intraorganellular fluid. A partial list of enzyme systems and the membranes with which they are associated is given below. Some of these systems (or members of them) have a demonstrated functional requirement for lipid.

     a. Electron transport and oxidative phosphorylation enzymes (bound to the inner mitochondrial membrane).

     b. Protein synthesis (polysomes are frequently attached to the rough endoplasmic reticulum and perhaps penetrate the membrane).

     c. Microsomal oxidase system (involved in heme catabolism and detoxication of other compounds; smooth endoplasmic reticulum).

     d. Enzyme system which synthesizes the polysaccharide portion of the mucopolysaccharides (smooth endoplasmic reticulum and Golgi apparatus).

     e. Miscellaneous

         i) Permeases (also called translocases or transporter proteins; used by the cell to transport molecules and ions across membrane permeability barriers; occur in most membranes; when energy for the transport is supplied by ATP hydrolysis, these are sometimes called ATPases; an example is the Na^+-K^+ ATPase or pump in the cell membranes of erythrocytes and other cells).

ii) <u>Adenylate cyclase</u> (located on the inside of a number of cell membranes; it is in communication with a <u>hormone receptor protein</u> attached to the outside of the cell membrane).

Fat Absorption

Fat is a major source of dietary calories and the intake ranges anywhere from 10 gm to 150 gm per day. The digestion and absorption of fat, which is accomplished in the small intestine, can be divided into three phases: intraluminal, intestinal and removal. The intraluminal phase is dependent upon pancreatic enzymes and bile and small-bowel secretions. The latter consists of both endocrine and exocrine secretions. The endocrine portion includes cholecystokinin-pancreozymin, secretin, gastric inhibitory peptide, etc. The functions of these have already been discussed (see protein digestion, Chapter 6). The smooth muscle contractions of the alimentary canal provide the motility for the mixing and propulsion of the intraluminal contents. Fat digestion and absorption require <u>a neutral pH</u> and the <u>action of a detergent</u>. The bile and pancreatic secretions (rich in bicarbonate) neutralize the acid chyme while the detergents (bile acids) are obtained from the bile. The gallbladder contracts under the influence of cholecystokinin-pancreozymin and empties its contents (bile) into the duodenum within twenty minutes of the beginning of a meal.

1. The steps involved in the absorption of dietary fat are outlined in Figure 106. Cholesterol is absorbed directly into the mucosal cells where most of it is first esterified with fatty acids and then incorporated into chylomicrons or, to a small extent, absorbed directly into the lymphatic system. The uptake of cholesterol from the intestine is dependent upon the presence of the bile salts (discussed later) which are presumably needed for emulsification of the sterol. Phospholipids are hydrolyzed by the phospholipases to free fatty acids and glycerylphosphorylcholine. The fatty acids are absorbed and the glycerylphosphorylcholine is either excreted in the feces or further degraded and absorbed.

2. The triglycerides are hydrolyzed by pancreatic lipase. As the name implies, it is secreted into the intestine by the pancreas, along with phospholipases, proteases, amylase, and other compounds. Anything which prevents this material from reaching the intestine can result in fatty stools (steatorrhea; discussed later in this section). The major products obtained from pancreatic lipase (an

Figure 106. Intestinal Absorption of Dietary Fats

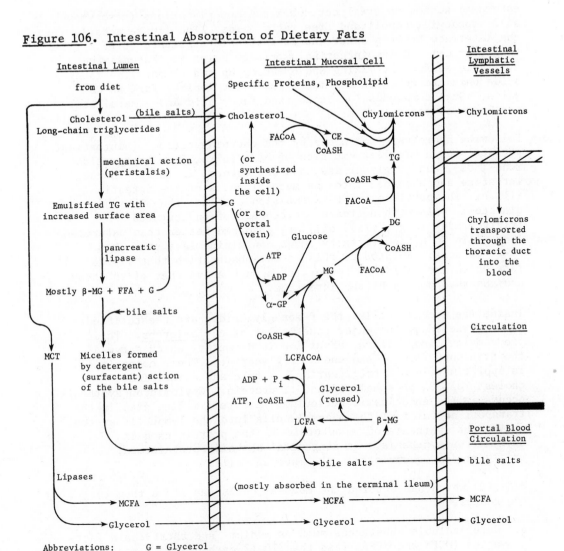

Abbreviations:
- G = Glycerol
- MCT = Medium chain triglycerides
- β-MG = β-Monoglyceride (2-monoacylglycerol)
- DG = Diglyceride
- TG = Triglyceride
- FFA = Free fatty acid
- LCFA = Long chain fatty acid (14 carbons or more)
- MCFA = Medium chain fatty acid (10 carbons or less)
- α-GP = α-Glycerolphosphate
- CE = Cholesterol ester
- FACoA = Fatty acyl CoA

esterase) action on triglycerides are β-monoglycerides (also known as 2-monoacylglycerol) and free fatty acids. The lipase acts at the interface between water and an insoluble substrate (e.g., triglyceride). Gastrointestinal motility helps to increase the available surface by mixing and breaking up the chyme. The fatty acids that are released are mostly in their ionized form as Ca^{+2} salts. Such compounds (of an ionized fatty acid and a metal ion) are known as soaps.

3. Following hydrolysis and aided by the churning action of intestinal peristalsis, a micellar emulsion of bile salts, free fatty acids, mono-, di-, and triglycerides, and cholesterol is formed in the intestine and absorbed into the mucosal cells of the intestinal villae. The particles of this emulsion are less than 1 μ in diameter. Oils with a greater percentage of unsaturated **triglycerides (e.g., olive oil) are absorbed better than saturated** long-chain TG's, as are fatty acids containing less than about 16-18 carbons. Related to this is the observation that lipids and fats which are solids at temperatures much above that of the body are absorbed poorly by the intestine.

4. Inside the mucosal cells, the β-monoglycerides are reesterified to triglycerides, the major constituent of chylomicrons. These are complex lipoproteins, about 1 μ in diameter, which serve as the transport form of exogenous triglyceride. Their composition is approximately 87% triglyceride, 4% free and esterified cholesterol, 8% phospholipid and 1% protein. Chylomicron synthesis occurs in the endoplasmic reticulum. A rate-limiting step in the transport of fat from the mucosal cells into the lymphatic system may be the synthesis of apoprotein-B, the protein used in chylomicron synthesis. In congenital abetalipoproteinemia, triglycerides accumulate in the mucosal cells.

5. Material which has entered the mucosal cells can leave them by two routes.

   a. Water-soluble substances such as medium- and short-chain fatty acids (MCFA and SCFA; less than 10-12 carbon atoms) and bile salts are absorbed directly into the enterohepatic portal circulation. The MCFA are transported as free (unesterified) fatty acids. The bile salts are removed from this blood by the liver and reinjected, along with the bile, into the duodenum. Because of this recycling, very little bile acid appears in the peripheral circulation and only about 200 mg/day are lost in the feces. The direct absorption of the smaller fatty acids is useful in the maintenance of patients who exhibit fatty acid malabsorption and steatorrhea due to bile salt deficiency. Medium-chain-length triglycerides (MCT) are provided in the diet in place of a large fraction

of the more usual long-chain triglycerides. Since neither
bile nor micellar formation is directly necessary for lipase
activity, the MCT can be hydrolyzed and the MCFA absorbed
despite this lack. Special MCT diets are commercially
available.

b. The chylomicrons and any remaining cholesterol are absorbed
into the intestinal lacteals (lymphatics originating in the
villi of the small intestine). Following a meal, the presence
of large amounts of chylomicrons in the lymph gives it a milky
appearance and endows it with the name <u>chyle</u>. From the lymph,
these materials enter the blood by way of the thoracic duct
and, to a lesser extent, accessory channels.

6. Lipid malabsorption, leading to the excessive loss of fat and other
lipids in the stool (i.e., steatorrhea), can be caused by several
things. It is important to note that, while the bile salts do not
activate the lipase, they do prevent its inactivation. Because
albumin can partially replace the bile salts in this function, it
is presumed that the solubilization of the fatty acids produced by
lipase action stops fatty acid inhibition of the enzyme. In the
laboratory assay of serum lipase (see Chapter 3), a much higher
value for lipase activity is obtained if sodium deoxycholate (a bile
salt) is added to the reaction mixture to emulsify the fatty acids
produced. The details of bile salt metabolism are discussed later
in this chapter.

a. Defective lipolysis in the lumen of the small intestine:

   i) Bicarbonate deficiency: due to pancreatic disease or
       gastric hypersecretion.

   ii) Bile salt deficiency: due to impaired liver function,
        obstruction of bile duct, administration of cholestyramine
        (binds to anions of bile salts and enhances their fecal
        loss--used as a hypocholesterolemic agent), or excessive
        loss of bile salt because of poor reabsorption due to
        ileal disease or resection.

   iii) Pancreatic lipase deficiency: due to malfunction or
         damage of pancreatic tissue, obstruction of the pancreatic
         duct (as in cystic fibrosis of the pancreas; see under
         protein digestion), or a genetic defect in enzyme synthesis.

b. Defective mucosal cell function:

   i) Impaired resynthesis of triglycerides: in nontropical
       sprue and adrenocortical hormone deficiency.

ii) Impaired apolipoprotein synthesis: abetalipoproteinemia.

iii) Blockage of the synthesis of the above proteins by certain drugs that inhibit protein synthesis (e.g., puromycin).

c. Defective transport of chylomicrons due to lymphatic obstruction (lymphangiectasia): occurs in lymphosarcoma, intestinal lipodystrophy, protein-losing enteropathy.

d. Abnormal connections between organs:

i) Chyluria: abnormal connection between the urinary tract and the lymphatic drainage system of the small intestine, causing the excretion of milky urine.

ii) Chylothorax: abnormal connection between the pleural space and lymphatic drainage system of the small intestine, giving rise to the accumulation of <u>milky pleural fluid</u>.

7. Defective fat absorption can also impair the uptake of other nutrients, notably the fat-soluble vitamins (A, D, E, and K). Since vitamin K is essential for blood clotting, care should be taken to avoid its deficiency in patients prior to gallbladder surgery.

8. Laboratory evaluation used in disorders of digestion and absorption of the intestines: The tests include fecal fat determination, xylose absorption test, carotene absorption, vitamin B_{12} absorption, bile salt absorption, x-ray examination of the small intestine after the ingestion of barium, and biopsy of the small intestine.

a. The fecal fat determination is the most sensitive measure of steatorrhea. The fecal lipids are derived from unabsorbed dietary lipids, cellular debris and excretions from the intestinal mucosa, other excretions from the glands other than the intestines, bacteria or other organisms. On a lipid-free diet, the fecal fat output is about 1-4 g per day. The test is performed on patients who have consumed at least 100 g of fat per day and the fecal fat is usually quantitated in a sample derived from a 3-5 day collection. The adult normal value (for a person ingesting 100 g of fat per day) for fecal fat is about less than or equal to 7 g per day (the normal range for children from the age of 2 months to 6 years is about 0.3-2 g of fat per day). Enhanced fat in the feces suggests maldigestion and/or malabsorption. Further tests are required to establish the etiology of steatorrhea.

b. The D-xylose (a pentose) absorption test consists of the administration of a known amount of D-xylose and the measurement of its concentration in the blood and urine at selected intervals. The absorption of xylose does not require pancreatic and bile secretions and therefore an impairment in its absorption is usually suggestive of a small intestinal disease.

c. Vitamin B_{12} (see Chapter 10) and the bile salts are absorbed in the ileum; their measurements are reflective of ileal absorption.

d. Histologic examination of biopsies from the small bowel is used in the diagnosis of abetalipoproteinemia, amyloidosis, Whipple's disease, gluten-sensitive enteropathy and mast-cell disease. The findings include flattening of the mucosa, loss of villi, loss of brush border, elongated crypts and an enhanced number of inflammatory cells in the lamina propria.

e. X-ray examination after a barium meal may show dilation of the small intestinal loops, thickening and edema of the intestinal wall, barium flocculation and segmentation. These and other specific patterns may be seen in blind-loop syndrome, fistula or Crohn's disease.

f. Steatorrhea occurs in diseases of the pancreas due to a deficiency of pancreatic lipase. Acute pancreatitis may be diagnosed by measuring the activity of amylase (see digestion of carbohydrates--Chapter 4) in the serum and urine. Plasma lipase is also measured, which shows a slower pattern of rise and fall than amylase: the peak activity is at about 48 hours and the level remains high for about a week. The tests for chronic pancreatic disorders consist of measuring the levels of bicarbonate and enzyme directly in the duodenal juice (pancreatic secretion). This is accomplished by duodenal intubation in an overnight-fasted patient and collecting the pancreatic juice before and after an injection of pancreatic stimulant (secretin and/or pancreozymin).

Fatty Acid Oxidation

1. Quantitatively, the most important pathway for fatty acid catabolism is β-oxidation (Figure 107). The name is derived from the oxidation of the activated fatty acid (fatty acyl-CoA) at the β-carbon, followed by removal of two carbon fragments as acetyl-CoA. The oxidation of fatty acyl-CoA takes place <u>entirely in the mitochondrial matrix</u>. Fatty acids are obtained either by the hydrolysis of exogenously derived triglycerides present in chylomicrons by the action of lipases (present on the walls of capillary beds of the target tissues) or by the action of hormone-sensitive lipase on triglycerides stored in adipose tissue cells (discussed later). The free fatty acids in the blood are transported bound to albumin.

Figure 107. Fatty Acid Activation, Transport and β-Oxidation

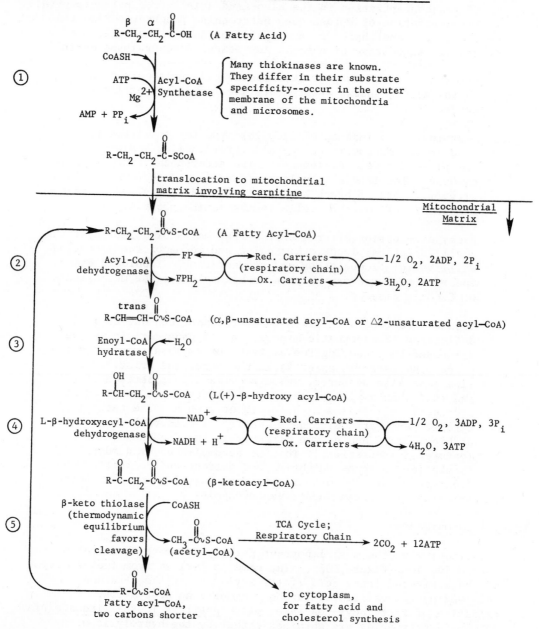

①

②

③

④

⑤

Note that the shortened fatty acyl–CoA from one cycle is further oxidized in successive passes until it is entirely converted to acetyl–CoA. Odd-chain fatty acids, discussed later, produce one molecule of propionyl-CoA. Ox. = oxidized, Red. = reduced; respiratory chain = oxidative phosphorylation and electron transport; "∿" = a high-energy bond (see Chapter 3); FP = flavoprotein.

The intracellular steps involved in fatty acid (e.g., palmitic acid) oxidation are: activation of the fatty acid in the cytoplasm, entry into the mitochondria, oxidation by the mitochondrial matrix enzymes to acetyl-CoA, oxidation of hydrogen atoms removed during β-oxidation in the mitochondrial respiratory chain, and oxidation of acetyl-CoA in the tricarboxylic acid cycle and respiratory chain. The latter two processes are associated with the mitochondrial inner membrane and matrix. The electrons are transported to molecular oxygen via the cytochrome system, generating ATP from ADP and P_i (oxidative phosphorylation); these aspects are discussed in Chapter 4.

2. Fatty acid β-oxidation consists of activation, transport into the mitochondrial matrix and oxidation.

   a. <u>Fatty acid activation</u>. The reaction is catalyzed by acyl-CoA synthetases and there are at least three different enzymes, each showing specificity based on the length of the carbon chain of a fatty acid. The enzymes and their substrates are as follows: acetyl-CoA synthetase: acetic, propionic and acrylic acids; medium-chain acyl-CoA synthetase: fatty acids with 4 to 12 carbon atoms; and long-chain acyl-CoA synthetase: fatty acids with more than 12 carbons. The latter two enzymes also catalyze the activation of unsaturated fatty acids and 2- and 3-hydroxy acids. The first two enzymes mentioned above are located in the outer membranes of the mitochondria and the third, the long-chain acyl-CoA synthetase, is found in the endoplasmic reticulum (microsomes). The overall reaction of activation is:

$$RCOOH + ATP + CoASH \rightleftharpoons \overset{\overset{\displaystyle O}{\displaystyle \|}}{RC}\text{-SCoA} + AMP + PP_i$$
$$\text{Fatty}$$
$$\text{acyl-CoA}$$
$$\text{(a thioester)}$$

The reaction is favored towards the formation of fatty acyl-CoA. The pyrophosphate formed is hydrolyzed to two inorganic phosphates by the enzyme pyrophosphatase: $PP_i + H_2O \rightarrow 2\ P_i$. The fatty acid activation apparently takes place in two steps:

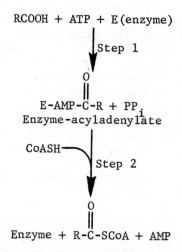

RCOOH + ATP + E(enzyme)

Step 1

$$E-AMP-\overset{\overset{\textstyle O}{\|}}{C}-R + PP_i$$

Enzyme-acyladenylate

CoASH

Step 2

$$Enzyme + R-\overset{\overset{\textstyle O}{\|}}{C}-SCoA + AMP$$

b. Transport of fatty acyl-CoA to mitochondrial matrix. This is accomplished by carnitine, whose chemical name is β-hydroxy-γ-trimethylammonium butyrate. The fatty acyl-CoA cannot penetrate the inner membrane of the mitochondria in the absence of carnitine. Thus, carnitine is required in catalytic amounts for the oxidation of fatty acids. It will be seen later in this chapter that carnitine also plays an important role in fatty acid synthesis, which takes place in the cytoplasm, by participating in the transport of acetyl-CoA. The role of carnitine in fatty acyl-CoA transport is shown in Figure 108. Note in the figure that there are two types of carnitine acyltransferases involved in the translocation of acyl-CoA: one located on the outer surface of the inner mitochondrial membrane and the other on the inner surface of the inner membrane. The overall translocation reaction is:

$$
\begin{array}{c}
\underset{\text{Acyl-CoA}}{R-\overset{\overset{\displaystyle O}{\|}}{C}-SCoA} +
\underset{\text{Carnitine}}{H_3C-\overset{\overset{\displaystyle CH_3}{|}}{\underset{\underset{\displaystyle CH_3}{|}}{N^+}}-CH_2-\overset{}{\underset{\underset{\displaystyle OH}{|}}{CH}}-CH_2-COOH}
\end{array}
$$

Carnitine Acyltransferase

$$
H_3C-\overset{\overset{\displaystyle CH_3}{|}}{\underset{\underset{\displaystyle CH_3}{|}}{N^+}}-CH_2-\overset{\overset{\displaystyle H}{|}}{\underset{\underset{\underset{\underset{R}{|}}{C=O}}{|}}{C}}-CH_2-COOH + CoASH
$$

$\left.\begin{array}{c} \\ \\ \end{array}\right\}$ acyl group

It should be noted that the standard free energy change of the reaction catalyzed by carnitine acetyltransferase is about zero. Therefore, the acyl-carnitine bond, an oxygen ester, may be considered as a high-energy linkage. Normally, oxygen-ester linkages are <u>not</u> high-energy bonds when compared with acid anhydrides, enol phosphates, thioesters, and other types of bonds.

Absence of carnitine palmitoyl transferase in biopsied muscle tissues from two brothers has been noted. These patients showed, consistent with the role of the enzyme carnitine transferase, impaired utilization of long-chain fatty acids along with elevated levels of plasma free fatty acids and triglycerides. Myoglobinuria was also found, indicating muscle destruction which may be due to inadequate production of ATP. Other clinical manifestations included renal failure and azotemia. It should be noted that fatty acid oxidation in man at rest on a regular diet is responsible for more than 50 percent of the energy required by the muscle tissue.

c. The pathway for β-oxidation in the mitochondria matrix. The general pathway is shown in Figure 107. The following aspects should be noted about this pathway.

   i) The fatty acyl-CoA is <u>dehydrogenated</u> by the removal of two hydrogen atoms from the α- and β-carbon atoms to yield the α,β-unsaturated acyl-CoA (also called

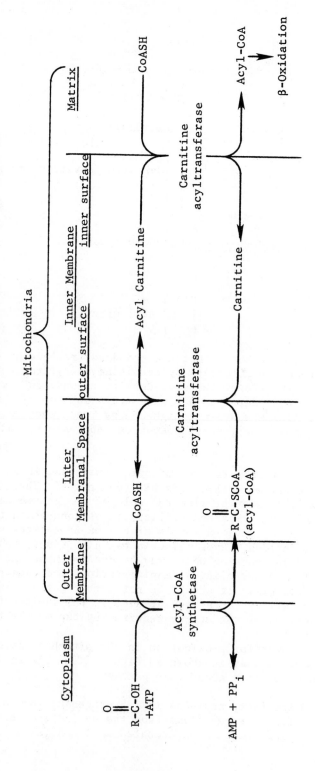

Figure 108. Transport of Fatty Acids Into the Mitochondrial Matrix--Role of Carnitine

Δ2-unsaturated acyl-CoA). The reaction is catalyzed by acyl-CoA dehydrogenase. There appears to be four different enzymes, each specific for a given range of fatty acid chain length. All of these enzymes are flavoproteins and contain a tightly linked prosthetic group of flavin adenine dinucleotide (FAD). The electrons of the acyl-CoA dehydrogenase-linked $FADH_2$ are transferred to the respiratory chain, probably at the CoQ level via a second flavoprotein (electron-transferring flavoprotein). It is of interest to point out that the Δ2 double bond formed in this reaction has a trans geometrical configuration. As noted earlier, the double bonds in the naturally occurring fatty acids are generally in the cis configuration. The oxidation of cis-fatty acids is discussed later, which requires two auxiliary enzymes (in addition to enzymes involved in the β-oxidation): an isomerase (enoyl-CoA isomerase) and an epimerase (3-hydroxyacyl-CoA epimerase).

ii) The hydration of the Δ2-unsaturated acyl-CoA is catalyzed by enoyl-CoA hydratase (reaction 3). This enzyme possesses broad specificity and can catalyze the hydration of α,β- (or Δ2) unsaturated CoA either in trans or cis configuration. The products formed in these reactions are as follows:

Δ2-trans enoyl-CoA $\longrightarrow$ L(+)-β-hydroxyacyl-CoA
                       (or L(+)-3-hydroxyacyl-CoA)

Δ2-cis enoyl-CoA $\longrightarrow$ D(-)-β-hydroxyacyl-CoA
                       (or D(-)-3-hydroxyacyl-CoA)

Note that the latter reaction is used in the oxidation of natural unsaturated fatty acids and that an epimerase converts it to the L-isomer which is the substrate for the next enzyme in β-oxidation.

iii) Oxidation of β-hydroxyacyl-CoA (reaction 4) is accomplished by a NAD^+-linked (broadly specific in terms of chain length) L-β-hydroxyacyl-CoA dehydrogenase. The enzyme is absolutely specific for the L-stereoisomer. The electrons from the NADH generated in this reaction are passed on to NADH dehydrogenase of the respiratory chain.

iv) The last step of β-oxidation (reaction 5) involves a thiolytic cleavage catalyzed by the enzyme β-ketothiolase (acetyl-CoA acetyltransferase). This enzyme has broad specificity and yields a molecule of acetyl-CoA and a fatty acyl-CoA shorter by two carbon atoms. The reaction is highly exergonic ($\Delta G^{o'}$ = -6.7 kcal/mole) and thiolysis (analogous to hydrolysis) is favored. The enzyme has a reactive -SH group on a cysteinyl residue which participates in the catalysis as follows:

$$\underset{\text{R-C-CH}_2\text{-C-SCoA}}{\overset{\text{O}\qquad\text{O}}{||\qquad||}} + \text{HS-Enz} \rightleftharpoons \underset{\text{RC-S-Enz}}{\overset{\text{O}}{||}} + \underset{\text{CH}_3\text{-C-SCoA}}{\overset{\text{O}}{||}}$$

β-ketoacyl-CoA thiolase   acetyl-CoA

$$\underset{\text{R-C-S-Enz}}{\overset{\text{O}}{||}} + \text{HS-CoA} \rightleftharpoons \underset{\text{R-C-SCoA}}{\overset{\text{O}}{||}} + \text{Enz-SH}$$

acyl-CoA   (regenerated
(with 2   enzyme)
less carbon
atoms)

Note that at each cycle of β-oxidation, the fatty acyl-CoA is shortened by a two-carbon fragment of acetyl-CoA. For example, after one activation, palmitic acid (C16) yields 8 acetyl-CoA molecules and 14 pairs of hydrogen atoms, requiring 7 cycles through the β-oxidation enzymatic system.

3. Energetics of β-oxidation. From Figure 107, it can be seen that one molecule of ATP is hydrolyzed to AMP and pyrophosphate for each fatty acid molecule oxidized (reaction 1). This is the only ATP consumed in these reactions. The energy from its hydrolysis is used to form the high-energy thioester bond in acyl-CoA.

a. Each time a mole of fatty acyl-CoA cycles through reactions 2-5, one mole of reduced flavoprotein (FPH_2), one mole of NADH, and at least one mole of acetyl-CoA are produced. On the last pass of an even-chain-length fatty acid, two moles of acetyl CoA are formed; and the final pass of an odd-chain-length molecule releases one mole of propionyl-CoA.

b. Based on the preceding paragraph, the amount of ATP formed in the β-oxidation of hexanoic acid can be calculated. In doing so, the hydrolysis of ATP to AMP + PP_i is considered as equivalent to two moles of ATP, since two high-energy phosphate bonds are lost. This calculation is shown below.

Reaction	Direct Consequences of the Reaction	Moles of ATP Gained or Lost Per Mole of Hexanoic Acid
1	hexanoic acid $\rightarrow$ hexanoyl-CoA	-2
2	dehydrogenation of acyl-CoA; $2(FP \rightarrow FPH_2)$	+4
3	hydration of α,β-unsaturated fatty acyl-CoA	0
4	dehydrogenation of β-hydroxy-acyl-CoA; $2(NAD^+ \rightarrow NADH + H^+)$	+6
5	formation of 3 moles of acetyl-CoA, followed by their oxidation to CO_2 and H_2O in the TCA cycle and electron-transport system	$+(3\times12) = +36$

$$\text{Total ATP} = +44$$

As was pointed out earlier (Chapter 4), fatty acid oxidation produces more moles of ATP per mole of CO_2 formed than does carbohydrate oxidation. In this case, oxidation of one mole of glucose (a six-carbon moiety as is hexanoic acid) produces, at most (assuming malate shuttle operation exclusively), 38 moles of ATP. This is six less than obtained from hexanoic acid.

c. By a similar type of calculation, it can be shown that the complete oxidation of one molecule of palmitic acid yields about 129 ATP molecules:

$$C_{15}H_{31}COOH + 8 \text{ CoASH} + ATP + 7 \text{ FAD} + 7 \text{ NAD}^+ + 7 \text{ H}_2O$$

$$\longrightarrow 8 \text{ CH}_3COSCoA + AMP + PP_i + 7 \text{ FADH}_2 + 7 \text{ NADH} + 7 \text{ H}^+$$

As noted earlier, each molecule of acetyl-CoA yields 12 ATP molecules (12 x 8 = 96), $FADH_2$ yields 2 ATP (7 x 2 = 14), NADH yields 3 (7 x 3 = 21), and two high-energy bonds are consumed (-2; $ATP \rightarrow AMP + PP_i$). Thus, net ATP production is about 129. The energy yield due to total combustion of palmitic acid in a bomb calorimeter (see Chapter 3) is -2380 kcal/mole:

$$C_{16}H_{32}O_2 + 23\ O_2 \longrightarrow 16\ CO_2 + 16\ H_2O$$

$$\text{at } 20^{\circ}C,\ \Delta H^{\circ} = -2380\ \text{kcal/mole}$$

In biological oxidation, the energy conserved as ATP is about 942 kcal/mole (129 x 7.3). Thus, the percent standard free energy of oxidation of palmitic acid conserved as high energy phosphate is about 40 (942/2380 x 100).

4. The propionyl-CoA produced by β-oxidation of odd-chain length fatty acids is metabolized as shown next.

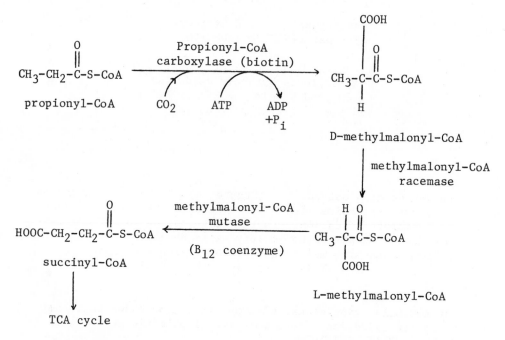

a. Propionyl-CoA is also produced in the oxidation of isoleucine, one of the branched-chain amino acids. The fermentation products of certain bacteria (especially strains of the genera <u>Propionibacterium</u> and <u>Veillonella</u>) contain propionic acid. This can be converted to propionyl-CoA by acetyl-CoA synthetase. Methylmalonyl-CoA is the chief product of valine oxidation.

b. There are two inheritable types of methylmalonic acidemia and aciduria. In young children, they are associated with mental retardation and failure to thrive. In one type (unresponsive to vitamin B_{12}), there is a lack of methylmalonyl-CoA mutase. The other type probably involves an inability to readily convert vitamin B_{12} to B_{12} coenzyme. It responds to large doses of vitamin B_{12}. Propionic and methylmalonic aciduria results from vitamin B_{12} deficiency in otherwise normal individuals.

5. The other oxidative pathways that are significant in fatty acid metabolism are as follows:

a. <u>Alpha-oxidation</u> (already mentioned in connection with Refsum's syndrome) is important in the catabolism of branched-chain and odd-chain-length fatty acids. The general reaction, shown below, is catalyzed by a monooxygenase and requires O_2, Fe^{+2}, and either ascorbate or tetrahydropteridine. It has been demonstrated in plants and in microsomes from brain and other tissues.

$$R\text{-}CH_2\text{-}CH_2\text{-}COOH + \text{reduced cofactor} + O_2$$

$$\xrightarrow{\text{monooxygenase}} \quad \underset{\overset{|}{OH}}{R\text{-}CH_2\text{-}CH\text{-}COOH} + \text{oxidized cofactor} + H_2O$$

α–hydroxy fatty acid

This is also one route for the synthesis of hydroxy fatty acids. The α-hydroxy fatty acids can be further converted in succeeding reactions to a fatty acid one carbon shorter than the original one. Thus, if an odd-chain-length compound is used initially, an even-chain-length acid is produced which can be further oxidized by β-oxidation.

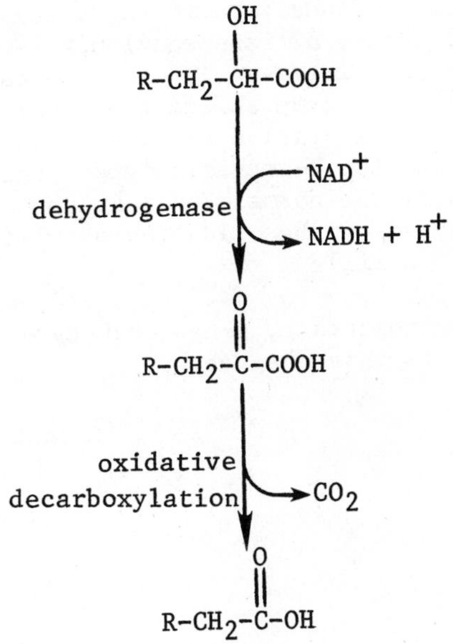

$$\underset{\overset{|}{OH}}{R\text{-}CH_2\text{-}CH\text{-}COOH}$$

dehydrogenase — NAD^+

↘ $NADH + H^+$

$$\underset{\overset{\|}{O}}{R\text{-}CH_2\text{-}C\text{-}COOH}$$

oxidative decarboxylation → CO_2

$$\underset{\overset{\|}{O}}{R\text{-}CH_2\text{-}C\text{-}OH}$$

In Refsum's disease, an inherited autosomal disorder, the biochemical lesion lies in the α-oxidation of phytanic acid (probably in the α-hydroxylation step). This is a 20-carbon, branched-chain fatty acid derived from the plant alcohol phytol which is present as an ester in chlorophyll. Thus, the origin of phytanic acid in the body is from dietary sources. The oxidation of phytanic acid is shown in Figure 109.

The clinical characteristics of the disease include peripheral neuropathy and ataxia, retinitis pigmentosa, and abnormalities associated with skin and bones. Significant improvement in clinical manifestations have been observed when patients are on prolonged, low phytanic acid diets (e.g., diets which exclude dairy and ruminant fat).

b. Omega-oxidation refers to oxidation of the carbon most remote from the carboxyl group in any fatty acid. The basic reaction, catalyzed by a monooxygenase which requires NADPH, O_2, and cytochrome P-450, is shown below. It has been demonstrated in liver microsomes and in some bacteria.

$$H_3C-(CH_2)_n-COOH \xrightarrow{\omega-ox.} HO-CH_2-(CH_2)_n-COOH$$

Further oxidation of the ω-hydroxy acids can produce the corresponding dicarboxylic acids which can be β-oxidized from either end.

6. Oxidation of mono- and polyunsaturated fatty acids. As noted earlier, the oxidation of unsaturated fatty acids requires two additional enzymes, Δ2cis, Δ2trans enoyl-CoA isomerase and D(-)-3-hydroxyacyl-CoA epimerase, in order to carry out β-oxidation. The first enzyme shifts the double bond into the proper position and configuration for the hydration step. The second enzyme converts the L-stereoisomer of the 3-hydroxyacyl-CoA to the D-stereoisomer, the normal β-oxidation intermediate. These aspects are illustrated by the oxidation of oleic and linoleic acids (Figures 110 and 111).

7. The metabolism of acetoacetate, β-hydroxybutyrate and acetone are discussed later in this chapter.

(Text continued on p. 892)

Figure 109. Oxidation of Phytol and Phytanic Acid

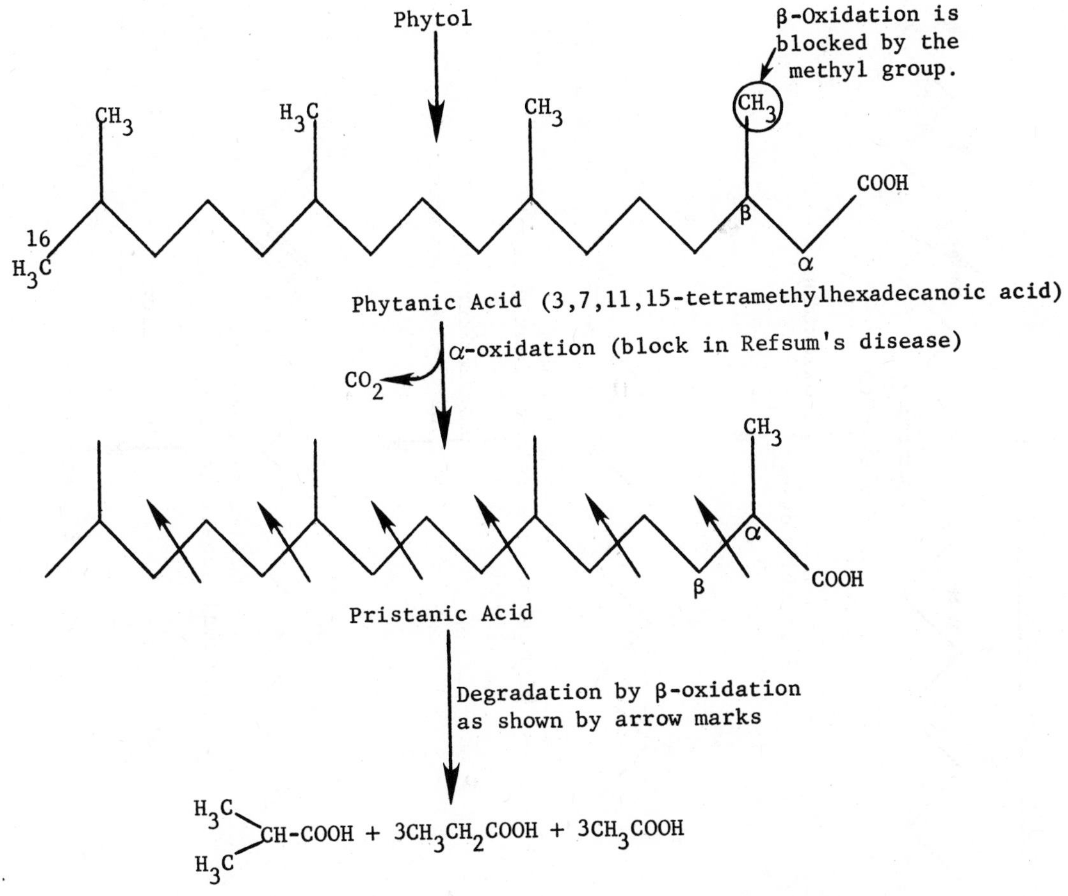

Phytol

β-Oxidation is blocked by the methyl group.

CH_3

H_3C

CH_3

CH_3

16

H_3C

COOH

β

α

Phytanic Acid (3,7,11,15-tetramethylhexadecanoic acid)

α-oxidation (block in Refsum's disease)

CO_2

CH_3

α

COOH

β

Pristanic Acid

Degradation by β-oxidation as shown by arrow marks

H_3C

CH-COOH + $3CH_3CH_2COOH$ + $3CH_3COOH$

H_3C

889

Figure 110. Oxidation of Oleic Acid

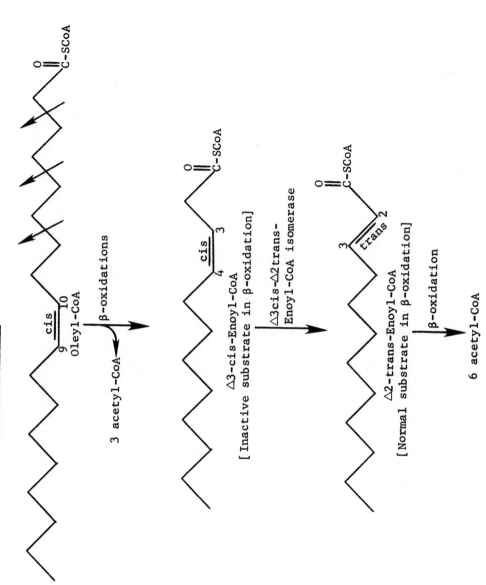

Figure 111. Oxidation of Linoleic Acid

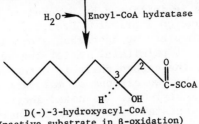

Δ3cis-Δ6cis-Dienoyl-CoA
(Inactive substrate in β-oxidation)

Δ3-cis-Δ2-trans-
Enoyl-CoA isomerase

Δ2-trans-Δ6-cis-Dienoyl-CoA
(Normal substrate in β-oxidation)

β-oxidations

2 acetyl-CoA

Δ2-cis-Enoyl-CoA
(Although the double bond is in cis-configuration, this is a
normal substrate for the enzyme enoyl-CoA hydratase.)

H_2O Enoyl-CoA hydratase

D(-)-3-hydroxyacyl-CoA
(Inactive substrate in β-oxidation)

3-hydroxyacyl-CoA
Epimerase

L(+)-3-hydroxyacyl-CoA
(Normal substrate in β-oxidation and yields 4 acetyl-CoA)

891

Fatty Acid Synthesis

1. The reactions of <u>de novo</u> fatty acid biosynthesis are outlined in
   Figure 112. The biosynthesis is catalyzed by <u>two enzyme systems</u>
   which function in sequence, namely acetyl-CoA carboxylase and
   fatty acid synthetase. Some important features of these
   reactions are:

   a. The <u>de novo</u> synthesis takes place in the <u>cytoplasm</u>--note that
      oxidation occurs in the mitochondria.

   b. All carbon atoms of the fatty acids are derived from
      acetyl-CoA (obtained from the oxidation of carbohydrates or
      amino acids), and palmitate (C16) is the predominant fatty
      acid produced. However, fatty acids with more than 16
      carbons, those that are unsaturated, and hydroxy fatty acids
      are obtained by separate processes such as chain elongation,
      desaturation and α-hydroxylations, respectively.

   c. The <u>first committed</u> (rate-controlling) step is the biotin-
      dependent carboxylation of acetyl-CoA to yield malonyl-CoA,
      catalyzed by acetyl-CoA carboxylase. The important allosteric
      effectors are citrate (<u>positive</u>) and long-chain acyl-CoA
      derivatives (<u>negative</u>).

   d. Although the initial and rate-controlling step requires CO_2
      fixation, the CO_2 is <u>not</u> incorporated into fatty acids.
      That is, no <u>net</u> CO_2 fixation occurs during fatty acid
      synthesis and, thus, the carbon of labelled $^{14}CO_2$ (as $H^{14}CO_3^-$)
      is not incorporated into the carbons of fatty acids synthesized.

   e. The synthesis is initiated by a required molecule of acetyl-CoA
      which functions as a primer. Eventually, the two carbons of
      the acetyl group end up as carbon atoms 15 and 16 (two terminal
      carbon atoms) of palmitate. After the addition of the starter
      acetyl group, the carbon chain is extended by successive
      additions of the two carbons of malonate which are originally
      derived from acetyl-CoA. As noted above, the third carbon of
      malonate, the unesterified carboxylic acid group, is removed
      as CO_2. It has been shown in mammalian liver and mammary
      gland that butyryl-CoA is a more active primer than acetyl-CoA.
      <u>Odd-chain-length</u> fatty acids found in some organisms are
      synthesized by priming the reaction with propionyl-CoA instead
      of acetyl-CoA.

   f. The final release of the finished fatty acid occurs when the
      chain length is C16. This length specificity is due to the
      substrate specificity of the enzyme β-ketoacyl synthetase
      (Figure 112; reaction 2). The enzyme accepts substrate acyl

groups up to 14 carbons in length. Thus, the synthesis terminates in palmitoyl-CoA, which functions as a feedback inhibitor of the acetyl-CoA synthetase enzyme.

g. The overall reaction for palmitate synthesis starting from acetyl-CoA is:

$$8 \text{ acetyl-CoA} + 14 \text{ NADPH} + 14 \text{ H}^+ + 7 \text{ ATP} + \text{H}_2\text{O}$$

$$\longrightarrow \text{Palmitate} + 8 \text{ CoASH} + 14 \text{ NADP}^+ + 7 \text{ ADP} + 7 \text{ P}_i$$

The reducing equivalents for the fatty acid synthesis are provided by NADPH, which are largely derived from the oxidation of glucose-6-phosphate by the hexose monophosphate shunt (or phosphogluconate) pathway.

2. The acetyl-CoA carboxylase reaction and the formation of malonyl-CoA.

   a. The enzyme catalyzes biotin-dependent carboxylation, the first committed step in fatty acid biosynthesis. In animal tissues the enzyme is found in the cytoplasm.

   b. The enzyme has been purified from a variety of cells; microorganisms, yeast, plants, and animals. In animal cells, the enzyme is present as an inactive protomer and an active polymer. Since the dissociation of protomers (from animal tissues and yeast) into subunits leads to inactivation, the function of each subunit is not understood. However, the enzyme isolated from E. coli can be readily dissociated into active components and hence has been used in the delineation of subunit functions.

   c. The protomeric form of the enzyme from avian (chicken) liver has a molecular weight of 470,000-500,000 daltons. It has four nonidentical subunits: two with molecular weights of 117,000--one of which contains biotin bound in an amide linkage to the ε-amino group of a specific lysine residue; one with a molecular weight of 129,000; and one with a molecular weight of 139,000. Each protomer contains binding sites for one molecule of the biotinyl prosthetic group, one acetyl-CoA, one bicarbonate, and one citrate. Citrate is responsible for shifting the equilibrium between inactive protomer to active polymer, in favor of the active polymer. The polymeric enzyme appears as long filaments of enzyme protomers in electron micrographic studies.

   d. As noted earlier, the details of the mechanisms of the carboxylation reactions have been elucidated by working with

(Text continues on p. 896)

Figure 112. Synthesis of Fatty Acid (Palmitic Acid)

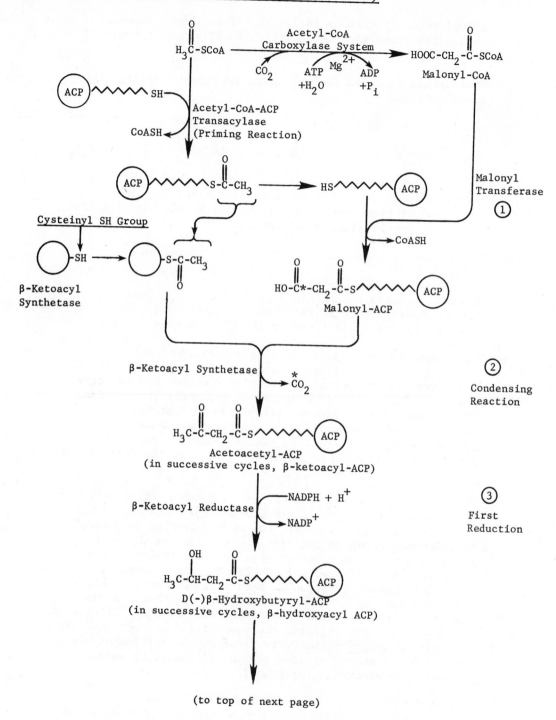

(to top of next page)

894

(from bottom of preceding page)

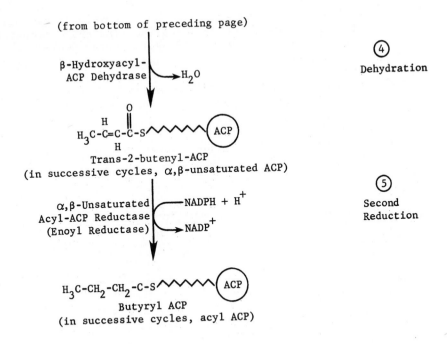

β-Hydroxyacyl-
ACP Dehydrase $\rightarrow H_2O$

④ Dehydration

Trans-2-butenyl-ACP
(in successive cycles, α,β-unsaturated ACP)

α,β-Unsaturated
Acyl-ACP Reductase
(Enoyl Reductase)

—NADPH + H⁺

$\rightarrow$ NADP⁺

⑤ Second Reduction

Butyryl ACP
(in successive cycles, acyl ACP)

Is the acyl group sixteen carbons long yet?

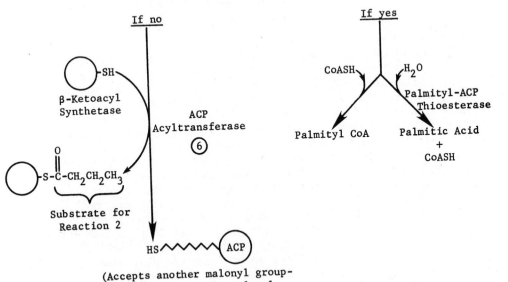

If no

If yes

β-Ketoacyl
Synthetase

ACP
Acyltransferase
⑥

CoASH

H_2O

Palmityl-ACP
Thioesterase

Palmityl CoA

Palmitic Acid
+
CoASH

Substrate for
Reaction 2

(Accepts another malonyl group-
reaction 1, and the malonyl-
ACP initiates the next cycle of
two carbon extension.)

the E. coli enzyme which can be successfully dissociated into its functionally active subunits. The enzyme consists of three proteins: biotin carboxyl carrier protein (BCCP, M.W. 22,500); biotin carboxylase (M.W. 98,000, with two subunits of identical size); and acetyl-CoA:malonyl-CoA transcarboxylase, also known as carboxyl transferase (M.W. 130,000, made up of nonidentical subunits of 30,000 and 35,000 daltons).

e. The steps involved in the carboxylation reaction consist of two half-reactions:

i) $ATP + HCO_3^- + BCCP \underset{Mg^{2+}}{\overset{\text{Biotin Carboxylase}}{\rightleftharpoons}} ADP + P_i + BCCP - COO^-$

ii) $BCCP - COO^- + Acetyl\text{-}CoA \rightleftharpoons BCCP + Malonyl\text{-}CoA$

The overall reaction:

$ATP + HCO_3^- + Acetyl\text{-}CoA \rightleftharpoons Malonyl\text{-}CoA + ADP + P_i$

The presumed reaction mechanisms are shown in Figure 113. The other biotin-dependent carboxylases discussed earlier are propionyl-CoA carboxylase (this chapter) and pyruvate carboxylase (see gluconeogenesis, Chapter 4). Recall that the latter enzyme, like acetyl-CoA carboxylase, is also subject to allosteric regulation. Pyruvate carboxylase, a mitochondrial enzyme, is activated by catalytic amounts of acetyl-CoA and catalyzes the conversion of pyruvate to oxaloacetate. The oxaloacetate is either converted to glucose via the gluconeogenic pathway or combines with acetyl-CoA to form citrate. Some of the citrate is transported to the cytoplasm where it not only activates the first committed step of fatty acid synthesis but provides acetyl-CoA and promotes fatty acid systems (discussed later in this section). There are other types of carboxylation reactions which use bicarbonate as a substrate but do not require biotin and are dependent upon vitamin K. In this vitamin K-dependent carboxylating enzyme system, the acceptor of the carboxyl groups are glutamyl residues of glycoprotein clotting factors II, VI, IX and X. The product formed is a γ-carboxyglutamyl residue, which is required for the function of these proteins (see Chapters 10 and 11).

Figure 113. The Chemical Reactions Involved in the Carboxylation of Acetyl CoA

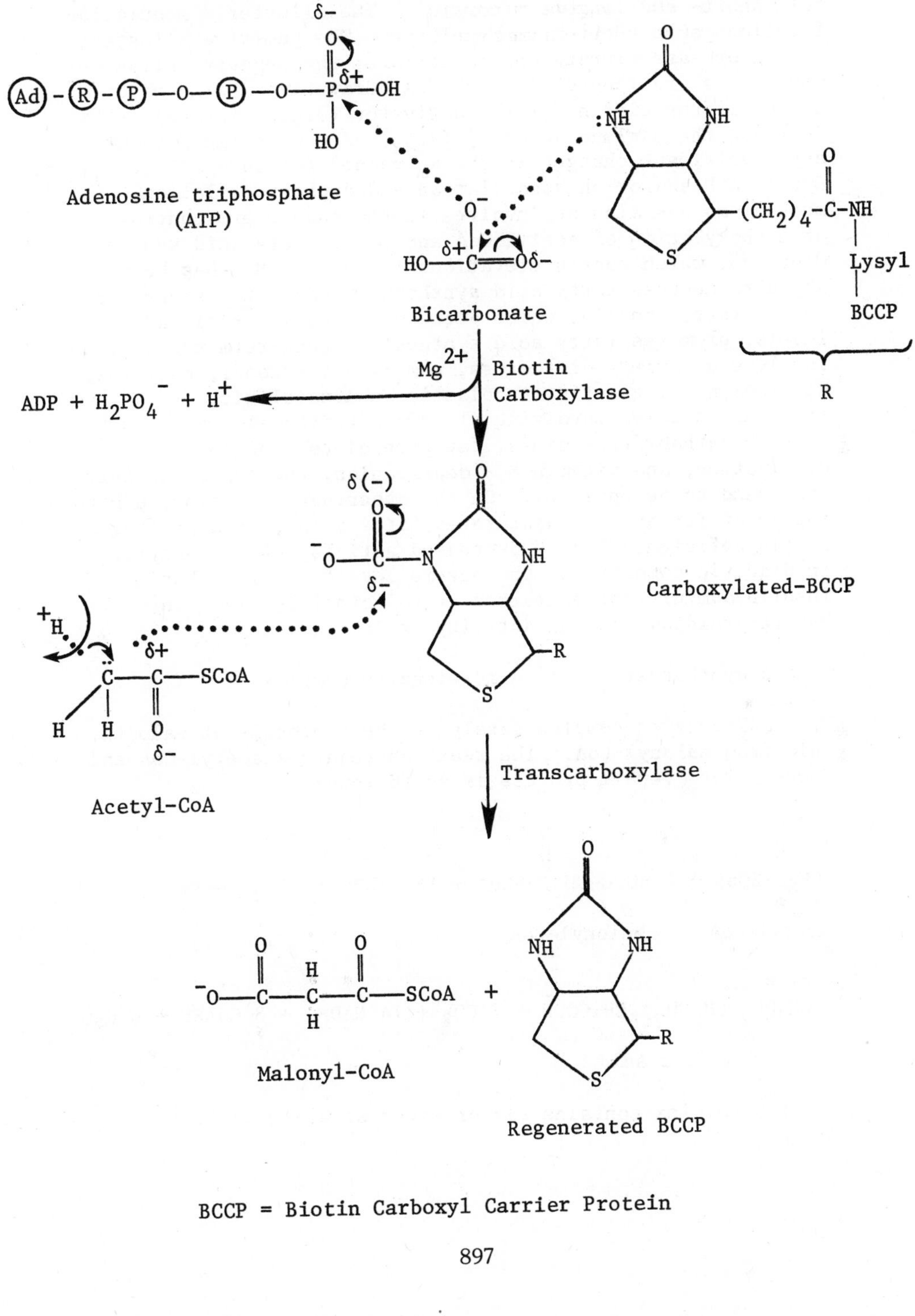

Adenosine triphosphate (ATP)

Bicarbonate

Lysyl

BCCP

R

$ADP + H_2PO_4^- + H^+$

Mg^{2+} | Biotin Carboxylase

Carboxylated-BCCP

Acetyl-CoA

Transcarboxylase

Malonyl-CoA

Regenerated BCCP

BCCP = Biotin Carboxyl Carrier Protein

897

f. The rate-limiting enzyme, acetyl-CoA carboxylase, is under both short- and long-term control. The allosteric modulation functions as a short-term regulator. The positive allosteric modulators are citrate and isocitrate; the negative allosteric modulators are long-chain acyl-CoA derivatives. The major effect of the binding of the activator (e.g., citrate) is to increase the maximum velocity (V_{max}) of the enzyme without any significant changes in the K_m values for substrates. Cyclic AMP and/or phosphorylation and dephosphorylation mechanisms may also be involved in the short-term control of carboxylation of acetyl-CoA and hence fatty acid synthesis. Glucagon, which causes elevations in cyclic AMP, has been shown to depress fatty acid synthesis in in vitro experiments. In contrast, insulin, which decreases hepatic cyclic AMP levels, promotes fatty acid synthesis. Long-term regulation involves a variety of factors, namely nutritional, hormonal, developmental, genetic, neoplastic, pharmacologic, etc. In animal studies involving altered nutritional states such as high carbohydrate diets, fat-free diets, choline deprivation, and vitamin B_{12} deprivation, the enzyme activity was found to be enhanced. On the other hand, fasting, a high intake of fat or polyunsaturated fatty acids, or a prolonged biotin deficiency led to decreased activity of the enzyme. In diabetic conditions, the enzyme levels are low, but insulin administration raises it to normal levels. In hyperthyroidism, enzyme activity is increased.

3. Fatty acid synthetase (FAS)--a multienzyme complex.

a. This multienzyme complex catalyzes the synthesis of palmitic acid from malonyl-CoA. The reaction requires acetyl-CoA and NADPH. The overall process is as follows:

$$CH_3\overset{\overset{\displaystyle O}{\|}}{C}\text{-SCoA} + 7\ HOOC\text{-}CH_2\overset{\overset{\displaystyle O}{\|}}{C}\text{-SCoA} + 14\ NADPH + 14\ H^+ \longrightarrow$$

* *
Acetyl-CoA   Malonyl-CoA

* *
$$CH_3CH_2(CH_2CH_2)_6CH_2COOH + 7\ CO_2 + 14\ NADP^+ + 8\ CoASH + 6\ H_2O$$

Palmitic Acid

The FAS complex contains six or seven separate catalytic sites.

b. Two types of FAS complexes are found: a tightly associated complex which resists dissociation into individual monofunctional components found in animals, yeast and certain higher bacteria, and an enzyme complex which can be dissociated into its individual functional components found in most procaryotic organisms (e.g., E. coli) and higher plant cells. It is of interest to point out that when Euglena, an organism which exhibits plant or animal characteristics, is grown under conditions conducive to chloroplast formation, an E. coli type of enzyme is synthesized. However, when the organism is grown under conditions suitable for the development of animal properties, a typical multienzyme complex characteristic of animal cells is produced. The elucidation of the various intermediate reactions has been accomplished by studies of purified component enzymes of the E. coli-FAS complex. It should be noted that the presence of tightly aggregated, multifunctional FAS in yeast and animal cells, and the dissociable FAS complexes found in procaryotes and plants, parallel the observations found with polyfunctional acetyl-CoA carboxylase systems.

c. The reason for the failure to dissociate yeast and animal FAS complexes into monomeric, monofunctional components is because these synthetases are made up of multifunctional polypeptides. That is, a given polypeptide chain contains a number of catalytic sites catalyzing different reactions, as opposed to each peptide chain catalyzing only one reaction (e.g., E. coli-FAS subunits). The yeast-FAS complex is composed of only two types of multifunctional polypeptide chains, and the sum of their catalytic activities accounts for the overall chemical synthesis. Each FAS complex, however, contains multiple copies of these two types of polypeptides. These two polypeptides are coded by no more than two genetically unlinked, polycistronic gene loci. The functional activities are divided between the two peptides as follows: polypeptide A consists of a 4'-phosphopantetheine binding region, β-ketoacyl synthetase and β-ketoacyl reductase; polypeptide B possesses activities of acetyl transacylase, malonyl transacylase, β-hydroxy-acyl dehydrase and enoyl reductases. The FAS complexes (rat liver FAS) appear to be assembled in three stages: initial synthesis of the two multifunctional polypeptide chains, the formation of the duplex-apoenzyme and, finally, the generation of the enzymatically active complex (the holoenzyme) by the attachment of the 4'-phosphopantetheine group by an enzyme-catalyzed reaction. This assembly process is under the influence of a variety of control mechanisms occurring with changes in developmental, hormonal and nutritional states. The remarkable evolution of the FAS complex with <u>covalently linked</u> catalytic sites appears to provide a great deal of catalytic efficiency. In this type of synthetic system there is no accumulation of free intermediates and, in a given multifunctional polypeptide chain, the stoichiometry of the individual catalytic activities are one to one.

d. The molecular weights of the FAS complexes vary from different sources and range from 400,000 to 2.5 million daltons. The largest synthetases are found in yeast and the smallest are present in mammalian cells (liver, mammary gland or brain).

e. The reactions of the fatty acid synthetase complex. In the reaction sequence, the <u>central role</u> of the acyl-carrier protein (ACP; homogeneous, small molecular weight ACP's have been isolated from bacteria and plants) and the acyl-carrier portion of the polypeptide (in yeast and animal cells), which contain the 4'-phosphopantetheine prosthetic group (20.02 nm long), should be noted. The prosthetic group, also part of coenzyme A (CoASH), is derived from it and is bound to ACP as a phosphodiester with the hydroxy group of a specific serine residue. The acyl intermediates are bound in thioester linkage to the SH group of the prosthetic group which serves as a swinging arm carrying the acyl groups from one catalytic site to the next.

Structure of 4'-phosphopantetheine attached to the serine residue of ACP:

$$\text{Serine Residue} - \text{O} - \underset{\underset{OH}{|}}{\overset{\overset{O}{\|}}{P}} - \text{O} - \underset{4'}{CH_2} - \underset{\underset{CH_3}{|}}{\overset{\overset{CH_3}{|}}{C}} - \underset{\underset{OH}{|}}{CH} - \overset{\overset{O}{\|}}{C} - \underset{H}{N} - CH_2 - CH_2 - \overset{\overset{O}{\|}}{C} - \underset{H}{N} - CH_2 - CH_2 - SH$$

The reaction sequence as delineated in purified <u>E. coli</u> systems is as follows:

i) The priming reaction:

$$H_3C-\overset{\overset{O}{\|}}{C}-SCoA + HS\text{-}ACP \rightleftharpoons CoASH + ACP\text{-}S\text{-}\overset{\overset{O}{\|}}{C}\text{-}CH_3$$

The reaction is catalyzed by acetyl-CoA-ACP transacylase.

ii) The acetyl group is then transferred to a specific cysteinyl-SH group of the β-ketoacyl synthetase enzyme (condensing enzyme) of the FAS complex.

$$ACP-S-\overset{\overset{\displaystyle O}{\|}}{C}-CH_3 + HS\text{-Condensing Enzyme}$$

$$\rightleftharpoons ACP-SH + H_3-\overset{\overset{\displaystyle O}{\|}}{C}-S\text{-Condensing Enzyme}$$

iii) Now the regenerated ACP-SH can accept the malonyl group, and the reaction is catalyzed by malonyl transferase (or malonyl-CoA-ACP transacylase).

$$HOOC-CH_2-\overset{\overset{\displaystyle O}{\|}}{C}-SCoA + HS-ACP \rightleftharpoons HOOC-H_2C-\overset{\overset{\displaystyle O}{\|}}{C}-S-ACP + CoASH$$

iv) Condensation of two carbon atoms of malonyl-CoA and two carbon atoms of acetyl-CoA, catalyzed by β-ketoacyl synthetase, occurs next.

$$ACP-S-\overset{\overset{\displaystyle O}{\|}}{C}-CH_2-COOH + H_3C-\overset{\overset{\displaystyle O}{\|}}{C}-S\text{-Condensing Enzyme}$$

$$\rightleftharpoons ACP-S-\overset{\overset{\displaystyle O}{\|}}{C}-CH_2-\overset{\overset{\displaystyle O}{\|}}{C}-CH_3 + HS\text{-Condensing Enzyme} + CO_2$$

This decarboxylation reaction is an exergonic one, and its equilibrium is pulled toward the formation of the product acetoacyl-ACP or β-ketoacyl-ACP.

v) The first reduction is catalyzed by NADPH-dependent β-ketoacyl reductase.

$$ACP-S-\overset{\overset{\displaystyle O}{\|}}{C}-CH_2-\overset{\overset{\displaystyle O}{\|}}{C}-CH_3 + NADPH + H^+$$

$$\longrightarrow ACP-S-\overset{\overset{\displaystyle O}{\|}}{C}-CH_2-\overset{\overset{\displaystyle OH}{|}}{C}H-CH_3 + NADP^+$$

The reaction is stereospecific and the product formed is D(-)β-hydroxyacyl ACP.

vi)   The dehydration reaction is catalyzed by β-hydroxyacyl-ACP
      dehydrase. The reaction is stereospecific, and the D(-)
      isomer (not the L(+) isomer) is converted to a trans-α,β-
      unsaturated acyl-ACP derivative.

$$\text{ACP-S-}\overset{\overset{\text{O}}{\|}}{\text{C}}\text{-CH}_2\text{-}\overset{\overset{\text{OH}}{|}}{\text{CH}}\text{-CH}_3 \;\rightleftharpoons\; \text{ACP-S-}\overset{\overset{\text{O}}{\|}}{\text{C}}\text{-}\overset{\overset{\text{H}}{}}{\text{C}}\text{=}\overset{}{\underset{\underset{\text{H}}{}}{\text{C}}}\text{-CH}_3 + \text{H}_2\text{O}$$

vii) The second reduction is catalyzed by enoyl reductase,
      and the electron donor is NADPH.

$$\text{ACP-S-}\overset{\overset{\text{O}}{\|}}{\text{C}}\text{-}\overset{\overset{\text{H}}{}}{\text{C}}\text{=}\underset{\underset{\text{H}}{}}{\text{C}}\text{-CH}_3 \;\rightleftharpoons\; \text{ACP-S-}\overset{\overset{\text{O}}{\|}}{\text{C}}\text{-CH}_2\text{-CH}_2\text{-CH}_3$$

      Note that the product formed is a saturated acyl-
      thioester of ACP.

viii) If the acyl group is less than sixteen carbons in
      length, it is transferred to a cysteinyl residue of
      β-ketoacyl synthetase, allowing ACP to accept a
      malonyl group from another molecule of malonyl-CoA
      (reaction 1). The next cycle of extension of two
      carbons then begins with the condensation reaction
      (Figure 112, reaction 2).

f. A schematic diagram of the FAS complex is shown in Figure 114.
    Each of the outer circles represents an enzyme (or a catalytic
    site in a polyfunctional polypeptide) which catalyzes the
    indicated reaction. The center circle is ACP or its
    equivalent. The dotted line with an -SH group is the swinging
    arm (4'-phosphopantetheine) in each of its successive
    positions. This swinging arm carries the acyl groups from
    one catalytic site to the next and accomplishes the six steps
    required for the addition of each two-carbon unit. The other
    -SH group of the acyl transferase functions as a temporary
    storage site for the acyl group before it is transferred to
    a specific cysteinyl group of β-ketoacyl synthetase.

g. The reducing agent for fatty acid synthesis is NADPH. The
    majority of the NADPH requirement is supplied by the reactions
    of the hexose monophosphate shunt pathway (for details see
    Chapter 4):

Figure 114. A Schematic Diagram of the Fatty Acid Synthetase Complex
(see text for details)

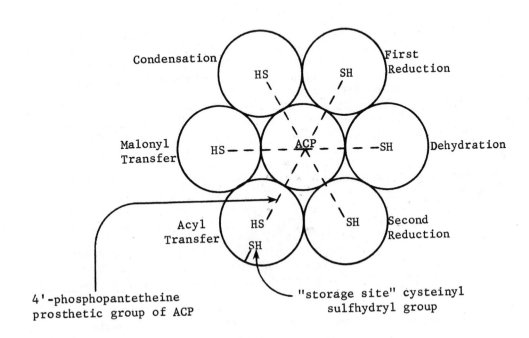

i)

$$\text{Glucose-6-Phosphate} \xrightarrow[\substack{NADP^+ \qquad NADPH + H^+}]{\substack{G-6-P \\ Dehydrogenase}} \text{6-}\textcircled{P}\text{-Gluconolactone}$$

ii)

$$\text{6-}\textcircled{P}\text{-Gluconate} \xrightarrow[\substack{NADP^+ \quad CO_2 \quad NADPH + H^+}]{\substack{\text{6-}\textcircled{P}\text{-Gluconate} \\ Dehydrogenase}} \text{Ribulose-5-}\textcircled{P}$$

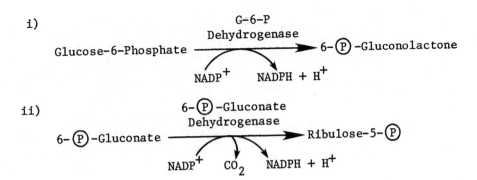

The oxidation of malate also provides NADPH.

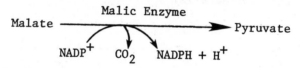

$$\text{Malate} \xrightarrow[\text{NADP}^+ \quad CO_2 \quad \text{NADPH} + H^+]{\text{Malic Enzyme}} \text{Pyruvate}$$

All of the above enzymes are located in the cytoplasm, where the FAS complex is also located. Active lipogenesis takes place in the liver, adipose tissue and lactating mammary glands. All of these tissues show a correspondingly high activity of the HMP shunt. Thus, lipogenesis is closely linked to carbohydrate oxidation. The rate of lipogenesis is high in well-fed humans whose diet is rich in carbohydrate. Restricted caloric intake, a high fat diet or an insulin deficiency decreases fatty acid synthesis (discussed later).

h. Source and transport of acetyl-CoA. Acetyl-CoA is synthesized in mitochondria by a number of reactions: oxidative decarboxylation of pyruvate (Chapter 4), catabolism of some amino acids (e.g., phenylalanine, tyrosine, leucine, lysine and tryptophan--see Chapter 6) and β-oxidation of fatty acids (already discussed in this chapter). Since acetyl-CoA cannot by itself be transported across the mitochondrial membrane to the cytoplasm, its carbon atoms are transferred by two transport mechanisms.

  i) Transport dependent upon carnitine: As previously discussed, carnitine participates in the transport of fatty acyl-CoA into the mitochondria. Carnitine plays a similar role in the transport of acetyl-CoA out of the mitochondria for its utilization in fatty acid synthesis (Figure 115).

  ii) Cytoplasmic generation of acetyl-CoA ("citrate shuttle"): This pathway is shown in Figure 116. Note that the citrate synthesized from oxaloacetate and acetyl-CoA is transported to the cytoplasm via the tricarboxylate anion carrier system and cleaved to yield acetyl-CoA and oxaloacetate.

$$\text{Citrate} + \text{ATP} + \text{CoA} \xrightarrow{\substack{\text{ATP-Citrate Lyase} \\ \text{(or citrate} \\ \text{cleavage enzyme)}}} \substack{\text{Acetyl-CoA} + \\ \text{Oxaloacetate} + \text{ADP} + P_i}$$

Thus, citrate not only modulates fatty acid synthesis as a positive effector but also provides carbon atoms for its synthesis. It should also be noted that the oxaloacetate

Figure 115. The Role of Carnitine in Mediating Acetyl and Acyl Transfers Across Mitochondrial Membranes

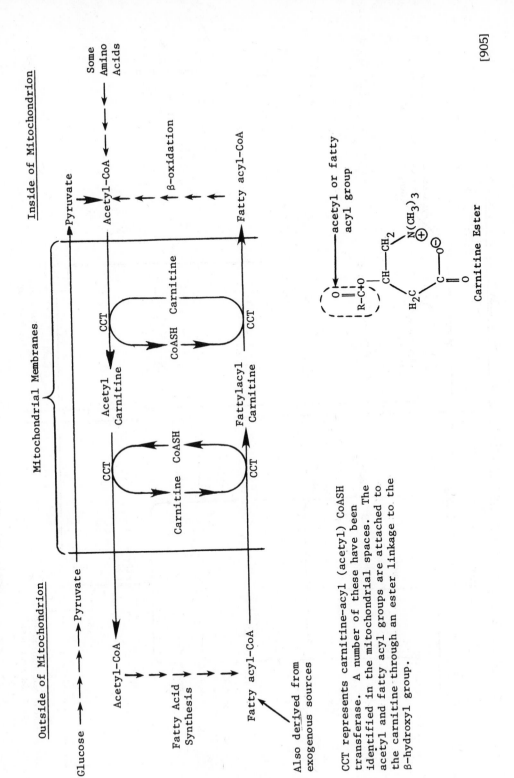

CCT represents carnitine-acyl (acetyl) CoASH transferase. A number of these have been identified in the mitochondrial spaces. The acetyl and fatty acyl groups are attached to the carnitine through an ester linkage to the β-hydroxyl group.

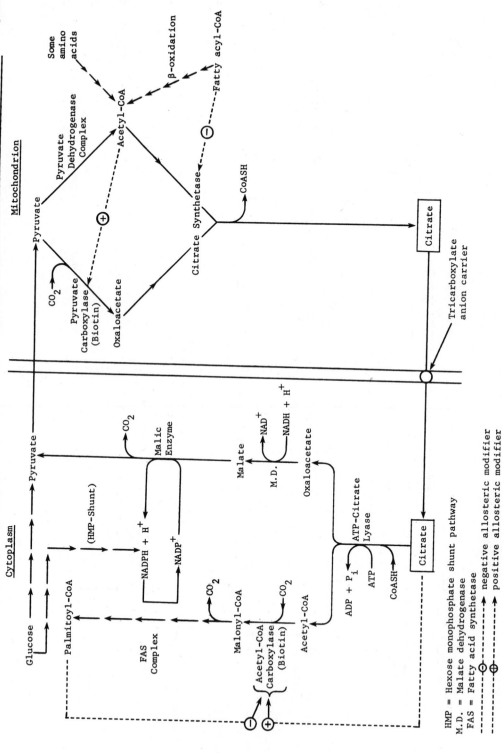

Figure 116. Cytoplasmic Generation of Acetyl-CoA Via Citrate Transport and Related Reactions   [906]

HMP = Hexose monophosphate shunt pathway
M.D. = Malate dehydrogenase
FAS = Fatty acid synthetase
――――――⊖―――――→ negative allosteric modifier
――――――⊕―――――→ positive allosteric modifier

formed from pyruvate (via malate) is eventually converted to glucose in the cytoplasm by the gluconeogenic pathway. The glucose is then oxidized via the HMP shunt pathway, thereby augmenting fatty acid synthesis by providing NADPH.

i. Regulation of fatty acid synthetase complex. Similar to acetyl-CoA carboxylase, FAS undergoes short- and long-term control. The short-term control is due to negative and positive modulation of preformed enzymes by allosteric modifiers and changes in the concentrations of substrate, cofactor and product. These aspects of FAS are not yet understood. The long-term (or adaptive) regulation usually consists of changes in the enzyme content as a result of protein synthesis and degradation.

Variations in hormone levels and the nutritional state can regulate fatty acid synthesis by affecting short-term as well as long-term control mechanisms. The hormones which influence fatty acid synthesis include insulin, glucagon, epinephrine, thyroid hormone, and prolactin. In the diabetic state, hepatic fatty acid synthesis is severely impaired and is corrected by administration of insulin. The impaired fatty acid synthesis in a diabetic is presumably due to defects in glucose metabolism, leading to a reduced level of inducer, increased level of repressor, or both. Glucagon and epinephrine inhibit fatty acid synthesis immediately and over longer periods of time and, since they both raise intracellular levels of cAMP, their action may in part be due to protein phosphorylation and dephosphorylation. It should be noted that these two hormones also promote lipolysis by stimulating a cAMP-sensitive triglyceride lipase (discussed later in this section under lipoprotein metabolism) and raising the intracellular levels of long-chain acyl-CoA. This molecule, as we have already seen, inhibits the acetyl-CoA carboxylase reaction and citrate synthetase, affecting fatty acid synthesis. The regulation of fatty acid synthesis by prolactin appears to be confined to the mammary gland where it stimulates the synthesis of the enzyme.

4. Although the pathway described above is the only route for de novo fatty acid synthesis, other routes exist for fatty acid elongation.

a. In the mitochondria, a slightly modified reversal of β-oxidation adds acetyl groups to saturated and unsaturated C_{12}, C_{14}, and C_{16} fatty acids to form primarily C_{18}, C_{20}, C_{22}, and C_{24} fatty acids. This system does not require CO_2, malonyl-CoA, or biotin.

b. Microsomes can elongate C_{10}-C_{16} saturated and C_{18} unsaturated fatty acids by successive addition of two carbon groups derived from malonyl-CoA. This route appears similar to de novo synthesis except that the intermediates are not bound to an acyl carrier protein. Recall that mammals are unable to synthesize fatty acids with multiple double bonds so that any multiple unsaturated fatty acids used by this pathway must be preformed.

5. Synthesis of branched, odd-chain-length, unsaturated, hydroxy, and dicarboxylic fatty acids has been mentioned a number of times. There are several other reactions which can be used to form such compounds that have not yet been described. They will be briefly summarized here.

a. Short, branched-chain acyl-CoA's derived from catabolism of valine, leucine, and isoleucine can be elongated, producing branched-chain fatty acyl-CoA's.

b. In addition to alpha- and omega-oxidation, hydroxy fatty acids can be formed by hydration of singly unsaturated fatty acids. These can then be further elongated by addition of acetyl residues.

c. As was pointed out under essential fatty acids, there are certain unsaturated fatty acids which the body cannot synthesize. The two ways in which mammals can satisfy their need for such compounds are shown below. These reactions occur primarily in the liver.

   i) The essential fatty acids (linoleic, linolenic, and arachidonic) can be lengthened by adding two-carbon units to them. The elongations may be performed by the microsomal-malonyl-CoA system described earlier. Additional double bonds can be introduced into this added material in a manner similar, but not identical, to the one described below for forming singly unsaturated acids.

   ii) Palmitic acid can be oxidized to palmitoleic acid containing one double bond and stearic acid can be oxidized to the singly unsaturated oleic acid. Both reactions simply remove one hydrogen each from carbons 9 and 10. Palmitoleic and oleic acids can be further elongated and oleic acid can have additional unsaturation added to the carbons used for this elongation.

<u>The Biosynthesis of Cholesterol</u> consists of the following five stages:

1. Synthesis of mevalonate (containing 6 carbons) from acetyl–CoA (2 carbons).

2. Loss of CO_2 from the mevalonate to form isoprenoid (5 carbon) units.

3. Condensation of six isoprenoid units to form squalene (30 carbons).

4. Conversion of squalene to lanosterol (30 carbons).

5. Transformation of lanosterol to cholesterol (27 carbons).

1. <u>Synthesis of Mevalonate</u>. Mevalonate is synthesized by condensation of three molecules of acetyl–CoA. The enzymes involved in its synthesis are extramitochondrial and found in the microsomal membrane fraction of cell homogenates.

   a. The steps for the formation of mevalonate are shown in Figure 117. The key step in cholesterol biosynthesis is the conversion of HMG–CoA to mevalonate, catalyzed by the enzyme HMG–CoA reductase (which is under complex regulatory control-- discussed later).

   b. A minor pathway for the synthesis of mevalonate has also been described. This pathway occurs in the cytoplasm and is mediated through an acyl carrier protein (ACP) and appropriate enzymes:

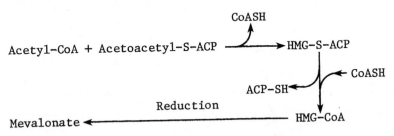

   This pathway is <u>not</u> quantitatively significant in the synthesis of cholesterol.

   c. HMG–CoA is also synthesized in the mitochondria by the same sequence of reactions described for cholesterol biosynthesis (see item 1a. above). However, the HMG–CoA in the mitochondria yields ketone bodies, namely, acetoacetate, D(-)β-hydroxybutyrate and acetone. The ketone bodies are not metabolized in the liver

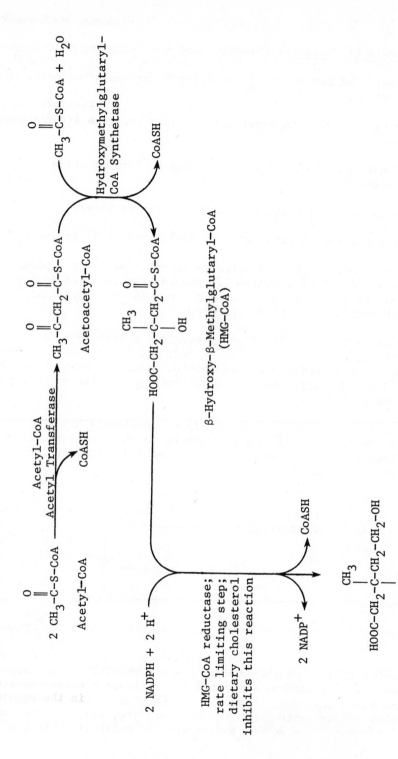

Figure 117. Synthesis of Mevalonic Acid

since it lacks the activating enzymes (the major one is succinyl-CoA-acetoacetate-CoA transferase). But this enzyme is present in extrahepatic tissues (e.g., skeletal muscle, cardiac muscle, kidney, brain, etc.) and, the ketone bodies, in particular acetoacetate and β-hydroxybutyrate, can serve as substrates for energy production. These aspects are discussed later in this section under <u>ketosis</u>.

As previously mentioned, since cholesterol biosynthesis is extramitochondrial, there are two distinct pools of HMG-CoA, one in the mitochondria concerned with the formation of ketone bodies and the other in the extramitochondrial pool involved with the synthesis of isoprenoid units and eventually cholesterol. These aspects are shown in Figure 118.

2. <u>Formation of Isoprenoid Units</u>. Isoprenoid units, the building blocks of the steroid skeleton, are synthesized by the successive phosphorylations of mevalonate with ATP to yield the esters of 5-monophosphate, 5-pyrophosphate, and 5-pyrophosphate-3-monophosphate. The latter compound, a very unstable intermediate, loses phosphoric acid and the carboxyl group to yield 3-isopentenyl pyrophosphate (IPPP) which isomerizes to <u>3,3-dimethylallyl pyrophosphate</u> (DMAPP). These reactions, catalyzed by soluble enzymes, are shown in Figure 119.

3. <u>Condensation of Isoprenoid Units to Form Squalene</u>. These reactions consist of the condensation of IPPP and DMAPP with the elimination of pyrophosphate to form a ten-carbon compound, <u>geranyl pyrophosphate</u>. The latter compound condenses with another molecule of IPPP with the release of pyrophosphate to yield a fifteen-carbon derivative, <u>farnesyl pyrophosphate</u>. Two molecules of this compound combine to form squalene, catalyzed by a microsomal enzyme, squalene synthetase. These chemical reactions are shown in Figure 120.

4. <u>Conversion of Squalene to Lanosterol</u>. These reactions require molecular oxygen and occur in the microsomes. Squalene, a molecule that closely resembles the steroid nucleus, is converted to lanosterol, the <u>first sterol</u> to be formed, by ring closures. The reactions (see Figure 121) consist of the formation of squalene 2,3-epoxide, catalyzed by a monooxygenase which requires molecular oxygen and NADPH, and cyclization to lanosterol by a series of concerted 1,2-methyl group and hydride shifts along the squalene hydrocarbon chain. In the conversion of squalene to lanosterol and eventually to cholesterol, the reactants are presumably bound to specific cytoplasmic carrier proteins (squalene or sterol carrier protein) in order to permit the reactions to occur in the aqueous phase of the cell.

<u>(Text continues on p. 915)</u>

Figure 118. Mitochondrial and Extramitochondrial Biosynthesis of HMG-CoA in the Liver and Its Utilization

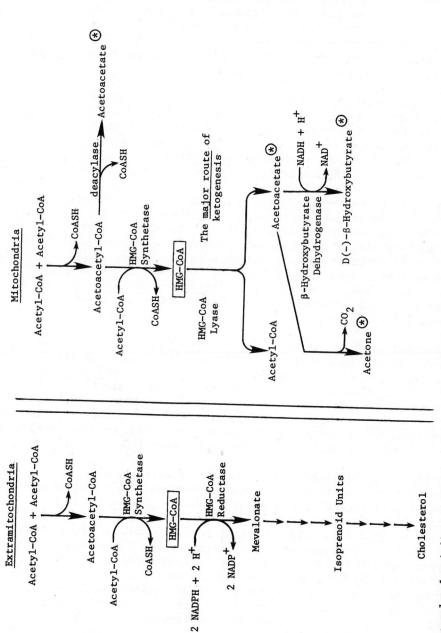

The molecules indicated by * are ketone bodies and once formed they cannot be reactivated and metabolized in the liver due to the absence of activating enzymes. But they diffuse out of hepatocytes into the blood stream and eventually are oxidized in the extrahepatic tissues. Note the major route of ketone body formation is the HMG-CoA pathway.

Figure 119. Synthesis of Isoprenoid Units From Mevalonate

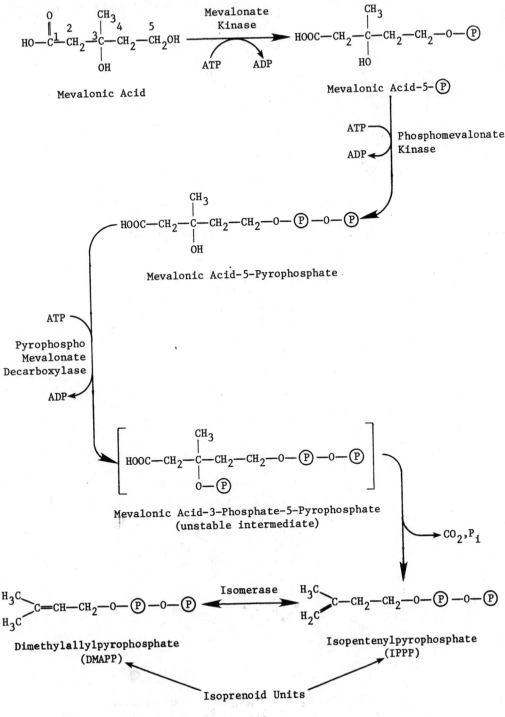

Mevalonic Acid

Mevalonic Acid-5-℗

Mevalonic Acid-5-Pyrophosphate

Mevalonic Acid-3-Phosphate-5-Pyrophosphate
(unstable intermediate)

Dimethylallylpyrophosphate
(DMAPP)

Isopentenylpyrophosphate
(IPPP)

Isoprenoid Units

913

Figure 120. Synthesis of Squalene From Isomeric Pentenyl Pyrophosphates

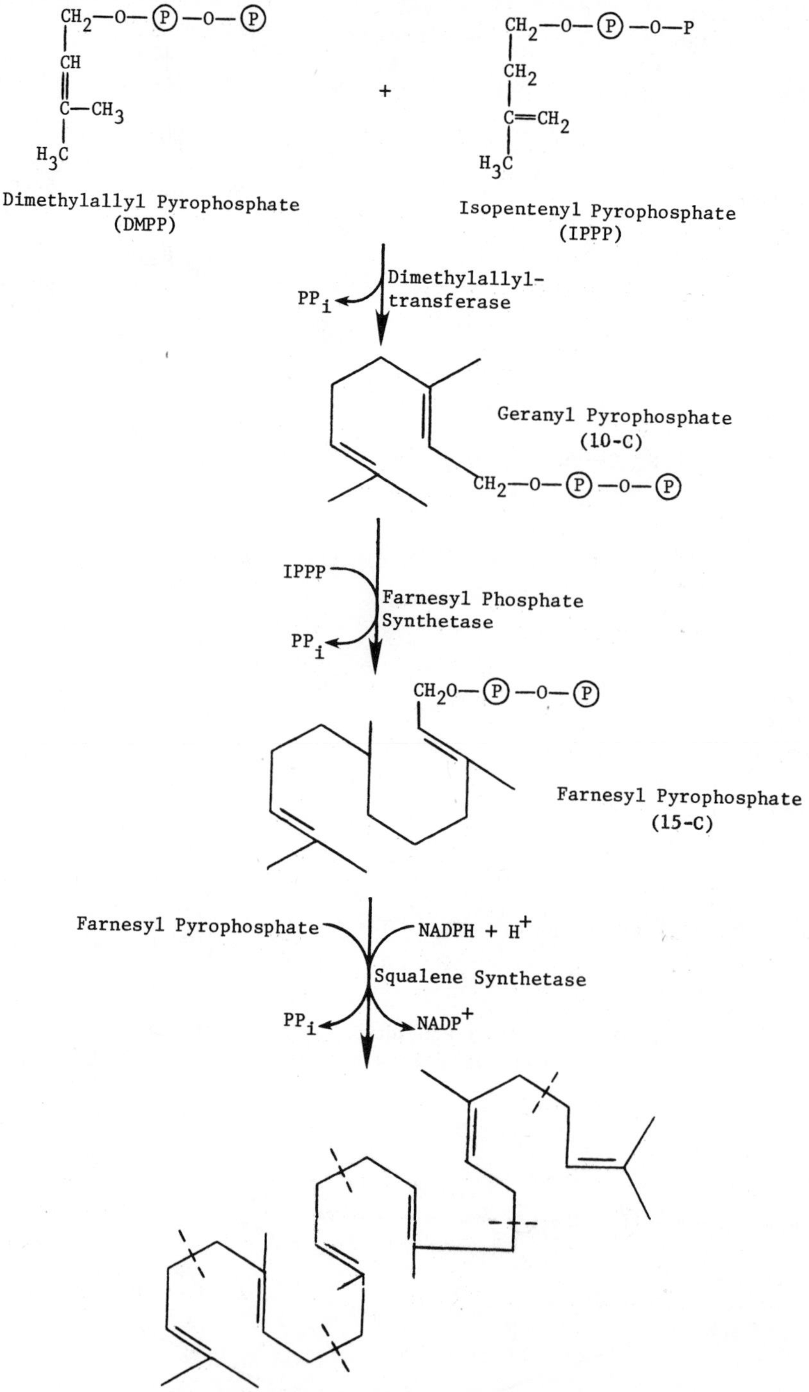

Squalene (the dashed lines indicate isoprene units)

5. Transformation of Lanosterol to Cholesterol. This is a complex
    process involving many steps. Most of the enzymes required in this
    sequence of conversions are embedded in the membranes of the
    endoplasmic reticulum (microsomes). As noted earlier, at least one
    cytoplasmic protein, known as the sterol carrier protein, is required
    which presumably functions as a carrier of reactants from one
    catalytic site to the next and perhaps affects the reactivity of the
    enzymes as well. The reactions in this series consist of the
    removal of three methyl groups, two attached to carbon 4 and one
    attached to carbon 14 (refer to the structure of lanosterol,
    Figure 121), the migration of the double bond from the 8,9 to the
    5,6 position, and the saturation of the double bond in the side
    chain. The conversion of lanosterol to cholesterol occurs via two
    pathways, one with 7-dehydrocholesterol as an intermediate and the
    other with desmosterol. It appears that the synthesis via
    7-dehydrocholesterol is quantitatively more significant. These
    reactions are shown in Figure 122.

Comments on Cholesterol Metabolism

1. A normal adult synthesizes 1.5-2.0 gm of cholesterol each day and
    consumes another 0.3 gm per day in the diet. Cholesterol is a
    product of animal metabolism and occurs primarily in food of animal
    origin. Egg yolk is rich in cholesterol, and meat, liver, brain,
    and shellfish contain appreciable amounts. As mentioned earlier,
    cholesterol is absorbed from the intestines along with other lipids.
    In the mucosal cells, most of it is esterified with fatty acids and
    incorporated into chylomicrons and VLDL which are transported to the
    blood by the lymphatics. Bile salts and cholesterol esterase
    (present in the pancreatic juice) are needed for cholesterol
    absorption.

2. The liver is the main site of cholesterol synthesis (1-1.5 gm per
    day) with the adrenal cortex, skin, intestine, testes, aorta, and
    perhaps other tissues forming the remainder of it. All of the
    carbon atoms of cholesterol are derived from acetyl-CoA, as can be
    seen in Figure 123.

(Text continues on p. 918)

Figure 121. Synthesis of Lanosterol From Squalene

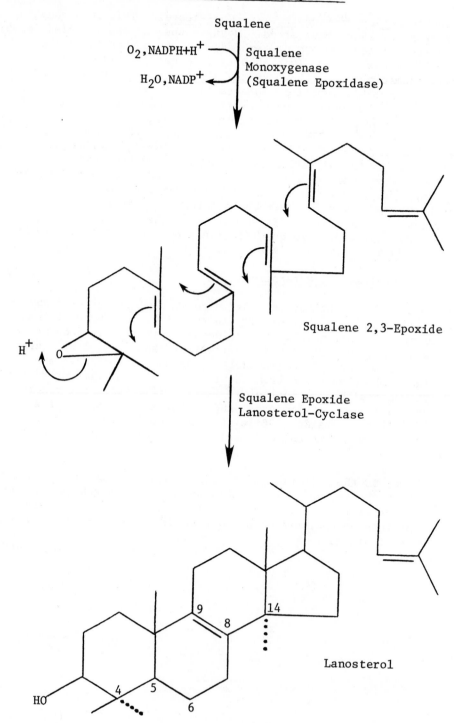

Squalene

O_2, NADPH+H$^+$ — Squalene Monoxygenase (Squalene Epoxidase)

H_2O, NADP$^+$

Squalene 2,3-Epoxide

H$^+$

Squalene Epoxide Lanosterol-Cyclase

Lanosterol

Figure 122. Conversion of Lanosterol to Cholesterol

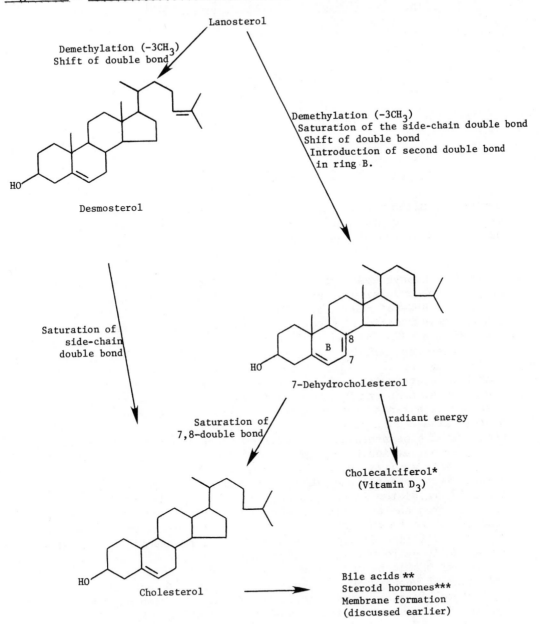

Lanosterol

Demethylation (-3CH₃)
Shift of double bond

Desmosterol

Demethylation (-3CH₃)
Saturation of the side-chain double bond
Shift of double bond
Introduction of second double bond
in ring B.

Saturation of
side-chain
double bond

7-Dehydrocholesterol

Saturation of
7,8-double bond

radiant energy

Cholecalciferol*
(Vitamin D₃)

Cholesterol

Bile acids **
Steroid hormones***
Membrane formation
(discussed earlier)

* See Chapter 9 for further discussion.
** Discussed later in this chapter.
*** See Chapter 13 for further information.

Figure 123. Origins of the Carbon Atoms in Cholesterol

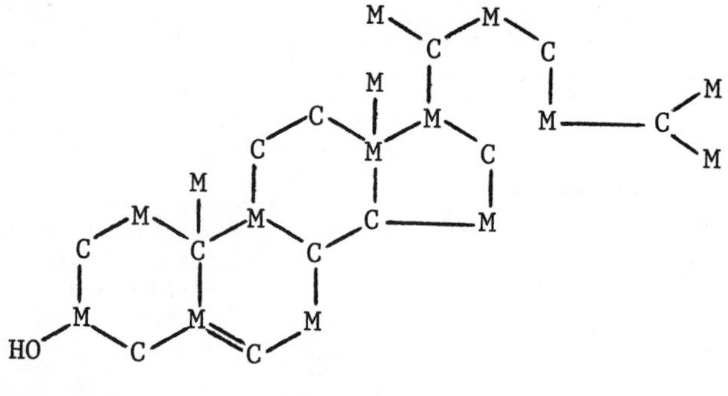

M = derived from the methyl carbon of the acetate; C = derived from the carbonyl carbon of acetate. The oxygen comes from molecular oxygen during the epoxidation of squalene.

3. Control of biosynthesis of cholesterol. The primary rate-limiting step is the conversion of HMG-CoA to mevalonic acid, catalyzed by the enzyme HMG-CoA reductase. High dietary cholesterol levels inhibit this reaction, affecting hepatic cholesterogenesis. The regulation of HMG-CoA reductase appears to be complex and can occur at three stages, namely, protein synthesis, modulation of enzyme activity and protein degradation. In rodents, it has been shown that HMG-CoA reductase exhibits a diurnal rhythm. In in vitro experiments, cAMP and Mg-ATP inhibit the HMG-CoA reductase step, although it has not been ascertained whether the cAMP-dependent protein kinase participates in the physiologic control process of cholesterol biosynthesis. In addition to cholesterol, there are two other potent inhibitors of HMG-CoA reductase present in the steroid intermediary metabolism. These inhibitors are 7α-hydroxycholesterol and 20α-hydroxycholesterol, which are the first intermediates in the production of bile acids and steroid hormones from cholesterol, respectively. It appears that cholesterol itself may not be the inhibitor. Instead, a cholesterol-containing lipoprotein formed during the absorption and transport of cholesterol may actually perform this function (discussed later). In this regard, it should be pointed out that mevalonic acid is also a precursor of coenzyme Q, although its formation is not affected by cholesterol feeding. A second rate-limiting reaction may be the cyclization of squalene into lanosterol. Fasting inhibits cholesterol biosynthesis while high-fat diets seem to accelerate the production of cholesterol. The thyroid gland also plays a role in the regulation of cholesterol metabolism. Thyroid hormone possibly increases the catabolism of cholesterol while hypothyroidism is

associated with hypercholesterolemia. Administration of thyroid
hormone to both normal and hypothyroid individuals reduces the serum
cholesterol level. It is of interest to point out that the
regulation of cholesterol synthesis is lost in cancerous hepatic
cells.

4. The routes of cholesterol elimination are indicated below. The
    role of the bile salts in lipid absorption from the intestines has
    already been mentioned, and this and other aspects of bile salt
    metabolism will be discussed later in more detail.

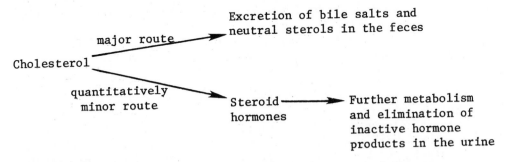

5. Other than as a precursor for bile salts and steroid hormones,
    the exact biochemical function of cholesterol is conjectural.
    Cholesterol esters may aid in the transport of fatty acid,
    particularly unsaturated ones. Brain tissue is rich in cholesterol.
    Whether it is involved in the production and propagation of impulses
    or serves merely as an insulating material is not known. A
    precursor of cholesterol (7,8-dehydrocholesterol) is converted to
    vitamin D_3 when exposed to ultraviolet light. The role of
    cholesterol in steroid hormone synthesis is covered in the chapter
    on hormones.

Role of Cholesterol in Arteriosclerosis and Atherosclerosis

1. Arteriosclerosis is a degeneration and hardening of the walls of the
    arteries, capillaries, and veins due to chronic inflammation which
    results in fibrous tissue formation and limits the blood supply to
    the vital organs. It can have several etiologies including defects
    in the basement membrane. Atherosclerosis is one type of
    arteriosclerosis, specifically involving an accumulation in the
    arterial intima of plaque-like lipids which undergo calcification.
    In atherosclerosis, cholesteryl esters and other lipids are
    deposited in the connective tissue of the arterial walls. It is
    important to notice that atherosclerosis is a subtype of
    arteriosclerosis and that the two words are not synonymous.
    Atherosclerosis of the coronary arteries, the vessels that supply

the heart muscle, is known as coronary heart disease. It should be noted that while occlusion of the blood vessel of the heart leads to heart attacks, occlusion of the blood vessels of the brain results in strokes. These aspects are discussed in greater detail along with lipoprotein metabolism in this chapter.

2. Diabetes mellitus, lipid nephrosis, hypothyroidism, and other hyperlipemic conditions, in which prolonged high levels of low density lipoproteins (LDL) and very low density lipoproteins (VLDL) occur in the blood, lead to atherosclerosis. High blood pressure (which is related to high salt ingestion and obesity), obesity, cigarette smoking, lack of exercise, and family history (genetics) also play a role in the development of atherosclerosis. Normal dietary fluctuations of cholesterol do not appear to cause this disease, although feeding of greater than usual amounts of cholesterol on a sustained basis may induce it.

3. There is a good correlation between high serum lipid levels and atherosclerosis. Total serum cholesterol levels have been used most extensively in such correlations. Most recently, cholesterol associated with high density lipoproteins (HDL) show a reciprocal correlation with coronary heart disease (that is, elevated levels of cholesterol in the HDL fraction may have a protective action against atherosclerosis--see lipoprotein metabolism). Hypertriglyceridemia also shows an association with the risk of coronary heart disease; however, it is a much weaker, independent risk factor than cholesterol.

4. Heparin, a polysaccharide normally found in serum (see Chapter 4), has a clearing effect in *vivo* on hyperlipemic blood. This may be due to stimulation of lipolytic activity by heparin and the rapid removal of the solubilized lipid from the bloodstream. There may be a deficiency of heparin in atherosclerosis. The ability of heparin, dextran sulfate, and other polyfunctional ions to complex to and, in the presence of Mg^{+2} or Mn^{+2}, precipitate all of the lipoproteins except the high density ones provides the basis of one technique for the separation of this class of lipoproteins and the measurement of cholesterol contained in the HDL fraction (discussed later).

5. The correlation of cholesterol levels with atherosclerosis suggests that the control of these levels may be useful in the management or prevention of the disease. However, convincing data showing that induced plasma cholesterol lowering will actually prevent atherosclerosis is still lacking. Blood cholesterol is susceptible to dietary influences and to the hypocholesterolemic drugs.

    a. The blood cholesterol can be lowered by replacing some of the saturated fatty acids in the diet by polyunsaturated fatty acids.

There is some evidence that atherosclerosis is due to a
<u>relative</u> deficiency of unsaturated fatty acids. On this basis,
safflower, corn, peanut, and cottonseed oils (in order of
decreasing efficacy) lower blood cholesterol while butter fat
and coconut oil raise it. The polyunsaturates probably exert
their effect by

   i)   stimulating cholesterol excretion into the intestine;

  ii)   stimulating the oxidation of cholesterol to bile acids;

 iii)   increasing the rate of metabolism of cholesterol esters,
since esters of cholesterol with unsaturated fatty acids
are more rapidly metabolized in the liver and other
tissues than are the esters with saturated fatty acids;

  iv)   shifting the distribution of cholesterol (decreasing it
in the plasma and increasing it in the tissues).

b.   The <u>hypocholesterolemic drugs</u> function in several ways to lower
serum cholesterol levels. Some examples of such drugs and
their actions are given below.

   i)   One group of drugs acts on cholesterol levels by
blocking the various stages of cholesterol synthesis.
<u>Clofibrate</u> (Atromid-S), shown below, is such a compound.

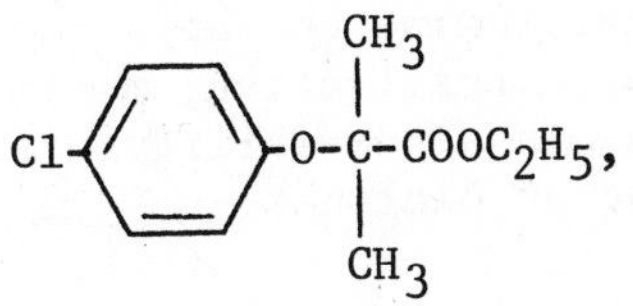

$$Cl-\!\!\left\langle\bigcirc\right\rangle\!\!-O-\underset{\underset{\displaystyle CH_3}{|}}{\overset{\overset{\displaystyle CH_3}{|}}{C}}-COOC_2H_5, \text{ Clofibrate}$$

Although the exact inhibitory mechanism is not understood,
clofibrate is known to inhibit hepatic cholesterol
synthesis. Clofibrate also lowers the VLDL,
presumably by inhibiting triglyceride and lipoprotein
synthesis. This compound is used in treating types III,
IV, and V hyperlipoproteinemias, but it is not of
value in the treatment of type I and II hyperlipoproteinemias (see later). Side-effects of clofibrate
include enhanced sensitivity to coumarin anticoagulants
and, in some cases, myositis with weakness, tenderness
of muscle (with an accompanying increase in CPK levels),
transient elevation of SGOT and SGPT, leukopenia, etc.

  ii)   <u>D-thyroxine</u> lowers the cholesterol concentration by
accelerating the catabolism of cholesterol and LDL.

Note that the normal thyroid hormone is L-thyroxine. The D-isomer is used to reduce cholesterol level because it has the highest ratio of hypocholesterolemic to calorigenic activity of any of the thyroid-like compounds tested. One side-effect of D-thyroxine is enhanced sensitivity to coumarin anticoagulants. This drug should not be used in individuals with organic heart disease or abnormal glucose tolerance, and possibly those with an elevated metabolic rate.

iii) **Cholestyramine**, a basic anion exchange resin, prevents reabsorption of bile acids and increases their fecal loss. The $Resin^+Cl^-$ exchanges its Cl^- ion for a bile $acid^-$ ion, giving rise to a Resin-bile acid complex which is eliminated in the feces. This interrupts the enterohepatic recycling and leads to increased catabolism of cholesterol to replace the lost bile acids. This compound has been found to be quite useful in type II hyperlipoproteinemia. Its side effects include nausea, constipation, possibly hyperchloremic acidosis in children, interference with absorption of fat-soluble vitamins (A, D, E, and K), etc. In patients with severe hypercholesterolemia or in patients who do not adhere reliably to dietary alterations or drug treatment, an ileal exclusion operation may be required. Since bile acids are reabsorbed almost entirely in the ileum, this operation, which results in bypassing the ileum for absorption purposes, has the same effects as that of cholestyramine therapy. Note that vitamin B_{12} is also absorbed in this region, making the development of macrocytic anemia a possibility. Some patients may also show persistent diarrhea.

iv) **Neomycin** blocks the intestinal absorption of cholesterol.

v) **Nicotinic acid** decreases the plasma lipids (both cholesterol and triglycerides) by interfering with the mobilization of FFA. It is useful in all of the hyperlipoproteinemias except type I. Side effects include flushing of the skin, pruritus, G.I. disturbances, hyperglycemia, hyperuricemia, and abnormal liver function.

vi) **β-Sitosterol**, a natural sterol of plant origin, prevents cholesterol absorption, causing a reduction in plasma cholesterol levels. The plant sterol itself is not absorbed to any significant extent.

Metabolism of Bile Acids (Salts)

1. The bile acids are 24-C steroid derivatives. The primary bile acids (cholic and chenodeoxycholic) are synthesized in the liver from cholesterol, as shown in Figure 124. The preferred substrate is newly synthesized, unesterified cholesterol. It has been shown that intravenously administered cholesterol primarily enters a pool of cholesterol which is inaccessible to 7α-hydroxylase, which catalyzes the first committed step in bile acid synthesis. Cholic acid is the most abundant of these compounds. The conjugation of bile acids with glycine and taurine precedes secretion into the bile. At the alkaline pH of the bile and in the presence of alkali cations (Na^+, K^+), the bile acids and their conjugates are actually present as salts (ionized forms), although the name bile acids and bile salts are used interchangeably.

2. The regulation of bile acid formation from cholesterol occurs at the 7α-hydroxylation step. The reaction is catalyzed by a microsomal monooxygenase system. It appears that this reaction requires the participation of cytochrome P-450, NADPH-cytochrome-P450 reductase and a phospholipid. The concentration of bile acids in the enterohepatic circulation regulates the activity of 7α-hydroxylase and thus bile acid synthesis. The other reactions that occur in the formation of bile acids after the 7α-hydroxylation step are: oxidation of the 3β-hydroxy group to a 3-keto group, isomerization of the Δ5 double bond to the Δ4 position, conversion of the 3-keto group to a 3α-hydroxyl group, reduction of the Δ4 double bond, 12α-hydroxylation in the case of cholic acid synthesis and finally, oxidation of the side chain. The 12α-hydroxylase, like 7α-hydroxylase, is associated with the microsomes and requires NADPH, molecular oxygen and participation of cytochrome P-450 (although it may not be an integral part of the 12α-hydroxylase complex). The 12α-hydroxylase, unlike the 7α-hydroxylase, does not exhibit diurnal variation (see later). The level of activity of this enzyme regulates the amount of cholic acid synthesized. The exact steps involved in the oxidation of the side chain and removal of three carbons from it are not clear. The reactions may be similar to β-oxidation of fatty acids. The substrates in the above-mentioned reactions are mostly water insoluble and, therefore, they appear to require a carrier protein(s) (namely sterol carrier protein(s)) during their synthesis and metabolism which occur in the aqueous interior of the hepatocyte. Recall that during the latter steps of cholesterol biosynthesis, similar carrier proteins (squalene and sterol carrier proteins) appear to be essential. The bile acids are normally synthesized in the liver at a rate of about 0.2-0.5 g per day. In instances where there is a loss of bile acids due to: drainage of bile through a biliary fistula, administration of bile acid-complexing resins (e.g., cholestyramine) or ileal exclusion (see cholesterol metabolism), the activity of 7α-hydroxylase is

Figure 124. Formation of Bile Acids

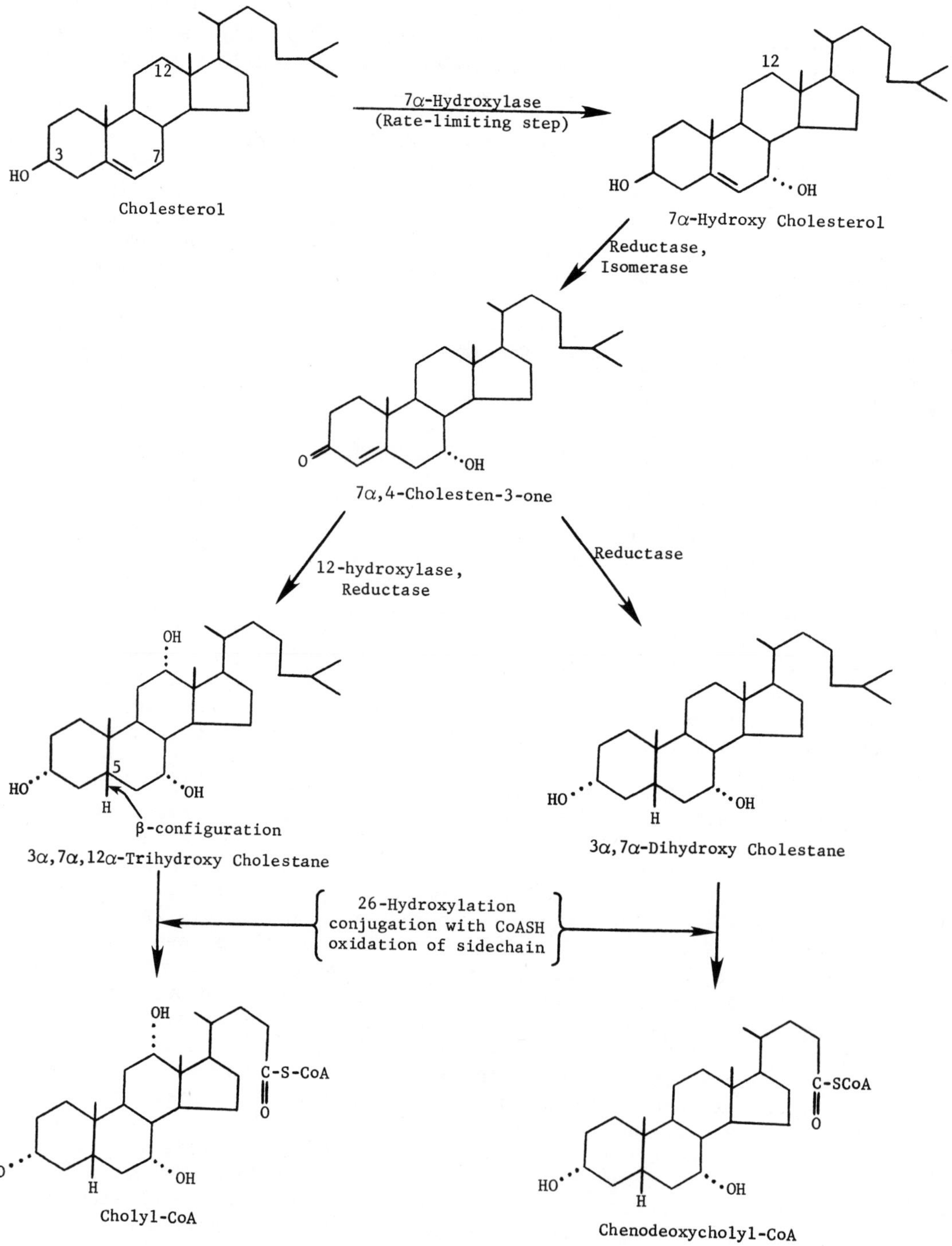

12
3
7
HO
Cholesterol
7α-Hydroxylase
(Rate-limiting step)
12
HO
OH
7α-Hydroxy Cholesterol
Reductase,
Isomerase
O
OH
7α,4-Cholesten-3-one
12-hydroxylase,
Reductase
Reductase
OH
HO
5
OH
H
β-configuration
3α,7α,12α-Trihydroxy Cholestane
HO
OH
H
3α,7α-Dihydroxy Cholestane
26-Hydroxylation
conjugation with CoASH
oxidation of sidechain
OH
C-S-CoA
O
HO
OH
H
Cholyl-CoA
C-SCoA
O
HO
OH
H
Chenodeoxycholyl-CoA

stimulated several-fold with consequent increases in bile acid formation. Since bile acid formation is one of the major pathways for the elimination of cholesterol, the latter two bile acid-losing processes are used in bringing about a reduction of cholesterol levels in hypercholesterolemic patients.

Another aspect of the regulation of bile acid formation involves the biosynthesis of cholesterol, the precursor of bile acids. As noted earlier in the section on cholesterol metabolism, the major rate-limiting step of cholesterol biosynthesis is the reduction of hydroxymethylglutaryl-CoA (HMG-CoA). 7α-Hydroxy-cholesterol, the first intermediate of bile acid formation, inhibits HMG-CoA reductase. The activities of both 7α-cholesterol hydroxylase and HMG-CoA reductase undergo parallel changes due to changes in bile acid levels. In the rat, both enzymes show similar patterns of diurnal variation with the highest activities occurring during the dark period. The inhibitory mechanism of bile acids or their intermediates on these two enzymes is not known, but it appears that they do <u>not</u> function directly as allosteric modifiers. Bile acid in the intestines, in addition to their role in cholesterol absorption, may also regulate cholesterol biosynthesis in the intestine. It is of interest to point out that in man the presence of excess cholesterol does not proportionately increase bile acid production, while it can suppress endogenous cholesterol synthesis and increase the excretion of neutral steroids in feces.

3. In mammals including man, the major bile acids synthesized belong to the 5β-type (see Figure 124). The bile acids with a 5α-configuration occur in small amounts. The bile acids are conjugated with taurine and glycine before they are excreted in bile. The conjugation with glycine takes place only in mammals. In man, the normal ratio of glycine- to taurine-conjugated bile acids is about 3:1. Sulfate esters of bile acids are also formed to a small extent. Bile produced in the hepatocytes is stored and concentrated in the gallbladder; a saccular, elongated, pear-shaped organ attached to the hepatic duct. Bile's constituents, in addition to bile acids, include bile pigments (e.g., bilirubin glucuronides--products of heme catabolism--see Chapter 7), cholesterol and lecithin. The pH of the gallbladder bile is about 6.9-7.7. It should be noted that cholesterol is normally water <u>insoluble</u>, but it is solubilized in the bile by the formation of micelles with bile acids and lecithin. Formation of cholesterol gallstones can occur due to excessive secretion of cholesterol into the bile or the presence of insufficient amounts of bile acids and lecithin relative to the cholesterol in the bile. The inadequate amounts of bile acids can result from decreased synthesis in the liver, decreased uptake of bile acids by hepatocytes from the portal blood (enterohepatic circulation), or increased loss of bile acids in the gastrointestinal tract. These aspects are discussed later in this section.

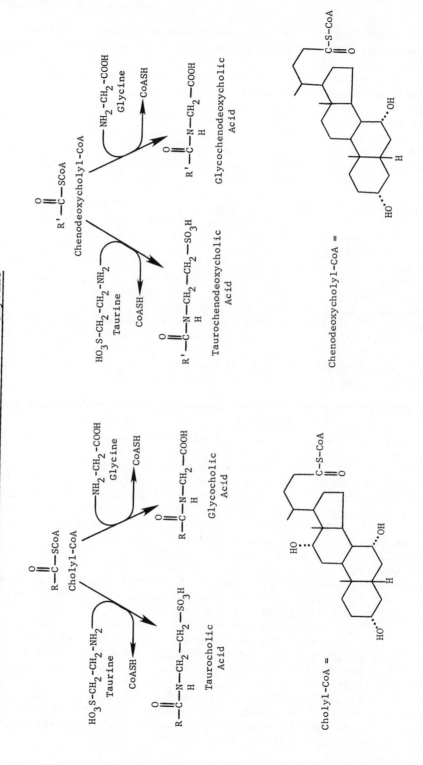

Figure 125. Conjugation of Bile Acids With Taurine and Glycine

With the ingestion of food, cholecytokinin-pancreozymin
(a gastrointestinal hormone--see Chapter 6) is released into the
blood and contracts the gallbladder. The contents of the gallbladder
are then rapidly emptied into the duodenum by way of the common bile
duct. In the duodenal wall, the bile duct fuses with the pancreatic
duct; this fused duct is known as the ampulla of Vater. The
functions of the bile system include the role of the bile salts in
the absorption of lipids and the lipid-soluble vitamins A, D, E
and K (for details see lipid absorption, this chapter);
neutralization of acid chyme; and excretion of toxic metabolites
(e.g., bile pigments, some drugs, toxins, etc.) in the feces.

4. The secondary bile acids, deoxycholic and lithocholic acids, are
    synthesized by 7-dehydroxylation of the deconjugated primary bile
    acids, cholic and chenodeoxycholic acids, respectively. The enzymes
    for these conversions are provided by bacteria residing primarily
    in the large intestine. The major portion (>90%) of the bile acids
    present in the intestines are reabsorbed by an active transport
    system into the portal circulation at the distal ileum. In the
    portal blood the bile acids are transported bound to albumin.
    The bile acids are taken up by the liver, promptly reconjugated
    with taurine and glycine, and resecreted into the bile. Both the
    ileal absorption and liver uptake of bile acids may be mediated
    by Na^+-dependent (carrier) transport mechanisms. This cyclic
    process of bile transport from the intestine to the liver and back
    to the intestine is known as the enterohepatic circulation (see
    Figures 126 and 127). During a single passage of portal blood
    through the liver, about 90% of the bile acids are extracted.
    The pool size of bile acids in the enterohepatic circulation is
    about 2 to 4 g and circulates about twice each meal. The amount
    of bile acids lost in the feces is about 0.5 g per day, consisting
    mostly of secondary bile acids (particularly lithocholic acid,
    the least soluble of the bile acids). This bile acid loss of
    about 0.5 g per day is made up by the synthesis of an equal amount
    in the liver.

5. Some clinical aspects of bile acid metabolism.

    a. In liver disorders, the serum levels of bile acids are
        elevated, and the measurement of serum bile acids is a sensitive
        indicator of liver disease. Bile acids are not normally found
        in the urine, due to efficient uptake by the liver and its
        excretion into the intestines. But in hepatocellular disease
        and obstructive jaundice (see Chapter 7), they are elevated.
        As noted earlier, lithocholic acid is toxic and can cause
        hemolysis, fever, etc. In general, the effects associated
        with increased levels of plasma bile acids due to liver disease
        include pruritis (see Chapter 7, under jaundice), steatorrhea,
                              (Text continues on p. 930)

Figure 126. Conversion of Primary to Secondary Bile Acids Catalyzed by Microbial Enzymes

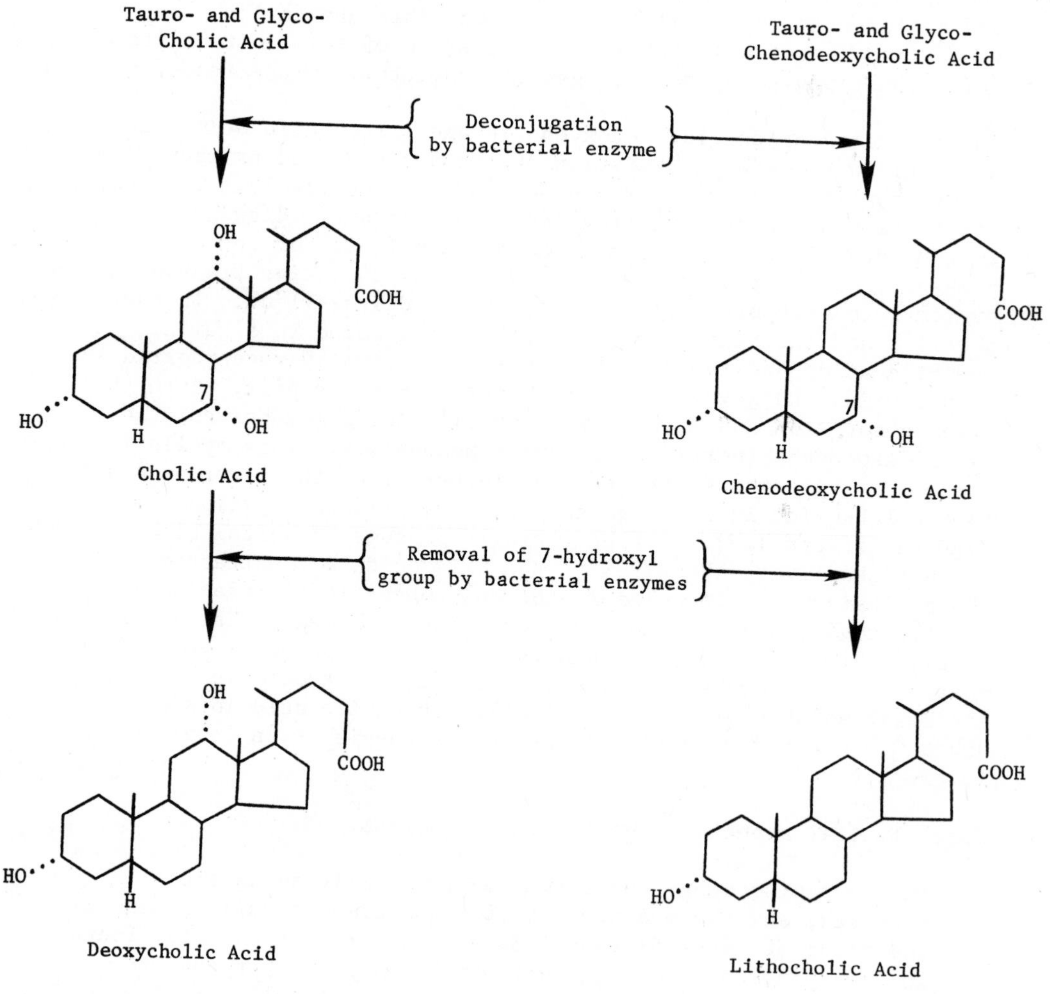

Tauro- and Glyco-
Cholic Acid

Tauro- and Glyco-
Chenodeoxycholic Acid

Deconjugation
by bacterial enzyme

OH

COOH

HO

H

7

OH

Cholic Acid

COOH

HO

H

7

OH

Chenodeoxycholic Acid

Removal of 7-hydroxyl
group by bacterial enzymes

OH

COOH

HO

H

Deoxycholic Acid

COOH

HO

H

Lithocholic Acid

Figure 127. <u>Formation and Disposition of the Bile Salts (Bile Acids)</u>

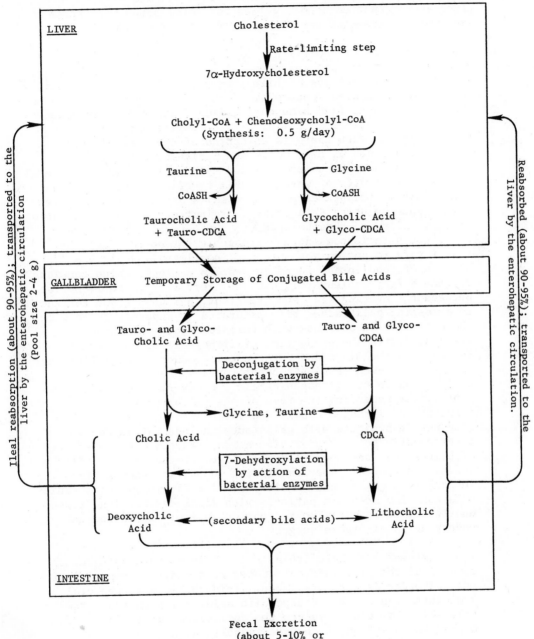

LIVER

Cholesterol

↓ Rate-limiting step

7α-Hydroxycholesterol

Cholyl-CoA + Chenodeoxycholyl-CoA
(Synthesis: 0.5 g/day)

Taurine   Glycine

CoASH   CoASH

Taurocholic Acid   Glycocholic Acid
+ Tauro-CDCA   + Glyco-CDCA

GALLBLADDER   Temporary Storage of Conjugated Bile Acids

Tauro- and Glyco-   Tauro- and Glyco-
Cholic Acid   CDCA

Deconjugation by
bacterial enzymes

Glycine, Taurine

Cholic Acid   CDCA

7-Dehydroxylation
by action of
bacterial enzymes

Deoxycholic   (secondary bile acids)   Lithocholic
Acid   Acid

INTESTINE

Ileal reabsorption (about 90-95%); transported to the
liver by the enterohepatic circulation
(Pool size 2-4 g)

Reabsorbed (about 90-95%); transported to the
liver by the enterohepatic circulation.

Fecal Excretion
(about 5-10% or
about 0.5 g/day--mostly secondary bile acids)

CDCA = Chenodeoxycholic Acid

hemolytic anemia, and further liver injury. The various
literature-reported normal values for total serum bile acids
are not in agreement but are usually below 2 μg per ml.

b. The major cause of <u>cholelithiasis</u> (presence or formation of
gallstones) is the precipitation of cholesterol in the bile.
Elevated concentrations of bile pigments (bilirubin glucuronides)
in the bile can also lead to the formation of concretions known
as pigment stones (see heme catabolism--Chapter 7). As noted
earlier, cholesterol in the bile is solubilized by bile acids
and lecithin. When there is an excess of cholesterol and/or
decreased amounts of bile acids and lecithin, the bile gets
supersaturated with cholesterol and eventually forms cholesterol
stones. The limits of water solubility of cholesterol in the
bile fluid in the presence of bile salts and lecithin has been
established by using triangular coordinates. If the mole ratio
of bile salts and phospholipids to cholesterol is less than
10:1, the bile is considered <u>lithogenic</u> (stone-forming), but
this figure is not an absolute one in all individuals. A
pattern of food intake found in western nations, which consists
of excess fat, cholesterol, and an interval of 12 to 14 hours
between the evening meal and breakfast, has been shown to lead
to a fasting gallbladder bile saturated or supersaturated with
cholesterol. In patients with gallstones, the problems
associated with the production of lithogenic bile appear to
reside with the liver and not the gallbladder. The gallbladder
may merely provide an environment for the precipitation of
cholesterol in the lithogenic bile. However, there is still
controversy regarding the role of the gallbladder in the
formation of lithogenic bile and gallstones. Characteristic
findings in subjects with gallstones include a reduced
enterohepatic pool size of bile acids, reduced hepatic secretion
of bile acids, and decreased half-lives of the primary bile
acids. In general, American Indian women, who show a high
incidence of gallstones, exhibit similar findings. Lithogenic
bile can also occur in patients with ileal dysfunction, those
undergoing cholestyramine therapy or in individuals who have
undergone ileal bypass surgery.

The treatment for cholelithiasis is cholecystectomy (surgical
removal of the gallbladder). However, a medical treatment of
this disorder is also available. This consists of the oral
administration of chenodeoxycholic acid, which enriches the
bile acid pool and dissolves the cholesterol gallstones. The
mechanism of action of chenodeoxycholic acid is not clear; it
decreases hepatic cholesterol secretion in bile but the effect
does not appear to be related to the expansion of the pool
size or enhanced secretion of the bile acid. For instance,
cholic acid administration does not dissolve cholesterol

gallstones, while it does expand the bile acid pool and causes increased secretion of bile acid. Chenodeoxycholic acid may have a specific action on one of the rate-limiting enzymes. The ratio of HMG-CoA reductase to cholesterol 7α-hydroxylase appears to be high in patients with gallstones and administration of chenodeoxycholic acid to these patients returned the ratios to normal levels. With regard to long-term management of gallstones with chenodeoxycholic acid, the following aspects should be noted: the effect of chenodeoxycholic acid is transient and, therefore, the therapy may have to be long-term; the treatment appears to be promising only in patients with radiolucent gallstones and functioning gallbladders; and since the therapy raises the serum transaminase levels, the significance of which is not clear, the measurement of other liver function tests may be required.

Lipoproteins

1. General Comments

   a. The lipids, which are generally insoluble in water, are largely transported within the body in an aqueous medium. To accomplish this and thereby make lipids available for metabolism, the lipids are first complexed with various specific proteins. The resultant lipoproteins (LP's) are globular particles and consist of varying amounts of neutral lipids (e.g., triglycerides or cholesterol esters) at the central core surrounded by a coat of unesterified cholesterol, phospholipid and protein (see later).

   b. Four major groups of plasma lipoproteins have been separated and identified based upon electrophoresis and ultracentrifugation. It should be emphasized that each group or class of lipoproteins is heterogeneous and can be further subdivided. The major groups of lipoproteins share several apolipoprotein (apoprotein) components (proteins found in association with lipids to form various lipoproteins) in common. Each group of apoproteins is heterogeneous and is made up of more than one polypeptide chain. No uniform nomenclature for these apoproteins has yet been accepted. In one terminology, they are identified by their carboxy-terminal amino acid. In another classification (adopted in this text and developed by Alaupovic and colleagues), the apoproteins are designated as A, B, C, D, and E. In addition to the usual chemical methods of identification (amino acid analysis, peptide fingerprinting, sequence determination, etc.), the apoproteins show immunologic specificity. This has been used in some studies for the detection and quantification of these molecules (e.g., elevated levels of apoprotein-E in type III hyperlipoproteinemia). The four major lipoproteins,

separable by ultracentrifugation, are the chylomicrons, very
low density lipoproteins (VLDL), low density lipoproteins (LDL),
and heavy density lipoproteins (HDL). Upon electrophoresis
at alkaline pH, where either paper or agarose is used as a
support medium, the lipoproteins migrate into different
positions. The nomenclature used is in reference to the
migration of serum proteins under comparable conditions (see
serum protein electrophoresis, Chapter 6). The lipoprotein
bands are identified by staining with appropriate lipid
stains (e.g., oil red o, fat red 7B, sudan black B). The
chylomicrons appear at the origin (point of application of
plasma sample), the VLDL appear at the α_2-region (more commonly
known as the pre-β region; hence they are called pre-β
lipoproteins), the LDL migrate to the β-globulin region
(designated as the β-lipoproteins), and the HDL appear at the
α_1-globulin region (designated as the α-lipoproteins). The
electrophoretic pattern of the lipoproteins separated on
paper (or on thin agarose films) is shown in Figure 128;

Figure 128.   Migration of Plasma Lipoproteins on Paper Electrophoresis

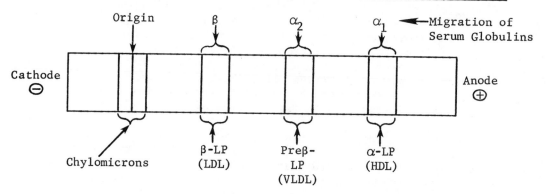

a correlation of the classification of lipoproteins by
electrophoresis and hydrated density is shown in Table 40.
The broad functions of the four major lipoproteins are as
follows: chylomicrons primarily transport dietary
triglycerides and cholesterol, VLDL transports endogenously
synthesized (primarily in the liver) triglycerides, and both
LDL and HDL participate in the transport of endogenously
synthesized cholesterol (see later for detailed aspects).
As previously mentioned, the apoproteins are heterogeneous
and differ from one another in a number of properties, namely,
biologic, immunologic and physicochemical. In addition to
their structural functions, some apoproteins have specific
roles as enzyme activators or inhibitors in lipoprotein

Table 40. Nomenclature and Some Physical Data on the Four Major Lipoprotein Types

Lipoprotein	Hydrated Density (g/ml)	S_f*	Position of Migration in Paper or Agarose Electrophoresis in Comparison With Serum Globulins
(1) Chylomicrons	< 0.95	> 400	origin
(2) Pre-β (very low density lipoprotein; VLDL)	0.95 - 1.006	20-400	α_2 (migrating ahead of β-globulin, and named as pre-β position)
(3) β-Lipoprotein (low density lipoprotein; LDL)	1.006 - 1.063	0-20	β (with β-globulin)
(4) α-Lipoprotein (high density lipoprotein, HDL)	1.063 - 1.210	---	α_1 (with α_1-globulin)

* The rate at which lipoprotein floats up through a solution of NaCl of specific gravity 1.063 is expressed in Svedberg, S_f units: one S_f unit = 10^{-13} cm/second/dyne/gm at 26°C. S_f can be thought of as a negative sedimentation constant.

metabolism. In order to understand the molecular 'organization of the lipoproteins, lipid-protein interactions, and mechanisms of normal and abnormal lipoprotein metabolism, knowledge of the molecular structure of apoproteins is essential. This is presently an active field of investigation. The amino acid sequences have been established for four apoproteins (human C-I, C-III, A-I and A-II), the phospholipid-binding segments have been identified in some, and the biologic role of a few apoproteins have been delineated. The major properties of the apoproteins are summarized in Table 41.

c. As noted under cholesterol metabolism, hypercholesterolemia is one of the major risk factors (the others are hypertension and cigarette smoking) for the development of premature cardiovascular disease. This has stimulated an intense investigation of the metabolism of lipoproteins in normal individuals and in patients with atherosclerosis and/or other disorders of lipid metabolism. In this section, the structure and metabolism of lipoproteins in normal and pathologic conditions are discussed. Regarding the four lipoprotein classes described above, it should be noted that they can be further subdivided, and in a number of clinical conditions there are plasma lipoproteins which do not fall under the four major lipoprotein categories. These disorders, which are discussed later, include obstructive liver disease, type III dyslipoproteinemia, abetalipoproteinemia, Tangier disease and lecithin:cholesterol acyltransferase.

2. Composition of Lipoproteins

a. <u>Chylomicrons</u>. Chylomicrons, carriers of exogenous triglycerides (dietary fat) consisting of long chain fatty acids, are synthesized in the intestinal epithelial cells and transported by the lymphatic system. They enter the systemic blood circulation via the thoracic duct. Note that the triglycerides, which consist of fatty acids of less than 10 carbons, are absorbed directly into the portal blood. The data on the size and composition of chylomicrons are as follows.

Size: heterogeneous (diameter range for lymph chylomicrons = 300-5000 Å)

Triglyceride: major constituent, about 86%

Phospholipids (lecithin and sphingomyelin are major constituents): about 8%

Cholesterol: about 5%

Protein: 0.5-2.0%

Table 41. The Properties of Human Plasma Apolipoproteins

Apoprotein	Molecular Weight and Number of Amino Acid (aa) Residues	Constituent of Lipoproteins		Biologic Role and Remarks
		in major quantity	in minor quantity	
A-I	28,000; single polypeptide chain of 245 aa residues; sequence is known.	HDL	Chylomicrons & VLDL	Activator of LCAT
A-II	17,000; two identical polypeptide chains of 77 aa residues; sequence is known.	HDL	Chylomicrons & VLDL	---
B	Not exactly known.	LDL, VLDL & Chylomicrons	---	---
C-I	6500; single polypeptide chain of 57 aa residues; sequence is known.	Chylomicrons & VLDL	HDL	Activator of LCAT and Postheparin-LPL
C-II	About 10,000; single polypeptide chain of about 95 aa residues.	Chylomicrons & VLDL	HDL	Activator of Postheparin-LPL
C-III	8800; single polypeptide chain of 79 aa residues; sequence is known; most abundant of C-proteins.	Chylomicrons & VLDL	HDL	Inhibitor of Postheparin-LPL
apo-D (also known as thin-line protein)	About 20,000.	---	HDL	Not exactly known. May be an activator of LCAT and may function as a specific carrier of lysolecithin generated due to the action of LCAT on HDL.
apo-E$_{1-3}$ (also known as arginine-rich protein)	32-39,000. Three components have been identified by isoelectric focusing.	VLDL	HDL	While the total apo-E is increased in the VLDL of type III hyperlipoproteinemia, apo E-III is grossly deficient or absent.

LCAT = Lecithin:Cholesterol Acyltransferase
LPL = Lipoprotein Lipase

The amount of protein found in lymph chylomicrons is small and variable; they are made up of about 66% apoC (apoC-I, apoC-II, and apoC-III), 22% apoB, and 12% apoA (apoA-I and apoA-II in about a 1:1 ratio).

b. Very low density lipoproteins (VLDL). VLDL is the major transport vehicle for endogenously synthesized triglycerides and is predominantly synthesized in the liver. The intestine also participates in its synthesis, but only to a small extent. The information concerning its size, density and composition are as follows.

Size: variable; range: 280-750 Å in diameter

The size of VLDL is directly proportional to its triglyceride content and inversely proportional to its phospholipid and protein content. The variation in size may be due to the different stages of degradation of VLDL by lipoprotein lipase (discussed later). The hydrated density range of VLDL is 0.95-1.006 g per ml; at the lower end of this range it overlaps with chylomicrons.

Composition (average values, expressed as % wt; 90-92% lipid, 8-10% protein)

Lipid composition (expressed as % of total lipids by wt)

Triglyceride: 55-56%

Phospholipid: 19-21% (lecithin and sphingomyelin are major constituents)

Cholesterol: 15-17% (the ratio of esterified cholesterol to unesterified cholesterol is about 1)

Apoprotein: 8-10%; heterogeneous--the major constituents are apoB (about 35% of the total VLDL protein), apoC (content varies in a given individual with the degree of lipidemia and with age; larger VLDL particles contain relatively greater quantities of apoC and lower quantities of apoB), and apoE.

c. Low density lipoproteins (LDL). LDL is the principal vehicle for the transport of cholesterol from the liver to the body cells. To a large extent it appears that LDL is derived from VLDL (discussed later). The size, density and composition of LDL are given below.

Size: relatively uniform, almost spherical; range: 210-250 Å in diameter

Hydrated density range: 1.019-1.063 g per ml

Overall composition: lipids--75% and protein--25%

Lipid composition

Esterified cholesterol: 50% (the major fatty acid is linoleic acid, an essential fatty acid)

Unesterified cholesterol: 10%

Phospholipids: 30%

Triglycerides: 10%

Apoprotein: apoB

d. <u>High density lipoproteins (HDL)</u>. These are synthesized in the liver (and in the intestine) and appear to play a significant role in the transport of cholesterol from peripheral cells to the liver, where the cholesterol is removed from the body via the bile. The HDL has been subfractionated into three density classes, namely, HDL_1 (1.050-1.063 g per ml), HDL_2 (1.063-1.120 g per ml), and HDL_3 (1.120-1.210 g per ml). Although the exact physiologic functions of each of these fractions is not known, it has been shown that the plasma ratio of HDL_2 to HDL_3 is altered in various physiologic and pathologic conditions. For example, estrogen has been implicated in affecting this ratio, as it has been observed that in premenopausal women the content of HDL_2 is three times greater than that found in men. The size and composition of HDL are as follows.

Size: 90-120 Å in diameter; smallest of the lipoprotein particles

Overall composition: about equal amounts of protein and lipids by weight

Composition of lipids (expressed as % total lipids) by weight:

Phospholipids--42-51%; predominantly lecithin

Cholesteryl esters--32%; the predominant fatty acid
of the cholesteryl ester is
linoleic acid

Apoprotein: Major apoproteins are apoA-I and apoA-II.
The ratios of these two proteins vary within
each HDL-subspecies. Minor apoproteins are
apoC, apoD and apoE.

e. <u>Other lipoproteins</u>. Two other lipoproteins, in addition to
the chylomicrons, VLDL, LDL and HDL, have been described.
These are lipoprotein (a) and lipoprotein-X.

i) Lipoprotein (a): It is a variant of LDL and appears in
the plasma of normal subjects in the density range of
1.050-1.120 g per ml. Lp(a) has a molecular weight
of 5×10^6 daltons, and on agarose electrophoresis it
migrates with VLDL (pre-β electrophoretic mobility).
The increased electrophoretic mobility of Lp(a) over
LDL may be due to its high content of sialic acid.
The apoprotein composition of Lp(a) is 65% apoB, 15%
albumin and an apoprotein designated as apoLp(a), which
is unique to Lp(a). The physiologic significance of
Lp(a) is not known. Its concentration (average
concentration: 14 mg per 100 ml of plasma) does not
appear to bear a relationship to age, sex, lipid levels
or coronary heart disease.

ii) Lipoprotein-X: Lp-X, an abnormal lipoprotein, is found
in patients with obstructive liver disease and in
individuals who lack the enzyme lecithin:cholesterol
acyltransferase (LCAT). Lp-X floats in the density
range of LDL and also exhibits the same electrophoretic
mobility as that of LDL. It is separated from LDL,
however, by hydroxyapatite chromatography or by zonal
centrifugation. The composition of Lp-X is different
from that of LDL, and it does not react with antisera
to LDL. The major apoproteins of Lp-X, isolated from
patients with LCAT deficiency, are albumin, apoC, and
apoA. It also contains small amounts of apoD and apoE.
It is of interest to point out that the Lp-X obtained
from patients with obstructive liver disease has been
reported to lack apoA-I, which is a powerful activator
of LCAT. The lipid constituents of Lp-X consist of
cholesterol (almost entirely in the <u>unesterified</u> form)
and phospholipids. The significance of these aspects
is made clear in the section on the metabolism of
lipoproteins. In electron microscopy the negatively

stained Lp-X preparations appear as stacks of disc-like
structures (rouleaux structures).

3. Molecular Organization of Lipoproteins

 a. Elucidating the spatial organization of lipids and proteins
    in lipoproteins involves results from low-angle x-ray
    scattering, nuclear magnetic resonance, calorimetry, electron
    microscopic observations, and lipid-apolipoprotein reconstitution
    experiments. The models that have been developed for the
    molecular architecture of lipoproteins are tentative and
    speculative.

 b. The model for the triglyceride-rich lipoprotein particles,
    chylomicrons and VLDL, consists of assigning the neutral
    lipids (triglycerides and cholesteryl esters) to the central
    core, which is surrounded by a layer of cholesterol,
    phospholipid and protein.

 c. The study of the molecular structure of LDL particles has
    been particularly difficult, because of its apoprotein (apoB)
    insolubility in an aqueous medium. This has prevented its
    full characterization. The suggested models for LDL include
    a bilayer or a trilayer structure. In the bilayer model,
    protein is present on both the inside and the outside of a
    phospholipid bilayer. The trilayer model consists of micelles
    of cholesteryl esters, cholesterol and triglycerides, with
    phospholipids projecting on both sides of the layer.

 d. Two major models have been suggested for HDL, the amphipathic
    helical model and the fluid mosaic model. In the former, the
    helical peptide segment is presumed to contain two sides, a
    polar or hydrophilic side and a nonpolar or hydrophobic side.
    These types of helices, called <u>amphipathic</u>, interact with the
    hydrophobic fatty-acyl chains of phospholipids at the nonpolar
    surface (with hydrophobic helical surfaces existing in
    parallel to the fatty acyl chains), while the polar surface
    of the helices interacts with the polar groups of phospholipids.
    Thus, in this model the apoproteins are organized into an
    amphipathic α-helical structure, with the hydrophilic surface
    interacting with the polar group of phospholipids, and a
    hydrophobic surface interacting with the fatty-acyl chains.
    The fluid mosaic model is very similar to the one described
    for membrane structure (see membrane structure, this chapter)
    in which the proteins are organized as "icebergs in a sea of
    lipid." It should be noted that in evolving an exact
    structure for HDL, these two models need not be mutually
    exclusive.

4. Metabolism of Lipoproteins

a. <u>General Comments</u>. The plasma lipoproteins are in a dynamic
state. They are continuously being synthesized and degraded
with rapid exchange of components (both lipid and protein)
among themselves. Two enzymes, lecithin:cholesterol
acyltransferase (LCAT) and lipoprotein lipase (also known as
triglyceride lipase), play a significant role in the catabolism
of the lipid fractions of lipoproteins. Some examples of lipid
and protein exchanges are: transfer of triglycerides from VLDL
to LDL and HDL, ready exchange of unesterified cholesterol
between various lipoproteins as well as membranes (in contrast,
cholesteryl esters exchange very slowly), phospholipid transfers
among lipoproteins and membranes, bidirectional transfer of
apoC between VLDL and HDL, and transfer of apoC from HDL to
chylomicrons. The major sites of plasma lipoprotein synthesis
are the intestine and liver. The synthesis takes place in
both rough and smooth endoplasmic reticulum. The components
that are necessary for the synthesis of lipoproteins are
triglycerides, cholesterol (and cholesteryl esters),
phospholipids, and apoproteins. The synthesis of each
component has already been described in the appropriate sections.

The fatty acids required for the synthesis of triglycerides
are derived from dietary lipids (transported in chylomicrons),
adipose tissue (regulated by hormone-sensitive, cAMP-mediated
triglyceride lipase--discussed later), and the liver (<u>de novo</u>
synthesis). A variety of factors (e.g., nutritional, hormonal)
regulate the availability of substrates for the synthesis of
lipid components. As noted previously, acetyl-CoA carboxylase
plays a major role in fatty acid synthesis. High dietary
lipid inhibits fatty acid synthesis, whereas a high carbohydrate
diet promotes fatty acid synthesis both in the liver and adipose
tissue. In the liver, whether the fatty acids are oxidized or
used for the synthesis of triglycerides presumably depends upon
the levels of the mitochondrial membrane enzyme, fatty acid-CoA-
carnityl transferase. Recall that this enzyme is essential
for the transport of fatty acyl-CoA into the mitochondria,
the major oxidation site of fatty acids. The fatty acyl-CoA
is diverted to triglyceride synthesis when the capacity of
this enzyme to transport fatty acyl-CoA across the mitochondrial
membrane is exceeded.

The major aspects of cholesterol metabolism consist of
absorption from the intestine, synthesis in the intestines and
liver, conversion to bile acids in the liver and excretion in
the bile, loss in the feces as neutral sterols and bile acids,
and transport from peripheral cells to liver and <u>vice versa</u>.
The synthesis of cholesterol is regulated by the <u>microsomal</u>

enzyme, HMG-CoA reductase. Cholesterol feeding has been shown
to inhibit the enzyme activity both by immediate inactivation
as well as long-term inhibition by affecting the synthesis of
new enzyme (see cholesterol biosynthesis, this chapter). In
the liver, the sterol or squalene carrier protein (SCP), which
transports water-insoluble intermediates in cholesterol
biosynthesis, has been postulated to play a role in lipoprotein
synthesis.

To a major extent, phospholipid synthesis and degradation
occurs in the liver. The synthesis takes place on the
endoplasmic reticulum; these aspects have been discussed
earlier.

b.   Chylomicrons. The major site of synthesis is the jejunal
mucosa (see lipid absorption for details). The synthesis
inside the cell occurs in the rough (apoprotein synthesis) and
smooth endoplasmic reticulum. For the synthesis and subsequent
transport of chylomicrons, apoprotein synthesis (e.g., apoB)
is essential. For example, subjects with an autosomal recessive
disorder known as abetalipoproteinemia (discussed later), who
lack the ability to synthesize apoB, have plasma which is
completely absent in chylomicrons, VLDL and LDL. Note that
the latter two lipoproteins also require apoB for their
synthesis and transport. The chylomicrons are released into
the intracellular space by reverse pinocytosis (exocytosis),
which consists of the fusion of chylomicron-containing vacuoles
with the plasma membrane at the lateral cell surface followed
by discharge of the chylomicrons. From the intercellular
spaces, chylomicrons enter the lymphatic system and eventually
gain access to the systemic blood circulation via the thoracic
duct.

The chylomicrons are rapidly cleared from the circulation with
a half-time of disappearance of less than 1 hour. They are
catabolized in the following steps. ApoC is transferred from
HDL to the nascent chylomicrons entering the circulation. In
particular, the transfer of apoC-II is important because it
activates lipoprotein lipase. The next step is the hydrolysis
of triglycerides by the action of apoC-II-activated lipoprotein
lipase. This process leads to the progressive delipidation of
chylomicrons. The free fatty acids and diglycerides are taken
up by the adipose tissue for further processing. The lipoprotein
lipase is located principally on the cells of the endothelial
surface (i.e., cells lining the blood capillaries) of muscle
and adipose tissue. Normally, this enzyme activity is not
detectable to any appreciable extent, but it is increased
significantly after intravenous administration of heparin
(known as post-heparin lipolytic activity--PHLA). The

presence of additional lipoprotein lipase has been detected
in normal post-heparin plasma and its function is not clearly
understood. These enzymes are an apoC-I-activated lipase
and a hepatic lipase. In a clinical disorder known as
familial-hyperchylomicronemia (type-I hyperlipoproteinemia),
the PHLA (C-II-activated adipose tissue lipase; deficient
activity is not due to an absence of apoC-II) is totally
deficient. The affected individuals have very high levels
of plasma triglycerides, most of which are in the chylomicrons.

The phospholipids of chylomicrons are presumably degraded by
both lipoprotein lipase and LCAT. During the catabolism of
chylomicrons, part of the apoC is retransferred to HDL and
the remnant chylomicron is presumably removed by the liver
for further catabolism. These aspects are shown in Figure 129.

c. The metabolism of very low density lipoproteins (VLDL) and
the formation of low-density lipoproteins (LDL). VLDL, the
carrier of endogenous triglycerides, is synthesized both in
the liver and intestine. It is secreted by the hepatic cells
by reverse pinocytosis into the hepatic sinusoids via the
space of Disse (see Chapter 15). In a series of steps, VLDL
is eventually converted to LDL and, thus, virtually all plasma
LDL is derived from VLDL. The sequence of events of stepwise
delipidation of VLDL leading to the synthesis of LDL is as
follows. Initially, like chylomicrons, apoB-containing,
triglyceride-rich particles are secreted. These particles
acquire apoC and apoE from HDL. Apparently, there is also
reversible exchange of cholesterol and phospholipids between
VLDL and HDL. The acquisition of apoC-II prepares the VLDL
for stepwise delipidation by lipoprotein lipase. During the
process of catabolism, VLDL gradually loses triglycerides and
apoC, and the ester content of cholesterol increases. The
source of cholesterol appears to be HDL, and the esterification
is mediated by LCAT, which catalyzes the transfer of a fatty
acid from the 2-position of lecithin to cholesterol. Note
that the apoA-I of HDL is an activator of LCAT. Thus, VLDL
is converted to LDL, which is rich in cholesteryl esters and
apoB. On the basis of studies using labelled VLDL particles,
the precursor-product relationship of VLDL and LDL has been
established. The conversion of VLDL to LDL results in an
increase in the hydrated density. An intermediate density
lipoprotein (IDL) has been characterized. The steps involved
in the conversion of VLDL to LDL are shown in Figure 130. A
metabolic defect involved in the catabolism of VLDL is presumed
to be the cause of primary familial dysbetalipoproteinemia
(type III hyperlipoproteinemia--discussed later). The abnormal
VLDL, which accumulates in this disorder, has reduced
electrophoretic mobility (discussed earlier). Its apoprotein

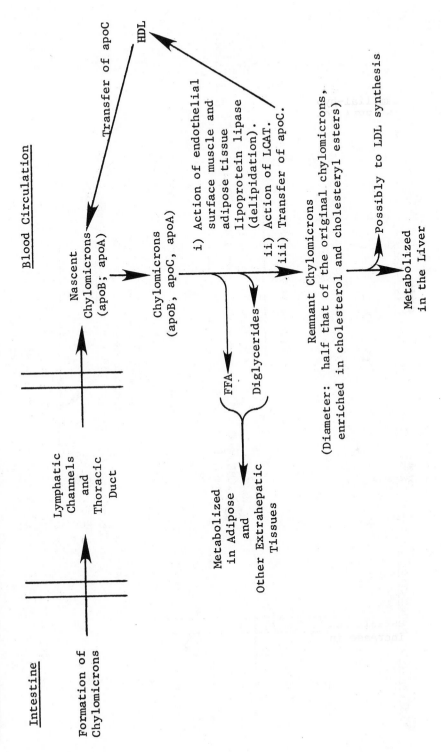

Figure 129. Steps Involved in the Metabolism of Chylomicrons

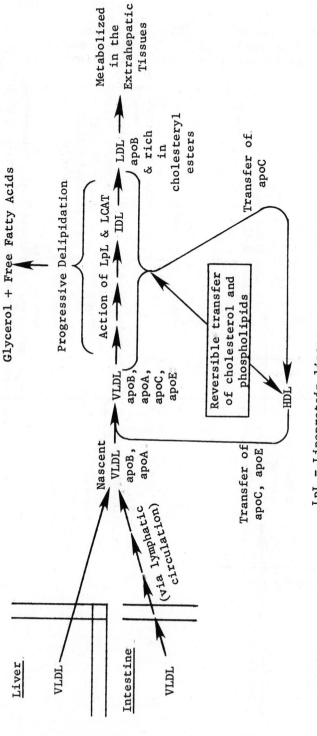

Figure 130. Schematic Representation of the Conversion of VLDL to LDL

composition consists of a lowered amount of apoC, and while the total amount of apoE is increased, one of the apoE components (apoE-III) is grossly deficient or absent. It should be noted that the pathway for the conversion of IDL to LDL is not fully understood at this time. For example, it is not certain in humans whether or not the liver plays any role in this conversion.

d. <u>Catabolism of LDL and the role of HDL</u>. Recent studies using hepatectomized animals have shown that the catabolism of LDL occurs in <u>extrahepatic</u> cells (e.g., arterial smooth muscle cells, lymphoid cells, fibroblasts, endothelial cells, etc.). In addition, a recent subject of extensive investigation (by Goldstein and Brown) has been the regulation of LDL catabolism (and hence cholesterol metabolism) in cells grown in tissue culture.

The metabolic events involved in the LDL catabolism as shown in tissue culture cells (e.g., human fibroblasts) are as follows. Note the <u>nonhepatic cells</u> acquire cholesterol used for the membrane synthesis and other metabolic reactions requiring steroid nucleus, from LDL.

i. Binding of LDL with a high affinity receptor located on the cell membrane. Plasma LDL binds specifically with <u>high affinity</u> receptors located on the cell membrane. The receptor is a protein and it presumably is a primary product of a single genetic locus. This hypothesis is supported in part by the observation that the fibroblasts obtained from patients with homozygous familial hypercholesterolemia lack LDL receptors and in heterozygotes the receptors are reduced by about 50%. The specificity of the binding of LDL to the receptor protein appears to reside with the apoB component of LDL. A lipoprotein, designated HDL_c which appears in the plasma of swine when fed with diets rich in cholesterol, binds with LDL receptors despite the fact that it does not contain any detectable apoB. However, it does contain a large amount of apoE and it is speculated that the apoE and apoB contain homologous peptide regions accounting for their similar function.

The LDL receptors on the cell membrane vary in number depending upon the needs of the cell for cholesterol and thereby regulating the entry of LDL (and hence cholesterol). Electron microscopic studies using ferritin-labelled LDL as probes has shown that the receptors are concentrated in the "coated regions" (constituting less than 2% of the total membrane surface) of plasma membrane of human fibroblasts.

ii. Internalization of LDL and its catabolism. The LDL particle bound to specific receptors enters the cell by endocytosis and it fuses with lysosomes. In the secondary lysosomes, the lysosomal hydrolases (e.g., acid lipase, protease) degrades the LDL to their monomeric constituent in the acidic environment and releases the hydrolyzed products (e.g., amino acids, cholesterol) into the cytoplasm. In particular, the cholesterol is incorporated into an intracellular pool where it has the following functions. The cell preferentially utilizes cholesterol contained in this pool for membrane synthesis and other synthetic reactions which require a sterol nucleus (e.g., steroid hormones). It suppresses the cellular cholesterol synthesis by inhibiting the HMG-CoA reductase enzyme, the rate-limiting enzyme of the cholesterol biosynthesis (see cholesterol biosynthesis, this chapter). As noted earlier, the cholesterol derived from LDL also regulates the number of receptor sites on the cell membrane and thereby regulates the cellular uptake of LDL (and therefore cholesterol). The unused (or excess) cholesterol in the cell is re-esterified (with oleate and palmitoleate) and stored as cholesteryl esters. This reaction is catalyzed by the microsomal enzyme, acyl-CoA:cholesterol acyltransferase (ACAT), and its activity is stimulated by the LDL-derived cholesterol. Note that the cholesteryl esters contained in the LDL are esters of linoleate whereas the esters formed in the cell catalyzed by the enzyme ACAT contain predominantly oleate and palmitoleate (monosaturated fatty acids). From the above, it should be noted that the receptor-mediated, LDL-derived cholesterol not only meets the requirements of the cell's cholesterol needs but also prevents its overaccumulation by inhibiting the de novo synthesis, suppressing the further entry of LDL and storing the unused cholesterol as cholesteryl esters. Despite the elaborate regulatory system, the cells do accumulate an excessive amount of cholesteryl esters, when the plasma levels of LDL exceeds saturating the high affinity receptor-mediated LDL uptake process. When this occurs, the LDL enters the cell by a nonspecific phagocytic process known as "bulk-phase pinocytosis." The LDL which enters by this mechanism apparently plays no role in the regulation of the de novo synthesis of cholesterol and thus leads to its excessive accumulation with pathologic consequences (e.g., atherosclerosis, discussed later). It should be noted that at low levels of LDL, the high affinity receptor process of LDL internalization predominates and only when the LDL receptors are saturated will the bulk-phase pinocytic process become operative in the LDL uptake.

e. Support for the above pathway for LDL catabolism comes from
    studies using fibroblasts from patients with genetic defects
    in lipid metabolism. These disorders, some of which are
    discussed in greater detail under coronary heart disease,
    are listed below with some of their characteristics.

    i.   Receptor-deficient disorder--familial hypercholesterolemia:
         LDL receptors are absent; the plasma level of LDL are very
         high (6 to 10 times above normal); the de novo synthesis
         of cholesterol in the peripheral cells proceeds unabated;
         the bulk-phase pinocytosis predominates; excessive
         accumulation of cholesteryl esters occurs and eventually
         leads to severe atherosclerosis with myocardial infarction
         occurring at a young age. The heterozygotes (who have
         receptors of about 50% normal) suffer from a less severe
         disease; the LDL levels are elevated about two- to
         four-fold above the normal level and are susceptible to
         coronary heart disease during the ages of 30 to 60 years.

    ii. A receptor-defective disorder: The fibroblasts obtained
         from these patients contain about 5-20% receptors as
         compared to normal. The characteristics of homozygotes
         and heterozygotes for this disorder are comparable with
         the homozygous and heterozygous patient of familial
         hypercholesterolemia.

    iii. LDL-internalization defect: Fibroblasts from these
         patients bind LDL but are unable to internalize it;
         the clinical manifestations are similar to patients
         with receptor-negative and receptor-defective disorders.

    iv. Complete absence of lysosomal acid lipase--Wolman
         Syndrome: The absence of acid lipase, which hydrolyzes
         cholesteryl esters and triglycerides, leads to the
         massive accumulation of cholesteryl esters and
         triglycerides in almost all cells (a lipid storage
         disease); the disease is fatal usually within the first
         year of life; autosomal recessive disease.

    v.   Reduced levels of lipoprotein lipase--known as
         cholesterol ester storage disease: In this disorder
         the acid lipase is reduced to about 1-5% of normal;
         this storage disease is relatively benign compared with
         Wolman Syndrome, a fatal lipid storage disease; the
         patients reach young adulthood and then death ensues;
         the disease is characterized by hepatosplenomegaly and
         hyperlipidemia (also present in type I-glycogen storage
         disease; therefore specific diagnosis requires the
         enzyme assay of peripheral blood leukocytes or cultured

skin fibroblasts; also see glycogen storage diseases, Chapter 4); the disorder is expressed as an autosomal recessive disease.

vi. Apoprotein B deficiency--known as abetalipoproteinemia: The apoB deficiency leads to the lack of VLDL and LDL in the plasma; the plasma cholesterol is mainly bound to HDL and this particle is not taken up by the cells; therefore the HMG-CoA reductase step is not regulated; an autosomal recessive disease. A schematic representation of LDL catabolism with some of its biochemical lesions is shown in Figure 131.

f. <u>Metabolism of high-density lipoprotein (HDL)</u>. HDL is synthesized in both the liver and the intestine. As previously discussed the functions of HDL include the source of apoC and apoE for the metabolism of triglyceride-rich lipoproteins, namely chylomicrons and VLDL; the substrate for the esterification of cholesterol catalyzed by LCAT; and the vehicle for transporting cholesterol from peripheral tissues to the liver. The role of HDL in the transport of cholesterol from the peripheral tissues to the liver is supported by the observations made in patients with familial HDL deficiency known as <u>Tangier disease</u> who show abnormal tissue deposits of cholesteryl esters. The plasma of these patients show markedly reduced levels of HDL (and LDL). The HDL exhibits an abnormal apoprotein composition with a apoA-I:apoA-II ratio of 1:12 (the normal ratio is about 3:1) and the HDL has been designated HDL_T. The molecular defect has not yet been elucidated. This is a rare autosomal recessive disorder and clinical manifestations include hepatosplenomegaly and enlarged yellowish-appearing tonsils (a pathognomonic feature of this disorder). The role of HDL in the cholesterol metabolism has recently received a considerable amount of attention. Elevated total plasma cholesterol has clearly been established as a major risk factor for coronary heart disease (CHD). However, recent retrospective analysis of epidemiological data of CHD has shown a <u>negative correlation</u> with HDL cholesterol, and the deficiency of HDL is considered as an independent risk factor. A working hypothesis for the negative correlation of HDL-cholesterol with CHD is that the lowered HDL levels (and hence lowered HDL cholesterol) implies an impaired or less effective removal of cholesterol from the peripheral tissues. In addition it has been shown that in cells grown in (fibroblasts, smooth muscle and endothelial cells) tissue culture, the HDL competitively inhibits the LDL binding. These observations suggest that the increased HDL prevents the cellular uptake of LDL whereas decreased HDL promotes LDL uptake. Further studies on the interrelationships between the major classes

Figure 131. Catabolism of LDL and Associated Biochemical Lesions

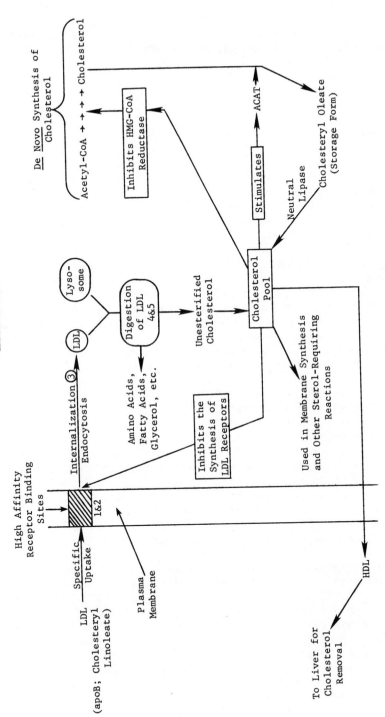

HMG-CoA = β-Hydroxy-β-methylglutaryl Coenzyme A reductase--Rate limiting enzyme of cholesterol biosynthesis
ACAT = Acyl-CoA:Cholesterol Acyltransferase
Biochemical Lesions (see text for details)
   1&2 = Receptor negative and defective disorders (familial hypercholesterolemia)
      3 = Internalization defect
      4 = Complete deficiency of acid lipase (Wolman Syndrome)
      5 = Partial acid lipase deficiency--cholesterol ester storage disease

[949]

of lipoproteins and the function of individual lipoproteins are required to elucidate the mechanisms involved in the development of premature heart disease.

The fate of HDL in the human circulation is not known and in rats the major site for catabolism is the liver.

5. Clinical Disorders of Lipoprotein Metabolism

These disorders consist of both hypo- and hyperlipidemias. Some aspects of these disorders have already been discussed. There are three inherited types of hypolipoproteinemia, all of which are usually detected initially by the presence of hypocholesterolemia. Their differentiation is based on a more detailed analysis of the serum lipoprotein pattern.

a. Tangier disease (familial HDL deficiency; α-lipoprotein deficiency) is characterized by hypocholesterolemia and high or (occasionally) normal serum triglyceride. Normal levels of HDL are markedly reduced from the serum, and a small amount of abnormal HDL (designated HDL_T) containing decreased quantities of apoA-I and an excessive amount of apoA-II is present instead. As noted previously, the ratio of apoA-I to apoA-II is 1:12 in these complexes, whereas the value in normal HDL is 3:1. The metabolic defect in this disorder appears to be one of catabolism rather than anabolism. The kinetic analysis of radiolabelled HDL administered to homozygously and heterozygously affected individuals has yielded results consistent with an enhanced rate of catabolism of HDL and with a relatively normal rate of synthesis of apoA-I and apoA-II. Clinical manifestations of this disorder include cholesteryl ester deposition in reticuloendothelial tissues (notably resulting in hepatosplenomegaly and yellow-orange, enlarged tonsils) and, in a number of patients, neurological abnormalities. The mode of inheritance is autosomal recessive. The disease does not appear to be generally fatal.

b. In abetalipoproteinemia (acanthocytosis; Bassen-Kornzweig syndrome), as the name implies, lipoproteins containing apoB (namely LDL, VLDL and chylomicrons) are absent. Thus, as noted earlier, this experiment of nature points to the fact that apoB is absolutely essential for the formation of chylomicrons and VLDL (and hence LDL). The important clinical findings are retinitis pigmentosa, malabsorption of fat and ataxic neuropathic disease. The erythrocytes are distorted with a "thorny" appearance due to protoplasmic projections of varying sizes and shapes (hence the term, acanthocytosis, from the Greek akantha, meaning thorn). Serum triglyceride levels are low. The genetic defect is probably one which completely shuts

off apoB synthesis, although other possibilities do exist.
It is transmitted as an autosomal recessive characteristic,
and the heterozygotes cannot be detected without an affected
offspring.

c. In hypobetalipoproteinemia, unlike abetalipoproteinemia, LDL
    is low (10-20% of normal) but not absent, HDL is normal, and
    VLDL is mildly lowered. In twenty-three affected individuals
    from the four known affected families, only one subject had
    central nervous system dysfunction and fat malabsorption.
    The other twenty-two known cases had only mild pathologic
    changes, if any were present at all. The symptoms and signs
    of abetalipoproeinemia are generally not shown in this
    disorder. The disease is inherited as an autosomal dominant
    trait. It is interesting to note the benignity of this
    reduction in β-lipoprotein compared to the relative seriousness
    of hyperbetalipoproteinemia. In the latter, LDL concentrations
    are two to six times normal, and such persons are predisposed
    to premature atherosclerosis.

6. Hyperlipidemias and Hyperlipoproteinemias

    a. The hyperlipidemias and hyperlipoproteinemias are, quantitatively,
        a much more significant group of lipoprotein disorders than the
        hypolipoproteinemias. The two are considered together because,
        as was pointed out earlier, all serum lipids are present as
        lipoproteins. Consequently, an increase in serum lipids
        (hyperlipidemia) must be accompanied by an elevation of the
        proteins needed to carry the lipid (hyperlipoproteinemia).
        There may be exceptions to this but certainly, at present, the
        most rational approach to these disorders is to treat them as
        being synonymous. It should be noted that while hyperlipidemia
        and hyperlipoproteinemia are considered synonymous, the term
        hyperlipemia is used to characterize the lactescence (milky
        appearance) in serum or plasma obtained from subjects in the
        post-absorptive state. The lactescence in the serum is due to
        increased concentrations of triglyceride-rich lipoproteins,
        namely chylomicrons and VLDL. Values of serum lipid levels
        are useful, however, for aiding in the evaluation of the
        lipoproteins which are abnormal and in the diagnosis of these
        diseases. This is especially true with respect to determinations
        of serum triglyceride and cholesterol levels as will be seen
        later. Most hyperlipidemias are initially detected by
        observation of elevated triglyceride or cholesterol, or both.

    b. The detection of hyperlipidemias is important because numerous
        studies have shown a correlation between elevated levels of
        lipids (in particular, cholesterol) and the development of
        coronary heart disease. Persons with hyperlipidemias also run

an increased risk of contracting atherosclerosis, regardless of the cause of the elevated lipid levels. Detection of the hyperlipidemia is only the beginning, however. In order to diagnose the specific disorder and to prescribe the most efficacious treatment to correct it, the hyperlipidemias have been divided into five (six, counting two subclasses of one type) classes or phenotypes, and a systematic procedure has been developed to aid in the proper classification of hyperlipidemic patients.

c. Although the term "phenotyping" implies a genetic origin of these diseases, hyperlipidemia can also be caused by other things. For example, hyperlipidemias are secondary to certain other diseases, including hypothyroidism, uncontrolled insulin-dependent diabetes mellitus, pancreatitis, nephrotic syndrome, dysglobulinemia, and biliary obstruction. Some of the secondary causes for each hyperlipidemia are indicated later in the descriptions of the phenotypes. Environmental factors, such as diet, alcohol consumption, and the use of certain drugs (e.g., estrogens, notably in birth control pills) can also result in elevated serum lipids. Thus, in addition to the phenotype, it is extremely important to consider the etiology of each hyperlipidemia. One factor which is useful in establishing the presence of a primary (genetic) hyperlipidemia is the identification of the same disorder in other members of the family. It should be emphasized that some of the phenotypes are genetically heterogeneous; furthermore, a given phenotype very rarely defines a specific genetic disorder (see Table 42). It is interesting that, unlike the hypolipoproteinemias, no abnormal apolipoproteins have been implicated as causative agents in the hyperlipoproteinemias, with the exception of apoE-III deficiency in subjects with primary dysbetalipoproteinemia (type III hyperlipoproteinemia--discussed later).

d. Screening for hyperlipidemia and phenotyping. Although the values of serum cholesterol and triglyceride which occur in a patient under his normal living conditions are useful in diagnosis, accurate phenotyping requires a standard "preparation" prior to the taking of the blood sample to be used.

   i) The individual should consume a normal diet, with no gain or loss of weight for the two weeks prior to sample collection. No medications which are known to affect plasma lipid or lipoprotein concentration should be administered to the patient during this period.

   ii) Blood should be taken 12-14 hours postprandially (after a meal) and the serum should be separated promptly.

Plasma can be used but a correction must be applied to account for the dilution caused by the addition of anticoagulants. The terms plasma and serum will be used interchangeably in this section. The plasma or serum sample should not be frozen. If normal plasma, collected as described above, is stored at 4°C for 18 hours, it should show no chylomicrons. However, the occurrence of a creamy, upper layer (supernatant) and/or uniform turbidity throughout the plasma indicates the presence of chylomicrons and VLDL due to an abnormality.

iii) The accurate measurement of serum cholesterol and serum triglycerides has already been discussed. The normal values for cholesterol vary with age, sex and diet. While the "normal" for one population may be associated with, say, a high risk of coronary disease, the "normal" for another population may be quite benign. This difference between the statistical normal for a population and the biological normal (i.e., that associated with good health) is important to recognize. Furthermore, the cholesterol values are continuously distributed in the population, and the values overlap between "normal" and abnormal" subjects. Thus, the normal values cannot be derived by calculating the mean and 2 standard deviations. In a given population, the cholesterol values falling in the upper 5 or 10 percent are considered abnormal. The designation hypercholesterolemia is, therefore, somewhat arbitrary. The relationship between plasma cholesterol and the development of atherosclerosis is a graded relationship, and no clearcut cutoff values have been established which can protect against coronary heart disease. The values below should be viewed as broad guidelines.

	Cholesterol (mg/100 ml)	Triglyceride (mg/100 ml)
Infants	45-170 (<10 at birth)	10-140
Children	less than 180	10-140
Adults	130-250	10-190 *

*In one recent study values greater than 250 mg/100 ml were considered abnormal.

iv) Post heparin lipolytic activity (PHLA) is a measure of the ability of the body to release lipoprotein lipase

in response to intravenous heparin administration.
Recall that this enzyme is synthesized within the major
cell types (e.g., muscle, adipocytes) and located on the
endothelial surface of the capillaries supplying the
tissues. The enzyme is responsible for the removal of
triglycerides as free fatty acids and glycerol from
chylomicrons (and VLDL) present in the blood perfusing
the tissue. A major portion of the free fatty acid
is taken up by the tissue cells where they are released.
The test is performed simply by doing two serum lipase
assays, one before and the other following heparin
administration. PHLA can also be qualitatively
assessed by noting the disappearance of chylomicrons in
the plasma ten minutes after the intravenous
administration of sodium heparin (10U/Kg body weight).

v)   The actual evaluation of the lipoprotein pattern is
usually done by electrophoresis although phenotyping
by this technique alone is not feasible. The analytic
ultracentrifuge can also be used for this purpose and
results from it are generally more accurate than those
obtained by electrophoresis. The ultracentrifuge is an
expensive instrument, however, and is therefore not
economically feasible for most laboratories to use.
From a practical standpoint, the phenotyping of
hyperlipidemic conditions are necessary in order to
place the patients into various major therapeutic groups.
For this purpose, all that is required is the measurement
of cholesterol and triglyceride in a serum obtained from
a 12-14 h fasting patient and the examination of
refrigerated serum for chylomicrons. Therefore for
screening purposes, lipoprotein electrophoresis is not
required.

vi) Other techniques which may be useful in lipoprotein
analysis include selective precipitation (heparin,
polyvinylpyrrolidine, dextran sulfate, etc.),
nephelometry (large molecules and particles, such as
chylomicrons and VLDL scatter light), and immunochemistry.
It is frequently helpful to use more than one method in
evaluating abnormal conditions.

7. Classification and Description of Phenotypes

The classification system for hyperlipoproteinemias was developed
by Fredrickson and his coworkers. The stepwise procedure for
phenotyping using methods described in the preceding section is
summarized in Figure 132 and described below.

Figure 132. Algorithm for the Differentiation of Lipoprotein Phenotypes*

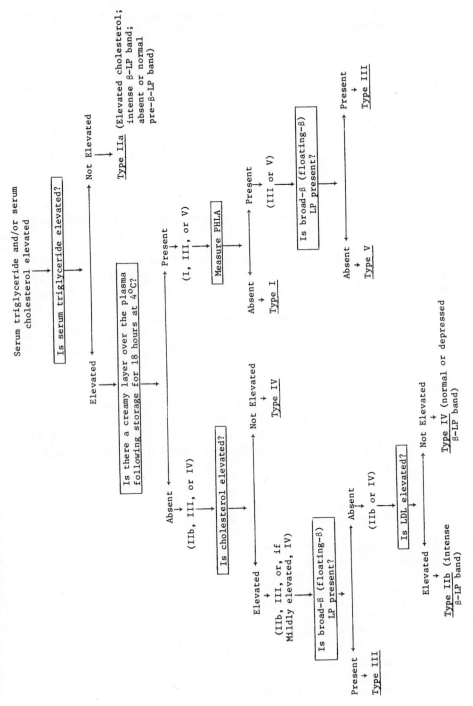

PHLA = postheparin lipolytic activity

*Modified from A.M. Gotto and L. Scott, J. Am. Dietetic Association, 62, 617, (1973), with permission.

a. <u>Type I</u> (familial fat-induced hyperlipemia; hyperchylomicronemia). Characteristics are:

   i) Plasma, collected 14 hours postprandially from a patient maintained on a normal diet, shows a creamy supernatant phase (due to excessive chylomicrons) following storage at $4^{\circ}C$ for 18 hours.

  ii) Triglyceride levels are increased with a modest rise in cholesterol, resulting in a ratio of cholesterol/triglyceride of less than 0.2. When the ratio declines to less than 0.1, it is characteristic of <u>only</u> Type I.

 iii) The electrophoretic pattern shows the presence of a thick chylomicron band at the origin, reduced beta and alpha bands, and the absence of a pre-beta band.

  iv) Clinical manifestations of this type include eruptive xanthomas, colic (abdominal pain), hepatosplenomegaly and lipemia retinalis, all of which appear at an early age.

   v) Type I is rare and is expressed as an autosomal recessive trait. Secondary causes include uncontrolled diabetes mellitus, acute alcoholism, dysglobulinemia, pancreatitis, hypothyroidism, and the use of oral contraceptives.

  vi) The biochemical lesion appears to be an absence of post-heparin lipolytic activity (PHLA). To confirm this type, it is essential to measure PHLA which is reduced in these individuals. This lipase aids in clearing the chylomicrons from the plasma. PHLA is normal in all of the rest of the types.

 vii) This type responds to low-fat diets supplemented with medium chain triglycerides. No particular drug therapy has been found to be of any value.

b. <u>Type II</u> (familial hyperbetalipoproteinemia; familial hypercholesterolemia).

   i) This is the most common phenotype, and it is subdivided into types IIa and IIb. Both show increased β-LP (LDL) while, in addition, IIb exhibits an elevated pre-β (VLDL) concentration which does not occur in IIa. Both phenotypes can occur in several forms of primary (genetic) hyperlipidemias (discussed later). Hypothyroidism, porphyria, dysglobulinemia, biliary obstruction, and dietary imprudence are other causes of this disorder which must be ruled out, in establishing a primary disorder.

ii) The clinical manifestations include xanthelasma, tendon and tuberous xanthomas, corneal arcus (juvenile), and premature and accelerated atherosclerosis. The treatment consists of restricted intake of cholesterol (less than 300 mg/day) and saturated fats, enhanced intake of polyunsaturated fats, and regulation of body weight by controlling caloric intake. Drugs such as cholestyramine, dextrothyroxine (D-thyroxine), and nicotinic acid are also used when needed.

iii) Characteristics of Type IIa:

    (a) Plasma maintained at $4^{\circ}C$ for 18 hours is clear (no visible chylomicrons).

    (b) Total plasma cholesterol values are elevated in association with a normal triglyceride level and a ratio of cholesterol/triglyceride greater than 1.5.

    (c) In this phenotype, the excess cholesterol is associated with the LDL fraction. In other situations, it is possible to have elevated total plasma cholesterol without an increase in the LDL cholesterol. For example, the total plasma cholesterol is elevated in women taking oral contraceptives but the cholesterol is associated with the HDL rather than the LDL fraction. An estimate of LDL cholesterol can be made using the relationship:

LDL-Cholesterol = Plasma Cholesterol

$$- \left[\text{VLDL-Cholesterol} + \text{HDL-Cholesterol}\right]$$

All values are expressed as mg/100 ml. The HDL-cholesterol is estimated by analysis of the supernatant obtained from the precipitation of the non-HDL lipoproteins with the heparin-Mn^{2+} or phosphotungastate-Mg^{2+} complex. The VLDL-cholesterol is calculated by dividing the plasma triglyceride by five. However, this latter approximation to obtain values for cholesterol associated with VLDL is only valid when serum triglyceride values are less than 400 mg/100 ml, no chylomicrons are present, and familial dyslipoproteinemia (phenotype

III) has been excluded. In the latter disorder the abnormal VLDL ratio for cholesterol:triglyceride is 1:1 instead of the normal value of about 1:5. Therefore,

LDL-Cholesterol = Plasma Cholesterol

$$- \left[\frac{\text{Plasma Triglyceride}}{5} + \text{HDL-Cholesterol} \right]$$

This formula should only be used when plasma triglyceride is less than 400 mg/100 ml and following exclusion of type III hyperlipoproteinemia as a possible cause. The upper limits for normal LDL cholesterol values (in mg/100 ml) are: 150 (0-29 years of age); 170 (30-49 years of age); and 190 (50-69 years of age).

(d) The electrophoretic pattern of type IIa plasma shows an absent or normal (not elevated) pre-β band. The β-band is heavily stained, chylomicrons are not visible, and the α-lipoprotein is normal.

iv) Characteristics of Type IIb:

(a) On maintaining plasma from a type IIb patient at 4°C for 18 hours, it is either clear or uniformly turbid (lipemic), in contrast to the clarity always seen with type IIa plasma treated similarly. No creamy supernatant of chylomicrons is present.

(b) Plasma triglyceride and LDL cholesterol are always increased and total plasma cholesterol is usually elevated.

(c) Both the beta- and pre-beta-lipoprotein electrophoretic bands are increased in intensity, while the α-lipoprotein band is usually normal and chylomicrons are not visible.

c. Type III (floating-β-lipoprotein; broad-β-lipoprotein; familial dysbetalipoproteinemia).

i) Plasma maintained at 4°C for 18 hours is usually turbid with a creamy layer above the plasma.

ii) Both plasma triglyceride and cholesterol values are nearly always elevated. The cholesterol/triglyceride ratio is most commonly around 1 but it can vary from 0.3 to greater than 2.

iii) In paper, agarose, or cellulose acetate electrophoresis, the plasma reveals a broad beta band encompassing both the beta and pre-beta regions. Sometimes a distinct pre-β band of abnormal intensity is seen. A faint chylomicron band is also frequently observed, even in a fasting sample. The α-lipoproteins are generally normal. Certain other aspects of the lipoprotein pattern are visible when polyacrylamide and starch are used as the electrophoretic support. This is helpful in the diagnosis of this disease.

iv) The definitive diagnosis of type III is made by demonstrating the presence of an abnormal lipoprotein with a density of less than 1.006 g/ml ("floating-β" lipoprotein). A decrease in one LDL subclass (density 0.010–0.163, S_f = 0–12) is seen with an increase in VLDL (100–400) and of the LDL having S_f = 12–20. The low density fraction also contains an unusually large amount of cholesterol. Recent studies have shown that the abnormal cholesterol-rich VLDL present in this disorder contains reduced amounts of apoC and increased amounts of apoE (values exceeding 40 mg/100 ml considered as diagnostic). Further, it has been shown that while the total apoE is increased, one of the polymorphic forms of apoE (apoE-III) is grossly deficient or absent.

v) Primary type III is relatively rare and is presumably inherited as an autosomal recessive trait. Secondary causes of type III are rare but myxedema and dysglobulinemia must be excluded. The biochemical lesion of this disorder has not yet been identified. The clinical manifestations of this type are tendon and palmar xanthomas, accelerated atherosclerosis of the coronary and peripheral vessels, and abnormal glucose tolerance curves.

vi) The treatment includes weight control by means of a restricted diet which is low in carbohydrate and cholesterol and which contains increased amounts of polyunsaturated fat. The drugs used in the treatment include clofibrate, D-thyroxine, and nicotinic acid.

d. Type IV (hyper-pre-beta-lipoproteinemia).

i) Plasma maintained at 4°C for 18 hours is either clear or turbid throughout, but no creamy chylomicron layer is present.

ii) Plasma cholesterol is normal or slightly elevated. Triglycerides are elevated.

iii) On electrophoresis, the plasma reveals a normal beta band and a pre-beta band with an increased intensity. The alpha-band appear normal, and there is an absence of a chylomicron band.

iv) Type IV is a common phenotype and is expressed in two of the genetic hyperlipidemia, namely familial hypertriglyceridemia and familial combined hyperlipidemia (discussed later). The triglyceridemia is frequently associated with an abnormal glucose tolerance. Estrogens, either alone or in oral contraceptives, seem to often cause triglyceridemia.

The lipoprotein abnormalities seen in this phenotype can also be due to a number of secondary causes, including pancreatitis, hypothyroidism, the nephrotic syndrome, alcoholism, diabetes mellitus, glycogen storage disease, dysglobulinemia, progestin administration, Werner's syndrome, Gaucher's disease, Niemann-Pick disease, and others.

Xanthomatosis and hepatosplenomegaly are characteristic clinical signs of this phenotype. Type IV hyperlipoproteinemia also carries with it a high risk of developing coronary disease.

v) The biochemical lesions leading to the hypertriglyceridemia are not clear although defects in the endogenous synthesis and clearance of triglyceride-rich particles are possible causes.

vi) The management of this type involves attaining an ideal body weight, the restriction of carbohydrates, and enhanced intake of polyunsaturated fats. Clofibrate and nicotinic acid are drugs which are sometimes useful in treatment.

e. Type V (familial hyper pre-beta-lipoproteinemia with hyperchylomicronemia).

i) When a plasma sample is maintained at $4^{\circ}C$ for 18 hours, it shows chylomicron as a creamy supernatant layer. The underlying plasma is turbid. This plasma turbidity distinguishes type V from type I.

ii) Elevated levels of both cholesterol and triglycerides occur. The plasma cholesterol/triglyceride ratio is usually greater than 0.15 and less than 0.6. The lighter and larger lipoprotein particles are increased (S_f 20-400; and S_f > 400).

iii) On electrophoresis, the pattern reveals a chylomicron band (at the point of application) and an intense pre-beta band while the α and β bands are generally decreased. This type does not show the presence of a "broad-beta" band encompassing both the beta and pre-beta regions as is observed in type III. This can be confirmed by the absence of any "floating-β" lipoprotein upon ultracentrifugation. Type V and type I are also distinguished from each other by the normal PHLA observed in the former type compared to the absence of PHLA in type I.

iv) The clinical manifestations of this type are abdominal pain, eruptive xanthomas, hepatosplenomegaly, hyperglycemia, hyperuricemia, and an abnormal glucose tolerance profile.

v) The disorders which can show the characteristics of type V include chronic pancreatitis, severe diabetes mellitus (insulin dependent), alcoholism, nephrosis, myeloma, and dysglobulinemia. In subjects with phenotype V, when no underlying cause is found, the disorder is considered as a primary hyperlipoproteinemia. Among the propositi, the prevalence of hyperuricemia, diabetes mellitus, pancreatitis and xanthomatosis are high. It is of interest to point out that surprisingly the prevalence of coronary heart disease is not significantly enhanced in the propositi and their relatives of type V disorder, despite the presence of highly elevated plasma triglyceride levels. This suggests that hypertriglyceridemia may not be an independent risk factor in the development of coronary heart disease. In type V, the lighter and the larger lipoproteins are increased, and they presumably have reduced atherogenic potential when compared with heavier and smaller lipoproteins that accumulate in types II and III disorders.

vi) The management of type V involves attaining an optimal body weight, enhanced intake of protein, and diminished intake of fat (less than 70 grams per day) and carbohydrate. Drugs used are nicotinic acid and clofibrate.

8. Genetic Hyperlipidemias

Six different entities have been identified. Their clinical
characteristics, risk factor for premature cardiovascular disease,
mode of inheritance, relationship to various phenotypes, and other
properties are summarized in Table 42. The familial lipoprotein
lipase deficiency and familial dysbetalipoproteinemia (broad beta
disease) have already been discussed (see phenotypes I and III
respectively). Following are brief descriptions of some of the
genetic hyperlipidemias.

a. Familial hypercholesterolemia. This disorder is most
extensively investigated. It is usually expressed as
phenotype IIa but rarely as IIb. The clinical manifestations
include tendon and tuberous xanthomas, corneal arcus and
xanthalasma. It is inherited as an autosomal dominant disorder
and shows increased incidence of premature cardiovascular
disease. The plasma cholesterol levels are usually greater
than 350 mg/100 ml (however, a normal level does not rule out
the primary disease), and, in particular, the cholesterol
associated with the LDL-fraction is elevated. The biochemical
lesions include deficient or defective high-affinity LDL
receptors and internalization of LDL in nonhepatic peripheral
cells (discussed earlier).

b. Polygenic hypercholesterolemia. The subjects in this group
show hypercholesterolemia and/or hypertriglyceridemia.
Multiple gene defects appear to be involved, and often it is
difficult to dissociate the influence of genetic and
environmental factors on the hyperlipidemia.

c. Familial hypertriglyceridemia. Expressed as phenotype IV
(and rarely as phenotype V), characterized by an elevated
triglyceride in VLDL and inherited as an autosomal dominant
trait. The disorder is often associated with mild diabetes
mellitus, obesity and hyperinsulinemia. The biochemical
lesion (or lesions) is not known.

d. Familial combined hyperlipidemia. The phenotypes observed
in this disorder are IIa, IIb, IV and V. The clinical
syndrome exhibits pleomorphic manifestations. Expressed as
an autosomal dominant trait.

9. Lecithin:Cholesterol Acyltransferase (LCAT) Deficiency

a. The LCAT catalyzes the following reaction:

Lecithin + Cholesterol → Cholesteryl ester + Lysolecithin

Table 42. The Genetic Hyperlipidemias

Type	Cholesterol Level	Triglyceride Level	Lipoprotein Type	Usual Appearance of Chilled Serum	Xanthomas	Premature Coronary Heart Disease	Genetic Mechanism	Frequency in Population	Frequency in Unselected Population with MI* < 60 Yr of Age	Comment
Familial hypercholesterolemia	Elevated	Usually normal	IIa (rarely IIb)	Usually clear	Frequent tendon xanthelasma	4+	Autosomal dominant	0.1-0.5%	3.6%	LDL-receptor defects; homozygotes have coronary heart disease in adolescence.
Polygenic hypercholesterolemia	Elevated	Normal	IIa or IIb	Clear	---	2+	Polygenic	5%	Increased	
Familial hypertriglyceridemia	Normal	Elevated	IV (rarely V)	Usually turbid (rarely creamy)	---	1+	Autosomal dominant	1%	5%	Probably heterogeneous.
Familial combined hyperlipidemia	Elevated in 66% of cases	Elevated in 66% of cases	IIa,IIb,IV (rarely V)	Usually turbid (rarely creamy)	Rare	3+	Autosomal dominant	1.5%	11-20%	About 33% have hypercholesterolemia; About 33% have hypertriglyceridemia; About 33% have both elevated cholesterol & triglyceride.
Broad beta disease	Elevated	Elevated	III	Turbid (often with creamy layer)	Frequent, tuberous, palmar creases	4+	Autosomal dominant	Rare	1%	Peripheral vascular disease common; requires tests on isolated VLDL.
Familial lipoprotein lipase deficiency	Slightly elevated	Very high	I	Creamy layer on top; clear below	Eruptive	0	Autosomal recessive	Very rare	---	Abdominal pain common.

* MI = myocardial infarct.
(Reprinted with permission from Motulsky, A.G.: The genetic hyperlipidemias. N. Engl. J. Med., 294: 823, 1976)

As noted earlier, this reaction is responsible for the formation of most of the cholesteryl ester in plasma. The preferred substrate is lecithin (phosphatidylcholine) which contains an unsaturated fatty acid residue on the 2-carbon of the glycerol moiety. HDL and LDL are the major sources of lecithin and cholesterol. The apoA-I, which is a part of HDL, is a powerful activator of LCAT. ApoC-I has also been implicated as an activator of this enzyme; however, the activation may be dependent upon the chemical nature of the phospholipid substrate. The major source of activity appears to be the liver. The blood level of LCAT is higher in males than in females. The role of LCAT in the lipoprotein metabolism has already been discussed. The enzyme is responsible for converting excess free cholesterol to cholesteryl ester with the simultaneous conversion of lecithin to lysolecithin. The products are subsequently removed from circulation. Thus, LCAT plays a significant role in the eventual removal of cholesterol and lecithin from the circulation, which is similar to lipoprotein lipase, and is responsible for the removal of triglycerides contained in the chylomicrons and VLDL. Since LCAT regulates the levels of free cholesterol, cholesteryl esters, and lecithin in plasma, it may play an important role in maintaining the normal membrane structure and fluidity of peripheral tissue cells.

b. The primary deficiency of LCAT, initially found in three Norwegian sisters, is a rare disorder and appears to be inherited as an autosomal recessive trait. The major clinical manifestations include corneal opacity, normocytic anemia due to both decreased erythropoiesis and increased erythrocyte destruction (the target cells are apparent in the blood smear and the erythrocytes contain increased amounts of cholesterol and lecithin), proteinuria and hematuria (the presence of foam cells in glomerular tufts have been noted). The abnormalities found with respect to plasma lipoprotein composition in patients with LCAT deficiencies is consistent with its function described above. In general, the lipoproteins contain abnormally high amounts of unesterified cholesterol and phospholipids. The LCAT-deficient plasma contains, in addition to small amounts of normal lipoproteins, an abnormal lipoprotein (known as Lp-X), VLDL with β-electrophoretic mobility, and two types of abnormal HDL particles. Some properties of Lp-X has already been described under lipoprotein metabolism.

10. Pathogenesis of Atherosclerosis

    a.   In the United States and western Europe, atherosclerosis, principally a disease of the coronary artery, is the leading cause of death due to myocardial and cerebral infarction and thrombosis. As previously noted, the three major risk factors for the development of premature atherosclerosis are hypercholesterolemia (and in particular hyperbetalipoproteinemia), hypertension, and cigarette smoking. Some of the secondary risk factors are diabetes mellitus and obesity. The disease progresses insidiously for many years before life-threatening situations develop. The major problem is the formation of discrete atherosclerotic plaques (a lumpy thickening of the arterial wall) which narrows the arterial lumen and initiates the formation of blood clots. These two events severely limit or completely shut off the blood supply to a vital organ, interfering with its normal function. Recent studies have shown that the atheromatous lesion consists of smooth muscle cells which can accumulate cholesteryl esters in large quantities. These cells have been grown in tissue culture and they possess an LDL pathway for cholesterol metabolism (discussed earlier). Thus, two major areas for discussion concerning the pathogenesis of atherosclerosis are causes for the smooth muscle cell proliferation and the accumulation of both extra- and intracellular cholesteryl esters.

    b.   In order to understand these aspects, the structure of a normal artery is described, and a schematic diagram of a cross-section of an artery is shown in Figure 133. The arteries are channels through which blood is transported under pulsating pressure. They are capable of maintenance and repair functions. The wall of the artery has three layers, namely the intima (the innermost layer), the media (the middle layer), and the adventitia (the outermost layer).

        i)   The intimal layer on the luminal side contains a <u>single layer of endothelial cells</u> (recall, that it is on the plasma membrane of these cells where lipoprotein lipase is located). These cells play an important role in the passage of water and other substances from the blood into the tissue cells. The intimal layer on the peripheral side is surrounded by a fenestrated sheet of elastic fibers (the principal component is elastin--see Chapter 5) known as internal elastic lamina. The middle portion of the intimal layer contains various extracellular components of connective tissue fibers. This portion may contain occasional smooth cells, depending upon the type of artery, and the age and sex of the subject.

Figure 133.   <u>A Schematic Diagram of a Cross-Sectional View of an</u>
<u>Artery</u> (see text for details)

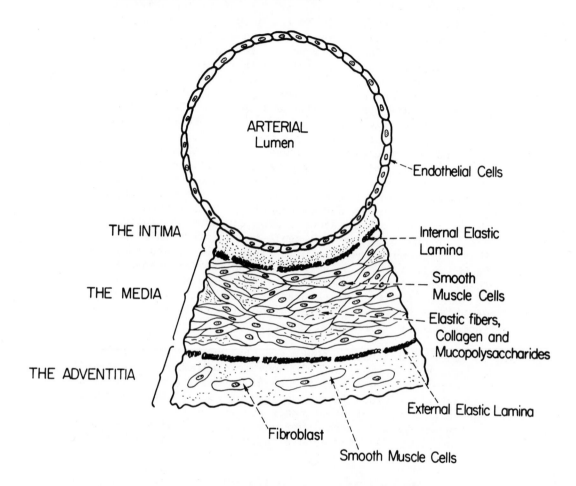

Atherosclerosis has its major effect in the intimal layer,
and the lesions occur due to proliferation of the smooth
muscle cells.

ii)   The media is composed entirely of smooth muscle cells,
which are diagonally oriented. They are surrounded by
collagen, small elastic fibers and mucopolysaccharides
(proteoglycans). Most of these cells are attached to
one another by specific junctional complexes, and these
are arranged as spirals between the elastic fibers in
order to support the arterial wall. The extracellular
products that are present in this region are all
secreted by the smooth muscle cells. In an elastic

artery, several units (lamellar units) described above are present, in order to increase elasticity of the blood vessel.

iii) The outermost layer of the artery, known as the adventitia, is separated from the media by a discontinuous sheet of elastic tissue (the external elastic lamina). This layer consists of smooth muscle cells and fibroblasts. These cells are mixed with bundles of collagen and mucopolysaccharides. This area of the artery is supplied with blood vessels in order to provide the nutrients.

c. There are three types of atherosclerotic lesions: <u>the fatty streak</u> (commonly found early in age; ubiquitously distributed among the world's population; a small number of smooth muscle cells aggregate surrounded by cholesteryl ester deposits; sessile lesion; does not obstruct; and presents no clinical symptoms), <u>the fibrous plaque</u> (found in atherosclerosis; lesion protrudes into the lumen; composed chiefly of cholesteryl ester-loaded smooth muscle cells proliferating in the intimal layer, surrounded by collagen, elastic fibers and proteoglycans; all of these are enclosed in a fibrous cap), and <u>the complicated lesion</u> (associated with the occlusive disease; in addition to the contents of the fibrous plaque, the material is calcified and altered due to hemorrhage, cell necrosis and mural thrombosis).

d. A number of hypotheses have been proposed to account for the formation of the atherosclerotic lesions. In one of these hypotheses, the initial process requires an injury to the endothelium (see Figure 133) leading to focal desquamation and thus exposing the underlying subendothelial region. The injury may be caused by the presence of excessive amounts of atherogenic lipoproteins (e.g., LDL), hypertension (causes increased shear stress), or abnormalities in the metabolism of some hormones. Once the subendothelial region is exposed, the platelets adhere to collagen, followed by aggregation and release of their granules which contain growth factors. The release platelet factors and the influx of plasma constituents, namely lipoproteins, hormones, etc., promote the smooth cell proliferation. The latter synthesizes more connective tissue elements and contributes to the deposition of cholesteryl esters both within the cells (these cells have been described as "foam cells") and in their surrounding area. Recall that the normal uptake of LDL (the cholesteryl ester-rich lipoprotein) by a cell is regulated by the high affinity receptors located on the plasma membrane. However, when the LDL levels exceed the capacity of the high affinity receptor process (i.e., when its concentration is not "normal" or when it is unphysiologic), the LDL is taken up a receptor-independent bulk-phase, phagocytic process, leading to its excessive accumulation. Normally, the injury can be self-limiting with re-endothelialization and suppression of further proliferation of smooth muscle cells. However, if the injury is repeated or present in a chronic manner, the pathologic process continues unabated resulting in a severe or fatal disease.

The exact mechanism involved in the eventual formation of the atherosclerotic lesion caused by the initial insult to the endothelial barrier, the role of mitogenic factors (e.g., factors released from platelets), and the role of atherogenic lipoproteins are not known. It should be noted that prostaglandins and thromboxanes play major roles in platelet adhesion and aggregation to damaged intima (see prostaglandins and platelet function, this chapter). Recall that the normal endothelium synthesizes PGI_2 (prostacyclin) which inhibits platelet aggregation, whereas the damaged endothelium causes the platelet to aggregate and causes the formation of thromboxanes. The latter compound along with endoperoxides (increased due to the lack of the utilization in the synthesis of PGI_2) accelerates the platelet aggregation--compounding the problem--with the formation of thrombus within the damaged blood vessels.

The other hypotheses proposed to explain the origin of atherosclerosis are: the monoclonal hypothesis and the clonal senescence hypothesis. The monoclonal hypothesis suggests that the smooth muscle cells in an atherosclerotic plaque are all derived from a single "mutated" cell. An experimental verification for this hypothesis comes from the observation that all of the smooth muscle cells obtained from a given atherosclerotic lesion contain only one type of isoenzyme of glucose-6-phosphate dehydrogenase (see for details G-6-PD and Lyon's hypothesis--Chapter 4), whereas cells in fatty streaks contained both types of isoenzymes, suggesting a heterogeneous cell population. Cholesterol, substances (or their yet unknown metabolites) present in the cigarette smoke and hypertension, presumably function as "mutagens."

The clonal-senescence hypothesis suggests that with increasing age there is a decrease in the number of smooth muscle cells in the media. Normally, the smooth muscle cell proliferation is regulated by substances known as "chalones" which are secreted by smooth muscle cells themselves. However, when their replication is diminished with advancing age, the supply of chalones is less abundant. It is speculated that this reduction may not by itself be a significant effect in the media, but in the intimal region, it may lead to smooth muscle cell proliferation leading to atherosclerotic plaques.

At the present time, additional experimental evidence is required to support one or the other of the hypotheses described above. It should be noted that these hypotheses may not be mutually exclusive.

e. Lipoproteins and coronary heart disease (CHD). As noted previously the relationship between the development of premature CHD and total plasma cholesterol and LDL-cholesterol is well established. Recent retrospective analyses of epidemiological data have suggested a negative correlation between HDL-cholesterol and premature CHD. Furthermore, it has been observed that at any given level of LDL-cholesterol, the probability of CHD increases as HDL-cholesterol decreases. These observations have led to the present concept that the lowered HDL-cholesterol level (or HDL deficiency) is an independent risk factor for premature CHD. In one study consisting of men aged 50-69, the prevalence (no. of cases/1000) of CHD at HDL-cholesterol levels (mg/100 ml of plasma) of less than 25, 25-34, 35-44, and 45-54 were 180, 123.6, 94.6, and 77.9, respectively. Note the decline of CHD prevalence with increasing levels of HDL-cholesterol. The proposed mechanism of this protective action of HDL may be due to its ability to remove cholesterol from the peripheral tissues and to transport it to the liver for catabolism. Thus lowered levels of HDL in the plasma can lead to less effective removal of cholesterol from the peripheral cells. In another mechanism based upon tissue culture studies, it has been suggested that the HDL may interfere with the uptake of LDL by the peripheral cells. HDL-cholesterol and triglyceride levels are negatively correlated across various populations. Plasma HDL levels are reduced in obesity, physical inactivity and cigarette smoking, and it is increased with increased physical activity and intake of moderate amounts of alcohol. The reasons for these variations in levels are not known. The influence of other factors, namely obesity, diabetes mellitus and other chronic disorders and the feeding of essential fatty acids and hypocholesterolemic drugs on plasma HDL-cholesterol, are not known at this time. It should be emphasized that in addition to environmental factors which influence the levels of HDL, there are genetic factors affecting the HDL levels. The familial aggregation of CHD is well recognized, and it has been shown that children of CHD patients have lower HDL concentrations. Extensive investigations are required to clarify these aspects.

Some of the Other Disturbances of Fat Metabolism

1. Obesity is a form of malnutrition. It is most prevalent in the
    lower socioeconomic groups. In obesity (abnormal increase in
    body weight due to fat deposition), the rate of lipid deposition
    is greater than the rate of utilization. Thus obesity is a
    disorder of caloric excess, and regardless of the origin of the
    disease, obesity develops when caloric intake exceeds the caloric
    needs of the body. It has a high association with maturity-onset
    diabetes mellitus, coronary heart disease, and hypertension.
    Diabetes mellitus and hypertension can be ameliorated with
    decreasing obesity.

    a. There are many types of obesity: juvenile-onset, adult-onset,
        hypothalamic (the hypothalamus is concerned with the regulation
        of appetite, and any injury namely tumors, inflammation,
        trauma, and increased cranial pressure to certain regions of
        this gland can cause obesity), hormone-related (Cushing's
        syndrome, insulinoma, hypothyroidism, Stein-Leventhal syndrome,
        castration, pregnancy), genetic (Prader-Willi syndrome,
        hyperostosis frontalis interna, Alstrom's syndrome, Laurence-
        Moon-Biedl syndrome), psychic, and drug induced (tricyclic
        depressants, phenothiazines, cycloheptadine).

        In the juvenile-onset obese individuals, the number of fat
        cells are increased when compared with the non-obese or
        adult-onset obese subjects. This type of obesity is known as
        hyperplastic obesity. However in the adult-onset obesity,
        the number of fat cells is not increased but they are enlarged
        to their increased fat content and termed as hypertrophic
        obesity. In the United States, by far the largest group of
        obesity (about 95%) come under the adult-onset category.
        Obesity most commonly results from simply eating more calories
        than are needed for the energy requirements of the body.
        This can occur because of failure to adjust one's caloric
        intake due to a decreased metabolic rate or a change from an
        active to a more sedentary life-style.

    b. The treatment of obesity includes several approaches, and its
        success depends upon the subject's attitude, the environment
        the subject comes from, and the physician's concerns and
        attitudes. In general, emphasis should be placed upon both
        the weight loss to seek an ideal body weight and its
        maintenance. It is in the latter that most individuals have
        problems. The treatment modalities include special diets,
        exercise, therapeutic fasting, medication to suppress appetite,
        and surgery. One surgical procedure, known as jejunoileal
        bypass, is performed on grossly obese individuals as a last
        resort. The procedure consists of connecting the jejunum to

the colon, thus severely decreasing the absorptive area.
Complications are many--steatorrhea; the deficiency of
calcium, potassium and magnesium; the impaired absorption of
vitamin B_{12}, fat soluble vitamins A, D and K, and bile acids;
and the development of severe liver disease--and the mortality
rate exceeds 9%. It is essential that, prior to surgery, the
knowledge of the patient's liver, cardiovascular and renal
function is required. A careful post-surgical follow-up is
also mandatory.

2. Xanthomatosis is the accumulation of excess lipids (often rich in
cholesterol) in multiple fatty tumors. The name is derived from
the yellow-orange color of the deposits (xanthos = yellow). This
is frequently accompanied by lipemia and hypercholesterolemia.
Hand-Schüller-Christian disease (chronic idiopathic xanthomatosis)
is a cholesterol lipidosis associated with diabetes insipidus and
lipid deposits in the liver, spleen, and the flat bones of the
skull.

3. Cachexia is the failure to ingest enough calories to maintain
normal lipid storage. A high rate of lipid mobilization from
fat depots occurs, relative to the rate of deposition. In severe
cases, adipose tissue may completely disappear. This may be
caused by anorexia (loss of appetite) due to damage to certain
portions of the hypothalamus (see above, under obesity) or to a
disorder known as anorexia nervosa. Anorexia nervosa occurs
almost exclusively in females between the ages of 10-30 years who
also develop amenorrhea. The majority of patients come from the
upper socioeconomic group and tend to be overachievers. The
other clinical manifestations include bradycardia, low blood
pressure, hypothermia, hyperactivity, cold sensitivity,
constipation (most common complaint), yellow palms with
hypercarotenemia (due to the excess consumption of carrots), to
some extent diabetes insipidus, occasionally growth of lanugo
hair all over the body, and edema (in severe cases). The
biochemical findings in patients of anorexia nervosa consist of
a low basal metabolic rate, decreased levels of the plasma
leutinizing hormone (LH), the follicle stimulating hormone (FSH)
and thyroxine, and increased levels of plasma corticoids. The
pathognomonic findings for this disorder, therefore, are decreased
levels of gonadotropins and elevated levels of cortisol in the
presence of cachexia. In many cases amenorrhea precedes weight
loss. This observation, along with deranged hormonal secretory
patterns, have been attributed presumably due to yet unknown
specific hypothalamus dysfunctions (e.g., responsiveness of
hypothalamus to gonadotropins and vice versa in starvation).
Attainment of the proper body weight corrects all of these
abnormalities and, if uncorrected, leads to death. Cachexia can
also be secondary to carcinoma, malnutrition, certain chronic

infectious diseases, hyperthyroidism, and uncontrolled diabetes. In the last disorder, fat is oxidized to provide energy, and the capacity to synthesize fat for storage is drastically reduced. This gives rise to the depletion of fat depots.

Functions of Some Tissues in Lipid Metabolism

Liver

1. The liver is actively concerned with lipid metabolism. Liver takes up plasma free fatty acids very actively and utilizes them for the synthesis of cholesterol, triglycerides and phospholipids, which are either stored or released into the circulation as VLDL and HDL. Note that the liver does not utilize FFA for energy purposes as much as muscle tissue does. About 30-60% of the circulating dietary triglycerides are also taken up by the liver.

2. The liver plays a major role in cholesterol metabolism by regulating the plasma cholesterol levels (maintaining cholesterol homeostasis), carrying on the biosynthesis of cholesterol and its conversion to the bile acids, etc.

3. The normal liver is 5% lipid, this figure being dependent on a number of factors. In fatty livers, due to certain pathological and physiological disturbances, the lipid content increases to 25-30%. In chronic lipid accumulation, the liver cells become fibrotic, leading to cirrhosis and impaired liver function. Lipids are normally found in the Kupffer cells in the form of droplets. In fatty livers, fatty acid droplets may replace the entire cytoplasm of the hepatic cell. In samples of fatty livers, chemical and histological findings correlate well.

4. Increased fat in the liver may result from decreased oxidation of fat, increased formation of fat, increased movement of fat from the depots to the liver, impaired removal of fat from the liver, or a combination of these factors. There are several etiologies for these four abnormal processes.

   a. Nutritional factors. Diets which are high in fat and low in carbohydrate; diets deficient in methionine and choline (lipotropic factors, needed for phospholipid synthesis); excessive amounts of thiamine and biotin; excessive alcohol ingestion (oxidation of the alcohol produces large amounts of NADH and acetyl-CoA. The NADH stimulates α-glycerol-phosphate synthesis from dihydroxyacetone phosphate and the acetyl-CoA contributes to fatty acid and cholesterol biosynthesis).

   b. Endocrine factors   Hormones which promote hepatic lipid

storage are ACTH, adipokinin, adrenal cortical hormones, sex steroids, insulin, and the thyroid hormones.

c. <u>Miscellaneous factors</u> include chemicals which inhibit protein synthesis (puromycin, ethionine) and other, less specific ones, such as CCl_4, $CHCl_3$, phosphorous, and bacterial toxins. Anoxia, due to anemia and respiratory or vascular congestion, has a similar effect.

5. Steps which are necessary for the transport of fat from the liver are summarized in Figure 134. The steps which are inhibited by some of the above agents are indicated.

Adipose Tissue

1. Adipose tissue functions as an area of caloric storage. When there is excess caloric intake, calories are stored there as triglycerides, and when there is stress or other energy deficit, free fatty acids are released. Glucose by itself can provide all the atoms needed for a triglyceride (TG) molecule. The reactions related to lipid metabolism which occur in the adipose cells are indicated in Figure 135.

2. The capillary lipoprotein lipase breaks down the TG in chylomicrons and VLDL to FFA and glycerol. The FFA then enters the adipose cells to become adipose FFA and glycerol returns to the blood. Adipose tissue cannot utilize glycerol due to the absence of the enzyme, glycerol kinase. In the formation of triglycerides in the adipose tissue, α-glycerolphosphate is obtained from G-6-P. Thus a supply of glucose must be available for TG synthesis.

3. Insulin plays several important roles in fatty acid metabolism in the adipose tissue: (a) It stimulates the uptake of blood glucose, and (b) It increases both fatty acid and triglyceride synthesis.

4. A number of other hormones and compounds also affect the adipose tissue. These are illustrated in Figure 136. This is another important example of hormonal effects mediated by cyclic AMP.

Ketosis

1. Under certain metabolic conditions (starvation, very high fat diets, severe diabetes mellitus and other states in which carbohydrate reserves are depleted, or unavailable for energy purposes), there is a high rate of fatty acid oxidation in the liver. This produces considerable quantities of the <u>ketone bodies</u>: acetoacetate, D(-)β-hydroxy butyrate, and acetone which are released into the circulation, causing <u>ketosis</u>. The name "ketone bodies" for these compounds is traditional but somewhat inaccurate.

(Text continues on p. 977)

Figure 134. Synthesis of VLDL and HDL Lipoproteins With Some of Their Inhibitors in the Liver

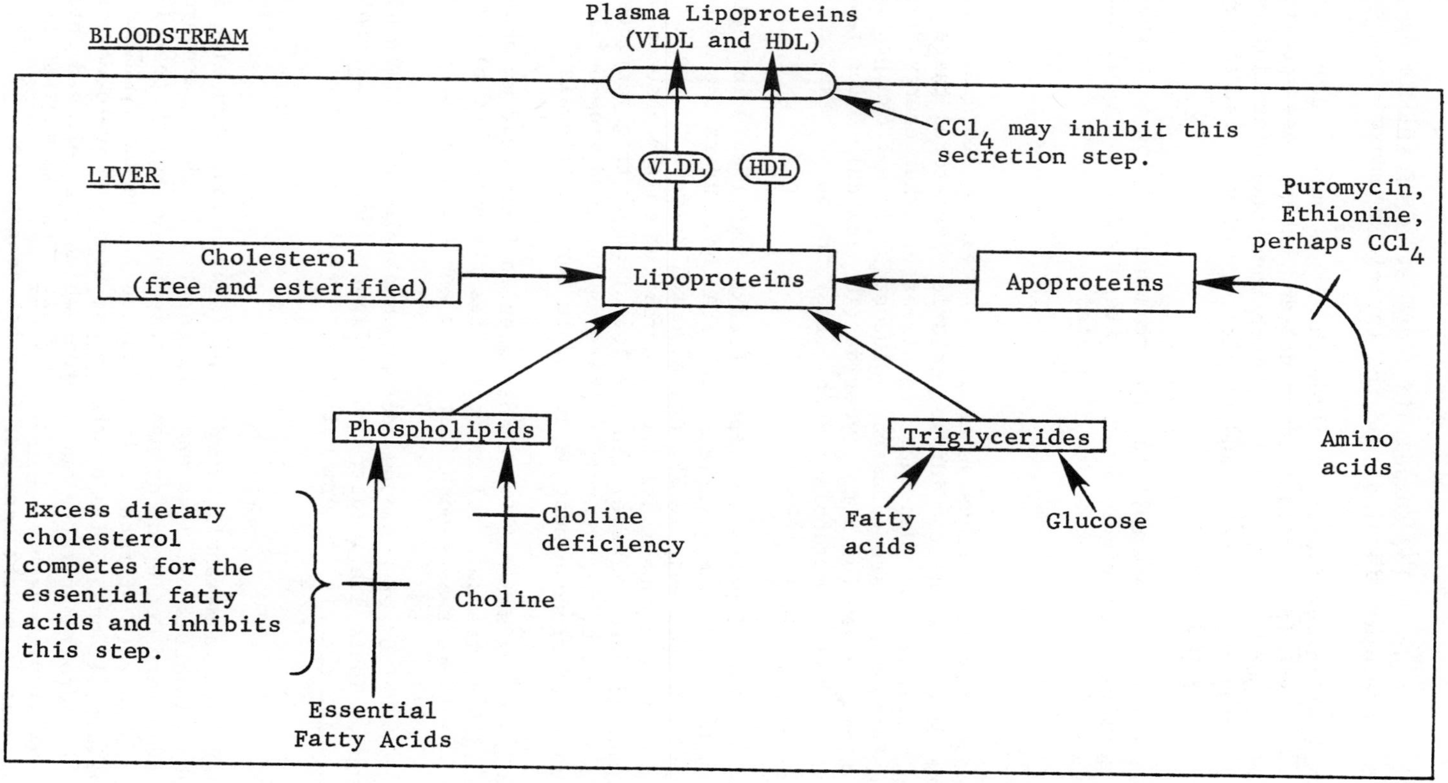

Figure 135. Lipid Metabolism in the Adipose Tissue Cells

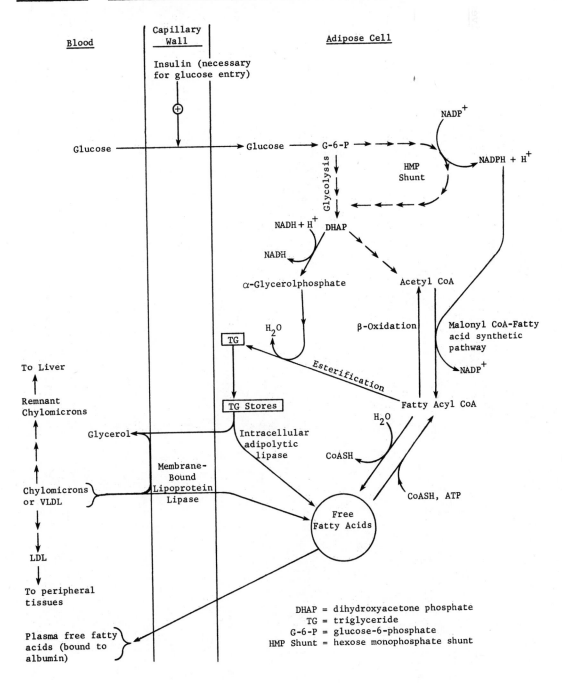

975

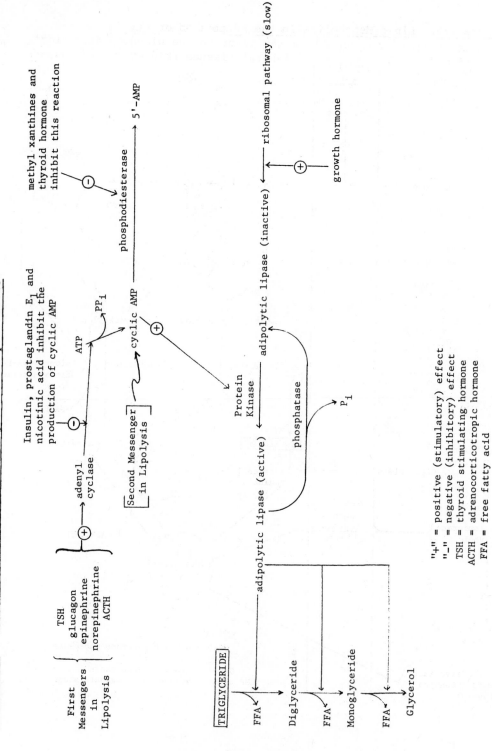

Figure 136. Hormonal Regulation of Lipolysis in Adipose Cells

[976]

2. Under normal conditions, only small amounts of the ketone bodies
    are formed. They pass from the liver to the blood by diffusion
    and are oxidized in the peripheral tissues (kidney, muscle, heart,
    brain and testes) by way of acetyl-CoA and the citric acid cycle.
    Only when the rate of production of ketone bodies increases much
    more than their utilization do the levels build up.

3. Acetoacetate and β-OH butyrate are moderately strong acids. They
    have to be buffered and, hence, deplete the alkali reserves,
    giving rise to ketoacidosis. This can be fatal, for example in
    uncontrolled diabetes mellitus.

4. The synthesis of ketone bodies in the liver proceeds by the
    pathways outlined in Figure 137. As previously noted, the enzymes
    required for their formation are associated mainly with the
    mitochondria. The major pathway for acetoacetic acid synthesis
    seems to be the one from HMG-CoA, catalyzed by HMG-CoA lyase.
    In addition to the fatty acids, three amino acids (tyrosine,
    phenylalanine, and leucine; see Chapter 6) are important
    contributors. The decarboxylation of acetoacetate to form acetone
    appears to be spontaneous in mammals. In bacteria, however, a
    specific enzyme, acetoacetate decarboxylase, has been shown to
    catalyze the reaction. The contribution of acetone to the total
    ketone bodies is minor.

5. The catabolism of acetoacetate and β-hydroxybutyrate, the two
    major ketone bodies, is essentially a reversal of the reactions
    in Figure 137, giving acetyl-CoA as an end product. The key
    enzyme in the process is β-ketoacid CoA transferase, needed to
    "activate" the acetoacetate by forming acetoacetyl CoA. The
    reactions involved are shown below.

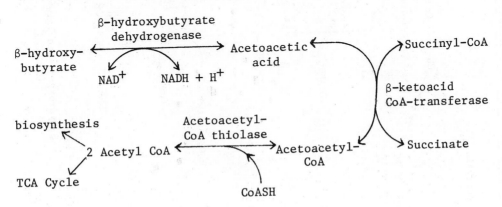

β-hydroxybutyrate dehydrogenase is an enzyme of the inner
mitochondrial membrane.

[978]

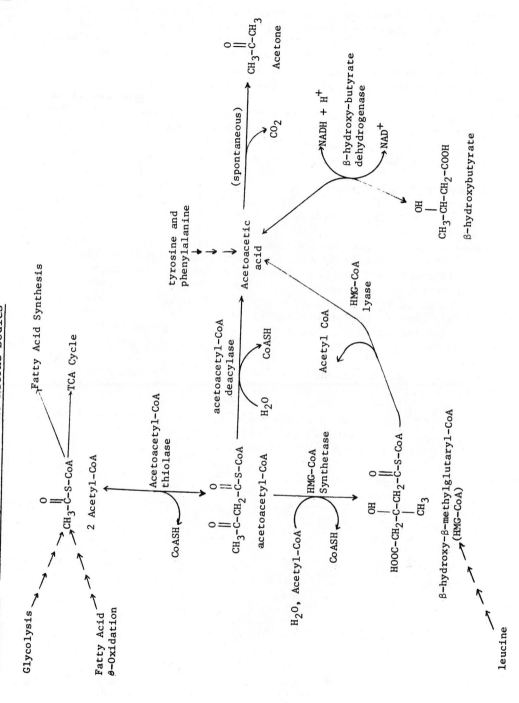

Figure 134. Hepatic Synthesis of the Ketone Bodies

6. β-Ketoacid CoA-transferase is absent in the liver which is one
    reason why the ketone bodies must be released into the blood.
    Tissues which can metabolize the ketone bodies (muscle, brain,
    testes, heart, kidney) have this enzyme. As was indicated earlier,
    the ability of the brain to metabolize the ketone bodies as an
    energy source is especially important in starvation. For some
    time it was thought that the brain had to have glucose and it is
    only fairly recently that the enzymes necessary for ketone body
    catabolism have been demonstrated to be present in this tissue.
    This capacity for ketone body utilization in the brain is also of
    significance in rat, canine, and human neonates. In these (and
    other) species, mother's milk is very rich in fats, producing a
    transient ketosis in suckling infants which lasts until weaning.
    In the rat, levels of most of the requisite enzymes increase
    following birth then decline again after weaning. The
    acetoacetyl-CoA thiolase follows a somewhat different pattern,
    with the final adult activity remaining fairly high, probably
    due to its role in other pathways. The brain cannot survive
    entirely without glucose, however, perhaps due to the need for
    succinyl-CoA in the β-ketoacid CoA-transferase reaction.
    Normally, succinyl-CoA hydrolysis (in the TCA cycle) is coupled
    to the formation of GTP, a compound needed for several tissue
    functions including protein synthesis and gluconeogenesis.
    The glucose may be necessary to supplement the succinyl-CoA
    supply to permit GTP synthesis.

9
Mineral Metabolism

Calcium and Phosphate Metabolism

1. Calcium is the fifth most abundant element, and it is the
    principal extracellular divalent cation in the human body. A
    healthy human adult (70 kg) contains about 1-1.25 kg of calcium
    (i.e., about 20-25 g/kg fat free tissue), and a newborn (3.5 kg)
    contains about 25 g of calcium. Most (about 80%) of the 25 g of
    calcium in the full-term fetal skeleton is acquired during the
    last trimester of pregnancy. The mechanism of placental transport
    of calcium from the mother to the fetus is an active process and
    takes place against a concentration gradient. Ninety-nine percent
    of the body calcium is in the skeleton, and the remaining one
    percent is in the extravascular, interstitial and intracellular
    fluids.

2. The level of calcium in the serum is maintained remarkably
    constant at a concentration of about 10 mg/100 ml (normal range:
    9-11 mg/100 ml <u>or</u> 4.5-5.5 mEq/liter; the normal level is about
    the same throughout life, with the exception of cord blood which
    shows a value of about 11.6 mg/100 ml of serum). In the serum
    the calcium is distributed into three forms: non-diffusible,
    protein-bound calcium (primarily bound to albumin); diffusible
    calcium complexes (calcium bound to lactate, bicarbonate,
    phosphate sulfate or citrate); and diffusible ionic calcium (Ca^{+2}).
    The ionized calcium is less than half of the total calcium, and
    it should be emphasized that it is the only form which is
    <u>physiologically active</u>. Furthermore, it is the ionic calcium
    fraction that is regulated by the parathyroid glands. A decrease
    in this fraction is responsible for the production of tetany,
    regardless of the changes in total calcium values. The calcium
    from a non-diffusible fraction is only slowly released to
    replenish the ionic form of the metal, should the levels of the
    free form fall. The ionized calcium and the protein-bound
    calcium are in equilibrium which is affected by changes in the
    pH (acid-base status) of the blood. As the pH increases
    (alkalosis), the ionic calcium decreases, and vice versa. For
    example, at pH 6.8 about 54.3% of the serum calcium is in the
    ionized form, whereas at pH 7.8 only 37.5% is in the ionized form.
    For each 0.1 pH unit increase, there is about a 4% decrease in
    the level of ionic calcium. The ionic calcium is quantitated
    after separation from protein-bound calcium by several different
    methods: dialysis, or ultrafiltration across a semipermeable

membrane, or ultracentrifugation, or dextran (e.g., Sephadex) gel
filtration. Ionic calcium can also be estimated by measuring
calcium in the cerebrospinal fluid (CSF), because CSF may be
considered as an ultrafiltrate of plasma containing only negligible
amounts of protein. Normal CSF calcium values are 4.2-5.4 mg/100 ml.
A knowledge of serum protein concentration and total calcium at
pH 7.35 and 25°C permits estimation of the ionized calcium by using
the following relationship which is based on the original McLean-
Hastings equation:

$$\text{Ionic Calcium (mg/100 ml)} = \frac{6[Ca] - (P/3)}{P + 6}$$

where [Ca] = mg total calcium/100 ml
       P = g protein/100 ml

Several studies have shown a good agreement between the values
for ionic calcium in normal sera measured by direct methods and
calculated by the above formula. However, the above relationship
used for calculating the concentration of ionic calcium from
total serum calcium and protein levels from abnormal patients
does not correlate well with the actual measurement of Ca^{+2}.
Ionized calcium can also be measured by an ion-selective electrode,
a procedure which is preferred over the others at this time. The
reported normal adult range for ionic calcium using an ion-
selective electrode procedure is 4.4-5.4 mg/100 ml (or 2.2-2.7
mEq/liter). The ionic calcium levels for childhood, adolescence
and old age do not significantly differ from the adult values.

3. a. Calcium functions as a principal component of the skeleton
        and teeth, required for blood coagulation (see Chapter 11),
        neuromuscular activities (membrane permeability, release of
        neurochemical transmitters at synaptic junctions, muscle
        contraction--see section on muscle contraction, Chapter 4),
        release and activation of enzymes, and transport function of
        cell membrane.

    b. Ca^{+2} is a component of many membranes, presumably being bound
        to the membrane proteins. Hormone binding to the plasma
        membrane receptor sites or depolarization of the membrane by
        electrical stimulation causes release of the bound Ca^{+2} into
        the cytoplasm by an unknown mechanism. This is accompanied
        by a local increase in membrane permeability. Passage of
        Ca^{+2} into the cell creates an unstable situation since, under
        resting conditions, the intracellular Ca^{+2} concentration is
        1/1000 of the extracellular value. This gradient is normally
        maintained by a Ca^{+2} pump located in the plasma membrane.
        Many of the enzymes activated by hormone interaction at the
        receptor site require Ca^{+2} for activity and frequently involve

the formation of cAMP by the membrane bound adenylate cyclase (see cyclic AMP, Chapter 4). Examples of processes which require Ca^{+2} and which produce cAMP include the release of neurotransmitter substances by the synapse in response to electrical stimuli, the release of growth hormone by the anterior pituitary in response to growth hormone releasing factor, the synthesis and release of glucose by the liver in response to glucagon, HCl secretion by the stomach in response to histamine, and a number of others.

A low plasma Ca^{+2} concentration due to dietary deficiency, inadequate absorption, vitamin D deficiency, hypoparathyroidism, or a thyroid tumor (of the parafollicular cells) has the immediate effect of perturbing several of these processes. Nerve fibers fire spontaneously, muscle cells show lowered threshold stimuli, and cardiac contractility is increased while relaxation is retarded, resulting in muscle spasms. This syndrome is known as tetany, and it may be either latent (triggered by stress) or overt. If uncorrected, it can lead to death, due to respiratory or cardiac failure due to impaired muscle function. In hypercalcemic states due to hyperparathyroidism, excessive intake of vitamin D or calcium, milk-alkali syndrome, etc., there is decreased membrane permeability and decreased cell excitability.

4. The absorption of calcium from the intestines occurs mainly in the duodenum against a concentration gradient and requires energy (i.e., active transport). The uptake requires 1,25-dihydroxy-cholecalciferol, a metabolite of vitamin D (discussed later). Calcium is also absorbed by a (presumably vitamin D-independent) passive ionic diffusion process, occurring in the jejunum and ileum. However, the contribution made by this process towards calcium absorption must be small in view of the fact that the deficiency of vitamin D results in significantly reduced calcium absorption and negative calcium balance. There appears to be a high degree of adaptability for absorption of calcium depending upon the amount of calcium present in the diet. Thus, a diet low in calcium increases its efficiency of uptake. In general, in both sexes beginning from about age 40, the calcium absorption decreases linearly with increase in age. There are a number of dietary factors which have an influence on calcium absorption. Citric acid, lactose and some amino acids increase, whereas the free fatty acids, phosphates, phytates and oxalates (combining with calcium to form insoluble salts within the intestinal lumen) decrease the intestinal absorption of calcium. It should be noted, however, that with the levels of these compounds normally found in most diets, it is doubtful that the calcium requirement is affected to any significant extent in order to maintain a proper calcium balance. It has been reported that a high intake of protein

increases the calcium loss from the body. For example, in young adult males, it has been observed that by raising the dietary protein consumption from 47 to 142 g/day, the urinary loss of calcium is approximately doubled.

5. Calcium requirements and sources

    a. The recommended daily allowances are as follows:

	mg/day
Infants	360–540
Children	800
Teenagers (pubertal growth period)	1200
Adults	800
Pregnancy & Lactation	1200

Concerning the dietary allowances for calcium during various stages of growth and physiological states, the following aspects should be noted.

    i) The rates of skeletal calcium deposition during growth are as follows:

	mg/day
Between birth and 10 years of age	80–150
During pubertal growth period:	
female	about 200
male	about 270

Under normal circumstances, with the cessation of growth any increase in the accumulation of calcium ceases, although there is a continual remodelling of the bone (discussed later).

    ii) In adults the calcium intake is primarily responsible for the making up of obligatory calcium losses in digestive juices, sweat and urine. This loss has been estimated to be about 300 mg/day. However, irrespective of nutritional and environmental factors, it has been shown (in many populations and racial groups) that after the age of 40, there is a decline in bone mass at an average rate of about 5 to 10 percent per every ten years. In women this bone loss appears to coincide or be enhanced at the time of menopause. Furthermore the rate of bone loss is greater in women than in men. It has been shown that this loss can be prevented or reduced in the post-menopausal women by estrogen replacement therapy or a diet supplemented with one gram of calcium per day. The latter may be of significance in calcium absorption; as

noted earlier it has also been observed in both sexes
that beginning from about age 40, the intestinal
absorption of calcium declines linearly with age. Thus,
it is important to recognize these aspects in arriving
at the calcium requirements for older subjects.
Prolonged calcium deficiency can cause osteoporosis
(discussed later).

iii) During pregnancy and lactation, the calcium needs are
increased. As previously noted about 80% of the 25 g
of calcium present in a full-term fetus is deposited
during the last trimester of pregnancy. About 250 mg
of calcium per day is lost in the milk in lactating
women.

iv) As previously noted, the calcium absorption is adaptable,
and with the reduction of dietary calcium, there is an
increased calcium absorption which is due to increased
production of the active form of vitamin D (see later).
Therefore, diets with calcium levels significantly lower
than the recommended allowances may not lead to a negative
calcium balance.

v) Calcium intake and urolithiasis (formation of stones
within the urinary tract). When calcium is excreted in
quantities greater than 300 mg per 24 hours, it can lead
to the formation of calcium phosphate or oxalate stones.
This problem, in the absence of any other abnormalities
(e.g., hyperparathyroidism, overdosage of vitamin D,
milk-alkali syndrome, bone diseases, neoplastic diseases),
appears to be related to increased intestinal absorption
of dietary calcium in the affected individuals. The
disorder is known as idiopathic urolithiasis. However,
recent observations have implicated that the hypercalciuria
in these patients is due to excessive production of
1,25-dihydroxyvitamin D, which is required for intestinal
calcium absorption. The mechanism for the excessive
production of the active metabolite of vitamin D is not
understood. Note in the normal individuals that the
intestinal calcium absorption is regulated with respect
to dietary calcium levels, and therefore if the dietary
calcium is excessive, the absorption decreases, with the
consequence of the absence of hypercalciuria. In the
management of the hypercalciuric patients in order to
prevent the formation of urinary calculi, a reduction
of dietary calcium to about 600 mg/24 hours and an
adequate fluid intake (about 50% as water) to maintain
a urine volume of about 2.5 liters/day are recommended.
In addition, an adequate amount of dietary phosphate

supplementation has also been reported to be effective in these patients. The beneficial effects of phosphate supplementation may be mediated by pyrophosphate, a potent inhibitor of crystal growth and aggregation, whose concentration is increased with increased dietary intake of phosphate. Additionally, phosphate also regulates the production of 1,25-dihydroxyvitamin D in the kidney. However, in patients with magnesium ammonium phosphate stones (struvite) associated with urinary tract infection, the supplementation of phosphate may be contraindicated.

In patients with oxalate urinary stones due to <u>enteric hyperoxaluria</u> and whose urine is characterized by <u>decreased</u> amounts of calcium, a dietary supplementation of 1 g of calcium per day and maintenance of a urinary volume of 2.5 liters per day have been suggested. Under normal circumstances dietary oxalate (present in rhubarb, citrus fruits and its juices, spinach, tea, cola drinks, etc.) is absorbed poorly due to formation of calcium oxalate complexes which are eliminated in the feces. However, in certain gastrointestinal disorders with steatorrhea, the fatty acids complex with intraluminal calcium, thus decreasing the available calcium for oxalate binding and thus causing an increased absorption of oxalate. The urinary excretion of oxalate in normal individuals is less than 40 mg per 24 hours and less than 4 mg of this quantity arises from dietary sources. Note the oxalate is also produced within the body in small amounts as the end products of glyoxalate and ascorbate (for glyoxalate metabolism see under glycine metabolism--Chapter 6).

b. The major sources of calcium are milk (cow's milk: 120 mg/100 ml; human breast milk: 30 mg/100 ml) and dairy products (e.g., cheese, yogurt). Sardines and other small fish in which bones are consumed and soybean products (e.g., soybean curd, known as tofu: 128 mg/100 g) also can provide significant amounts of calcium. Other calcium-containing foods are leafy vegetables, legumes and nuts, and whole grain cereal products.

6. Phosphate metabolism. Along with calcium, it is the chief inorganic constituent of skeleton (contains about 80% of body phosphorus). Its role as organic esters, acid anhydrides and other derivatives in the macromolecular synthesis (DNA, RNA, protein glycogen, phospholipids, etc.) and intermediary metabolism is discussed throughout this book. An average adult contains about 12 g of phosphorus per kg fat free tissue (i.e., males = 670 g; females = 630 g). Phosphates are widespread in natural foods and the body

phosphate is efficiently conserved (disucssed later). Thus its
deficiency under normal circumstances is not common. Phosphate is
absorbed along with calcium as its counterion and also by an
independent active transport process mediated by 1,25-dihydroxy-
vitamin D. The normal ranges for serum phosphorus levels are as
follows:

	mgP/100 ml of serum
Children (in the first year of life)	4-7 (2.32-4.06 mEq/L)
Adults	2.5-4.8 (1.45-2.76 mEq/L)

There are no appreciable differences between sexes. However, there
is a considerable variation in plasma phosphate during the day,
particularly following meals. The normal ranges are therefore
obtained under fasting conditions. The phosphate homeostasis and
disorders of phosphate metabolism are discussed along with calcium
in this section.

7. Bone structure, composition formation and remodelling

   a. The skeleton not only functions as a support and serves as a
       structural framework for the body but also plays a vital role
       in the calcium homeostasis. It serves as a ready reservoir
       of calcium (recall, 99% of the body calcium is in the bones).
       Bone is comprised of a cellular component (occupying about
       2-3% of the total volume), an organic matrix, and an inorganic
       phase (the mineral component). The organic matrix consists of
       collagen, mucopolysaccharides, sialoproteins and lipids. The
       mineral component is largely made up of a form of calcium
       phosphate which is generally considered to possess the crystal
       structure of hydroxyapatite: $Ca_{10}(PO_4)_6(OH)_2$. Bone also
       contains an amorphous phase, a hydrated tricalcium phosphate
       which is present in larger quantities in early life but matures
       to an apatitic crystalline form with increase in age. Thus,
       the initial deposition of the mineral in the bone formation
       occurs in the amorphous form which is eventually transformed
       into a crystalline state.

   b. The formation of bone requires <u>osteoblasts</u>. These cells are
       evenly distributed and connected with one another by long
       cytoplasmic arms known as canaliculi. This interconnecting
       network provides the means for transporting a variety of
       molecules and provides the surface for the formation of bone
       along with the surface of osteoblasts. These cells are
       situated close to capillaries (no cell is more than 100 μ from
       a capillary) in order to obtain raw materials for bone formation,
       nutrients and oxygen. The bone formation begins with the
       secretion of organic intercellular substances (e.g., collagen,
       mucopolysaccharides, etc.). This material surrounds the

osteoblasts and their canalicular system. The secreted soluble collagen molecules aggregate to form an insoluble fibrillar network (see Chapter 5 for details on collagen biosynthesis), and it is on this matrix that the mineralization occurs. However, before mineralization can take place, the matrix undergoes complex biochemical alterations with regard to its mucopolysaccharide and lipid contents. The impregnation of the organic matrix with the salt calcium phosphates begins with the initial crystal nucleation followed by extraction of minerals from the surrounding fluid. The further deposition and crystal growth is made possible presumably due to the fact that the solubility product of hydroxyapatite is significantly lower than the Ca X P ion product in the extracellular fluid. Alkaline phosphatase has been associated with bone formation. However, its <u>exact</u> role is unclear. The enzyme is present in the osteoblasts as well as in the extracellular layers of a newly synthesized matrix. The osteoblasts may achieve a high local concentration of phosphate by the action of alkaline phosphatase on the sugar phosphates and pyrophosphates. In many disorders with significant osteoblastic activity, namely rickets, osteomalacia and hyperparathyroidism, the serum levels of alkaline phosphatase may be elevated. The osteoblasts, now completely surrounded by the intercellular substances they have secreted, are known as <u>osteocytes</u>. These cells, although encrusted in the solid surroundings of calcium phosphate crystals, are supplied with oxygen and nutrients through their interacting passageways (canaliculi) and their close proximity to capillaries. Thus bone is a dynamic, well-vascularized tissue.

c. Bone remodelling (or bone turnover) may be visualized as a cyclic process and consists of <u>resorption</u> (dissolution of both mineral and matrix with the loss of bone tissue) and formation. These processes occur continually in order to maintain calcium homeostasis, to meet the changing structural needs of the bone, and to renew the aging interstitial material. The remodelling process is an orderly process and involves the following sequence of events. Undifferentiated <u>mesenchymal cells</u> present on the bone surface are activated to become <u>osteoprogenitor cells</u> which by further division are transformed to <u>preosteoclasts</u>. These cells fuse to become <u>osteoclasts</u>, which are large multinucleated cells. Osteoclasts are short lived, and they are responsible for the dissolution of the bone (resorption). Osteoclasts can also arise from osteocytes via the formation of an intermediate known as osteoclastic osteocyte. The resorption requires the action of collagenase and several other lysosomal proteases. The solubilization of the mineral requires H^+ ions which are derived from organic acids and H_2CO_3. The latter is synthesized from the hydration of CO_2 catalyzed by carbonic anhydrase, an enzyme localized in the osteoclasts. After completing their function, the osteoclasts undergo division and transformation eventually to osteoblasts via the preosteoblasts. The osteoblasts after finishing their synthetic function become mature osteocytes locked within the bone matrix or transformed to undifferentiated mesenchymal cells. These processes at a given bone remodelling locus are summarized as follows:

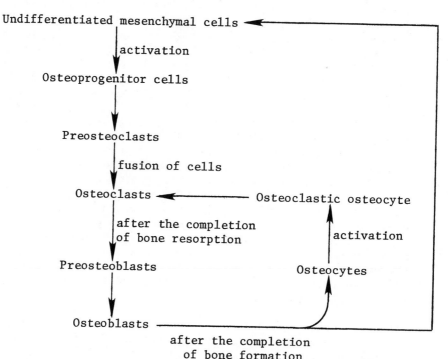

The bone remodelling process is controlled by a variety of factors. The transformation of osteoprogenitor cells to osteoclasts is stimulated by parathyroid hormones (PTH-- discussed later), thyroxine, pituitary growth hormone and vitamin D, whereas calcitonin, estrogens and glucocorticoids inhibit the formation of osteoclasts. The conversion of an osteoclast to an osteoblast is promoted by calcitonin, estrogen, pituitary growth hormone, inorganic phosphate and mechanical stress but the process is antagonized by PTH and vitamin D. The mechanism of action in some of these agents are mediated by cyclic AMP.

Mechanism of Calcium Homeostasis

As noted previously, the plasma calcium level is maintained at a constant level due to the participation of a number of important regulatory mechanisms. These aspects are discussed below.

1. Vitamin D and its effect on calcium and phosphate metabolism

    a. Vitamin D, a lipid soluble vitamin, can be obtained from the diet or can be synthesized from 7-dehydrocholesterol (an immediate precursor of cholesterol--see cholesterol biosynthesis, Chapter 8) when exposed to the ultraviolet rays of the sun. Radiant energy striking the skin is used in the opening of a "B" ring of 7-dehydrocholesterol, producing cholecalciferol (vitamin D_3). Ergosterol (24-methyl-21-dehydro-7-dehydrocholesterol), a common plant steroid which is present in the diet, may also serve as a substrate in the photometabolism reaction, producing ergocalciferol (vitamin D_2, also called calciferol). Irradiation of ergosterol is an important commercial method for the synthesis of vitamin D_2, which is used for enriching cow's milk. Vitamin D deficiency, once an important public health problem of infancy and childhood, has become rare with the advent of vitamin D fortification of milk and its products. It should be noted that the metabolism of vitamin D_2 and D_3 are analogous, and they evoke the same physiologic response. The conversion of 7-dehydrocholesterol to cholecalciferol is shown below.

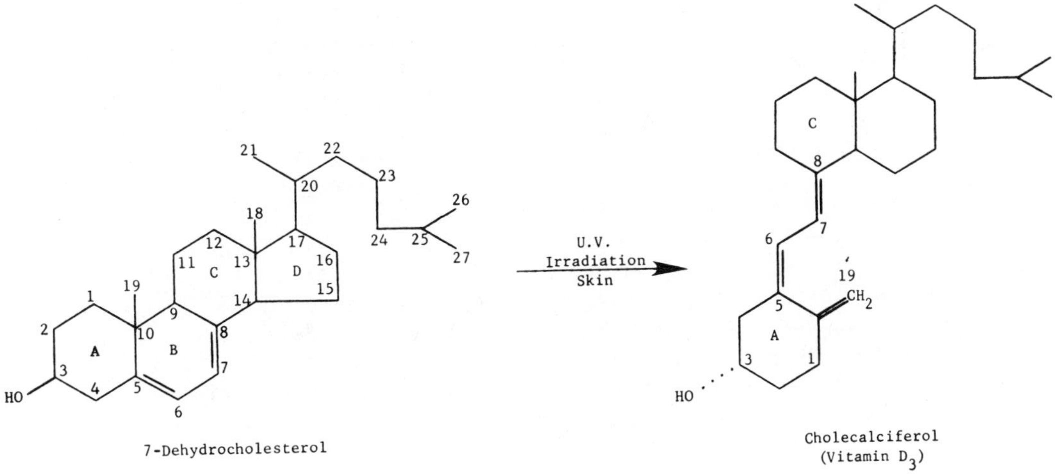

7-Dehydrocholesterol   U.V. Irradiation Skin   Cholecalciferol (Vitamin D_3)

b. The function of vitamin D includes the absorption of
    calcium and phosphate from the intestine which is required
    for the sketal mineralization and other biochemical
    functions of these minerals and, in concert with
    parathyroid hormone, the mobilization of calcium from bone
    in order to maintain plasma calcium levels. However,
    neither vitamin D_2 or D_3 is the active form of the vitamin.
    Both substances undergo hydroxylation at carbon 25 and
    carbon 1 before they are biologically active (discussed
    below). With regards to phosphate transport in the
    intestine, it should be noted that phosphate can accompany
    the translocation of calcium as its counterion. However,
    the active metabolite of vitamin D stimulates a phosphate
    transport in the intestine which is independent of calcium
    transport. The translocation of calcium and phosphate
    occurs against an electrochemical gradient, appears to
    require specific carriers, and is an active transport
    process (discussed later).

c. The recommended dietary allowance for vitamin D is 400 IU.
    Deficiency of vitamin D leads to rickets in children and
    osteomalacia in adults (discussed later). As noted
    earlier, in the United States the vitamin D deficiency is
    rare because of vitamin D fortification of milk and its
    products. But vitamin D deficiency can occur in
    individuals who do not consume adequate amounts of
    fortified milk (in many countries milk is not enriched
    with vitamin D), or who have insufficient exposure to
    sunlight (due to prolonged illness, cultural patterns,
    decreased intensity of sunlight in the more northerly
    lattitude), or who have an increased requirement for
    vitamin D due to intestinal malabsorption problems
    associated with steatorrhea (see lipid absorption,
    Chapter 8). The latter problems also cause the depletion
    of an important metabolite of vitamin D synthesized in the
    liver, namely 25-hydroxyvitamin D_3 which undergoes an
    active enterohepatic circulation. Vitamin D toxicity can
    result when it is consumed for several weeks in large
    quantities of 2000-5000 units/kg/day. The clinical
    manifestations include hypercalcemia (12 mg% or more),
    hypercalciuria (results in nephrolithiasis and renal
    injury), polyuria, polydipsia, anorexia, nausea, vomiting,
    constipation, and hypertension. Extreme hypercalcemia may

lead to drowsiness and coma. Management consists of the correction of fluid and electrolyte balance. Cortisone is administered in order to reduce the calcium level in the blood to normal.

d. The metabolism of vitamin D. The liver and the kidney are the major sites of the metabolic activation of vitamin D in the body.

   i) The initial transformation of vitamin D_3 to 25-hydroxyvitamin D_3 (25-(OH)D_3) takes place predominantly in the liver catalyzed by 25-hydroxylase.

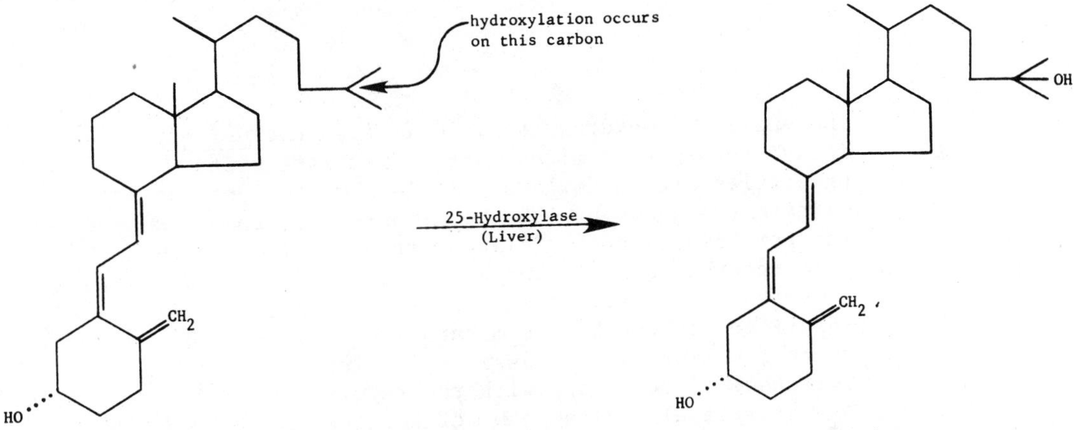

hydroxylation occurs on this carbon

25-Hydroxylase (Liver)

D3

25-(OH)D_3

The properties of 25-hydroxylase are as follows: the enzyme is located on the endoplasmic reticulum and requires NADPH and molecular oxygen; the enzyme is not inhibited by carbon monoxide, inhibitors of lipid peroxidation, and inhibitors of cytochrome P-450; the enzyme is not induced by phenobarbital. The exact mechanism of this reaction and the regulatory aspect of this enzyme are not known with certainty.

   ii) The 25-(OH)D_3 is converted to the active metabolite by hydroxylation in the kidney to 1,25-dihydroxy-vitamin D_3 (1,25-(OH)$_2D_3$) catalyzed by 1-hydroxylase.

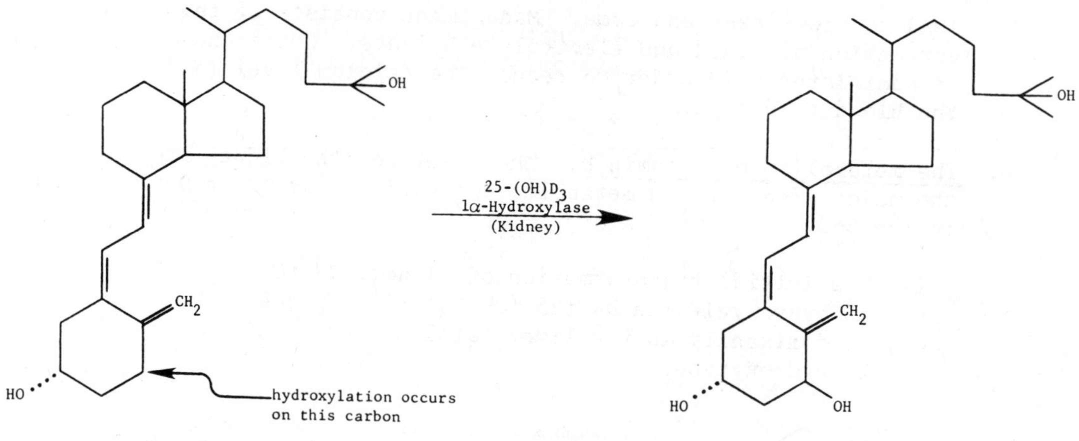

25-(OH)D_3

hydroxylation occurs on this carbon

1α-Hydroxylase (Kidney)

25-(OH)D_3

1α,25-(OH)$_2D_3$
(active metabolite)

The enzyme, 1α-hydroxylase, is located <u>exclusively</u> in the renal inner mitochondrial membrane. Its properties are as follows: it is a mixed function oxidase (as shown by ^{18}O experiments); it requires the participation of a flavoprotein, an iron-sulfur protein (ferridoxin), and a cytochrome P-450. Using a variety of experiments, 1,25-(OH)$_2D_3$ has been identified as the <u>active metabolite</u> of vitamin D, which functions in a classic endocrine function (discussed later along with the regulation of 1-hydroxylase). Vitamin D metabolites are transported in blood, bound to specific plasma proteins. A protein (molecular weight: 50,000-60,000, an α-globulin), which binds to both D_3 and 25-(OH)D_3, has been identified and characterized from human plasma. Under normal circumstances, about 1 to 3% of this protein is saturated with the vitamin D metabolites. The level of the binding protein is not altered in a number of disorders of calcium and phosphate metabolism; however in pregnancy its level is increased. The plasma levels of vitamin D metabolites have been measured in humans by competitive protein-binding procedures. The normal range for plasma 1,25-(OH)$_2D_3$ is 2.1-4.5 ng/100 ml.

iii) Formation of other metabolites of vitamin D_3

24,25-Dihydroxyvitamin D_3 (24,25-(OH)$_2D_3$) is also synthesized in the kidney and catalyzed by the 24-hydroxylase located in the mitochondria. However, the kidney is not the only site of 24-hydroxylation, as it has been shown that nephrectomized animals produce $24,25-(OH)_2$ upon administration of large quantities of $25-(OH)D_3$. It is of interest to point out that of the two possible stereoisomers (R and S), the product of the 24-hydroxylase reaction is the R epimer, namely $24R,25-(OH)_2D_3$. The exact role of this compound is not known; it has a <u>limited</u> biologic activity; no specific target organs have been identified; in animals with a sufficient supply of calcium and phosphate, it has been shown that the 1-hydroxylase is inhibited and the $24-(OH)D_3$, instead of being converted to $1,25-(OH)_2D_3$, is transformed to $24,25-(OH)_2D_3$. Thus, the 24-hydroxylase reaction may play an important role in the calcium homeostasis. In this regard it should be pointed out that this reaction is repressed by the parathyroid hormone, and the presence of $1,25-(OH)_2$ is required for the appearance of 24-hydroxylase.

$1,24R,25$-Trihydroxyvitamin D_3 $(1,24,25-(OH)_3D_3)$ has also been detected, primarily in <u>in vitro</u> incubation experiments. This product can be synthesized either from 1-hydroxylation of $24,25-(OH)_2D_3$ or 24-hydroxylation of $1,25-(OH)_2D_3$. The exact role of this trihydroxylated derivation of vitamin D_3 in humans is not clear.

$25,26$-Dihydroxyvitamin D_3 has been isolated from pig plasma. This metabolite is relatively biologically inactive and its exact role is not known.

iv) The primary excretory route for vitamin D metabolites is via the bile into the feces, and urinary excretion appears to be negligible.

e. Regulation of vitamin D metabolism (vitamin D endocrine system). $1,25-(OH)_2D_3$ functions as a hormone, and its metabolic production from $25-(OH)D_3$ is modulated according to the calcium and phosphorus need of the body. The endocrine feedback step, where stringent control is applied, is at the 1-hydroxylase reaction, which occurs exclusively in the kidney. As stated previously, the liver 25-hydroxylase does not appear to be under any specific regulatory mechanism. Thus, the $25-(OH)D$ level is primarily dependent upon vitamin D intake or the extent to which skin is exposed to sunlight. The primary regulators of 1-hydroxylase are parathyroid hormone (PTH), phosphate and $1,25-(OH)_2D_3$. Some of these aspects are summarized in Figure 138. Recent studies have also implicated other

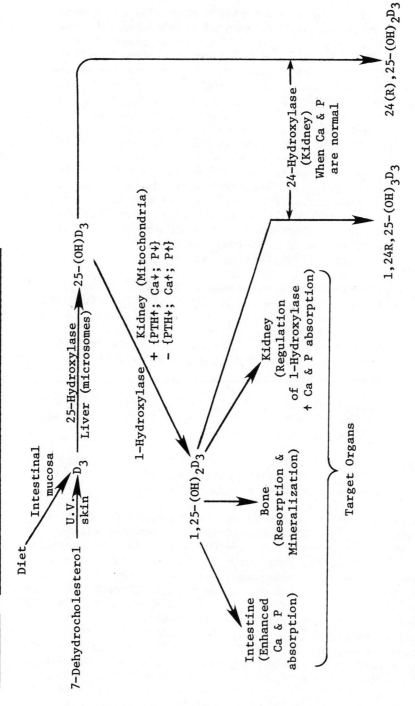

Figure 138. Schematic Representation of Vitamin D Metabolism

See text for details; Ca = Calcium, P = Phosphate, PTH = Parathyroidhormone;
↑ = increase, ↓ = decrease--these refer to plasma levels;
+ = the enzyme activity increased, - = enzyme activity decreased.

hormones, namely growth hormones, estrogen, prolactin, and
cortisol, as modulators of this process.

The two primary stimuli for the appearance of 1-hydroxylase
in the kidney are low levels of serum calcium and phosphate.
Low phosphate directly stimulates the appearance of the
enzyme, whereas low calcium does this indirectly via the
parathyroid glands. The low serum calcium level <u>stimulates</u>
the parathyroid glands to secrete PTH, which in turn
<u>stimulates</u> the 1-hydroxylase and inhibits the 24-hydroxylase
reaction (see Figure 138). Thus, PTH functions as a
tropic hormone in the synthesis of $1,25-(OH)_2D_3$. These
events lead to the enhanced production of $1,25-(OH)_2D_3$
which is then released into the blood circulatory system
to reach the target organs (intestine and bone) in order
to mobilize calcium and phosphate. Once the hypophosphatemia
and hypocalcemia are corrected, the production of $1,25-(OH)_2D_3$
ceases, terminating its action with closure of the feedback
loop. The vitamin D endocrine system for the correction of
deficits in calcium and phosphate is shown in Figures 139
and 140 respectively. Note that the restoration of
normocalcemia involves the production of PTH, which is
responsible for the mobilization of calcium along with
$1,25-(OH)_2D_3$ and is also responsible for the elimination of
excess mobilized phosphate in the urine by preventing its
reabsorption (Figure 139). Also, note that $1,25-(OH)_2D_3$
acts on the intestines to increase the absorption of dietary
calcium and phosphate. The normalization of hypophosphatemia
does <u>not</u> involve PTH, and it is corrected by the action of
$1,25-(OH)_2D_3$ on the intestines. The excess calcium mobilized
in this process is lost in the urine, due to the absence of
PTH which occurs under this circumstance (Figure 140). It
has also been observed that $1,25-(OH)_2D_3$ decreases the
excretion of filtered phosphate, sodium and calcium by
causing an increase in the absorption of these ions at the
proximal renal tubule. Thus with respect to phosphate
reabsorption, $1,25-(OH)_2D_3$ may be antagonistic to the
phosphaturic action of PTH. Also note that it is possible
for $1,25-(OH)_2$ to govern its own synthesis by functioning
as a feedback regulator either at the parathyroid glands,
or at the kidney, or both (not shown in the figures).
Additional studies are required to clarify these aspects.

As discussed above, $1,25-(OH)_2D_3$ plays a significant role
in the maintenance of blood calcium levels by increasing
the intestinal absorption of the mineral and resorption
(dissolution) of the bone. The latter action of the
active metabolite vitamin D on the bone may appear as
antagonistic to its well-established role in the normal

(Text is continued on p. 998)

Figure 139. Interrelationships of Vitamin D Endocrine System and Parathyroid Hormone (PTH) in Calcium Homeostasis

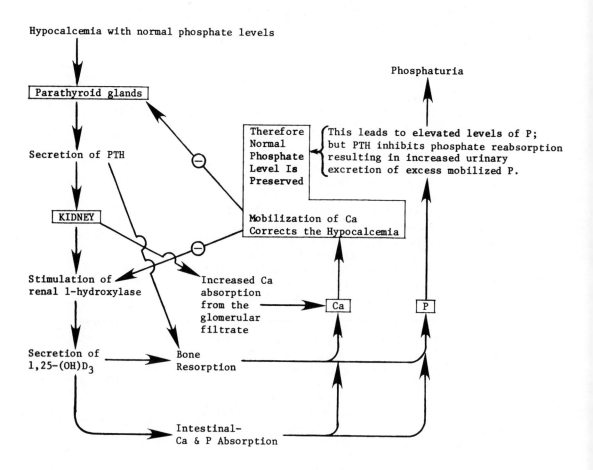

Hypocalcemia with normal phosphate levels

Phosphaturia

Parathyroid glands

Secretion of PTH

Therefore Normal Phosphate Level Is Preserved

This leads to elevated levels of P; but PTH inhibits phosphate reabsorption resulting in increased urinary excretion of excess mobilized P.

KIDNEY

Mobilization of Ca Corrects the Hypocalcemia

Stimulation of renal 1-hydroxylase

Increased Ca absorption from the glomerular filtrate

Ca

P

Secretion of 1,25-(OH)D$_3$

Bone Resorption

Intestinal- Ca & P Absorption

Ca = Calcium; P = Phosphate

──○➤ = negative feedback (inhibitory step)

996

Figure 140. Vitamin D Endocrine System and Phosphate Homeostasis

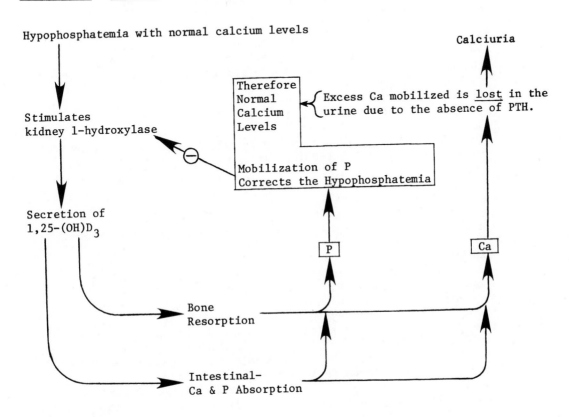

P = Phosphate; Ca = Calcium

———⊖——▶ = negative feedback (inhibitory step)

process of mineralization of bone. It should be noted, however, that the initial bone dissolution (osteoclastic osteolysis) may be required to supply the minerals at the appropriate concentrations for the formation of new bone (initiate mineralization of osteoid and/or activation of mesenchymal cell differentiation--see bone formation, discussed earlier).

As noted previously the absorption of dietary calcium is adaptable depending upon the availability of calcium in the intestinal lumen. When there are low amounts of calcium, the absorption increases, and vice versa. These changes in absorption have been attributed to the regulation that occur with the $1,25-(OH)_2D_3$ synthesis.

The effects of calcitonin (a hypocalcemic hormone--discussed later) on vitamin D metabolism in humans is unclear at this time.

f. Mechanism of action of $1,25-(OH)_2D_3$. The current working hypothesis of the mode of action of $1,25-(OH)_2D_3$ is analogous to that of steroid hormone action (see Chapter 13). The principal steps involved in this process are: the binding of the hormone to a specific receptor in the cytoplasm, the translocation of this hormone-receptor protein complex to the nucleus, the binding of the complex to specific non-histone proteins of chromatin, the stimulation of unique genes leading to the synthesis of new mRNA's (transcription), and finally the translation of mRNA on the polysomes to specific functional proteins. In the intestine, two proteins which presumably play a role in the transport of phosphate and calcium have been identified, due to the action of $1,25-(OH)_2D_3$. They are, a phosphatase (described as alkaline phosphatase or calcium-ATPase) and a calcium-binding protein. In other tissues, it has been speculated that this mode of action of $1,25-(OH)_2D_3$ is similar to that of the intestine but specific details are not yet known.

g. Therapeutic use of vitamin D metabolites and analogues. The delineation of vitamin D metabolism has led to the synthesis of a number of metabolites and analogues. These compounds are now undergoing active investigation in order to establish their therapeutic value and/or mode of action in a variety of disorders affecting calcium and phosphate homeostasis.

i) Dihydrotachysterol (DHT). This is an analog of vitamin D and may be considered as the reduction product of the vitamin (D_2 or D_3):

DHT

It has been shown that DHT is not highly active when
compared with vitamin D (1/450 as active as vitamin D)
in the antirachitic assays. However, when used in
high doses, it has been shown to be more effective
than high doses of vitamin D in calcium mobilization
from bone or absorption from the intestines. Thus,
DHT has been used in the treatment of hypocalcemia
(e.g., hypoparathyroidism) and a vitamin D-resistant
form of rickets. DHT undergoes metabolic activation
by 25-hydroxylation in the liver by the same system
as vitamin D_3. But, unlike 25-(OH)D_3, it appears
that 25-(OH)DHT exerts no feedback inhibition on the
hydroxylation process; furthermore, DHT is active
in nephrectomized rats, suggesting that the
1-hydroxylation reaction of 25-(OH)DHT is not
required for its function. This aspect is clarified
by the comparison of the structures of 1,25-(OH)$_2D_3$
and 25-(OH)DHT. The 3β-hydroxyl group of DHT is
situated approximately in the same position as the
1α-OH group of 1,25-(OH)D_3. Thus, 25-(OH)DHT may be
considered as an <u>effective analogue</u> of 1,25-(OH)$_2D_3$.

ii) Other metabolites or analogues that have been
     synthesized are 24-(OH)D_3, 1-(OH)D_3, 1,25-(OH)$_2D_3$
     and 5,6-trans-25-(OH)D_3. As expected, the therapeutic

administration of either $1,25-(OH)_2D_3$ or $1-(OH)D_3$ effectively bypasses the block in the renal 1α-hydroxylation, and hence they are useful in the treatment of hypoparathyroidism, renal osteodystrophy and other renal disorders (see parathyroid hormone, this section), and vitamin D-dependent rickets. The latter is a genetic disorder characterized by either a complete absence of the renal 1-hydroxylase or a defect in its regulation. In one study, it has been shown that $1,25-(OH)_2D_3$ oral administration of as low a quantity as 1 µg/day was found to be beneficial. $25-(OH)D_3$ has shown to be beneficial in patients with hepatic disease (recall that 25-hydroxylation occurs in the liver) and patients receiving anticonvulsant drugs.

iii) Glycosides of $1,25-(OH)_2D_3$ have been isolated from the plants, Solanum malacoxylon and Cestrum diurnum. The cattle grazing on these plants develop calcinosis which can lead to death. The extract from S. malacoxylon has been used in the treatment of uremic bone disease.

2. Parathyroid hormone (PTH)

a. The primary function of PTH is to maintain the constancy of the concentration of the calcium ion in the extracellular fluid. There are usually four (sometimes five or rarely six) parathyroid glands in humans. They are so named because they are located by the side of the thyroid gland, two embedded in the superior poles of the thyroid and two in its inferior poles. The hormone-secreting cells of the glands are known as chief cells. The function of the oxyphil cells, which are presumably derived from chief cells, is unknown.

b. PTH from different sources (human, bovine and porcine) have been chemically characterized; they are all single polypeptide chains of 84 amino acid residues (molecular weight: 9500 daltons). The full biologic activity of the entire molecule (84 amino acid residues) resides in the amino terminal segment, consisting of amino acid residues 1 to 34. This portion of the peptide of 34 amino acid residues, with the sequence present in the human and bovine PTH, has been chemically synthesized. It possesses the characteristic biologic activity. It should be noted that, although PTH from different sources have similar biologic activity (but not the same biologic potency), they can be

distinguished immunologically. It is of interest to point
out that the amino terminal sequence of the 34 amino acid
residues of human PTH differs from the porcine sequence
at two amino acid positions (i.e., at position 16, Asn in
human PTH is substituted with Ser; similarly at position 18,
Met → Leu) and from the bovine sequence at three positions
(1 Ser → Ala; 7 Leu → Phe; 16 Asn → Ser). The porcine and
human PTH are biologically less active than bovine PTH;
this is presumably due to the presence of the N-terminal
alanine residue in bovine PTH (both human and porcine PTH
contain serine in this position).

c. The synthesis of PTH in the chief cells proceeds in a
sequential manner from a larger, biologically less active
molecule to a smaller (consisting of 84 amino acid residues),
biologically active form. The prohormone (pro-PTH) has been
identified in parathyroid glands of many sources, namely
bovine, porcine, chicken, rat, human, and in human
parathyroid adenomas. The bovine and human prohormone has
six additional amino acid residues (Lys-Ser-Val-Lys-Lys-Arg)
linked to the amino terminus of PTH. Pro-PTH synthesized
on the ribosomes enters the cisternal space of the endoplasmic
reticulum to be transferred to the Golgi complex, where it is
converted to PTH by a peptidase and stored in the granules.
A precursor for pro-PTH (pre-pro-PTH), containing 115 amino
acid residues (molecular weight: 13,000 daltons), has also
been identified. These aspects of PTH formation are similar
to insulin biosynthesis (see Chapter 4) and other polypeptide
hormones. The general theme appears to be that the
biologically active substances, initially formed as _inactive_
precursors, are converted to biologically _active_ products by
specific enzymic modification of the precursors. This
phenomenon is not only restricted to polypeptide hormones
but also occurs in numerous other processes such as the
activation of pancreatic zymogens, the complement system,
the blood clotting process, the kininogen system, etc.

d. The secretion of PTH is regulated by the levels of ionized
calcium in the blood, and it is inversely related. When
the concentration of ionized calcium is low, the secretion
of PTH is increased (chronic hypocalcemia leads to
hypertrophy and hyperplasia of the glands); when the ionized
calcium is high, the secretion of PTH is diminished (chronic
hypercalcemia results in hypoplasia of the gland). It
should be noted that magnesium also regulates PTH secretion
in a similar fashion to that of calcium; PTH secretion is
enhanced with _decreases_ in magnesium concentration at normal
levels of calcium. At extremely low levels of magnesium,

the release of PTH stops to negligible levels presumably due to inhibition of a magnesium-dependent adenylate cyclase. This leads to decreased synthesis of cAMP which is required for PTH secretion.

The calcium ion regulates a number of processes involved in the PTH synthesis and secretion. These include amino acid uptake by the chief cells, synthesis of pro-PTH, conversion of pro-PTH to PTH, maturation of secretory granules and finally secretion of PTH from the cell. All of the above processes are augmented in the hypocalcemic state, whereas they are inhibited in the hypercalcemic state. Magnesium regulates only the PTH secretion, but has no influence in any of the other processes concerned with the PTH synthesis.

e.   As noted earlier, the function of PTH is to maintain the constancy of ionic calcium in the extracellular fluid. When the level of calcium is low, the hormone is secreted into the bloodstream to restore the normal concentration by acting on target cells located in the various peripheral tissues in order to mobilize calcium into the extracellular fluid. Thus PTH acts on bone, intestinal and renal cells.

   i)   In the bone, PTH promotes the dissolution of bone (resorption) by increasing the rate of conversion of undifferentiated mesenchymal cells and osteocytes to osteoclasts (osteolytic cells) and delaying their conversion to osteoblasts (see bone formation, this section). The bone matrix undergoes destruction due to the stimulation of collagenase activity by PTH. This is reflected in the enhanced excretion of hydroxyproline and hydroxylysine, the hydrolytic products of collagen in the urine. PTH has also been shown to inhibit the collagen synthesis in the osteoblasts. During resorption, lysosomes which contain hydrolases are released into extracellular spaces by osteoclasts. These enzymes aid in the resorptive process. Furthermore, PTH may promote glycolysis and inhibit the oxidation of substrates in the tricarboxylic cycle, resulting in the accumulation of lactate, citrate and other organic acids. This can lead to a decline in the pH providing a suitable condition for the dissolution of hydroxyapatite.

   The intracellular mechanism of PTH action in the bone appears to involve cAMP elevation within the target cells. Agents which lower intracellular cAMP concentrations, either by activation of the phosphodiesterase (as does imidazole) or by inhibition of the adenylate cyclase (the mechanism of action of 2-thiophene carboxylic acid), cause a fall in plasma Ca^{+2} even in the presence of PTH. Agents which raise cAMP levels (such as dibutyryl cAMP, or caffeine and theophylline, both phosphodiesterase inhibitors) will raise plasma Ca^{+2}. In addition to cAMP, Ca^{+2} itself has been implicated as an intracellular mediator in bringing about various metabolic alterations.

ii) In the kidney, PTH increases the renal tubular reabsorption of calcium and the excretion of phosphate (phosphate principally absorbed in the proximal tubule is inhibited by PTH). These effects lead to the elevation of extracellular calcium ion concentration. In the absence of PTH, rapid loss of Ca^{+2} in the urine occurs, a situation which can be reversed by administration of the hormone. Under normal conditions, 97% to 99% of the Ca^{+2} filtered by the glomerulus is actively reabsorbed along the entire length of the nephron, with 65% to 80% of the reabsorption occurring in the proximal tubule. Under the same conditions, 85% to 90% of the phosphate is also actively reabsorbed. Thus, there is normally a slight continual phosphate loss (amounting to about 700–800 mg/day; this loss is made up of a dietary phosphate intake of about 1000 mg/day) but almost complete Ca^{+2} retention. Any increase in the PTH level beyond the normal range results in progressively smaller amounts of phosphate reabsorption. PTH also influences the excretion of other ions which are involved with bone metabolism. It decreases the renal clearance of magnesium and hydrogen ions and increases the clearance of Na^+, K^+, HCO_3^-, H_2O, amino acids, citrate, chloride, and sulfate. The renal effects of PTH appears to be mediated by cAMP, whose synthesis increases in response to PTH, preceding other physiologic actions. Under normal circumstances, the urine contains micromolar quantities of cAMP, one hundred times the levels found within cells or in plasma. This urinary cAMP is formed by the action of PTH on the Ca^{+2} pumping mechanism of the cells of the collecting tubules. It is not known how the cAMP is secreted into the urine in this unusual manner. In some cases of hypocalcemia, calcium reabsorption from the collecting tubules is impaired because the tubule cells do not respond to PTH. In this situation, called pseudohypoparathyroidism, serum levels of PTH are high but the urine contains very little cAMP. Presumably, PTH receptor sites are lacking in the cells of such patients or some other defect in the cAMP mechanism is present (end-organ resistance).

iii) The effect of PTH on the increased calcium absorption in the gastrointestinal tract is mediated by $1,25-(OH)_2D_3$ (PTH promotes 1-hydroxylase activity-- already discussed). Note the latter also participates in bone resorption. Thus in the vitamin D-deficient animals, the effect of PTH upon calcium mobilization is attenuated.

As noted earlier, there exists a complex relationship among serum Ca^{+2} and phosphate, PTH, and $1,25-(OH)_2D_3$, which functions to maintain calcium homeostasis. In a typical cycle, PTH is released in response to low serum Ca^{+2} which mobilizes calcium by direct action on the bone and by increasing the synthesis of the renal hormone, $1,25-(OH)_2D_3$. The vitamin D derivative proceeds to the bone (further mobilizing skeletal Ca^{+2}) and to the intestine (causing increased Ca^{+2} absorption). The feedback loop is completed when the serum Ca^{+2} rises sufficiently to shut off PTH secretion. Phosphate levels are also related to this process. Hypophosphatemia, by itself with no increase in PTH, can stimulate $1,25-(OH)_2D_3$ synthesis. This also suggests a mechanism whereby PTH may exert its influences on the endocrine function of the kidney (see Figures 139 and 140). The effects of PTH on the kidney occur within minutes (best suited for emergency adjustment) and on the bone within hours (recall, bone has the largest reserve of calcium). However, the intestinal response to PTH requires days to become apparent (note, this is mediated via the formation of $1,25-(OH)_2D_3$, and thus the delayed response).

f. Hyperparathyroidism. Primary hyperparathyroidism is the condition resulting from excess PTH production due to hyperplasia, adenoma or carcinoma (rarely) of the parathyroid glands. Tumors originating in non-parathyroid tissues, such as squamous cell carcinoma of the lung and adenocarcinoma of the kidney, may also secrete PTH-like polypeptides giving rise to hyperparathyroidism (ectopic production of the hormone).

Secondary hyperparathyroidism develops whenever there is

a negative calcium balance, leading to hypocalcemia.
These include chronic renal failure and malabsorption.
In chronic renal insufficiency, the increased secretion
of PTH is due to two causes: with a decrease in renal
mass, there is an increase in phosphate retention which
leads to elevated serum phosphate concentration resulting
in decreased serum calcium levels and promoting the
secretion of PTH to restore normal serum calcium; the
second cause involves a decreased amount of $1,25-(OH)_2D_3$
synthesis due to a decreased amount of the enzyme
1-hydroxylase leading to decreased intestinal calcium
absorption and resulting in low serum calcium levels
which again promptly stimulates the release of PTH.
Subjects who are undergoing long-term hemodialysis therapy
develop progressive hyperparathyroidism, which can be
arrested by the presence of adequate amounts of ionized
calcium in the dialyzing fluid. The methods used in the
prevention of secondary hyperparathyroidism due to chronic
renal disease include decreasing the dietary intake of
phosphate, the prevention of intestinal phosphate
absorption by the presence of phosphate binding agents
(aluminum hydroxide, calcium carbonate or calcium
lactate), and the administration of a vitamin D analogue
which does not require 1-hydroxylation in order to be
functional (e.g., $1\alpha-(OH)D_3$, dihydrotachosterol) or the
active metabolite of vitamin D itself, namely $1,25-(OH)_2D_3$.

The clinical manifestations in uncomplicated primary
hyperparathyroidism are due to hypercalcemia with
hypercalciuria and hyperphosphaturia. These are
hypotonicity of the muscle with general skeletal muscle
weakness, dysfunction of the smooth muscle resulting in
constipation, flatulence, anorexia, nausea and vomitting,
occasional cardiac irregularities, a high incidence of
peptic ulcers and pancreatitis, neuropsychiatric
manifestations, and a high incidence of urolithiasis
(which may lead to diffuse nephrolithiasis progressing
to severe renal insufficiency). Furthermore, exaggerated
PTH secretion due to any cause may lead to a bone disorder
known as osteitis fibrosa (also known as von Recklinghausen's
bone disease), in which bone is replaced by a highly
cellular, fibrous tissue. These changes are due to the
stimulation of resorptive activities of the bone
(accumulation and activation of osteoclasts). In osteitis
fibrosa, bone pain and pathologic fractures are the
principal clinical manifestations.

In the primary hyperparathyroidism the initial diagnostic
laboratory finding is hypercalcemia with low serum

phosphate. Measurements of the urinary cAMP level (which
is elevated due to the action of PTH on the renal cells--
already discussed) have also been used. Direct measurements
of plasma PTH concentration by radioimmunoassay (RIA) is
also employed in confirming the diagnosis. This test has
been found to be particularly useful in pre-operative
localization of abnormal glands, which is accomplished by
obtaining blood samples through selective catheterization
of venous drainage from the parathyroid glands. It should
be noted that the RIA for serum PTH has been complicated
by the fact that the hormone with 84 amino acid residues
released into the thyroid vein is rapidly degraded into
at least two major fragments by undergoing cleavage between
the 33 and 34 amino acid residues. The biologically active
amino terminus portion has a short half-life and shows
poor immunoreactivity with the antibody, whereas the
carboxy terminus portion (the larger fragment) is
biologically inactive but reacts strongly with the antibody
produced against the complete hormone. Despite these
limitations of RIA due to biologic, chemical and
immunological heterogeneity of circulating PTH peptides,
the assay has yielded useful information in the diagnosis
of the disorders of the parathyroid glands, because
immunoreactive, biologically inactive fragments are
derived predominantly from complete PTH molecules. With
the availability of the synthetic human biologically active
peptides, a specific RIA has been developed which should
aid considerably in the laboratory diagnosis of parathyroid
disorders. The treatment of primary hyperparathyroidism
consists of removal of the hyperplastic or adenomatous
glands by surgery. In subjects, when it has been clearly
established that the overactivity of the parathyroid
glands is due to chief cell hyperplasia (non-neoplastic
hyperfunctioning gland), a subtotal resection is performed,
in order to prevent the development of hypoparathyroidism.
In one study patients with primary parathyroid hyperplasia
were treated by total parathyroidectomy but with
autotransplantation of a portion of the gland into the
forearm muscle. It was shown that these grafted pieces
secreted PTH and were able to maintain normocalcemia.

g. Hypoparathyroidism can occur in subjects who have undergone
total thyroidectomy or parathyroidectomy for a variety of
reasons. Congenital forms of hypoparathyroidism are also
known. The disorder can also occur due to congenital
dysplasia of the structures of the third and fourth
pharyngeal pouches (DiGeorge Syndrome). Sporadic forms
are also known. As previously noted, pseudohypoparathyroidisms
are due to end-organ resistance to PTH. These patients,

despite the elevated levels of PTH, show <u>hypoc</u>alcemia and hyperphosphatemia, and the urinary excretion of cAMP is diminished suggesting the unresponsiveness of renal cells to PTH. Exogenous administration of PTH will not affect these parameters. <u>Albright hereditary osteodystrophy type I</u> is a type of pseudohypoparathyroidism. This is a sex-linked disorder. The clinical characteristics of the disorder are short stature, short fourth and fifth metacarpal and metatarsal bones, round facies, metastatic calcification, mental retardation, strabismus, cataracts, and tetany. The effects of tetany have already been discussed.

In general, hypocalcemia of hypoparathyroidism is managed with the administration of vitamin D and dietary supplementation of calcium.

3. Calcitonin

a. Calcitonin is a hypocalcemic peptide hormone of 32 amino acids and has a molecular weight of 3,590 daltons. It is produced by the parafollicular cells (known as "C" cells) scattered throughout the thyroid gland. Its presence has also been demonstrated in the parathyroid and thymus glands of humans. The C-cells are of neural crest origin, which migrated to the last branchial pouch during early embryogenesis. Total thyroidectomy may, therefore, not remove all cells capable of producing this hormone. Normal serum Ca^{+2} levels appear to cause continuous calcitonin secretion and the level of this hormone is directly responsive to the concentration of Ca^{+2} in the blood. Glucagon also stimulates calcitonin release and gastrin is reported to act similarly. The half-life of calcitonin in the blood is about 2-15 minutes, somewhat shorter than that of PTH.

b. The only site of action of calcitonin appears to be in the bone, where it can cause deposition of calcium and phosphate (as hydroxyapatite) and thereby rapidly reduce plasma concentrations of <u>both</u> ions. The effect is most profound in young or growing animals and in other conditions where bone remodelling is active. The effect decreases in the post-adolescent human and its homeostatic role in the adult man remains unclear. It is conceivable that the main function of calcitonin may be to prevent the occurrence of hypercalcemia during childhood.

c. Calcitonin appears to inhibit bone resorption by regulating both the number and activity of osteoclasts, making this hormone directly antagonistic to the action of PTH. The mechanism of action of calcitonin is unclear, but it does involve either protein synthesis or activation of adenylate cyclase. The parafollicular cells are capable of storing large amounts of calcitonin, in marked contrast to the parathyroid cells which store very little PTH. Both hormones are present in the plasma of adults (despite a marked decrease in the effect of calcitonin in the adult) and both have a linear response to changes in Ca^{+2} concentration. Calcitonin varies <u>directly</u> with Ca^{+2}, while PTH varies <u>inversely</u> with Ca^{+2}. An <u>increase</u> in the concentration of this ion <u>elevates</u> the calcitonin level while <u>decreasing</u> the PTH concentration.

d.   Elevated plasma levels of calcitonin (measured by radioimmunoassays using antisera to natural or synthetic human calcitonin) are found in patients with medullary carcinoma of the thyroid gland. The carcinoma originates from the parafollicular cells; this disorder occurs sporadically but in some instances it manifests as a familial disease transmitted as an autosomal dominant trait; the treatment consists of total thyroidectomy. In relatives who show hyperplasia of the parafollicular cells, basal and stimulated levels of calcitonin are increased. Increased plasma levels of calcitonin have also been noted in some patients with hyperparathyroidism, pseudohypoparathyroidism, Zollinger-Ellison syndrome (gastrin-stimulated secretion of calcitonin), carcinoid tumors, pernicious anemia, Paget's disease and chronic renal insufficiency.

e.   Since it decreases hypercalcemia and hyperphosphatemia and inhibits bone resorption, calcitonin is effective in the treatment of disorders which give rise to these abnormalities (e.g., vitamin D intoxication, hyperparathyroidism, idiopathic hypercalcemia, osteolytic bone metastasis, Paget's disease). In particular calcitonin has been useful in the treatment of chronic disorders characterized by continual rapid bone formation and resorption (i.e., increased skeletal remodelling). <u>Paget's disease</u> or <u>osteitis deformans</u> is such a disorder, which is characterized by deformities in the contours of the bones and internal skeletal architecture. The biochemical lesion is not known. The patients show normal plasma levels of calcium and phosphate with frequent hypercalciuria which leads to stone formation. The serum alkaline phosphatase level is significantly elevated (occasional elevation of acid phosphatase has also been noted). Increased urinary excretion of hydroxyproline is also observed in patients with the active disease. Calcitonin administered subcutaneously, either alone or with oral phosphate, has been found to be beneficial in the treatment of Paget's disease. For this purpose salmon calcitonin, which is most active among the calcitonins isolated from different species, is used. The two other drugs which are used in the control of disease activity are mithramycin and ethane-1-hydroxy-1,1-diphosphonate (EHDP). The diphosphonates contain a P-C-P bond and are structurally similar to inorganic pyrophosphate (P-O-P), which has been suggested to be involved in the regulation of bone calcium deposition and dissolution. However, unlike pyrophosphate, the diphosphonates are completely resistant to enzymatic (pyrophosphatase) breakdown. The therapeutic usefulness of diphosphonate is presumably due to its inhibition of bone resorption. They also, like pyrophosphate, can inhibit calcium and phosphate deposition.

4. Other hormones

The effect of glucocorticoids and thyroxine on the calcium and phosphorous homeostasis is rather perplexing. Both are required in low concentrations for normal bone growth. Additionally, administration of cortisone is clinically useful in correcting hypercalcemia and the hyperosteolytic activity of malignant bone disease, thyrotoxicosis, and vitamin D intoxication. Yet both agents in large amounts cause bone deterioration, an effect seen in Cushing's syndrome and thyrotoxicosis. The effects of hypercorticolism include inhibition of intestinal (antagonizes the action of $1,25-(OH)_2D_3$) and renal absorption of calcium, and augmentation of the resorptive effect of PTH resulting in the loss of bone mass (cortisol enhances protein catabolism and causes the dissolution of skeletal matrix). These effects lead to osteoporosis and secondary hyperparathyroidism. Table 43 summarizes information pertaining to various substances that participate in the calcium and phosphate homeostasis.

Some Disorders of Calcium and Phosphate Homeostasis

A number of disorders have already been discussed. A few of the major remaining disorders are discussed in this section.

1. In general, hypocalcemia (total serum calcium less than 7.0 mg/100 ml and the ionized calcium below 2.5 mg/100 ml) occurs due to various types of hypoparathyroidism (congenital, acquired, end-organ resistance), vitamin D deficiency or biochemical lesions in its metabolism, or other causes (hypoproteinemia, hypernatremia with potassium deficiency, calcitonin excess, dietary calcium deficiency, renal tubular defects, acute pancreatitis with extensive fat necrosis, gastrointestinal biliary or pancreatic disease). Early neonatal hypocalcemia (which leads to neonatal tetany) can occur due to the above causes but also is a result

Table 43. Substances Which Participate in the Homeostatic Control of Calcium and Phosphate. Some of these are discussed elsewhere in this chapter and in other parts of the book.

Substance	Source	Chemical Nature	Site of Action	Effect
Vitamin D	a) Diet b) UV irradiation of 7,8-dehydro-cholesterol, a precursor of cholesterol; or of ergosterol, a plant sterol	Steroid derivative; active metabolite: $1,25-(OH)_2D_3$; $25-(OH)$: Liver $1-(OH)$: Kidney	Intestinal mucosa; bone; kidney	Increases calcium and phosphate adsorption from intestine by increasing synthesis of calcium binding protein and calcium dependent ATPase in intestinal mucosa. Acts on bone to cause resorption; in the kidney enhances reabsorption of calcium and phosphate.
Parathyroid Hormone (hypercalcemic factor)	Parathyroid	Small protein; molecular weight is 9500 daltons	1) Bone: activates osteoclasts, causing removal of calcium and phosphate from bone into plasma (bone resorption) 2) Kidney: increases reabsorption of calcium (decreases urinary excretion of calcium); decreases phosphate reabsorption, thereby increasing urinary phosphate excretion 3) Intestinal mucosa: action mediated by $1,25-(OH)_2D_3$	Increases serum calcium and enhanced urinary phosphate excretion; hydroxyproline in the urine increases, due to increased collagenase action
Calcitonin (hypocalcemic factor)	Parafollicular cells (thyroid, parathyroid, thymus)	Small proteins; molecular weight is 3500-4000 daltons	Bone: increase deposition of calcium	Decreases serum calcium, and urinary phosphate and hydroxyproline (decreased collagenase action)
Growth Hormone	Anterior Pituitary	Protein; molecular weight is 21,500 daltons	Bone: activates osteoblasts to increase synthesis of collagen and chondroitin sulfate	Increased utilization of calcium during growth
Estradiol	Ovary	Steroid	Bone: affects osteoblasts	Specific action on epiphyseal cartilage of bone during puberty; causes ends of epiphyses to close and thereby stops bone elongation
Thyroid Hormone	Thyroid	Iododerivatives of tyrosine	Bone: affects osteoblasts	In growing animal it is synergic to growth hormone in cartilage growth; in adults, prevents bone resorption
Glucocorticoids	Adrenal Cortex	Steroid	Bone	Aids in calcium and phosphate deposition in bone; depresses osteolytic action of osteoclasts
Androgens	Testes	Steroid	Bone	Aids in calcium and phosphate deposition in bone

Primary Calcium Homeostatic Agents (Vitamin D, Parathyroid Hormone, Calcitonin)

of maternal complications (e.g., diabetes mellitus, toxemia, deficiency of dietary calcium and vitamin D, hyperparathyroidism), complications of delivery (e.g., prematurity, cerebral injury, etc.), and post-natal complications (e.g., respiratory distress syndrome, sepsis). Severe magnesium deficiency (due to malabsorption or dietary lack) or high phosphate diet (leading to hyperphosphatemia) can also lead to hypocalcemia, particularly after the fourth or fifth day of life. Note, the plasma levels of ionic magnesium also regulate PTH secretion in addition to ionic calcium in the similar direction, but severe hypomagnesemia is associated with inhibition of PTH release (due to inhibition of adenylate cyclase, a magnesium-dependent enzyme required for secretion of PTH), and decreased sensitivity to the calcium mobilizing effect of PTH (with the execution of its effect on the renal tubules).

2. Causes of hypercalcemia (values greater than 11 mg/100 ml of serum) include hyperparathyroidism, hypervitaminosis D, vitamin A intoxication, disuse osteoporosis, neoplasms (osseous metastases, ectopic secretion of PTH, solid tumors which secrete prostaglandin-like substances--inhibitors of prostaglandin synthesis such as aspirin reduces the hypercalcemia--see prostaglandin metabolism, Chapter 8), or other causes (e.g., granulomatous disease, milk alkali syndrome, hypoadrenocorticism, thyroid disorders, hyperproteinemia, thiazide diuretics, etc.). In addition to the above, neonatal hypercalcemia can occur due to maternal hypoparathyroidism or familial neonatal parathyroid hyperplasia.

Hypercalcemia in general causes hypercalciuria, which may lead to the formation of calcium containing renal stones. However, it should be noted that hypercalciuria is also associated with normocalcemia. The normocalcemic hypercalciuria may be the result of increased intestinal calcium absorption and excretion-absorptive calciuria (due to enhanced production of $1,25-(OH)D_3$-- discussed earlier) or an impairment in the renal tubular reabsorption of calcium-renal hypercalciuria. Attempts have been made to identify the various causes of hypercalciuria by measuring the levels of serum calcium and urine calcium, creatinine and cAMP. Alteration in these parameters are shown in Table 44. The creatinine measurement is performed in order to normalize calcium and cAMP levels in the urine.

3. Hypophosphatemia (levels less than 2.5 mg/100 ml of serum in adults; normal range for adults = about 2.5-4.8 mg/100 ml; normal range for infants during the first year of life = about 4-7 mg/100 ml; note all values are expressed as mg of elemental phosphorus/100 ml).

Table 44. Alterations of Biochemical Parameters in Different Types of Hypercalciurias

Type of Hypercalciuria	Fasting Serum Calcium	Urinary Parameters			
		Fasting		After Oral Feeding of Calcium (1g)	
		Calcium	cAMP*	Calcium	cAMP*
Absorptive (Increased intestinal absorption)	→	↑	↑	↑	↑
Renal (Decreased renal tubular reabsorption)	→	↑**	↑**	↑	↓ or ↑
Primary Hyperparathyroidism	↑	↑	↑	↑	↑

→ = normal; ↑ = increased; ↓ = decreased

* Urinary cAMP levels are taken as a measure of parathyroid function.

** A high urinary calcium loss produced secondary hyperparathyroidism resulting in increased cAMP excretion.

a. Phosphate depletion due to dietary deficiency is essentially unknown, because phosphates are widely distributed in natural foods. However, hypophosphatemia can occur essentially because of three types of problems: increased renal phosphate loss; preferential intracellular shift of phosphate into muscle, adipose tissue and liver from extracellular fluids, depriving other cells of phosphate (e.g., erythrocytes, nervous system); and diminished intestinal absorption and/or enhanced intestinal loss of phosphate. Conditions which give rise to any of the above abnormalities include genetic hypophosphatemia; starvation; malabsorption problems; liver disease; acute gout; acute myocardial infarction; pregnancy; hypothyroidism; hyperparathyroidism; vitamin D deficiency or excess; rickets and osteomalacia; alkalosis or acidosis; hypokalemia or hypomagnesemia; sepsis (gram negative and positive organisms); administration of sex hormones, catecholamines and thiazides; hyperventilation; renal tubular defects; intake of phosphorus binding agents (antacids); hemodialysis with phosphate poor solutions; chronic or acute alcoholism; diabetes; hyperalimentation with phosphate deficient solution; and glucose administration (particularly intravenous administration).

b. As noted earlier, phosphate participates in innumerable reactions, and they are discussed throughout this book. Thus, its deficiency affects all cells. In general, the numerous malfunctions are results of diminished ATP levels in phosphate-deprived cells and a reduction of both 2,3-diphosphoglycerate (2,3-DPG) and ATP in the red blood cells. Recall, 2,3-DPG affects the binding of oxygen to hemoglobin; lowering of 2,3-DPG levels causes higher affinity of hemoglobin for oxygen, and thus oxygen is less readily released to the tissues (see hemoglobin, Chapter 7) causing generalized tissue hypoxia. The drop in the red blood cell ATP also leads to their membrane alteration, as ATP is required for the maintenance of the erythrocyte membrane integrity. These cells assume a spherical shape (spherocytes) losing their normal biconcave shape. These disturbances affect the erythrocyte life span (eventually causing anemia) and microcirculation. Severe hypophosphatemia is potentially a fatal syndrome. The subjects develop paresthesia, muscular weakness and lethargy which may lead to convulsions and coma.

c. A point which should be emphasized is that in many clinical situations, well-intentioned iatrogenic maneuvers may drive a phosphate-depleted (but may be normophosphatemic) patient to severe hypophosphatemia, perhaps with its fatal consequence. Some of the clinical conditions which have the potentiality of developing profound hypophosphatemia during treatment are

diabetic ketoacidosis, acute alcoholism, carbohydrate loading,
hyperalimentation, etc. Note that a patient with diabetic
ketoacidosis is dehydrated, obtunded, hyperventilating and
acidotic. In the management of this disorder, the immediate
therapeutic goals are to provide insulin to normalize glucose
metabolism and to restore fluid and electrolyte balance.
Both of these maneuvers may lead to hypophosphatemia: a
shift of phosphate from extracellular fluids to intracellular
fluids for the metabolism of glucose and other substrates,
in particular muscle, liver and adipose tissue, depriving
phosphate from the rest of the cells of the body (e.g., the
central nervous system, red blood cells); and fluid replacement
with phosphate poor solutions.

The treatment of an acute alcoholic, who to begin with is
generally phosphate-depleted due to poor dietary habits and
complicated by vomiting and/or diarrhea, includes rehydration
and hyperalimentation. Both of these procedures can cause
hypophosphatemia. It should also be noted that a concurrent
hypomagnesemia can cause increased urinary excretion of
phosphates. Furthermore, an overzealous administration of
alkali in the management of acidosis in the above clinical
situation to hypophosphatemic patients, may be hazardous.
One important reason for this appears to be related to the
shifting of the hemoglobin-oxygen saturating curve by
alkalosis in the same direction (i.e., to the left of the
normal curve) as that produced of 2,3-DPG. Thus, the combined
effects of alkalosis and 2,3-DPG severely limits the capacity
of hemoglobin to release oxygen to the tissues. In these
patients, the plasma phosphate should be monitored carefully,
and if the hypophosphatemia were to develop, it should be
corrected promptly.

4. Disease of the Bone

    a. Osteoporosis (porosity of the bone). In this condition,
        normal bone material (both matrix elements and hydroxyapatite)
        are lost. This can proceed to a state where frequent,
        multiple fractures occur. Other manifestations include
        vertebrate compression, loss of vertical height and back pain.
        There are multiple causes for this disorder, giving rise to
        several different patterns of bone loss. The most common
        cause is the loss of anabolic hormones with advancing age
        (senile osteoporosis). While osteoporosis does occur in old
        men, it is more common in post-menopausal women where decreased
        levels of estrogens have a bone-wasting effect. The normalcy
        of the serum alkaline phosphate levels is useful in
        distinguishing this disease from rickets and osteomalacia.
        The plasma calcium and phosphate levels are normal. In post-

menopausal and senile osteoporosis, prolonged calcium
deficiency can be contributory to the development and
progression of the disorder (see calcium absorption, this
section). The treatment includes enhanced dietary calcium
intake (1.5 g/day), the presence of an adequate amount of
vitamin D, administration of estrogens and/or androgens, and
oral intake of fluoride. The fluoride is responsible for
rendering the bone more resistant to resorption. This is
made possible because the fluoride ions displace the hydroxyl
groups in the calcium hydroxyapatite crystals, and the
resulting fluoroapatite is more resistant to dissolution.
It should be noted, however, that optimal concentrations of
fluoride stabilizes the bone (and tooth enamel), whereas at
high concentrations it can cause osteosclerosis and
osteomalacia. This has been termed as osteofluorosis. In
some geographic areas (e.g., Punjab, India; Arabia), this is
a common form of bone disorder and is a result of water-mediated
excess fluoride intake.

Osteoporosis can occur due to nutritonal disorders such as
protein and caloric malnutrition, Kwashiorkor, chronic
alcoholism, scurvy, malabsorption syndromes, and calcium
deficiency.

As previously noted the endocrine disorders, namely
hyperthyroidism and hypercortisolism (e.g., Cushing's
syndrome), can cause osteoporosis.

Bed-ridden patients or normal individuals with prolonged
minimal physical activity develop negative calcium balance
(with loss mostly from weight-bearing bones) which leads to
osteoporosis (known as disuse osteoporosis). Prolonged
weightlessness, as in space flight, can give rise to a
similar disorder. These instances also result in
hypercalciuria.

b.   Rickets and osteomalacias (softening of the bone). In these
disorders there is a failure of deposition of calcium and
phosphate (mineralization) leading to unmineralized osteoid.
Rickets is primarily a disorder of children (thus the
mineralization defects during periods of growth involves
epiphyseal cartilage; ends of the large bones become enlarged)
whereas osteomalacia is a disorder of adults. These disorders
are results of inadequate amounts of calcium and phosphorus
in extracellular fluid and abnormalities in the matrix that
does not permit normal mineralization. These disorders can
occur due to vitamin D deficiency (dietary deficiency or
problems related to malabsorption), metabolic lesions
associated with the formation of the active metabolite of

vitamin D (see vitamin D metabolism, this section), phosphate deficiency, renal tubular defects leading to urinary loss of phosphate (Fanconi's syndrome), familial hypophosphatemia (a sex-linked dominant disorder--renal phosphate clearance is greatly increased; biochemical lesion not known), and renal osteodystrophy.

In rickets and osteomalacia caused by vitamin D deficiency, the characteristic biochemical parameters are normal or low plasma calcium, low plasma phosphate, elevated alkaline phosphatase, and very low levels of urinary calcium.

c. Osteogenesis imperfecta is an inherited disorder of defects in collagen (type I) biosynthesis. For details see under collagen metabolism, Chapter 5.

Magnesium Metabolism

1. Magnesium is the second most abundant cation (after K^+) within the cellular fluids and the fourth most abundant cation in the body. An average normal 70 Kg adult contains about 25 grams of magnesium. More than 50% of this magnesium is found in bone complexed with calcium and phosphate, the rest being present in the soft tissues and body fluids. The normal plasma levels of magnesium vary between 1.3 and 2.1 mEq/liter (or 1.6-2.3 mg/100 ml), about 30% of which is protein bound. Serum levels may not accurately estimate the total body stores of magnesium. The mechanism for the control of plasma magnesium levels are not completely understood, but parathyroid hormone (see PTH and calcium homeostasis, this chapter) and aldosterone play a role in this process.

2. Magnesium is indispensable to normal health; however, its exact daily requirement is not known. The recommended adult male dietary allowance is about 350 mg/day; an adult female is about 300 mg/day. During pregnancy and lactation, an intake of 450 mg/day has been found to be satisfactory in maintaining a positive balance. Foods of vegetable origin supply adequate amounts of magnesium (note chlorophyll contains magnesium). Other edible substances which contain magnesium are whole grains, raw dried beans, peas, cocoa, various nuts, soybeans, and some seafoods (100-400 mg/100 g).

3. Magnesium is principally absorbed in the small intestine (small amounts are also absorbed in the colon). $1,25-(OH)_2D_3$ does <u>not</u> aid in the absorption of the magnesium. However, calcium affects the magnesium absorption presumably by a competition of the absorptive pathway. However, the occurrence of a rare, genetic, selective, non-absorption of magnesium disorder suggests that

they are absorbed into the body by different mechanisms.
Approximately 33% of the absorbed magnesium is excreted in the
urine, which varies with absorption (diminished absorption with
decreased excretion). Fecal magnesium, which represents the
major portion of metals excreted daily, is unabsorbed magnesium.

4. Magnesium is required in many enzymatic reactions, particularly
    those in which ATP participates. As indicated under the comments
    on glycolysis (Chapter 4), the complex $ATP^{-4}\text{-}Mg^{+2}$ serves as the
    substrate in these reactions. In some instances, it can be
    replaced by Mn^{+2}. Magnesium and calcium have interdependent
    influences on the neuromuscular irritability. When the plasma
    magnesium levels fall below their normal values (hypomagnesemia),
    tetany, a functional disturbance also seen in hypocalcemia
    (discussed earlier), can result. Antagonism between magnesium
    and calcium can be observed when they are administered in higher
    concentrations. When the plasma magnesium level is increased to
    about 20 mg/100 ml, animals become anesthetized with a paralysis
    of peripheral neuromuscular activity. This action can be
    reversed by the intravenous administration of a suitable amount
    of calcium. There is no biochemical explanation available for
    this reversal of magnesium narcosis by calcium. That magnesium
    is an intracellular cation is shown by the observation that the
    magnesium concentration within a muscle cell is about ten times
    greater than the value for plasma magnesium. This unequal
    distribution between extracellular and intracellular compartments
    is comparable to that seen for K^+. The relationship between
    Mg^{+2} and Ca^{+2} is somewhat analogous to that between K^+ and Na^+.
    Aldosterone regulates the excretion of both K^+ and Mg^{+2} in the
    kidney.

5. <u>Hypermagnesemia</u> can occur in acute or chronic renal failure, in
    patients undergoing hemodialysis, and in patients who have
    received excessive amounts of magnesium salt during replacement
    therapy. The clinical manifestations are due to impaired
    neuromuscular transmission and resembles a curare effect. At
    serum magnesium levels of 5–10 mEq/liter, the cardiac conduction
    is affected, and at levels greater than 25 mEq/liter cardiac
    arrest occurs in diastole.

6. <u>Hypomagnesemia</u> is observed in many disorders or clinical
    situations, namely steatorrhea, alcoholism, diabetic ketoacidosis,
    thyrotoxicosis, hyperaldosteronism, hyperparathyroidism, following
    starvation and trauma after surgery, with intravenous fluid
    replacement, loss via gastric fluids, in infants with protein-
    calorie malnutrition (Kwashiorkor), in hypercalcemic states
    associated with malignant osteolytic metastases, in congestive
    heart failure after diuresis with mercurials, ethacrymic acid and
    furosemide. As noted earlier, the most distinctive clinical

manifestation of hypomagnesemia due to neuromuscular hyperexcitability (with normal serum calcium and blood pH) is tetany which usually occurs of serum levels less than 1 mEq/liter. Other clinical characteristics of hypomagnesemia are weakness, tremors, stupor, nausea, vomiting, behavioral disturbances, etc.

Other Inorganic Ions of Biological Importance

The importance of other major inorganic substances in human metabolism has already been discussed. These include iron (Chapter 7), iodine (Chapter 6), sodium, potassium and chloride (Chapter 12).

Trace Elements

1. This group of inorganic elements found in living systems are so named because they are present only in minute quantities (picograms to micrograms per gram of wet tissue or cells). These include aluminum, antimony, arsenic, barium, boron, bromine, cadmium, chromium, cobalt, copper, fluorine, gallium, iodine, lead, lithium, manganese, mercury, molybdenum, nickel, rubidium, selenium, silicon, silver, strontium, tin, vanadium, and zinc. Of these the following are believed to be essential for the maintenance of normal metabolism: chromium, copper, fluorine, iodine, manganese, molybdenum, nickel, selenium, silicon, tin, vanadium, and zinc. It should be noted, however, that many of these inorganic elements both essential and non-essential in higher concentrations are toxic to the living systems. Their toxicity is due to alterations in the membrane permeability, inhibition of protein synthesis, binding to nucleic acids and affecting their normal function, or inhibitions of enzyme-catalyzed processes (e.g., oxidative phosphorylation). While the acute toxic manifestations of a trace metal is known, the <u>chronic</u> effect and potential hazards to normal health of trace metals are largely <u>unknown</u>.

2. ## Copper Metabolism

   a. An extremely important trace element and is necessary for the normal function of many enzymes. Copper is an element which occurs widely. The suggested daily allowance for copper in adults is about 2.5 mg/day (0.05 mg/Kg body weight for infants and children). The average daily diet contains anywhere from 2.5 to 5.0 mg of copper, for exceeding normal requirements. Good dietary sources of copper are liver, kidney, shellfish, nuts, raisins and dried legumes. The predominant absorption of copper occurs in the stomach and upper small intestine, bound to ligands, such as amino acids and small peptides. A small protein (molecular weight: 10,000 daltons) appears to be involved in the mucosal transport; the role of a protein known as metallothionein with similar weight present in the mucosal cells which strongly binds to copper, zinc and cadmium in the intestinal absorption is not clear. Transition elements (cadmium, mercury, zinc, silver) antagonize copper absorption by competing for common binding sites. An adult individual contains a total of 100–150 mg of copper. This is distributed so that about 50% of the copper is in the muscles and the bones, about 10% is in the liver, and the remainder is distributed among the red blood cells, the plasma, and other tissues. As in the case of iron, most of the copper in the diet is lost in the feces and only a small part of it reaches the blood. Copper metabolism is also similar to that of iron in that, once copper is absorbed, it is almost completely retained by the body, and of the excretory routes, the <u>biliary system</u> is the predominant one. Thus, liver plays a major role in the copper metabolism.

b. The role of copper in human metabolism includes mitochondrial energy production, crosslinking of elastin and collagen, melanin formation, hematopoiesis, development of nervous and bone tissue.

   i) It is a constituent of a number of proteins, some of which are important enzymes. These include cytochrome oxidase, catalase, tyrosinase, monoamine oxidase (e.g., lysyl oxidase--see collagen and elastin metabolism, Chapter 5), ascorbic acid oxidase, superoxide dismutase, δ-aminolevulinic acid dehydrase, dopamine-β-hydroxylase, galactose oxidase, uricase, and ceruloplasmin. Ceruloplasmin, the major copper protein in serum, exhibits oxidase activity toward some polyphenols and toward amines, such as epinephrine and serotonin. Plasma levels of ceruloplasmin is determined by measuring the rate at which it catalyzes the oxidation of p-phenylenediamine by molecular oxygen. It has been demonstrated that ceruloplasmin is essential in promoting erythropoiesis, and this may be related to its ability to catalyze the oxidation of ferrous iron to the ferric state (i.e., it functions as a ferroxidase). Recall, iron is transported in the blood in the ferric state bound to transferrin and made available to bone marrow in the synthesis of red blood cells (see Chapter 7). However, patients with Wilson's disease, who have very low levels of ceruloplasmin, show a <u>normal</u> iron metabolism. Thus, the role of copper in the erythropoiesis is not understood.

   ii) Superoxide dismutase (contains both copper and zinc) catalyzes the inactivation of superoxide free radical anions and thus plays an important role in protecting the cells from the deleterious effects of this ion (see Chapter 4).

$$O_2^- + O_2^- + 2 H^+ \rightarrow O_2 + H_2O_2$$

This copper-containing enzyme is found in many tissues or cells such as erythrocytes, brain and liver, and they were named erythrocuprein, cerebrocuprein and hepatocuprein, respectively.

c.   The copper in the blood is distributed into two different fractions: plasma and erythrocyte. The plasma copper is contained as ceruloplasmin, complexed to albumin and amino acids. In the erythrocytes, it is present in superoxide dismutase and is bound to other proteins. In the serum, about 98% (60-99% depending on the species) of the copper is bound to ceruloplasmin, an α_2-globulin having a molecular weight of 160,000 daltons. It is synthesized in the liver. It contains 8 atoms of copper per molecule (about half in the form of Cu^{+2}) bound probably to histidyl and either lysyl or tyrosyl residues. The normal plasma level of ceruloplasmin is about 30 mg per 100 ml. Ceruloplasmin does not participate in the transport of copper in the blood. In the blood, copper in ceruloplasmin does not exchange with non-ceruloplasmin copper. However, the remainder of the copper present in the serum is loosely bound to the histidyl and perhaps the carboxyl groups of albumin. This may be a metabolically available source of copper for tissue use. The albumin-bound copper is known as direct-reacting copper because it is readily available for reaction with chromogens. This is in contrast to the ceruloplasmin-bound metal which requires prior treatment with HCl for its release.

The colorimetric methods for serum copper determination are based on treatment of the sample to release protein-bound copper, precipitation of the proteins, and reaction of the copper in the filtrate with a chromogen. In order of increasing sensitivity, diethyldithiocarbamate (yellow complex with Cu^{+2}), cuprizone (blue complex with Cu^{+2}), and oxalyldihydrazide (lavender color with Cu^{+2} in the presence of NH_4OH and acetaldehyde) are used as chromogens. Copper can also be determined by atomic absorption spectrophotometry, following its separation from the serum proteins. The principle of this method is the selective absorption of light by a flame containing ions of a particular metal. A solution of the metal being measured is drawn into a flame, where it is ionized. Light passing through this flame will have one or several wavelengths (energies) absorbed, corresponding to the electronic transitions which can occur in the metal ions. This method is both sensitive and selective (though subject to some interferences), permitting accurate measurement of a number of metals including most or all of the alkali metals (Na^+, K^+, Li^+, etc.), alkaline earths (Ca^{+2}, Mg^{+2}, etc.), and transition metals (copper, zinc, iron, cobalt, etc.).

d. <u>Wilson's disease</u> (also known as hepatolenticular degeneration)

    i)   This is a rare, autosomal recessive disturbance of copper metabolism, with a prevalence of about 1:200,000. The precise biochemical lesion is not known. In this disease, there are increased amounts of copper in various tissues, particularly in the liver, brain, kidneys, and the cornea of the eye.

    ii)   The disease is characterized by abnormalities in the brain, cirrhosis, excessive excretion of copper in the urine, and aminoaciduria. These patients also show greenish-brown Kayser-Fleischer rings at the limbus of the cornea. Usually the serum ceruloplasmin levels are markedly decreased (used in the diagnosis) with elevated non-ceruloplasin-bound copper. However, in some patients <u>normal</u> levels of serum ceruloplasmin have been observed indicating that the primary defect in Wilson's disease may not involve ceruloplasmin. The basic defect appears <u>not</u> to be involved with absorption of copper from the gastrointestinal tract or its uptake by the liver. The problem appears to reside in the disposition of copper in the liver. Abnormalities in lysosomal handling of copper and/or in a storage protein leading to its increased affinity and, therefore, decreased elimination of copper via the bile, may account for the metabolic lesions in this disorder.

    iii) Penicillamine (β,β-dimethylcysteine) is the drug of choice for treatment of Wilson's disease. The structure of penicillamine is:

$$\begin{array}{c} CH_3 \\ | \\ H_3C-C-CH-COOH \\ || \\ HS\ \ NH_2 \end{array}$$

It is a chelating agent for copper (as well as for mercury, zinc, and lead), which solubilizes these metals and promotes their elimination from the body in the urine. Its action is thus similar to that of deferoxamine on iron (see Chapter 7).

e. Menke's Kinky Hair Syndrome (also see under some of the
disorders of collagen and elastin metabolism--Chapter 5).
This is a rare X-linked, recessive copper-deficiency disorder.
The disorder is a result of a defect in the intestinal
transport of copper. The clinical manifestations are
progressive cerebral degeneration, abnormalities of the bone,
hair with less crimp and steely structure (pili torti),
arterial degeneration and hypothermia. Most of these effects
can be explained based upon the defect in the copper-dependent
enzyme due to copper deficiency. The patients show low serum
copper and ceruloplasmin levels, low levels of hepatic and
cerebral copper levels. Although therapeutic administration
of copper by intravenous route restores the serum copper
and ceruloplasmin levels, it does not bring about any
significant clinical changes or prevent early death.

f. Acquired copper deficiency in adults is very uncommon; however,
in infants the deficiency can occur due to malnourishment
with chronic diarrhea. These infants may show varying degrees
of anemia, pallor, retarded growth, edema and anorexia. In
general it should be pointed out that intravenous
hyperalimentation with preparations lacking in sufficient
amounts of copper may lead to copper deficiency and its
manifestations. These problems are responsive to copper
therapy.

3. Zinc Metabolism

This element is essential to growth, development and normal
function. It is a component of a number of enzymes (carbonic
anhydrase, alkaline phosphatase, alcohol dehydrogenase, lactic
dehydrogenase, carboxypeptidase, DNA-dependent RNA polymerase,
DNA polymerase, etc.) involved in the metabolism of nucleic
acids, proteins, carbohydrates and lipids. The recommended
dietary allowance is about 15 mg/day. An average diet in the
United States contains about 10-15 mg of zinc. Foods which are
good sources of zinc are meat, liver, eggs, seafoods (especially
oysters), milk and whole grains.

A clinical syndrome, found in Iranian and Egyptian boys whose
diets contain inadequate amounts of zinc, consists of small
stature, hypogonadism and mild anemia. Their zinc levels in the
plasma, red blood cells and hair are all decreased. After an
oral supplementation of zinc, these subjects showed clinical
improvements (i.e., puberty development and growth rates
acceleration).

Acrodermatitis enteropathica is an autosomal recessive disorder, and its clinical manifestations appear to be related to zinc deficiency. The clinical characteristics are severe chronic diarrhea, alopecia, wasting, and thickened and ulcerated skin around body orifices and extremities. The subjects with this disorder show markedly low levels of serum and hair zinc. The administration of zinc has been found to be beneficial. An acquired form of this disorder has been observed in subjects who have been maintained on long-term total parenteral nutrition with zinc-poor preparations. Wound healing appears to be delayed in patients with zinc deficiency which can be normalized with increased zinc intakes.

Zinc appears to play a role in the mobilization of vitamin A from the liver and maintain optimal concentrations in the plasma. Zinc deficiency in apparently healthy individuals may cause hypogeusia (abnormally diminished acuteness of the sense of taste), who respond to zinc therapy. In these patients, the salivary zinc has been reported to be low (serum levels may be normal) and upon administration of oral zinc (as zinc sulfate), the salivary zinc levels return to normal with the return of normal taste acuity.

4. There are several other elements which are required for normal body functions, are toxic in some way, or are used as drugs. Several examples are cited below.

    a. Lithium, in the form of its salts, has found particular use in the control of severe mania. This alkali metal has many similarities to Na^+ and K^+, being almost completely absorbed from the gastrointestinal tract and eliminated by renal excretion. Lithium is not treated entirely like sodium and potassium, however, being much more evenly distributed between the intracellular and extracellular fluids. Some of its diverse effects on different organ systems are probably related to this partial replacement of Na^+ and K^+ in bodily processes. Because Li^+ apparently can move across cell membranes more freely than can other electrolytes, it may also act by alteration of the chemical microenvironment of the cell. Its effects probably include increasing the osmolarity and thereby altering the water distribution, and modifying the pH and the ionic atmosphere of proteins and other biomolecules. Cyclic AMP-mediated processes which are regulated by peptide hormones also seem to be particularly disturbed by lithium, perhaps by one of the mechanisms described above. This suggests the usefulness of lithium salts in the management of thyrotoxicosis and other disorders in which there is excessive activation of adenylate cyclase by peptide hormones. Cases of lithium-induced diabetes insipidus have been reported where the metal apparently inhibits the ability of antidiuretic hormone (vasopressin) to synthesize cAMP, needed as a mediator in the control of water metabolism. There have been reports of lithium intoxication, some resulting in death.

b. Fluorine (as fluoride) is necessary for the prevention of dental caries (decalcification of the tooth enamel resulting from the action of microorganisms on carbohydrates during formation of the teeth). A fluoride concentration in the drinking water of about 0.9-1.0 mg/liter appears to be effective as a prophylactic. On the other hand, mottled enamel (dull, chalky patches on the teeth with pitting, corrosion, and deficient calcification) results from the consumption of water having fluoride in excess of 1.5 mg/liter during the years of tooth development (see osteofluorosis under calcium metabolism, this chapter). Fluoride has also been shown to stimulate adenylate cyclase by an unknown mechanism.

c. Selenium is an analogue of sulfur and can replace it in cysteine, methionine, etc. Plants grown in soil which has a high selenium content concentrate this element and can cause selenium toxicity when used as food by sheep and cattle. Such selenium-rich plants seem to not be harmful to humans. Other actions attributed to selenium include carcinogenesis in laboratory animals, enhancement of tooth decay (if ingested during tooth development), and antagonism of magnesium, copper and manganese levels in rats. This last effect was magnified when cobalt was also present in the diet. On the other hand, smaller amounts of selenium (as selenate, SeO_4^{-2}) appear to be necessary for prevention of hepatic necrosis in laboratory mice. Several seleno-enzymes (ones which require selenium for activity) have been identified. These include glutathione peroxidase in higher animals and bacteria and formate dehydrogenase in bacteria. This duality (requirement of low levels for normal growth but toxicity with slightly greater amounts) seems relatively common among a number of "trace nutrients."

d. Depending on the local environment, at least 43 different elements are normally incorporated into developing teeth and another 25 are seen in some instances. The remaining elements (notably the heavy metals) have never been detected. At least 14 different trace elements are demonstrably necessary for normal human health. The list of elements which are toxic in

some circumstances is also quite long. <u>Cadmium</u>, a waste-product
from certain industrial processes, was implicated as the cause
of itai-itai ("ouch-ouch") disease, a syndrome characterized
by severe, painful decalcification of bone and other tissues.
In the 1950's, it afflicted many residents of Japan's Jinzu
River basin. <u>Lead</u> poisoning has resulted from the ingestion
of alcohol prepared illegally in vessels partly fabricated
from lead. Air pollution and the consumption of flakes of
lead-based paint by children have been implicated as causes
of lead poisoning in the poorer sections of several cities.

A number of plants concentrate trace elements from the soil
and water. These include selenium (see above), strontium
(mesquite beans), sulfate (cabbage), and lithium (wolfberries,
used by Indians in the Southwest United States for making jam).

The application of trace metal analysis to clinical problems
has generally been very limited, but several observations
have been made which may prove useful in this regard. The
serum zinc concentration is elevated by physical trauma such
as operations and burns. The degree of elevation correlates
well with the severity of the injury and the extent of
recovery, returning to normal when the stress abates.

e. As levels of metals in the environment continue to rise and
as more examples of metal toxicity present themselves to
physicians, a greater knowledge of the metabolism of the
trace elements will develop. Additionally, there is the
problem of radioactive isotopes which are also increasing in
the environment.

10
Vitamins

with John H. Bloor, M.S.

General Considerations

1. The name <u>vitamin</u> arose from vita (= life, in Greek) + amine because
    thiamine (antiberiberi factor; vitamin B_1), the first essential
    food factor to be prepared in relatively pure form, was an amine.
    The word is now used to describe any of a rather loosely-defined
    collection of organic molecules which are needed by the body in
    small amounts for normal human growth, reproduction and homeostasis,
    but which it is unable to synthesize in adequate amounts. These
    include the <u>fat-soluble</u> vitamins (A, D, E and K) and the <u>water-</u>
    <u>soluble</u> vitamins (the B-complex and vitamin C). Note, there are
    other organic compounds which are not synthesized in the body and
    are required for the maintenance of normal metabolism. These
    comprise essential amino acids (see Chapter 6) and essential fatty
    acids (see Chapter 8). A number of other substances have been
    found to be essential food factors in various non-human species
    and a group of compounds exists which are classified as "vitamin-
    like." The latter substances will be discussed again later in
    this chapter.

    a. The B-complex vitamins are a group of about twelve chemically
        unrelated compounds which commonly occur together in dietary
        sources. Many are also synthesized by the intestinal flora.
        Of these, eight are distinct compounds (or groups of closely
        related compounds) known to be of dietary significance in man.
        They are thiamine, riboflavin, niacin, pyridoxine, biotin,
        pantothenic acid, folic acid, and cyanocobalamin. Deficiencies
        of these generally cause disorders of the nervous system,
        mucous membranes, skin, hematopoiesis, etc. Synonyms and
        properties of these compounds have been indicated throughout
        the book and will be summarized later in this chapter.

    b. Despite the fact that individual members. of the fat- and
        water-soluble vitamins are chemically dissimilar, these two
        general classifications have been retained because the
        compounds in each group have some properties in common. The
        fat-soluble vitamins are usually absorbed in the intestine,
        along with other lipids, into the intestinal lymphatics by
        processes facilitated by bile salts, and they are all
        originally derived from isoprenoid building blocks.
        Examination of the structures of vitamins A, E and K reveals

the presence of isoprene units. In the case of vitamin D, the structure itself does not show the isoprene units, but recall that the synthesis of its precursor requires isoprene building blocks (see cholesterol biosynthesis, Chapter 8). Significant amounts of the fat-soluble compounds are stored in the liver (perhaps as a consequence of their lipid solubility), while water-soluble vitamins are not generally stored in the body to any great extent.

c. It is sometimes convenient to use one name to designate a group of closely related substances. "Vitamin D" refers to ergocalciferol (D_2), cholecalciferol (D_3), their 25-hydroxy derivatives (25-(OH)D_2 and 25-(OH)D_3), and 1,25-dihydroxy-cholecalciferol (see vitamin D metabolism, Chapter 9). Similarly, pyridoxine (vitamin B_6) is actually a mixture of pyridoxol (for which pyridoxine is an accepted synonym), pyridoxal, and pyridoxamine while niacin (nicotinic acid) is really a combination of nicotinic acid and nicotinamide. Frequently, there is only one (or at most two or three) active form, however, to which the several members of the group are converted by the body. Thus, 1,25-dihydroxycholecalciferol is considered to be the final active metabolite of vitamin D, while the coenzyme forms of vitamin B_6 and niacin are, respectively, pyridoxal phosphate and the two related dinucleotides, NADH and NADPH.

2. Many vitamins were discovered through the observation of patients who were suffering from vitamin deficiency diseases. A substance was generally considered to be a vitamin if it was shown to be a normal dietary constituent absent in the diet of the afflicted individual which, when administered in pure form, would relieve the symptoms seen in the patient.

a. For example, scurvy (scorbutus) used to be observed frequently among sailors who ate no fresh fruit or vegetables during long sea voyages. Citrus fruits (for example, limes, hence the name "limey" or "lime-juicer" for British sailors) were known to relieve this disorder. It was later shown that large amounts of ascorbic acid (vitamin C) were present in these foods and that pure ascorbate was effective as an antiscorbutic (hence the name ascorbic acid). It is rare to encounter the deficiency of a single vitamin, except in instances such as those discussed below, where the dietary supply is adequate but utilization of the vitamin is impaired. Frequently, if one vitamin is deficient, others are also lacking and the intake of protein, trace elements, and other nutrients is probably insufficient. This is particularly true of the members of the B-complex.

b. Vitamin deficiencies <u>are not always caused by dietary insufficiency</u>. In addition to nutritional inadequacy, other causes which can give rise to vitamin deficiency are malabsorption, pharmacological agents, and biochemical lesions affecting the metabolism of vitamins. Malabsorption syndromes are known where the vitamins are ingested but largely excreted in the feces. In disorders which cause steatorrhea (e.g., biliary obstruction, pancreatic disorders), the fat-soluble vitamins are generally poorly absorbed. Cases of congenital vitamin B_{12} deficiency are known which are due to the absence of a protein factor needed for B_{12} uptake in the intestine (discussed later). The intestinal bacteria are capable of synthesizing some vitamins (notably K and those of the B-complex). Consequently, loss of these organisms (as in sulfonamide and antibiotic treatment) may cause the symptoms of vitamin K deficiency disease. Certain drugs may antagonize vitamin metabolism resulting in an inadequate supply for the body's needs. For example, isoniazid (used in the treatment of tuberculosis) is reported to produce the symptoms of pyridoxine deficiency, presumably due to interference with the metabolism of this vitamin. Isoniazid is a competitive inhibitor of pyridoxal kinase which catalyzes the conversion of pyridoxal to pyridoxal phosphate, the active coenzyme form of the vitamin. Another example, already discussed, is the use of folate antagonists (methotrexate, etc.) in the treatment of neoplasia (see Chapter 5). Anticonvulsant drugs (e.g., phenobarbital and dilantin) occasionally produce rickets or osteomalacia. This has been attributed to the stimulation of the production of non-specific drug-metabolizing microsomal enzymes by these pharmacological agents which also inactivate vitamin D (and its metabolites).

Like any other metabolite (e.g., amino acids, carbohydrates), the participation of vitamins in the normal metabolism also requires a number of transformations mediated by membrane-binding proteins, transport proteins, and enzymes (including enzymes involved in the activation of the vitamin and the processes concerned with the synthesis of apoenzymes). The synthesis of all of these proteins is under genetic control and therefore a wide variety of different mutations can interfere with the activity of a vitamin. An appreciation of these aspects will become apparent as one examines the metabolism of vitamin B_{12} (discussed later), folate (Chapters 5 and 6), vitamin D (Chapter 9), and others discussed throughout this book.

As opposed to a <u>nutritional</u> vitamin deficiency which is acquired and correctable with physiologic doses, the deficiency

diseases with a genetic etiology, while exhibiting the same
biochemical and clinical manifestations, require pharmacological
doses (i.e., administration of large doses, 10-1000 times the
normal dose) or specific metabolites of vitamins (e.g.,
administration of $1,25-(OH)_2D_3$ in vitamin D-dependent rickets--
see Chapter 9) overcoming the genetic block. The group of
disorders with genetic etiology and which requires pharmacological
doses of the vitamin for the correction of deficiency are known
as <u>vitamin-dependent</u> or vitamin-responsive conditions. In some
of these disorders, the route of administration may have to be
changed from oral to parenteral. There are more than two
dozen vitamin-dependent (or responsive) metabolic disorders
which have been reported. Some of these are discussed under
the respective vitamins. Note (discussed in Chapter 9) that
the two vitamin D-responsive disorders are familial
hypophosphatemia (also known as vitamin D-resistant rickets;
inherited as a sex-linked dominant character with complete
penetrance but with varying expressivity; the nature of the
genetic defect is not known; unresponsive to the treatment
with physiologic doses of $1,25-(OH)_2D_3$; and requires 50,000-
200,000 units of vitamin D daily) and vitamin D-dependent
rickets (renal 1-hydroxylase deficiency; treatable with large
doses of vitamin D or physiologic doses of $1,25-(OH)D_3$).

c. A variety of techniques or procedures are used in the
assessment of nutritional status with respect to vitamins.
These include direct measurement of vitamins in blood, urine
or other biologic samples, measurement of a metabolite of
the particular vitamin in consideration, measurement of a
metabolite accumulating due to the deficiency of the vitamin,
and functional tests (loading tests or the determination of
enzyme activity which is dependent upon the presence of
vitamin in its appropriate active form). The examples of
these are given in this chapter and others.

d. To plan diets which contain adequate supplies of the vitamins
and to aid in diagnosing vitamin deficiency diseases, it is
necessary to know how much of these materials are needed by
members of the population. For this purpose, two sets of
standards have been established.

  i) The <u>Minimum Daily Requirement</u> (MDR) is defined as the
  smallest amount of a substance needed by a person to prevent
  a deficiency syndrome. This is also considered to represent
  the basic physiological requirement of the material by the
  body. These values are established by the United States
  Food and Drug Administration. The MDR is not known for
  all vitamins.

ii) The <u>Recommended Daily Allowance</u> (RDA) is the daily amount
of a compound needed to maintain good nutrition in most
healthy people. These values are intended to serve as
nutritional goals, not as dietary requirements. They are
defined by the Food and Nutrition Board of the National
Academy of Science of the United States.

RDA values have also been established for mineral elements,
energy (calories, carbohydrates, and fats), protein, and
electrolytes and water. The MDR for some of these substances
has also been defined. Of these two, the RDA is used more
widely today than is the MDR. The RDA's for the vitamins are
given later, with the individual compounds. Many of these
values are given in International Units (I.U.), which are
empirically defined for each substance. It is appropriate to
note here that, in general, <u>vitamins are needed only in small
quantities</u>. As was pointed out earlier (see coenzymes,
Chapter 3), many vitamins serve as coenzymes or coenzyme
precursors and, as such, play a <u>catalytic</u> role in body
function. Many examples of this have already been noted
among members of the B-complex.

e. In the case of vitamins A and D, overdosage produces
<u>hypervitaminosis</u>. In recognition of this, the United States
Food and Drug Administration recently ruled that vitamin A
in doses greater than 10,000 I.U. and vitamin D in doses
greater than 400 I.U. will be available on a prescription-only
basis. Additionally, all products which contain quantities
of a vitamin in excess of 150 percent of its RDA will be
classified as drugs. This allows closer regulation of the
vitamins and will permit a more rapid response to any future
cases of toxicity which may result from chronic or acute
overdosage. A and D hypervitaminosis will be discussed in
the sections on these vitamins.

3. As was implied above, certain vitamins can be synthesized by man
in limited quantities.

a. Niacin (nicotinamide), one of the B-complex vitamins, can be
formed within the body from tryptophan (see Chapter 6). This
pathway is not active enough to satisfy all of the body's
needs but, in calculating the recommended allowance for
niacin, 60 mg of dietary tryptophan can be considered equivalent
to one mg of dietary niacin. In Hartnup's disease (see
Chapter 6), a rare hereditary disorder in the transport of
monoamino monocarboxylic acids (e.g., tryptophan), a pellagra-
like rash may appear. This suggests that, over a long period
of time, dietary intake of niacin is insufficient for metabolic

needs. This pattern also occurs in <u>carcinoid syndrome</u> in which much tryptophan is shunted into the synthesis of 5-hydroxytryptamine.

b. As described in Chapter 9, vitamin D can be synthesized by the body provided radiant energy is available for the conversion:

$$\text{7-dehydrocholesterol} \xrightarrow{\text{photons}} \text{cholecalciferol (vitamin } D_3).$$

Since dietary vitamin D is necessary, the amount supplied by this pathway is apparently not adequate for the body's needs.

Fat-Soluble Vitamins

1. Vitamin A

   a. This vitamin is needed for vision, normal growth, reproduction (in both males and females), and normal embryonic development.

   i) The earliest sign of vitamin A deficiency is night blindness (discussed later). Keratinization and cornification of the epithelial cells of all tissues occur, the best-known changes in humans being the development of xerophthalmia and "toad skin." The latter lesion may also involve B-complex deficiencies. The two visual disorders occur independently of each other. Males become sterile and, although deficient females are able to conceive, the offspring seldom survive to term due to placental defects. Failure of skeletal growth is also seen early in vitamin A deficiency. Because the most characteristic of these symptoms manifest themselves only during growth and because of significant liver storage of this vitamin, it is less common to see the deficiency disease in adults than in children. Additionally, several years' supply of vitamin A is stored in the liver of a normal adult. This supply is absent in neonates and very young children.

   ii) Although its role in vision is fairly well understood, the mechanisms by which vitamin A performs its other functions remain obscure. The RDA for vitamin A is 5,000 I.U., one I.U. of vitamin A being equivalent to 0.3 µg of all-<u>trans</u> β-carotene.

   b. Eggs, butter, cod liver oil, and the livers of other fish are the principal dietary sources of retinol (vitamin A_1). In these materials, it is present as the free alcohol (shown below) and as retinyl esters of fatty acids (primarily palmitic acid).

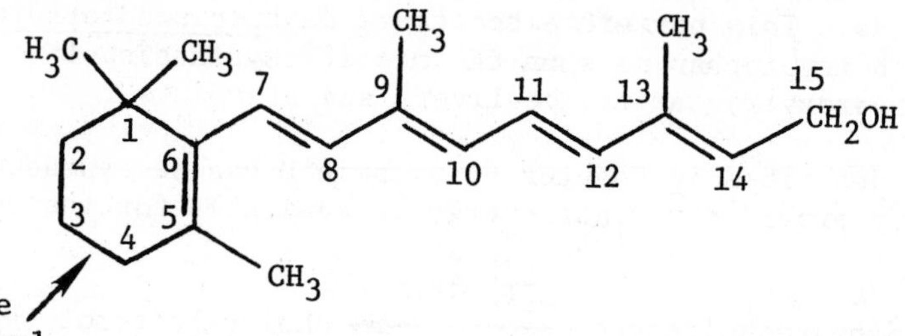

β−ionone
(trimethyl−
cyclohexenyl)
ring

All-<u>trans</u> retinol (Vitamin A$_1$)

If the alcohol at carbon 15 is oxidized to an aldehyde or an
acid, the resulting compounds are, respectively, retinal and
retinoic acid. Retinal is the metabolically active form of
vitamin A in some processes and can be converted to forms
which entirely fulfill the need for the vitamin. The acid,
while perhaps being the form needed for growth, is probably a
catabolic product, unable to entirely replace the vitamin.

i) In most of the dietary retinol, the four sidechain double
    bonds are in the <u>trans</u> conformation as shown above. <u>Trans</u>
    refers to the fact that bonds to adjacent atoms of the
    chain are on opposite sides of the planar double bond.
    This type of isomerism is important in vitamin A chemistry.
    It is further illustrated below, with 2-butene as an
    example.

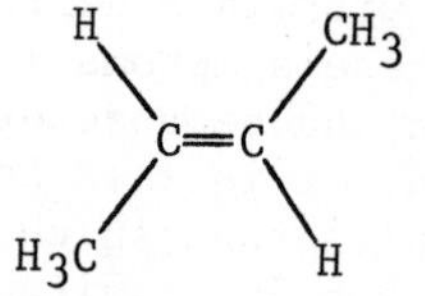

trans−2−butene 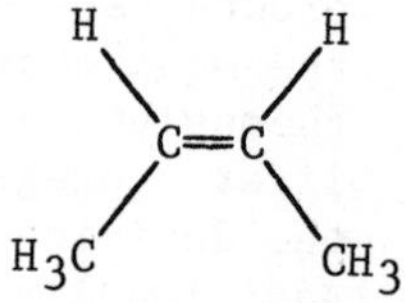   cis−2−butene

In this example, the methyl groups can be on <u>opposite</u>
sides of the planar double bond (<u>trans</u> to each other) or on
the <u>same</u> side (<u>cis</u> to each other). The double bonds of
retinol are readily oxidized by atmospheric oxygen,
inactivating the vitamin, unless protected by the presence
of antioxidants such as vitamin E.

ii) The retinyl esters are hydrolyzed by one or more specific
    hydrolases (derived from pancreatic juice-pancreatic
    retinyl ester hydrolase or intestinal brush border
    hydrolase) in the lumen and brush border of the intestine
    prior to their absorption. In the mucosal cells, they
    are re-esterified, incorporated into the chylomicrons,

and transferred to the lymph. After the delipidation of
the chylomicrons in the blood, the remnant chylomicrons
are taken up by the liver and stored as retinyl esters.
In an adult human receiving the RDA of vitamin A, a
year's supply or more may be retained by the liver,
making it very difficult for vitamin A deficiency to occur.

c. Vitamin A can also be derived from a group of plant pigments
    known as <u>carotenes</u>. These are present in carrots, mangoes,
    cantaloupe, and other fruits and vegetables. One-half cup of
    canned carrot juice, for example, contains about 20,000 I.U.
    of vitamin A (four times the RDA). The structure of all-<u>trans</u>
    β-carotene, the most important of these, is as follows:

β-carotene is cleaved
here to produce retinol

All-<u>trans</u> β-carotene

This molecule is cleaved at the indicated bond by β-carotene-
15,15'-dioxygenase, a soluble enzyme present in the intestinal
mucosa, producing two molecules of retinal. This retinal is
then reduced to retinol by NADH or NADPH and absorbed by the
route described previously.

d. As indicated in figure 141, vitamin A is mobilized from the
    liver as the free alcohol. Zinc appears to be involved in
    the mobilization of vitamin A from the liver. In the plasma,
    it is normally bound to a specific <u>retinol binding</u> protein
    (molecular weight: 21,000 daltons), each protein molecule
    being able to bind one molecule of retinol. This complex
    is associated, in turn, with prealbumin (55,000 daltons), so
    that the retinol transport complex is actually about 76,000
    daltons in mass. It migrates electrophoretically as an
    α-globulin. The interaction with prealbumin is highly
    specific, with an association constant of 10^6–10^7. In normal
    individuals, fewer than 5% of the total circulating vitamins
    are present as retinyl esters.

e. The serum concentration of vitamin A appears to depend on the
    amount of retinol binding protein available and hence, on the

Figures 141. Sources and Metabolism of Vitamin A in the Human Body

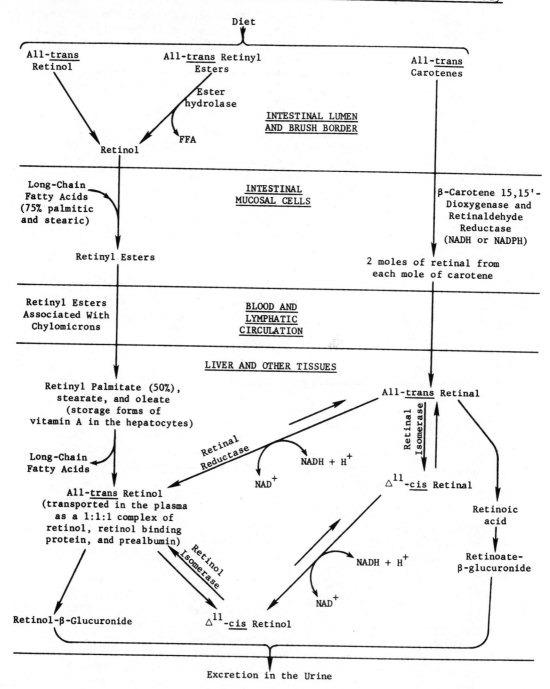

Note: The relative lengths of the arrows in the reversible reactions indicate the direction of the equilibrium.

rates of synthesis and degradation of this carrier. Patients
suffering from parenchymal liver disease and protein-calorie
malnutrition have reduced amounts of both the binding protein
and vitamin A in the circulation. When an adequate diet is
provided to malnourished children, plasma retinol binding
protein and retinol concentrations both increase, even without
vitamin A supplementation. This indicates the presence of
vitamin A stores which could not be mobilized on the inadequate
diet.

At least in rats, the converse is also true. That is, in
vitamin A deficient animals, the binding protein accumulates
in the liver. Repletion with vitamin A results in its rapid
release. The amount released is proportional to the dose of
vitamin A. The rapidity indicates the release of pooled
protein rather than new synthesis.

The normal rate of production of retinol binding protein
appears near to the maximum rate. In normal individuals,
there are 25.3-67.1 µg of it per milliliter of plasma. In
patients with hypervitaminosis A (up to seven times the
normal concentration of total plasma vitamin A), values were
within or only slightly above this normal range.

f. As was indicated above, retinoic acid is at least partly a
    catabolic product of retinol and retinal. It can be converted
    to its β-glucuronide and excreted in the urine or it can be
    decarboxylated. It has also been shown that retinol is
    excreted in the urine as its β-glucuronide. These and other
    metabolic routes of vitamin A are shown in Figure 141.
    Although it is known that the 11-cis isomers of both retinol
    and retinal are converted to the all-trans forms in the liver,
    the isomerase enzymes have not actually been studied. Retinal
    reductase is also found in the retina. The conversion of
    all-trans retinal to the 11-cis isomer can also be accomplished
    by 380 nm radiation but the process is slow and inefficient,
    requiring high levels of illumination.

g. Retinal is the form of vitamin A involved in the visual
    process. Rhodopsin (visual purple) is a complex of opsin
    (a protein) with 11-cis-retinal (details discussed later).
    The 11-cis-isomer is regenerated and rhodopsin forms again.
    Rhodopsin is localized in the retinal rods of the eye which
    are responsible for the basic detection of light (as opposed
    to color vision). Night blindness (nyctalopia), due to a
    vitamin A deficiency, is thus detected by noting an increase
    in the minimum amount of illumination which can be sensed
    by the eye (the visual threshold).

h. The visual process.

    i) This is a very complex process and consists of many
parts. The retina contains photoreceptor cells which
convert the light signals to electrical pulses (process
known as transduction); these pulses are transmitted
to the brain via the optic nerve. The other components
of the vertebrate eye, namely the cornea, iris and lens,
provide the necessary optical system for the proper
focusing of light on the retina. Thus, the retina is
a thin layer of light sensitive cells (photoreceptors)
and interconnecting nerve cells.

    ii) The photoreceptor system. There are two types of cells:
rods and cones. Rods are extremely sensitive, function
in dim illumination (designed for nighttime black and
white vision), and are concentrated around the periphery
of the retina; in the human eye, there are about 100
million cells. Cones, on the other hand, are involved
in color vision, are less sensitive (i.e., they function
under bright conditions), show different spectra due to
the presence of three different types of cone cells, and
are abundant in the center of the retina; there are only
about 5 million cones in the human eye. In general,
several rods are connected to a common neuron, whereas
each cone is connected to a single neuron.

    iii) In order to understand the molecular basis of the visual
process, a description of the structure of these cells
is essential. These cells are named rod cells because
of their cylindrical shape (length: 50 µM and diameter:
1 µM). The outer segment of each rod contains 500–2000
flattened sacs known as disks, which are about 160 Å
thick. The plasma membrane surrounding the outer segment
of the rod cell apparently is <u>not</u> attached to the disks.
In other words, the plasma membrane is electrically
isolated from the disks. The outer segment is only
concerned with the photoreception and transduction
process, while the inner segment is relegated to the
role of the maintenance of normal cell metabolism (e.g.,
ATP production, protein synthesis, etc.). The rod cell
in itself has a long life; however, the disks in the
outer segment are <u>continually replenished</u> by replacement
of new disks at the end nearest to the nucleus and
removal at the other end. This process of regeneration
is essential to the maintenance of the normal visual
process. Hereditary defects in regeneration of disks
which lead to blindness have been observed in rats.
Cone cells, in contrast to rods, have no disks; their

outer membrane is invaginated and appear to have no
rejuvenation system.   Figure 142 is a schematic
representation of a rod cell.

iv)   The disks are composed essentially of proteins and
phospholipids in about equal amounts.   Approximately 40%
of the protein is rhodopsin, the photoreceptor protein
of the rod cells.   Rhodopsin is a lipoprotein insoluble
in water (contains about 50% hydrophobic amino acids),
has a molecular weight of about 35,000 daltons, and is
bound to the lipid bilayer of the disk membrane.   Recent
studies have suggested that rhodopsin can move about
freely on the surface of the bilayer membrane showing
both rotational and translational motions (see fluid-
mosaic structure of membrane, Chapter 8).   This may also
occur with photoreceptor proteins of the cone (known as
iodopsin) located on the invaginated membranes.   The
photoreceptor portion of the molecule of the rhodopsin
is 11-cis-retinal.   The retinals are linear polyene
aldehydes and contain delocalized π electrons, which
determine many of their physical properties.   Rhodopsin
shows a broad absorption band with a maximum at 500 nm
and has a very high extinction coefficient at this
wavelength, making it ideal for black and white reception.
It should be noted that opsin, by itself, does not absorb
visible light but varies in its chemical nature from one
cell type to the other.   It is remarkable that the
chromophore (vitamin A_1 aldehyde or vitamin A_2 aldehyde-
3 dehydroretinal) is the same as that found in all visual
systems (molluscs, anthropods, and vertebrates).   However,
the visual pigment in differnet species exhibits different
absorption spectra which are attributed to the differences
in the chemical nature of the protein (opsins) and their
interaction with the chromophore.

There are three different pigments present in the
retinal cones which are responsible for color vision
("cone vision").   Each has a different absorption
maximum: 430 nm (blue), 540 nm (green), and 575 nm
(red), in accord with the trichromatic theory of color
vision.   The pigments all contain 11-_cis_-retinal but
appear to differ with respect to the opsin (protein
portion) which they possess.   Each of the three types
of hereditary color-blindness is due to the absence of
a specific one of these opsins, presumably due to
mutation of one of the three genes coding for these
proteins.

The 11-cis-retinal is attached to the opsin through a

Figure 142. <u>A Schematic Representation of a Mammalian Rod Cell</u>.
Note there are about 500-1500 discs in the outer
segment.

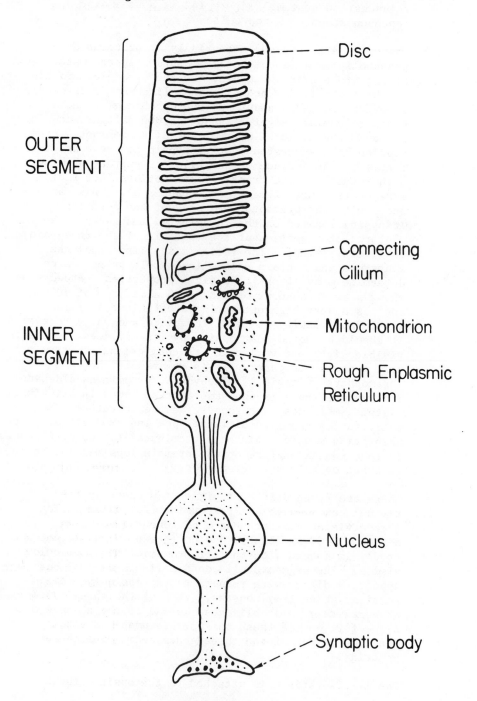

OUTER
SEGMENT

INNER
SEGMENT

Disc

Connecting
Cilium

Mitochondrion

Rough Enplasmic
Reticulum

Nucleus

Synaptic body

protonated Schiff base linkage involving the aldehyde
group of the chromophore and the ε-amino group of a
specific lysine residue of the protein.

$$
\overset{\overset{\displaystyle H}{\displaystyle |}}{C_{19}H_{27}C}=O + H_2N-(CH_2)_4-\text{opsin}
$$

11-cis-retinal   lysyl side chain

H_2O

$$
\overset{\overset{\displaystyle H}{\displaystyle |}}{C_{19}H_{27}C}=N-(CH_2)_4-\text{opsin}
$$

H^+

$$
\overset{\overset{\displaystyle H}{\displaystyle |}}{C_{19}H_{27}C}=\overset{+}{N}-(CH_2)_4-\text{opsin}
$$

Protonated Schiff Base Opsin

It is of interest to point out that of the several
possible stable isomers of retinal in nature, only the
11-cis compound has been found.

v) The primary photochemical process consists of
    isomerization of 11-cis-retinal to all-trans (see
    Figure 143) configuration which finally is responsible
    for the generation of a nerve impulse. The isomerization
    results in the hydrolysis of the Schiff base linkage with
    the chromophore being <u>detached</u> from the opsin. When
    rhodopsin is exposed to light, its color changes from
    red to yellow; this change is termed bleaching. The
    whole process occurs within a time period of a few
    milliseconds and involves many intermediates:

Figure 143. The Light-Induced Isomerization of the 11-cis-Retinaldehyde Group of Rhodopsin to all-trans-Retinal

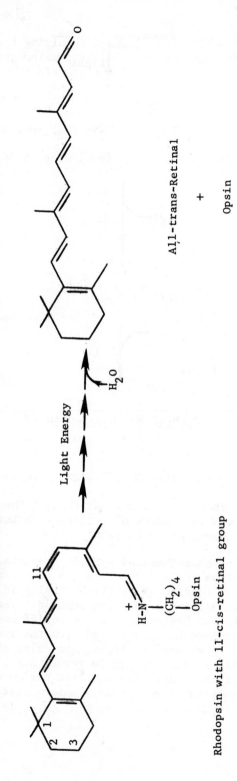

Rhodopsin with 11-cis-retinal group

Light Energy

H_2O

All-trans-Retinal

+

Opsin

Note the substantial change in the shape of the retinal molecule when 11-cis-retinal is isomerized to all-trans compound--the primary photochemical process.

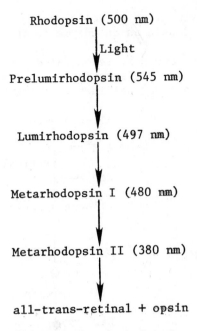

Rhodopsin (500 nm)

Light

Prelumirhodopsin (545 nm)

Lumirhodopsin (497 nm)

Metarhodopsin I (480 nm)

Metarhodopsin II (380 nm)

all-trans-retinal + opsin

vi) The change in conformation of rhodopsin accompanying the conversion of 11-cis-retinal to all-trans-retinal is responsible for the generation of the nerve impulse. This process is explained as follows. In the absence of light, the outer segment of the rod is permeable (which is unusual) to Na^+ ions. Therefore Na^+ ions enter the cell at the outer segment, and it is pumped out of the cell from the inner segment by an ATP-mediated process, in maintaining the dark current. In the presence of light, the permeability of the membrane to Na^+ decreases with the result that the membrane becomes more negative on the inside. The change in the decrease of membrane permeability is proportional to the intensity of impinging light. Thus the effect of light is to decrease the permeability of Na^+ with an increase in the polarization of membrane (i.e., hyperpolarization). The hyperpolarization is then transmitted via the synapse at the base of the rod cell through the neuron network, for ultimate processing in the visual cortex of the brain.

The mechanism by which the action of light on rhodopsin brings about a decrease in the outer membrane permeability to Na^+ is not completely known. In one mechanism calcium ions released from the interdiskal space due to conformational changes of rhodopsin upon photon absorption have been suggested to block the sodium pores, causing hyperpolarization. Other mechanisms include the involvement of a cyclic nucleotide and the functioning of rhodopsin as an enzyme in the hyperpolarization process.

vii) The rhodopsin is regenerated from the combination of opsin and 11-cis-retinal. The latter is obtained from all-trans-retinal in the following reactions:

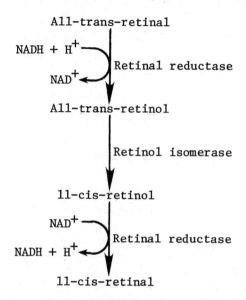

11-cis-Retinal can also be derived from other tissues (e.g., liver). The dissociation and the regeneration of rhodopsin are summarized as follows:

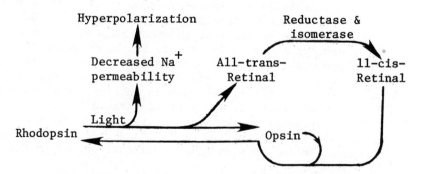

i. Hypervitaminosis A, mentioned earlier, is characterized by skeletal decalcification (in chronic cases), tenderness over the long bones, dermatitis (redness, drying, and scaling, resembling sunburn), headaches (due to elevated cerebrospinal fluid pressure--symptoms may be mistaken for cerebral neoplasm), and sometimes a yellow dyspigmentation of the skin caused by

carotene deposition. A chronic dosage of 25,000–50,000 units
per day in children or over 100,000 units per day in adults
for a period of several months is probably necessary for
manifestation of the more severe symptoms. The Food and Drug
Administration of the United States has imposed a ceiling of
10,000 I.U. on the amount of vitamin A that can be present in
the various multivitamin preparations available without
prescription.

Total serum vitamin A is elevated in cases of vitamin A
toxicosis, with most of the elevation being due to an increase
in retinyl esters. Unesterified retinol in the transport
complex is generally increased less than two-fold while total
serum vitamin A is raised two- to eight-fold. Recent work
has shown (in rats and humans) that the retinyl esters are
associated with the plasma lipoproteins. As indicated before,
retinol binding protein is normal or only slightly elevated
in this disorder.

Vitamin A is a surfactant and its toxicity appears due to its
ability to labilize and disrupt biological membranes. Excessive
amounts of retinol increase the synthesis and release of
lysosomal hydrolases. The destructive action of these enzymes
on tissues may account for many of the effects of
hypervitaminosis A. Only free retinol, or retinol found in
association with plasma lipoproteins, is harmful rather than
retinol specifically bound to retinol binding protein.
Apparently the specific binding protein transport complex is
stable enough to preclude the release of sufficient retinol
for the manifestation of toxic effects. In in vitro systems
designed to evaluate the toxicity, addition of the retinol-
retinol binding protein complex is harmless and the binding
protein alone is able to protect against the prior addition
of free retinol. Thus it would appear that the symptoms of
vitamin A toxicosis would not appear until the available
binding protein capacity had been exceeded.

2. Vitamin D

   a. The structure, function, and metabolism of this vitamin are
      discussed in Chapter 9 (Mineral Metabolism). It is needed
      for the proper intestinal absorption of calcium and is
      thereby intimately involved in the actions of parathyroid
      hormone (PTH) and calcitonin on calcium metabolism (see
      Chapter 9). It also aids in the regulation of phosphate
      levels by promoting its absorption in the intestines. In
      the Fanconi Syndrome, a renal tubular reabsorption disorder,
      the calcium and phosphate losses are treated with large
      doses of Ca^{+2} and vitamin D.

Several antirachitic agents (compounds which have vitamin D
activity and are thereby able to prevent rickets) are known,
especially ergocalciferol, cholecalciferol, and
dihydrotachysterol which are described in Chapter 9. Fish-
liver oils (e.g., cod liver oil) contain a ketonic sterol,
derived from an unknown source, which has about twice the
potency of the calciferols. The RDA for vitamin D is 400 I.U.
per day. One international unit is the activity present in
0.01 ml of average medicinal cod liver oil and is equivalent
to approximately 0.05 µg of ergocalciferal (vitamin D_2) or
0.025 µg (65 picomoles) of cholecalciferol (vitamin D_3). The
principal dietary sources in the United States are milk
(which is usually supplemented with vitamin D) and fish liver.
An unknown but significant amount of cholecalciferol is also
synthesized within the body from cholesterol. There is
appreciable storage of vitamin D in the liver.

b. <u>Hypervitaminosis D</u>, mentioned earlier, presents clinically with
demineralization and increased fragility of the bones. Serum
Ca^{+2} is elevated, resulting in soft-tissue calcification and
formation of renal calculi. These symptoms are seen in
patients who have received, on a prolonged basis, ten times the
amount of vitamin normally required. All forms of vitamin D
are toxic in this regard. The adult MDR is not actually known
for this vitamin, however, partly due to the difficulty in
measuring the amount of active vitamin formed within the body
by the action of light striking the skin.

3. <u>Vitamin E</u> (tocopherol)

a. The name tocopherol is derived from the Greek <u>tokos</u> (childbirth)
+ <u>pherein</u> (to bear) + <u>ol</u> (alcohol) and the classic manifestation
of vitamin E deficiency is infertility of both male and female
animals.

i) Seven tocopherols, all derivatives of <u>tocol</u>, have been
isolated from natural sources. The most common and most
active of these is <u>α-tocopherol</u> (5,7,8-trimethyltocol),
shown below.

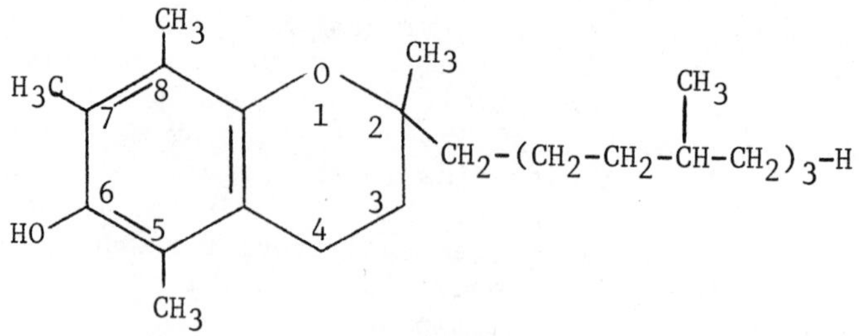

α-tocopherol

Other tocopherols (β, γ, δ) differ in the number of methyl groups on the aromatic ring.

ii)   The principal sources of the tocopherols are vegetable oils, especially wheat germ oil; legumes, nuts, cereals, and green vegetables are also important sources. Although rich in vitamins A and D, the fish liver oils are devoid of vitamin E. The RDA for vitamin E is 10-20 I.U. This is a decrease from the value first recommended in 1965. One international unit is equivalent to 1 mg of synthetic d,l-α-tocopherol acetate. The potency of d-α-tocopherol acetate, the natural form, is 1.36 I.U. per mg while d-α-tocopherol, the free alcohol, has 1.49 I.U. per mg.

iii) Although the metabolism of vitamin E is not well understood, it can be oxidized in man to the product shown below.

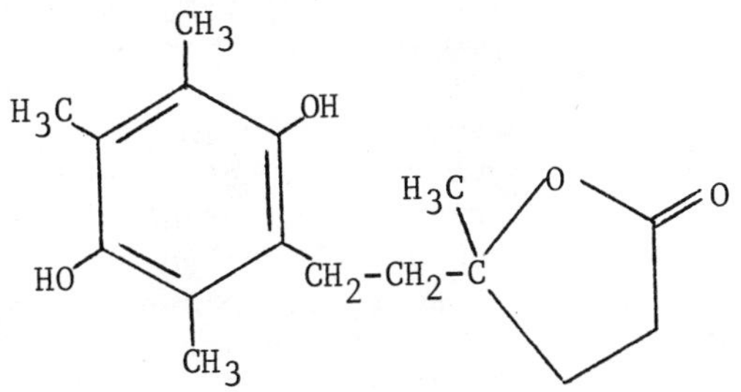

This material has been found in human bile secretions conjugated with two moles of glucuronic acid via the two hydroxyl groups.

b. Despite intensive investigation, the exact biological role(s) of vitamin E remain unknown. This is at least partly due to the multiplicity of disorders observed in vitamin E deficiency (in animals) and to the way in which various non-vitamin agents can apparently correct some of them. There is also considerable tissue storage of vitamin E (as there is of all the fat-soluble vitamins) making it very difficult or impossible to produce a deficiency in adults. In the only long-term study, adult men were depleted for three years without showing any symptoms of the deficiency, although serum tocopherol concentrations became very low. Patients with defects in dietary fat absorption may have very low serum levels of vitamin E without manifesting any symptoms which can be corrected by administration of the vitamin.

i)   The role of vitamin E as an antioxidant was referred to in the discussion of vitamin A. The latter material is

readily oxidized to an inactive form, a process which
can be prevented by the presence of vitamin E.
Similarly, vitamin E protects dietary essential fatty
acids (see Chapter 8) from oxidative degradation, the
amount of the vitamin needed being proportional to the
quantity of unsaturated fatty acid in the diet.
Synthetic antioxidants will replace vitamin E in these
processes. It is also interesting that the tissues of
vitamin E deficient animals consume oxygen at a greater
rate than do those of normal animals.

ii) Certain disorders which are prevented only by the
tocopherols can be produced in experimental animals by
vitamin E deficiency. These include nutritional
muscular dystrophy, irreversible degeneration of rat
testicular tissue, and anemia in monkeys.

iii) For some time it has been known that selenium could
correct some but not all of the symptoms of vitamin E
deficiency. It is now known that selenium serves as
an "antioxidant" in its role as a cofactor for
glutathione (GSH) peroxidase, an enzyme present in
erythrocytes and a number of other tissues (see
Chapter 9). In selenium-deficient animals, increased
hemolysis results from a lack of GSH peroxidase activity.
It has been hypothesized that the overlap in the efficacy
of selenium and vitamin E is related to their respective
abilities to prevent oxidative and peroxidative damage
to tissues. There may also be a direct interaction
between selenium and α-tocopherol, but if so, it remains
obscure.

iv) Membrane changes which are reversed by tocopherol
administration have been noted in cases of vitamin E
deficiency. These include increased fragility of the
erythrocytes in the presence of peroxides and dialuric
acid (5-hydroxy barbituric acid) and changes in
membranes of the intestinal mucosal cells. In the
latter situation, several types of cellular and
subcellular membranes lost their ability to be stained
by osmium tetroxide. This compound is thought to bind
to sites of unsaturation (e.g., unsaturated fatty
acids) in organic molecules.

v) The only clear cut cases of vitamin E deficiency
reported in humans are among premature infants. In
these infants, the hemolysis and the anemia caused by
the oxidant-induced injury to membranes respond to
vitamin E therapy. It has also been reported that in

vitamin E-deficient premature infants, the administration of this vitamin may ameliorate oxygen-induced tissue injury (e.g., retrolental fibroplasia, bronchopulmonary dysplasia). Vitamin E has been shown to lower the platelet hyperaggregability, and this action is presumably somehow related to the inhibition of synthesis of endoperoxides from essential fatty acids (see prostaglandin metabolism, Chapter 8).

vi) Efforts to find a relationship between vitamin E and any specific enzyme activity have produced ambiguous results. Vitamin E deficiency may decrease the activity of the microsomal hydrolase system in rats, but the regulation of this system is very complex and the effects observed may have nothing directly to do with vitamin E. It has also been reported that vitamin E deficiency in rabbits <u>increases</u> the rate of xanthine oxidase synthesis. Other claims for effects of vitamin E on specific enzymes have also been made. Many of the observed effects of vitamin E have not been explained, however, and future studies may well identify systems which are directly affected by vitamin E.

c. There seem to be two major hypotheses regarding the part which vitamin E plays in nutrition. One maintains that it is largely a non-specific antioxidant. The other concedes the vitamin's role as an antioxidant but believes that another, more specific function (perhaps as a coenzyme) exists.

d. For some time claims have been made that massive doses of vitamin E (10-20 times the RDA) are beneficial in man. Although many people continue to consume large quantities of the tocopherols--to the enrichment of "health-food" dealers--controlled studies have failed to substantiate these claims. It appears, however, that the ingestion by humans of up to 300 I.U. per day is innocuous.

4. <u>Vitamin K</u>

a. This group of compounds (whose name derives from the initial letter of the German word Koagulation) is needed for normal blood clotting and no characteristics other than impaired blood clotting have been noted in vitamin K deficiency. Although they are related to coenzyme Q and have been shown in the mycobacteria to play a role similar to that of CoQ in man, the K vitamins do not appear to be involved in oxidative phosphorylation in higher organisms.

b. There are three series of K vitamins: the phyloquinones
    (exemplified by K_1); the menaquinones (for example K_2); and
    the menadiones (e.g., K_3). All are derivatives of
    1,4-naphthaquinone, shown below, which itself has some
    vitamin K activity.

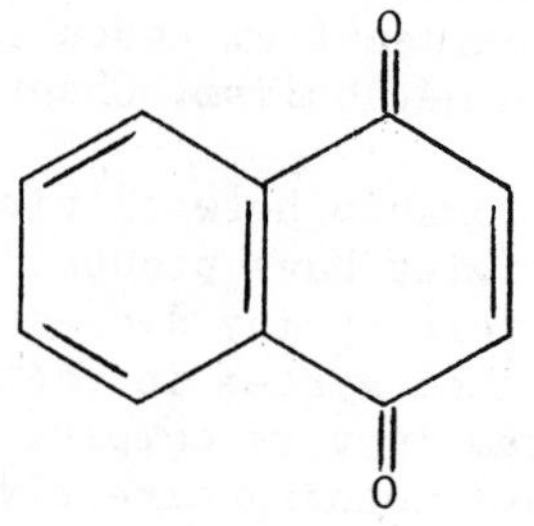

1,4-naphthoquinone

i) The <u>phyloquinones</u> all have the basic structure

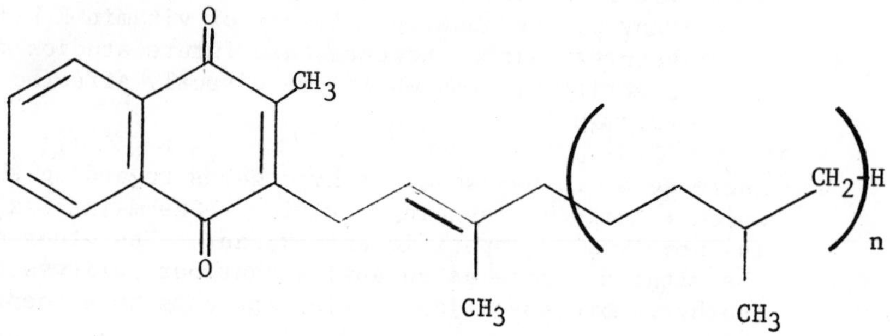

Vitamin K_1, found in alfalfa, spinach, cabbage, and
other green vegetables, has n=4. In this compound,
the unsaturated sidechain is known as a phytyl group.

ii) The <u>menaquinones</u> are variations on the structure

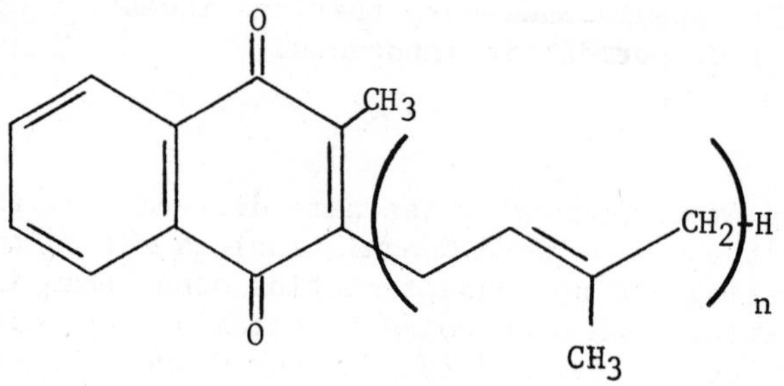

Vitamin K_2 (n=5) is synthesized by intestinal bacteria
and other microorganisms and is found in animal tissue.

The intestinal flora are an important source of this vitamin (see earlier in this chapter) and may be instrumental in preventing deficiency diseases caused by low dietary intake. Human milk contains 15 µg/liter while cow's milk has 60 µg/liter, making milk an important source of this vitamin especially in infants.

iii) The menadiones are synthetic compounds. Of these, vitamin K_3 (menadione), shown below, is the most important. It is three to four times as potent as vitamin K_1.

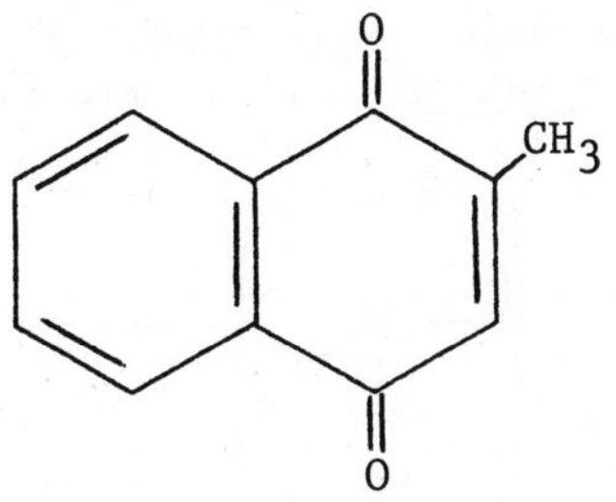

menadione

Note that all of the vitamins K actually contain menadione as a nucleus.

c. An RDA for vitamin K has not been established because of inadequate information regarding daily human intake and the contribution of the intestinal bacteria to this supply. As with the other fat-soluble vitamins, anything which causes steatorrhea or poor lipid absorption in the intestine may also cause some degree of vitamin K deficiency, but few cases of this have been reported in humans. The symptoms center around clotting abnormalities, notably hypoprothrombinemia and hemorrhagic disorders. Menadione and its derivatives, given in excess, reportedly produced kernicterus in premature infants and hemolytic anemia in rats. The mechanism for this is not understood. Consequently, vitamin K_1 is the preferred form for neonates and pregnant women.

d.   i) As noted earlier, the biologic function of vitamin K is exclusively associated with blood coagulation. The coagulation process consists of a cascade (or waterfall) of enzymatic reactions that sequentially convert zymogens of proteolytic enzymes to their corresponding active forms (see Chapter 11). Vitamin K is <u>required</u> for the normal formation of four of the zymogens, namely factor II (prothrombin), factor VII (proconvertin), factor IX (Christmas factor) and factor X (Stuart factor). The active forms of these zymogens are serine

endopeptidases. These factors are synthesized in the
liver and are <u>activated</u> upon binding to phospholipids
of certain biologic membranes (e.g., aggregated platelets,
damaged cells). The binding of the vitamin K-dependent
zymogens to membranes require a prior complexation of
calcium to the zymogen proteins. Thus calcium binding
is essential for the expression of biologic activity of
these proteins. Vitamin K is involved in the structural
modification of these proteins in order to provide
calcium-binding sites. This vitamin-mediated process
of the synthesis of calcium-binding sites on the proteins
occurs at a <u>posttranslational</u> step. The modification
consists of <u>carboxylation</u> of glutamic residues:

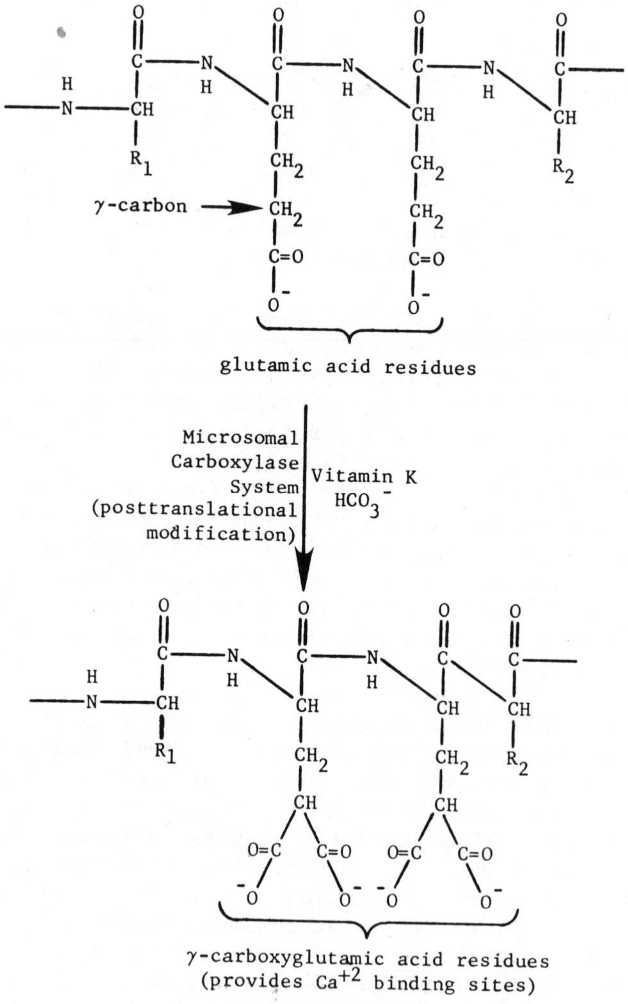

glutamic acid residues

Microsomal
Carboxylase System
(posttranslational modification)

Vitamin K
HCO_3^-

γ-carboxyglutamic acid residues
(provides Ca^{+2} binding sites)

ii)   The detailed mechanism of the vitamin K-dependent
      carboxylase process is not yet known. The in vitro
      system has been studied with washed microsomes as well
      as its solubilized preparation. The carboxylation
      requires the presence of precursor protein (recently
      a pentapeptide, phe-leu-glu-glu-val, has been shown to
      serve as a substrate), vitamin K (hydroquinone is the
      active form; however, if the quinone form is used, a
      reducing agent--NADH or NADPH--is required for activity),
      O_2 and bicarbonate. In the washed microsomal preparations,
      the vitamin K-antagonists (warfarin, 2-chloro-3-phytyl-
      1,4-naphthoquinone) inhibit the carboxylase activity
      which can be overcome by high concentrations of vitamin K.
      In the solubilized system, however, warfarin does not
      inhibit the activity, whereas the 2-chloro analogue of
      the vitamin is still a powerful inhibitor. In normal
      prothrombin, the first ten glutamic acid residues present
      in the amino terminal portion of the molecule are
      carboxylated. These carboxylations account for the
      calcium ion binding and biologic activity.

iii) It is of interest to point out that the clues for
      vitamin K action came from studies done on abnormal
      prothrombin found in the plasma of patients treated
      with vitamin K-antagonists (e.g., coumarin
      anticoagulants--discussed later) and in the plasma of
      coumarin-fed cows. The abnormal prothrombin does not
      bind calcium, does not adsorb to barium citrate, and
      does not participate in the normal coagulation process.
      Supporting the present thinking of vitamin K action,
      the abnormal prothrombin lacks the γ-carboxyl groups
      of glutamic acid. It should be noted that the
      abnormal and normal prothrombin are immunologically
      indistinguishable, as they contain the same main
      antigenic determinants. Thus, the functional defect
      in the abnormal prothrombin resides in the calcium
      binding sites located on the pro-portion of the
      molecule. The thrombin part of the molecule is
      unaffected, suggesting the entire vitamin K-dependent
      modification consists of γ-carboxylations in the
      pro-portion of the molecule. This can be demonstrated
      by converting the abnormal prothrombin to biologically
      active thrombin by using unphysiologic activators,
      namely the enzymes from the venom of the snake, Echis
      carinatus.

Recently, proteins containing γ-carboxyglutamic acid
residues have been found in bone and in renal tissue.
The implication of these observations is not clear at

this time; perhaps vitamin K may be involved in bone metabolism as well.

e.   As noted previously, a number of compounds antagonize the action of vitamin K thereby acting as <u>anticoagulants</u>. The most important of these are the coumarin-related materials, including warfarin (an important rodenticide) and dicumarol. The structures of these compounds are shown below.

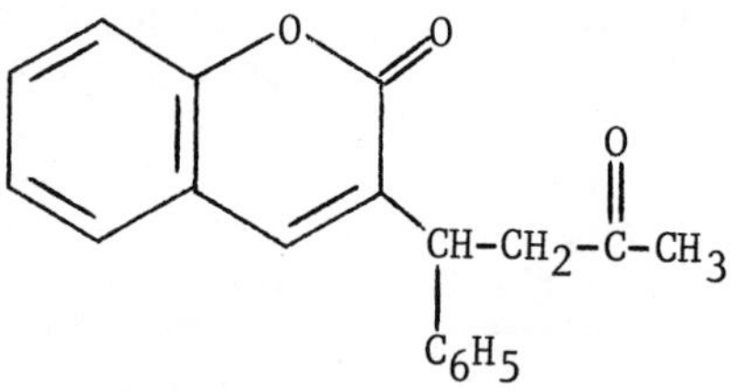

Warfarin

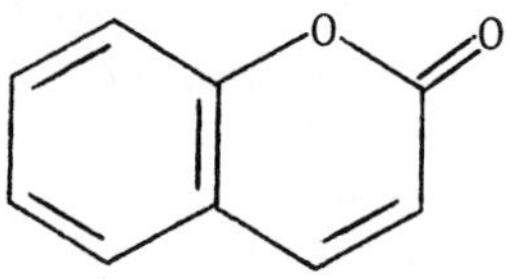

Coumarin

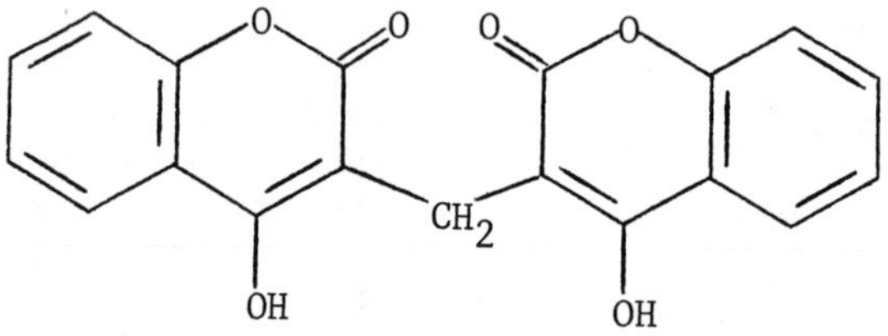

Dicumarol (Bishydroxycoumarin)

Dicumarol is the toxic substance in "spoiled sweet clover" disease, a hemorrhagic condition which occurs in cows.

i)   Although the mechanism of action of the coumarins is not known in detail, it is clear that it interferes with the vitamin K-mediated γ-carboxylation of the prothrombin precursor which forms the Ca^{+2} binding sites on prothrombin. It seems likely that it similarly alters the other affected clotting factors. Abnormal forms of factors II (prothrombin), VII, IX, and X (clotting factors; see Chapter 11) have been isolated from bovine and human plasma and rat liver following dicumarol treatment. The abnormal prothrombin has been shown to lack the γ-carboxyglutamic acid residues. Since normal factor X has also been shown to contain these modified residues, presumably the abnormal one lacks them. In vivo, the activity of factor X is most severely depressed while factors II, VII, and IX are less severely affected.

ii) The antagonism between vitamin K and the coumarins seems to be competitive, since it can be totally reversed by sufficiently high doses of the vitamin. There is evidence that they both bind to the same site on albumin so that coumarins may act by decreasing the plasma transport of the vitamin. This is also consistent with the observed competitive inhibition. It has recently been shown that a newly discovered metabolite of vitamin K, the 2,3-epoxide, accumulates in the livers of warfarin-treated rats. A reductase which normally converts the epoxide back to the vitamin is inhibited by warfarin. The original theory--that the epoxide is the actual vitamin K antagonist--has now been proven incorrect, so that the significance of this observation. is unclear. It suggests, however, that the effect of the coumarins not be in the transport step after all.

iii) Many drugs apparently modify the activity of the coumarins and, in turn, the coumarins can alter the metabolism of other compounds. Substances which potentiate the anticoagulant effect may decrease the albumin binding of the coumarins, thereby increasing the concentration of the free molecules in the plasma (e.g., phenylbutazone); decrease the rate of coumarin metabolism by inhibiting the liver microsomal oxidase system (e.g., chloramphenicol and phenyramidol) and perhaps, in some cases, decrease the synthesis of normal clotting factors. The last mechanism has yet to be clearly demonstrated. Compounds which antagonize the coumarins, thereby enhancing coagulation, generally seem to function by accelerating coumarin metabolism. Most notable and potent among this group are the barbiturates, which apparently stimulate the hepatic microsomal oxidase system, thereby increasing the rate of coumarin hydroxylation. Other coumarin antagonists are glutethimide (a hypnotic) and griseofulvin (an antifungal agent). Coumarins potentiate the activity of diabinese (chlorpropamide, a hypoglycemic agent), dilantin (diphenylhydantoin, an anticonvulsant), and orinase (tolbutamide, a hypoglycemic). They are also reported to potentiate the activity of phenobarbital, although both compounds are catabolized by the microsomal oxidase system in the liver. An unknown mechanism of interaction is seemingly involved in this instance.

iv)   The area of drug-drug interactions is of great practical
importance in medicine. It is discussed generally in
Chapter 14 and other examples are cited in appropriate
places in this book.

Water-Soluble Vitamins

The majority of these compounds have already been covered in appropriate
places throughout the book. Reference to these discussions will be
given below with each vitamin. All apparently serve as coenzymes or
coenzyme-precursors in one or more biological reactions. It is
important to notice that the B-complex vitamins are generally obtained
from the same sources (whole grain cereals, meats, yeast, wheat germ)
and that deficiency of any one of these is usually accompanied by
deficiency of the entire group as well as of protein.

1. Vitamin B$_1$ (thiamine; aneurine; cocarboxylase) is the antiberiberi
(antineuritic) factor referred to earlier in this chapter. The
active coenzyme form is thiamine pyrophosphate (TPP; see comments
on citric acid cycle, Chapter 4). Thiamine triphosphate (TTP; not
to be confused with thymidine triphosphate) which is synthesized
from TPP and ATP by TPP-ATP phosphoryl transferase may be the form
of thiamine which is neurophysiologically active. Because TPP and
TTP are rapidly interconverted, however, all that is certain is
that some form of thiamine is important for the normal functioning
of the nervous system. Of the total thiamine in the body, 10% is
TTP, 80% is TPP, and 10% is TMP (thiamine monophosphate; formed by
the action of thiamine pyrophosphatase on TPP). Thiamine
pyrophosphate is synthesized from dietary thiamine and ATP by
thiamine pyrophosphokinase, in a single step.

   a.   It is required for a variety of reactions such as:

   i)   oxidative decarboxylations of α-keto acids (Chapter 4):
   pyruvic acid → acetyl CoA
   α-ketoglutarate → succinyl CoA
   branched-chain amino acid catabolism (Chapter 6).

   ii)   transketolase reaction (Chapter 4) and related processes.

   iii)   non-oxidative decarboxylations of α-keto acids:
   pyruvate → acetaldehyde + CO_2 (in yeast).

   b.   Principal dietary sources of this vitamin are fish, lean meat
   (especially pork), milk, poultry, dried yeast, and whole grain
   cereals. Bread, cereals, and flour-based products are
   frequently enriched with this material. The requirement of
   thiamine is related to caloric intake, and the RDA is about
   1.5 milligrams.

c. Thiamine deficiency is common in the countries of Asia where polished rice is the staple foodstuff in the diet. In the United States and Europe thiamine deficiency is seen primarily in association with <u>chronic alcoholism</u>. Several syndromes in man due to thiamine deficiency have been reported, namely beriberi, Wernicke's Syndrome and Korsakoff's Syndrome. The severity of the syndrome is dependent upon the degree and duration of deprivation of the vitamin. Wernicke's (mortality is 90% without therapy; common cause of death is sudden heart failure) and Korsakoff's syndromes are the result of severe deficiency. The less severe deficiency of the vitamin leads to beriberi and a milder form of this disorder is known as polyneuritis (or dry beriberi). An infantile form of beriberi has also been reported in breast-fed infants (usually between 2 and 5 months of age) whose mothers have a history of poor diet. In general cardiovascular and neurologic findings are the hallmarks of thiamine deficiency. The neurologic manifestations of fully-developed deficiency syndromes are parasthesias, foot drop, ataxia, confusion, extraocular muscle palsies, psychosis, coma and death. In most subjects the cardiac disturbances develop at some stage of thiamine deficiency. It should be emphasized that the cardiac and neurological derangements respond completely to thiamine replacement.

The thiamine-responsive metabolic disorders (i.e., those disorders which require pharmacologic doses for treatment) are branched-chain ketoaciduria (increased plasma levels of branched-chain amino acids and their keto acids; mutant enzyme: oxidative decarboxylase of branched-chain keto acids), pyruvicacidemia (elevated level of plasma pyruvate, lactate and alanine; mutant enzyme: pyruvate decarboxylase), and subacute necrotizing encephalomyelopathy (elevated level of plasma pyruvate and lactate).

d. The significant biochemical parameters that aid in the diagnosis of thiamine deficiency are elevated blood pyruvate and lactate concentrations and the <u>reduced</u> erythrocyte transketolase (a TPP-dependent enzyme--see hexose monophosphate shunt pathway, Chapter 4) activity. The latter parameter is reliable and sensitive to mild degrees of depletion. The blood levels of the vitamin are decreased in the deficiency states; however, they are not very useful in making a definitive diagnosis.

2. Vitamin B_2 (<u>riboflavin</u>) is incorporated into flavin mononucleotide (FMN) and flavin adenine dinucleotide (FAD). Both of these function, as part of flavoproteins, in electron transfer (redox) reactions (Chapter 6). As was pointed out earlier, these are not

truly flavin nucleotides since the bond between the sugar and the flavin ring is not a glycosidic one. Enzymes which contain FAD as a prosthetic group include D-amino acid oxidase, glucose oxidase, xanthine oxidase, NADH and NADPH dehydrogenases (such as NADPH-glutathione reductase), succinate dehydrogenase, and acyl CoA dehydrogenase. FMN is part of the L-amino acid oxidase found in the kidney, "old yellow enzyme" (involved in glucose-6-phosphate oxidation; first flavoprotein discovered), and other dehydrogenases. FAD appears to be more commonly used than FMN as a cofactor. Although much riboflavin occurs in man and many other organisms as noncovalently bound FAD, an appreciable amount also is found in covalent linkage to amino acid residues in proteins. Succinate dehydrogenase is an FAD-histidyl protein while monoamine oxidase contains FAD attached to a cysteinyl residue. Other proteins contain FAD linked to these same residues by different types of bonds.

The riboflavins are ubiquitous in plant and animal tissues. Particularly good dietary sources are liver, yeast, and wheat germ; while milk and eggs are important when consumed in sufficient quantities. Leafy green vegetables also provide this vitamin, and cereals and bread are frequently enriched with B_2. The RDA is 1.7 milligrams.

No specific symptoms are associated with riboflavin deficiency (ariboflavinosis), partly because it is difficult to produce uncomplicated B_2 deficiency. Although cheilosis (red, swollen, cracked lips), a magenta-colored tongue, seborrheic dermatitis, and congestion of the conjunctival blood vessels have been observed apparently as a result of riboflavin deficiency, some or all of these are also caused by other dietary inadequacies, notably of niacin and iron. The most sensitive bodily indicators for B_2 deficiency are the erythrocytes and a normal person contains 20 µg/100 ml of the vitamin in the blood. The measurement of the activity of erythrocyte glutathione reductase and the extent of its stimulation to the in vitro addition of FAD have been used in the assessment of riboflavin deficiency.

3. Vitamin B$_6$ (<u>pyridoxine</u>) is actually three closely related
    compounds.

Pyridoxol	Pyridoxal	Pyridoxamine
(for the alco<u>hol</u> group in position four)	(for the al<u>dehyde</u> group in position four)	(for the <u>amine</u> group in position four)

Pyridoxine designates the entire group as well as being synonymous
with pyridoxol. The active form in the body is pyridoxal-5-phosphate
which seems to be transiently converted to pyridoxamine-5-phosphate
in functioning as a coenzyme (see Chapter 6). This coenzyme is
generally important in reactions involving α-amino acids. Examples
of this include the formation of N^5,N^{10}-methylene FH$_4$ from FH$_4$ and
serine (Chapter 5), nicotinic acid synthesis (tryptophan metabolism,
Chapter 6), cysteine metabolism (Chapter 6), and heme synthesis
(formation of δ-aminolevulinic acid from glycine and succinyl CoA,
Chapter 7). Its role in transaminase reactions is discussed in
Chapter 6. Pyridoxol is the form usually present in plants while
the aldehyde and amine are animal products. The three compounds
are readily interconverted by the body.

a. The RDA for vitamin B$_6$ is 2.0 milligrams in adults and 0.3
    milligrams for infants, although the actual requirement
    increases with an increase in dietary protein. The studies
    upon which these are based were very inadequate, however, and
    accurate information on human requirements for this vitamin
    are practically nonexistent. Oral contraceptives alter
    tryptophan metabolism, increasing the need for pyridoxine.
    The estrogen in the pills seems to be the cause of this
    abnormality, with progesterone having little effect. There
    is evidence, also, that the need for B$_6$ may be greater in
    the elderly than in young or middle-aged adults. An
    abnormality of B$_6$ metabolism may be involved with mongolism.

    The principal dietary sources are whole grain cereals, wheat
    germ, yeast, meat, and egg yolk. Although no deficiency
    syndrome specifically due to a lack of pyridoxine is known
    in man, rats and monkeys maintained on a B$_6$-free diet develop
    dermatitis and exhibit neuropathological changes. Other
    manifestations include lassitude, weakness, anorexia, and
    iron-resistant hypochromic, microcytic anemia. Infants having
    an inherited requirement for increased amounts of the vitamin

have convulsions if the supply is inadequate. This symptom can be relieved by pyridoxine. The cerebral dysfunction has been ascribed to reduced activity of glutamic acid decarboxylase. This is a B_6-dependent enzyme which catalyzes the conversion of glutamic acid to γ-aminobutyric acid (GABA), an inhibitory neurotransmitter. The reaction involved is:

$$
\begin{array}{ccc}
\text{COOH} & & \text{COOH} \\
| & & | \\
\text{CH}_2 & & \text{CH}_2 \\
| & \text{Glutamate} & | \\
\text{CH}_2 & \text{decarboxylase} & \text{CH}_2 \\
| & \text{(}B_6\text{-phosphate)} & | \\
\text{CH-NH}_2 & \longrightarrow & \text{CH}_2\text{NH}_2 \\
| & & \\
\text{COOH} & \text{CO}_2 &
\end{array}
$$

glutamic acid

γ-aminobutyric acid
(a neurotransmitter)

b. Pyridoxine antagonists such as deoxypyridoxine and isonicotinoyl hydrazide (isoniazid, a tuberculostatic drug) are capable of producing B_6-deficiency symptoms when administered to humans.

isoniazid   deoxypyridoxine

Although most of these symptoms (e.g., nausea, seborrheic dermatitis, glossitis, and polyneuritis) are associated with other disorders, deoxypyridoxine also causes oxaluria (due to impairment of glycine metabolism, Chapter 6) and excretion of xanthurenic acid (caused by a lack of kynureninase activity, Chapter 6). Roughening of the skin and a pyridoxine-responsive anemia have also been reported in some cases of pyridoxine deficiency.

c. The pyridoxine-responsive disorders are infantile convulsions (mutant enzyme: probably glutamic acid decarboxylase), cystathioninuria (elevated levels of urinary cystathionine; mutant enzyme: cystathionase), xanthurenic aciduria (elevated

level of urinary xanthurenic acid; mutant enzyme: kynureninase),
homocystinuria (elevated levels of plasma and urinary methionine
and homocystine; mutant enzyme: cystathionine synthetase), and
hyperoxaluria (elevated levels of urinary oxalate and glyoxylate;
mutant enzyme: glyoxylate:α-ketoglutarate carboligase).

d. The pyridoxine deficiency can be assessed by direct measurement
of the vitamin in the blood, by the tryptophan load test
(consists of administering tryptophan and measuring xanthurenic
acid in the urine--see tryptophan metabolism, Chapter 6), or
by measurements of transaminase activities in the plasma.

4. Niacin (the generic name for nicotinic acid and nicotinamide as
well as a synonym for nicotinic acid) is a B-vitamin by virtue of
its occurrence in meat, eggs, yeast, and whole-grain cereals in
conjunction with other members of the B-complex. The RDA is
20 milligrams but, since a limited amount can be synthesized from
tryptophan by the human body (see Chapter 6), the actual
requirements depend on the amount and type of dietary protein.
In addition, it has recently been found that about 20% of the
niacin in cereals is unavailable for use by the body. There
appear to be a number of such bound forms of niacin. Two that
have been characterized are the niacinogens (peptides of molecular
weight 12,000 to 13,000 daltons) and a carbohydrate complex with
a molecular weight of about 2,370 (the one isolated from wheat
brain is named niacytin). A better understanding of these
compounds is necessary in order to accurately assess the amount
of niacin available in different foods.

The principal known role of niacin in the body is as part of
NADH and NADPH, both dinucleotide coenzymes involved in glycolysis,
fatty acid metabolism, and a number of other oxidation-reduction
reactions. The structure and functions of these compounds are
given in Chapter 4 and their synthesis from nicotinic acid is
outlined under tryptophan metabolism. The structures of nicotinic
acid and nicotinamide are shown below.

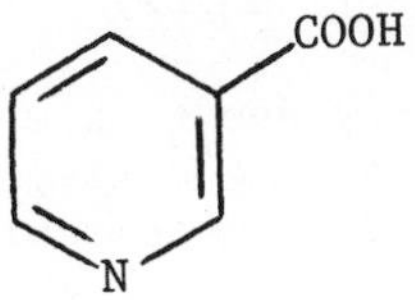

niacin 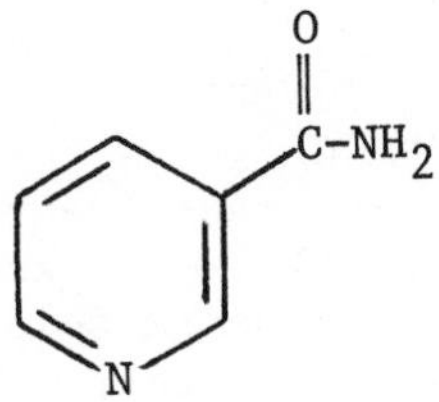   nicotinamide

As with most of the other B-vitamins, it is difficult to pinpoint
symptoms due specifically to a deficiency of niacin. Pellagra,

generally attributed to a lack of this vitamin, probably reflects an insufficiency of several members of the B-complex. Symptoms include dermatitis (pellagra rash), stomatitis, abnormalities of the tongue and of digestion, and diarrhea. These are difficult to correlate with the known function of the vitamin as a precursor to redox coenzymes, suggesting some other roles for niacin. In Hartnup disease, an inherited disorder of amino acid transport (see Chapter 6), a pellagra-like rash may be seen. Presumably this occurs because the dietary supply of niacin is inadequate for the body's needs in the absence of tryptophan. Carcinoid syndrome, in which tryptophan is diverted to the formation of large quantities of 5-hydroxytryptamine, presents a similar clinical picture.

5. Pantothenic acid (Pantoyl-β-alanine) is ubiquitous in plant and animal tissues, being especially abundant in those materials which are rich in other B-vitamins. There is no recommended RDA for pantothenic acid. A daily intake of 5-10 milligrams is probably adequate for adults. Pregnant women should use about 10 milligrams per day. In the body, it is an essential precursor to the synthesis of coenzyme A (CoA, CoASH), used to carry acetyl groups (acetyl CoA) in a wide variety of two-carbon-requiring reactions. CoASH is also the carrier of fatty acyl groups (Chapter 8), succinyl groups, and other carboxyl-containing materials (as in amino acid metabolism). Pantothenic acid is also part of the "swinging sulfhydryl arm" of the fatty acid synthetase complex. These aspects are discussed under carbohydrate, amino acid and lipid metabolism. The structure of pantothenic acid is shown below.

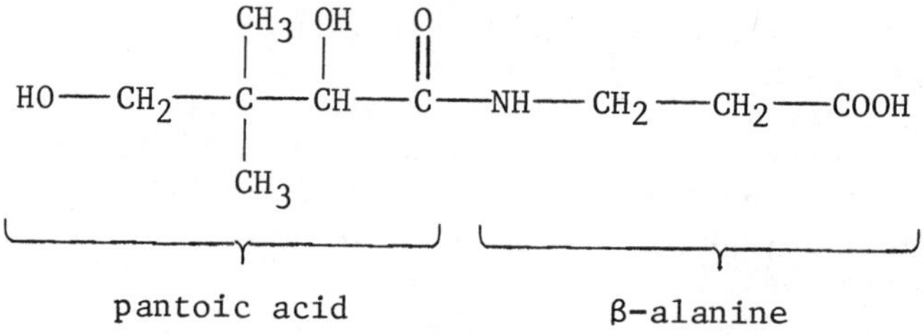

$$HO-CH_2-\underset{\underset{CH_3}{|}}{\overset{\overset{CH_3}{|}}{C}}-\underset{\underset{}{|}}{\overset{\overset{OH}{|}}{CH}}-\overset{\overset{O}{||}}{C}-NH-CH_2-CH_2-COOH$$

pantoic acid   β-alanine

Pantothenic Acid (Pantoyl-β-Alanine)

Uncomplicated pantothenic acid deficiency due to dietary restriction alone is probably not known in man. When the antivitamin ω-methylpantothenic acid is administered, symptoms similar to those seen in animals occur. The significant biochemical changes are insensitivity of the adrenal cortex to ACTH (resulting in adrenal insufficiency), increased insulin sensitivity, and the acceleration of erythrocyte sedimentation. Neuromuscular degeneration is also observed and antibody synthesis in response to an antigen to which the individual is known to be immune is sluggish. If the person is deficient in both pyridoxine and pantothenic acid, antibody formation is completely abolished.

6. Biotin is another member of the B-complex. It is widely distributed in natural products with beef liver, yeast, peanuts, kidney, chocolate, and egg yolk being especially rich in it. The RDA is 0.3 milligrams although no MDR has been established. As with the other B-vitamins, the intestinal flora synthesize biotin and, in fact, the total daily urinary and fecal excretion of the vitamin usually exceeds the daily dietary intake.

   a. Biotin is the coenzyme in a number of carbon dioxide fixation reactions. Examples of this are given below.

       i) Acetyl CoA + CO_2 → Malonyl CoA (Chapter 8)

       ii) Propionyl CoA + CO_2 → Methylmalonyl CoA (Chapter 8)

       iii) Pyruvate + CO_2 → Oxaloacetate (Chapter 4)

   The mechanism by which biotin is believed to mediate the CO_2 attachment is discussed under fatty acid biosynthesis, Chapter 8. An unusual carboxylation, for which biotin is apparently not required, is the addition of carbon 6 to the purine ring (see Chapter 5) and γ-carboxylations of glutamic residues of the zymogen of blood coagulation, which is a vitamin K-dependent process (already discussed, this chapter).

   b. The structure of biotin is shown below. In oxybiotin, which is capable of substituting for biotin in most species; the sulfur is replaced by oxygen.

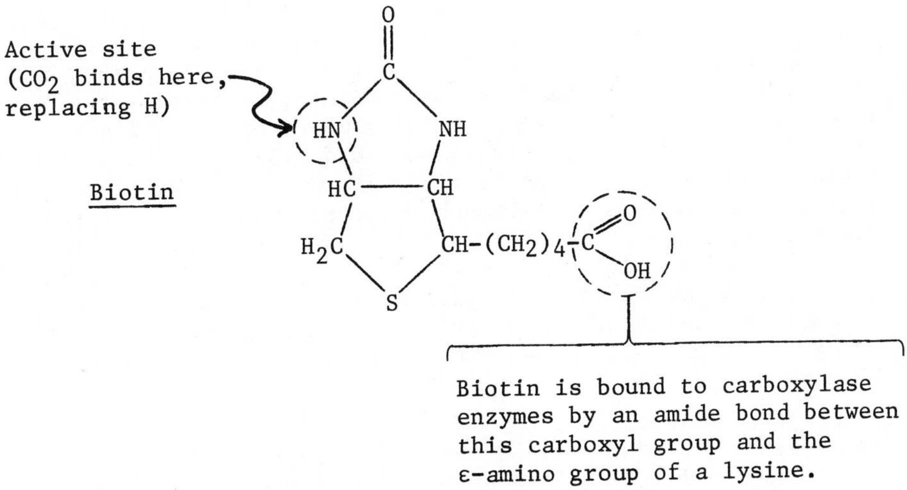

Active site (CO_2 binds here, replacing H)

Biotin

Biotin is bound to carboxylase enzymes by an amide bond between this carboxyl group and the ε-amino group of a lysine.

The biotin-carboxylase (biotin-enzyme) is formed in two steps:

$$\text{biotin} + \text{ATP} \longrightarrow \text{biotinyl-5'-adenylate} + PP_i$$

$$\text{biotinyl-5'-adenylate} + \text{apocarboxylase} \longrightarrow \text{holocarboxylase} + \text{AMP}$$

Proteolysis (in vivo and in vitro) releases biocytin, also known as ε-biotinyllysine. Biocytinase, in the blood and liver, cleaves this molecule to biotin and lysine. In the rat (and probably in other mammals) the biotin ring system is not usually degraded prior to excretion. The side-chain is largely β-oxidized, leaving bisnorbiotin. This compound, a small amount of sulfoxide, and traces of the methyl ketone derivatives, together with the unaltered free vitamin, are excreted in the urine.

c. Biotin deficiency occurs in man when large amounts of raw egg white are consumed. Avidin, a protein of about 70,000 molecular weight, is present in egg white and binds biotin strongly ($K_{assoc.} = 10^{15}$), preventing its absorption by the body. Avidin is readily inactivated by heating. It contains four identical subunits, each having 128 amino acid residues and each capable of binding one molecule of biotin. This binding is abolished by heat and other denaturing influences. Sterilization of the intestine (as with sulfonamide treatment) has also been known to cause the biotin deficiency syndrome and biotin antimetabolites have a similar effect. The symptoms of biotin deficiency initially include a scaling dermatitis, lassitude, anorexia, muscle pains, and depression. As the condition progresses, nausea, anemia, hypercholesterolemia, and changes in the electrocardiogram are observed.

7. Folic acid was discussed extensively in the section on one-carbon metabolism (Chapter 5). The major clinical manifestations of folate deficiency are megaloblastic anemia, hypersegmented polymorphonuclear leukocytes, and thrombocytopenia. Its active coenzymic form is folacin (FH_4, tetrahydrofolic acid). It is found in many foods, especially the glandular meats, leafy green vegetables, and yeast. Although folic acid is stored in the liver (unlike most of the water-soluble vitamins), the RDA for folate is 400 milligrams. This reflects the fact that, while the MDR is probably about 50 milligrams, the amount of utilizable vitamin in food varies, absorption from the gastrointestinal tract may be incomplete, and some of the folic acid is destroyed by cooking.

a. Most dietary sources contain pteroylpolyglutamic acid which is hydrolyzed to pteroylglutamic acid (folic acid) by "conjugases." These enzymes are peptidases which are ubiquitously distributed in nature. They are poorly characterized. In man, they occur in the lysosomes of the intestinal epithelial cells, with no detectable activity being observed in the intestinal lumen. The uptake of the polyglutamates by these cells is probably an energy-requiring, carrier-mediated process. Failure of the body to perform this hydrolysis may produce folate deficiency. Other causes of the deficiency state may be increased demands by the body (as during pregnancy, hemolytic anemia, leukemia, and Hodgkins disease) or excessive excretion of the vitamin. Folacin antimetabolites (e.g., methotrexate) are also well known. Sulfonamides function as antibacterials by interfering with folate synthesis in susceptible organisms. Sprue, a syndrome characterized by a sore mouth and gastrointestinal disturbances including periodic diarrhea and steatorrhea, is associated with folate deficiency. Megaloblastic anemia (appearance of the red cell precursor in the bone marrow changes) frequently results from either folate or vitamin B_{12} deficiency but the relationship between these two vitamins is not clear. This point is further discussed in the next section on vitamin B_{12}.

b. As was suggested previously, folacin functions as a carrier of one-carbon fragments in biosynthetic reactions. Some important examples are summarized below.

i)   Glycinamide ribosyl-5-$\textcircled{P}$ + N^5,N^{10}-methenyl FH_4

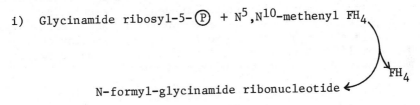

FH_4

N-formyl-glycinamide ribonucleotide

(see purine biosynthesis, Chapter 5)

ii) 5-Aminoimidazole-4-carboxamide + N^{10}-formyl FH_4
        ribonucleotide

FH_4

5-formamidimidazole-4-carboxamide ribonucleotide

(see purine biosynthesis, Chapter 5)

iii)   dUMP + N^5,N^{10}-methylene FH_4 $\longrightarrow$ Thymidylic Acid

$\downarrow$

FH_2

(see pyrimidine biosynthesis, Chapter 5)

iv)   Homocysteine + N^5-methyl FH_4 $\longrightarrow$ Methionine

$\downarrow$

FH_4

(see sulfur-containing amino acid metabolism, Chapter 6)

8. Vitamin B_{12} (cyanocobalamin) is the last member of the B-vitamin complex. Animals and higher plants cannot synthesize B_{12} although a number of animal tissues are able to concentrate it, making lean meat, liver, sea food, and milk important dietary sources of this vitamin. Microorganisms are responsible for the de novo synthesis of the B_{12}. Most cooking does not destroy B_{12} since it is stable to 250°C. The RDA for cyanocobalamin is 3 micrograms. There are reports (as yet unconfirmed) that very high doses of vitamin C (1 g or more with each meal) destroy substantial amounts of the B_{12} in the food consumed. Until further data are available, it is probably wise to occasionally check the B_{12} status of anyone consuming large amounts of ascorbate. A serum B_{12} concentration of less than 100 pg per milliliter is usually diagnostic for vitamin deficiency. Exceptions are conditions which cause high concentrations of B_{12} binding proteins in the circulation (e.g., the myeloproliferative disorders), reducing the amount of free vitamins available for uptake by the tissues.

a. The structure of cobalamin is shown in Figure 144. The corrin ring system, composed of four pyrrole rings linked by three methene groups and one direct carbon-carbon bond, is similar to the porphyrin system discussed in Chapters 4 and 7. In the commercially available form of the vitamin, the R-group (indicated in the structure by (R)) is usually CN^-, hence the name cyanocobalamin. This is probably an artifact of the isolation procedure, however, and cyanocobalamin does not occur naturally in microorganisms. In B_{12}-coenzyme (cobamide coenzymes), (R) is 5'-deoxyadenosine (adenosyl B_{12}) or a methyl group (methyl B_{12}), depending on the reaction in which the coenzyme participates. Both of these compounds are unusual among biomolecules because they contain carbon bonded to a metal (in this case cobalt). This is termed an organo-metallic bond. In the coenzymes, the adenosyl group is donated by ATP while the methyl group probably comes from N^5-methyl FH_4. The 5,6-dimethylbenzimidazole ring is

Figure 144. The Structure of Vitamin B₁₂

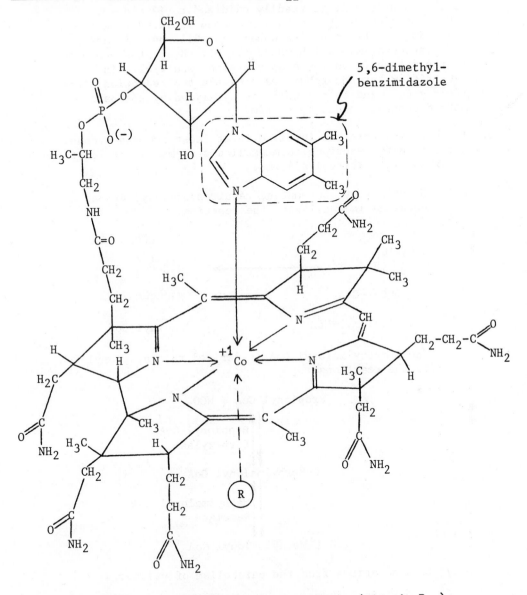

5,6-dimethyl-benzimidazole

If R = -CN, then the molecule is cyanocobalamin (vitamin B_{12});
  R = 5'-deoxyadenosine, the molecule is adenosyl B_{12};
  R = CH_3, the molecule is methyl B_{12}.

sometimes replaced by 5-hydroxybenzimidazole, adenine, or other similar groups. Although the cobalt is shown in a +1 oxidation state, it is readily oxidized to +2 and +3. Hydroxocobalamine ($\boxed{R}$ = OH$^-$; has been isolated from mammalian tissue) contains cobalt +3 while methyl, adenosyl, and cyanocobalamin all have cobalt +1. B_{12s}, the precursor to the coenzymic forms, also contains cobalt +1. In Clostridium tetanomorphum, each of the two reduction steps ($Co^{+3} \rightarrow Co^{+2}$ and $Co^{+2} \rightarrow Co^{+1}$) is catalyzed by a separate, NADH-dependent reductase system.

b. The B_{12} coenzymes are involved in at least two and probably three or more reactions in mammalian systems and at least two reactions in bacterial systems.

    i) Methylmalonyl CoA-mutase requires adenosyl B_{12} as a coenzyme in catalyzing the reaction.

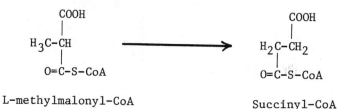

      L-methylmalonyl-CoA   Succinyl-CoA

The L-methylmalonyl CoA is derived from propionyl CoA in the reactions:

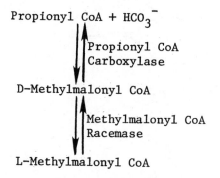

It also arises from the catabolism of valine.

    ii) The remethylation of homocysteine to methionine requires methyl B_{12} as a coenzyme. The reaction uses N^5-methyl FH_4 as the methyl donor.

    iii) Adenosyl B_{12} was originally identified as a cofactor for the enzyme catalyzing the conversion of glutamic acid to β-methylaspartic acid in C. tetanomorphum. This reaction is similar to the methylmalonyl CoA mutase reaction but it has not been shown to occur in mammals.

iv) In various species of Lactobacillus, the deoxynucleoside triphosphates are formed by reduction of the nucleoside triphosphates in a reaction catalyzed by an adenosyl-B_{12} coenzyme-requiring enzyme. This is different from the nucleoside diphosphate reductase system (see biosynthesis of deoxyribonucleotides, Chapter 5) which does not require a B_{12} coenzyme. The triphosphate reduction system has been demonstrated only in the lactobacilli.

c. In man, B_{12} deficiency as the direct result of inadequate dietary intake is known to occur only among strict vegetarians. Since plants are unable to synthesize B_{12}, it is present in vegetables only as the result of contamination by microorganisms. Thus, vegetarians slowly develop B_{12} deficiency over many years. Although the colonic bacteria make large amounts of the vitamin, it is only absorbed when it is complexed with the intrinsic factor synthesized in the stomach, so that this production is of no dietary use. Note that the intestinal absorption of vitamin B_{12} requires both gastric and ileal components. The fecal excretion of B_{12} is about 5 μg per day, both in normal persons and those with pernicious anemia. Disorders have also been described which seem to result from B_{12} malabsorption, from defective plasma transport, or from an inability of the body to utilize the vitamin once it has been absorbed.

The most characteristic result of B_{12} (and folate) deficiency is megaloblastic anemia. In this disorder, the red cell precursors in the bone marrow are mildly enlarged and the nuclear chromatin is altered in appearance (megaloblastic change). Erythroid hyperplasia is also apparent in the bone marrow. Advanced cases also show leukopenia and thrombocytopenia of the peripheral blood. Vitamin B_{12} deficiency also results in neurological damage which is not seen in persons lacking only folate (spinocerebellar neurologic dysfunction).

Megaloblastic anemia is diagnosed from a bone marrow aspirate. When this disorder is caused specifically by a deficiency of gastric intrinsic factor (IF, needed for B_{12} absorption; discussed below), it is termed pernicious anemia. In classic pernicious anemia, the megaloblastic changes and IF absence are accompanied by a complete lack of gastric HCl (achlorhydria) and the occurrence of atrophic gastritis. This name is somewhat misleading since "pernicious" is a general term meaning "tending to a fatal conclusion." As more is understood about the mechanism of B_{12} absorption, it now appears that some cases of pernicious anemia may also result from a defective intrinsic factor or other heritable errors in the absorption process (see below).

Pernicious anemia is diagnosed by the <u>Schilling test</u> and by analysis of gastric aspirates. In the Schilling test, vitamin B_{12} containing radioactive cobalt is administered orally and the amount of radioactivity appearing in a 24-hour urine specimen is determined. A normal person will excrete about 1/3 of the label in this period while less than 8% is found in the urine of a patient with classic pernicious anemia (a positive Schilling test). The test is repeated several days later, this time including intrinsic factor with the B_{12}. If the defect is one of factor synthesis or an abnormal IF, the B_{12} uptake should be normal in the second test. The megaloblastic anemia of folate deficiency is usually indistinguishable from that of B_{12} deficiency except with respect to the Schilling test. Serum B_{12} can now be measured directly by competitive protein binding methods (normal range: 200-1600 pg/ml).

i) Two types of <u>methylmalonic aciduria</u> (see Chapter 6) have been described. One type, responsive to the administration of adenosyl B_{12} or large doses of vitamin B_{12}, is believed to be caused by a defect in deoxyadenosyl transferase, the enzyme which transfers a deoxyadenosyl group from ATP to cobalamin during adenosyl B_{12} synthesis. The other type, which is <u>not</u> relieved by vitamin B_{12}, is probably due to an abnormal methylmalonyl CoA mutase enzyme. Both defects are apparently inherited as autosomal recessive traits but there are insufficient data to state this definitely. <u>Propionicacidemia</u> (misleadingly called ketonic hyperglycinemia because of its clinical picture) is a defect of propionate metabolism distinct from those producing methylmalonic aciduria. It is caused by a defect in propionyl-CoA carboxylase.

ii) Another case of methylmalonic aciduria, accompanied by defects in sulfur-containing amino acid metabolism, has been described. Although definitive data are lacking, this could be due to an abnormality in an enzyme catalyzing the formation of a precursor common to both methyl-B_{12} and adenosyl-B_{12}. The reductase enzymes have been suggested as possible sites for the mutation. An alternative hypothesis which has not been ruled out is that the error is one in folate metabolism, since these two coenzymes are closely related (see below).

iii)   Several hereditary B_{12} malabsorption disorders are also known. These are characterized by megaloblastic anemia, an inability to take up orally administered vitamin B_{12}, and responsiveness to parenteral B_{12}. The normal route of B_{12} absorption is shown in Figure 145.

Gastric juice intrinsic factor (IF), isolated from humans, is a mucoprotein of molecular weight about 50,000. It usually contains about 13% carbohydrate and is secreted by the gastric parietal cells along with HCl. Transcobalamin II is a β-globulin of about 35,000 molecular weight. Its primary function appears to be the transport of newly absorbed B_{12} into the circulation. It appears that transcobalamin II is the more important among the transport proteins in the delivery of vitamin B_{12} to the tissues. Its absence causes severe vitamin B_{12} abnormality. Transcobalamin I has a molecular weight of roughly 121,000 and migrates with the α-globulins during serum electrophoresis. Most of the circulating B_{12} is bound to this protein, suggesting that transcobalamin I is principally a carrier molecule for B_{12} within the body. The binding constant of transcobalamin I for B_{12} is also greater than that of transcobalamin II for the vitamin. The transport of B_{12} across the ileal mucosal membranes is an active, energy-requiring process.

Defects have been reported in which no IF is secreted into the gastric juice (pernicious anemia; see above). Recently, another type of pernicious anemia, caused by secretion of a defective IF, has been described. The IF in this patient will bind B_{12} normally but it is unable, even when administered to normal individuals, to attach to the ileal mucosal receptor site and transfer the B_{12} to the circulation. Still a third etiology for B_{12}-malabsorption pernicious anemia appears to involve defective mucosal cells. In these cases, the B_{12}-IF complex is formed and binds to the mucosal cell receptor sites but the B_{12} cannot pass through the cells and into the circulation. Certain disorders and blockages of the terminal ileum can also cause B_{12}-malabsorption and lead to the symptoms of pernicious anemia. These can usually be distinguished from a true lack of IF by the occurrence of a positive Schilling test even with added factor. This is not absolute, however, and a Schilling test should not be used alone for diagnosis.

d. Folate and B_{12} metabolism appear closely related, although just how this is so is not clear. A deficiency of either vitamin

Figure 145.

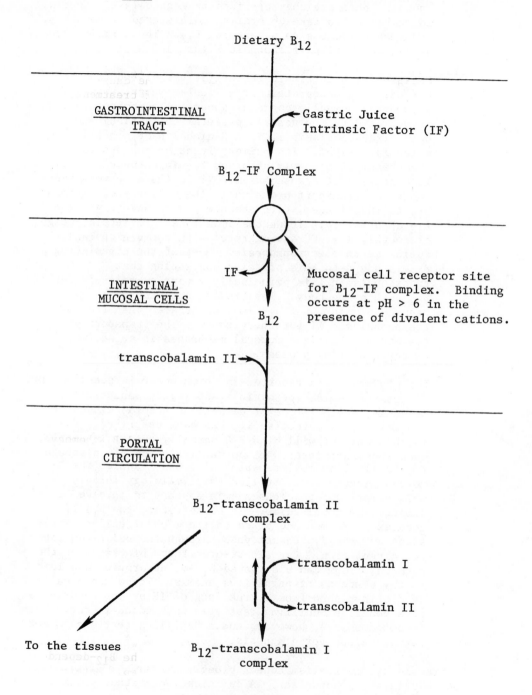

produces megaloblastic anemia and the symptoms of B_{12}-deficiency anemia are reversed by large doses of folate while B_{12} ameliorates folate-deficiency anemia. (Folate does <u>not</u> reverse the degeneration of the long tracts of the spinal cord which occurs in B_{12} deficiency, however, and may even worsen it. Consequently, folate should not be used clinically to treat B_{12} deficiency and care should be taken to accurately determine the cause of megaloblastic anemia prior to the initiation of treatment.)

Several other observations also indicate the interrelationship of B_{12} and folate.

i) A lack of either vitamin causes increased urinary excretion of formiminoglutamic acid (FIGLU; see Chapter 6) in the histidine-loading test. Urinary formate and amino-imidazolcarboxamide are also increased. All three compounds are substrates for folate-requiring reactions, and B_{12} is not known to be directly involved in their metabolism. This suggests that B_{12} deficiency impairs folate metabolism, making it unavailable as a cofactor for the reactions needed to metabolize these compounds. The role of B_{12} in this case may be in the synthesis of methionine from homocysteine. N^5-Methyl FH_4 is a substrate for the methylation reaction.

ii) Patients with B_{12} deficiency have elevated plasma concentrations of N^5-methyl FH_4. The ratio of this compound to FH_4 is also greatly increased. Both methionine and B_{12} decrease the relative and absolute amounts of the methyl derivative. Since FH_4 is the form used in most reactions requiring folate, the demethylation reaction, catalyzed by methyl B_{12}:homocysteine methyl transferase, is again implicated as the affected step. The ability of methionine to decrease N^5-methyl FH_4 is not clear but it may be related to its conversion to S-adenosyl-methionine (SAM). This is known as the "methyl-trap" theory.

iii) Vitamin B_{12} deficiency decreases hepatic folate stores. The defect appears to be one of retention of folate by the liver, rather than of uptake, although the latter may also be somewhat affected. Synthesis of pteroylpolyglutamates (recall that folate is pteroylmonoglutamate), which are non-transportable, hence, retained by the cells, is decreased. Since it is the non-methylated folate derivatives which serve as substrates for polyglutamate formation, the B_{12}-dependent transmethylation is once again the key step. Methionine again can replace B_{12}. The mechanism for its action is again not clear. Some workers find that methyl folates can be converted to polyglutamates directly, however, so there may be other pathways involved here, as well.

In summary, then, the B_{12}-dependent transmethylation of homocysteine is the common point between B_{12} and folate metabolism. Impairment of this reaction can explain most of the similarities between the symptomatology of deficiencies of either vitamin. The ability of methionine to replace B_{12}, the significance of pteroylpolyglutamates in folate metabolism, and the effect, if any, of B_{12} on folate transport across membranes require further investigation.

9. Vitamin C (ascorbic acid) was discussed earlier in this chapter. It is a vitamin in man, other primates, and guinea pigs because these species lack the enzyme which converts L-gulonolactone to 2-ketogulonolactone (see Chapter 4). All other species investigated are able to synthesize ascorbate. This pattern suggests that the ancestors of both primates and guinea pigs had the enzyme but lost it through two independent mutations. The need for separate mutations, one in a human ancestor, the other in the guinea pig line, arises because of the genetic isolation of the two species relative to each other.

The structure of L-ascorbic acid is as follows:

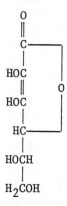

The principal dietary sources of this vitamin are fresh fruits (notably citrus fruits), tomatoes, leafy green vegetables, and new potatoes. Cooking (but not freezing) destroys (by oxidation) a portion of the ascorbate in these foods. The RDA for vitamin C is 45 milligrams, an amount which is probably well in excess of the MDR.

The vitamin C deficiency disease, <u>scurvy</u>, presents a variety of symptoms, almost all of which can be related to the formation of

defective collagen. The anemia of scurvy may result from a
defect in iron or folate metabolism as a result of the vitamin C
deficiency. The role of ascorbate as a cofactor for protocollagen
hydroxylase is discussed in Chapter 5 and the collagen formed in
scorbutic patients is reportedly low in hydroxyproline. Despite
this, it has yet to be definitively shown that ascorbate is
absolutely required as a cofactor in this or any other reactions
in the human body. Vitamin C is also involved in the conversion
of p-hydroxyphenylpyruvic acid to homogentisic acid (phenylalanine
metabolism; see Chapter 6) and probably in other hydroxylations
and in the microsomal electron transport system.

Other bodily processes for which there is some evidence that
vitamin C is required are:

a. Wound healing, blood clotting, and hematopoiesis. The data
    are primarily phenomenological although it is known that
    ascorbate aids in the synthesis of N^5 and N^{10} folate
    derivatives which are necessary for DNA synthesis, hence for
    red cell formation.

b. Neurotransmitter synthesis. The synthesis of norepinephrine
    from dopamine and the hydroxylation of tryptophan which is
    the first step in serotonin synthesis, both involve vitamin C
    in some way. Other hydroxylations may also require vitamin C.

c. Lipid metabolism. Although the mechanism is not known, there
    are several reports that both serum cholesterol and
    triglycerides are decreased by the consumption of pharmacologic
    doses of ascorbate (about 1 gm per day). Data to support this
    observation are not extensive, however, and it remains to be
    confirmed that this is a real effect.

Recently, there have been extravagant claims that massive doses
of vitamin C (1-2 or more grams per day) can prevent or cure the
common cold as well as other diverse ailments. After careful
evaluation of all available material, there seems to be very
little if any data to support this position. In fact, there is
some evidence suggesting that a daily intake of 2-4 gm may
actually be harmful, causing reproductive failure, vitamin B_{12}
inactivation, and perhaps nephrolithiasis. Self-treatment with
large quantities of vitamin C is almost certainly neither useful
nor wise.

Vitamin-Like Substances

As was mentioned early in this chapter, several compounds are
apparently required in the diet despite the fact that pathways for
their synthesis are known in the body. Such a situation could

conceivably arise if the pathways do not provide an adequate supply for the needs of the cell (due to low substrate concentrations, rapid degradative processes, or an inefficient enzyme system) or if the material cannot be readily transported from the site of synthesis to the place of action.

a. Choline (N,N,N-trimethyl-β-hydroxyethylamine) is an important constituent of phospholipids (lecithin is phosphatidyl choline) and of acetylcholine, a neurotransmitter substance. It can be completely synthesized from serine (see phospholipid metabolism, Chapter 8). Note, however, that this synthesis only occurs when the serine is present as phosphatidyl serine and is dependent on an adequate supply of amino acids (and, hence, of protein). Betaine (N,N,N-trimethylglycine) readily replaces choline in the diet for all species. Choline is utilized by a salvage pathway. In the lung, this salvage route is the principal one for the synthesis of the lecithin needed as a surfactant to prevent alveolar collapse (see Chapter 8). It has not yet been demonstrated that choline is an essential dietary nutrient in man although it is necessary in some other species.

b. Inositol (1,2,3,4,5,6-hexahydroxycyclohexane) can occur in a number of different isomeric forms. Myo-inositol (meso-inositol), is an important constituent of phospholipids. It is the only one of the inositols with biological activity. Inositol hexaphosphate (phytic acid) is found in avian erythrocytes where it binds to hemoglobin, thereby regulating the oxygen-capacity of the blood. Although inositol deficiency has been shown in laboratory animals, the dietary importance of this compound in humans has not been demonstrated.

c. Lipoic acid and ubiquinone (coenzyme Q) are sometimes classified as vitamin-like materials. In man, however, it appears that synthesis of these compounds by the body is adequate to meet all normal needs of the cells.

11
Immunochemistry

with Yoshitsugi Hokama, Ph.D.*

This section on the immune system of host defense is an integral part
of human biochemistry and physiology. Its consideration in this text
will be limited, and more extensive presentations are available in
the textbooks of immunology. This chapter will encompass the
immunoglobulins of serum proteins and the cells and tissue systems
associated with host defense or surveillance against deleterious
agents, particularly those of an infectious nature. It includes:

1. The immune system: the central and peripheral lymphoid systems,
    including the thymus, thymus-dependent (T) and thymus-independent
    (B) lymphocytes, spleen, lymph nodes and their relationships to
    T- and B-lymphocytes (the cell-mediated and humoral systems,
    respectively);

2. Immunoglobulins (IgG, IgA, IgM, IgD, IgE): isolation and
    properties, structure, nomenclature, and functions in disease
    processes;

3. Phagocytes of the peripheral blood and the reticulo-endothelial
    system;

4. The complement, kinin, blood coagulation and fibrinolytic systems
    and their functions in host responses;

5. Clinical considerations relative to the immune system.

The Immune System

The immune system, as we shall see, is highly complex and
involves a network of interacting cells and their molecular products.
One of the working hypotheses of the immune system, supported by a
variety of experimental observations, postulates that a host contains
a certain number of antigen-reactive cells as determined by the
genetic make-up. When a particular antigen is encountered by the
host, it reacts with a specific cell carrying the specific receptor;
this interaction leads to its proliferation of that cell (clonal
expansion).

---

* Professor of Pathology, John A. Burns School of Medicine,
  University of Hawaii at Manoa.

Among these cells, there are lymphocytes which are divided into two classes: thymus-derived cells (T-cells) and bursa-equivalent (or bone-marrow-derived; B-cells) cells. The B-cells following antigenic stimulation transform into immunoglobulin-secreting plasma cells. This is known as humoral immunity (i.e., acquired immunity mediated through circulating immunoglobulins present in the various body fluids). The T-cells, while participating in the cell-mediated immunity (i.e., acquired immunity mediated by cell-cell interactions), play an important role in the regulation of the humoral immune response. Some T-cells act as helper cells and others as suppressor cells of the B-cell activity. It should be noted that these two different groups of T-cells are genetically committed to function either as suppressor or helper cells. Thus the immune system is finely tuned and counter-balanced with stimulatory and inhibitory effects of different types of cell-cell interactions and their molecular products.

1. The Central Lymphoid System

The immune system consists of two major central lymphoid systems and their associated peripheral components. As noted above, in mammals, these are the thymus-dependent (cell-mediated) system and the thymus-independent or bursa-equivalent (humoral) system. Unlike birds (Aves), which have the bursa of Fabricius (located near the rectal region of the gastrointestinal tract and composed of B-cell lymphocytes associated with immunoglobulin synthesis), man has no such specific organ. Instead, in mammalian systems, one speaks of thymus-independent or bursa-equivalent lymphocytes. Since these cells perform the same functions in mammals as do the bursal lymphocytes in birds, the term B-lymphocyte is retained. The bone marrow, liver, appendix, tonsils, and lymphoid tissues of the intestines (GALT: gut associated lymphoid tissues) have been suggested as possible bursa-equivalent tissues or sources of B-lymphocyte regulator in mammals.

The genesis and development of immunocompetent T- and B-lymphocytes, and their relationship to each other and to various tissues of the body are shown in Figure 146. As indicated, both T- and B-cells originate from the same sources (yolk sac, fetal liver, and bone marrow) during embryogenesis. The distinguishing features of these two types of cells are summarized in Table 45.

Genesis of the Lymphoid System

a. Thymus

The thymus is formed from the III and IV pharyngeal pouches, which contribute the mesenchymal (epithelial) cells, and the

Figure 146. Genesis of the Lymphoid Immune System

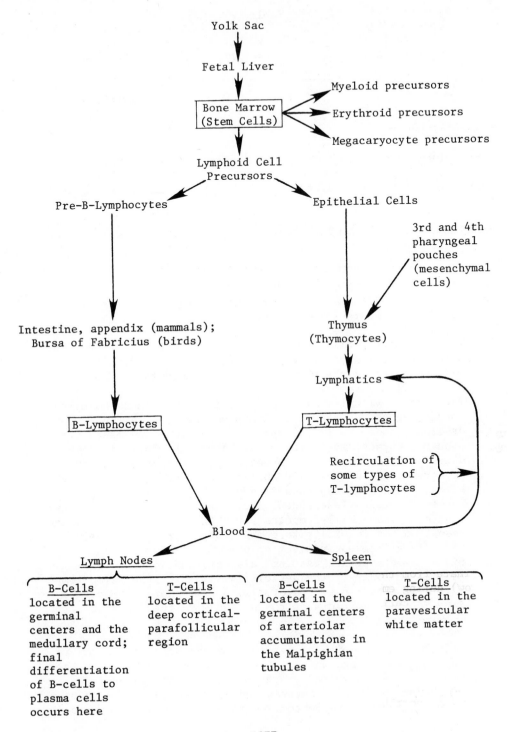

Table 45

Major Functions of the B-Cells and T-Cells,
the Primary Members of the Two-Component Lymphoid System

B-Cells	T-Cells
Humoral immunity	Cell-mediated immunity
Immunoglobulin synthesis	Homograft rejection
Production of the wide diversity of antibodies (immunoglobulins)	Delayed hypersensitivity reactions
Immediate hypersensitivity reactions	Graft versus host (GVH) reactions

---

stem cells of the bone marrow, which constitute the
lymphoid elements (pre-thymocytes).

The epithelial cells are the source of the hormone thymosin,
which is necessary for the maturation of thymocytes to
T-lymphocytes of the peripheral lymphoid system. The
congenital occurrence of rare thymic abnormalities (aplasia)
in newborns with T-cell immunodeficiency disorders have led
to the suggestion that the thymus was related to the
development of the immune system. Additional evidence from
neonatal thymectomy in mouse have shown the following: a
decrease in circulating lymphocytes (lymphopenia); severe
impairment of the animal's ability to reject graft; reduced
humoral antibody responses to some (those antigens requiring
T-helper cells - T-dependent antigens) but not all antigens;
wasting disease at 1-3 months after thymectomy, probably the
result of the inability to combat infections effectively,
since germ free environment or use of antibiotics can
minimize or prevent "wasting" disease; and loss of cell-
mediated immunity. The deficit initiated by the thymectomy
of the neonate can be restored with transplant of whole
thymus, crude extracts of thymus and with the purified
hormone thymosin.

Impairment of the immune system also can be demonstrated in
adult thymectomized mouse. This has been attributed to the
loss of thymosin which is synthesized by the residual
epithelial cells of hypertrophied adult thymus. The changes
observed in adult thymectomized mouse can be divided into

two phases, early and later changes in the immune system.
Early changes include loss of T-cells; loss of RNA synthesis;
loss of serum thymosin; and decrease in T-suppressor cell
which regulates B-cells, hence tendency to the formation of
autoantibodies and development of autoimmune diseases,
especially in the older individuals. Later changes include
loss of surface antigens (θ antigen in mouse T-cell) or the
decrease in number of these cells (due to loss of thymosin);
loss of mixed lymphocyte toxicity and mitogen (e.g.,
phytohemagglutinin-PHA, concanvalin A) responses,
graft versus host reactions, and humoral immune responses
to antigens dependent upon T-helper cells tend to decrease.

b.   T-Lymphocytes

T-Lymphocyte precursors from bone marrow, known as pre-thymic
cells, seed the thymus. There they become thymocytes and
trigger the development of the thymus. The thymocytes
proliferate from the cortex of the thymus at a rapid rate and
move inward to the medullary region where they undergo
maturation. These mature thymocytes enter the circulating
pool of lymphocytes and are then called T-lymphocytes. They
have a relatively long half-life. In addition to comprising
70-80% of the circulating lymphocytes of the blood, T-cells
are found in the perivascular region of the white matter of
the spleen (Malpighian corpuscles), the parafollicular and
deep cortical regions of the lymph node, the thoracic duct,
and the bone marrow. The T-lymphocytes are a heterogeneous
group of cells with diverse functions. They differ in
turnover rates, circulating routes, and surface antigenic
markers. In the blood and the thoracic duct lymph, there
are two populations of small lymphocytes. One population
is formed at a slower rate, is long-lived (having a half-life
of 100-200 days), and moves back and forth between blood and
lymph. The other is produced rapidly (2 to 3 mitotic
divisions per day) and has a circulating life span of less
than two weeks ($t_{\frac{1}{2}}$ of 3 to 4 days). Although the thymus and
bone marrow are the major sites for these short-lived
lymphocytes, other lymphoid centers can also produce them.

c.   B-Lymphocytes

B-Lymphocyte precursors are also of bone marrow origin.
They are either distributed directly to spleen, lymph nodes,
and other lymphoid tissues or may first traverse to or be
regulated by the liver, bone marrow, appendix, intestines
(GALT), or tonsils in order to be functional B-lymphocytes.
Evidence for B-cell control similar to that shown for birds
is not as convincing in the mammalian system as yet,

although the intestine, appendix, tonsils, liver and bone
marrow have been suggested as the possible bursa-equivalent
tissues. In addition to the B-lymphocytes in the peripheral
blood (20 to 30% of circulating lymphocytes of varying sizes),
these cells are also found in the germinal centers of lymph
nodes, the lamina propria of the gastrointestinal tract,
secretory glands, bone marrow, appendix, and other lymphoid
tissues associated with the reticulo-endothelial system.

2. The Peripheral Lymphoid System

    a. B-Lymphocytes

        i) The peripheral lymphoid system which is associated with
the immune system in the mature individual (capable of
immune responses) resides primarily in the peripheral
lymphoid tissues. These consist of the lymphatics, the
spleen, lymph nodes, lymphoid tissues of the
gastrointestinal tract, the respiratory tract,
secretory glands, and the lymphocytes of the blood.

        ii) As indicated earlier, the B-lymphocytes are situated in
the germinal centers of lymphoid follicles within the
lymph nodes and spleen and the plasma cells situated in
the medullary regions of the lymph nodes and red pulp
of the spleen. The distribution of B- and T-cells
within a lymph node is illustrated in Figure 147.
Plasma cells, the final cellular product of B-lymphocyte
development, are involved in antibody (Ab) or specific
immunoglobulin synthesis for the elaboration into blood
and serous fluids. These cells may be found scattered
throughout the lymphoid systems, including the bone
marrow. The sequence of B-cell transformation to plasma
cells via lymphoblast, following antigenic stimulation,
is shown later, in section 3. Plasma cells found in
endothelial areas of the respiratory tract, nasal
cavities, secretory glands, and gastrointestinal tract
are associated generally with the IgA and IgE syntheses.
These immunoglobulins are consequently referred to as
secretory immunoglobulins. Secretory IgA (S-IgA), a
constituent of the humoral immune system, is responsible
for the surface immunity of mucous membranes. Note
that this is one of the first immunologic defenses
that a microorganism has to encounter while attempting
to penetrate a mucous membrane.

       iii) B-Lymphocytes and the process of their conversion to
plasma cells play a significant role in humoral
immunity. B-Lymphocytes contain the light chain and

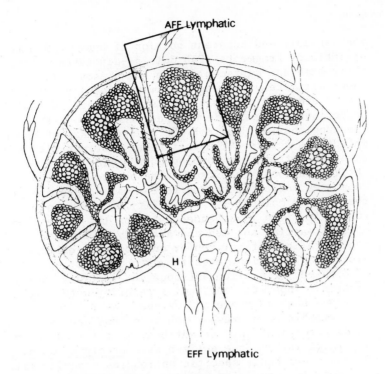

AFF Lymphatic

EFF Lymphatic

Figure 147. Detailed Diagram of Lymph Node
F(G)C-Follicular (Germinal Center) Cells, H-Hilus;
LS-Lymphatic Sinus; MC-Medullary Cord; P-Para-
follicular Cells; RC-Reticular Cells; T-Trabeculae; AFF=afferent;
EFF=efferent

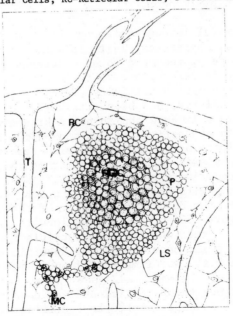

the variable and constant portion of heavy chain regions
of antibody fragments (Fab domain--discussed later) on
their membrane surfaces, while the plasma cells, in
turn, ultimately secrete the immunoglobulin. This
system provides defenses against viral and bacterial
infections by synthesis of specific antibodies
(immunoglobulins) against the antigens (Ag) of the
infectious agents. The antibodies interact with their
specific antigens (a homologous reaction) to give
antigen-antibody complexes. The nature of these
complexes varies according to the size and shape of
the antigen molecules. Visibility of the complexes
formed in vitro is dependent also on the concentrations
of the reacting antigen and antibody. Extreme
concentrations of antigen or antibody may obscure
visible aggregation, even though complexes (soluble)
have formed or antigen-antibody interactions have
occurred. Specific antibodies reacting with soluble
antigens such as proteins, carbohydrates, and viruses,
form visible complexes. Such reactions are referred
to as precipitation, the antibody is referred to as
a precipitin, and the antigen as a precipitinogen.
On the other hand, specific antibodies reacting with
antigens in suspensions (for example, bacteria or red
blood cells), results in an agglutination reaction.
The antibody is termed an agglutinin and the antigen
is an agglutinogen.

iv)   The humoral immune system has been associated with
immediate hypersensitivity in man. In many instances,
this type of response to an exogenous or endogenous
antigen is deleterious to the host. One type of
immediate hypersensitivity, referred to below as an
atopic allergy, has been attributed to the immunoglobulin
IgE. This immunoglobulin has a high affinity for mast
cells in connective tissues and basophils of blood.
The release of histamine and slow-reacting substance
of anaphylaxis (SRS-A) and other mediators of immediate
hypersensitivity reactions results when an antigen
interacts with an IgE molecule which is bound to a mast
cell via the Fc portion (a part of the immunoglobulin,
see Figure 153) of the immunoglobulin. The activation
of the enzymes which are responsible for the degranulation
of mast cells occurs when the specific antigen forms
bridges with two or more cell bound IgE on the membrane
surface of the mast cell (see Figure 148). The
redistribution of the IgE-antigen complexes on the cell
membrane is also related to the bridging phenomenon.
The histamine and SRS-A released initiate a classical

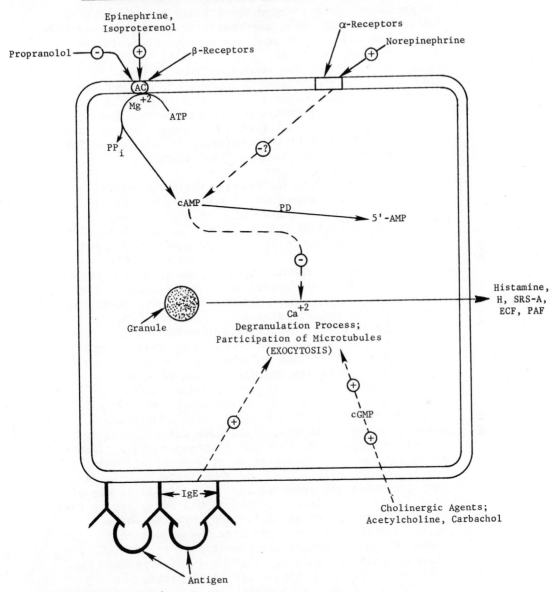

----▶ = Multistep reaction.
  − = Inhibitory (or suppressive) effect on the indicated process.
  + = Stimulatory effect on the indicated process.
  H = Heparin.
SRS-A = Slow-reacting substance of anaphylaxis.
  ECF = Eosinophic chemotactic factor.
  PAF = Platelet activating factor.
  AC = Adenylate cyclase.
  PD = Phosphodiesterase (inhibited by methylxanthines).

systemic anaphylaxis or an immediate skin reaction.
The immediate type of hypersensitivity associated with
IgG and exogenous antigens (resulting in antigen-antibody
complexes) can also trigger the release of histamine,
acetylcholine, serotonin, anaphylatoxin, and bradykinin.
These agents affect the smooth muscles of blood vessels,
collagen-containing tissues, leukocytes, and various
other tissues, producing histologic patterns characteristic
of early acute inflammation.   The release of the
pharmacologically active substances occur in some instances
through the activation of complement by the IgG-antigen
or IgM-antigen complexes (discussed later).

v)   A variety of factors, including both immunological
     (atopic) and non-immunological (non-atopic) stimuli,
     appears to activate attacks of bronchial asthma.   To
     explain this clinical entity involving catecholamines,
     the β-adrenergic theory has been proposed, which is
     depicted in Figure 148.   The bronchial system is
     affected by the catecholamines.   Those affecting the
     α-receptor (for a discussion on α- and β-receptors of
     catecholamines, see Chapter 6), such as norepinephrine,
     induce broncho-constriction and those affecting the
     β-receptor (e.g., epinephrine) induce bronchial
     relaxation.   Under normal circumstances, homeostasis
     exists between these two influences.   The β-receptor
     is linked to adenylate cyclase, an enzyme which
     catalyzes synthesis of cyclic-3',5'-AMP (cAMP) from
     ATP in the presence of Mg^{+2} ions (see Chapter 4).   It
     has been postulated that patients with bronchial
     asthma, eczema, rhinitis, etc., have deficiencies in
     adenylate cyclase in the cells of these various tissues,
     since the key to suppression or release of histamine and
     SRS-A is the cyclic nucleotide cAMP (suppresses) and
     cGMP (stimulates), respectively.   Diminished levels of
     cAMP due to insufficient synthesis via adenylate cyclase
     or hyperactive degradation (to AMP) via the enzyme
     phosphodiesterase, would increase the amount of
     histamine (H), eosinophilic chemotactic factor (ECF),
     platelet activating factor (PAF) and SRS-A release in
     response to the presence of IgE-Ag complexes, by
     infections, or by other agents.

Experimentally, suppression of adenylate cyclase by propranolol (a β-adrenergic blocking agent; antagonizes the action of epinephrine) along with an increase in α-receptor stimulation will suppress cAMP production and increase cGMP leading to increased histamine and SRS-A release following either atopic or non-atopic stimulation. Isoproterenol, which stimulates β-receptors, together with aminophylline, a methylxanthine (which inhibits phosphodiesterase) would increase the levels of cAMP and thus cause suppression of histamine and SRS-A release. Conversely, the level of cGMP would essentially be diminished.

In summary, it appears that the intracellular levels of cAMP and cGMP modulate the release of histamine, SRS-A, etc. β-Adrenergic agents, prostaglandin E_2 and histamine itself via the H_2-receptors (for a discussion on histamine receptors, see Chapter 6) inhibit degranulation and also enhance intracellular levels of cAMP; whereas α-adrenergic compounds lower cAMP levels and may augment degranulation. Cholinergic agents promote degranulation presumably by increasing the cGMP levels by stimulating guanylate cyclase. The degranulation is prevented by atropine, a cholinergic antagonist.

vi) The role of cyclic nucleotides, cAMP and cGMP in immunobiology, in addition to that discussed in the preceding paragraphs, is presently being investigated. It appears that cAMP suppresses antibody synthesis when given in large doses prior to antigen administration, whereas antibody stimulation is seen when cAMP is administered after antigen injection. The effect of both cAMP and cGMP appear to be on lymphocyte division. cAMP generally suppresses lymphocyte division with induction of tolerance and cGMP stimulates cell division. Both cyclic nucleotides are considered as "second messengers."

vii) The B-lymphocytes of the lymph nodes and the peripheral blood of man can be functionally differentiated from T-lymphocytes by their response to non-specific mitogens (substances which induce blastogenesis). This is summarized below. A positive response (indicated by a +) consists of an increased rate of mitosis and, consequently, of DNA synthesis.

Mitogens	B-Lymphocytes	T-Lymphocytes
Pokeweed mitogen (PWM)	+	+ (insoluble PWM)
Phytohemagglutinin (PHA)	±	+
Concanavalin A (ConA)	−	+
Purified Lipopolysaccharide	+ (no response in man)	−

T-Lymphocytes respond to PHA (enhanced by adherent cells--accessory cells--macrophages) and ConA, with no responses to LPS and to soluble PWM. B-Lymphocytes respond only to PWM, LPS (mouse B-cells) and possibly to PHA stimulation. B-Lymphocytes of man show no responses to LPS. Both T- and B-lymphocytes respond to insoluble PHA, PWM and ConA. These responses are generally assayed by determining the amount of ^{3}H-thymidine incorporated into the new DNA synthesized in the lymphocyte culture following stimulation by the mitogen in question. This is done using cells in short-term cultures (generally 2 to 7 days). The ^{3}H-thymidine uptake is determined in the scintillation spectrometer following precipitation of the cellular material with trichloroacetic acid, appropriate washing with alcohol and ethyl ether, and solubilization of the precipitate in the scintillation solution.

b. T-Lymphocytes

   i) T-Lymphocyte function is associated with cell-mediated-immunity (CMI). The distribution of these cells has been discussed in earlier paragraphs. They have been implicated in primary transplantation rejection, delayed hypersensitivity, surveillance against neoplastic transformation, and protection against viruses and bacterial infections. T-Lymphocytes do not transform to plasma cells nor do they elaborate immunoglobulins, but they do have specific antigens on their membrane surfaces (the θ antigens and Ly antigens of mouse T-lymphocytes for example) and MHC (major histocompatibility complexes) antigens.

   ii) Sensitized T-lymphocytes must be in direct contact with target antigens for their activation. This interaction initiates the release of a variety of factors which affect macrophages, polymorphonuclear leukocytes, lymphocytes and other cells. Release of these factors is not restricted to T-lymphocytes only. B-Lymphocytes have been shown to release some of these lymphokines (Table 46).

Mediators affecting macrophages:

i) Migration-inhibition-factor (MIF): stabilizes or retains macrophages in the inflamed area (this observation is currently used for the <u>in vitro</u> determination of cell-mediated-immunity).

ii) Macrophage activating factor: this factor is indistinguishable from (MIF).

iii) Macrophage (monocyte) chemotactic factor: attracts macrophage (monocytes) to area of injury (delayed hypersensitivity).

iv) Monocyte growth factor(s): factors in lymphocyte ConA cultures which induces and maintains monocytic growth in continuous culture.

Mediators affecting polymorphonuclear leukocytes:

i) Leukocyte chemotactic factors (LCF): for neutrophils, eosinophil, and basophil, they play a role in chronic inflammation and cutaneous delayed hypersensitivity.

ii) Leukocyte inhibitory factor (LIF): inhibits migration of polymorphonuclear leukocytes, retains cells in area of inflammation.

Mediators affecting lymphocytes:

i) Factors affecting antibody production (B-cells): a factor in mouse which triggers IgM production in B-cell to sheep erythrocyte; factors which induce IgG and IgE antibody production and factors associated with suppression of antibody synthesis have also been reported.

ii) Mitogenic factors: factors released by sensitized lymphocyte reacting with specific antigens, which stimulate other non-sensitized lymphocytes to blastogenesis.

iii) Transfer factor: a low molecular weight dialyzable
fraction obtained from sensitized T-lymphocytes,
either by direct extraction or release from cells
following reaction with specific antigen. Transfers
specific delayed hypersensitivity in man.

Mediators affecting other cells:

i) Lymphotoxin and growth inhibitory factors (cytotoxic
factors): lymphotoxin is a factor released by
sensitized lymphocytes which is cytotoxic to certain
target cells. Growth inhibitory factor(s) inhibits
growth of cells, rather than causing lysis of cells,
also referred to as cloning inhibition factor. Effect is
only on certain target cells such as Hela cells.

ii) Interferon: factor released in culture medium and
produced by normal lymphocytes when they are stimulated
with viruses, antigens or mitogens; interferon inhibits
or interferes with virus proliferation.

iii) Other factors: osteoclast activating factor, formed
by lymphocytes stimulated with specific antigens or
mitogens; tissue factor (pro-coagulant factor activity),
act similarly to coagulant blood factor VIII in decreasing
clotting time in blood factor VIII deficiency plasma.
This factor is produced by lymphocytes stimulated with
mitogens or antigens. Colony stimulating factor induces
bone marrow stem cells to differentiate into
granulocytes and mononuclear cells in vitro. Table 46
summarizes some of the physical and chemical properties
of these mediators produced by lymphocytes (T- and B-
cells).

These lymphocytic factors have collectively been called
"lymphokines." Examples of the classical type of delayed
hypersensitivity are the tuberculin reaction and the contact
dermatitis caused by poison ivy. With the possible exception
of the transfer factor and interferon, most of the factors
released in the cell-mediated immune responses have not been
completely characterized. In part, this is due to the small
amounts of these factors released at any one time. Evidence
for their occurrence is derived primarily from biological
assays and observations. Thus, whether a factor contributes
to one or more of the observed functions of the lymphocytes
remains to be determined.

Further characteristics and biological properties of T- and B-
cell lymphocytes, polymorphonuclear cells and macrophages are
summarized in Table 47.

Table 46. Some Physical and Chemical Characteristics of Mediators of Human Lymphocytes

Mediators	Heat Stability $56°$ for 30'	Produced by Cell Type	M.W. $\times 10^3$	Mobility in Polyacrylamide Gel-Electrophoresis	Effect of Enzymes[†] A	B
MIF protein	Stable	T & B	25-50	Albumin	R	S
MCF --	Stable	T & B	12-25	Albumin	--	S
MAF protein	Stable	T & B	25	Albumin	R	S
LIF protein	Stable	T & B	68	Albumin	R	S
LCF --	--	--	24-55	Variable	--	--
LMF --	Stable	T	20-30	--	--	--*
FAAP Enhancement --	Stable	--	--	--	--	--
Suppression --	--	--	25-55	--	--	--
TF	Stable	T	4	--	--	--**
CF protein	Stable	T	80-90	Post-Albumin	S	S
Interferon --	--	T & B		--	--	--***
OAF --	Liable		13-25	--	--	[+]S
CSF glycoprotein	Stable	T	40-60	--	--	--

  * Resistant to RNASE, DNASE and proteolytic enzymes.

 ** Resistant to RNASE and DNASE.

*** Not affected by DNASE, RNASE but destroyed by trypsin.

 † Inactivated by proteolytic enzymes other than chymotrypsin.

A = Neuraminidase

B = Chymotrypsin

R = Resistant

S = Sensitive

See text for the explanation of other abbreviations.

Table 47. Membrane Characteristics and Biological Properties of
Various Cell Types

Cell Type	Membrane Receptors
T-cell	For sheep RBC (E-rosettes) (Immunoglobulins)? Clq
B-cells	(Immunoglobulins), Fc, C3d, C4b, Clq, C3b (EAC-rosettes)
Macrophage	Fc, C3b
Granulocytes	C3b, Fc

T- and B-cells are distributed in the tissues and blood in the
following percentages:

Tissue	T-Cells	B-Cells
Thymus	> 75	< 25
Bone marrow	< 25	> 75
Tonsil	50	50
Spleen	50	50
Lymph node	75	25
Lymph	> 75	< 25
Blood, peripheral	55-75	15-30

3. Lymphocyte Interactions and Functional Relationship

   a. Recent evidence suggests that, for some antigens (<u>thymus-dependent</u>), antibody synthesis occurs only after T-B lymphocyte interactions. Aggregated human IgG (AHGG) is an example of a natural antigen which requires T-cell (T-helper lymphocyte) intervention. This can be depicted as follows:

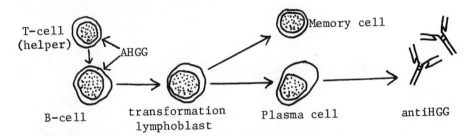

   The memory cells encode for future reference the information to synthesize antiHGG. On the other hand, antigens such as lipopolysaccharide (LPS) of gram-negative organisms like <u>Escherichia coli</u> require no T-cell intervention. They can act directly on B-cells resulting in plasma cell formation with subsequent antiLPS synthesis.

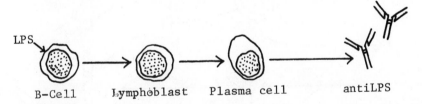

   Pneumococcal type-specific capsular polysaccharides behave in a manner similar to LPS. These antigens are referred to as <u>T-independent antigens</u>. Whether any interaction of T-lymphocytes with other cells is required for cell-mediated responses has not been thoroughly examined. However, regulatory T-lymphocytes such as T-suppressor cells have been demonstrated in control of antibody synthesis and cell-mediated immunity. Deficiency of these T-suppressor cells can lead to autoimmune diseases associated with increase in autoantibody synthesis as seen in diseases such as systemic lupus erythematosus.

   b. Particulate antigens, such as sheep erythrocytes (SRBC) and bacteria, must initially be digested to appropriate antigenic subunits ("activated" antigens) before T-B-lymphocyte interaction can occur for antibody synthesis. This requires

the intervention of a third cell type called the accessory-
cell (A-cell). These cells are also referred to as adherent
cells because of their ability to attach themselves to
surfaces such as glass. They are macrophages and have been
shown to enhance antibody synthesis by providing "activated"
antigens. This can be depicted as follows:

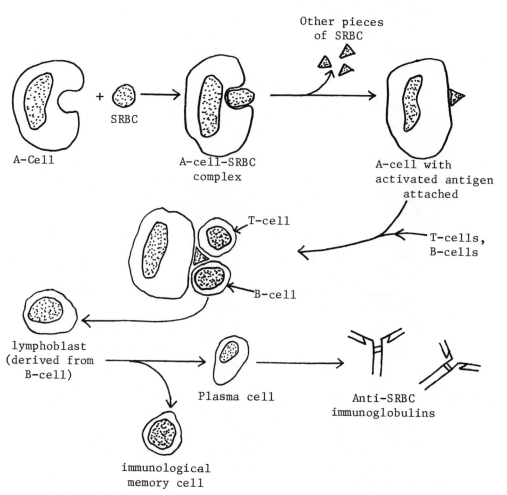

Though the illustrations depict direct contact between cells,
it has been shown that specific and non-specific soluble
factors obtained from lymphocyte extracts or substances
elaborated by lymphocytes could give similar effects.

Chemistry of the Immunoglobulins

Immunoglobulins are elaborated by plasma cells following the transformation of antigen-stimulated B-lymphocytes. As indicated earlier, this constitutes the humoral immune system. The genesis of the humoral immune system and the lymphoid tissues involved in this development has been presented earlier in this chapter. Discussions here will be confined to (1) the isolation and properties of immunoglobulins, (2) their biological functions, (3) nomenclature and classification based on the recommendations of the WHO (World Health Organization), and (4) the structures of these molecules and the theories which account for antibody diversity.

1. The Isolation and Physicochemical Characterization of Immunoglobulins

   a. The solubility properties of the immunoglobulins in salt solutions (especially the salting-out effect; see Chapter 2) were utilized in early procedures for their isolation from serum proteins. The solubility of any non-electrolyte solute in an aqueous solvent is decreased by the addition of a neutral salt. This effect is observed only when the dielectric constant of the solute is less than that of water. Ammonium sulfate has been most frequently used, though other salts such as sodium sulfate, sodium chloride, potassium chloride, potassium phosphate, and calcium chloride have also been employed. Two factors account for the popularity of ammonium sulfate: (1) its salting-in action at low ionic strength is less than that of calcium, potassium, and sodium chlorides, and (2) the salting-out effect of the sulfate ions is far greater than that of chloride ions. In addition, the greater solubility of ammonium sulfate over broad ionic concentrations affords a wider range of salting-out conditions than is attainable with the other salts. The disadvantages of ammonium sulfate are: ammonium ions interfere with direct nitrogen analysis and hence must be removed by exhaustive dialysis; it is not a good buffer and hence the control of pH below 8.0 is difficult; and its concentration is strongly dependent on temperature. For temperatures of 20 to 25°C, a half-saturated ammonium sulfate solution may be taken as equivalent to a concentration of 2.05 molal. Sodium sulfate gives a better salting-out effect than ammonium sulfate at comparable ionic concentrations, but is less popular than the latter because it is less soluble in water. This limits the range of conditions available for serum protein fractionations.

   Immunoglobulins of all classes have been precipitated by one-third saturated ammonium sulfate at room temperature.

At best, these are mixtures of IgG, IgA, IgM, and at times
other globulins. Nonetheless, with careful reprecipitation
and washing with one-third ammonium sulfate solution at
pH 7.4, immunoglobulins of reasonable purity (85 to 90%)
which show a major peak in the γ-region in cellulose acetate
electrophoresis (γ-globulins) may be obtained (see Chapter 6).

b. With the advent of gel filtration and chromatography using
anionic and cationic adsorbents conjugated to cellulose or
other inert polymers and gels (agarose, polyacrylamide),
immunoglobulins of high purity have been obtained. For
example, chromatography of serum proteins equilibrated with
0.0175 M phosphate buffer, pH 6.3, on diethylaminoethanol-
cellulose (DEAE-cellulose), an anionic exchanger, will give
relatively homogeneous immunoglobulin-G in the first major
eluate fraction. IgA and IgM appear in later fractions with
increasing ionic concentration, but these fractions also
contain other serum proteins. A typical fractionation
pattern on DEAE-cellulose, with the distribution of the
immunoglobulins and other serum proteins, is shown in
Figure 149.

Isolation of immunoglobulins has also been achieved by gel
filtration chromatography. In this technique, the separation
is based on molecular size and shape of the proteins. Gel
filtration chromatography has been most useful in the
separation of IgM from IgG and of heavy chains (H-chain) from
light chains (L-chain), to be discussed later. These
chromatographic methods have been widely used for
immunoglobulin isolation and characterization because of
their simplicity and less severe denaturation or alteration
of the immunoglobulins.

c. Another method utilized in the characterization and, in many
instances, isolation of purified immunoglobulin products is
zonal electrophoresis. This utilizes a semi-solid medium
that serves as a bridge between the electrodes and as a
support on which the sample is separated. The migration of
the proteins depends on their net electrical charge at the
pH being used, and, hence, on their isoelectric point (pI).

Zonal electrophoresis is not the recommended procedure for
isoelectric point determination with most support materials.
At the alkaline pH used (8.6), the usual supports possess a
negative charge. Since the support is a solid and therefore
unable to migrate when the electric field is applied, the
solvent (water) will move instead, in a direction <u>opposite</u>
to that which the support would take if it were able to.
Thus, for a negatively charged support (attracted to the

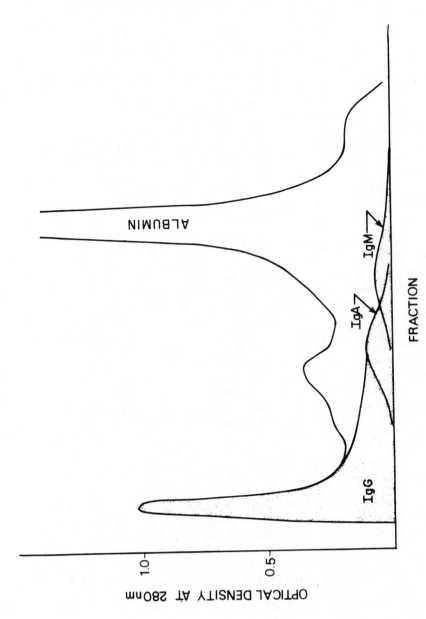

Figure 149. <u>DISTRIBUTION OF I_GG I_GA AND L_GM ON DEAE-CELLULOSE
COLUMN CHROMATOGRAPHY.</u> The upper line outlines the
fractionation profile of whole serum, while the stippled
regions indicate where the three major immunoglobulin
fractions occur within this profile. IgE and IgD are so
small quantitatively that they cannot readily be shown
in this diagram.

anode), water will migrate toward the cathode, creating a decreasing concentration gradient from the anode to the cathode. This partially counters the electrical force and thereby alters the mobility of serum proteins. This phenomenon of water migration is termed the <u>electro-osmotic effect</u>. It is discussed in more detail in most books on physical chemistry. The influence of electro-osmosis is especially significant with the serum proteins which have isoelectric points ranging from 4.8 (albumin) to 7.4 (γ-globulin). All of these proteins bear a negative charge at pH 8.6 and will therefore migrate <u>toward</u> the anode, <u>against</u> the water migration. The isoelectric points of serum proteins have generally been determined by the free boundary electrophoresis method of Tiselius, utilizing the mobilities of the proteins in buffer systems of different pH's. Here, the problem is one of convection and, hence, temperature control is a critical factor. To minimize this effect, the electrophoresis is carried out at $0^{o}C$.

Some of the more common media used in zonal electrophoresis are cellulose acetate, agar, agarose, gelatin, starch, and polyacrylamide. Zonal electrophoresis is performed at low ionic concentrations (0.05 to 0.075) and the buffer systems used (at pH 8.6) include sodium barbital, borate, and tris-glycine. Starch and polyacrylamide media add a second dimension to electrophoresis in that proteins are separated on the basis of molecular shape and size in addition to their net electrical charges. This is called the sieving effect and is controlled by altering the concentration of the gels and thus the sizes of the pores in them. Typical zonal electrophoretic patterns of normal human serum carried out on cellulose acetate are shown in Figure 150. Five major components can be seen: near the anode ($\oplus$) are the albumin peak, followed by α_1-, α_2-and β-globulin peaks and finally the γ-globulin or immunoglobulin peak. The contributions of the individual immunoglobulin are also indicated.

d. Immunoelectrophoresis is a useful procedure for the characterization of immunoglobulins with specific antibodies which are prepared against the Ig's (see later). This procedure combines the principles of zonal electrophoresis and immunodiffusion for the analysis of antigens. The method consists of first separating the immunoglobulins in serum by zonal electrophoresis in agar, agarose or gelatin, and subsequently adding specific antiserum to the electrophoresed antigens in wells cut parallel to the electrophoretic migration. Figure 151 illustrates how this procedure is used in recognizing the five major classes of immunoglobulin.

Figure 150. <u>Cellulose Acetate Electrophoresis of Serum Proteins and of the Immunoglobulins</u>

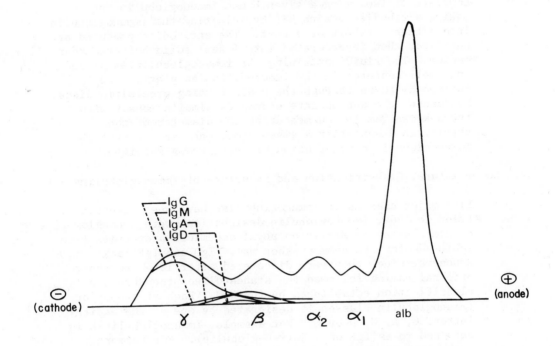

Figure 151. <u>Immunoelectrophoresis of the Immunoglobulins</u>

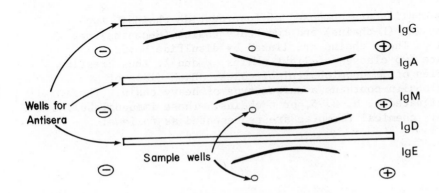

e. Immunoglobulins are found mostly in the γ-globulin fraction and to a small extent in the β-globulin fraction of serum. Quantitative determination of human immunoglobulins can be made by a radial immunodiffusion procedure. This method consists of isolating a given human immunoglobulin and making antibodies against it by injecting the immunoglobulin into either a rabbit or a goat. The antibodies produced are purified, then incorporated into a semi-solid buffered agar medium. The fluid containing the immunoglobulin is placed in a well centered in the immunodiffusion plate. The antigen diffuses through the plate forming precipitin discs. By running the appropriate standards simultaneously with the unknown sample and measuring the diameter of the precipitin discs after a given time, one can estimate the immunoglobulin concentration of the unknown specimen.

2. Nomenclature, Classification and Structure of Immunoglobulins

a. Five major classes of immunoglobulins have been recognized. Prior to 1964, immunoglobulin designations were a conglomerate of names based primarily on physicochemical properties (see Table 48 for the common synonyms). IgD and IgE lack synonyms, since they were discovered after 1964 and their original naming followed the standardized procedure for classification established by the WHO. Each of the major immunoglobulin classes is designated by one of the capital letters G, A, M, D or E. For example, Immunoglobulin-E is referred to as IgE or γ-E (γ-E-globulin). Furthermore, each class can be referred to as a member of the kappa (κ) or lambda (λ) type, based on the nature of the L-chain constituent. Thus, the designation of the immunoglobulins depends upon the antigenic determinants of the polypeptide chains which constitute the complete molecule (see below).

b. Immunoglobulins are composed of four peptide chains: two light chains (L-chains) and two heavy chains (H-chains) per molecule. These chains are linked by disulfide bonds. There are two classes of light chains, κ and λ, thus creating two series of immunoglobulin molecules. Each class of immunoglobulin contains a unique type of heavy chain. These are designated γ, α, μ, δ, or ε chains. These immunoglobulins and their chemical formulae are represented as follows:

	H-Chain	κ-Type	λ-Type
IgG	γ	$\kappa_2\gamma_2$	$\lambda_2\gamma_2$
IgA	α	$\kappa_2\alpha_2$	$\lambda_2\alpha_2$
IgM	μ	$\kappa_2\mu_2$	$\lambda_2\mu_2$
IgD	δ	$\kappa_2\delta_2$	$\lambda_2\delta_2$
IgE	ε	$\kappa_2\epsilon_2$	$\lambda_2\epsilon_2$

c. The structural analysis of immunoglobulins began with Porter's early studies using the proteolytic enzyme papain on rabbit immunoglobulin. Since then, in addition to papain, pepsin (another proteolytic enzyme; see Chapter 6) and the reducing agents, mercaptoethanol and dithiothreitol, have been widely used for this work. The reducing agents are necessary for the breakage of the disulfide linkages. Immunoglobulin fragments obtained by proteolysis or reduction have subsequently been analyzed and separated by adsorption and gel filtration chromatography and by zonal electrophoresis on starch or polyacrylamide gel supports. Diagrammatic representations of the fragments obtained following cleavage with papain and pepsin are shown in Figure 152. Papain hydrolysis, under proper pH and time conditions, cleaves IgG into three fragments. These fragments consist of two similar fragments designated Fab, and each contain the antibody (ab) combining sites. The third fragment is called Fc because it is crystallizable, being obtained as flat, thromboid crystals from aqueous solution. Pepsin, on the other hand, produces one large fragment (designated F(ab')$_2$), which has both ab sites, and eight smaller peptides (four per chain). The small peptides are derived from the region corresponding to the Fc fragment of papain digestion. The principal site of both papain and pepsin action is in the "hinge" region of the molecule (see Figures 152 and 153). This is a segment of the heavy chain peptides which is linear (or randomly coiled) due to the presence of three proline residues which prevent helical folding (see Chapter 2). This "openness" makes the hinge region susceptible to enzymatic attack.

d. Reduction of IgG with mercaptoethanol or dithiothreitol, followed by alkylation of the exposed sulfhydryls, results in the cleavage of the interchain disulfide linkages. L-Chains and H-chains are obtained by gel filtration

(Text continued on p. 1102)

Figure 152. Papain and Pepsin Cleavage of 1 G A solid line between chains indicates an interchain disulfide bond while "SS" represents an intrachain cystinyl group.

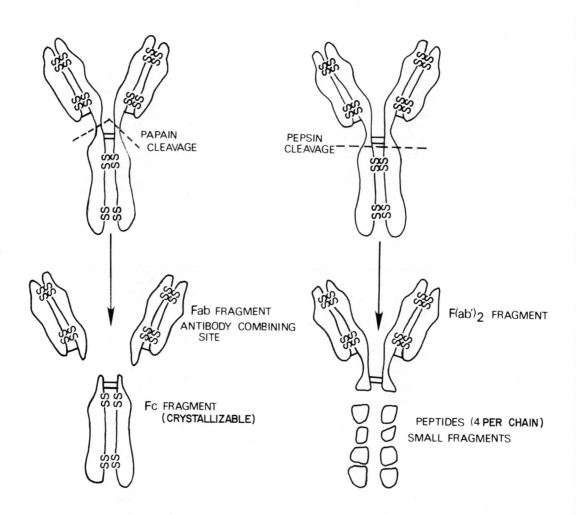

PAPAIN CLEAVAGE

PEPSIN CLEAVAGE

Fab FRAGMENT
ANTIBODY COMBINING
SITE

F(ab')₂ FRAGMENT

Fc FRAGMENT
(CRYSTALLIZABLE)

PEPTIDES (4 PER CHAIN)
SMALL FRAGMENTS

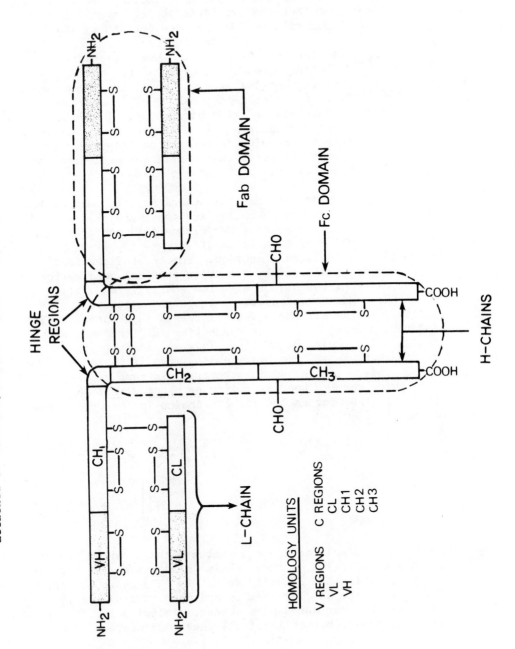

Figure 153. Diagrammatic Structure of IgG. CHO represents carbohydrate attached to the heavy chain constant regions.

chromatography on Sephadex following reduction and alkylation
of the interchain -S-S- bonds. This is illustrated in
Figure 154. Under mild conditions, the intrachain -S-S
bonds remain intact and retain the configuration and
biological properties of the chains. These enzymatic
cleavages and chemical reductions have led to the
understanding of the spatial orientation of the subunits
and the structure-function relationships of immunoglobulins,
which will be discussed in later paragraphs. In addition,
amino acid sequence studies and immunological analysis have
contributed much to the elucidation of the heterogeneity,
genetics, antibody diversity, and configurational arrangement
of the five major classes of immunoglobulins.

e. To reiterate, monomeric immunoglobulins are composed of four
polypeptide chains (two H-chains and two L-chains), which
loop and twist to form the three dimensional configuration.
The configuration of each Ig is of extreme importance for
its ultimate biological function. The three-dimensional
structure is maintained by two major physical bonding forces:
(i) noncovalent bonds (non-electron sharing), and (ii)
disulfide bonds formed by two neighboring cysteine (half-
cystine) residues. These were discussed in Chapter 2. The
interchain disulfide bridge connecting the light and heavy
chains in molecules of subclass IgG_1 is closer to the
bridges between the two heavy chains than it is in the other
Ig's. This is reflected in schematic diagrams of the Ig's
by drawing the L-H bond as a line <u>perpendicular</u> to both
chains in IgG_1 but <u>angled</u> to both chains in the other Ig's,
as shown below.

IgG_1 Other Igs

IgA_2 has no disulfide linkages between the L- and H-chains
and the orientation of the L-chains relative to the H-chains
is opposite to that normally seen. In this instance, the
H-L bridges are attributable to hydrogen bonding, salt
linkages, and van der Waal's forces. Individually, these
are weak bonds, but collectively they constitute a strong
bonding force (see Chapter 2).

f. The light (L) chain consists of about 214 amino acids. The
NH_2-terminal region, about 107 to 115 amino acids long, is
called the L-chain <u>variable (VL) region</u>. The remaining

Figure 154. <u>Reduction of Alkylation of IgG.</u>

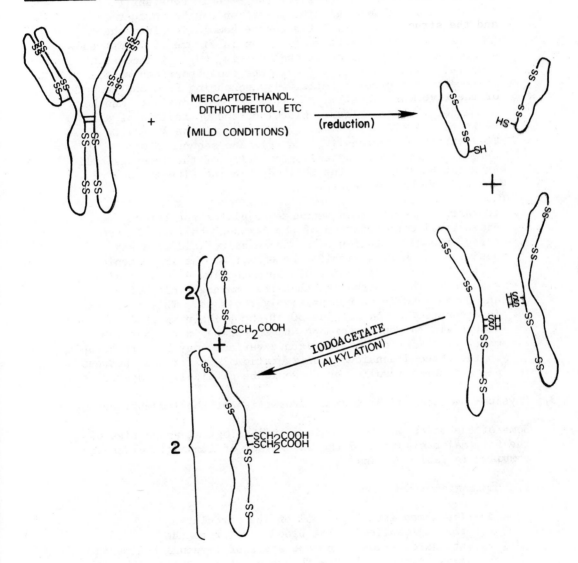

(carboxyl) portion of the L-chain has about 107 to 110 amino acids and is termed the L-chain <u>constant (CL) region</u>. The heavy (H) chain has about 450 amino acid residues. As in the L-chain, the 107 to 115 NH_2-terminal residues compose the variable H-chain (VH) region while the remaining (310 to 330) COOH-terminal residues are the H-chain constant (CH) region. For IgG and IgA, the CH-region can be further subdivided, on the basis of sequence homologies, into the CH_1, CH_2, and CH_3 sequences as shown in Figure 153. IgM and IgE appear to have an additional domain, CH_4. The constant regions (CH_1, CH_2, CH_3, and CH_4) are those portions whose amino acid sequence is relatively unchanging from one member of a particular class (γ, μ, etc.) of heavy chain to another member of the same class. Likewise, the CL regions of all κ-chains are closely alike as are those of the λ-chains. The "variable" regions (VL and VH) have sequences which depend on the immunological specificity of the immunoglobulin. As might be expected, the antibody binding site is located in the variable sequences.

g. In part, the amino acid sequence regulates the three dimensional configuration of the immunoglobulins. In some cases, the substitution of a single amino acid at a key position may have a significant effect on the immunogenicity or the biological function of the molecule. A classical example is the observation that the immunogenicity of pig and human insulin is not precisely the same. This difference is accounted for by residue 30 of the B-chain which is alanine in the human hormone and threonine in the porcine material. Part of this antigenic variance may be attributable to the three-dimensional configurational differences between pig and human insulin, which would be difficult to assess.

3. Physico-Chemical and Biological Properties of the Immunoglobulins

Some of the major physico-chemical and biological properties of the five major classes and the subclasses of immunoglobulins are compiled in Tables 48 and 49.

a. Immunoglobulin-G (IgG; γG)

i) IgG comparises 80 to 85% of the circulating immunoglobulins in the blood. Prior to the standardization of nomenclature of immunoglobulins in 1964 by the World Health Organization, these globulins were designated by a variety of names such as γ, $7s\gamma$, $\gamma\gamma$, 6.6γ, γ_1, and γss. These designations were derived from electrophoretic mobilities and ultra-centrifugal studies. The approximate molecular weights,

(Text continued on p. 1108)

Table 48. Physicochemical Properties and Other Biochemical Characteristics of the Immunoglobulins of Man

Immunoglobulin Class / Immunoglobulin Subclasses	IgG				IgA		IgM	IgD	IgE
	IgG$_1$	IgG$_2$	IgG$_3$	IgG$_4$	IgA$_1$	IgA$_2$	---	---	---
1. Synonyms	γ, 7Sγ, 6.6γ, γ$_2$, γss				β$_2$A, γ$_1$A		γ$_1$M, β$_2$M, 19Sγ, γ-macroglobulin	none	none
2. Approximate M.W.*	149,000–153,000				150,000–600,000		900,000	180,000	188,000–209,000
3. Approximate S$_{20,w}$*	6.6–7.2				7,9,11,13		18–32	7	7–8.2
4. Diffusion Coefficient (1 × 10^{-7} cm^2/sec)	4.0				3.0–3.6		1.71–1.75	---	3.71
5. Extinction Coefficient (E$^{1\%}_{280\ nm}$)	13.8				13.4–13.9		13.3	---	15.3
6. Partial Specific Volume ($\overline{V}_{20}$, cm^3/gm)	0.739				0.729–0.723		0.723	---	0.713
7. Isoelectric Point in pH units	5.8–7.3				---		5.1–7.8	---	---
8. Valence (Ag binding sites)	2				2		5	2	2
9. Total Carbohydrate Content (wt. %)	3.0				12.0		12.0	13.0	12.0
10. Mean Survival in Serum (T/2 in days)	11–12	11–12	6–7	20–21	5–6	4.5	5	2.8	1.5
11. Mean Concentration in Serum: (mg/dl) (range in parentheses)	Total for IgG: 1240 (900–2000)				280 (200–350)		120 (75–150)	3.0 (<0.3–40)	0.03 (0.007–0.18)
	820	286	90	50					
12. Heavy Chain Class (M.W.)	γ (50,000)				α (65,000)		μ (65,000–70,000)	δ	ε (72,500)
13. Light Chain	κ or λ				κ or λ		κ or λ	κ or λ	κ or λ

Table 48. (continued)

Immunoglobulin Class / Immunoglobulin Subclasses	IgG				IgA		IgM	IgD	IgE
	IgG$_1$	IgG$_2$	IgG$_3$	IgG$_4$	IgA$_1$	IgA$_2$			
14. κ to λ ratio (range)	1.42 to 2.41	0.96 to 1.10	1.25 to 1.12	7 to 5	1	2	1, 1-4	1/20	---
15. Number of Inter-chain Disulfide-Bonds	4	6	6	4	5?	5?	15	3	3?
16. Allotypes: H-Chain Gm	+	+	+	0	-	-	-	-	-
Am_2	-	-	-	-	-	+	-	-	-
L-Chain Inv	++	++	++	++	++	++	++	++	++
OZ	++	++	++	++	++	++	++	++	++
17. Electrophoretic Mobility in Agar	γ_2	γ_2	γ_2	γ_1	$\gamma_1-\beta$		$\gamma_2-\beta_2$	$\gamma-\beta$	γ_1
18. Intrinsic Viscosity		0.06			---		0.162	---	---
J-Chain	-	-	-	-	+	+	+	-	-
Secretory piece	-	-	-	-	+	+	+ (secretory) IgM	-	-

* M.W. = molecular weight;

$S_{20,w}$ = sedimentation coefficient corrected to water at 20°C;

blank spaces = data unknown.

In some cases, data are given for each subclass while in others, average values for the class are indicated.

Table 49. The Biological Properties of Human Immunoglobulins

Immunoglobulin Class — Subclasses	IgG				IgA		IgM	IgD	IgE
	IgG₁	IgG₂	IgG₃	IgG₄	IgA₁	IgA₂	IgM	IgD	IgE
1. mg/ml (serum)	5-12	2-6	0.5-1	0.2-1	0.5-2	0-0.2	0.5-2	0-0.4	0-0.002
2. Specific antibodies									
a. anti-RH	+		+				+		+
b. anti-dextran	+	+							
c. anti-levan		+							
d. anti-factor VIII				+					
3. In Secretions	0	0	0	0	+++	+++	++	0	+
4. Complement-fixation Classical	+	+	+	0	0	0	+++	0	0
Alternate pathway	0	0w	0	+	+	+	0	0	+
5. Passive cutaneous Anaphylaxis (PCA)	+	+	+	+	0		0	0	0
6. Placental transfer (Pt)	+++	+	+	++	0		0	0	+
7. Reacts with Rheumatoid factor (antigen)	+	+	0	+	+	+	+	0	0
8. Rheumatoid factor (antibody)	+	+	+	+					
9. Reacts with Staphylococcus Protein A	+	+	0	+					
10. Fc Receptor on:									
Macrophage	+	0	+	0	0		0	0	0
Basophils	0	0	0	0	0		0	0	++
Neutrophils	+	±	+	+	0	0	0	0	0
Platelets	+	±	+	±	0		0	0	0
Lymphocytes	0	0	0	0	0		0	0	++
11. Prausnitz-Kustner P-K reaction	(positive, (+) in hyper-immunized individuals)				+				(0.1-1.0 ng AbN/ml)
12. Isohemagglutinin	+		+	+	+	rare	+	0	0
13. Synthesis: mg/Kg/day	25		3.4		24	21.3	7	0.4	0.02
14. Cryoglobulins	+	0	+	+	+	rare	+	0	
15. Reverse P.C.A.	+	0	+	+	0	0	0		0

+ = reacts (antibody present); 0 = does not react; w = weakly; AbN = antibody nitrogen; ± = possible reaction; Blanks indicate data not known.

sedimentation, extinction and diffusion coefficients, isoelectric points, and partial specific volume are data compiled for IgG without reference to subclass distinction. This may account for the range found especially in the sedimentation coefficient and isoelectric points. Subclasses IgG_1, IgG_2, and IgG_3 have similar mobilities in agar immunoelectrophoresis migrating with the γ_2 globulins, while IgG_4 appears to have slightly faster mobility (moving with the γ_1 fraction). This, in part, may account for the variation in the isoelectric points for the mixture of subclasses. The valence or binding capacity (number of binding sites per molecule) for IgG is 2 although, in some instances, a valence of 1 has been reported. This may be attributable to a weaker binding capacity rather than a true valence of 1, since this so-called univalent antibody has been shown to have all four polypeptide chains (two H- and two L-chains). Molecules in all four of the IgG subclasses contain γ-H chains and either κ or λ L-chains. The total carbohydrate content of IgG molecules is approximately one-fourth that of the other Ig classes.

ii) Kappa/Lambda ratios in the subclasses increase in the order $IgG_2 < IgG_3 < IgG_1 < IgG_4$, while serum concentrations of the subclasses decrease in the order $IgG_1 > IgG_2 > IgG_3 > IgG_4$. The mean total serum concentration of IgG is 1240 mg/dl with a normal range of 900-2000 mg/dl. The mean half-life of the IgG subclasses vary from a low of 6-7 days for IgG_3 to a high of 20-21 days for IgG_4. Those for IgG_1 and IgG_2 are intermediate, at 11 to 12 days. A mean of 23 days has been indicated for total IgG. IgG_1 and IgG_4 have four interchain disulfide bridges, two between L-H chains and two between H-H chains. IgG_2 and IgG_3 possess six interchain disulfide covalent bonds with four bonds between the H-H chains rather than two.

iii) Immunoglobulins are excellent antigens or immunogens and antisera to them can be raised in a variety of animals (monkey, horse, sheep, donkey, goat, and rabbit). Using these antisera, specific antigenic sites on the Ig's can be correlated with structural features and with specific amino acids in the polypeptide chain. Antigens are conveniently categorized into three major groups: isotype antigens are present in all normal sera and can be differentiated into the five major classes and their subgroups (types);

allotype antigens are present in some (but not all)
normal sera and are governed by allelic genes;
idiotype antigens are unique to one particular antibody
molecule, hence by inference, they are related to
the V-region or antibody combining site.

iv) The isotype and allotype antigens are related to the
C-region of the H-chains. The class-specific antigens
of the C-region represent one group of isotypes.
Cross-reactions between the major immunoglobulin
classes have not been demonstrated, though regions of
homology based on sequence studies have been shown.
Within subclasses, however, many shared common
antigenic sites have been found. These analyses have
been carried out by precipitation and hemagglutination
inhibition tests. Antibodies to some subclass-antigens
are difficult to raise in animals.

v) The allotype antigens for the H-chains are referred to
as Gm (gamma-markers). The Gm markers are restricted
to the CH region of the H-chains of IgG and IgA (see
IgA section). Gm 1, 22, 4, and 17 are restricted to
IgG_1; Gm 24 is specific for IgG_2; and Gm 3, 5, 6, 13,
14, 15, 16, and 21 are specific for IgG_3. No regular
Gm for IgG_4 has been demonstrated as yet. IgG_4 has a
non-marker, Gm-variant antigen designated 4a and 4b on
the CH_3-homology region. Other Gm antigens are shared
among the subclasses. There are at present a total
of 27 Gm antigens. The allotype marker InV is found
on the kappa chain. Valine at residue position 191 on
the κ-chain represents a positive test for InV (b+);
whereas substitution at this position with leucine
results in InV (a+). Amino acid substitution at
position 190 in the λ-chain controls the presence or
absence of the isotype Oz.

The biological activities attributed to the IgG
subclasses are shown in Table 49. The major subclass
associated with specific antibodies is IgG_1. Specific
antibodies to dextran, levan, Rh, and factor VIII
reside in the different subclasses of IgG. IgG_4 does
not fix complement, while IgG_2 does so weakly.
Passive-cutaneous anaphylaxis (PCA) in guinea pig is
negative for IgG_2. IgG_3 does not react with rheumatoid
factor and staphylococcal protein A. No reaction with
macrophage receptor has been shown for IgG_2 and IgG_4.
Isohemagglutinin related to IgG antibody is found in
individuals following hyperimmunization with ABO
antigens. IgG antibodies are generally related to the

later-occurring antibodies following antigenic
stimulation. IgG antibodies do not participate in the
P-K (Prausnitz-Kutzner) reaction. Cryoglobulins and
rheumatoid factor (antibody) in some diseases may be
associated with IgG. With the exception of IgG_3, all
other IgG subclasses have been shown as antigens for
rheumatoid factor (antibody).

b. Immunoglobulin-A (IgA; γA)

   i) IgA has two subclasses designated IgA_1 and IgA_2. There
are two major differences between these subclasses:
the allotype Am_2 is present in the C region of the
H-chain of IgA_2 (Am_2+) and is absent in IgA_1 (Am_2-);
the L-chains in IgA_2 are not linked to the H-chains by
disulfide bonds. In addition, as was indicated
earlier, the orientation of the L-chains in IgA_2 is
opposite to that of the L-chains shown in the other
immunoglobulins.

   ii) The significant physicochemical and biological
properties of IgA are listed in Tables 48 and 49.
Some of these properties are for the secretory IgA
which is a dimer of IgA_1 held together by a secretory
piece or component (SP, SC) and a polypeptide "J"-
(juncture) chain. Properties of both SC and "J"-chain
are shown in Table 50.

Table 50.   Properties of Secretory Component and J-Chain

Properties	SC	J-Chain
Molecular weight	70,000	15,000
Carbohydrate	9.5-15.0%	8.0%
Sedimentation coefficient, $S_{20,w}$	4.20	1.28
Partial Specific Volume (cm^3/gm)	0.726	--
$E_{278}^{1\%}$	12.7	6.3-7.0
Electrophoresis: (relative mobility)	β-globulin	fast-moving component in gel
Associated with H-chains	α and μ	α and μ

Thus, the polymeric IgA contains other components in addition to H- and L-chains. The secretory piece and J-chain have also been noted in IgM. Skeleton diagrams of each of the major classes and subclasses of immunoglobulins with inter-chain disulfide (S-S) linkages are shown below. The locations of the J-chain and SC in IgM and secretory IgA_1 is still speculative although it is postulated to occur, as shown below, somewhere near the carboxyl termini of the heavy chains. The secretory component is antigenic and is present in its free form in bodily secretions.

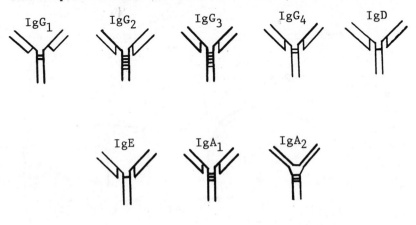

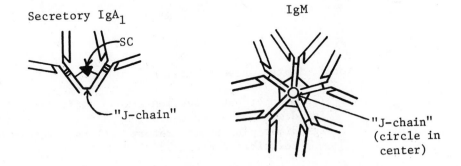

iii) It appears that there are no chemical differences between circulating forms of IgG, IgD, and IgE and those found in the secretions of mucous membrane cells. As was just indicated, however, a portion of IgM and almost all of IgA have a secretory component (SC) attached to them when they occur in external secretions. An additional connecting glycoprotein

(J-chain) has also been identified which is associated with dimeric and polymeric forms of IgA and IgM molecules.

In humans, it has been shown that the secretory component is synthesized in the mucosal cells and is joined by disulfide linkages to the H-chains of IgA molecules before they are secreted. IgA molecules themselves are synthesized in response to a proper stimulus in the plasma cells that are present in the submucosa. This can be diagrammatically represented as follows:

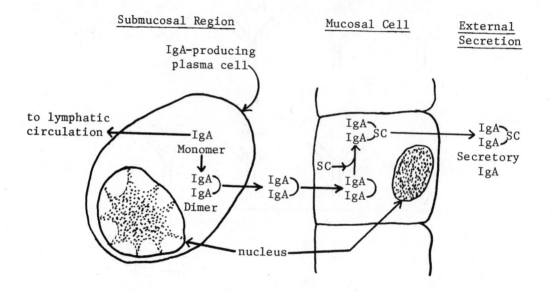

Note: in the above diagram

   IgA monomer has a sedimentation coefficient of 7S.
   IgA dimer has a sedimentation coefficient of 10S.
   IgA dimer plus SC (secretory IgA) has a sedimentation coefficient of 11S.

A question arises as to the need for the SC component that is synthesized and attached to IgA dimers in the mucosal cells. It has been suggested that the SC component may facilitate transport of the dimer through the mucosal cell by protecting the molecule from intracellular degradative processes (e.g., proteolysis). It may also participate in the functioning of IgA as an antibody in an external environment. A similar

mechanism for synthesis of secretory IgM containing SC
seems likely.

iv) The primary function of IgA has not been elucidated.
Recent studies in orally vaccinated or spontaneously
infected human subjects appear to confirm the protective
value of exocrine IgA antibodies. Previous reports on
the absence of complement-fixation by IgA antibodies
may have to be revised, since it has been observed that
colostral-type IgA to Escherichia coli was able to lyse
these bacteria in the combined presence of lysozyme and
complement. The tendency of IgA to polymerize probably
accounts for the various sedimentation coefficients and
molecular weights observed. It is likely that this
tendency also contributes to the activation of
complement via the alternate complement activation
route. This will be discussed later, in the section
on complement.

It has also been reported that porcine milk IgA enhances
the ability of phagocytes to destroy certain types of
encapsulated bacteria, such as pneumococcus. This is
known as opsonization and substances, such as IgA, which
have this ability are known as opsonins. It has also
been suggested that one of the major functions of IgA
is to react with various harmful antigens to form
unabsorbable antibody-antigen complexes. This would
prevent entrance of these deleterious materials via the
gut and other endothelial surfaces such as the respiratory
tract. This is compatible with the high circulating
antibody titers to food antigens found in individuals
with IgA deficiency. IgA deficiency also correlates with
a high incidence of autoimmune disease. Nonetheless,
complete absence of IgA is not incompatible with good
health, since in many instances of secretory IgA
deficiency, IgM appears to assume IgA functions. IgA is
negative for PCA, placental transfer, and P-K reaction,
while several naturally occurring serum isohemagglutinins
are IgA molecules. Some IgA cryoglobulins may occur,
but rarely in certain diseases.

Other possible functions of secretory immunoglobulins
(in particular IgA) are as follows: prevention of
viral and bacterial infections that may have access to
the external secretions such as in the lungs, mouth,
intestines, ear canals, etc.; and participation as
opsonins in the phagocytic process (involving
principally monocytes in the submucosal region) in
cases where the infective agent has penetrated the
mucosal barriers.

v) With diseases produced by respiratory viruses such as
   influenza, local immunization appears to augment the
   effectiveness of immunization by parenteral routes.
   In fact, in those viral infections affecting primarily
   the mucosal regions, local immunization may be the
   preferred method. The local immunization obtained,
   for example, by administration of a vaccine via the
   alimentary route, may have an additional advantage in
   control of viruses that produce systemic infections.
   This advantage is the elimination of the carrier state.
   For example, in poliomyelitis immunization by the Salk
   method (systemic introduction of viral antigens), one
   may develop neutralizing antibodies systemically but
   not in the secretory IgA. Therefore, these individuals
   may be immune to the disease, but may become carriers
   by harboring the virus in their intestinal tract. On
   the other hand, the Sabin method of immunization
   (attenuated virus by oral route) not only produces
   systemic antibodies but also secretory antibodies.
   The latter may prevent local infections by neutralizing
   the virus particles.

c. Immunoglobulin-M (IgM; γM)

i) IgM is the largest immunoglobulin with a molecular
   weight of 900,000 or greater. Evidence for the
   occurrence of subclasses of different allotypes has
   not been demonstrated, although they have been suggested.
   The significant physicochemical properties of IgM are
   indicated in Table 48. The valence number is of
   particular interest. Though 10 would be anticipated by
   structural analysis, only 5 binding sites have been
   shown in reactions with antigens. This may, in part,
   be due to steric interferences at binding sites and
   thus may not represent the true potential valence.
   Like IgA and IgE, the total carbohydrate content of
   the molecule is four times that of IgG. Because of
   its size, the intrinsic viscosity and diffusion
   coefficients are strikingly different from the other
   immunoglobulins.

ii) The significant biological characteristics of IgM are
    shown in Table 49. IgM fixes complement and it is
    probably the major isohemagglutinin (anti-A or anti-B).
    One of the cardinal features of IgM is its early
    appearance following antigenic stimulation. It
    possesses strong agglutination or precipitation
    properties, perhaps due to its pentavalency. It can
    react with other Ig's as a rheumatoid factor, and is

considered to be a factor itself. Rheumatoid factors
with both antigenic and antibody roles have also been
associated with IgG and IgA. It has been suggested
that IgM is one of the more primitive of the
immunoglobulins, since it is synthesized early in
neonatal life and even, in some species by the fetus
in utero. IgM does not traverse the placental barrier
under normal conditions. "J" chain is also a part of
the IgM molecule, probably bound to some region at or
near the carboxyl terminus of the H-chains. IgM also
participates as a secretory immunoglobulin containing
secretory piece.

d.   Immunoglobulin-D (IgD; γD)

Information on IgD is limited and no biological function has
yet been attributed to these molecules. Of interest is its
low level in serum (only IgE is lower) and the high lambda/kappa
ratio. It is not a secretory Ig like IgA or IgE and its P-K
response is negative. The study of surface IgD as receptors
on B-cells have been of interest in the immune response. In
general the biological activity of IgD appears to be limited.
The alternate pathway of C activation seems to be initiated
by IgD. The available physicochemical data for IgD is
shown in Table 48.

e.   Immunoglobulin-E (IgE; γE)

The physicochemical properties of IgE are summarized in
Table 48. It is the least concentrated and has the shortest
survival time in serum of any of the immunoglobulins. The
affinity of IgE for mast cells and its relationship to
immediate hypersensitivity and to the homocytotropic
antibodies of other animals are among the most studied
aspects of this immunoglobulin. IgE has been equated to the
reaginic antibody of man associated with the release of
histamine and SRS-A (slow-reacting-substance of anaphylaxis)
following interaction with specific antigens. The term
reagin is used in man to refer to cytotropic antibodies.
These antibodies are bound to target cells (e.g., mast cells)
in individuals sensitized to an antigen. When presented
again with the antigen, the cytotropic antibodies bind it,
thereby triggering the release of vasoactive amines from the
target cells. IgE does not bind complement in the classical
pathway but activates complement via the alternate pathway.
IgE can passively sensitize monkey and human skin (P-K
reaction) and leukocytes, but gives no PCA response in the
guinea pig. Addition of specific antigen to IgE-sensitized
leukocytes induces release of histamine and SRS-A. This has

been utilized as a means of detecting allergies in man by
examining the morphological changes of the leukocytes or
measuring histamine release after addition of the antigen.
Similar changes, with release of histamine and SRS-A, occur
with monkey lung tissues. The Fc region of the IgE molecule
binds to basophilic leukocytes and mast cells, which contain
the membrane Fc receptors for IgE.

Though much is known of the role of IgE in allergic
manifestations of the atopic variety, there is little
knowledge of its real function in normal metabolic processes.
IgE is considered a secretory immunoglobulin with a
distribution similar to that of IgA in areas such as the
respiratory and gastrointestinal tract mucosal tissues.
However, IgE does not have secretory component or J-chain.
That IgE is an important anthelmintic antibody has been
recently demonstrated against some parasites.

4. Antibody Diversity

   a. Multivalent antigens, such as bacteria, have many antigenic
   determinants. Since each determinant can, theoretically,
   induce the synthesis of a different immunoglobulin molecule,
   the response of the humoral system to even the simplest
   antigen can be quite heterogeneous. When a B-cell transforms
   to a plasma cell, each member of the clone which develops
   from that cell is capable of producing only one molecular
   species of immunoglobulin. Thus, the number of different
   Ig molecules reflects the number of B-lymphocyte clones which
   synthesized them.

   b. A typical antibody response of Coturnix coturnix (Japanese
   quail) to Brucella antigen (containing 1 x 10^6 bacteria/ml)
   given intraperitoneally is shown in Figure 155. The initial
   administration of 0.5 ml of the bacterial suspension
   initiates a primary response. The following sequence of
   events accompanies the primary response.

      i)   An initial induction period (2 to 4 days for Brucella
           antigen in Coturnix);

      ii) A rapid synthesis of antibody and its appearance in
           the serum (note that IgM appears earlier than IgG and
           has a significantly higher titer);

      iii) Attainment of a maximum concentration on about the
           seventh day;

      iv) A gradual decline in the level of the antibodies in
           the serum.

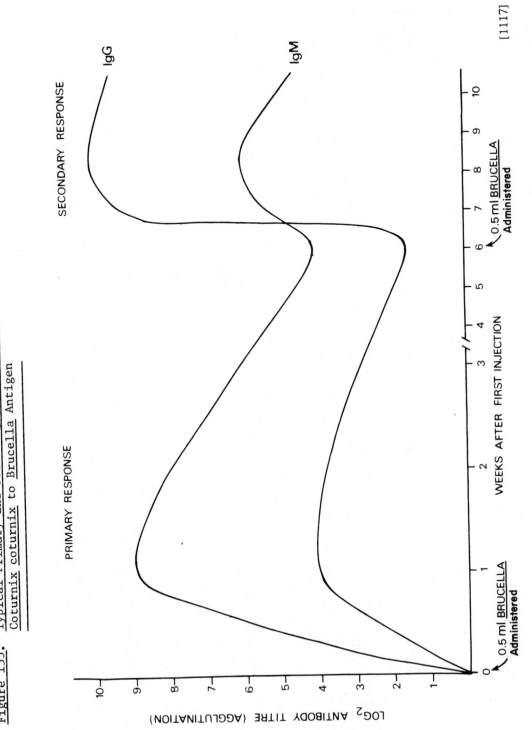

Figure 155. Typical Primary and Secondary Immune Responses of Coturnix coturnix to Brucella Antigen

[1117]

This sequence will vary according to the antigen and animal used. A second administration of the same antigen when antibody levels are decreasing following the primary response results in a secondary response. This time, the rate of synthesis is greater, the induction period is shorter, a greater peak titer occurs, and the antibodies disappear less rapidly from the blood. The decreased rate of disappearance could be accounted for by the observations that (1) IgG has a longer serum-half-life than IgM, and (2) in the secondary response, the amount of IgG synthesized is greater than the amount of IgM. The differentiation of IgG and IgM in the agglutination reaction with Brucella antigen in vitro is accomplished by the addition of a dilute solution of mercaptoethanol. This reducing agent inactivates IgM by degrading the large (polymeric) IgM molecules to their monomeric forms, which lacks agglutinin activity. IgG is essentially resistant to mild mercaptoethanol treatment. IgM can also be readily separated from IgG by gel filtration chromatography.

c. The biochemistry of protein synthesis, transcription and translation using DNA, mRNA, tRNA, polyribosomes, and the appropriate enzymes was discussed in Chapter 5. Like albumin and hemoglobin, there is a constant rate of immunoglobulin degradation and de novo synthesis. The rates depend on the nature of the immunoglobulins. Specific immunoglobulin (antibody) synthesis occurs when the appropriate clone is stimulated by a specific antigenic determinant as was indicated earlier. Although much is known of the fundamentals of protein synthesis and of the immunoglobulin structures, the manner in which H-chains and L-chains are put together to form the tetrapeptide has not been elucidated. Of interest also is the manner in which a single B-cell switches from IgM to IgG synthesis. Some mechanism of feedback inhibition by IgG has been suggested but information in these areas are limited.

d. Figure 153 is a schematic diagram of the basic structure of human immunoglobulin. The important features of this molecule are:

   i) Two identical L-chains (about 22,500 daltons each) and two identical H-chains (approximately 53,000 to 70,000 daltons each) are present, linked by disulfide bonds and non-covalent interactions;

   ii) The molecule is folded into three domains (2 Fab and 1 Fc), which are separated by the H-chain hinge region;

iii) The polypeptide chains of both H and L can be divided into an NH_2-terminal (variable) region and a COOH-terminal (constant) portion. The V- and C-regions are defined by amino acid sequence homology; for example, in the C-region of the predominant L-chain, only one amino acid substitution occurs and it behaves as a single Mendelian allele. In the V-region of the same L-chain, as many as 15 to 40 amino acid substitutions can occur. The V-regions of H- and L-chains from the same Ig show striking sequence homologies with one another. Such homologies would also be anticipated between the V-region of the H- and L-chains of antibodies from different species following stimulation of both species by the same antigenic determinant (i.e., the binding site sequence is relatively independent of species when the Ig's are against the same antigen). Sequence homologies are also found in the C-regions of H- and L-chains.

e. Present evidence strongly supports the concept of a single immunoglobulin polypeptide chain encoded by two germline (heritable) genes, a V-gene and a C-gene, which are joined somatically. Thus, three major families of immunoglobulin polypeptides could be envisioned based on amino acid sequence homology and genetic linkage studies. These are the families of the H-chain and the two L-chains (κ and λ). The constant regions of some of these families can be further divided into classes and subclasses. For example, classes and subclasses of C H-chains are $C\gamma_1$, $C\gamma_2$, $C\gamma_3$, $C\gamma_4$, $C\alpha_1$, $C\alpha_2$, $C\mu$, $C\delta$, and $C\epsilon$. Classes and subclasses of C L-chains are C_{arg}, C_{lys} for λ; and C_k for κ. Each of these are encoded by a germline gene. V-regions are also divided into subgroups controlled by separate genes, but they are of a more complex nature.

f. The structural features of the immunoglobulins are intimately associated with their biological function. Thus, antibody diversity is encoded in the V-regions while the general functions of antibody molecules (placental transfer; complement and macrophage binding) are mediated by the various C-regions of the H-chain (Fc portion).

g. Numerous theories have been advanced to explain antibody production in the past. With the tremendous advances over the past decade in the knowledge of antibody molecules, immunologists have now generally accepted Burnet's clonal selection theory. This theory postulates that clones of potential immunologically competent cells, each containing the genetic codes (C-gene and V-gene, one of each for H- and L-chains) of the immunoglobulin complementary to one antigenic determinant, are present from birth. A sufficient number of different clones would provide an immune potential against all existing antigens that an individual might encounter. The immunogen thus selects or affects its specific complementary clone, which is provided with the appropriate, specific receptor site, and stimulates it to proliferate and synthesize the corresponding specific antibody. This concept stresses the role of the immune cells. Since antibody diversity reflects a corresponding immune cell gene diversity, there is the question of how this gene diversity arose in the immune cell population in the first place. Three generally acceptable theories have been formulated for explaining this.

    i)   The <u>somatic mutation theory</u> postulates that antibody diversity results for hypermutation during somatic differentiation of the immune cells.

    ii) The <u>germline theory</u> proposes that antibody diversity is encoded by a large number of germline genes (i.e., multiple genes encode the V-regions, while a separate gene encodes the C-region in a given immune cell).

    iii) The <u>somatic recombination theory</u> suggests that somatic recombinations occur in a limited number of antibody genes.

Of these three hypotheses, the germline and somatic mutation theories have greatest support from amino acid sequence analysis of the immunoglobulins of various species.

<u>Immunogen (Antigen)</u>

1. General Considerations

   a. It is obvious that any discussion of antibody is intimately associated with reference to antigen. This term, synonymous with immunogen, refers to any substance, exogenous or endogenous, which, when administered to a host, induces or stimulates the production of specific antibodies (any of the five major immunoglobulins discussed earlier). These antibodies can combine specifically with the immunogen as demonstrated by various immunological tests. Antigenicity is dependent on molecular weight, size and shape, and the physicochemical properties of the substances.

   b. There are many varieties of immunogens, including bacteria

and their subunits, viruses, proteins, carbohydrates, nucleic
acids and in some cases, lipids. It is estimated that there
are up to 1×10^6 different kinds of antigens. Immunogens
are designated complete and incomplete. The former group can
induce production of antibody in the host and also react with
the specific antibody formed against it. Incomplete
immunogens, called haptens, are incapable of antibody
stimulation in the host unless coupled with a carrier
antigen. Nevertheless, they can react with antibody as
determined by special procedures such as the blocking test
and passive hemagglutination reactions. Tolerogens are a
third class of "antigens" which induce unresponsiveness in
individuals. Persons showing no response to a tolerogen are
said to be tolerant toward it. The tolerant state can be
natural or induced. The latter form is brought about by
introduction of an antigen which is qualitatively inadequate
(as is readily demonstrated in rats, monomeric γ-globulin is
tolerogenic, while aggregated γ-globulin is highly antigenic)
or quantitatively insufficient to produce an immune response.
Tolerogens usually affect the T-cells so that they lose
their ability to interact with the B-cells to bring about
antibody synthesis. However, since tolerance refers to a
general inability of the host to respond to an antigen via
either the cell-mediated or humoral system, an animal can be
rendered tolerant by blocking either B- or T-cells or both.

c. Proteins of high molecular weights are usually good antigens
capable of stimulating antibody production. Some examples in
order of decreasing molecular weight are viruses,
thyroglobulins, immunoglobulins, and serum albumins.
Carbohydrates, which are basically made of hexoses, hexosamine,
and sialic acid residues, are generally less satisfactory
immunogens than proteins. Examples of carbohydrate antigens
are type-specific pneumococcal polysaccharides, the
lipopolysaccharides of gram-negative bacteria, and the
blood-group mucopolysaccharides of erythrocytes. Lipids
generally act as haptens and are poor complete antigens.
Nucleoproteins have been shown to be complete immunogens,
while nucleic acids and the base residues (pyrimidines and
purines) of DNA and RNA show good haptenic properties.

2. Isoantigens (Homologous Antigens)

a. ABO and Rh Blood Groups

All proteins and protein-lipopolysaccharide complexes of
human serous fluids and tissues have immunogenic properties.
Of interest in this section are the isoantigens, which have
been extensively investigated and which play significant

roles in transfusion, transplantation, and certain
immunological disorders (isoimmunization; hemolytic disease
of the newborn). Isoantigens are expressions of different
alleles which occur at a single genetic locus within
different individuals of one interbreeding population
(genetic polymorphism). They can be considered as the
factors which make a person's tissues and blood cells
immunologically unique (i.e., unlike those found in another
individual).

i) In man, the ABO and Rh red cell isoantigens are
medically important in transfusions and, to some
extent, in tissue transplantation. They are found on
the surface of red blood cell membranes and on various
cells (especially those of epithelial origin) scattered
throughout the body. Some ABO isoantigens are also
related to bacterial antigens such as the somatic
lipopolysaccharides of gram-negative bacteria. Rh
isoantigens, on the other hand, have been demonstrated
only in the erythrocytes of man and monkey (the term
Rh is derived from Rhesus monkey). The phenotypes and
genotypes of the ABO and Rh isoantigens are summarized
in Tables 51 and 53, respectively. The ABO antigens
have reciprocal antibodies called isohemagglutinins
present in the blood. As indicated in the earlier
sections, these appear soon after birth and are
primarily IgM and IgA antibodies. Rh antigens have
no reciprocal natural antibodies, but are so highly
immunogenic that improper transfusion from an Rh-
positive (Rh^+) to an Rh-negative (Rh^-) individual
results in anti-Rh^+ antibody production in the Rh^-
recipient. Similarly, transmission of red cells
during parturition from an Rh^+ infant to an Rh^- mother
(across the placental barrier, for example) can stimulate
anti-Rh^+ antibodies in the parent. An Rh^+ fetus
conceived in subsequent pregnancies can be seriously
affected by the anti-Rh^+ IgG_1 antibodies carried by
the mother since these are able to cross the placenta
and enter the fetal circulation. This disorder is
known as erythroblastosis fetalis or hemolytic disease
of the newborn. Severe destruction of fetal erythrocyte
occurs with release of excessive bilirubin which causes
jaundice and brain tissue damages. Lack of glucuronyl
transferase in infant's immature liver contributes to
this phenomenon (see Chapter 7).

ii) The ABO blood group is genetically controlled by three
alleles at one genetic locus (three allelic genes).
These are the A, B, and O alleles, where O is an amorph

(Text continued on p. 1127)

Table 51. ABO Blood Groups

Genotype	Phenotype	Isoantigen In Red Cells	Isohemagglutinin Present in Plasma	Frequency, United States, (%)		
				Caucasians	Negroes	Chinese
A_1O A_1A_1 A_1A_2	A_1	A and A_1	Anti-B	41	27	28
A_2A_2 A_2O	A_2	A	Anti-B			
BO BB	B	B	Anti-A and Anti-A_1	10	21	23
OO	O	H	Anti-A and Anti-B	45	48	36
A_1B	A_1B	A_1 and B	None	4	4	13
A_2B	A_2B	A_2 and B	None			

Table 52. Antigenic Determinants of the ABO and Lewis Blood Groups

Specificity	Allele	Non-Reducing Terminal Residues of the Isoantigen
None	"precursor"	$Gal\overset{\beta1,3}{-----}NacGlu\overset{\beta1,3}{-----}Gal\overset{\beta1,3}{-----}NacGal\cdots\cdots$
Type XIV (pneumococcus)	"precursor"	$Gal\overset{\beta1,4}{-----}NacGlu\overset{\beta1,3}{-----}Gal\overset{\beta1,3}{-----}NacGal\cdots\cdots$
H	H	$\beta1,3;$ $Gal\overset{\beta1,4}{-----}NacGlu\overset{\beta1,3}{-----}Gal\overset{\beta1,3}{-----}NacGal\cdots\cdots$ $\mid\alpha1,2$ Fuc
Le[a]	Le	$Gal\overset{\beta1,3}{-----}NacGlu\overset{\beta1,3}{-----}Gal\overset{\beta1,3}{-----}NacGal\cdots\cdots$ $\mid\alpha1,4$ Fuc
Le[b]	H & Le	$Gal\overset{\beta1,3}{-----}NacGlu\overset{\beta1,3}{-----}Gal\overset{\beta1,3}{-----}NacGal\cdots\cdots$ $\mid\alpha1,2 \quad \mid\alpha1,4$ Fuc $\quad$ Fuc
A	A	$\beta1,3;$ $NacGal\overset{\alpha1,3}{-----}Gal\overset{\beta1,4}{-----}NacGlu\overset{\beta1,3}{-----}Gal\overset{\beta1,3}{-----}NacGal\cdots\cdots$ $\mid\alpha1,2$ Fuc
B	B	$\beta1,3;$ $Gal\overset{\alpha1,3}{-----}Gal\overset{\beta1,4}{-----}NacGlu\overset{\beta1,3}{-----}Gal\overset{\beta1,3}{-----}NacGal\cdots\cdots$ $\mid\alpha1,2$ Fuc

Note: a) The first two isoantigens, labeled as "precursor" are the basic carbohydrate sequences to which are added specific residues to form the ABO antigenic determinants. The first one is non-specific while the second one is, as indicated, related to the antigen of type XIV (pneumococcus).

b) Le[a] and Le[b] are the Lewis blood group isoantigens. They were discovered after the original ABO grouping was formed but are chemically closely related to the ABO isoantigens.

c) Gal = galactose; NacGlu = N-acetylglucosamine; NacGal = N-acetylgalactosamine; Fuc = fucose (6-deoxy-L-galactose).

d) When more than one type of linkage ($\beta1,3$; $\beta1,4$) is shown, both are known to occur without altering the antigenic specificity.

1124

Table 53. Rh Blood Groups: Rh-Hr Phenotypes, Genotypes and Frequencies in White Population

RH POSITIVE = 83.2%

Reactions with anti-						Designation		Genotypes		Frequencies	
Rh_0 D	rh' C	rh" E	hr' c	hr" e	hr f	Wiener	Fisher-Race	Wiener	Fisher-Race	White %	Total
+	-	-	+	+	+	Rh_0rh	cDe/c-e	R^0r R^0R^0	cDe/cde cDe/cDe	1.9950 0.0659	2.1
+	+	-	+	+	+	Rh_1rh	cDe/c-e	R^1r, R^1R^0, R^0r'	CDe/cde CDe/cDe cDe/Cde	32.6808 2.1585 0.0505	35.0
+	+	-	-	+	-	Rh_1Rh_1	CDe/C-e	R^1R^1 R^1r'	CDe/CDe CDe/Cde	17.6803 0.8270	18.5
+	-	+	+	+	+	Rh_2rh	cDE/c-e	R^2r, R^2R^0, R^0r''	cDE/cde cDE/cDe cDe/cdE	10.9657 0.7243 0.0610	11.8
+	-	+	+	-	-	Rh_2Rh_2	cDE/c-E	R^2R^2 R^2r''	cDE/cDE cDE/cdE	1.9906 0.3353	2.3
+	+	+	+	+	-	Rh_1Rh_2	CDe/c-E	R^1R^2, R^1r'', R^2r'	CDe/cDE CDe/cdE cDE/Cde	11.8648 0.9992 0.2775	13.1
+	+	+	+	+	+	$Rh_z rh$	CDE/c-e	$R^z r, R^z R^0, R^0 r^y$	CDE/cde CDE/cDe cDe/CdE	0.1893 0.0125 0.0003	0.2
+	+	+	-	+	-	$Rh_z Rh_1$	CDE/C-e	$R^z R^1, R^z r', R^1 r^y$	CDE/Cde CDE/Cde CDe/CdE	0.2048 0.0048 0.0042	0.21
+	+	+	+	-	-	$Rh_z Rh_2$	CDE/c-E	$R^z R^2, R^z r'', R^2 r^y$	CDE/cDE CDE/cdE cDE/CdE	0.0687 0.0058 0.0014	0.07
+	+	+	-	-	-	$Rh_z Rh_z$	CDE/C-E	$R^z R^z$ $R^z r^y$	CDE/CDE CDE/CdE	0.0006 0.00001	0.0006

Table 53. (continued)   Rh Blood Groups: Rh-Hr Phenotypes, Genotypes and Frequencies in White Population

RH NEGATIVE = 16.8%

Rh0 D	rh' C	rh" E	hr' c	hr" e	hr f	Designation Wiener	Designation Fisher-Race	Genotypes Wiener	Genotypes Fisher-Race	Frequencies White %
–	–	–	+	+	+	rh	cde/cde	rr	cde/cde	15.102
–	+	–	+	+	+	rh'rh	Cde/cde	r'r	Cde/cde	0.7664
–	+	–	–	+	–	rh'rh'	Cde/Cde	r'r'	Cde/Cde	0.0097
–	–	+	+	+	+	rh"rh	cdE/cde	r"r	cdE/cde	0.9235
–	–	+	+	–	–	rh"rh"	cdE/cdE	r"r"	cdE/cdE	0.0414
–	+	+	+	+	–	rh'rh"	Cde/cdE	r'r"	Cde/cdE	0.0234
–	+	+	+	+	+	$rh_y rh$	CdE/cde	$r^y r$	CdE/cde	0.0039
–	+	+	–	+	–	$rh_y rh'$	CdE/Cde	$r^y r'$	CdE/Cde	0.0001
–	+	+	+	–	–	$rh_y rh"$	CdE/cdE	$r^y r"$	CdE/cdE	0.0001
–	+	+	–	–	–	$rh_y rh_y$	CdE/CdE	$r^y r^y$	CdE/CdE	0.000001

Reactions with anti-

(i.e., an uncharacterized antigen). The O allele is expressed as the <u>absence</u> of the A and B alleles and is also involved in the expression of the H-isoantigen. Subgroups (notably A_1 and A_2) have been demonstrated in the A group. The secretory (Se) genes, distinct from the ABO locus (though closely related to them), regulate the elaboration of the A, B, and H isoantigens in body fluids (especially saliva). Although the ABO isoantigens have not been completely characterized structurally, the terminal residues are known and are indicated in Table 52.

iii) The isoantigens of the Rh system are genetically complex. Two acceptable genetic schema have been suggested for the explanation of the expression of the Rh antigens on the erythrocyte surfaces. These are the <u>Fisher-Race</u> and <u>Wiener</u> theories of Rh isoantigen inheritance. The Fisher-Race theory suggests the presence of five distinct determinants designated D, C, E, c, and e, which are products of genes situated at three distinct but closely linked chromosomal loci. Crossing over is infrequent due to the close proximity of the genes to each other. It is thought that one set of three alleles is inherited from each parent. This is similar to the concept of one cistron regulating the product of one antigenic determinant, the Rh phenotype, and it can be illustrated as follows.

Chromosomal loci	Antigenic determinant	Rh Phenotype
D	D	
C	C	Rh^+
E	E	
D	D	
C	C	Rh^+
e	e	
d	none	
c	c	Rh^-
e	e	

The gene regulating D is significant since this expresses the Rh^+ isoantigen important in clinical medicine. This isoantigen is responsible for <u>Erythroblastosis fetalis</u> in newborns. It accounts for nearly all (90%) of the hemolytic diseases of the newborn resulting from maternal isoimmunization. The alternative allele, small d, is hypothetical. It represents the <u>absence</u> of D-isoantigen since the presence of a d-isoantigen has not been shown to date.

The second theory, that of Wiener, suggests that one chromosomal locus is responsible for the regulation of several antigenic determinants (allele-multiple determinants). This can be shown as follows.

<u>Chromosomal locus</u>	<u>Antigenic determinant</u>	<u>Rh Phenotype</u>
R^1	Rh_o rh' rh"	Rh^+
r'	rh' hr"	Rh^-

In the Wiener scheme R^0, R^1, R^2 and R^z relate to Rh^+ isoantigens and r, r', r", and r^y represent Rh^-. This is seen in Table 53 in greater detail.

iv) There are approximately twelve other blood group isoantigens of the human erythrocytes. They have been primarily useful in medico-legal and anthropological problems. With the exception of rare transfusion difficulties, the contribution of these groups to diseases has not been ascertained.

b. Major Histocompatibility Complex (MHC)

i) The biological significance of the chromosomal region, defined as the major histocompatibility complex (MHC) which controls the synthesis of transplantation antigens, is not only restricted to allograft survival, but is also involved in the relationships to a large array of other biological phenomena, such as the immune responses and susceptibility to diseases.

With the advent of tissue transplantation, this major group of isoantigens have become of tremendous importance in clinical medicine. These isoantigens are found on the surfaces of white blood cells, particularly lymphocytes and designated as HLA (HLA: human leukocyte antigen). They are also found in platelets, serum and most of the fixed tissues of the body. These antigens can be detected serologically by means of cytotoxicity assays (discussed later) and utilized for donor selection in tissue transplantation.

ii) The HLA gene complex (or MHC) is located on the short arm of chromosome 6. It consists of four closely linked but distinct loci (Locus: the position of a gene on a chromosome) designated as A, B, C and D. These loci are highly polymorphic. Thus, loci are given latter names and the alleles (i.e., alternate forms of a gene which occur at the same locus) are designated numerically. For example, HLA-B8 refers to B locus and 8th allele. The prefix w (workshop) is used between the letter and the number to indicate tentative assignments: HLA-Bw17, etc. (see Table 54).

The HLA antigenic determinants have not been thoroughly characterized chemically because of the difficulty of separating them from the other components of the cell membrane and because they usually are of rather heterogeneous molecular composition. Nevertheless, recent work suggests that the HLA alloantigenicity probably resides in a protein (unlike the ABO determinants which are carbohydrates) of molecular weight about 31,000 daltons. Samples of HLA antigens have been prepared by methods which remove non-covalently bound lipids and carbohydrates. These purified isoantigens are immunogenic proteins containing less than 1% carbohydrate or lipid indicating that it is probably not a glyco- or lipoprotein. These findings are still quite controversial, however, and it may yet be shown that carbohydrate or lipid is at least partially necessary for the complete identity of these antigens.

iii) The genetic control of HLA is accomplished by a complex of linked genes at two chromosomal loci, termed segregants or subloci. The HLA system is similar to that of the Rh system in that the linked genes are transmitted in a group from one generation to the next.

The group of allelic genes (pair of HLA allelic determinants) contributed by each parent is referred to

Table 54. Major Human Histocompatibility Loci (1975*) and the Ethnic Distributions in HLA-A and HLA-B Loci

Frequency (%) in:

Loci	Caucasian (Dutch, Swiss)	Mongoloid (Chinese)	Negroid (Zanbian)
HLA-A(LA)			
A1	16	1	2
A2	28	31	14
A3	16	1	5
A9	11	17	11
A10	4	5	5
A11	7	28	0
Aw19	7	16	43
Aw28	6	0	9
Blank	5	1	11
HLA-B(4)			
B5	6	6	1
B13	3	12	1
Bw17	4	9	17
Bw27	4	2	0
B12	13	1	19
Bw35	9	3	8
Bw15	8	19	2
Bw16	3	7	2
Bw21	1	0	2
Blank	7	15	27
B7	15	1	10
B8	11	0	2
Bw40	10	19	0
Bw14	2	0	5
Bw18	1	1	4
Bw22	3	5	0
HLA-C			
−Cw1			
−Cw2			
−Cw3			
−Cw4			
−Cw5			

HLA-D (Formally the Mix Lymphocyte Culture Locus (MLC))
 −Dw1
 −Dw2
 −Dw3
 −Dw4
 −Dw5
 −Dw6

* As announced by the WHO-IUIS Terminology Committee for the HLA systems in Cellular Immunol. 21, 382-389, 1976.

as the <u>haplotype</u>. These are shown schematically as segregated, linked genes according to Mendelian principles of genetics (see Figure 156).

The genes have many alleles within the population and are therefore polymorphic. There are approximately 51 known HLA antigens. Phenotypically, no individual can have more than four major HLA antigens (two from each segregant series) since there are only four subloci (two on each member of the chromosomal pair). The segregant series of HLA antigens and their distribution in some of the ethnic groups thus far examined are shown in Table 54.

iv) The loci HLA-A, HLA-B, and HLA-C have been established through use of antibodies (anti-leukocyte) from pregnant women and from blood transfused individuals. These antibodies define the cell surface antigens of lymphocytes. The locus HLA-D has been defined via the mixed lymphocyte culture procedure (MLC), in which cell surface differences are examined by proliferative response of peripheral lymphocytes <u>in vitro</u> (the LD or lymphocyte-defined approach). These approaches define the differences in the biological role of and in the determinants of the MHC.

v) The induction of an immune response is noted <u>in vivo</u> as rejection of the tissue or graft. The primary (or first set) rejection, which occurs approximately ten days after transplant, is generally related to the cell-mediated (T-lymphocyte) system. Secondary (or second set) graft reaction and hyperacute rejection involve the humoral system and all the attendant amplification processes (complement, fibrinolysis, kinins, etc.). In the secondary response, the graft is generally rejected within four to five days after transplantation in a sensitized host. Induction of an immune response can also be detected <u>in vitro</u> by cytotoxicity and leukoagglutination tests. Serum from sensitized individuals is incubated with lymphocytes in the presence of complement and the viability of the cells is estimated by the dye exclusion method. Since viable cells do not stain with dyes such as trypan blue or eosin, the number of surviving cells may thus be determined. This is the general procedure utilized in tissue typing. Recipient and donor lymphocytes are treated, in the presence of complement, with a number of standard cytotoxic sera capable of reacting with the major antigens of the two segregant series of HLA

<u>**Figure 156.**</u> <u>Schematic Diagram of the Inheritance of the HLA Haplotypes</u>

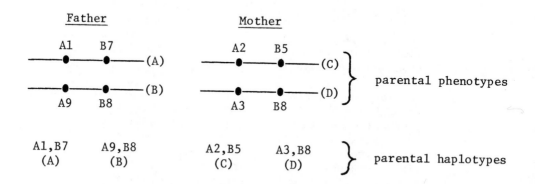

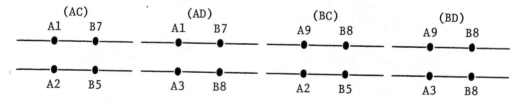

The four possible phenotypes of the offsprings:

It is a common practice to label the paternal haplotypes A,B and the maternal C,D. Since there can be only two paternal and two maternal haplotypes, there are only four possible combinations in the offsprings as indicated in this figure. It thus follows that there is a 25% chance that two siblings will be HLA-identical and share two haplotypes. An additional 50% will show one haplotype, and the remaining 25% will share no haplotype. Parents share one haplotype with all children and almost always differ for the other.

(see earlier) and the percentage of viable cells is determined. The matching grade of the recipient and donor are classified as shown in Table 55.

vi)   In any discussion of transplantation immunity, the terminologies listed in Table 56 are of importance. Relative to genetic identity, the terms syngeneic and isogeneic (isologous), allogeneic (homologous) and xenogeneic (heterologous) are also used.

Information on the HLA of man and its possible influence on the immune responses is limited compared with the information available on the murine H-2 complex. The HLA complex of man and its relationship to Ir (immune response gene) can be illustrated diagrammatically as follows:

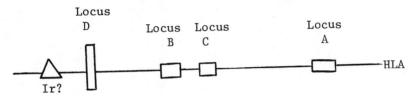

Ir: Immune response gene regulates the ability to respond to various immunogenic determinants.

vii) Relationship of HLA to disease. Different disease states have been examined relative to the ABO isoantigens with no clearcut patterns. Similar studies are presently being extensively examined in attempts to relate HLA antigens to disease states. Some of these findings for the HLA are summarized in Table 57.

viii) In summary, significant properties of the three known human isoantigen systems are compared in Table 58. Soluble forms of HLA occur in the sera of normal individuals, in a manner similar to that of the ABO system. The soluble HLA may have significant implications in the tolerance states of individuals.

(**Text continued on p. 1137**)

Table 55. Tissue Typing on the Basis of HLA Antigen

Match Type	Hypothetical Recipient Genotype		Hypothetical Donor Genotype		Remarks
A	A2,B5	A3,B8	A2,B5	A3,B8	Identical HLA's in both donor and recipient (e.g., identical twins or inbred strains-isologous)
B	A2,B5	A3,B8	A2,B5	A3,B5	No group mismatches but donor is missing one antigen found in the recipient
C	A2,B5	A3,B8	A2,B5	A3,B7	One group mismatch; donor has one antigen which recipient does not have (i.e., one unrelated antigen)
D	A2,B5	A3,B8	A10,B12	A3,B8	Two group mismatches, both on the same chromosome (i.e., one haplotype completely different between donor and recipient)
E	A2,B5	A3,B8	A2,B12	A3,B7	Two or more group mismatches involving both chromosomes (i.e., both donor haplotypes are at least partially different from recipient haplotypes)
F	---		---		Match is classified as type F if the recipient has an antibody to one of donor's HLA antigens. This can occur even in identical twins if one of them has an autoimmunity to one of his own HLA antigens.

Table 56. Transplantation Immunology: Terminology

Nature of Graft	Tissue	Definition
Autograft	autologous	Graft from one part of the body to another in the same individual
Isograft	isologous	Graft from genetically identical individuals (identical twins, inbred strains of animals)
Allograft	homologous	Graft from a genetically dissimilar donor of the same species
Xenograft	heterologous	Graft from a donor of another species

Table 57. HLA and Disease

Diseases	HLA Antigen	Findings (Frequency)
1. Leukemia	A2, B12 haplotype	increase
	A1 or A11	decrease
Acute myelogenous	A2 or B12	increase
	A11	decrease or normal in some studies
2. Hodgkin's disease	B5, Bw35, Bw15	50% increase
	B7, A11	increase
3. Lymphoma	B12, B13	increase
4. Idiopathic ankylosing spondylitis	B27	over 90% of patients
5. Other diseases:		
glomerulonephritis	A2	increase
infectious mononucleosis	B5	increase
systemic lupus erythematosus	Bw5	increase
myasthenia gravis	B8	increase

1135

Table 58. Comparison of ABO, Rh, and HLA Isoantigens

Characteristics	Blood-Group System		HLA
	ABO	Rh	
Clinical significance	Transfusion reaction Hemolytic disease of the newborn	Erythroblastosis fetalis	Transplantation and Transfusion Reactions
Techniques for the immunosuppression of the humoral and cell-mediated responses	None	Specific anti-Rh antibody given to Rh⁻ mother just prior to Rh exposure	Steroids; antilymphocyte serum (ALS)
Genetic polymorphism	Moderate	Extensive	Extensive
Chromosome loci	Single	Complex	Complex

The Complement System

Complement is an essential part of the normal host defense mechanism and functions as the main effector pathway of the humoral immune and inflammatory process. As we will see, the complement system is involved in the phagocytosis and lysis of invading cells (e.g., virus, bacteria). The system can be activated by one of two pathways: the classical or alternate (or properdin) pathway. The classical pathway is activated by antibody (IgG or IgM) interacting with the antigen present on the invading cell whereas the alternate pathway (first described in 1954 by Pillemer and his coworkers) is activated by yeast and bacterial cell-wall components (e.g., lipopolysaccharides; zymosan--a yeast cell-wall polysaccharide) and aggregated IgA or IgD. It is important to emphasize that the activation of the alternate pathway is not necessarily dependent upon a specific antigen-antibody interaction and therefore may represent a natural defense system against the invading organisms in the unimmunized host. Of the five classes of immunoglobulins, the only immunoglobulin which is not involved in the complement activation either in the classical or alternate pathway is IgE.

The complement system is a double-edged sword. On the one hand it protects the host against the invading organism, and it is becoming increasingly apparent that various types of complement deficiency (acquired or hereditary) in man are associated with an enhanced predisposition to infections and certain diseases. On the other hand, in some instances the activation of the complement system by the soluble antigen-antibody complexes can initiate a pathologic process. The antibody may be directed against a foreign antigen or a host antigen itself (an autoimmune process). The disorders of this type are known as immune-complex diseases, in which the soluble antigen-antibody complexes are deposited on the surface of vascular endothelial cells. These deposits activate the complement system leading to deleterious effects consisting of acute inflammation and intravascular coagulation (vasculitis).

The overall complement system consists of a group of about 19 plasma proteins (mainly synthesized in the macrophages) and most of these are present as inactive precursors until their activation. The activation process is similar to blood clotting reactions in that it involves a cascade of limited proteolytic reactions in which a series of activated enzymes are formed from their precursors. The cleavage products participate in a variety of biologic reactions. These include a provision of binding sites for other complement components to allow assembly of these molecules leading to lysis, chemotaxis of leukocytes, enhancement of phagocytosis by opsonization, platelet aggregation, and increased vascular permeability.

This section will include discussions relative to the chemical composition of the components of the system, the molecular events of complement action leading to cytolysis, a brief mention of complement deficiencies, and the relationship of this to other inflammatory amplification systems.

1.   Chemical Characteristics

   a.   The overall complement (C) comprises a group of about 19 distinct but functionally interrelated plasma proteins. These molecules are found in the fresh sera of animals of many species. Complement levels tend to fluctuate during certain diseases. In immunological diseases, complement consumption occurs so that serum levels generally decrease. In some inflammatory states where immune systems are not involved, an increase in complement activity may occur. Differences in complement activity between species, which are attributable to the variation in concentrations of one or the other of the 19 components, have been demonstrated. Complement components of man have been linked to the major histocompatibility complex.

   b.   Examination of fresh serum has demonstrated that certain components of C are readily destroyed by oxidizing agents and by heat. Heat treatment of fresh serum at 56°C irreversibly inactivates C1- and C2-activities, while treatment of fresh serum with zymosan (a complex containing mostly protein and carbohydrate; prepared from yeast cell walls; also called anticomplementary factor) removes C3-component. Dilute ammonium hydroxide treatment of fresh serum inactivates C4-component.

   c.   With the advent of chromatographic procedures utilizing anionic and cationic cellulose derivatives, gel filtration chromatography, and zonal electrophoresis, rapid advances in the characterization of many of the components have been achieved. The significant, known physicochemical characteristics of the 11 components of C that participate in the classical pathway are shown in Table 59. Molecular weights range from a low of 80,000 daltons (C1s and C9) to a high of 400,000 daltons (C1q). Concentrations of C-components in fresh serum range from a low of 2.5 mg/100 ml (C2) to a high of 138 mg/100 ml (C3). In immunological assays by radial immunodiffusion, C3 and C4 levels in serum have been utilized for recognition of complement action in immunological diseases. Complement fixed to an antigen-antibody complex in tissue has been detected with fluorescence labeled anti-C3 in immunofluorescence analysis using fluorescence-microscopy. Complement component levels have also been estimated by

Table 59. Physicochemical Properties of the Components of Complement

Properties	Components, in order of reaction										
	C1			C4	C2	C3	C5	C6	C7	C8	C9
	C1q	C1s	C1r								
Molecular Weight (daltons)	400,000	90,000	170,000	206,000	120,000	180,000	180,000	90,000	110,000	163,000	80,000
Sedimentation Coefficient ($S_{20,w}$)	11.1	4.0	7.0	10.0	5.5	9.5	8.7	5-6	5-6	8.0	4.5
Serum Concentration (mg/dl)	19.0	11.0	20.0	60.0	2.5	138.0	8.0	7.5	5.5	5-10	5-10
Relative Electrophoretic Mobility	γ_2	α_1	β_1	β_1	$\beta_1-\alpha_2$	$\beta_2-\beta_1$	β_1	β_2	β_2	γ_1	α_2
Carbohydrate Content (%)	15.0	-	-	14.0	-	2.7	19.0	-	-	-	-
Functional Sulfhydryl Content (number of SH groups)	-	-	-	-	2	1-2	+	-	-	-	-

+ = present

- = unknown or absent

measuring the concentrations of C1, C2, C3, C4 and C5 in serum by radial immunodiffusion. The majority of the C-components migrate in the β-globulin region in zonal electrophoresis, although C1q, C8, and C9 migrate in the γ, γ_1, and α regions, respectively. C1q is an interesting molecule containing large amounts of carbohydrate, hydroxyproline, hydroxylysine, and glycine. It appears to be chemically related to the collagenic proteins. This is the component which binds to the Fc portion of the antibody in an antibody-antigen complex.

2. Activation of the Complement System by the Classical Pathway

The pathway consists of 11 C-components which interact sequentially in a cascade of limited proteolytic reactions leading to the conversion of precursors to activated enzymes. The classical C-system is activated by the interaction of an antibody with its specific antigen (present either on a cell surface such as an erythrocyte, parasitic cell, or malignant cell or in an aqueous system as soluble antigens). The sequence has been thoroughly defined using the lysis of sheep erythrocytes (SRBC-antigen) by their specific antibody (hemolysin) produced in rabbit by administration of SRBC. The general scheme of the complement reactions is shown in Figure 157. The classical pathway has been divided into three functional components: the recognition unit (the interaction of the C1 complex with the antibody), the activation unit (involves C4, C2 and C3), and finally the membrane attack unit (consists of C5b, C6, C7, C8, C9 and results in cytolysis).

a. The recognition unit. This unit, the C1 component (a trimolecular complex), consists of three subunits--C1q, C1r and C1s--which are held together by calcium ions. The specific sites are located on the C1q for combining with specific receptors on the Fc portion of the antibody molecules (IgG_1, IgG_2, IgG_3 and IgM) which have aggregated on the surface of the antigen-containing cells. Specific antibodies, when complexed to antigen residing on the cell surface, fix and activate C1q (it undergoes conformational change). The changes in C1q unmasks an enzymatic activity in C1r, which in turn cleaves a peptide from C1s, exposing its active site and thereby creating the first major enzyme of the system (known as C1 esterase). The activated complex is shown as $\overline{C1}$ (where the bar over the number indicates activated component). The initial activation of C1q requires at least two binding sites which are provided by two Fc fragments. Thus two IgG molecules or one IgM molecule are essential for activation. Note IgM, a pentamer, can provide up to 5 Fc binding sites.

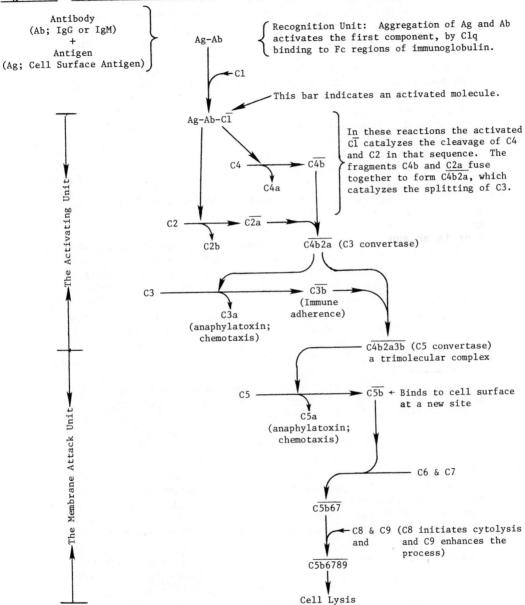

Figure 157. Schematic Representation of Classical Pathway of Complement

Antibody
(Ab; IgG or IgM)
+
Antigen
(Ag; Cell Surface Antigen)

Ag-Ab

Recognition Unit: Aggregation of Ag and Ab activates the first component, by C1q binding to Fc regions of immunoglobulin.

← C1

This bar indicates an activated molecule.

Ag-Ab-C̄1

In these reactions the activated C̄1 catalyzes the cleavage of C4 and C2 in that sequence. The fragments C4b and C2a fuse together to form C4b2a, which catalyzes the splitting of C3.

C4 → C̄4b

C4a

C2 → C̄2a →

C2b

C̄4b2a (C3 convertase)

C3 → C̄3b →
(Immune adherence)

C3a
(anaphylatoxin; chemotaxis)

C̄4b2a3b (C5 convertase) a trimolecular complex

C5 → C̄5b ← Binds to cell surface at a new site

C5a
(anaphylatoxin; chemotaxis)

C6 & C7

C̄5b67

C8 & C9 (C8 initiates cytolysis and C9 enhances the process)
and

C̄5b6789

Cell Lysis

The Activating Unit

The Membrane Attack Unit

1141

b. The activation unit. The activation is initiated by C̄1
which cleaves C4 and then C2 (the adsorption of C2 is
promoted by Mg^{+2}) into their major cleavage products,
namely C4b and C2a. The C̄1 amplifies the cleavage process
and produces a shower of these fragments. However, the
subsequent reactions occur only with those fragments bound
to the cell membrane. The C4b and C2a form a bimolecular
complex (C4b2a) on the cell membrane which exhibits an
enzymatic activity towards C3. C4b2a, also known as C3
convertase, splits the C3 into C3a (smaller fragment) and
C3b (larger fragment). The attachment of C3b to C4b2a
leads to a trimolecular complex (C4b2a3b) which has the
enzymatic activity towards C5. The C3a released into the
fluid phase is an anaphylatoxin and functions in the
inflammatory processes. The C3b fragment, in addition to
its role in the enzyme complex C4b2a3b, promotes phagocytosis
by PMN leukocytes, monocytes and macrophages by binding to
several sites on the cell membrane.

c. The membrane attack system. The C4b2a3b trimolecular complex
(also known as C5 convertase) catalyzes the cleavage of C5
into C5a and C5b. C5a is an anaphylatoxin and participates
in the inflammatory processes like C3a. The C5b attaches
to a new site on the cell membrane alongside the C4b2a3b
complex which decays rapidly. Following the binding of C5b,
the remainder of the terminal components of the sequence is
self assembled on the membrane. Initially a trimolecular
complex (C5b,6,7) is formed followed by C8 and C9 attachments
to the complex. These processes are non-enzymatic. The
actual cytolysis is initiated by C8 and C9, and the exact
mechanisms are not understood. Electron microscopic
observations have revealed "hole-like" lesions on the membrane.
Through these channels K^+ leaves the cell, and water and Na^+
enter the cell, eventually resulting in osmotic lysis.

d. As noted above the complement system not only mediates
cytolysis but also participates in the inflammatory processes.
The biologic activities of various complement components are
indicated in Figure 157 and Table 60. The cells possessing
receptors for the complement component are shown in Table 61.

e. Regulation of complement system. This is accomplished by
two sets of processes: the presence of specific inhibitors
(or inactivators) of the enzymes that participate in the
complement system and the instability of some complexes that
are produced in the sequential reactions.

Table 60. Biological Action of Complement Components

Complement Component

C2 and C4	Virus neutralization.
C3a and C3b	Enhances phagocytosis of PMN leukocytes, with lysosomal release, immune adherence, anaphylactoid reaction.
C3a, C5a and C5b67	Increases chemotaxis, C3a exhibits kinin-like activity.
C3a and C5a	Anaphylatoxins release histamine from mast cells; C5a causes lysosomal enzyme release.
C1-7	Facilitates lymphocytotoxicity.
C1-9	Cytotoxic to specific sensitized target cell or adjacent cells; activates tissue lysosomal enzymes.

Table 61. Cells Possessing Receptors for Complement Fragments

Cells With Receptors	Complement Component
B lymphocytes	C3b, C4b
Neutrophils	C3b
Monocytes	C3b, C3d, C4b
Macrophages	C3b, C4b
Erythrocytes (primates)	C3b, C4b
Platelets	C3b, C4b

i)   C1-inhibitor (C1INH)(105,000 daltons) is an α_2-globulin
     which inhibits the enzymatic activities of C1 activity
     by combining with the enzyme in stoichiometric
     proportions. A congenital deficiency of C1INH, known
     as hereditary angioedema, gives rise to recurrent
     episodes of subepithelial edema of the skin and the
     upper respiratory and gastrointestinal tracts. It is
     inherited as an autosomal dominant trait.

ii) The inactivation of C3b (the fragment of C3 that
     attaches to C4b2a giving C4b2a3b) is
     accomplished by an enzyme (also known as C3 inactivator)
     that catalyzes the cleavage of C3b to two fragments,
     C3c and C3d.

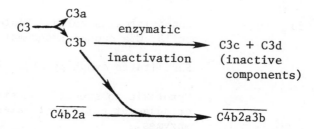

     The deficiency of C3 inactivator has been observed in
     two subjects. The clinical manifestations in this dis-
     order are due to the accumulation of C3b which causes
     the activation of the alternate pathway (discussed
     later) with severe depletion of factor B and C3, which
     may be responsible for the recurrent pyogenic infections
     seen in these patients. In addition the increased
     production of C3a may be responsible for the observed
     allergic reactions. Infusion of purified C3 inactivator
     (or fresh plasma) ameliorated the symptoms by inhibiting
     the C3 hypercatabolism.

iii) A protein substance (β_1-globulin) has been isolated
     which inactivates C6.

iv) Inactivation of C3a and C5a (anaphylatoxins) is done
     by cleavage of the carboxyl-terminal arginine from
     these molecules by carboxypeptidases (300,000 daltons,
     α-globulin). Formation of unstable complexes which
     undergo spontaneous decay also provides a control
     mechanism by preventing the accumulation of
     intermediates.

3. The Alternate (or Properdin) Pathway of Complement Activation

a. As noted earlier, this pathway can be activated by bacterial polysaccharides, endotoxin aggregated IgA and others (see Table 62). This pathway is considered a phylogenetically older system in complement activation. Some of the properties of the components of this system are presented in Table 63.

Table 62. Activators of the Alternate Complement Pathway

Substances	Specifically Identified Components
1. Immunoglobulins (aggregated)	IgG_1, IgG_2, IgG_3, IgG_4 IgA_1, IgA_2, IgD
2. Polysaccharides	Endotoxin, lipopolysaccharides of gram negative organisms, zymosan, inulin, and agar
3. Others	Cobra venom factor (CVF), trypsin, plasmin

b. The recognition unit of the pathway is a protein known as initiating factor (IF) or factor I. This factor binds to the appropriate site (repetitive sugar residues of polysaccharides present on a bacterial or viral cell) on the cell membrane and undergoes activation. The activated IF by interacting with C3 and factors B and D becomes catalytically active and cleaves C3 to C3b and C3a. The C3b formed, along with factors B and D, is transferred to a <u>new location</u> on the cell membrane. Factor D is presumably involved in the cleavage and activation of factor B. The resulting complex, C3bB, brings about additional cleavage of C3 to C3b (i.e., C3bB functions like a classical C3 convertase). The generated C3b molecules bind with the activated B (B) to form a complex which in turn combines with properdin (P) to form C3bBP, a polymolecular complex with C5 convertase activity. This complex converts C5 to C5b. In a new location on the cell membrane, C5b initiates the process of membrane attack and cell lysis with the terminal components of the complement. These reactions are identical with the classical pathway. It should be noted that the complement activation can also be initiated by proteolytic enzymes,

Table 63. Physicochemical Characteristics of Factors in the Alternate Pathway

Component	Synonym	Molecular Weight	Electrophoretic Mobility	Approximate Concentration in Serum (µg/ml)
Initiating Factor	Nephritic Factor	150,000	γ_1	---
Factor B	C3 Proactivator	93,000	β	200
Factor $\overline{B}$	C3 Activator	63,000	γ	---
Factor $\overline{D}$	C3 Proactivator Convertase	24,000	α	---
Properdin	---	220,000	γ_2	25

Figure 158. Activation of the Alternate Pathway of the Complement

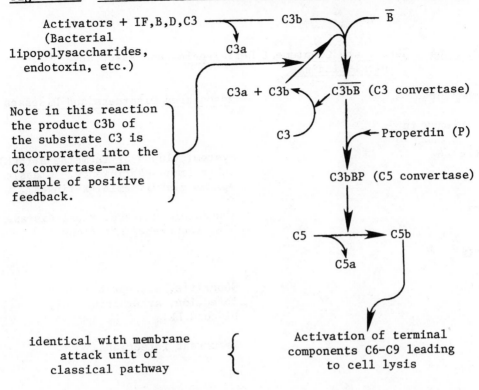

Activators + IF,B,D,C3 ───── C3b ───── $\overline{B}$
  (Bacterial
lipopolysaccharides,
 endotoxin, etc.)

plasmin or thrombin by converting C3 to C3b. C3 inactivator is an inhibitor of this pathway. A simplified scheme of the alternate pathway is shown in Figure 158.

4. Complement Deficiency and in Disease

Inherited abnormalities of the complement system of man have been described for 9 of the 11 classical components and two inhibitors. Table 64 summarizes the information available regarding hereditary complement deficiencies. There are also _acquired_ abnormalities that lead to alterations in the serum complement levels. The elevated levels may be observed in many acute inflammatory states or infections. These disorders may be found in association with diabetes, amyloidosis, thyroiditis, pregnancy, ulcerative colitis, polyarteritis, rheumatoid arthritis, rheumatic fever, spondylitis, hepatic disease, surgery, myocardial infarction, cancer, sarcoidosis, etc. Low levels are seen in immune complex disorders, namely systemic lupus erythematosus, rheumatoid arthritis, acute post-streptococcal nephritis, serum sickness, cryoglobulinemia, etc.

Table 64. Hereditary Complement Deficiencies and Associated
Clinical Findings

Component	Associated Disorders and Diseases
Deficiency in:	
C1q	Systemic lupus erythematosus (SLE), other immuno-deficiency. Classic Agamma globulinemia (X-linked).
C1r	Rheumatoid disease, renal disease, SLE, recurrent infections.
C1s	SLE.
C2	Nephritis, susceptibility to infection, arthralgia, SLE, Discoid LE.
C3	Recurrent infections.
C4	SLE.
C5	SLE, Leiner's disease, repeated infections.
C6	Infections.
C7	Raynaud's phenomenon.
C8	Gonorrhea, SLE.
C̄Ī inactivator (also known as C̄Ī esterase inhibitor)	Hereditary Angioedema, SLE.
C3b inactivator	Recurrent infections.

The Inflammatory Process

The inflammatory process involves interlocking networks of the
complement, clotting, fibrinolysis and kinin systems. All of these
systems are dependent upon proteolytic, enzymatic reactions which
are regulated by activation of inactive precursors by selective
proteolysis, positive feedback, stoichiometric inhibition, multistep
amplification and enzymatic degradation of the active products.
The inflammatory process is initiated in response to injury, and
it is essential to the survival of the host. The inflammatory
reaction often becomes a part of the immune process and the final
mediator of expression of immunologic reactions (both deleterious
and advantageous). The process comprises a series of biochemical
and microanatomical changes of the terminal vascular bed and of the
connective tissues. These processes are intended to eliminate the
injurious agents and to repair the damaged tissues. Inflammation
involves a great many responses of the host. Discussions in this
section will be limited to a brief review of the biochemical events
associated with coagulation, fibrinolysis, and kinin generation in
their relationships to the inflammatory state. The complement system,
a part of the inflammatory process, has already been discussed.
Discussions of the inflammatory cells, peripheral blood leukocytes,
and tissue macrophages will be presented later. The sequence of
events resulting from tissue injury which make up the inflammatory
response are as follows:

1. Tissue injury results in an increased permeability and dilation
    of blood vessels (erythema) in the injured region. Initiation
    of fibrin deposition via the clotting system occurs with the
    activation of Hageman factor (HF; clotting factor XII).

2. The deposited fibrin aids in trapping the deleterious agents
    and platelets within the coagulum. This also enhances
    phagocytosis by neutrophils and macrophages.

3. HF subunits (formed by the action of plasmin on activated HF)
    activate prekallikrein by converting it to kallikrein
    (chemotaxis factor). This stimulates the emigration of blood
    leukocytes and plasma cells, and these cells contribute to
    phagocytosis and localized antibody synthesis.

4. Increases in blood plasma and tissue fluids at the injured site
    occurs (as shown by swelling-edema) with the elevation of the
    local levels of the bactericidal serum factors (specific
    antibody, complement factors, kinin, opsonins, etc.). At the
    same time, the increase in fluids contributes to the dilution
    of the toxins formed.

5. The appearance of non-specific acute phase proteins such as
    C-reactive protein may occur, which could stimulate localized
    phagocytosis and activation or suppression of lymphocytes.

6. The physiological and biochemical alterations (e.g., an increase
    in CO_2, decrease in O_2, increase in body temperature, and
    accumulation of organic acids) which occur at the site of injury
    may be deleterious to the invading bacteria.

7. Finally, tissue repair occurs following macrophage action with
    the appearance of fibroblasts and deposition of collagen.

The Blood Clotting Process

1. As noted previously (see vitamin K metabolism, Chapter 10), the
    clotting process consists of the sequential conversion of inactive
    enzymes into active form. Table 65 summarizes some of the
    properties of the blood clotting factors. There are two
    coagulation pathways: intrinsic and extrinsic. The intrinsic
    pathway derives all of its factors from plasma whereas the
    extrinsic system is dependent upon factors derived from plasma
    as well as tissue extracts.

2. The coagulation factors are all proteins with the exception of
    Ca^{+2} (also labelled as a factor) which is <u>required</u> in a number
    of reactions. The various steps involved in both intrinsic and
    extrinsic pathways are shown in Figure 159. Note the intrinsic
    system begins with the activation of factor XII which is thought
    to occur due to its interaction with collagen (exposed when
    endothelium is injured). Activated factor XII (XIIa) also
    mediates the conversion of prekallikrein to kallikrein and the
    eventual formation of kinins (discussed later). The eventual
    formation of a clot is a result of the conversion of soluble
    fibrinogen by the action of thrombin (a serine protease) into
    insoluble gel fibrin. The fibrin peptides undergo polymerization
    by forming a linkage between glutamine side chains in one fibrin
    monomer and a lysine side chain of another fibrin molecule.
    This reaction is catalyzed by factor XIIIa (see under lysine,
    Chapter 2).

3. There are a number of inherited deficiencies of clotting factors
    that lead to excessive or uncontrolled bleeding. Some examples
    of these are: hemophilia A (deficiency of factor VIII),
    hemophilia B (deficiency of factor IX), hemophilia C (deficiency
    of factor XI), congenital parahemophilia (deficiency of
    factor V), afibrinogenemia, dysfibrinogenemia, etc.

Table 65. Nomenclature and Some Physicochemical Properties of the Blood Coagulation Factors and Members of the Kinin System

International Classification (number)	Synonyms	Molecular Weight		Electrophoretic mobility
		Bovine	Human	
Factor I	Fibrinogen	340,000	341,000	β_1
Factor II	Prothrombin	68,500	69,000	α_2
Factor IIa	Thrombin	33,700	35,000	β
Factor III	Tissue thromboplastin	1.7×10^7	---	---
Factor IV	Calcium	---	---	---
Factor V	Ac-globulin, labile factor, proaccelerin prothrombin accelerator	290,000	70,000-350,000	β_2,β
Factor VII	Stable factor, autoprothrombin I	35,000-63,000	50,000-100,000	β,α
Factor VIII	Antihemophilic globulin	180,000	300,000-400,000	β_2,β
Factor IX	Christmas factor, autoprothrombin II, plasma thromboplastin component	49,900	100,000-200,000	β
Factor X	Stuart-Prower factor	86,000	---	α
Factor XI	Plasma thromboplastin antecedent	---	100,000-200,000	$\beta-\alpha$
Factor XII	Hageman factor	---	60,000-100,000	γ_1
Factor XII (activated)	---	---	100,000	γ_1
Factor XII (fragments; subunits)	Prekallikrein activators (PKA)	---	30,000-70,000	prealbumin
Factor XIII	Fibrin-stabilizing factor	---	350,000	globulin
Prekallikrein	Fletcher factor	---	127,000	γ_2
Kallikrein	---	---	108,000	γ_2
Plasminogen	pro-fibrinolysin	---	81,000	β
Plasmin	fibrinolysin	---	75,400	β
Kininogen	α_2-globulin	---	50,000-200,000	pretransferrin
Bradykinin	Kallidin I	---	1,060	---
Kininases	Peptidases	---	---	---

Figure 159. Pathway for Blood Coagulation

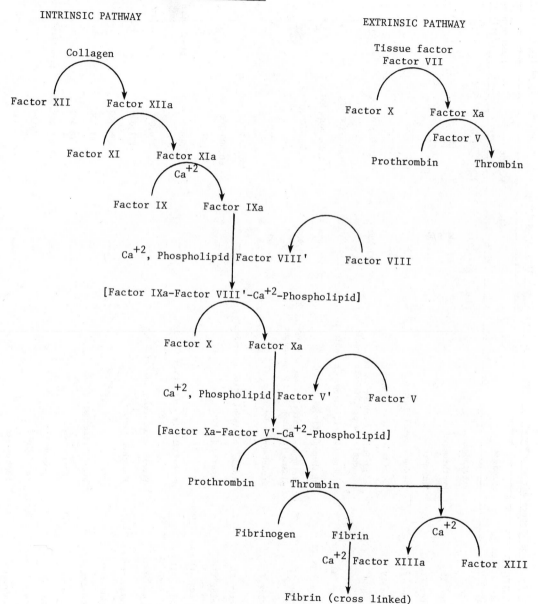

The Fibrinolytic System

Lysis of a fibrin clot, part of the tissue repair process, is
mediated by the plasminogen-plasmin system. Plasmin, a proteolytic
enzyme, degrades fibrin (and fibrinogen) to polypeptides of varying
molecular weights known as fibrinogen degradation products (FDP).
The reactions that lead to the generation of plasmin are as follows:

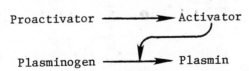

The plasminogen conversion to plasmin can also be catalyzed by
thrombin or enzymes found in certain bacteria (e.g., streptokinase,
staphylokinase) or urokinase. The proactivator is converted to an
activator by proteolytic fragments of factor XIIa known as
prekallikrein activators (PKA). PKA itself is generated from XIIa
by the action of plasmin (an example of positive feedback). These
interrelationships are shown in Figure 160.

Kinin System

1. The kinins are a group of highly active peptides and are derived
    from plasma precursors known as kininogens (see Table 65).
    Kinins produce most of the signs of acute inflammation; these
    include generalized dilatation of the peripheral arterioles,
    increased capillary permeability, production of pain, and migration
    of leukocytes to the injured area. These highly potent biologic
    substances are rapidly inactivated in the plasma and in the lungs.
    There are two enzymes (kininases) that catalyze the inactivation
    of kinins. Kininase I is a carboxypeptidase and kininase II is
    a dipeptidyl carboxypeptidase. It is of interest to point out
    that the kininase II also catalyzes the conversion of angiotensin I
    to angiotensin II by removal of terminal histidylleucine moiety
    (see Chapter 12). Angiotensin II is a powerful vasoconstrictor;
    thus, kininase II, while inactivating powerful vasodilators (the
    kinins), is also able to generate a potent vasoconstrictor.

2. The major kinins are bradykinin (a nonapeptide: H-Arg-Pro-Pro-
    Gly-Phe-Ser-Pro-Phe-Arg-OH), lysylbradykinin (known as kallidin)
    and methionyl-lysylbradykinin.

3. The kinin-forming system and its relationship to blood coagulation,
    complement and fibrinolytic processes are shown in Figure 160.
    Note the central role of factor XIIa and the positive feedback
    of plasmin and kallikrein over the conversion of XIIa to
    prekallikrein activators (PKA). The PKA's are proteases which
    convert prekallikrein to kallikrein and proactivator to activator.

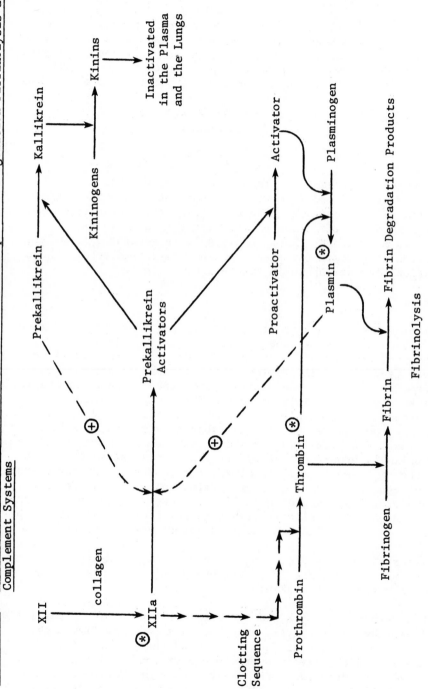

Figure 160. Kinin-Generating System and Its Interrelationships With Coagulation Fibrinolysis and Complement Systems

4. Kallikrein, the proteolytic enzyme responsible for the conversion of kininogen to kinin, is inhibited by two inhibitors: C1 inhibitor and α_2-macroglobulin. Recall, in the hereditary angioneurotic edema the C1 inhibitor is deficient; thus the subject with this disorder not only show problems due to uncontrolled complement activation but also to excessive production of kinins. Other disorders which can give rise to severe problems due to excessive generation of kinins are gram-negative septicemia, carcinoid syndrome, coronary artery disease, inflammatory-joint disease, certain transfusion reactions, and dumping syndrome.

Peripheral Leukocytes and Macrophages

The previous sections dealt with the contribution of the humoral system to the exudation associated with inflammation and of the many biological peptides stimulating emigration of the cells making up the exudates. The present section will briefly describe the pertinent morphological characteristics, functions, and the deficiencies or dysfunctions of these cells in man.

1. Characteristics of Leukocytes

    a. Neutrophils

        i) Polymorphonuclear neutrophilic leukocytes (PMN-leukocytes; see Chapter 4) are granulocytes which possess a number of functional properties important in host resistance. These typical inflammatory cells, associated with acute inflammation, are essential for the pathogenesis of acute, necrotic reactions of immunological diseases, such as the vasculitis of the immune-complex disease (serum sickness), Arthus' reaction, the severe glomerulonephritis of nephritoxic nephritis, and many other immune-complex diseases. PMN-leukocytes are, nevertheless, primarily involved in the continuous defense against non-immunologically related inflammatory infections caused by bacteria and other parasites. The major route by which PMN-leukocytes exert their defenses is phagocytosis which will be discussed in detail later in this section. Neutrophils vary in size from 10 to 15 μm diameter.

        ii) Neutrophils are the major granulocytes in peripheral blood (2 to 3×10^{10} cells in the total blood volume and 1.5 to 3.0×10^{12} in bone marrow). A mature neutrophil has a half-life of 6.6 hours as measured by use of ^{32}DFP-labeled PMN-leukocytes (DFP = diisopropylfluorophosphate; see Chapter 3). The nucleus of neutrophils is multilobed and in stained smears many pink-staining granules can be seen in the cytoplasm. When examined with the electron microscope, these granules are of varying density and size and have been designated α, β, and γ. They constitute the major source of the enzymes used for the killing of bacteria following phagocytosis. Some of the enzymes found in the cytoplasm of neutrophils are summarized in **Table 66.**

iii) Receptors for Fc of IgG and IgE, and C3b have been demonstrated on the membrane of **neutrophils.**

b. Eosinophils

i) The eosinophilic leukocytes are similar in size and in nuclear structure to the neutrophils. The major distinguishing characteristics of eosinophils are their cytoplasmic granules (0.2 to 1.0 µ), which are much larger than neutrophilic granules and which have a strong affinity for acidic aniline dyes such as eosin. The **granules contain large quantities of stable peroxides,** lipids, and proteins. Crystalloid structures of various patterns can be seen in these granules by electron-microscopic examinations. In human eosinophils, the crystalloid structures have various shapes (oval, squarish, two or more cores, etc.) and are generally localized in the central region of the granule. The crystalloids appear to stain strongly for peroxidase activity when analyzed by electron-microscope histochemical staining using 3,3'-diaminobenzidine tetrachloride. Eosinophil granules **also contain enzymes similar to those in neutrophils, but** appear to lack lysozyme and phagocytin. Nevertheless, these granules are considered to be lysosomes. Like the neutrophils, eosinophils have a short life span in peripheral circulation. They constitute 2 to 5% of the total circulating leukocytes of peripheral blood. Bone marrow contains 200 times more eosinophils than blood, whereas, the tissues have 500 times more than the blood. The eosinophils of the tissues are primarily found in the intestinal walls, skin, external genitalia, and lungs. Blood eosinophils are regulated by the adrenal corticosteroids. An increase in these hormones due to stress or therapeutic administration results in decreased numbers of eosinophils in the circulating blood.

ii) The functions of eosinophils are unknown, but they do have limited phagocytic capabilities. When phagocytosis occurs, the cells can degranulate and in this respect eosinophils appear to be similar to neutrophils. It has been suggested that eosinophils contain a factor(s) which neutralizes histamine, serotonin, and perhaps kinins. This could be

Table 66. Some of the Enzymes Associated with the Cytoplasmic Granules of Granulocytes, Macrophages (Monocytic Origin), and Lymphocytes

Cell Type	Organelle	Enzymes
1. Granulocytes		
a. Neutrophil	Lysosome	Acid phosphatase, acid lipase, aryl sulfatases A and B, acid ribonucleases and deoxyribonucleases, cathepsin B,C,D,E, collagenase, phosphoprotein and phosphatidic acid phosphatases, phospholipase, organophosphate-resistant esterase, β-glucuronidase, β-galactosidase, β-N-acetyl-glucosaminidase, α-L-fucosidase, α-1,4-glucosidase, α-mannosidase, α-N-acetylglucosaminidase, α-N-acetyl-galactosaminidase, hyaluronidase, lysozyme
	Peroxisome-related	D-amino acid oxidase, L-α-OH acid oxidase, catalase, myeloperoxidase
b. Eosinophil	Lysosome	Acid protease, β-glucuronidase, aryl sulfatase, nucleases and phosphatase, peroxidase
c. Basophil (mast cell)	Lysosome	Proteases, phospholipase A, glucuronidase, acid and alkaline phosphatases, ATPase
	Granules	Histamine, serotonin, heparin, dopamine, acid mucopoly-saccharides, (dopa, tryptophan and histidine decarboxylases), heparin-synthesizing enzymes of cytoplasm, SRS-A (Note that not all of these substances are enzymes.)
2. Lymphocytes	Peroxisome-related	D-amino acid oxidase, L-α-OH acid oxidase, peroxidase catalase
	Lysosome	Arylsulfatase, β-glucuronidase, β-galactosidase, N-acetyl-β-glucosaminidase, N-acetyl-α-galactosaminidase, α-mannosidase, α-arabinosidase, β-xylosidase, β-cellobiosidase, β-fucosidase, cathepsin D
3. Macrophage	Lysosome and cytoplasmic mitochondria	Acid-phosphatase, β-glucuronidase, cathepsin, esterase, lysozyme, α-glycerolphosphate dehydrogenase, DPN- and TPN-diaphorases (isocitric, lactic, malic, succinic dehydrogenases), uridine diphosphate glucose-glycogen transglycosylase

the means by which these mediators of vascular changes are controlled. This concept helps to explain the appearance of large numbers of eosinophils during an allergic manifestation due to an immediate type of hypersensitivity associated with the release of the vasoamines. However, evidence for the existence of the neutralizing factor(s) has not been presented as yet. The enzymes of eosinophils are summarized in Table 66.

iii) **Eosinophils contain receptors for the Fc portion of IgG.**

c. Basophils (Mast Cells)

i) **Basophils are specialized inflammatory cells which are** chemically, structurally, and functionally identical to the mast cells encountered in connective tissues. These cells, together with platelets, probably play a significant role in the release of vasoamines such as histamine and serotonin. Basophils constitute 0.5% of the circulating blood leukocytes. They are characterized by the presence of large electron dense cytoplasmic granules which have a predilection for basic aniline dyes suggesting that they contain acid mucopolysaccharides. It has been known for some time that these granules contain heparin which is responsible for the metachromasia. Degranulation of mast cells releases histamine (see Chapter 6) in addition to heparin. In some species, such as the rat, mast cells also contain serotonin (5-hydroxytryptamine) in their cytoplasmic granules. About 30% of the dry weight of mast cell granules is heparin and approximately 35% is basic protein. The histamine content of mast cell granules varies from species to species and from tissue to tissue (7 to 40 µg/cell). Packed peritoneal mast cells of the rat contain 630 to 700 µg serotonin/ml of cells. It has been suggested that mast cell granules are composed of polysaccharide-protein-ion complexes to which vasoamines (histamine, serotonin, dopamine) are bound electrostatically. Other constituents of mast cell granules are shown in Table 66.

ii) Substances which can initiate mast cell or basophil degranulation with the concomitant release of heparin and vasoactive amines, are IgE-Ag complexes, anaphylatoxin (C3A, C5A), basic protein (neutrophil lysosomal protein), and antimast cell antibody plus complement; chemicals such as dextrans, polyvinylpyrrolidine, bee venom, rose thorn; surface active agents such as bile salts, lysolecithin, and Tween 20; and physical agents (heat, ultraviolet radiation, x-rays, and radioisotopes). The mechanisms by which these agents produce degranulation differ, although vasoactive amines are released in all cases. The degranulation process is an active one and requires energy. The primary physiological functions of basophils and mast cells are not clear. Their characteristic perivascular distributions suggest a possible role in the regulation of the permeability of the terminal vascular beds via the release of vasoactive amines. Abnormal release of the vasoactive amines by mast cells contributes to the inflammatory patterns observed in immediate and delayed hypersensitivities and in non-immunological disorders.

iii) Membranes of mast cells and basophils contain receptors for Fc of IgG and IgE. Eosinophil chemotactic factors and platelet activating factor are released by activated mast cells or basophils.

d. Platelets

i) The significance of platelets in the intrinsic pathway for coagulation of blood is well known. They are the fundamental formed elements involved in the creation of the hemostatic plug (thrombus) of flowing blood. The brief discussion presented here will pertain to the vasoactive amines of platelets involved in the immunological reactions. Like mast cells and basophils, platelets are affected by antigen-antibody complexes in the presence of complement. In some species, vasoactive amines are also released in this process. The reaction in rabbit platelets can be summarized by

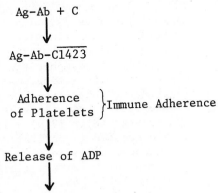

Ag-Ab + C

Ag-Ab-C$\overline{1423}$

Adherence of Platelets } Immune Adherence

Release of ADP

Further platelet aggregation; release of vasoactive amines.

(As mentioned earlier, in immune adherence, neutrophils, eosinophils, and macrophages also adhere to the Ag-Ab-C$\overline{1423}$

complex.) These amines are contained within the platelets in granules distinct from lysosomes. Energy (presumably as ATP) is required for the release of the vasoactive amines and platelet lysis does not usually accompany the release.

ii) The concentrations of vasoactive amines are lower in human platelets than in those of the rabbit. Also, instead of adhering to Ag-Ab-C1423 complexes, they aggregate directly with the antibody globulin of the AgAb complexes. Thus, the release of ADP and vasoactive amines occurs without the intervention of complement. Platelets of ruminants and pigs are similar to those of man. However, although the mechanisms are different in all of the species mentioned here, the end result (release of vasoactive amines) is the same.

iii) Collagen and several non-immunological materials can also initiate release of vasoactive amines from platelets. Other immunological factors which can trigger platelet release of amines include aggregated IgG, antigen plus sensitized leukocytes, antiplatelet antibody, and the enzymes thrombin and trypsin.

iv) For a discussion on the relationship between platelet aggregation and prostaglandin derivatives, see Chapter 8.

e. Lymphocytes and Plasma cells

i) Lymphocytes of the peripheral blood comprise nearly 40 to 50% of circulating leukocytes and are the major immunological cells. They are a heterogeneous population in both size and function. Both B-cells and T-cells can be found in the circulation. The T-cells constitute the major group of lymphocytes in a normal individual (70 to 80% of the circulating small lymphocytes). B-cell lymphocytes constitute the remainder of the lymphocytes (20 to 30%) varying in size from small to large lymphocytes (8 to 12 μ). The functions of lymphocytes have been discussed in earlier sections. These cells are the pivotal cells in immunology. The intracellular enzyme profile, which is very limited, of these lymphocytes is shown in Table 66.

ii) The major morphologic characteristics of lymphocytes are their large homogeneous, non-lobed nuclei, and light-staining, non-granular cytoplasm. Plasma cells are rarely found in peripheral blood in normal individuals. These cells, as indicated earlier, are associated with antibody synthesis and are the end cells of B-lymphocyte development. Their major morphologic characteristics are their eccentric cartwheel-like nuclei. The cytoplasmic area is much larger and stains much more intensely with basic dyes (greyish blue) than does that of a lymphocyte. Electron micrographs show numerous channels of endoplasmic reticulum lined with polyribosomes in the cytoplasm. Few or no granules characteristic of granulocytes are evident.

f. Monocytes and Macrophages

i) Monocytes of the peripheral blood measure 12 to 15 μ in diameter when examined in stained blood smears. The nucleus is generally kidney shaped and the cytoplasm stains greyish blue with aniline dyes such as crystal violet. The cytoplasm also contains azurophil granules which are reddish blue in color. Blood monocytes are considered immature macrophages and their primary source is the bone marrow. There is general agreement that monocytes migrate from the circulation and mature into typical tissue and inflammatory macrophages at the site of injury. This has been substantiated with radioautographic procedures using ^{3}H-thymidine. The source of the macrophages in the lung, peritoneal cavity, and inflammatory exudate is the bone marrow. Macrophages may also arise by the mobilization and proliferation of existing tissue macrophages, migration of Kupffer cells from the liver into the lungs, the transformation of lymphocytes to macrophages, and (possibly) differentiation of specialized endothelial cells to macrophages. Differences and similarities are evident between macrophage cells from different tissue sites. This can be shown as follows.

Source of Macrophage	Mitotic Rate	Ability to Adhere and Spread on Glass
1. Sessile or fixed tissue (bone marrow, spleen, and liver)	high	slow
2. Alveolar tissue	high	readily
3. Peritoneal and blood monocytes	low	readily

ii) Tissue macrophages are larger than blood monocytes (15 to
80 μ) and the number of azurophil granules varies. Their
nuclei vary in size and shape and multinucleated forms
may be present. The epithelioid cells of chronic
inflammation (granulomous) are macrophages with ovoid
nuclei resembling epithelial cells. The presence of
diffuse lipids in the cytoplasm gives these cells a pale
appearance. Lipid droplets may also be shown in the
cytoplasm of macrophages. Electron micrographs of
monocytes and unstimulated peritoneal macrophages show
great similarity. Unlike peripheral granulocytes, these
cells have a well-defined Golgi apparatus composed of
flattened sacs, small vesicles, and a few granules;
with a moderate amount of rough endoplasmic reticulum,
varying numbers of electron dense granules, and many
small vesicles in the cytoplasm.

iii) The major function of macrophages, like that of
neutrophils, is the phagocytosis and the degradation of
the ingested material. Significant numbers of neutrophils
are also ingested by macrophages following tissue damage
or an acute infection. The half-life of monocytes has
been estimated to be three days. Some of the enzymes
found in macrophages are listed in Table 66.

iv) Macrophages, again like neutrophils, are influenced by
complement factor (AgAbC1423 in immune adherence) and by
sensitized T-cell lymphokines such as MIF (migration
inhibitory factor). The ability of certain antibodies (IgG$_1$
and IgG$_3$) to adhere to macrophage membrane (macrophage
receptor - MR) permits the IgG to react with its antigen and
may play a significant role in macrophage-mediated protection
against certain infections. These macrophage-adherent
immunoglobulins are known as the cytophilic antibodies. The
adherence occurs through the Fc portion of the IgG molecule.
Similar binding of IgG to macrophages which react with
erythrocytes (sheep RBC) give the typical rosette appearance
attributed to macrophages. The role of the macrophage as an
accessory or auxiliary cell (A-cell) in the processing of
antigens (which is sometimes needed for T-cell--B-cell
interactions) has been discussed previously.

2. Phagocytosis

a. Phagocytosis - the process by which a phagocytic cell destroys
invading bacteria - can be viewed as a sequence of three events.
As discussed earlier, the major phagocytic cells are the
neutrophils and the macrophages.

i) <u>Chemotaxis</u>, the attraction of the phagocytes to the site of injury, has been discussed. It is related to the peptides released during complement activation, kallikrein activation, and antigen reaction with specific sensitized T-lymphocytes. In addition, chemotactic factors released by bacteria and from tissue breakdown products have been demonstrated. These seem to be small glycoprotein fragments and polysaccharide.

ii) The term <u>phagocytosis</u> is used both for the overall process and for the second step, the engulfment (ingestion) of the bacteria. This occurs when the outer cytoplasmic membrane of the cell surrounds the bacteria, with the formation of a vacuole within the cytoplasm of the phagocyte.

iii) Soon after engulfment, the actual killing of the bacteria begins. Lysosomal (and perhaps peroxisomal) granules fuse with the vacuole and release their enzymatic contents into it. The resultant packet is now called a phagolysosome or phagosome, within which the killing and digestion of the bacteria occurs.

The subunits or fragments formed are egested (eliminated from the cell) and, in many instances, passed on to the appropriate lymphocytes for cell-mediated or humoral immune responses. These subunits or antigens coming from macrophages have been designated as "super-immunogens" by some investigators. Figure 161 is a schematic representation of the phagocytic process.

Phagocytosis is enhanced by naturally occurring opsonins (immunoglobulins) against bacteria and by specific antibody immunoglobulins. These effects can be amplified through the participation of complement and the immune adherence phenomenon as discussed earlier. Fibrin clots also assist phagocytosis by creating surfaces and by restricting bacterial migration, both of which make entrapment easier.

<u>Pinocytosis</u> is used to describe a process similar to phagocytosis. It involves the intake of soluble material (proteins) and colloidal suspensions (such as virus particles).

b. Biochemistry of Phagocytosis (see also carbohydrate metabolism and phagocytosis, Chapter 4)

i) The significant biochemical events associated with phagocytosis are increases in anaerobic glycolysis and

[1164]

Figure 161. Phagocytosis of Bacteria by PMN Leukocyte
(N–Nucleus, M–Mitochondria, ER–Endoplasmic
Reticulum)

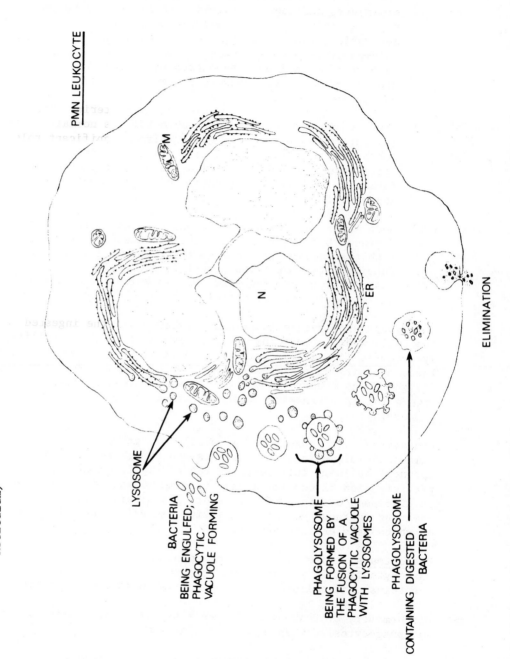

PMN LEUKOCYTE

M

N

ER

M

LYSOSOME

BACTERIA
BEING ENGULFED;
PHAGOCYTIC
VACUOLE FORMING

PHAGOLYSOSOME
BEING FORMED BY
THE FUSION OF A
PHAGOCYTIC VACUOLE
WITH LYSOSOMES

PHAGOLYSOSOME
CONTAINING DIGESTED
BACTERIA

ELIMINATION

in the hexose monophosphate-shunt pathway (HMP-shunt; see Chapter 4). These increases are necessary for the ingestion step. Increases in glucose utilization and in the formation of lactic acid and hydrogen peroxide occur following phagocytosis. Energy increases of from 5 to 40% over those in a resting leukocyte have been shown to occur in a leukocyte which is actively ingesting bacteria. Thus, suppression of glycolysis or the HMP will result in decreased ingestion of bacteria. The H_2O_2 (which increases to four to six times normal levels following active phagocytosis) plays a significant role in killing the ingested organisms. Superoxide anion (O_2^-), hydrogen peroxide plus peroxidase and chloride ions have been shown to play a significant role in killing of ingested bacteria. In some instances, halogenation (involving Cl^-, hydrogen peroxide, and peroxidase) of the bacterial membrane resulted in the death of the organism. In other instances, the aldehydes and ketones formed through peroxidatic action have been shown to be the toxic substances.

ii)   Hydrogen peroxide, in the presence of ascorbic acid, results in the formation of free radicals which could alter membrane surfaces such as those of the ingested bacteria. This effect would increase the vulnerability of the bacterial surface to the lysosomal enzymes present in the phagolysosome. For example, it has been shown that H_2O_2 plus ascorbic acid mixtures can render <u>Salmonella</u> membrane surfaces susceptible to lysozyme attack by exposing the muramic acid (2-amino-3-0-(1-carboxyethyl)-2-deoxy-D-glucose) residues previously protected by the membrane capsules of the bacteria. Evidently, the free radicals formed depolymerize the protective covering of the plasma membrane, since any reducing chemical, such as thiosulfite, which inhibits free radical formation, reverses this phenomenon. In this context, it is interesting to note that macrophages have a high ascorbic acid content.

iii) Once the ingested bacteria are killed, the complete destruction follows, catalyzed by the numerous hydrolases found in the lysosomes. The enzymes contained in the phagocytes have been summarized in Table 66. A congenital absence of any of these enzymes causes one of the so-called <u>lysosomal disorders</u> discussed in Chapter 8. The synthesis of H_2O_2 as it occurs in PMN leukocytes is presented in Chapter 4. The oxidases (which produce H_2O_2 from O_2) and the peroxidases (which degrade it to H_2O) are probably those enzymes which are found in the peroxisomes. The aldehydes and ketones are perhaps formed by $NADP^+$-dependent dehydrogenases but this process, as it pertains to phagocytosis, is not well understood in man.

3. Leukocyte Deficiencies

Defects in the non-specific immune system associated with neutrophils have been demonstrated. These changes include quantitatively, a decrease in the number of neutrophils (neutropenia), and qualitatively, defects in some metabolic or biochemical function of these cells, but with normal morphologic appearance and number.

a. Neutropenias associated with decreased production of myeloid precursors due to bone marrow defects have been observed. There are acquired and inherited forms of neutropenias. The acquired types can be caused by drugs, pollutants, radiation, endotoxin, overwhelming infections, neoplasia, and disorders of the bone marrow and the spleen. The inherited forms are x-linked recessive or autosomal dominant characteristics.

Increased susceptibility to infections is common in both acquired and inherited neutropenias. Micrococcus pyogenes, Streptococcus pyogenes, and Diplococcus pneumoniae are the common causes of infection in the neutropenic individual.

b. Qualitative changes in neutrophils have been associated with chronic granulomatous disease (CGD; see Chapter 4). Transmission is familial, presumably as an x-linked recessive trait. However, both males and females have been affected, although at different ages. Defects in the biochemical pathways have been implicated in the inability to kill low-grade pathogens such as Escherichia coli, Micrococcus pyogenes, Serratia marcescens, Paracolon hafnia, and Klebsiella enterobacter. CGD has been associated with a deficiency of glucose-6-phosphate dehydrogenase, a decrease in NADPH oxidase, deficient myeloperoxidase, or defects in the glutathione peroxidase and reductase systems. Since all of these defects have not been found in any one patient, the defect in CGD has yet to be completely elucidated.

c. Myeloperoxidase deficiency in leukocytes has been reported which is associated with an inability of the leukocytes to kill ingested bacteria. The generation of H_2O_2 is severely diminished and the ability to break down H_2O_2 is defective in leukocytes with this disorder.

Abnormalities of the Immune System

1. Three major categories of immune diseases can be considered.

    a. Diseases characterized by abnormal proliferation of the lymphoid
        cells (B-lymphocytes and plasma cells) which normally synthesize
        immunoglobulins. Abnormal increases in plasma cells are
        associated with excessive immunoglobulin synthesis and the Ig
        produced is characteristic of the disorder;

    b. Diseases characterized by immunodeficiencies, either congenital
        or acquired, with a decrease or lack of Ig synthesis,
        agammaglobulinemia (B-lymphocyte deficiency), or decrease in
        cell-mediated immunity (T-lymphocyte deficiency or defect);

    c. Diseases associated with lymphocytic proliferation with or
        without an increase in immunoglobulin synthesis and with
        general deficiencies in the T-cell system.

2. Diseases in the first and third categories are immunoproliferative
    disorders which include multiple myeloma, macroglobulinemia, heavy
    and light chain diseases. The third also includes lymphomas,
    lymphosarcomas, lymphatic leukemias, and Hodgkins disease. These
    two groups are also synonymous with neoplasia. By
    immunofluorescence procedures, using fluorescein-labeled anti-
    bodies specific for IgM, IgG, IgA, IgD, and IgE and kappa and
    lambda chains, the specific nature of the lymphoproliferative dis-
    order in lymphatic leukemias has been ascertained in some studies.

3. The multiple myelomas are also called gammopathies and are
    classified according to the immunoglobulin whose level is increased
    in the circulating blood. For example, a myeloma with an increase
    in IgG and decrease in IgM and IgA is called a monoclonal
    gammopathy-IgG. Increases in two or more immunoglobulins are called
    polyclonal gammopathies. For example, if IgG and IgA are increased,
    the disorder is referred to as a polyclonal gammopathy-IgG,A.
    Figure 162 shows normal immunoglobulin synthesis compared to the
    situation in a monoclonal gammopathy-IgG.

4. The major criteria for the diagnosis of multiple myelomas are

    a. abnormal quantitative increase in serum immunoglobulin;

    b. changes in the x-ray patterns of the bones, especially the skull,
        vertebrae, and long bones (osteoporosis; see Chapter 9);

    c. a marked increase in number of pre-plasma and plasma cells in
        the bone marrow.

Figure 162. Synthesis of Immunoglobulins in Normal Individuals and in an IgG Monoclonal Gammopathy Normal Immunoglobulin Synthesis

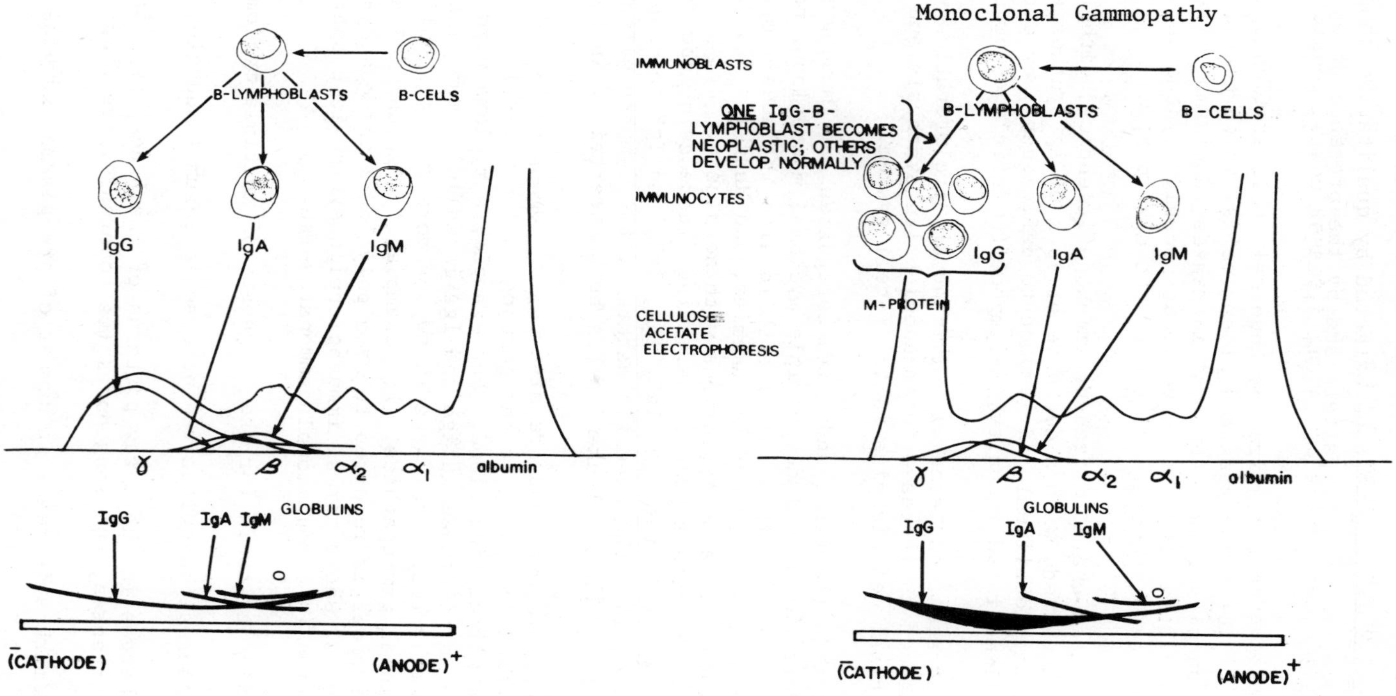

The changes in serum can be detected by qualitative changes of the immunoelectrophoretic pattern and by the presence of an M-protein peak in cellulose acetate electrophoresis. This M-protein is the immunoglobulin (or Ig-chain) which is being synthesized by the abnormally proliferating B-lymphocyte. It can be a myeloma globulin (IgG, A, D, or E), a Bence-Jones protein (free kappa-or free lambda-chains) or a combination of the two. An increase in total serum protein is suggestive of but not diagnostic for a multiple myeloma. Quantitation of the immunoglobulin by radial immunodiffusion with specific anti-Ig is highly important in the diagnosis of multiple myeloma.

5. Because an increase in serum immunoglobulins is normally an expression of stimulation by specific antigens, the immunoproliferative disorders such as multiple myeloma - a classical monoclonal gammopathy - are an enigma, since no specific antigens have been implicated in any of these diseases. Recently, however, the IgM of macroglobulinemia in man and the IgA of multiple myeloma of the experimental mouse have been shown to interact in an antigen-antibody type reaction with, respectively, lecithin and pneumococcal C-polysaccharide. The antigenic determinant in the C-polysaccharide-IgA reaction was shown to be phosphorylcholine. The immunoglobulins of monoclonal gammopathies have contributed immeasurably to the elucidation of structure and thus to structural-functional relationships of the immunoglobulins.

6. The significant immunoproliferative and immunodeficiency diseases are summarized in Tables 67 and 68. Figure 163 summarizes the areas of the immune system which are affected (indicated by a heavy bar) by the diseases associated with immunodeficiency. Pertinent remarks related to these diseases are also included. More details are available in pathology texts and in specific reviews pertaining to these diseases.

7. A significant increase in the incidence of cancer has been observed in patients with immunodeficiency disorders such as the Wiskott-Aldrich syndrome and ataxia-telangiectasia. A high incidence of cancer is also more common in patients undergoing long periods of immunosuppressive therapy following kidney transplants. These findings strongly support the concept of "immune surveillance" against cancer as expressed by Burnet and others. In addition, patients with lymphoproliferative disorders tend to have depressed T-lymphocyte functions.

Table 67. Immunoproliferative Disorders

Diseases	Immunoglobulin					Characteristics
	IgG	IgA	IgM	IgD	IgE	
1. Multiple Myeloma						Bone marrow changes; osteoporosis; increase in plasma cells; increase in a single immunoglobulin (M-protein in cellulose acetate electrophoresis); general decrease (D) in other Ig; 30 to 40% of the patients have Bence-Jones protein in urine
Monoclonal Gammopathy-G	I	D	D	D	D	
Monoclonal Gammopathy-A	D	I	D	D	D	
Monoclonal Gammopathy-D	D	D	D	I	D	
Monoclonal Gammopathy-E	D	D	D	D	I	
2. Macroglobulinemia						Increase in IgM with general decrease in the other Ig; changes are in lymphoid tissue, hyperplasia, M-protein peak
Monoclonal Gammopathy-M	D	D	I	D	D	
3. Heavy Chain Disease	H-chains in serum and urine; abnormal γ, α, μ H-chains have been reported					Lymphoid cell hyperplasia; other Ig generally decrease; M-protein peak
4. Light Chain Disease	L-chains in serum and in urine					Bone marrow changes; increase in plasma cells; generally no M-protein peak in cellulose acetate electrophoresis
5. Hodgkin's disease, lymphoma, lymphocytic leukemias	Variable Ig levels, dependent on whether T-or B-cell hyperplasia occurs					Changes in lymphoid tissues; hyperplasia; anaplasia; variable sizes of cells; specific cell characteristics (Hodgkins-Reed-Sternberg giant cells)

I = increase; D = decrease.

Table 68.   Immune Deficiency Disorders

Nature of Deficiency	Inheritance	Characteristic Defects and Pattern	Susceptibility
1. Cell-Mediated			
a. Congenital thymic aplasia (Di Georges' Syndrome)	---	Defect in embryogenesis; lack thymus and parathyroid; defect of the third and fourth pharyngeal pouches	Viruses, fungi, bacteria
b. Congenital thymic dysplasia (Nezelof Syndrome)	Autosomal recessive	Faulty dysgenesis of thymus; normal immunoglobulin synthesis; defect in cell-mediated immune system	Candida, Herpes, Pneumocystis carinii, Pneumococcus, Streptococcus, Staphylococcus
2. Humoral Immunity			
a. Agammaglobulinemia			
i) Congenital (Bruton Type)	X-linked (males); autosomal recessive (males and females)	Congenital; generally normal until nine months of age or older (some until age 5 to 6 years); immunoglobulin 0 to less than 100 mg/100 ml; plasma cells absent in lymph nodes	Micrococcus pyogenes, Diplococcus pneumoniae, Streptococcus, Pneumocystis carinii
ii) Acquired	---	Acquired; seen at any age, "Primary"--no underlying disease; "Secondary-- associated with disease	---

[1171]

Table 68. Immune Deficiency Disorders (continued)

Nature of Deficiency	Inheritance	Characteristic Defects and Pattern*					Susceptibility
			IgG	IgA	IgM	IgE	
iii) Transient hypogamma-globulinemia	---	Gamma-globulin synthesis delayed until 18-30 months of age; normal at three months; cell-mediated response intact; infections of skin, meninges, and respiratory tract					Gram positive bacteria
b. Dysgammaglobulinemia (Selective Ig deficiencies)							
i) Type I	X-linked (males)	Manifest autoimmune disorders	D	D	I (150-1000 mg per 100 ml)	---	Increased pyogenic infections of lung and skin
ii) Type II	---	Nodular lymphoid hyperplasia	N	D	D	---	Increased
iii) Type III	---	Associated with autoimmune disease	N	D	N	I	Recurrent respiratory infections
iv) Type IV	(Familial?)		D	N	N	N	---
v) Type V	(Familial?)	Low levels of isohemagglutinin	N	N	D	---	Gram-negative bacteria

Table 68. Immune Deficiency Disorders (continued)

Nature of Deficiency	Inheritance	Characteristic Defects and Pattern*					Susceptibility
		IgG	IgA	IgM	IgE		
vi) Type VI	---	N	N	N	---		Staphylococcus
vii) Type VII	---	Deficiency in IgG subclass					Pneumonias
		D	I	N	---		
c. Wiskott-Aldrich Syndrome	X-linked recessive (males)	N	N	D	---	Absence of isohemagglutinin; thrombocytopenia; eczema; inability to process polysaccharide antigens; some reconstitution with transfer factor also occurs in some T-cell deficiencies	Viruses: Vaccinia, and Herpes; Gram- and Gram+ bacteria; Fungi: C. albicans; Protozoa: P. carinii.
d. Ataxia-telangiectasia	Autosomal recessive	N?	D	N?	D	T-cell defects also; degeneration of cerebellum with ataxia	Increased; respiratory infections.
3. Cell-Mediated and Humoral Swiss type, thymic dysplasia, agamma-globulinemia	X-linked recessive, autosomal recessive					Peripheral lymphoid; lack B- and T-cells; defect at lymphoid stem cell; no humoral or cell-mediated responses; lymphopenia and agammaglobulinemia	Increase: viral fungal, bacterial (Gram+ and Gram-) Protozoan infections.

Table 68. Immune Deficiency Disorders (continued)

Nature of Deficiency	Inheritance	Characteristic Defects and Patterns*	Susceptibility
4. Reticular dysgenesis	---	Stem cell defect; infants die within the first week of life; thymic dysplasia; lymphopenia; lymphoid depletion; neutropenia; depletion of myeloid cell precursors	Severe sepsis due to Staphylococcus infection

* N = normal; D = decreased; I = increased; all refer to the serum levels of the indicated immunoglobulin.

Figure 163. Summary of the Steps Affected in the Immune Deficiency Diseases

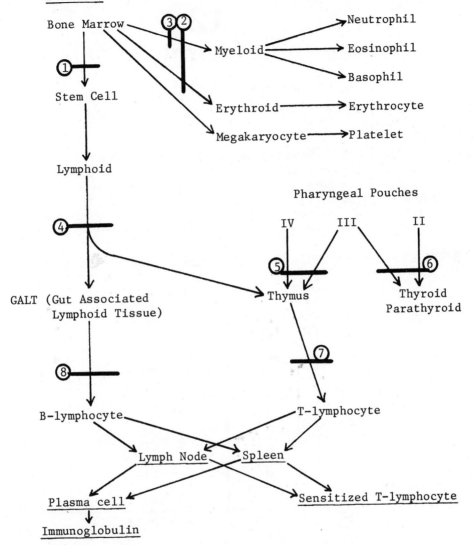

Note: The numbered bars on the diagram indicate the step or steps affected in each of the following disorders:

1. Reticular Dysgenesis
2. Fanconi's Syndrome
3. Chronic Granulomatous Disease and Myeloperoxidase deficiency
4. Thymic Dysplasia and Agammaglobulinemia (Swiss Type)
5. Thymic Aplasia (III)

   and IV; Di Georges' Syndrome)
6. Thymic Aplasia (II and III)
7. Thymic Dysplasia (Nezelof Syndrome)
8. Agammaglobulinemia (Bruton Type)

12
Water and Electrolyte Balance

with John C. McIntosh, Ph.D.*

Water Metabolism

1. Water is the most abundant body constituent, comprising 45 to 60% of the total body weight (see Figure 164). In a lean person, water accounts for a larger fraction of the body mass than it does in a fat person, accounting for this wide variation. Since biochemical reactions take place in a predominantly aqueous environment, control of water balance is an important requirement for homeostasis.

Figure 164. Proportional Distribution of Solids and Water in a Healthy Adult**

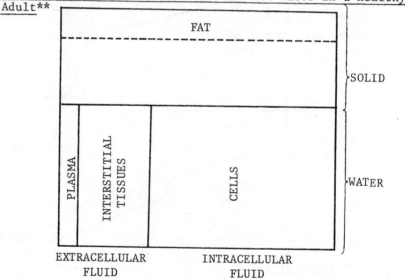

2. Although water permeates freely throughout the body, the various solutes are less mobile due to barriers in the form of the various membrane systems. This gives rise to various fluid pools or compartments of differing composition (see Figure 165). The composition of each compartment is kept constant by a combination of active and passive processes.

---

* Lecturer in Clinical Biochemistry, Massey University, New Zealand.

** From Baron, D.N.: A Short Textbook of Clinical Biochemistry, Philadelphia, 1969, J.B. Lippincott, page 1.

Figure 165. Composition of Body Fluids

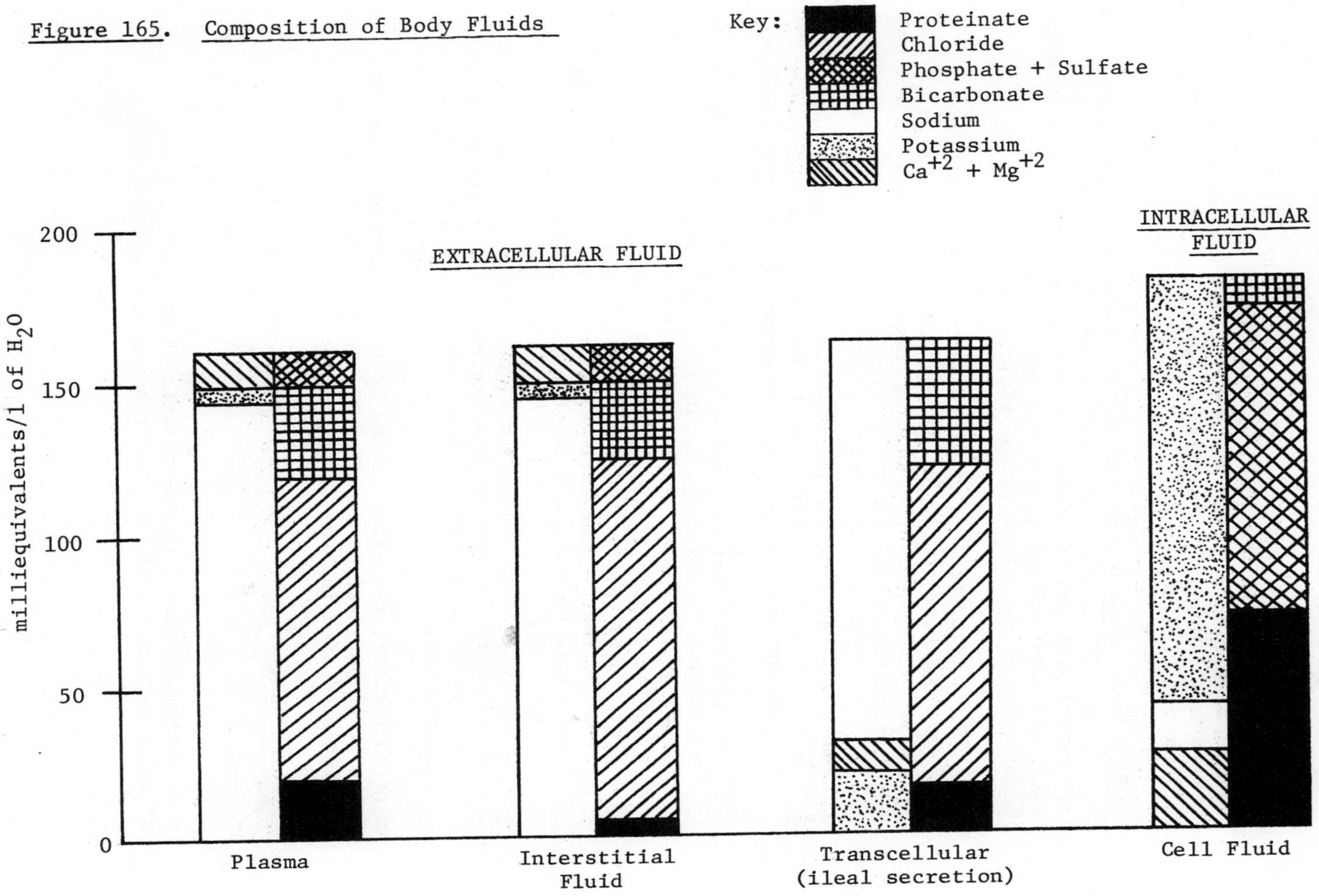

[1177]

a. <u>Intracellular fluid</u> comprises 30 to 40% of the body weight
    (about two-thirds of the total body water). Potassium is the
    predominant cation with magnesium being the next most common.
    The anions are mainly protein residues and organic phosphates,
    chloride and bicarbonate being low in intracellular fluid.

b. <u>Extracellular fluid</u> has sodium as the predominant cation and
    accounts for 20 to 25% of the body weight (one-third of the
    body water). It can be subdivided into four pools   (1) vascular
    fluid; (2) interstitial fluid; (3) transcellular fluid;
    (4) bone-dense connective tissue fluid.

    i)   The vascular pool is the circulating portion of the
         extracellular fluid. It is characterized by a high
         protein content which does not readily pass the
         endothelial membranes.

    ii) Interstitial fluid is the solution that actually surrounds
         the cells. It accounts for 18 to 20% of the body water.
         It exchanges with the vascular space via the lymph system.

    iii) The transcellular pool includes the fluid present in
         secretions of the body, including the digestive juice,
         intra-ocular fluid, cerebrospinal fluid (CSF) and
         synovial (joint) fluid. These solutions are actively
         secreted by specialized cells and can differ considerably
         from the rest of the extracellular fluid. Under normal
         conditions they rapidly recycle with the main pool.

    iv) Dense connective tissue (bone, cartilage) fluid exchanges
         slowly with the rest of the extracellular fluid so it is
         considered separately. It accounts for 15% of the total
         body water.

c. As noted earlier, the water in the body constitutes about 60%
    of the total body weight and thus the normal body water may
    be calculated by using the relationship:

    Normal body weight = 0.6 x normal body weight in Kg

                                              ...Equation 1

    In an abnormal situation, the existing body water can be
    estimated by knowing the subject's serum sodium value and
    weight in kg:

$$\text{Existing body water} = \frac{\text{ideal serum Na}^+ \text{ x normal body water}}{\text{measured serum Na}^+}$$

...Equation 2

Note in the above formula normal body water = 0.6 x normal body weight in kg and Na$^+$ values are expressed in mEq/L.

The body water deficit or excess can be calculated from values derived from equations 1 and 2; for example,

Body water deficit = normal body water (equation 1)
— existing body water (equation 2)

The body water deficit can also be calculated by using the following formula:

$$\text{Body water deficit} = \frac{\text{measured Na}^+ - \text{ideal Na}^+}{\text{ideal Na}^+} \text{ x } \begin{array}{l}\text{normal}\\\text{body water}\end{array}$$

In these calculations, a value of 138 mEq/L may be used for ideal Na$^+$.

3. Water Movements

Unlike ions, water is not actively secreted. Its movements are due to osmosis and filtration. In osmosis water will move to the area of highest solute concentration. Thus the active movement of salts into an area creates a concentration gradient down which water will flow following the salt movement. In filtration, hydrostatic pressure in the arterial blood moves water and nonprotein solutes through specialized membranes to give a protein free filtrate. The formation of the renal glomerular filtrate is such a process. Filtration also accounts for the transfer of water from the vascular space into the interstitial compartment. This fluid movement is opposed by the osmotic pressure (oncotic pressure) caused by the proteins present in the plasma (Starling's Law; see discussion on albumin, Chapter 5).

4. Ionic Pumps

Cells have the capacity to move ions (especially Na$^+$ and K$^+$) against a concentration gradient. The postulated "sodium pump" actively transports sodium across cellular membranes, usually from the interior to the exterior of the cell. It is an active process requiring metabolic energy and is stopped by cooling and by metabolic inhibitors. This carrier system may be associated with a magnesium-, sodium-, and potassium-dependent ATPase which is inhibited by ouabain (a digitalis derivative). This enzyme has been isolated from cellular membranes. The sodium pumping action is frequently coupled to the cellular uptake of potassium as shown below. As is indicated, it appears that two potassium ions are pumped into the cell for each three sodium ions secreted.

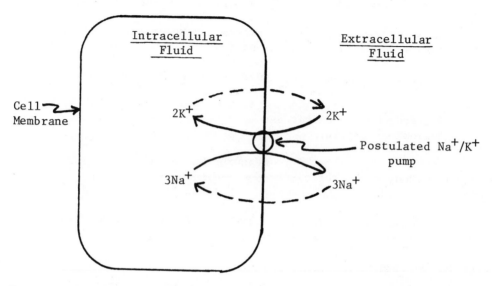

Note: <u>Passive</u> processes (diffusion) are represented by dashed lines while <u>active</u> (energy requiring) processes are shown by solid lines. Diffusion occurs <u>down</u> a concentration gradient (from more concentrated to more <u>dilute</u>); <u>active transport</u> moves ions <u>up</u> a concentration gradient.

This is the postulated mechanism by which the difference in cation concentrations between intra- and extracellular fluid is explained. Other postulated pumps include a K^+-H^+ exchange by cells while the renal tubular cells have an active H^+ pump which results in the formation of an acid urine.

Renal Physiology

As the kidney is the major organ regulating extracellular composition and volume, a brief explanation of renal physiology is needed to understand its function. Further details can be obtained from standard physiology texts.

1. Three main processes occur in the renal nephron. (1) Formation of a virtually protein-free ultrafiltrate at the glomerulus; (2) Active reabsorption (principally in the proximal tubule) of solutes from

the glomerular filtrate; (3) Active excretion of substances such as hydrogen ions into the tubular lumen, usually in the distal portion of the tubule (see Figure 166).

2. Under normal circumstances, the amount of sodium reabsorbed from the glomerular filtrate exceeds 99% of that originally filtered by the glomerulus. Most (65-85%) of this reabsorption occurs in the proximal tubule. It is thought that entry of sodium from the filtrate into the tubule cells is a passive process. The cation is then actively transported to intercellular channels which line the tubule. This produces a gradient and results in the passive diffusion of further salt and water from the tubular lumen. It has been calculated that one-third of the sodium is transported actively and the remainder is passively transported. In the loop of Henle and the distal tubule, the remaining sodium is absorbed by active processes, possibly coupled with H^+ and K^+ secretory mechanisms.

3. The amount of sodium delivered to the tubules varies with the glomerular filtration rate (GFR). When the filtered load of sodium is increased, the readsorption of sodium and water increases and when the load falls, the readsorption of sodium falls. This means that, in spite of GFR variation, the excretion of sodium does not change. This phenomenon is called glomerulotubular balance and implies that the nephron responds to a varying sodium load. This is more important in animals like the dog where the GFR is variable, usually rising after a meal due to the increased urea load.

4. K^+ ion is thought to be totally reabsorbed in the proximal tubule. The potassium appearing in the urine is due to additional secretion in the distal tubule.

5. Reabsorption of Water

   a. The normal GFR is 100-120 ml/min which means that about 150 liters of fluid pass through the renal tubules each day. Ninety-nine percent of this is reabsorbed, giving an average daily urine volume of 1-1.5 liters. Approximately 80% of the water is reabsorbed in the proximal tubule, a consequence of the active absorption of solutes in this region. The filtrate is still isoosmotic with the blood, however, and water reabsorption in the rest of the tubule can be varied according to the water balance of the individual as opposed to the obligatory water reabsorption in the proximal tubule.

   b. The facultative absorption of water depends on the loop of Henle establishing an osmotic gradient by the secretion of sodium ions from the ascending and the uptake of Na^+ by the descending loop. As a result, the tip of the loop is

Figure 166. A Summary of Renal Physiology As It Relates to Electrolytes
(Na+ and K+) and Water

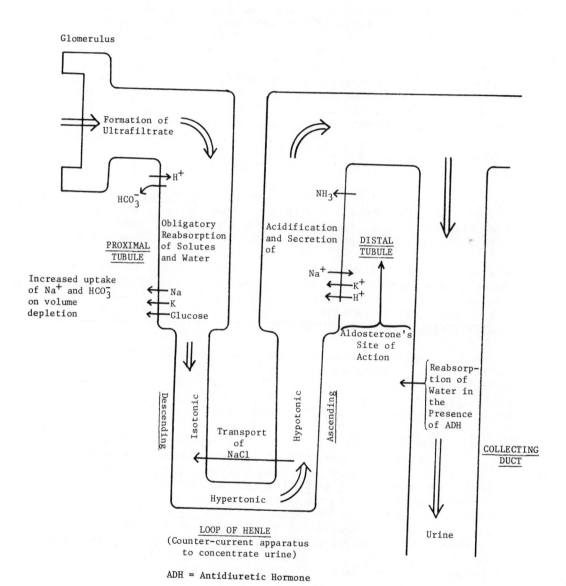

LOOP OF HENLE
(Counter-current apparatus
to concentrate urine)

ADH = Antidiuretic Hormone

hyperosmotic (1200 mOsmol) and the distal end is hypoosmotic
with respect to blood. The collecting ducts run through the
hyperosmotic region. In the absence of antidiuretic hormone
(ADH; vasopressin), the membranes of the cells comprising
the ducts are relatively impermeable to water. They become
permeable in the presence of ADH, water is withdrawn, and
the urine becomes hyperosmotic with respect to the blood.

Homeostatic Controls

1. The composition and volume of extracellular fluid is regulated by
    a number of complex hormonal and nervous mechanisms which are
    only partially understood. They can be divided into three groups
    although there is interaction among all three: (1) osmolarity;
    (2) volume control; (3) pH control.

2. The osmolarity of the body fluid is kept within narrow limits
    (285-295 mOsmol/kg) through regulation of water intake (via a
    thirst center?) and water excretion by the kidney under the
    influence of an antidiuretic hormone. The osmolarity of
    extracellular fluid is mainly due to sodium ion and its
    accompanying anions.

3. The volume of extracellular fluid is kept relatively constant,
    provided an individual's weight remains constant to within $\pm$ 1 kg.
    Volume receptors measure the effective circulating blood volume
    and a decreased blood volume leads to stimulation of the renin-
    angiotensin-aldosterone system with retention of sodium ions
    (discussed later). The increased sodium ion levels lead to rises
    in osmolarity and stimulation of ADH, resulting in increased
    water retention. An antagonistic system has been postulated but
    no definitive evidence has been accumulated. This system would
    result in increased Na^+ excretion when extracellular fluid volume
    was increased.

4. The pH of extracellular fluid is kept within very narrow units
    (7.35-7.45) by means of buffering mechanisms and regulation by
    lungs and kidneys. The buffers in the blood were discussed in
    Chapters 1 and 7.

5. The three systems do not act independently. For example, acute
    blood loss results in the release of ADH and aldosterone to restore
    blood volume and renal regulation of the pH leads to ionic shifts
    in potassium and sodium.

Water and Osmolarity Control

1. In spite of considerable variations in fluïd intake, an individual
    maintains water balance and a constant composition of body fluids.
    The homeostatic regulation of water is described in this section.
    Figure 167 summarizes the factors which will be covered and their
    interactions.

Figure 167. Regulation of Osmolarity Within the Body

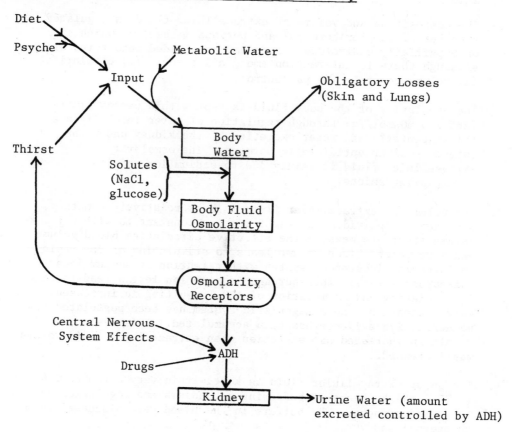

a. Sources of Water

    i) Two to four liters of water a day are consumed in food and
        drink.

    ii) 300 ml of metabolic water are formed each day by
        oxidation of lipids and carbohydrates (see Chapter 4).

b. Loss of Water

   i) Perspiration ⎱ 1 liter/day; increased by fever or
   ii) Lungs   ⎰ hot climatic conditions.

   iii) Gastro-intestinal tract. Although a large volume of fluid
        is secreted during digestion, most of the water is
        reabsorbed and losses in the feces are small. However,
        disturbances in the gastro-intestinal tract can lead to a
        considerable loss of salts and water with profound
        effects on the homeostasis of these materials (see later).

   iv) One to two liters of water are excreted each day in the
        urine. This volume can be regulated by the body to
        maintain homeostasis.

2. The water balance of the body is regulated to maintain constant
    osmolarity of the body fluids.

   a. If two solutions of unequal concentration are separated by a
       semi-permeable membrane, solvent (and any solute particles small
       enough to pass through the membrane) will move through the
       membrane, by osmosis, from the low-concentration side into the
       solution of high concentration. Osmotic pressure is defined as
       the hydrostatic pressure which must be applied (to the solution
       of higher concentration) in order to prevent this solvent flow.
       When one speaks of the osmotic pressure of a solution, it is
       the value relative to pure water (i.e., the pressure differential
       across the membrane when the solution is on one side and water
       is on the other).

   b. Osmotic pressure is directly related to the number of particles
       present per unit weight of solvent. Thus it is a colligative
       property, as are boiling point elevation (and vapor pressure)
       and freezing point depression. A solution containing one mole
       of particles in 22.4 kg of water (22.4 l at 4°C) exerts an
       osmotic pressure of one atmosphere and is said to have an
       osmolality of 0.0446. Conversely, the osmotic pressure of a
       one osmolal solution (one mole of particles in one kilogram
       of water) is 22.4 atmospheres. In this sense, "number of
       particles" is roughly defined as the quantity of non-interacting
       molecular or ionic groups which are present. Since glucose
       does not readily dissociate, one mole of this compound
       dissolved in one kilogram of water (a one molal solution)
       produces one mole of "particles" and the solution has an
       osmolality of one. Sodium chloride, on the other hand,
       dissociates completely in water, forming two particles from
       each molecule of NaCl. Thus a one molal solution of NaCl is a
       two osmolal solution. A one molal solution of Na_2SO_4 or $(NH_4)_2SO_4$

is a three osmolal solution, and so on. The number of moles of
a substance which, when added to one kg of water, produces a
one osmolal solution, is referred to as an <u>osmol</u>. One osmol of
glucose is one mole, while one osmol of NaCl is 0.5 moles and
one osmol of Na_2SO_4 is 0.333 moles. In practice, a milliosmol
(mOsmol), equal to one-one thousandth of an osmol, is the unit
usually used.

c.   When working with aqueous solutions, <u>osmolarity</u> is sometimes used
     interchangeably with (or instead of) <u>osmolality</u>. Although this
     is not strictly correct (since one is moles of particles per
     <u>liter</u> of <u>solution</u> and the other is moles of particles per
     <u>kilogram</u> of <u>solvent</u>), in water at temperatures of biological
     interest the error is fairly small unless solute concentrations
     are high (i.e., when an appreciable fraction of the solution is
     not water). Thus, with urine the approximation is acceptable
     while with serum it is not, because of the large amount of
     protein present. Although osmolarity is more readily measured,
     it is temperature dependent, unlike osmolality.

d.   Osmolality is commonly measured by freezing point or vapor
     pressure depression.

     i)   In terms of vapor pressure, (P^v), the osmotic pressure (Π)
          is defined as

$$\Pi = P^v_{\text{pure solvent}} - P^v_{\text{solution}}$$

          As mentioned before,

$$\text{Osmolality} = \Pi/22.4$$

          where the osmotic pressure is in atmospheres. In one
          commercially available instrument, the solution and solvent
          vapor pressures are measured by using sensitive thermistors
          to detect the difference in temperature decrease caused by
          the evaporation of solvent from a drop of pure solvent and
          a drop of solution. Because the rate of evaporation
          (vapor pressure) of the solution is lower, the temperature
          change will be less and the vapor pressure difference can
          be calculated. This change in temperature due to a
          difference in vapor pressure is related to the phenomenon
          of boiling point elevation.

     ii) The freezing point of a solution is always lower than that
          of the pure solvent. The exact value of the depression
          depends on the solvent and the osmolality of the solution.
          For water,

$$\text{Osmolality} = \frac{\Delta T}{1.86}$$

where ΔT is the freezing point depression in degrees centigrade. Instruments which measure the freezing of a sample are used in clinical laboratories to determine serum and urine osmolality.

e. Since water passes freely through most biological membranes, all fluids in the body are in osmotic equilibrium. The osmolarity of a plasma sample is thus roughly representative of the osmolarity of the other body fluids.

i) The osmotic pressure of extracellular fluid is due primarily to Na^+ and its accompanying anions, Cl^- and HCO_3^-. Since Na^+ is the principal osmotically-active cation, doubling the sodium concentration provides a good estimation of serum osmolality. Thus, the normal range of plasma Na^+ values is 135–145 mEq/l (about 3.1–3.3 g/l) and the normal plasma osmolality is about 270–290 mOsmol/kg (corresponding to an osmotic pressure of 6.8–7.3 atmospheres and a freezing point depression of 0.50–0.54°C).

ii) Although glucose is present to the extent of one gram/liter it does not dissociate and so provides only 5–6 mOsmols of particles per kg (contributing about 0.1 atmospheres to the osmotic pressure). For similar reasons, although there are 60–80 grams of protein per liter of plasma, the contributions of these molecules to the osmolarity is small (about 10.8 mOsmols/kg). Because these proteins are large and generally unable to pass through biological membranes, they are important in determining fluid balance between the intravascular and extravascular spaces. Hence, that portion of the osmotic pressure due to proteins is often referred to as the oncotic pressure.

iii) Since many of the molecules in plasma interact with each other, plasma is not an ideal solution. Consequently, the measured osmolarity of a plasma sample is an effective osmolality and is lower than that which would be obtained by adding up the concentrations of all ions and molecules present. A solution which has the same effective osmolality as plasma is said to be isotonic. Examples include 0.9% saline, 5% glucose, and the more complex Ringer's and Locke's solutions. It should be noted that, if a solute is able to permeate a membrane freely, then a solution of that solute will behave as if it were pure water, with respect to the membrane. Thus, a urea solution will cause red cells to swell and burst as does pure water, because urea moves freely across erythrocyte membranes.

iv) The osmolality of urine can differ markedly from that of plasma, as pointed out earlier, because of active (energy-requiring) concentrative processes going on in the renal tubules. Also, the membranes of the collecting ducts show a varying degree of water permeability (see later), permitting the removal of certain solutes from the urine without simultaneous loss of water.

v) As noted above, the serum osmolality can be estimated from serum Na^+, glucose and blood urea nitrogen (BUN) concentrations:

$$\text{Estimated osmolality} = 1.86(Na^+ \text{ mEq/L})$$

$$+ \frac{\text{Glucose in mg/100 ml}}{20}$$

$$+ \frac{\text{BUN mg/100 ml}}{3}$$

Note that the concentrations of glucose and BUN, expressed in milligrams per 100 ml, are converted to moles with the appropriate conversion factors. The estimated osmolality is usually 5 to 8 mOsm less than the osmolality as determined by colligative measurements (discussed earlier). However, if the latter value is much greater than the estimated (or calculated) value, it indicates that there are molecules other than Na^+, glucose and BUN accounting for the appreciable elevation of serum osmolality. Such discrepancies are seen in subjects with drug toxicity (alcohol, barbiturates and slicylates), acute poisoning due to unknown toxic substances, liver failure, etc. Thus, the determination of osmolality by both methods is useful in evaluating drug intoxication. In these instances a prompt hemodialysis may be required. It has been reported that in these situations, if the measured value exceeds the calculated value by more than 40 mOsm/Kg, the prognosis is poor. In general, the determination of osmolality is helpful in the management of patients with a wide range of fluid and electrolyte disorders, namely chronic renal disease, non-ketotic diabetic coma, hypo- and hypernatremia, hyperglycemia, hyperlipemia, burns, after major surgery or severe trauma (particularly following serious head injuries), hemodialysis, diabetes insipidus, etc.

3. Changes in osmolality on an order of 2% or more are detected by
    osmoreceptors, probably located in the hypothalamic region (see
    Chapter 13). These receptors have properties similar to a
    semipermeable membrane which permits passage of urea but restricts
    Na^+ and glucose. Thus, changes in the plasma levels of Na^+ and
    glucose trigger the receptors while urea has no effect.

    a. The physiological responses to an <u>increase</u> in osmolality are
        a sensation of thirst and restriction of urinary water loss
        (production of hypertonic urine). Under conditions of fluid
        restriction, urine osmolality can rise to between 800 and
        1200 mOsmol/kg (normal is 390-1090 mOsmol/kg), three to four
        times the plasma levels of 270-290 mOsmol/kg. A <u>decrease</u> in
        plasma osmolarity (as in excessive water intake) can cause
        the excretion of hypotonic urine with an osmolality as low
        as 40 to 50 mOsmol/kg. Other water losses, from the skin
        and lungs, are not subject to any controls of this sort.

    b. The biochemical response to hypertonicity of the body fluids
        is the release of antidiuretic hormone (ADH; also called
        vasopressin) from the neurohypophysis. As indicated, this
        is mediated by the osmoreceptors via neuronal impulses. The
        synthesis and storage of this hormone is discussed in
        Chapter 13. In brief, ADH is synthesized by neurons located
        principally in the supraopticoneurohypophyseal nucleus and
        transported down their axons to the posterior pituitary. The
        principal physiologic effect of ADH is the retention of water
        by the kidney.

        i) ADH is a nonapeptide having (in man and most mammals)
            the sequence shown below.

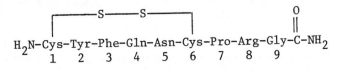

Arginine Vasopressin

            The "arginine" in the name refers to the amino acid at
            position 8. Lysine vasopressin, with a lysine residue
            in place of arginine, is the ADH in pigs and hippopotami.
            Vasopressin closely resembles oxytocin, another
            neurohypophyseal hormone (see Chapter 13). Both are
            nonapeptides and have some overlap in their activities.
            ADH also has vasoconstrictive activity (as the synonym
            vasopressin implies) but this and its weak oxytocic
            action are probably not physiologically important.

        ii) ADH is released into the bloodstream where it is

transported apparently not bound to any proteins. Being small enough to be filtered by the glomerulus, it is rapidly cleared by the kidney and excreted in the urine. It is also destroyed in the liver. One milligram of vasopressin is defined as having 450 units of biological activity. Under conditions of normal water intake (no restrictions) 11 to 30 milliunits (about 70 ng) of ADH are secreted each day. When water intake is restricted, this secretion rate can double.

iii) The site of action of ADH is believed to be the renal collecting ducts which become permeable to water in the presence of ADH. A partial interchange (equilibration) can then occur between the urine and hypertonic regions of the nephron causing the urine to become hypertonic relative to the plasma. The action of ADH is mediated by cAMP and the urinary levels of cAMP are increased in the presence of ADH. The disulfide ring (cystine group) is necessary for action and may be involved in the binding of the hormone to the receptor site.

c. Although the primary stimulus for ADH secretion is a rise in the osmolarity of the blood perfusing the receptors, other factors may also be involved.

i) In acute hemorrhage where there is an abrupt drop in extracellular fluid volume, there is apparent coordination between the two systems controlling blood volume and osmolarity. Under these circumstances, ADH is released to increase the volume even at the expense of a drop in osmolarity. Central nervous stimuli are probably involved in this reaction.

ii) In some situations, ADH is released even in the presence of a water overload and a decline in plasma sodium and osmolarity. This is called inappropriate ADH secretion (IADH) and results in the formation of a hypertonic urine. It can be caused by fear and pain through cortical influence on the hypothalamus. In addition ADH is synthesized and secreted from diseased malignant and nonmalignant lung tissue, carcinomas of the pancreas and duodenum, lymphosarcoma, reticulum cell sarcoma, Hodgkin's disease and thymoma. The IADH secretion leads to hyponatremia and water retention. Various drugs such as morphine and barbiturates cause release of ADH from the pituitary. It is of interest to point out that narcotic antagonists have been useful in the treatment of patients with IADH secretion of central (nervous system) origin. It has been speculated that the narcotic antagonists bring about this action by inhibiting the endogenous opiates, known as endorphins (see Chapter 13) found in the central nervous system. Ethanol inhibits the ADH release.

4. Man's intake of water (thirst) is predominantly influenced by psychic and cultural influences. Hyperosmolarity will cause thirst but even under usual conditions man consumes more water than required. In certain situations, however, abnormal water loss can occur, resulting in pathological conditions.

   a. Excretion of water alone will be accompanied by an increase in osmolarity and a contraction of both the extracellular and intracellular fluid compartments. Cations are usually lost along with water but this depletion may be more restricted to the extracellular compartment. Unless corrected by water and cation replacement, this loss can lead to circulatory collapse, failure of body heat control, and death.

   b. Several factors can increase the amount of urine formed (polyuria), resulting in elevated water excretion by this route.

      i) In diabetes insipidus the renal tubules fail to concentrate the urine and recover the water from the glomerular filtrate. This can be due to defective ADH receptors in the nephrons (nephrogenic diabetes insipidus; unresponsiveness of the receptors to ADH); or to diminished or absent ADH secretion. In patients of diabetes insipidus due to deficiency or lack of ADH, the hormone replacement therapy is needed. 8-Lysine vasopressin is such a compound. A recent synthetic analogue of vasopressin (1-deamino-8-D-arginine vasopressin) which has very little pressure activity has been found to be very effective. These compounds may be administered as a nasal spray or subcutaneously (note oral administration is ineffective as with other peptide hormones).

      ii) In osmotic diuresis, as is seen in diabetes mellitus with severe glycosuria, a large solute load increases the osmolality of the glomerular filtrate and impairs the ability of the kidney to concentrate the urine.

   c. Excessive amounts of water may also leave the body during sweating, vomiting, and diarrhea. Some of the mechanisms by which enteric pathogens cause water and electrolyte losses from the intestines have recently been investigated.

i)   There are two "classes" of pathogens which can cause diarrhea. One group, exemplified by Shigella dysenteriae, must invade the epithelial cells of the intestinal mucosa for its toxins to work. The other type, which includes E. coli and Vibrio cholerae, grow in the intestinal lumen or on the surface of the epithelial cells. The toxin which they secrete is sufficient, by itself, to cause the disease.

ii)   Cholera toxin (from Vibrio cholerae) has been studied most extensively. It has been obtained in crystalline form and shown to be a protein with a molecular weight of 84,000 daltons. It is heat- and acid-labile and is hydrolyzed by pronase but not trypsin.

iii) The water and electrolyte loss associated with cholera seems to result from the ability of the toxin to stimulate chloride secretion into and to inhibit sodium absorption from the intestinal lumen. This raises the osmolarity of the intestinal contents and causes diarrhea. The mechanism by which cholera toxin accomplishes this is not completely understood but there seems to be an increase in the adenylate cyclase activity (and cAMP levels) of the epithelial cells of the entire small intestine (duodenum, jejunum and ileum) and a decrease in the activity of a Na^+-K^+ ATPase. The elevated cAMP synthesis may be due to the formation of additional molecules of adenylate cyclase. The rate of lipolysis in isolated fat cells, a process mediated by cAMP, is also stimulated by the toxin. Cycloheximide, which inhibits protein synthesis, inhibits the fluid loss from the intestine but does not block the effect of cholera toxin on fat cells. This somewhat obscures just how cholera toxin interacts with the cAMP system. Further, the possibility has been raised that cholera toxin has a phospholipase activity and that it performs its functions by altering the phospholipids in the cell membrane where the cyclase and the ATPase are located. Both cyclase and ATPase are known to require phospholipids for activity. There is also evidence that the membrane binding site for toxin may be a ganglioside. Much remains to be learned about this process.

iv)   Currently, the most reliable treatment for cholera is the parenteral and oral administration of solutions containing glucose and electrolytes. The latter fluid replacement therapy is based on the observation that the glucose- and amino acid-stimulated Na^+ and water absorption is not affected by cholera toxin. Antibiotics such as tetracycline and chloramphenicol, used in conjunction with this therapy, reduce the duration of the diarrhea and **hence the amount of fluid needed. Adenylate cyclase** inhibitors such as ethacrynic acid and cycloheximide have also been tried. Although ethacrynic acid has been successful in relieving diarrhea in dogs with experimentally induced cholera, this approach is not yet ready for use in humans.

v)   As indicated above, some strains of E. coli produce an enterotoxin which is immunologically related to cholera toxin and which causes symptoms similar to those of cholera. Enterotoxins of Clostridium perfringens (an organism involved in food infection) has been implicated in the production of a mild diarrhea of short duration in the disease produced by this organism. The toxin has been **shown to be heat-labile and stimulates adenylate cyclase** activity in the frog erythrocyte membrane.

d. Water excess gives rise to vague cerebral symptoms of headaches, malaise, and, in more severe cases, muscular weakness and cramps. It is associated with low plasma sodium and hypo-osmolarity. Abnormal water intake may be iatrogenic (e.g., administration of excessive I.V. fluids) or may stem from psychological causes. Inappropriate ADH secretion, as is sometimes seen due to postoperative pain, causes increased water retention. Ectopic ADH production, as in bronchiogenic carcinoma, has a similar effect.

Extracellular Volume Control

1. The volume of extracellular fluid (ECF) in a normal adult is kept surprisingly constant. This can be seen from the observation that a person's body weight does not vary by more than a pound per day despite fluctuations in food and fluid intake. Decreases in extracellular fluid result in a lowering of the effective blood volume, and a compromised circulatory system. An increase in extracellular fluid may lead to hypertension, edema, or both. The homeostatic mechanism for volume control centers upon the renal regulation of the sodium balance. When the ECF volume decreases, less sodium is excreted, while increased ECF volume increases sodium loss. Increased sodium retention leads to an expansion of the extracellular space. This is because sodium is confined to this region and the regulation of extracellular osmolarity (discussed earlier) causes increased water retention in order to keep the sodium concentration constant. Sodium movement in the kidneys is under hormonal influence from the aldosterone-angiotensin-renin system and the postulated natriuretic hormone. These controls are summarized in Figure 168.

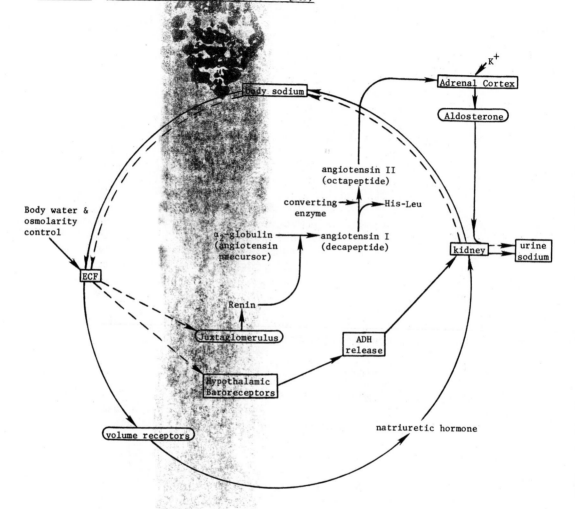

Figure 168. Regulation of Extracellular Fluid (ECF)

1194

2. Volume receptors which perceive changes in extracellular volume and then transduce this information into hormonal changes for ECF regulation must exist, but their nature is controversial. They are probably stretch receptors which monitor the vascular compartment and respond to the degree of filling of some area of the cardiovascular system. Increases in effective blood volume result in increased sodium excretion as is seen after assumed supinity or exposure to low temperatures. Both conditions cause peripheral vascular constriction which increases the effective blood volume because there is less pooling of blood in various part of the circulation. Decreases in effective blood volume lead to sodium retention. This occurs in hemorrhage or prolonged standing with pooling of blood in the limbs.

3. <u>Aldosterone</u> is the principal mineral corticoid known. It is the hormone which produces increased sodium retention and expansion of the extravascular space. It is formed in the zona glomerulosa of the adrenal cortex. The structure and biosynthesis of aldosterone is given in Chapter 13.

   a. There is no specific binding protein for aldosterone in the plasma and it circulates bound non-specifically to albumin with a certain amount "free" in solution. Plasma levels are 30–130 ng/1 with a secretion rate of ____ per 24 hours. The hormone is rapidly inactivated by both the liver and kidney, with 80–90% of it being removed in one pass through the liver. The hormone is reduced to the dihydro derivative, then conjugated with glucuronic acid and excreted in the urine. 10% of the hormone is removed by the kidney as the 18 oxyglucuronide. A small percentage (0.1–0.5%) of aldosterone is excreted unchanged, with the rest occurring as inactive products.

   b. Aldosterone stimulates the sodium pump causing the retention of sodium ions and an increased excretion of potassium or hydrogen ions. This could be due to a coupled exchange of ions via a common carrier or may be secondary to the increased sodium uptake (i.e., a passive H^+-K^+ loss in response to Na^+ uptake). The principal site of action is the distal renal tubule, and aldosterone has no effect on sodium absorption by the proximal tubules. Other areas which have an active sodium pump that is sensitive to aldosterone are the gastrointestinal tract, sweat glands, and muscle. All show a potassium loss and net sodium gain under the influence of aldosterone.

   c. Aldosterone acts by stimulating the production of messenger RNA, thereby increasing the production of specific proteins. The exact nature of these proteins is unknown. A sodium permease has been proposed as one candidate, but the only definite

increase noticed has been in citrate synthetase. It is possible
that the limiting factor in the activity of the sodium pump is
the availability of energy-providing substrates. Then increased
levels of enzymes, such as citrate synthetase, that provide
these substrates will have the effect of stimulating sodium
transport.

d. Several factors cause the release of aldosterone from the
adrenal cortex.

  i) Angiotensin II is a potent stimulator and is the main
  physiological influence.

  ii) Elevated plasma potassium will cause aldosterone levels
  to rise. This is part of the homeostatic regulation of
  potassium levels.

  iii) ACTH plays a "permissive" role. It is necessary for
  the effect of angiotensin or potassium ions.
  Pharmacological doses cause a release of aldosterone but
  its physiological role may be slight.

  iv) The kidney produces several prostaglandins which have a
  hypotensive and natriuretic effect. One PGA causes the
  release of aldosterone independently of angiotensin.

e. Two controlling steps in the biosynthesis of aldosterone exist
(see Chapter 13).

  i) The conversion of cholesterol to pregnenolone is stimulated
  by ACTH, angiotensin, and K^+. This step is common to
  the formation of all steroid hormones.

  ii) The last step in the conversion of corticosterone to
  aldosterone, catalyzed by an 18-ol-dehydrogenase also
  appears to be a control point, although little is known
  about it. The enzyme is found only in the zona
  glomerulosa. Since this 18-aldehyde formation is unique
  to aldosterone synthesis, control at this point is more
  specific in regulating this process.

4. Renin is a proteolytic enzyme secreted by the juxtaglomerular
apparatus of the kidney. The substrate for renin is an alpha-2
globulin, present in plasma, called angiotensinogen.
Angiotensinogen is synthesized in the liver and its production
is enhanced by glucocorticoids, angiotensin II, and estrogens.
The increased levels of angiotensinogen have also been observed
after nephrectomy, ureteral ligation, hemodilution and hypoxia.
Renin splits off a decapeptide (angiotensin I) from

angiotensinogen. Angiotensin I is converted to angiotensin II by loss of a terminal histidylleucine group, leaving an octapeptide. This reaction is catalyzed by a "converting enzyme" present in the pulmonary vascular bed. Converting enzyme is also found in the blood, kidneys, and other tissues. Angiotensin II is the physiological active form of the hormone, having a much greater activity than angiotensin I, its precursor. Recent studies have shown that the haptapeptide (des-Asp^1-angiotensin II or angiotensin III, see Figure 169) derived from angiotensin II by the action of aminopeptidase, is biologically active and may mediate some of the actions of angiotensin II (e.g., aldosterone stimulation). Renin has also been found in non-renal tissue such as the brain, submaxillary glands and uterus. Angiotensin has been found in amniotic fluid, presumably formed by uterine renin. The function of renin in these tissues is unknown. The formation and structure of angiotensins are shown in Figure 169.

a. The release of renin is caused by:

  i) A <u>decrease in renal perfusion pressure</u> is a measure of the effective blood volume. Renin levels <u>increase</u> in prolonged standing (where pooling of blood in the extremities occurs) and <u>decrease</u> upon lying down. The juxtaglomerular apparatus acts as a local volume receptor.

  ii) A <u>decrease in sodium transport across the tubular membrane</u> (decrease in tubular sodium) also stimulates renin release, a factor which is of obvious importance in the regulation of sodium balance* A fall in the filtered load of sodium reaching the tubule results in an increase in aldosterone and an increase in sodium retention.

  iii) <u>Increased adrenergic nervous activity</u>, independent of the other two factors, causes renin release. It is apparently mediated by cAMP.

  iv) The release of renin is <u>inhibited</u> by angiotensin II, thus completing the feedback loop. ADH also inhibits renin release.

b. Angiotensin II has several important actions. It participates in the regulation of blood pressure and of the volume and composition of the extracellular fluid. The physiological actions of angiotensin II include arteriolar constriction

---

*The macula densa appears to be the receptor in this process.

Figure 169. The Formation, Degradation and Structures of Angiotensins

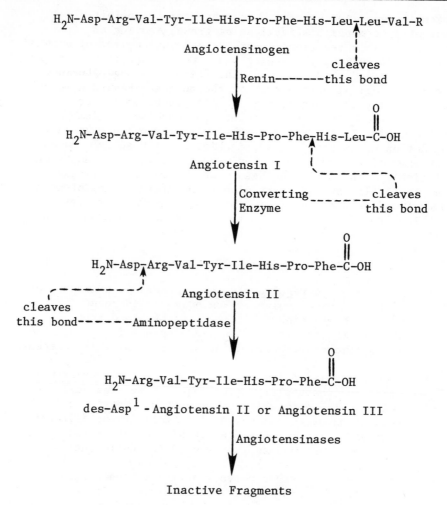

Inhibitors of Renin-Angiotensin System

1. Structural analogue of angiotensin II. An example of this class of inhibitors is saralasin (sarcosine-1, alanine-8, angiotensin II). It reduces blood pressure in hypertensive patients with elevated plasma renin activity and has been useful diagnostically for identifying angiotensin-dependent hypertension.

2. Inhibitors of angiotensin converting enzyme: Teprotide (a nonapeptide) and D-3-mercapto-2-methylpropanoyl-L-proline (an orally active compound). Note the converting enzyme is identical to kininase II that degrades bradykinin (see Chapter 11) and thus inhibition of this enzyme may lead to accumulation of vasodilator bradykinin.

(most potent vasoconstrictor known; pressor activity diminished in sodium depleted individuals and in patients with cirrhosis), stimulation of the secretion of aldosterone (which affects water and electrolyte balance), increased secretion of catecholamines from the adrenals, increased blood pressure by directly acting on the brain, stimulation of thirst (leading to increased water intake), and the stimulation of ACTH and ADH.

5. Third factor or <u>natriuretic hormone</u> has not been isolated but evidence has accumulated to substantiate its existence. The hormone antagonizes aldosterone and mediates increased excretion of sodium ions. It may act on the proximal tubule portion of the nephron. The stimulus for the hormone's release is an expansion of the extracellular volume. Natriuretic hormone is possibly a polypeptide and it may be identical to melanocyte stimulating hormone from the pituitary.

6. A <u>low extracellular fluid volume</u> (hypovolemia) leads to a decrease in the effective blood volume. This is best detected by measuring <u>postural hypotension</u>. A fall in blood pressure of over 10 torr in moving from a supine to sitting position indicates hypovolemia. The commonest cause of low ECF is the loss of sodium-containing body fluids from the skin by sweating or from the gastro-intestinal tract through vomiting or diarrhea. There are frequently attendant acid-base disorders due to losses of hydrogen or bicarbonate ions at the same time. Deficiency of sodium can also result from increased renal losses as in cases of renal disease or adrenal insufficiency. Treatment consists of the replacement of sodium ion, correction of accompanying potassium deficiency, and correction of the acid-base disorder.

7. An <u>increase in extracellular fluid volume</u> results first in a weight gain followed by edema. There are a number of causes of edema (see albumin synthesis, Chapter 5).

    a. Hypoalbuminemia (nephrosis and cirrhosis); low plasma albumin results in a lower oncotic pressure and an increase in fluid transfer to the interstitial space from the vascular system.

    b. Interference with lymphatic drainage;

    c. Increased salt retention (as in cardiac failure and hyperactivity of the adrenal gland).

Treatment consists in eliminating (where possible) the cause of the edema. Excess fluid can be removed by diuretics which interfere with the reabsorption of sodium by the kidneys. There is increased loss of potassium in diuretic therapy which can cause a deficiency of this ion.

Some forms of hypertension are associated increased aldosterone levels. It has even been suggested that as many as one quarter of all hypertensive patients suffer from hyperaldosteronism. Primary aldosteronism results in high blood pressure, sodium retention, and low renin levels. This can be ameliorated by spironolactone, an aldosterone antagonist.

Sodium

The average sodium content of the human body is 60 mEq/kg of which 50% is in extracellular fluids, 40% is in the bone, and 10% is intracellular. The chief dietary source of sodium is the salt added in cooking. Excess sodium is largely excreted in the urine although some sodium is lost in sweat. Gastrointestinal losses are small except in diarrhea. The sodium deficit in a clinical situation can be estimated by using the following formula:

$$\text{Sodium deficit} = (\text{ideal serum Na}^+ - \text{measured serum Na}^+)$$

$$\text{x } 0.2 \text{ x body weight in kg}$$

In the above relationship, the term 0.2 x body weight in kg represents extracellular fluid volume which contains most of the Na^+. Most aspects of sodium metabolism have already been considered in earlier sections of this chapter.

Potassium

1. The average human body contains 40 mEq/kg of potassium, distributed mainly in the intracellular space. Potassium is required for carbohydrate metabolism and there is increased cellular uptake of potassium ions during glucose catabolism. Potassium is widely distributed in plant and animals foods. The human requirement is about 4 grams a day. Excess potassium is excreted in the urine, the process being regulated by aldosterone (see previously).

2. Plasma potassium plays a role in the irritability of excitable tissue. A high plasma potassium leads to electrocardiogram (EKG) abnormalities and possibly to cardiac arrhythmia. This is possibly due to a lowering of the membrane potential. Low potassium in the plasma increases the membrane potential, resulting in decreased irritability, EKG abnormalities of another sort, and muscle paralysis.

3. Hyperkalemia (potassium retention with elevated plasma potassium) can occur in renal disease and adrenal insufficiency, where the normal secretory mechanisms are impaired. Metabolic acidosis, in particular diabetic acidosis, causes K^+ to be released from the cells as does the catabolism of cellular protein which occurs in starvation or fever. Treatment consists of correction of the

acidosis and the promotion of cellular uptake of potassium by insulin administration. Insulin enhances glucose uptake and therefore glycolysis, a process which requires K^+. In severe cases, ion exchange resins are given orally to bind potassium in the intestinal secretions.

4. Hypokalemia (potassium deficiency) can be caused by the loss of gastrointestinal secretions which contain significant amounts of potassium and by excessive losses of potassium in the urine due to an elevated aldosterone level or diuretic therapy. Low potassium levels are usually associated with alkalosis (discussed later). Potassium deficit can be estimated by using the following relationship:

Potassium deficiency = (ideal potassium - measured potassium)

x 0.4 x body weight in kg

Note that the K^+ is primarily present in the intracellular fluid and its volume is about twice the extracellular fluid, hence the term 0.4 x body weight in kg.

Acid-Base Balance

1. The blood pH is usually regulated within very narrow limits. The normal pH range is 7.35-7.45 (corresponding to 35-45 nmoles of H^+/1) and values below 6.80 (160 nmoles of H^+/1) or above 7.70 (20 nmoles of H^+/1) are seldom compatible with life. At the same time, the body generates a large quantity of acid as a by-product of metabolism. Fourteen thousand milliequivalents of CO_2 are removed by the lungs each day. On a diet of 1-2 g/Kg of protein per day, the kidneys remove 40-70 meq of acid each day. This is in the form of sulfate (from the oxidation of sulfur-containing amino acids), phosphate (from phospholipid and phosphoamino acid catabolism), and organic acids (e.g., lactic, β-hydroxybutyric and acetoacetic). The organic acids are produced by incomplete oxidation of carbohydrate and fats. Under some conditions (e.g., ketosis; see Chapter 8), considerable amounts of these materials may form.

2. The major buffer systems of the blood were discussed previously (see Chapter 1). The most important <u>extracellular</u> buffer is the carbonic acids-bicarbonate system, shown below.

$$CO_2 + H_2O \rightleftharpoons H_2CO_3 \rightleftharpoons HCO_3^- + H^+$$

As discussed in Chapter 1, at the normal blood pH of 7.4, the ratio of HCO_3^- to H_2CO_3 is 20 to 1. Thus the system is not at its maximum buffering capacity but it is capable of neutralizing a large amount

of acid. The carbonic acid-bicarbonate system is independently
regulated by two organs. The kidneys control the HCO_3^- levels in
the blood while the lungs (via the respiratory rate) regulate the
P_{CO_2} (partial pressure of CO_2). Other buffer systems (proteins,
phosphates) also operate in the plasma and erythrocytes.
Proteins are especially important in the buffering of
the intracellular fluid. Lastly, the hydroxyapatite of the bone also
acts as a buffer.

a.   The respiratory center in the medulla responds to the pH of the
blood perfusing it and is the source of pulmonary control. P_{CO_2}
and perhaps P_{O_2} also influence the center along with nervous
impulses from higher centers of the brain. A fall in pH results
in an increased rate and deeper breathing with a consequent
increase in the respiratory exchange of gases and a lowering
of the P_{CO_2}. The lower P_{CO_2} (and hence H_2CO_3) brings about an
elevation of the pH. Similarly, a fall in respiratory rate
leads to an accumulation of CO_2, a rise in P_{CO_2}, and a fall in
pH. The pulmonary responses to pH fluctuations are rapid while
the renal compensatory mechanisms (below) are slower to develop.

b.   The kidneys actively secrete hydrogen ions derived from the
dissociation of H_2CO_3, into the urine. The formation of carbonic
acid is a slow reaction whose rate is increased by carbonic
anhydrase.

$$H_2O + CO_2 \underset{(slow)}{\overset{\text{carbonic anhydrase}}{\rightleftharpoons}} H_2CO_3 \underset{(fast)}{\rightleftharpoons} H^+ + HCO_3^-$$

This is a zinc-containing enzyme found in renal tubular cells
and erythrocytes. The dissociation of carbonic acid is rapid
with no need for catalysis. The H^+ ion released is available
for secretion into the lumen of the nephron while the bicarbonate
ion passively diffuses into the blood stream. Thus, every H^+
secreted into the urine is associated with production of a
bicarbonate ion in the extracellular fluid.

c.   The hydrogen ion secretion is associated with three processes
in the nephron: bicarbonate reabsorption, acidification of
the urine, and ammonium ion production.

i)   The proximal tubule is responsible for the major reabsorption
of the 4500 mEq of bicarbonate filtered through the
glomerulus each day. Hydrogen ion secreted into the tubule
lumen combines with HCO_3^- to form CO_2 and water. The CO_2
diffuses into the tubular cells where it is rehydrated to
H_2CO_3 by carbonic anhydrase. It then dissociates to
bicarbonate and H^+ again. The HCO_3^- diffuses into the

blood stream, resulting in net reabsorption of bicarbonate (see Figure 170). This proximal tubular mechanism becomes saturated at approximately 26 mEq/1. At higher levels HCO_3^- appears in the distal tubule and can be excreted in the urine. Under conditions of elevated P_{CO_2}, H^+ secretion is more active, possibly due to intracellular acidosis.

In proximal renal tubular acidosis, bicarbonate reabsorption is impaired and saturation occurs at a lower value (about 16-18 mEq/1). The plasma bicarbonate is low while the urine pH is frequently high due to the presence of bicarbonate.

ii) The distal tubule cells can actively secrete hydrogen ions even against a concentration gradient until the urine attains a pH of 4.4. Beyond this point, any additional hydrogen ions that are to be excreted must be buffered as the tubule cells cannot work against a lower pH. This buffering comes from phosphate ions and ammonia. These two processes, described below, are illustrated in Figure 171.

Any phosphate ions in the glomerular filtrate that are not reabsorbed are converted to the dihydrogen form. The amount of $H_2PO_4^-$ present (termed the <u>titratable acidity</u>) can be measured by titrating the urine to pH 7 (pK_2 for H_3PO_4 = 6.15). Normal values for titratable acidity are 16-60 mEq /24 hours, depending on the phosphate load.

Phosphate is a valuable body resource, however, and is mostly reabsorbed, so additional buffering is provided by ammonia. NH_3 readily diffuses into the lumen of the nephron where the addition of a hydrogen ion to it forms an ammonium ion. Because of the positive charge on NH_4^+, it is no longer able to freely pass through the membranes and remains "trapped" in the urine. One hydrogen ion is absorbed for each ammonia molecule produced. The source of the ammonia is the hydrolysis of glutamine catalyzed by <u>glutaminase</u>, a specific renal enzyme. The reaction is shown below.

$$H_2O + glutamine \xrightarrow{\text{glutaminase}} glutamic\ acid + NH_3$$

$$NH_3 + H^+ \longrightarrow NH_4^+$$

This enzyme is induced under acidotic conditions and is responsible for the increased capacity of the kidney to remove hydrogen ions in renal compensation. On a normal protein diet, 0.2 to 1.0 gm of ammonia are excreted each day by this route.

(Text continued on p. 1206)

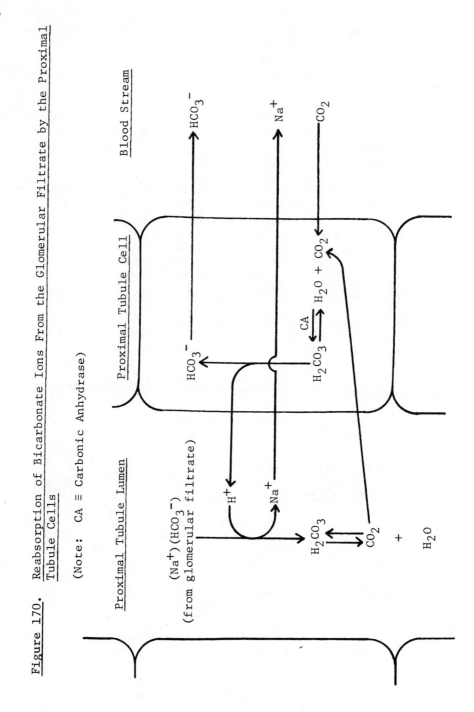

Figure 170. Reabsorption of Bicarbonate Ions From the Glomerular Filtrate by the Proximal Tubule Cells

(Note: CA ≡ Carbonic Anhydrase)

Figure 171. Buffering of the Hydrogen Ions Secreted into the Distal Tubule Lumen

(Note: CA ≡ Carbonic Anhydrase)

a. Excretion of Titratable Acid

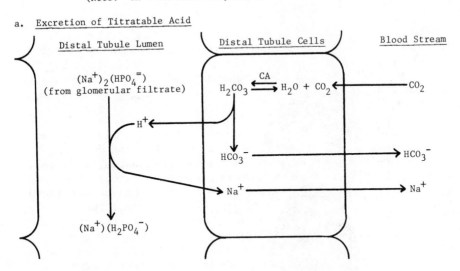

b. Secretion of Ammonia

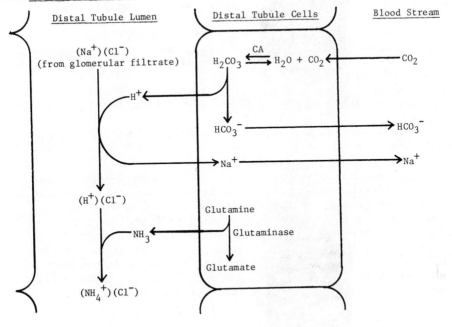

3. In any derangement [of acid-]base balance, compensatory changes occur
   to restore homeostasis. [Ac]idosis due to respiratory failure will
   lead to compensatory [chan]ges in the kidney which can give rise to a
   metabolic alkalosis.

   Disorders of acid-base balance are classified into four major
   groups based on their cause and the direction of the pH change. They
   are (1) respiratory acidosis, (2) metabolic acidosis, (3)
   respiratory alkalosis, and (4) metabolic alkalosis. Alterations in
   the plasma electrolytes in each of these conditions is indicated
   in Table 69.

   In the assessment [of these] disorders, the commonly measured
   serum electrolytes [are Na+,] H$^+$ (as pH), Cl$^-$, and HCO$_3$$^-$.
   However, there a[re other anions] (sulfates, phosphates, proteins,
   etc.) and cations [calcium,] magnesium, proteins, etc.) which are
   present and not meas[ured rou]tinely. These ions can be estimated
   indirectly. The [concentrati]on of underlined unmeasured anions, for example,
   can be assessed as [follows.] We know, according to the principle
   of electrical neu[trality tha]t the sum of the positive charges
   must equal the su[m of the neg]ative charges. That is, the
   number of positive [charges = t]he number of negative charges.
   It should be emph[asized that a]lthough the individual anionic or
   cationic componen[ts vary] (discussed later), their sum must
   always be equal. [In plasma, N]a$^+$ and K$^+$ account for 95% of the
   cations and Cl$^-$ a[nd HCO$_3$$^-$ acc]ount for about 85% of the anions.
   The other anions, [i.e., phos]phate, sulfate and proteins,
   contribute towar[d the remain]der of the negative charges (about
   15%). The concen[tration of] this group of anions can be calculated
   from the following[:]

$$\text{Unmeasured anions} = (Na^+ + K^+) - (Cl^- + HCO_3^-)$$

   The unmeasured anion is commonly known as anion gap (AG). The
   normal range for AG is 12 ± 4.

   As we will see, AG is useful in assessing the acid-base status
   of a patient and aids in particular in the diagnosis of metabolic
   acidosis. In general, the disorders which cause a high AG are
   metabolic acidosis, dehydration, therapy with sodium salts of
   strong acids, therapy with certain antibiotics (e.g.,
   carbenicillin--its anion contributing to the increased gap) and
   alkalosis. A decrease in normal AG can occur in various
   dilutional states, hypocalcemia, hypermagnesemia, hypernatremia,
   hypoalbuminemia, [conditions ass]ociated with hyperviscosity, in
   some paraproteine[mias and bromi]de toxicity.

   a. Respiratory a[cidosis is de]fined as a decrease in blood pH
      caused by an [accumulation] of CO$_2$ in the lungs with a consequent

Table 69. Changes in Plasma Electrolytes During Acid-Base Disorders

Parameter	Normal Values	Acidosis Resp. U.	Acidosis Resp. C.	Acidosis Metabolic U.	Acidosis Metabolic C.	Alkalosis Resp. U.	Alkalosis Resp. C.	Alkalosis Metabolic U.	Alkalosis Metabolic C.
Plasma bicarbonate (mEq/1)	22-30	-	$\uparrow$	$\downarrow$	$\downarrow$	-	$\downarrow$	$\uparrow$	$\uparrow$
P_{CO_2} (torr)	38-42	$\uparrow$	$\uparrow$	-	$\downarrow$	$\downarrow$	$\downarrow$	-	$\uparrow$
pH	7.35-7.45	$\downarrow$	-	$\downarrow$	-	$\uparrow$	-	$\uparrow$	-
[illegible] (mEq/1)	3.5-5.0	$\uparrow$	$\uparrow$	$\uparrow$	$\uparrow$	$\downarrow$	$\downarrow$	$\downarrow$	$\downarrow$

U = uncompensated

C = compensated

$\uparrow$ = elevated

$\downarrow$ = lowered

\- means no change

rise in P_{CO_2} (hypercapnia or hypercarbia). It is associated with emphysema, pneumonia, and other disorders which interfere with the pulmonary removal of CO_2. The increased carbonic acid in the blood is partly buffered by the uptake of H^+ by cells with a corresponding loss of intracellular K^+. In acute hypercapnia, the primary compensatory mechanism consists of tissue buffering.

The kidneys respond to the elevated P_{CO_2} by increasing the amount of HCO_3^- formed by the carbonic anhydrase system in the renal tubules and by excreting more hydrogen ions (producing a more acid urine). These renal compensatory mechanisms attempt to neutralize the respiratory acidosis by producing a metabolic alkalosis.

The primary goal of the treatment in this disorder is to remove the cause of the disturbed ventilation. Immediate intubation and assisted ventilation will also aid in improving the gas exchange.

b. Metabolic acidosis. This class of disorders may be grouped into two categories: metabolic acidosis associated with increased AG and metabolic acidosis with normal AG. The disorders of metabolic acidosis with increased AG include diabetic ketoacidosis, alcoholic ketoacidosis, lactic acidosis (some causes of lactic acidosis are hypoxia, shock, severe anemia, alcoholism, malignancies, etc.), toxicity of salicylates, methanol, paraldehyde and ethylene glycol, and renal failure. The metabolic acidosis with normal AG are seen in renal tubular acidosis, carbonic anhydrase inhibition, diarrhea, ammonium chloride administration, chronic pyelonephritis and obstructive uropathy.

In both groups of metabolic acidosis bicarbonate levels are decreased. The tissue buffering occurs with the exchange of extracellular H^+ ions with the intracellular K^+ ions. Thus, in metabolic acidosis with the exception of the disorders caused due to diarrhea, carbonic anhydrase inhibition and renal tubular acidosis, the plasma levels of potassium are usually increased. The metabolic acidosis with normal AG is associated with elevated levels of Cl^- (hyperchloremia). This is due to, for example when HCl is added, the H^+ ion neutralizing HCO_3^- (leading to acidosis) and replacing it with Cl^- ions. In the case of renal tubular acidosis, the hyperchloremic acidosis is a result of the kidney's inability to either reabsorb or regenerate bicarbonate ions in the proximal tubule.

i) Compensation initially consists of the prompt stimulation of respiratory rate leading to a decrease in P_{CO_2} and respiratory alkalosis. This cannot be sustained, however, due to tiring of the respiratory muscles. A slower renal compensation also occurs which can be maintained for an extended period because of the induction of glutaminase. In this process, the excess acid is excreted and the production of HCO_3^- is increased, correcting the acidosis.

ii) Metabolic acidosis is treated by first correcting the cause of the acidosis (e.g., insulin administration in diabetic ketoacidosis). The acid is then neutralized by treatment with a base such as $NaHCO_3$ (the treatment of choice as it directly replaces the lost bicarbonate and has no disadvantages), sodium lactate, or tris (trishydroxymethylaminomethane) buffer. It should be emphasized that certain problems may occur following the alkali replacement therapy in the management of metabolic acidosis: development of post-treatment respiratory alkalosis particularly if the low CO_2 tension (which has occurred due to respiratory compensation) persists after alkali therapy and a paradoxical further decline in cerebrospinal fluid (CSF) pH which may cause a deterioration of consciousness in some acidotic patients. The alkaline "overshoot" is a result of the resumption of normal oxidation of anions (e.g., lactate, acetoacetate) leading to the eventual production of bicarbonate from CO_2. The deterioration of the CSF pH has been attributed to a relative impermeability of the blood-CSF membranes to bicarbonate ions and permeability to CO_2. Thus, it has been suggested that the treatment with alkali should be restricted to patients with severe metabolic acidosis (an arterial pH of 7.1 or less) or a serum bicarbonate level of less than 5 mEq/L. The severe acidosis is corrected in small increments and over a period of several hours. The quantity of bicarbonate that is required in order to raise plasma bicarbonate by a given amount is estimated from the following formula:

Amount of HCO_3^- required in milliequivalents =

(desired HCO_3^- - measured HCO_3^-)

x 0.4 x body weight in kg

The volume of distribution of bicarbonate is assumed to be approximately 40% of body weight. Potassium replacement therapy is also frequently needed because of the transport of the intracellular K^+ to extracellular fluid during cellular buffering and its subsequent loss in the urine.

c. <u>Respiratory alkalosis</u> occurs when the respiratory rate increases abnormally (hyperventilation) leading to a decrease in P_{CO_2} and a rise in pH.

   i) Hyperventilation can be caused by hysteria, by pulmonary irritation (as in pulmonary emboli), or by a head injury with damage to the respiratory center. If the alkalosis is the result of hysteria, the usual treatment is simply sedation.

   ii) The pH increase is buffered by a lowering of plasma HCO_3^- (probably due to the law of mass action) and, to some extent, by the exchange of plasma K^+ for intracellular H^+. Renal compensation seldom occurs because this type of alkalosis is usually transitory.

d. <u>Metabolic alkalosis</u> is used to describe a decrease in hydrogen ion or an elevated plasma HCO_3^- level. It can be subdivided into three types.

   i) The <u>administration of excessive amounts of alkali</u> (as during $NaHCO_3$ treatment of a peptic ulcer) increases the HCO_3^-:H_2CO_3 ratio with a consequent rise in plasma pH. Acetate, citrate, lactate, and other substrates which can be oxidized to HCO_3^- can also cause alkalosis.

   ii) <u>Vomiting</u>, with the concomitant loss of H^+ and Cl^-, produces an acid-losing alkalosis.

   iii) If there is <u>excessive loss of extracellular potassium</u> (and extracellular fluid) in the kidneys, K^+ diffuses out of the cells and is replaced by Na^+ and H^+ from the extracellular fluid. Additionally, K^+ and H^+ are normally secreted by the distal tubule cells to balance Na^+ uptake during Na^+ reabsorption (see Figure 172). If extracellular K^+ is depleted, more H^+ is lost to permit reabsorption of the same amount of Na^+. Loss of H^+ by both routes causes <u>hypokalemic</u> alkalosis. The use of excessive amounts of some diuretics or elevated aldosterone production can cause the hypokalemia which initiates this type of alkalosis.

   iv) To compensate for a metabolic alkalosis, the respiratory rate decreases, raising P_{CO_2} and lowering the pH. This mechanism is limited, however, because if the respiratory rate falls too low, P_{O_2} decreases to the point where respiration is once again stimulated. Renal compensation involves decreased reabsorption of bicarbonate and the consequent formation of an alkaline urine. Unfortunately,

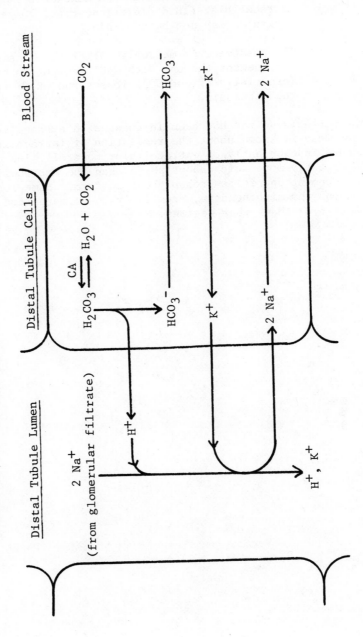

Figure 172. Secretion of Potassium and Hydrogen Ions by the Distal Tubule Cells During Sodium Ion Reabsorption

(Note: CA ≡ Carbonic Anhydrase)

[1211]

the bicarbonate excreted is accompanied by a loss of cations (Na^+, K^+) and if the alkalosis is accompanied by extracellular fluid depletion, renal compensation by this mechanism may not be possible.

v)   The treatment of metabolic alkalosis consists of fluid replacement therapy with NaCl and KCl to correct electrolyte (especially K^+ and Cl^-) losses and NH_4Cl to counteract the alkalosis.

4.   The emphasis so far has been in changes in extracellular fluid. Much less is known about the regulation of intracellular pH which occurs independently of extracellular control. The cell interior is probably more acid (about pH 7) than the extracellular environment and it is probably not uniform. Some localized areas, such as the mitochondria, are likely to be more acid (with a pH of about 6.8) than other regions.

13
Hormones

General Aspects of Hormones

1. The roles of hormones in the metabolism of a number of substances have been discussed in appropriate places throughout this text. A list of the most important instances of this are given below.

	Previously discussed in:
Cyclic AMP and hormones	Chapter 4
Erythropoietin	Chapters 4 and 7
Effects of glucagon and other hormones on carbohydrate metabolism	Chapter 4
Insulin, glucagon, somatostatin and their relationship to diabetes	Chapter 4
Serotonin and melatonin	Chapter 6
Influence of hormones on protein metabolism	Chapter 6
Gastrointestinal hormones	Chapter 6
Epinephrine and norepinephrine	Chapter 6
Thyrotropin releasing hormone (TRH), thyrotropin, thyroid hormones (T_3 and T_4)	Chapter 6
Cholesterol (as a precursor of the steroid hormones)	Chapter 8
Prostaglandins	Chapter 8
General remarks on steroids	Chapter 8
Parathyroid hormone, 1,25-dihydroxy-vitamin D, calcitonin, and the effect of other hormones on mineral metabolism	Chapter 9
Effects of renin-angiotensin system, aldosterone and antidiuretic hormone on water and electrolyte metabolism	Chapter 12

This chapter concerns itself with some general aspects of hormones and with those hormones (notably the steroids) which have not already been discussed.

2. Hormones are elaborated and secreted by the <u>endocrine systems</u> of the body under the direction of higher centers of the brain and of chemical stimuli transmitted in the blood stream. The actions of hormones include the regulation of normal levels of various substances within the body (homeostasis), integration of bodily activities with one another, and morphogenesis (both during development and in changes occurring after maturation). Some aspects of the interrelationship of the nervous and endocrine systems (neuroendocrinology) are discussed in the section on the hypothalamus.

3. Hormones produced by specialized cells are released into the circulation in order to reach their target tissues and exert their action. A few hormones (notably those of the hypothalamus) are also transported at least partially by secretory neurons. A given hormone may have a specific effect on a particular target cell (testosterone acts on seminiferous tubule epithelial cells in the formation of sperm) as well as the rest of the body cells (testosterone stimulates biosynthetic processes in non-testicular tissue). It is important to point out that, from an evolutionary point of view, the development of an efficient circulatory system was a prerequisite to the development of an endocrine system. Endocrine (as opposed to exocrine) glands are those tissues which secrete hormones <u>into</u> the body. Note that, although albumin is synthesized in the liver and secreted into the circulation, it is not classified as an endocrine secretion or hormone because it has no specific physiological action or target tissue.

4. Some of the properties of hormones are summarized below.

    a. Hormones are synthesized and released in extremely minute quantities and exert their biological effects at quite low concentrations (10^{-6} g/100 ml for steroids; somewhat less for peptide hormones).

    b. The rate of secretion of a hormone by an endocrine gland is established by the need for the hormone and its rate of inactivation. Several types of self-regulatory feedback mechanisms exist to control the secretion of a given hormone and to thus maintain it at a required physiologic level.

    c. Although not all hormones are stored (e.g., some steroid hormones), those which are may be kept in the endocrine cells as inactive precursors in the form of granules (pituitary cells; β-cells of the islets of Langerhans in the pancreas); or as a colloid in the lumen (cavity) of the gland, surrounded by secretory cells (thyroid acini). Examples of inactive precursors and their associated active forms (indicated in parentheses) include: thyroglobulin, (T_3 and T_4); proinsulin, (insulin); testosterone, (dihydrotestosterone); and progesterone, (dehydroprogesterone).

d. The stimuli which cause the release of the stored hormones vary. Insulin, for example, is secreted when blood glucose levels increase above the physiologic "fasting" level. In the case of the adrenal medullary chromaffin cells (cells of neural crest origin which have lost their ability to participate in the conduction of nerve impulses), the hormones epinephrine and norepinephrine (catecholamines) are secreted only in response to neural stimulation. The hypothalamic neurosecretory cells can participate both in nerve impulse conduction and the elaboration and release (in response to nervous and chemical stimuli)of neurohormones.

e. Hormones normally do not exert any significant effects on the cells (tissues) that produce them.

f. Hormones may have very specific effects on a particular target organ (the action of luteinizing hormone on the interstitial cells of the male and female gonads) or may influence several organs (recall the action of insulin on muscle, adipose tissue, etc.). Hormones can also show multiple effects (i.e., different responses to one hormone occur in different tissues).

g. The physiologic response of a tissue to a hormone is dependent upon other variables such as age, genetic makeup, etc.

h. Hormonally initiated responses may persist long after the hormone is no longer measurable by biochemical and biological assays.

Mechanisms of Actions of Hormones

1. The initial encounter of any hormone with a target cell is mediated by a specific receptor molecule. This receptor molecule can be located on the plasma membrane or in the cytoplasm or nucleus of the cell. Some aspects of receptor proteins were discussed in Chapter 4 under cyclic AMP. In general, the peptide and glycoprotein hormones have receptors on the plasma membrane, steroid hormones bind to cytosol receptors, and triiodothyronine (T_3) binds receptors in the nucleus. Recently, specific receptors have also been detected in various other intracellular membranes, namely microsomes, and Golgi apparatus for catecholamines prolactin, growth hormone, somatomedin, etc.

Some of the properties of the hormone receptors are that they exhibit a strict structural specificity, the number of receptors are finite and saturable, they are present only in the hormone responsive cells, and the hormone-receptor interaction is rapid, reversible and has affinity constants consistent with the physiologic levels of the hormone. Recent studies made on the hormone receptors have led to the dilineation of some endocrine abnormalities that are related to receptor defects, namely normal receptors deficiency (e.g., pseudohypoparathyroidism, ADH resistant diabetes insipidus, pseudohermaphroiditism) or the development of antibodies to receptors (e.g., Graves disease, myasthenia gravis, in some types of insulin-resistant diabetes). In a number of disorders the measurement of the level of receptors (by using radioreceptor assays--RRA) has been useful in the management of the clinical entity. An example of RRA is the estradiol receptor assay which is used in the treatment of breast cancer. It has been shown that breast cancers which contain receptors to estradiol are responsive to endocrine management (antiestrogen therapy). The RRA's have also been used in the determination of hormones.

2. There are two general types of mechanism which have been described to explain the action of hormones.

   a. A number of (non-steroid) hormones activate the adenylate cyclase system leading to the production of cAMP, which appears to stimulate intracellular metabolism by releasing inhibitory constraints on these pathways. The cAMP produced functions as an allosteric modulator and promotes the phosphorylation of proteins, in addition to altering membrane permeability and calcium flux (see Chapters 4 and 9). The locus of specific receptor sites for these hormones is on the cell plasma membrane. The details of cAMP production have already been discussed in Chapter 4. Intracellular changes in the steady-state levels of cyclic GMP (cGMP) due to the action of hormones have also been observed. In many instances, the cAMP and cGMP levels show a reciprocal relationship with opposite biological effects. However, it should be noted that the precise regulatory role of cGMP is not understood, and this subject is of considerable controversy at this time.

   b. The action of steroid hormones does not invoke the production of cAMP. They stimulate general and specific protein synthesis in the target cells, acting at the levels of transcription and translation.

3. Estrogens, androgens, progesterone, ecdysone (an insect steroid hormone; causes visible puffing of fruit fly salivary cell chromosomes), aldosterone and other steroid hormones have been

shown to stimulate messenger RNA synthesis in their respective target tissues. Studies using actinomycin D inhibition of mRNA production and DNA hybridization techniques with newly formed RNA have shown that the steroid hormones are capable of causing transcription of unique portions of the DNA as well as stimulating the generalized production of RNA to enhance overall protein synthesis.

4. Following is a possible sequence of events leading to effects of steroid hormones on transcription:

    a. Entry of steroid hormone into the target cell.

    b. Binding of the hormone or its metabolite with a specific protein receptor molecule in the <u>cytoplasm</u> and the concentration of the hormone in the target cell. These cytoplasmic steroid binding proteins show specificity for particular steroids, as well as serving as carriers to facilitate the entry of the steroids into the nucleus. The hormone binding alters the conformation of the receptor protein complex. The hormone receptor is a dimer compound of two subunits, A and B (4S and 8S). Each subunit binds one molecule of hormone and therefore each receptor contains two hormone molecules. The translocation of the receptor-hormone complex requires a temperature-dependent activation.

    c. The hormone-receptor protein complex enters the nucleus where it becomes bound to the chromosome through a specific acceptor protein associated with chromatin. The chromosomal acceptor protein appears to be an acidic, non-histone protein. In the case of progesterone, it has been shown, working with the chick oviduct system, that the B subunit of the hormone-receptor complex has specific binding affinity for a particular acidic chromosomal protein. In contrast, the A subunit binds DNA only, and does not exhibit any specificity or requirement for chromosomal proteins. The molecular details that lead to the activation of the transcription process is not clear.

5. With the exception of testosterone, which is converted to its reduced derivative, 5α-dihydrotestosterone, before binding with its specific receptor, there is little to suggest that steroid hormones undergo metabolic conversion in order to bind to their specific receptors. The conversion of testosterone to its 5α-dihydro derivative is shown below:

OH

5α–reductase

NADPH
+H⁺

NADP⁺

O

Testosterone (T)

OH

O

H

5α–Dihydrotestosterone (DHT)

The 5α–reductase appears to be associated with the nuclear membrane as well as with the endoplasmic reticulum.

6. Pretreatment of a progesterone target tissue (guinea pig uterus, mouse vagina, chick oviduct) with estrogen enhances the uptake of progesterone by cells of these tissues, thereby demonstrating that there can be an interdependence (synergism) between steroid hormones in some tissues. Estrogen has been shown to induce the production of progesterone-receptor protein.

The Hypothalamus

1. This endocrine gland is located below (ventral to) the thalamic region of the forebrain. Under the control of chemical and nervous stimuli from higher centers of the brain, it regulates the secretion of hormones (see Table 71) from the adenohypophysis (and hence many other hormonal secretions in the body) and provides the neurohypophyseal hormones, vasopressin and oxytocin. These processes are summarized in Figure 173. It is interesting to note that, since the nervous impulses are at least partially mediated by neurotransmitter substances, hormones play a role in yet one more step of the control sequence.

2. Vasopressin and oxytocin, the neurohypophyseal hormones (see Table 71), are synthesized by cells in the supraoptic and para-ventricular nuclei of the anterior hypothalamus. As granules, in association with the neurophysins, they migrate along secretory neurons and accumulate at the ends of these nerves in the neurohypophysis. The neurophysins are a class of "carrier" proteins (about 10,000 molecular weight) synthesized in the hypothalamus. In the bovine system, neurophysin I is primarily associated with oxytocin in the neurosecretory granules while neurophysin II apparently carries vasopressin, but this is known to differ in other species. Oxytocin is released from the neurohypophysis in response to the suckling of an infant, contraction of the uterus during labor, vaginal stimulation, and cervical dilation. Vasopressin (antidiuretic hormone) secretion is stimulated by small changes (a few percent) in the osmolarity of the blood

Figure 173. Hypothalamic and Hypophyseal Hormonal System

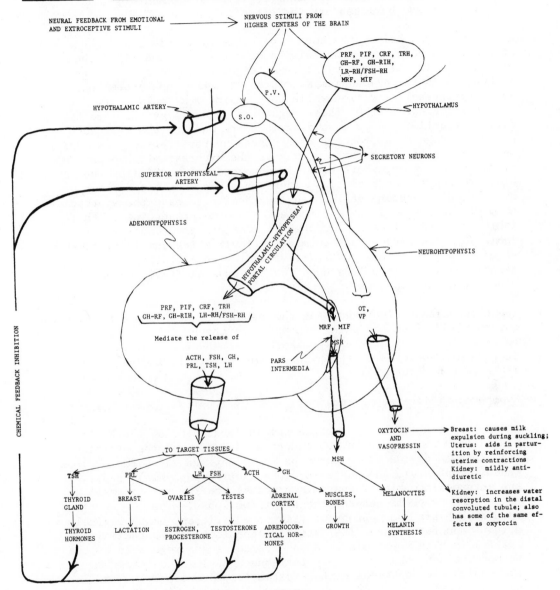

Key: P.V. = paraventricular nucleus; S.O. = supraoptic nucleus;
OT = oxytocin; VP = vasopressin; PRL = prolactin.
For explanation of other abbreviations, see Tables 70 and 71.

1219

(an effect mediated by osmoreceptors in the hypothalamus), hemorrhage, adrenalcortical insufficiency, and alterations in the renin-angiotensin system (see Chapter 12). The physiological effects of these hormones are indicated in Figure 173.

3. As is discussed below, the adenohypophysis synthesizes and secretes at least six hormones. The release of these compounds into the bloodstream is regulated by at least eight hormones and factors elaborated in the hypothalamus. Three of these substances inhibit adenohypophyseal secretion while the remainder stimulate it. It is not yet clear whether these regulators are all synthesized by different cell-types or if two or more are made by the same sort of cell. Regardless, once formed, these neurosecretions pass down nerve fibers to the primary plexus of the hypothalamic-hypophyseal portal circulation, where they are absorbed and carried by the blood to the adenohypophysis. There they act upon the adeno-hypophyseal secretory cells, causing them to release the appropriate hormones into the systemic circulation. Some properties of these inhibitory and releasing hormones and factors are summarized in Table 70. The nervous and chemical stimuli which can initiate this sequence of events are quite diverse. In general, they are associated with the body's need for the physiological response produced by the hormone.

4. The hypothalamus also regulates the release of melanocyte stimulating hormone (MSH) from the pars intermedia of the hypophysis (see below). In this instance also, both releasing and inhibitory factors are secreted by the hypothalamus. This is indicated in Table 70 and Figure 173.

The Hypophysis (Pituitary Gland)

1. The hypophysis (meaning undergrowth) is located below the brain and plays a major role in the endocrine function of the organism. The gland is divided into three parts. They are the anterior lobe (adenohypophysis), the posterior lobe (neurohypophysis), and the intermediate "lobe" (pars intermedia), really just a small collection of cells between the other two lobes. The adenohypophysis and neurohypophysis have different embryological origins and functions. The anterior lobe (of oral epithelial origin) is highly vascularized and its cells are regulated by the hormones elaborated by the hypothalamus and released into the hypothalamic-hypophyseal portal system (see previously). The neurohypophysis (derived from the neural epithelium), on the other hand, is connected with the hypothalamus through many nerve fibers as well as by vascular interaction through the hypophyseal stalk. The neurohypophysis also contains non-neural connective tissue cells (neuroglial cells) known as pituicytes.

Table 70. Hypothalamic Releasing and Inhibitory Hormones and Factors

Hormone	Chemical Nature	Target Organ	Effect Produced
A. Corticotrophin-releasing (CRF)	peptide (not isolated in pure form)	adenohypophysis	Adrenocorticotrophic hormone (ACTH) release
B. Follicle-stimulating hormone releasing hormone (FSH-RH) (same as LH-RH, below)	decapeptide	adenohypophysis	Follicle-stimulating hormone (FSH) release
C. Thyrotropin releasing hormone (TRH)	tripeptide	adenohypophysis	Thyrotropin (TSH) release
D. Luteinizing hormone releasing hormone (LH-RH)	decapeptide	adenohypophysis	Luteinizing hormone (LH) release
E. Growth hormone releasing factor (GH-RF)	polypeptide	adenohypophysis	Growth hormone (GH) release
F. Growth hormone release-inhibiting hormone (GH-RIH) or somatostatin	tetradecapeptide	adenohypophysis	Inhibition of GH release; suppresses the secretions of other endocrine and exocrine glands
G. Prolactin releasing factor (PRF)	not yet known	adenohypophysis	Prolactin release
H. Prolactin release inhibiting factor (PIF)	not yet known	adenohypophysis	Inhibition of prolactin release
I. Melanocyte stimulating hormone (MSH) releasing factor (MRF)	not yet certain	intermediate lobe of the pituitary	Melanocyte stimulating hormone (MSH) release
J. MSH release-inhibiting factor	not yet certain (may be a tripeptide)	intermediate lobe of the pituitary	Inhibition of MSH release

2. The adenohypophysis contains two main groups of cells which are histologically distinct. They are designated chromophobes (cells that do not stain very well; small, with granular cytoplasm) and chromophils (cells that stain readily; large with granular epithelium). The chromophils are further subdivided into basophils and acidophils, depending on the chemical nature of the dyes which are capable of staining them. By the use of a variety of dyes, additional divisions of these classes of cells are possible. Functionally, the adenohypophysis contains at least six different cell types, each secreting a specific hormone.

3. Table 71 lists the hormones of the hypophysis along with their target organs and functions. As can be seen, FSH, LH, and TSH are glycoproteins and are secreted by (morphologically distinct) basophilic cells. There is some overlap in the secretions from these cells. All three contain one each of two types of subunits, designated α and β. The α-subunits of TSH, LH, and FSH are identical in amino acid sequence, responsible for the immunological cross-reactivity of these three hormones, and have no significant hormonal activity. The β-subunits of these glycoprotein hormones are different and confer biological specificity and hormonal activity. It should be noted that neither α-chains nor β-chains, by themselves, possess biologic potency. But a biologically active molecule can be generated in vitro by combining an α- and a β-chain, whose specificity is determined by the β-chain. Human chorionic gonadotropin (HCG), also a glycoprotein, bears structural similarity to FSH, LH, and TSH. The α-chain of HCG is immunologically indistinguishable from those of LH, FSH or TSH. However, the amino acid sequence of the α-chain of HCG shows a two amino acid inversion and a three amino acid deletion at the NH_2 terminus when compared with the α-chain of other glycopeptide hormones. As noted above the β-chain of each of these four glycopeptide hormones is structurally unique and confers biologic and immunologic specificity. HCG exhibits primarily luteinizing and luteotropic activity. An early pregnancy test (as early as 14 days after conception) is based upon the detection of HCG in the urine.

Human placental lactogen (HPL), also known as human chorionic somatomammotropin, is another placental protein hormone synthesized, stored and released from the synctiotrophoblast cells. It consists of 199 amino acid residues held together by two intrachain disulfide bonds and has a molecular mass of 21,500 daltons. The structural similarities between HPL and human growth hormone suggest that these two proteins (and prolactin) evolved from a common progenitor polypeptide. The biologic activity of HPL in general falls into two general categories—namely lactogenic and somatotropic (resembling the action of prolactin and growth hormone of the anterior pituitary).

Table 71. Actions and Some Characteristics of the Pituitary (Hypophyseal) Hormones

Hormone	Chemical Nature	Target Tissues	Effect Produced
A. Neurohypophysis (Posterior Pituitary)			
i) Oxytocin	cyclic nonapeptide	Mammary glands; uterus	Ejection of milk from the breasts; smooth muscle contraction; pressor efferent
ii) Vasopressin (antidiuretic hormone ADH)	cyclic nonapeptide	Kidneys, arteries	Reabsorption of water; smooth muscle contraction
B. Pars intermedia (intermediary lobe)			
i) Melanocyte-stimulating hormone (MSH)	peptide α-MSH: 13 amino acids β-MSH: 18 amino acids	Skin, melanophores	Dispersal of skin in melanophores, leading to darkening of skin
C. Adenohypophysis (Anterior pituitary)			
i) Prolactin (leuteotropin) secreted by basophil cells	Protein	Mammary gland	Initiation of production of milk in a prepared gland
ii) Adrenocorticotrophin (ACTH; corticotrophin) secreted by chromophobe	44 residue peptide	Adrenal cortex	Synthesis and/or secretion of adrenal cortical steroids
iii) Thyrotrophin(see pages 492-493) (Thyroid-stimulating hormone, TSH) secreted by basophil cells	Glycoprotein; molecular weight about 28,000	Thyroid gland	Synthesis and secretion of T_3 and T_4
iv) Growth hormone (somato-trophin; GH); secreted by acidophil cells	188 residue peptide	Adipose tissue	Lipolysis and release of lipid Release of lipids (mobilization of fat) Anabolic hormone; growth of long bones
		Adipose tissue; Rest of the tissues of the body	
v) Follicle-stimulating hormone (FSH); secreted by the basophil cells	Glycoprotein; molecular weight about 16,000-17,000	Follicle of ovary in female; seminiferous tubules in males	Maturation of follicles; sperm production
vi) Luteinizing hormone (LH), (interstitial cell-stimulating hormone; ICSH) secreted by basophil cells	Glycoprotein; molecular weight about 16,000-17,000	Interstitial cells of the ovary in females. Interstitial cells of the testis in males	Estrogen and progesterone secretion; maturation of follicle; formation of corpus luteum Androgen secretion; interstitial tissue cell maturation

[1223]

The somatomedins, a family of peptides, appears to mediate some of the actions of growth hormone, for example stimulation of sulfate uptake by cartilage, an insulin-like action (non-suppressible insulin-like activity), and growth promotive actions. These peptides are synthesized under the influence of growth hormone in the liver, kidneys and possibly other tissues.

4. Endorphins, a group of peptides present in the pars intermedia of the pituitary, bind to opiate receptors in the brain and exhibit opiate (morphine and its analogues, see Chapter 15)-like properties, such as analgesia. It is of interest to point out that the amino acid sequence of β-lipotropin, a polypeptide hormone of anterior pituitary gland, contains the sequences for α, β, γ, and δ endorphins. The amino acid residues of β-lipotropin which corresponds to α, β, γ, and δ endorphins respectively are 61-76, 61-91, 61-77, and 61-87. It has been suggested that the β-lipoprotein is a prohormone (it itself lacks the opiate-like activity) for all the endorphins. In support of this is the observation that the in vitro incubation of β-lipotropin with brain extracts yields peptide with opiate-like activity. The endorphin is also present in the brain and perhaps in other organs. Enkephalins are smaller (pentapeptides) than endorphins and also show morphine-like activity by combining with opiate receptors. These have been isolated from the brain tissue and characterized. There are two enkephalins: Met^5- and Leu^5-enkephalin. The amino acid sequence of Met-enkephalin is: Tyr-Gly-Gly-Phe-Met. The endorphins contain the Met-enkephalin sequence in their structure. The physiologic role of endorphins and enkephalins is yet unknown.

5. Several disorders of the pituitary have been described, producing both under and oversecretion of the hypophyseal hormones.

   a. Overproduction of growth hormone (GH) causes gigantism if it occurs in children and acromegaly if it is present in adults. Gigantism results from a delay in epiphyseal fusion due to hypogonadism resulting from the excessive GH. It may be accompanied by mild acromegalic features. In acromegaly, the bulk of bone and soft tissue increases causing an increase in the size of the hands and other parts of the body. Changes in facial appearance and growth of excessive hair (hirsuteness) also frequently occur. The usual cause is a secreting acidophil tumor of the pituitary gland.

   b. Elevated ACTH levels are found in Addison's disease and are one of the common causes of Cushing's syndrome. Both of these disorders are discussed later.

c. <u>Panhypopituitarism</u> (see Chapter 6) is, as the name implies, a general undersecretion of the pituitary hormones. It is considered to be a chronic condition which poses no threat to life except in times of stress. Causes include pituitary tumors and infarctions, granulomatous lesions due to tuberculosis and other diseases, and pituitary surgery or irradiation. The degree to which each hormone is decreased depends on the duration of the disorder and on the individual patient.

d. A specific deficiency of growth hormone is the cause of about 10% of the cases of dwarfism. When GH is demonstrably decreased, the disorder is known as pituitary dwarfism. Dwarfism with this specific etiology is responsive to treatment with GH.

Adrenals

1. In an adult, the adrenals lie on the superior pole of the kidneys (ad-renal meaning next to the kidney). The glands are triangular and each is enclosed by a capsule of fibrous connective tissue. In a cross section of a fresh gland, the cortex region appears yellow in color and is readily distinguishable from the brownish medullary region.

2. Each adrenal actually consists of two endocrine glands which differ in their embryological origins. The adrenal medulla (the inner region) is of ectodermal (neural crest) origin while the cortex is derived from the cephalic end of the mesodermal region.

3. The cortex and medulla secrete chemically distinct hormones which produce differing effects. The medullary hormones are catecholamines which have already been discussed (see Chapter 6), while the cortical hormones are steroids (adrenocorticosteroids). The medullary hormones indirectly (through the adenohypophysis) affect the secretory activity of the cortex.

4. Over fifty different steroids have been isolated and characterized. They can be grouped into three classes based upon their functional characteristics: mineral corticoids (affect electrolyte (Na^+, K^+, Cl^-) concentrations and water balance), glucocorticoids (affect the metabolism of carbohydrates, proteins, and lipids), and a few sex steroids. The principal mineral corticoid is aldosterone and the principal glucocorticoid is cortisol. There is some overlap in the activities of these hormones, however, as can be seen in Table 72.

Table 72.   Activities of Some Natural and Synthetic Corticoids

	Glucocorticoid Activity (units/mg)	Mineralcorticoid Activity (units/mg)
Aldosterone	0.4	3000
Cortisol (hydrocortisone)	1	1
Cortisone	0.8	0.8
Prednisone (Δ^1-cortisone)	3.5	0.8
Dexamethasone (9α-fluoro-16α-methylprednisolone)	25	0
Fluorocortisone (9α-fluorocortisol)	15	125

Cortisol is defined as having unit glucocorticoid and mineralcorticoid activity. The first three compounds occur naturally while the last three are synthetic.

The reason that aldosterone does not exhibit a physiologically noticeable effect as a glucocorticoid, even though it has 40% of the activity of cortisol on a units per milligram basis, is because its normal, physiological blood level is low compared to that of the glucocorticoids.

5. The cortex is further differentiated into three concentrically arranged layers or zones of cells. The outermost layer, known as the zona glomerulosa, is thin and is primarily responsible for secretion of aldosterone. The middle layer is the zona fasciculata and is the widest of the three concentric layers. The zona reticularis is the innermost layer and it surrounds the adrenal medulla. The two inner layers are responsible for the synthesis and secretion of glucocorticoids (in particular cortisol) and small amounts of adrenal androgens. Under normal circumstances, in man, the amounts of the principal corticoids secreted during a 24 hour period are as follows: cortisol (10-30 mg); aldosterone (0.3-0.4 mg); and corticosterone (2-4 mg).

6. The zonae fasciculata and reticularis are regulated by ACTH in a negative-feedback type of control system. ACTH maintains the structure and function of these cells and in hypophysectomy the zonae fasciculata and reticularis undergo atrophy. On the other hand, the zona glomerulosa maintains normal structure and function upon hypophysectomy or may even show hypertrophy. Thus the zona glomerulosa is not dependent upon ACTH for its secretion and function, although its activity may be influenced indirectly by ACTH. Recall that ACTH secretion is regulated by ACTH-releasing factor produced by the hypothalamus. Stress stimuli which bring about hypothalamic secretions originate at higher centers of the central nervous system.

Figure 174a. **Biosynthesis of Cortisol and Aldosterone**
(For explanatory notes, see Figure 174b)

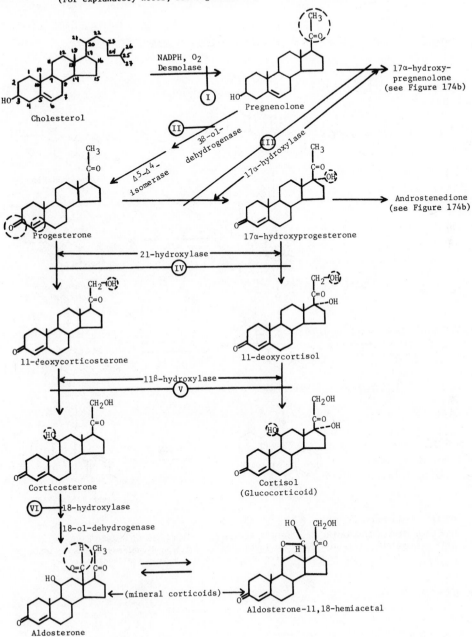

<u>Figure 174b</u>. <u>Biosynthesis of Estrogen, Testosterone, and Other Sex Steroids</u>

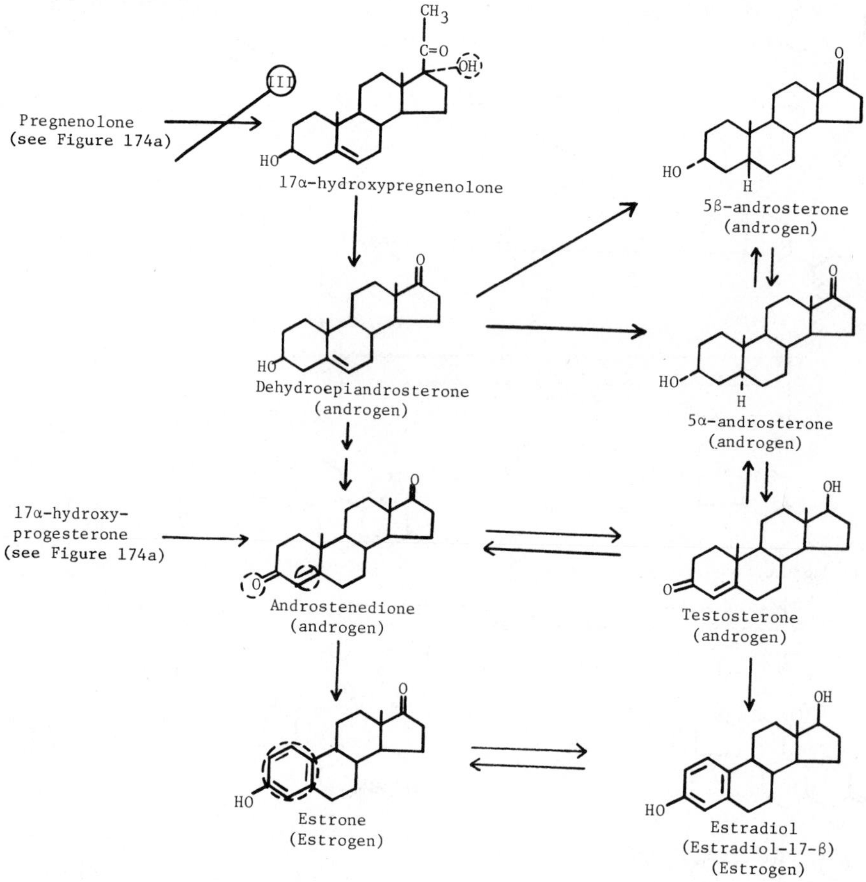

In both **Figures** 174a and b, the Roman numerals (I-VI) refer to biochemical lesions of steroidogenesis. They are described in the text. The groups enclosed by dashed lines are the regions of the molecules which are altered in the reaction leading to the structure indicated.

7. Aldosterone secretion is regulated by a variety of factors including the renin-angiotensin system, the plasma electrolyte (Na^+, K^+) levels, growth hormone, etc. The metabolism of aldosterone is discussed in the chapter on water and electrolyte balance (see Chapter 12).

Biosynthesis of Adrenal Cortical Hormones

1. The precursor to the adrenal steroids is cholesterol which in turn is synthesized from acetate (see Chapter 8). More cholesterol is present in the adrenal cortex than in any other structure of the body except for the nervous tissue. The adrenal cortex cholesterol is esterified with long chain unsaturated fatty acids.

2. In the biosynthetic pathways, a number of specific hydroxylations of the steroid ring system take place. These reactions are catalyzed by specific hydroxylases which are mixed function oxidases requiring O_2, NADPH, cytochrome P-450 and adrenodoxin. Except for the 11- and 18-hydroxylating systems (which are mitochondrial), these reactions occur in the microsomes. The cortex also contains large amounts of vitamin C.

3. The synthetic pathways for some of the steroids are shown in Figure 174. Also shown in this figure are the biochemical lesions of steroidogenesis which cause various congenital adrenal cortical hyperplasias. In these syndromes, there generally are elevations in levels of testosterone, estrogens (along with 17-ketosteroids), and (except for type II) pregnanetriol.

4. The first step in steroidogenesis is the conversion of cholesterol to pregnenolone, catalyzed by a mixed function oxidase enzyme complex system known as desmolase (cholesterol oxidase). The reaction takes place in the mitochondria and requires NADPH, cytochrome P-450, and either Mg^{+2} or Ca^{+2}. This step is the rate-limiting reaction in steroidogenesis and pregnenolone functions as a feedback inhibitor of this enzyme.

5. The basis for ACTH stimulation of steroidogenesis is its activation of the adrenal cortical adenylate cyclase, causing a rise in cAMP and activating glycogen phosphorylase (see Chapter 4). This enzyme breaks down glycogen to produce glucose-6-phosphate, which then is oxidized via the hexose monophosphate shunt pathway yielding NADPH. The NADPH is a co-substrate in the hydroxylase reactions and an increase in its availability promotes overall steroid synthesis. The action of ACTH also involves stimulation of protein synthesis, presumably by increasing mRNA synthesis and, hence, growth of the cortex.

6. Transport and catabolism of corticosteroids: albumin is a general carrier of steroid hormones, particularly aldosterone. Cortisol, however, is bound to an α_1-globulin known as transcortin or corticosteroid-binding protein. Each molecule of transcortin binds one molecule of cortisol (or corticosterone). Transcortin has a molecular weight of 52,000 daltons and is synthesized in the liver. Estrogen and thyroid hormone increase the transcortin levels while liver diseases and nephrosis decrease them. Catabolism of steroids occurs through the various reduction and conjugation reactions that occur in the liver **(see Chapters 4 and 14)**.

7. Cortisol has several biological actions

    a. It exerts a catabolic effect on muscle, epidermal, and connective tissue leading to negative nitrogen balance, release of amino acids, and atrophy of these tissues.

    b. Most of these tissues also exhibit reduced amino acid uptake. Hence, any unutilized amino acids released from protein breakdown enter the liver where they are converted to glucose via gluconeogenesis. The glucose produced is released into the blood, causing hyperglycemia. A prolonged hyperglycemia due to excess glucocorticoids can lead to diabetes mellitus.

    c. The immediate (acute) effect of cortisol is to transfer amino acids to the liver where they are converted to glucose and glycogen. Cortisol induces the synthesis of tyrosine transaminase, pyruvic-glutamic transaminase, tryptophan oxygenase, and ornithine decarboxylase. The long term effect also leads to induction of synthesis of enzymes which enhance the production of glucose: glucose-6-phosphatase, fructose-1,6-diphosphatase, PEP carboxykinase, and pyruvate carboxylase. The net effect of all of these is to produce glucose in the liver and hyperglycemia in the blood. Recall that insulin suppresses the activity of these gluconeogenic enzymes while inducing glycolytic and glycogenic ones.

    d. Cortisol causes increased release of glycerol and fatty acids from the adipose tissue. The glycerol is converted to glucose in the liver while the increased fatty acid levels cause an inhibition of key glycolytic enzymes in the liver. Administration of cortisol increases peripheral lipogenesis. This action is partially explained by the observation that cortisol produces hyperglycemia which increases insulin secretion and thereby promotes lipid synthesis. The fat synthesized has the characteristic distribution seen in Cushing's Syndrome (moon face, buffalo hump, pendulous abdomen).

e. Cortisol has anti-allergic and anti-inflammatory activities. These properties are therapeutically useful and are employed widely. There is also enhanced lymph node lysis resulting in lymphocytopenia and decreased antibody production. Erythropoiesis is increased. The antiallergic reactions may be due to decreased production of histamine and histamine-like substances and protection of the surface of the affected cells from the antigen and antibody interaction.

f. Excess cortisol can cause increased gastric acidity due to elevated HCl secretion. Enhanced pepsin secretion is also observed, with production of a watery mucous in the stomach. These can cause ulcers or aggravate an already existent ulcer.

8. The interrelationships between the adrenal cortex, adenohypophysis, the hypothalamus, and the higher centers of the central nervous system are shown in Figure 175.

9. The proper functioning of the feedback mechanisms at various levels (outlined in Figure 175) can be assessed by the following methods.

a. The <u>administration of ACTH</u> to a normal person should cause the adrenal cortex to respond by releasing cortisol, resulting in an increase in blood cortisol levels. Hypofunctioning of this gland should be suspected when this response does not occur.

b. <u>Dexamethasone administration</u> will indicate whether or not the hypothalamic feedback centers are capable of turning off CRF release and thus ACTH release. If this does not occur, then cortisol cannot function as a feedback inhibitor of its own synthesis.

c. The response of the hypothalamic feedback centers to low levels of cortisol is assessed by inhibiting the synthesis of cortisol. This is accomplished by administering metyrapone which inhibits the 11-hydroxylation step (final hydroxylation step) causing a fall in circulating cortisol levels and stimulating the hypothalamic feedback center to release CRF and hence ACTH.

d. The adrenal response of the patient to stress is assessed by the insulin hypoglycemic test. This consists of reducing the blood glucose level of the patient to 35 mg/100 ml by intravenous administration of insulin. Normal response to this stressful situation is the rapid elevation of plasma cortisol levels.

<u>Figure 175</u>. <u>Interactive Controls Among the Brain, Hypothalamus,</u>
<u>Anterior Pituitary, and Adrenal Cortex</u>

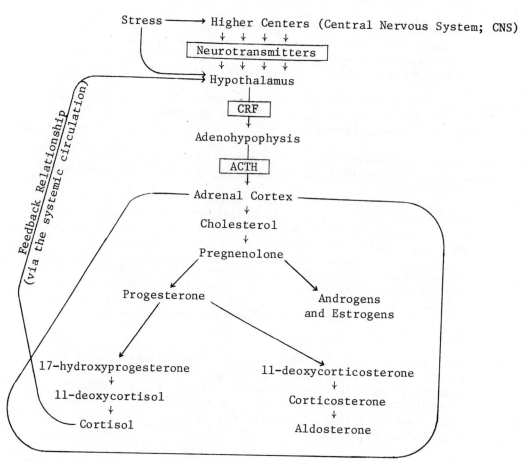

CRF = corticotrophin releasing factor.

Disorders of Adrenal Steroidogenesis

Six disorders of steroid synthesis in the adrenal gland are indicated in Figures 174a and b. The numbers below (I-VI) refer to this diagram.

1. <u>Desmolase deficiency</u> is also known as lipoid adrenal hyperplasia. No steroid hormone synthesis takes place in the adrenals, leading to the excessive accumulation of cholesterol and other lipids. Adrenal insufficiency (salt and water loss) occurs and an early death ensues unless the disease is treated extensively. No urinary steroids are detectable.

2. In <u>3β-ol-dehydrogenase deficiency</u>, progesterone is not formed from pregnenolone resulting in deficient production of mineral and glucocorticoids and elevation of dehydroepiandrosterone and related products. Affected individuals show salt loss, hypospadias (male), and mild virilism (female). High urinary excretion of Δ^5-3β-hydroxysteroids is observed.

3. <u>17α-hydroxylase deficiency</u> is a defect in the production of androgens, estrogens, and cortisol with an accompanying elevation of corticosterone and deoxycorticosterone. These individuals show immature sexual development and hypertension. Urinary 17-ketosteroids and estrogens are absent, while elevated levels of pregnanediol and corticosterone are present.

4. <u>21-hydroxylase deficiency</u> is the most common of these lesions. It is transmitted as an autosomal recessive trait and is found particularly in Eskimos. It is a deficiency of production of cortisol and aldosterone. Elevated levels of androgens are seen and affected individuals show masculinization with increased urinary excretion of 17-ketosteroids, pregnanediol, and C-21 deoxysteroids.

5. Patients with <u>11β-hydroxylase deficiency</u> show depressed production of cortisol and aldosterone, elevated levels of androgens, and to a lesser extent, an increase in 11-deoxycorticosterone and 11-deoxycortisol. These individuals show masculinization and hypertension with elevated urinary 17-ketosteroids, 11-deoxycorticosterone, and 11-deoxycortisol.

6. <u>18-hydroxylase deficiency</u> is a defect in the production of aldosterone. There is an increase in corticosterone. Affected individuals show the salt-losing syndrome and urinary levels of corticosterone and related products are elevated.

As a general consequence of these lesions, the plasma cortisol levels are low and a feedback mechanism is triggered to compensate

for this. ACTH production is enhanced until normal levels of
cortisol are attained. This situation eventually leads to
adrenal hyperplasia.

Structural and Chemical Aspects of Steroid Hormones (see Figures 174a and b)

1. Cholesterol (cholestane is the name of the parent ring system) is
   the precursor of the following five principal classes of steroid
   hormones. These are listed below with the parent ring system
   indicated in the parentheses: progestogen (pregnane), estrogen
   (estrane), androgen (androstane), mineral corticoid (pregnane),
   and glucocorticoid (pregnane).

2. Glucocorticoids (cortisol), mineral corticoids (aldosterone), and
   progestogens (progesterone) all contain 21 carbon atoms and are
   referred to as C-21 steroids. The androgens (testosterone) are
   C-19 steroids while the estrogens (estradiol) are C-18 steroids.
   Note that estrogen is unique because it contains an aromatic
   (ring A is benzenoid) ring system and does not contain the methyl
   group (C-19) attached to C-10. Consequently, in estrogens, the
   -OH group attached to the benzene ring is phenolic and weakly
   acidic, making these compounds soluble in dilute basic solutions.
   This property is employed in the extraction procedures used in
   the identification of estrogens.

3. Androgens and estrogens contain a potential keto ($>$C=O) group in
   the C-17 position and hence are called 17-ketosteroids. It should
   be pointed out that, in man, 17-ketosteroids are not only secreted
   by the adrenal cortex, but also by the ovaries and testes. About
   one-third of the total 17-ketosteroids are of testicular origin.
   Androgens generally appear in the urine as 17-ketosteroids.
   Urinary 17-ketosteroids can be separated into three classes:
   acidic compounds (derived from the bile acids), phenolic compounds
   (which are slightly acidic and are metabolites of estrogens), and
   neutral compounds (derived from the androgens of the adrenal
   cortex and gonads). These metabolites are present in the urine
   as glucuronides. Prior to their separation for purposes of
   identification, free steroid derivatives are obtained by hydrolysis
   of these glucuronides. This is done by treatment with concentrated
   H_2SO_4 or HCl, or by use of the enzyme glucuronidase. The steroids
   are extracted into ether, the acidic (bile acid derived) fraction
   is separated with $NaHCO_3$, and the phenolic portion with 2N NaOH.

A group of steroids which have a sidechain at the 17 position
which can be oxidized by sodium bismuthate to give a 17-keto group
are known as the 17-ketogenic steroid. This name has nothing to
do with the metabolism of these compounds, referring only to their
reaction in vitro during their determination in the clinical
laboratory. They include cortisol, cortisone, Kendall's compound S
(11-deoxycortisol), pregnanetriol, and 17α-hydroxyprogesterone.

This oxidation permits the determination of these compounds by the Zimmerman reaction, which is also used to determine the natural 17-ketosteroids (dehydroepiandrosterone, androsterone, etiocholanolone, androstendione, estrone, and others).

4. The attachment of an -OH group to a saturated ring system can give rise to two isomers. In the α-configuration, the -OH group extends <u>below</u> the plane of the ring and the substitution is shown with a dashed line (---). The β-isomer, in which the hydroxyl bond is indicated with a solid line, has the -OH group <u>above</u> the plane of the ring. The 17-OH groups present in testosterone and estradiol are β-substitutions, while cortisol has an α-OH group (indicated by ---) attached to C-17.

5. An oxy compound has one <u>additional</u> oxygen atom relative to its parent compound while a deoxy compound has one <u>less</u> oxygen. Similarly, a dihydro compound contains two <u>additional</u> hydrogen atoms while dehydro designates a compound having two <u>less</u> hydrogen atoms than the molecule from which the name is derived.

Cushing's Syndrome

1. This problem is due to the presence of excess circulating cortisol. It can result from (a) excessive secretion of ACTH by the adenohypophysis (the most common form), (b) ectopic ACTH production by a tumor of non-endocrine origin (as seen in bronchiogenic carcinoma) with the excess ACTH production leading to adrenal hyperplasia, (c) carcinoma or adenoma of the adrenal cortex, (d) administration of cortisone and its synthetic substitutes in a variety of disorders.

2. The clinical features consist of red cheeks, bruisability, thin skin, obesity with a characteristic distribution of fat (moon face, pendulous abdomen, buffalo hump), hypertension, muscular weakness, osteoporosis, hirsutism, menstrual disturbances, poor wound healing, hyperacidity, peptic ulcers, etc.

3. Laboratory Aspects

    a. The plasma ACTH and, hence, the cortisol level normally shows a diurnal variation with the highest level occurring in the morning and the lowest level around midnight. In Cushing's Syndrome, the cortisol level does <u>not</u> show a diurnal change, remaining constantly elevated.

    b. Urinary 11-hydroxycorticosteroids (cortisol) are elevated. Note any stressful situation can result in both (a) and (b).

    c. Dexamethasone suppression test. Dexamethasone is a synthetic steroid which produces the same effect as cortisol. When this compound is administered to normal individuals, there is a fall in the adenohypophysis ACTH secretion due to an effect on the hypothalamus. This reduction in ACTH output can be monitored by plasma cortisol or urinary 17-hydroxycorticosteroid levels and both should drop in normal individuals. In Cushing's Syndrome, due to adrenal hyperplasia there is an impairment in the suppression of the cortisol level at low dosages of dexamethasone, while with higher levels of the drug, the suppression can be accomplished. In adrenal and ectopic ACTH tumors, however, there is no suppression even after the administration of high doses of dexamethasone.

Addison's Disease

1. This disease is caused by the <u>hypofunctioning</u> of the adrenal cortex, as a result of tuberculosis (the most common cause), idiopathic atrophy of the cortex (presumably due to an autoimmune mechanism), secondary malignancy, mycotic infections, and amyloidosis.

2. The severity of the disease depends upon the amount of functional tissue which remains available to synthesize and secrete the steroid hormones. The most important consequence of the deficiency is the lack of aldosterone, leading to sodium depletion. The clinical characteristics include dehydration, hemoconcentration, hyponatremia, elevated plasma potassium, metabolic acidosis, low blood pressure, muscular weakness, hypoglycemia, gastro-intestinal disturbances, and increased circulating levels of ACTH due to the deficiency of cortisol (lack of feedback suppression of ACTH). The excess ACTH leads to brown pigmentation since the ACTH molecule also has some MSH activity. The two are both peptide hormones and they share part of their amino acid sequence in common. An Addisonian crisis, which needs immediate treatment, is a condition in which the patient is in a dehydrational shock.

3. The laboratory investigation shows decreased excretion of 17-hydroxycorticosteroids and aldosterone. The most important diagnostic step is the demonstration of the non-responsiveness of the adrenal cortex to ACTH administration. This can be accomplished by measuring 17-hydroxycorticosteroids before and after the administration of synthetic ACTH peptide (which contains 24 of the 39 amino acid residues yet possesses the same biological activity). A rise in plasma values of these hormones excludes Addison's disease.

Gonadal Hormones
===

Androgens

1. The name androgen means "producing male characteristics." Any
    substance which has this effect may be referred to as an androgen.
    The Leydig cells (interstitial cells) of the testes synthesize
    two androgens: testosterone and androstenedione. Testosterone is
    the principal one of these and it is responsible for the observed
    endocrine effects of testicular activity. The ovaries also secrete
    small amounts of testosterone and the adrenal cortex in both
    sexes secretes androgens (mentioned earlier).

2. The gonadotrophic hormones FSH and LH (also called ICSH, see
    Table 71) stimulate the production of spermatozoa in the
    seminiferous tubules and of androgens in the Leydig cells,
    respectively. The circulating level of testosterone regulates
    the secretion of the gonadotrophic hormones by a feedback
    mechanism.

3. Biosynthesis of testosterone in the Leydig cells is similar to
    that described in adrenal cortex (see Figure 174b). It has been
    shown that the active form of testosterone is dihydrotestosterone
    (discussed earlier). The major catabolic transformations of the
    androgens take place in the liver. Testosterone is converted to
    5α- and 5β-androsterone (5β-androsterone is also known as
    etiocholanolone) which is then conjugated with glucuronic acid
    or sulfate and excreted as a 17-ketosteroid.

4. The basic molecular mechanism by which steroid hormones (including
    the androgens) function were discussed earlier. The specific
    effects of androgens include

    a. Development of primary and secondary male sexual characteristics
        (masculinization). The effects of testosterone are manifested
        in both sexes. In males, its influence predominates while in
        females, it has a minimal effect partly because it has to
        overcome feminization caused by the estrogens.

    b. A major response to androgens is the anabolism of nitrogen
        and calcium compounds, reflected (especially in males) by an
        increase in skeletal muscle mass and growth of the long bones
        prior to closure of the epiphyseal cartilage. Castration
        results in a delay of epiphyseal closure, while excessive
        secretion of androgens can cause premature closing of the
        epiphyses.

c. Some of the other effects of testosterone are an increase in the basal metabolic rate (BMR), promotion of Na^+, Cl^- and water reabsorption by the kidney tubules, and stimulation of erythrocyte production in the bone marrow.

5. Attempts have been made to synthesize androgens with anabolic activity but without their masculinizing effects. Testosterone is used therapeutically in carcinoma of breast. In this regard, it is highly desirable to have a compound with androgenic activity which does not cause masculinization. Following are some examples of synthetic androgens.

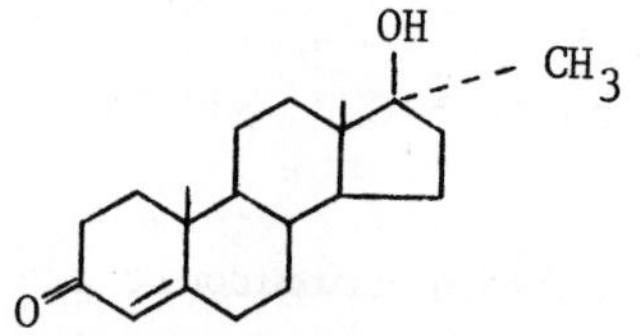

17α-Methyltestosterone
(a potent androgen, orally effective)

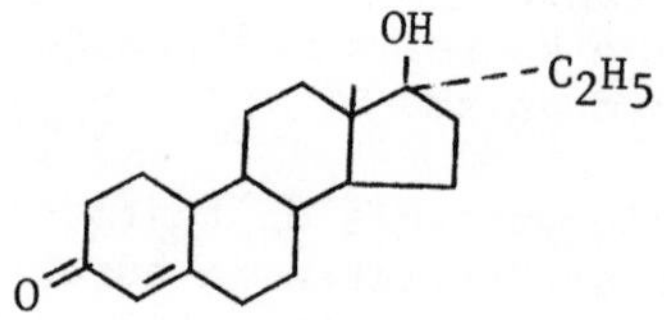

17α-Ethyl-19-nortestosterone

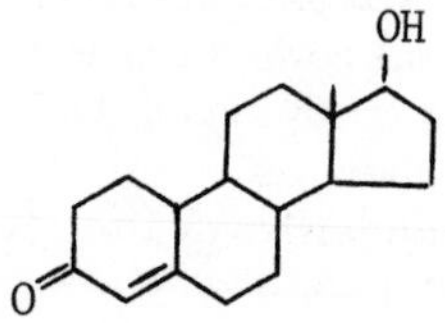

19-Nortestosterone

The above compounds are named as <u>norsteroids</u> because they lack an angular methyl group (C-19) attached to C-10. Estrogens also lack this C-19 angular methyl group. These norsteroids have been shown by bioassay procedures to have an anabolic activity twenty times greater than their masculinizing activity.

Estrogens

1. The term estrus (or oestrus) from which the word estrogen is derived comes from the Greek <u>oistros</u> (gadfly). It refers to the state of sexual excitability or receptivity in females of many species which occurs with varying periodicity. Apparently, the frenzy caused **by the irritation of the gadfly is likened to this state.** Estrogen is now used to refer to any of a group of compounds which cause feminization as seen by primary and secondary sexual characteristics. The natural estrogens are produced in the

Graafian follicles of the ovaries under the stimulus of FSH. Three
types of estrogens have been isolated from ovarian tissue and
human urine. These are

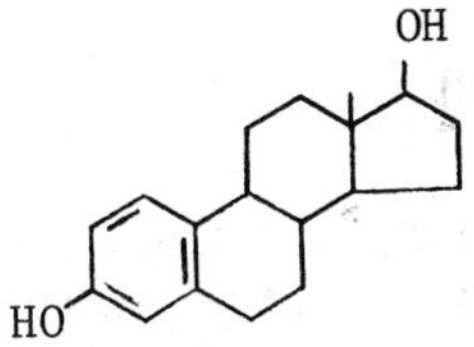

β-Estradiol 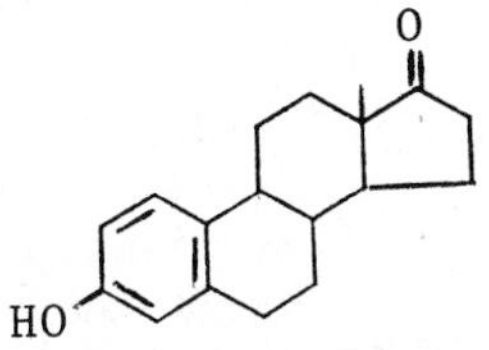   Estrone 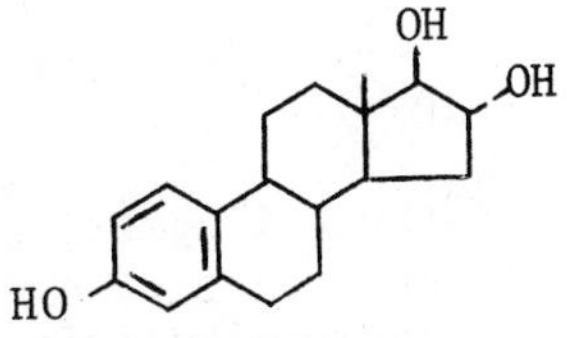   Estriol

2. Estradiol is the major hormone of the ovarian secretion and is
   the most potent of the three. The other two are metabolic products
   of estradiol. Estriol is produced mainly in the liver from the
   other two hormones. It undergoes conjugation with glucuronic acid
   and sulfate and is eliminated in the bile and urine. Measurement
   of urinary estriol provides a useful index of estrogen production
   (see below).

3. Estrogens are C-18 steroids, lacking the angular methyl group at
   C-10, and ring A is aromatic in these compounds. Androgens are
   C-19 steroids. The biosynthesis of the estrogens is given in
   Figure 174b.

4. The target organs of the estrogens are the uterus, the vagina, the
   adenohypophysis, the hypothalamus, and the mammary gland. At these
   sites, these hormones have the actions summarized next.

   a. They promote the development, maturation, and function of the
      internal and external female reproductive organs. The
      proliferative phase of the uterine cycle is caused by the
      cyclic production of estrogen.

   b. Their action on the hypothalamus and the adenohypophysis is
      to suppress FSH secretion and promote LH secretion.

   c. Other effects include enhancement of the activity of alkaline
      phosphatase, an increase in the amounts of glycogen in the
      endometrium and the vagina, increased RNA and protein synthesis,
      activation of transhydrogenase, and increased accumulation of
      lipids in muscle and other tissues.

5. Natural and synthetic estrogens are used clinically in ovarian
   agenesis, other estrogen deficiencies, and in treating the
   symptoms associated with menopause. One of the synthetic estrogens
   is diethylstilbestrol whose structure is shown below.

H5C2

OH

OH

C=C

or

HO

HO

C2H5

Diethylstilbestrol therapy has been initiated in the first trimester of pregnancy if bleeding occurs or if there is risk of a miscarriage. A tragic side effect of this treatment has recently been reported. The adolescent female offspring of women who had received this non-steroidal estrogen during the indicated period of pregnancy show an increased incidence of adenocarcinoma of the vagina.

6. Measurement of estriol, a metabolite of estrogen, in a 24-hour urine sample has been useful in the assessment of placental and fetal functions in pregnant women. After fertilization, the ovum gets implanted in the uterine wall and the developing placenta secretes chorionic gonadotrophin (CG) which continues the secretion of estrogen and progesterone from the corpus luteum. This maintains the corpus luteum in the luteal phase and prevents the onset of menstruation. The CG secretion reaches a peak at about 13 weeks of gestational life. At this time, the fetus and placenta together take over the production of estrogens and progesterone and chorionic gonadotrophin secretion declines. The estrogens produced by the integrated feto-placental unit appear in the maternal circulation and the estriol formed appears in the urine. Urinary estriol levels can reach sufficiently high values for accurate laboratory measurements to be made. For a particular individual under normal conditions, urinary estriol increases throughout pregnancy. In monitoring normal feto-placental development, it is essential to perform assays on 24-hour urine specimens and to make periodic determinations. Single determinations are misleading unless they are extremely low (indicating non-secretion), because the normal range varies greatly.

Progesterone

1. Progesterone is an intermediate in the biosynthesis of the adrenal corticosteroids, testesterone, and estradiol (see Figures 174a and b). Thus it is present in the adrenal cortex, the placenta, and the ovary. It is also produced in the corpus luteum, and this tissue may therefore be considered as a transitory endocrine gland.

2. In a non-pregnant female, the corpus luteum is the principal source of progesterone, while the placenta provides most of this hormone during the second and third trimesters of pregnancy. During the follicular phase of a normal menstrual cycle, progesterone is present in the blood in very low concentrations. This follicular

phase progesterone is of adrenal origin. The progesterone
rises in the luteal phase of the cycle if and only if ovulation
occurs. Thus, a blood progesterone measurement during the second
half of the cycle can be used to determine if ovulation did occur
and if, in fact, a corpus luteum capable of secreting progesterone
is present in the ovary.

3. The major metabolite of progesterone is pregnanediol which is
    excreted in the urine as both the glucuronide and the sulfate.

Progesterone   Pregnanediol

4. The major biological effects of progesterone are

    a. Preparation of the uterus for implantation of the ovum to
        occur, a process initiated by estrogens. The actions of
        progesterone occur only <u>after</u> prior estrogen action. The
        preparation consists of developing a secretory type of
        endometrium and of diminishing the muscular contractions of
        the Fallopian tubes and of the uterus.

    b. Stimulation of the glandular development of the breasts by
        inhibition of the effects of estrogens.

    c. Progesterone is reponsible for stimulating the basal metabolic
        rate and thus causing a periodic rise in the body temperature
        (about 0.5 to 1.0°F) during the luteal phase (second half)
        of the normal menstrual cycle.

    d. Suppression of ovulation, hence progesterone and its
        derivatives, are used to prevent conception.

The Menstrual Cycle

The reproductive system of the female, unlike that of the male, shows
regular cyclic changes. During each cycle, the organs in the female
which are needed for reproduction are prepared for fertilization and
pregnancy. Although the duration of a menstrual cycle is extremely

variable, an average figure of 28 days is generally used. The first day of the cycle corresponds with the first day of vaginal menstrual bleeding. The period from then until ovulation is the <u>follicular phase</u> of the cycle, during which time the follicle is developing. At the end of this phase, the follicle cell bursts open and releases the ovum. From then until onset of the next menses, the corpus luteum is developing, hence the name <u>luteal phase</u> for the second half of the cycle.

The hormonal changes associated with the menstrual cycle are illustrated in Figure 176. To start the cycle, a hypothalamic-releasing factor (LRH) stimulates secretion of the pituitary gonadotropins which in turn stimulate the secretion of estrogen. Rising estrogen levels stimulate a sharp rise in releasing factor and a rise in LH that triggers ovulation and formation of the corpus luteum. This structure secretes high levels of progesterone and estrogen which inhibit further output of gonadotropins. As the gonadotropin levels decline, the corpus luteum ceases to function, steroid (estrogen and progesterone) levels drop, and gonadotropin levels once again begin to rise, starting the hormonal cycle over again. If pregnancy occurs, chorionic **Gonadotropin (CG)**, secreted by the corion of the placenta, keeps the corpus luteum competent (replacing the pituitary gonadotropins) so that estrogen and progesterone continue to be secreted at high levels. Thus the cycle is broken until the pregnancy terminates. Some birth control pills function in a similar fashion to break the rhythm and prevent ovulation.

For many years, it was thought that the hypothalamic-hypophyseal system contained inherent cyclic capabilities. More recent studies on negative and positive feedback modulations by the ovarian steroids have shown this to be incorrect. Rather, the cyclic gonadotropin released during the reproductive years is a <u>secondary</u> phenomenon dependent on ovarian function. Operationally, the hypothalamic-hypophyseal system can be viewed as a provider and the cyclic ovarian steroid output as a modulator. This output exerts an overlapping, interrelated sequence of negative and positive feedback action and produces a differential effect on the release of FSH and LH.

Figure 176. Composite Diagram of the Variation of Certain Hormones in the Normal Menstrual Cycle
(note: ng = nanograms; pg = picograms; mIU = milli-international units)

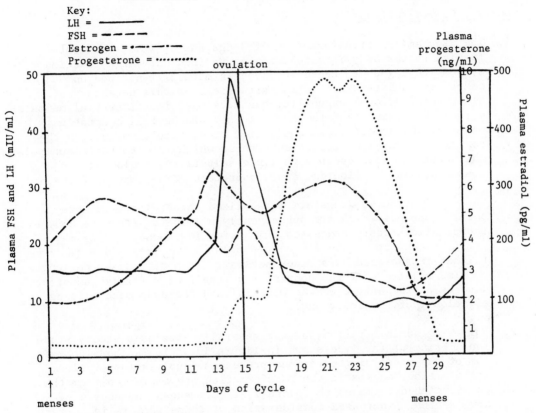

Key:
LH = ————
FSH = — — — —
Estrogen = •—•—•—•—
Progesterone = •••••••••••

The hypothalamic-releasing factor (LRH) stimulates pituitary gonadotropins, which in turn stimulate the secretion of estrogen. Rising estrogen levels stimulate a sharp rise in releasing factor and a rise in LH that triggers ovulation and the formation of the corpus luteum. High levels of progesterone and estrogen inhibit further output of gonadotropins, and the levels decline. The corpus luteum ceases to function, steroid levels drop.

14
Biochemical Transformation of Drugs

with Pushkar N. Kaul, Ph.D. *

General Considerations

1. The biochemical transformation of drugs and of other chemicals foreign to the body is a function of various enzymatic reaction systems analogous to those involved in metabolizing food and endogenous substances. It is fortunate that this necessary biotransformational capability has developed in terrestrial mammals, perhaps as a result of the evolutionary biochemical adaptation process. Without it, man would find it difficult to survive the toxicity of foreign organic substances which, due to their non-polar character and consequent high lipid solubility, would tend to accumulate in the lipoidal and lipoprotein compartments of the body.

   In most instances, drugs are metabolized into relatively more polar compounds which are subsequently eliminated from the body by the usual excretory processes. In a few instances, a drug may first be converted into a more non-polar metabolite which is then further metabolized by another pathway. In most cases, the length of time required for the body to rid itself of a foreign substance depends on the nature of the material and the drug-metabolizing efficiency of the individual.

2. Following administration, a drug generally attains a steady state equilibrium of the type shown in Figure 177. It is the algebraic sum of the rates of the various processes involved (absorption, distribution, enzymatic biotransformation, and excretion from the body) which determines the overall effectiveness and course of drug action. Increased consideration of these factors in current clinical practice has improved the rationale of therapy and thus the quality of patient care.

3. a. Drugs, unlike vitamins and amino acids, are absorbed from the gastrointestinal tract by passive diffusion. Only unionized drug molecules are believed to cross the biological membranes, except in rare cases when an active or a facilitated transport

---

* Professor of Pharmacodynamics and Toxicology and Adjunct Professor of Pediatrics and Research Medicine, University of Oklahoma, Norman, Oklahoma.

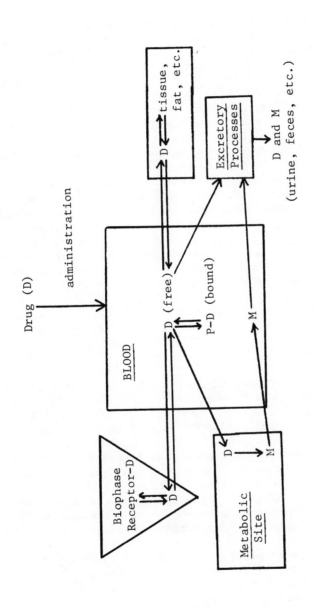

Figure 177. Processes Which Contribute to the Steady State Equilibrium of a Drug In Vivo

Drug (D)

administration

BLOOD

D (free)

P–D (bound)

M

Biophase
Receptor–D

D

Metabolic
Site

D → M

tissue,
fat, etc.

Excretory
Processes

D and M
(urine, feces, etc.)

D = drug

P–D = protein-bound drug

M = metabolite of D

[1245]

may be involved. Acidic drugs do not ionize in the acidic medium of the stomach and are therefore absorbed through the gastric mucosa. Basic drugs ionize in the stomach and are hence absorbed later, in the intestinal segment.

b. Not only is the extent of absorption of a drug important but also the rate at which its absorption occurs. In general, the faster a drug is absorbed, the higher and more rapidly attained will be the maximum blood level. It is possible to have two dosage forms of the same drug which differ in efficacy simply because of unequal rates of absorption. Although both forms may be completely absorbed, the one which enters the circulation more rapidly will result in a higher peak blood level and, hence, a greater clinical effect. This can be critical in the administration of antibiotics where it is desirable to attain the highest possible peak level.

c. In the actual process of absorption of a drug from any site of administration other than intravascular, the drug molecules must cross the capillary membrane to enter the blood. The unidirectional flow of drug molecules into the circulation continues until no more of the drug remains outside of the capillary wall at the site of absorption. Like a dialysis system, this entry of molecules can be considerably affected by the rate of blood flow inside the capillary lumen. Any changes in the blood flow due to normal diurnal variations, spontaneous over-exertion, or a pathological state affecting the cardio-vascular system can significantly influence the rate of absorption of a drug.

4. Having entered the circulation, a drug molecule has several possible fates. It may undergo elimination via the urine or bile, escape into the interstitial fluid and then into the cells of the various organs, or be metabolically altered while passing through the liver, kidney, lungs, or other organs. All these possibilities, however, are open only to the free molecules (see Figure 177). Molecules which are bound to plasma or cellular proteins are not available for these processes unless they first dissociate from their binding site. The extent of this dissociation increases (by the law of mass action) when the quantity of free drug in the blood decreases as a result of the various dispositional processes described above.

5. The binding of drugs to blood proteins, as well as to the cellular or subcellular structures of the biophase (site of action), involves an interaction between the drug molecule and a small portion of the biological macromolecule known as the receptor. This binding may lead to no discernable pharmacological response, as in the case of plasma protein-binding of imipramine. In other cases, a definite response may be seen as when acetazolamide binds to and inhibits

carbonic anhydrase in the kidney tubular cells, producing diuresis.
In both situations, however, the binding involves similar types
of forces.

6. There are factors other than protein-binding which influence the
    distribution of a drug. Preferential uptake of iodides by the
    thyroid, of tetracyclines by bone, of chlorpromazine by hair and
    skin, and of thiopental by adipose tissues are only some of the
    examples of how specific tissues may affect the distribution and
    blood level of a drug following its administration. A drug may be
    distributed in all the fluid compartments of the body and consequently
    develop only a low circulating concentration. Such a drug is said
    to have a large volume of distribution.

7. Those drugs which enter the cerebrospinal fluid (CSF) are believed
    to have crossed the so-called blood-brain barrier. This barrier
    is not an anatomical structure but rather a concept devised to
    explain why some drugs enter the CSF while others do not. From
    the evidence available, this property of drugs is probably due to
    their physico-chemical characteristics and to the presence of glial
    cells in the brain. Generally speaking, highly lipid-soluble drugs
    cross the blood-brain barrier whereas highly polar and ionized
    drugs do not seem to enter the CSF.

Protein-Binding of Drugs

1. The bonding forces involved in the drug-protein binding include
    those encountered earlier in the discussion of protein structure
    (see Chapter 2). They are generally classified into primary and
    secondary bonds. The primary bonds are mostly ionic in nature,
    involving coulombic (electrostatic) attractions between opposite
    charges on the drug and the receptor molecules. They operate at
    relatively long distances (< 4 Å) and aid in the initial
    juxtaposition of the drug and the receptor site so that the
    secondary bonds can be easily formed. The secondary bonds include
    hydrogen bonds and van der Waal's (London) forces.

2. Hydrogen bonds were discussed earlier in Chapter 2. They involve
    the sharing of a proton by two electronegative atoms having
    unbonded electrons. The two atoms sharing the H^+ are usually
    about 3 Å apart. An example of hydrogen bonding is illustrated
    by the hypothetical binding of warfarin to albumin shown in
    Figure 178.

3. Another secondary bonding force is the van der Waal's bond or
    London dispersion force. These are relatively weak (0.1 to 1
    kcal/mole), but in multiples they can reinforce the binding in-
    teraction significantly. These bonds usually occur between carbon
    atoms of the two interacting molecules (drug and protein) when they

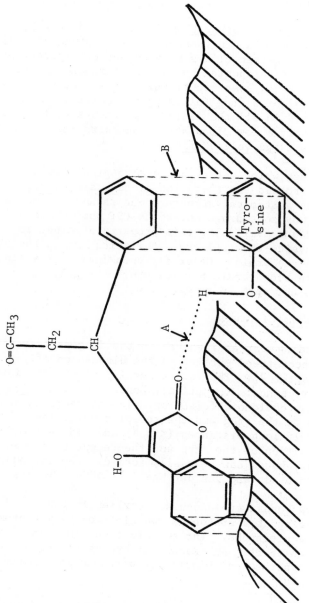

Figure 178. Postulated Hydrogen and Van der Waal's Bond Formation Between Warfarin (Coumadin) and a Tyrosine Residue of Albumin

A = hydrogen bond

B = van der Waal's bond

approach each other closely enough for the outermost electrons of
the carbon of one molecule to be within the range of the electrostatic field of the nucleus belonging to the carbon atom of the
other molecule. Such interactions usually occur in multiples and
involve either flat aromatic rings or long aliphatic carbon chains
of the drug molecule and a comparable structural region of the
protein molecule. The sum of the primary and secondary forces
determines the strength of the drug-protein interaction.

4. The majority of drugs which undergo plasma-protein binding are
    attached to albumin. Some drugs may also be bound to erythrocyte
    membranes or to cytoplasmic proteins of erythrocytes. For example,
    it is postulated that chlorpromazine is bound to the erythrocyte
    membrane, whereas its metabolites are most likely bound to sites
    in the red cell interior (other than carbonic anhydrase).
    Acetazolamide, on the other hand, is largely bound to carbonic
    anhydrase.

5. The binding of drugs to blood proteins may have important clinical
    implications. If the drugs are strongly bound, as are the
    dicumarol anticoagulants (see Chapter 10) and suramin (a
    sulfonamide), the circulating level of <u>free</u> drug will be low and
    the drug will be only slowly removed from the blood stream. As was
    indicated earlier, the amount of free drug in the blood is crucial
    because it is only these molecules which are active. The slow
    elimination of a tightly bound drug is another manifestation of the
    same effect, since only free drug is actively metabolized and
    excreted.

    a. Protein-bound drug molecules act as a storage form of the drug.
        This can be either advantageous or dangerous. Suramin is used
        to treat African sleeping sickness which is caused by the
        parasite <u>Trypanosoma gambiense.</u> Because of its binding to
        plasma proteins, a single injection of the drug provides a
        parasiticidal effect for up to a week, a distinct advantage.

    b. Warfarin, one of the coumarol anticoagulants, is normally
        distributed in the blood with 97% of it bound to albumin and only
        3% free. Since it is a potent drug, this free level is clinically
        adequate for anticoagulant therapy. A number of acidic drugs
        (penicillin, sulfonamides, salicylates, etc.) are able to
        displace warfarin from albumin, thus increasing the level of
        free anticoagulant. A displacement of only 3% would double the
        free level and the danger of internal hemorrhage is imminent in
        such a situation. Many drug-drug interactions (discussed
        later) are based on this type of competitive binding and
        displacement of bound drugs.

6. Although both acidic and basic drugs are capable of binding to
    blood proteins, most studies carried out so far deal mainly with
    the acidic drugs. In general, all drugs possessing one or more
    homocyclic aromatic rings (benzene, naphthalene, etc.), one or
    more potential sites for hydrogen bonding, and an atom capable of
    acquiring a charge at physiological pH, are potential protein
    binders.

Pathways of Metabolism

1. It is frequently true that the activity of a drug ends when it is
    metabolized, but it is <u>not</u> correct to believe that drug metabolism
    is <u>synonymous</u> with detoxication. In some cases, the metabolites
    have greater activity than the drug itself and may be more toxic
    or even better therapeutic agents. Phenacetin and acetanilid are
    both converted in the body to acetaminophen which is also an
    antipyretic analgesic agent. Recently it was shown that
    chlordiazepoxide (Librium) yields two metabolites which are at least
    as active as the parent drug. One of these metabolites is already
    being used clinically as a minor tranquilizer.

2. Drugs undergo four major types of reactions in the body. These
    are <u>hydrolysis</u>, <u>reduction</u>, <u>oxidation</u>, and <u>conjugation</u> (synthesis).
    The major site of these reactions is the liver, but recent evidence
    suggests that the lungs may also have a rich store of drug-
    metabolizing enzymes. These enzymes are generally located in the
    endoplasmic reticulum, a membrane located inside the cell.
    During disruption of the cell and fractionation of the subcellular
    components, it disintegrates yielding small particles or fragments
    known as <u>microsomes</u>. Enzymes which are associated with this cell
    fraction are termed microsomal enzymes. There are also many drug-
    metabolizing enzyme systems present in the cytoplasm.

3. <u>Hydrolysis</u>

    Drugs which are either esters or amides are usually hydrolyzed.
    For example, acetylcholine, an endogenous ester, is hydrolyzed by
    acetylcholinesterase to choline and acetic acid. Several other
    choline esters are handled similarly by plasma cholinesterases.
    Local anesthetics such as procaine and analgesics such as aspirin
    are also hydrolyzed by the body. These reactions are catalyzed
    by hydrolases (esterases and amidases) found in blood plasma and
    in the soluble fraction of the cells of tissues such as the liver.
    Substrate specificity of the hydrolases may vary from tissue to
    tissue and from species to species.

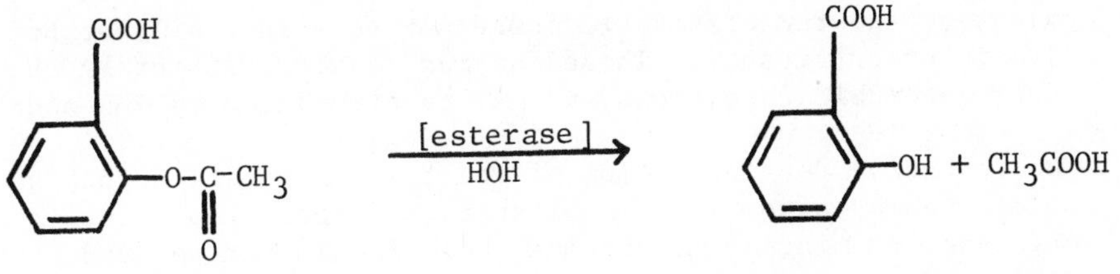

Acetyl-salicylic Acid
(aspirin)
→ [esterase] / HOH →
Salicylic Acid + Acetic Acid

Hydrolytic biotransformations are relatively simple in that the
products are usually predictable and can therefore be traced and
studied with ease. Not all ester-containing drugs are hydrolyzed
to a significant extent in man. For example, only about 2% of a
given dose of atropine is degraded in this manner in man. Some
strains of rabbits, however, have atropinesterase, an enzyme not
found in human tissues, which effectively metabolizes this drug by
hydrolysis.

4. Reduction

a. Only a few drugs have structures capable of being reduced
    in vivo. Azo reduction of prontosil rubrum (sulfamidochrysoidine),
    an azo dye and the precursor of the sulfonamide family
    of drugs, is perhaps the first (1935) and best-known
    example of this type of transformation. Though inactive by
    itself, the dye is reduced **in vivo** to an active antibacterial
    metabolite, sulfanilamide, as shown below.

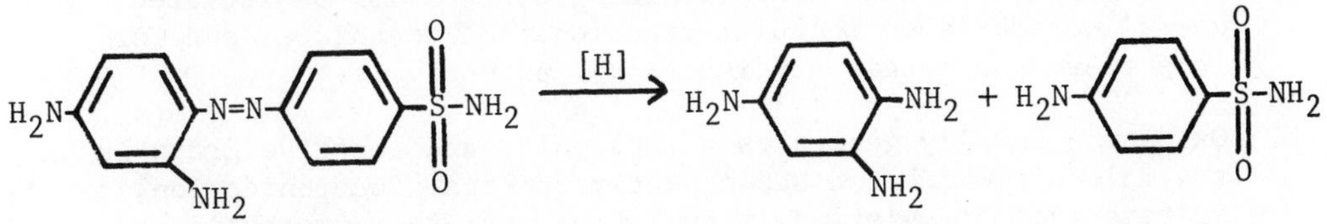

Prontosil rubrum → [H] → Triaminobenzene + Sulfanilamide

Reduction of a nitro group can also occur if one is present in
the drug. Chloramphenicol, a broad-spectrum antibiotic, is an
example. Both the bacterial and the mammalian nitroreductase
systems are capable of catalyzing this metabolic reaction.

b. The azo and nitro reductase activities are present in the light
    microsomal fraction which sediments more slowly than that
    containing the drug-oxidizing enzyme systems. Unlike

oxidases, the reductases are found not only in the liver but also in other tissues. These enzymes require NADH or NADPH under anaerobic conditions and may be stimulated by FMN and FAD.

c. The reductive dehalogenation of volatile anesthetics such as halothane may also be of importance. In these reactions, again catalyzed by microsomal enzymes, chlorine is removed and replaced by hydrogen. Chlorpromazine, a major tranquilizer, appears likely to undergo a similar transformation, but this has yet to be confirmed experimentally.

d. Non-microsomal reductions of aldehydes are generally catalyzed by alcohol dehydrogenase in the presence of NADH. (Note that this is the <u>reverse</u> of the reaction which occurs when an <u>alcohol</u> is the substrate.) This enzyme, under the name retinal reductase, plays an important role of converting retinal to retinol during the process of vision (see Chapter 10). Chloral hydrate, a classical sedative, is another aldehyde which is reduced in the body by alcohol dehydrogenase, in this case to trichloroethanol. It is this metabolite which possesses the sedative property and not the parent drug. Dimethyl sulfoxide, disulfiram (the inhibitor of alcohol dehydrogenase used in alcoholism; see Chapters 4 and 6), and compounds of pentavalent arsenic and antimony are all reduced by non-microsomal enzymatic systems.

5. <u>Oxidations</u>

a. The greatest proportion of drugs undergo oxidation in the body. The oxidases are present both in the endoplasmic reticulum and the cytoplasm. **These drug-oxidizing systems can be isolated** by well-established cellular fractionation techniques and for this reason have been studied fairly extensively.

b. Oxidation usually generates a more polar and reactive group in the molecule which can subsequently undergo glucuronidation, sulfatation, acetylation, methylation, or other synthetic or conjugative reactions.

6. <u>Nonmicrosomal Oxidations</u> (Cytoplasmic and Mitochondrial)

a. <u>Alcohols</u> are oxidized by NAD-linked alcohol dehydrogenase (see Chapter 4) and found in liver, kidney and lung. Recall that this same enzyme, in the presence of NADH, also catalyzes aldehyde reductions. Certainly the major pathway of metabolism of ethanol is oxidation by this enzyme, though recent evidence suggests that a small proportion may be metabolized by the microsomal ethanol oxidizing system.

b. <u>Aldehydes</u> are oxidized by an NAD-dependent aldehyde dehydrogenase and by aldehyde oxidase. Both of these enzymes have been isolated from liver. Purines and xanthines are oxidized by xanthine oxidase (see Chapters 3 and 5) which is very similar to aldehyde oxidase.

c. <u>Monoamines</u> such as catecholamine, tryptophan, tyramine, and norepinephrine, having alkylamine chains with no α-substitution, are oxidized by MAO present in mitochondria. Alpha-carbon substituted short-chain amines such as amphetamine are metabolized by the microsomal oxidizing system. Non-specific MAO and diamine oxidase, present in plasma, can also oxidize the monoamines mentioned, but the reaction rates may be relatively slow.

d. <u>Steroidal hormones</u> such as deoxycorticosterone, are oxygenated by a mixed function oxygenase system which, as an exception to the rule, is present in the mitochondria of the adrenal cortex. Unlike the rest of the non-microsomal oxygenase systems, the adreno-cortical mitochondrial oxygenase contains cytochrome P-450, the heme protein otherwise found only in the endoplasmic reticulum of the liver, kidney and lung cells **(see Chapter 4).**

7. <u>The Microsomal Oxidizing System</u>

a. The supernatant fraction of a cell homogenate centrifuged at 9000 gravities contains microsomes and several co-factors capable of the oxidative transformation of drugs. The microsomes can be isolated as a pellet by centrifuging at about 110,000 gravities for one hour. When reconstituted, in the presence of oxygen, with the cell supernate, $NADP^+$, Mg^{++}, and nicotinamide, these microsomes regain their ability to perform oxidations. The supernate provides the enzymes required to maintain $NADP^+$ in its reduced (NADPH) form.

b. The microsomal oxygenation system differs from the mitochondrial and cytoplasmic systems in that it fixes <u>molecular</u> oxygen to the organic substrate (drug) and does not produce any useful energy. Briefly, NADPH (or NADH) reduces (activates) molecular oxygen. The "active oxygen", [0], is carried by cytochrome P-450 to the drug molecule where it is attached, usually as a hydroxyl group. The microsomal system is referred to as a <u>mixed function</u> oxygenase because it <u>reduces</u> O_2 while <u>oxidizing</u> the drug or other organic substrate. It is capable of oxidizing carbon, sulfur, nitrogen, and other groups.

c. Although the $NADP^+ \leftrightarrow$ NADPH and flavoprotein dependence of the oxidation is common to both the mitochondrial and the microsomal oxidative systems, there is an established difference between them. Unlike mitochondria, microsomes normally contain no cytochrome C which is part of the electron-flow pathway in the mitochondrial oxidative chain. Instead, this function is carried out in a somewhat modified fashion by cytochrome P-450. This heme protein, in its reduced form, is capable of combining with carbon monoxide to form a complex having a characteristic absorption maximum at 450 nm. According to current data, cytochrome P-450 appears to be the oxygen-activating enzyme. It reduces one atom of oxygen to water and introduces the other into the drug molecule. The electron carriers involved and their relationship to each other was shown earlier in **Figure 93, Chapter 7.**

d. Because it has been difficult to solubilize and purify cytochrome P-450 (the systematic name is B-420; **see Chapter 4**), a vast amount of questionable data has been gathered on this hepatic microsomal enzyme. It has been established, however, that it is present on the smooth endoplasmic reticulum.

e. Oxidation, frequently followed by conjugation, is the primary pathway for the biotransformation of drugs. The oxidation usually occurs at a carbon, nitrogen, or sulfur atom and many times involves removal of a group, such as the demethylation of an alkylated oxygen, ring nitrogen, or amino group. Some examples of these reactions are given below. Conjugation (synthetic) reactions are discussed in the section following this one.

Oxidized drug may further be transformed into a relatively more polar derivative by any one of many synthetic routes described in the next section. Since oxidations usually occur on either a carbon atom of the drug molecule or on nitrogen or sulfur, the examples will be discussed in that order. In addition, many oxidative transformations involve removal of a group such as a methyl from a nitrogen or an oxygen, or from an amine group.

i) <u>Aliphatic</u> (alkyl) sidechain oxidation results in the attachment of an oxygen atom to the carbon to produce a hydroxyl group. <u>Pentobarbital</u> in man is transformed into an alcohol by this sort of reaction.

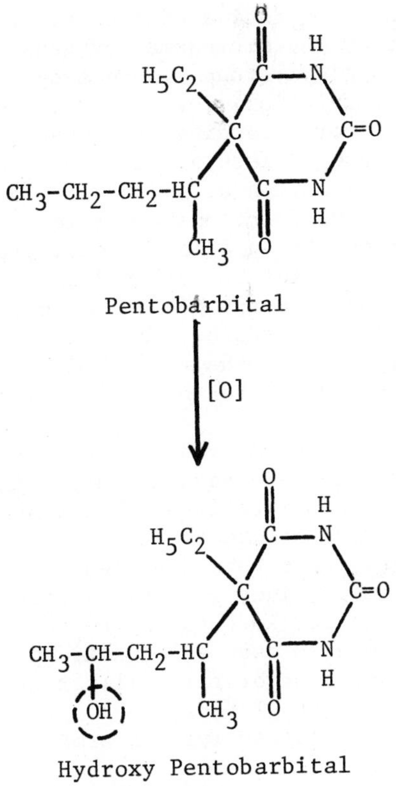

Pentobarbital

[O]

Hydroxy Pentobarbital

Sidechain oxidation of phenylbutazone, a drug used in arthritis, generates the corresponding hydroxy derivative which is a potent uricosuric. For this reason, phenylbutazone is also useful in treating acute gout.

ii) <u>Ring hydroxylation</u> usually occurs with both alicyclic and aromatic compounds. Phenobarbital, for example, is hydroxylated on the phenyl ring to p-hydroxyphenobarbital

Phenobarbital   Hydroxy Phenobarbital

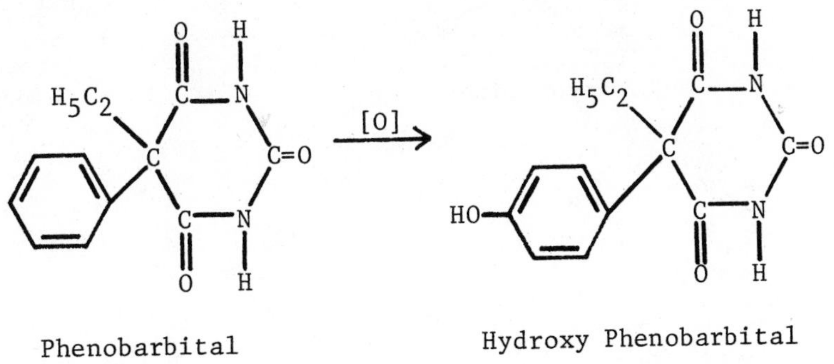

iii)   Dealkylation by oxidation of a carbon can also be
       accomplished by the microsomal oxygenase system. Oxidative
       removal of an alkyl group from nitrogen (N-dealkylation),
       oxygen (O-dealkylation), or sulfur (S-dealkylation)
       generates the corresponding hydrogenated ($>$NH, -OH or -SH)
       group in the drug molecule. Each of these groups is then
       capable of conjugation or synthesis. During the removal
       process, the alkyl functions are oxidized to an aldehyde
       and then further to an acid. Methyl groups generally
       are finally converted to carbon dioxide. The
       N-demethylation of morphine was one of the first reactions
       of this type to be studied. N-(^{14}C-methyl)-morphine was
       given to subjects and the radioactivity was partially
       recovered from their breath as $^{14}CO_2$.

iv)   N-dealkylation of primary amines occurs at a faster rate
       than that of secondary and tertiary amines. Demethylation
       of tertiary amine drugs is of importance, however, because
       in several cases the demethylated metabolites have turned
       out to be active and have thus led to the discovery of
       new drugs. Imipramine, a tertiary amine used as an
       antidepressant, is demethylated to desipramine, an active
       metabolite now also used as an antidepressant.
       Chlorpromazine, a major tranquilizer is also demethylated
       to yield monodesmethyl (Nor_1) chlorpromazine which has
       many pharmacological activities similar to the parent
       drug.

Chlorpromazine Nor_1 Chlorpromazine

v) O-dealkylation, yielding a phenol, usually occurs in drugs
which have an alkylaryl ether structure. In man, nearly
10 percent of a dose of codeine is dealkylated to morphine.

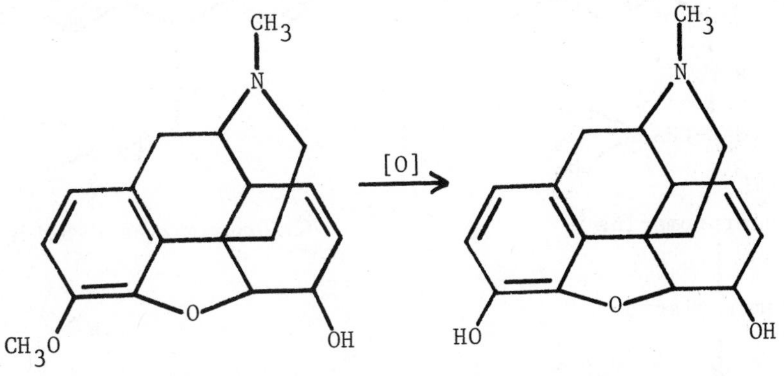

Codeine   Morphine

Acetophenetidin (phenacetin) is an analgesic and
antipyretic drug which is O-deethylated to generate an
active antipyretic metabolite, N-acetyl-p-aminophenol.

vi) S-dealkylation of various thioethers has been found to
occur *in vitro* with liver, kidney, and spleen microsomal
preparations in the presence of NADPH and oxygen.
An *in vivo* example is the S-demethylation of 6-methyl
thiopurine to 6-mercaptopurine, an antitumor compound
(see Chapter 5).

vii) Tertiary amine drugs (e.g., tremarine, imipramine,
chlorpromazine) are also oxygenated on the nitrogen to
form N-oxides. Primary and secondary amines, on the
other hand, undergo N-hydroxylation to form hydroxylamines.
In some cases, the N-hydroxylated metabolites are more
toxic than the parent drug. Of considerable interest,
relative to chlorpromazine metabolism, is the recently reported
identification of large amounts of N-hydroxyl-Nor$_1$
Chlorpromazine in the red cells of schizophrenic
patients who had been receiving daily chlorpromazine
doses. Using chlorpromazine as an example, these two
types of transformation are shown below.

Chlorpromazine   [O] ⟶   Chlorpromazine N-oxide

demethylation ↓

Nor$_1$ Chlorpromazine   [O] ⟶   N-hydroxyl-Nor$_1$ Chlorpromazine

viii) Oxidative deamination (of a primary amino group) is
analogous to dealkylation. Amphetamine and other alpha
methyl amine drugs are metabolized by oxidative
deamination catalyzed by the mixed function microsomal
oxygenase system of the endoplasmic reticulum. This
reaction is different from the non-microsomal oxidative
deamination of short-chain, non-alpha-substituted amines,
such as norepinephrine. Those reactions are carried out
by mitochondrial monoamine oxidase (MAO). The
deamination of amphetamine to phenylacetone was the
first microsomal oxidation discovered.

Amphetamine   [O] ⟶   Phenylacetone

ix) The S-oxidations of phenothiazine derivatives such as
     chlorpromazine are good examples of this type of reaction.
     Sulfoxides of this major tranquilizer as well as of its
     various demethylated and ring hydroxylated metabolites
     have been recognized in man. The same oxygenase system
     may further oxidize the sulfoxides to sulfone.

Chlorpromazine   Chlorpromazine Sulfoxide

x) Oxidative desulfuration of the thiobarbiturates and
    thioureas can be accomplished by the microsomal oxygenase
    system, replacing the sulfur with an atom of oxygen.
    Thiopental is biotransformed into pentobarbital by a
    reaction wich requires NADH (instead of the usual NADPH)
    and which is stimulated by Mg^{++}.

Thiopental (enol)   Pentobarbital (enol)

xi) In conclusion, the oxidative pathways of drug metabolism
     occur frequently and include a wide variety of biochemical
     transformations. In most of the reactions, both the
     microsomal and the non-microsomal, the product is usually
     more polar (water soluble) than the reactant. Often a
     hydroxyl or a primary or secondary amine is produced,
     all of which are capable of undergoing further synthetic
     reactions.

8. Synthesis or Conjugation

These are reactions in which a number of endogenous small molecules react with polar functions of drugs or their metabolites to yield various derivatives. The reactions are catalyzed by specific enzymes and require energy. The conjugations include glucuronidation, O- and N-acetylation, O-, S-, and N-methylation, glycine conjugation, ethereal sulfate formation, and (occasionally) glutathione conjugation. The commonest and probably most-studied conjugations are those producing glucuronides.

a. Glucuronidation

    i)   In those tissues capable of this type of conjugation, the enzymatic machinery necessary is present in the soluble fraction and the microsomes. The active donor of glucuronic acid, uridine diphosphoglucuronic acid (UDPGA), is synthesized from glucose-1-phosphate and uridine triphosphate (UTP) by the cytoplasmic enzymes of the uronic acid pathway (see Chapter 4). The transfer of the glucuronic acid moiety to the drug molecule containing a hydroxyl, amine, or a carboxyl function is carried out by glucuronyl transferases present in the endoplasmic reticulum (microsomes). Typical reactions of this sort were shown previously in Chapter 4.

   ii)   Glucuronidation is essentially a nucleophilic substitution in which the drug molecule acts as a nucleophile and launches a back-side attack on the electron-deficient carbon atom 1 of the glucuronic acid part of UDPGA. Inversion of configuration is an inherent characteristic of this type of reaction. Thus, although the configuration of glucuronic acid in UDPGA is alpha, the final reaction product (the glucuronidated drug) has the beta configuration. The glucuronide products of hydroxylated compounds are known as O- or ether glucuronides, those of carboxylic compounds are called ester glucuronides, and the ones formed from amines are termed N-glucuronides.

  iii)   Although many drugs (and other compounds) are metabolized by glucuronidation, the true significance of these derivatives is still open to question. Previously, it had been believed that it was simply a way of rendering drugs and their metabolites more polar and water soluble, thereby increasing their rate of renal excretion. It may also, however, be an important mechanism for enabling these compounds to cross biological membranes, including the blood-brain barrier. There is one report (yet to be

confirmed or refuted) that the glucuronide of a drug is
in fact able to enter the brain while the unconjugated
drug cannot.

Another consideration is the ubiquitousness in tissues of
β-glucuronidase, an enzyme which specifically hydrolyzes
the β-glucuronides of drugs and other compounds. A
conceivable sequence of events in the action of a drug
might be administration, glucuronidation, entry of the
glucuronide into the cell (or wherever the site of action
for the drug is), and hydrolysis of the glucuronide by
β-glucuronidase, releasing the free drug molecule. The
biochemical machinery for the active transport of glucuronides is known to be present in the body. Bilirubin glucuronide is actively excreted into the bile by hepatocytes
(see Chapter 7) and other glucuronides are generally secreted
by the tubular cells into the lumen of the renal tubules
by an active (energy-requiring) process. Further studies
may well show that glucuronidation is required for the
transport of many drug molecules to their site of action.

iv) Research into the role of glucuronic acid conjugation is
hindered by the poor availability of glucuronides. These
compounds are difficult to synthesize in vitro and their
water solubility makes them hard to isolate from natural
sources. Any efforts to generate workable quantities of
glucuronides of drugs and their metabolites would certainly
be worthwhile.

b. Acylation

i) Various carboxylic acids such as acetic or benzoic acid
are activated by coenzyme A (CoA) to form acetyl or
benzoyl∿CoA. In this form, they can react with primary
amines and amides to form acylated metabolites.
Derivatives of aniline (such as the sulfonamide drugs)
and hydrazines (e.g., isoniazid, an antitubercular drug)
are metabolized in this fashion.

Isoniazid   Acetylated Isoniazid

ii) There are also examples in man where the carboxylic group
of the drug is activated by conjugation with CoA. This
compound then reacts with an amino acid such as glycine
or glutamic acid to form peptide acylates. The ability
of the liver to conjugate glycine with benzoic acid
reflects the condition of this organ and may be used as
a liver function test in humans. Glycine conjugation of
the bile acids (see Chapter 8) is an important step
in their excretion.

iii) Acetylation can occur in lung, liver, spleen, and many
other tissues in the presence of appropriate acyl
transferase. All reactions, however, are dependent on CoA.
A genetic deficiency of acetyl transferase enzyme has been
found to exist in nearly half the Caucasian population
of the United States. When administered to this group,
the usual therapeutic dose of isoniazid is toxic,
producing symptoms such as peripheral neuritis.

c. Methylation

i) Transmethylation reactions are brought about by a variety
of transmethylase enzymes present in microsomal and
soluble fractions of the liver cell. Drugs containing
aromatic hydroxyl, sulfhydryl, or amine functions undergo
methylation. S-adenosylmethionine (SAM), synthesized
from methionine (the actual methyl donor) and ATP,
serves as the methyl carrier. The synthesis and structure
of this compound were shown previously (see Chapter 6).
The methylation reaction is summarized below.

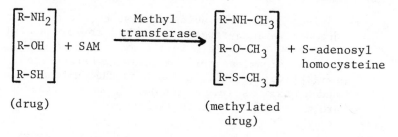

(drug)   (methylated
                               drug)

ii) Some of the drugs methylated in the body include
norepinephrine which is both N- and O-methylated;
norcodeine and normorphine both of which are N-methylated;
and dimercaprol (**British Anti-Lewisite**) which is
S-methylated. Of these, norepinephrine is one example
of a series of catecholamines with a phenylethylamine
structure. Such compounds are O-methylated in the presence
of catechol O-methyl transferase (COMT; see Chapter 6).

iii) The adenosylmethionine-methyl transferase system has no relationship to the oxidative demethylation discussed earlier. These two types of biochemical transformations of drugs are not the reverse of each other, but are two distinctly different reactions. Both are, however, catalyzed by microsomal enzymes.

d. Sulfate Ester Conjugation

Various alcoholic and phenolic drugs are sulfated to form ethereal sulfates. Sulfate is first activated by ATP to adenosine-5'-phosphosulfate, then to 3'-phosphoadenosine-5'-phosphosulfate (PAPS) and is finally transferred to the oxygen function of the drug in the presence of sulfokinases (also known as sulfate transferases) found in the cytoplasm of mammalian livers. The formation of PAPS and the structures of the compounds involved were shown previously (see Chapter 6). A typical sulfatation is shown below.

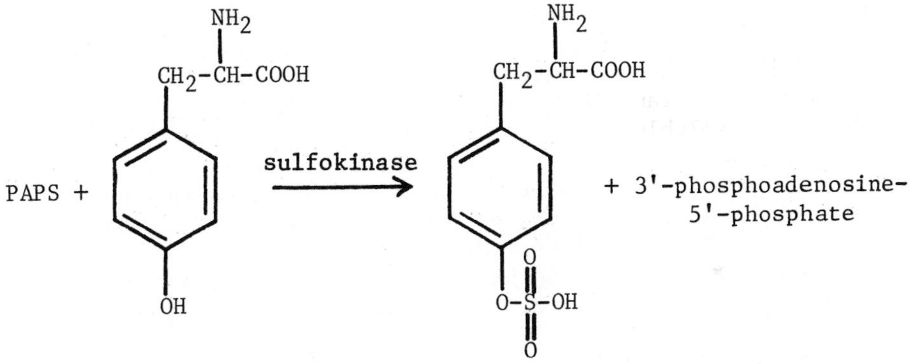

PAPS + [Tyrosine] —sulfokinase→ [Tyrosine-O-Sulfate] + 3'-phosphoadenosine-5'-phosphate

Tyrosine   Tyrosine-O-Sulfate

e. Drug-Induced Alkylations

Several sulfur and nitrogen mustards spontaneously change into an electrophilic intermediate and then attack important endogenous molecules such as RNA and DNA, forming alkylated products within the functioning cells. This interferes with the cellular activities, since the alkylated molecules behave differently than do the normal ones. This is the basis for the antineoplastic and mutagenic activity (see Chapter 5) of the nitrogen mustards. Alkylating agents of this sort are degraded in the body by hydrolysis.

9. Predicting the Metabolic Fate of Drugs

    From what has been described about the biotransformations of drugs,
    it is clear that the route by which a drug is metabolized depends
    on its chemical structure. Based on this, some predictions
    concerning the fate of a drug in the body can be made.

    a. Aromatic and aliphatic cyclic structures are usually hydroxylated
        and subsequently conjugated with either glucuronic acid or
        sulfate. Aliphatic sidechains are also hydroxylated.

    b. Aromatic hydroxyl groups are generally glucuronidated or
        sulfated, but occasionally may be methylated. Aliphatic
        hydroxyl groups undergo oxidation or glucuronide conjugation.

    c. Aliphatic carboxyl groups are oxidized and decarboxylated, or
        conjugated with glucuronic acid. Aromatic carboxyl groups are
        conjugated with either glucuronic acid or an amino acid.

    d. Primary aliphatic amines are glucuronidated or oxidatively
        deaminated, whereas aromatic primary amines may be methylated,
        acetylated, or glucuronidated. Secondary and tertiary amines
        are dealkylated or sometimes methylated.

    Although these are generalities and may not hold in all cases, they
    may aid in the synthesis of derivatives of drugs which have a
    desired duration of action on the basis of their ability or lack
    of it to get metabolized. Highly polar drugs such as strong acids
    and bases are excreted unchanged while extremely non-polar substances
    such as liquid paraffin do not get absorbed in the first place.

Clinical Significance of Drug Metabolism

1. One thing common to all biotransformational reactions of drugs is
    that they involve specific enzymes. In recent years, it has been
    observed that these drug-metabolizing enzymes can be induced or
    inhibited by drugs and organic air pollutants which find entry into
    the body. In the current practice of polypharmacy (simultaneous
    prescribing of more than one drug), there is always a chance that
    one of the administered drugs may affect the enzymes involved in
    metabolizing the other coadministered drugs. Since
    biotransformation may either terminate or enhance drug action, the
    implications of the effect of drugs on the drug-metabolizing enzymes
    are obvious. This is one type of drug-drug interaction (discussed
    below).

2. A number of compounds can inhibit the oxidation of drugs. An
    example of a widely studied experimental agent having this effect
    is beta-diethylaminoethyl diphenylpropylacetate, known commonly as

SKF 525A. Monoamine oxidase inhibitors such as iproniazid (see Chapter 6) and pargyline are more familiar examples of agents which have clinical applications. Estrogens are also capable of inhibiting the metabolism of certain drugs.

3. a. Similarly, there are a number of compounds, many used commonly as drugs, which are capable of inducing drug-metabolizing enzymes. Among the most widely investigated of these are phenobarbital, the steroidal hormones, and 3,4-benzpyrene, a carcinogenic hydrocarbon. These agents induce the microsomal enzymes which metabolize a large number of drugs and endogenous waste products (e.g., bilirubin).

    b. The induction is believed to involve enhanced enzyme synthesis via an increased synthesis of mRNA. So far, however, in experiments designed to prove induction, only the end results such as the increased rate of metabolism of a drug and increased synthesis of liver proteins have been measurable. By inference, it is believed that induction of drug metabolism is a result of increased protein (enzyme) synthesis.

4. Drug-Drug Interactions

    As was pointed out at the start of this chapter, the level of any compound in the body, including drugs from exogenous sources, depends on a balance among several factors. Any factor which alters this balance, including the presence of another drug, is likely to affect the systemic response to the compound.

    a. In the past decade, a large number of adverse drug reactions or paradoxical drug responses were discovered to have occurred as a result of inhibition or induction of the metabolic pathway of one drug by another. One of the simplest clinical examples of induction is the development of tolerance to a drug. Many **drugs on repeated administration induce their own metabolizing** enzymes, with the result that larger doses are required to produce the desired therapeutic effect as time passes.

        i) Although inhibition of drug metabolism has not been studied as much as induction, toxicity of a normal therapeutic dose of one material may be due to the inhibition of its metabolism by another drug which had been previously given to the patient for some other purpose. Estrogens are known to inhibit oxidative and conjugational metabolism of some drugs. Occasionally, the inability of a female patient to tolerate a dose of a drug, which has no adverse effect on a male patient, may be attributed to this.

        ii) There are many clinical examples and uses of the phenomenon of enzyme induction. One well-known clinical application is the use of phenobarbital in patients suffering from a deficiency of glucuronyl transferase. In such cases, the circulating levels of bilirubin, which is normally glucuronidated and secreted into bile, increase to the point of producing jaundice. Administration of phenobarbital results in the lowering of bilirubin levels to near normal, since this barbiturate induces the enzyme glucuronyl transferase. Another example is the inhibition of the anticoagulative effect of the coumarine by barbiturates and glutathimide. These compounds both induce the hepatic microsomal oxidase system which catabolizes the coumarins.

b. There are other ways in which one drug may influence the biological activity of another. These include alterations in absorption, transport, and function at the receptor site. Some examples are given below.

    i) Protamines (salmin , clupeine, etc.; see Chapter 5) act as heparin antagonists. These are basic peptides which are thought to neutralize the relatively acidic heparin and prevent it from acting as an anticoagulant.

    ii) Gentamicin is inactivated by carbenicillin or ampicillin when they are combined in the same infusion fluid. Once in the serum, however, this does not occur, so that administration of these antimicrobials by separate routes is desirable.

    iii) Most drugs are carried to their site of action bound to plasma proteins. Low levels of a specific binding protein (e.g., a deficiency of cortisol-binding globulin) or a general decrease in plasma proteins (as in hypoalbuminemia) will thus affect this transport. One compound may compete with another for binding sites on the plasma proteins. In hyperbilirubinemia, the ability of albumin to bind other compounds (drugs) is limited because it is heavily loaded with bilirubin. Aspirin (acetyl-salicylic acid) behaves similarly. The ability of some substances to potentiate the activity of the coumarin anticoagulants by displacing them from protein-binding sites and thereby increasing the level of the free compound was cited previously (see Chapter 10). Chloral hydrate, a sedative and hypnotic, has this effect.

    iv) The potentiation of 6-mercaptopurine (6MP) by allopurinol was discussed earlier (see Chapter 5). Allopurinol inhibits xanthine oxidase, the enzyme which catabolizes 6MP to

6-thiouric acid, an inactive metabolite. This also
increases the toxic side effects (bone marrow degeneration
leading to pancytopenia) of 6MP, however.

   v) Disulfiram, the inhibitor of aldehyde oxidase (see Chapter 4)
       used in many alcoholism treatment programs, greatly
       increases the toxicity of ethanol due to the accumulation
       of acetaldehyde. Deaths have been reported due to
       massive alcohol ingestion among alcoholic patients
       participating in mandatory disulfiram programs.

   c. As was briefly indicated above, drug-drug interactions are
       potentially useful as well as dangerous. As more is understood
       about these processes, a more rational approach to the control
       of the magnitude and duration of drug effects should arise.
       Also, much of the estimated three billion dollars currently spent
       each year on the treatment of adverse drug reactions in patients
       could be put to better use if such reactions were avoided in
       the first place.

5. Pharmacogenetics is a relatively new term used to refer to the
    genetically based variation among individuals in their response to
    drugs. In many instances, these differences are related to a
    decreased (or sometimes increased) ability of some persons to
    metabolize a drug. This in turn governs the intensity and duration
    of action of the compound. Pharmacogenetics is a quite logical
    extension of the growing interest and awareness of enzyme mutants
    and inborn metabolic errors. It was mentioned briefly in connection
    with hemoglobin variants.

   a. Genetic variants of various esterases including carbonic
       anhydrase, acetylcholinesterase, and serum cholinesterase have
       been recognized in man. Not only do these enzymes behave
       differently in various physico-chemical analytical systems, but
       they also clearly exhibit varying degrees of specificity and
       activity toward their substrates.

       i) Clinically, this may be witnessed dramatically following
           the administration of succinylcholine. In normal
           individuals, this drug is hydrolyzed by the plasma
           cholinesterase at a rate such that doses of 50 to 80 mg
           do not produce any toxic symptoms (prolonged paralysis).
           In individuals with a genetic deficiency of the active
           enzyme (fluoride resistant or silent variants), these
           doses produced paralysis lasting from 25 to 300 minutes
           and requiring extended artificial respiration.

      ii) Individuals having the cynthiana serum cholinesterase
           variant show a resistance to succinylcholine. This seems

to be due to an increase in the serum level of cholinesterase, resulting in abnormally rapid hydrolytic inactivation of the molecule.

   iii) The significance of esterase variation is not known in terms of the basic physiological function of the body and the genetic variants of esterases have been found in healthy individuals. Whether or not these variants evolved in a certain generation or over several generations as a result of constant exposure to some inhibitor cannot be answered unequivocally, but a possibility for this does exist.

b. Isoniazid, a drug used in the treatment of tuberculosis, is inactivated by acetylation catalyzed by the liver microsomal enzyme N-acetyl transferase. In individuals with a slow acetylation capacity this drug may cause toxic effects such as peripheral neuritis. This can be prevented by administration of excess pyridoxine, which does not interfere with the antitubercular activity.

c. Diphenylhydantoin is inactivated by the liver microsomal hydroxylase system. Three genetic types have been demonstrated with regard to the rate of inactivation: (1) slow metabolizers show toxic effects (nystagmus, ataxia, etc.) with normal doses; (2) normal metabolizers; and (3) rapid metabolizers who require an increased dose of the drug to produce the expected biological response.

d. Individuals who are deficient in liver microsomal dealkylase enzyme develop methemoglobinemia and hemolysis upon the administration of acetophenetidin.

e. G-6-P dehydrogenase deficient individuals show methemoglobinemia and hemolysis upon administration of a variety of drugs (primaquine, etc.; see Chapter 4).

f. Other enzymes involved in drug metabolism which exhibit genetic variation in man include glucuronyl transferase (discussed earlier) and various aromatic hydroxylases and oxidases. A reported lack of sulfite oxidase in children is interesting but may represent an example of developmental enzymology rather than a pharmacogenetic characteristic. This newly evolving concept recognizes that infants and children are deficient in certain enzyme activities which develop as they approach or attain maturity. Another example, discussed earlier is neonatal jaundice resulting from delayed development of hepatic glucuronyl transferase in neonates.

g. The clinical implications of the genetic differences in man's ability to metabolize drugs is currently evolving with reference to the blood levels of drugs obtained following administration of a fixed dose. A tenfold difference in blood levels of imipramine has been found to occur in different patients receiving the same dose, on a body weight basis, of this drug. Likewise, antipyrine, chlorpromazine, meperidine (demerol) and many other important drugs develop varying blood levels in different individuals. Most of these differences can be attributed to inherent variations in the drug-metabolizing enzymes of the patients. Those who are totally or partially deficient in the enzymes may exhibit very high blood levels of drugs which can at times result in undue toxicity.

h. Practically, it would add to the quality of clinical practice if the physician tested his subjects for pharmacogenetic anomalies before administering certain drugs known to be metabolized by any one of the enzymes which exist as genetic variants. For example, a small test dose of succinylcholine could be injected to see that no paralysis developed, prior to embarking on the usual dosage schedule of this drug.

6. Another topic which is currently being considered is the role of the intestinal microflora in the metabolism of exogenous and endogenous substances.

a. In patients with a decreased NH_3 tolerance due to advanced liver disease, neomycin may be administered to suppress the formation of this compound by the action of bacterial urease.

b. Salicylazosulfopyridine, an anti-inflammatory agent used to treat ulcerative colitis and other bowel inflammations, is cleaved at the diazo bond by intestinal bacteria. The products, 5-aminosalicylate and sulfopyridine, have respectively anti-inflammatory and antibacterial actions. The effectiveness of the parent (uncleaved) compound can be potentiated by the concomitant administration of antibiotics such as neomycin.

c. Caffeic acid, a constituent of most of the normal dietary vegetables, seems to be metabolized exclusively by the intestinal flora. Of the four reactions involved in this process, none of eleven species of human fecal bacteria tested was able to carry out more than one of them. This suggests a way of investigating in situ the functioning of these organisms. Future studies will hopefully reveal other examples of this sort of selective (species-specific) metabolism.

d. L-dihydroxyphenylacetic acid (L-DOPA; see Chapter 6), a drug used to treat Parkinsonism, is apparently converted to m-tyramine and m-hydroxyphenylacetic acid by bacterial enzymes of the intestines. This may contribute to the variability of patients to L-DOPA treatment.

e. Quite little is known about other transformations catalyzed by bacterial enzyme systems in the gut. In vitro studies are difficult because the conditions of pH, osmolarity, and oxygen supply found in this region are hard to reproduce in the laboratory. The use of fecal isolates has only questionable value for understanding how these microorganisms function in their actual environment. Germ-free strains of animals have provided some information but its applicability to human systems is not clear. This is a new and relatively unexplored area of study whose importance may be much greater than was hitherto expected.

7. Drug and nutrient interaction is another important consideration which should be taken into account in the maintenance or restoration of normal health. Many drug-nutrient interactions lead to clinically observable side effects and adverse reactions. There are two aspects to this problem, namely the effect of nutritional status on the drug metabolism and the effect of drugs on the nutritional status. Several examples of the latter has already been discussed in the book (e.g., the production of vitamin D deficiency by anticonvulsant drugs, vitamin B_1 deficiency in alcoholics). In general, the nutritional status can be altered by drugs by affecting nutrient absorption (by interference with absorption--e.g., laxatives, alcohol, penicillin), nutrient availability (by binding with nutrients--e.g., hypocholesterolemic agents--cholestyramine decreases absorption of vitamin B_{12}, iron, etc.), nutrient metabolism and excretion (effects of alcohol on folate, B_6 and mineral metabolism, anticonvulsants on vitamin D metabolism, oral contraceptives and estrogens on B_6 and folate metabolism), and destruction or injury to intestinal epithelial cells (cytotoxic drugs, colchicine, etc.). It should also be noted that some drugs alter taste acuity (e.g., penicillamine, griseofulvin and lincomycin). The effects of some of these drugs on taste and appetite may be related to zinc deficiency. The subject drug-nutrient interaction is in need of extensive investigation.

15
Liver Function

Anatomical Considerations (with Virgil L. Jacobs, Ph.D.*)

1. The liver represents the largest and heaviest gland in the body (3.5 pounds or 1500-1600 grams in the adult). It is situated in the upper abdomen directly beneath the diaphragm and can be palpated mainly in the right hypochondriac and epigastric regions.

2. The four lobes of the liver are further subdivided into lobules by partitions or septa of connective tissue that are continuous with the peritoneal covering (Glisson's capsule).

3. The arterial supply is derived from branches of the celiac artery called the right and left hepatic arteries. These arteries further subdivide and eventually supply oxygenated blood to the liver sinusoids.

4. The portal vein is formed from the venous drainage of the intestines, spleen, and pancreas. Branches of the portal vein in the liver follow the hepatic arterial route and drain into the sinusoids. Thus both arterial and venous blood mixes in the liver sinusoids. About 75% of the blood supply of the liver is from the portal vein and about 25% from the hepatic artery. The function of the arterial blood is to raise both the oxygen content and the pressure of the sinusoidal blood.

5. Blood in the sinusoids passes slowly through the rows or cords of liver cells where the exchange of oxygen, metabolites, and nutrients occur (see Figure 179). A group of sinusoids anastomose to form the central vein. This vein represents the beginning of the venous drainage of the liver. The central veins of many surrounding liver lobules join to form larger veins and these further anastomose to form the hepatic vein which joins the inferior vena cava.

6. Bile is both a secretory and an excretory product of liver cells. The biliary system begins as the bile canaliculi running between two apposed rows of liver cells. As more canaliculi anastomose, larger bile ducts with a simple cuboidal epithelium are formed. Bile ducts from the various liver lobes join to form the common bile duct which opens into the duodenum.

---

* Associate Professor of Anatomy, College of Medicine, Texas A & M University, College Station, Texas.

Figure 179. Schematic Drawing of the Anatomy and Circulation in the Liver

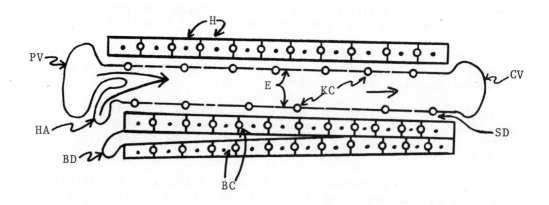

The portal area (triad) is to the left. It consists of a branch of
the portal vein (PV), a branch of the hepatic artery (HA), and a bile
duct (BD). The arterial and venous blood enter the sinusoid which
runs between two rows (cords) of hepatocytes (H). The sinusoid is
lined by endothelial cells (E) which are periodically studded with
Kupffer cells (KC). The endothelium is not continuous but has openings
permitting the fluid to contact the surface of the hepatocytes. The
space of Disse (SD) is between the endothelium and the hepatocyte
surface. After passing through the sinusoids, the blood enters the
central vein (CV) which joins with the general venous drainage
of the liver. Bile is secreted by the hepatocytes into the bile
canaliculi (BC) which collectively drain into the bile ducts (BD).
The bile ducts transport it to the gallbladder where it ultimately
enters the intestinal tract at the duodenum.

7. The parenchyma of the liver consists of one type of cell, the
    hepatic cell. These are arranged in rows or cords, usually as
    doublets, with a great deal of branching. The general histological
    appearance of the cells and sinusoids is that of a sieve.

8. The classical or hepatic liver lobule has the central vein in its
    center from which liver cords and sinusoids radiate peripherally
    like spokes of a wheel. The rim or edge of the lobule contains
    connective tissue and several portal areas, each of which contains
    a portal triad (hepatic artery branch, portal vein branch, and a
    bile duct) and sometimes a lymphatic vessel. The shape of the
    hepatic lobule is usually hexagonal in cross-section (see Figure 180).

9. Alternatively, the liver can be subdivided into portal lobules which
    have the portal canals at their centers. The peripheral boundary
    of this type of lobule is marked by three central veins. This forms
    a triangular lobule with the biliary system draining bile into
    the centrally placed bile duct.

10. A so-called functional unit of the liver can also be defined based
    on the blood supply to the lobules. The unit is diamond shaped
    with two central veins at opposite ends and two portal canals in
    the corners. Pathologically, liver damage is frequently related to
    the blood supply. Normally those parenchymal cells nearest the
    central vein will receive less oxygen and food material than the
    liver cells nearest to the hepatic artery branches.

11. Liver cells are polygonal in shape with one surface related to a
    sinusoid, another applied to other liver cells, and one other
    surface minutely separated from an adjacent liver cell to form a
    bile canaliculus. The nuclei are highly variable in size but
    somewhat spherical in shape. The cytoplasm has a slightly basophilic
    appearance. Glycogen in the liver is first deposited in the more
    peripheral cells of a lobule. As more glucose is brought in
    through the portal system, the liver cells closer to the central
    vein begin to show deposits. The reverse occurs when a glycogen
    demand is placed on the liver (i.e., the cells nearest the central
    vein release their glycogen first).

12. The hepatic cell membrane is thrown into folds or microvilli
    where the cell surface is applied to the sinusoidal lining.
    This cell surface is separated from the sinusoid by the space of
    Disse (a very narrow extravascular or interstitial fluid space
    between the sinusoidal endothelial lining and the hepatocytes).
    Near the midportion of where two liver cells meet, the two
    membranes of an interface separate slightly to form a bile
    canaliculus. This side or surface of the cell is called the
    biliary pole.

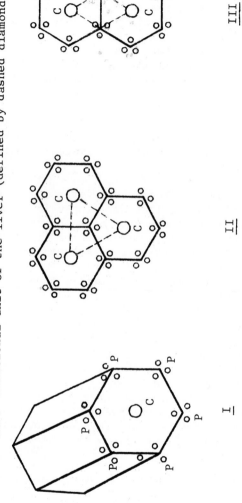

Figure 180. Three Types of Hypothetical Subunits (Lobules) of the Liver

I: Classical hepatic liver lobule (hexagonal)

II: Portal lobule (defined by dashed triangle)

III: Functional unit of the liver (defined by dashed diamond)

Note: C = central vein (large circle)

P = portal triad (portal area containing branches of the portal vein, hepatic artery, and bile ducts, represented by three small circles)

13. Sinusoids of the liver are larger than capillaries and lined by a reticuloendothelium. The cell types of this lining are an endothelial cell and a phagocytic cell (Cell of Kupffer). These cells do not form a continuous lining of the sinusoids but are fenestrated to present a sieve-like appearance. This endothelial barrier selectively allows blood plasma a direct access to the microvilli of liver cells for selective absorption. At death, the liver shows a widened edematous expansion of the space of Disse that is plasma filled. The phagocytic cells lining the sinusoids are active and may contain degenerating erythrocytes, pigment granules, droplets of fatty material and iron-containing granules.

Functions of the Liver

1. The liver performs a multitude of diverse functions which are discussed throughout this text at the appropriate places. It is important to stress that this organ is placed in a strategic position where it receives all of the nutrients, with the exception of lipids, from the portal vein which drains the capillary beds of the stomach, intestines, spleen, and pancreas. The portal blood goes through the columns of liver cells and enters the systemic circulation through the hepatic vein. This propitious intervention of the liver between the portal and systemic circulations not only permits the liver to absorb, transform, and store the nutrients resulting for the intestinal digestion and absorption but also protects the rest of the body from toxic metabolites (such as NH_3) produced in the intestines or present in the diet, by converting them into non-toxic substances before they enter the systemic blood.

2. The numerous functions of liver can be categorized into the following groups. More details on these can be found in other parts of this book.

   a. Metabolic functions: involving carbohydrates, proteins, lipids, minerals, and vitamins.

   b. Detoxication and protective functions: conversion of ammonia to urea; formation of bilirubin glucuronides; drug detoxication; steroid inactivation; phagocytic activity of Kupffer cells in removing foreign substances; others.

   c. Excretory functions: formation of bile and the excretion of bilirubin glucuronides (bile pigments), bile salts (cholesterol metabolites), bromosulfalein (BSP), and other foreign compounds.

   d. Storage functions: storage of glycogen, iron, several vitamins (A, D, B_{12}), and other materials.

e. Hematological functions: hematopoiesis during embryonic
development; formation of clotting factors (VII, IX, X, and
prothrombin, all of which are dependent upon vitamin K for
their synthesis); synthesis of many plasma proteins except for
immunoglobulins (recently, though, it has been shown that
hepatocytes do synthesize some immunoglobulins and that this
may play an important role in the overall immune mechanism);
special carrier proteins for various metabolites; etc.

f. Circulatory functions: participates in blood storage; acts as
an intermediary between the portal and systemic blood circulation;
reticulo-endothelial activities of Kupffer cells lining the
sinusoids.

Tests of Liver Function

1. Since the liver plays a central role in many diverse metabolic
processes, no single test or combination of tests will clearly
establish the diagnosis of a specific disease. Often the results
of a test indicate the nature of the pathological process but not
its cause. In order to assess a liver disease, a spectrum of tests,
commonly known as a liver profile, is performed.

2. Hepatic disorders can cause parenchymal cell dysfunction or damage
and/or cholestasis. Consequently, the most common laboratory tests
performed measure the secretory and excretory capabilities of the
hepatocytes and liver cell dysfunction and damage.

3. Tests used to evaluate the secretory and excretory functions

a. Measurements of plasma bilirubin and bilirubin glucuronides

An increase in unconjugated serum bilirubin is associated with
excessive erythrocyte destruction, or defects in the uptake or
conjugation of bilirubin by the liver cells. In jaundice, both
the conjugated and unconjugated forms are often increased. In
cholestasis, conjugated bilirubin is the major species present.
In general, the presence of bilirubin glucuronide in the plasma
is suggestive of a pathological condition. These diseases
are discussed in Chapter 7.

b. Measurement of urobilinogen in the urine (see also Chapter 7).

The presence of excess amounts of urinary urobilinogen are
associated with excessive erythrocyte destruction and failure
of the liver to re-excrete urobilinogen due to liver cell
damage, as in hepatitis. Excessive concentration of normal urine
can also cause apparently elevated urobilinogen levels. Since
urobilinogen gets converted to urobilin on standing, a more

accurate way of assessing the amount of urobilinogen involves converting the urobilinogen to urobilin and measuring the level of this compound.

A total absence of urinary urobilinogen is indicative of complete obstruction of bile flow (cholestasis). In cholestasis, the feces have a very pale coloration due to a lack of bile pigments.

c. The bromosulfalein (BSP) excretion test

When BSP (a dye) is administered intravenously liver cells remove the dye from the circulation, conjugate it with the cysteine portion of a glutathione molecule, and excrete it into the bile. A 45-minute post-BSP injected (in one procedure 5 mg/kg body weight is used) blood sample normally contains about 3% of the total dose administered. Since excretion of the BSP depends upon the integrity of the liver cells, adequacy of the blood circulation, and the patency of the biliary tract, any abnormalities in these cause elevated retention of BSP. Loss of BSP in the urine is negligible and normally is discounted. This is a sensitive test and is only performed when other tests have produced negative results.

d. Serum alkaline phosphatase measurement

   i) There are a number of enzymes present in the serum that hydrolyze mono-esters of phosphates. The acid and alkaline phosphatases show a lack of substrate specificity and show optimal activities at alkaline and acid pH's, respectively. 5'-Nucleotidase, which catalyzes the hydrolysis of nucleoside-5'-phosphates (such as AMP), does have a substrate specificity permitting its distinction from the other two phosphatases.

   ii) Serum alkaline phosphatase is derived from several sources including the liver (cells lining the sinusoids and the bile canaliculi), bone (where it is needed for osteoblastic activity; see Chapter 9), kidney, intestine, and placenta (in pregnant women). Therefore, elevated levels of serum alkaline phosphatase are observed in a variety of disorders such as rickets and osteomalacia (see Chapter 9), Paget's disease, hyperparathyroidism, during the healing of fractures, osteoblastic carcinoma, and intra- and extrahepatic cholestasis (hepato-biliary involvement).

   The elevation of serum alkaline phosphatase occurs in biliary obstruction. Other enzymes used in the assessment of biliary tract integrity are 5'-nucleotidase, Gamma glutamyltransferase, and leucine aminopeptidase.

iii) The various tissues have different isoenzymes of alkaline phosphatase. These isoenzymes can be distinguished by electrophoretic, chemical, and physical means (the enzyme of each permitting identification of the source of the elevated alkaline phosphatase). Electrophoresis (of lactic acid dehydrogenase) was discussed in Chapter 4. The relative susceptibility of the bone isozyme to heat inactivation permits its distinction from the liver form. The intestinal and placental isozymes are inhibited by L-phenylalanine, unlike the liver and bone forms. A combination of heating and inhibitory methods can be used to quantitate all four isozymes. (Serum acid phosphatase activity, which may also arise from several sources, notably the prostate, can also be fractionated by selective inhibition with L-tartrate which inhibits only prostatic acid phosphatase.)

iv) The level of serum 5'-nucleotidase is not enhanced in individuals with bone diseases but is elevated in hepatic disorders. Serum 5'-nucleotidase measurement is thus useful in confirming hepatic involvement when alkaline phosphatase level are elevated and liver disease is suspected.

4. Tests based on liver cell dysfunctions and damage

   a. Measurement of plasma proteins: Since the liver is responsible for the synthesis of albumin (and other plasma proteins), the plasma albumin level is decreased in liver cell dysfunction such as that caused by viral hepatitis. This can be assessed by plasma protein electrophoresis (see Chapter 6).

   b. Alterations in the coagulation factors: In the presence of vitamin K, liver parenchymal cells are responsible for the synthesis of clotting factors II (prothrombin), VII, IX, and X. The plasma levels of these molecules can be assessed by measuring prothrombin time. In addition to the above four factors, deficiencies of factors I (fibrinogen) and V occur in severe forms of liver disease. Recall that vitamin K is a fat-soluble vitamin which requires bile salts for its absorption. Consequently, cholestasis results in a lack of absorption of vitamin K giving rise to abnormal clotting time despite the normalcy of the liver parenchymal cells. In order to establish parenchymal damage using tests involving clotting factors, it is essential to administer vitamin K parenterally.

c. As has already been mentioned, liver cells are rich in a wide variety of enzymes, several of which are to some extent tissue-specific. An increase in the plasma levels of these enzymes is strongly suggestive of extensive damage of the hepatic cells and the release of their contents into the circulation. Enzymes whose levels are frequently measured for this purpose include the transaminases SGOT and SGPT (see Chapter 6) and LDH (isozymes LD-4 and LD-5; see Chapter 4).

d. Recent work indicates that the detection of antimitochondrial antibody in the circulation is indicative of primary biliary cirrhosis (cirrhosis resulting from the chronic retention of bile in the liver). This type of autoimmunity is also found in chronic active hepatitis and cryptogenic cirrhosis, but with a much lower frequency. The assay is performed by the fluorescent antibody technique. This test may be useful in confirming a diagnosis of primary biliary cirrhosis since the other clinical and laboratory findings are not pathognomic.

Appendix I

Ranges of Normal Values for Some Biochemical Parameters
(for others refer to the text)

The normal ranges vary with procedure. Every clinical laboratory or a group of laboratories in a given geographic location should establish normal values for each constituent, derived from persons unique to that location. It should be noted that normal values may vary with sex, age, race, administration of drugs, nutritional status, time of specimen collection and others. B = blood, P = plasma, S = serum, U = urine, dl = deciliter (100 ml).

Acetoacetate plus acetone (S).....0.5-3.0 mg/dl

Adrenocorticotrophic hormone
        (ACTH)(P)...................8-10 AM: 20-100 pg/ml

Albumin (S).......................3.2-5.6 g/dl

        spinal fluid..............10-30 mg/dl

Aldolase (S)......................1.0-7.5 U/1 at 30°C

Aldosterone (P)...................Recumbent (2h): 0.3-2.5 ng/dl
                                  upright (4h): 4.0-19.6 ng/dl

δ-Aminolevulinic acid (S)........15-23 μg/dl (lower in children)

                        (U)........1.5-7.5 mg/day

Ammonia (P).......................56-122 μg/dl

Amylase (S).......................60-180 Somogyi units/dl

        (U).......................70-275 Somogyi units/h

α_1-Antitrypsin (S)................160-350 mg/dl

Ascorbic acid (P).................0.6-2.0 mg/dl (fasting)

Bicarbonate (P)...arterial........21-28 mEq/1

            (P)...venous..........22-29 mEq/1

Bilirubin (S)...total.............0.2-1.0 mg/dl
                conjugated........0-0.20 mg/dl
                                  (For other values, see Chapter 7)

Calcium (S)...ionized............4.4-5.4 mg/dl (2.2-2.7 mEq/1)
              total..............9-11 mg/dl (4.5-5.5 mEq/1)

Carbon dioxide content (S).......24-30 mEq/1

β-Carotene (S)...................Adults: 60-200 µg/dl
                                 Newborns: about 70 µg/dl

Catecholamines
    (P)...epinephrine...........18-26 ng/dl
          norepinephrine........47-69 ng/dl

    (U)...random................0-14 µg/dl
          24h...................less than 100 µg/day

Ceruloplasmin (S)................25-43 mg/dl

Chloride (S or P)................96-106 mEq/1

Cholesterol (S)..................150-280 mg/dl (See also Chapter 8)

Complement.......................See Chapter 11

Copper (S).......................70-160 µg/dl

Cortisol (P).....................9-10 AM: 6-26 µg/dl
                                 4 PM: 2-18 µg/dl

Creatine kinase (S)..............Males: 12-65 U/1 at 30°C
                                 Females: 10-50 U/1 at 30°C

Creatinine (S or P)..............Males: 0.9-1.4 mg/dl
                                 Females: 0.8-1.2 mg/dl

           amniotic fluid........Greater than 2.0 mg/dl suggests
                                 fetal maturity when maternal serum
                                 creatinine level is normal.

Creatinine clearance.............See Chapter 6

Electrophoresis...serum..........See Chapter 6
                  hemoglobin......See Chapter 7

Fat (on a diet of about
    100 g of fat/day)
    feces, 72h collection........< 7 g/day (See Chapter 8)

Gastrin (S).....................Fasting: < 100 pg/ml
                                Borderline elevated: 100–200 pg/ml
                                Elevated fasting gastrin: > 200 pg/ml
                                Elders: 200–800 pg/ml

Globulins (S)...total............2.3–3.5 g/dl (For individual fractions,
                                        see Chapters 6 and 11)

Glucose (S or P)................70–100 mg/dl (See Chapter 4)

Glucose-6-phosphate dehydrogenase,
     red cells..................3.4–8.0 U/g Hb

Hemoglobin, hematocrit and
     other related parameters.....See Chapter 7

Immunoglobulins.................See Chapter 11

Insulin.........................See Chapter 4

Iron (S)...total................Males: 60–150 µg/dl
                                Females: 50–130 µg/dl

Iron binding capacity (S)........250–410 µg/dl

Lactate (B)...arterial...........3–7 mg/dl

       (B)...venous.............4–12 mg/dl

Lactate dehydrogenase (S)........85–190 U/1 at 30°C

Lactate dehydrogenase isoenzymes..See Chapter 4

Lactose tolerance...............See Chapter 4

Lecithin/Sphingomyelin ratio
     (L/S ratio)...
     amniotic fluid.............$\geq$ 2 indicates probable fetal lung
                                maturity

Lipase (S)......................28–280 U/1 at 37°C

Magnesium (S)...................1–2 mg/dl (1.5–2.5 mEq/1)

Osmolality (S)..................275–295 mOsmol/kg

            (U)...random..........> 600 mOsmol/kg

            (U)...after 12h fluid
                restriction......> 900 mOsmol/kg

P_{CO_2} (B)...arterial...............35–45 Torr (or mm Hg)

P_{O_2} (B)...arterial...............80–104 Torr (or mm Hg)

pH (B)...........................7.35–7.45

Phenylalanine (S)................Premature or underweight newborns:
2.0–7.5 mg/dl
Full-term normal weight newborns:
1.2–3.4 mg/dl
Adults: 0.8–1.8 mg/dl

Phosphatase, acid (S)...total.....Males: 0.13–0.63 Sigma units/ml
Females: 0.01–0.56 Sigma units/ml

Phosphatase, alkaline (S).........2.0–4.5 Bodansky units/ml
(values for children are higher)

Phospholipids (S)................145–225 mg/dl

Phosphorous, inorganic (S).......Birth—1 year: 4.0–7.0 mg/dl
(1.29–2.26 mmol/l)
Children: 4.5–6.5 mg/dl
(1.45–2.09 mmol/l)
Adults: 3.0–4.5 mg/dl
(0.96–1.45 mmol/l)

(U)...24h..0.4–1.3 g/day (12.9–42.0 mmol/day)
(varies with the diet)

Potassium (S)....................3.5–5.0 mEq/l

Protein (S)...total..............6.0–8.0 g/dl
spinal fluid.......15–45 mg/dl

(U)...random.............< 10 mg/dl
24h...............50–100 mg/day

Pyruvic acid (S)................0–0.11 mEq/l

Sodium (S)......................135–145 mEq/l

Thyroxine and related parameters..See Chapter 6

Transaminase (S)...SGOT..........Males: 7–21 U/l at 30°C
Females: 6–18 U/l at 30°C

SGPT..........Males: 6–21 U/l at 30°C
Females: 4–17 U/l at 30°C

Triglycerides (S)................10–190 mg/dl (See Chapter 8)

Urea nitrogen (S or P)...........8–26 mg/dl (varies with the amount
                                        of protein intake)

Uric acid (S or P)..............Males: 3.5–7.0 mg/dl
                                Females: 2.5–6.0 mg/dl
                                Children: 2.0–5.5 mg/dl

         (U)...24h..............Average diet: 250–750 mg/day

Vitamin A (S)...................30–65 µg/dl

Vitamin B$_{12}$ (S)................200–1600 pg/ml

Vitamin C (P)...................0.6–2.0 mg/dl

Appendix II

Food and Nutrition Board, National Academy of Sciences–National Research Council Recommended Daily Dietary Allowances,[a] Revised 1974[*]

Designed for the maintenance of good nutrition of practically all healthy people in the USA

	Age (yrs)	Weight (kg)	Weight (lbs)	Height (cm)	Height (in.)	Energy (Cal)[b]	Protein (g)	Vitamin A activity (RE)[c]	Vitamin A (IU)	Vitamin D (IU)	Vitamin E activity[e] (IU)	Ascorbic acid (mg)	Folacin[f] (μg)	Niacin[g] (mg)	Riboflavin (mg)	Thiamin (mg)	Vitamin B6 (mg)	Vitamin B12 (μg)	Calcium (mg)	Phosphorus (mg)	Iodine (μg)	Iron (mg)	Magnesium (mg)	Zinc (mg)
Infants	0.0–0.5	6	14	60	24	kg X 117	kg X 2.2	420[d]	1400	400	4	35	50	5	0.4	0.3	0.3	0.3	360	240	35	10	60	3
	0.5–1.0	9	20	71	28	kg X 108	kg X 2.0	400	2000	400	5	35	50	8	0.6	0.5	0.4	0.3	540	400	45	15	70	5
Children	1–3	13	28	86	34	1300	23	400	2000	400	7	40	100	9	0.8	0.7	0.6	1.0	800	800	60	15	150	10
	4–6	20	44	110	44	1800	30	500	2500	400	9	40	200	12	1.1	0.9	0.9	1.5	800	800	80	10	200	10
	7–10	30	66	135	54	2400	36	700	3300	400	10	40	300	16	1.2	1.2	1.2	2.0	800	800	110	10	250	10
Males	11–14	44	97	158	63	2800	44	1000	5000	400	12	45	400	18	1.5	1.4	1.6	3.0	1200	1200	130	18	350	15
	15–18	61	134	172	69	3000	54	1000	5000	400	15	45	400	20	1.8	1.5	1.8	3.0	1200	1200	150	18	400	15
	19–22	67	147	172	69	3000	54	1000	5000	400	15	45	400	20	1.8	1.5	2.0	3.0	800	800	140	10	350	15
	23–50	70	154	172	69	2700	56	1000	5000		15	45	400	18	1.6	1.4	2.0	3.0	800	800	130	10	350	15
	51+	70	154	172	69	2400	56	1000	5000		15	45	400	16	1.5	1.2	2.0	3.0	800	800	110	10	350	15
Females	11–14	44	97	155	62	2400	44	800	4000	400	12	45	400	16	1.3	1.2	1.6	3.0	1200	1200	115	18	300	15
	15–18	54	119	162	65	2100	48	800	4000	400	12	45	400	14	1.4	1.1	2.0	3.0	1200	1200	115	18	300	15
	19–22	58	128	162	65	2100	46	800	4000	400	12	45	400	14	1.4	1.1	2.0	3.0	800	800	100	18	300	15
	23–50	58	128	162	65	2000	46	800	4000		12	45	400	13	1.2	1.0	2.0	3.0	800	800	100	18	300	15
	51+	58	128	162	65	1800	46	800	4000		12	45	400	12	1.1	1.0	2.0	3.0	800	800	80	10	300	15
Pregnant						+300	+30	1000	5000	400	15	60	800	+2	+0.3	+0.3	2.5	4.0	1200	1200	125	18+	450	20
Lactating						+500	+20	1200	6000	400	15	80	600	+4	+0.5	+0.3	2.5	4.0	1200	1200	150	18	450	25

[*]Recommended Dietary Allowances, 8th ed. National Academy of Sciences, 1974, p 128.

[a]The allowances are intended to provide for individual variations among most normal persons as they live in the United States under usual environmental stresses. Diets should be based on a variety of common foods in order to provide other nutrients for which human requirements have been less well defined. See text for more detailed discussion of allowances and of nutrients not tabulated

[b]Kilojoules (kJ) = 4.2 X Cal

[c]Retinol equivalents

[d]Assumed to be all as retinol in milk during the first 6 months of life. All subsequent intakes are assumed to be half as retinol and half as β-carotene when calculated from international units. As retinol equivalents, three-fourths are as retinol and one-fourth as β-carotene

[e]Total vitamin E activity, estimated to be 80% as α-tocopherol and 20% other tocopherols

[f]The folacin allowances refer to dietary sources as determined by *Lactobacillus casei* assay. Pure forms of folacin may be effective in doses less than one-fourth of the recommended dietary allowance

[g]Although allowances are expressed as niacin, it is recognized that on the average 1 mg niacin is derived from each 60 mg dietary tryptophan

[b]This increased requirement cannot be met by ordinary diets; therefore, the use of supplemental iron is recommended

Suggested Readings

General References

Metabolic Basis of Inherited Diseases, 4th ed., Stanbury, J.B., Wyngaarden, J.B., & Fredrickson, D.S. (eds.), McGraw-Hill, New York, 1978.

Text Book of Medicine, 14th ed., Beeson, P.B., & McDermott, W. (eds.), Saunders, Philadelphia, 1975.

Clinical Diagnosis by Laboratory Methods, 15th ed., Davidsohn, I., & Henrey, J.B. (eds.), Saunders, Philadelphia, 1974.

Histology, 7th ed., Ham, A.W., Lippincott, 1974.

Nutritional Support of Medical Practice, Schneider, H.A., Anderson, C.E., & Coursin, D.B. (eds.), Harper & Row, Hagerstown, Maryland, 1977.

Nutrition Reviews: Present Knowledge in Nutrition, 4th ed., Hegsted, D.M. (ed.), The Nutrition Foundation, Inc., New York, 1976.

Clinical Chemistry, 2nd ed., Henrey, R.J., Cannon, D.C., & Winkelman, J.W. (eds.), Harper & Row, Hagerstown, Maryland, 1974.

Fundamentals of Clinical Chemistry, 2nd ed., Tietz, N. (ed.), Saunders, Philadelphia, 1976.

The Pharmacological Basis of Therapeutics, 5th ed., Goodman, L.S., & Gillman, A. (eds.), Macmillan, New York, 1975.

Physiology and Biophysics (in three volumes), 20th ed., Ruch, T.C., & Patton, H.D., Saunders, Philadelphia, 1974.

Chapter 1
Acids, Bases and Buffers

ABC of Acid-Base Chemistry, 5th ed., Davenport, H.W., University of Chicago Press, Chicago, 1969.

Quantitative Problems in Biochemistry, 4th ed., Dawes, E.A., Williams & Wilkins, Baltimore, 1969.

Quantitative Problems in Biochemical Sciences, Montgomery, R., & Swenson, C.A., Freeman, San Francisco, 1969.

Biochemical Calculations, Segal, I.H., Wiley, New York, 1968.

A Biologist's Physical Chemistry, Morris, J.G., Addison-Wesley, Reading, Massachusetts, 1968.

The Structure and Properties of Water, Eisenberg, D., & Kauzmann, W., Oxford University Press, Fairlawn, New Jersey, 1969.

Acid Base Regulation, 2nd ed., Masoro, E.J., & Siegal, P.D., Saunders, Philadelphia, 1977.

Chapter 2
Amino Acids and Proteins

The Proteins, Volumes 1-5, Neurath, H. (ed.), Academic Press, New York, 1963-1970.

Chemistry of the Amino Acids, Three Volumes, Greenstein, J.P. & Winitz, M., Wiley, New York, 1961.

Chemical Synthesis of Peptides and Proteins, Marglin, A., & Merrifield, R.B., Ann. Rev. Biochem. 39:841, 1970.

Stereochemistry and Its Application to Biochemistry, Alworth, W.L., Wiley-Interscience, New York, 1972.

The Structure and Action of Proteins, Dickerson, R.E., & Geis, I., Harper & Row, New York, 1969.

The Total Synthesis of an Enzyme with Ribonuclease A Activity, Gutte, B., & Merrifield, R.B., J. Am. Chem. Soc. 91:501, 1969.

Chemical Structures of Pancreatic Ribonuclease and Deoxyribonuclease, Moore, S., & Stein, W.H., Science 180:458, 1973.

Evolution of Structure and Function of Proteases, Neurath, H., Walsh, K.A., & Winter, W.P., Science 158:1638, 1967.

The Amino Acid Sequence in the Glycyl Chain of Insulin, Sanger, F., & Thompson, E.O.P., Biochem. J. 53:353, 1963.

The Amino Acid Sequence in the Phenylalanine Chain of Insulin, Sanger, F., & Tuppy, H., Biochem. J. 49:463, 1961.

The Structure and Action of Proteins, Dickerson, R.E., & Geis, I., Benjamin, Menlo Park, California, 1969.

The Hydrophobic Effect, Tanford, C., Wiley-Interscience, New York, 1973.

Principles That Govern the Polypeptide Chain, Anfinsen, C.B., Science 181:223, 1973.

X-Ray Studies of Protein Mechanisms, Dickerson, R.E., Ann. Rev. Biochem. 41:815, 1972.

Three-Dimensional Pictures of Molecular Models, Harte, R.A., & Rupley, J.A., J. Biol. Chem. 243:1664, 1968.

The Three Dimensional Structure of a Protein Molecule, Kendrew, J.C., Scientific American 205:96, December 1961.

Protein Denaturation, Tanford, C., Adv. Prot. Chem. 23:121, 1968 and 24:1, 1970.

Acquisition of Three Dimensional Structure of Proteins, Wetleufer, D.B., & Ristow, S., Ann. Rev. Biochem. 42:135, 1973.

Chapter 3
Enzymes

Enzyme-Catalysed Reactions, Gray, C.J., Van Nostrand Reinhold Co., New York, 1971.

Methods in Enzymology, Colowick, S.P., & Kaplan, N.O. (eds.), Academic Press, New York, A Continuing Series.

Structure and Function of Proteins at the Three Dimensional Level, Cold Spring Harbor Symposia on Quantitative Biology, Volume XXXVI, Cold Spring Harbor Laboratory, Cold Spring Harbor, New York, 1972.

The Enzymes, 3rd ed., Boyer, P.D. (ed.), Academic Press, New York, 1973. Ten Volumes.

Enzymes: Physical Principles, Gutfreund, H., Wiley-Interscience, New York, 1972.

Enzymic Catalysis, Westley, J., Harper & Row, New York, 1969.

Enzyme Kinetics, Plowman, K., McGraw-Hill, New York, 1972.

Advances in Enzymology, Meister, A. (ed.), Academic Press, New York, A Continuing Series.

Enzymes, 2nd ed., Dixon, M., & Webb, E.C., Longmans, London, 1964.

Enzyme Nomenclature, American Elsevier, New York, 1973.

The Enzymes, Haldane, J.B.S., MIT Press, Cambridge, Massachusetts, 1965.

The Chemical Kinetics of Enzyme Action, 2nd ed., Laidler, K.J., Oxford, New York, 1973.

Dissociation, Enzyme Kinetics, Bioenergetics, Christensen, H., Saunders, Philadelphia, 1975.

Principle and Practice of Diagnostic Enzymology, Wilkinson, H., Year Book, Chicago, 1976.

Concepts and Perspectives in Enzyme Stereochemistry, Hanson, K.R., Ann. Rev. Biochem. 45:307, 1976.

Serine Proteases: Structure and Mechanism of Catalysis, Kraut, J., Ann. Rev. Biochem. 46:331, 1977.

Enzyme Topology of Intracellular Membranes, DePierre, J.W., & Ernster, L., Ann. Rev. Biochem. 46:201, 1977.

Enzymes as Drugs, Holcenberg, J.S. and Roberts, J., Ann. Rev. Pharmacol. Toxicol. 17:97, 1977.

Current Concepts About the Treatment of Selected Poisonings: Nitrite, Cyanide, Sulfide, Barium, and Quinidine, Smith, R.P., & Gosselin, R.E., Ann. Rev. Pharmacol. Toxicol. 16:189, 1976.

Relations Between Structure and Biological Activity of Sulfonamides, Maren, T.H., Ann. Rev. Pharmacol. Toxicol. 16:309, 1976.

Enzyme Polymorphism & Function During Embryonic Development, Netzloff, M.L., & Rennert, O.M., Ann. Clin. & Lab. Sci. 7:216, 1977.

Subcellular Localization of Enzymes, Zaki, F.G., Ann. Clin. & Lab. Sci. 7:222, 1977.

Chapter 4
Carbohydrates

General

Carbohydrate Digestion and Absorption, Gray, G., N. Engl. J. Med. 292:1225, 1975.

Pathophysiology of Diseases Involving Intestinal Brush Border Proteins, Alpers, D.H., & Seetharam, B., N. Engl. J. Med. 296:1047, 1977.

Recognition of Lactose Intolerance, Bayless, T.M., Hosp. Pract. 11:97, October 1976.

Muscle Amino Acid Metabolism and Gluconeogenesis, Ruderman, N.B., Ann. Rev. Med. 26:245, 1975.

Glycerol Utilization and Its Regulation in Mammals, Lin, E.C.C., Ann. Rev. Biochem. 46:765, 1977.

The Control of Glycogen Metabolism in the Liver, Hers, H.G., Ann. Rev. Biochem. 45:167, 1976.

Low Activities of the Pyruvate and Oxoglutarate Dehydrogenase Complexes in Five Patients With Friedreich's Ataxia, Blass, J.P., et al., N. Engl. J. Med. 295:62, 1976.

Glucose-Lactate Interrelations in Man, Kreisberg, R.A., N. Engl. J. Med. 287:132, 1972.

Caloric Homeostasis and Disorders of Fuel Transport, Havel, R.J., N. Engl. J. Med. 287:1186, 1972.

Fuel Homeostasis in Exercise, Felig, P., & Wahren, J., N. Engl. J. Med. 293:1078, 1975.

Myocardial Ischemia (in three parts), Hillis, L.D., & Braunwald, E., N. Engl. J. Med. 296:971, 1034 and 1093, 1977.

Disorders of Galactose Metabolism, Rennert, O.W., Ann. Clin. Lab. Sci. 7:443, 1977.

The Genetic Defect in Galactosemia, Tedesco, T.A., et al., N. Engl. J. Med. 292:737, 1975.

Galactokinase Deficiency as a Cause of Cataracts, Beutler, E., et al., N. Engl. J. Med. 288:1203, 1973.

Diagnosis of Myocardial Infarction by Serum Isoenzyme Analysis, Roe, C.R., Ann. Clin. Lab. Sci. 7:201, 1977.

Creatine Phosphokinase-MB (CPK-MB) and Diagnosis of Myocardial Infarction, Guzy, P.M., West J. Med. 127:455, December 1977.

Clinical Applications of Lactate Dehydrogenase Isoenzymes, Papadopoulos, N.M., Ann. Clin. Lab. Sci. 7:506, 1977.

Current Concepts: Principles of Medical Genetics (in two parts),
Erbe, R.W., N. Engl. J. Med. 294:381 and 480, 1976.

Amylase-Creatinine Clearance Ratio in the Diagnosis of Acute
Pancreatitis, Scottolini, A.G., & Bhagavan, N.V., Hawaii Med. J.
36:391, 1977.

Specificity of Increased Renal Clearance of Amylase in Diagnosis of
Acute Pancreatitis, Warshaw, A.L., & Fuller, A.F., N. Engl. J. Med.
292:325, 1975.

Enhancement of the Amylase-Creatinine Clearance Ratio in Disorders
Other Than Acute Pancreatitis, Levin, R.J., Galuser, F.L., &
Berk, J.E., N. Engl. J. Med. 292:329, 1975.

Differentiation of Pancreatitis From Common Bile Duct Obstruction
With Hyperamylasemia, Lesser, P.B., & Warshaw, A.L., Gastroenterology
68:636, 1975.

Amylase--Its Clinical Significance: A Review of the Literature,
Salt, W.B., & Schenker, S., Medicine 55:269, 1976.

Mechanism of Increased Renal Clearance of Amylase/Creatinine in Acute
Pancreatitis, Johnson, S.G., Ellis, C.J., & Levitt, M.D., N. Engl. J.
Med. 295:1214, 1976.

Diagnosis of Pancreatitis Masked by Hyperlipemia, Lesser, P.B., &
Warshaw, A.L., Ann. Intern. Med. 82:795, 1975.

Electron Transport and Oxidative Phosphorylation

Electron Transport Phosphorylation, Baltscheffsky, H., & Baltscheffsky, M.,
Ann. Rev. Biochem. 43:871, 1974.

Oxidative Phosphorylation and Photophosphorylation, Boyer, P.D., et al.,
Ann. Rev. Biochem. 46:955, 1977.

The Biosynthesis of Mitochondrial Proteins, Schatz, G., & Mason, T.L.,
Ann. Rev. Biochem. 43:51, 1974.

Structure and Function of Cytochromes C, Salemme, F.R., Ann. Rev.
Biochem. 46:299, 1977.

Biological Applications of Ionophores, Pressman, B.C., Ann. Rev.
Biochem. 45:501, 1976.

Coupling of "High-Energy" Phosphate Bonds to Energy Transductions,
Boyer, P.D., et al., Fed. Proc. 34:1711, 1975.

Bioenergetics: Past, Present and Future, Slater, E.C., in "Reflections on Biochemistry," in honor of Severo Ochoa, Kornberg, A., et al. (eds.), Pergamon Press, New York, 1976, p. 45.

Reconstitution, Mechanism of Action and Control of Ion Pumps, Racker, E., The Tenth Hopkins Memorial Lecture, Biochemical Society Transactions, Volume 3, 1975, p. 785.

Cyclic Nucleotides

Cyclic Nucleotides in Cell Growth and Differentiation, Friedman, D.L., Physiol. Rev. 56:652, 1976.

Role of Cyclic Nucleotides in Growth Control, Pastan, I.H., Johnson, G.S., & Anderson, W.B., Ann. Rev. Biochem. 44:491, 1975.

Selective Cyclic Nucleotide Phosphodiesterase Inhibitors as Potential Therapeutic Agents, Weiss, B., & Hait, W.N., Ann. Rev. Pharmacol. Toxicol. 17:441, 1977.

Protein Phosphorylation, Rubin, C.S., & Rosen, O.M., Ann. Rev. Biochem. 44:831, 1975.

Clinical Cyclic Nucleotide Research, Broadus, A.E., in "Adv. Cyclic Nucleotide Res.," Volume 8, Greengard, P., & Robison, G.A. (eds.), Raven Press, New York, 1977, p. 509.

Hypoglycemia and Hyperglycemia

Fasting Hypoglycemia in Adults, Fajans, S.S., & Floyd, Jr. J.C., N. Engl. J. Med. 294:766, 1976.

Neonatal Hypoglycemia, Senior, B., N. Engl. J. Med. 289:790, 1973.

Jamaican Vomitting Sickness, Tanaka, K., Kean, E.A., & Johnson, B., N. Engl. J. Med. 295:461, 1976.

The Unrike Akee--Forbidden Fruit, Bressler, R., N. Engl. J. Med. 295:500, 1976.

Insulin, Glucagon, and Somatostatin Secretion in the Regulation of Metabolism, Unger, R.H., Dobbs, R.E., & Orci, L., Ann. Rev. Physiol. 40:307, 1978.

Somatostatin: Physiological and Clinical Significance, Guillemin, R., & Gerich, J.E., Ann. Rev. Med. 27:379, 1976.

Metabolic Effects of Somatostatin in Maturity-Onset Diabetes, Tamborlane, W.V., et al., N. Engl. J. Med. 297:181, 1977.

"Somatistatinoma":   A Somatostatin Containing Tumor of the Endocrine Pancreas, Ganda, O.P., et al., N. Engl. J. Med. 296:963, 1977.

Somatostatinoma, Unger, R.H., N. Engl. J. Med. 296:998, 1977.

Physiology and Pathophysiology of Glucagon, Unger, R.H., & Orci, L., Physiol. Rev. 56:778, 1976.

The Role of Glucagon in the Endogenous Hyperglycemia of Diabetes Mellitus, Unger, R.H., & Orci, L., Ann. Rev. Med. 28:119, 1977.

Familial Hyperglucagonemia--An Autosomal Dominant Disorder, Boden, G., & Owen, O.E., N. Engl. J. Med. 296:534, 1977.

Glucagon and Its Clinical Significance, Lawrence, A.M., & Abraira, C., Lab. Med. 8:21, January 1977.

Hyperglycemia, Polyol Metabolism, and the Complications of Diabetes Mellitus, Gabbay, K.H., Ann. Rev. Med. 26:521, 1975.

Diabetic Polyneuropathy: The Importance of Insulin Deficiency, Hyperglycemia and Alterations in Myoinositol Metabolism in Its Pathogenesis, Winegrad, A.I., & Greene, D., N. Engl. J. Med. 295:1416, 1976.

The Glycosylation of Hemoglobin: Relevance to Diabetes Mellitus, Bunn, H.F., Gabbay, K.H., & Gallop, P.M., Science 200:21, 1978.

The Sorbitol Pathway and the Complications of Diabetes, Gabbay, K.H., N. Engl. J. Med. 288:831, 1973.

Biochemistry of the Renal Glomerular Basement Membrane and Its Alterations in Diabetes Mellitus, Spiro, R.G., N. Engl. J. Med. 288:1337, 1973.

Diet and Diabetes, West, K., Postgraduate Medicine 60:209, 1976.

Diabetes Mellitus, 4th ed., Sussman, K., & Metz, R., American Diabetes Assn, 1975 (several excellent articles).

Inheritance of Diabetes Mellitus, Zonana, J., & Rimoin, D.L., N. Engl. J. Med. 295:603, 1976.

Insulin Resistance: New Insights, Martin, D.B., N. Engl. J. Med. 294:778, 1976.

Familial Hyperproinsulinemia: An Autosomal Dominant Defect, Gabbay, K.H., et al., N. Engl. J. Med. 294:911, 1976.

Errors in Insulin Biosynthesis, Steiner, D.F., N. Engl. J. Med. 294:952, 1976.

Decreased Insulin Binding in Congenital Generalized Lipodystrophy, Oseid, S., et al., N. Engl. J. Med. 296:245, 1977.

Hormone Resistance and Hormone Sensitivity, Roth, J., et al., N. Engl. J. Med. 296:277, 1977.

Glucose Production in New Borns of Insulin Dependent Diabetic Mothers, Kalhan, S.C., Savin, S.M., & Adam, P.A., N. Engl. J. Med. 296:375, 1977.

Pathophysiology of Insulin Secretion, Brooks, M., Lab. Med. 8:15, January 1977.

The Syndrome of Insulin Resistance and Acanthosis Nigricans, Kahn, C.R., et al., N. Engl. J. Med. 294:739, 1976.

Lipoatrophic Diabetes in a Young Woman--Case Records of Massachusetts General Hospital, Arry, R.A., & McCully, K.S., N. Engl. J. Med. 292:35, 1975.

Molecular Basis of Insulin Action, Czech, M.P., Ann. Rev. Biochem. 46:359, 1977.

Neuropharmacology of the Pancreatic Islets, Smith, P.H., & Porte, Jr. D., Ann. Rev. Pharmacol. Toxicol. 16:269, 1976.

Oral Antidiabetic Agents (in two parts), Shen, S.-W., & Bressler, R., N. Engl. J. Med. 296:493 and 787, 1977.

Islet Cell Transplantation in Treatment of Diabetes, Najarian, J.S., Hosp. Pract. 12:63, December 1977.

Phagocytosis

Phagocytosis (in three parts), Stossel, T.P., N. Engl. J. Med. 290:717, 774 and 833, 1974.

Defective Leukocyte Superoxide Production in Chronic Granulomatous Disease, Curnutte, J.T., Whitten, D.M., & Babior, B.M., N. Engl. J. Med. 290:593, 1974.

Superoxide Radical and the Bactericidal Action of Phagocytes, Fridovich, I., N. Engl. J. Med. 290:624, 1974.

Neutrophil Actin Dysfunction and Abnormal Neutrophil Behavior, Boxer, L.A., Hedley-White, E.T., & Stossel, T.P., N. Engl. J. Med. 291:1093, 1974.

Leukocyte Muscular Dystrophy, Ward, P.A., N. Engl. J. Med. 291:1134, 1974.

Superoxide Dismutases, Fridovich, I., Ann. Rev. Biochem. 44:147, 1975.

Oxygen-Dependent Microbial Killing by Phagocytes (in two parts), Babior, B.M., N. Engl. J. Med. 298:659 and 721, 1978.

Defect in Superoxide Production in Chronic Granulomatous Disease, Carnutte, J.T., Kipnes, R.S., & Babior, B.M., N. Engl. J. Med. 293:628, 1975.

Contractile Proteins

Chemistry of Muscle Contraction, Taylor, E.W., Ann. Rev. Biochem. 41:577, 1972.

The Cooperative Action of Muscle Proteins, Murray, J.M., & Weber, A., Scientific American 230:58, February 1974.

The Protein Switch of Muscle Contraction, Cohen, C., Scientific American 233:36, November 1975.

Calcium-Binding Proteins, Kretsinger, R.H., Ann. Rev. Biochem. 45:239, 1976.

Proteins of Contractile Systems, Mannherz, H.G., & Goody, R.S., Ann. Rev. Biochem. 45:427, 1976.

Nonmuscle Contractile Proteins: The Role of Actin and Myosin in Cell Motility and Shape Determination, Clarke, M., & Spudich, J.A., Ann. Rev. Biochem. 46:797, 1977.

Biochemical Studies on the Regulation of Myocardial Contractility, Morkin, E., & La Raia, P.J., N. Engl. J. Med. 290:445, 1974.

Congestive Heart Failure: Role of Altered Myocardial Cellular Control, Katz, A.M., N. Engl. J. Med. 293:1184, 1975.

Microtubules: Structure, Chemistry and Function, Stephens, R.E., & Edds, K.T., Physiol. Rev. 56:709, 1976.

Immotile Cilia, Chronic Airway Infections and Male Sterility, Eliasson, R., et al., N. Engl. J. Med. 297:1, 1977.

What Makes Cilia and Sperm Tails Beat?, Fawcett, D.W., N. Engl. J. Med. 297:46, 1977.

Biochemistry and Physiology of Microtubules, Snyder, J.A., & McIntosh, J.R., Ann. Rev. Biochem. 45:699, 1976.

Pharmacology of Drugs That Affect Intracellular Movement,
Samson, F.E., Ann. Rev. Pharmacol. Toxicol. 16:143, 1976.

Use of Drugs in Myopathies, Grob, D., Ann. Rev. Pharmacol. Toxicol.
16:215, 1976.

Red Blood Cell Enzymes

Genetic Disorders of Human Red Blood Cells, Beutler, E., J. Am. Med.
Assoc. 233:1184, 1975.

Red Cell Enzymes, Paniker, N.V., CRC Critical Reviews in Clinical
Laboratory Sciences, page 469, April 1975.

Biochemical Studies in a Patient With Pyroglutamic Acidemia
(5-Oxoprolinemia), Marstein, S., N. Engl. J. Med. 295:406, 1976.

Genetic Implications of G-6-PD Deficiency, Desforges, J.F., N. Engl.
J. Med. 294:1438, 1976.

G-6-PD Deficiency, Keller, D.F., CRC Press, Cleveland, 1971.

Metabolism of Alcohol

Metabolic and Hepatic Effects of Alcohol, Isselbacher, K.J.,
N. Engl. J. Med. 296:612, 1977.

Behavioral and Biochemical Interrelations in Alcoholism,
Mendelson, J.H., & Mello, N.K., Ann. Rev. Med. 27:321, 1976.

The Metabolism of Alcohol, Lieber, C.S., Scientific American 234:25,
March 1976.

Chapter 5
Nucleic Acids

Nucleic Acids and Protein Synthesis

Translational Control of Protein Synthesis, Lodish, H.F.,
Ann. Rev. Biochem. 45:39, 1976.

Recognition Mechanisms of DNA-Specific Enzymes, Jovin, T.M.,
Ann. Rev. Biochem. 45:889, 1976.

Transfer RNA: Molecular Structure, Sequence, and Properties,
Rich, A., & RajBhandary, U.L., Ann. Rev. Biochem. 45:805, 1976.

Soluble Factors Required for Eukaryotic Protein Synthesis,
Weissbach, H., & Ochoa, S., Ann. Rev. Biochem. 45:191, 1976.

Processing of RNA, Perry, R.P., Ann. Rev. Biochem. 45:605, 1976.

Structure of Chromatin, Kornberg, R.D., Ann. Rev. Biochem. 46:931, 1977.

The Mechanism of Action of Inhibitors of DNA Synthesis, Cozzarelli, N.R., Ann. Rev. Biochem. 46:641, 1977.

Recombinant DNA, Sinsheimer, R.L., Ann. Rev. Biochem. 46:415, 1977.

Eukaryotic Messenger RNA, Brawerman, G., Ann. Rev. Biochem. 43:621, 1974.

Animal RNA Viruses: Genome Structure and Function, Shatkin, A.J., Ann. Rev. Biochem. 43:643, 1974.

Methods of Gene Isolation, Brown, D.D., & Stern, R., Ann. Rev. Biochem. 43:667, 1974.

Replication of Circular DNA in Eukaryotic Cells, Kasamatsu, H., & Vinograd, J., Ann. Rev. Biochem. 43:695, 1974.

The Selectivity of Transcription, Chamberlin, M.J., Ann. Rev. Biochem. 43:721, 1974.

Peptide Synthesis, Fridkin, M., & Patchornik, A., Ann. Rev. Biochem. 43:419, 1974.

Structure and Function of the Bacterial Ribosome, Kurland, C.G., Ann. Rev. Biochem. 46:173, 1977.

Eukaryotic DNA Polymerases, Weissbach, A., Ann. Rev. Biochem. 46:25, 1977.

Biosynthesis of Small Peptides, Kurahashi, K., Ann. Rev. Biochem. 43:445, 1974.

Diphtheria Toxin, Pappenheimer, Jr. A.M., Ann. Rev. Biochem. 46:69, 1977.

Eukaryotic Nuclear RNA Polymerases, Chambon, P., Ann. Rev. Biochem. 44:613, 1975.

DNA Replication, Gefter, M.L., Ann. Rev. Biochem. 44:45, 1975.

Restriction Endonucleases in the Analysis and Restructuring of DNA Molecules, Nathans, D., & Smith, H.O., Ann. Rev. Biochem. 44:273, 1975.

Enzymatic Repair of DNA, Grossman, L., et al., Ann. Rev. Biochem. 44:19, 1975.

Chromosomal Proteins and Chromatin Structure, Elgin, S.C.R., & Weintraub, H., Ann. Rev. Biochem. 44:725, 1975.

Chemical Carcinogenesis, Heidelberger, C., Ann. Rev. Biochem. 44:79, 1975.

Biological Methylation: Selected Aspects, Cantoni, G.L., Ann. Rev. Biochem. 44:435, 1975.

The Biochemistry of Mutagenesis, Drake, J.W., & Baltz, R.H., Ann. Rev. Biochem. 45:11, 1976.

DNA Repair and Human Disease, Lambert, B., & Ringborg, U., Acta Med. Scand. 200:433, 1976.

Oncogenic Implications of Chromosomal Instability, German, J., Hosp. Pract. 8:93, February 1973.

Gene De-Repression, Frenster, J.H., & Herstein, P.R., N. Engl. J. Med. 288:1224, 1973.

The Manipulation of Genes, Cohen, S., Scientific American 233:25, July 1975.

Gene Manipulation, Cohen, S., N. Engl. J. Med. 294:883, 1976.

Carcinogenesis and Aging--Two Related Phenomena?, Pitot, H.C., Am. J. Pathol. 87:444, 1977.

DNA Insertions and Gene Structure, Williamson, B., Nature 270:295, 1977.

Unconventional Viruses and the Origin Disappearance of Kuru, Gajdusek, D.C., Science 197:943, 1977.

DNA Sequencing Techniques, Salser, W.A., Ann. Rev. Med. 43:923, 1974.

A New Method for Sequencing DNA, Maxam, A.M., & Gilbert, W., Proc. Natl. Acad. Sci. U.S.A. 74:560, 1977.

The Nucleotide Sequence of Viral DNA, Fiddes, J.C., Scientific American 237:54, December 1977.

Collagen, Elastin and Albumin

The Biosynthesis of Collagen (in three parts), Grant, M.E., & Prockop, D.J., N. Engl. J. Med. 286:194, 242 and 291, 1972.

Collagen Biology: Structure, Degradation, and Disease, Gross, J., Harvey Lect. 68:351, 1973.

The Biochemistry of Collagen, Veis, A., Ann. Clin. Lab. Sci. 5:123, 1975.

Molecular Defects in Collagen and the Definition of "Collagen Disease," Uitto, J., & Prockop, D.J., in "The Molecular Pathology," Good, R.A., Day, S.B., & Unis, Y. (eds.), Charles C Thomas, Publisher, Springfield, Illinois, 1975.

Defects in the Biochemistry of Collagen in Diseases of Connective Tissue, Uitto, J., & Lichtenstein, J.R., J. Invest. Dermatol. 66:59, 1976.

Biosynthesis of Collagen and Its Alterations in Pathological States, Kivirikko, K.I., & Risteli, L., Med. Biol. 54:159, 1976.

Intracellular Steps in the Biosynthesis of Collagen, Prockop, D.J., et al., in "Biochemistry of Collagen," Ramachandran, G.N., & Reddi, A.H. (eds.), Plenum Publishing, New York, 1976.

Collagen Diseases and the Biosynthesis of Collagen, Prockop, D.J., & Guzman, N.A., Hosp. Pract. 12:61, December 1977.

The Biosynthesis of Collagen, Bornstein, P., Ann. Rev. Biochem. 43:567, 1974.

Collagenases (in three parts), Harris, Jr. E.D., & Krane, S.M., N. Engl. J. Med. 291:557, 605 and 632, 1974.

A Heritable Disorder of Connective Tissue: Hydroxylysine-Deficient Collagen Disease, Pinnell, S.R., et al., N. Engl. J. Med. 286:1013, 1972.

Posttranslational Protein Modifications, With Special Attention to Collagen and Elastin, Gallop, P.M., & Paz, M.A., Physiol. Rev. 55:418, 1975.

Arterial Elastin, Rucker, R.B., & Tom, K., Am. J. Clin. Nutri. 29:1021, 1976.

Albumin Synthesis (in two parts), Rothschild, M.A., Oratz, M., & Schreiber, S.S., N. Engl. J. Med. 286:748 and 816, 1972.

Regulation of Albumin Metabolism, Rothschild, M.A., Oratz, M., & Schreiber, S.S., Ann. Rev. Med. 26:91, 1975.

Purine and Pyrimidine Metabolism

Folate Absorption and Malabsorption, Rosenberg, I.H., N. Engl. J. Med. 293:1303, 1975.

Inborn Errors of Folate Metabolism (in two parts), Erbe, R.W.,
N. Engl. J. Med. 293:753 and 807, 1975.

Pyrimidine Metabolism in Man, Smith, L.H., N. Engl. J. Med. 288:764,
1973.

Control of Pyrimidine Biosynthesis in Human Lymphocytes, Ito, K., &
Uchino, H., J. Biol. Chem. 251:1427, 1976.

Purine Nucleoside Phosphorylase Deficiency and Cellular
Immunodeficiency, Stoop, J.W., et al., 296:651, 1977.

Combined Immunodeficiency Disease Associated With Adenosine Deaminase
Deficiency--A Special Article and Combined Report, J. Pediatr.
86:169, 1975.

Adenosine-Deaminase Deficiency in Two Patients With Severely Impaired
Cellular Immunity, Gilbert, E.R., et al., Lancet, page 1067,
November 18, 1972.

Enzyme Replacement Therapy for Adenosine Deaminase Deficiency and
Severe Combined Immunodeficiency, Polmar, S.H., et al., N. Engl. J.
Med. 295:1337, 1976.

Abnormal Purine Metabolism and Purine Overproduction in a Patient
Deficient in Purine Nucleoside Phosphorylase, Cohen, A., et al.,
N. Engl. J. Med. 295:1449, 1976.

Complete Deficiency of Adenine Phosphoribosyltransferase,
Van Acker, K.J., et al., N. Engl. J. Med. 297:127, 1977.

Gout

Genetic Aspects of Gout, Becker, M.A., & Seegmiller, J.E., Ann. Rev.
Med. 25:15, 1974.

Urate Excretion, Fanelli, Jr. G.M., Ann. Rev. Med. 28:349, 1977.

The Physiologic Approach to Hyperuricemia, Rastegar, A., & Thier, S.O.,
N. Engl. J. Med. 286:470, 1972.

Current Therapy of Gout and Hyperuricemia, Kelley, W.N., Hosp. Pract.
11:69, May 1976.

Molecular Basis of Gouty Inflammation: Interaction of Monosodium Urate
Crystals With Lysosomes and Liposomes, Weissmann, G., & Giuseppe, A.R.,
Nature (New Biology) 240:167, 1972.

Hyperuricemia, Renal Disease, and Abnormal Electrophoretic Findings, Case Records of Massachusetts General Hospital, Ellman, L., & Galdabini, J.J., N. Engl. J. Med. 294:998, 1976.

Urinary-Tract Stones Resulting From the Excretion of Oxypurinol, Landgrebe, A.R., Nyhan, W.L., & Coleman, M., N. Engl. J. Med. 292:626, 1975.

Antineoplastic Agents

The Clinical Pharmacology of Antineoplastic Agents (in two parts), Chabner, B.A., et al., N. Engl. J. Med. 292:1107 and 1159, 1975.

Neurotoxicity of Commonly Used Antineoplastic Agents (in two parts), Weiss, H.W., Walker, M.D., & Wiernik, P.H., N. Engl. J. Med. 291:75 and 127, 1974.

Second Neoplasm--A Complication of Cancer Chemotherapy, Chabner, B., N. Engl. J. Med. 297:213, 1977.

Antineoplastic Agents From Plants, Wall, M.E., & Wani, M.C., Ann. Rev. Pharmacol. Toxicol. 17:117, 1977.

The Clinical Applications of Cell Kinetics in Cancer Therapy, Livingston, R.B., & Hart, J.S., Ann. Rev. Pharmacol. Toxicol. 17:529, 1977.

Immunologic Aspects of Cancer Chemotherapy, Haskell, C.M., Ann. Rev. Pharmacol. Toxicol. 17:179, 1977.

Selective Suppression of Humoral Immunity by Antineoplastic Drugs, Heppner, G.H., & Calabresi, P., Ann. Rev. Pharmacol. Toxicol. 16:367, 1976.

Chapter 6
Amino Acid Metabolism

General

Amino Acid Metabolism in Man, Felig, P., Ann. Rev. Biochem. 44:933, 1975.

Glutathione and Related γ-Glutamyl Compounds: Biosynthesis and Utilization, Meister, A., & Tate, S.S., Ann. Rev. Biochem. 45:559, 1976.

Regulation of Amino Acid Decarboxylation, Morris, D.R., & Fillingame, R.H., Ann. Rev. Biochem. 43:303, 1974.

A Family of Protein-Cutting Proteins, Stroud, R.M., Scientific American 231:74, July 1974.

Aminoacidurias Due to Inherited Disorders of Metabolism (in two parts), Frimpter, G.W., N. Engl. J. Med. 289:835 and 895, 1973.

Routine Newborn Screening for Histidinemia, Levy, H.L., Shih, V.E., & Madigan, P.M., N. Engl. J. Med. 291:1214, 1974.

Disorders of Renal Amino Acid Transport, Segal, S., N. Engl. J. Med. 294:1044, 1976.

Nephrolithiasis, Williams, H.E., N. Engl. J. Med. 290:33, 1974.

The Pharmacology of Renal Lithiasis, Steele, T.H., Ann. Rev. Pharmacol. Toxicol. 17:11, 1977.

Research in Cystic Fibrosis (in three parts), DiSant'Agnese, P.A., & Davis, P.B., N. Engl. J. Med. 295:481, 534 and 597, 1976.

Cystic Fibrosis, Bowman, B.H., & Mangos, J.A., N. Engl. J. Med. 294:937, 1976.

Enzyme Replacement Therapy of Exocrine Pancreatic Insufficiency in Man, Graham, D.Y., N. Engl. J. Med. 296:1314, 1977.

Fate of Orally Ingested Enzymes in Pancreatic Insufficiency, DiMagno, E.P., et al., N. Engl. J. Med. 296:1318, 1977.

The Ins and Outs of Oral Pancreatic Enzymes, Meyer, J.H., N. Engl. J. Med. 296:1347, 1977.

1,4-Diaminobutane (Putrescine), Spermidine, and Spermine, Tabor, C.W., & Tabor, H., Ann. Rev. Biochem. 45:285, 1976.

Putative Peptide Neurotransmitter, Otsuka, M., & Takahashi, T., Ann. Rev. Pharmacol. Toxicol. 17:425, 1977.

"Second Messengers" in the Brain, Nathanson, J.A., & Greengard, P., Scientific American 237:108, August 1977.

Modification of Brain Serotonin by the Diet, Fernstrom, J.D., Ann. Rev. Med. 25:1, 1974.

Neurologic Disorders in Renal Failure (in two parts), Raskin, N.H., & Fishman, R.A., N. Engl. J. Med. 294:143 and 204, 1976.

Familial Azotemia: Impaired Urea Excretion Despite Normal Renal Function, Hsu, C.H., et al., N. Engl. J. Med. 298:117, 1978.

The Use of Amino Acid Precursors in Nitrogen-Accumulating Diseases, Close, J.H., N. Engl. J. Med. 290:663, 1974.

Serum Proteins: Diagnostic Significance of Electrophoretic Patterns, Larson, P.H., Human Pathology 6:629, 1974.

Plasma Protein Measurements as a Diagnostic Aid, Alper, C.A., N. Engl. J. Med. 291:287, 1974.

Basic and Clinical Aspects of the $Alpha_1$-Antitrypsin, Talamo, R.C., Pediatrics 56:91, 1975.

$Beta_2$ Microglobulin and Lymphocyte Infiltration in Sjogren's Syndrome, Michalski, J.P., et al., N. Engl. J. Med. 293:1228, 1975.

Structure and Function of Alpha-Fetoprotein, Tomasi, Jr. T.B., Ann. Rev. Med. 28:453, 1977.

Oncofetal Antigens, Shuster, J., Freedman, S.O., & Gold, P., Am. J. Clin. Pathol. 68:679, 1977.

Protease Inhibitor Phenotypes and Obstructive Lung Disease, Morse, J.O., et al., N. Engl. J. Med. 296:1190, 1977.

Hopi Indians, Inbreeding and Albinism, Woolf, C.M., & Dukepoo, F.C., Science 164:30, 1969.

Gastrointestinal Hormones

Secretin, Cholecystokinin and Newer G-I Hormones (in two parts), Rayford, P.L., Miller, T.A., & Thompson, J.C., N. Engl. J. Med. 294:1093 and 1157, 1976.

Gastrin (in two parts), Walsh, J.H., & Grossman, M.I., N. Engl. J. Med. 292:1324 and 1377, 1974.

Circulating Gastrin, Walsh, J.H., Ann. Rev. Physiol. 37:81, 1975.

The Newer Gut Hormones, Pearse, A.G.E., Polak, J.M., & Bloom, S.R., Gastroenterology 72:746, 1977.

Molecular Evolution of Gut Hormones: Application of Comparative Studies on the Regulation of Digestion, Dockray, G.J., Gastroenterology 72:344, 1977.

Gastrointestinal Hormones and Their Functions, Johnson, L.R., Ann. Rev. Physiol. 39:135, 1977.

Gastrointestinal Pharmacology, Burks, T.F., Ann. Rev. Pharmacol. Toxicol. 16:15, 1976.

Hyperammonemia

Urea Cycle Disorders, Shih, V.E., & Efron, M.L., in "The Metabolic Basis of Inherited Disease," Stanbury, J.B., Wyngaarden, J.B., & Fredrickson, D.S. (eds.), McGraw-Hill, New Jersey, 1972, p. 370.

Pathogenesis of Hepatic Coma, Zieve, L., & Nicoloff, D., Ann. Rev. Med. 26:143, 1975.

Treatment of Carbamyl Phosphate Synthetase Deficiency With Keto Analogues of Essential Amino Acids, Batshaw, M., Brusilow, S., & Walser, M., N. Engl. J. Med. 292:1085, 1975.

Hepatic Pre-Coma and Coma, 5th ed., Sherlock, S., in "Diseases of the Liver and Biliary System," Blackwell Scientific Publications, 1975, p. 84.

Lethal Neonatal Deficiency of Carbamyl Phosphate Synthetase, Gelehrter, T.D., & Snodgrass, P.J., N. Engl. J. Med. 290:430, 1974.

Inherited Hyperammonemic Syndromes, Hsia, Y.E., Gastroenterology 67:347, 1974.

Current Diagnosis and Treatment of Hepatic Coma, Conn, H.O., Hosp. Pract., page 65, February 1973.

Urea-Cycle Enzyme Deficiencies and an Increased Nitrogen Load Producing Hyperammonemia in Reye's Syndrome, Snodgrass, P.J., & DeLong, G.R., N. Engl. J. Med. 294:855, 1976.

Transiently Reduced Activity of Carbamyl Phosphate Synthetase and Ornithine Transcarbamylase in Liver of Children With Reye's Syndrome, Brown, T., et al., N. Engl. J. Med. 294:861, 1976.

Ammonia Disposal in Reye's Syndrome, Smith, A.L., N. Engl. J. Med. 294:897, 1976.

Reye's Syndrome, Schiff, G.M., Ann. Rev. Med. 27:447, 1976.

Histamine

Histamine H_2-Receptor Antagonists--A Symposium, Fed. Proc. 35:1923, 1976.

Reflections on Histamine, Gastric Secretion, and H_2 Receptor, Code, C.F., N. Engl. J. Med. 296:1459, 1977.

Histamine (in two parts), Beaven, M.A., N. Engl. J. Med. 294:30 and 320, 1976.

Histaminergic Mechanisms in Brain, Schwartz, J.C., Ann. Rev. Pharmacol. Toxicol. 17:325, 1977.

Catecholamines

Catecholamines, Axelrod, J., & Weinshilboum, R., N. Engl. J. Med. 287:237, 1972.

Interrelations Between β-Adrenergic Receptors, Adenylate Cyclase and Calcium, Steer, M.L., Atlas, D., & Levitzki, A., N. Engl. J. Med. 292:409, 1975.

β-Adrenergic Receptors: Recognition and Regulation, Lefkowitz, R.J., N. Engl. J. Med. 295:323, 1976.

Catecholamines--A Symposium, Calif. Med. 117:32, 1972.

Clinical Features and Laboratory Diagnosis of Functional Paraganglioma (Pheochromocytoma), Markel, S.F., & Johnson, R.M., Lab. Med. 6:39, 1975.

In Vitro Study of β-Adrenergic Receptors, Wolfe, B.B., Harden, T.K., & Molinoff, P.B., Ann. Rev. Pharmacol. Toxicol. 17:575, 1977.

Psychopharmacological Implications of Dopamine and Dopamine Antagonists: A Critical Evaluation of Current Evidence, Hornykiewicz, O., Ann. Rev. Pharmacol. Toxicol. 17:545, 1977.

The Effects of Psychopharmacological Agents on Central Nervous System Amine Metabolism in Man, Maas, J.W., Ann. Rev. Pharmacol. Toxicol. 17:411, 1977.

Catecholamines and Neurologic Diseases (in two parts), Maskovitz, M.A., & Wurtman, R.J., N. Engl. J. Med. 293:274 and 332, 1975.

Multiple Endocrine Neoplasia Type II, Wells, Jr. S.A., & Ontjes, D.A., Ann. Rev. Med. 27:263, 1976.

Phenylketonuria

Phenylketonuria and Its Variants, Kauffman, S., & Milstien, S., Ann. Clin. Lab. Sci. 7:178, 1977.

Unrecognized Phenylketonuria, Perry, T.L., et al., N. Engl. J. Med. 289:395, 1973.

Phenylketonuria Due to Deficiency of Dihydropteridine Reductase, Kauffman, S., et al., N. Engl. J. Med. 293:785, 1975.

A Comparison of Effectiveness of Screening for Phenylketonuria in the United States, United Kingdom and Ireland, Starfield, B., & Holtzman, N.A., N. Engl. J. Med. 293:118, 1975.

Thyroid Hormones

Impaired Peripheral Conversion of Thyroxine to Triiodothyronine, Cavalieri, R.R., & Rapoport, B., Ann. Rev. Med. 28:57, 1977.

The Thyroid and Its Control, Sterling, K., & Lazarus, J.H., Ann. Rev. Physiol. 39:349, 1977.

Active Form of the Thyroid Hormone and Somatomedin, Ingbar, S.H., & Braverman, L.E., Ann. Rev. Med. 26:443, 1975.

Serum Triiodothyronine in Man, Utiger, R.D., Ann. Rev. Med. 25:289, 1974.

Thyroidal and Peripheral Production of Thyroid Hormones, Schimmel, M., & Utiger, R.D., Ann. Intern. Med. 87:760, 1977.

Initiation of Thyroid Hormone Action, Oppenheimer, J.H., N. Engl. J. Med. 292:1063, 1975.

Thyroid Thermogenesis, Edeman, I., N. Engl. J. Med. 290:1303, 1974.

Revised Nomenclature for Tests of Thyroid Hormones in Serum, Solomon, D.H., et al., J. Clin. Endocrin. Metab. 42:595, 1976.

Self-Tolerance and Autoimmunity in the Thyroid, Allison, A.C., N. Engl. J. Med. 295:821, 1976.

Clinical Application of Thyroid-Releasing Hormone, Hershman, J.M., N. Engl. J. Med. 290:886, 1974.

Low Thyroid Hormones and Respiratory-Distress Syndrome of the New Born, Cuestas, R.A., et al., N. Engl. J. Med. 295:297, 1976.

Neonatal Detection of Hypothyroidism, Fisher, D.A., J. Pediatr. 86:822, 1975.

Preliminary Report on a Mass Screening Program for Neonatal Hypothyroidism, Dussault, J.H., et al., J. Pediatr. 86:670, 1975.

Improved Prognosis in Congenital Hypothyroidism Treated Before Age Three Months, Klein, A.H., Meltzer, S., & Kenny, F.M., J. Pediatr. 81:912, 1972.

Screening for Congenital Hypothyroidism, Fisher, D.A., Hosp. Pract. 12:73, December 1977.

Hyperthyroidism During Pregnancy, Burrow, G.N., N. Engl. J. Med. 298:150, 1978.

A New In-Vitro Blood Test for Determining Thyroid Status--The Effective Thyroxine Ratio, Mincey, E.K., Thorson, S.C., & Brown, J.L., Clin. Biochem. 4:216, 1971.

Hypothyroidism Due to Thyroid-Hormone Binding Antibodies, Karlsson, A.F., Wibell, L., & Wide, L., N. Engl. J. Med. 296:1146, 1977.

Chapter 7
Hemoglobin and Porphyrin Metabolism

General

Regulation of Hemoglobin Synthesis (in three parts), Nienhuis, A.W., & Benz, Jr. E.J., N. Engl. J. Med. 297:1318, 1371 and 1430, 1977.

Cooperative Interactions of Hemoglobin, Edelstein, S.J., Ann. Rev. Biochem. 44:209, 1975.

Hemoglobin Function, Oxygen Affinity, and Erythropoietin, Adamson, J.W., & Finch, C.A., Ann. Rev. Physiol. 37:351, 1975.

Regulation of Respiration in Man, Guz, A., Ann. Rev. Physiol. 37:303, 1975.

Mechanism of Action of Hemoglobin, Arnone, A., Ann. Rev. Med. 25:123, 1974.

2,3-DPG and Erythrocyte Oxygen Affinity, Brewer, G.J., Ann. Rev. Med. 25:29, 1974.

New Aspects of the Structure, Function, and Synthesis of Hemoglobins, Huisman, T.H.J., & Schroeder, W.A., CRC Press, Cleveland, 1971.

Stereochemistry of Cooperative Effects in Hemoglobin, Perutz, M.F., Nature 228:726, 1970.

Intracellular Organic Phosphates as Regulators of Oxygen Release by Haemoglobin, Benesch, R., & Benesch, R.E., Nature 221:618, 1969.

Erythrocyte Metabolism: Interaction With Oxygen Transport, Brewer, G.J., & Eaton, J.W., Science 171:1205, 1971.

Regulation of Fetal Hemoglobin Production, Kazazian, Jr. H.H., Seminars in Hematol. 11:525, 1974.

Androgens and Erythropoiesis, Shahidi, N., N. Engl. J. Med. 289:72, 1973.

The Glycosylation of Hemoglobin: Relevance to Diabetes Mellitus, Bunn, F.H., Gabbay, K.H., & Gallop, P.M., Science 200:21, 1978.

Hemoglobin Components in Diabetes Mellitus: Studies in Identical Twins, Tattersall, R.B., et al., N. Engl. J. Med. 293:1171, 1975.

Exposure of Humans to Lead, Hammond, P.B., Ann. Rev. Pharmacol. Toxicol. 17:197, 1977.

Vulnerability of Children to Lead Exposure and Toxicity (in two parts), Lin-Fu, J.S., N. Engl. J. Med. 289:1229 and 1289, 1973.

Iron, Iron Chelation Therapy, Hemochromatosis

Iron in Biochemistry and Medicine, Jacobs, A., & Worwood, M. (eds.), Academic Press, New York, 1974.

Iron Deficiency: Pathogenesis, Clinical, Aspects, Therapy, Hallberg, L., Harwerth, H.-G., & Vannotti, A. (eds.), Academic Press, New York, 1970.

Who Needs Iron?, Crosby, W.H., N. Engl. J. Med. 297:543, 1977.

Hemochromatosis: The Unsolved Problems, Crosby, W.H., Seminars in Hematol. 14:135, 1977.

Porphyrin and Iron Metabolism in Sideroblastic Anemia, Bottomley, S., Seminars in Hematol. 14:169, 1977.

Sideroblasts, Siderocytes and Sideroblastic Anemia, Cartwright, G.E., & Deiss, A., N. Engl. J. Med. 292:185, 1975.

Hemochromatosis: Course and Treatment, Sherlock, S., Ann. Rev. Med. 27:143, 1976.

Ferritin in Serum: Clinical and Biochemical Implications, Jacobs, A., & Worwood, M., N. Engl. J. Med. 292:951, 1975.

Normal Serum Ferritin in Precirrhotic Hemochromatosis, Crosby, W.H., N. Engl. J. Med. 296:1116, 1977.

Normal Serum Ferritin Concentrations in Precirrhotic Hemochromatosis, Wands, J.R., et al., N. Engl. J. Med. 294:302, 1976.

Physiologic Studies in Precirrhotic Stage of Familial Hemochromatosis, Feller, E.R., et al., N. Engl. J. Med. 296:1422, 1977.

Continuous Subcutaneous Administration of Deferoxamine for Iron Overload, Propper, R.D., et al., N. Engl. J. Med. 297:418, 1977.

Iron Overloading and Thalassemia--Experimental Success and Practical Realities, Weatherall, D.J., Pippard, M.J., & Callender, S.T., N. Engl. J. Med. 297:445, 1977.

Thalassemia

Thalassemia, Forget, B.G., & Nathan, D.G., Ann. Rev. Med. 26:345, 1975.

The Thalassemias, Orkin, S.H., & Nathan, D.G., N. Engl. J. Med. 295:710, 1976.

Molecular Basis of Thalassaemia, Clegg, J.B., & Weatherall, D.J., Br. Med. Bull. 32:262, 1976.

Screening for Thalassemia Trait by Electronic Measurement of Mean Corpuscular Volume, Pearson, H.A., O'Brien, R.T., & McIntosh, S., N. Engl. J. Med. 288:351, 1973.

Molecular Basis for Acquired Haemoglobin H Disease, Old, J., et al., Nature 269:524, 1977.

Prenatal Diagnosis of α-Thalassemia: Clinical Application of Molecular Hybridization, Kan, Y.W., Golbus, M.S., & Dozy, A.M., N. Engl. J. Med. 295:165, 1976.

Antenatal Diagnosis of Hemoglobinopathies: A Exquisite Molecular Brew, Nathan, D.G., N. Engl. J. Med. 295:1196, 1976.

Prenatal Diagnosis in a Pregnancy at Risk for Prenatal Homozygous β-Thalassemia, Kan, Y.W., et al., N. Engl. J. Med. 292:1096, 1975.

Prenatal Diagnosis of Hemoglobinopathies, Kazazian, Jr. H.H., N. Engl. J. Med. 292:1125, 1975.

Diagnosis of Homozygous α-Thalassemia in Cultured Amniotic-Fluid Fibroblasts, Wong, V., et al., N. Engl. J. Med. 298:669, 1978.

Prenatal Diagnosis of Hemoglobinopathies: A Review of 15 Cases, Alter, B.P., et al., N. Engl. J. Med. 295:1437, 1976.

Sickle-Cell Anemia

Prenatal Diagnosis of Sickle-Cell Anemia, Kan, Y.W., Golbus, M.S., & Trecartin, R., N. Engl. J. Med. 294:1039, 1976.

Prenatal Diagnosis of Sickle Cell Anemia and Alpha-G Philadelphia, Alter, B.P., et al., N. Engl. J. Med. 294:1040, 1976.

Effects of Urea and Cyanate on Sickling In Vitro, Segal, G.B., et al., N. Engl. J. Med. 287:59, 1972.

Sickle Hemoglobin Aggregation: A New Class of Inhibitors, Votano, J.R., Gorecki, M., & Rich, A., Science 196:1216, 1977.

Porphyrin Metabolism and the Porphyrias

Enzymatic Formation and Cellular Regulation of Heme Synthesis, Gidari, A., & Levere, R.D., Seminars in Hematol. 14:145, 1977.

Interaction of Proteins With Porphyrins, Koskelo, P., & Muller-Eberhard, U., Seminars in Hematol. 14:221, 1977.

Experimental Porphyrias as Models for Human Hepatic Porphyrias, De Matteis, F., & Stonard, M., Seminars in Hematol. 14:187, 1977.

The Porphyrias: A Review, Elder, G.H., Gray, C.H., & Nicholson, D.C., J. Clin. Path. 25:1013, 1972.

The Hepatic Porphyrias: Pathogenesis, Manifestations and Management, Bloomer, J.R., Gastroenterology 71:689, 1976.

Acute Intermittent Porphyria: Clinical and Selected Research Aspects-- NIH Conference Moderator: Tschudy, D.P., Ann. Intern. Med. 83:851, 1975.

Effects of Hematin in Hepatic Porphyria, Dhar, G.J., et al., Ann. Intern. Med. 83:20, 1975.

Hematin and Porphyria, Watson, C.J., N. Engl. J. Med. 293:605, 1975.

Erythropoietic Protoporphyria, Methews-Roth, M.M., N. Engl. J. Med. 297:98, 1977.

The Neurologic Manifestations of Porphyria: A Review, Becker, D.M., & Kramer, S., Medicine 56:411, 1977.

Erythropoietic Porphyrias: Current Mechanistic and Therapeutic Considerations, Poh-Fitzpatrick, M.B., Seminars in Hematol. 14:211, 1977.

Porphyrin Metabolism in Porphyria Cutanea Tarda, Elder, G.H., Seminars in Hematol. 14:227, 1977.

Hepatic Siderosis and Porphyria Cutanea Tarda: Relation of Iron Excess to the Metabolic Defect, Felsher, B., & Kushner, J.P., Seminars in Hematol. 14:243, 1977.

Treatment of Porphyria Cutanea Tarda by Phlebotomy, Ippen, H., Seminars in Hematol. 14:253, 1977.

Modern Spectroscopic and Chromatographic Techniques for the Analysis of Porphyrin on a Microscale, Jackson, A.H., Seminars in Hematol. 14:193, 1977.

Clinical Aspects of Porphyrin Measurement Other Than Lead Poisoning, Lamon, J.M., Clin. Chem. 23:260, 1977.

History and Background of Protoporphyrin Testing, Labbe, R.F., Clin. Chem. 23:256, 1977.

Bilirubin Metabolism

Bilirubin Metabolism in Man, Schmid, R., N. Engl. J. Med. 287:703, 1972.

Endogenous Production of Carbon Monoxide, Coburn, R.F., Hosp. Pract. 8:89, July 1973.

Development of Bilirubin Transport and Metabolism in the Newborn Rhesus Monkey, Gartner, L.M., et al., J. Pediatr. 90:513, 1977.

Liver Disease in Infants, Part II: Hepatic Disease States, Mathis, R.K., J. Pediatr. 90:864, 1977.

Physiologic Approach to Jaundice, Badley, B.W.D., Clin. Biochem. 9:144, 1976.

Neonatal Nonhemolytic Jaundice, Johnson, J.D., N. Engl. J. Med. 292:194, 1975.

Jaundice in Older Children and Adults: Algorithm for Diagnosis, Ostrow, J.D., J. Am. Med. Assoc. 234:522, 1975.

Chronic Nonhemolytic Unconjugated Hyperbilirubinemia With Glucuronyl Transferase Deficiency, Arias, I.M., et al., Am. J. Med. 47:395, 1969.

Pigment Gallstones, Soloway, R.D., Trotman, B.W., & Ostrow, J.D., Gastroenterology 72:167, 1977.

Chemistry and Physiology of Bile Pigments: An International Symposium, Fogarty International Center Proceedings No. 35, Berk, P.D., & Berlin, N.I. (eds.), DHEW Publication No. (NIH) 77-1100, U.S. Dept. Health, Education and Welfare, NIH, Bethesda, Maryland, 1977.

Chapter 8
Lipids

General

Essential Fatty Acids Reinvestigated, Alfin-Slatter, R.B., &
Aftergood, L., Physiol. Rev. 48:758, 1968.

Current Studies on Relation of Fat to Health, Kummerow, F.A.,
J. Am. Oil Chem. Soc. 51:255, 1974.

Lipids and Lipid Metabolism, Masoro, E.J., Ann. Rev. Physiol. 39:301,
1977.

Mechanisms and Regulation of Biosynthesis of Saturated Fatty Acids,
Volpe, J.J., & Vagelos, R.P., Physiol. Rev. 56:339, 1976.

Control Mechanisms in the Synthesis of Saturated Fatty Acids,
Bloch, K., & Vance, D., Ann. Rev. Biochem. 46:263, 1977.

Multifunctional Proteins, Kirschner, K., & Bisswanger, H.,
Ann. Rev. Biochem. 45:143, 1976.

Metabolic Alterations of Fatty Acids, Fulco, A.J., Ann. Rev. Biochem.
43:215, 1974.

A Disorder of Muscle Lipid Metabolism and Myoglobinuria--Absence of
Carnitine Palmityl Transferase, Bank, W.J., et al., N. Engl. J. Med.
292:443, 1975.

Regulation of HMG-CoA Reductase, Rodwell, V.W., Nordstrom, J.L., &
Mitschelen, J.J., Adv. Lipid Res. 14:1, 1976.

Cholesterol Methodologies: A Review, Zak, B., Clin. Chem. 23:1201, 1977.

Phospholipids in Biology and Medicine (in two parts), Jackson, R.L., &
Gotto, Jr. A.M., N. Engl. J. Med. 290:24 and 87, 1974.

Phosphoglyceride Metabolism, van den Bosch, H., Ann. Rev. Biochem.
43:243, 1974.

The Molecular Organization of Membranes, Singer, S.J., Ann. Rev. Biochem.
43:805, 1974.

Abnormalities of Cell-Membrane Fluidity in the Pathogenesis of Disease,
Cooper, R.A., N. Engl. J. Med. 297:371, 1977.

Genetic Modification of Membrane Lipid, Silbert, D.F., Ann. Rev. Biochem.
44:315, 1975.

Prenatal Therapy of a Patient With Vitamin B_{12}-Responsive Methylmalonic Acidemia, Ampola, M.G., et al., N. Engl. J. Med. 293:313, 1975.

Prenatal Treatment of Methylmalonic Acidemia, Nyhan, W.L., N. Engl. J. Med. 293:346, 1975.

Dietary Fiber and Human Health, Mendeloff, A.I., N. Engl. J. Med. 297:811, 1977.

Applications and Hazards of Total Intravenous Hyperalimentation, Dudrick, S.J., & Long, III J.M., Ann. Rev. Med. 28:517, 1977.

Total Parenteral Nutrition, Law, D.H., N. Engl. J. Med. 297:1104, 1977.

Metabolism of Ketone Bodies by the Brain, Sokoloff, L., Ann. Rev. Med. 24:271, 1973.

Diabetic Ketoacidosis, Felig, P., N. Engl. J. Med. 290:1360, 1974.

Carnitine Metabolism in Human Subjects (in two parts), Mitchell, M., Am. J. Clin. Nutri. 31:293 and 481, 1978.

Prostaglandins

An Enzyme Isolated From Arteries Transforms Prostaglandin Endoperoxides to an Unstable Substance That Inhibits Platelet Aggregation, Moncada, S., et al., Nature 263:663, 1976.

Coronary Arterial Smooth Muscle Contraction by Substance From Platelets: Evidence That It Is Thromboxane A_2, Ellis, E.F., et al., Science 193:1135, 1976.

Metabolism and Physiologic Roles of Vitamin E, Oski, F.A., Hosp. Pract. 12:79, October 1977.

Influence of Vitamin E on Prostaglandin Biosynthesis in Rat Blood, Hope, W.C., et al., Prostaglandins 10:558, 1975.

Vitamin E Deficiency and Enhanced Platelet Function: Reversal Following E Supplementation, Lake, A.M., Stuart, M.J., & Oski, F.A., J. Pediatr. 90:722, 1977.

Thromboxanes: The Power Behind the Prostaglandins--Research News, Science 190:770, 1975.

Aspects of Prostaglandin Function in the Lung (in two parts), Mathe, A.A., et al., N. Engl. J. Med. 296:850 and 910, 1977.

Blood Clotting: The Role of Prostaglandins--Research News, Science 196:1072, 1977.

Advances in Prostaglandin and Thromboxane Research, Volume I, Samuelsson, B., & Paoletti, R. (eds.), Raven Press, New York, 1976.

Prostaglandin Endoperoxides: Novel Transformations of Arachidonic Acid in Human Platelets, Hamberg, M., & Samuelsson, B., Proc. Nat. Acad. Sci. U.S.A. 71:3400, 1974.

Prostaglandins as Mediators of Hypercalcemia Associated with Cancer, Seyberth, H.W., et al., N. Engl. J. Med. 293:1278, 1975.

Prostaglandins, Hypercalcemia & Cancer, Tashijian, Jr. A.H., N. Engl. J. Med. 293:1317, 1975.

Prostaglandins, Samuelsson, B., Granstrom, E., Green, K., Hamberg, M., & Hammarstrom, S., Ann. Rev. Biochem. 44:669, 1975.

Basic Mechanisms of Prostaglandin Action on Autonomic Neurotransmission, Hedqvist, P., Ann. Rev. Pharmacol. Toxicol. 17:259, 1977.

Lecithin/Sphingomyelin Ratio

Diagnosis of the Respiratory Distress Syndrome by Amniocentesis, Gluck, L., et al., Am. J. Obstet. Gynecol. 109:440, 1971.

Assessment of the Risk of the Respiratory-Distress Syndrome by a Rapid Test for Surfactant in Amniotic Fluid, Clements, J.A., et al., N. Engl. J. Med. 286:1077, 1972.

A Prospective Evaluation of the Lecithin Sphingomyelin Ratio and Rapid Surfactant Test in Relation to Fetal Pulmonary Maturity, Keniston, R.C., et al., Am. J. Obstet. Gynecol. 121:324, 1975.

Assessment of Fetal Maturation by the Foam Test, Roux, J.F., Nakamura, J., & Brown, E., Am. J. Obstet. Gynecol. 117:280, 1973.

Lecithin Sphingomyelin Ratio and a Rapid Test for Surfactant in Amniotic Fluid, Boehm, F.H., Srisupundit, S., & Ishii, T., Obstet. Gynecol. 41:829, 1973.

Amniotic Fluid Shake Test Versus Lecithin/Sphingomyelin Ratio in the Antenatal Prediction of Respiratory Distress Syndrome, Mukherjee, T.K., et al., Am. J. Obstet. Gynecol. 119:648, 1974.

A Comparison Between the Lecithin Sphingomyelin Ratio and Other Methods of Assessing the Presence of Pulmonary Surfactant in Amniotic Fluid, Parkinson, C.E., & Harvey, D.R., Br. J. Obstet. Gynecol. 80:406, 1973.

Lecithin/Sphingomyelin Ratios in Amniotic Fluid in Normal and Abnormal Pregnancy, Gluck, L., & Kulovich, M.V., Am. J. Obstet. Gynecol. 115:539, 1973.

The Lecithin/Sphingomyelin Ratio in High-Risk Obstetric Population, Aubrey, R.H., et al., Obstet. Gynecol. 47:21, 1976.

Clinical Experience with Amniotic Fluid Lecithin/Sphingomyelin Ratio, Donald, I.R., et al., Am. J. Obstet. Gynecol. 115:547, 1973.

Prediction of Respiratory Distress Syndrome by Estimation of Surfactant in the Amniotic Fluid, Dewhurst, C.J., et al., Lancet 1:1475, 1973.

Phosphatidyl Inositol and Phosphatidyl Glycerol in Amniotic Fluid: Indices of Lung Maturity, Hallman, M., et al., Am. J. Obstet. Gynecol. 125:613, 1976.

Prediction of Respiratory-Distress Syndrome by Shake Test on New Born Gastric Aspirate, Evans, J.J., N. Engl. J. Med. 292:1113, 1975.

Tests for Maturity of Fetal Lung, Gross, T.L., & Cook, W.A., N. Engl. J. Med. 297:671, 1977.

Amniotic-Fluid Cortisol and Human Fetal Lung Maturation, Sharp-Cageorge, S.M., et al., N. Engl. J. Med. 296:89, 1977.

Lipid Storage Diseases

Absence of Hexosaminidase A and B in Normal Adult, Dreyfus, J.C., Poenaru, L., & Svennerholm, L., N. Engl. J. Med. 292:61, 1975.

Replacement Therapy for Inherited Enzyme Deficiency: Use of Purified Ceramidetrihexosidase in Fabry's Disease, Brady, R.O., et al., N. Engl. J. Med. 289:9, 1973.

Replacement Therapy for Inherited Enzyme Deficiency: Use of Purified Glucocerebrosidase in Gaucher's Disease, Brady, R.O., et al., N. Engl. J. Med. 291:989, 1974.

Gaucher's Disease, A Review, Peters, S., Lee, R.E., & Glew, R.H., Medicine 56:425, 1977.

Enzymological Approaches to the Lipidoses, Brady, R.O., Ann. Clin. Lab. Sci. 7:105, 1977.

Enzyme Replacement Therapy in Gaucher's and Fabry's Disease, Pentcher, P.G., Ann. Clin. Lab. Sci. 7:251, 1977.

The Carrier Potential of Liposomes in Biology and Medicine (in two parts), Gregoriadis, G., N. Engl. J. Med. 295:704 and 765, 1976.

Toward Enzyme Therapy for Lysosomal Storage Diseases, Desnick, R.J., Thorpe, S.R., & Fiddler, M.B., Physiol. Rev. 56:57, 1976.

Enzymes in Amniotic Fluid, Watkins, B.F., & Bermes, E.W., Ann. Clin. Lab. Sci. 7:231, 1977.

Inherited Disorders of Lysosomal Metabolism, Neufeld, E.F., Timple, W.L., & Shapiro, L.J., Ann. Rev. Biochem. 44:357, 1975.

Lysosomal Storage Diseases, Kolodny, E.H., N. Engl. J. Med. 294:1217, 1976.

GM_3 (Hematoside) Sphingolipodystrophy, Max, R.S., et al., N. Engl. J. Med. 291:929, 1974.

Synthetic Defect in Ganglioside Metabolism, O'Brien, J.S., N. Engl. J. Med. 291:975, 1974.

Bile Acids

Bile Acid Metabolism, Danielsson, H., & Sjovall, J., Ann. Rev. Biochem. 44:233, 1975.

Synthesis and Enterohepatic Circulation of Bile Salts, Hanson, R.F., & Pries, J.M., Gastroenterology 73:611, 1977.

Metabolism of Ursodeoxycholic Acid in Man, Fedorowski, T., et al., Gastroenterology 73:1131, 1977.

Enzymatic Analysis of Bile Acids, Macdonald, I., Clin. Biochem. 9:153, 1976.

Chenodeoxycholic Acid Treatment of Gallstones, Iser, J.H., et al., N. Engl. J. Med. 293:378, 1975.

Desaturation of Bile and Cholesterol Gallstone Dissolution With Chenodeoxycholic Acid, Hoffman, A.F., Am. J. Clin. Nutri. 30:993, 1977.

Effect of Chenodeoxycholic Acid and Phenobarbital on the Rate-Limiting Enzymes of Hepatic Cholesterol and Bile Acid Synthesis in Patients With Gall Stones, Coyne, M.J., et al., J. Lab. Clin. Med. 87:281, 1976.

Bile-Acid Metabolism and the Liver, Williams, C.N., Clin. Biochem. 9:149, 1976.

Bile Acid Pools, Kinetics and Biliary Composition Before and After Cholecystectomy, Almond, H.R., et al., N. Engl. J. Med. 289:1213, 1973.

Efficacy and Specificity of Chenodeoxycholic Acid Therapy for Dissolving Gallstones, Thistle, J.L., & Hoffmann, A.F., N. Engl. J. Med. 289:655, 1973.

Gallbladder Bile Composition in Different Ethnic Groups, Oviedo, M.A., et al., Arch. Pathol. Lab. Med. 101:208, 1977.

Radioimmunoassay of Bile Acids in Serum, Demers, L.M., & Hepner, G., Clin. Chem. 22:602, 1976.

Lipoproteins

Lipoprotein Structure and Metabolism, Jackson, R.L., Morrisett, J.D., & Gotto, Jr. A.M., Physiol. Rev. 56:259, 1976.

Lipoproteins: Structure and Function, Morrisett, J.D., Jackson, R.L., & Gotto, Jr. A.M., Ann. Rev. Biochem. 44:183, 1975.

Interconversions of Apolipoprotein Fragments, Blum, C.B., & Levy, R.I., Ann. Rev. Med. 26:365, 1975.

The Common Hyperlipoproteinemias, Fisher, W.R., & Truitt, D.H., Ann. Internal. Med. 85:497, 1976.

Classification of the Hyperlipidemias, Havel, R.J., Ann. Rev. Med. 28:195, 1977.

The Genetic Hyperlipidemias, Motulsky, A.G., N. Engl. J. Med. 294:823, 1976.

The Low-Density Lipoprotein Pathway and Its Relation to Atherosclerosis, Goldstein, J.L., & Brown, M.S., Ann. Rev. Biochem. 46:897, 1977.

The Effect of Hypolipidemic Drugs on Plasma Lipoproteins, Levy, R.I., Ann. Rev. Pharmacol. Toxicol. 17:499, 1977.

It's Time To Be Practical, Fredrickson, D.S., Circulation 51:209, 1975.

HDL Cholesterol and Other Lipids in Coronary Heart Disease, Castelli, W.P., et al., Circulation 55:767, 1977.

Distribution of Triglyceride and Total, LDL and HDL Cholesterol in Several Populations: A Cooperative Lipoprotein Phenotyping Study, Castelli, W.P., et al., J. Chron. Dis. 30:147, 1977.

Plasma-High-Density-Lipoprotein Concentration and Development of Ischaemic Heart Disease, Miller, G.J., & Miller, N.E., The Lancet, page 16, January 4, 1975.

The Tromsø Heart Study, High-Density Lipoprotein and Coronary Heart Disease: A Prospective Case-Control Study, Miller, N.E., et al., The Lancet, page 965, May 7, 1977.

Diet-Heart: End of an Era, Mann, G.V., N. Engl. J. Med. 297:644, 1977.

Serum Lipoproteins and Coronary Heart Disease, Rhoads, G.G., Gulbrandsen, C.L., & Kagan, A., N. Engl. J. Med. 294:293, 1976.

Atherosclerotic Cardiovascular Disease: Current Perspectives, Gotto, Jr. A.M., Yeshurun, D., & DeBakey, M.E., J. Chron. Dis. 29:677, 1976.

Disorders of Cholesterol Metabolism: A Symposium, Am. J. Clin. Nutri. 30:966, 1977.

Cholesterol Ester Storage Disease: Clinical, Biochemical and Pathologic Studies, Beaudet, A.L., et al., J. Pediatr. 90:910, 1977.

Primary Type V Hyperlipoproteinemia, Greenberg, B.H., et al., Ann. Intern. Med. 87:526, 1977.

Type III Hyperlipoproteinemia: Diagnosis in Whole Plasma by Apolipoprotein-E Immunoassay, Kushwaha, R., et al., Ann. Intern. Med. 87:509, 1977.

Type III Hyperlipoproteinemia: Paradoxical Hypolipidemic Response to Estrogen, Kushwaha, R.S., Ann. Intern. Med. 87:517, 1977.

Rapid Method for the Isolation of Lipoproteins From Human Serum by Precipitation With Polyanions, Burstein, M., Scholnick, H.R., & Morfin, R., J. Lipid Res. 11:583, 1970.

Cholesterol Determination in High-Density Lipoproteins Separated by Three Different Methods, Lopes-Virella, M.F., et al., Clin. Chem. 23:882, 1977.

Quantitation of Apolipoprotein A-1 of Human Plasma High Density Lipoprotein, Albers, J.J., et al., Metabolism 25:633, 1976.

The Pathogenesis of Atherosclerosis (in two parts), Ross, R., & Glomset, J.A., N. Engl. J. Med. 295:369 and 420, 1976.

The Origin of Atherosclerosis, Benditt, E.P., Scientific American, page 74, February 1977.

Cellular Mechanisms for Lipid Deposition in Atherosclerosis (in two parts), Small, D.M., N. Engl. J. Med. 297:873 and 924, 1977.

Obesity

Influence of Obesity on Health (in two parts), Mann, G.V., N. Engl. J. Med. 291:178 and 226, 1974.

Diet and Weight Loss, Van Itallie, T.B., & Yang, M.-U., N. Engl. J. Med. 297:1158, 1977.

Metabolic, Hormonal, and Neural Regulation of Feeding and Its Relationship to Obesity: A Symposium, Am. J. Clin. Nutri. 30:739, 1977.

Therapy of Obesity With Hormones, Rivlin, R.S., N. Engl. J. Med. 292:26, 1975.

Pharmacologic Control of Feeding, Hoebel, B.G., Ann. Rev. Pharmacol. Toxicol. 17:605, 1977.

Metabolic Complications of Jejunoileal Bypass Operations for Morbid Obesity, Scott, Jr. H.W., et al., Ann. Rev. Med. 27:397, 1976.

Protein Nutrition and Liver Disease After Jejunoileal Bypass for Morbid Obesity, Moxley, III R.T., Pozefoky, T., & Lockwood, D.H., N. Engl. J. Med. 290:921, 1974.

Jejunoileal Shunt for Obesity, Hirsch, J., N. Engl. J. Med. 290:962, 1974.

Anorexia Nervosa

Anorexia Nervosa and the Hypothalamus, Vande Wiele, R.L., Hosp. Pract. 12:45, December 1977.

Clinical and Metabolic Features of Anorexia Nervosa, Warren, M.P., & Vande Wiele, R.L., Am. J. Obstet. Gynecol. 117:435, October 1973.

The Significance of Weight Loss in the Evaluation of Pituitary Response to LH RH in Women With Secondary Amenorrhea, Warren, M.P., et al., J. Clin. Endocrinol. Metab. 40:601, 1975.

Anorexia Nervosa: Behavioural and Hypothalamic Aspects, Vigersky, R.A., et al., Clin. Endocrinol. Metab. 5:517, July 1975.

Hypothalamic Dysfunction in Patients With Anorexia Nervosa, Mechlenburg, R.S., et al., Medicine 53:147, March 1974.

Cortisol Secretion and Metabolism in Anorexia Nervosa, Boyar, R.M., et al., N. Engl. J. Med. 296:190, 1977.

Anorexia Nervosa: Immaturity of the 24-Hour Luteinizing Hormone Secretory Pattern, Boyar, R.M., et al., N. Engl. J. Med. 291:861, 1974.

Critical Body Weight in Anorexia Nervosa, Johanson, A., N. Engl. J. Med. 291:904, 1974.

Toward an Elucidation of the Psychoendocrinology of Anorexia Nervosa, Katz, J.L., et al., in "Hormones, Behavior, and Psychopathology," Sachar, E.J. (ed.), Raven Press, New York, 1976, pp. 263-283.

Chapter 9
Mineral Metabolism

Calcium and Phosphate

Basic and Clinical Concepts Related to Vitamin D Metabolism and Action (in two parts), Haussler, M.R., & McCain, T.A., N. Engl. J. Med. 297:974 and 1041, 1977.

Metabolism and Mechanism of Action of Vitamin D, DeLuca, H.F., & Schnoes, H.K., Ann. Rev. Biochem. 45:631, 1976.

Recent Advances in Calcium Metabolism (in two parts), Root, A.W., & Harrison, H.E., J. Pediatr. 88:1 and 177, 1976.

Recent Advances in Vitamin D: A Symposium, Am. J. Clin. Nutri. 29:1257, November 1976.

Nutrition and Bone Disease, Gallagher, J.C., & Riggs, B.L., N. Engl. J. Med. 298:193, 1978.

Paget's Disease of Bone, Ryan, W.G., Ann. Rev. Med. 28:143, 1977.

The Effects of Calcium and Phosphorus in Hemodialysis, Goldsmith, R.S., Ann. Rev. Med. 27:181, 1976.

Calcium Oxalate Renal Stones, Prien, Jr. E.L., Ann. Rev. Med. 26:173, 1975.

Medical Management of the Hypercalcemia of Malignancy, Deftos, L.J., & Neer, R., Ann. Rev. Med. 25:323, 1974.

Nutrition and Urolithiasis, Smith, L.H., Van Den Berg, C.J., & Wilson, D.M., N. Engl. J. Med. 298:87, 1978.

Evidence for the Secretion of an Osteoclast Stimulating Factor in Myeloma, Mundy, G.R., et al., N. Engl. J. Med. 291:1041, 1974.

Osteoclastic Bone Resorption and the Hypercalcemia of Cancer,
Brewer, B.H., N. Engl. J. Med. 291:1081, 1974.

Serum 25-OHC and Bone Mass in Children Given Anticonvulsants,
Hahn, T.J., et al., N. Engl. J. Med. 292:550, 1975.

The Cellular Basis of Metabolic Bone Disease, Rasmussen, H., &
Bordier, P., N. Engl. J. Med. 289:25, 1973.

A Simple Test for the Diagnosis of Absorptive, Resorptive and Renal
Hypercalciurias, Pak, Y.C.C., et al., N. Engl. J. Med. 292:497, 1975.

Selective Deficiency of 1,25-Dihydroxycholecalciferol, Metz, S.A.,
et al., N. Engl. J. Med. 297:1084, 1977.

Parathyroid Autotransplantation in Primary Parathyroid Hyperplasia,
Wells, Jr. S.A., et al., N. Engl. J. Med. 295:57, 1976.

Hypophosphatemia--Medical Staff Conference, U.C.S.F., West J. Med.
122:482, 1975.

Phosphorous Deficiency and Hypophosphatemia, Kreisberg, R.A.,
Hosp. Pract. 12:121, March 1977.

Hyperphosphatemia in Lactic Acidosis, O'Connor, R.L., Klein, K.L., &
Bethune, J.E., N. Engl. J. Med. 297:707, 1977.

Magnesium

Essential Metals in Man: Magnesium, Schroeder, H.A., & Nason, A.P.,
J. Chron. Dis. 21:815, 1969.

Experimental Production of Magnesium Deficiency in Man, Shils, M.E.E.,
Ann. N.Y. Acad. Sci. 162:847, 1969.

Magnesium Metabolism (in three parts), Wacker, W.E.C., & Paris, A.F.,
N. Engl. J. Med. 278:658, 712 and 772, 1968.

Magnesium Deficiency and Cardiac Disorders, Iseri, L.T., et al.,
Am. J. Med. 58:837, 1975.

Hypomagnesemia, Hall, R.C., & Joffe, J.R., JAMA 244:1749, 1973.

Pharmacology of Magnesium, Massry, S.G., Ann. Rev. Pharmacol. Toxicol.
17:67, 1977.

Serum Magnesium Levels in Patients With Acute Myocardial Infarction,
Abraham, A.S., et al., N. Engl. J. Med. 296:862, 1977.

Copper

Copper Homeostasis in the Mammalian System, Evans, G.W., Physiol. Rev. 53:535, 1973.

Steely Hair, Mottled Mice and Copper Metabolism, Danks, D.M., N. Engl. J. Med. 293:1147, 1975.

Miscellaneous

Trace Elements, Ulmer, D.D., N. Engl. J. Med. 297:318, 1977.

Mechanisms of Lithium Action, Singer, I., & Rotenberg, D., N. Engl. J. Med. 289:254, 1973.

Arsenic Toxicology and Industrial Exposure, Pinto, S.S., & Nelson, K.W., Ann. Rev. Pharmacol. Toxicol. 16:95, 1976.

Pharmacology and Toxicology of Lithium, Schou, M., Ann. Rev. Pharmacol. Toxicol. 16:231, 1976.

Effect of Acute Disease and ACTH on Serum Zinc Proteins, Falchuk, K.H., N. Engl. J. Med. 296:1129, 1977.

Chapter 10
Vitamins

Vitamin K, Prothrombic and γ-Carboxyglutamic Acid, Stenflo, J., N. Engl. J. Med. 296:624, 1977.

Vitamin Homeostasis in the Central Nervous System, Spector, R., N. Engl. J. Med. 296:1393, 1977.

Special Issue on the Chemistry of Vision, Menger, E.L. (ed.), Accounts of Chemical Research 8:81, March 1975.

Metabolism and Physiologic Roles of Vitamin E, Oski, F.A., Hosp. Pract. 12:79, October 1977.

Vitamin C and Acute Illness in Navajo School Children, Coulehan, J.L., N. Engl. J. Med. 295:973, 1976.

Circulating Antibody to Trans-Cobalamin II Causing Retention of Vitamin B_{12} in the Blood, Carmel, R., Tatsis, B., & Baril, L., Blood 49:987, 1977.

Human B_{12} Transport Proteins, Brown, E.B. (ed.), Progress in Hematology 9:57, 1975, Grune & Stratton, New York.

Biosynthesis of Water-Soluble Vitamins, Plaut, G.W.E., Smith, C.M., & Alworth, W.L., Ann. Rev. Biochem. 43:899, 1974.

Biotin Enzymes, Wood, H.G., & Barden, R.E., Ann. Rev. Biochem. 46:385, 1977.

Vitamin K-Dependent Formation of γ-Carboxyglutamic Acid, Stenflo, J., & Suttie, J.W., Ann. Rev. Biochem. 46:157, 1977.

Vitamin Toxicity, DiPalma, J.R., & Ritchie, D.M., Ann. Rev. Pharmacol. Toxicol. 17:133, 1977.

Vitamin K and the Oral Anticoagulant Drugs, O'Reilly, R.A., Ann. Rev. Med. 27:245, 1976.

Vitamin A Transport in Human Vitamin A Toxicity, Smith, F.R., & Goodman, D.S., N. Engl. J. Med. 294:805, 1976.

Changes in Circulating Transcobalamin II After Injection of Cyanocobalamin, Donaldson, Jr. R.M., Brand, M., & Serfilippi, D., N. Engl. J. Med. 296:1427, 1977.

Chapter 11
Immunochemistry

General

Laboratory Diagnosis of Immune and Autoimmune Reactions: Basic Principles and Practical Applications--A Symposium, Am. J. Clin. Pathol. 68:633, 1977.

Three-Dimensional Structure of Immunoglobulins, Davies, D.R., Padlan, E.A., & Segal, D.M., Ann. Rev. Biochem. 44:639, 1975.

The Biological Origin of Antibody Diversity, Williamson, A.R., Ann. Rev. Biochem. 45:467, 1976.

Thymic Hormones: Biochemistry, and Biological and Clinical Activities, Bach, J.F., Ann. Rev. Pharmacol. 17:281, 1977.

Thymosin Activity in Patients With Cellular Immunodeficiency, Wara, D.W., et al., N. Engl. J. Med. 292:70, 1975.

Thymosin Therapy Creates a "Hassall"(?), Hong, R., & Horowitz, S., N. Engl. J. Med. 292:104, 1975.

Structure and Biological Activity of Immunoglobulin E, Ishizaka, K., Hosp. Pract. 12:57, January 1977.

Alterations in T- and B-Cells in Human Disease States, Williams, Jr. R.C., & Messner, R.P., Ann. Rev. Med. 26:181, 1975.

Bence-Jones Proteins and Light Chains of Immunoglobulins (in two parts), Solmon, A., N. Engl. J. Med. 294:17 and 91, 1976.

Some Impacts of Clinical Investigation on Immunology, Franklin, E.C., N. Engl. J. Med. 294:531, 1976.

Intestinal Antibodies, Walker, W.A., & Isselbacher, K.J., N. Engl. J. Med. 297:767, 1977.

The Association of Genes in the Major Histocompatibility Complex and Disease Susceptibility, Sasazuki, T., Grumet, F.C., & McDevitt, H.O., Ann. Rev. Med. 28:425, 1977.

HLA and Disease Susceptibility: A Primer, Rosenberg, L.E., & Kidd, K.K., N. Engl. J. Med. 297:1060, 1977.

Formation of Human Plasma Kinin, Colman, R.W., N. Engl. J. Med. 291:509, 1974.

Suppressors in the Network of Immunity, Siegal, F., N. Engl. J. Med. 298:102, 1978.

Myasthenia Gravis (in two parts), Drachman, D.B., N. Engl. J. Med. 298:136 and 186, 1978.

Secretory Function of Mononuclear Phagocytes, Unanue, E.R., Am. J. Pathol. 83:396, 1976.

Acute Inflammation, Ryan, G.B., & Majno, G., Am. J. Pathol. 86:185, 1977.

Immunopathogenesis of Systemic Lupus Erythematosus, Koffler, D., Ann. Rev. Med. 25:149, 1974.

IgA-Associated Glomerulonephritis, McPhaul, Jr. J.J., Ann. Rev. Med. 28:37, 1977.

The Antiglobulin Test in Autoimmune Hemolytic Anemia, Rosse, W.F., Ann. Rev. Med. 26:331, 1975.

Immuno-Prevention of Rh Hemolytic Disease of the Newborn, McConnell, R.B., & Woodrow, J.C., Ann. Rev. Med. 25:165, 1974.

Bronchodilator Therapy (in two parts), Webb-Johnson, D.C., & Andrews, Jr. J.L., N. Engl. J. Med. 297:476 and 758, 1977.

Immunotherapy of Cancer, Oettgen, H.F., N. Engl. J. Med. 297:484, 1977.

Complement

Complement, Muller-Eberhard, H.J., Ann. Rev. Biochem. 44:697, 1975.

The Value of Complement Assays in Clinical Chemistry, Whicher, J.T., Clin. Chem. 24:7, 1978.

Complement Deficiency States, Agnello, V., Medicine 57:1, 1978.

Chemistry and Function of the Complement System, Muller-Eberhard, H.J., Hosp. Pract. 12:33, August 1977.

Complement Genes on Chromosome 6, Bruns, G.A.P., N. Engl. J. Med. 296:510, 1977.

Complement and Host Defense Against Infection, Johnston, R.B., & Stroud, R.M., J. Pediatr. 90:169, 1977.

Linkage of the Gene Controlling C4 Synthesis to the Histocompatibility Complex, Ochs, H.D., et al., N. Engl. J. Med. 296:470, 1977.

Evaluation of Thrombogenic Effects of Drugs, Zbinden, G., Ann. Rev. Pharmacol. Toxicol. 16:177, 1976.

Blood Coagulation

Basic Mechanisms in Blood Coagulation, Davie, E.W., & Fujikawa, K., Ann. Rev. Biochem. 44:799, 1975.

Endothelial Cells and the Biology of Factor VIII, Jaffe, E.A., N. Engl. J. Med. 296:377, 1977.

Detection of the Carrier State of Classic Hemophilia, Klein, H.G., et al., N. Engl. J. Med. 296:959, 1977.

Genetic Counselling in Classic Hemophilia A, Graham, J.B., N. Engl. J. Med. 296:996, 1977.

Factor VIII Inhibitors, Deykin, D., N. Engl. J. Med. 291:205, 1974.

Platelets: Physiology and Abnormalities of Function (in two parts), Weiss, H.J., N. Engl. J. Med. 293:531 and 580, 1975.

Von Willebrand's Disease, Holmberg, L., & Nilsson, I.M., Ann. Rev. Med. 26:33, 1975.

Molecular Immunology of Factor VIII, Zimmerman, T.S., & Edgington, T.S., Ann. Rev. Med. 25:303, 1974.

<u>Phagocytosis</u>--See Chapter 4.

Chapter 12
Water and Electrolyte Balance

<u>Kidney, Renin and Angiotensin</u>

Kidney, Lassiter, W.E., Ann. Rev. Physiol. 37:371, 1975.

Angiotensin-Converting Enzyme and the Regulation of Vasoactive Peptides, Soffer, R.L., Ann. Rev. Biochem. 45:73, 1976.

Proximal Tubular Reabsorption and Its Regulation, Jacobson, H.R., & Seldin, D.W., Ann. Rev. Pharmacol. Toxicol. 17:623, 1977.

Renin and the Therapy of Hypertension, Guthrie, Jr. G.P., Genest, J., & Kuchel, O., Ann. Rev. Pharmacol. Toxicol. 16:287, 1976.

Relationship of Renal Sodium and Water Transport to Hydrogen Ion Secretion, Arruda, J.A.L., & Kurtzman, N.A., Ann. Rev. Physiol. 40:43, 1978.

Mechanisms Regulating Renin Release, Davis, J.O., & Freeman, R.H., Physiol. Rev. 56:1, 1976.

Nonosmolar Factors Affecting Renal Water Excretion (in two parts), Schrier, R.W., & Berl, T., N. Engl. J. Med. 292:81 and 141, 1975.

Acute Oliguria, Harrington, J.T., & Cohen, J.J., N. Engl. J. Med. 292:89, 1975.

Renal Function During and Following Obstruction, Wilson, D.R., Ann. Rev. Med. 28:329, 1977.

Interrelations of Renin, Angiotensin II, and Sodium in Hypertension and Renal Failure, Gavras, H., Oliver, J.A., & Cannon, P.J., Ann. Rev. Med. 27:485, 1976.

Low Renin Hypertension, Gunnells, Jr. J.C., & McGuffin, Jr. W.L., Ann. Rev. Med. 26:259, 1975.

Solving the Adrenal Lesion(s) of Primary Aldosteronism, Melby, J.C., N. Engl. J. Med. 294:441, 1976.

Mechanics of Glomerular Ultrafiltration, Brenner, B.M., & Humes, H.D., N. Engl. J. Med. 297:148, 1977.

Renin-Angiotensin System, Peart, W.S., N. Engl. J. Med. 292:302, 1975.

Location of Aldosterone-Producing Adenomas With ^{131}I-19-Iodocholesterol, Hogan, M.J., et al., N. Engl. J. Med. 294:410, 1976.

The Renin-Angiotensin System, Reid, I.A., Morris, B.J., & Ganong, W.F., Ann. Rev. Physiol. 40:377, 1978.

Osmotic Diuresis, Gennari, F.J., & Kassirer, J., N. Engl. J. Med. 291:714, 1974.

Drug-Induced Dilectional Hyponatremia, Moses, A.M., & Miller, M., N. Engl. J. Med. 291:1234, 1974.

Angiotensin Antagonist in Diagnosing Angiotensinogenic Hypertension, Streeten, D.H.P., et al., N. Engl. J. Med. 292:657, 1975.

Angiotensin Blockade: New Pharmacologic Tools for Understanding and Treating Hypertension, Laragh, J.H., N. Engl. J. Med. 292:695, 1975.

Renin, Aldosterone and Glucagon in the Natriuresis of Fasting, Spark, R.F., et al., N. Engl. J. Med. 292:1335, 1975.

Brain Edema, Fisman, R.A., N. Engl. J. Med. 293:706, 1975.

Osmolality

Osmometry, Dormandy, T.L., The Lancet, page 267, February 4, 1967.

Ethanol Ingestion--Commonest Cause of Elevated Plasma Osmolality?, Robinson, A.G., & Loeb, J.N., N. Engl. J. Med. 284:1253, 1971.

The Hyperosmolar State, Loeb, J.N., N. Engl. J. Med. 290:1184, 1974.

Osmolality, Warhol, R.M., Eichenholz, A., & Mulhausen, R.O., Arch. Intern. Med. 116:743, 1965.

Osmometry: A New Bedside Laboratory Aid for the Management of Surgical Patients, Boyd, D.R., & Baker, R.J., Surgical Clinics of North America 51:241, 1971.

Hyperosmolality and Trauma, Hallberg, D., Acta Anaesth. Scand. Suppl. 55:21, 1974.

Diabetes Insipidus

Vasopressin in Osmotic Regulation in Man, Robertson, G.L., Ann. Rev. Med. 25:315, 1974.

DDAVP in the Treatment of Central Diabetes Insipidus, Robinson, A.G., N. Engl. J. Med. 294:507, 1976.

Diabetes Insipidus and ADH Regulation, Moses, A.M., Hosp. Pract. 12:37, July 1977.

Anion Gap, Acid-Base Disorders

The Anion Gap, Oh, M.S., & Carroll, H.J., N. Engl. J. Med. 297:814, 1977.

Clinical Use of the Anion Gap, Emmett, M., & Narins, R.G., Medicine 56:38, 1977.

Acid-Base Status of the Blood, 4th ed., Siggard-Andersen, O., Williams & Wilkins, Baltimore, 1974.

Illustrated Manual of Fluid and Electrolyte Disorders, Collins, R.D., Lippincott, Philadelphia, 1976.

Serious Acid-Base Disorders, Kassirer, J.P., N. Engl. J. Med. 291:773, 1974.

Measurement of Urinary Electrolytes--Indications and Limitations, Harrington, J.T., & Cohen, J.J., N. Engl. J. Med. 293:1241, 1975.

Diabetic Ketoacidosis, Felig, P., N. Engl. J. Med. 290:1360, 1974.

Acid-Base Balance in Cerebrospinal Fluid, Posner, J.B., Swanson, A.G., & Plum, F., Arch. Neurol. 12:479, 1965.

Cerebrospinal Fluid, Fisher, R.G., Mayo Clinic Proc. 50:482, 1975.

Acid-Base Balance of Cisternal and Lumbar Cerebrospinal Fluid in Hospital Patients, Plum, F., & Price, R.W., N. Engl. J. Med. 289:1346, 1973.

Cerebrospinal-Fluid Acid-Base and Electrolyte Changes Resulting From Cerebral Anoxia in Man, Kalin, E.M., et al., N. Engl. J. Med. 293:1013, 1975.

Chapter 13
Hormones

Mechanism of Action of the Sex Steroid Hormones (in three parts), Chan, L., & O'Malley, B.W., N. Engl. J. Med. 294:1322, 1372 and 1430, 1976.

Steroid Hormone Receptors in the Regulation of Differentiation, McCarty, Jr. K.S., & McCarty, Sr. K.S., Am. J. Pathol. 86:705, March 1977.

Hypothalamic Hormones: The Link Between Brain and Body, Schally, A.V., Kastin, A.J., & Arimura, A., American Scientist 65:712, 1977.

Hormone Receptors, Jarett, L., & McDonald, J., Arch. Pathol. Lab. Med. 101:156, 1977.

Localization and Release of Neurophysins, Sief, S.M., & Robinson, A.G., Ann. Rev. Physiol. 40:345, 1978.

Antidiuretic Hormone, Bleich, H.L., & Boro, E.S., N. Engl. J. Med. 295:659, 1976.

Endorphins, Brain Peptides That Act Like Opiates, Guillemin, R., N. Engl. J. Med. 296:226, 1977.

Opiate Receptors in the Brain, Snyder, S.H., N. Engl. J. Med. 296:266, 1977.

Peptide Hormones, Tager, H.S., & Steiner, D.S., Ann. Rev. Biochem. 43:509, 1974.

Regulation of Steroid Biosynthesis, Dempsey, M.E., Ann. Rev. Biochem. 43:967, 1974.

Steroid Receptors: Elements for Modulation of Eukaryotic Transcription, Yamamoto, K.R., & Alberts, B.M., Ann. Rev. Biochem. 45:721, 1976.

Cyclic GMP Metabolism and Involvement in Biological Regulation, Goldberg, N.D., & Haddox, M.K., Ann. Rev. Biochem. 46:823, 1977.

Common Mechanisms of Hormone Secretion, Trifaro, J.M., Ann. Rev. Pharmacol. Toxicol. 17:27, 1977.

Clinical Pharmacology of Systemic Corticosteroids, Melby, J.C., Ann. Rev. Pharmacol. Toxicol. 17:511, 1977.

The Pharmacology of the Pineal Gland, Minneman, K.P., & Wurtman, R.J., Ann. Rev. Pharmacol. Toxicol. 16:33, 1976.

The Effects of Light on Man and Other Mammals, Wurtman, R.J.,
Ann. Rev. Physiol. 37:467, 1975.

Neural Control of the Posterior Pituitary, Hayward, J.N.,
Ann. Rev. Physiol. 37:191, 1975.

Membrane Receptors, Cautrecasas, P., Ann. Rev. Biochem. 43:169, 1974.

Hormonal Regulation of the Reproductive Tract in Female Mammals,
Brenner, R.M., & West, N.B., Ann. Rev. Physiol. 37:273, 1975.

The Regulation of Growth by Endocrines, Daughaday, W.H.,
Herington, A.C., & Phillips, L.S., Ann. Rev. Physiol. 37:211, 1975.

Peripheral Actions of Glucocorticoids, Leung, K., & Munck, A.,
Ann. Rev. Physiol. 37:245, 1975.

Estrogens and the Human Male, Marcus, R., & Korenman, S.G.,
Ann. Rev. Med. 27:357, 1976.

Circadian Rhythms and Episodic Hormone Secretion in Man,
Weitzman, E.D., Ann. Rev. Med. 27:225, 1976.

Relation Between Growth Hormone and Somatomedin, Van Wyk, J.J., &
Underwood, L.E., Ann. Rev. Med. 26:427, 1975.

The Role of Hormone Receptors in the Action of Adrenal Steroids,
Feldman, D., Ann. Rev. Med. 26:83, 1975.

Endocrine Therapy of Breast Cancer, McGuire, W.L., Ann. Rev. Med.
26:353, 1975.

Gonadotropin-Releasing Hormone, Yen, S.S.C., Ann. Rev. Med. 26:403,
1975.

Isolated Growth Hormone Deficiency and Related Disorders, Merimee, T.J.,
Ann. Rev. Med. 25:137, 1974.

ACTH and the Regulation of Adrenocortical Secretion, Ganong, W.F.,
Alpert, L.C., & Lee, T.C., N. Engl. J. Med. 290:1006, 1974.

Neuroendocrine Neoplasms and Their Cells of Origin, Tischler, A.S.,
et al., N. Engl. J. Med. 296:919, 1977.

The Pineal Organ (in two parts), Wurtman, R.J., & Moskowitz, M.A.,
N. Engl. J. Med. 296:1329 and 1383, 1977.

Luteinizing-Hormone-Releasing Hormone, McCann, S.M., N. Engl. J. Med.
296:797, 1977.

Glycopeptide Hormones and Neoplasms, Odell, W.D., N. Engl. J. Med. 297:609, 1977.

Nerve Growth Factor (in three parts), Chalmers, T.C., et al., N. Engl. J. Med. 297:1096, 1149 and 1211, 1977.

Cell Receptors in Disease, Jacobs, S., & Cuatrecasas, P., N. Engl. J. Med. 297:1383, 1977.

Prolactin, Frantz, A.G., N. Engl. J. Med. 298:201, 1978.

Action of Hormones on the Kidney, Katz, A.I., & Lindheimer, M.D., Ann. Rev. Physiol. 39:97, 1977.

Biosynthesis of Polypeptide Hormones, Hew, C.-L., & Yip, C.C., Can. J. Biochem. 54:591, 1976.

Chapter 14
Biochemical Transformation of Drugs

Drug Metabolism and Disposition: The Biological Fate of Chemicals, in "Proceedings of the Second International Symposium on Microsomes and Drug Oxidations," Leibman, K.C. (ed.), Williams & Wilkins, Baltimore, 1973.

Pharmacogenetics, in "Annals of the New York Academy of Science," Volume 151, LaDu, B.N., & Kalow, W. (eds.), New York Academy of Science, New York, 1968.

Drug Metabolism in Man, in "Annals of the New York Academy of Science," Volume 179, Vessell, E.S. (eds.), New York Academy of Science, New York, 1971.

Principles of Drug Action, Goldstein, A., Aronow, L., & Kalman, M., Hoeber, Harper & Row, New York, 1968.

Factors Influencing Drug Metabolism, Conney, A.H., & Burns, J.J., Adv. Pharmacol. 1:31, 1962.

Mechanisms of Drug Absorption and Excretion, Rall, D.P., & Zubrod, C.G., Ann. Rev. Pharmacol. 2:109, 1962.

Metabolic Factors Controlling Duration of Drug Action, in "Proceedings of the First International Pharmacological Meeting," Volume 6, Brodie, B.B., & Erdos, E.G. (eds.), McMillan, New York, 1962.

Pharmacogenetics: Defective Enzymes in Relation to Reaction to Drugs, LaDu, B.N., Ann. Rev. Med. 23:453, 1972.

Clinical Pharmacology of Drug Interactions, Raisfeld, I.H.,
Ann. Rev. Med. 24:385, 1973.

Biochemistry of Drug Dependence, Takemori, A.E., Ann. Rev. Biochem.
43:15, 1974.

Cardiovascular Drug Interactions, Brater, D.C., Morrelli, H.F.,
Ann. Rev. Pharmacol. Toxicol. 17:293, 1977.

Pharmacokinetic Consequences of Aging, Richey, D.P., & Bender, A.D.,
Ann. Rev. Pharmacol. Toxicol. 17:49, 1977.

Developmental Aspects of the Hepatic Cytochrome P450 Monooxygenase
System, Neims, A.H., et al., Ann. Rev. Pharmacol. Toxicol. 16:427, 1976.

Therapeutic Implications of Bioavailability, Azarnoff, D.L., &
Huffman, D.H., Ann. Rev. Pharmacol. Toxicol. 16:53, 1976.

Behavioral Pharmacology and Toxicology, Bignami, G., Ann. Rev.
Pharmacol. Toxicol. 16:329, 1976.

Chemicals, Drugs, and Lipid Peroxidation, Plaa, G.L., & Witschi, H.,
Ann. Rev. Pharmacol. Toxicol. 16:125, 1976.

Effects of Anesthesia on Intermediary Metabolism, Brunner, E.A.,
Cheng, S.C., & Berman, M.L., Ann. Rev. Med. 26:391, 1975.

Fatal Reactions to Drug Therapy, Koch-Weser, J., N. Engl. J. Med.
291:302, 1974.

Drug Therapy: Bioavailability of Drugs (in two parts), Koch-Weser, J.,
N. Engl. J. Med. 291:233 and 503, 1974.

Handling of Bioactive Materials by the Lung (in two parts),
Fishman, A.P., & Pietra, G.G., N. Engl. J. Med. 291:884 and 953, 1974.

Anticonvulsant Drugs and Calcium Metabolism, Anast, C.S., N. Engl. J.
Med. 292:587, 1975.

Clinical Pharmacokinetics (in two parts), Greenblatt, D.J., &
Koch-Weser, J., N. Engl. J. Med. 293:702 and 964, 1975.

Binding of Drugs to Serum Albumin (in two parts), Koch-Weser, J., &
Sellers, E.M., N. Engl. J. Med. 294:311 and 526, 1976.

Nephrotoxicity of Antimicrobial Agents (in three parts), Appel, G.B.,
& Neu, H.C., N. Engl. J. Med. 296:663, 722 and 784, 1977.

Drug Disposition in Liver Disease, Shand, D.G., N. Engl. J. Med.
296:1527, 1977.

Food and Drug Interactions, Hartshorn, E.A., J. Am. Dietet. A 70:15, 1977.

Drug Bioavailability Studies, Wagner, J.G., Hosp. Pract. 12:119, January 1977.

How the Liver Metabolizes Foreign Substances, Kappas, A., & Alvares, A.P., Scientific American 232:22, June 1975.

Chapter 15
Liver Function

Intravascular Coagulation in Liver Disease, Verstraete, M., Vermylen, J., & Collen, D., Ann. Rev. Med. 25:447, 1974.

Mechanisms of Hepatic Bile Formation, Forker, E.L., Ann. Rev. Physiol. 39:323, 1977.

Liver Disease in Infants (in two parts), Mathis, R.K., Andres, J.M., & Walker, A.W., J. Pediatr. 90:686 and 864, 1977.

Liver Function, Burke, D.M., Human Pathol. 6:273, 1975.

Diseases of the Liver and Biliary System, 5th ed., Sherlock, S., Blackwell, Oxford, 1975.

Index

Numbers in italics indicate a figure; "t" following a page number indicate a table.

1335